AF702179

Clove (*Syzygium aromaticum*)

Clove (*Syzygium aromaticum*)

Chemistry, Functionality and Applications

Edited by

Mohamed Fawzy Ramadan
Professor of Biochemistry and Food Chemistry,
Umm Al-Qura University, Saudi Arabia

ELSEVIER

ACADEMIC PRESS
An imprint of Elsevier

Academic Press is an imprint of Elsevier
125 London Wall, London EC2Y 5AS, United Kingdom
525 B Street, Suite 1650, San Diego, CA 92101, United States
50 Hampshire Street, 5th Floor, Cambridge, MA 02139, United States
The Boulevard, Langford Lane, Kidlington, Oxford OX5 1GB, United Kingdom

Notices

Knowledge and best practice in this field are constantly changing. As new research and experience broaden our understanding, changes in research methods, professional practices, or medical treatment may become necessary.

Practitioners and researchers must always rely on their own experience and knowledge in evaluating and using any information, methods, compounds, or experiments described herein. In using such information or methods they should be mindful of their own safety and the safety of others, including parties for whom they have a professional responsibility.

To the fullest extent of the law, neither the Publisher nor the authors, contributors, or editors, assume any liability for any injury and/or damage to persons or property as a matter of products liability, negligence or otherwise, or from any use or operation of any methods, products, instructions, or ideas contained in the material herein.

ISBN: 978-0-323-85177-0

For information on all Academic Press publications
visit our website at https://www.elsevier.com/books-and-journals

Publisher: Nikki P. Levy
Acquisitions Editor: Megan R. Ball
Editorial Project Manager: Li Emerald Production
Project Manager: Bharatwaj Varatharajan
Cover Designer: Victoria Pearson

Typeset by STRAIVE, India

To the soul of my father, Professor Fawzy Ramadan Hassanien,
and to my beloved family

Contents

Contributors xxv
About the editor xxxi
Preface xxxiii
Description xxxv

CHAPTER 1 Introduction to clove: Chemistry, functionality, and techno-applications 1
Mohamed Fawzy Ramadan
1 United Nations sustainable development goals and applications of medical plants 1
1.1 Clove (*S. aromaticum*): Chemistry, functionality, and techno-applications 2
References 7

SECTION 1 Clove (*Syzygium aromaticum*): Cultivation, composition and applications

CHAPTER 2 Composition of clove (*Syzygium aromaticum*) buds 11
Tossaton Charoonratana
1 Introduction 11
2 Production of clove buds in the Pacific Ocean 14
3 Development stages of clove flowers 15
4 Description of clove buds 18
5 Clove genetics 20
6 Chemical compositions of clove buds 21
7 Application of clove buds 24
7.1 Food applications 25
7.2 Traditional medicine 27
7.3 Kretek cigarettes 30
7.4 Animal feeds 31
7.5 Air freshener 31
7.6 Source for eugenol extraction 32
8 Conclusion 32
Acknowledgment 32
References 33

CHAPTER 3 Clove (*Syzygium aromaticum*) leaves 37
Tossaton Charoonratana
1 Introduction 37
2 Description of clove leaves 38
3 Chemical composition of clove leaves 39
4 Application of clove leaves 40
4.1 Source of eugenol extraction 40
4.2 Traditional medicine 40
5 Conclusion 40
Acknowledgment 40
References 41

CHAPTER 4 Clove (*Syzygium aromaticum*) polysaccharides 43
Ahmad Cheikhyoussef, Natascha Cheikhyoussef, Ahmed A. Hussein, and Mohamed Fawzy Ramadan
1 Introduction 43
2 Clove polysaccharides structure and composition 44
3 Clove polysaccharides' biological activities 45
4 Conclusions 45
References 46

CHAPTER 5 Clove (*Syzygium aromaticum*) oleoresins 49
Tossaton Charoonratana
1 Introduction 49
2 Clove oleoresins characteristics and compositions 50
3 Clove oleoresins from different plant varieties 51
4 Clove oleoresins extraction 52
5 Detection of compounds in clove oleoresins 53
6 Clove oleoresins application 54
7 Prospect in clove oleoresins application 55
7.1 Clove oleoresins safety 55
7.2 Food preservative 56
7.3 Pharmaceuticals 57
7.4 Biopesticides 59
7.5 Animal feeds 60
7.6 Fish farming 60
7.7 Miscellaneous products 60
8 Conclusion 60
Acknowledgment 61
References 61

CHAPTER 6 Cloves (*Syzygium aromaticum*) cultivars: Convenient source of eugenol and its role in commercially important formulations 67
Abdul Hafeez Laghari and Salman Tariq Khan
1 Introduction 67
2 Useful compounds of clove 69
3 Eugenol and its derivatives 74
4 Eugenol-based formulations 76
5 Conclusion 78
References 78

CHAPTER 7 Chemistry and functionality of clove oil nanoemulsions 81
Bimlesh Mann, Richa Singh, S. Athira, Rajesh Kumar, and Rajan Sharma
1 Introduction 81
2 Clove oil and eugenol 81
3 Biofunctional properties of clove oil 82
4 Challenges for using the clove oil in the food matrix 83
5 Encapsulation of clove oil 83
6 Nanoemulsions for delivery of clove oil 84
7 Importance of emulsifiers/surfactants in an emulsion system 85
7.1 Low molecular weight surfactants 88
7.2 High molecular weight surfactants 88
8 Stability of clove oil nanoemulsions 89
8.1 Gravitational separation 89
8.2 Flocculation and coalescence 90
8.3 Ostwald ripening 91
9 Characterization of clove oil nanoemulsions 91
9.1 Particle size 91
9.2 Particle charge 92
9.3 Morphological characterization 92
10 Functionalities of clove oil nanoemulsions 93
11 Application of clove oil/eugenol nanoemulsions as an antimicrobial agent in food 96
12 Conclusion 96
References 96

CHAPTER 8 Eugenol-based nutraceuticals 103
Mohamad Taleuzzaman, Rishabh Verma, and Jabeen Farhana
1 Introduction 103
2 Source, extraction, and chemistry of eugenol 105

2.1 Solvent extraction ... 105
2.2 Hydrodistillation ... 106
2.3 Microwave-assisted extraction of eugenol ... 106
2.4 Supercritical carbon dioxide extraction ... 106
2.5 Ultrasound-assisted extraction of eugenol ... 106
3 Nanoformulation-based nutraceuticals ... 107
4 Eugenol's pharmacology and toxicological effects ... 109
4.1 Pharmacology effects ... 109
4.2 Toxicological studies ... 111
5 Conclusion ... 112
Acknowledgment ... 112
References ... 112

CHAPTER 9 Clove (*Syzygium aromaticum*) volatiles ... 117
Tossaton Charoonratana
1 Introduction ... 117
2 Clove volatile oil characteristics and compositions ... 118
3 Clove volatile oil from different plant varieties ... 128
4 Clove volatile oil from different plant ages ... 129
5 Clove volatile oil from different maturity of plant organs ... 130
6 Clove volatile oil from different cultivation areas ... 132
7 The volatile oil in cloves with different post-harvesting processes ... 133
8 Clove volatile oil extraction and storage ... 134
8.1 Clove volatile oil distillation ... 135
8.2 Clove volatile oil pre- and post-distillation processes ... 137
8.3 Clove oil extraction using alternative methods ... 139
8.4 Clove volatile oil storage ... 140
9 Clove volatile oil composition detection ... 140
10 Clove volatiles application ... 141
10.1 Clove volatile oil in food processing ... 141
10.2 Clove volatile oil in medicines ... 142
10.3 Clove volatile oil in perfumery ... 145
10.4 Clove volatile oil in biopesticides ... 145
10.5 Clove volatile oil as clearing agents in tissue histology ... 148
10.6 Clove volatile oil as a source of eugenol and β-caryophyllene ... 149
11 Prospect in clove volatiles application ... 149
11.1 Clove volatile oils safety ... 150
11.2 Food preservative ... 151
11.3 Pharmaceuticals ... 155
11.4 Cosmetics ... 158

11.5 Bacteria-resistant fabric 158
11.6 Animal feeds 158
11.7 Fish anesthesia 159
11.8 Mosquito control 160
12 Conclusion 160
Acknowledgment 161
References 161
Further reading 174

CHAPTER 10 Secondary metabolites of clove (*Syzygium aromaticum*) 175
Niranjan Das, Akash Dey, Sudip Kumar Mandal, Debanjan Chatterjee, Rajan Logesh, and Hari Prasad Devkota
1 Introduction 175
2 Economic value of clove 176
3 Phytochemistry 177
4 Biological activities 184
4.1 Antibacterial activity 184
4.2 Antioxidant activity 185
4.3 Antifungal activity 185
4.4 Anti-inflammatory activity 186
4.5 Anticancer activity 186
4.6 Nematicidal activity 187
4.7 Anesthetic effect 187
4.8 Herbicidal activity 188
4.9 Insecticidal activity 188
5 Conclusions and future prospective 188
References 189

CHAPTER 11 Health-promoting activities of clove (*Syzygium aromaticum*) 195
Hari Prasad Devkota, Rajan Logesh, and Niranjan Das
1 Introduction 195
2 Health-promoting and other pharmacological activities of clove extracts 197
2.1 Antimicrobial activities 197
2.2 Antioxidant activities 197
2.3 Anti-inflammatory activities 198
2.4 Analgesic activities 198
2.5 Hepatoprotective activities 198
2.6 Antidiabetic activities 199
3 Conclusions 199
References 199

CHAPTER 12 Nanocomposites using clove (*Syzygium aromaticum*) chemical constituents **203**
Nusrat M. Begum, Abdulrahman M. Elbagory, Ateeq Rahman, Ahmed A. Hussein, Natascha Cheikhyoussef, and Ahmad Cheikhyoussef
1 Introduction........ 203
2 Nanocomposites........ 204
3 Conclusions........ 206
References........ 207
Further reading 208

CHAPTER 13 Food applications of clove (*Syzygium aromaticum*) buds **209**
Rajan Logesh, Niranjan Das, and Hari Prasad Devkota
1 Introduction........ 209
2 Food applications (culinary uses)........ 209
3 Traditional medicinal uses 209
4 Other uses 211
References........ 212
Further reading 213

CHAPTER 14 Clove (*Syzygium aromaticum*) phenolics: Extraction, compositions, and biological activities **215**
Ahmad Cheikhyoussef, Natascha Cheikhyoussef, Ateeq Rahman, and Ahmed A. Hussein
1 Introduction........ 215
2 Clove phenolics extraction, structure, and composition........ 216
2.1 Clove phenolics extraction........ 216
2.2 Clove phenolics structure and composition........ 217
3 Clove phenolics biological activities 223
4 Conclusions........ 226
References........ 226

CHAPTER 15 Clove (*Syzygium aromaticum*) in poultry feed **235**
Jet Saartje Mandey
1 Introduction........ 235
2 Clove as alternative of phyto-additive 236
3 Clove in poultry diet........ 238
4 Essential oil of clove 242
5 Essential oil in diet of poultry........ 244
6 Clove as phyto-additive in drinking water........ 247

7 Phytochemicals of clove for intestinal health 247
8 Conclusion 249
References 250

CHAPTER 16 Clove (*Syzygium aromaticum*) and its derivatives in fish feed 257
Muhammad Saeed, Muhammad Sajjad Khan, Umair Younas, Faisal Siddique, Kamal Niaz, Zahid Farooq, Mayada R. Farag, Shaaban S. Elnesr, Mahmoud Alagawany, and Mohamed Fawzy Ramadan

1 Introduction 257
1.1 Pharmacological properties of clove and its derivatives 258
1.2 Chemical structure of cloves 260
1.3 Health benefits and beneficial uses of clove and its derivatives in fish 260
2 Conclusion 263
References 263

CHAPTER 17 Clove (*Syzygium aromaticum*) and eugenol toxicity 267
Zeynep Aksoylu Özbek and Pelin Günç Ergönül

1 Introduction 268
2 Bioactive compounds of clove and their toxic effects 268
2.1 Toxicity of clove against fish 282
2.2 Toxicity of clove against pests 283
3 Eugenol 286
4 Uses areas of eugenol 287
5 Toxicity of eugenol and its effects 288
6 Acute toxicity of eugenol 290
7 Subacute and sub-chronic toxicity of eugenol 293
8 Chronic toxicity of eugenol 296
9 Ecotoxicity of eugenol 299
10 Regulatory requirements 299
11 Conclusion 300
Acknowledgment 300
References 300

CHAPTER 18 Anticancer effects of the active fraction from clove in vitro and in vivo 315
Gan Qiao and Xiukun Lin

1 Introduction 315
2 Anticancer effect of liposoluble fraction of clove 315
3 The anticancer effect of the aqueous fraction of clove 320

4 Discussion 330
References 331

CHAPTER 19 Clove and its constituents against urban pests: Examples from ants and cockroaches 335
Lekhnath Kafle and Suwimol Chinkangsadarn

1 Urban pests 335
2 Control methods of urban pests 335
3 Application of clove and its constituents to control ants and cockroaches 336
4 Mortality and repellency of clove products against ants 336
4.1 Mortality 337
4.2 Repellency 338
5 Mortality and repellency of clove products against cockroaches 338
5.1 Mortality 340
5.2 Repellent 341
6 Mode of action of bioactive compounds of cloves 342
7 Conclusion 343
References 343

CHAPTER 20 Clove (*Syzygium aromaticum*) and clove extracts applications in the meat industry 347
Oscar Abel Sánchez-Velázquez, Alan Javier Hernández-Álvarez, and Guiomar Melgar-Lalanne

1 Introduction 347
2 Clove oil production 349
3 Chemical composition and mechanisms of action of clove 350
3.1 Eugenol 350
4 Quality control of meats with clove and clove eugenol 352
5 Antibacterial activity of clove and clove eugenol in meat and meat products 353
6 Antioxidant activity of clove in meat and meat products 354
7 Conclusion and perspectives 357
References 358

CHAPTER 21 Nanostructured pharmaceutical formulations for topical application of clove oil and eugenol 363
Talita Nascimento da Silva, Stephani Araujo Cardoso, and Thaís Nogueira Barradas

1 Introduction 363
2 Clove oil as an excipient in pharmaceutical formulations for topical drug delivery 364

2.1 Clove oil as a solvent for poorly soluble drugs 365
2.2 Clove oil as a penetration enhancer 366
3 Clove oil as a bioactive molecule in pharmaceutical formulations: Biological and pharmacological role 369
3.1 Antiinflammatory, antioxidant, and antinociceptive properties 369
3.2 Antimicrobial applications 372
3.3 Anticancer and antiproliferative properties 374
4 Wound healing properties 376
5 Nanostructured systems used in the encapsulation of clove oil 377
5.1 Nanoemulsions 378
6 Lipid-based nanoparticles: Liposomes, SLN, and NLC 380
7 Polymeric nanoparticles 383
8 Electrospun nanofibers 386
9 Hydrogel-based nanomaterials 387
10 Other vesicle-derived carriers 388
11 Inclusion complexes and cyclodextrins 389
12 Conclusion 390
References 390

CHAPTER 22 Extraction of bioactive compounds from clove (*Syzygium aromaticum*) 405
Muhammad Kamran Khan, Sadia Hassan, Muhammad Imran, and Muhammad Haseeb Ahmad

1 Introduction 405
2 Extraction of bioactive compounds 405
2.1 Solvent extraction (maceration) 409
2.2 Soxhlet extraction 410
2.3 Ultrasound-assisted extraction 411
2.4 Microwave-assisted extraction 412
2.5 Supercritical fluid extraction 412
3 Applications of clove extracts 413
4 Conclusion 414
References 414

SECTION 2 Clove (*Syzygium aromaticum*) fixed oil: Chemistry, functionality and applications

CHAPTER 23 Composition and functionality of clove (*Syzygium aromaticum*) fixed oil 421
Mohamed Fawzy Ramadan

1 Introduction 421
2 Fatty acid profile of CFOil 422

3 Lipid classes and subclasses of CFOil ... 422
4 Tocol profile of CFOil ... 423
5 Antiradical potential of CFOil ... 423
6 Antimicrobial activity of CFOil ... 423
7 Hepatoprotective effect of CFOil against CCl_4-induced toxicity in animals ... 423
8 Application of CFOil in meat products ... 424
9 Application of CFOil in edible oil blends ... 424
10 Conclusion ... 424
References ... 424

CHAPTER 24 Food applications of clove (*Syzygium aromaticum*) fixed oil ... 427
Muhammad Imran and Mohamed Fawzy Ramadan
1 Introduction ... 427
2 Applications of clove oil in edible oil blends ... 428
3 Applications of microencapsulated clove oil ... 428
4 Applications of clove oil in meat products ... 428
5 Clove oil, baked product, and vegetable plant ... 430
6 Applications of clove oil in edible coating ... 430
7 Conclusion ... 432
References ... 432

CHAPTER 25 Health-promoting activities of clove (*Syzygium aromaticum*) fixed oil ... 435
Zarina Mushtaq, Muhammad Imran, Muhammad Haseeb Ahmad, Ayesha Rasheed, Nosheen Asghar, Marwa Waheed, Muhammad Bilal Hussain, and Mohamed Fawzy Ramadan
1 Introduction ... 435
2 Chemical composition of clove fixed oil ... 436
3 Extraction of clove oil ... 437
4 Pharmacological activities of clove ... 438
4.1 Antioxidant/free radical scavenging activity ... 438
4.2 Anticarcinogenic activity ... 438
4.3 Antibacterial activity ... 438
4.4 Antiinflammatory activity ... 439
4.5 Antithrombotic activity ... 439
4.6 Antidiabetic activities ... 439
4.7 Other pharmacological activities ... 440
5 Toxicity doses ... 440
6 Conclusion ... 440
References ... 440

CHAPTER 26 Clove oil as bio-additives in biodiesel-diesel fuel 445
Elaine Elaine and Kar Lin Nyam

1 Introduction 445
1.1 Clove oil 445
1.2 Extraction of clove oil 446
1.3 Biodiesel-diesel fuel 447
1.4 Fuel additives 448
2 The effect of clove oil on diesel-biodiesel fuel 448
2.1 General overview 448
2.2 Fuel properties 449
2.3 Engine performance 450
2.4 Exhaust emission 452
2.5 Cost 455
2.6 Future trends 455
3 Conclusion 455
References 455

SECTION 3 Clove essential oil: Chemistry, functionality and applications

CHAPTER 27 Composition and functionality of clove (*Syzygium aromaticum*) essential oil 461
Sonia Kaushal and Rashmi

1 Introduction 461
2 Distribution 461
3 Morphology 462
4 Chemical constituents of clove essential oil 462
4.1 Clove bud essential oil 462
4.2 Clove leaves essential oil 463
4.3 Clove stem oil 463
5 The functionality of clove essential oil 463
5.1 Antibacterial activity 474
5.2 Antioxidant activity 475
5.3 Antifungal activity 476
5.4 Anti-inflammatory activity 477
5.5 Acaricidal activity 477
5.6 Leishmanicidal activity 478
5.7 Anesthetic activity 478
5.8 Anticancer activity 478
5.9 Nematicidal activity 478

5.10 Herbicidal activity 479
5.11 Larvicidal activity 479
References 479

CHAPTER 28 Effect of extraction techniques on the yield, composition, and quality of clove (*Syzygium aromaticum*) essential oil 485
Jyotimala J. Jadhav, Girirajsinh C. Jadeja, and Meghal A. Desai

1 Introduction 485
2 Methods for extraction of essential oil 486
2.1 Classical methods 486
2.2 Advanced/innovative methods 491
3 Comparative studies 495
4 Conclusion 498
References 498

CHAPTER 29 Health-promoting activities of clove (*Syzygium aromaticum*) essential oil 501
Eman T. El Kenany

1 Introduction 501
2 Health-promoting biological activities 501
2.1 Antibacterial activities 501
2.2 Antifungal activities 507
2.3 Other biological activities and applications 508
References 509
Further reading 515

CHAPTER 30 Encapsulation of clove (*Syzygium aromaticum*) essential oil 517
Onur Özdikicierler and Pelin Günç Ergönül

1 Introduction 517
2 Clove oil encapsulation methods 518
2.1 Encapsulation using spray drying or lyophilization (freeze-drying) 518
2.2 Encapsulation by emulsification 519
2.3 Encapsulation by entrapment in liposomes 521
3 Biologic activities of encapsulated clove essential oil 522
4 Potential uses of encapsulated clove essential oil in food matrices 524
5 Conclusion 526
References 527

CHAPTER 31 Supercritical carbon dioxide extraction of clove essential oil: Optimization and characterization 531
Moncef Chouaibi
1 Introduction 531
2 Supercritical SC-CO_2 apparatus 532
3 Optimization of essential oil yield 533
3.1 Effect of extraction pressure 535
3.2 Effect of extraction temperature on the essential oil yield 535
3.3 Optimal conditions of essential oil extraction 536
4 Chemical characteristics 536
5 Conclusion 538
References 538
Further reading 540

CHAPTER 32 Clove essential oil nanoemulsion: Optimization using artificial neural network 541
Moncef Chouaibi
1 Introduction 541
2 Optimization techniques 542
2.1 Response surface methodology 543
2.2 Artificial neural network 544
2.3 Adaptive neuro fuzzy inference system 548
3 Techniques of essential oil nanoemulsions 549
4 Comparison between RSM, ANN, and ANFIS 552
5 Optimal conditions of stable O/W nanoemulsion 552
6 Conclusion 553
References 553

CHAPTER 33 Preparation of oil-in-water (O/W) clove essential oil nanoemulsion: Characterization and stability 559
Moncef Chouaibi
1 Introduction 559
2 Clove oil nanoemulsions 560
3 Emulsification process 560
4 Nanoemulsion morphology 562
5 Stability studies of clove nanoemulsion 563
5.1 Effect storage period on droplet size 563
5.2 Effect storage period on PDI 565
5.3 Effect storage period on ζ-potential 565

5.4 Effect storage period on dynamic viscosity 566
5.5 Effect of storage time on antibacterial activity 566
6 Conclusion 567
Acknowledgments 568
References 568

CHAPTER 34 Storage quality of clove (*Syzygium aromaticum*) oil 573
Muhammad Imran, Muhammad Kamran Khan, Muhammad Nadeem, Muhammad Haseeb Ahmad, Rabia Shabir Ahmad, and Muhammad Abdul Rahim
1 Introduction 573
2 Extraction of clove bud essential oil 573
3 Chemistry of clove bud essential oil 575
4 Storage stability 575
5 Storage quality of food products supplemented with clove bud essential oil 575
6 Biological activities 577
7 Multidisciplinary applications 578
8 Conclusions 578
References 579

SECTION 4 Clove extracts: Chemistry, functionality and applications

CHAPTER 35 Composition and functionality of clove (*Syzygium aromaticum*) extracts 585
Kevser Karaman and Mahmut Kaplan
1 Clove (*Syzygium aromaticum*) 586
2 Extraction of clove (*Syzygium aromaticum*) 586
3 Composition of clove extracts 589
4 Functionality of clove extracts 592
4.1 Antimicrobial effect 592
4.2 Antioxidant and antiradical effect 594
4.3 Anticancer effect 595
4.4 Anti-inflammatory effect 596
4.5 Antidiabetic effect 596
4.6 Anti-stress effect 597
4.7 Effect on testicular function 597
4.8 Food practices 597
4.9 Encapsulation applications of clove extract 598
4.10 Other applications 599
References 600

CHAPTER 36 Food applications of clove (*Syzygium aromaticum*) extracts 607
Mustafa Kiralan and Onur Ketenoglu
1 Introduction 607
2 Antioxidant and antimicrobial properties of clove extracts 608
3 Food applications of clove extracts 611
3.1 Meat and meat products 611
3.2 Fish products 612
3.3 Other food products 613
References 614

CHAPTER 37 Health-promoting activities of clove (*Syzygium aromaticum*) extracts 619
A. Tomas, A. Maroyi, Natascha Cheikhyoussef, Ahmed A. Hussein, and Ahmad Cheikhyoussef
1 Introduction 619
2 Ethnobotanical knowledge and traditional uses 620
3 Biological activities 620
3.1 Antioxidant activity 621
3.2 Antibacterial activity 621
3.3 Antiviral activity 622
3.4 Anti-inflammatory activity 623
3.5 Antifungal activity 623
3.6 Anticancer activity 624
3.7 Antimycobacterial and antiplasmodial activities 625
3.8 Antidiabetic activity 625
3.9 Anticonvulsant activity 626
3.10 Hepatoprotective activity 626
3.11 Skin protective effect 627
3.12 Neuroprotection effect 627
3.13 Immunomodulation effects 628
4 Other health benefits 628
5 Conclusions 629
References 629

CHAPTER 38 Clove (*Syzygium aromaticum*)-mediated metallic nanoparticles: Synthesis, characterization, and possible pharmacological and industrial applications 639
Abdulrahman M. Elbagory, Ateeq Rahman, Natascha Cheikhyoussef, Ahmad Cheikhyoussef, Nusrat M. Begum, and Ahmed A. Hussein
1 Introduction 640
2 Role of clove phytochemicals in the synthesis 641

2.1 AgNPs 646
2.2 AuNPs 649
2.3 FeNPs 650
2.4 PdNPs 651
2.5 ZnSNPs/ZnONPs 653
2.6 Other mono-MNPs 653
2.7 Multi-MNPs 654
3 Conclusions 655
References 655

CHAPTER 39 Toxicity of clove (*Syzygium aromaticum*) extract 663
M. Joyce Nirmala, P.J. Shiny, Uppada Sumanth Raj, Narasa Saikrishna, and R. Nagarajan
1 Introduction 663
1.1 A brief description of clove 663
1.2 Biological properties of the clove extract 664
2 Extraction types and active compounds identified 665
2.1 Aqueous extract of clove flower buds 665
2.2 Solvent extraction 665
2.3 Supercritical extraction 666
2.4 Identification of chemical compounds 667
3 Toxicity studies 667
3.1 Toxic effect of clove extract at cellular level 667
3.2 Toxic effect of clove extract on reproduction in mice 669
3.3 Sub-chronic assessment of Wistar rats 669
3.4 Toxic effect of clove extract on aquatic species 670
3.5 Other toxic effects 671
4 Conclusion 672
Acknowledgments 672
References 672

CHAPTER 40 Antiviral properties of clove (*Syzygium aromaticum*) 675
Raya Soltane, Mtat Dalila, Abdelwaheb Dhahri, Ahlam Alasiri, Rami Adel Pashameah, Nuha Almulla, Amani Chrouda, and Ahmed Mostafa
1 Introduction 675
2 Antiherpes simplex virus (HSV) activity of *S. aromaticum* 676
3 Antiviral activity of green silver nanoparticles produced using *Syzygium aromaticum* aqueous buds extract 677

4 Antiviral activity of eugenol ... 677
4.1 *In vitro* and *in vivo* antiviral activity of eugenol against human herpesvirus ... 677
4.2 Eugenol inhibitor of the Ebola virus *in vitro* ... 678
4.3 Inhibition of dengue virus protease by eugenol ... 678
5 Antiviral activities of clove oil ... 678
6 Antiviral effects of *Syzygium aromaticum* against severe acute respiratory syndrome coronavirus 2 (SARS-CoV-2) ... 679
7 Conclusion ... 680
References ... 680

Index ... 683

Contributors

Muhammad Haseeb Ahmad
Department of Food Science; Institute of Home and Food Sciences, Faculty of Life Sciences, Government College University, Faisalabad, Punjab, Pakistan

Rabia Shabir Ahmad
Department of Food Science, Faculty of Life Sciences, Government College University, Faisalabad, Punjab, Pakistan

Zeynep Aksoylu Özbek
Department of Food Engineering, Faculty of Engineering, Manisa Celal Bayar University, Yunusemre, Manisa, Turkey

Mahmoud Alagawany
Poultry Production Department, Faculty of Agriculture, Fayoum University, Fayoum, Egypt

Ahlam Alasiri
Department of Basic Sciences, Adham University College, Umm Al-Qura University, Makkah, Saudi Arabia

Nuha Almulla
Department of Basic Sciences, Adham University College, Umm Al-Qura University, Makkah, Saudi Arabia

Nosheen Asghar
Institute of Home and Food Sciences, Faculty of Life Sciences, Government College University, Faisalabad, Punjab, Pakistan

S. Athira
Kerala Veterinary and Animal Sciences University, Pookode, Kerala, India

Thaís Nogueira Barradas
Department of Pharmaceutical Sciences, Federal University of Juiz de Fora, Juiz de Fora, Brazil

Nusrat M. Begum
Department of Environmental Health, Water and Health Research Centre, University of Johannesburg, Doornfontein, South Africa

Stephani Araujo Cardoso
Postgraduate Program in Nanobiossystems, Federal University of Rio de Janeiro, University City, Rio de Janeiro, Brazil

Tossaton Charoonratana
Department of Pharmacognosy, College of Pharmacy, Rangsit University, Lak-Hok, Mueang Pathum Thani, Thailand

Debanjan Chatterjee
Department of Natural Products, National Institute of Pharmaceutical Education and Research, Ahmedabad, Gandhinagar, Gujarat, India

Ahmad Cheikhyoussef
Science and Technology Division, Multidisciplinary Research Unit, Centre for Research Services, University of Namibia, Windhoek, Namibia

Natascha Cheikhyoussef
Ministry of Higher Education, Technology and Innovation, Windhoek, Namibia

Suwimol Chinkangsadarn
Department of Tropical Agriculture and International Cooperation, National Pingtung University of Science and Technology, Pingtung, Taiwan, ROC; Faculty of Animal Sciences and Agricultural Technology, Silpakorn University, Cha-Am, Phetchaburi, Thailand

Moncef Chouaibi
Higher School of Food Industries of Tunisia, Food Engineering Department, University of Carthage, Tunis, Tunisia

Amani Chrouda
Department of Chemistry, College of Science Al-Zulfi, Majmaah University, Al-Majmaah, Saudi Arabia

Talita Nascimento da Silva
Postgraduate Program in Pharmaceutical Sciences, Federal University of Rio de Janeiro, University City, Rio de Janeiro, Brazil

Mtat Dalila
Department of Chemistry, Preparatory Institute for Engineering Studies of Nabeul (IPEIN), University of Carthage, Tunis, Tunisia

Niranjan Das
Department of Chemistry, Iswar Chandra Vidyasagar College, Belonia, Tripura, India

Meghal A. Desai
Department of Chemical Engineering, Sardar Vallabhbhai National Institute of Technology, Surat, Gujarat, India

Hari Prasad Devkota
Graduate School of Pharmaceutical Sciences, Kumamoto University, Kumamoto, Japan

Akash Dey
Department of Pharmaceutical Chemistry, Dr. B.C. Roy College of Pharmacy and Allied Health Sciences, Durgapur, West Bengal, India

Abdelwaheb Dhahri
Department of Polymer Materials Engineering, University of Lyon, UMR CNRS 5223, Lyon, Villeurbanne, France

Eman T. El Kenany
Botany and Microbiology Department, Faculty of Science, Alexandria University; Oral Biology Department, Faculty of Dentistry, Pharos University in Alexandria, Alexandria, Egypt

Elaine Elaine
Faculty of Applied Sciences, UCSI Heights, UCSI University, Kuala Lumpur, Malaysia

Abdulrahman M. Elbagory
Chemistry Department, Cape Peninsula University of Technology, Cape Town, Western Cape, South Africa

Shaaban S. Elnesr
Poultry Production Department, Faculty of Agriculture, Fayoum University, Fayoum, Egypt

Mayada R. Farag
Forensic Medicine and Toxicology Department, Veterinary Medicine Faculty, Zagazig University, Zagazig, Egypt

Jabeen Farhana
Faculty of Pharmacy, BRIG, Hyderabad, Telangana, India

Zahid Farooq
Cholistan University of Veterinary and Animal Sciences Bahawalpur, Bahawalpur, Pakistan

Pelin Günç Ergönül
Department of Food Engineering, Faculty of Engineering, Manisa Celal Bayar University, Yunusemre, Manisa, Turkey

Sadia Hassan
Department of Nutritional Sciences, Faculty of Science and Technology, Government College Women University, Faisalabad, Pakistan

Alan Javier Hernández-Álvarez
School of Food Science and Nutrition, University of Leeds, Leeds, United Kingdom

Muhammad Bilal Hussain
Institute of Home and Food Sciences, Faculty of Life Sciences, Government College University, Faisalabad, Punjab, Pakistan

Ahmed A. Hussein
Chemistry Department, Cape Peninsula University of Technology, Cape Town, Western Cape, South Africa

Muhammad Imran
Department of Food Science; Institute of Home and Food Sciences, Faculty of Life Sciences, Government College University, Faisalabad, Punjab, Pakistan

Girirajsinh C. Jadeja
Department of Chemical Engineering, Sardar Vallabhbhai National Institute of Technology, Surat, Gujarat, India

Jyotimala J. Jadhav
Department of Chemical Engineering, Sardar Vallabhbhai National Institute of Technology, Surat, Gujarat, India

Lekhnath Kafle
Department of Tropical Agriculture and International Cooperation, National Pingtung University of Science and Technology, Pingtung, Taiwan, ROC

Mahmut Kaplan
Faculty of Agriculture, Department of Field Crops, Erciyes University, Kayseri, Turkey

Kevser Karaman
Faculty of Agriculture, Department of Agricultural Biotechnology, Erciyes University, Kayseri, Turkey

Sonia Kaushal
Department of Chemistry, Punjab Agricultural University, Ludhiana, Punjab, India

Onur Ketenoglu
Faculty of Agriculture, Department of Food Engineering, Eskisehir Osmangazi University, Eskisehir, Turkey

Muhammad Kamran Khan
Department of Food Science, Faculty of Life Sciences, Government College University, Faisalabad, Punjab, Pakistan

Muhammad Sajjad Khan
Cholistan University of Veterinary and Animal Sciences Bahawalpur, Bahawalpur, Pakistan

Salman Tariq Khan
Pakistan Council of Scientific and Industrial Research, Laboratories Complex, Karachi, Pakistan

Mustafa Kiralan
Faculty of Engineering, Department of Food Engineering, Balikesir University, Balikesir, Turkey

Rajesh Kumar
ICAR-National Dairy Research Institute, Karnal, Haryana, India

Abdul Hafeez Laghari
Pakistan Council of Scientific and Industrial Research, Laboratories Complex, Karachi, Pakistan

Xiukun Lin
Delisi Group, Weifang, Shandong, China

Rajan Logesh
TIFAC-CORE in Herbal Drugs, Department of Pharmacognosy and Phytopharmacy, JSS College of Pharmacy (JSS Academy of Higher Education and Research), Rockland's, Ooty, Tamil Nadu, India

Sudip Kumar Mandal
Department of Pharmaceutical Chemistry, Dr. B.C. Roy College of Pharmacy and Allied Health Sciences, Durgapur, West Bengal, India

Jet Saartje Mandey
Department of Animal Nutrition, Animal Husbandry Faculty, University of Sam Ratulangi, Manado, North Sulawesi, Indonesia

Bimlesh Mann
ICAR-National Dairy Research Institute, Karnal, Haryana, India

A. Maroyi
Medicinal Plants and Economic Development (MPED) Research Centre, Department of Botany, University of Fort Hare, Alice, South Africa

Guiomar Melgar-Lalanne
Instituto de Ciencias Básicas, Universidad Veracruzana, Xalapa, Veracruz, Mexico

Ahmed Mostafa
Center of Scientific Excellence for Influenza Viruses, Water Pollution Research Department, Environmental and Climate Change Research Institute, National Research Centre, Giza, Egypt

Zarina Mushtaq
Institute of Home and Food Sciences, Faculty of Life Sciences, Government College University, Faisalabad, Punjab, Pakistan

Muhammad Nadeem
Department of Dairy Technology, University of Veterinary and Animal Sciences, Lahore, Punjab, Pakistan

R. Nagarajan
Department of Chemical Engineering, Indian Institute of Technology Madras, Chennai, Tamil Nadu, India

Kamal Niaz
Cholistan University of Veterinary and Animal Sciences Bahawalpur, Bahawalpur, Pakistan

M. Joyce Nirmala
Department of Chemical Engineering, Indian Institute of Technology Madras, Chennai, Tamil Nadu, India

Kar Lin Nyam
Faculty of Applied Sciences, UCSI Heights, UCSI University, Kuala Lumpur, Malaysia

Onur Özdikicierler
Faculty of Engineering, Food Engineering Department, Ege University, İzmir, Turkey

Rami Adel Pashameah
Department of Basic Sciences, Adham University College, Umm Al-Qura University, Makkah, Saudi Arabia

Gan Qiao
Central Nervous System Drug Key Laboratory of Sichuan Province, School of Pharmacy, Southwest Medical University, Luzhou, China

Muhammad Abdul Rahim
Department of Food Science, Faculty of Life Sciences, Government College University, Faisalabad, Punjab, Pakistan

Ateeq Rahman
Department of Physics, Chemistry & Material Science, Faculty of Agriculture, Engineering & Natural Sciences; Department of Chemistry and Biochemistry, Faculty of Science, University of Namibia, Windhoek, Namibia

Uppada Sumanth Raj
Department of Chemical Engineering, Indian Institute of Technology Madras, Chennai, Tamil Nadu, India

Mohamed Fawzy Ramadan
Department of Clinical Nutrition, Faculty of Applied Medical Sciences, Umm Al-Qura University, Makkah, Saudi Arabia

Ayesha Rasheed
Institute of Home and Food Sciences, Faculty of Life Sciences, Government College University, Faisalabad, Punjab, Pakistan

Rashmi
Department of Chemistry, Punjab Agricultural University, Ludhiana, Punjab, India

Muhammad Saeed
Cholistan University of Veterinary and Animal Sciences Bahawalpur, Bahawalpur, Pakistan

Narasa Saikrishna
Department of Chemical Engineering, Indian Institute of Technology Madras, Chennai, Tamil Nadu, India

Oscar Abel Sánchez-Velázquez
Facultad de Ciencias Químico-Biológicas, Universidad Autónoma de Sinaloa, Culiacán, Sinaloa, Mexico

Rajan Sharma
ICAR-National Dairy Research Institute, Karnal, Haryana, India

P.J. Shiny
Department of Chemical Engineering, Indian Institute of Technology Madras, Chennai, Tamil Nadu, India

Faisal Siddique
Cholistan University of Veterinary and Animal Sciences Bahawalpur, Bahawalpur, Pakistan

Richa Singh
ICAR-National Dairy Research Institute, Karnal, Haryana, India

Raya Soltane
Department of Basic Sciences, Adham University College, Umm Al-Qura University, Makkah, Saudi Arabia; Department of Biology, Faculty of Sciences, Tunis El Manar University, Tunis, Tunisia

Mohamad Taleuzzaman
Department of Pharmaceutical Chemistry, Faculty of Pharmacy, Maulana Azad University, Jodhpur, Rajasthan, India

A. Tomas
Department of Biochemistry, Microbiology, and Biotechnology, Faculty of Agriculture, Engineering & Natural Sciences, University of Namibia, Windhoek, Namibia

Rishabh Verma
Department of Pharmaceutics, School of Pharmaceutical Education and Research, Jamia Hamdard, New Delhi, India

Marwa Waheed
Institute of Home and Food Sciences, Faculty of Life Sciences, Government College University, Faisalabad, Punjab, Pakistan

Umair Younas
Cholistan University of Veterinary and Animal Sciences Bahawalpur, Bahawalpur, Pakistan

About the editor

Mohamed Fawzy Ramadan is Professor of Biochemistry and Food Chemistry at Umm Al-Qura University, Makkah, Saudi Arabia. He is a consultant of international publishing at the Deanship of Scientific Research (Umm Al-Qura University). Since 2014, Ramadan has been Professor in the Agricultural Biochemistry Department, Faculty of Agriculture at Zagazig University, Egypt.

Prof. Ramadan obtained his PhD (Dr. rer. nat.) in food chemistry from the Berlin University of Technology, Germany (2004). He continued his postdoctoral research in ranked universities such as the University of Helsinki, Finland; the Max-Rubner Institute, Germany; the Berlin University of Technology, Germany; and the University of Maryland, United States. In 2010, he was appointed as Visiting Professor (100% research) at King Saud University. In 2012, he was appointed as Visiting Professor (100% teaching) in the School of Biomedicine, Far Eastern Federal University in Vladivostok, Russian Federation.

Prof. Ramadan has published more than 300 research papers and reviews in international peer-reviewed journals. He has edited and published several books and book chapters (with a Scopus *h*-index of 42 and more than 5500 citations). In addition, he has been an invited speaker at several international conferences. Since 2003, Prof. Ramadan has been a reviewer and an editor of several highly cited international journals such as the *Journal of Medicinal Food* and the *Journal of Advanced Research*.

Prof. Ramadan has received several prizes, including the Abdul Hamid Shoman Prize for Young Arab Researcher in agricultural sciences (2006); the Egyptian State Prize for encouragement in agricultural sciences (2009); the European Young Lipid Scientist Award (2009); the AU-TWAS Young Scientist National Awards (Egypt) in basic sciences, technology, and innovation (2012); the TWAS-ARO Young Arab Scientist (YAS) Prize in scientific and technological achievement (2013); and the Atta-ur-Rahman Prize in chemistry (2014).

Preface

Aromatic and medical plants have been globally used as nutraceuticals and pharmaceuticals due to their biosafety. According to the World Health Organization, about 80% of the world's population depends on conventional medicine, which uses plant extracts or pure bioactive phytochemicals to treat diseases. Thus, there is great interest in novel medical plants as phytoconstituent-rich sources for functional food and drugs.

Syzygium aromaticum (clove, family Myrtaceae) is a traditional medicinal plant that has been used in food preservation and exhibits pharmacological effects. Clove buds contain up to 18% essential oil (CEO), while CEO composition and yield differences depend on agroecological conditions, pretreatments, and extraction methods. Eugenol ($C_{10}H_{12}O_2$), eugenyl acetate, carvacrol, and β-caryophyllene are the main CEO phytochemicals used to treat wounds and burns and as pain relievers in dental care. CEO is also used in nonfood applications (i.e., perfumes and soaps). Studies report CEO and eugenol's antifungal, antiviral, antimicrobial, antioxidant, anticancer, and anti-inflammatory traits. CEO also exhibits antipyretic, appetizer, aphrodisiac, anxiolytic, hypnotic, antiemetic, decongestant, and antiepileptic properties. Moreover, *in vitro* tests have revealed the significant antiradical effect of CEO. When CEO compared with common antioxidants, the antioxidant potential declined as follows: CEO > BHT > α-tocopherol. On the other hand, cold-pressed clove oil is a rich source of bioactive lipids, including essential fatty acids, polar lipids, and tocols. The oil exhibits high antiradical and antimicrobial properties. It could be blended with common vegetable oils to produce healthy oil blends with health-promoting properties.

Clove bud and leaf extracts are rich sources of phytochemicals that show high antioxidant, antipiroplasmic, antimicrobial, and hepatoprotective efficacy. *S. aromaticum* aqueous extracts show antiviral impact against herpes simplex virus type 1 and influenza A virus. Eugeniin, a bioactive compound in *S. aromaticum* extracts, exhibits antiviral potential toward herpes viruses and hepatitis C virus.

Given the numerous reports on the functional, nutritional, and pharmacological traits of clove, this book aims to create a multidisciplinary forum for discussion on the phytochemical profile and biological activities as well as the food and nonfood applications of clove buds, clove oils, clove bioactive compounds, and clove extracts. The outcomes will explore their beneficial uses for developing pharmaceuticals, nutraceuticals, novel food, natural drugs, and feed. The book contains chapters focusing on the following sections:

Section 1. Clove (*Syzygium aromaticum*): Cultivation, composition and applications
Section 2. Clove (*Syzygium aromaticum*) fixed oil: Chemistry, functionality and applications
Section 3. Clove essential oil: Chemistry, functionality and applications
Section 4. Clove extracts: Chemistry, functionality and applications

Intending to provide reference work for those of the scientific community involved in horticulture, food science, nutrition, health, and pharmacology as well as for undergraduate and graduate students, this book presents a comprehensive review of the aspects and results that have led to the advancements in clove chemistry, applications, and functionality. I hope that the book will be a valuable resource for people involved in related disciplines.

I sincerely thank all the authors and contributors for their valuable contributions and their cooperation during the preparation of the book. In addition, the Elsevier staff's help and support, especially that of Megan Ball and Emerald Li, was essential for completing the task and is highly appreciated.

Mohamed Fawzy Ramadan
Makkah, Saudi Arabia

Description

Clove (*Syzygium aromaticum*, family Myrtaceae) is a medicinal plant that has been used as a spice and possesses pharmacological traits. Clove is used for medicinal purposes, in the perfume industries and as food preservatives due to its antioxidant and antimicrobial potential. Studies on the antioxidant traits revealed that clove was the most effective compared to other common spices and herbs.

Clove is rich in phytoconstituents, including monoterpenes, sesquiterpenes, hydrocarbons, and phenolics. Clove has been reported as a rich source of phenolics such as eugenol, flavonoids (i.e., quercetin and kaempferol), phenolic acids, hydroxyphenyl propane, and gallic acid derivatives. The Food and Drug Administration confirmed the safety of clove oils, clove buds, eugenol, and oleoresins as food ingredients. Besides, dentistry studies reported that eugenol has an allergic effect.

Clove (Syzygium aromaticum): Chemistry, Functionality, and Applications addresses the cultivation, composition, and applications of clove, along with the chemistry, functionality, and applications of clove fixed oil; clove essential oil; and clove extracts as well as their role in food, health, and medicine. Specifically, the book delves into the functional, nutritional, and pharmacological traits of clove by demonstrating the phytochemical profile, biological activities, and food and nonfood applications of clove buds, clove oils, clove bioactive compounds, and clove extracts. This reference will be of use to food scientists, technologists, chemists, nutritionists, and pharmacists in developing new pharmaceuticals and novel food products.

Key features

- Explores the chemistry and functionality of clove buds, clove oils, and clove extracts.
- Discusses clove bioactive phytochemicals and their health-promoting potential.
- Presents the functional applications of clove buds, clove oils, and clove extracts.
- Includes literature reviews and coverage of recent developments.
- Authored by international scientists and industry experts.
- Addresses the growing application areas, including novel food, pharmaceuticals, nutraceuticals, and cosmetics.

Readership

- Researchers and students with an interest in the area (phytochemistry, food chemistry, pharmacology, food technology, dentistry, economy, and horticulture).
- Developers of functional food, pharmaceuticals, cosmetics, and R&D managers working in sectors that use natural products.

CHAPTER

Introduction to clove: Chemistry, functionality, and techno-applications

1

Mohamed Fawzy Ramadan
Department of Clinical Nutrition, Faculty of Applied Medical Sciences, Umm Al-Qura University, Makkah, Saudi Arabia

1. United Nations sustainable development goals and applications of medical plants

The United Nations Sustainable Development Goals (UNSDG) offer a sustainable world vision in 2015 (https://sustainabledevelopment.un.org). Regarding food, the way it is cultivated, manufactured, stored, marketed, transported, and consumed is where the essential link between people and the path to a sustainable economy is found. The good health and well-being goal is aimed at enhancing a healthy life, and is linked to the use of health-enhancing medicinal plants and environmentally friendly methods in food manufacturing (Ramadan, 2021).

Food chemists are searching for novel spices and herbs with health-enhancing properties that might be designed to enhance health. The recent investigations being carried out will be essential to the way we eat (McClements, 2019; Ramadan, 2021). Novel spices, herbs, and medicinal plants have been used to formulate pharmaceuticals and nutraceuticals. Recently, the world's attention has been directed to the outbreak of COVID-19. Vicidomini et al. (2021) discussed the role of *S. aromaticum* in the framework of COVID-19 therapy, focusing on the antiviral, antithrombotic, and antiinflammatory impacts of *S. aromaticum* and its bioactive compounds. The World Health Organization (WHO) reported that about 80% of the global population depends on traditional medicine, which uses spices, herbs, medicinal plant extracts, and phytochemicals to treat diseases. With the developments in nutraceuticals, there is great interest in medicinal plants and spices as a rich source for phytochemicals, nutraceuticals, functional foods, and drugs. The extraction of specialty oils, phytoconstituents, and bioactive constituents has increased due to their fruitful roles. Recent research has studied bioactive constituents and therapeutic effects as well as the mode of action and toxicological impacts of phytoextracts, oils, and bioactive constituents (Ramadan and Moersel, 2003, 2004; Ramadan, 2013, 2020; Hassanien et al., 2015; Carvalho et al., 2021). The WHO is giving importance to identifying bioactive phytochemicals and exploring medicinal plants for health benefits. Emphasis has been given to the safety, efficacy, quality control, quality assurance, toxicity, clinical trials, dosage, therapeutic uses, and drug interactions.

Clove (*Syzygium aromaticum*). http://doi.org/10.1016/B978-0-323-85177-0.00003-3

1.1 Clove (*S. aromaticum*): Chemistry, functionality, and techno-applications

Medical and aromatic plants have been widely used as pharmaceuticals and nutraceuticals due to their biosafety. There is great interest in novel medical plants as phytochemical-rich sources for functional food and drugs. *S. aromaticum* (family Myrtaceae) is a traditional medicinal plant that has been used as a spice in food preservation and possesses pharmacological traits. Clove is a flower bud native to the Maluku islands in Indonesia and is currently farmed in different parts of the world. *S. aromaticum* is used for medicinal purposes, in the perfume industries, and as a food preservative due to its antioxidant and antimicrobial potential. Studies on the antioxidant traits revealed that clove was the most effective when compared to other common spices and herbs (Ramadan et al., 2013; El-Hadary and Ramadan, 2016; Abo El-Maati et al., 2016; Osman et al., 2019; Saeed et al., 2021).

Clove is rich in phytoconstituents, including monoterpenes, sesquiterpenes, hydrocarbons, and phenolics. Clove has been reported as a rich source of phenolics such as eugenol, flavonoids (i.e., quercetin and kaempferol), phenolic acids, hydroxybenzoic acids, hydroxyphenyl propens, and gallic acid derivatives. The US Food and Drug Administration (FDA) confirmed the safety of clove buds, clove oil, eugenol, and clove oleoresins as a food ingredient. Some dentistry studies reported that eugenol has allergic effects. The WHO stated that the acceptable daily amount of clove for humans is 2.5 mg/kg bodyweight (El-Hadary and Ramadan, 2016; Abo El-Maati et al., 2016; El-Saber Batiha et al., 2020; Saeed et al., 2021).

Cloves reduced in vivo tissue injury in the liver, lens, and cardiac muscles in experimental animals. Cloves are used to treat vomiting, nausea, flatulence, and liver, bowel, and stomach disorders as well as provide stimulants for nerves. Clove inhibited foodborne pathogens and was used to treat candida, viruses, worms, and bacterial infections. *S. aromaticum* exhibited activity toward hepatitis C viruses. Clove also possessed a boosting effect on nerve stimulation and sexual behavior in mice (El-Hadary and Ramadan, 2016; El-Saber Batiha et al., 2020).

Clove buds contain up to 18% essential oil (CEO) while differences in composition and yield depend on agro-ecological conditions, pretreatments, and extraction methods. Eugenol, eugenyl acetate, carvacrol, and β-caryophyllene are the main CEO phytochemicals used to treat wounds and burns as pain relievers in dental care. CEO is also used in industrial applications (i.e., perfumes and soaps). Studies reported the antifungal, antimicrobial, antiviral, antioxidant, anticancer, and antiinflammatory traits of CEO and eugenol. CEO is used to treat acne, asthma, allergy disorders, and rheumatoid arthritis; it also showed acaricidal and antispasmodic impacts. CEO also exhibited antipyretic, appetite, aphrodisiac, anxiolytic, hypnotic, antiemetic, decongestant, and antiepileptic activities. Moreover, in vitro tests revealed the significant antiradical activity of CEO. When CEO is compared with other common antioxidants, the antioxidant effect decreased as follows: CEO > butylated hydroxytoluene (BHT) > α-tocopherol > butylated hydroxyanisole (BHA) > Trolox. On the other hand, cold-pressed *S. aromaticum* oil is a rich source of bioactive lipids, including fatty acids, polar lipids, and tocols. The oil exhibited high antiradical and antimicrobial activities. The oil could be blended with common vegetable oils to produce healthy blends with health-promoting potential (Ramadan, 2013; Ramadan et al., 2013; Osman et al., 2019).

Eugenol ($C_{10}H_{12}O_2$), the main bioactive compound in CEO, has been utilized in dentistry because it can penetrate dental tissues and enter the blood. Eugenol donates hydrogen and fixes the phenoxyl radicals. Studies reported the antiseptic, analgesic, anticancer, antidepressant, antiviral, antispasmodic, antifungal, antiinflammatory, and hepatoprotective traits of eugenol (Carvalho et al., 2021).

Clove buds and leaf extracts are rich sources of bioactive phytochemicals (i.e., biflorin, kaempferol, myricetin, rhamnocitrin, oleanolic acid, gallic acid, and ellagic acid). *S. aromaticum* ethanol and methanol extracts showed high antioxidant, antipiroplasmic, antimicrobial, and hepatoprotective efficacy. Tannins, gallic acid, ellagic acid, and flavonoids from alcohol and water bud extracts exhibited antithrombotic, antimicrobial, antiobesity, hypoglycemic, antiprotozoal, antiinflammatory, and gastroprotective effects. *S. aromaticum* aqueous extract showed an antiviral impact against influenza A virus and herpes simplex virus type 1. Eugeniin, a bioactive compound in *S. aromaticum* extracts, also exhibited antiviral potential toward herpes viruses and hepatitis C virus (Abo El-Maati et al., 2016; Osman et al., 2019; Saeed et al., 2021).

1.1.1 Clove (S. aromaticum) market

Clove is globally traded. In the 14th century, the clove trade acted as a stimulant in the establishment of commerce at Asian and European ports. During the following centuries, cloves were traded through phases such as increased prices, warfare, decreased prices, and smuggling seedlings for planting (Zheng et al., 1992; Bulbeck et al., 1998; Chomchalow, 1996; Kamatou et al., 2012).

Indonesia, Malaysia, India, and Sri Lanka are the main clove producers, and high quantities are produced in Madagascar, the West Indies, and Zanzibar (Kamatou et al., 2012). Bud harvesting is done using hand-picking, which is labor-intensive; the bud dry weight is one-third the fresh weight (Chomchalow, 1996). It was estimated in 2001 that about 2000t of clove oils are annually manufactured. In 2006, a gross market value of $30–70 million per year was estimated through clove use as a fragrance ingredient and flavor (Bohnert et al., 2008; Kamatou et al., 2012).

Increasing demand for nutraceuticals and pharmaceuticals rich in bioactive constituents owing to their health impacts is likely to increase the *S. aromaticum* market size (www.gminsights.com). The clove trade and the search for valuable spices have stimulated economic development (Cortés-Rojas et al., 2014).

Fig. 1 presents global export and import quantities (tonnes) from 2015 to 2019.

Clove (*S. aromaticum*) greatly attracts international scientific research. A survey and scientific literature search with the keywords "clove and *S. aromaticum*" in the database PubMed (May 2021) revealed 1500 published documents about *S. aromaticum* bioactive phytoextracts, oils, fatty acids, bioactive constituents, and techno-applications. When *S. aromaticum* was searched for in the Web of Science (through May 2021), almost 600 documents were found, including articles (540), proceedings papers (30), and review articles (17).

A careful search on *S. aromaticum* in Scopus (www.scopus.com) showed that the number of documents is high (approx. 2000 through May 2021).

Fig. 2 presents the document counts on *Syzygium aromaticum* from 2000 to 2020. The documents annually published on *Opuntia* significantly increased from 48 contributions in 2000 to 328 contributions in 2020. This measurable indicator reflects the importance and recent interest in *S. aromaticum* as a topic in the scientific community.

Fig. 3 presents the distribution of types of documents on *Syzygium aromaticum* from 2000 to 2020, which includes research articles (1856), conference papers (83), reviews (70), and book chapters (9). The documents are related to the subject fields of agricultural and biological science (35%), biochemistry, genetics and molecular biology (12%), pharmacology, toxicology and pharmaceutics (10%),

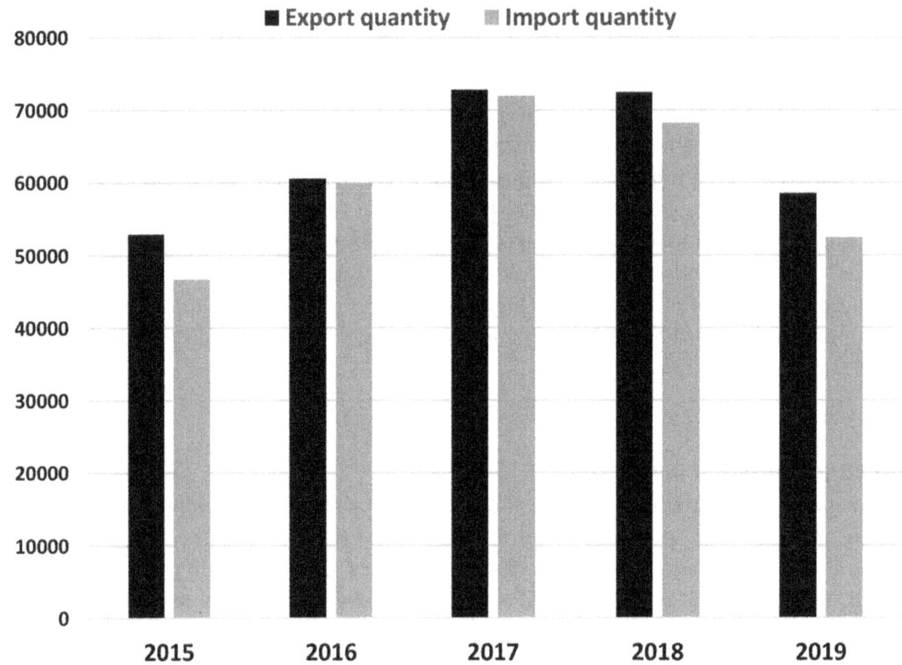

FIG. 1

Global export and import quantities (tonnes) of clove (*Syzygium aromaticum*) from 2015 to 2019 (FAOSTAT 2021).

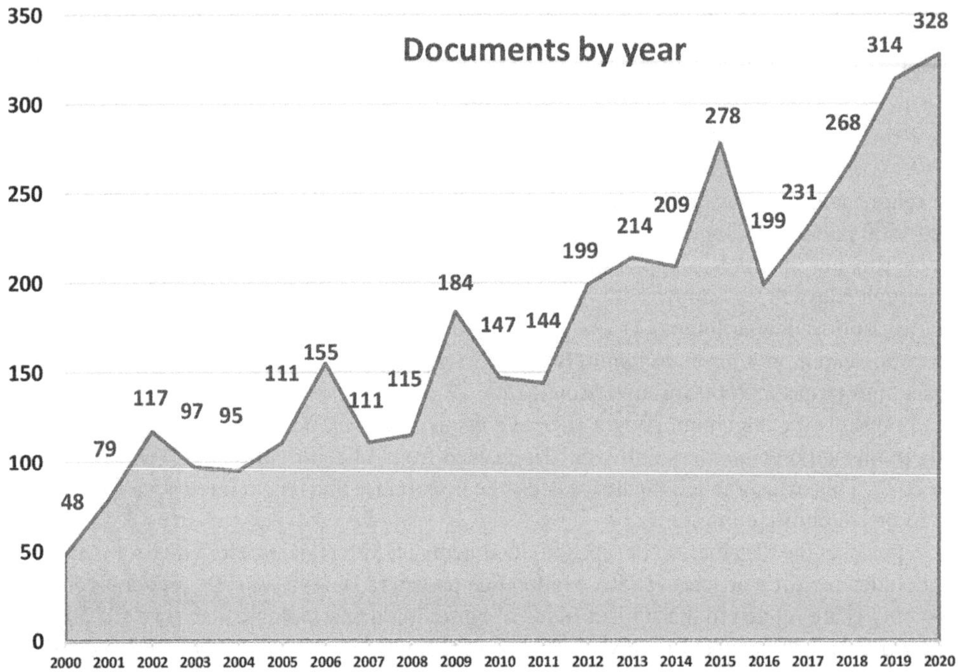

FIG. 2

Scholarly output on *Syzygium aromaticum* from 2000 to 2020 (www.scopus.com).

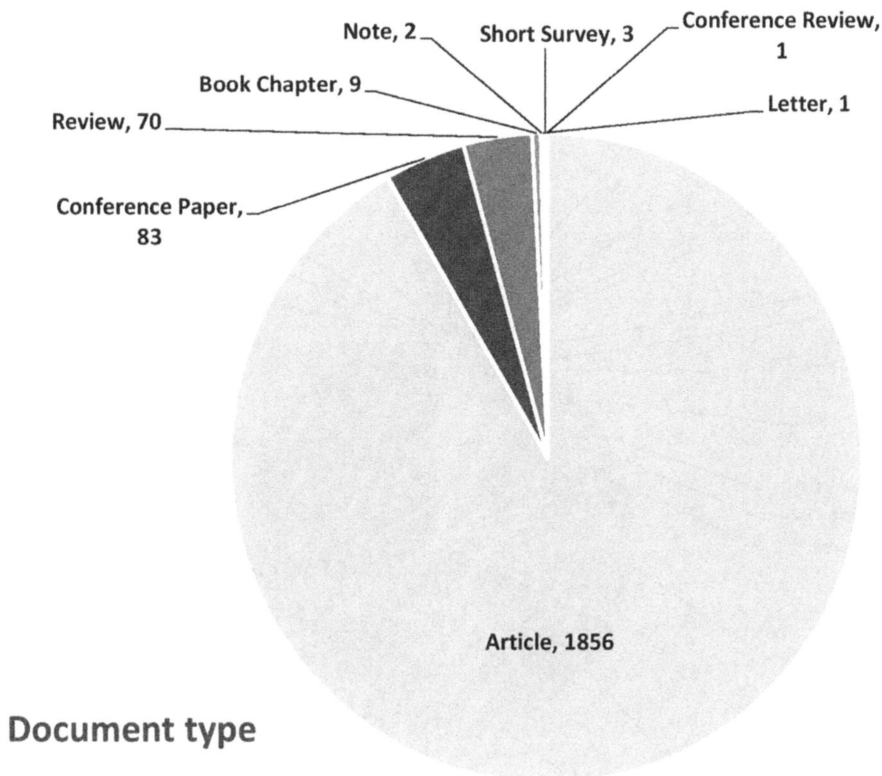

FIG. 3

Distribution by types of documents on *Syzygium aromaticum* from 2000 to 2020 (www.scopus.com).

medicine (9%), chemistry (8%), and immunology and microbiology (6%) (Fig. 4). Scientists from India, the United States, Brazil, Egypt, South Korea, Iran, China, Spain, Japan, and Indonesia were the principal authors (Fig. 5). Acta Horticulturae, the Journal of Agricultural and Food Chemistry, Aquaculture, Aquaculture Research, Food Chemistry, the Journal of Ethnopharmacology, Industrial Crops and Products, LWT-Food Science and Technology, Phytotherapy Research, and the Journal of Food Science and Technology are the leading journals that published scientific research on the clove (*S. aromaticum*).

Many scientific books have been published on the structural composition and functional traits of herbs, spices, and aromatic plants. It is difficult to find a scientific book focused on the agricultural practices, structural profile, and functional traits of clove. This project has chapters regarding clove that contain discussion on the unique phytochemicals that led to it being considered a health-enhancing plant. The goal of editing this book was to discuss the phytochemical profile, structural composition, therapeutic traits, and functional properties of phytochemicals, high-value oils, nutrients, extracts, and volatiles of clove (*S. aromaticum*) as well as to explore their beneficial uses in pharmaceuticals, nutraceuticals, novel food, natural drugs, and feed.

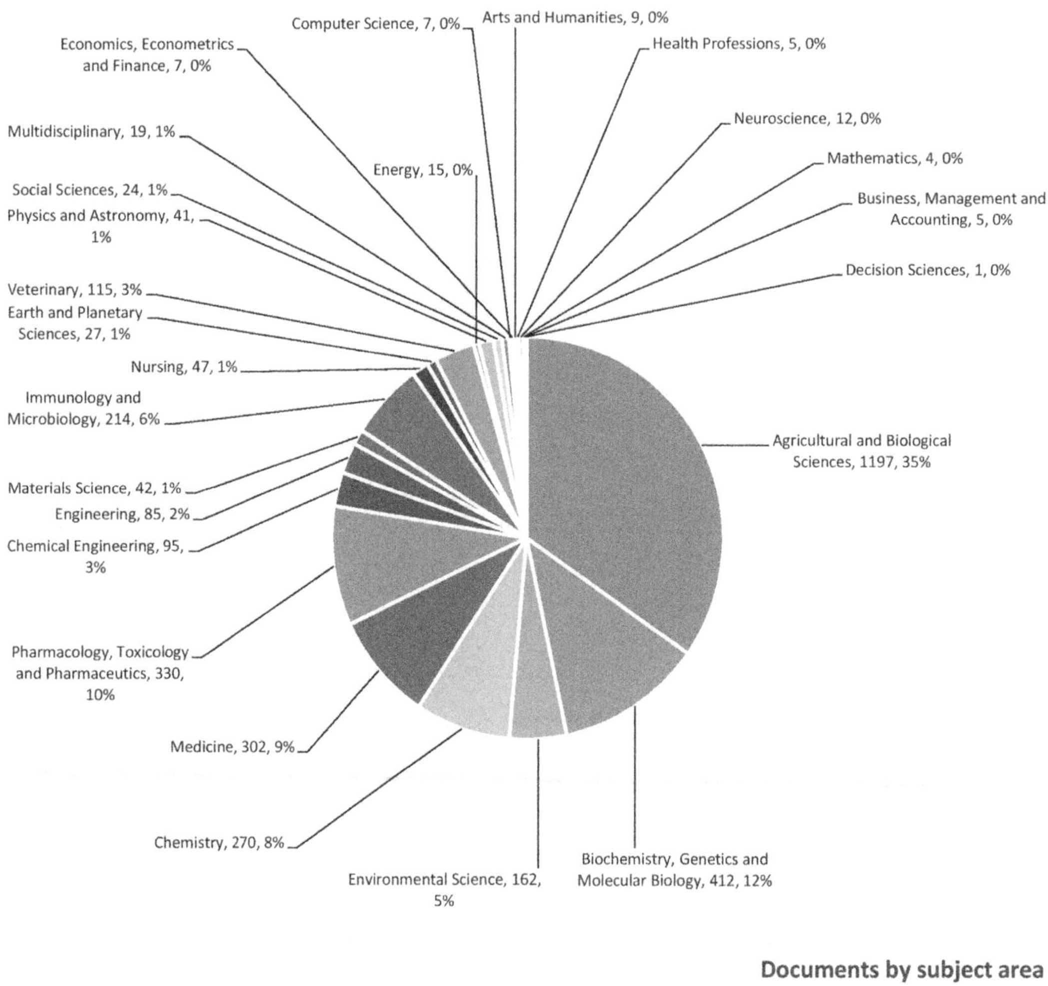

FIG. 4

Distribution by subject area of documents on *Syzygium aromaticum* from 2000 to 2020 (www.scopus.com).

The chapters provide a diversity of developments in food science, phytotherapy, and horticultural research. The editor invited highly cited researchers to write high-quality chapters with the significant purpose of letting the readers know more about this specific area of science. The book contains comprehensive chapters under these main sections:

Section 1. Clove: cultivation, composition, and applications
Section 2. Clove fixed oil: chemistry, functionality, and applications
Section 3. Clove essential oil: chemistry, functionality, and applications
Section 4. Clove extracts: chemistry, functionality, and applications

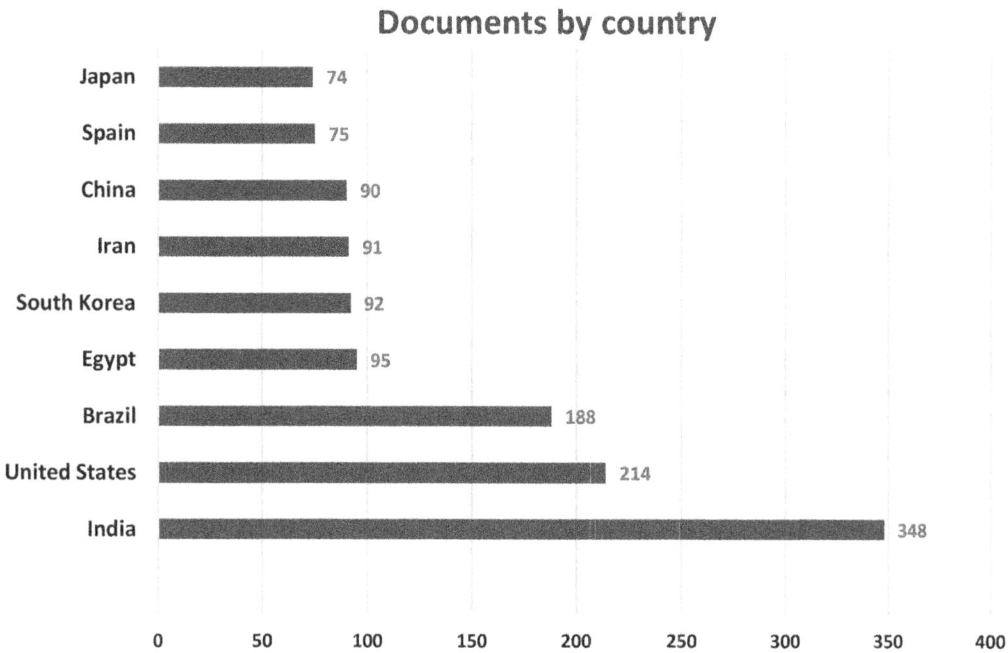

FIG. 5

Distribution by country of documents on *Syzygium aromaticum* from 2000 to 2020 (www.scopus.com).

References

Abo El-Maati, M.F., Mahgoub, S.A., Labib, S.M., Al-Gaby, A.M.A., Ramadan, M.F., 2016. Phenolic extracts of clove (*Syzygium aromaticum*) with novel antioxidant and antibacterial activities. Eur. J. Intern. Med. 8, 494–504. https://doi.org/10.1016/j.eujim.2016.02.006.

Bohnert, H.J., Nguyen, H.R., Lewis, N.G., 2008. Bioengineering and Molecular Biology of Plant Pathways. Elsevier, San Diego, CA, USA.

Bulbeck, D., Reid, A., Tan, L.C., Wu, Y., 1998. Southeast Asian Exports since the 14th Century: Cloves, Pepper, Coffee, and Sugar. Singapore, Institute of Southeast Asian Studies.

Carvalho, R.P.R., de Lima, G.D.A., Machado-Neves, M., 2021. Effect of eugenol treatment in hyperglycemic murine models: A meta-analysis. Pharmacol. Res. 165, 105315.

Chomchalow, N., 1996. Spice production in Asia–an overview. In: Presented at IBC Asian Spice Markets Conference, Singapore, 27–28 May 1996.

Cortés-Rojas, D., Souza, C., Oliveira, W., 2014. Clove (*Syzygium aromaticum*): a precious spice. Asian Pac. J. Trop. Biomed. 4, 90–96.

El-Hadary, A.E., Ramadan, M.F., 2016. Hepatoprotective effect of cold-pressed *Syzygium aromaticum* oil against carbon tetrachloride (CCl_4)-induced hepatotoxicity in rats. Pharm. Biol. 54, 1364–1372. https://doi.org/10.3109/13880209.2015.1078381.

El-Saber Batiha, G., Alkazmi, L.M., Wasef, L.G., Beshbishy, A.M., Nadwa, E.H., Rashwan, E.K., 2020. *Syzygium aromaticum* L. (Myrtaceae): traditional uses, bioactive chemical constituents, pharmacological and toxicological activities. Biomolecules 10, 202. https://doi.org/10.3390/biom10020202.

Hassanien, M.F.R., Assiri, A.M.A., Alzohairy, A.M., Oraby, H.F., 2015. Health-promoting value and food applications of black cumin essential oil: an overview. J. Food Sci. Technol. 52, 6136–6142.

Kamatou, G.P., Vermaak, I., Viljoen, A.M., 2012. Eugenol-from the remote Maluku Islands to the international market place: a review of a remarkable and versatile molecule. Molecules 17 (6), 6953–6981. https://doi.org/10.3390/molecules17066953.

McClements, D.J., 2019. The science of foods: Designing our edible future. In: McClements, D.J. (Ed.), Future Foods: How Modern Science Is Transforming the Way we Eat. Springer International Publishing, Cham.

Osman, A., Mahgoub, S.A.M., Wahdan, K.M.M., Ramadan, M.F., 2019. Antimicrobial and antioxidant infuence of *Syzygium aromaticum* oil supplementation on minced beef quality during cold storage. J. Food Saf. Food Qual. 70, 31–56. https://doi.org/10.2376/0003-925X-70-40.

Ramadan, M.F., 2013. Healthy blends of high linoleic sunflower oil with selected cold pressed oils: functionality, stability and antioxidative characteristics. Ind. Crop Prod. 43, 65–72.

Ramadan, M.F., 2020. Chapter 1: Introduction to cold pressed oils: green technology, bioactive compounds, functionality, and applications. In: Ramadan, M.F. (Ed.), Cold Pressed Oils. Academic Press, pp. 1–5.

Ramadan, M.F., 2021. Introduction to black cumin (*Nigella sativa*): chemistry, technology, functionality and applications. In: Ramadan, M.F. (Ed.), Black Cumin (*Nigella sativa*) Seeds: Chemistry, Technology, Functionality, and Applications. Food Bioactive Ingredients. Springer, Cham, https://doi.org/10.1007/978-3-030-48798-0_1.

Ramadan, M.F., Asker, M.M.S., Tadros, M., 2013. Lipid profile, antiradical power and antimicrobial properties of *Syzygium aromaticum* oil. Grasas Aceites 64, 509–520. https://doi.org/10.3989/gya.011713.

Ramadan, M.F., Moersel, J.-T., 2003. Analysis of glycolipids from black cumin (*Nigella sativa* L.), coriander (*Coriandrum sativum* L.) and niger (*Guizotia abyssinica* Cass.) oilseeds. Food Chem. 80, 197–204.

Ramadan, M.F., Moersel, J.-T., 2004. Oxidative stability of black cumin (*Nigella sativa* L.), coriander (*Coriandrum sativum* L.) and niger (*Guizotia abyssinica* Cass.) upon stripping. Eur. J. Lipid Sci. Technol. 106 (1), 35–43.

Saeed, M., Khan, M.S., Alagawany, M., Farag, M.R., Alqaisi, O., Aqib, A.I., Qumar, M., Siddique, F., Ramadan, M.F., 2021. Clove (*Syzygium aromaticum*) and its phytochemicals in ruminant feed: an updated review. Rend. Lincei Sci. Fis. Nat. 32, 273–285.

Vicidomini, C., Roviello, V., Roviello, G.N., 2021. Molecular basis of the Therapeutical potential of clove (*Syzygium aromaticum* L.) and clues to its anti-COVID-19 utility. Molecules 26, 1880. https://doi.org/10.3390/molecules26071880.

Zheng, G.Q., Kenney, P.M., Lam, L.K.T., 1992. Sesquiterpenes from clove (*Eugenia caryophyllata*). J. Nat. Prod. 55, 999–1003.

SECTION 1

CLOVE (*SYZYGIUM AROMATICUM*): CULTIVATION, COMPOSITION AND APPLICATIONS

CHAPTER

Composition of clove (*Syzygium aromaticum*) buds

2

Tossaton Charoonratana

Department of Pharmacognosy, College of Pharmacy, Rangsit University, Lak-Hok, Mueang Pathum Thani, Thailand

1. Introduction

Flos Caryophylli is the dried, unopened flower buds of the clove tree, *Syzygium aromaticum* (L.) Merr. & L.M.Perry, in Myrtaceae family. Clove, or Cengkeh in the Indonesian language, is native to Indonesia, which is now extensively growing in Sulawesi, Maluku, Java, and Sumatra Islands (Fig. 1). The major clove-producing countries are Indonesia, Madagascar, Tanzania, Sri Lanka, Brazil, and Comoros. The main importing countries are India, Singapore, Indonesia, United Arab Emirates, and Saudi Arabia. The world production is estimated to be 161,504 tons, and Indonesia alone accounts for 80% of the world production (Danthu et al., 2020).

In Indonesia, four cultivated aromatic clove varieties are popularly recognized; Siputih, Sikotok, and Tuni, which are native varieties, and Zanzibar, believed to derive from a few seeds reintroduced from Zanzibar Island in 1932 (Hadiwijaya, 1979(. Other varieties such as Afo, Air mata, Cengkeh hutan, Dokiri, Indari, Raja, Sibela, and Tauro are also recognized but are of little commercial value (Pool et al., 1986(. The clove is an evergreen tree reaching a height of 10–20 m (Fig. 2). Crop improvement of clove is restricted to a selection of high-yielding mother cloves. The propagation of clove trees is through seeds. The fruit is a one-seeded drupe that contains one oblong seed 1.5 cm long. Clove flowers are bisexual, and self-pollination is common (Nurdjannah and Bermawie, 2012).

Clove trees start flowering from the fourth year after planting. Thus, the full-bearing stage is reached after 15 years. For aromatic cloves, the period of flowering at the full-budding stage generally ranges from October to December. Clove buds are harvested when they turn pink-red (Fig. 3). Harvesting clove at different stages yields different clove compositions such as water, oil, fiber, and silicon. Clove buds are sun-dried for 5–6 days or oven-dried for 1 day (Fig. 4). On average, 3–4 kg of dried buds can be harvested from 15-year-old clove trees. However, the yields vary significantly with 1 year of high production and 2–3 years of lower production (Nurdjannah and Bermawie, 2012; Mahulette et al., 2020).

The quality required for clove buds varies between countries. Indonesia has three standard grades for clove quality; their characteristics are shown in Table 1 (BSN, 1998). The specifications of clove buds in the United States are also described in the same table (Tainter and Grenis, 1993).

The clove buds can accumulate chemical constituents, mostly odoriferous volatiles, in the oil glands. The dried buds yield 0.6%–20.0% oil from the clove volatiles chapter, which contains

Clove (*Syzygium aromaticum*). https://doi.org/10.1016/B978-0-323-85177-0.00004-5

FIG. 1

Clove garden in the mountain of Java.

FIG. 2

The bushy evergreen clove trees.

FIG. 3

Clove tree with flowers at full-budding stage.

FIG. 4

Drying of clove buds in Java.

47.4%–89.6% eugenol, 0.9%–21.3% eugenyl acetate, 2.3%–25.3% β-caryophyllene, and other minor compounds.

Since ancient times, humans have used clove buds for food flavoring, food preservation, and traditional medicine. Interestingly, in Indonesia, most clove bud production is for kretek cigarette manufacturers. This chapter provides the botanical aspect of clove buds, their compositions, and their applications. Brief information about clove bud production in the Pacific Ocean region is also included.

Table 1 Indonesian Clove Quality Standard (SNI 01-3392-1998) and US specifications of clove.

Standard parameter	Grade quality I	Grade quality II	Grade quality III	Analysis method	US specifications
Size	Even	Even	Uneven	Organoleptic	–
Color	Brownish	Black brown	Brown	Organoleptic	–
Aroma	Specific	Specific	Specific	Organoleptic	–
Foreign matter[a] (%w/w)	≤0.5	≤1.0	≤1.0	SP-SMP-32-1975	≤1.0
Clove twig (%w/w)	≤1.0	≤3.0	≤5.0	SP-SMP-32-1975	≤5.0
Inferior clove[b] (%w/w)	≤2.0	≤2.0	≤5.0	SP-SMP-32-1975	≤1.0
Molded clove[c]	Negative	Negative	Negative	SP-SMP-32-1975	Negative
Moisture content (%w/w)	≤14.0	≤14.0	≤14.0	SP-SMP-32-1975	≤8.0
Volatile content (%v/w)	≥20.0	≥18.0	≥16.0	SP-SMP-37-1975	≥16.0

[a]*Foreign matter = all material not from clove bud.*
[b]*Inferior clove = shrunken, broken, and fertilized flowers.*
[c]*Molded clove = clove which is already molded.*

2. Production of clove buds in the Pacific Ocean

The Molucca Sea and the Banda Sea are the seas in Sulawesi Island and the Maluku Islands, Indonesia, and they are part of the western South Pacific Ocean. In Maluku, the distribution of clove plantations is in Central Maluku Regency, Seram Island, Buru Island, and Ambon Island. Cloves grow well on the soil rich in organic matter with optimum pH of 5.5–6.5. The appropriate environmental conditions are optimal rainfall of around 500–3500 mm per year, the temperature of 22–32°C, and humidity of 60%–80%. On Ambon Island, clove trees are found at an altitude of 100–250 m, while on Seram Island, at an altitude of 100–500 m (Mahulette et al., 2019a,b).

Tuni species generally dominate clove cultivation in the Maluku Islands; however, the farmers' interest in developing forest cloves increases. Although the forest clove contains a low amount of eugenol, this kind of clove tree is strong, resistant to pests and diseases, continues production every year, and has less fluctuation, heavier flower weights, and rapid growth (Rehatta et al., 2019).

The productivity of Indonesian cloves is low, which is around 260–360 kg ha^{-1}. In the southern Halmahera district, Maluku Islands, although clove plantations have increased, the production of cloves in this district has decreased by 32% from 2013 to 2017. The reduction of clove production is influenced by the age of old plants and the declining quality of land (Pusdatin, 2014). Various approaches can be used to fix the problem. Firstly, an analysis of the suitable land for clove trees is needed; secondly, the land which is insufficiently suitable class but had limiting factor must be improved; and lastly, replacement of old trees with the new good quality trees can be done (Laban et al., 2020; Umasugi et al., 2019; Rehatta et al., 2019; Ermiati and Ardana, 2020). Moreover, the data about the influence of climate on clove buds productivity in these regions should be helpful since there are reports about the effect of temperature and rainfall on clove production (Martin et al., 1988; Fitri, 2020). For example, if excessive rainfall occurs, the farmers are suggested to use paclobutrazol, which blocks the biosynthesis of gibberellin, one of the plant growth regulators that accelerates vegetative growth in clove trees. This chemical retardant was also reported to improve clove bud quality (Runtunuwu et al., 2016).

Since March and October are transition seasons between the dry and rainy seasons, sometimes it is rainy when cloves are harvested; sun-drying of clove buds is not possible. Therefore, besides using the curing method to preserve fresh clove buds until the sun rises, the farmers can use a drying machine for clove drying. The drying machine with 100kg/h capacity was invented to provide convenience (Insyirah, 2019).

Dried clove buds are usually packed in burlap sacks and stored in a clean, dry room with good ventilation. Storage in this condition should not cause any significant changes except the loss of brightness. Clove buds lose oil by evaporation, the rate depends mainly on the moisture content of the products, temperature, and relative humidity during storage (Purseglove et al., 1981).

Ground clove is produced by milling or grinding dried clove buds. The process is usually conducted at the temperature of 25–35°C to prevent the loss of much of the volatiles. Several methods, such as pre-chilling, water cooling, or refrigeration of the grinding chambers, have been used to minimize the heat generated during processing. Ground clove is produced with varying degrees of fineness; for distillation and extraction, coarsely ground clove is accepted, while a finer product is needed for direct use in food flavoring (Nurdjannah and Bermawie, 2012). From the United States, the specific requirements for ground clove are adulteration with stem by weight 5% and minimum volatile oil 14%. The maximum moisture, ash, and acid-insoluble ash are 8%, 5%, and 0.5%, respectively (Tainter and Grenis, 1993).

3. Development stages of clove flowers

Clove flowers develop in small clusters initiated from mature shoots. From the literature, the whole development stage generally takes around 8 months, divided into eight stages, starting from young buds to the full-fruit stage. At the young bud stage, they are initially pale with a shiny rotund appearance (Fig. 5). Then at stage-1, they turn green and possess a long, narrow ovary with four small triangular sepals projecting to one end. The sepals surround a small ball of four overlapping petals that protect the developing flower parts inside (Fig. 6). They are in this form, and the sizes are bigger at stage-2 and stage-3, respectively (Fig. 7). Finally, the buds have reached their full size in the full-budding stage; the hypanthium has also turned pink-red (Fig. 8). The clove buds are harvested by hand or thresher machine and dried for use as a spice (Razafimamonjison et al., 2013).

The full-flowering stage starts when the buds develop to mature; the petals fall off by force from white-yellow stamens, which are then revealed. These stamens surround a single narrow stigma (Fig. 9). Clove buds are not flowering simultaneously, and flowering times vary in different parts of the world. The fruits are initiated 1 month after the flowering stage, and they are mature 2 months later (Fig. 10) (Panco, 2013).

FIG. 5

Young bud stage of cloves.

FIG. 6

Budding stage-1 of cloves.

FIG. 7

Budding stage-3 of cloves.

FIG. 8

Full-budding stage of cloves.

FIG. 9

Full-flowering stage of cloves.

FIG. 10

Red color clove fruits.

It was reported that the flower development cycle of the Raja variety and the forest clove needed a shorter time. This is because they grow wild in the forest of Maluku and flower twice a year, while Zanzibar, Sikotok, Siputih, and Tuni flower once a year (Wael et al., 2018).

4. Description of clove buds

Various clove varieties possess different bud characteristics. For example, the Tuni variety has an average flower length of 1.95 cm, a width of 0.49 cm, and a weight of 0.37 g; when blooming, the color is deep pink (52C). While the Zanzibar variety has an average flower length of 1.89 cm, a diameter of 0.50 cm, and a weight of 0.38 g, when blooming, the color is intense red (53D). On the other hand, forest clove flower, when blooming, is generally light yellow-green (154D). It has an enormous size with an average flower length of 2.41 cm, a width of 0.64 cm, and a weight of 0.82 g. However, they all share the same funnel-shaped flower buds (Hariyadi et al., 2020; Mahulette et al., 2019a,b).

Dried clove buds are 10–20 mm long and blackish-brown; the hypanthium solid, cylindrical, flattened, tapering toward the base and bearing at the apex, thick, triangular, divergent sepals, alternating with four rounded, fragile, unexpanded, membranous, imbricated petals forming a pale, nearly spherical head that encloses numerous stamens, curved inward, and inserted on small disk (Fig. 11) (World Health Organization, 2003).

The hypanthium shows an epidermis with a thick cuticle and numerous, scattered, ranunculaceous stomata on the surface in the transverse section. The outer cortex contains two to three rows of large oil reservoirs lined by a secretory epithelium and embedded in the parenchyma; beneath is an area comprising vascular bundles accompanied by a few fibers containing rosette calcium oxalate crystals. Beneath the bundles is a zone of parenchyma, in which some cells possess thickened walls and some cells are around intercellular air spaces. Next is a location of small-celled parenchyma followed by an inner zone of fibrovascular bundles, along with a few thick-walled, pericyclic fibers (Wilson, 2016).

FIG. 11

Dried clove bud.

The sepals have an epidermis similar to the hypanthium, with numerous stomata on the outer surface, a mesophyll with a few vascular strands and multiple oil glands, and rosette crystals of calcium oxalate. The petals have straight-walled epidermal cells, an undifferentiated mesophyll containing oil glands, rosette calcium oxalate crystals, and small vascular strands. The filament of the stamen bears oil glands beneath the epidermis and the connectives bear a large oil gland at the apex (Wilson, 2016).

The powdered material of *S. aromaticum* is brownish with a characteristic aromatic odor and spicy taste. For microscopical, the clove powder consists of an oil gland (Fig. 12); fragment of vascular tissue with a narrow spiral vessel (Fig. 13); fiber is abundant, isolated, often whole with the narrow lumen and thickened wall (Fig. 14); pollen grain is numerous (Fig. 15) (World Health Organization, 2003).

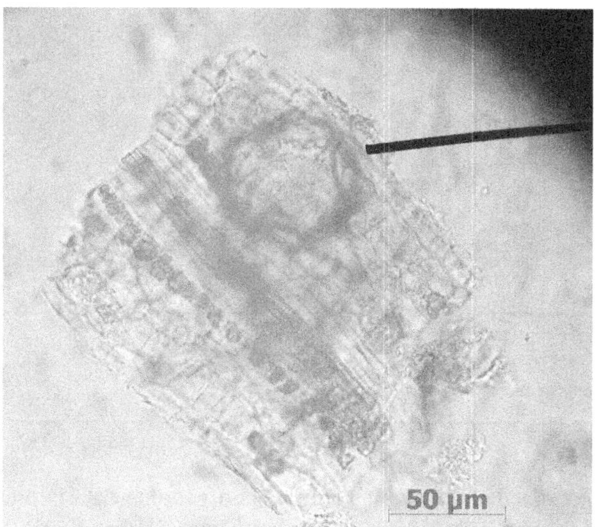

FIG. 12

Oil gland.

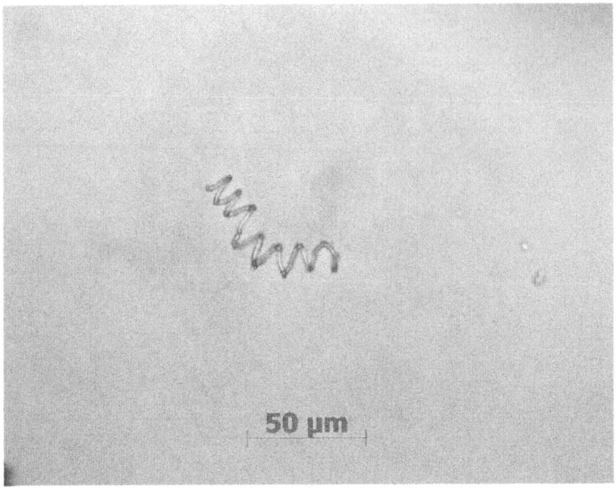

FIG. 13

Spiral vessel.

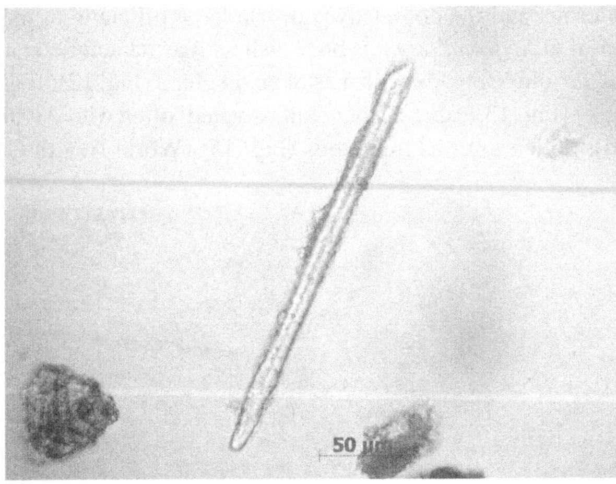

FIG. 14

Fragment of fiber.

5. Clove genetics

There is no whole genomics data available; only plastid genome data is presented in GenBank by Chinese researchers. All genes involved in the biosynthesis of eugenol and other major compounds in this plant are not clarified. Developing genomics and metabolomics are now under-investigated by Prof. Dr. Felix Kessler.

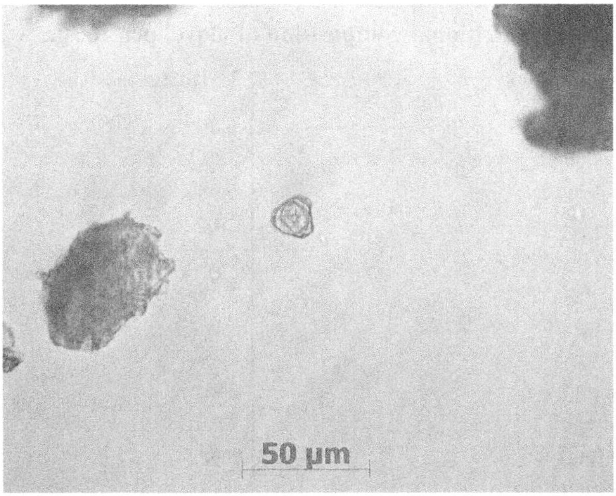

FIG. 15

Pollen grain.

The genotyping markers were created, and it was found that ribulose-1,5-biphosphate carboxylase had been successfully amplified from one sample (MB 1) of clove plant in Ternate Island, North Maluku (Nurhasanah et al., 2019). Alanine/arginine aminopeptidase and malate dehydrogenase 2 were used as genetic markers for Zanzibar variety (Bermawie and Pool, 1991). The noncoding region in the nuclear genome suggested that Afo cloves from Ternate Island and Sibela cloves from Bacan Island are from one ancestor (Afo 1) in North Maluku (Sundari et al., 2019). In India, significant levels of polymorphism among six genotypes of clove collected from different altitudes were found using random amplification polymorphic DNA (Saranya and Ravi, 2018).

6. Chemical compositions of clove buds

The United States Department of Agriculture provides the nutritional composition of clove buds (Table 2). It is showed that clove buds have relatively high food energy. They contain a high level of fat and carbohydrate (Tainter and Grenis, 1993). Linoleic acid and oleic acid are the primary fatty acids found in clove buds (Ramadan et al., 2013). Phenylpropanoids, terpenoids, chromone glycosides, tannins, and acetophenones are the groups of secondary metabolites mainly found in clove buds. In addition, simple benzoic acids, cinnamic acids, ketones, flavonoids, and sterols were found in trace amounts.

Phenylpropanoids; eugenol, eugenyl acetate, *trans*-coniferyl aldehyde, 3-(4-hydroxy-3-methoxyphenyl) propane-1,2-diol, 1-omethyl-guaiacylglycerol, epoxiconiferyl alcohol, benzyl acetate, chavicol, and isoeugenol are shown in Fig. 16 (Blascheck et al., 2008; Ryu et al., 2016).

Table 2 Nutritional composition of clove, per 100 g.

Composition	USDA (ground)
Water	6.86
Food energy (kcal)	323
Protein (g)	5.98
Fat (g)	20.0
Carbohydrate (g)	61.2
Ash (g)	5.88
Ca (g)	0.64
P (mg)	105
Na (mg)	243
K (mg)	1102
Fe (mg)	8.68
Thiamin (mg)	0.11
Riboflavin (mg)	0.26
Niacin (mg)	1.45
Ascorbic acid (mg)	80.8
Vitamin A activities (RE)	53

Source: Tainter, D.R., Grenis, A.T., 1993. Cloves, in Spices and Seasonings. VCH Publishers, New York, pp. 64-67.

Cinnamic acid; ferulic aldehyde is presented in Fig. 16 (Nassar et al., 2007).

Simple benzoic acid; gallic acid is presented in Fig. 16 (Nonaka et al., 1980).

Tannins; eugeniin, ellagic acid, and 3,3′,4′-tri-omethylellagic acid are shown in Fig. 17 (Ryu et al., 2016; Nonaka et al., 1980).

Flavonoids; quercetin, tamarixetin 3-*O*-β-D-glucopyranoside, and ombutin 3-*O*-β-D-glucopyranoside are shown in Fig. 18 (Nassar et al., 2007).

Terpenoids; β-caryophyllene, caryophyllene oxide, α-humulene, humulene epoxide II, caryolane-1,9β-diol, clovane-2,9-diol, oleanolic acid, maslinic acid, limonin, calacorene, calamenene, α-cadinene, δ-cadinene, α-copaene, α-cubebene, farnesol, germacrene D, α-guaiene, linalool, and α-ylangene are shown in Fig. 19 (Ryu et al., 2016; Nassar et al., 2007; Blascheck et al., 2008).

Sterol; β-sitosterol-3-*O*-β-D-glucoside is presented in Fig. 19 (Ryu et al., 2016).

Chromone glycosides; biflorin, isobiflorin, 6-C-β-D-(6′-*O*-galloyl)glucosylnoreugenin, and 8-C-β-D-(6′-*O*-galloyl)glucosylnoreugenin are shown in Fig. 20 (Ryu et al., 2016).

Acetophenones; 2,4,6-trihydroxy-3-methylacetophenone-2-*O*-β-D-glucoside and 2,4,6-trihydroxy-3-methylacetophenone-2-C-β-D-glucoside are shown in Fig. 20 (Ryu et al., 2016; Nasomyon and Charoonratana, 2021).

Ketone; 2-heptanone is presented in Fig. 20 (Blascheck et al., 2008).

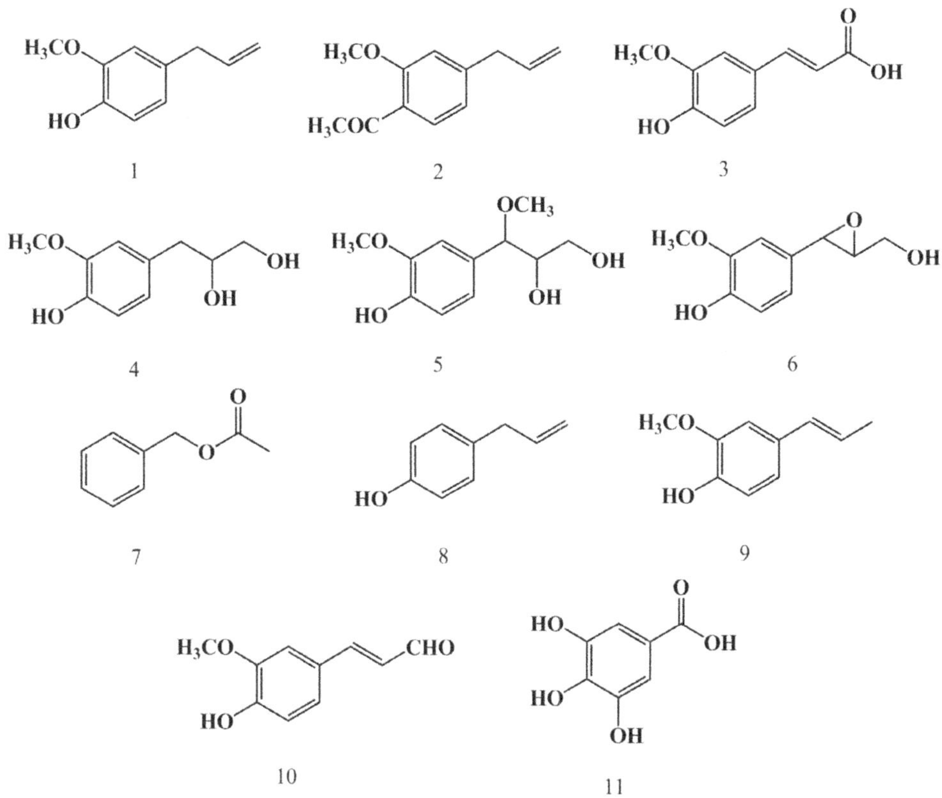

1 = eugenol
2 = eugenyl acetate
3 = trans-coniferyl aldehyde
4 = 3-(4-hydroxy-3-methoxy-phenyl)propane-1,2-diol
5 = 1-omethyl-guaiacylglycerol
6 = epoxiconiferyl alcohol
7 = benzyl acetate
8 = chavicol
9 = isoeugenol
10 = ferulic aldehyde
11 = gallic acid

FIG. 16

Phenylpropanoids, cinnamic acid, and simple benzoic acid in clove buds.

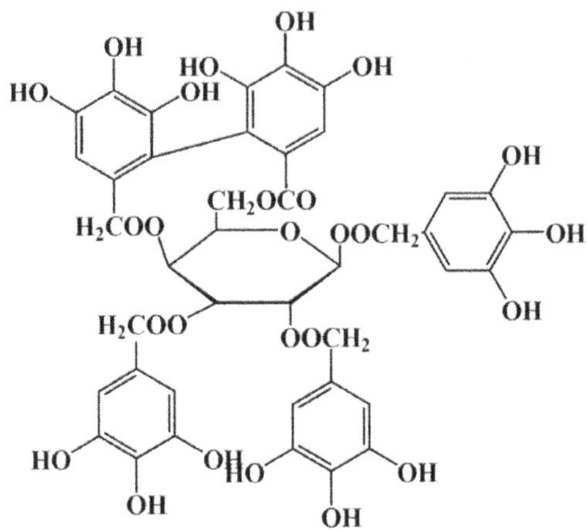

1

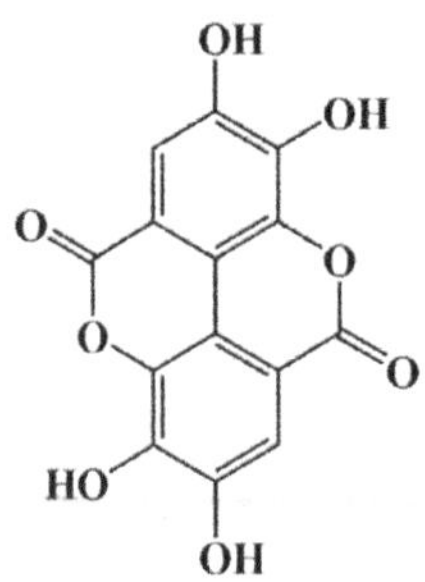

2

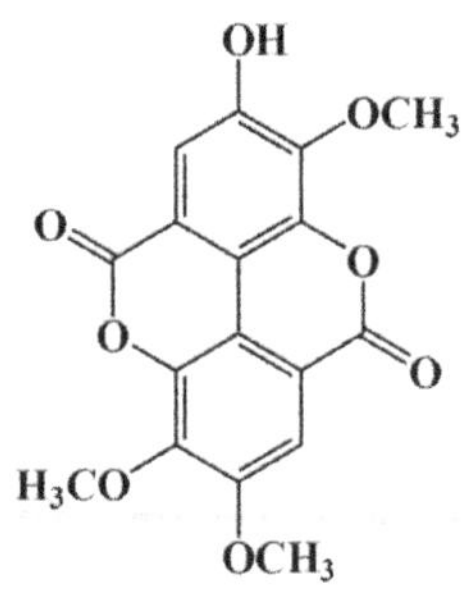

3

1 = eugeniin
2 = ellagic acid
3 = 3,3',4'-tri-omethylellagic acid

FIG. 17

Tannins in clove buds.

7. Application of clove buds

Generally, the primary use of clove buds is for domestic culinary purposes and as a flavoring agent in the food industry. It is also used as the key ingredient in many traditional medicine recipes. However, in Indonesia, 90% of clove buds production is for kretek cigarette manufacture (Nurdjannah and Bermawie, 2012). Other applications such as animal feeds, air freshener, and source for eugenol extraction were also included.

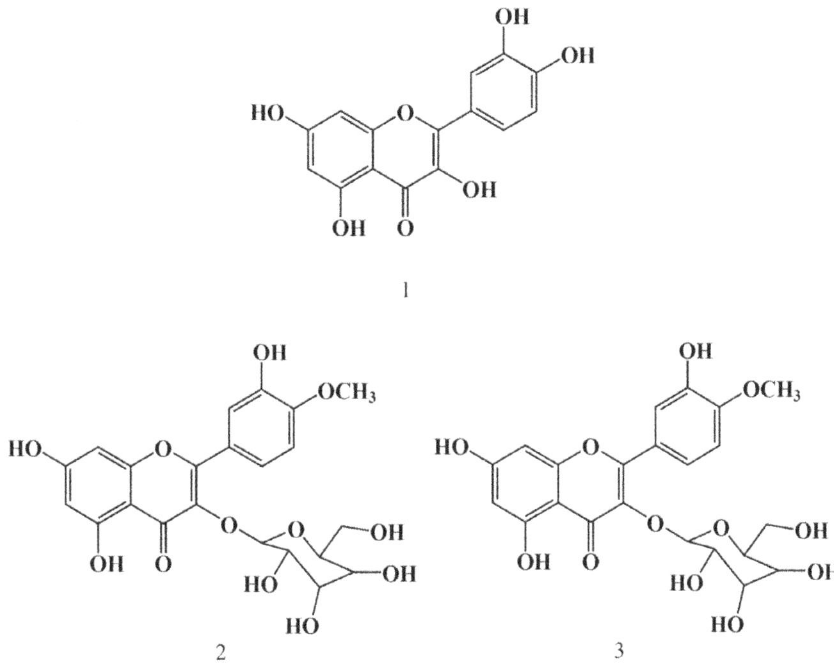

1 = quercetin
2 = tamarixetin 3-O-beta-D-glucopyranoside
3 = ombutin 3-O-beta-D-glucopyranoside

FIG. 18

Flavonoids in clove buds.

7.1 Food applications

Several domestic culinary recipes using clove buds are present all around the world. Bandrek is a hot beverage made of a mixture of ginger, palm sugar, and cinnamon. Other ingredients such as clove buds and star anise are sometimes added (Fig. 21). It is a traditional beverage native to the Sundanese of West Java, Indonesia. While in Yogyakarta, Central Java, Wedang uwuh is a typical drink. It has a sweet and spicy taste with bright red color and fragrant aroma. The spicy taste is from ginger, clove, and other ingredients (Zain, 2014). Masala chai, a spiced sweetened black tea mixed with milk, uses ground cloves. This beverage is originally from India and gains worldwide popularity (Movsowitz, 2020).

Clove is one of the significant spices used in Chettinad cuisine from Tamil Nadu state in South India (Rajagopalan, 2017). For Thai cuisine, clove bud is used in the Kai Palo recipe, in which boiled eggs, tender pork belly, and tofu are simmered in a sweet-salty five-spice soup (Fig. 22) (Chongchitnant, 2019).

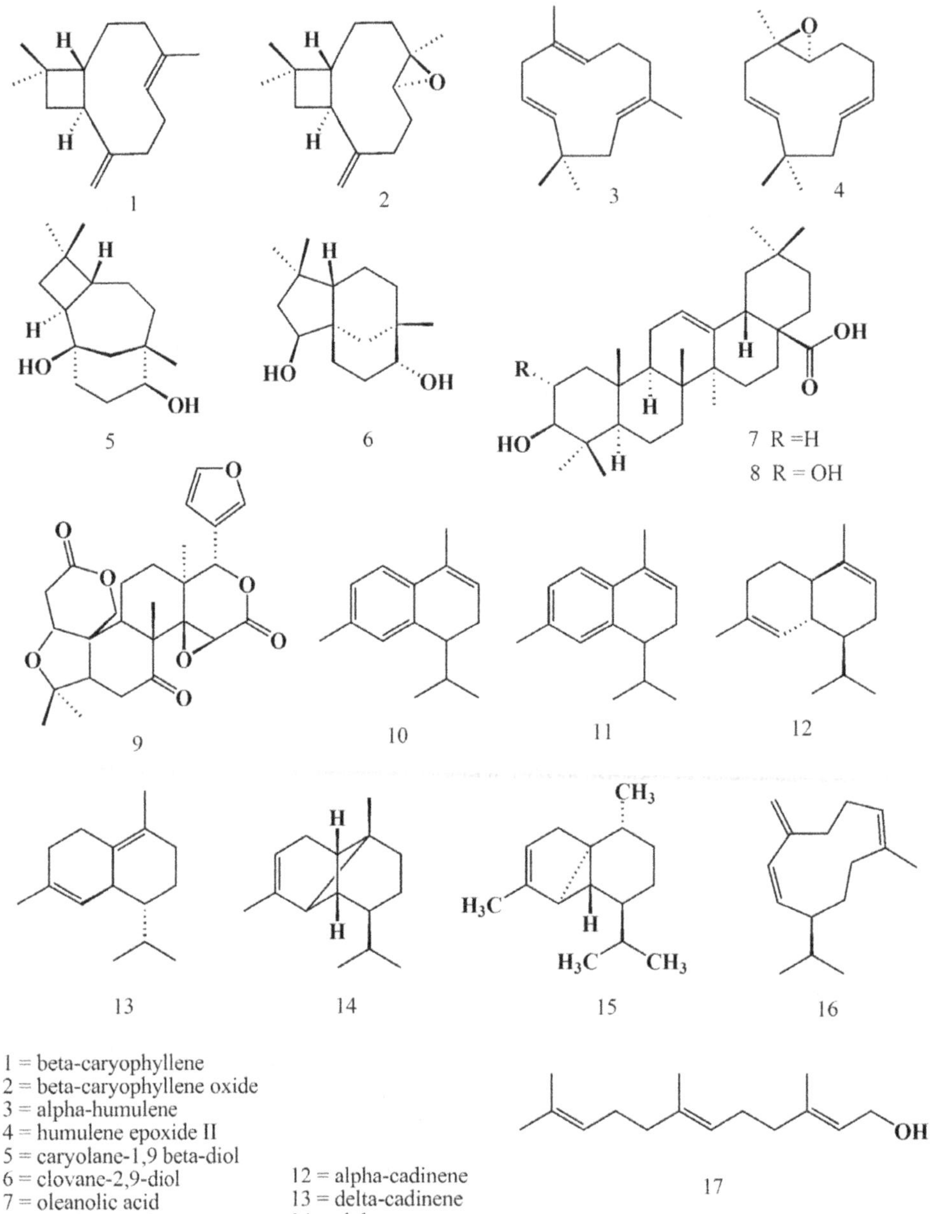

1 = beta-caryophyllene
2 = beta-caryophyllene oxide
3 = alpha-humulene
4 = humulene epoxide II
5 = caryolane-1,9 beta-diol
6 = clovane-2,9-diol
7 = oleanolic acid
8 = maslinic acid
9 = limonin
10 = calacorene
11 = calamenene
12 = alpha-cadinene
13 = delta-cadinene
14 = alpha-copaene
15 = alpha-cubebene
16 = germacrene D
17 = farnesol

FIG. 19

Terpenoids and sterol in clove buds.

(Continued)

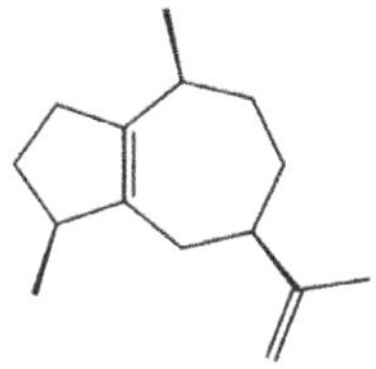

18

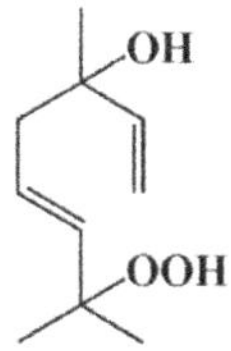

19

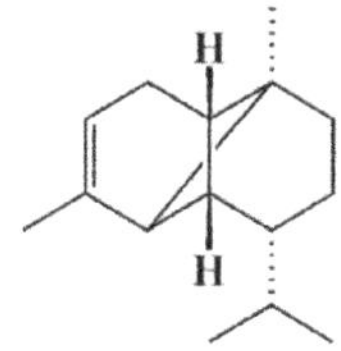

20

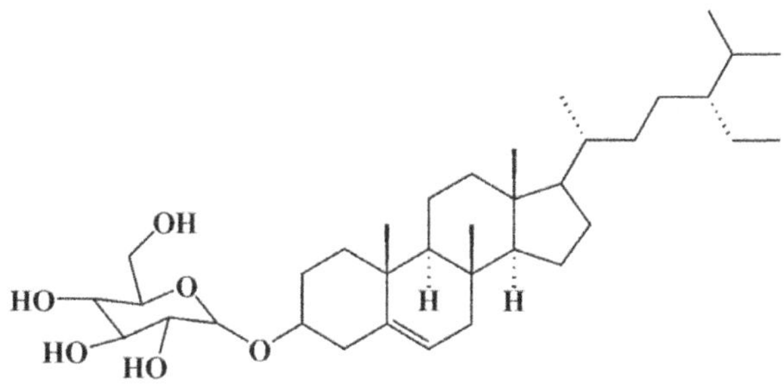

21

18 = alpha-guaiene
19 = linalool
20 = alpha-ylangene
21 = beta-sitosterol-3-O-beta-D-glucoside

FIG. 19, CONT'D

For dessert, clove bud is sometimes added in Canjica, a typical Brazilian sweet dish usually served at a winter festival in June (Fig. 23) (Parés, 2013).

Food products that use clove bud are curry powder, sauces, sausages, and cheese. For example, curry powder and meat sauces use 2%–3% and 0.37% by weight of ground clove buds, whereas Sweet Italian sausages use 0.111% ground clove. Moreover, Frisian clove cheese was developed in the Frisian Islands of the Netherlands. It is a gouda-style cheese made from skimmed milk, cumin, and dried clove buds (Fig. 24). This cheese possesses a firm texture, intense flavor, and low fat (Farrel, 1990).

7.2 Traditional medicine

Clove bud has been used in traditional knowledge developed over generations in many regions such as Asia and Africa. For example, Oke sou, a traditional herbal drink at Lako Akediri village in North Maluku, is made from various plants such as cloves, pepper, lemongrass, etc. It is believed to have the ability to maintain the health of reproductive function and remove women's body odor. This drink is consumed when women get their first menstruation (Wakhidah et al., 2017).

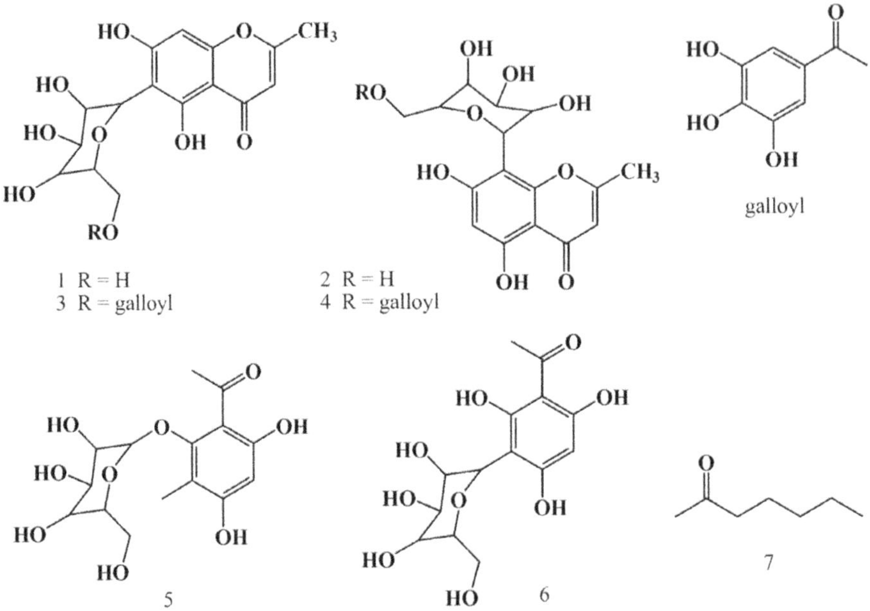

1 = biflorin
2 = isobiflorin
3 = 6-C-beta-D-(6'-O-galloyl)glucosylnoreugenin
4 = 8-C-beta-D-(6'-O-galloyl)glucosylnoreugenin
5 = 2,4,6-trihydroxy-3-methylacetophenone-2-O-beta-D-glucoside
6 = 2,4,6-trihydroxy-3-methylacetophenone-2-C-beta-D-glucoside
7 = 2-heptanone

FIG. 20

Chromone glycosides, acetophenones, and ketone in clove buds.

FIG. 21

Bandrek, traditional beverage native to Sundanese, Indonesia.

FIG. 22

Kai Palo, a Thai childhood favorite food.

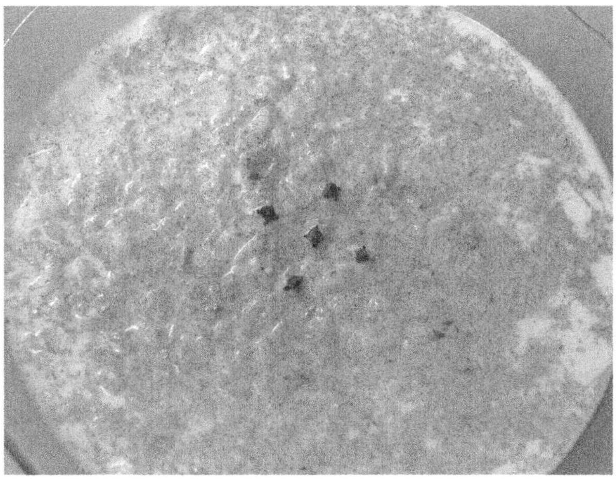

FIG. 23

Brazilian dessert sweet Canjica of white corn.

FIG. 24

Frisian clove cheese.

FIG. 25

Tolak Angin, Indonesian herbal supplement.

Tolak Angin is a famous herbal supplement from Indonesia. The supplement is honey infused with herbs like ginger, fennel, cloves, cinnamon, etc. (Fig. 25). It is believed to boost the immune system, improve the digestive system, and help symptoms associated with colds (Dharmana and Susilaningsih, 2007; Saptaningtyas and Indrahti, 2020).

Boreh is a type of herbal medicine in Bali, a body scrub composed of natural ingredients such as herbs, rice, and spices. The common herbs and spices are cloves, nutmeg, and pepper. All the ingredients are dried and then ground with water using a mortar and pestle to create a yellowy-brown paste. This body scrub is most commonly administered to the skin of people who have a fever, headache, or complaints about joint stiffness, aches, and pains (Hartayu and Widiasih, 2012). Clove bud is the ingredient in 23 Thai traditional medicine recipes on the National List of Essential Medicines, which can be prescribed in the hospital. It is a major component in the Ya Pra Sa Kan Plu recipe for stomachache relief (Urumarudappa et al., 2020).

Clove bud is used in Ayurveda and is called Lavang in India. It is one of the ingredients in Tambul, a mouth freshener. Ayurveda mentions chewing Tambul daily to maintain oral hygiene and prevent oral cavity and throat (Raghavan and Baruah, 1958). Moreover, ground clove is reported to be used as an ingredient or coating material of Mercury pills that treat Frank's disease, which is likely correlated to syphilis (Wujastyk, 2015). In Chinese medicine, clove bud is considered acrid, warm, and aromatic, inflowing the kidney and stomach meridians. It is noteworthy to warm the middle, direct stomach qi downward, fortify the kidney yang, and relieve hiccough (Bensky et al., 2004).

In Nigeria, clove bud is an ingredient used in herbal medicine for neonates and infants less than 6 months old. The recipes treat common ailments, including abdominal spasms and diarrhea (Nwaiwu and Oyelade, 2016).

7.3 Kretek cigarettes

Kretek cigarettes are tobacco mixed with cloves and other ingredients. In 1880, it was firstly manufactured in Kudus, Central Java, for sore throat and asthma treatment. Later, kretek became massively

FIG. 26

Kretek production using a traditional rolling machine.

popular as cigarettes, and the reason behind this happening remains unclear (Sellato, 2000). In kretek production, the quality and variety of tobacco play an essential role. First, crushed dried clove buds weighing about one-third of the tobacco blend are added. Then, the process can be performed using machine-made or hand-rolled kreteks (Fig. 26). The evidence on the health risks related to smoking kretek is reported, and it was found that kretek is likely to be as harmful as a regular cigarette (Nuryunarsih et al., 2021).

7.4 Animal feeds

Clove bud is frequently supplemented into feedstuff since it can improve the intake and growth of animals. Many studies were reported about its effect on poultry. It was found that feed supplemented with clove bud powder and agrimony extract significantly boosted chickens' body weight compared to the control group within 14 days of fattening. Dry matter and crude protein of thigh muscle were also improved (Marcincak et al., 2011). Another study optimized the effect of clove bud powder on Japanese quails. It was found that feed supplemented with 0.5% clove can increase carcass yield and breast weight of Japanese quails without harm to quails (Tariq et al., 2015).

7.5 Air freshener

Cloves have a great aroma and can be used to remove offensive smells rather than artificial air fresheners. For example, orange pomander stuck with cloves can be hung in various places to have continual aromatherapy and improve respiratory problems (Fig. 27) (Bhowmik et al., 2012).

FIG. 27

Orange pomander stuck with clove buds.

7.6 Source for eugenol extraction

Eugenol was reported to be extracted directly using fine ground clove buds. The optimum yield of eugenol was 129.8mg/g dry clove buds using supercritical carbon dioxide extraction. The condition was 20g of clove powder with a particle diameter of 0.5mm at a temperature of 60°C and a pressure of 250bar. The extraction time was 90min at a constant flow rate of 2$\text{L}\,\text{min}^{-1}$ of CO_2 (Chatterjee and Bhattacharjee, 2012). Moreover, a patent related to eugenol isolation from clove buds such as CN101781178A presents using the refluxing method (Yan, 2009).

8. Conclusion

This chapter provides comprehensive information about clove buds. Cloves have funnel-shaped, triangular sepals, alternating with four unexpanded petals. The color and size of cloves are different among varieties. Several oil glands are scattered in hypanthium, sepals, petals, and stamens. Cloves contain high carbohydrate and fat levels; the primary fatty acids are linoleic acid and oleic acid. The aromatic odor of cloves is from its secondary metabolites, mainly phenylpropanoids, terpenoids, acetophenones, and ketones. However, genetic and enzymatic information in biosynthetic pathways of secondary metabolites needs clarification.

Acknowledgment

The author thanks Asst. Prof. Dr. Apirak Sakunpak and Mr. Nirun Vipunngoun for providing support materials.

References

Bensky, D., Clavey, S., Stöger, E., 2004. Chinese Herbal Medicine: Materia Medica. Eastland Press, Seattle.

Bermawie, N., Pool, P.A., 1991. The use of isozyme analysis to determine the status of Zanzibar type clove (*Syzygium aromaticum* L.). Ind. Crops Res. Rep. 17 (1), 1–9.

Bhowmik, D., Kumar, K.P.S., Yadav, A., Srivastava, S., Paswan, S., Dutta, A.S., 2012. Recent trends in Indian traditional herbs *Syzygium aromaticum* and its health benefits. J. Pharmacogn. Phytochem. 1 (1), 13–22.

Blascheck, W., Ebel, S., Hackenthal, E., Holzgrabe, U., Keller, K., Reichling, J., Schulz, V., 2008. *Syzygium Caryophylli* Aetheroleum (Nelkenöl). Springer Verlag, Heidelberg.

BSN, 1998. SNI (Standar Nasional Indonesia) Cengkeh (*Eugenia caryophyllus* (Sprengel), Bullock et Harrison). SNI 01-3392-1998.

Chatterjee, D., Bhattacharjee, P., 2012. Supercritical carbon dioxide extraction of eugenol from clove buds. Food Bioprocess Technol. 6, 2587–2599.

Chongchitnant, P., 2019. Five-Spice Egg & Pork Belly Stew (Instant Pot & Stovetop). https://hot-thai-kitchen.com/kai-palo/.

Danthu, P., Simanjuntak, R., Fawbush, F., Leong Pock Tsy, J.M., Razafimamonjison, G., Abdillahi, M.M., et al., 2020. The clove tree and its products (clove bud, clove oil, eugenol): prosperous today but what of tomorrow's restrictions? Fruits 75 (5), 224–242.

Dharmana, E., Susilaningsih, N., 2007. Pengaruh Pemberian Tolak Angin Cair terhadap Jumlah Sel T, Kadar IFN-γ, dan IL-4 serta Fungsi Hati dan Ginjal Orang Sehat. Fakultas Kedokteran Universitas Diponegoro, Semarang.

Ermiati, Ardana, I.K., 2020. Model of seedling institution development to improve sustainability of clove farming system in West Java. In: Proceeding of 1st ICSP 2020: IOP Conference Series: Earth and Environmental Science. IOP Publishing, Bristol.

Farrel, K.T., 1990. Spices, Condiments and Seasoning, second ed. Springer US, New York.

Fitri, K.A.N., 2020. Analisis Agihan Iklim terhadap Tanaman Cengkeh di Bagian Lereng Timur Gunung Merapi dan Merbabu Tahun 2017. Universitas Muhammadiyah, Surakarta.

Hadiwijaya, T., 1979. Cengkeh, Data dan Petunjuk ke arah Swasembada. Gunung Agung, Jakarta.

Hariyadi, S., Mahulette, A.S., Yahya, S., Wachjar, A., 2020. Agro-morphologies and physicochemical properties of flower bud, stem and leaf oils in two clove varieties (*Syzygium aromaticum* L. Merr. and Perry.) originated from Ambon Island. CMU J. Nat. Sci. 19 (3), 516–530.

Hartayu, T.S., Widiasih, K.A., 2012. Pemahaman masyarakat desa pererenan, kecamatan mengwi kabupaten badung, Bali tentang Boreh-anget. J. Farm. Sains Dan Komun. 9 (2), 71–74.

Insyirah, N.A., 2019. Design of Clove Drying Machine With Capacity 100 kg/hour. University of Muhammadiyah Malang, Malang.

Laban, S., Solle, M.S., Ahmad, A., Jayadi, M., 2020. Land suitability evaluation for clove plants in Bacan Island. In: Proceeding of IC-FSSAT 2020: IOP Conference Series: Earth and Environmental Science. IOP Publishing, Bristol.

Mahulette, A.S., Hariyadi, S., Yahya, S., Wachjar, A., Alfian, A., 2019a. Morphological traits of Maluku native forest clove (*Syzygium aromaticum* L. Merr & Perry). J. Trop. Crop Sci. 6 (2), 105–111.

Mahulette, A.S., Hariyadi, S., Yahya, S., Wachjar, A., Marzuki, I., 2019b. Morpho-agronomical diversity of forest clove in Moluccas, Indonesia. Hayati J. Biosci. 26 (4), 156–162.

Mahulette, A.S., Hariyadi, S., Yahya, S., Wachjar, A., 2020. Physico-chemical properties of clove oil from three forest clove accession groups in Maluku. In: Proceeding of the 1st International Conference on Sustainable Plantation.

Marcincak, S., Popelka, P., Zdolec, N., Martonova, M., Simkova, J., Marcincakova, D., 2011. Effect of supplementation of phytogenic feed additives on performance parameters and meat quality of broiler chickens. Slov. Vet. Res. 48 (1), 27–34.

Martin, P.J., Butler, D.R., Dabek, A.J., 1988. Cause of irregular clove production in the Islands of Zanzibar and Pemba. Exp. Agric. 24, 105–114.

Movsowitz, J., 2020. The History of Chai: A Superfood Before Superfoods Were Cool. https://www.gr8nola.com/blogs/health-wellness/the-history-of-chai-a-superfood-beforesuperfoods-were-cool.

Nasomyon, T., Charoonratana, T., 2021. Comparative chemical composition and *in vitro* cytotoxic activity of *Syzygium aromaticum* against selected cancer cell lines. Trop. J. Nat. Prod. Res. 5 (2), 331–335.

Nassar, M.I., Gaara, A.H., El-Ghorab, A.H., Farrag, A.H., Shen, H., Hug, E., et al., 2007. Chemical constituents of clove (*Syzygium aromaticum*, Fam. Myrtaceae) and their antioxidant activity. Rev. Latinoam. Quím. 35 (3), 47–57.

Nonaka, G., Harada, M., Nishioka, I., 1980. Eugeniin, a new ellagitannin from cloves. Chem. Pharm. Bull. 28 (2), 685–687.

Nurdjannah, N., Bermawie, N., 2012. Cloves. Woodhead Publishing, Cambridge.

Nurhasanah, Sundari, Papuangan, N., 2019. Amplification and analysis of Rbcl gene (ribulose-1,5-bisphosphate carboxylase) of clove in Ternate Island. In: Proceeding of International Conference on Life Sciences and Technology: IOP Conference Series: Earth and Environmental Science. IOP Publishing, Bristol.

Nuryunarsih, D., Lewis, S., Langley, T., 2021. Health risks of kretek cigarettes: a systematic review. Nicotine Tob. Res. 23 (8), 1274–1282.

Nwaiwu, O., Oyelade, O.B., 2016. Traditional herbal medicines used in neonates and infants less than six months old in Lagos Nigeria. Niger. J. Paediatr. 43 (1), 40–45.

Panco, M., 2013. Assessment of Local Knowledge and Agricultural Practices in the Clove Tree Based Cropping Systems in the East Fénérive District, Madagascar. Retrieve fro https://afs4food.cirad.fr/en/content/download/4469/33986/version/2/file/Panco%2Cclove+Fenerive+2013+synth%C3%A8se.pdf.

Parés, L.N., 2013. The Formation of Candomble: Vodun History and Ritual in Brazil (Latin America in translation/en traducción/em tradução). University of North Carolina Press, p. 288.

Pool, P.A., Eden-green, S.J., Muhammad, M.T., 1986. Variation in clove (*Syzygium aromaticum*) germplasm in the Moluccan Islands. Euphytica 35, 149–159.

Purseglove, J.W., Brown, E.G., Green, C.L., Robbins, S.R.J., 1981. Spices. vol. 1 Longman, New York, pp. 229–285.

Pusdatin, 2014. Outlook komoditi cengkeh Pusat Data dan Informasi Pertanian. Sekretaris Jenderal-Kementerian Pertanian, Jakarta.

Raghavan, V., Baruah, H.K., 1958. Arecanut: India's popular masticatory -history, chemistry and utilization. Econ. Bot. 12, 315–345.

Rajagopalan, A., 2017. Chettinad Food: 10 Ingredients That Make It a Lip-Smacking Affair. NDTV. Retrieved 16 August 2020 https://food.ndtv.com/food-drinks/chettinad-food-10-ingredients-that-make-it-a-lip-smacking-affair-1253214.

Ramadan, M.F., Asker, M.M.S., Tadros, M., 2013. Lipid profile, antiradical power and antimicrobial properties of *Syzygium aromaticum* oil. Int. J. Fats Oils 64 (5), 509–520.

Razafimamonjison, G., Jahiel, M., Ramanoelina, P., Fawbush, F., Danthu, P., 2013. Effects of phenological stages on yield and composition of essential oil of *Syzygium aromaticum* buds from Madagascar. Int. J. Basic Appl. Sci. 2 (4), 312–318.

Rehatta, H., Marasabessy, D.A., Sopalauw, S.H., 2019. Productivity of forest clove (*Syzygium obtusifolium* L.) in Leihitu Sub-District, Central Maluku District. J. Budid. Pertan. 15 (1), 31–37.

Runtunuwu, S.D., Mamarimbing, R., Tumewu, P., Rengkung, R.M.N., 2016. The effect of paclobutrazol on flower quality of clove (*Syzygium aromaticum* L). J. Bioslogos 6 (2), 33–41.

Ryu, B., Kim, H.M., Woo, J.H., Choi, J.H., Jang, D.S., 2016. A new acetophenone glycoside from the flower buds of *Syzygium aromaticum* (cloves). Fitoterapia 115, 46–51.

Saptaningtyas, A.I., Indrahti, S., 2020. Dari industri Jamu tradisional ke industri Jamu modern: perkembangan industri Jamu Sido muncul dalam mempertahankan eksistensi perusahaan tahun 1951–2000. Historiografi 1 (2), 172–180.

Saranya, D., Ravi, R., 2018. *Syzygium aromaticum*: a subtropical tree scrutinized for the genetic differentiation among the altitudes of Nilgiris. Ann. Plant Sci. 7 (4), 2134–2138.

Sellato, B., 2000. In: Hanusz, M. (Ed.), Kretek. The Culture and Heritage of Indonesia's Clove Cigarettes. Equinox Publishing, Jakarta, p. 203.

Sundari, Nurhasanah, Mas'ud, A., Amin, M., Arumingtyas, E., Azrianingsih, R., 2019. Update phylogenetic information of the local varieties of cloves (*Syzygium aromaticum*) from North Maluku, Indonesia based on ITS sequences data. Biodiversitas 20 (6), 1510–1515.

Tainter, D.R., Grenis, A.T., 1993. Cloves, in Spices and Seasonings. VCH Publishers, New York, pp. 64–67.

Tariq, H., Rao, P.V.R., Raghuvanshi, R.S., Mondal, B.C., 2015. Effect of Aloe vera and clove powder supplementation on carcass characteristics, composition and serum enzymes of Japanese quails. Vet. World 8, 664–668.

Umasugi, B., Prijono, S., Soemarno, Ariffin, 2019. Potential for land and lengths available to water needs on the clove land in Ternate Island. In: Proceeding of 5th International Conference on Food, Agriculture and Natural Resource. Atlantis Press SARL, Paris.

Urumarudappa, S.K.J., Tungphatthong, C., Prombutara, P., Sukrong, S., 2020. DNA metabarcoding to unravel plant species composition in selected herbal medicines on the National List of Essential Medicines (NLEM) of Thailand. Sci. Rep. 10, 18259.

Wael, S., Nuringtyas, T.R., Wijayanti, N., Astuti, P., 2018. Secondary metabolites production in clove (*Syzygium aromaticum*): chemical compounds. J. Biol. Sci. 18, 399–406.

Wakhidah, A.Z., Silalahi, M., Nisyawati, 2017. Ethnobotanical study of Oke Sou: traditional herbal drink from Lako Akediri village in West Halmahera, Indonesia. In: Proceeding of 3rd Asia Future Conference Environment & Coexistence. Japan Book Inc, Tokyo.

Wilson, L., 2016. Spices and flavoring crops: leaf and floral structures. In: Caballero, et al. (Eds.), Encyclopedia of Food Sciences and Nutrition. Academic Press, pp. 84–92.

World Health Organization, 2003. WHO Monographs on Selected Medicinal Plants. vol. 2 WHO Press, Geneva. [monograph on the internet]. [cited 2021 May 5]. Available fro http://apps.who.int/medicinedocs/en/d/Js4927e/7.html.

Wujastyk, D., 2015. Mercury as an antisyphilitic in Ayurvedic medicine. De Gruyter 69 (4), 1043–1067.

Yan, Z., 2009. Extraction and Separation Method of Eugenol. China Patent. CN101781178A.

Zain, M.Y., 2014. "Torehan Sejarah dalam Segelas Bandrek" (in Indonesian). Kompas.com (Retrieved 20 August 2014).

CHAPTER

Clove (*Syzygium aromaticum*) leaves 3

Tossaton Charoonratana

Department of Pharmacognosy, College of Pharmacy, Rangsit University, Lak-Hok, Mueang Pathum Thani, Thailand

1. Introduction

Most people recognize clove and clove oil from the flower buds of *Syzygium aromaticum* (L.). However, clove leaves are very mean to rural Madagascan and Indonesian people since they are the primary material for clove leaf oil production. Dry leaves, which fall naturally, are used for oil distillation in the household industry, where the workforce is low cost (1.5$ or 3.5$ per day in Madagascar or Indonesia, respectively) (Lutfi et al., 2020; Danthu et al., 2020). These raw materials are available in the dry season for only 6 months a year (Khozali et al., 2012).

In Indonesia, the price of 1 kg dried clove leaves is around 0.11$ to 0.14$. It was reported that women farmers in Larewa, South Sulawesi, can collect up to 100 kg clove leaves per day (Mustadjar and Ismail, 2018). In addition, clove oil household enterprises give significant extra income to the farmers. The monthly average income of clove leaf oil during the dry season in Palau village, Donggala, Central Sulawesi was about 77.0$ (Arizona and Lamusa, 2016). In 2007, the Indonesian tropical plant research center reported that Java Island had around 500,000 km^2 of clove cultivation area. It was estimated that the potential of deciduous clove leaves was about 27,600 kg per day or equivalent to 4000 kg of clove leaf oil per day (Yuliani, 2007). Besides oil production, it was reported that clove leaves had been used as mosquito repellents for generations by the people of Maluku by burning and smoking them in the house (Nindatu and Noya, 2018).

The clove leaves accumulate various chemical constituents, both volatiles and nonvolatiles. The dried leaves yield only 0.34%–4.80% oil, which contains 60.5%–94.4% eugenol, 0%–2.80% eugenyl acetate, 2.91%–32.2% β-caryophyllene, and other minor volatile compounds. The reported nonvolatiles are flavonoids, chromone glycosides, and acetophenones; however, more informative data is required to identify the major nonvolatiles in clove leaves. This chapter provides the botanical aspect of clove leaves, their compositions, and their applications. Brief information about major compounds' biosynthesis in the leaves is also proposed.

Clove (*Syzygium aromaticum*). https://doi.org/10.1016/B978-0-323-85177-0.00014-8

2. Description of clove leaves

Generally, the mature clove leaves are evergreen, alternate to mostly opposite, simple, glabrous; petiole 1–3 cm long, somewhat thickened at base; blade obovate-oblong to elliptical, base very acute, apex acuminate, coriaceous, shining, gland-dotted. Numerous oil glands are present on the lower surface. In the Myrtaceae family, the character of plants the phloem locations on both sides of the xylem (Board, 2010).

Various clove varieties possess different leaf characteristics. Firstly, the forest cloves have an enormous leaf size among all varieties. It can be 15.2–20.0 cm in length and 7.0–9.0 cm in width with dark green color (Mahulette et al., 2019). Secondly, the Raja variety is suggested to be a cross-variety between forest clove and cultivated clove. It possesses elliptic to elliptic-oblong leaf with a size of 11.0–16.5 × 4.4–7.5 cm in dark green color (Alfian et al., 2019). Thirdly, it was reported that the Madagascar variety possesses leaves up to 13.5 cm in length (Razafimamonjison et al., 2016). Lastly, other researchers noted that the Tuni variety has a bigger leaf size than Zanzibar; the size is around 8.9 × 4.1 cm and 8.2 × 3.3 cm, respectively. Both varieties have an elliptical shape with dark green but different top colors (Hariyadi et al., 2020). Other remarks about clove leaves from other varieties were suggested (Wael et al., 2018). The morphology of clove leaf from the Siputih variety is large and dark green. The Sikotok have small leaf blades, and the leaves are reddish; the old leaves are small, green, and slightly shiny.

Researchers separate clove leaves into four development stages in Madagascar, young leaves, expanded leave-1, expanded leave-2, and mature leaves. They are initially red at the young leaf stage with 2–4 cm in length. At the expanded-1 stage, they turn pink and possess an elliptical shape with 6–8 cm long. At the expanded-2 stage, the leaf color is pale green with 8–10 cm long, then they change to dark green with 11–13 cm in length at the last stage (Fig. 1). The moisture content of mature leaves (57%) was the highest, while the others were around 47%–53% (Razafimamonjison et al., 2016).

FIG. 1

Clove leaves at different stages; left to right, two expanded leave-1 samples, two expanded leave-2 samples, and a mature leaf.

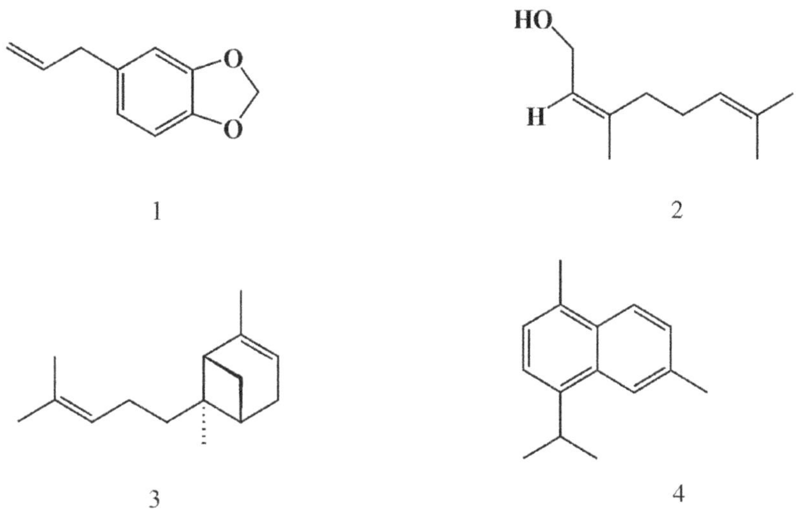

1 = safrole
2 = nerol
3 = alpha-bergamotene
4 = cadalene

FIG. 2

Volatile compounds in clove leaves that are not reported in clove buds.

3. Chemical composition of clove leaves

Generally, the chemical compositions of volatiles in clove leaves are almost similar to clove buds. However, some volatile compounds in leaves are not reported in clove buds, such as safrole, nerol, α-bergamotent, and cadalene (Fig. 2). Moreover, different chromatographic fingerprints were found between clove leaves and buds. Thus, it was suggested that other major nonvolatiles are presented in clove leaves (Nasomyon and Charoonratana, 2021).

Phenylpropanoids, terpenoids, chromone glycosides, and acetophenones are secondary metabolites found in clove leaves. Phenylpropanoids reported in clove leaves are eugenol, eugenyl acetate, chavicol, isoeugenol, and safrole (Blascheck et al., 2008). Terpenoids reported in clove leaves are β-caryophyllene, caryophyllene oxide, α-humulene, humulene epoxide II, δ-cadinene, calamenene, α-copaene, α-cubebene, farnesol, germacrene D, linalool, nerol, α-bergamotent, and cadalene (Blascheck et al., 2008). Chromone glycosides reported in clove leaves are biflorin and isobiflorin (Singh and Bharate, 2006; Nasomyon and Charoonratana, 2021). Acetophenone reported in clove leaves is 2,4,6-trihydroxy-3-methylacetophenone-2-C-β-D-glucoside (Nasomyon and Charoonratana, 2021).

Eugenol is the primary compound in the volatile part of clove leaves. The biosynthetic pathway of eugenol is well studied in other plant species' leaves and occurs via the shikimate pathway. For example, in *Ocimum* species, it was reported that eugenol was synthesized by eugenol synthase using coniferyl acetate as a substrate. Moreover, eugenol synthase gene expression was positively related to eugenol contents (Anand et al., 2016). Therefore, for eugenyl acetate, it is predicted that this compound is synthesized by eugenol acyltransferase using eugenol as a substrate. Furthermore, since eugenyl acetate is at a high level in young clove buds and leaves, it is proposed that eugenol acyltransferase gene expression is also high at this time.

β-Caryophyllene, a major sesquiterpene in various plants, is reported to be synthesized in the cytoplasm via the mevalonate pathway. The β-caryophyllene synthase gene is not cloned yet in *S. aromaticum*; however, it is predicted to be the responsible gene in β-caryophyllene biosynthesis in clove. Twenty-four plants are reported to contain this gene (Dewick, 2002).

4. Application of clove leaves

The primary use of clove leaves is for oil production. However, another way to increase the value of clove leaves is to isolate the valuable compound, eugenol, from the oil. Moreover, clove leaves used as the ingredient in traditional medicine recipes are also reported.

4.1 Source of eugenol extraction

A patent related to eugenol isolation from clove leaves such as CN104628542A presents decompression and distillation (Senxiong, 2013). Moreover, cellulose acetate membrane was suggested to use as an add-on in eugenol purification. It was reported that the filtration process using the cellulose acetate membrane could increase the eugenol content by up to 7.8% (Kusworo et al., 2018).

4.2 Traditional medicine

The people in the Sultanate of Jailolo, North Maluku, consist of four original tribes; Gamkonora, Tobaru, Sahu, and Wayoli, who still use traditional medicine to treat diseases. Their tradition is called Bakera, and clove leaves are the primary ingredient in the recipes for postnatal care and to cure liver disease, malaria, and pain (Syahdar et al., 2019).

5. Conclusion

Clove leaves are simple, bright green, and glossy. The lower surface contains oil glands. Elliptical leaves with petioles grow up to 20 cm long depending on the varieties. Clove leaves contain high eugenol, eugenyl acetate, and β-caryophyllene. The proposed eugenol and eugenyl acetate synthesis in clove occurred via the shikimate pathway using eugenol synthase and eugenol acyltransferase. However, genetic and enzymatic information in the biosynthesis of clove metabolites needs clarification.

Acknowledgment

The author thanks Asst. Prof. Thanapat Songsak and Ms. Emerald Li for their kindly supports.

References

Alfian, A., Mahulette, A.S., Zainal, M., Hardin, Bahrun, A.H., 2019. Morphological character of raja clove (*Syzygium aromaticum* L. Merr & Perry.) native from Ambon Island. In: Proceeding of IOP Conference Series Earth and Environmental Science 2019. IOP Publishing, Bristol.

Anand, A., Jayaramaiah, R.H., Beedkar, S.D., Singh, P.A., Joshi, R.S., Mulani, F.A., et al., 2016. Comparative functional characterization of eugenol synthase from four different *Ocimum* species: implications on eugenol accumulation. Biochim. Biophys. Acta 1864 (11), 1539–1547.

Arizona, M.N.H., Lamusa, A., 2016. Analisis Pendapatan Industri Rumah Tangga Penyulingan Minyak Daun Cengkeh di Desa Palau Kecamatan Balaesang Tanjung Kabupaten Donggala. Agrotekbis 4 (4), 461–467.

Blascheck, W., Ebel, S., Hackenthal, E., Holzgrabe, U., Keller, K., Reichling, J., Schulz, V., 2008. *Syzygium Caryophylli* Aetheroleum (Nelkenöl). Springer Verlag, Heidelberg.

Board, N., 2010. Handbook on Spices, Reprinted ed. Asia Pacific Business Press, Delhi.

Danthu, P., Simanjuntak, R., Fawbush, F., Leong Pock Tsy, J.M., Razafimamonjison, G., Abdillahi, M.M., et al., 2020. The clove tree and its products (clove bud, clove oil, eugenol): prosperous today, but what of tomorrow's restrictions? Fruits 75 (5), 224–242.

Dewick, P.M., 2002. Medicinal Natural Products: A Biosynthetic Approach. Wiley, Chichester, West Sussex, England.

Hariyadi, S, Mahulette, A.S., Yahya, S., Wachjar, A., 2020. Agro-morphologies and physicochemical properties of flower bud, stem and leaf oils in two clove varieties (*Syzygium aromaticum* L. Merr. and Perry.) originated from Ambon island. CMU J. Nat. Sci. 19 (3), 516–530.

Khozali, A., Supardi, S., Hastuti, D., 2012. Analisa Usaha Penyulingan Minyak Daun Cengkeh (*Syzygium aromaticum*, syn. *Eugenia aromaticum*) (Studi Kasus di Kecamatan Sukorejo Kabupaten Kendal). J. Med. 8 (2), 32–42.

Kusworo, T.D., Soetrisnanto, D., Widayat, W., Budiyono, B., Utomo, D.P., 2018. Study of polymeric membranes potential for eugenol purification from crude clove leaf oil. Asian J. Chem. Eng. 18 (2), 81–92.

Lutfi, M., Andajani, K., Ilhamuddin, Utami, H.N., Affifah, F.N., 2020. Appropriate technology application of traditional clove oil production: effort to upgrade quality. Adv. Food Sci. Sustain. Agric. Agroind. Eng. 3 (2), 75–80.

Mahulette, A.S., Hariyadi, S., Yahya, S., Wachjar, A., Alfian, A., 2019. Morphological traits of Maluku native forest clove (*Syzygium aromaticum* L. Merr & Perry). J. Trop. Crop Sci. 6 (2), 105–111.

Mustadjar, M., Ismail, A., 2018. Working pattern of women farmers in increasing family income. In: Witjaksono, et al. (Eds.), Advances in Social Science, Education and Humanities Research. vol. 226. Atlantis Press, Zhengzhou, pp. 1095–1099.

Nasomyon, T., Charoonratana, T., 2021. Comparative chemical composition and *in vitro* cytotoxic activity of *Syzygium aromaticum* against selected cancer cell lines. Trop. J. Nat. Prod. Res. 5 (2), 331–335.

Nindatu, M., Noya, Y., 2018. Effectiveness of clove leaves (*Syzygium aromaticum* L.) steeping repellent to *Anopheles* sp. J. Biol. Eduk. Edisi 10 (2), 17–20.

Razafimamonjison, G., Boulanger, R., Jahiel, M., Ramanoelina, P., Fawbush, F., Lebruu, M., et al., 2016. Variations in yield and composition of leaf essential oil from *Syzygium aromaticum* at various phases of development. Int. J. Basic Appl. Sci. 5 (1), 90–94.

Senxiong, Z., 2013. Production Process of Eugenol for Flavoured Product. China Patent CN104628542A.

Singh, I.P., Bharate, S.B., 2006. Phloroglucinol compounds of natural origin. Nat. Prod. Rep. 23, 558–591.

Syahdar, S., Tamalene, M.N., Hasan, S., 2019. Bakera: tradition of medicinal plants utilization for therapy, prevention and recovery of diseases in Jailolo Sultanate custom society, Indonesia. Asian J. Ethnobiol. 2 (2), 41–47.

Wael, S., Nuringtyas, T.R., Wijayanti, N., Astuti, P., 2018. Secondary metabolites production in clove (*Syzygium aromaticum*): chemical compounds. J. Biol. Sci. 18, 399–406.

Yuliani, S., 2007. Vanilin dari Limbah daun Cengkeh. Media: Sinar Tani. 22 August 2007. Available from: www.litbang.pertanian.go.id/artikel/173/.

CHAPTER

Clove (*Syzygium aromaticum*) polysaccharides

4

Ahmad Cheikhyoussef[a], Natascha Cheikhyoussef[b], Ahmed A. Hussein[c], and Mohamed Fawzy Ramadan[d]

Science and Technology Division, Multidisciplinary Research Unit, Centre for Research Services, University of Namibia, Windhoek, Namibia[a] Ministry of Higher Education, Technology and Innovation, Windhoek, Namibia[b] Chemistry Department, Cape Peninsula University of Technology, Cape Town, Western Cape, South Africa[c] Department of Clinical Nutrition, Faculty of Applied Medical Sciences, Umm Al-Qura University, Makkah, Saudi Arabia[d]

1. Introduction

Clove (*Syzygium aromaticum*) belongs to the family of Myrtaceae (130–150 genera and around 3000 species); one of the most common plant families in terms of food uses and application in the food industry, thanks to their unique organoleptic and therapeutic properties (Devkota and Adhikari-Devkota, 2020). It is native to the Maluku islands in eastern Indonesia; however, recently, it has been cultivated in different regions in the world (Hussein et al., 2019). Clove tree is composed of leaves and buds, and it has a single flower of 1–2 cm long (Saeed et al., 2021). Three different stages of clove flowering bud have been reported by Kasai et al. (2016), which include phase 1 that is characterized with the green color and budding stage, phase II with the pink color and flowering stage, and the last phase with the fruiting stage and distinctive red color. They further emphasized that the clove bud's full-budding stage is the most economic value of the clove phases.

Plant-based polysaccharides are a group of bioactive natural products with a high molecular weight with increasing interest in recent years and have several attractive biological activities (Mohan et al., 2020). The approved polysaccharides for food products formulation with industrial applications are referred as food polysaccharides, with the main function of boosting the textural attributes of the foods. Naturally, these biomolecules tend to form complexes with other types of biomolecules, such as proteins and lipids (Lu et al., 2019; Yang et al., 2020; Beaumont et al., 2021). Food polysaccharides are well known for their remarkable rheological properties (thickening, stabilizing, and gelling) and for their emulsifying properties. Furthermore, their use at a very low concentration may significantly impact the textural properties of the food products (Bernaerts et al., 2019; Yang et al., 2020; Guimarães et al., 2020). Unfortunately, very little information could be found in the literature, and few studies were reported regarding the polysaccharides from clove extracts; therefore, this chapter provides an overview of the clove (*S. aromaticum*) polysaccharides composition, structure, and biological activities, as well as their reported mode/mechanism of action.

Clove (*Syzygium aromaticum*). https://doi.org/10.1016/B978-0-323-85177-0.00039-2

2. Clove polysaccharides structure and composition

Carbohydrates were reported to be one of the chemical components of clove plant part extract with different concentrations based on the nature of the used plant parts and the solvent used whereby Sulieman et al. (2007) reported 51.5% content in clove bud powder. Meanwhile, Bello and Jimoh (2012) reported lower content of carbohydrates with 30.9% in clove bud, and almost the same percentage (31%) was reported by Ereifej et al. (2015). Other studies revealed carbohydrates in clove extracts on a qualitative basis only (Jimoh et al., 2017; Kaur and Kaushal, 2019).

Lee et al. (2000) compared the chemical composition of clove extracts from different regions and reported high total sugar of 26.5%–27.4% and 21.6%–21.7% for the cloves extracts from Singapore and Holland, respectively, with high arabinose (27.1% and 15.1%) and galactose (6.60% and 4.97%) for the clove extracts from Holland and Madagascar. Meanwhile, uronic acid was high (25.3%–30.3% and 25.3%–27.8%) for clove extracts from Singapore and Holland, respectively, compared to Madagascar and China clove extracts (15.2%–18.5% and 14.8%–15.2%, respectively).

In terms of structure, polysaccharides from clove belong to sulfated polysaccharides. This type of polysaccharides are extracted from natural resources or synthesized by modifying the sugar chain with sulfates (Wang et al., 2018; Lin et al., 2019; Ray et al., 2020). Two polysaccharides with relatively high antithrombotic activity were isolated from the flower buds of *S. aromaticum* (L.) Merr. & Perry (Lee et al., 2000) by anion-exchange chromatography, hydrophobic interaction column chromatography, and size exclusion chromatography with molecular weight ranging from ca. 34,000 for the low molecular weight polysaccharide (LMWPS) to the high molecular weight polysaccharide (HMWPS) of ca. 103,000 Da through the usage of butyl-toyopearl 650 M chromatography to eliminate the impurities from proteins due to the hydrophobic characteristics of resin (Lee et al., 2001).

The salt concentration (ca. 1.0 M) reported in the sulfated polysaccharide from cloves might be necessary for elution from the ion-exchange column is slightly lower than ca. 1.5 M, the concentration of algal sulfated polysaccharide (Pavão et al., 1998). This can be partly explained by the relatively low sulfate/sugar ratio in the clove-based polysaccharides. Therefore, Lee et al. (2001) assumed that the polysaccharides from cloves have distinctive structural characteristics compared to algal polysaccharides. The methylation studies on the two clove sulfated polysaccharides confirmed their polymeric complexity and heterogeneity, whereby Lee et al. (2001) reported that both clove polysaccharides are rich in pectic polysaccharides as can be deduced from the natural occurrence of variously linked residues of rhamnosyl, galactosyl, and arabinosyl (Selvendran and O'Neill, 1987). Lee et al. (2001) further added that as most of the uronic acid was shown to be galacturonic acid (GalA), very high amounts of rhamnogalacturonan were likely to be present. Remarkably, significant amounts of 4-linked Glc residues in the HMW fraction suggest the existence of appreciable cellulose in the structure.

Lee et al. (2001) detected the IR absorption bands at 818 cm^{-1} which is regarded as the C-O-S stretching vibration by sulfated polysaccharides (Perlin and Casu, 1982). The sulfates in clove polysaccharides are located at position 6 on the 3-linked Gal, and that is confirmed with the work of Nishino et al. (1994), who reported that the absorption bands at 820 cm^{-1} were caused by 6-sulfated-D-Gal.

In terms of sugar composition (molar %), LMWPS is mainly composed of 24.1% Rha, 18.9% Gal, 18.0% GalA, and 17.9% Ara with 10.8% sulfate and 18.2% protein. Meanwhile, the HMWPS consists of 26.0% Ara, 23.7% Gal, 17.5% Glc, and 12.4% Rha with 15.4% sulfate and 8.0% protein (Lee et al., 2001). Type I rhamnogalacturonan was found to be the backbone of the clove LMW and HMW polysaccharides structures with repeated units of 4-linked GalA, 2-linked Rha, and a side chain of

2,4-branched Rha with arabinan or galactan on position 4, whereby only HMW polysaccharide is likely to contain arabino-3,6-galactan moiety and this assumption is supported by the observation of a strong reactivity of the b-D-glucosyl-Yariv antigen, which reacts with arabinogalactan (Lee et al., 2001).

3. Clove polysaccharides' biological activities

Sulfated polysaccharides possess several potential biomedical applications, such as anticoagulant activity (Mourão, 2004; Fonseca et al., 2009; Zheng et al., 2021), antioxidant activity (Ahmad, 2021), antithrombotic activity (Mourão, 2004; Carvalhal et al., 2019; Chagas et al., 2020), antivirals (Andrew and Jayaraman, 2021; Hans et al., 2021), immunomodulation (Patel, 2012; Chen and Huang, 2018; Barbosa et al., 2019), and tissue engineering (Dinoro et al., 2019). For example, Lee et al. (2000) compared the anticoagulant activity of clove extracts with other herbs such as thyme, sage, oregano, and mugwort and recorded the clotting time (s) with the shortest time of $\pm$40 s for the oregano and clove hot water extracts.

Sulfated polysaccharides from clove extracts are known to possess antithrombotic activity (Lee et al., 2001). Therefore, they presumed that the mode of action for the clove LMW polysaccharide is likely to be a low molecular weight heparin type and the heparin type for the clove HMW polysaccharide is based on the obtained results of the activated partial thromboplastin time (APTT) and thrombin time (TT) assays that are widely used to monitor the therapeutic level of heparin-induced anticoagulation (Chen et al., 2003; Eikelboom and Hirsh, 2006; Undas, 2017; McRae et al., 2021). Although polysaccharides from cloves have an antithrombotic ability, their in vitro activities are lower than that of heparin, considering that the efficacy of antithrombotic agents' comparison with heparin cannot be entirely explained in a single antithrombotic parameter (Lee et al., 2001). The antithrombotic activity of the LMW polysaccharide displayed a low activity (plasma clotting time of 90 s in APTT assay) in comparison to the HMW fraction (145 s).

The animal studies indicated that crude LMW polysaccharides did not show acute toxicity, while the acute LD_{50} of the HMW polysaccharide was approximately twofold lower than that of heparin (Lee et al., 2001). Therefore, the anticoagulation action with no toxicity in vivo of low molecular weight sulfated polysaccharide may represent a valuable and safe antithrombotic agent. The clove LMW polysaccharide has shown less in vitro antithrombotic potency than the HMW polysaccharide. It did not show the acute toxic effect in vivo while further pre-clinical studies are needed to investigate its advantages over the clove HMW polysaccharide to verify its potential application (Lee et al., 2001).

4. Conclusions

Carbohydrates form an essential component of the clove chemical structure; however, data on polysaccharides from clove are lacking, and few studies could be found in the literature. Clove polysaccharides belong to sulfated polysaccharides with molecular weight ranging from 34 to 103 kDa with promising antithrombotic activity. However, only two polysaccharides have been reported from clove extracts with approximately twofold lower acute toxicity for the HMW polysaccharide than that one reported for heparin. Therefore, further studies are recommended on clove polysaccharides to purify, characterize, and unravel their interesting structures-functions relationship using state-of-the-art analytical techniques and instrumentation to explore their potential applications.

References

Ahmad, M.M., 2021. Recent trends in chemical modification and antioxidant activities of plants-based polysaccharides: a review. Carbohydr. Polym. Technol. Appl. 2. https://doi.org/10.1016/j.carpta.2021.100045, 100045.

Andrew, M., Jayaraman, G., 2021. Marine sulfated polysaccharides as potential antiviral drug candidates to treat Corona Virus disease (COVID-19). Carbohydr. Res. 505. https://doi.org/10.1016/j.carres.2021.108326, 108326.

Barbosa, J., Costa, M., Melo, L., Medeiros, M., Pontes, D.L., Scortecci, K.C., Rocha, H., 2019. *Caulerpa Cupressoides* Var. *Flabellata*. Mar. Drugs 17 (2), 105. https://doi.org/10.3390/md17020105.

Beaumont, M., Tran, R., Vera, G., Niedrist, D., Rousset, A., Pierre, R., Shastri, V.P., Forget, A., 2021. Hydrogel-forming algae polysaccharides: from seaweed to biomedical applications. Biomacromolecules 22, 1027–1052.

Bello, M.O., Jimoh, A.A., 2012. Nutrients composition of seed, chemical characterization and fatty acid composition of oil of *Syzygium aromaticum*. Appl. Chem. 42, 6065–6068.

Bernaerts, T.M.M., Gheysen, L., Foubert, I., Hendrickx, M.E., Van Loey, A.M., 2019. The potential of microalgae and their biopolymers as structuring ingredients in food: a review. Biotechnol. Adv. 37. https://doi.org/10.1016/j.biotechadv.2019.107419, 107419.

Carvalhal, F., Cristelo, R.R., Resende, D., Pinto, M., Sousa, E., Correia-da-Silva, M., 2019. Antithrombotics from the sea: polysaccharides and beyond. Mar. Drugs 17 (3), 170. https://doi.org/10.3390/md17030170.

Chagas, F., Lima, G.C., Dos Santos, V., Costa, L., de Sousa, W.M., Sombra, V.G., de Araújo, D.F., Barros, F., Marinho-Soriano, E., de Andrade Feitosa, J.P., de Paula, R., Pereira, M.G., Freitas, A., 2020. Sulfated polysaccharide from the red algae *Gelidiella acerosa*: anticoagulant, antiplatelet and antithrombotic effects. Int. J. Biol. Macromol. 159, 415–421. https://doi.org/10.1016/j.ijbiomac.2020.05.012.

Chen, F., Huang, G., 2018. Preparation and immunological activity of polysaccharides and their derivatives. Int. J. Biol. Macromol. 112, 211–216. https://doi.org/10.1016/j.ijbiomac.2018.01.169.

Chen, C.C., You, J.Y., Ho, C.H., 2003. The aPTT assay as a monitor of heparin anticoagulation efficacy in clinical settings. Adv. Ther. 20 (5), 231–236. https://doi.org/10.1007/BF02849851.

Devkota, H.P., Adhikari-Devkota, A., 2020. Cold pressed clove (*Syzygium aromaticum*) oil. In: Ramadan, M.F. (Ed.), Cold Pressed Oils. Academic Press, pp. 273–276 (ISBN9780128181881) https://doi.org/10.1016/B978-0-12-818188-1.00023-2.

Dinoro, J., Maher, M., Talebian, S., Jafarkhani, M., Mehrali, M., Orive, G., Foroughi, J., Lord, M.S., Dolatshahi-Pirouz, A., 2019. Sulfated polysaccharide-based scaffolds for orthopaedic tissue engineering. Biomaterials 214. https://doi.org/10.1016/j.biomaterials.2019.05.025, 119214.

Eikelboom, J.W., Hirsh, J., 2006. Monitoring unfractionated heparin with the aPTT: time for a fresh look. Thromb. Haemost. 96 (5), 547–552.

Ereifej, K., Feng, H., Rababah, T., Tashtoush, S., Al-U'datt, M., Al-Rabadi, G., Torley, P., Alkasrawi, M., 2015. Microbiological status and nutritional composition of spices used in food preparation. Food Nutr. Sci. 6, 1134–1140. https://doi.org/10.4236/fns.2015.612118.

Fonseca, R.J., Santos, G.R., Mourão, P.A., 2009. Effects of polysaccharides enriched in 2,4-disulfated fucose units on coagulation, thrombosis and bleeding. Practical and conceptual implications. Thromb. Haemost. 102 (5), 829–836. https://doi.org/10.1160/TH08-11-0773.

Guimarães, J.T., Balthazar, C.F., Silva, R., Rocha, R.S., Graça, J.S., Esmerino, E.A., et al., 2020. Impact of probiotics and prebiotics on food texture. Curr. Opin. Food Sci. 33, 38–44. https://doi.org/10.1016/j.cofs.2019.12.002.

Hans, N., Malik, A., Naik, S., 2021. Antiviral activity of sulfated polysaccharides from marine algae and its application in combating COVID-19: mini review. Bioresour. Technol. Rep. 13. https://doi.org/10.1016/j.biteb.2020.100623, 100623.

Hussein, M., Abd El-Hack, M.E., Mahgoub, S.A., Saadeldin, I.M., Swelum, A.A., 2019. Effects of clove (*Syzygium aromaticum*) oil on quail growth, carcass traits, blood components, meat quality, and intestinal microbiota. Poult. Sci. 98 (1), 319–329. https://doi.org/10.3382/ps/pey348.

Jimoh, S.O., Arowolo, L.A., Alabi, K.A., 2017. Phytochemical screening and antimicrobial evaluation of *Syzygium aromaticum* extract and essential oil. Int. J. Curr. Microbiol. App. Sci. 6, 4557–4567.

Kasai, H., Shirao, M., Ikegami-Kawai, M., 2016. Analysis of volatile compounds of clove (*Syzygium aromaticum*) buds as influenced by growth phase and investigation of antioxidant activity of clove extracts. Flavour Fragr. J. 31, 178–184. https://doi.org/10.1002/ffj.3299.

Kaur, K., Kaushal, S., 2019. Phytochemistry and pharmacological aspects of *Syzygium aromaticum*: a review. J. Pharmacogn. Phytochem. 8 (1), 398–406.

Lee, J.I., Lee, H.S., Jun, W.J., Yu, K.W., Shin, D.H., Hong, B.S., Cho, H.Y., Yang, H.C., 2000. Screening of anticoagulant activities in extracts from edible herbs. J. Korean Soc. Food Sci. Nutr. 29, 335–341.

Lee, J.I., Lee, H.S., Jun, W.J., Yu, K.W., Shin, D.H., Hong, B.S., Cho, H.Y., Yang, H.C., 2001. Purification and characterization of antithrombotics from *Syzygium aromaticum* (L.) MERR. & PERRY. Biol. Pharm. Bull. 24 (2), 181–187. https://doi.org/10.1248/bpb.24.181.

Lin, L., Yu, Y., Zhang, F., Xia, K., Zhang, X., Linhardt, R.J., 2019. Bottom-up and top-down profiling of pentosan polysulfate. Analyst 144 (16), 4781–4786. https://doi.org/10.1039/c9an01006h.

Lu, X., Chen, J., Guo, Z., Zheng, Y., Rea, M.C., Su, H., Zheng, X., Zheng, B.D., Miao, S., 2019. Using polysaccharides for the enhancement of functionality of foods: a review. Trends Food Sci. Technol. 86, 311–327. https://doi.org/10.1016/j.tifs.2019.02.024.

McRae, H.L., Militello, L., Refaai, M.A., 2021. Updates in anticoagulation therapy monitoring. Biomedicines 9 (3), 262. https://doi.org/10.3390/biomedicines9030262.

Mohan, K., Muralisankar, T., Uthayakumar, V., Chandirasekar, R., Revathi, N., Ramu Ganesan, A., Velmurugan, K., Sathishkumar, P., Jayakumar, R., Seedevi, P., 2020. Trends in the extraction, purification, characterisation and biological activities of polysaccharides from tropical and sub-tropical fruits—a comprehensive review. Carbohydr. Polym. 238. https://doi.org/10.1016/j.carbpol.2020.116185, 116185.

Mourão, P.A., 2004. Use of sulfated fucans as anticoagulant and antithrombotic agents: future perspectives. Curr. Pharm. Des. 10 (9), 967–981. https://doi.org/10.2174/1381612043452730.

Nishino, T., Takabe, Y., Nagumo, T., 1994. Isolation and partial characterization of a novel β-D-galactan sulfate from the brown seaweed *Laminaria angustata* var. *longissima*. Carbohydr. Polym. 23 (3), 165–173. https://doi.org/10.1016/0144-8617(94)90099-X.

Patel, S., 2012. Therapeutic importance of sulfated polysaccharides from seaweeds: updating the recent findings. 3 Biotech 2 (3), 171–185. https://doi.org/10.1007/s13205-012-0061-9.

Pavão, M.S., Aiello, K.R., Werneck, C.C., Silva, L.C., Valente, A.P., Mulloy, B., Colwell, N.S., Tollefsen, D.M., Mourão, P.A., 1998. Highly sulfated dermatan sulfates from Ascidians. Structure versus anticoagulant activity of these glycosaminoglycans. J. Biol. Chem. 273 (43), 27848–27857. https://doi.org/10.1074/jbc.273.43.27848.

Perlin, A.S., Casu, B., 1982. In: Aspinall, G.O. (Ed.), The Polysaccharides. vol. 1. Academic Press, New York, pp. 133–192.

Ray, B., Schütz, M., Mukherjee, S., Jana, S., Ray, S., Marschall, M., 2020. Exploiting the amazing diversity of natural source-derived polysaccharides: modern procedures of isolation, engineering, and optimization of antiviral activities. Polymers 13 (1), 136. https://doi.org/10.3390/polym13010136.

Saeed, M., Khan, M.S., Alagawany, M., et al., 2021. Clove (*Syzygium aromaticum*) and its phytochemicals in ruminant feed: an updated review. Rend. Lincei Sci. Fis. Nat. 32, 273–285. https://doi.org/10.1007/s12210-021-00985-3.

Selvendran, R.R., O'Neill, M.A., 1987. Isolation and analysis of cell walls from plant material. Methods Biochem. Anal. 32, 25–153. https://doi.org/10.1002/9780470110539.ch2.

Sulieman, A.M.E., El Boshra, I.M.O., El Khalifa, A.A., 2007. Nutritive value of clove (*Syzygium aromaticum*) and detection of antimicrobial effects of its bud oil. Res. J. Microbiol. 2, 266–271.

Undas, A., 2017. Determination of fibrinogen and thrombin time (TT). Methods Mol. Biol. 1646, 105–110. https://doi.org/10.1007/978-1-4939-7196-1_8.

Wang, Z.J., Xie, J.H., Shen, M.Y., Nie, S.P., Xie, M.Y., 2018. Sulfated modification of polysaccharides: synthesis, characterization and bioactivities. Trends Food Sci. Technol. 74, 147–157. https://doi.org/10.1016/j.tifs.2018.02.010.

Yang, X., Li, X., Sun, L., Guo, Y., 2020. An overview of classifications, properties of food polysaccharides and their links to applications in improving food textures. Trends Food Sci. Technol. 102 (2020), 1–15. https://doi.org/10.1016/j.tifs.2020.05.020.

Zheng, J., Li, B., Ji, Y., Chen, Y., Lv, X., Zhang, X., Linhardt, R.J., 2021. Prolonged release and shelf-life of anticoagulant sulfated polysaccharides encapsulated with ZIF-8. Int. J. Biol. Macromol. 183, 1174–1183. https://doi.org/10.1016/j.ijbiomac.2021.05.007.

CHAPTER

Clove (*Syzygium aromaticum*) oleoresins

5

Tossaton Charoonratana
Department of Pharmacognosy, College of Pharmacy, Rangsit University, Lak-Hok, Mueang Pathum Thani, Thailand

Abbreviations

CFR	Code of Federal Regulations
CO	clove oleoresins
DAD	diode array detector
DENV	dengue virus
FCC	Food Chemical Codex
FEMA	The Flavor and Extract Manufacturers Association
FID	flame ionization detector
GC	gas chromatography
GRAS	Generally Recognized As Safe
HPLC	high-performance liquid chromatography
JECFA	The Joint FAO/WHO Expert Committee on Food Additives
MS	mass spectrometer
NMR	nuclear magnetic resonance
SFE	supercritical fluid extraction
TTC	the conservative threshold of toxicological concern
UAE	ultrasound-assisted extraction

1. Introduction

Spice oleoresins are resin-like viscous materials obtained by solvent extraction and possessing the full character of natural spices. They comprise volatile and nonvolatile components such as terpenes, phenolics, alcohols, fatty acids, sterols, and waxes. It is known that eugenol, eugenyl acetate, and β-caryophyllene are the major compounds which strongly influence clove characteristic. All mentioned compounds are readily soluble in an organic solvent but slightly water-soluble. Thus, clove oleoresins (CO) are semisolid material likely obtained from *Syzygium aromaticum* using extraction of the solvents except for water. The mixture of water and alcohol as the extractive solvent can be considered as clove extracts discussed in another chapter.

Moreover, supercritical fluid extraction (SFE) using CO_2 is an alternative method for CO extraction. Extraction pressure ≥ 10 MPa and temperature $\geq 40°C$ are suitable for CO extraction (Yazdani

Clove (*Syzygium aromaticum*). https://doi.org/10.1016/B978-0-323-85177-0.00001-X

et al., 2005; Ivanovic et al., 2013). On the other hand, at lower pressure and temperature, volatile oil is selectively extracted from clove (Reverchon and Marrone, 1997).

The implication from the Code of Federal Regulations (CFR) title 21 section 582.20 CO can be obtained by extracting flower buds, twigs, and leaves (FDA, 2020b). However, the yield and composition of CO are significantly varied depending on solvent uses and extraction methods. For example, it was reported that CO from buds prepared by ethanol extraction yielded about 22% to 31%, while using benzene provided 18% to 22% yield (Weiss, 1997).

CO is generally used as a flavoring agent and adjuvant in food products. Recently, according to CFR title 21 section 184.1257(a), CO has been listed as GRAS (Generally Recognized As Safe) (FDA, 2021). It can be used as a substance added to impart or help impart a taste or aroma in food like clove volatile oil (FDA, 2020a). Even some extraction processes of CO are more expensive than the distillation of clove volatile oil; CO contains low contamination, mainly if produced organically. Thus it may fit niche markets such as cosmetics and perfumery. Moreover, CO also possessed unique chemical constituents that affected its biological activity, which can be a new source of compounds in the medical field.

The characteristic of CO is affected by factors like plant varieties which will be discussed. The influences of plant ages, plant organ maturity, cultivation area, and postharvesting process were discussed entirely in the chapter on clove volatiles. It was found that all factors affected clove volatile oil constituents; thus, it was implied to relate to CO. This chapter also provides information about CO characteristics, composition, extraction method, the present application, and possible application according to its safety and efficacy data.

2. Clove oleoresins characteristics and compositions

The specification of CO is recommended in CFR title 21 section 184.1257(b)(1)-(5). It was noted that CO had specific gravity between 1.036 and 1.060 g mL^{-1}, the angular rotation was between −2 and 0 degrees, and the refractive index was between 1.527 and 1.538 at 20°C (FDA, 2021). The eugenol content was not less than 85% (Committee on Codex Specifications, 1996). Moreover, the residual solvents are limited for spice oleoresins, and they are listed in CFR title 21 section 173 subpart C.

CO is rich in valuable compounds, and the composition is varied since various solvents extract it. Eugenol, eugenyl acetate, oleanolic acid, biflorin, isobiflorin, and gallic acid were presented when clove buds were extracted with dichloromethane mixture methanol (7:13) by ultrasonication (Ladurner et al., 2017). Moreover, it was reported that clove bud residues extracted with ethanol by maceration had biflorin, isobiflorin, quinic acid, gallic acid, and quercetin (Fathoni et al., 2017). The presence of those compounds along with acetophenone derivatives was proved by Nuclear Magnetic Resonance (NMR) spectroscopy (Ryu et al., 2016). Another publication showed the successive extraction of CO from buds using various solvents. It was found that eugenol and eugenyl acetate were the major compounds in hexane CO. Low amount of limonin and ferulic aldehyde was presented in dichloromethane fraction. In contrast, a trace amount of tamarixetin 3-*O*-β-D-glucopyranoside, ombutin 3-*O*-β-D-glucopyranoside, and quercetin was shown in ethanolic oleoresins (Dwivedi et al., 2011). Vitamin E, sitosterol, diethyl phthalate, and long-chain hydrocarbons were detected when clove materials were extracted by SFE (Wenqiang et al., 2007; Frohlich et al., 2019).

The chemical composition of CO was not deeply studied. Generally, the major compounds were almost similar to clove volatile oil; it was found that eugenol was the most abundant following by β-caryophyllene, α-humulene, and eugenyl acetate. Other compounds that are not present in volatile oil can be found in high amounts, such as oleanolic acid, biflorin, isobiflorin, and 2,4,6-trihydroxyacetophenone-3-*C*-β-D-glucoside.

Even the extraction is the most crucial process-affected characteristic of CO. Other factors also play an essential role because of the variable nature of the constituent profile of plants.

3. Clove oleoresins from different plant varieties

The plant genotypes affected its chemical compositions. It was reported that cold-pressed rice bran oil from different cultivars possessed different levels of oryzanol, tocopherol, and fatty acids (Charoonratana et al., 2015). Several CO from various Indonesian clove varieties were investigated for the chemical constituents.

A group of Indonesian researchers reported a yield and major compound content of leaves oleoresin from Zanzibar, Sikotok, Siputik, and Raja varieties. The yield of CO using hexane extraction was 2.57%. The chemical composition was analyzed by GC-MS, revealing eugenol, β-caryophyllene, α-humulene, and caryophyllene oxide. Only Zanzibar had tetratetraconten (3.12%), while Sikotok had β-isopropylidene glycerol (16.5%). The eugenol content was in this order Siputih > Raja > Zanzibar > Sikotok. The β-caryophyllene content was conversely to eugenol (Sikotok > Zanzibar > Raja > Siputih) (Wael et al., 2018a,b). The amount of compounds CO from *Syzygium aromaticum* buds, twigs, and leaves from different varieties is reported in Table 1.

Table 1 Chemical composition (%) of CO from *Syzygium aromaticum* buds, twigs, and leaves from different varieties.

Clove varieties	Plant parts	Extraction solvent	Compound content			
			Eugenol	β-Caryophyllene	Caryophyllene oxide	α-Humulene
Zanzibar	Leaves	Hexane	70.43	16.79	2.07	3.05
Sikotok	Leaves	Hexane	48.33	28.80	3.40	2.07
Siputih	Leaves	Hexane	80.15	13.44	4.90	1.50
Raja	Leaves	Hexane	77.24	16.15	5.08	1.52
Afo	Buds	Methanol	56.57	15.43	1.08	–
	Twigs	Methanol	85.50	9.68	2.68	–
	Leaves	Methanol	92.51	4.23	1.78	–
Gorontalo	Buds	Methanol	62.02	10.60	0.81	–
	Twigs	Methanol	88.96	7.50	1.26	–
	Leaves	Methanol	92.51	2.60	n.d.	–
Tuni	Buds	Methanol	36.43	13.09	1.82	–
	Twigs	Methanol	88.93	n.d.	2.01	–
	Leaves	Methanol	91.18	n.d.	4.01	–

n.d., not detected; –, not reported.

Another group from Java studied comparing the primary composition of methanolic oleoresins from Afo, Gorontalo, and Tuni clove varieties. It was found that the average eugenol content in CO was highest in Gorontalo (81.1%) and lowest in Tuni (72.1%). On the other hand, the β-caryophyllene content was highest in Afo (9.78%) and lowest in Gorontalo (6.90%). From Table 1, it was shown that the Afo clove bud was suitable for β-caryophyllene extraction, while Afo and Gorontalo clove leaves were appropriate materials for eugenol extraction (Taher et al., 2019).

It is difficult to conclude the effect of variety on CO since only two studies were presented. However, it was suggested that CO from leaves of Zanzibar, Siputih, Raja, Afo, Gorontalo, and Tuni was a good source for eugenol extraction. In contrast, CO from the leaves of Sikotok was rich in β-caryophyllene.

4. Clove oleoresins extraction

The extraction of oleoresins from spices has been the subject of research interest because of the possible uses in various applications. Extraction is the step to isolate the desired chemical compounds from spice materials. Solvent extraction is the most extensively used process. The extraction process proceeds through four phases; firstly, the solvent penetrates the spice matrix. Secondly, the compound dissolves in the solvents. Thirdly, the compound is diffused out of the spice matrix. Finally, the extracted compounds are collected. Various methods enhancing the diffusivity and solubility in the above steps will assist the extraction. Many extraction methods include maceration, ultrasound-assisted extraction (UAE), Soxhlet extraction, and SFE to extract oleoresins from spices. Besides the extraction method, it was found that the types of solvent, the particle size of spice materials, the solvent to solid ratio, the extraction temperature, and the duration affect the extraction efficiency (Zhang et al., 2018). The solvent selection is crucial for solvent extraction; for example, it was found that using different solvents provided different chemical profiles in the same herbal recipe (Charoonratana et al., 2014).

There was a study of ethanolic CO extraction from buds using the maceration method by varying the particle size, temperature, time, and solvent to solid ratio. It was found that the optimal particle size for extraction of eugenol from clove at room temperature is 250 μm. In comparison, the optimal eugenol extraction temperature for particle size of 250 μm is 50°C. Furthermore, the solvent to solid ratio of 9.4:1 with 60 min extraction time provided 37.8% CO yield (Myint et al., 1995).

UAE uses acoustic waves that travel through the solvent, generating cavitation bubbles. When the cavitation bubbles collapse, which produces a shockwave of solvent toward spice materials, shockwave-induced damage to the plant cell wall boosts the mass transfer of compounds across cellular membranes into the solvent (Vinatoru et al., 2017). Using UAE in CO extraction from buds, a study reported that the optimal condition of UAE was at the temperature of 60°C for 45 min using 5% clove concentration; the CO yield was 28.2%. Eugenol was found to be the major compound (64.0%). Other compounds were β-caryophyllene, α-humulene, caryophyllene oxide, and 2-methoxy-4-(2-propenyl) phenol acetate (Tekin et al., 2015).

For efficient extraction, the increasing temperature may liberate more compounds from spice materials. Extraction at a high temperature can be performed using the Soxhlet apparatus. The Nigerian researchers optimized extraction time and solvent ratio for CO extraction from buds by the Soxhlet apparatus. It was found that the optimal extraction time of 120 min and hexane/ethanol ratio of 1.00 at 90°C provided a 30.2% CO yield. The major compounds in CO were eugenyl acetate (46.8%) and 16-Octadecenoic acid, methyl ester (16.4%) (Galadima and Muhammad, 2016). Another study from China performed the Soxhlet extraction using 30 g of clove buds in ethanol for 6 h. The yield of the extraction was 41.8%. The major compounds in CO were eugenol (57.2%), eugenyl acetate (19.3%), and β-caryophyllene (17.5%). In this experiment, the CO yield was high because ethanol is considered a universal solvent in plant extraction (Wenqiang et al., 2007). However, noting from the previous study, the CO yield was dropped when hexane was added to ethanol as the cosolvent for extraction (Galadima and Muhammad, 2016). Since hexane is a solvent specifically for oil extraction, it was suggested that a high amount of nonvolatile phenolic compounds is also presented in cloves. Besides this point, it was apparent that eugenol was absent in CO from Nigeria (Galadima and Muhammad, 2016). The data in the chapter on clove volatiles suggested that clove buds from Madagascar (which is expected to export to Nigeria) naturally contained a high amount of eugenyl acetate. Moreover, if the farmer collected the clove buds younger than the full budding stage, the eugenyl acetate content would be very high, while eugenol content was low.

The observation was also done in clove leaves using hexane as the extraction solvent. The researcher performed the Soxhlet extraction using 10 g of clove leaves for 6 h. The yield of the extraction was 1.90%. The CO contained eugenol 1.12 g/kg leaves (Frohlich et al., 2019).

SFE method is another alternative method that can be used in CO extraction from spices. It provides oleoresins without contamination of organic solvents. Several studies evaluated the extraction with supercritical CO_2 of cloves. Firstly, the study assessed the effects of temperature and pressure on the yield and composition of CO from buds. It was found that the increase of temperature from 30°C to 40°C increased the CO yield and high eugenol content, while the rise of temperature from 40°C to 50°C did not increase the CO yield, but the eugenol content increased.

Furthermore, the extraction yield boosted notably with the increase of pressure. However, more fatty acids, sterol, and wax were coextracted; it is attributed to enhancing the CO_2 density, increasing their dissolving ability. The best condition was at 50°C and 10 MPa, which provided a CO yield of 19.6% and eugenol plus eugenol acetate content of 58.8% (Wenqiang et al., 2007).

Another study was performed on clove leaves. The best result for SFE was 40°C and 22 MPa, which provided the highest CO yield (1.08%) and eugenol content (29.8%) (Frohlich et al., 2019). Furthermore, it was noticed that the CO yield from clove buds was higher than leaves; the same result happened in clove volatile oil extraction. Thus, it was suggested that the volatiles part was richer in buds than in leaves.

CO obtained from Soxhlet extraction was compared to CO acquired from SFE. It was found that Soxhlet extraction provided more CO yield than the latter; however, the content of eugenol and eugenol acetate in CO from Soxhlet was about two times lower than CO from SFE. The oleoresins from Soxhlet were brown ointment which referred to undesired impurities (if the purpose was to extract eugenol). SFE had the advantage of a shorter extraction time but required more investment. The user can choose the method depending on the meaning of extraction. The composition of CO obtained from different extraction methods was shown in Table 2.

5. Detection of compounds in clove oleoresins

The specification of eugenol in CO is recommended in CFR title 21 section 184.1257(b)(1). Furthermore, it is referred to the Food and Chemical Codex (FCC) that the eugenol content is determined as phenols using the method in Appendix VI, and the value is not less than 85.0% (Committee on Codex Specifications, 1996).

Table 2 Oil yield (%w/w) and major constituents (%) of CO from *Syzygium aromaticum* buds and leaves from different extraction methods.

Plant part	Extraction method	Conditions	Time (min)	Oil yield (%w/w)	Compounds content (%)		
					Eugenol	Eugenyl acetate	β-Caryophyllene
Buds	Maceration	50°C; 9.4:1 solvent to solid ratio	60	37.80	–	–	–
	UAE	60°C; 5% clove concentration	45	28.20	64.00	–	–
	Soxhlet	>78°C; 30g clove	360	41.80	57.24	19.34	17.50
	SFE	50°C; 10MPa	120	19.60	57.36	22.34	13.99
Leaves	Soxhlet	>69°C; 10g clove	360	1.90	5.67	–	–
	SFE	40°C; 22MPa	80	1.08	29.84	–	–

–, *not reported.*

In CO, both volatiles and nonvolatiles are mixed. However, gas chromatography (GC) coupling with flame ionization detector (FID) or mass spectrometer (MS) is only used for volatiles detection. According to AOAC International, the chemical composition of CO was analyzed by high-performance liquid chromatography (HPLC) coupling with a diode array detector (DAD). The CO was prepared by reflux ground clove sample (0.3g) in ethanol (30mL) for 2h. It was filtrated through a 0.45μm syringe filter before analysis. The stationary phase was the reverse phase (C18) for the chromatographic condition, and the mobile phase was methanol:water (60:40). The analysis was performed at a flow rate of 0.8mL/min, the column temperature was kept at 30°C, and the detection wavelength was recorded at 280nm. Eugenol was the main target, and it was eluted within 10min. The limit of quantification was 2.47ng/mL (Yun et al., 2010). This method can be modified and applied for other major compounds detection in CO, such as eugenyl acetate, oleanolic acid, biflorin, isobiflorin, and 2,4,6-trihydroxyacetophenone-3-*C*-β-D-glucoside.

6. Clove oleoresins application

The odor and flavor of CO are regarded as superior to distilled oil and much closer to the natural spice. The advantages of CO over dry cloves are little risk of bacterial contamination and standard strength and quality. It is gradually being used in the food and perfume industries in place of whole or ground spices. The recent annual usage of clove bud oleoresin in the United States was 390μg/person/day (Harman and Murray, 2018).

CO used as flavoring agents is often standardized to contain a specific percentage of volatile oil by diluting food-grade ingredients. For oil-based applications, CO may be diluted with edible vegetable oil. In addition, CO may be diluted with a polysorbate ester, resulting in a water-soluble standardized oleoresin (Reineccius, 1994). It was found that clove bud oleoresins standardized to contain volatile oil ranging from 82% to less than 25% may be used according to the Flavor and Extract Manufacturers Association (FEMA) no. 2324 (Gooderham et al., 2020). The literature also reported using alcoholic beverages in flavoring (Leung, 1980).

7. Prospect in clove oleoresins application

This section presented an overview of prospects in CO application based on the safety and efficacy data of the oleoresins and their components.

7.1 Clove oleoresins safety

According to CFR title 21 section 184.1257(a), CO has been listed as GRAS (FDA, 2021). It can be used as a flavoring agent and adjuvant in food products. Even it was revealed that the estimated intake of unidentified compounds from clove bud oleoresin (FEMA 2324) in the US population was 12 to 35 μg/person/day, which was lower than the conservative threshold of toxicological concern (TTC) for Structural Class III (Kroes et al., 2000; Munro et al., 1996). It was interesting to evaluate the safety of CO based on the major or potent constituents (besides eugenol) expected in volatile oil, resin, and standardization ingredient content. Methyl eugenol, eugenyl acetate, β-caryophyllene, α-humulene, oleanolic acid, biflorin, isobiflorin, and 2,4,6-trihydroxyacetophenone-3-*C*-β-ᴅ-glucoside are selected for safety consideration.

For clove bud oleoresin (FEMA 2324), the range for the estimated intake of methyl eugenol from the use as a flavoring is from 0.04 to 0.13 μg/person/day. Thus, estimated intakes from the natural occurrence of methyl eugenol in clove bud oleoresin (FEMA 2324) are below the TTC of 0.15 μg/person/day for compounds with structural alerts for genotoxicity (Kroes et al., 2004).

The Committee in 65th and 78th meetings of the Joint FAO/WHO Expert Committee on Food Additives (JECFA) included eugenyl acetate and β-caryophyllene in no safety concern list (Additives, 65th report of the Joint FAO/WHO Expert Committee on Food Additives, 2005; Additives, 78th report of the Joint FAO/WHO Expert Committee on Food Additives, 2014). It is recommended to use 1 to 5 mg/kg of eugenyl acetate for food products (European Food Safety Authority, 2011).

Cytotoxicity was observed in human acute promyelocytic leukemia cell line HL-60 treated with oleanolic acid at a concentration twice of its IC_{50} value (loss of more than 20% viable cells) (Wang et al., 2006). The acute toxicity of oleanolic acid was performed consistently with the OECD guideline no. 425. The LD_{50} value in mice, higher than 2.0 g/kg, indicated that this compound was low toxic (Mshelia et al., 2019).

The toxicological information of several compounds such as α-humulene, biflorin, isobiflorin, and 2,4,6-trihydroxyacetophenone-3-*C*-β-ᴅ-glucoside have not been significantly investigated. It was reported that α-humulene may cause oxidative damage to the cell membrane (Agus, 2021). Biflorin, isobiflorin, and 2,4,6-trihydroxyacetophenone-3-*C*-β-ᴅ-glucoside were only said to have no cytotoxic activity against human ovarian cancer cells (A2780) (Ryu et al., 2016).

7.2 Food preservative

The increasing consumer awareness of the potential hazard of synthetic food preservatives has provoked the search for alternative natural preservatives to improve safety and food quality. CO had antioxidant and antimicrobial activities, which means that it can be considered a food preservative.

7.2.1 Antioxidant

Oxidation can cause deterioration in food during cooking and storage, leading to off-flavor, loss of color and texture, and decreased nutritional value. Natural antioxidants are fascinating since the concern for the potential health hazard of synthetic antioxidants is increasing. The study found that CO possessed antioxidant activity compared with butylated hydroxyanisole and *tert*-butylhydroquinone (synthetic antioxidants), inhibiting lipid oxidation and extension shelf life of a cooked chicken and chevon meat during storage at 4°C for 21 days (Sultana et al., 2018). Moreover, it was reported that hydroethanolic clove extract had higher antioxidant activities than CO using ethyl acetate extraction in all experiments, DPPH· radicals scavenging assay, $ABTS^+$ radical scavenging activity, and β-carotene/linoleic acid bleaching assay (Abo El-Maati et al., 2016).

7.2.2 Antimicrobial activity against foodborne pathogens

CO possesses inhibitory activity against food pathogens such as *Escherichia coli*, *Pseudomonas aeruginosa*, *Salmonella enterica*, and *Staphylococcus aureus* (Pandey and Singh, 2011; El-Shawaj et al., 2012; Hoque et al., 2007). It was reported that the antimicrobial activity of CO is due to the presence of phenolic compounds. Hydroethanolic clove extract and ethyl acetate CO contained a high amount of phenolics showed relatively high antimicrobial activity. In addition, hydroethanolic followed by ethyl acetate extract showed inhibition zone diameter with tested Gram-positive and Gram-negative bacteria (*Escherichia coli*, *Listeria monocytogenes*, *Salmonella enterica*, *Serratia marcescens*, and *Staphylococcus aureus*) (Abo El-Maati et al., 2016).

It was found that soaking poultry meat in 0.5% CO could prevent the growth of *Pseudomonas aeruginosa*. The product can be stored at 4°C for more than 1 week without the formation of slime, off-odors, and off-colors (Carlos and Harrison, 1999). CO mixing with sodium alginate was used as a coating material to improve breast fillet stability. It was found that meat coating with CO had the best shelf life and lasted till the 18th day of chiller storage (4°C). CO at 3% to 5% can be used for prolonging the shelf life of alginate-coated breast (James, 2018).

Ghee is a clarified butter commonly used in the cuisine of the Indian subcontinent. In a study, CO was used to improve the stability of ghee. The samples added with 0.5% CO was found to be improved in peroxide value, conjugated dienes, thiobarbituric acid value, fatty free acid, and radical scavenging activity. Moreover, 0.5% CO had higher antioxidant properties during deep-frying than synthetic antioxidant butylated hydroxyanisole (Shende et al., 2014).

The mechanism of antimicrobial activity of CO is due to eugenol. It was found that eugenol can induce cellular protein and lipid leakage on bacterial cell walls (Oyedemi et al., 2009). In *E. coli* and *L. monocytogenes*, eugenol inhibited adhesion genes responsible for bacterial adhesion (Baskaran et al., 2016; Upadhyay et al., 2013). In *Salmonella* spp., it was suggested that eugenol expressed the inhibition against efflux pump, disrupt the bacterial membrane, and caused cytoplasm leakage (Miladi et al., 2017).

Ethanolic CO seems more appropriate in food preservation than CO using ethyl acetate as an extractive solvent since it is an edible solvent. In addition, it possesses high antioxidant and antifungal activities.

7.3 Pharmaceuticals

Few investigations have described the therapeutic potential of CO comparing to clove volatile oil. The pharmacological activities of CO have also been thought to be due to the existence of its volatiles and nonvolatile compounds such as biflorin, isobiflorin, and oleanolic acid. This section will summarize the potentials of CO to develop into preparations to cure some diseases.

7.3.1 Dental formularies and related products

Porphyromonas gingivalis is a human pathogen that causes oral inflammation that destroys the tissues supporting the tooth, eventually leading to tooth loss. Crude methanol oleoresin of clove exhibited inhibitory activity against *Porphyromonas gingivalis* (Cai and Wu, 1996). It was found that eugenol could destroy the integrity of the cell membrane of *Porphyromonas gingivalis*. Moreover, eugenol inhibited biofilm formation and reduced the preformed biofilm of *Porphyromonas gingivalis*. The down-regulation of genes related to biofilm (*fimA*, *hagA*, *hagB*, *kgp*, *rgpA*, *rgpB*) suggested that eugenol suppressed biofilm formation at the initial stage (Zhang et al., 2017). Therefore, herbal mouthwash containing CO can be formulated to prevent the disease.

Candida albicans causes superficial infection such as oral candidiasis, which causes loss of taste and pain while eating or swallowing. CO (0.4%) dispersed in a concentrated sugar solution was reported to have fungicidal activity (99.6% reduction) on *C. albicans* after 60 min (Nunez et al., 2001). The fungicidal activity against *C. albicans* was from eugenol, which causing cell leakage and increase cell permeability. The structure and cell surface morphology were also altered by eugenol (Latifah-Munirah et al., 2015).

Various concentrations of CO were prepared to investigate its potential for denture cleansing. The denture base material is a part of the artificial tooth attached to the supporting tissue and where the adhesive of the artificial tooth is applied. Thermoplastic nylon is a material that has many advantages, such as good aesthetics and high flexibility; however, the discoloration can occur by pigment dyes from food. It was found that 0.8% CO could prevent discoloration of thermoplastic nylon compared to artificial denture cleansers (Wibawaningtyas et al., 2017).

7.3.2 Cancer drugs

Few studies have been investigated the anticancer activity of CO using in vitro experiments. It was reported that CO possesses anticancer activities against MCF-7 and MDA-MB-231 breast cancer, HeLa cervical cancer, HT-29 colorectal cancer, TE-13 esophageal cancer, BEL-7402, and SKOV-3 liver cancer, PANC-1 pancreatic cancer, and DU-145 prostate cancer (Liu et al., 2014; Dwivedi et al., 2011). Eugenol was proposed to be the compound responsible for anticancer activity against MCF-7, HeLa, TE-13, and DU-145.

The in vivo study showed that ethyl acetate CO exhibited anticancer activity against HT-29 xenograft in mice. It was suggested that oleanolic acid, one of the major active compounds in CO, was responsible for this activity (Liu et al., 2014).

The researchers were preparing CO in the nanoparticle to improve its efficacy against cancer. One study found that methanolic CO loaded with polyvinylpyrrolidone and Iron oxide nanoparticles induced apoptosis in MCF-7 breast cancer cell lines. The nanoparticle also activated caspase-3, caspase-8, and caspase-9 response and induced oxidative stress through ROS formation (Thenmozhi, 2020). From another study, the nanoparticle CO was synthesized using chitosan and calcium alginate. It could improve the IC_{50} against HeLa cervical cancer cell lines from 22.5 to 6.2 μg/mL compared to ethanolic CO (Fadilah et al., 2019). However, to conclude that CO can be used as anti-cancer drugs, all evidence still needs more support data from clinical trials.

7.3.3 Diabetes drugs

Diabetes and its complications cause morbidity and mortality in many countries and contribute significantly to health care costs. Long-term administration of synthetic diabetes drugs can cause weight loss, heart failure, and intestinal disorders. Therefore, effective alternative therapy could be developed (Nathan, 2015).

It was reported that CO could upregulate Takeda G protein-coupled receptor 5, which is a membrane-type receptor for bile acid expressing in a wide variety of tissues. Activation of this receptor had been shown to enhance insulin secretion in the pancreas (Ladurner et al., 2017). Moreover, CO and its constituents, dehydrodieugenol and dehydrodieugenol B, had potent PPAR-γ ligand-binding activities which stimulate 3T3-L1 preadipocyte differentiation (Kuroda et al., 2012). This data may explain the therapeutic potential for CO against diabetes.

From in vivo study, the ethanolic CO significantly suppressed an increase in blood glucose level in type 2 diabetic KK-A^y mice (Kuroda et al., 2012). In the metaanalysis from the studies in rodents, it was concluded that eugenol, which is the major compound in cloves, affected diabetes by inhibiting carbohydrate-metabolizing enzymes (Carvalho et al., 2021).

7.3.4 Immunomodulators

These are natural compounds that can modulate any aspect of the immune system, including the immune system's adaptive and innate arms. A study reported the possibility of CO utility as an immunomodulator. It was found that oleoresins from clove leave extracted with hexane significantly increased the number of lymphocytes, lymphoblasts, and ROI secretion of macrophages in Balb/c mice infected with *Salmonella typhimurium* at a dose of 150 mg/kg BW (Wael et al., 2018b). However, a lot more researches are needed to prove its immunomodulatory activity.

7.3.5 Antimalarial agents

The study was performed both in vitro and in vivo. Regarding in vitro study, methanolic CO extracted from buds and twigs had the IC_{50} against *Plasmodium falciparum* at 1 to 10 μg/mL. The effect was not expected to be from eugenol alone since pure eugenol showed weak activity in the same test. Regarding in vivo study, methanolic CO extracted from buds and twigs were fed to mice infected with *Plasmodium berghei*. It was found that CO from buds and twigs suppressed the growth of *P. berghei* after the first date and second date, respectively. Furthermore, both CO showed inhibitory activity at low doses (25 mg/kg BW).

Moreover, this study indicated that methanolic CO extracted from buds and twigs had LD_{50} 19,743 and 47,304 mg/kg BW, respectively. It was nontoxic in mice (Taher et al., 2018). Further study finding the compounds responsible for the activity was fascinating.

7.3.6 Hangover supplements

Acetaldehyde is a metabolite of alcohol formed by the action of alcohol dehydrogenase. The current knowledge is that the blood levels of acetaldehyde and its metabolite play an essential role in mediating alcohol-related side effects, including hangover. It was found that polyphenol in clove buds extracted using ethanol can alleviate the oxidative stress and inflammation caused by the accumulation of acetaldehyde after heavy drinking. Thus, the result supported CO as an effective herbal supplement to reduce hangover severity (Mammen et al., 2018).

7.3.7 Antiviral agents

Identification of lead molecules to fight COVID-19 was performed by computational approach. Ligands from the natural compounds library at Selleck Inc. (L1400) were screened for their capability to inhibit the main protease (3CLpro) of SARS-CoV-2. It was found that biflorin from cloves showed high affinities for Mpro (−8.5 kcal/mol), suggesting its potential inhibitory activity (Rehman et al., 2020).

Biflorin and isobiflorin were also found to inhibit the protease of the Dengue virus (DENV). The IC_{50} values of biflorin and isobiflorin against DENV2 protease were found to be 89.6 and 58.9 μM, respectively, while the values against DENV3 protease were found to be 219 and 337 μM, respectively. Furthermore, atomic-level details of the binding of these two phytochemicals to DENV serotype-2, -3 were obtained by computational approach and saturation transfer difference NMR spectroscopy, which showed that the molecular recognition at the active site of the DENV protease-mediated by a network of hydrogen bonds and hydrophobic contacts (Saleem et al., 2019).

7.3.8 Hepatoprotective supplements

Oleanolic acid can be found in a high amount in CO. Oleanolic acid at low concentration could exhibit hepatoprotective activity, as demonstrated in different works of literature (Liu, 2005; Chen et al., 2014). However, prolonged exposure to increased doses of oleanolic acid resulted in cholestasis (Lu et al., 2013; Vilas-Boas et al., 2020).

From pharmacokinetic studies in rodents, the hepatoprotective activity of oleanolic acid was suggested to link to its low total plasma clearance after the i.v. injection (Jeong et al., 2007). Also, a high concentration of oleanolic acid was determined in the liver and kidney after a 4 weeks consumption of a diet with 0.5% of this triterpene (Yin et al., 2012). However, from pharmacokinetic studies in humans, the elimination rate of oleanolic acid was not low when taken as a compound or ingredient in the food (Song et al., 2006; De la Torre et al., 2020). The hepatoprotective of oleanolic acid occurred by introducing metallothionein synthesis to prevent glutathione emptying and inhibiting several subtypes of Cytochrome P450 (Liu, 1995; Liu et al., 1993).

7.4 Biopesticides

Developing new plant protection materials is needed since synthetic pesticides possess high toxicity and nonbiodegradable properties. Biopesticide from natural products is a good option for artificial pesticide replacement. CO was reported to use as biopesticides in some studies. For example, CO could be used in the control of cocoa fruit rot caused by *Phytophthora palmivora*. It was found that 1% methanolic CO from leaves completely prevented the damage of cocoa fruit caused by *Phytophthora palmivora* (Dewi et al., 2019). Moreover, it was found that ethanolic CO could reduce the acceleration of the ripening and fungal spoilage in navel oranges. Thus, it could alleviate fruit deterioration during postharvest storage, but it has little effect on weight loss (Zeng et al., 2012).

7.5 Animal feeds

Due to human health concerns about antibiotics use and their effects on humans' antibiotic resistance, the antibiotic growth promoter has already been banned in some countries. Therefore, the use of phytogenic plants in poultry feeds has been examined relative to their stimulation of digestion (Basmacioğlu Malayoğlu et al., 2010). It was found that 400ppm supplementation of CO to diets improved the digestibility of the feeds for broilers (Dalkilic and Guler, 2009). Moreover, ethanolic CO could increase testosterone which may help to enhance reproductive performance in West African Dwarf goats (Omotayo et al., 2018).

7.6 Fish farming

Anesthesia is helpful in fish farming in such practices as egg extrusion, fish biometric, and transportation. It can decrease fish motility, the intensity of stress, and reduce physical damage, thus protecting the vital physiological functions of fish. Ethanolic CO was reported to be used as fish anesthesia. The optimal concentration producing total loss of balance and response to external stimuli with shorter induction time was CO from buds followed by CO from leaves at 50 and 100mg/L with a duration time of 20.1 and 22.3min, respectively (Wongphaosakul et al., 2017).

In aquaculture farming, fish are susceptible to microbial pathogens, especially when subjected to poor water quality and overstocking. Infectious diseases are the primary cause of commercial losses in the aquaculture industry. In African sharp tooth catfish, the infection is regularly caused by *Aeromonas* spp., *Pseudomonas* spp., *Vibrio* spp., and *Staphylococcus* spp. It was found that CO showed a strong inhibitory effect against all microbes. CO could reduce the mortality rate of catfish infected with *Aeromonas sobria* from 83% to 16%. Productive performance parameter was also improved in the group treated with CO (Ghaly et al., 2015).

7.7 Miscellaneous products

In South Korea, reports were utilizing CO in different ways. Firstly, methanolic CO was used as a natural colorant in silk fabrics. Dyeing silk fabrics in CO at 60°C for 60min had good wet-fastness. CO dyed brownish-black on the Fe-mordanted silk while dyed yellow on other mordants (Lee et al., 1997). Secondly, CO was reported to use as the ingredient in eco-friendly biocide for biofilm removal on outdoor stone monuments. Excellent antifungal activity was confirmed against *Penicillium* sp., *Trichoderma* sp., *Aspergillus* sp., and the others isolated from the stone monument. Moreover, the biocide was applied to the stonework in the Korea National University of Cultural Heritage. The observed ATP reduction indicated that the treatment with the biocide with CO reduced the number of fungus biofilm (Jeong et al., 2018).

8. Conclusion

CO is the mixture of volatiles and nonvolatiles from cloves proven to be used safely as flavoring agents. Moreover, since its specification is almost similar to clove bud oil (CAS number 8000-34-8), it was suggested to use nontoxic biopesticides if the cost of production is less than clove bud oil distillation. Different extractive solvents yield other CO characteristics. Even eugenol is the primary compound in

CO, oleanolic acid and biflorin were presented in a high amount when the extractive solvents are ethyl acetate and alcohol. The investigation can be performed using various extraction methods and pharmacological activity testing to obtain lead compounds for drugs. It can also be added the value of clove wastes after distillation.

Acknowledgment

The author thanks Ms. Urasa Charoonratana and Ms. Rapiya Charoonratana for kindly providing support.

References

Abo El-Maati, M.F., Mahgoub, S.A., Labib, S.M., Al-Gaby, A.M.A., Ramadan, M.F., 2016. Phenolic extracts of clove (*Syzygium aromaticum*) with novel antioxidant and antibacterial activities. Eur. J. Integr. Med. 8, 494–504.

Agus, H.H., 2021. Toxicology: oxidative stress and dietary antioxidants. In: Patel, V., Preedy, V. (Eds.), Terpene Toxicity and Oxidative Stress. Academic Press, Cambridge, pp. 33–42.

Baskaran, S.A., Kollanoor-Johny, A., Nair, M.S., Venkitanarayanan, K., 2016. Efficacy of plant-derived antimicrobials in controlling enterohemorrhagic *Escherichia coli* virulence in vitro. J. Food Prot. 79, 1965–1970.

Basmacioğlu Malayoğlu, H., Baysal, Ş., Misirlioğlu, Z., Polat, M., Yilmaz, H., Turan, N., 2010. Effects of oregano essential oil with or without feed enzymes on growth performance, digestive enzyme, nutrient digestibility, lipid metabolism and immune response of broilers fed on wheat soybean meal diets. Br. Poult. Sci. 51 (1), 67–80.

Cai, L., Wu, C.D., 1996. Compounds from *Syzygium aromaticum* possessing growth inhibitory activity against oral pathogens. J. Nat. Prod. 59 (10), 987–990.

Carlos, A.M.A., Harrison, M.A., 1999. Inhibition of selected microorganisms in marinated chicken by pimento leaf oil and clove oleoresin. J. Appl. Poult. Res. 8, 100–109.

Carvalho, R.P.R., de Almeida Lima, G.D., Machado-Neves, M., 2021. Effect of eugenol treatment in hyperglycemic murine models: a meta-analysis. Pharmacol. Res. 165, 105315.

Charoonratana, T., Songsak, T., Monton, C., Saingam, W., Bunluepuech, K., Suksaeree, J., et al., 2014. Quantitative analysis and formulation development of a traditional Thai antihypertensive herbal recipe. Phytochem. Rev. 13, 511–524.

Charoonratana, T., Songsak, T., Sakunpak, A., Pathompak, P., Charoenchai, L., 2015. Using liquid chromatography-mass spectrometry based metabolomics to discriminate between cold pressed rice bran oils produced from two different cultivars of *Oryza sativa* L. ssp. Indica in Thailand. Chin. J. Chromatogr. 33 (9), 966–973.

Chen, P., Zeng, H., Wang, Y., Fan, X., Xu, C., Deng, R., et al., 2014. Low dose of oleanolic acid protects against lithocholic acid-induced cholestasis in mice: potential involvement of nuclear factor-E2-related factor 2-mediated upregulation of multidrug resistance-associated proteins. Drug Metab. Dispos. 42 (5), 844–852.

Committee on Codex Specifications, Food and Nutrition Board, Division of Biological Sciences, Assembly of Life Sciences, National Research Council, 1996. Food Chemicals Codex. National Academy Press, Washington, DC.

Dalkilic, B., Guler, T., 2009. The effects of clove extract supplementation on performance and digestibility of nutrients in broilers. Firat Univ. J. Health Sci. 23 (3), 161–166.

De la Torre, R., Carbo, M., Pujadas, M., Biel, S., Mesa, M., Covas, M., et al., 2020. Pharmacokinetics of maslinic and oleanolic acids from olive oil—effects on endothelial function in healthy adults. A randomized, controlled, dose-response study. Food Chem. 322, 126676.

Dewi, N.L.P.S.S., Suprapta, D.N., Suada, I., 2019. Effectiveness of clove leaf extract (*Syzygium aromaticum* L.) against *Phytophthora palmivora* the cause of cocoa fruit rot (*Theobroma cacao* L.). J. Agroekoteknologi Tropika 8 (4), 458–467.

Dwivedi, V., Shrivastava, R., Hussain, S., Ganguly, C., Bharadwaj, M., 2011. Comparative anticancer potential of clove (*Syzygium aromaticum*)—an Indian spices—against cancer cell lines of various anatomical origin. Asian Pac. J. Cancer Prev. 12, 1989–1993.

El-Shawaj, A.M., El-zainy, A.R.M., El-Zamzamy, F.M., Nabila Awad, M.S.A.M., 2012. *In vitro* antibacterial effect of some natural plant extracts against *Escherichia coli* O157:H7, *Escherichia coli* NRRL and *Salmonella* sp. J. Prod. Dev. 17 (3), 367–384.

European Food Safety Authority, 2011. Scientific opinion on the safety and efficacy of allyhydroxybenzenes (chemical group 18) when used as flavourings for all animal species. EFSA J. 9.

Fadilah, F., Widodo, V.A., Priawan, R.P., Joyo, E.O., Abdurrohman, H., Fatmawaty, et al., 2019. Synthesis nanoparticle extract of clove (*Syzygium aromaticum* L.) and characterization by differential scanning colorimetry profiling and its activities as inhibitor of HeLa cell lines. Res. J. Pharm. Technol. 12 (7), 3355–3358.

Fathoni, A., Saepudin, A., Cahyana, A.H., Rahayu, D.U.C., Haib, J., 2017. Identification of nonvolatile compounds in clove (*Syzygium aromaticum*) from Monado. AIP Conf. Proc. 2017, 1862.

FDA, 2020a. Food Additives. 21 C.F.R. 170.3(o)(12),.

FDA, 2020b. Substances Generally Recognized as Safe. 21 C.F.R. 582.20,.

FDA, 2021. Clove and Its Derivatives. 21 C.F.R. 184. 1257(a),.

Frohlich, P.C., Santos, K.A., Palu, F., Cardozo-Filho, L., da Silva, C., da Silva, E.A., 2019. Evaluation of the effects of temperature and pressure on the extraction of eugenol from clove (*Syzygium aromaticum*) leaves using supercritical CO_2. J. Supercrit. Fluids 143, 313–320.

Galadima, M.S., Muhammad, A., 2016. Modellong of binary solvent extraction of clove (*Syzygium aromaticum*) oil using response surface methodology. Bayero J. Eng. Technol. 11 (2), 20–26.

Ghaly, F.M., Baraka, D.M., Mahmoud, S.H., Abd EL-Salam, S.S., Awad, S.M., El-Makhzangy, A.A., 2015. Efficacy of ciprofloxacin and clove extract on bacterial infection of *Clarias gariepinus*. Middle East J. Appl. Sci. 5 (1), 1–9.

Gooderham, N.J., Cohen, S.M., Eisenbrand, G., Fukushima, S., Guengerich, P., Hecht, S.S., et al., 2020. FEMA GRAS assessment of natural flavor complexes: clove, cinnamon leaf and West Indian bay leaf-derived flavoring ingredients. Food Chem. Toxicol. 145, 111585.

Harman, C.L., Murray, I.J., 2018. 2015 Poundage and Technical Effects Survey. Flavor and Extract Manufacturers Association of the United States (FEMA), Washington, DC, USA.

Hoque, M.M., Bari, M.L., Juneja, V.K., Kawamoto, S., 2007. Antimicrobial activity of cloves and cinnamon extracts against food borne pathogens and spoilage bacteria, and inactivation of *Listeria monocytogenes* in ground chicken meat with their essential oils. J. Food Sci. Technol. 72, 9–21.

Ivanovic, J., Zizovic, I., Ristic, M., Stamenic, M., Skala, D., 2013. The analysis of simultaneous clove/oregano and clove/thyme supercritical extraction. J. Supercrit. Fluids 55, 983–991.

James, L., 2018. Effect of Sodium Alginate Edible Coating With Clove (*Syzigium Aromaticum*) Bud Oleoresin on Quality of Broiler Chicken Breast Fillets (M.V.Sc. thesis). Kerala Veterinary and Animal Sciences University, Pookode.

Jeong, D.W., Kim, Y.H., Kim, H.H., Ji, H.Y., Ji, H.Y., Yoo, S.D., et al., 2007. Dose-linear pharmacokinetics of oleanolic acid after intravenous and oral administration in rats. Biopharm. Drug Dispos. 28, 51–57.

Jeong, S.H., Lee, H.J., Kim, D.W., Chung, Y.J., 2018. New biocide for eco-friendly biofilm removal on outdoor stone monuments. Int. Biodeterior. Biodegradation 131, 19–28.

Joint FAO/WHO Expert Committee on Food Additives (2005: Geneva, Switzerland), World Health Organization & Food and Agriculture Organization of the United Nations, 2005. Evaluation of Certain Food Additives and

Flavoring Agents: Sixty-Fifth Report of the Joint FAO/WHO Expert Committee on Food Additives. World Health Organization.

Joint FAO/WHO Expert Committee on Food Additives (2014: Geneva, Switzerland), World Health Organization & Food and Agriculture Organization of the United Nations, 2014. Evaluation of Certain Veterinary Drug Residues in Food: Seventy-Eighth Report of the Joint FAO/WHO Expert Committee on Food Additives. World Health Organization.

Kroes, R., Galli, C., Munro, I., Schilter, B., Tran, L., Walker, R., et al., 2000. Threshold of toxicological concern for chemical substances present in the diet: a practical tool for assessing the need for toxicity testing. Food Chem. Toxicol. 38, 255–312.

Kroes, R., Renwick, A.G., Cheeseman, M., Kleiner, J., Mangelsdorf, I., Piersma, A., et al., 2004. Structure based thresholds of toxicological concern (TTC): guidance for application to substances present at low levels in the diet. Food Chem. Toxicol. 42, 65–83.

Kuroda, M., Mimaki, Y., Ohtomo, T., Yamada, J., Nishiyama, T., Mae, T., et al., 2012. Hypoglycemic effects of clove (*Syzygium aromaticum* flower buds) on genetically diabetic KK-Ay mice and identification of the active ingredients. J. Nat. Med. 66, 394–399.

Ladurner, A., Zehl, M., Grienke, U., Hofstadler, C., Faur, N., Pereira, F.C., et al., 2017. Allspice and clove as source of triterpene acids activating the G protein-coupled bile acid receptor TGR5. Front. Pharmacol. 8, 468.

Latifah-Munirah, B., Himratul-Aznita, W.H., Zain, N.M., 2015. Eugenol, an essential oil of clove, causes disruption to the cell wall of *Candida albicans* (ATCC 14053). Front. Life Sci. 8 (3), 231–240.

Lee, H.S., Chang, J.H., Kim, I.H., Nam, S.W., 1997. Dyeing of silk with clove extract. J. Korean Soc. Dyers Finish. 9 (5), 19–29.

Leung, A.Y., 1980. Encyclopedia of Common Natural Ingredients Used in Food, Drugs and Cosmetics. Wiley, New York, pp. 131–132.

Liu, J., 1995. Pharmacology of oleanolic acid and ursolic acid. J. Ethnopharmacol. 49 (2), 57–68.

Liu, J., 2005. Oleanolic acid and ursolic acid: research perspectives. J. Ethnopharmacol. 100, 92–94.

Liu, Y., Kreppel, H., Liu, J., Choudhuri, S., Klaassen, C.D., 1993. Oleanolic acid protects against cadmium hepatotoxicity by inducing metallothionein. J. Pharmacol. Exp. Ther. 266 (1), 400–406.

Liu, H., Schmitz, J.C., Wei, J., Cao, S., Beumer, J.H., Strychor, S., et al., 2014. Clove extract inhibits tumor growth and promotes cell cycle arrest and apoptosis. Oncol. Res. 21, 247–259.

Lu, Y.F., Wan, X.L., Xu, Y., Liu, J., 2013. Repeated oral administration of oleanolic acid produces cholestatic liver injury in mice. Molecules 18 (3), 3060–3071.

Mammen, R.R., Mulakal, J.N., Mohanan, R., Maliakel, B., Madhavamenon, K.I., 2018. Clove bud polyphenols alleviate alterations in inflammation and oxidative stress markers associated with binge drinking: a randomized double-blinded placebo-controlled crossover study. J. Med. Food, 1–9.

Miladi, H., Zmantar, T., Kouidhi, B., Chaabouni, Y., Mahdouani, K., Bakhrouf, A., et al., 2017. Use of carvacrol, thymol, and eugenol for biofilm eradication and resistance modifying susceptibility of *Salmonella enterica* serovar *Typhimurium* strains to nalidixic acid. Microb. Pathog. 104, 56–63.

Mshelia, H.E., Ugwah-Oguejiofor, C.J., October, N., Abubakar, K., 2019. Isolation of oleanolic acid from *Parinari curatellifolia* (Planch Ex. Benth) stem bark and evaluation of its anticonvulsant and sedative activities in rodents. Trop. J. Nat. Prod. Res. 3 (1), 17–21.

Munro, I.C., Ford, R.A., Kennepohl, E., Sprenger, J.G., 1996. Correlation of structural class with no-observed-effect levels: a proposal for establishing a threshold of concern. Food Chem. Toxicol. 34, 829–867.

Myint, S., Daud, W.R.W., Mohamad, A.B., 1995. Determination of optimal conditions for extraction of alcohol-soluble eugenol containing material from cloves. Pertanika J. Sci. Technol. 3 (1), 99–106.

Nathan, D.M., 2015. Diabetes: advances in diagnosis and treatment. J. Am. Med. Assoc. 314, 1052–1062.

Nunez, L., Aquino, M.D., Chirife, J., 2001. Antifungal properties of clove oil (*Eugenia caryophylata*) in sugar solution. Braz. J. Microbiol. 32, 123–126.

Omotayo, O., Ikechukwuka, O.M., Oyebola, T.O., Ogundimu, O.A., 2018. Effect of *Syzygium aromaticum* (L.) extract on the reproductive functions of West African dwarf bucks. J. For. Res. Manage. 15 (3), 137–143.

Oyedemi, S.O., Okoh, A.I., Mabinya, L.V., Pirochenva, G., Afolayan, A.Y., 2009. The proposed mechanism of bactericidal action of eugenol, α-terpineol and γ-terpinene against *Listeria monocytogenes*, *Streptococcus pyogenes*, *Proteus vulgaris* and *Escherichia coli*. Afr. J. Biotechnol. 8 (7), 1280–1286.

Pandey, A., Singh, P., 2011. Antibacterial activity of *Syzygium aromaticum* (clove) with metal ion effect against food borne pathogens. Asian J. Plant Sci. Res. 1 (2), 69–80.

Rehman, M., Alajmi, M.F., Hussain, A., 2020. Natural compounds as inhibitors of SARS-CoV-2 main protease (3CLpro): a molecular docking and simulation approach to combat COVID-19. Curr. Pharm. Des. 16, 3577–3589.

Reineccius, G., 1994. Natural flavouring materials. In: Sourcebook of Flavours, second ed. Chapman & Hall, New York.

Reverchon, E., Marrone, C., 1997. Supercritical extraction of clove bud essential oil: isolation and mathematical modeling. Chem. Eng. Sci. 52 (20), 3421–3428.

Ryu, B., Kim, H.M., Woo, J.H., Choi, J.H., Jang, D.S., 2016. A new acetophenone glycoside from the flower buds of *Syzygium aromaticum* (cloves). Fitoterapia 115, 46–51.

Saleem, H.N., Batool, F., Mansoor, H.J., Shahzad-ul-Hussan, S., Saeed, M., 2019. Inhibition of dengue virus protease by eugeniin, isobiflorin, and biflorin isolated from the flower buds of *Syzygium aromaticum* (cloves). ACS Omega 4, 1525–1533.

Shende, S., Patel, S., Arora, S., Sharma, V., 2014. Oxidative stability of ghee incorporated with clove extracts and BHA at elevated temperatures. Int. J. Food Prop. 17 (7), 1599–1611.

Song, M., Hang, T., Wang, Y., Jiang, L., Wu, X., Zhang, Z., et al., 2006. Determination of oleanolic acid in human plasma and study of its pharmacokinetics in Chinese healthy male volunteers by HPLC tandem mass spectrometry. J. Pharm. Biomed. Anal. 40, 190–196.

Sultana, K., Jayathilakan, K., Pandey, M.C., 2018. Evaluation of antioxidant activity, radical scavenging, and reducing power of clove oil and clove oleoresin in comparison with natural and synthetic antioxidants in chevon (*Capra aegagrus hircus*) and chicken meat. Def. Life Sci. J. 3 (1), 51–58.

Taher, D.M., Solihin, D.D., Cahyaningsih, U., Sugita, P., 2018. Methanol extract of cloves (*Syzygium aromaticum* (L.) Merry & Perry) of Tuni Buru Selatan varieties as antimalarial. Acta Vet. Indones. 6 (2), 38–47.

Taher, D.M., Solihin, D.D., Cahyaningsih, U., Sugita, P., 2019. Chemical components analysis of clove (*Syzygium aromaticum* (L.) Merr. & Perry). Global Sci. J. 7 (4), 379–391.

Tekin, K., Akalin, M.K., Seker, M.G., 2015. Ultrasound bath-assisted extraction of essential oils from clove using central composite design. Ind. Crop. Prod. 77, 954–960.

Thenmozhi, T., 2020. Functionalization of iron oxide nanoparticles with clove extract to induce apoptosis in MCF-7 breast cancer cells. 3 Biotech 10, 82.

Upadhyay, A., Upadhyaya, I., Kollanoor-Johny, A., Venkitanarayanan, K., 2013. Antibiofilm effect of plant derived antimicrobials on *Listeria monocytogenes*. Food Microbiol. 36, 79–89.

Vilas-Boas, V., Gijbels, E., Jonckheer, J., Waele, D.E., Vinken, M., 2020. Cholestatic liver injury induced by food additives, dietary supplements and parenteral nutrition. Environ. Int. 136, 105422.

Vinatoru, M., Mason, T.J., Calinescu, I., 2017. Ultrasonically assisted extraction (UAE) and microwave assisted extraction (MAE) of functional compounds from plant materials. Trends Anal. Chem. 97, 159–178.

Wael, S., Nuringtyas, T.R., Wijayanti, N., Astuti, P., 2018a. Secondary metabolites production in clove (*Syzygium aromaticum*): chemical compounds. J. Biol. Sci. 18, 399–406.

Wael, S., Watuguly, T.W., Arini, I., Smit, A., Matdoan, N., Prihati, D.R., et al., 2018b. Potential of *Syzygium aromaticum* (clove) leaf extract on immune proliferation response in Balb/c mice infected with *Salmonella typhimurium*. Case Rep. Clin. Med. 7, 613–627.

Wang, F., Hua, H., Pei, Y., Chen, D., Jing, Y., 2006. Triterpenoids from the resin of *Styrax tonkinensis* and their antiproliferative and differentiation effects in human leukemia HL-60 cells. Nat. Prod. 69, 807–810.

Weiss, E.A., 1997. Essential Oil Crops. CAB, Wallingford, pp. 235–239.

Wenqiang, G., Shufen, L., Ruixiang, Y., Shaokun, T., Can, Q., 2007. Comparison of essential oils of clove buds extracted with supercritical carbon dioxide and other three traditional extraction methods. Food Chem. 101, 1558–1564.

Wibawaningtyas, N., Kristiana, D., Probosari, N., 2017. The effect of the thermoplastic nilon enterprises (valplast) on clove extract (*Syzygium aromaticum*) in various concentrations on the color change. J. Dentomaxillofacial Sci. 2 (3), 180–182.

Wongphaosakul, S., Posri, S., Suphan, S., Buddhakala, N., 2017. Effect of clove flower and leaf extracts on sedation in hybrid catfish. J. Sci. Technol. 7 (1), 24–33. Rajamangala University of Technology Thanyaburi.

Yazdani, F., Mafi, M., Farhadi, F., Tabar-Heidar, K., Aghapoor, K., Mohsenzadeh, F., Darabi, H.R., 2005. Supercritical CO_2 extraction of essential oil from clove bud: effect of operation conditions on the selective isolation of eugenol and eugenyl acetate. Z. Naturforsch. 60b, 1197–1201.

Yin, M., Lin, M., Mong, M., Lin, C., 2012. Bioavailability, distribution, and antioxidative effects of selected triterpenes in mice. J. Agric. Food Chem. 60, 7697–7701.

Yun, S., Lee, M., Lee, K., Ku, H., Son, S., Joo, W., 2010. Quantitative analysis of eugenol in clove extract by a validated HPLC method. J. AOAC Int. 93 (6), 1806–1810.

Zeng, R., Zhang, A., Chen, J., Fu, Y., 2012. Postharvest quality and physiological responses of clove bud extract dip on 'Newhall' navel orange. Sci. Hortic. 138, 253–258.

Zhang, Y., Wang, Y., Zhu, X., Cao, P., Wei, S., Lu, Y., 2017. Antibacterial and antibiofilm activities of eugenol from essential oil of *Syzygium aromaticum* (L.) Merr. & L. M. Perry (clove) leaf against periodontal pathogen *Porphyromonas gingivalis*. Microb. Pathog. 113, 396–402.

Zhang, Q., Lin, L., Ye, W., 2018. Techniques for extraction and isolation of natural products: a comprehensive review. Chin. Med. 13, 20.

CHAPTER

6

Cloves (*Syzygium aromaticum*) cultivars: Convenient source of eugenol and its role in commercially important formulations

Abdul Hafeez Laghari and Salman Tariq Khan

Pakistan Council of Scientific and Industrial Research, Laboratories Complex, Karachi, Pakistan

1. Introduction

Syzygium aromaticum (L.) Merr. & L. M. Perry (Syn. *Eugenia caryophyllus*) is a tree in the family Myrtaceae, an evergreen plant of 10–20cm with spear-shaped leaves and racemiferous leaves; yellowish flowers are famous for the production of cloves. Clove is a nail-shaped dried flower bud of *Syzygium aromaticum* are among the oldest spices and drugs in history (Öztürk and Özbek, 2005). The clove is harvested primarily for the unopened flower buds borne in clusters, which are dried to produce whole clove buds (Fig. 1). Name of cloves in local languages in different regions is *Kabshqarunfil, Kabshqaranful* (Arabic), *Ding xiang, Ting hsiang* (Chinese), *Mikhak* (Persian), Clou de girofle (French), *Nelke* (German), *Laung* (Hindi, Urdu, Sindhi), *CengkihGraambu* (Tamil). It naturally grew in some islands of Indonesia (Malay Archipelago, especially the Moluccas) and is now cultivated in many parts of the world, including Tanzania (Zanzibar and Pemba), Islands near Africa Seychelles, Mauritius, Madagascar, Sri Lanka, India (Tamil Nadu and Kerala), Indonesia, Malaysia, Brazil, Jamaica, Guinea, and Brazil as well as in some parts of the mediterranean region such as Turkey as per recent reports (Mbaveng and Kuete, 2017). The plants have a strong phenolic smell and a sharp acrid taste. Clove is used as a spice stimulant by chewing tobacco and a betel chew, flavoring agent. The oil obtained by the distillation of cloves is used in cooking, food processing, pharmacy, perfumery, and cosmetics (Ntamila and Hassanali, 1976; Öztürk and Özbek, 2005; Karmakar et al., 2012; Devkota and Adhikari Devkota, 2020).

Eugenol, IUPAC name, 4-allyl-2-methoxyphenol with different synonyms, 1,3,4-1-hydroxy-2-methoxy-4-allylbenzene, 1-hydroxy-2-methoxy-4-prop-2-enylbenzene, 1-hydroxy-2-methoxy-4-propenyl benzene, and 2-methoxy-1-hydroxy-4-allylbenzene has Chemical Abstracts Service Registry Number 97 53 0 while molecular formula is $C_{10}H_{12}O_2$ (Sellamuthu, 2014).

Clove (*Syzygium aromaticum*). https://doi.org/10.1016/B978-0-323-85177-0.00017-3

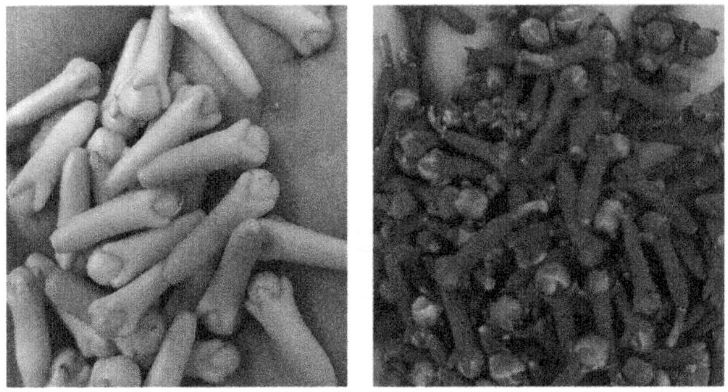

FIG. 1

Clove bud (A) and dried form (B).

The major source of eugenol is clove oil, while synthesis methods have also been reported, it is still considered that clove is a convenient and economical source of eugenol. Isolation of eugenol from clove oil and its main components (i.e., eugenol and acetyl eugenol) separation has been reported (Ntamila and Hassanali, 1976), which involves steam distillation of cloves and then extraction using chloroform and basic aqueous solvents. The advantage of the method is its convenience and not requiring a very high quantity of cloves.

Eugenol belongs to phenylpropanes such as anethol, estragole, and cinnamaldehyde biosynthesized via the shikimate pathway. Eugenol is mainly isolated from numerous aromatic plants (Bendre et al., 2016). Eugenol (4-allyl-2-methoxyphenol), a methoxyphenol with a short hydrocarbon chain, is a main constituent of the essential oil obtained from many angiospermic plants consumed as spices such as *Pimenta racemose* (bay leaves), *Cinnamomum verum* (cinnamon leaf), and *Syzygium aromaticum* (clove) that possess eugenol as the major component (80%–95%) of its oil. Eugenol shows aromatic flavor; that is why it is called aromatic phytochemical making clove a choice for spice and fragrance in ancient China for more than 2000 years. Eugenol has been used as a spice because of its strong odor.

It has been extensively used as a therapeutic agent in dentistry for sedation in patients with toothache arising from various sources, including pulpitis and dentinal hypersensitivity, because of its detergent-like effect. Moreover, mixing with zinc oxide, it forms cement for temporary fillings of the teeth. Recently, much attention has been paid to eugenol derivatives due to their diverse biological activities such as antiseptic and analgesic, anesthetic, antispasmodic, antipyretic, and antioxidant (Rahim et al., 2017).

Pharmacologic studies have demonstrated that eugenol has anticonvulsant, local anesthetic and antistress, bacteriostatic and bactericidal, anticandidal, and antifungal properties. In addition, eugenol has antigenotoxic, insecticidal, antioxidant, and anti-inflammatory activities and anticarcinogenic potential. It has also shown positive effects on neuro disorders. Therefore, it is also a good choice for the food industry, using eugenol as a flavoring agent (Carrasco et al., 2008). Furthermore, eugenol and methyl eugenol are used as insect attractants. As such, eugenol is commercially available in large quantities, and the specific structure and ready availability of eugenol have turned this natural product into an interesting starting material. Therefore, the synthetic chemists have been chossing Eugenol as a convenient starting material to make useful complex structures to synthesize a valuable substrate for various biotransformations (Bendre et al., 2016).

Eugenol itself is believed to be toxic in high doses; hence its modification has been done from time to time by the synthesis of its derivatives to cure particular diseases. Of those, major are esters and eugenol ethers, which show antispasmodic activity. Besides, biphenyl was also believed to enhance antiproliferative and pro-apoptotic activities against malignant melanoma activity (Rahim et al., 2017; Daniel et al., 2009).

In recent studies, the mode of action of eugenol is also the focus of many researchers. For example, eugenol's mode of antibacterial action is also attributed to its ability to disrupt the cytoplasmic membrane, thereby increasing its nonspecific permeability. In addition, recent evidence reveals that the hydroxyl group on eugenol is thought to bind to proteins, preventing enzyme action in *Enterobacter aerogenes* (Devi et al., 2010). Unfortunately, however, safety data of most synthesized compound formulations containing eugenol still lacks in many cases.

2. Useful compounds of clove

Though the high content of clove is eugenol and its derivatives, it has been learned through the literature that many bioactivities of eugenol itself are weaker than eugenol extracts such as essential oil mixture. It proves the synergetic effect of other compounds present in cloves, which must be listed with all the details available in the published literature.

Different compounds of clove and eugenol have been reported from time to time using analytical instrumentation techniques and partial confirmation of structures, such as essential oil compositions (Jirovetz et al., 2006), volatile chemical constituents (Oliveira et al., 2009) including glucosidically bound volatiles (Sgorbini et al., 2015). Besides, the isolation and purification of new and previously known compounds have also been carried out and summarized in Table 1.

Table 1 Compounds isolated from clove and eugenol.

Sr. no.	Name of compound	Group/class	Isolation/ detection method	Molecular structure numbers (Fig. 2)	Importance	References
1.	α-Caryophyllene β-Caryophyllene γ-Caryophyllene	Sesquiterpene	Steam distillation as well as extraction by organic solvents, i.e., benzene, ethanol, ether	**1–3**	Help in cold tolerance for mammals	Walter (1972)
2.	Syzyginin A, 1,2,3-tri-*O*-galloyl-4,6-(*S*)-tergalloyl-β-D-glucopyranose Syzyginin B, A novel hydrolyzable tannin	Tannins	Extraction in organic solvents and purification using Sephadex LH 20 column	**4–5**		Tanaka et al. (1996)

Continued

Table 1 Compounds isolated from clove and eugenol—cont'd

Sr. no.	Name of compound	Group/class	Isolation/ detection method	Molecular structure numbers ()	Importance	References
3.	Dehydrodieugenol *Trans*-coniferyl aldehyde (Methyl ethers of both are also synthesized by reaction with diazomethane)	Phenylpropanoids	Extraction by organic solvents and purification using silica gel column	**10–11**	Antimutagenic activity (Umu test method)	Miyazawa and Hisama (2003)
4.	Gallic acid (783.5, 2375.8) Eugenol (9381.7) Acetyl eugenol (2075.1) Quercetin (28.4) Kaempferol (63.9) (mg/100 g of dry weight)	Phenolics	Extracted in 80% methanol	**12–16**	Antioxidant	Shan et al. (2005)
5.	Rhamnetin 3-*O*-β-D-glucuronide Rhamnazin 3-*O*-β-D-glucuronide Rhamnazin-3-*O*-β-D-glucuronide-6″-methyl ester Rhamnocitrin-3-*O*-β-D-glucuronide-6″-methyl ester (along with 15 known flavonoids)	Flavonoids	Ethanol extract, fractionation, and column chromatography	**17–20**	Cytotoxicity against human 26 ovarian cancer cells (Known compounds were stronger than new compounds)	Ryu et al. (2016)
6.	Kumatakenin	Flavonoid	As per Ryu et al. (2016)	**21**	Anticancer	Woo et al. (2017)

It is important to note that many compounds are isolated from different plants, but most are explored until their identification and publication. However, it is essential to check their bioactivity, which is seldom done later on by the same or other research groups. For example, in the case of clove constiuent Kumatakenin is notable, which was isolated earlier by Ryu et al. (2016) and evaluated for cytotoxicity assay of ovarian cancer cells and found active with three times higher MIC value as compared to control cisplatin. However, its other aspects of bioactivity were later reported (Woo et al., 2017) in which anti-cancer activities by inducing apoptosis of ovarian cancer cells and inhibiting the alternative activation of tumor-associated macrophage was found prominant. Thus, the same compounds may be either tested for other bioactivity assays or modified/derivatized to make them more effective.

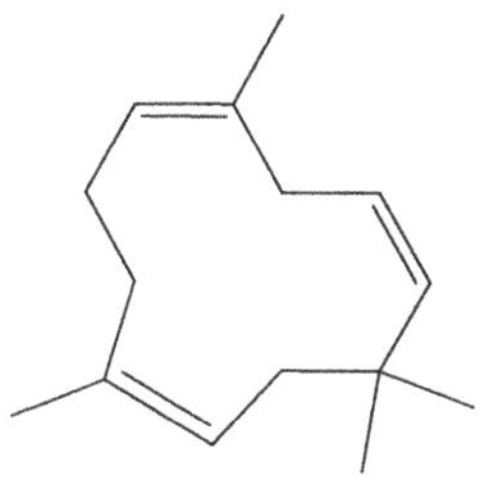

1

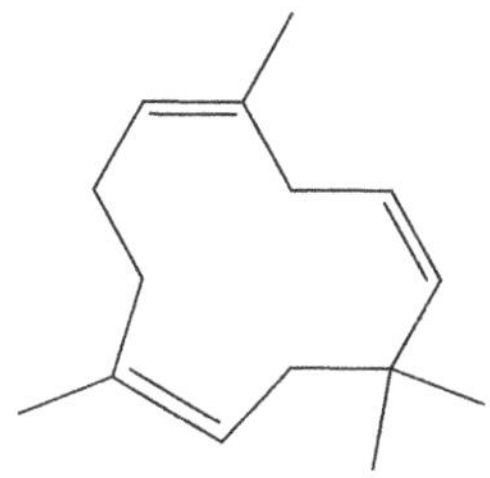

2

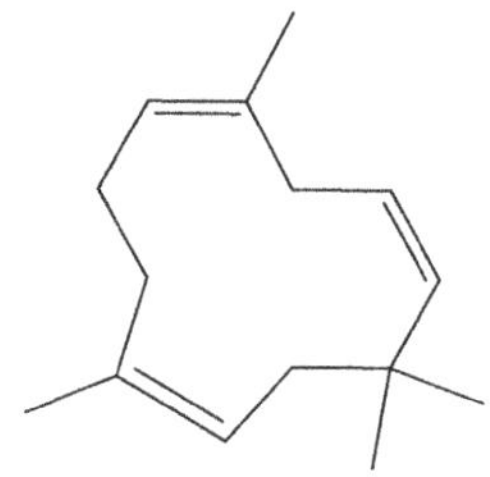

3

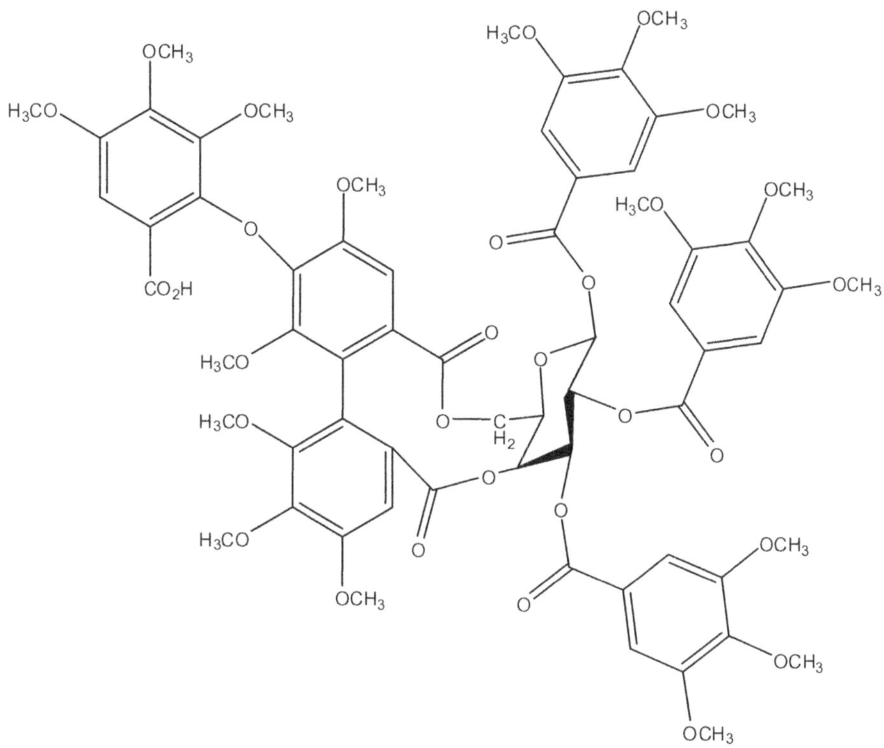

4

FIG. 2

Molecular structure of compounds isolated from clove.

(Continued)

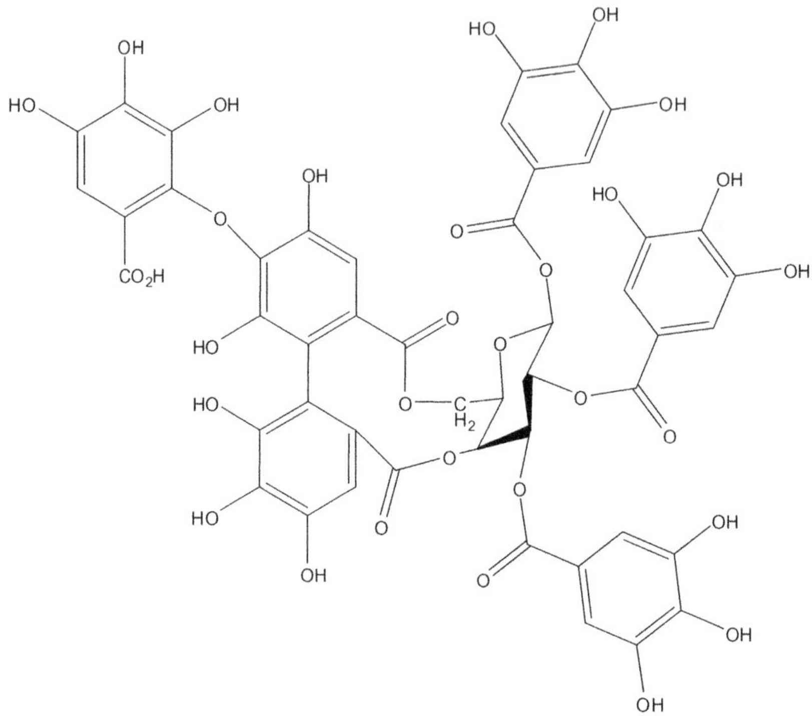

5

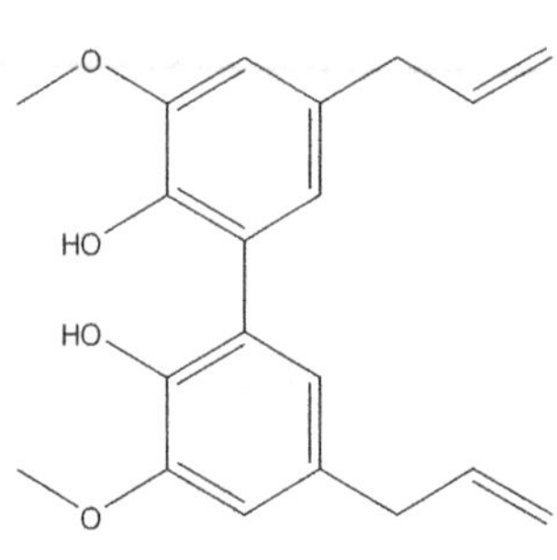

6

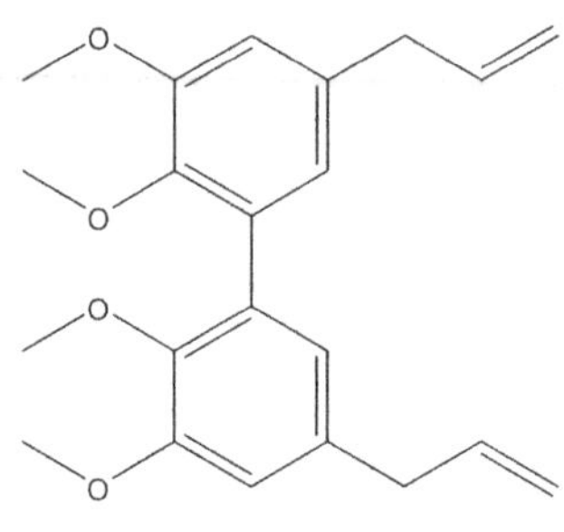

7

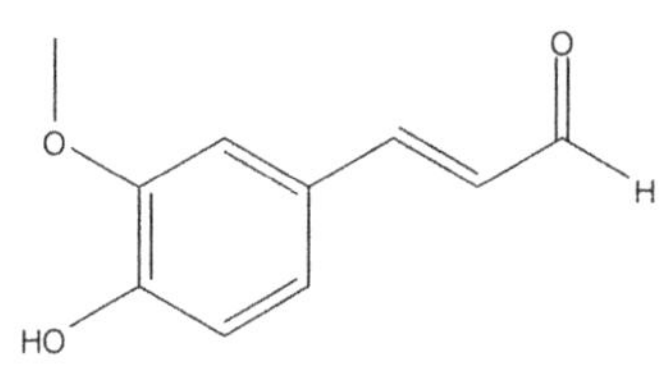

8

FIG.2, CONT'D

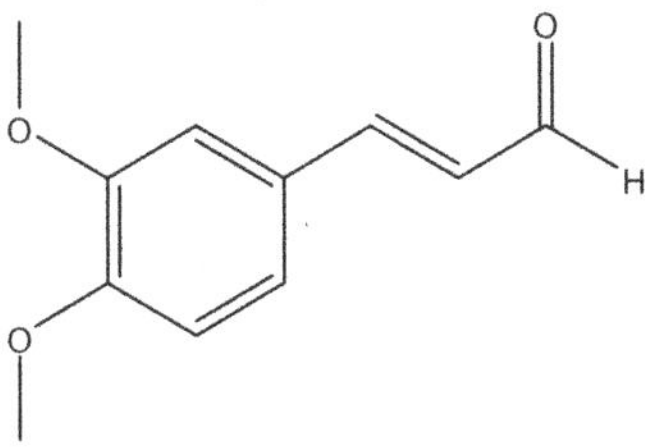

9

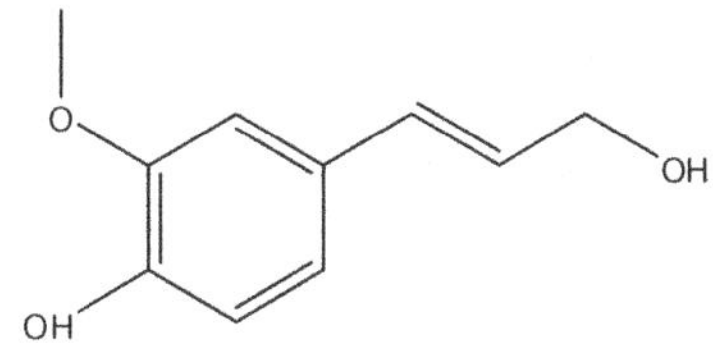

10

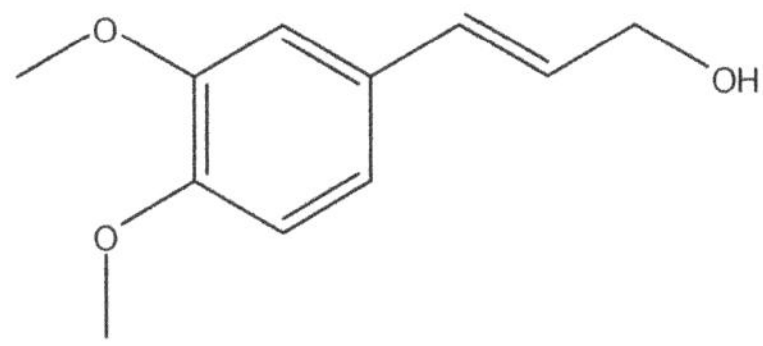

11

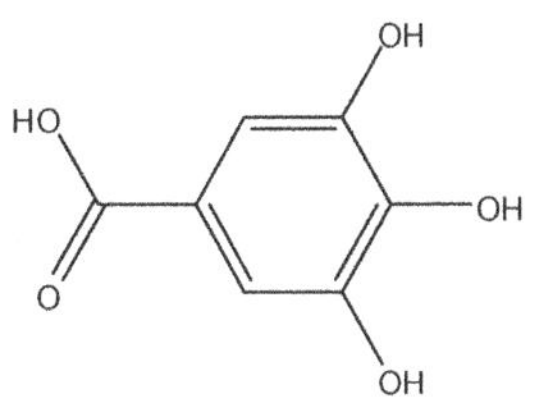

12

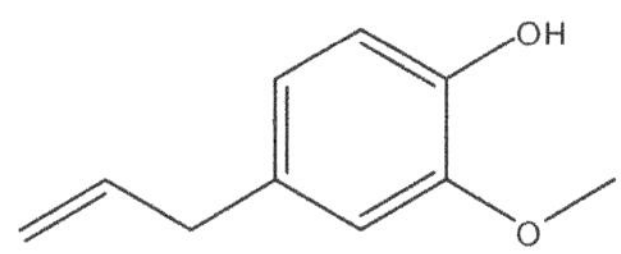

13

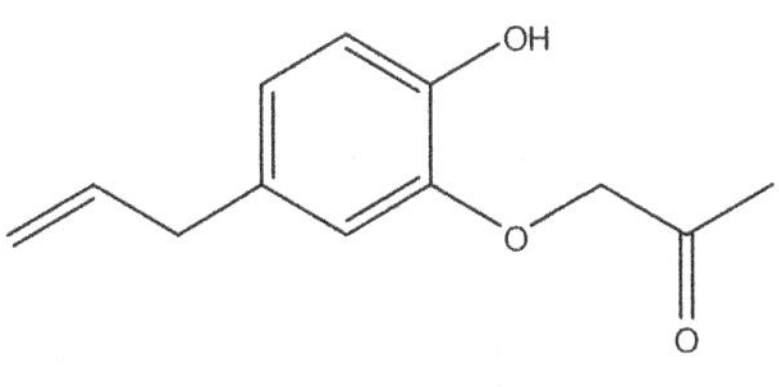

14

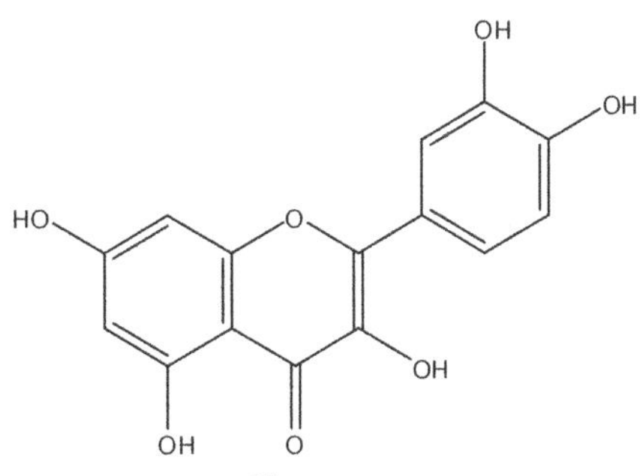

15

FIG.2, CONT'D

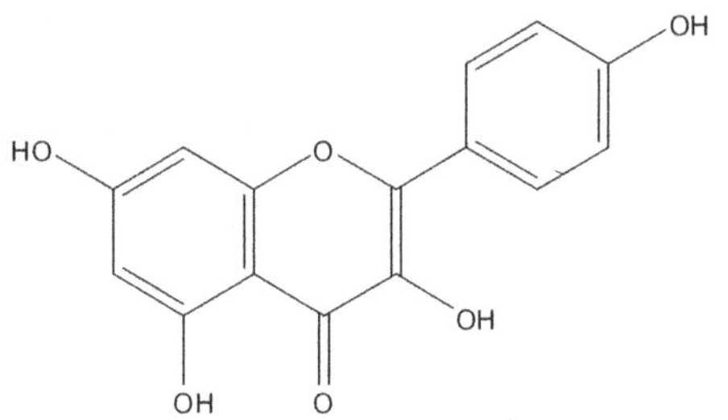

16

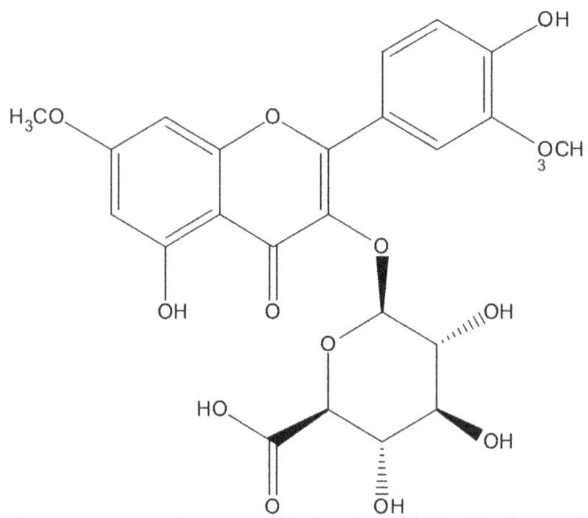

17

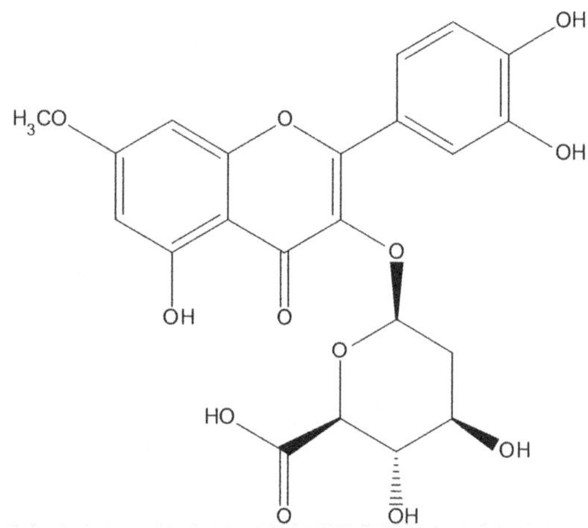

18

FIG.2, CONT'D

3. Eugenol and its derivatives

3,4-Dimethoxy isoamylcinnamic is the condensation product of eugenol modified to veratraldeyde and isoamyl acetic synthesized by acetylation of isoamyl alcohol obtained from fusel oil. It showed a promising in vitro SPF value even at low concentrations (Wahyuningsih et al., 2010).

In an attempt to study and enhance the anticancer properties of eugenol, acylated and nitrated derivatives of eugenol were prepared. Two out of six prepared analogues, i.e., 5-allyl-3-nitrobenzene-1,2-diol and 4-allyl-2-methoxy-5-nitrophenyl acetate, were found more active than eugenol itself against cell growth activity. This study may assist in finding some future chemotherapeutics (Carrasco et al., 2008).

In a study based on steam distillation extraction products of clove, eugenol, methyl eugenol, and acetyl eugenol were identified by GCMS and quantified by HPLC. These compounds were used to evaluate their antioxidant and anti-inflammatory properties. Eugenol showed the strongest DPPH·

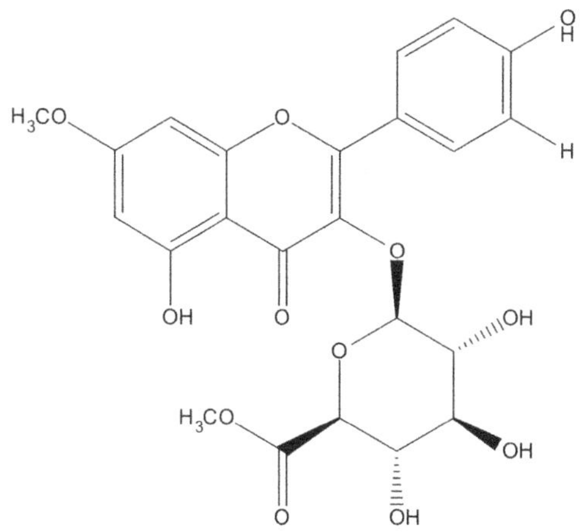

19

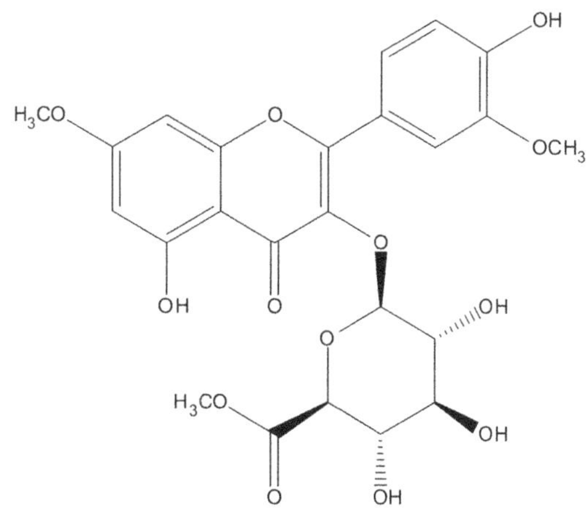

20

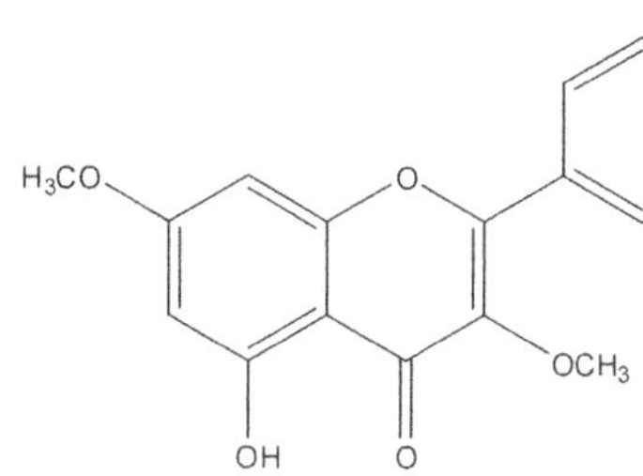

21

FIG.2, CONT'D

radical scavenging activity and cyclooxygenase (COX 2) inhibitory activity, whereas methyl eugenol exhibited the strongest lipoxygenase (15 LOX) inhibitory activity (Leem et al., 2011).

Eugenol extracted from clove oil was subjected to acetylation and nitration by Hidalgo et al., searching for potential compounds to use in pharmaceutical, cosmetic, and/or food industries. Their findings concluded that the eugenol and a nitrate derivative (5-allyl-3-nitrobenzene-1,2-diol) with more phenolic groups contributed most as antioxidants among all prepared derivatives. In addition, 5-allyl-3-nitrobenzene-1,2-diol showed the lowest toxicity (Hidalgo et al., 2009). Hidalgo et al. (2009) have synthesized a series of eugenol derivatives seconding to claim of Carrasco et al. (2012), Devi et al. (2010), Chang et al. (2001), and Gayoso et al. (2005) that eugenol ethers and ester derivatives are simple to synthesis with stability and the broad spectrum of biological activity. All of those were evaluated for their antibacterial activity against Gram-positive bacteria (*Bacillus subtilis, Staphylococcus aureus,* and *Staphylococcus epidermidis*) and two Gram-negative bacteria (*Escherichia coli* and *Salmonella typhimurium*) using well diffusion method. Though none of the derivatives (or modified eugenol) could beat the activity of control, i.e., Streptomycin, few derivatives were found promising. On the contrary (Carrasco et al., 2012), nitro group-containing compound showed lower activity attributed to its deactivating para position. On the other hand, bromine derivative 4-allyl-2-methoxyphenyl-4-bromobenzoate was best against all selected bacteria. However, this lead could be more authentic if the compounds were tested for toxicity and other side effects before further trials.

4. Eugenol-based formulations

Five derivatives (coded as E2 to E6) have been synthesized by acetylation and nitration of eugenol (E1), obtained from clove oil, to evaluate their ability to decrease cell growth of human cancer cells DU 145 and KB tumor cell lines using MTT assay. Though moderate activity was observed, all derivatives were more active than eugenol (Carrasco et al., 2008). Later on, the same group further evaluated these compounds for toxicity and effectiveness as an antioxidant (ORAC assay and DPPH·). Of those, 5-allyl-3-nitrobenzene-1,2-diol (E5) was found promising with antioxidant more excellent than eugenol while all others were less active than eugenol. The activity is attributed to two OH groups of the synthesized compound, one in eugenol and none in any other derivative, decreasing the antioxidant activity of all derivatives. It is always necessary to see other aspects of active compounds, including toxicity, as the most important factor. Two methods were used in this study to check the toxicity of eugenol and all synthesized derivatives. Fortunately, the least toxic compound showed the greatest activity among all compounds (eugenol and five derivatives synthesized in this study). The most toxic compound was 4-allyl-2-methoxyphenyl acetate (E3) due to the nitro group. Based on the results, the authors suggested using E1 and E5 in the pharmaceutical, cosmetic, and food industries for formulations of products (Shan et al., 2005).

Eugenol-like molecules were prepared as an alternative to synthetic antioxidant compounds to prevent undesired actions such as affecting the liver, which included γ diisoeugenol, hydroxymethyl eugenol, dihydro eugenol, and 4-(1,3-dioxanyl)phenol. It was done during a medicinal program to get small natural or synthetic molecules for drug delivery. Dihydroeugenol was synthesized by reduction of eugenol, while isoeugenol produced different compounds in the same conditions with anticipation that all compounds would show antioxidant activity by possessing phenolic hydroxyl and methoxy groups in *ortho* position as a functional moiety. The synthesized compounds showed higher activity than common synthetic antioxidants such as BHT and BHA (Trolox equivalent antioxidant capacity (TEAC)). Obtained compound r-1-ethyl-5-hydroxy-c-3-(4-hydroxy-5-methoxyphenyl)-6-methoxy-t-2-methylindane (γ isoeugenol) showed the most prominent activity among all while dihydroeugenol was the least active. However, novel compound 4-(1,3-dioxanyl)phenol was suggested as a molecular model in pharmacological research for antioxidant formulation due to its convenient synthesis and considerable activity. However, toxicity study of all compounds has not been reported, making its choice more authentic (Merchán Arenas et al., 2011).

The practical application of antiparasitic agents has been a challenge due to drug resistance, environmental pollution, and undesired action of commercially available sources. Hence, the development of new effective and safe alternatives is being tried from time to time. In this regard, Shang et al. (2019),

Shang et al. (2016), and Guo et al. (2017) had already tried to check antiparasitic (i.e., acaricidal) activity of essential oils, oregano oil, and essential oil from *Rhododendron nivale* against a parasite. However, the mechanism of action was not known until this study, in which eugenol has been chosen as an antiparasitic agent, and systemic investigation of its targets by combined analysis of surface plasmon resonance (SPR) and RNA Seq is done against *P. cuniculi* (Shang et al., 2021). In molecular docking assay, 12 possible sites were found unbind with eugenol, suggesting binding to Mitochondrially Encoded NADH: Ubiquinone Oxidoreductase Core Subunit 2 (MTND2), one of the 12 genes encoding subunits of complex I in insects and parasites.

On the contrary, eugenol did not bind a specific peptide to inhibit it. It proved the difference in sensitivity to eugenol between insects and mammals. Hence, it is proved that eugenol targets complex I of the mitochondrial respiratory chain in parasites. Besides that, the safety study revealed exciting facts about the 2.5 mg/kg dose of eugenol and the positive control that rotenone gave to rats, and histopathological changes in the liver were observed. Rotenone caused loss of nigral and other neurological disabilities in rats, leading to the death of all rats in the test, but eugenol did not degenerate TH immunoreactive neurons in the striatum of rats and therefore did not cause Parkinson's disease (PD) or related symptoms. This study suggests eugenol as the best alternative to available commercial products. However, eugenol volatile property is disadvantage which should be overcome to make it a final choice for above mentioned activities (Shang et al., 2021).

Contact dermatitis occurs by exposure of skin to irritants, which is one of the most common skin diseases. Eugenol has been reported to possess many bioactivities; however, it is allergic to skin, leading to a bad choice for topical formulations. Nevertheless, its importance is needed to overcome that problem to get its therapeutic benefit. Polymeric nanocarriers of eugenol were for targeted drug delivery to retain its penetration in the superficial layers of the skin. To meet dermatitis's desire, the polymer's negative charge reduces its interaction with skin cells and the drug penetration (deAraújo Lopes et al., 2018).

Vulvovaginal candidiasis is a vaginal yeast or fungal infection that causes irritation, discharge, and intense itchiness of the vagina and the vulva. One-third of women face this infection, and 5% of those get it again while reason is yet not finalized, but few known are diabetes mellitus, antibiotherapy, corticotherapy, and pregnancy (Fidel and Sobel, 1996; Sheehan et al., 1999). To substitute the available curing agents (i.e., polyenes and azole) carvacrol and eugenol have been tried as natural products following their antifungal activity reported earlier. The anticandidal activity of carvacrol and eugenol was evaluated using prophylactic and therapeutic treatment in rats inoculated intravaginally with *Candida albicans* and compared with nystatin (used as a positive control) by microbiological and histological techniques. In a careful in vivo study, results were obtained by taking a swab test of available pathogenic yeast, i.e., *Candida albicans*. The immunosuppressed group applied with agar only remained positive till 10 days, while the untreated was completely negative. Treatment with standard medicine Nystatin cured 100% animals, almost similar response by carvacrol; however, eugenol-treated animals 7/9 were positive even on the 10th day proving it weak among all. Though authors claim carvacrol and eugenol are naturally occurring antifungal agents for patients with AIDs, more studies on eugenol and its enormous derivatives may be carried out with their safety data to get a more suitable solution curing agent of the disease (Chami et al., 2004).

Clove has been used to relieve tooth pain, and eugenol is considered the functional compound for this property. However, eugenol also has a promising role in the dental sector. Recently, a patent has been filed, which is based on a composition possessing eugenol for safely removing a cured dental cement or resin from metal brackets, metal braces, and enamel surfaces of teeth. The composition is based on 0.5%–10% eugenol (product weight base) with one or more cross-linked polymers of acrylic acid or carbopol 934 or 940 as a gelling agent. The product has sufficient power to reduce the Vickers hardness of a cured or hardened orthodontic bonding adhesive cement to remove a metal bracket from an enamel surface of a tooth (Eugenol Dental Composition, 2021).

5. Conclusion

Though eugenol has been evaluated in vitro and in vivo for many assays to confirm its ability to cure different diseases or get relief from chronic conditions, very limited formulation has been marketed so far. At the same time, the potential of eugenol makes it a perfect choice for the health industry. Kamatou et al. (2012) summarized different efforts for the above-mentioned potential of eugenol, such as treating toothache in 1600 in "Practice of Physic" in France. Since then, many researchers have been trying eugenol as a safe substitute for treating patients with different ailments. Including those, the remaining literature has been cited in this chapter as much as available at different scientific published data sources. Nevertheless, more safety studies of eugenol, its derivatives, and synergetic effect-giving combinations are necessary to get maximum health benefits from this valuable compound.

References

Bendre, R.S., Rajput, J.D., Bagul, S.D., Karandikar, P., 2016. Outlooks on medicinal properties of eugenol and its synthetic derivatives. Nat. Prod. Chem. Res. 4 (3), 1–6.

Carrasco, A.H., Espinoza, C.L., Cardile, V., Gallardo, C., Cardona, W., Lombardo, L., Catalán, M.K., Cuellar, F. M., Russo, A., 2008. Eugenol and its synthetic analogues inhibit cell growth of human cancer cells (Part I). J. Braz. Chem. Soc. 19, 543–548.

Carrasco, H., Raimondi, M., Svetaz, L., Liberto, M.D., Rodriguez, M.V., Espinoza, L., Madrid, A., Zacchino, S., 2012. Antifungal activity of eugenol analogues. Influence of different substituents and studies on mechanism of action. Molecules 17 (1), 1002–1024.

Chami, F., Chami, N., Bennis, S., Trouillas, J., Remmal, A., 2004. Evaluation of carvacrol and eugenol as prophylaxis and treatment of vaginal candidiasis in an immunosuppressed rat model. J. Antimicrob. Chemother. 54 (5), 909–914.

Chang, S.T., Chen, P.F., Chang, S.C., 2001. Antibacterial activity of leaf essential oils and their constituents from *Cinnamomum osmophloeum*. J. Eethnopharmacol. 77 (1), 123–127.

Daniel, A.N., Sartoretto, S.M., Schmidt, G., Caparroz Assef, S.M., Bersani Amado, C.A., Cuman, R.K.N., 2009. Anti-inflammatory and antinociceptive activities A of eugenol essential oil in experimental animal models. Rev. Bras. Farmacogn. 19, 212–217.

deAraújo Lopes, A., da Fonseca, F.N., Rocha, T.M., de Freitas, L.B., Araújo, E.V.O., Wong, D.V.T., Lima Júnior, R.C.P., Leal, L.K.A.M., 2018. Eugenol as a promising molecule for the treatment of dermatitis: antioxidant and anti-inflammatory activities and its nanoformulation. Oxid. Med. Cell. Longev., 2018. 8194849.

Devi, K.P., Nisha, S.A., Sakthivel, R., Pandian, S.K., 2010. Eugenol (an essential oil of clove) acts as an antibacterial agent against *Salmonella typhi* by disrupting the cellular membrane. J. Ethnopharmacol. 130 (1), 107–115.

Devkota, H.P., Adhikari Devkota, A., 2020. Cold pressed clove (*Syzygium aromaticum*) oil. In: Ramadan, M.F. (Ed.), Cold Pressed Oils. Elsevier, pp. 273–276, https://doi.org/10.1016/B978-0-12-818188-1.00023-2.

Anon., 2021. Eugenol Dental Composition. A61C 7/02 (20060101); A61K 47/10 (20060101); A61K 31/085 (20060101).

Fidel Jr., P.L., Sobel, J.D., 1996. Immunopathogenesis of recurrent vulvovaginal candidiasis. Clin. Microbiol. Rev. 9 (3), 335–348.

Gayoso, C., Lima, E., Oliveira, V., Pereira, F., Souza, E., Lima, I., Navarro, D., 2005. Sensitivity of fungi isolated from onychomycosis to *Eugenia cariophyllata* essential oil and eugenol. Fitoterapia 76 (2), 247–249.

Guo, X., Shang, X., Li, B., Zhou, X.Z., Wen, H., Zhang, J., 2017. Acaricidal activities of the essential oil from Rhododendron nivale Hook. f. and its main compund δ cadinene against *Psoroptes cuniculi*. Vet. Parasitol. 236, 51–54.

Hidalgo, M.E., De la Rosa, C., Carrasco, H., Cardona, W., Gallardo, C., Espinoza, L., 2009. Antioxidant capacity of eugenol derivatives. Quim. Nova 32, 1467–1470.

Jirovetz, L., Buchbauer, G., Stoilova, I., Stoyanova, A., Krastanov, A., Schmidt, E., 2006. Chemical composition and antioxidant properties of clove leaf essential oil. J. Agric. Food Chem. 54 (17), 6303–6307.

Kamatou, G.P., Vermaak, I., Viljoen, A.M., 2012. Eugenol-from the remote Maluku Islands to the international market place: a review of a remarkable and versatile molecule. Molecules 17 (6), 6953–6981.

Karmakar, S., Choudhury, M., Das, A.S., Maiti, A., Majumdar, S., Mitra, C., 2012. Clove (*Syzygium aromaticum* Linn) extract rich in eugenol and eugenol derivatives shows bone preserving efficacy. Nat. Prod. Res. 26 (6), 500–509.

Leem, H.H., Kim, E.O., Seo, M.J., Choi, S.W., 2011. Antioxidant and anti inflammatory activities of eugenol and its derivatives from clove (*Eugenia caryophyllata* Thunb.). J. Korean Soc. Food Sci. Nutr. 40 (10), 1361–1370.

Mbaveng, A., Kuete, V., 2017. Syzygium aromaticum. In: Medicinal Spices and Vegetables From Africa. Elsevier, pp. 611–625.

Merchán Arenas, D.R., Muñoz Acevedo, A., Vargas Méndez, L.Y., Kouznetsov, V.V., 2011. Scavenger activity evaluation of the clove bud essential oil (*Eugenia caryophyllus*) and eugenol derivatives employing $ABTS^{+\bullet}$ decolorization. Sci. Pharm. 79 (4), 779–792.

Miyazawa, M., Hisama, M., 2003. Antimutagenic activity of phenylpropanoids from clove (Syzygium aromaticum). J. Agric. Food Chem. 51 (22), 6413–6422.

Ntamila, M., Hassanali, A., 1976. Isolation of oil of clove and separation of eugenol and acetyl eugenol. An instructive experiment for beginning chemistry undergraduates. J. Chem. Educ. 53 (4), 263.

Oliveira, R.A.D., Reis, T.V., Sacramento, C.K.D., Duarte, L.P., Oliveira, F.F.D., 2009. Constituintesquímicosvoláteis de especiariasricasem eugenol. Rev. Bras. Farmacogn. 19, 771–775.

Öztürk, A., Özbek, H., 2005. The anti-inflammatory activity of Eugenia caryophyllata essential oil: an animal model of anti-inflammatory activity. Eur J Gen. Med. 2 (4), 159–163.

Rahim, N.H.C.A., Asari, A., Ismail, N., Osman, H., 2017. Synthesis and antibacterial study of eugenol derivatives. Asian J. Chem. 29 (1), 22.

Ryu, B., Kim, H.M., Lee, J.S., Lee, C.K., Sezirahiga, J., Woo, J.H., Choi, J.H., Jang, D.S., 2016. New flavonol glucuronides from the flower buds of *Syzygium aromaticum* (clove). J. Agric. Food Chem. 64 (15), 3048–3053.

Sellamuthu, R., 2014. Eugenol. Academic Press/Elsevier, Cambridge, MA, pp. 539–541.

Sgorbini, B., Cagliero, C., Pagani, A., Sganzerla, M., Boggia, L., Bicchi, C., Rubiolo, P., 2015. Determination of free and glucosidically bound volatiles in plants. Two case studies: L menthol in peppermint (Mentha x piperita L.) and eugenol in clove (*Syzygium aromaticum* (L.) Merr. & LM Perry). Phytochemistry 117, 296–305.

Shan, B., Cai, Y.Z., Sun, M., Corke, H., 2005. Antioxidant capacity of 26 spice extracts and characterization of their phenolic constituents. J. Agric. Food Chem. 53 (20), 7749–7759.

Shang, X.F., Dai, L.X., Yang, C.J., Guo, X., Liu, Y.Q., Miao, X.L., Zhang, J.Y., 2021. A value added application of eugenol as acaricidal agent: the mechanism of action and the safety evaluation. J. Adv. Res. 34, 149–158.

Shang, X., Wang, Y., Zhou, X., Guo, X., Dong, S., Wang, D., Zhang, J., Pan, H., Zhang, Y., Miao, X., 2016. Acaricidal activity of oregano oil and its major component, carvacrol, thymol and p cymene against *Psoroptes cuniculi* in vitro and in vivo. Vet. Parasitol. 226, 93–96.

Shang, X.F., Dai, L.X., Liu, Y.Q., Zhao, Z.M., Li, J.C., Yang, G.Z., Yang, C.J., 2019. Acaricidal activity and enzyme inhibitory activity of active compounds of essential oils against *Psoroptes cuniculi*. Vet. Parasitol. 267, 54–59.

Sheehan, D.J., Hitchcock, C.A., Sibley, C.M., 1999. Current and emerging azole antifungal agents. Clin. Microbiol. Rev. 12 (1), 40–79.

Tanaka, T., Orii, Y., Nonaka, G.I., Nishioka, I., Kouno, I., 1996. Syzyginins A and B, two ellagitannins from *Syzygium aromaticum*. Phytochemistry 43 (6), 1345–1348.

Wahyuningsih, T.D., Raharjo, T.J., Tahir, I., 2010. Synthesis of 3, 4 dimethoxy isoamyl cinnamic as the sunscreen compound from clove oil and fusel oil. Indian J. Chem. 2 (1), 55–63.

Walter, R., 1972. β Caryophyllene in native clove bud oil. Phytochemistry 11 (1), 405–406.

Woo, J.H., Ahn, J.H., Jang, D.S., Lee, K.T., Choi, J.H., 2017. Effect of kumatakenin isolated from cloves on the apoptosis of cancer cells and the alternative activation of tumor associated macrophages. J. Agric. Food Chem. 65 (36), 7893–7899.

CHAPTER

Chemistry and functionality of clove oil nanoemulsions

7

Bimlesh Mann[a], Richa Singh[a], S. Athira[b], Rajesh Kumar[a], and Rajan Sharma[a]

ICAR-National Dairy Research Institute, Karnal, Haryana, India[a] Kerala Veterinary and Animal Sciences University, Pookode, Kerala, India[b]

1. Introduction

Syzygium aromaticum (synonym: *Eugenia cariophylata*), commonly known as clove, is a median-sized tree (8–12 m) from the *Myrtaceae* family native to North Molucca islands in east Indonesia (Cortés-Rojas et al., 2014). The clove, *Syzygium aromaticum*, is one of the most precious spices that has been used for centuries as a flavoring agent, food preservative, and active ingredient in traditional therapies (Cortés-Rojas et al., 2014). The main products obtained from clove include whole or ground clove buds, essential oils produced from clove buds, stem and leaf, and clove oleoresins. In the food industry, this spice is generally used in a small quantity due to their intense flavor in form of powdered, extracted essential oils, or oleoresin. Besides flavoring properties, it also contains nutrients, such as proteins, vitamins, minerals, etc. (Nurdjannah and Bermawie, 2006). The antimicrobial properties of the clove oil are well known because they are effective against a large number of food spoilage and pathogenic microorganisms like *Escherichia coli, Listeria monocytogenes,* and *Salmonella enteric* (Chaieb et al., 2007). Clove oil has antiseptic and analgesic, antifungal, anticarcinogenic, antiallergic, and antimutagenic and can be used as alternative medicine. The major component of clove oil is eugenol which constitutes about 70%–95%, and it acts as a critical factor for all the bio-functional properties of clove oil. Still, clove oil or its active ingredient eugenol has limited use in the food system due to poor water solubility, high volatility, less stability, and high loss during food processing. Therefore, a strategy is essential to incorporate this into the food matrix without compromising the biological property of the active ingredient and sensory and nutritional quality of the targeted food. Encapsulation of these lipophilic components is the best platform as a delivery system to overcome most of the mentioned challenges (Sharma et al., 2017). This chapter describes the use of emulsification techniques to protect the bio-functional properties of clove oil for its wider applications such as food, pharma, etc.

2. Clove oil and eugenol

Clove comprises mainly volatile and non-volatile components. Volatile clove oil can be mainly extracted from leaves, stems, buds, and fruits. The yield and quality of the oils are influenced by origin, variety, quality of raw materials, pre-treatment before distillation, distillation method, and post-

Clove (*Syzygium aromaticum*). https://doi.org/10.1016/B978-0-323-85177-0.00012-4

distillation treatment (Nurdjannah and Bermawie, 2006). Clove oil is a colorless to pale yellow liquid with a sharp, spicy odor and taste. Eugenol ($C_{10}H_{12}O_2$) is the major component of clove oil comprising 83%–95% of the oil, an allyl chain-substituted guaiacol, i.e., 2-methoxy-4-(2-propenyl) phenol. It is a colorless or yellowish liquid that is slightly water-soluble but easily soluble in organic solvent (Nurdjannah and Bermawie, 2012).

Clove bud oil contains 15%–20% essential oil with eugenol (70%–85%), eugenyl acetate (15%), and β-caryophyllene (5%–12%) (Fig. 1). Clove leaves yield 3.0%–4.8% essential oil, and with the maturity of leaves, eugenol content increased from 38.3% to 95.2%. Clove stem yields 6% volatile oil containing 80.2% eugenol and 6.6% β-caryophyllene. At the same time, ripe fruits yield 2% of oil, which comprises 50%–55% eugenol (Milind and Deepa, 2011). Both clove oil and eugenol have GRAS (Generally Regarded as Safe) status by the Food and Drug Administration and are also considered non-carcinogenic and non-mutagenic.

3. Biofunctional properties of clove oil

Besides the flavoring properties, clove oil has a broad spectrum of biological properties like antimicrobial, antioxidant, and anti-inflammatory, making it a suitable choice as an ingredient in food. Clove oil or eugenol has well-known antimicrobial properties. The ability of eugenol to disrupt the cytoplasmic membrane of the bacteria, which further leads to an increase in non-specific permeability, is the basic mechanism of antibacterial property (Gill and Holley, 2006). The hydrophobic eugenol easily penetrates into the lipopolysaccharide of the Gram-negative bacterial cell membrane and alters its cell structure, leading to the leakage of intracellular constituents (Cui et al., 2018). The hydroxyl group of eugenols also interacts with the protein, preventing enzyme action in different pathogenic and spoilage bacteria like *Enterobacter aerogenes* (Burt, 2004). Clove oil dispersed in a concentrated sugar solution (0.4%, v/v) had a germicidal effect against various bacteria like *S. aureus, Klebsiella pneumoniae, Pseudomonas aeruginosa, Clostridium perfringens, E. coli*, and *Candida albicans* (Briozzo et al., 1989). Eugenol at a concentration of 0.05% has a bactericidal action against *Bacillus tuberculosis* (Lopez et al., 2005). Benencia and Courrèges (2000) studied the antiviral activity against HSV-1 and HSV-2 viruses, and 50% inhibition was obtained with a concentration of 25.6 and 16.2 μg/mL eugenol, respectively. The inhibitory effect of clove oil on *L. monocytogenes* and its effect on the cell

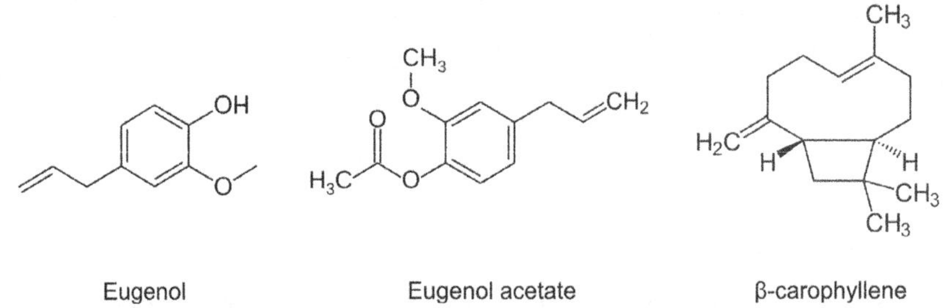

FIG. 1

Chemical structure of eugenol, eugenol acetate, and carophyllene.

membrane, respiratory metabolism, and its interaction with DNA were investigated by Cui et al. (2018). The results showed that the clove oil could cause the leakage of biological macromolecules like protein, ATP, and DNA and the reduction in the activity of two intracellular enzymes (β-galactosidase and AKP), finally altering cell membrane permeability (Cui et al., 2018).

Clove oil also has a comparable antioxidant property with potent synthetic antioxidants like BHA (butylated hydroxyl toluene), BHT (butylated hydroxy toluene), and pyrogallol (Dorman et al., 2000). Nagababu and Lakshmaiah (1994) reported the ability of eugenol to prevent the formation of thiobarbituric acid reactive substances (TBARS) at concentrations of 4–15 μM. Eugenol also can scavenge hydroxyl-free radicals, thereby protecting the skin damaged by UV light (Taira et al., 1992). The basic mechanism underlying the antioxidant activity of clove oil is due to its ability to suppress lipid peroxidation and act as scavengers of free radicals and chelators of transition metal ions (Jung et al., 2006; Masae et al., 2005). Fujisawa et al. (2002) showed that electron spin resonance spectroscopic studies showed that eugenol (18.50 mU per 100 μM sol) significantly scavenged reactive oxygen species.

4. Challenges for using the clove oil in the food matrix

Besides several bioactivities, the direct use of clove oil in foods is not encouraged due to its sensitivity to oxygen, light, temperature, and pH. In the presence of oxygen and light, the eugenol may get oxidized, thus leading to the formation of free radicals and finally resulting in the reduced biofunctional property. As the clove essential oil is a highly volatile component, high-temperature storage may cause loss. Moreover, it is highly unstable with varying pH (Dima and Dima, 2015). So, the stability issues hinder using these bioactive components in the food matrix as a natural source of food preservatives and flavoring agents. Using essential oil in food is increasing day by day due to the rise of green consumerism and the health consciousness of consumers regarding synthetic food additives. However, incorporating essential oil in the aqueous food system is limited due to its lipophilic nature. The unsaturated ring structure of the essential oil tends to interact with food components (Burt, 2004). This may lead to a reduction in the nutritional value of the food and the biofunctional property of the active ingredient. Besides, the addition of essential oil based on the flavor or biological activity threshold value may alter the flavor characteristics of the product. This may reduce the overall acceptability of the product in the market as every consumer are habitual and particular about the sensory quality, especially the flavor of each food product (Chon et al., 2020). Thus, there is a broad scope for a technique that can protect essential oils from the above hurdles, preserve their flavor and biological activity, and convert them into a water-soluble powder form.

5. Encapsulation of clove oil

Encapsulation technology is a method of enclosing molecules of interest into capsules before delivery into a system. This process is beneficial in preventing the degradation and volatilization of active compounds in essential oil, preserving the active component against harsh and unpredicted processing and storage conditions. It provides a longer shelf life and bioavailability for the active compounds and can be used for the controlled release of entrapped materials on the target site (Zuidam and Shimoni, 2010). Encapsulation widens the application area of clove oil in food formulations. Several encapsulation methods are used for these purposes. The use of nanotechnology in encapsulation is a revolutionary technology that provides the exclusive property of nanoscale structures (size ≤ 100 nm) with the main advantage of homogeneity, which leads to improved encapsulation performance and improved physical and chemical properties. The major advantages of nanotechnology for encapsulating bioactive food ingredients are improved accessible surface area and solubility, which improve the bioavailability of bioactive ingredients. Nanoencapsulation also protect early metabolization of bioactive ingredient after ingestion before reaching the target site. Nanoencapsulation of clove oil has added advantage of allowing its addition to aqueous products, masking its unpleasant taste when used in excess amount, and protecting its loss due to volatility during harsh processing treatments like heating (Woranuch and Yoksan, 2013).

Nanoencapsulation of bioactive ingredients can be achieved by different mechanisms divided into approaches: "top-down," "bottom-up," and combination of both. In the "top-down" method, particle size is reduced by employing various mechanical forces, while in the "bottom-up" method, methods such as self-assembly are used to increase particle size. Various nanoencapsulation techniques like nanoemulsions, nanoliposomes, and nanostructured lipid carriers are used to encapsulate essential oils (Fig. 2).

6. Nanoemulsions for delivery of clove oil

Among the different techniques mentioned in Fig. 2, emulsification is a practical approach to improve physicochemical properties of clove oil, such as increased water solubility and lower volatility of essential oils (Ezhilarasi et al., 2013). Emulsions are colloidal dispersions prepared by mixing an oil phase, aqueous phase, surfactant, and co-surfactant in a specific ratio (Quintanilla-Carvajal et al., 2010).

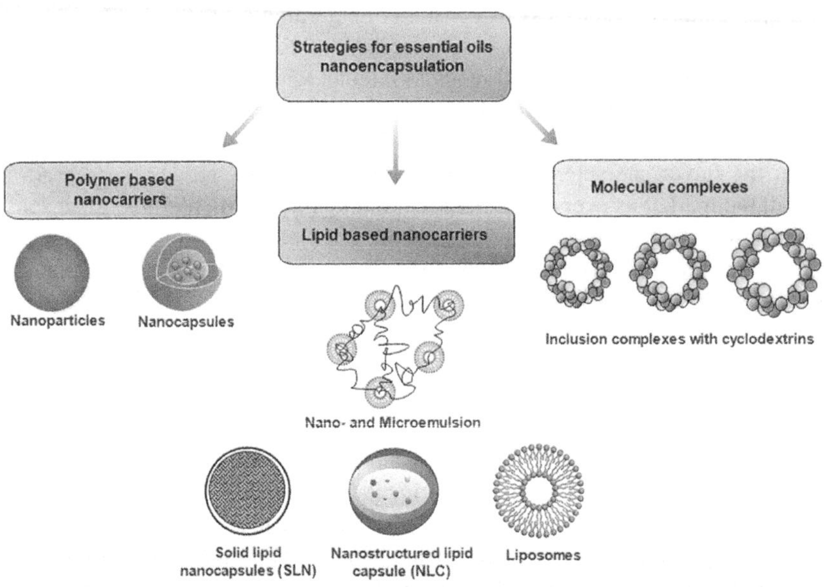

FIG. 2

Schematic representation of nanoencapsulation techniques used for essential oils (Rai et al., 2017).

Both micro- and nano-emulsions of clove oil are considered to have bioactive properties (mainly antimicrobial) due to the small size of oil particles with a high surface tension that can fuse with and disrupt the membrane of microorganisms. Nanoemulsion of essential oil attracts over microemulsion because of their smaller size, higher stability, and better bioavailability (McClements and Rao, 2011).

For the delivery of water-insoluble essential oils like clove oil, oil-in-water nanoemulsions are valuable vehicles. Different workers have used different methods for the preparation of clove oil nanoemulsion. Nanoemulsions are prepared using low-energy (high-speed homogenization, phase inversion) or high-energy emulsification methods (high-pressure homogenization). In literature, clove oil nanoemulsions used for food applications are majorly prepared using high-energy inputs (such as ultrasonication and microfluidization) (Håkansson and Rayner, 2018; Akhavan et al., 2018). These techniques are widely used due to their non-toxicity, reproducibility, and environmental friendliness. However, high-speed homogenization is not a preferred method to prepare O/W nanoemulsion but can be used for the preparation of coarse emulsions and these coarse emulsions can be converted into nanoemulsions using high-energy methods (Espitia et al., 2019). Fig. 3 depicts the concepts of high-energy techniques used to make essential oil nanoemulsions.

Ultrasonication to prepare oil-in-water emulsion is an ancient and powerful technique in which ultrasonic waves induce cavitation effect (Abismaıl et al., 1999). Wood and Loomis were the first to report ultrasonic-assisted emulsification, and the first patent was issued in Switzerland in 1944. The ultrasonic waves induce emulsification in two steps. The eruption of dispersed phase droplets into continuous phase is caused by a combination of interfacial waves and instability followed by breaking of droplets by cavitation near the interface (Mahdi Jafari et al., 2006). The shear force produced by ultrasound generates temperature and pressure that help in achieving homogenization (generation of smaller particles) in mixed dispersions (Leong et al., 2017). As a result, the nanoemulsions prepared by ultrasonication have a smaller size, great stability, and low polydispersity. In ultrasonication, the properties of prepared nanoemulsions are affected by different factors like the ultrasonic cycle and intensity of radiation. Shahavi et al. (2019) reported that with an increase in duty cycle, the average droplet size of clove oil nanoemulsion decreased, whereas pulsed ultrasound with proper intervals was more effective than continuous ultrasonication. The prolonged ultrasonic treatment effectively disrupts large particles in emulsion. According to Kentish et al. (2008), ultrasonic treatments decreased the size of nanoemulsion droplets to a minimum at intermediate speed but increased the size at higher power. As per the observation of Shao et al. (2018), higher ultrasonic power (750 W) is not effective as a significant increase in clove oil nanoemulsion particle size was observed that might be attributed to the tendency to recoalescence. This is also termed as over-processing of ultrasonic treatment. In microfluidization, high-pressure stream is forced to pass through fine-sized microchannels, creating a high shearing action that provides an excellent emulsion. The size of emulsion prepared by microfluidization is of smaller size than the product of high-pressure homogenization; therefore, the prepared oil-in-water emulsion is even stable at very low emulsifier concentration (Mahdi Jafari et al., 2006) (Table 1).

7. Importance of emulsifiers/surfactants in an emulsion system

Emulsifiers/surfactants are amphiphilic molecules that could adsorb on the interface, thereby reducing the interfacial tension at the oil-water interface. They have hydrophobic/lipophilic parts, typically aliphatic chains, hydrophilic parts, and are not highly soluble in water or organic solvent/oil with the least repulsive force from these solvents at an interface (Walstra, 1996). Therefore, they can form a

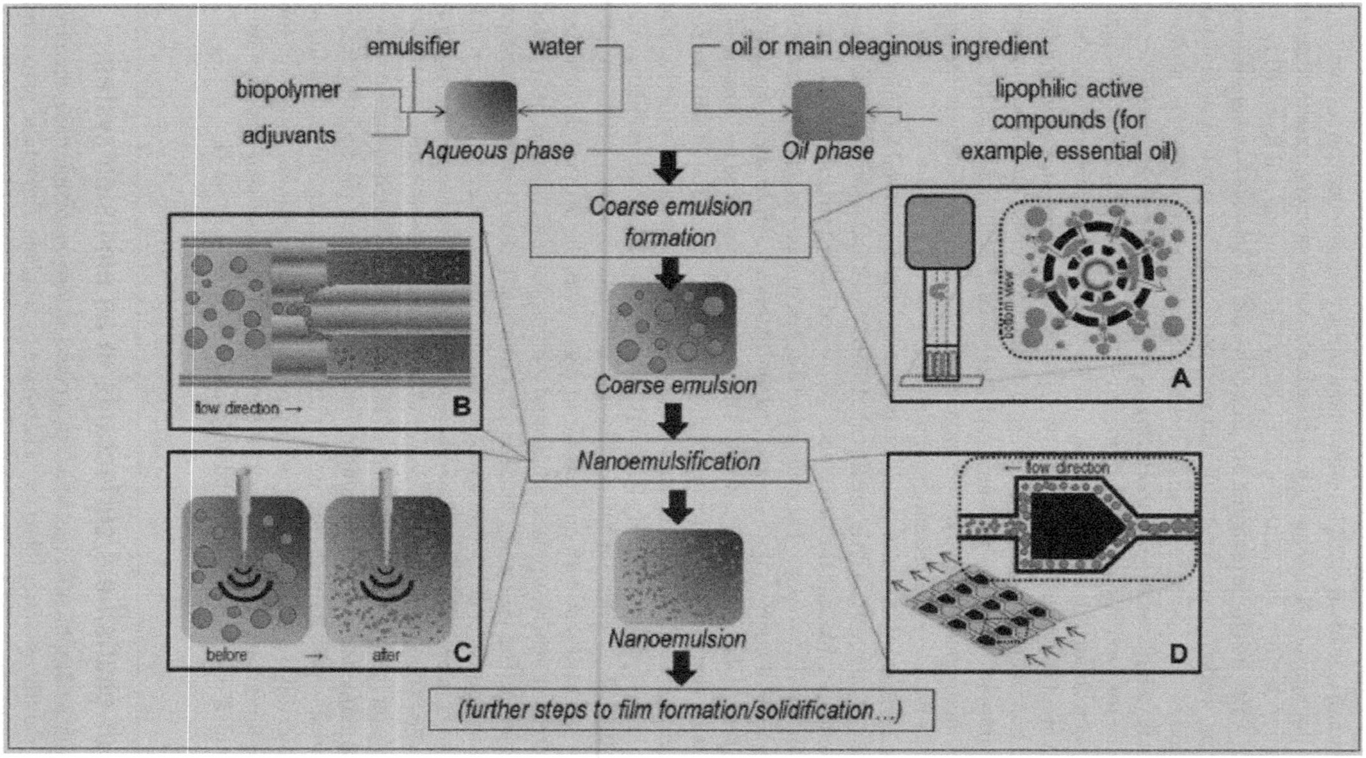

FIG. 3

Schematic representation of different techniques used for the preparation of essential oil nanoemulsions. (A) The high-speed homogenizer is used to form coarse oil-in-water emulsion before emulsification by (B) high-pressure homogenization, (C) ultrasonication, (D) microfluidization (Espitia et al., 2019).

Table 1 Different techniques used for the preparation of clove oil/eugenol nanoemulsion.

Technique	Conditions	Materials	Observations	References
Ultrasonication	400 W, 50% amplitude	Clove oil with crodamol as co-stabilizer	Size: 59–134 nm PDI: 0.196–0.363 Stable for 60 days	de Meneses et al. (2019)
Ultrasonication	30 KHz, 50 W, 40% amplitude	Eugenol, Tween 80, Labrasol	Size: 89.98 ± 6.48 nm PDI: 0.238 ± 0.001 Zeta: −10.05 ± 0.11 Stable for 14 weeks	Ahmad et al. (2018)
Ultrasonication	450 W	Eugenol, chitosan, Tween 20	Size: 432.3 ± 5.2 nm PDI: 0.372 ± 0.010 Zeta: 2.25 ± 0.16 mV Stable for 30 days at 4–50°C	Shao et al. (2018)
Ultrasonication	20 kHz, 750 W	Eugenol, sesame oil, Tween 80	Size: 20–80 nm PDI: 0.092–0.30 Stable for 4 weeks	Ghosh et al. (2014)
Microfluidization High-pressure homogenization	150 MPa	Clove oil, sodium alginate, Tween 80	Size: 20.88 ± 5.1 nm PDI: 0.82 ± 0.07 Size: 23.23 ± 10.51 nm PDI: 0.94 ± 0.09	Salvia-Trujillo et al. (2015)
Microfluidization	11,000 psi for five passes	Eugenol, bean oil, Tween 80, SDS	Stable for 4 days	Li et al. (2015)
Microfluidization	15,000 psi	Clove oil, MCT, Tween 80	Size: 85.64 ± 3.46 30 days at 4 and 25°C	Wan et al. (2019)
High-pressure homogenizer	15,000 psi	Eugenol, MCT, and Tween 80	Size: 80–300 nm	Terjung et al. (2012)
Ultrasonication	10% amplitude		Size: 500–3000 nm	
High-speed homogenization followed by microfluidization	18,000 rpm for 3 min followed by 100 MPa with five processing cycles	Eugenol, flaxseed oil, gum	Size: 200–240 nm PDI: 0.07 ± 0.006 Zeta potential: −22 to −35	Sharif et al. (2017)
High-speed homogenization followed by microfluidization	13,500 rpm for 2 min followed by 50–150 MPa with 1–20 passes	Clove oil, canola oil, gum	Size: 151.3 ± 0.56 PDI: 0.14 ± 0.002 Stable for 30 days at room temperature	Majeed et al. (2016)
High-speed homogenization	13,500 rpm for 3 min	Eugenol, gum arabic, lecithin	Size: 319.2 PDI: 0.47 Zeta potential: −56.5 mV Stable for 14 days at room temperature	Hu et al. (2016)
High-speed homogenizer	15,000 rpm for 3 min	Eugenol, WPI-maltodextrin conjugate	Size:127–255 nm	Shah et al. (2013)

viscoelastic film around the dispersed droplets (water/oil) to keep the emulsion more stable (Lam and Nickerson, 2013).

7.1 Low molecular weight surfactants

They are classified based on the polar nature of the hydrophilic portion, i.e., non-ionic, anionic, and cationic, respectively (Table 2). HLB (hydrophilic lipophilic balance) value is the basic strategy to determine the suitability of emulsifiers to prepare various emulsion systems. HLB values are calculated from the ratio of the hydrophilic head group to the lipophilic tail group (McClements, 2005). The value ranges from 1 to 40. If the value is 7, then it indicates equal solubility of surfactant in water and oil. The lower value implies their high affinity toward the oil phase, and they can be used for the W/O emulsion system. The surfactants with HLB value of more than 7 are suitable for making O/W emulsions (Walstra, 1996).

Non-ionic surfactants influence the stability of protein-stabilized emulsion in several different ways: (i) competitive adsorption, (ii) specific protein-surfactant interactions, and (iii) depletion flocculation (McClements, 1998). Thus, low molecular weight surfactants can be effectively used as a co-surfactant to prepare emulsion systems along with high molecular weight surfactants, which provide more stability to the dispersed phase (Gupta et al., 2010a,b). The factors that affect the non-ionic surfactants' effectiveness as a good emulsifier are pH, temperature, and concentration. Cloud point is the term used to denote the temperature at which non-ionic surfactant solution becomes turbid. This value varies with the non-ionic surfactant structure (i.e., length of the polyoxyethylene groups) and the non-polar compound (Gaysinksy, 2004).

7.2 High molecular weight surfactants

Proteins are amphiphilic molecules with hydrophilic tails and hydrophobic heads, which help stabilize an emulsion by adsorbing at the interface. The hydrophobic head region aligns itself with the lipid phase and the hydrophilic tail region with the water phase (Foegeding and Davis, 2011).

Table 2 Small molecular weight surfactants generally used for emulsion formulation (Walstra, 1996).

Type	Example of surfactant	HLB value
Non-ionic		
Aliphatic alcohol	Hexadecanol	1
Monoacyl glycerol	Glycerol monostearate	3.8
Esters of monoacylglycerols	Lactoylmonopalmitate	8
Spans	Sorbitanmonostearate	4.7
	Sorbitanmonooleate	7
	Sorbitanmonolaurate	8.6
Tween 80	Polyoxyethylenesorbitanmonooleate	16
Anionic		
Soap	Na oleate	18
Lactic acid esters	Na stearoyl-2-lactoyl lactate	21
Phospholipids	Lecithin	Fairly large

While comparing with small molecular emulsifiers that rapidly diffuse to the interface to achieve excellent emulsion-forming abilities, protein tends to be bulkier at the interface by slow diffusion (Lam and Nickerson, 2013). Proteins generally undergo a partial denaturation to form a stable viscoelastic film around the oil dispersed droplet by exposing buried hydrophobic amino acids to the surface/within the oil phase and hydrophilic amino acids within the aqueous phase. This mechanism of adsorption helps to resist all the mechanical stresses and provides electrostatic and steric stabilization for the formation of stable emulsions (Tcholakova et al., 2006).

Food proteins generally used as emulsifiers include whey protein isolate, casein, ovalbumin, soy gluten, and bovine serum albumin. The emulsifying ability of the protein mainly depends on its physicochemical characteristics. Surface hydrophobicity influences the ability of the protein to adsorb to the oil side at the interphase, leading to greater disintegration, ultimately giving higher emulsion capacity (Kim et al., 2005). At the same time, the surface charge of the protein enhances their solubility in the aqueous phase, which enables them for greater diffusion at the interphase (Lam and Nickerson, 2013). The emulsion pH is also an important determinant for emulsion stability as it affects the structure of the protein molecule. If the emulsion pH is above or below the isoelectric point of the specific food protein used in the particular emulsion formulation, the droplets will get either positive or negative. The charged droplets may execute high electrostatic repulsive force between them, which provides greater emulsion stability. However, if the pH condition is close to the isoelectric point, the oil droplets will flocculate/aggregate, leading to coalescence and instability of the emulsion (McClements, 2005).

Polysaccharides/hydrocolloids are generally used in the food systems as a stabilizer for emulsions, suspensions, and foams, including gum arabic, guar gum carboxy methylcellulose carrageenan, agar, starch, and pectin. Gelatin, a protein derived from collagen, is one of the few non-carbohydrate stabilizers used extensively. They are hydrophilic and dispersed in solution as colloids, which leads to their designation as hydrocolloids (Walstra, 1996). These hydrocolloids could stabilize oil-in-water emulsions by increasing the aqueous phase's viscosity and acting as a barrier against destabilization by slowing down flocculation and coalescence between dispersed droplets (Dickinson, 1988; Akhtar and Dickinson, 2003).

8. Stability of clove oil nanoemulsions

The prepared oil-in-water nanoemulsions must be physically stable during their storage for their applicability. The physical instability in these nanoemulsions due to undesirable molecular interactions at the interface causes a hydrophobic effect that induces thermodynamic instability. Due to their smaller size, nanoemulsions are more physically stable in comparison to the conventional emulsion. However, despite their smaller size, nanoemulsions will eventually degrade due to different mechanisms, e.g., gravitational separation, flocculation, coalescence, Ostwald ripening, and phase separation. Therefore, the designing of nanoemulsions is always done by keeping these instabilities in mind. Following is the physical instability (Fig. 4) that commonly occurs in oil-in-water nanoemulsions:

8.1 Gravitational separation

Gravitational separation is the mechanism by which nanoemulsion droplets travel downward (sedimentation) or upward (creaming) since their density exceeds or falls below that of the liquid around them. Most essential oils are less dense than water; they tend to migrate upward, while water tends to migrate

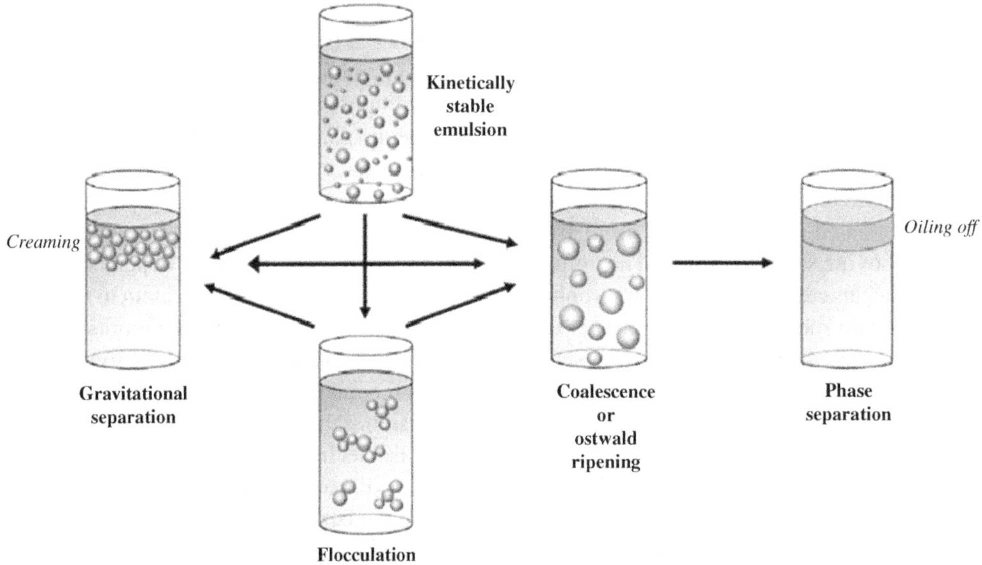

FIG. 4

Schematic representation of the mechanisms responsible for physical instability in nanoemulsion (Liu et al., 2019).

downward. Therefore, essential oil nanoemulsions (including clove oil) exhibit more cases of creaming. When there is a low concentration of emulsifiers during the preparation of nanoemulsions that is insufficient to cover the droplet surface completely, extensive creaming is observed (Liu et al., 2019). However, thick and dense coats while preparing oil-in-water nanoemulsion make them more prone to sedimentation. Clove oil nanoemulsions prepared using sodium caseinate and pectin showed sedimentation and creaming after storage for 20 days at 25°C. The sedimentation in clove oil nanoemulsion was visible with a progressively increased particle size of clove oil nanoemulsion from 172 nm at 0 day to 415 nm on 20th day at 25°C (Sharma et al., 2017).

8.2 Flocculation and coalescence

Coalescence and flocculation are two forms of droplet accumulation that are determined by colloidal contact between the droplets. Flocculation is the mechanism of two or more droplets attracting each other to form clusters. The use of amphiphilic polysaccharides as emulsifiers above certain concentration promotes emulsion instability due to flocculation. Coalescence is the mechanism by which many droplets collide and amalgamate to create a larger droplet. Processing conditions such as acidic pH (pH 2–4) or high ionic pressure (200–500 mM NaCl) cause the droplets to coalesce due to decreased electrostatic repulsion. When protein is used as wall material for nanoencapsulation of clove oil, droplet flocculation is observed mainly at the pH (4.6) near to pI of protein (Surh et al., 2006).

8.3 Ostwald ripening

Dispersed phase droplets diffuse through the continuous phase, allowing disappearance of smaller droplets by dissolution and deposition on the larger droplets is called Ostwald ripening, which is considered to be the main instability mechanism for such nanoemulsions, driven by the curvature differences of the particles. Nanoemulsions containing essential oil are particularly prone to Ostwald ripening due to a considerable oil phase solubility in the aqueous phase, despite its primarily hydrophobic nature (Ryu et al., 2018).

Besides physical instability, the clove oil nanoemulsions also suffer from chemical instability like flavor loss due to oxidation. Since the interfacial areas of nanoemulsions are relatively broad, water-soluble pro-oxidants like metals (Cu and Fe) contact essential oils and accelerate their oxidation (McClements and Decker, 2000). As a result, in nanoemulsions, various chemical reactions occur, such as loss of taste, color, oxidation, and hydrolysis of lipid, causing them to lose their desirable properties.

Nanoemulsions of essential oils need to be engineered for higher kinetic stability, which is accomplished by careful selection of coat material with stabilizers and emulsifiers to meet the precise requirements of commercial applications (Zhang and McClements, 2018). The physicochemical properties of nanoemulsions, such as composition, particle size, charge, physical and aggregation state, rheology property, etc., are typically referred to as the stabilization mechanism (McClements and Jafari, 2018). The addition of thickeners and weighting agents that increase viscosity and reduce density can minimize gravitational separation. Droplet aggregation (flocculation and coalescence) is controlled by balancing repellent (e.g., steric and electrostatic) and attractive (e.g., hydrophobic, van der Waals, and depletion) interactions of droplets by changing the type and amount of emulsifier (Zhang and McClements, 2018). Ostwald ripening can be slowed down by adding ripening inhibitors or using an oil phase with low water solubility (Wooster et al., 2008).

9. Characterization of clove oil nanoemulsions

The techno-functional attributes (size, shape, charge, and interfacial properties) of nanoemulsion depend on the raw materials (e.g., oil, water, emulsifier, and thickening agent both on their type and concentration) and the method of manufacturing (McClements, 2004). The small size of nanoemulsions is responsible for better stability to gravitational separation due to the dominance of Brownian motion over gravitational forces (McClements, 2015) and decreased attractive forces between droplets. Following are the different important properties of nanoemulsions.

9.1 Particle size

Measurement of particle size in nanoemulsions is essential as the stability, appearance, rheology, release profile, and bioavailability of active ingredient are affected by its size (Assadpour et al., 2016). Particle size can be measured using different analytical techniques like microscopy, light scattering, electric pulse counting, sedimentation, ultrasonic spectroscopy, and NMR. However, in literature, light scattering and microscopy techniques characterize the particle size of clove oil nanoemulsion. Dynamic light scattering (DLS) is one of the most popular techniques for determining the size of nanoemulsion. The hydrodynamic diameter and polydispersity index are typically expressed as an intensity-weighted mean of the particle size (PDI). The PDI measures the nanoemulsion's homogeneity, with a value of 0.2 indicating a narrow droplet size distribution. DLS is a susceptible technique that presents several advantages like simplicity, automation, and time-saving. However, adequate sample dilution is required to avoid multiple scattering in the emulsion as DLS measurement is based on fluid's Newtonian behavior (Egerton and Tooley, 2014). The interfacial structure of particles in nanoemulsion is also associated with nanoemulsion stability and can also be determined using light scattering. Generally, the surface of nanoemulsion droplets is heterogeneous and complex in structure due to the organization of surfactant molecules at the particle surface which causes changes in physicochemical and sensory properties of nanoemulsions (McClements, 2015). All researchers have used particle size to observe physical instability in clove oil nanoemulsion (as mentioned in Table 1).

9.2 Particle charge

The particle charge is measured by surface charge density, surface potential, and zeta potential (Rao and McClements, 2011). Out of all the above-mentioned, zeta potential is mainly used to evaluate the electric properties of the droplets in nanoemulsion due to its strong representation in terms of electric properties and ease of determination. Between 30 and 60 mV, zeta potential of particles represents good stability of nanoemulsion and minimizes chances of flocculation, which occurs mainly when zeta potential is <15 mV (Silva et al., 2011). The zeta potential of particles in nanoemulsion is influenced by the pH of the medium, particle morphology, and state of hydration and surfactants used during the preparation of nanoemulsions (Simunkova et al., 2009). This is also one of the prime techniques used to characterize nanoemulsions (as mentioned in Table 1).

9.3 Morphological characterization

Microscopy can be used as a direct imaging technique for nanoemulsion that will provide information about nanoemulsion morphology and structure and their size that may complement the above-mentioned DLS technique for measuring the size of nanoemulsion (Hosseini et al., 2015). Various microscopy techniques like electron microscopy, laser scanning confocal microscopy, and atomic force microscopy are popular for the morphological characterization of nanoemulsions. The most popular electron microscopy is used for morphological characterization of clove oil nanoemulsions. Transmission electron microscopy (TEM), which produces a 2D cross-section of the sample, and scanning electron microscopy, which produces a 3D image, are the most commonly used electron microscopies (Klang et al., 2012). These images predict the shape, distribution, and size of nanoemulsions to illustrate the structural changes in nanoemulsions during processing and storage. The fine and evenly distributed oil droplets indicate that the emulsification process is efficient, and further coalescence and flocculation types of physical instability can be predicted using these images. However, proper care needs to be presented while analyzing nanoemulsions with microscopy, and it is advised to analyze different regions within a sample for statistically reliable data. Fig. 5 represents the SEM and TEM micrographs of 5% clove oil nanoemulsions prepared using sodium caseinate and pectin.

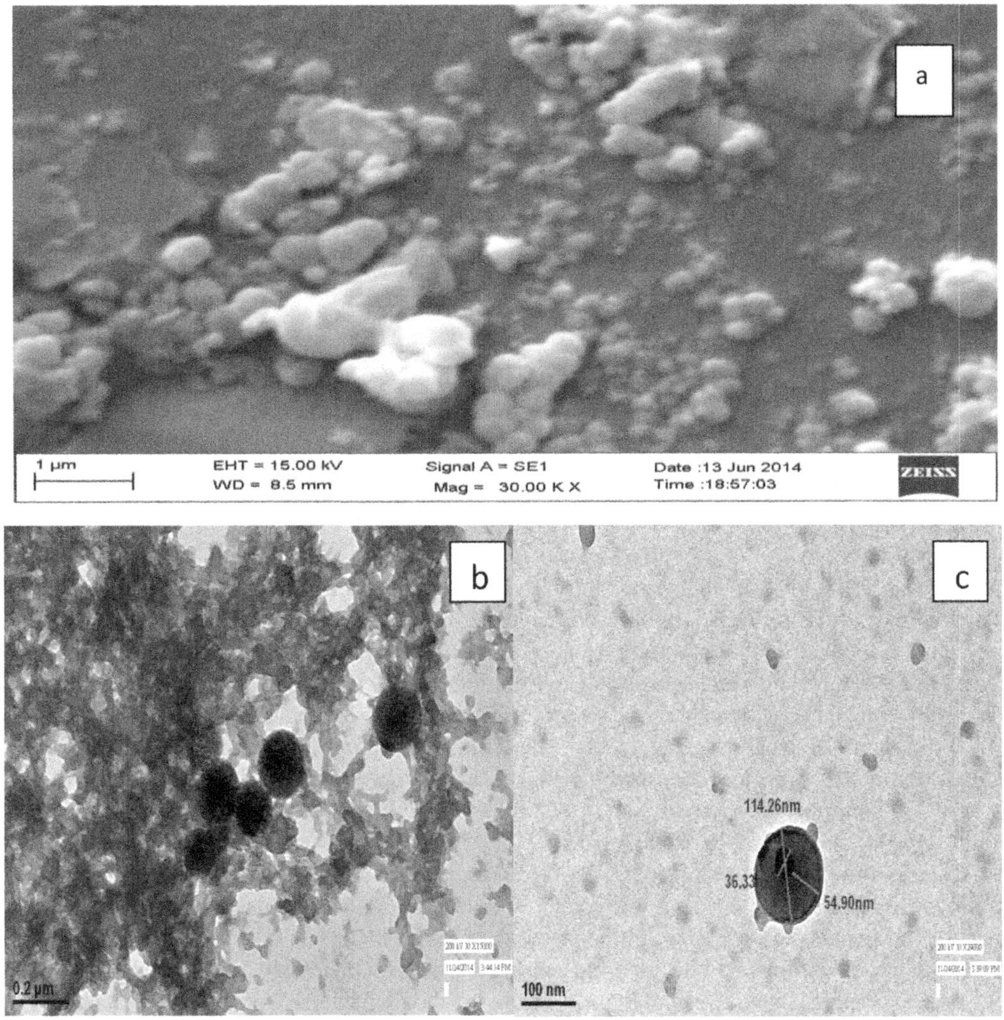

FIG. 5

(A) SEM, (B) TEM of stable 5% clove oil nanoemulsions prepared by using sodium caseinate (5%) and pectin (0.1%). (C) The core ingredient, clove oil, has a particle size of 36 nm and is encapsulated inside the coating material, NaCas and pectin, which has a particle size of 54 nm (Sharma et al., 2017).

10. Functionalities of clove oil nanoemulsions

Clove oil is known for its potent antimicrobial, anti-inflammatory, and antioxidant activities. The known activity of antimicrobial action of clove oil relies on its hydrophobic nature, which enables it to migrate into microbial cell membrane that causes damage to membrane structure and makes them more permeable, which leads to loss of cytoplasmic constituents that causes the loss of microbial cell

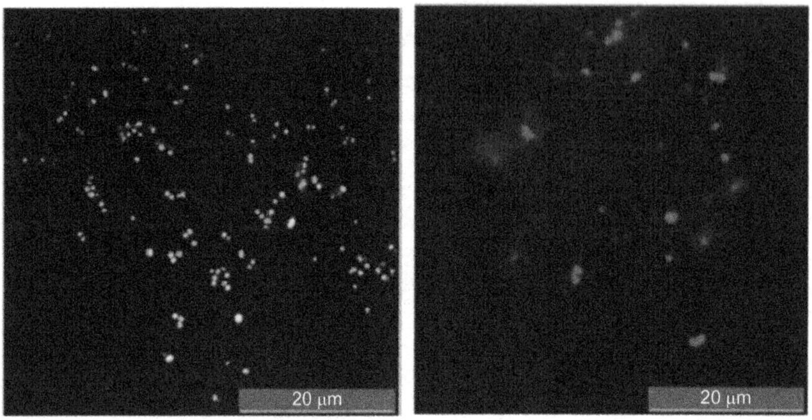

FIG. 6

Fluorescence micrograph of *S. aureus* live (control) and dead cells (after treating with 10% eugenol nanoemulsion for 150 min) (Ghosh et al., 2014).

(Bakkali et al., 2008). Fig. 6 represents a fluorescence micrograph of live cells (untreated, appeared as green) and nanoemulsion-treated dead cells (eugenol, appeared as red) of *S. aureus*.

Eugenol is the major bioactive component of clove oil, and bacterial enzyme activity is negatively affected by the OH free radical group on eugenol, leading to bacterial cell death (Woranuch and Yoksan, 2013). Nanoemulsions of clove oil/eugenol protected its bioactivity and emerged as a promising technique for its application (Donsì et al., 2011). After emulsification, the clove oil/eugenol has better solubility, stability, and enhanced antimicrobial activities due to the controlled/slow release of clove oil/eugenol. The solubility of clove oil nanoemulsion is also positively correlated with its permeability through the cell membrane into bacteria (Li et al., 2015). Different studies have suggested that antimicrobial and antifungal activities of essential oil (Eos) are considerably improved over that of bulk essential oils due to increased surface area as a result of smaller size (Wan et al., 2019; Donsì et al., 2011). The non-ionic surfactants present in the coating layer when comes in contact with microbial cell membrane destabilize due to their interaction with some constituents present over there, which causes the release of clove oil and explains the delivery of clove oil from its nanoemulsions (Pérez-Conesa et al., 2011).

Clove oil exhibits outstanding antibacterial activity against a broad group of microorganisms like *L. monocytogenes, E. coli, S. aureus,* and *P. aeruginosa*. The antimicrobial activity is presented in terms of minimum inhibitory concentration (MIC), and different researchers have reported that emulsification of clove oil/eugenol did not significantly interfere with the antimicrobial activity of clove oil/eugenol; even the MIC reported for nanoemulsion is lower than free clove oil/eugenol that suggests encapsulation by nanoemulsion improved antimicrobial activity (Shao et al., 2018). Ghosh et al. (2014) reported that dilution of nanoemulsion did not affect the antimicrobial activity of eugenol as 10% and 1% eugenol nanoemulsion solution provided similar antimicrobial activity. In contrast, it was reported that microemulsions lose their bioactivity on dilution due to significant changes in the structure of their droplets (Teixeira et al., 2007). Salvia-Trujillo et al. (2015) compared the effect of method of preparation (ultrasonication and microfluidization) on antimicrobial activity of essential oil and reported that in ultrasonication, the temperature reaches to as high as 50°C during preparation that leads to loss of antimicrobial activity of essential oils. A similar effect of ultrasonication on the decreased antimicrobial activity of essential oils is due to oxidation at high temperatures. Microfluidization is much more gentle than ultrasonication, where sample temperature is below 20°C and preserves heat-sensitive volatile compounds. Different functional activities of clove oil/eugenol nanoemulsions are presented in Table 3.

Table 3 Functional activity of clove oil/eugenol nanoemulsions.

Functional activity	References
Antimicrobial activity	
Effective against *S. aureus* (MIC 195.3 µg/mL), *E. coli* (MIC 97.6 µg/mL), *P. aeruginosa* (MIC 195.3 µg/mL), and *Salmonella* (MIC 390.6 µg/mL)	Shao et al. (2018)
2–3-log reduction in *S. aureus* after treatment with 1% eugenol nanoemulsion for 60 and 120 min, respectively	Ghosh et al. (2014)
Complete inhibition of growth of *Listeria monocytogenes* for up to 48 h in skimmed milk. Viable counts of *E. coli* O157:H7 were reduced below detectable levels in 1 h	Gaysinsky et al. (2007)
Effective against *S. aureus*, *B. cereus*, *E. coli* O157:H7, *L. monocytogenes*, *P. aeruginosa*, and *Salmonella typhi*. Larger zone of inhibition against Gram-positive strains than Gram-negative strains	Hamed et al. (2012)
Effective against *Staphylococcus* spp.	Joseph and Sujatha (2011)
Effective against *L. monocytogenes* (MIC 0.1%), *S. aureus* (MIC 0.14%), and *E. coli* (MIC 0.1%)	Majeed et al. (2016)
Effective against *L. monocytogenes*. MBC and MIC similar to free eugenol	Hu et al. (2016)
Potent inhibitory effect against *Candida* spp. (MIC 327 µg/mL)	de Siqueira et al. (2019)
Effective against *E. coli*, *Salmonella choleraesuis*, *Klebsiella pneumoniae*, *S. aureus*, *Staphylococcus epidermidis*, and *S. mutans* both Gram-positive and Gram-negative bacteria	de Meneses et al. (2019)
Clove-loaded nanoemulsions showed bactericidal against *Escherichia coli* with 3.6 log cycle reduction after 30 min of contact time	Salvia-Trujillo et al. (2015)
Antimicrobial activity against *E. coli* and *S. aureus*	Li et al. (2015)
Nanoemulsions are effective. Larger droplet sizes were more effective at inhibiting growth and inactivating cells (*E. coli*, *L.innocua*) than smaller ones	Terjung et al. (2012)
Effective against *B. subtilis*, *S. aureus*, *P. vulgaris*, *P. aeruginosa*, and *K. pneumonia*	Anwer et al. (2014)
Antimicrobial activity against *E. coli* O157:H7	Shah et al. (2013)
Other functional properties	
Possess antioxidant activity	Shao et al. (2018)
Wound healing effect was comparable to antibiotic gentamycin proven in vivo using a rat model	Alam et al. (2017)
Improved anti-inflammatory effect of eugenol nanoemulsion compared to piroxicam proven in vivo using a rat model	Esmaeili et al. (2016)
More effective than pure eugenol in the treatment of wound healing and anti-inflammatory activity in the Wistar rat	Ahmad et al. (2018)
Potential antibacterial, analgesic, and anesthetic properties to treat periodontal disease	Ahmad et al. (2019)
Antifungal agent against *Fusarium oxysporum* and plays a pivotal role in sustainably controlling plant pests and diseases	Sharma et al. (2018)
Antifungal against *Fusarium graminearum*	Wan et al. (2019)
Effective against mycotoxin (Fumonisins)	Singh et al. (2020)

Table 4 Application of clove oil/eugenol nanoemulsion in real food products as antimicrobial agent.

Product	Added value by clove oil/eugenol nanoemulsion	References
Milk	Higher antimicrobial activity against *E. coli, Listeria monocytogenes*, and *Salmonella enteritidis* than free essential oil due to the improved dispersion in the food matrix above the solubility limit	Shah et al. (2013), Xue et al. (2015)
Orange juice	Enhanced antimicrobial activity against native bacteria due to alteration in membrane permeability	Ghosh et al. (2014)
Fresh lettuce/ spinach leaves	Applied via spraying/immersion and found to be effective against *E. coli, L. monocytogenes, S. typhimurium,* and *S. enterica*	Bhargava et al. (2015), Ruengvisesh et al. (2015)
Sliced bread	Increased antibacterial and antifungal properties against endogenous flora	Otoni et al. (2014)
Chicken frankfurters	Significant decrease in *L. monocytogenes* during storage at 5°C and 15°C	Mytle et al. (2006)

11. Application of clove oil/eugenol nanoemulsions as an antimicrobial agent in food

The oil-in-water nanoemulsions of clove oil exhibit the unusual property of efficiently penetrating biological barriers to deliver active molecules. In recent years, consumer demand for health promotion by diet has driven the food industry to develop functional food containing beneficial bioactive compounds. The use of natural antimicrobial essential oils as a food preservative against pathogenic and spoilage bacteria has gained significant interest in the list of functional ingredients. Clove oil nanoemulsions have been extensively studied against various microorganisms, yielding promising results. However, only a few studies have looked into the use of clove oil nanoemulsions in real food matrixes, and even fewer have looked into endogenous flora inactivation. The use of clove oil nanoemulsions for food preservation is defined in Table 4.

12. Conclusion

The recent research on nanoencapsulation proved to be an efficient delivery system of clove oil for food and pharma applications. However, research is required to establish its stability in the food system regarding its interaction with food matrices. Moreover, further sufficient scientific evidence is also required to establish the bioavailability and safety of the nanoencapsulated clove oil in the human system, enhancing its application for health benefits.

References

Abismaıl, B., Canselier, J.P., Wilhelm, A.M., Delmas, H., Gourdon, C., 1999. Emulsification by ultrasound: drop size distribution and stability. Ultrason. Sonochem. 6 (1–2), 75–83.

Ahmad, N., Alam, M.A., Ahmad, F.J., Sarafroz, M., Ansari, K., Sharma, S., Amir, M., 2018. Ultrasonication techniques used for the preparation of novel eugenol-nanoemulsion in the treatment of wounds healings and anti-inflammatory. J. Drug Deliv. Sci. Technol. 46, 461–473.

Ahmad, N., Ahmad, F.J., Bedi, S., Sharma, S., Umar, S., Ansari, M.A., 2019. A novel nanoformulation development of eugenol and their treatment in inflammation and periodontitis. Saudi Pharm. J. 27 (6), 778–790.

Akhavan, S., Assadpour, E., Katouzian, I., Jafari, S.M., 2018. Lipid nano scale cargos for the protection and delivery of food bioactive ingredients and nutraceuticals. Trends Food Sci. Technol. 74, 132–146.

Akhtar, M., Dickinson, E., 2003. Emulsifying properties of whey protein-dextran conjugates at low pH and different salt concentrations. Colloids Surf. B: Biointerfaces 31, 125–132.

Alam, P., Ansari, M.J., Anwer, M.K., Raish, M., Kamal, Y.K., Shakeel, F., 2017. Wound healing effects of nanoemulsion containing clove essential oil. Artif. Cells Nanomed. Biotechnol. 45 (3), 591–597.

Anwer, M.K., Jamil, S., Ibnouf, E.O., Shakeel, F., 2014. Enhanced antibacterial effects of clove essential oil by nanoemulsion. J. Oleo Sci. 63 (4), 347–354.

Assadpour, E., Maghsoudlou, Y., Jafari, S.M., Ghorbani, M., Aalami, M., 2016. Optimization of folic acid nano-emulsification and encapsulation by maltodextrin-whey protein double emulsions. Int. J. Biol. Macromol. 86, 197–207.

Bakkali, F., Averbeck, S., Averbeck, D., Idaomar, M., 2008. Biological effects of essential oils—a review. Food Chem. Toxicol. 46 (2), 446–475.

Benencia, F., Courrèges, M.C., 2000. In vitro and in vivo activity of eugenol on human herpesvirus. Phytother. Res. 14, 495–500.

Bhargava, K., Conti, D.S., da Rocha, S.R., Zhang, Y., 2015. Application of an oregano oil nanoemulsion to the control of foodborne bacteria on fresh lettuce. Food Microbiol. 47, 69–73.

Briozzo, J., Nunez, L., Chirife, J., Herszage, L., D'Aquino, M., 1989. Antimicrobial activity of clove oil dispersed in a concentrated sugar solution. J. Appl. Bacteriol. 66 (1), 69–75.

Burt, S., 2004. Essential oils: their antibacterial properties and potential applications in foods—a review. Int. J. Food Microbiol. 94 (3), 223–253.

Chaieb, K., Hajlaoui, H., Zmantar, T., Kahla-Nakbi, A.B., Rouabhia, M., Mahdouani, K., Bakhrouf, A., 2007. The chemical composition and biological activity of clove essential oil, *Eugenia caryophyllata* (*Syzigium aromaticum* L. Myrtaceae): a short review. Phytother. Res. 21 (6), 501–506.

Chon, J.W., Seo, K.H., Bae, D., Kim, B., Jeong, D., Song, K.Y., 2020. Antibacterial activity of clove oil against foodborne pathogenic bacteria and sensory attributes in clove oil-enriched dairy products: a preliminary study. J. Dairy Sci. Biotechnol. 38 (4), 197–206.

Cortés-Rojas, D.F., de Souza, C.R.F., Oliveira, W.P., 2014. Clove (*Syzygium aromaticum*): a precious spice. Asian Pac. J. Trop. Biomed. 4 (2), 90–96.

Cui, H., Zhang, C., Li, C., Lin, L., 2018. Antimicrobial mechanism of clove oil on *Listeria monocytogenes*. Food Control 94, 140–146.

de Meneses, A.C., Sayer, C., Puton, B.M., Cansian, R.L., Araújo, P.H., de Oliveira, D., 2019. Production of clove oil nanoemulsion with rapid and enhanced antimicrobial activity against gram-positive and gram-negative bacteria. J. Food Process Eng. 42 (6), e13209.

de Siqueira, L.B.D.O., dos Santos Matos, A.P., da Silva Cardoso, V., Villanova, J.C.O., Guimarães, B.D.C.L.R., Dos Santos, E.P., Vermelho, A.B., Santos-Oliveira, R., Junior, E.R., 2019. Clove oil nanoemulsion showed potent inhibitory effect against Candida spp. Nanotechnology 30 (42), 425101.

Dickinson, E., 1988. The structure and stability of emulsions. In: Blanshard, J.M.V., Mitchell, J.R. (Eds.), Food Structure—Its Creation and Evaluation. Butterworths Publishing, London, UK, pp. 41–57.

Dima, C., Dima, S., 2015. Essential oils in foods: extraction, stabilization, and toxicity. Curr. Opin. Food Sci. 5, 29–35.

Donsì, F., Annunziata, M., Sessa, M., Ferrari, G., 2011. Nanoencapsulation of essential oils to enhance their antimicrobial activity in foods. LWT Food Sci. Technol. 44 (9), 1908–1914.

Dorman, H.D., Surai, P., Deans, S.G., 2000. In vitro antioxidant activity of a number of plant essential oils and phytoconstituents. J. Essent. Oil Res. 12 (2), 241–248.

Egerton, T.A., Tooley, I.R., 2014. Physical characterization of titanium dioxide nanoparticles. Int. J. Cosmet. Sci. 36 (3), 195–206.

Esmaeili, F., Rajabnejhad, S., Partoazar, A.R., Mehr, S.E., Faridi-Majidi, R., Sahebgharani, M., Syedmoradi, L., Rajabnejhad, M.R., Amani, A., 2016. Anti-inflammatory effects of eugenol nanoemulsion as a topical delivery system. Pharm. Dev. Technol. 21 (7), 887–893.

Espitia, P.J., Fuenmayor, C.A., Otoni, C.G., 2019. Nanoemulsions: synthesis, characterization, and application in bio-based active food packaging. Compr. Rev. Food Sci. Food Saf. 18 (1), 264–285.

Ezhilarasi, P.N., Karthik, P., Chhanwal, N., Anandharamakrishnan, C., 2013. Nanoencapsulation techniques for food bioactive components: a review. Food Bioprocess Technol. 6 (3), 628–647.

Foegeding, E.A., Davis, J.P., 2011. Food protein functionality: a comprehensive approach. Food Hydrocoll. 25, 1853–1864.

Fujisawa, S., Atsumi, T., Kadoma, Y., Sakagami, H., 2002. Antioxidant and prooxidant action of eugenol-related compounds and their cytotoxicity. Toxicology 177, 39–54.

Gaysinksy, S.S., 2004. Physicochemical and Antimicrobial Properties of Antimicrobials Encapsulated in Surfactant-Based Nanoparticles. A thesis submitted for the Master of Science degree Program to The University of Tennessee-Knoxville. http://trace.tennessee.edu/utk_gradthes/2193.

Gaysinsky, S., Taylor, T.M., Davidson, P.M., Bruce, B.D., Weiss, J., 2007. Antimicrobial efficacy of eugenol microemulsions in milk against *Listeria monocytogenes* and *Escherichia coli* O157: H7. J. Food Prot. 70 (11), 2631–2637.

Ghosh, V., Mukherjee, A., Chandrasekaran, N., 2014. Eugenol-loaded antimicrobial nanoemulsion preserves fruit juice against, microbial spoilage. Colloids Surf. B: Biointerfaces 114, 392–397.

Gill, A.O., Holley, R.A., 2006. Disruption of *Escherichia coli*, *Listeria monocytogenes* and *Lactobacillus sakei* cellular membranes by plant oil aromatics. Int. J. Food Microbiol. 108 (1), 1–9.

Gupta, S., Chauhan, D., Mehla, K., Sood, P., Nair, A., 2010a. An overview of nutraceuticals: current scenario. J. Basic Clin. Pharm. 1 (2), 55.

Gupta, J., Pandit, K., Kumar, A., Swaroop, P., Gupta, S., 2010b. Pharmaceutical nanotechnology novel nanoemulsion-high energy emulsification preparation, evaluation and application. J. Pharm. Res. 3, 117–138.

Håkansson, A., Rayner, M., 2018. General principles of nanoemulsion formation by high-energy mechanical methods. In: Jafari, S.M., McClements, D.J. (Eds.), Nanoemulsions. Academic Press, pp. 103–139.

Hamed, S.F., Sadek, Z., Edris, A., 2012. Antioxidant and antimicrobial activities of clove bud essential oil and eugenol nanoparticles in alcohol-free microemulsion. J. Oleo Sci. 61 (11), 641–648.

Hosseini, A., Jafari, S.M., Mirzaei, H., Asghari, A., Akhavan, S., 2015. Application of image processing to assess emulsion stability and emulsification properties of Arabic gum. Carbohydr. Polym. 126, 1–8.

Hu, Q., Gerhard, H., Upadhyaya, I., Venkitanarayanan, K., Luo, Y., 2016. Antimicrobial eugenol nanoemulsion prepared by gum arabic and lecithin and evaluation of drying technologies. Int. J. Biol. Macromol. 87, 130–140.

Joseph, B., Sujatha, S., 2011. Bioactive compounds and its autochthonous microbial activities of extract and clove oil (*Syzygium aromaticum* L.) on some food borne pathogens. Asian J. Biol. Sci. 4 (1), 35–43.

Jung, B.O., Chung, S.J., Lee, S.B., 2006. Preparation and characterization of eugenol-grafted chitosan hydrogels and their antioxidant activities. J. Appl. Polym. Sci. 99, 3500–3506.

Kentish, S., Wooster, T.J., Ashokkumar, M., Balachandran, S., Mawson, R., Simons, L., 2008. The use of ultrasonics for nanoemulsion preparation. Innov. Food Sci. Emerg. Technol. 9 (2), 170–175.

Kim, H.J., Decker, E.A., McClements, D.J., 2005. Influence of protein concentration and order of addition on the thermal stability of betalactoglobulin stabilized n-hexadecane oil-in-water emulsions at neutral pH. Langmuir 21, 134–139.

Klang, V., Matsko, N.B., Valenta, C., Hofer, F., 2012. Electron microscopy of nanoemulsions: an essential tool for characterisation and stability assessment. Micron 43 (2–3), 85–103.

Lam, R.S., Nickerson, M.T., 2013. Food proteins: a review on their emulsifying properties using a structure-function approach. Food Chem. 141 (2), 975–984.

Leong, T.S., Martin, G.J., Ashokkumar, M., 2017. Ultrasonic encapsulation—a review. Ultrason. Sonochem. 35, 605–614.

Li, W., Chen, H., He, Z., Han, C., Liu, S., Li, Y., 2015. Influence of surfactant and oil composition on the stability and antibacterial activity of eugenol nanoemulsions. LWT Food Sci. Technol. 62 (1), 39–47.

Liu, Q., Huang, H., Chen, H., Lin, J., Wang, Q., 2019. Food-grade nanoemulsions: preparation, stability and application in encapsulation of bioactive compounds. Molecules 24 (23), 4242.

Lopez, P., Sanchez, C., Batle, B., Nerin, C., 2005. Solid and vapour phase antimicrobial activities of six essential oils: susceptibility of selected food borne bacterial and fungal strains. J. Agric. Food Chem. 53, 6338–6346.

Mahdi Jafari, S., He, Y., Bhandari, B., 2006. Nano-emulsion production by sonication and microfluidization—a comparison. Int. J. Food Prop. 9 (3), 475–485.

Majeed, H., Liu, F., Hategekimana, J., Sharif, H.R., Qi, J., Ali, B., Bian, Y.Y., Ma, J., Yokoyama, W., Zhong, F., 2016. Bactericidal action mechanism of negatively charged food grade clove oil nanoemulsions. Food Chem. 197, 75–83.

Masae, I.T.O., Keiko, M., Masataka, Y., 2005. Antioxidant action of eugenol compounds: role of metal ion in the inhibition of lipid peroxidation. Food Chem. Toxicol. 43, 461–466.

McClements, D.J., 1998. Food Emulsions: Principles, Practice and Techniques. CRC Press, Boca Raton, FL.

McClements, D.J., 2004. Protein-stabilized emulsions. Curr. Opin. Colloid Interface Sci. 9 (5), 305–313.

McClements, D.J., 2005. Food Emulsions: Principles, Practice, and Techniques. CRC Press, Boca Raton.

McClements, D.J., 2015. Food Emulsions: Principles, Practices, and Techniques. CRC Press.

McClements, D.J., Decker, E.A., 2000. Lipid oxidation in oil-in-water emulsions: impact of molecular environment on chemical reactions in heterogeneous food systems. J. Food Sci. 65 (8), 1270–1282.

McClements, D.J., Jafari, S.M., 2018. General aspects of nanoemulsions and their formulation. In: Jafari, S.M., McClements, D.J. (Eds.), Nanoemulsions: Formulation, Applications, and Characterization. Academic Press, pp. 3–20.

McClements, D.J., Rao, J., 2011. Food-grade nanoemulsions: formulation, fabrication, properties, performance, biological fate, and potential toxicity. Crit. Rev. Food Sci. Nutr. 51 (4), 285–330.

Milind, P., Deepa, K., 2011. Clove: a champion spice. Int. J. Res. Ayurveda Pharm. 2 (1), 47–54.

Mytle, N., Anderson, G.L., Doyle, M.P., Smith, M.A., 2006. Antimicrobial activity of clove (*Syzgium aromaticum*) oil in inhibiting Listeria monocytogenes on chicken frankfurters. Food Control 17 (2), 102–107.

Nagababu, E., Lakshmaiah, N., 1994. Inhibition of microsomal lipid peroxidation and monooxygenase activities by eugenol. Free Radic. Res. 20, 253–266.

Nurdjannah, N., Bermawie, N., 2006. Clove. In: Peter, K.V. (Ed.), Handbook of Herbs and Spices. vol. 3. Woodhead Publishing, pp. 154–161.

Nurdjannah, N., Bermawie, N., 2012. Clove. In: Peter, K.V. (Ed.), Handbook of Herbs and Spices, second ed. Woodhead Publishing, pp. 197–250.

Otoni, C.G., Pontes, S.F., Medeiros, E.A., Soares, N.D.F., 2014. Edible films from methylcellulose and nanoemulsions of clove bud (*Syzygium aromaticum*) and oregano (*Origanum vulgare*) essential oils as shelf life extenders for sliced bread. J. Agric. Food Chem. 62 (22), 5214–5219.

Pérez-Conesa, D., Cao, J., Chen, L., McLandsborough, L., Weiss, J., 2011. Inactivation of *Listeria monocytogenes* and *Escherichia coli* O157: H7 biofilms by micelle-encapsulated eugenol and carvacrol. J. Food Prot. 74 (1), 55–62.

Quintanilla-Carvajal, M.X., Camacho-Dıaz, B.H., Meraz-Torres, L.S., Chanona-Perez, J.J., Alamilla-Beltran, L., Jimenez-Aparicio, A., Gutierrez-Lopez, G.F., 2010. Nanoencapsulation: a new trend in food engineering processing. J. Food Eng. Rev. 2, 39–50.

Rai, M., Paralikar, P., Jogee, P., Agarkar, G., Ingle, A.P., Derita, M., Zacchino, S., 2017. Synergistic antimicrobial potential of essential oils in combination with nanoparticles: emerging trends and future perspectives. Int. J. Pharm. 519 (1–2), 67–78.

Rao, J., McClements, D.J., 2011. Food-grade microemulsions, nanoemulsions and emulsions: fabrication from sucrose monopalmitate & lemon oil. Food Hydrocoll. 25 (6), 1413–1423.

Ruengvisesh, S., Loquercio, A., Castell-Perez, E., Taylor, T.M., 2015. Inhibition of bacterial pathogens in medium and on spinach leaf surfaces using plant-derived antimicrobials loaded in surfactant micelles. J. Food Sci. 80 (11), M2522–M2529.

Ryu, V., McClements, D.J., Corradini, M.G., McLandsborough, L., 2018. Effect of ripening inhibitor type on formation, stability, and antimicrobial activity of thyme oil nanoemulsion. Food Chem. 245, 104–111.

Salvia-Trujillo, L., Rojas-Graü, A., Soliva-Fortuny, R., Martín-Belloso, O., 2015. Physicochemical characterization and antimicrobial activity of food-grade emulsions and nanoemulsions incorporating essential oils. Food Hydrocoll. 43, 547–556.

Shah, B., Davidson, P.M., Zhong, Q., 2013. Nanodispersedeugenol has improved antimicrobial activity against *Escherichia coli* O157: H7 and *Listeria monocytogenes* in bovine milk. Int. J. Food Microbiol. 161 (1), 53–59.

Shahavi, M.H., Hosseini, M., Jahanshahi, M., Meyer, R.L., Darzi, G.N., 2019. Evaluation of critical parameters for preparation of stable clove oil nanoemulsion. Arab. J. Chem. 12 (8), 3225–3230.

Shao, Y., Wu, C., Wu, T., Li, Y., Chen, S., Yuan, C., Hu, Y., 2018. Eugenol-chitosan nanoemulsions by ultrasound-mediated emulsification: formulation, characterization and antimicrobial activity. Carbohydr. Polym. 193, 144–152.

Sharif, H.R., Goff, H.D., Majeed, H., Shamoon, M., Liu, F., Nsor-Atindana, J., Haider, J., Liang, R., Zhong, F., 2017. Physicochemical properties of β-carotene and eugenol co-encapsulated flax seed oil powders using OSA starches as wall material. Food Hydrocoll. 73, 274–283.

Sharma, M., Mann, B., Sharma, R., Bajaj, R., Athira, S., Sarkar, P., Pothuraju, R., 2017. Sodium caseinate stabilized clove oil nanoemulsion: physicochemical properties. J. Food Eng. 212, 38–46.

Sharma, A., Sharma, N.K., Srivastava, A., Kataria, A., Dubey, S., Sharma, S., Kundu, B., 2018. Clove and lemongrass oil based non-ionic nanoemulsion for suppressing the growth of plant pathogenic *Fusarium oxysporum* f. sp. lycopersici. Ind. Crop. Prod. 123, 353–362. https://doi.org/10.1016/j.indcrop.2018.06.077.

Silva, H., Cerqueira, M., Vicente, A., 2011. Nanoemulsions for food applications: development and characterization. Food Bioprocess Technol. 5 (3), 854–867.

Simunkova, H., Pessenda-Garcia, P., Wosik, J., Angerer, P., Kronberger, H., Nauer, G.E., 2009. The fundamentals of nano- and submicro-scaled ceramic particles incorporation into electrodeposited nickel layers: zeta potential measurements. Surf. Coat. Technol. 203 (13), 1806–1814.

Singh, P., Dasgupta, N., Singh, V., Mishra, N.C., Singh, H., Purohit, S.D., Srivastava, N., Ranjan, S., Yadav, N.P., Mishra, B.N., 2020. Inhibitory effect of clove oil nanoemulsion on fumonisin isolated from maize kernels. LWT 134, 110237.

Surh, J., Decker, E.A., McClements, D.J., 2006. Influence of pH and pectin type on properties and stability of sodium-caseinate stabilized oil-in-water emulsions. Food Hydrocoll. 20 (5), 607–618.

Taira, J., Ikemoto, T., Yoneya, T., Hagi, A., Murakami, A., Makino, K., 1992. Essential oil phenyl propanoids. Useful as OH scavengers? Free Radic. Res. 16, 197–204.

Tcholakova, S., Denkov, N.D., Ivanov, I.B., Campbell, B., 2006. Coalescence stability of emulsions containing globular milk proteins. Adv. Colloid Interface Sci. 123–126, 259–293.

Teixeira, P.C., Leite, G.M., Domingues, R.J., Silva, J., Gibbs, P.A., Ferreira, J.P., 2007. Antimicrobial effects of a microemulsion and a nanoemulsion on enteric and other pathogens and biofilms. Int. J. Food Microbiol. 118 (1), 15–19.

Terjung, N., Löffler, M., Gibis, M., Hinrichs, J., Weiss, J., 2012. Influence of droplet size on the efficacy of oil-in-water emulsions loaded with phenolic antimicrobials. Food Funct. 3 (3), 290–301.

Walstra, P., 1996. Dispersed systems: basic considerations. In: Fennema, O.R. (Ed.), Food Chemistry. Food Science and Technology. Marcel Dekker, New York, pp. 96–129.

Wan, J., Zhong, S., Schwarz, P., Chen, B., Rao, J., 2019. Physical properties, antifungal and mycotoxin inhibitory activities of five essential oil nanoemulsions: impact of oil compositions and processing parameters. Food Chem. 291, 199–206.

Wooster, T.J., Golding, M., Sanguansri, P., 2008. Impact of oil type on nanoemulsion formation and Ostwald ripening stability. Langmuir 24 (22), 12758–12765.

Woranuch, S., Yoksan, R., 2013. Eugenol-loaded chitosan nanoparticles: I. Thermal stability improvement of eugenol through encapsulation. Carbohydr. Polym. 96 (2), 578–585.

Xue, J., Davidson, P.M., Zhong, Q., 2015. Antimicrobial activity of thyme oil co-nanoemulsified with sodium caseinate and lecithin. Int. J. Food Microbiol. 210, 1–8.

Zhang, Z., McClements, D.J., 2018. Overview of nanoemulsion properties: stability, rheology, and appearance. In: Jafari, S.M., McClements, D.J. (Eds.), Nanoemulsions: Formulation, Applications, and Characterization. Academic Press, pp. 21–49.

Zuidam, N.J., Shimoni, E., 2010. Overview of microencapsulates for use in food products or processes and methods to make them. In: Zuidam, N.Z., Nedovic, V.A. (Eds.), Encapsulation Technologies for Active Food Ingredients and Food Processing. Springer, New York, NY, pp. 3–29.

CHAPTER

Eugenol-based nutraceuticals

8

Mohamad Taleuzzaman[a], Rishabh Verma[b], and Jabeen Farhana[c]

Department of Pharmaceutical Chemistry, Faculty of Pharmacy, Maulana Azad University, Jodhpur, Rajasthan, India[a]
Department of Pharmaceutics, School of Pharmaceutical Education and Research, Jamia Hamdard, New Delhi, India[b]
Faculty of Pharmacy, BRIG, Hyderabad, Telangana, India[c]

1. Introduction

Herbal remedies based on traditional systems of medicine have been widely accepted as essential components of the healthcare system (Abushouk et al., 2017a,b). Herbal plants are widely accepted as they have better efficacy over their synthetic counterparts and fewer side effects (Abushouk et al., 2017a,b). Currently, herbal and natural medicines reach about 80% of the world's population as a significant form of primary healthcare (Ekor, 2014). Many plant species have been reported to have antiinflammatory, antimicrobial, spasmolytic, sedative, analgesic, and anesthetic activities (El-Saber Batiha et al., 2018). Medicinal plants have been reported to play a vital role in the discovery of novel pharmaceutical molecules that are breakthroughs in the therapeutic world. Phytochemicals identified from these plants have been considered remarkable leading compounds in the search for efficacious new drugs (Beshbishy et al., 2019).

People take traditional medicine because it is more affordable, safer than allopathic medicine, acts upon people's psyche of having more personalized healthcare, and greater public access to health information. Traditional medicine is primarily used as a healthy supplement as well as therapy for chronic and life-threatening conditions. The use of traditional medicine also increases when conventional or allopathic medicine seems ineffective, as seen in the emergence of new infections. Moreover, herbal medicine is perceived as natural, safe, and non-toxic. However, this is not true, especially when these medicines are taken with other allopathic medicines or adulterated herbs (Wachtel-Galor and Benzie, 2011).

Clove, *Syzygium aromaticum*, is a herb that has been widely used in various traditional systems of medicine. It is a dried bud that belongs to the Myrtaceae family; it is indigenous to the Maluku islands in Indonesia, but is now cultivated worldwide (Cortés-Rojas et al., 2014; Batiha et al., 2019). The commercial parts of the clove tree are its leaves and buds. About 4 years after planting, flowering buds appear. Commercially, clove is used for medical purposes in various traditional medicine systems. It is also used in the perfume industry and as a spice and preservative in food such as meat because of its antioxidant and antimicrobial activities (Chomchalow, 2001; Cortés-Rojas et al., 2014). Cloves are utilized in the Indian and Chinese traditional systems of medicine as a warming and stimulating agent, and clove essential oil is used in the treatment of burns and wounds and as a pain reliever in

Clove (*Syzygium aromaticum*). https://doi.org/10.1016/B978-0-323-85177-0.00025-2

dental anomalies (Batiha et al., 2019; Chomchalow, 2001). The pharmacological and therapeutic effects of clove oil are depicted in Fig. 1.

Clove is a significant vegetation source of phenolic compounds such as flavonoids, hydroxybenzoic acids, hydroxycinnamic acid, and hydroxyphenyl propene (Sarrami et al., 2002; Neveu et al., 2010). A yield of 18% essential oil can be obtained from clove buds; approximately 89% is eugenol, and the remaining 5%–15% are eugenol acetate and β-cariofileno. Another major molecule found in the oil is α-humulen, which makes up about 2.1% of the essential oil (Jirovetz et al., 2006). Another major component of clove oil is gallic acid and its derivatives. The presence of phenolic structures in eugenol and phenolic acid makes clove oil a potent antioxidant (El Ghallab et al., 2019) (Fig. 2).

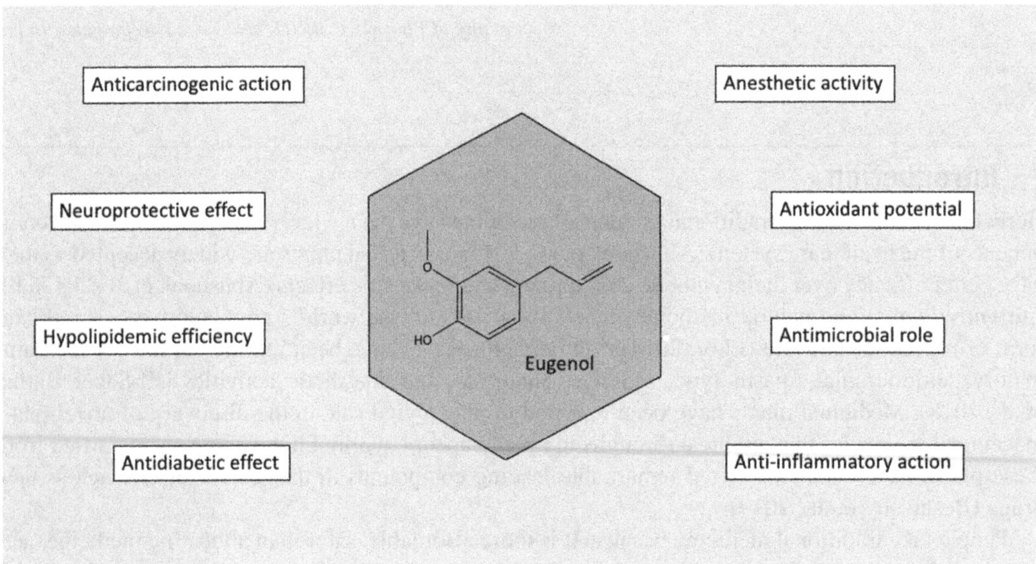

FIG. 1

Pharmacological and therapeutic potential of eugenol (Batiha et al., 2020).

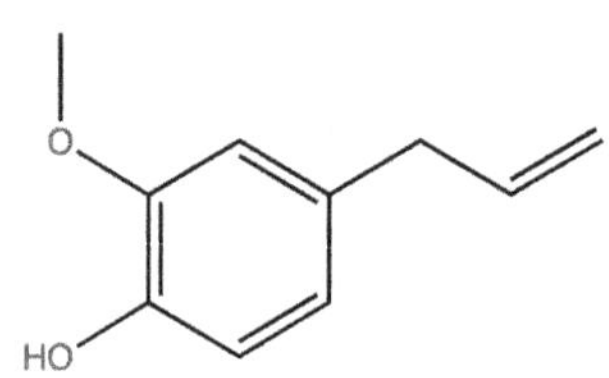

FIG. 2

Structure of eugenol.

2. Source, extraction, and chemistry of eugenol

Besides clove buds, other plants such as cinnamon, tulsi, turmeric, pepper ginger, oregano, and thyme contain eugenol. The highest concentration of eugenol is found in the clove buds and flower (180 mg/g) (Khalil et al., 2017). Clove and cinnamon are considered abundant sources of eugenol, but a significant drawback is the high cultivation and commercial extraction of eugenol from them. The aerial parts of plants such as leaves, bark, and flowers have a considerable amount of essential oils, thus they are a significant eugenol source (Prakash and Gupta, 2005; Kamatou et al., 2012). Studies reported that the concentration of eugenol varies with season, with the maximum yield during autumn (Yadav et al., 2015). This herbal ingredient has an influential health-promoting role that makes it versatile. Besides its natural source, it can be synthesized in the laboratory from the allylation of guaiacol with allyl chloride, and has nearly similar properties. The World Health Organization (WHO) declared this molecule as "generally recognized as safe" (GRAS) and non-mutagenic. This bioactive compound belongs to the phenylpropanoid class with the IUPAC name 4-allyl-2-methoxy phenol. It has a molecular weight of 164 g/mol, and is partially soluble in water and highly soluble in organic solvent (Charan Raja, 2015). Steam distillation is the most common method of extraction for eugenol. To extract eugenol, first essential oils from parts of the plant need to be extracted. This essential oil is further mixed with a 3% sodium or potassium hydroxide solution to isolate eugenol. Phenolic alkali salt is formed when sodium or potassium hydroxide is added. By solvent extraction or steam distillation, this insoluble portion is separated. Further, the remaining solution is acidified at refrigeration temperature to release eugenol. For better separation, various fractional distillation and chromatographic techniques such as high-performance liquid chromatography (HPLC) and thin-layer chromatography (TLC) are used. The purity of the isolated compound is confirmed by different spectroscopic instrumental techniques such as Fourier transform infrared spectroscopy (FTIR), Fourier transform near-infrared spectroscopy (FTNIR), mass spectrometry (MS), and nuclear magnetic resonance (NMR) (Kar Mahapatra et al., 2009). Some important extraction methods are listed below.

2.1 Solvent extraction

The solvent extraction method is considered one of the most common and widely used forms of extraction of essential oils from plants. Various solvents are used to extract eugenol, including methanol, ethanol, petroleum ether, and *n*-hexane. Significant drawbacks of using solvent extraction are changes in the flavor profile and the inclusion of other soluble residues (Guan et al., 2007). To extract eugenol from clove, the clove buds are first ground and wrapped in filter paper. This filter paper is subjected to an extraction thimble. Further extraction is carried out using a suitable organic solvent in the Soxhlet apparatus (Quan et al., 2004). The final extract is obtained by concentrating it on a rotary vacuum evaporator at 50°C. Several modifications have been made to the conventional solvent extraction method to increase the efficiency of extraction. One such method includes the batch extraction process. This method uses a reactor equipped with an agitator that has four blades and a motor with a 1200 rpm speed. This method has been reported to have satisfactory extraction results for eugenol extraction from tulsi (Garkal et al., 2012).

2.2 Hydrodistillation

Hydrodistillation is also a widely used method of extracting essential oils (Jeyaratnam et al., 2016). For this method, part of the plant is first dried and ground. This powdered sample is then taken into a 500 mL volumetric flask and soaked in water. This is further subjected to hydrodistillation for 4–6 h. The volatile distillate is collected and saturated with sodium chloride following the addition of a suitable organic solvent. After this, the hydro and solvent layer are separated and dehydrated using anhydrous sodium sulfate. Later, the sample/extract is recovered and concentrated by heating over a water bath at 60°C. This method yields an average of 11.5% yield with 50.5%–53.5% of eugenol. This can further be optimized using different additions (Guan et al., 2016).

2.3 Microwave-assisted extraction of eugenol

The conventional method of extraction has a few drawbacks such as hydrolysis, thermal degradation, and leaching of some fragrance components (Khalil et al., 2017). Several modern techniques have been incorporated to counter these losses to lead to a maximum yield with reduced processing time and energy demands. One such approach is microwave-assisted extraction (MWAE), also regarded as a green extraction approach. MWAE can produce eugenol and other essential oils with the same sensorial attributes and qualities as the traditional methods. This method is also rapid and economical (Li et al., 2014).

The MWAE approach has been configured with many other extraction processes such as microwave-assisted hydrodistillation (MWHD) (Golmakani and Rezaei, 2008), coaxial microwave-assisted hydrodistillation (coaxial MWHD), microwave-assisted hydrodistillation and gravity (MWHG) (Vian et al., 2008), and microwave steam distillation (MWSD) (Chemat et al., 2006). The MWHD is considered superior, with high savings in heating time (400%) and energy demands (30%). Furthermore, this method is safe, economical, and easy to scale up. A team of scientists also compared coaxial MWHD to conventional hydrodistillation and reported that eugenol and other essential oils extracted using coaxial MWHD have higher thermal stability and less extraction time than the traditional method.

2.4 Supercritical carbon dioxide extraction

Another efficient extraction method for eugenol is the supercritical carbon dioxide (SC-CO_2) method (Chatterjee and Bhattacharjee, 2013). Several studies have shown the efficiency of the SC-CO_2 method. For example, a team of scientists led by Gopalakrishnan compared the SC-CO_2 extraction method with liquid extraction and reported that SC-CO_2 has a superior extraction yield of clove oil (Gopalakrishnan et al., 1990). Extraction also was performed using SC-CO_2 method and extracts from other conventional methods (Guan et al., 2007; Yu et al., 2009). Further studies are still needed to validate using supercritical carbon dioxide to extract eugenol from clove bud oil.

2.5 Ultrasound-assisted extraction of eugenol

Another green method of extraction for eugenol is ultrasound (US)-assisted extraction. This method has significantly increased the yield and decreased energy demand. Another advantage of using the US extraction method is easy handling, rapid execution, no residue, high yield, prevention of degradation (Chemat et al., 2011). Various scientists assessed the use of US extraction for clove oil and compared it with the traditional method. The studies reported that the extractive yield of essential oil is highly influenced by temperature in the US-assisted extraction method. The studies concluded that the US-assisted extraction method can be influential on a commercial scale (Alexandru et al., 2013; Tekin et al., 2015).

3. Nanoformulation-based nutraceuticals

Presently, natural food ingredients are highly beneficial over artificial and synthetic additives for preventing and curing diseases. Bioactive compounds are formulated to protect from environmental conditions and safely delivered the bioactive compounds to the target organs and cells. The strategy for the development of nutraceuticals is perfect for this situation. The nanoencapsulation of food components has better therapeutic effects and safety. Nanoencapsulation is classified based on the nature of the ingredients such as lipid-based nanocarriers, nature-inspired nanocarriers, special equipment-based nanocarriers, biopolymer nanocarriers, and others (Assadpour and Mahdi Jafari, 2019; Taleuzzaman et al., 2020; Jahangir et al., 2020).

Nutraceuticals product developed in the food industry having eugenol as active constituents prepared nanoemulsion dosage form that provides the better quality, nutritional value, functional properties, and self-life used as a coloring agent, flavoring agent, and pharmacological activity as antimicrobials. The advantage of nanoemulsions is that they encapsulate the active compound, control the release, have better solubility, and absorb and minimize the oxidation of the active component by preventing moisture and gas exchange. The antimicrobial activity of eugenol against foodborne pathogens is enhanced by adopting nanoformulations, which overcome the hydrophobicity of the active component (Aswathanarayan and Vittal, 2019; McClements and Rao, 2011; Salvia-Trujillo et al., 2017; Alexandre et al., 2016).

The pharmacology of eugenol as an antifungal agent founded more than 95% when formulated as eugenol-casein nanoparticles (EC-NPs). Experimented against the spore germination fungus and 100% prevention of anthracnose incidence in postharvest pear. In the food industry, it is used frequently for the preparation of nutraceuticals. The stability and bioactivity of eugenol were observed to be better when nanoparticles were used as the carrier, but the limitation is applying high-energy techniques. The amphiphilic properties of casein make it suitable for the encapsulation of hydrophobic drugs such as eugenol and give better-controlled release (Xue et al., 2019; Woranuch and Yoksan, 2013; Wang and Zhang, 2017).

The hydrophobic properties of eugenol decrease the therapeutic effects of nutraceuticals because of their poor solubility, low absorption and bioavailability, chemical stability, and metabolism. Overall, the therapeutic effects are enhanced by nanoformulation, improving eugenol bioavailability (Surve et al., 2019). Nanodelivery gives better pharmacology effects such as antifungal, antibacterial, antioxidant, preservative, and germicide. Thermal stability against heat, light, and oxygen is also a challenge for the formulation development, the thermal stability increases by using electrospun nanofibers where incorporated cyclodextrin inclusion of eugenol (Kayaci et al., 2013).

Presence of β-carotene and eugenol in nutraceuticals decline the bioavailability, which affects the pharmacological activity. The physical and chemical stabilities as well as the bioavailability were improved by the development of food-grade nanoemulsions that overcome these limitations and significantly impact the stability of the formulation.Advantage was observed in this formulation: the degradation of β-carotene decreased at ambient storage and other ingredients such as whey protein, lecithin, and soybean oil. Overall, the therapeutic effects of eugenol improved with nanoemulsion delivery nutraceutical formulation. The chemical stability increased by adding lipophilic antioxidants (eugenol, co-enzyme Q10, and α-tocopherol) in carotenoid emulsions (Guan et al., 2016; Kim and Choi, 2020).

A mayonnaise nutraceutical on the market was studied. Its two types—mayonnaise with eugenol lean clove extract and mayonnaise with mustard—were compared. The antioxidant therapeutic value of the nutraceutical with eugenol was better (Chatterjee and Bhattacharjee, 2015). Dietary supplements with eugenol are used by humans, and the effects showed prevention of liver damage and inhibition of glutathione-*S*-transferase (GST) (Rompelberg et al., 1996). The researcher reported male infertility developed by organophosphate insecticide chlorpyrifos (CPF) that change physiology such as oxidative stress, structural destruction, and hemostasis of testicular tissue. These symptoms were examined in male rats by aerobic exercise and eugenol supplement. CPF developed in male rats; the experiment indicates that the CPF effects were reduced by improving the antioxidant capacity and the secretion of a sex hormone by aerobic exercise with a eugenol supplement (Nikbin et al., 2020).

Scientists studied the relationship between diet and reactive oxygen species (ROS) responsible for cancer and age-related clinical complications. Food containing natural antioxidant compounds such as eugenol were used as spices; this compound has a dual effect on oxidative stress as an antioxidant and pro-oxidant agent (Bezerra et al., 2017).

This bioactive compound influences cell division in the sub-G1 phase, including apoptosis in a regular time-dependent manner. p53 activation and proline-rich acidic protein cleavage by eugenol were confirmed. So, food supplements with such bioactive compounds and their derivatives can treat colon cancer (Kuppusamy et al., 2014). The antioxidant property of eugenol is beneficial for cancer treatment. Hepatocellular cancer is regulated by proprotein convertase subtilisin/kexin type 9 (PCSK9), a lipid regulator that alters the low density lipoprotein (LDL) metabolism that causes dyslipidemia. The use of nutraceuticals with eugenol, which regulates LDL levels, showed antioxidant and antihypercholesterolemic properties. LDL oxidations are inhibited by eugenol and prevent dyslipidemia formation (Zia et al., 2020).

Eugenol has always been used for better color, taste, and preservation in food. Several studies confirmed the therapeutic effects, which indicated that it could also be used to treat many diseases. Eugenol can alter the various transcription factors, growth factors, protein kinases, and inflammatory mediators. Scientists confirmed the cause of cancer is the dysregulation of about 500 gene products; eugenol can target multiple gene products (Sung et al., 2012).

In cancer cells, the anticancer agent's primary mechanisms are the induction of apoptosis, DNA damage, G2/M arrest, inhibition of proliferation, migration, and invasion of cancer cells. Nutraceuticals with phytomolecules such as eugenol, piperine, isoflavones, capsaicin, crocetin, and crocin are used to treat cancer (Siddiqui et al., 2018). Nutraceuticals have rich in eugenol are used to treat a neurological disorder, inhibit the 6-hydroxydopamine-induced reduction in the dopamine level by adopting the mechanism of decline 6-hydroxydopamine-induced lipid peroxidation and increased GSH level (Kabuto et al., 2007).

Arsenic trioxide, a chemotherapeutic agent, affects the treatment of promyelocytic leukemia, but its toxicity is a concern. Eugenol-contained nutraceuticals were used to control the toxicity. In experiments with rats, arsenic trioxide (4 mg/kg/day) was administered alone and in combination with eugenol (5 mg/kg/day) for 30 days. Results indicated that the thiol group was protected in the myocardium from oxidative stress in the combination dose (Binu et al., 2018). The antioxidant and antimicrobial activities of clove bud essential oil (CEO) were evaluated using a DPPH· free radical scavenging assay and the agar disk dilution method. Compared to the pharmacological effects of individual eugenol microemulsions and CEO microemulsions, the result indicated that individual eugenol emulsions are more effective. Eugenol microemulsions had promising pharmacological effects and could be used as preservatives and flavoring agents (Hamed et al., 2012).

Eugenol industrial application is limited because of its sensitivity to light, heat, and oxygen, which degrade its efficacy. This limitation can be overcome by the microencapsulation of eugenol by the double emulsion solvent evaporation technique. By surface methodology, the developed method was characterized. The reported encapsulation efficiency and product yield were 82.0%–94.7%. The study was beneficial in developing eugenol-based nutraceuticals as well as the encapsulation of eugenol in ethyl cellulose polymer-based microparticles to potentiate the therapeutic effects in nutraceuticals and food supplements (Paulo and Santos, 2018). The iota-carrageenan and eugenol models of hydrocolloid and nutraceuticals in the presence of Na and Ca ions were studied. Carrageenan maintains its network organization after the incorporation of eugenol molecules. The study indicated that cations have a role in encapsulation efficiency and release profiles of hydrocolloid-based nutraceutical carriers (Akhavan et al., 2018). Bioactive encapsulation through nanotechnology is a very innovative idea; its application in the health sector has been a revolution. With nanoencapsulation delivery, volatile substances, phenols, and lipophilic ingredients, which are unstable in traditional dosage forms, can overcome such challenges and provide better therapeutic effects in nutraceuticals, pharmaceuticals, and food supplements. The advancement of microencapsulation is a platform through which several natural substances are used to better health (Patra et al., 2018).

4. Eugenol's pharmacology and toxicological effects

4.1 Pharmacology effects

Clove and eugenol are used for various purposes such as medicines, spices, and perfumes. Clove in the Middle East and Asia region has been used as a spice since ancient times. It can be used as preservatives in many food items to replace synthetic ones so that it has more health benefits. Research has documented the pharmacological activities such as antiviral, anticarcinogenic, antibacterial, and antifungal effects of aromatic herbs such as cinnamon, oregano, clove, thyme, and mint. Among these, clove has gained more attention because of its practical antimicrobial and antioxidant activities. The antioxidant activity of clove is due to a high concentration of eugenol; it is used to treat different types of degenerative disease such as burns and wounds and as a pain reliever in dental care. In addition, tooth's infection and toothache cured using clove essential oil (CEO). Traditionally clove used in various parts of the world, including in Indian and Chinese traditional medicine as a warming and stimulating agent and for the treatment of vomiting: flatulence; nausea; liver, bowel and stomach disorder, and nerve's stimulant. Cloves are used to fight various microorganisms such as scabies, cholera, malaria, and tuberculosis in tropical Asia. They have been used for the treatment of foodborne pathogens, viruses, worms, candida, and different bacterial and protozoan infections (Batiha et al., 2020). The pharmacological activities of eugenol are presented in Table 1.

Table 1 Pharmacological activities of eugenol.

No.	Pharmacology activity	Mechanism	References
1	Antibacterial	Eugenol interacts with the bacterial cell membrane, resulting in leakage of intracellular content of bacteria such as proteins and lipids; the action depends on the duration of contact and the concentration of eugenol. Inhibits the growth of bacteria	Oyedemi et al. (2009)
2	Antiinflammatory	Macrophages inhibit the expression of cytokines and also decrease the activity of cyclooxygenase-II enzymes. Eugenol suppressed nuclear factor-kappa B (NF-κB) that prevented cell proliferation, which is a mechanism for treating gastric tumors. Inflammation of ankle joint is reduced by eugenol via decreasing cytokines such as TNF-α, interferon-gamma, and TNF-β	Leem et al. (2011), Grespan et al. (2012)
3	Antipyretic	Eugenol inhibits the production of prostaglandins by inhibiting cyclooxygenase (COX) and reducing fever when given intravenously and intragastrically	Feng and Lipton (1987)
4	Anticancer	Eugenol reduced the pro-inflammatory cytokine levels (i.e., IL-6 and TNF-α), and the enzyme marker responsible for inflammation such as (COX and iNOS) indicates anticancer activity	Kaur et al. (2010)
5	Antifungal	The concentration level of H_2O_2 and CO_2 increased by eugenol in the cytoplasm; this results in destabilization and disruption of the cell membrane, which leads to its permeability. Eugenol interferes with the synthesis of ergosterol, which affects the cell integrity and functionality of the cell membrane	de Oliveira Pereira et al. (2013)
6	Antiviral	Inhibits the replication of viruses such as HSV-1 and herpes simplex-2 (HSV-2)	Reichling et al. (2009)
7	Antiobesity	They have shown antiobesity effects by inhibiting transcription factors integral to adipogenesis and lipogenesis	Ali et al. (2011)
8	Antiulcer	The mechanism of action of eugenol is mediated by opening K (ATP) channels, scavenging free radicals, decreasing acid-pepsin secretion, increasing mucin production, and preventing the deleterious rise in nitric oxide level	Morsy and Fouad (2008)
9	Hypotensive efficacy	Hypotensive effects of eugenol are because of vasodilator and smooth muscle relaxant action as well as inotropic activities in heart muscle. This is dose-dependent	Pulikottil and Nath, (2015)
10	Antidiabetic	The level of carbohydrate metabolism enzymes and liver markers in the blood increased in diabetic patients. Experiments in rats induced diabetes by streptozotocin. By administration of eugenol, the level of all carbohydrate enzymes returned to normal	Srinivasan et al. (2014)

4.2 Toxicological studies

In Chinese traditional medicine, clove oil's pharmacological properties such as antimicrobial, antiseptic, and antispasmodic effects were reported. Several studies indicated the pharmacological properties of eugenol and its uses in household products such as fragrances in soaps and cosmetics; skin products; and additives in food, pharmaceutical, and dental products (Mohammadi Nejad et al., 2017). In the food and cosmetic industry, the use of clove oil as a flavoring and fragrance gent has been approved by the US Food and Drug Administration as a natural analgesic. The pharmacokinetics of eugenol suggest that a minor amount is absorbed in the lungs after inhalation of clove cigarettes, and in vitro studies indicated it undergoes biotransformation in hepatocytes (Barnes et al., 2007). The absorption and metabolism of eugenol after oral intake in male and female volunteers was examined. Within 24 h, it was wholly excreted in the urine as conjugates of eugenol. Less than 1% of the administrated doses were found in urine that were non-metabolized, whereas more than 90% of the metabolic products were phenolic conjugates. In conjugated products, eugenol-glucuronide and eugenol-sulfate was 50%. Metabolic routes apart from this were the epoxide-diol pathway, the synthesis of a thiophenol, a substituted propionic acid, allylic oxidation, and migration of the double bond (Oyedemi et al., 2009). Three different formulations of eugenol (Tween 80, ethoxylate, and lecithin) were studied in vivo for efficacy against pathogens that affect apples (*Phlyctema vagabunda, Penicillium expansum, Botrytis cinerea,* and *Monilinia fructigena*). Formulations (ethoxylate and Tween 80) of eugenol tested at room temperature were ineffective to control the disease, but when heated to 50°C, both formulations developed phytotoxicity on the apple surface and caused cuticle damage. The formulation of eugenol (2 mg/mL) and soy lecithin (50 mg/mL) suppressed the phytotoxicity and reduced the disease incidence (Amiri et al., 2008). Table 2 presents the toxicity studies of eugenol.

Table 2 Toxicity studies of eugenol.

No.	Dose volume/route of administration	Facts	References
1.	5–10 mL of clove oil administered to a 2-year-old boy	After ingestion affects at: 3 h: Distress and crying 8 h: Generalized seizure 24 h: Liver failure Toxicity was similar to paracetamol poisoning	Hartnoll et al. (1993)
2.	4 and 8 μL eugenol through intravenous infusion in the rat	Respiratory disorder and pulmonary edema	Wright et al. (1995)
3.	Exposure of eugenol to isolated rat hepatocytes in in vitro experiment	More than 85% of the cells were damaged and indicated hepatotoxicity	Soundran et al. (1994)
4.	Na+/K+ATPase is inhibited in the kidney and intestine by spice (cinnamon and clove)	Alters the mitochondrial function in cells that leads to a decrease in ATP level. Ultimately, it affects cell growth, viability, and aging	Usta et al. (2002)
5.	Experimented on African clawed frogs, given anesthetic doses	Toxicity effects on the organ kidney, morphological change and cell apoptosis in renal cells	Goulet et al. (2011)
6.	Intake of eugenol through the mouth for 15 days of different doses	In blood, the level of aspartate aminotransferase, alanine, aminotransferase, and total bilirubin increase	Li et al. (1980)

5. Conclusion

Eugenol is a bioactive compound found in various plants and herbs. Clove buds are where the highest concentration is obtained. Scientists reported several therapeutic effects of eugenol. Various formulations of eugenol are available on the market. Globally, the demand for nutraceuticals increases day by day. In the international market, several eugenol-based nutraceuticals are available worldwide. Scientists have developed nanoformulation-based nutraceuticals focused on bioactive molecule bioavailability and stability with better therapeutic effects.

Acknowledgment

The authors thank the pharmacy faculty of Maulana Azad University, Jodhpur, Rajasthan, India.

References

Abushouk, A.I., Ismail, A., Salem, A.M.A., Afifi, A.M., Abdel-Daim, M.M., 2017a. Cardioprotective mechanisms of phytochemicals against doxorubicin-induced cardiotoxicity. Biomed. Pharmacother. 90, 935–946.

Abushouk, A.I., Negida, A., Ahmed, H., Abdel-Daim, M.M., 2017b. Neuroprotective mechanisms of plant extracts against MPTP induced neurotoxicity: future applications in Parkinson's disease. Biomed. Pharmacother. 85, 635–645.

Akhavan, S., Assadpour, E., Katouzian, I., Jafari, S.M., 2018. Lipid nano scale cargos for the protection and delivery of food bioactive ingredients and nutraceuticals. Trends Food Sci. Technol. 74, 132–146.

Alexandre, E.M.C., Lourenço, R.V., Bittante, A.M.Q.B., Moraes, I.C.F., Sobral, P.J.A., 2016. Gelatine based films reinforced with montmorillonite and activated with nanoemulsion of ginger essential oil for food packaging applications. Food Packag. Shelf Life 10, 87–96.

Alexandru, L., Cravotto, G., Giordana, L., Binello, A., Chemat, F., 2013. Ultrasound-assisted extraction of clove buds using batch- and flow-reactors: a comparative study on a pilot scale. Innov. Food Sci. Emerg. Technol. 20.

Ali, S., Prasad, R., Naime, M., Zafar, H., Mahmood, A., Routray, I., Yalniz, M., Bahcecioglu, İ.H., Sahin, K., 2011. Dried peel fraction of Citrus sinensis partially reverses pathological changes in rat model of liver cirrhosis. Mediterr. J. Nutr. Metab. 4, 57–67.

Amiri, A., Dugas, R., Pichot, A.L., Bompeix, G., 2008. In vitro and in vivo [corrected] activity of eugenol oil (*Eugenia caryophylata*) against four important postharvest apple pathogens. Int. J. Food Microbiol. 126 (1–2), 13–19.

Assadpour, E., Mahdi Jafari, S., 2019. A systematic review on nanoencapsulation of food bioactive ingredients and nutraceuticals by various nanocarriers. Crit. Rev. Food Sci. Nutr. 59 (19), 3129–3151.

Aswathanarayan, J.B., Vittal, R.R., 2019. Nanoemulsions and their potential applications in food industry. Front. Sustain. Food Syst. 3 (95), 1–21.

Barnes, J., Anderson, L., Phillipson, D., 2007. Herbal Medicine. Pharmaceutical Press, London.

Batiha, G.E.S., Beshbishy, A.M., Tayebwa, D.S., Shaheen, H.M., Yokoyama, N., Igarashi, I., 2019. Inhibitory effects of *Syzygium aromaticum* and *Camellia sinensis* methanolic extracts on the growth of Babesia and Theileria parasites. Ticks Tick Borne Dis. 10 (5), 949–958.

Batiha, G.E., Alkazmi, L.M., Wasef, L.G., Beshbishy, A.M., Nadwa, E.H., Rashwan, E.K., 2020. *Syzygium aromaticum* L. (Myrtaceae): traditional uses, bioactive chemical constituents, pharmacological and toxicological activities. Biomolecules 10 (2), 202.

Beshbishy, A.M., El-Saber Batiha, G., Adeyemi, O.S., Yokoyama, N., Igarashi, I., 2019. Inhibitory effects of methanolic Olea europaea and acetonic *Acacia laeta* on growth of Babesia and Theileria. Asian Pac. J. Trop. Med. 12 (9), 425–434.

Bezerra, D.P., Militão, G.C.G., De Morais, M.C., De Sousa, D.P., 2017. The dual antioxidant/prooxidant effect of eugenol and its action in cancer development and treatment. Nutrients 9, 1367.

Binu, P., Gifty, K., Vineetha, R.C., Abhilash, S., Arathi, P., Nair, R.H., 2018. Eugenol, a plant-derived phenolic nutraceutical, protects thiol (SH) group in myocardium from ROS-mediated oxidation under chemotherapeutic stress induced by arsenic trioxide—a in vivo model study. Drug Chem. Toxicol. 41 (3), 352–357.

Charan Raja, M.R., 2015. Versatile and synergistic potential of eugenol: a review. Pharm. Anal. Acta 6 (5).

Chatterjee, D., Bhattacharjee, P., 2013. Supercritical carbon dioxide extraction of eugenol from clove buds. Food Bioprocess Technol. 6, 2587–2599.

Chatterjee, D., Bhattacharjee, P., 2015. Use of eugenol-lean clove extract as a flavoring agent and natural antioxidant in mayonnaise: product characterization and storage study. J. Food Sci. Technol. 52 (8), 4945–4954.

Chemat, F., Lucchesi, M.E., Smadja, J., Favretto, L., Colnaghi, G., Visinoni, F., 2006. Microwave accelerated steam distillation of essential oil from lavender: a rapid, clean and environmentally friendly approach. Anal. Chim. Acta 555 (1), 157–160.

Chemat, F., Zill-E-Huma, Khan, M.K., 2011. Applications of ultrasound in food technology: processing, preservation and extraction. Ultrason. Sonochem. 18 (4), 813–835.

Chomchalow, N., 2001. Spice production in Asia - an overview*. AU J. Technol. 5, 1–3.

Cortés-Rojas, D.F., de Souza, C.R.F., Oliveira, W.P., 2014. Clove (*Syzygium aromaticum*): a precious spice. Asian Pac. J. Trop. Biomed. 4 (2), 90–96.

de Oliveira Pereira, F., Mendes, J.M., de Oliveira, L.E., 2013. Investigation on mechanism of antifungal activity of eugenol against *Trichophyton rubrum*. Med. Mycol. 51 (5), 507–513.

Ekor, M., 2014. The growing use of herbal medicines: issues relating to adverse reactions and challenges in monitoring safety. Front. Pharmacol. 10 (4), 177.

El Ghallab, Y., Al Jahid, A., Jamal Eddine, J., Ait Haj Said, A., Zarayby, L., Derfoufi, S., 2019. *Syzygium aromaticum* L.: phytochemical investigation and comparison of the scavenging activity of essential oil, extracts and eugenol. Adv. Tradit. Med. 20, 153–158.

El-Saber Batiha, G., Magdy Beshbishy, A., Stuart Tayebwa, D., et al., 2018. Inhibitory effects of *Uncaria tomentosa* bark, *Myrtus communis* roots, *Origanum vulgare* leaves and *Cuminum cyminum* seeds extracts against the growth of Babesia and Theileria in vitro. Jpn. J. Vet. Parasitol. 17 (1), 1–6.

Feng, J., Lipton, J.M., 1987. Eugenol: antipyretic activity in rabbits. Neuropharmacology 26, 1775–1778.

Garkal, D.J., Taralkar, S.V., Kulkarni, P., Jagtap, S., Nagawade, A., 2012. Kinetic model for extraction of eugenol from leaves of *Ocimum Sanctum* Linn (Tulsi). Int. J. Pharm. Appl. 3 (1), 267–270.

Golmakani, M.T., Rezaei, K., 2008. Comparison of microwave-assisted hydrodistillation withthe traditional hydrodistillation method in the extraction of essential oils from *Thymus vulgaris* L. Food Chem. 109 (4), 925–930.

Gopalakrishnan, N., Shanti, P.P.V., Narayanan, C.S., 1990. Composition of clove (*Syzygium aromaticum*) bud oil extracted using carbon dioxide. J. Sci. Food Agric. 50 (1), 111–117.

Goulet, F., Vachon, P., Helie, P., 2011. Evaluation of the toxicity of eugenol at anesthetic doses in African clawed frogs (*Xenopus laevis*). Toxicol. Pathol. 39, 471–477.

Grespan, R., Paludo, M., Lemos Hde, P., Barbosa, C.P., Bersani-Amado, C.A., Dalalio, M.M., Cuman, R.K., 2012. Anti-arthritic effect of eugenol on collagen induced arthritis experimental model. Biol. Pharm. Bull. 35, 1818–1820.

Guan, W., Li, S., Yan, R., Tang, S., Quan, C., 2007. Comparison of essential oils of clove buds extracted with supercritical carbon dioxide and other three traditional extraction methods. Food Chem. 101, 1558–1564.

Guan, Y., Wu, J., Zhong, Q., 2016. Eugenol improves physical and chemical stabilities of nanoemulsions loaded with β-carotene. Food Chem. 194, 787–796.

Hamed, S.F., Sadek, Z., Edris, A., 2012. Antioxidant and antimicrobial activities of clove bud essential oil and eugenol nanoparticles in alcohol-free microemulsion. J. Oleo Sci. 61 (11), 641–648.

Hartnoll, G., Moore, D., Douek, D., 1993. Near fatal ingestion of oil of cloves. Arch. Dis. Child. 69, 392–393.

Jahangir, M.A., Taleuzzaman, M., Beg, S., Verma, S., Gilani, S.J., Alam, P., 2020. A review of eugenol-based nanomedicine: recent advancements. Curr. Bioact. Compd. 16, 1.

Jeyaratnam, N., Nour, A.H., Kanthasamy, R., Nour, A.H., Yuvaraj, A.R., Akindoyo, J.O., 2016. Essential oil from Cinnamomum cassia bark through hydrodistillation and advanced microwave assisted hydrodistillation. Ind. Crop. Prod. 92, 57–66.

Jirovetz, L., Buchbauer, G., Stoilova, I., Stoyanova, A., Krastanov, A., Schmidt, E., 2006. Chemical composition and antioxidant properties of clove leaf essential oil. J. Agric. Food Chem. 54 (17), 6303–6307.

Kabuto, H., Tada, M., Kohno, M., 2007. Eugenol [2-methoxy-4-(2-propenyl) phenol] prevents 6-hydroxydopamine-induced dopamine depression and lipid peroxidation inductivity in mouse striatum. Biol. Pharm. Bull. 30 (3), 423–427.

Kamatou, G.P., Vermaak, I., Viljoen, A.M., 2012. Eugenol—from the remote Maluku Islands to the international market place: a review of a remarkable and versatile molecule. Molecules 17 (6), 6953–6981.

Kar Mahapatra, S., Chakraborty, S.P., Majumdar, S., Bag, B.G., Roy, S., 2009. Eugenol protects nicotine-induced superoxide mediated oxidative damage in murine peritoneal macrophages in vitro. Eur. J. Pharmacol. 623 (1–3), 132–140.

Kaur, G., Athar, M., Alam, M.S., 2010. Eugenol precludes cutaneous chemical carcinogenesis in mouse by preventing oxidative stress and inflammation and by inducing apoptosis. Mol. Carcinog. 49 (3), 290–301.

Kayaci, F., Ertas, Y., Uyar, T., 2013. Enhanced thermal stability of eugenol by cyclodextrin inclusion complex encapsulated in electrospun polymeric nano fibers. J. Agric. Food Chem. 61, 8156–8165.

Khalil, A.A., Rahman, U.U., Khan, M.R., Sahar, A., Mehmood, T., Khan, M., 2017. Essential oil eugenol: sources, extraction techniques and nutraceutical perspectives. RSC Adv. 7, 32669–32681.

Kim, J., Choi, S.J., 2020. Improving the stability of lycopene from chemical degradation in model beverage emulsions: impact of hydrophilic group size of emulsifier and antioxidant polarity. Foods 9 (8), 971.

Kuppusamy, P., Yusoff, M.M., Maniam, G.P., Ichwan, S.J., Soundharrajan, I., Govindan, N., 2014. Nutraceuticals as potential therapeutic agents for colon cancer: a review. Acta Pharm. Sin. B 4 (3), 173–181.

Leem, H.H., Kim, E.O., Seo, M.J., Choi, S.W., 2011. Antioxidant and anti-inflammatory activities of eugenol and its derivatives from clove (*Eugenia caryophyllata* Thunb.) Korean. J. Food Sci. 40, 1361–1370.

Li, Y., Fabiano-Tixier, A.S., Chemat, F., 2014. Essential oils as green solvents. In: Essential Oils as Reagents in Green Chemistry. Springer Briefs in Molecular Science. Springer, Cham. https://doi.org/10.1007/978-3-319-08449-7_6.

Li, J., Yu, Y., Yang, Y., Liu, X., Zhang, J., Li, B., Zhou, X., Niu, J., Wei, X., Liu, Z., 1980. A 15-day oral dose toxicity study of aspirin eugenol ester in Wistar rats. Food Chem. Toxicol. 50, 1980–1985.

McClements, D.J., Rao, J., 2011. Food-grade nanoemulsions: formulation, fabrication, properties, performance, biological fate, and potential toxicity. Crit. Rev. Food Sci. Nutr. 51, 285–330.

Mohammadi Nejad, S., Özgüneş, H., Başaran, N., 2017. Pharmacological and toxicological properties of eugenol. Turk. J. Pharm. Sci. 14 (2), 201–206.

Morsy, M.A., Fouad, A.A., 2008. Mechanisms of gastroprotective effect of eugenol in indomethacin-induced ulcer in rats. Phytother. Res. 22 (10), 1361–1366.

Neveu, V., Perez-Jiménez, J., Vos, F., et al., 2010. Phenol-explorer: an online comprehensive database on polyphenol contents in foods. Database (Oxford) 2010, bap024.

Nikbin, S., Derakhshideh, A., Hozouri, T.M., et al., 2020. Synergic effects of aerobic exercise and eugenol supplement on germ cell development and testicular tissue structure in chlorpyrifos-treated animal model. Environ. Sci. Pollut. Res. 27, 17229–17242.

Oyedemi, S.O., Okoh, A.I., Mabinya, L.V., Pirochenva, G., Afolayan, A.J., 2009. The proposed mechanism of bactericidal action of eugenol α-terpineol and terpinene against *Listeria monocytogenes*, *Streptococcus pyogenes*, *Proteus vulgaris* and *Escherichia coli*. Afr. J. Biotechnol. 8, 1280–1286.

Patra, J.K., Das, G., Fraceto, L.F., Campos, E.V.R., Rodriguez-Torres, M.D.P., Acosta-Torres, L.S., Diaz-Torres, L.A., Grillo, R., Swamy, M.K., Sharma, S., Habtemariam, S., Shin, H.S., 2018. Nano based drug delivery systems: recent developments and future prospects. J. Nanobiotechnol. 16 (1), 71.

Paulo, F., Santos, L., 2018. Double emulsion solvent evaporation approach as a novel eugenol delivery system—optimization by response surface methodology. Ind. Crop. Prod. 126, 287–301.

Prakash, P., Gupta, N., 2005. Therapeutic uses of *Ocimum sanctum* Linn (Tulsi) with a note on eugenol and its pharmacological actions: a short review. Indian J. Physiol. Pharmacol. 49, 125–131.

Pulikottil, SJ, Nath, S., 2015. Potential of clove of Syzygium aromaticum in development of a therapeutic agent for periodontal disease: a review. S. Afr. Dent. J. 70, 108–115.

Quan, L., Li, S., Tian, S., Xu, H., Lin, A., Gu, L., 2004. Determination of organochlorine pesticides residue in ginseng root by orthogonal array design soxhlet extraction and gas chromatography. Chromatographia 59, 89–93.

Reichling, J., Schnitzler, P., Suschke, U., Saller, R., 2009. Essential oils of aromatic plants with antibacterial, antifungal, antiviral, and cytotoxic properties—an overview. Forsch. Komplementmed. 16, 79–90.

Rompelberg, C.J., Vogels, J.T., de Vogel, N., Bruijntjes-Rozier, G.C., Stenhuis, W.H., Bogaards, J.J., Verhagen, H., 1996. Effect of short-term dietary administration of eugenol in humans. Hum. Exp. Toxicol. 15 (2), 129–135.

Salvia-Trujillo, L., Soliva-Fortuny, R., Rojas-Graü, M.A., McClements, D.J., Martín-Belloso, O., 2017. Edible nanoemulsions as carriers of active ingredients: a review. Annu. Rev. Food Sci. Technol. 8, 439–466.

Sarrami, N., Pemberton, M.N., Thornhill, M.H., Theaker, E.D., 2002. Adverse reactions associated with the use of eugenol in dentistry. Br. Dent. J. 193 (5), 257–259.

Siddiqui, Z.H., Hareramdas, B., Abbas, Z.K., Parween, T., Khan, M.N., 2018. Use of plant secondary metabolites as nutraceuticals for treatment and management of cancer: approaches and challenges. In: Akhtar, M., Swamy, M. (Eds.), Anticancer Plants: Properties and Application. Springer, Singapore.

Soundran, V., Namagiri, T., Manonayaki, S., Vanithakumari, G., 1994. Hepatotoxicity of eugenol. Anc. Sci. Life 13, 213–217.

Srinivasan, S., Sathish, G., Jayanthi, M., Muthukumaran, J., Muruganathan, U., Ramachandran, V., 2014. Ameliorating effect of eugenol on hyperglycemia by attenuating the key enzymes of glucose metabolism in streptozotocin-induced diabetic rats. Mol. Cell. Biochem. 385 (1–2), 159–168.

Sung, B., Prasad, S., Yadav, V.R., Aggarwal, B.B., 2012. Cancer cell signaling pathways targeted by spice-derived nutraceuticals. Nutr. Cancer 64 (2), 173–197.

Surve, D.H., Paul, A.T., Jindal, A.B., 2019. Nanotechnology based delivery of nutraceuticals. In: Dasgupta, N., Ranjan, S., Lichtfouse, E. (Eds.), Environmental Nanotechnology. Environmental Chemistry for a Sustainable World. Springer, Cham, p. 21.

Taleuzzaman, M., Imam, S.S., Gilani, S.J., 2020. Clove oil/eugenol as the nanotechnological perspective for healthcare applications. In: Rahman, M., Beg, S., Kumar, V., Ahmad, F. (Eds.), Nanomedicine for Bioactives. Springer, Singapore.

Tekin, K., Akalin, M.K., Şeker, M.G., 2015. Ultrasound bath-assisted extraction of essential oils from clove using central composite design. Ind. Crop Prod. 77, 954–960.

Usta, J., Kreydiyyeh, S., Bajakian, K., Nakkash-Chmaisse, H., 2002. In vitro effect of eugenol and cinnamaldehyde on membrane potential and respiratory chain complexes in isolated rat liver mitochondria. Food Chem. Toxicol. 40, 935–940.

Vian, M.A., Fernandez, X., Visinoni, F., Chemat, F., 2008. Microwave hydrodiffusion and gravity, a new technique for extraction of essential oils. J. Chromatogr. A 1190, 14–17.

Wachtel-Galor, S., Benzie, I.F.F., 2011. Herbal medicine: an introduction to its history, usage, regulation, current trends, and research needs. In: Benzie, I., et al. (Eds.), Herbal Medicine: Biomolecular and Clinical Aspects, second ed. CRC Press/Taylor & Francis.

Wang, L., Zhang, Y., 2017. Eugenol nanoemulsion stabilized with zein and sodium caseinate by self-assembly. J. Agric. Food Chem. 65, 2990–2998.

Woranuch, S., Yoksan, R., 2013. Eugenol-loaded chitosan nanoparticles: I. Thermal stability improvement of eugenol through encapsulation. Carbohydr. Polym. 96, 578–585.

Wright, S.E., Baron, D.A., Heffner, J.E., 1995. Intravenous eugenol causes hemorrhagic lung edema in rats: proposed oxidant mechanisms. J. Lab. Clin. Med. 125, 257–264.

Xue, Y., Zhou, S., Fan, C., Du, Q., Jin, P., 2019. Enhanced antifungal activities of eugenol-entrapped casein nanoparticles against anthracnose in postharvest fruits. Nanomaterials (Basel) 9 (12), 1777.

Yadav, M.K., Chae, S.W., Im, G.J., Chung, J.W., Song, J.J., 2015. Eugenol: a phyto-compound effective against methicillin-resistant and methicillin-sensitive Staphylococcus aureus clinical strain biofilms. PLoS One 10 (3), e0119564. https://doi.org/10.1371/journal.pone.0119564.

Yu, H., Wu, K., Wu, T., Zhong, S., Xiang-hua, C., Zehong, L., 2009. Studies on supercritical CO_2 extraction-molecular distillation of essential oil from clove (*Eugenia caryophyllata* Thunb.) Bud. cabdirect.org. 29 (5), 74–78.

Zia, S., Batool, S., Shahid, R., 2020. Could PCSK9 be a new therapeutic target of eugenol? In vitro and in silico evaluation of hypothesis. Med. Hypotheses 136, 109513.

CHAPTER

9 Clove (*Syzygium aromaticum*) volatiles

Tossaton Charoonratana

Department of Pharmacognosy, College of Pharmacy, Rangsit University, Lak-Hok, Mueang Pathum Thani, Thailand

Abbreviations

CFR	Code of Federal Regulations
CVO	clove volatile oil
EFSA	The European Food Safety Authority
FCC	Food Chemicals Codex
FDA	US Food and Drug Administration
FEEDAP	The Panel on Additives and Products or Substances used in Animal Feed
FEMA	The Flavor and Extract Manufacturers Association
FID	flame ionization detector
FIFRA	The Federal Insecticide, Fungicide, and Rodenticide Act
GC	gas chromatography
GLP	good laboratory practice
GMP	good manufacturing practice
GRAS	generally recognized as safe
ISO	International Organization for Standardization
JECFA	The Joint FAO/WHO Expert Committee on Food Additives
LC	liquid chromatography
MIC	the minimum inhibitory concentration
MS	mass spectrometer
NTP	The National Toxicology Program
OECD	The Organization for Economic Co-operation and Development
QSAR	quantitative structure-activity relationship
SNI	Indonesian National Standard
TTC	the conservative threshold of toxicological concern

1. Introduction

Clove volatiles are chemical compounds from the clove tree, *Syzygium aromaticum* in the Myrtaceae family, which quickly evaporate at an average temperature. A mixture of these compounds usually appears in a liquid form as a volatile oil. Clove volatile oil (CVO) has a hot and robust aroma that varies in color from colorless to yellow or dark brown; its main composition is eugenol. CVO can be obtained

Clove (*Syzygium aromaticum*). https://doi.org/10.1016/B978-0-323-85177-0.00022-7

by distillation of various parts, including flower buds, twigs, and leaves with different yields as 0.60%–20.00%, 0.40%–8.72%, and 0.34%–4.80%, respectively (Santin et al., 2011; Nurdjannah et al., 2001; Mahulette et al., 2020; Hariyadi et al., 2020a,b; Raina et al., 2001). From the statistics available, the leading producer countries of CVO are Indonesia (4000 tons/year) and Madagascar (2000 tons/year), while other countries play a minor role, for example, CVO in Thailand is imported from China (Alighiri et al., 2017; Danthu et al., 2020).

CVO has been used for a long time in food processing. The Code of Federal Regulations (CFR) title 21 section 170.3(o)(12) included the use of CVO as flavoring agents and adjuvants (FDA, Food Additives, 2020a). Since the 1910s, it has been used as the flavor note in perfumery such as Mitsouko, Coco, Vitriol d', and Ysatis. Moreover, CVO from buds can be used in aromatherapy since it contains less eugenol than CVO from twigs, while the high content of eugenol can cause irritation (Danthu et al., 2020). CVO was used in mouthwash preparation to treat oral and pharyngeal mucosa inflammation, and its authorized product in the United Kingdom was used to relieve pain in the tooth (Blumenthal, 1998). CVO also possessed various biological activities such as antibacterial, antifungal, antiviral activities, and others; so, there is a space for developing additional use in the medicinal field (Danthu et al., 2020; European Scientific Cooperative on Phytotherapy, 2003).

The quality of CVO is affected by many factors such as plant varieties, plant age, plant organ maturity, cultivation area, post-harvesting process, distillation process, and post-distillation process. Different CVO types are used for other purposes. This chapter provides information about CVO composition, the factors that affected its composition, the present application, and possible application according to its safety and efficacy data.

2. Clove volatile oil characteristics and compositions

Generally, to obtain volatile oil with good quality for food processing and traditional uses, the physical properties and total eugenol content of CVO must meet the requirements. The physicochemical properties of CVO are specified in many trustable standards such as International Organization for standardization (ISO), Food Chemical Codex (FCC), and Indonesian National Standard (SNI); noticeably, the values in each property are different among those standards. Only FCC is freely available. However, there is some information from secondary resources reported about some properties of CVO from other standards. From ISO standard, it was shown that CVO from buds should meet the criteria of the total phenol between 85.0% and 93.0%, specific gravity between 1.04 and 1.05 $g\,mL^{-1}$, the angular rotation between -1.50 and 0 degrees, the refractive index between 1.52 and 1.58 at 20°C, and the solubility in 70% alcohol was 1:2 by volume (Isa and Nazarifah, 1986).

It was indicated that CVO from buds presented colorless to pale yellow color (Fig. 1; Reineccius, 1994). From FCC 4th edition, it was noted that clove bud oil possessed a specific infrared absorption spectrum. It had specific gravity between 1.038 and 1.060 $g\,mL^{-1}$, the angular rotation between -1.50 and 0 degrees, and the refractive index between 1.527 and 1.535 at 20°C. The solubility in 70% alcohol was 1:2 by volume. The eugenol content was not less than 85.0% (Committee on Codex Specifications, 1996).

From SNI 06-4267-1996, it had specific gravity between 1.04 and 1.07 $g\,mL^{-1}$, the angular rotation between -1.35 and 0 degrees, and the refractive index between 1.529 and 1.537 at 20°C. The solubility in 70% alcohol was 1:2 by volume. The eugenol content was between 80% and 95% (Badan Standardisasi Nasional Standar Standar Indonesia, 2011).

FIG. 1

Clove buds volatile oil.

Moreover, in herbal medicine production, the consistency of plant materials is crucial since it affects the efficacy and toxicity of the products. However, the chemical composition of plants varies, naturally, according to genetics, environmental conditions, and agricultural processes; the specification of materials or products from plants is needed to evaluate the qualified samples (Danthu et al., 2020; Charoonratana et al., 2021). The specification of CVO from buds for medical uses was indicated in monographs. According to the monographs, the primary compounds in CVO are eugenol (75%–88%), eugenyl acetate (4%–15%), and β-caryophyllene (5%–14%). The minor compounds are benzyl acetate, calacorene, calamenene, β-caryophyllene oxide, caryophyllene epoxide, α-cadinene, δ-cadinene, chavicol, α-copaene, α-cubebene, ethyl hexanoate, farnesol, 2-heptanone, α-humulene, humulene epoxide II, β-humulene, isoeugenol, and α-ylangene (less than 1%). The contamination of heavy metals was not more than 0.004%; for specific heavy metals, it was indicated in CAS number 8000-34-8 that Arsenic, Cadmium, Mercury, and Lead were not more than 3, 1, 1, and 10 ppm, respectively (European Scientific Cooperative on Phytotherapy, 2003; The United States Pharmacopeia, 2006; Blascheck et al., 2008; Chaieb et al., 2007).

The chemical composition of CVO was observed in clove buds from different areas in the world (Table 1). According to the monograph mentioned in the previous paragraph, it was found that there was only one literature in which CVO composition meets the requirement for herbal medicine production; it was commercial samples from some Indonesian and Madagascar exporting companies (Razafimamonjison et al., 2014).

Almost reported that CVO in the literature failed at least in one criterion. Some sources provided low-quality CVO, and the causes of substandard were from either plant sources or distillation methods. For example, in Indonesia, CVO from forest clove in Maluku contained a meager amount of eugenol (<30%) while obtained CVO using only steam distillation from Java, Toli Toli, and Bali contained eugenol less than 75% (Mahulette et al., 2020; Amelia et al., 2017; Sulistyoningrum et al., 2017). A report of CVO from growing clove in Cuba contained eugenol 70%, which was lower than the accepted level (Pino et al., 2001). CVO from cloves growing in south Turkey, Tunesia, and Algeria had high eugenol content (>75%) but β-caryophyllene was lower than 5%, while CVO from Brazil contained a low amount of eugenyl acetate (1.7%) (Alma et al., 2007; Boughendjioua, 2018; Hajlaoui et al., 2010; Santin et al., 2011). On the other hand, some sources provided CVO with appropriate major compounds amount (eugenol, eugenyl acetate, β-caryophyllene) but failure in minor compounds amount. For example, in China, CVO from Guangdong province contained dodecane 1.3% (Huang et al., 2013), while CVO from Guangxi province contained caryophyllene oxide 4.29%. CVO from

Table 1 Chemical composition (relative percentage) of CVO from *Syzygium aromaticum* buds from different sources.

	Exporting companies		Plantations				
			Indonesia				
Compound	Indonesia	Madagascar	Java	Bali	Toli Toli	Monado	Ambon
Eugenol	77.32–82.36	72.08–80.71	55.61	72.34	66.37	74.65	47.40–67.90
Eugenyl acetate	8.61–10.55	11.68–21.32	20.55	5.34	12.99	8.71	6.20–19.10
β-Caryophyllene	5.34–8.64	2.76–6.38	14.85	12.51	15.38	12.80	21.10–25.30
Benzyl acetate	–	–	–	0.04	0.02	–	–
Calacorene	–	–	–	0.04	0.02	–	–
Caryophyllene oxide	0.06–0.32	0.11–0.37	0.71	0.18	0.47	0.48	–
α-Cadinene	–	–	0.05	0.03	0.01	n.d.	1.90
δ-Cadinene	–	–	0.05	0.03	n.d.	0.11	–
Chavicol	0.13–0.18	0–0.24	0.31	0.25	0.18	0.17	–
α-Copaene	0.17–0.27	0–0.11	0.81	0.52	0.46	0.46	1.50
α-Cubebene	–	–	0.03	0.03	0.03	0.03	–
Eucalyptol	–	–	–	–	–	–	–
Ethyl hexanoate	–	–	–	–	–	–	–
Farnesene	–	–	0.17	0.25	0.07	0.07	–
Germacrene D	–	–	–	–	–	–	–
α-Guaiene	–	–	–	–	–	–	–
2-Heptanone	–	–	–	–	–	–	–
α-Humulene	0.65–1.04	0.34–1.04	2.76	2.34	1.97	1.54	1.30–3.30
α-Humulene epoxide	–	–	n.d.	–	–	0.05	–
Isoeugenol	0.02–0.24	0–0.20	0.02	–	–	n.d.	–
Limonene	–	–	–	–	–	–	–
Linalool	–	–	–	–	–	–	–
Methyl eugenol	0.04–0.08	0–0.06	–	–	–	–	–
Methyl salicylate	0.04–0.16	0–0.32	0.03	0.07	0.04	0.05	–
Pathchouli alcohol	–	–	–	–	–	–	1.20
Ylangene	–	–	n.d.	–	–	0.03	–

Table 1 Chemical composition (relative percentage) of CVO from *Syzygium aromaticum* buds from different sources—cont'd

Compound	Plantations						
	China		India				Bangladesh
	Guangdong	Guangxi	Sippighat	Ram Krishna Pur	Katchal	Chennai	Chittagong
Eugenol	76.80	76.23	59.90	61.06	76.06	74.80	49.71
Eugenyl acetate	9.50	1.76	4.06	3.66	3.77	0.94	–
β-Caryophyllene	6.00	11.54	21.07	24.86	13.19	17.45	18.94
Benzyl acetate	–	–	–	–	–	–	0.07
Calacorene	–	–	–	–	–	–	–
Caryophyllene oxide	0.10	4.29	–	–	–	–	–
α-Cadinene	–	–	0.06	0.05	0.03	0.04	–
δ-Cadinene	–	0.12	1.17	1.68	1.24	2.17	0.17
Chavicol	–	0.09	0.11	0.05	0.08	0.03	0.30
α-Copaene	–	0.05	–	–	–	–	0.39
α-Cubebene	–	–	3.45	3.06	0.84	0.14	–
Eucalyptol	–	0.14	–	–	–	–	0.06
Ethyl hexanoate	–	–	–	–	–	–	–
Farnesene	0.10	–	–	–	–	–	1.11
Germacrene D	–	–	n.d.	0.02	0.01	n.d.	0.08
α-Guaiene	–	–	–	–	–	–	0.02
2-Heptanone	–	–	–	–	–	–	–
α-Humulene	0.70	0.64	2.32	2.58	1.67	2.08	–
α-Humulene epoxide	–	–	–	–	–	–	–
Isoeugenol	0.10	–	–	–	–	–	–
Limonene	–	–	–	–	–	–	0.05
Linalool	0.10	–	–	–	–	–	–
Methyl eugenol	0.10	–	–	–	–	–	–
Methyl salicylate	–	0.06	–	–	–	–	0.25
Pathchouli alcohol	–	–	–	–	–	–	–
Ylangene	–	–	0.33	0.25	0.12	0.11	–

Compound	Plantations					
	Sri Lanka	Algeria	Tunisia	Turkey	Brazil	Cuba
	Colombo	Skikda	Mahdia	Hatay	Santa Catarina	Havana
Eugenol	79.20	80.00	88.58	87.00	89.60	69.80
Eugenyl acetate	10.40	5.01	5.62	8.01	1.70	16.10

Continued

Table 1 Chemical composition (relative percentage) of CVO from *Syzygium aromaticum* buds from different sources—cont'd

	Plantations					
	Sri Lanka	Algeria	Tunisia	Turkey	Brazil	Cuba
Compound	**Colombo**	**Skikda**	**Mahdia**	**Hatay**	**Santa Catarina**	**Havana**
β-Caryophyllene	7.50	2.27	1.38	3.56	8.60	13.00
Benzyl acetate	–	0.30	<0.10	–	–	–
Calacorene	–	–	<0.10	–	–	–
Caryophyllene oxide	–	–	<0.10	0.10	–	n.d.
α-Cadinene	–	–	<0.10	0.04	–	–
δ-Cadinene	–	–	n.d.	–	–	–
Chavicol	–	–	–	–	–	–
α-Copaene	–	–	<0.10	0.10	–	–
α-Cubebene	–	0.30	<0.10	–	–	–
Eucalyptol	–	0.50	–	–	–	–
Ethyl hexanoate	–	–	0.66	–	–	<0.10
Farnesene	–	0.99	–	–	–	<0.10
Germacrene D	–	0.60	n.d.	–	–	–
α-Guaiene	–	0.71	–	–	–	–
2-Heptanone	–	–	0.93	0.04	–	<0.10
α-Humulene	–	–	0.19	0.40	–	0.60
α-Humulene epoxide	–	–	–	–	–	<0.10
Isoeugenol	–	–	–	–	–	–
Limonene	–	0.30	n.d.	–	–	<0.10
Linalool	–	0.75	<0.10	0.01	–	<0.10
Methyl eugenol	–	–	<0.10	–	–	–
Methyl salicylate	–	–	–	0.07	–	<0.10
Pathchouli alcohol	–	–	–	–	–	–
Ylangene	–	–	–	–	–	–

n.d. = not detected, – = not reported.

Nicobar Island in India also possessed high content of α-humulene (1.6%) and δ-cadinene (1.2%) (Waman, 2020). In this case, less precise criteria can be applied; only contents of major compounds, eugenol, eugenyl acetate, and β-caryophyllene, should be considered in CVO sample quality control for medical purposes. CVO from other sources besides big companies or CVO gained from in-house distillation can also be used in herbal medicine production (Ranasinghe et al., 2002). Moreover, observation of any toxic or very potent compounds must be done and included in the criteria.

CVO content in leaves and twigs is less than in buds and they contain some other compounds. If the aim is to use them for medicine production, well-established monographs are required. However, they are acceptable for different applications. For example, it was mentioned that CVO from twigs is suitable for food applications. CVO from leaves usually contains a low amount of eugenyl acetate, making it a harsher note and not ideal for food flavoring. However, it has high eugenol and caryophyllene, so it is a good source of eugenol and caryophyllene extraction (Nurdjannah and Bermawie, 2012).

CVO from twigs is in yellow to dark brown color (Reineccius, 1994). The general specification of CVO from twigs was available in FCC and SNI. From FCC 4th edition, it was indicated that clove stems (twigs) oil possessed a specific infrared absorption spectrum. It had specific gravity between 1.048 and 1.056 g mL^{-1}, the angular rotation between −1.50 and 0 degrees, and the refractive index between 1.534 and 1.538 at 20°C. The solubility in 70% alcohol was 1:2 by volume. The eugenol content was between 89.0% and 95.0% (Committee on Codex Specifications, 1996). From SNI 06-4374-1996, CVO from twigs possessed specific gravity between 1.033 and 1.063 g mL^{-1}, the angular rotation between −1.30 and 0 degrees, and the refractive index between 1.510 and 1.520 at 20°C. The solubility in 70% alcohol was 1:2 by volume. The eugenol content was between 78% and 95% (Badan Standardisasi Nasional Standar Standar Indonesia, 2011). According to literature survey, the major compounds in CVO from twigs are eugenol (68%–97%), eugenyl acetate (0.1%–2.5%), β-caryophyllene (1.7%–24.8%), and α-humulene (0.2%–2.7%). The minor compounds are β-caryophyllene oxide, chavicol, α-copaene, isoeugenol, methyl eugenol, methyl salicylate, and patchouli alcohol (less than 1%) (Razafimamonjison et al., 2014; Hariyadi et al., 2020a,b). The data of CVO from twig were collected from 27 Madagascar samples, 7 Indonesian samples, and 20 Tanzania samples. Samples from Madagascar contained the richest eugenol, and only some of them contained chavicol. The chemical composition of CVO from clove twigs from different locations was shown in Table 2.

Table 2 Chemical composition (relative percentage) of CVO from *Syzygium aromaticum* twigs from different sources.

	Exporting companies			Plantations
				Indonesia
Compounds	Indonesia	Madagascar	Tanzania	Ambon
Eugenol	88.76–89.28	91.81–96.65	87.52–89.47	67.50–80.60
Eugenyl acetate	0.07–0.17	0.37–2.53	0.55–0.88	1.80–2.50
β-Caryophyllene	7.40–7.75	1.66–4.48	7.19–9.70	13.00–24.80
Caryophyllene oxide	0.20–0.26	0.14–0.60	0.25–0.68	1.00–1.20
Chavicol	n.d.	0–0.22	n.d.	–
α-Copaene	0.07–0.16	0–0.20	0–0.27	–
α-Humulene	0.93–1.31	0.22–0.79	0.75–1.08	1.60–2.70
Isoeugenol	0.01–0.03	0–0.80	0.01–0.10	–
Methyl eugenol	0.03–0.11	0–0.15	0–0.15	–
Methyl salicylate	0.06–0.20	0–0.56	0–0.27	–
Pathchouli alcohol	–	–	–	1.50

n.d. = not detected, – = not reported.

FIG. 2

Clove leaves volatile oil.

CVO from leaves is in pale yellow to ashen yellow color (Fig. 2; Reineccius, 1994). From FCC 4th edition, it was indicated that clove leaves oil possessed a specific infrared absorption spectrum. It had specific gravity between 1.036 and 1.046 $g\,mL^{-1}$, the angular rotation between −2 and 0 degrees, and the refractive index between 1.531 and 1.535 at 20°C. The solubility in 70% alcohol was 1:2 by volume. The eugenol content was between 84.0% and 88.0% (Committee on Codex Specifications, 1996). The general specification of CVO from leaves was also presented in the updated Indonesian National Standard (SNI 06-2387-2006). It had specific gravity between 1.025 and 1.049 $g\,mL^{-1}$ and the refractive index between 1.528 and 1.535 at 20°C. The solubility in 70% alcohol was 1:2 by volume. The eugenol content was not less than 78%, while β-caryophyllene was not more than 17% (Badan Standardisasi Nasional Standar Standar Indonesia, 2011). For CVO from leaves, according to literature survey, the major compounds are eugenol (61%–94%), eugenyl acetate (0%–2.8%), β-caryophyllene (2.9%–32.2%), and α-humulene (0.4%–3.5%). The minor compounds are benzyl *n*-octane, α-bergamotene, β-bisabolene, cadalene, δ-cadinene, γ-cadinene, cadinol, carvone, β-caryophyllene oxide, caryophyllene alcohol, cedr-9-ene, chavicol, methyl chavicol, cineole, α-clovene, α-copaene, α-cubene, cubenol, γ-elemene, isoeugenol, methyl eugenol, methyl isoeugenol, α-farnesene, germacrene D, globulol, guaiene, heptadecane, hexadecyl acetate, humulene epoxide II, humulane-1,6-dien-3-ol, hymachalene, ledol, limonene, limonene oxide, linalool, linalyl acetate, myristic acid, isopropyl myristate, nerol, nerolidol, oleic acid, terpinene-4-ol, terpinyl acetate, methyl salicylate, and safrole (less than 1%) (Razafimamonjison et al., 2016a,b; Srivastava et al., 2005; Wei and Shibamoto, 2010; Bhuiyan et al., 2010; Hariyadi et al., 2020a,b; Tahir et al., 2020).

The data of CVO from leaves were collected from 46 samples in four continents. The chemical composition of CVO from clove leaves from different locations was shown in Table 3. CVO from

Table 3 Chemical composition (relative percentage) of CVO from *Syzygium aromaticum* leaves from different sources.

	Exporting/importing companies				Plantations		
					Indonesia		
Compound	Indonesia	Madagascar	United States	Germany	Java	Palu	Ambon
Eugenol	75.04–77.54	80.87–83.58	76.51	76.80	65.03	64.81	60.50–78.30
Eugenyl acetate	0–0.06	0.29–1.45	n.d.	1.20	–	–	0.47–2.80
β-Caryophyllene	17.04–19.53	11.65–15.02	17.16	17.40	20.94	28.01	17.37–32.20
α-Bergamotene	–	0.40	n.d.	–	–	–	–
Cadalene	–	–	–	–	–	–	–
γ-Cadinene	–	0.30	–	–	–	–	–
δ-Cadinene	–	n.d.	0.34	–	0.31	–	–
Calamenene	–	n.d.	0.19	–	0.13	–	–
Caryophyllene oxide	0.37–0.43	0.05–0.55	0.46	0.40	–	–	0.87–1.00
Chavicol	0.07–0.13	0–0.12	–	0.10	–	–	–
α-Copaene	0.16–0.24	0–0.10	0.61	–	0.30	–	–
α-Cubebene	–	–	–	–	–	–	–
Farnesene	–	–	–	0.10	–	–	–
Germacrene D	–	–	–	–	–	–	–
α-Humulene	1.93–2.17	1.39–1.67	2.05	2.10	3.04	4.42	2.02–3.50
α-Humulene epoxide	–	0.10	–	0.10	–	–	–
Isoeugenol	0.08–0.13	0–0.24	–	<0.10	–	–	–
Limonene	–	–	n.d.	0.10	–	–	–
Linalool	–	–	n.d.	–	–	–	–
Methyl eugenol	0.15–0.24	0–0.10	–	<0.10	–	–	–
Methyl salicylate	<0.10	0.10	–	0.10	–	–	–
Nerol	–	–	–	–	–	–	–
Safrole	–	–	n.d.	–	–	–	0.96

	Plantations		
	India	Bangladesh	Cuba
Compounds	Nicobar	Chittagong	Havana
Eugenol	94.41	74.28	78.10
Eugenyl acetate	–	–	n.d.
β-Caryophyllene	2.91	3.85	20.50
α-Bergamotene	–	–	–
Cadalene	0.18	–	–
γ-Cadinene	0.18	–	–

Continued

Table 3 Chemical composition (relative percentage) of CVO from *Syzygium aromaticum* leaves from different sources—cont'd

	Plantations		
	India	Bangladesh	Cuba
Compounds	Nicobar	Chittagong	Havana
δ-Cadinene	–	0.21	–
Calamenene	–	–	–
Caryophyllene oxide	0.67	0.78	0.10
Chavicol	–	0.08	–
α-Copaene	0.04	0.17	–
α-Cubebene	–	0.02	–
Farnesene	0.06	n.d.	–
Germacrene D	–	0.38	–
α-Humulene	0.36	1.52	1.10
α-Humulene epoxide	–	–	<0.10
Isoeugenol	–	–	–
Limonene	–	2.08	<0.10
Linalool	0.08	0.14	<0.10
Methyl eugenol	–	–	–
Methyl salicylate	–	0.20	n.d.
Nerol	0.79	–	–
Safrole	–	–	–

n.d. = not detected, – = not reported.

leaves usually contained a low amount of eugenyl acetate but samples from Ambon Island, Indonesia, contain a high amount (>2%). The representative from Little Andaman in India had the most prosperous eugenol amount, but the sample from Ambon Island contained the lowest eugenol amount (Hariyadi et al., 2020a,b; Raina et al., 2001). The variation of β-caryophyllene amount occurred in samples from Indonesia. While the representative from Ambon Island contained the highest amount of β-caryophyllene, the representative from Palu in Sulawesi Island contained the lowest amount (Hariyadi et al., 2020a,b; Tahir et al., 2020). The only sample from Ambon Island in Indonesia contained safrole, and representatives from Bangladesh and Germany contained cineole and limonene (Sohilait et al., 2018; Bhuiyan et al., 2010; Jirovetz et al., 2006).

The composition of CVO from buds, twigs, and leaves was compared in Table 4. Each major compound's content in buds, twigs, and leaves is presented in the same manner. Interestingly, only eugenyl acetate content in buds is offered in a broader range than in twigs and leaves. It may be due to different harvesting times. If the collector harvested the younger buds than the full budding stage, the eugenyl acetate content was significantly additional from its full stage. While the eugenyl acetate contents between mature leaves and expanded-2 leaves were not that different, the range in leaves was smaller.

Table 4 Chemical composition (relative percentage) of CVO from *Syzygium aromaticum* buds, twigs, and leaves.

Compounds	Plant parts		
	Buds	**Twigs**	**Leaves**
Eugenol	47.40–89.60	67.50–96.65	60.50–94.41
Eugenyl acetate	0.94–21.32	0.07–2.50	0–2.80
β-Caryophyllene	2.27–25.30	1.66–24.80	2.91–32.20
α-Humulene	0.34–3.30	0.22–2.70	0.36–3.50
Caryophyllene oxide	0.06–4.29	0.14–1.20	0.05–1.00
α-Copaene	0–1.50	0–0.27	0–0.61
Isoeugenol	0–0.24	0–0.80	0–0.24
Methyl eugenol	0–0.10	0–0.15	0–0.24
Methyl salicylate	0–0.32	0–0.56	0–0.20
Chavicol	0–0.31	0–0.22	0–0.13
Pathchouli alcohol	1.20	1.50	–
α-Cubebene	0.03–3.45	–	0.02
δ-Cadinene	0–2.17	–	0–0.34
Farnesene	0.07–1.11	–	0–0.10
Germacrene D	0–0.60	–	0.38
Limonene	0.05–0.30	–	0–2.08
Linalool	0.01–0.75	–	0–0.14
α-Humulene epoxide	<0.1	–	≤0.10
α-Cadinene	0–1.90	–	–
2-Heptanone	0.04–0.99	–	–
α-Guaiene	0.02–0.71	–	–
Eucalyptol	0.06–0.50	–	–
Benzyl acetate	0.02–0.30	–	–
Ylangene	0–0.33	–	–
Ethyl hexanoate	<0.1	–	–
Calacorene	0.02–0.04	–	–
Safrole	–	–	0–0.96
Nerol	–	–	0.79
α-Bergamotene	–	–	0–0.40
γ-Cadinene	–	–	0.18–0.30
Calamenene	–	–	0–0.19
Cadalene	–	–	0.18

– = not reported.

3. Clove volatile oil from different plant varieties

Plant volatile oil content and composition have proven to be related to plant varieties, which possessed different genotypes. In the study of the composition of nine *Curcuma longa* varieties, only the Pitamber variety had cis-sesquisabinene hydrate, curzerenone, β-bisabolol, and farnesol (Mishra et al., 2018). In the Myrtaceae family, it was reported that only leaves of *Psidium guajava* Palumba variety contained high 1,8-cineole and α-terpineol when compared to the other three varieties (De Lima et al., 2010). In *Syzygium aromaticum*, few publications specified the variety of cloves used for CVO production in their studies. Since Indonesia is the motherland of cloves, wide varieties of clove exist there. The information was gathered from the primary producer islands, Ambon, Central Sulawesi, and Java.

The physical properties were investigated in Tuni, Zanzibar, and forest clove varieties from Ambon Island (Hariyadi et al., 2020a,b; Mahulette et al., 2020). First, Tuni CVO color was in the yellow group, which are brilliant yellow (buds), brilliant greenish-yellow (twigs), and light greenish-yellow (leaves). Tuni CVO from buds, twigs, and leaves had a specific gravity of 1.0443, 1.0541, and 1.0323 $g\,mL^{-1}$, respectively. Tuni CVO from buds, twigs, and leaves had a refractive index of 1.5350, 1.5350, and 1.5286. All Tuni CVO achieved 70% ethanol-soluble properties (1:2). Second, Zanzibar CVO color from buds was in the grayed orange group, orange-brown, while twigs and leaves were a brilliant yellow and light greenish-yellow, respectively. Zanzibar CVO from buds, twigs, and leaves had a specific gravity of 1.0327, 1.0382, and 1.0372 $g\,mL^{-1}$, respectively. Zanzibar CVO from buds, twigs, and leaves had a refractive index of 1.5263, 1.5317, and 1.5299, respectively. All Zanzibar CVO achieved 70% ethanol-soluble properties (1:2). Third, the forest CVO color from buds and leaves was in the yellow group, while twigs belonged to the grayed orange group. The whole forest CVO had a specific gravity between 0.9559 and 1.101 $g\,mL^{-1}$. Similarly, the index of refraction had a particular value in the range of 1.5075–1.5467. However, all forest CVO was poorly soluble in 70% ethanol (1:10).

The specific gravity can roughly predict whether CVO contains high or low eugenol β-caryophyllene. Since the specific gravity of eugenol is high (1.0651) and β-caryophyllene's is low (0.9075). CVO containing the highest eugenol and lowest β-caryophyllene showed the most increased specific gravity of 1.0541, while CVO holding less eugenol and tall β-caryophyllene showed low specific gravity of 1.0323 (Hariyadi et al., 2020a,b).

A group of researchers from Ambon Island reported the volatile oil yield and major compound amount of CVO from Tuni, Zanzibar, and forest clove varieties in 2018. The oil yield of the Zanzibar variety was superior; it was shown that CVO yield from buds, twigs, and leaves were 15.87%, 8.72%, and 2.20%, respectively. The volatile oil yield of the Tuni variety was average; it was shown that the volatile oil yields from buds, twigs, and leaves were 13.08%, 3.40%, and 1.60%, respectively. The forest cloves contained minimum volatile oil yield; it was found that the volatile oil yields from buds, twigs, and leaves were 1.33%–3.00%, 0.40%–1.00%, and 0.44%–0.93%, respectively. According to SNI 06-4267-1996, it was shown that only Tuni buds volatile oil contained appropriate total eugenol content (84%) determined by the Cassia method. The Zanzibar twigs volatile oil held total eugenol content between 78% and 95%, which is acceptable for general use according to SNI 06-4374-1996. However, all leaves volatile oil samples did not pass the criteria (SNI 06-2387-2006). β-Caryophyllene from the Tuni and Zanzibar leaves volatile oil exceeded 17%, while total eugenol in the forest cloves was very low (around 24%) (Hariyadi et al., 2020a,b; Mahulette et al., 2020). The refined process can be applied to adjust the compound amount to the appropriate level.

Table 5 Oil yield (%w/w), total eugenol (%), and β-caryophyllene (%) of CVO from *Syzygium aromaticum* buds, twigs, and leaves from different varieties.

			Compounds content	
Clove varieties	**Plant parts**	**Oil yield**	**Total eugenol**	**β-Caryophyllene**
Zanzibar	Buds	13.74–15.87	76.00–88.39	19.03
	Twigs	8.72	82.00	15.38
	Leaves	2.20	82.00	18.26
Tuni	Buds	13.08	84.00	12.68
	Twigs	3.40	98.00	6.88
	Leaves	1.60	78.00	20.93
Forest I	Buds	1.33	20.00	5.30
	Twigs	1.00	20.00	1.70
	Leaves	0.44	21.00	0.74
Forest II	Buds	1.67	26.00	11.79
	Twigs	0.40	22.00	3.32
	Leaves	0.34	26.00	6.16
Forest III	Buds	3.00	23.00	6.40
	Twigs	0.77	28.00	3.29
	Leaves	0.93	24.00	5.88
Siputih	Buds	13.13	81.05	–

– = not reported.

The Zanzibar clove variety was also popular in Central Sulawesi Island. In the year 2012, it was reported that eugenol content from buds reached 80%. However, solubility in alcohol was 1:1, which was failed from SNI 06-4267-1996 criteria (Pencarian Koleksi Perpustakaan Universitas Islam Indonesia, 2012).

In Java Island, there was a study comparing the volatile oil yield and total eugenol from Siputih and Zanzibar clove varieties in the year 2011. It was found that CVO yields in Siputih and Zanzibar buds were 13.13% and 13.74%, respectively. The total eugenol from both bud varieties was 81.05% and 88.39%, which was acceptable according to SNI 06-4267-1996. The author also mentioned that total eugenol from Afo clove variety was 70.65%–73.19% (Dan Syafaruddin, 2011).

From Indonesia, the Zanzibar variety contained the highest oil yield and eugenol content (Table 5). However, Tuni and Siputih varieties can become the best varieties if there is development in plant breeding and growing conditions.

4. Clove volatile oil from different plant ages

The genetics of plant in the same variety is identical. However, it does not express the same way throughout the plant's life. The change may cause variation in the chemical composition and content. Studies showed that the aging process of plants might cause either promotion or reduction in the volatile oil yield and alter the chemical composition (Innecco, 2004; Shiferaw et al., 2019). *Eucalyptus citriodora*, which belonged to Myrtaceae, showed that the volatile oil yield was highest in 5-year-old trees and declined with tree age. Citronellal and citronellol content also fluctuated with tree age (Gu et al., 2019). However, a study showed that CVO yield in young (3–4 years old) and mature (45 years old) clove trees of Zanzibar variety was not different (around 14.80%) (Alfikri et al., 2020). In the full-budding stage, it was found that mature trees contained higher eugenol content than young trees, while eugenyl acetate and β-caryophyllene were higher in young trees. The composition of both trees was different; for example, only the young tree possessed 3-ethyl-o-xylene, while only the mature tree contained mesitylene. Since their compositions were different, they affected the physical properties of CVO. It was found that only CVO from mature trees has the appropriate optical rotation, specific gravity, refractive index, and solubility in 70% ethanol, according to SNI 06-4267-1996.

Generally, clove oil is initially obtained from buds, twigs, and leaves of 7- to 8-year-old clove trees, and the tree reaches its total bearing capacity after 20 years (Alfikri et al., 2020; Purseglove et al., 1981). Even 3- to 4-year-old clove trees produced high yield CVO, the physical property (optical rotation) did not meet the criteria. Moreover, to conclude whether the ages of clove tree have influenced CVO yield, further studies with more variation of samples are required such as the number of samples, the ages of the tree (clove tree life span is up to 100 years old), and the variety of selections.

5. Clove volatile oil from different maturity of plant organs

The metabolic regulation of the volatile oil in plant organs with different maturities is different. A study on olives showed that different ripening degrees of fruit affected the volatile oil profiles (Karagoz et al., 2017). In flowers, some volatiles potentially active in the attraction of pollinators are short-life compounds. For example, β-ionone level in *Boronia megastigma* (Nees) reached the peak at the open flower stage, then it dropped after that (Mactavish and Menary, 1997). The report showed that the volatile oil yield in *Pogostemon cablin* leaves declined significantly from vegetative to fully mature stages. It may be due to the evaporation of volatiles from leaves or the catabolism of volatiles to other compounds (Tripathi and Hazarika, 2014).

Some exciting publications reported about effect of buds maturity on CVO yield, physical properties, and composition. One study observed that an unidentified clove variety showed an increase in the volatile oil yield from young bud to full budding stages (14.9% v/b to 16.1% v/b). However, no statistical analysis was performed. The change of chemical composition was detected in different buds' maturities. Eugenol was accumulated in young buds (55.6%); the level was gradually higher when buds grew more prominent and peaked in the full budding stage (72.2%). Eugenyl acetate content presented oppositely; it was high in the young bud stage (36.5%) and drastically reduced in the full budding stage (9.4%; Nurdjannah and Bermawie, 2012).

Another study was observed in the Zanzibar clove variety, and it showed no difference in CVO yield of budding-3, full budding, and flowering stages from mature clove trees. For physical properties, buds' maturities had most influenced the angular rotation of CVO from mature trees. It was found that CVO at the budding-3 stage possessed the highest angular rotation (1.63 degrees). The result may be due to the difference in chemical composition affecting the viscosity and the polarization angle of CVO. For chemical composition, it was found that eugenol content increased from the budding-3 stage (71.8%) to the full budding stage (77.6%) and flowering stage (82.3%). In comparison, eugenyl acetate content was decreased from the budding-3 stage (15.47%) to the full budding stage (6.54%) and flowering stage (3.93%). β-Caryophyllene level was highest in budding-3 stage (5.57%) and lowest in full budding stage (2.53%). Other minor compounds highly accumulated in the full budding stage, such as Psi-cumene, hemimellitene, 4-ethyl-o-xylene, 1,2,3,4-tetramethylfulvene, and prehnitene, while mesitylene and p-cymene were detected only in this stage (Alfikri et al., 2020).

Last, in the Madagascar clove variety, the researcher divided clove buds into eight development stages, young bud, budding-1, budding-2, budding-3, full budding, flowering, initial fruiting, and full fruiting. It was found that CVO yield was around 18% from young bud to full budding stages, then it sharply dropped to 12% in the flowering stage, 10% in the initial fruiting stage, and 2% in full fruiting stage. For physical properties, specific gravity gradually increased from 1.021 to 1.056, while the refractive index also acted in the same way (1.5126 to 1.5423) in young bud to the full fruiting stage. Considering only budding-3, full budding, and flowering stages, the results of specific gravity and refractive index from this study were correlated to the survey of Zanzibar variety. For chemical composition, it was found that eugenol content increased from young bud (39.6%) to the full budding stage (87.7%) and the flowering stages (93.6%), then it was steady until the full fruiting stage (94.89%). In comparison, eugenyl acetate content was decreased from young bud stage (56.0%) to the full budding stage (10.68%) and from the flowering stage (4.22%) to the full fruiting stage (2.01%). However, β-caryophyllene level fluctuated, it was highest in budding-3 stage (2.27%) and lowest in flowering stage (0.37%). Caryophyllene oxide was most increased in the full fruiting stage (1.01%) and lowest in the full budding stage (0.06%) (Razafimamonjison et al., 2013).

Leaves maturity also affected CVO yield and composition. In the year 1988, a report showed that during leaves growth (between 2 and 41 days), the content of eugenol increased from 38.3% to 95.2%, but the content of eugenyl acetate and β-caryophyllene decreased from 51.2% to 1.5% and 6.3% to 0.2%, respectively (Gopalakrishnan and Narayanan, 1988). In Madagascar clove variety, the researcher separated clove leaves into four development stages, young leaves, expanded leave-1, expanded leave-2, and mature leaves. Each stage of leaves was initially collected from July 2012 to January 2013 bimonthly. It was found that CVO yield was 5.1% in young leaves, then it significantly dropped to 4.5%, 4.1%, and 3.8% in each development stage. Like in the previous study, the eugenol content was 25.43%–30.38% in young leaves and upsurged to 58.29%–61.53% and 84.00%–90.48% in expanded leaves-1 and -2, respectively, then it remained steady around 90% in mature leaves. The eugenyl acetate content acted oppositely; high accumulation was detected in young leaves (61.44%–65.52%), then sharply dropped to 26.67%–32.65% and 0.96%–7.16% in expanded leaves-1 and -2, respectively. Finally, it was accumulated in a low amount in mature leaves (0.36%–1.64%). However, β-caryophyllene content was not different in all development stages of leaves. In this study, up to 2% hymachalene was detected in mature leaves, and it was the first study to report this compound in clove (Razafimamonjison et al., 2016a,b).

CVO accumulates in high amounts from young buds until mature buds; then, the yield declines after the full budding stage. The specific gravity and refractive index increased per clove bud maturity. Eugenol content increased along with the development of buds to flowers and fruits, but eugenyl acetate decreased. β-Caryophyllene amount fluctuated; however, it was highest in the budding-3 stage of Zanzibar and Madagascar varieties. In leaves, CVO yield was increased in young leaves and decreased in mature leaves. However, the mature leaves contained eugenol dramatically higher than in young leaves. The eugenyl acetate was high in young leaves and low in mature leaves (Table 6).

Table 6 Oil yield (%w/w) and major constituents (%) of CVO from *Syzygium aromaticum* buds and leaves from different maturities.

Plant part	Maturities	Oil yield	Compounds content				
			Eugenol	Eugenyl acetate	β-Caryophyllene	α-Humulene	Caryophyllene oxide
Buds	Young	17.58	39.66–55.60	36.50–56.07	1.77	–	0.48
	Stage 1	~17.6	42.16	54.44	2.01	–	0.74
	Stage 2	~17.6	49.97	48.50	0.80	–	0.27
	Stage 3	~17.6	52.13–71.89	15.47–44.00	2.27–5.57	n.d.	0.25
	Full bud	17.94	68.50–87.78	6.54–10.68	0.62–2.53	n.d.	0.06
	Flowering	12.25	82.38–93.68	3.93–4.22	0.37–3.21	0.39	0.27
	Fruit	9.46	94.48	2.53	1.03	–	0.60
	Full fruit	2.52	94.89	2.01	1.99	–	1.01
Leaves	Young	5.10	25.43–30.38	61.44–65.52	5.64–6.97	0.68–0.85	0.06–0.15
	Expand 1	4.50	58.29–61.53	26.67–32.65	5.66–8.63	0.61–1.02	0.03–0.08
	Expand 2	4.10	84.00–90.48	0.96–7.16	5.21–5.70	0.63–0.74	0.06–0.10
	Mature	3.80	88.36–90.22	0.36–1.64	4.63–6.42	0.57–0.78	0.13–0.24

n.d. = not detected, – = not reported.

6. Clove volatile oil from different cultivation areas

In diverse plant cultivation areas, the climate is different. Temperature is a climate factor that influences plant metabolic processes. Surface temperature is controlled by altitude (elevation above sea level). The altitude affected the quality and quantity of the volatile oil of several plants such as *Musa acuminata* AAA group, *Nigella sativa* L., etc. (Brat et al., 2004; Herlina et al., 2017). Further investigation studied the impact of different altitudes on the volatile oil yield and total eugenol of clove in Indonesia. Since the 1980s, it was reported that CVO yield and total eugenol from buds increased following an increase in altitude. However, the yield dropped when it was higher than 800 m above sea level (Nurdjannah and Rusli, 1988).

The study on the influence of altitude on CVO from leaves was performed in North Sulawesi Island. The samples were divided into seven groups according to the heights. The elevation ranges started at 100 m above sea level and increased every 100 m to achieve 800 m above sea level. The CVO yield increased from 100 to 400 m above sea level (2.80% to 3.08%). However, it sharply dropped below 2.00% at 401–500 m above sea level, then gradually decreased to 1.27% at 701–800 m above sea level. The eugenol content was expressed in the same manner; it was highest at 301–400 m above sea level (68.3%) and lowest at 701–800 m above sea level (59.5%) (Arrijani and Kapahang, 2017).

The responses to an altitude of CVO yield and total eugenol in buds and leaves are different at 401–800 m above sea level. Since the data are from two studies, more investigations are needed to draw the relationship between altitude and CVO quality and quantity.

7. The volatile oil in cloves with different post-harvesting processes

After collecting the plant parts from the tree, post-harvesting is a process such as drying and storage. Drying is the most common preservative method which protects plant materials from microbial overgrowth and enzymatic degradation. However, heat can also cause some alterations in plant cells and their composition. After drying, plant materials were stored for a while before using in manufacturing processes. Longer or harsher storage conditions may cause the loss of more volatile compounds leading to an increased relative content of harder volatiles. Different drying methods and storage conditions affected volatile oil yield and compositions in various plants. In *Lippa citriodora* Kunth, oven drying provided more volatile oil yield than shade drying, while freeze-drying preserved the highest amount of citral (Ebadi et al., 2015).

On the other hand, shade drying was more profitable in the volatile oil yield than oven drying in *Origanum vulgare* L. and *Origanum onites* L. All compositions of the volatile oil were different related to drying methods, except carvacrol in *Origanum vulgare* (Ozdemir et al., 2017). About storage time, after 1 year of storage of *Piper nigrum* L. in a glass bottle at 25°C, the volatile oil yield and terpenes content decreased, while oxygenated terpenes increased (Orav et al., 2004). Different storage temperatures affected the volatile oil composition in *Thymus officinalis* L. samples; it was shown that the frozen plant sample kept under −20°C showed the best preservation of thymol γ-terpinene and carvacrol (Usai et al., 2011).

Drying and storage were essential processes in preserving a good quality of clove. A group of researchers from Indonesia reported that the CVO yield of fresh cloves significantly increased after sun drying and oven drying. The duration of the drying process was not indicated, but it was stopped when the buds reached a moisture content of 13%. The yield was boosted from around 0.8% to 6.0% for oven drying, while the yield was 5.0% under sun drying. Oven drying at 50°C yielded more CVO than sun drying (6.0% w/w and 5.0% w/w, respectively). However, considering each major compound, oven and sun-drying influences on CVO composition were different. The eugenol content in fresh clove (56.69%) increased after oven drying (63.7%) and sun drying (60.9%), while the eugenyl acetate content in fresh clove (39.0%) decreased after oven drying (33.1%) and sun drying (34.4%). The α-humulene content in fresh clove (0.38%) decreased after oven drying (0.29%) and sun drying (0.33%). The β-caryophyllene content in fresh clove (2.45%) was close to that of in the sun drying clove (2.47%), but it is lower in the oven drying clove (2.25%). The caryophyllene oxide content in fresh clove (0.13%) decreased after oven drying (0.07%) but increased after sun drying (0.29%). The increase in caryophyllene oxide after sun drying may be due to the oxidation of caryophyllene in open-air conditions, and it might indicate CVO of inferior quality (Argyropoulos and Mullera, 2011). In this condition, α-humulene epoxide can also occur, however, it was not observed in the study (Hastuti et al., 2017; Murni et al., 2017).

In clove leaves, the yield of CVO increased depending on drying, then dropped to steady from the eighth day of drying. The maximum CVO yield was obtained on the sixth day (5.58%) with a humidity of around 17%, corresponding to an optimal drying time between 6 and 8 days. The influence of drying on CVO composition was found in leaves. After 15 days, the eugenol and β-caryophyllene content remained approximately stable. The eugenyl acetate content decreased, but the caryophyllene oxide content increased (Razafimamonjison et al., 2016a,b).

After 4 months of storage of clove buds, CVO yield increased from 5.0%–6.0% to around 7.0%. Eugenol content increased around 8%, while eugenyl acetate decreased around 10%.

The caryophyllene oxide content in oven-dried clove (0.07%) and sun-dried clove (0.29%) increased after 4 months of storage to 0.14% and 0.98%, respectively (Hastuti et al., 2017; Murni et al., 2017). For minor compounds, it was found that isoeugenol and farnesene disappeared during storage. It was noticeable that some volatiles in buds disappeared after 4 months of storage, and it was related to data in Table 4, in which some compounds were in the range of zero to some amount. Besides clove varieties and maturities, the storage condition of dried cloves was a crucial factor influencing CVO composition.

Another study reported about effect of the clove curing method on CVO yield. Clove curing is the method of preserving clove buds in a plastic bag on a rainy day and drying it in the sunshine. It was found that CVO decreased from 23.6% to 19.5% after 6 days of curing. The best curing period was 2 days, yielding 82.5% eugenol and 17.4% eugenyl acetate (Apriati et al., 2018).

In clove buds, CVO yield and eugenol content increased, but eugenyl acetate decreased after drying. After storage for a period (4 months), eugenol and caryophyllene oxide content increased but eugenyl acetate decreased. The β-caryophyllene and α-humulene content fluctuated. It was necessary to extract the oil from leaves approximately 1 week after their harvest in clove leaves since CVO yield would be maximum.

8. Clove volatile oil extraction and storage

Distillation is the primary method used in the isolation of CVO because of its simplicity. The distillation was used for volatile oil extraction in the local area and industry (Figs. 3 and 4). Other methods such as solvent extraction and cold-pressed extraction were excluded since the oil also possessed a high

FIG. 3

Distillation equipment for volatile oil production in rural Indonesia.

FIG. 4

Distillation equipment for volatile oil production in industry.

amount of nonvolatile compounds. The solvent extraction method using an organic solvent or in combination with water yields clove oleoresin or clove extract. Simultaneously, cold-pressed extraction provides long-chain hydrocarbon and other compounds, which yield a product called oil or fixed oil. Supercritical fluid extraction using CO_2 is an alternative method for CVO extraction from cloves. This method provides either volatile clove oil or fixed oil, depending on the setting condition of the instrument. However, it is not economically friendly for local people, and there was no report about the physical properties of CVO from supercritical fluid extraction compared to the standard criteria. There are generally three distillation methods which are steam distillation, hydrodistillation, and steam and water distillation. Primary manufacturers are Indaroma, Indesso, Kautilya Commodities, Biolandes, Divehi Holding, Fuzhou Farwell, and Takasago.

8.1 Clove volatile oil distillation

In steam distillation, cloves were placed on a mesh through which steam will pass. Before CVO was volatilized, cloves had to attain the maximum temperature at the operating pressure. It needed sufficient contact time between steam and cloves being stripped of the volatile oil. In hydrodistillation, cloves were boiled in water for enough time until they had been exhausted from the volatile oil. The drawback of this method was the inability to tell when cloves were exhausted. The runtimes will also be different for clove buds, twigs, and leaves, because of differences in plant matrices. In steam and water distillation, cloves were sunk in the water while steam passed through the still charge until it started to volatilize the volatile oil in the cloves. CVO was evaporated through the water as a vapor to be cooled in the condenser.

The distillation method and condition are essential factors yielding different CVO quality. For example, a study on clove leaves found that steam and water distillation provided more CVO yield and eugenol content than steam distillation. Steam and water distillation for 4h provided 78.2% eugenol content (Nurdjannah et al., 1993).

In steam distillation and hydrodistillation of clove buds, it was found that using 100g of buds, hydrodistillation provided higher CVO yield (11.5%) than steam distillation (10.1%). The contents of β-caryophyllene and α-humulene in CVO from hydrodistillation (36.9% and 4.41%, respectively) were higher than those of steam distillation (20.5% and 2.61%, respectively) (Wenqiang et al., 2007). Another group of researchers found that hydrodistillation provided a significantly higher CVO yield (12.9%) than steam distillation (11.54%). The contents of eugenol, β-caryophyllene, and α-humulene in CVO from hydrodistillation were 87.26%, 1.35%, and 0.19%, respectively. They were higher than steam distillation (82.65%, 0.91%, and 0.13%, respectively). In this study, the author applied microwave power at 1000 watts to assist CVO extraction. It was found that it can reduce the total extraction of hydrodistillation from 240 to 80min since the rate of temperature was increased. CVO yield also increased to 13.9%, but major compounds content was not different from hydrodistillation. The physical properties of CVO (specific gravity, refractive index, and color) were not significantly influenced by the different distillation methods (Golmakani et al., 2017).

Microwave radiation can disrupt the structure of plant cells, so it was used in compound extraction. Another study observed the optimum condition for microwave-assisted hydrodistillation of clove leaves using Response Surface Methodology. It was found that the best condition was extraction for 30 min with 599.5 watts of microwave power and using 65.86 g of clove leaves; it yielded 1.18% CVO (Handayani et al., 2020).

The multi-response optimization was used to predict the optimum condition for hydrodistillation of CVO buds. Four critical factors were included, cloves loading, cloves particle size, distilled water volume, and distillation time. Firstly, as the amount of cloves loading increased from 10 to 50g, CVO yield decreased after 20g of cloves. It is because the energy distribution per clove particle was reduced. The highest CVO yield was around 17% for 20g of cloves. Eugenol content was similar to CVO yield, but it decreased after 30g of cloves loading. Secondly, as the particle size of buds decreased, CVO yield and eugenol content increased. Because the smaller size of particles provided more surface area leading to better heat and mass transfer. The highest CVO yield was 12.07% at the 565μm particle. Thirdly, within 45min extraction time of 20g buds, it was found that 300mL of water provided the highest CVO yield (~17%). An inadequate amount of water can lead to overheating of clove buds, while an excessive amount of water required more energy to reach boiling point. Lastly, measuring the time needed for complete extraction, the experiment was performed for 50min. The highest CVO yield was achieved at 45min of extraction time. A decrease in the yield of eugenol was observed after 35min of extraction (Kapadiya and Desai, 2017).

Besides using microwave energy, ohmic heating can be used to assist volatile oil extraction. Ohmic heating is a process that the electrical resistance of plant material itself generates heat as electric current is passed through it, causing the eruption of the plant cells. A study used a Central Composite Design to optimize the ohmic heating assisted hydrodistillation method to extract CVO from clove buds. The appropriate condition which provided high CVO yield with less energy consumption was found as 12.5 V/cm voltage gradient, 97.9 min for process time, and 40 g of clove buds in 240 mL water. CVO yield of the ohmic heating assisted hydrodistillation method (13.1%) was higher than the hydrodistillation method (8.23%). However, the eugenol and β-caryophyllene contents in the hydrodistillation process were 87.1% and 1.01%, respectively, which were higher than those in the ohmic-assisted method (82.6% and 0.81%, respectively) (Tunça and Koca, 2019).

It was found that CVO yield from 25 g of buds was 7.04% at 6-h extraction time for steam distillation. However, the author did not report whether it will be increased after 6 h. The chemical composition of obtained CVO was as same as that of commercial CVO (Ratri et al., 2020). Another study reported that CVO yield from clove leaves was 1.6% using steam distillation for 7 h. The contents of eugenol and β-caryophyllene were 75.6% and 21.0%, respectively (Jayanti, 2011).

The yield and composition of obtained CVO from buds using different distillation methods were compared in Table 7. Hydrodistillation was among the best approaches, providing high volatile oil yield and rich in active compounds with low energy consumption.

8.2 Clove volatile oil pre- and post-distillation processes

Pretreatment processes were an add-on to improve the yield and quality of distilled CVO from leaves. Several pretreatment methods had been used, such as fermentation, delignification, and crushing. The pretreatment purpose was to degrade leaves tissue, so trapped volatile oil was quickly released during the distillation process. The first study optimized the appropriate time used for leaf fermentation with *Trichoderma harzianum* before steam distillation. *Trichoderma harzianum* is a cellulase enzyme-producing fungus that is easy to find in nature and produces many conidia. It was found that pretreatment with fermentation increased CVO yield up to 2.8 times. The highest yield was obtained at a fermentation time of 8 days with a 1.62% CVO yield. The eugenol and β-caryophyllene compositions were 75.28% and 21.61%, respectively (Nurhadianty et al., 2017).

Secondly, CVO from leaves was extracted by steam distillation with delignification, fermentation, and a combination of delignification and fermentation pretreatment. Delignification of clove leaves using 0.25% NaOH solution causes the lignin molecule to degrade due to the breaking of the aryl-ether, aryl-aryl, and alkyl-alkyl bonds. From the distillation process of dignified clove leaves, CVO was clear yellowish color, and the yield was 2.36%. It possessed a refractive index of 1.5185, a specific gravity of 1.031, and 1:2 solubility in 70% ethanol. Fermentation of clove leaves using *Trichoderma harzianum* was carried out for 8 days. From the distillation process of fermented clove leaves, CVO was clear yellowish color and the yield was 2.66%. It possessed a refractive index of 1.5190, a specific gravity of 1.030, and 1:1.5 solubility in 70% ethanol. The combination of delignification and fermentation processes before distillation yielded only 1.36% CVO with a slightly yellow color. It possessed a refractive index of 1.5190, a specific gravity of 1.025, and 1:2 solubility in 70% ethanol. Based on the physical properties, all pretreated CVO from leaves was per the range specified in SNI 06-2387-2006. It was concluded that the fermentation process gave the best result following delignification. However, the reduction in yield from the combined delignification and fermentation was presented. The eugenol and β-caryophyllene content also met the established standard (Wijaya et al., 2015).

Table 7 Oil yield (%w/w) and major constituents (%) of CVO from *Syzygium aromaticum* buds and leaves from different distillation methods.

Distillation method	Assisted conditions	Time (min)	Yield	Compounds content				
				Eugenol	Eugenyl acetate	β-Caryophyllene	α-Humulene	Caryophyllene oxide
Hydrodistillation	No	240	12.98	87.26	10.43	1.35	0.19	0.20
40 g bud/400 mL water	Microwave 1000 W	80	13.94	88.80	7.46	2.65	0.40	0.19
20 g bud/300 mL water	No	45	~17.0	–	–	–	–	–
40 g bud/240 mL water	No	97.9	8.23	87.12	–	1.01	–	0.15
	Ohmic heat 12.5 V/cm	97.9	13.18	82.68	–	0.81	–	0.15
100 g bud	No	360	11.50	48.82	3.89	36.94	4.41	0.32
Steam distillation	No	240	11.54	82.65	15.63	0.91	0.19	0.17
40 g bud/400 mL water	Microwave 1000 W	80	12.71	83.39	14.34	1.37	0.21	0.22
25 g bud	No	360[a]	7.04	85.01	13.06	1.14	0.18	0.32
100 g bud	No	600	10.10	58.20	13.89	20.59	2.61	0.49
Hydrodistillation 65.86 g leaves waste	Microwave 599.5 W	30	1.18	–	–	–	–	–
Steam distillation 500 kg leaves	No	420	1.60	75.64	–	21.03	–	–

[a]*According to the literature, the yield may be increased if the distillation time is prolonged.*
– = not reported.

Lastly, a simple pretreatment method is crushing. The process of leaves crushing is to open plant tissue making it easier for volatile oil to flow when leaves were distilled. When the leaves were crushed with low rotation operation, smooth crushed and roughly crushed leaves were obtained. Non-crushed, smooth crushed, and rough crushed were extracted using steam and water distillation to compare the volatile oil yield and composition. The highest CVO yield was obtained from rough crushed leaves distillation for 3h, and it was 33.40%, followed by smooth crushed leaves distillation (14.2%) and non-crushed leaves distillation (2.11%). For non-crushed leaves distillation, the yield reaches its peak at 4h of distillation (6.86%). The author also reported that changing the boiling chamber from iron to stainless steel provided better brightness of CVO (Lutfi et al., 2020).

The materials used to construct the boiling chamber and other equipment contribute to the dark color of CVO since metal ions from them react with eugenol resulting in a colored metal complex. Besides the material being replaced by stainless steel, other post-distillation methods that are more economically friendly can be applied to improve volatile oil quality. Adsorption materials such as bentonite and zeolite were used to refine CVO. Bentonite and zeolite comprise aluminum hydrosilicate, which can exchange cation (Mg and Fe). It was found that refined CVO with 10% bentonite changes the color of raw CVO from brownish-black to yellow. The clarity increased from 4.5% to 91.7%, while the brightness rose from 18.83 to 48.42. Also, bentonite can adsorb Mg and Fe; the value of those metal ions reduced by almost 50% in refined CVO. The content of eugenol, β-caryophyllene, and α-humulene in refined CVO (81.5% and 10.8%, respectively) was also higher than those of in raw CVO (76.1% and 2.05%, respectively). However, it was noted that β-caryophyllene was absent in refined CVO (Marwati et al., 2017). Another study reported that using 25 g of zeolite in a 60-min adsorption process, the eugenol content in CVO increased to 89.94% (Suci et al., 2013). A chelating agent such as citric acid was also used to refine CVO from leaves. It was found that the highest Fe bounding percentage condition was 3% citric acid at 60°C (Widayat et al., 2014).

8.3 Clove oil extraction using alternative methods

Supercritical fluid extraction is the optional method for CVO extraction. It possesses adjustable characteristics due to its density, which can be controlled by altering the pressure or temperature. The volatile oil is selectively extracted at lower pressure and temperature, while other constituents such as wax, dye, resin, and high molecular-weight compounds have low solubility under this condition. From the literature reviews, it was found that at extraction pressure $\geq$ 10 MPa and temperature $\geq$ 40°C, wax and long-chain hydrocarbons were coextracted with CVO (Yazdani et al., 2005; Wenqiang et al., 2007; Ivanovic et al., 2011, 2013).

The best condition for supercritical fluid extraction of CVO was as follows: the pressure may not exceed 10 MPa and the temperature must be lower than 40°C. In 1997, a group of researchers extracted CVO using supercritical CO_2 extraction with two separators. The first separator condition was 9 MPa and −10°C, the second separator condition was 1.5 MPa and 10°C, and CVO yield from this method was 20.8% (Reverchon and Marrone, 1997). Later in 2013, other researchers used two separators plus ultrasonication to extract CVO. It was found that ultrasonication-assisted supercritical fluid extraction yielded more CVO, eugenol, eugenyl acetate, β-caryophyllene, and α-humulene with the value of 22.0%, 59.1%, 18.6%, 15.3%, and 1.93% (Yang et al., 2014). The simple scale-up condition (8.06 MPa, 39.85°C) was successfully used for scaling-up extraction of clove residue on a 5.15 L scale. The CVO compositions were similar and the yield was 20% higher on the pilot scale than on the laboratory scale (Hatami et al., 2010; Prado et al., 2011).

Besides volatile oil extraction, fixed oil was extracted from clove buds using the cold-pressed method. It was found that linoleic acid and oleic acid were the primary fatty acids (42.1%) in cold-pressed clove buds oil. In contrast, the primary saturated fatty acids were palmitic acid and stearic acid, which comprised together around 15.0%. Tocopherols and tocotrienols were also found in cold-pressed clove buds oil (Ramadan et al., 2013).

8.4 Clove volatile oil storage

The stability of volatile oil is one of the most critical issues. It is reported that the composition of volatile oil changed upon storage, while factors such as light, temperature, and oxygen available are involved in alteration processes. However, the individual character of volatile oil was proved to influence the response to temperature, light, and the promptness of the particular volatile oil to oxidize. A study reported that α-terpinene in rosemary volatile oil decreased to 7% within 12 weeks of storage at 38°C under daylight but did not change during the same period at room temperature in the dark. While α-terpinene in thyme volatile oil seemed to undergo only minor changes upon different storage conditions (Turek and Stintzing, 2012). The volatile oil of *Cuminum cyminum* L. kept its primary quality, especially monoterpene constituent when stored in refrigerator and freezer (Mehdizadeh et al., 2017).

In CVO, eugenol, eugenyl acetate contents, and β-caryophyllene were observed after 90 days of storage at room temperature (25°C) and 0 ± 1°C. In both experiments, eugenol content increased but eugenyl acetate and β-caryophyllene decreased; room temperature was sufficient for CVO storage. The result was similar to that of clove buds (Gopalakrishnan, 1994). Moreover, there was a report about CVO stability to metal contact. For reactivity to Zn, CVO from buds, twigs, and leaves were precipitated within 2 weeks and changed in color within 4 months. For reactivity to Sn, CVO from buds changes in color within 8 months. For reactivity to Al, CVO from twigs changes in color while CVO from leaves was altered in the viscosity within 8 months (Jayawardee and Balachandran, 1979).

From the product specification of Sigma-Aldrich, the shelf life of clove bud oil (CAS number 8000-34-8) is 5 years.

9. Clove volatile oil composition detection

Specific instruments measure the physiological properties of CVO. However, CVO chemical composition and content can be quantified using different techniques. Conventionally, modified phenol determination was used to determine total eugenol in CVO. This method can be called the Cassia flask method. The volatile oil is shaken with an alkali solution such as 4% KOH in the Cassia flask for 5 min at room temperature and 10 min in a steam bath. Suddenly, after removing the flask from the steam bath, a quantity of alkali is added to drive the unreacted volatile oil into the neck of the flask. It is necessary to add while the flask is still hot since the nonphenolic part may partially solidify. Measure the quantity of volatile oil that does not dissolve in the alkali (Hariyadi et al., 2020a,b). The percentage of total eugenol content is calculated from the following formula:

$$P = (10 - V) * 10$$

P = The percentage of total eugenol content.

V = Volume of undissolved oil (mL).

Flame ionization detector (FID) and mass spectrometer (MS) are commonly used for volatiles detection. During FID operation, the volatiles in the carrier gas are mixed with hydrogen, and a flame burns volatiles in the oxygen-rich environment to become ions. The ions are detected by a collector electrode which is located above the flame. MS also comprises many types of ionization units. Atmospheric Pressure Chemical Ionization is more suitable for volatiles. It contains a heated nebulizer that volatiles can mix with N_2 gas and convert to a gas stream. The gas stream is ionized at corona discharge; then, ions pass to the mass analyzer.

Moreover, MS measures the mass-to-charge ratio of charged ions, allowing to determine the molar mass. The molar mass of different volatiles can be gathered for database construction. It is helpful for the identification of compounds in volatile oil. As described in the previous section, CVO comprises many volatiles constituents; the separation technique is needed to isolate each volatile before detection is performed. Chromatography is a separating technique whose basis is the different distribution ability of compounds between moving streams, called a mobile phase and an adjacent stationary phase. If the mobile phase is gas, the technique is called Gas Chromatography (GC). If the mobile phase is liquid, the method is Liquid Chromatography (LC). GC is more famous for volatiles detection since they are easily vaporized without decomposition. GC can combine with FID or MS detectors; it was used to analyze CVO composition in almost all studies. For example, compounds in CVO from different extraction methods were analyzed using GC-FID and GC-MS (Wenqiang et al., 2007). LC can be combined with MS since the liquid mobile phase must be nebulized to fine droplets before converting to the gas stream. Typically, it was used to analyze nonvolatile constituents in cloves; few studies reported CVO analysis using LC (Higashi and Fujii, 2010).

10. Clove volatiles application

CVO has been currently used for many purposes such as food processing, medicines, perfumery, biopesticides, clearing agents in tissue histology, and source of eugenol and β-caryophyllene. Different CVO types can be used for various purposes. This section provides that information.

10.1 Clove volatile oil in food processing

CVO is commonly used in cooking and as a flavoring agent in food products. Previously, according to CFR title 21 section 184.1257(a), CVO has been listed as GRAS (Generally Recognized As Safe) for human food ingredients under production with Good Manufacturing Practice (GMP). GMP includes the requirement that the human food ingredient is of appropriate food grade. The quantity of the element added to food is within the amount rationally required to achieve the proposed physical, nutritional, or other technical effect in food (FDA, 2020a,b,c). From CFR title 21 section 170.3(o)(12), CVO can be used as flavoring agents and adjuvants (substances added to impart or help impart a taste or aroma in food) (FDA, Food Additives, 2020a).

According to a cooking recipe, meat sauce, chili sauce, and tomato ketchup used 0.11%, 0.025%, and 0.139% clove buds oil, respectively (Farrel, 1990). Food seasoning such as Bologna seasoning B and C used 0.07% and 0.45% clove buds oil, respectively (Fig. 5). Clove stems oil can also be used in flavoring alcoholic beverages (Fig. 6; Leung, 1980).

FIG. 5

Clove volatiles as a seasoning.

In 2015, the Economic Research Service/United States Department of Agriculture reported that in the United States, the intake of clove oil from the use of cloves in food is estimated to be 1600 μg/person/day. The estimated intake of clove oil from the consumption of clove spice is significantly higher than the estimated intake for clove leaves oil (430 μg/person/day) and clove stems oil (190 μg/person/day) but considerably lower than the estimated intake of clove buds oil (2350 μg/person/day) (Gooderham et al., 2020).

10.2 Clove volatile oil in medicines

Traditionally, CVO has been used for centuries to prepare some herbal remedies; it primarily has been used by traditional folk healers. Nowadays, several preparations contain CVO, such as toothpaste and dental analgesic, used by people, pharmacists, and dentists to lessen the symptoms associated with toothache and dental decay (Figs. 7 and 8; Barnes et al., 2007). The German commission E proposes

FIG. 6

Clove in Alchermes liqueur.

FIG. 7

Kan Plu toothpaste containing clove oil.

1%–5% CVO as a mouthwash to treat oral inflammation and pharyngeal mucosa. Moreover, undiluted CVO or solutions with a strength of a minimum of 50% or gels with a strength of 20% are proven for topical anesthesia in dentistry (Blumenthal, 1998). The relief of toothache by CVO is only a provisional measure. Dental attention should be sought as soon as possible. The administration can be repeated after 20 min, then every 2 h after that if necessary. It was suggested that the analgesic and antimicrobial

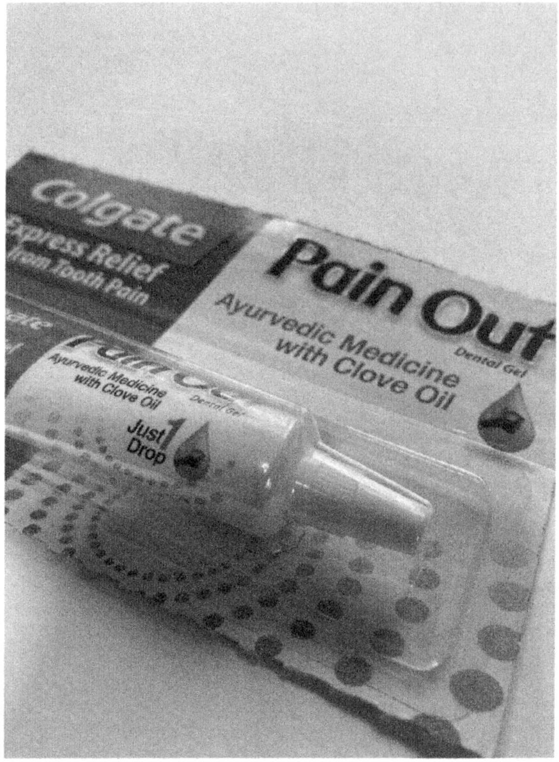

FIG. 8

Clove oil in dental analgesic.

properties of CVO make dental and oral mucosal use plausible (Park et al., 2011). The topical use of CVO is contraindicated in hypersensitivity to clove oil and hypersensitivity to Peru balsam due to possible cross-reactivity (Niinimäki, 1984).

CVO is used as a carminative, occasionally used in the treatment of flatulent colic. Dosage for oral administration for adults is 0.05–0.2 mL (Barnes et al., 2007). Moreover, it can be used in chronic anal fissure treatment. Dosage for using is 1% CVO in cream, applied three times daily (Elwakeel et al., 2007). Other applications treat a minor infection on the skin, dressing the slight wound, sore throat, and cough associated with the common cold, muscle pain, rheumatism, and insect bites (Blascheck et al., 2008; WHO, 2002).

According to application in the mouthwash, a study evaluated several phenolic compounds to develop the quantitative structure-activity relationship (QSAR). The observation was made in phenolic compounds in plants that demonstrated antibacterial activity. Among them, eugenol showed vigorous activity against oral bacteria. The QSAR approach found that lipophilicity and steric effects were key factors responsible for antibacterial activity. The optimum lipophilicity, measured by the logarithm of the octanol/water partition coefficient (log P), was 7.90 for *Streptococcus mutans*, the oral bacteria that causes tooth decay. The optimum log P was found to be 5.50 for *Fusobacterium nucleatum*, the oral bacteria that cause bad breath. The steric effect of substituents ortho to the phenolic group was related to log P of phenolic compounds, which was crucial in reducing antibacterial activity. However, log P of eugenol is 2.27; hence, the antibacterial activity of eugenol was likely exerted by multiple mechanisms; one probably came from its capability to act as a nonionic surfactant disrupting the lipid-protein interface (Greenberg et al., 2008).

10.3 Clove volatile oil in perfumery

In perfumery, the flavor notes presented by CVO are used as a critical factor for many well-established names such as Chypre, created by perfumer François Coty in 1917 and marketed until 1950, or Mitsouko created by Jean Paul Guerlain in 1919, which is still in production. In addition, rent perfumes such as Coco by Chanel (Fig. 9), Ysatis by Givenchy, and Vitriol d'O Eillet by Serge Lutens also used clove flavor notes (Danthu et al., 2020).

10.4 Clove volatile oil in biopesticides

Fungicides and insecticides are both pesticides used in plant protection. Moreover, insect repellants are also pesticides since they can be used to prevent insect bites. Ecological problems caused by the overuse of pesticides have been a matter of concern since pesticides possess high toxicity and nonbiodegradable properties. The residues in soil, water, and plants affect public health. The move toward green chemistry and the need for developing new plant protection tools discover natural products as biopesticides. The biopesticides refer to all types of innocuous natural materials that can reduce the pest population and increase food production (Koul et al., 2008).

FIG. 9

Coco by Chanel with clove flavor.

CVO is among the most effective pesticides compared with other volatile oils. It is easily transformed from a liquid to a gaseous state at room temperature without undergoing degradation. Moreover, clove, clove bud oil, and eugenol are safe for use as pesticides since they are listed as the active ingredients which can be used in pesticides products that are exempt from the Federal Insecticide, Fungicide, and Rodenticide Act (FIFRA) under the Minimum Risk Exemption regulations in 40 CFR 152.25(f) (FDA, 2011, Exemptions for pesticides of a character not requiring FIFRA regulation). There are current uses of CVO as pesticides which are described in the next section.

10.4.1 Fungicides

CVO was reported to have antifungal activity against plant pathogens such as *Fusarium* spp., *Erisyphe cruciferarum*, *Peronospora belbahrii*, *Podosphaera xanthii*, and *Plasmopara viticola*. CVO inhibited vanilla stem rot disease caused by *Fusarium oxysporum* sp. *vanillae* (Tombe et al., 1992). It was found that CVO can inhibit mycelial growth and conidial germination of *F. oxysporum* sp. *lycopersici* and reduce the severity of Fusarium wilt of tomato under in vitro and greenhouse conditions (Torre et al., 2015). Powdery mildew, *Erisyphe cruciferarum*, can infect specific crucifer plants, such as radish. It was found that 20% CVO at a maximum rate of 1.5 qt/acre can be used for disease control (radish). The powdery mildew, *Podosphaera xanthii*, is also a common disease of cucurbits. It was found that 20% of CVO possessed antifungal activity, and the area under the disease progress curve (AUDPC) was 96.9 (Lange et al., 2020).

Sclerotinia sclerotiorum causes white mold disease in Jerusalem artichoke tubers, resulting in considerable losses for tuber production. It was found that CVO at a concentration of 2% showed complete inhibition against *Sclerotinia sclerotiorum*. As the experiment continued to 120 days of storage, although deterioration in the control tubers was presented due to physiological changes and infection by decay organisms, treatment of these tubers with CVO significantly reduced their deterioration compared to control. Treatment of the tubers with CVO completely inhibited tuber sprouting and reduced weight loss. Moreover, CVO can reduce the degradation of the tubers' nutrients such as carbohydrate, inulin, and protein contents compared to control (Ghoneem et al., 2016).

Basil downy mildew, caused by the fungus-like oomycete pathogen, *Peronospora belbahrii*, becomes a destructive sweet basil disease. It was found that 20% of CVO showed moderate antifungal activity with the area under the disease progress curve value of 2577 (Homa et al., 2014). In grape fungus infection caused by *Plasmopara viticola*, copper oxychloride and copper hydroxide were used to control it. However, since copper is incompatible with the organic farming philosophy, some biopesticides were introduced instead. There was a study that compared the efficacy of copper and CVO against *Plasmopara viticola*. It was found that the inhibitory activity of 20% CVO, although the effect was lower than that of copper, controlled the development of downy mildew (Scalzo et al., 2012). Examples of CVO-based commercial products as fungicides are Sporatec and Bioxeda.

There was a report about limitations in the use of CVO. Even CVO can inhibit *Alternaria* spp. (Alternaria fruit rot), *Botrytis cinerea* (gray mold), and *Colletotrichum* spp. (ripe rot), which causes post-harvest disease on blueberry fruit. It negatively impacted sensory qualities and failed to increase blueberries' antioxidant level during post-harvest fumigation under refrigerated holding conditions (Mehra et al., 2013).

The QSAR of eugenol and antifungal activity were studied. It was suggested that enzymatic oxidation reaction played a crucial role in the mechanism of antifungal activity. The competitive reaction to the binding of the inhibitors to the thiol groups of cysteine residues in the fungal enzymes' active sites was possibly inactivating phenolic inhibitors, such as in the *Trametes versicolor* laccase. Moreover, it was found that oxygenated aromatic compounds which contained oxygen in the second position of the aromatic ring such as eugenol possessed more antifungal activity than compounds that had oxygen at a fourth position such as vanillin. However, the activity decreased when two or more oxygen-containing groups presented in the aromatic ring (Voda et al., 2004).

10.4.2 Insecticides

The primary pest occurring in rubber tree plantations in Brazil is *Tenuipalpus heveae* Baker. This species causes significant damage to the plant resulting in a decrease in latex yield. CVO showed insecticidal activity against this pest in a dose-dependent manner. The highest mortality was observed in the pest exposed to a concentration of 100 μL/L (De Melo et al., 2019).

CVO delayed the development of bean weevil, *Acanthoscelides obtectus* (Jumbo et al., 2014). Moreover, CVO had been tested for insecticidal activity on the bean weevil. CVO containing 92.94% eugenol possessed the LD_{50} at 43.6 μL/kg in bean weevil and steadily decreased its growth rate in a dose-dependent manner. The insecticidal activity decreased over time. Similarly, CVO was tested at doses of 35, 17.9, 8.9, 3.6, 1.8, 0.4, and 0.2 μL/g for maize weevil *Sitophilus zeamais* under laboratory conditions. CVO caused mortality of 100% for a species 48 h after treatment with concentrations of 17.9 μL/g. The LD_{50} for *Sitophilus zeamais* was 10.15 μL/g (Jairoce et al., 2016). An example of CVO-based commercial products as insecticides is Pest Out.

10.4.3 Insect repellants

CVO and its components had been tested on mites, permethrin-sensitive *Sarcoptes scabei* var. *suis*, and permethrin-resistant *Sarcoptes scabei* var. *canis*, separately at concentrations of 1.56%, 3.12%, 6.25%, 12.5%, and 25%. All insects contacted with CVO resulted in 100% mortality of permethrin-sensitive mites after 0.25 h. Permethrin-resistant mites died at the same time but required higher concentrations ($\geq$6.25%) of volatile oils. The acaricidal activities of eugenol, eugenyl acetate, isoeugenol, and methyl eugenol had been tested in contact bioassays against the two scabies mites. There was no significant difference between the activity observed for the positive control acaricide (benzyl benzoate) and the test compounds eugenol, eugenyl acetate, and isoeugenol. However, methyl eugenol had no acaricidal effect in sensitive mites after 1 h, but it showed the activity in resistant mites after 24 h at the highest concentration (100 mM) (Pasay et al., 2010).

The elimination of house dust mites, *Dermatophagoides pteronyssinus*, and direct contact using CVO were evaluated in three experiments. Firstly, synthetic fibers had been immersed in 2% CVO for 30 min and dried in a hot-air oven at 60°C for 2 h. Then, 0.5 g of dust mites were exposed to the coated fibers placed in the Siriraj chamber, restricting the mites to a specified area to experiment. Other experiments were performed using 10 mites placed in the chamber, and 10 μL of CVO was pipetted or sprayed onto them. These two procedures were carried out for three consecutive days at 0, 1, 3, and 6 months. The effectiveness of pipetting and spraying was 99% and 81%, respectively, while the placebo mortality was <5%. The results derived from inspecting the microscope suggested that coating synthetic fibers with CVO may also be an effective procedure for eliminating mites and their eggs. CVO may be applied to the mattress to enhance protection from mites and their allergens, thereby reducing the risk of developing allergic symptoms (Mahakittikun et al., 2013).

Evaluation of the repellency of CVO against *Periplaneta americana* had shown complete repellency against first instar nymphs at the concentration of 2% after 48 h. CVO ultimately repelled all fourth instar nymphs after 24h at a concentration of 8%. For the adult stage, the most significant repellency percentage was 90% after 48h at a concentration of 10%. CVO also provided fumigant toxicity against nymphs and adults of *Periplaneta americana* after 24 and 48h, respectively. Total mortality was recorded at a concentration of 7.5μL/L of air for first instar nymphs, 10μL for fourth instar nymphs, and 17.5μL for adults after 48h of fumigation (Omara et al., 2015).

CVO from leaves at the concentrations of 1%, 5%, 10%, and 15% had been evaluated in vitro for acaricidal activity against soft ticks, *Argas* spp., collected from poultry flocks. CVO at 15% concentration killed 80% of ticks in 24h, with the LD50 of 3.15% (Hasson and Al-Zubaidi, 2015). Eugenol had been tested against the poultry red mite, *Dermanyssus gallinae*. The undiluted solution provided 100% mortality in toxicity test, or at 1% of this dose, eugenol was 20% effective. LD_{50} values confirmed the superior efficacy of eugenol over both citral and geraniol (Sparagano et al., 2013). Examples of CVO-based commercial products as insect repellant are Auntie's and Pure Planet.

10.4.4 Mechanisms

The mechanism of eugenol for insecticidal activity was studied in *Periplaneta americana*, *Blattella germanica*, and *Camponotus pennsylvanicus* De Geer. It was reported that changes in the frequency of heartbeat, changes in the cAMP level in the nervous system, and decreased binding to octopamine receptors in *Periplaneta americana* in response to treatment with eugenol supported the involvement of octopamine receptors in the toxicity of eugenol (Enan, 2001).

In silico identification of protein target involving the insecticidal activity of eugenol was investigated. Eugenol derivatives were docked into each target with six different docking programs (PLP, ASP, ChemScore, GoldScore, Vina, and LeDock). Odorant-binding proteins were the target with the highest affinity toward eugenol derivatives, followed closely by acetylcholinesterase, chitinases, and octopamine receptors (Vieira et al., 2020).

The insect tolerance behavior to CVO occurred in the maize weevil *Sitophilus zeamais* population that resists phosphine and pyrethroid insecticides. From the study, the population growth rate of *Sitophilus zeamais* was less affected by CVO. Insects from this population reduced their respiratory rates after being exposed to CVO and changing their walking behavior to volatile oils treated surfaces. The physiological actions of volatile oils are related to interrupting the GABAergic receptor (Priestley et al., 2003). Although the primary molecular target of pyrethroids is voltage-gated sodium channels, a high amount of pyrethroids also target the GABAergic receptor (Soderlund, 2012; Clark and Symington, 2012; Casida and Durkin, 2013), which could help to explain why the specific *Sitophilus zeamais* population was significantly less susceptible to CVO action (Correa et al., 2015).

10.5 Clove volatile oil as clearing agents in tissue histology

In tissue histology, the step between dehydration and infiltration with embedding medium is referred to as clearing. It is a process that requires the use of solvents known as clearing agents that are miscible with both absolute alcohol and the embedding medium. CVO from buds is generally used as a clearing agent in animal and plant tissue histology. Compare to xylene, CVO is safer, nonflammable, and minimal in toxicity. However, CVO can also remove safranin stain; thus, it cannot be used in all cases (Lapin and Yakovleva, 1957; Adeniyi et al., 2016).

10.6 Clove volatile oil as a source of eugenol and β-caryophyllene

It was mentioned in Section 2 about the utilization of CVO from buds, twigs, and leaves for different purposes, such as CVO from leaves is a good source for eugenol extraction or CVO from buds and twigs can be used in food processing. However, the data from Tables 1 to 6 have provided us more exciting information. From Table 1, it was found that CVO from Chennai may not be suitable for food flavoring since it contained a low amount of eugenyl acetate, making it a harsher note. Moreover, CVO from buds and twigs in some locations (Ambon, Sippighat, and Ram Krishna Pur) was also a good source for β-caryophyllene extraction since the amount was closed to CVO from leaves. In contrast, CVO from leaves in places such as Nicobar and Chittagong possessed a low amount of β-caryophyllene. Although Zanzibar clove variety is the famous one, from Table 5, it was found that Tuni variety may be a better source of eugenol and β-caryophyllene. It was shown that CVO from twigs in the Tuni variety possessed the highest eugenol content (98%), while CVO from its leaves contained the highest β-caryophyllene content (21%). For specific compound extraction, it was concluded from Table 6 that clove buds were collected from full budding to full fruit stages, and clove leaves were collected from expanding-2 to mature leaves stages for eugenol extraction. For β-caryophyllene, stage-3 buds and expanding-1 leaves stages are considered as the best time to harvest the clove materials for extraction. For eugenyl acetate, it was appropriated to collect the buds and leaves when they are in young buds or stage-1 buds stages, or young leaves stage.

Firstly, CVO is extracted from the clove materials using distillation in the eugenol extraction and isolation processes. The CVO is subsequently mixed with a 3% sodium or potassium hydroxide solution to change the compounds in phenolic alkali salt form. The insoluble portion was discarded, while the alkaline solution is then acidified at a low temperature (4°C). The isolation of eugenol was performed by various techniques such as fractional distillation or liquid chromatography. There is a patent related to eugenol isolation from CVO, such as CN103819317A. The invention provides a method of producing eugenol from CVO from twigs with a simple and efficient extracting process (Xijun and Haijun, 2014).

β-Caryophyllene is used for designating the food flavor. It is obtained through vacuum distillation, and then eugenol is removed from clove twigs oil afterward. The high purity β-caryophyllene obtained from this method was reported as 94% (Alias, 2013; Yun and Jilin, 2012).

To the best of my knowledge, there is no large-scale method to isolate eugenyl acetate from CVO. Suitable materials for eugenyl acetate isolation are young buds and leaves, but the farmers typically harvest clove materials in mature stages when the eugenyl acetate level is naturally low. A specific supercritical CO_2 extraction method provided the mixture of eugenol and eugenyl acetate from clove buds (Yazdani et al., 2005). Generally, eugenyl acetate has been prepared following either biocatalytic or chemical approaches. The use of supported catalysts was optimized to develop alternative cleaner, safer, and less environmentally aggressive processes (Lerin et al., 2015; Tischer et al., 2019). Clinical study of eugenyl acetate has been limited. Nevertheless, there were reports about its users, which may be insight for future applications (Musthafa and Voravuthikunchai, 2016; Saraphanchotiwitthaya et al., 2019).

11. Prospect in clove volatiles application

This section presented an overview of potential or future uses of CVO based on the safety and efficacy data of the volatile oil and its composition.

11.1 Clove volatile oils safety

The Organization for Economic Co-operation and Development (OECD) classified eugenol as a mild skin irritant R38 UN GHS cat.3 (OECD, 2010). The irritation of human skin was found in 2 of 25 normal subjects using 20% CVO in Vaseline in a closed-patch test, but the erythema was not found in 65 subjects using 0.2%–2% CVO (Opdyke, 1975).

Although the low toxicity of CVO against humans and the relative harmlessness of the product residues are represented, the US Food and Drug Administration (FDA) noted that CVO might not be GRAS if they are used outside the scope described in the CFR title 21 section 184.1. Historically, CVO and isoeugenol have been used in foods, and eugenol has been used in animal foods. However, concerns regarding these compounds led to the submission of eugenol, isoeugenol, and methyl eugenol for investigation under the National Toxicology Program (NTP). The NTP conducts studies on compounds and extracts to determine their possibility to cause cancer. NTP concluded that eugenol is not clear as a carcinogen, but isoeugenol and methyl eugenol are carcinogenic to rodents (National Toxicology Program, 2010). Although the NTP concluded that the evidence of carcinogenicity in mice for eugenol was not clear, the lack of a dose-response in male mice, the lack of statistical significance in female mice, and a high level of background hepatocellular neoplasms in mice strain used in the study indicated that the hepatocellular neoplasms in the study were likely not related to administration of eugenol (Gooderham et al., 2020). The hepatotoxicity of isoeugenol in tested mice was also raised, and it was later concluded that the result was a secondary biological response to dose-dependent hepatotoxicity. Thus, it was irrelevant to humans who consume isoeugenol at a low level (<0.1 mg/kg body weight per day) from intended use in food flavoring. The toxicity occurrence was presented at a high-dose isoeugenol for histiocytic sarcoma, which was 3,750,000 times greater than the dose levels in estimated daily per capita intake of isoeugenol as a flavoring agent in the United States (0.00002 mg/kg BW/day) (Smith et al., 2009).

The genotoxicity assay results were reported, and it was found that CVO from leaves showed a negative result for mutagenicity in a good laboratory practice (GLP) compliant Ames assay. Moreover, CVO from buds was absent in clastogenic and aneugenic activities in the in vitro chromosomal aberration assay and the in vivo micronucleus assay (DeGraff, 1983; Ishidate et al., 1988). These negative results were consistent with analogous studies for eugenol (FDA, Scientific Literature Review of Eugenol and Related Substances. PB283-501/AS, 1978; JECFA, 1982; Rietjens et al., 2014).

An estimated acceptable daily intake of eugenol is up to 2.5 mg/kg body weight for humans (Reynolds, 1989). Limitation of methyl eugenol intake was concluded in the Expert Panel report of the Flavor and Extract Manufacturers Association (FEMA). Considering the substances containing methyl eugenol, the estimated intakes of CVO from twigs and leaves are below the conservative threshold of toxicological concern (TTC) of 0.15 μg/person/day. T estimated intakes exceed the TTC of 0.15 μg/person/day (Gooderham et al., 2020). The criteria were related to different contents of methyl eugenol in buds, twigs, and leaves, as shown previously in Table 4. From Tables 1–3, it was found that CVO from buds originated from Indonesia and Madagascar possessed the same methyl eugenol content (around 0.1%). Methyl eugenol content was higher in CVO from leaves from Indonesia than Madagascar, but the occurrence was the opposite in CVO from twigs.

From Table 4, it was found that CVO from buds, twigs, and leaves contained a high amount of eugenyl acetate and β-caryophyllene. Moreover, caryophyllene oxide, α-humulene, α-copaene, methyl salicylate, and chavicol can be found in all clove plant parts. Farnesene, α-cadinene, and δ-cadinene contents were more than 1% in CVO from buds. Thus, the author takes them into account for safety consideration.

Methyl salicylate in minute amounts caused no toxicity (FDA, 2020b,c). The committee in 57th meeting of the Joint FAO/WHO Expert Committee on Food Additives (JECFA) retained the previously established daily intake of 0–0.5 mg/kg of body weight for methyl salicylate (Additives, 57th report of the Joint FAO/WHO Expert Committee on Food Additives, 2002). From Table 4, CVO contained a low level of methyl salicylate (less than 0.6%). From the 63rd meeting of JECFA, β-caryophyllene, α-cadinene, and δ-cadinene were considered as no safety concern (Additives, 63rd report of the Joint FAO/WHO Expert Committee on Food Additives, 2004). Later in the 65th meeting of the JECFA, eugenyl acetate and caryophyllene oxide were included in the no safety concern list (Additives, 65th report of the Joint FAO/WHO Expert Committee on Food Additives, 2005). According to the JECFA and the European Food Safety Authority (EFSA), it is recommended to use 1–5 mg/kg of eugenyl acetate for food products (European Food Safety Authority, 2011). Re-evaluation of farnesene safety was also confirmed that it is safe for a food additive. Chavicol, or 4-allylphenol, was listed as no safety concern in the 69th meeting of the JECFA (Additives, 69th report of the Joint FAO/WHO Expert Committee on Food Additives, 2009). In the 78th meeting of the JECFA, additional data about β-caryophyllene were added. No evidence of mutagenicity was observed in Ames assays when β-caryophyllene up to 150,000 μg/plate was incubated with *Salmonella typhimurium* with and without metabolic activation. Moreover, β-caryophyllene did not induce acute or subchronic toxicities in rats (Additives, 78th report of the Joint FAO/WHO Expert Committee on Food Additives, 2014).

The toxicological properties of many compounds have not been thoroughly investigated. Some product datasheets, α-humulene or α-caryophyllene, can cause eye and skin irritation. It may be associated with reactive oxygen species overproduction and subsequent oxidative damage to proteins and lipids (Agus, 2021). For α-copaene that was not previously assessed for use in food, the QSAR analysis using Toxtree was made for potential genotoxicity and mutagenicity. No alerts were identified (FEEDAP, 2015). The experiment also confirmed that α-copaene did not show a genotoxic effect against neuron and N_{2a} neuroblastoma cell lines (Turkez et al., 2014).

From clinical reports, there are safety data in special populations and situations. CVO is contraindicated in case of hypersensitivity to CVO and Peru balsam due to cross-reactivity. It is not recommended to use in women with pregnancy and lactation since no data is available; however, in amounts commonly found in food is acceptable (monograph ESCOP). Overdosing at about 500 mg/kg in 7-month-old boy caused metabolic acidosis, leukocytosis, proteinuria, and ketonuria (Lane et al., 1991). A 15-month-old boy developed a fulminant hepatic failure after ingestion of 10 mL of CVO (Janes et al., 2005). For herb-drug interactions, taking cloves along with diabetes medications might cause your blood sugar to go too low. Moreover, the antiplatelet effect of CVO may increase the risk of bleeding if taken with antiplatelet drugs. Clove may also result in a false increase in phenytoin levels (Gruenwald et al., 2004).

The safety data are not entirely conclusive, whereby chronic toxicity, cytotoxicity, and genotoxicity of CVO still require more research in specific organisms. Furthermore, effective and safe concentrations and the quality standards of CVO for each future application must be established.

11.2 Food preservative

Although most food preservatives are classified as GRAS, there are problems concerning the safety of some chemicals, such as carcinogenicity of nitrosamines and allergic trigger of benzoic acid and sulfites. Concerning the safety of food preservatives is increasing, research has been performed to find safe food preservatives. Natural antimicrobial agents derived from plant sources such as volatile oil have been used for a long time. CVO possesses antioxidant, antibacterial, and antifungal activities, which means that it can be considered a food preservative, increasing the shelf life of foods (Nurdjannah and Bermawie, 2012).

11.2.1 Antioxidant

Oxidation of lipids in the process causes rancidity in food products. The search for natural antioxidants has been intensive. CVO possesses strong antioxidant property. Antioxidant activity of CVO was higher than the standard butylated hydroxytoluene; eugenol and eugenyl acetate were identified as the major antioxidants (Kramer, 1985). The antioxidant activity of eugenol and eugenyl acetate was similar to that of α-tocopherol (Lee and Shibamoto, 2001). Eugenol and isoeugenol also inhibit the peroxidation of lecithin induced by the Fe^{2+}-H_2O_2 system (Toda et al., 1994). The clove/cinnamon oil combination showed synergistic antioxidant potential in the volatile oils combination test, leading to more potency in antimicrobial activity, thus increasing food stability (Purkait et al., 2020). The presence of various aroma compounds may explain the improvement of food shelf life.

11.2.2 Antibacterial activity against foodborne pathogens

CVO was reported to have antibacterial activity against common food pathogens such as *Bacillus subtili*, *Campylobacter jejuni*, *Escherichia coli*, *Halomonas* spp., *Klebsiella pneumoniae*, *Listeria monocytogenes*, *Proteus* spp., *Pseudomonas aeruginosa*, *Salmonella enterica*, *Staphylococcus aureus*, and *Yersinia enterocolitica*. Moreover, it possessed activity against rare foodborne bacteria strains such as *Chromobacterium violaceum*. The reviewed literature found that CVO is among the most inhibitory volatile oils, owning a bacteriostatic concentration of not more than 0.075% against five food pathogens (*Campylobacter jejuni*, *E. coli*, *L. monocytogenes*, *Salmonella enterica*, and *Staphylococcus aureus*) (Palmer et al., 1998). Moreover, CVO from buds displayed effectiveness against another food pathogen such as *Bacillus subtili* (Ame et al., 2007). CVO also possessed antibacterial activities, with IC_{50} values of 1.5 μL/mL against *Klebsiella pneumoniae* and *Proteus* spp., and 6.25 μL/mL against *Yersinia enterocolitica*, so it can be used to prevent foodborne diseases (Al-Mariri and Safi, 2014).

Concerning *L. monocytogenes*, contamination in the food has gained much consideration, especially for food preserved in refrigerators, because this bacterium remains able to grow at low temperatures (Juntila et al., 1988). It was found that CVO was very effective against *L. monocytogenes*. It was recently reported that CVO could destroy the dense membrane structure of *L. monocytogenes* biofilm and reduce the bacteria's metabolic activity by 60.2% (Zhang et al., 2020). A study based on various *L. monocytogenes* strains revealed that the minimum inhibitory concentration (MIC) of CVO to 10 strains of *L. monocytogenes* ranged from 1.2 to 2.5 μg/mL (Hu et al., 2018). The activity was investigated in many kinds of food. The antibacterial activity of CVO against *L. monocytogenes*, added to bovine ground meat stored under refrigeration for 3 days was reported, and MIC was 1.56% (De Oliveira et al., 2013).

Moreover, CVO reduced the bacterial population to 1 mL/L in hotdogs with different levels of fat content. Raising the concentration to 10 mL/L had no significant effect on the full-fat hotdog, indicating that the antimicrobial activity was related to the fat content (Singh et al., 2003). Not only at low temperatures but it was also reported that clove oil at 1% concentration could inhibit the growth of *L. monocytogenes* at 30°C (Menon and Garg, 2001). It was also said that the MICs of CVO to different

Salmonella enterica strains ranged from 0.6 μg/mL to 2.5 μg/mL. Therefore, it should be noted that strain biodiversity has an essential impact on the antibiotic activity of CVO to foodborne microorganisms (Mazzarrino et al., 2015).

Psychrophilic bacteria, which are harmful to humans, can grow at low temperatures causing egg spoilage. It was reported that CVO at 40 μg/g could be used to prevent psychrophilic bacteria growth in eggs after 45 days of storage in the refrigerator. The flavor and taste of the boiled egg after storage were as same as the control. In addition, the physicochemical properties of the eggs during storage met the standard criteria. The primary constituent identified from the eggshell when kept in CVO was eugenol (Phothisuwan et al., 2018).

The shelf life of salt-cured fish was limited mainly due to red halophilic bacteria. CVO can be used to control the growth of red halophile in cured fish. It was reported that treatment with CVO at 0.02% and 0.5% concentration caused a reduction in the counts of red halophile bacteria and complete elimination of *Halomonas* spp. Also, CVO at a concentration of 0.02% exposed to brine overnight resulted in the complete elimination of red halophiles (Prasad and Seenayya, 2000).

Infection caused by *Chromobacterium violaceum* from food is very uncommon. However, it was reported to recover from many kinds of food, such as salad (Alfaro et al., 2019). Quorum sensing is a communication system correlating to population density. Many species of bacteria use quorum sensing to coordinate gene expression for biofilm formation and drug efflux pumps. CVO showed promising anti-quorum sensing activity against *Chromobacterium violaceum* (Khan et al., 2009).

Moreover, the three major quorum-sensing systems that control the virulence and biofilm formation in *Pseudomonas aeruginosa* are LasI-LasR, RhlI-RhlR, and PQS-MVFR. It was found that CVO significantly affects the activity of signal molecules of the Las and Rhl pathways. Furthermore, molecular modeling demonstrated that eugenol bonded to the amino acid residues Arg61 and Tyr41 of LasR receptor by hydrogen bond and hydrophobic interactions, which played a significant role in the activity (Jayalekshmi et al., 2016).

Eugenol possessed antibacterial activity in various foodborne pathogens; there were extensively studied in *E. coli*, *Salmonella* spp., and *Staphylococcus* spp. It was found that *E. coli* was sensitive to eugenol. The highest bactericidal activity of eugenol was detected when *E. coli* was exposed to eugenol for the first 10 min. The biofilm morphology and the cell viability detection indicated that a sufficiently high concentration of eugenol inhibited the biofilm growth and boosted the number of dead cells (Pérez-Conesa et al., 2011). Eugenol treatment significantly reduced the number of *Salmonella enteritidis*, *Salmonella heidelberg*, and *Salmonella typhimurium* on turkey cutlets in a concentration-dependent manner (Nair et al., 2014). Moreover, eugenol can significantly inhibit the growth of *Staphylococcus aureus*, *taphylococcus carnosus*, and *Staphylococcus epidermidis* (Morán et al., 2014; Manrique et al., 2017).

The mechanism of the antibacterial activity of eugenol was studied. It was found that eugenol possessed the ability to induce cellular protein and lipid leakage in both Gram-negative such as *E. coli* and *Proteus vulgaris* and Gram-positive such as *L. monocytogenes* and *Streptococcus pyogenes*. Eugenol had a damaging effect on cell walls and caused protein leakage in the range of 120–325 μg/mL at two × MIC concentrations. Treatment of bacteria with eugenol showed lipid leakage ranging from 120 to 220 μg/mL at two × MIC and 80 to 170 μg/mL at 1 × MIC (Oyedemi et al., 2009).

Moreover, in *E. coli*, it was found that eugenol significantly inhibited the expression of migration-related genes *fimA*, *fliC*, *hcpA*, and *lpfA*, which encode type 1 fimbriae, flagellin A, hemorrhagic coli pilus, and long polar fimbriae, respectively. Eugenol also inhibited adhesion genes *eae* and *ler*, which are the genes responsible for bacterial adhesion (Baskaran et al., 2016). The mechanism of eugenol in the inhibition of migrating gene expression may be linked with the constant and competitive binding of eugenol to the quorum sensing receptor and leading to significant suppression of bacterial quorum sensing (Rathinam et al., 2017).

Eugenol inhibited the biofilm synthesis and inactivated the fully formed biofilm of *L. monocytogenes*, affecting the downregulation of biofilm-associated gene *dnaK*, bacterial adhesive related genes *flaA*, *fliP*, *flgE*, and *motA*, transcriptional activator gene *prfA*, and quorum sensing related genes (Upadhyay et al., 2013).

In *Salmonella* spp., the antibacterial mechanism of eugenol was revealed by ethidium bromide determination using Spectrophotometry. It was shown that 500 mg/mL concentration considerably inhibited the accumulation of ethidium bromide, which suggested that eugenol expressed the inhibition against efflux pump of *Salmonella typhimurium*. Moreover, eugenol concentration in the range of 128–256 μg/mL can destroy biofilm integrity, disrupt the bacterial membrane, and caused cytoplasm leakage (Miladi et al., 2017).

11.2.3 Antifungal activity against foodborne pathogens

Many studies suggested that CVO exhibited antifungal activity against foodborne fungi such as *Aspergillus* spp., *Penicillium* spp., and *Eurotium* spp., which often contaminate bakery products. Although aatoxin is a toxic compound produced by foodborne fungi such as *Aspergillus flavus* and *Aspergillus parasiticus*, CVO at 0.6 mg/mL showed inhibitory activity against aflatoxin (Farag et al., 1989). Moreover, it was found that CVO possessed complete inhibition against *Aspergillus niger* (Ame et al., 2007). For *Penicillium* spp., it was found that the inhibition against mycelia of *P. digitalis* and *P. italicum* happened upon direct contact with the oil or vapor of CVO. Another fungus from food contamination, namely *Mucor* spp., was found susceptible to CVO (Rana et al., 2011).

Eugenol possessed an MIC value of 600 μg/mL against *spergillus ochratus* (Hua et al., 2014). Moreover, it significantly inhibited *Aspergillus parasiticus* growth at a 150 μg/mL concentration and completely inhibited the radial growth of *Aspergillus parasiticus* after incubation for 20 days (Pillai and Ramaswamy, 2012). The result of the mycelia dry weight of this fungus showed that eugenol inhibited its development in a dose-dependent manner in the concentration range of 19.16%–95.83% (Jahanshiri et al., 2015).

The mechanism of antifungal activity of eugenol was strongly correlated to suppression of gene expression in aflatoxin biosynthesis, including *aflD*, *aflM*, *aflR*, *aflP*, *aflT*, *omtA*, *pksA*, *nor-1*, and *ver-1*. In *Aspergillus parasiticus*, *omtA* and *pksA* were the most affected genes, which showed significant reduction at mRNA level when exposed to eugenol. In the concentration range of 15.07% to 98.0%, eugenol can inhibit the biosynthesis of aflatoxin B1 in this fungus (Jahanshiri et al., 2015). In *Aspergillus flavus*, eugenol at 0.80 mM completely inhibited the aflatoxin B1 biosynthesis with a rate of 95.4%. The mechanism was related to the downregulation of the transcription level of *aflD*, *aflM*, *aflR*, *aflP*, and *aflT* genes, and among these genes, *aflP* has highly downregulated about 2061 folds (Liang et al., 2015).

Foodborne yeasts, excluding yeasts used as starter cultures, commonly cause food spoilage. Along with *Candida albican*, it was reported that *andida krusei*, *Candida lusitaniae*, *Candida famata*, *Candida colliculosa*, *Candida parapsilosis*, and *Candida tropicalis* were in a group at risk of potential foodborne pathogens (Rajkowska and Kunicka-Styczniska, 2018). Furthermore, it was found that CVO at

68.4%–84.2% concentration showed a significant reduction in yeast biofilm formation. Therefore, CVO can be used as effective antibiofilm agent to prevent materials colonization by *Candida* spp. (Rajkowska et al., 2019).

It was suggested that a most promising approach would be combining CVO with other preservation methods such as low temperature, low O_2, or high-pressure techniques to develop a synergistic alternative for food preservation (Nurdjannah and Bermawie, 2012). However, based on antibacterial and antifungal activities, it was suggested that CVO and eugenol alone could also be used as a food preservative. Only some topics needed to strengthen, such as precise antimicrobial spectrum and effective concentration, safety concentration, and the quality standards of CVO and eugenol used for food preservation.

Instead of CVO fortified directly into food products, CVO can be incorporated into packaging materials. For example, chitosan film mixed with CVO at 2.5% w/w was in good physical properties and possessed high antioxidant and antimicrobial activities (Lekjing and Chinarak, 2018). On the other hand, the polyvinyl alcohol film containing 9% CVO showed the best quality protection. In addition, t microbiological shelf life could be extended for 2 days, and a 28.07% reduction of malonaldehyde (product from lipid oxidation) was obtained on day 7 (Chen et al., 2017).

11.3 Pharmaceuticals

Several investigations have documented the therapeutic potential of CVO. The pharmaceutical functions of CVO have also been thought to be due to the existence of its major compounds such as eugenol, β-caryophyllene, and eugenyl acetate. This section will summarize the possibilities of CVO to develop into preparations to encounter some diseases.

11.3.1 Pediculicidal shampoo

Head lice, or *Pediculus humanus capitis*, are an ectoparasite of humans that causes pediculosis capitis, with a high prevalence in children. *Pediculus humanus capitis* has been treated by methods that include physical removal and prescription medication. However, head lice resistance to conventional insecticides such as malathion has increased. Moreover, absorption of synthetic chemicals from the skin has the potential risk of chronic diseases. Thus, new topical treatments are needed. *Syzygium aromaticum* herbal shampoo against head lice has been tested on children to assess an in vitro efficacy. The results showed that the herbal shampoo at $6\,\mu L/cm^2$ is more effective than malathion shampoo at $3\,\mu L/cm^2$. The highest pediculicidal activity was the herbal shampoo at a dose of $6\,\mu L/cm^2$, with 100% mortality at 10min, an LT_{50} value of 0.9min, and an LC_{50} value of $2.3\,\mu L/cm^2$. However, malathion shampoo showed low toxicity, with 34.5% mortality at 10min and an LT_{50} value of 35.9min during the experiment periods. Thus, it was established that the herbal shampoo based on *Syzygium aromaticum* has the potential to alternative pediculicides for head lice treatments in children (Soonwera and Phasomkusolsil, 2016).

11.3.2 Dental formularies

Candida albicans is a well-known human pathogen that causes superficial infections such as oral candidiasis. It has been reported that approximately 90% of patients with AIDS are being affected with oral candidiasis (Musthafa and Voravuthikunchai, 2016). CVO from buds had antibiofilm activity against the potent biofilm-forming strains of *Candida albicans* at half MIC value and showed better activity than nystatin, as determined by a well diffusion method (Khan and Ahmad, 2012; Mansourian et al., 2014). CVO from leaves also showed effectiveness against *Candida albicans* at a concentration of 17.86% (Musta and Nurliana, 2019). Eugenol and eugenyl acetate contribute to the activity (He et al., 2007; Musthafa et al., 2016). Pastille or mouthwash containing CVO can be formulated for oral candidiasis treatment.

A canker sore is a painful and inflamed sore in the mouth. It may occur with viral infections and may also relate to an immune system problem. CVO provides an anti-inflammatory action and analgesic benefit. The anti-inflammatory activity of CVO is mainly because of eugenol. The eugenol mechanisms are primarily due to its inhibitory effect on prostaglandin synthesis and neutrophils/macrophages chemotaxis (Kim et al., 2003). Moreover, eugenol exerts analgesic activity due to its antinociceptive capacity via TRPV1 receptors (Kumarswamy, 2016). Therefore, a herbal gel containing CVO may be suitable for the treatment.

It was found that eugenol is more effective at reducing pain, inflammation, and infection than 0.2% chlorhexidine gel. In addition, t patients with alveolar osteitis who used the eugenol-based paste also had better wound healing than patients who used the other treatment or no treatment (Jesudasan et al., 2015).

11.3.3 Ear preparation drugs

Acute external otitis refers to inflammation of the external ear canal which may end up in loss of hearing (Flint et al., 2010). *Pseudomonas aeruginosa* is the most common pathogen responsible for about 40% of illnesses, followed by *Staphylococcus aureus*, which accounts for 9% of the cases. Other pathogens such as *S. epidermidis*, *Proteus* spp., and other Gram-negative *bacilli* and *Candida albicans* have also been reported from acute external otitis cultures (Osguthorpe and Nielsen, 2006; Van Balen et al., 2005). The first-line drugs in otitis treatment are the use of topical antibiotics. Moreover, alternative herbal preparation was also reported to be effective in curation. A combination herbal drop composed of volatile oils from *Syzygium aromaticum*, *Lavandula angustifolia*, and *Geranium robertianum* exhibited good efficacy in reducing the burden of infection and alleviating acute external otitis symptoms in patients (Panahi et al., 2014).

11.3.4 Cancer drugs

Several studies have been investigated on the anticancer activity of CVO. It was reported that CVO and eugenol possess anticancer activities against breast, cervical, colorectal, leukemia, liver, and lung cancer cell lines in vitro (Kouidhi et al., 2010; Kumar et al., 2014; Jaganathan and Supriyanto, 2012; Vidhya and Devaraj, 2011; Al-Sharif et al., 2013; Jaganathan et al., 2011; Hussain et al., 2012; Majeed et al., 2014). However, to conclude that CVO can be used for cancer treatment, all evidence still needs more support data from a clinical study. The mechanism of anticancer properties of CVO was investigated by mathematical modeling. It was predicted that the activities are involving many enzymes such as dehydrocholesterol reductase in cholesterol biosynthesis, dehydrofolate reductase in carbon metabolism, and serine palmitoyl-transferase long chain in sphingolipid biosynthesis (Azadi et al., 2020). There was an idea to prepare CVO in the nanoscale emulsion to improve its efficacy against cancer (Nirmala et al., 2019). Eugenyl acetate was recommended as a new filler biomaterial to reconstruct the tissue damage occurring after the resection in case of bone cancer (Vladu et al., 2020).

Cervical cancer is a significant health problem since it is the most frequent female genital tract cancer. The primary etiology is infection by Human Papillomavirus. Cervical cancer prevention using vaccination is one strategy to reduce its burden. However, worldwide shares a high commitment to this disease, vaccination is not fully available (Ferlay et al., 2015; Parkin et al., 2014). Alternative topical drugs such as 5-Fluorouracil, imiquimod, cidofovir, and retinoic acid can be applied to the cervix to induce regression of the lesions. Moreover, it was reported that AV2, a mixture of FDA-approved natural volatiles (eugenol, carvone, geraniol, nerolidol), was effective in patients with cervical lesions. Therefore, V2 was proposed for topical preparation for cervical cancer prevention with only mild local side effects such as burning, irritation, and itching (Mutombo et al., 2019).

11.3.5 Antibiotics for urinary tract infections

Proteus mirabilis is a Gram-negative bacteria naturally related to human urinary tract infections (Dayton et al., 1996). It is one of the most common causes of urinary tract infection in individuals with long-term indwelling catheters and bacteremia in the elderly (Burall et al., 2004). Urinary tract infection is generally managed with antibiotics, which are quite effective without complications. However, bacteria can resist antibiotics. Recently, the acceptance of herbal medicine for health care and the development of microbial resistance to antibiotics have led researchers to investigate the potential of medicinal plants for infectious treatment (Sosa and Zunino, 2009). Eugenol has potential antibacterial activity against *Proteus mirabilis*. The MIC and MBC experiments showed that 0.125 and 0.25% (v/v) were the bacteriostatic and bactericidal concentrations of eugenol against *Proteus mirabilis*. Furthermore, it inhibited the virulence factors of *Proteus mirabilis*, interacted with the cell membrane rapidly, changing the membrane permeability, and releases the intracellular components resulting in cell death (Devi et al., 2013).

11.3.6 Diabetes drugs

Diabetes mellitus is one of the most critical health problems of this century (Zimmet et al., 2014). It is often associated with liver dysfunction, diabetic nephropathy, and dyslipidemia (Taskinen, 2003; Chapman et al., 2011; Scholze et al., 2016). Although the drugs used in diabetes treatment are remarkably effective, they cannot be administered to all patients. For example, iulin treatment can cause a high degree of glucose variability, enhancing the risk of hypoglycemia (Frandsen et al., 2016). Likewise, long-term administration of drugs such as biguanides, meglitinides, and sulfonylureas can cause hypoglycemia, weight loss, heart failure, and intestinal disorders. Therefore, developing a more effective therapy for the prevention and treatment of diabetes is essential (Nathan, 2015).

A clinical study in patients with type 2 diabetes concluded that clove could be used as an antihyperglycemic agent (Khan et al., 2006). Also, diabetic animals with eugenol reduced blood glucose levels and attenuated oxidative damage (Singh et al., 2016; Al-Trad et al., 2019). In the meta-analysis from animal model studies, it was summarized that eugenol affected diabetes by inhibiting carbohydrate-metabolizing enzymes and the possible inhibition of glucose production by the liver and the glucose uptake by skeletal muscle cells. Moreover, eugenol could have potent effects on diabetic oxidative stress, restoring redox status and slowing the disease progression. In addition, it may regulate lipid profile and subsequently modulated body weight (Carvalho et al., 2021).

11.3.7 Neuroprotective agents

Neuroprotection is a term used to refer to approaches that defend the central nervous system from neuronal injuries caused by protein aggregation, inflammation, oxidative stress, and neurotoxicity. Many drugs are approved for neuroprotection, such as donepezil, memantine, benztropine, and nortriptyline (Da Conceicao Machado et al., 2018). Moreover, β-caryophyllene was proposed as a neuroprotective agent. It was reported that β-caryophyllene inhibited iNOS in Wistar rats and nitric oxide production in BV2 microglial cells (Javed et al., 2016; Hu et al., 2017). Moreover, β-caryophyllene protected dopaminergic neurons and decreases microglia and astrocyte activation (Ojha et al., 2016). It can also upregulate the expression of Nrf2 and heme oxygenase 1 to ameliorate oxidative damage and neuronal apoptosis in Wistar rats (Lou et al., 2016). Thus, the available data suggest that β-caryophyllene affects the nervous system and can be helpful for the development of neuroprotective agents.

11.4 Cosmetics

The topical formulation is among the famous preparation for insect repellent. Even CVO is an ingredient in pesticides that are exempt from FIFRA under the Minimum Risk Exemption regulations, the skin irritation test of preparation must be evaluated in humans. It was reported that *Syzygium aromaticum* based repellent lotion with 7% CVO from leaves possessed good color and smelled with a value for the color of 6.8 and aroma of 6.6, which the consumer prefers. No irritation has occurred (Kurniasari, 2016).

11.5 Bacteria-resistant fabric

Clothing materials are in contact with human skin; however, they are also carriers of odor-generating bacteria and fungi because of the adhesion of microorganisms on the fabric surface. These bacteria cause staining, discoloration, degradation, and unpleasant odor of fiber. The fabric can be protected from microbial attack; it can be given finishes with antimicrobial agents. The synthetic antimicrobial agents used for fabric are organometallics and triclosan, toxic and nonbiodegradable (Sun and Worley, 2005). It was found that low concentration CVO with cross-linker helped develop the fabric with desired functional utility. In cotton fabric, a result showed that 0.5% CVO gave a 17 mm zone of inhibition against *Staphylococcus aureus* and *Klebsiella pneumoniae*, whereas 1% CVO with cross-linker, dimethylol dihydroxy ethylene urea-based in-built catalyst, showed a 47 mm zone of inhibition against *Staphylococcus aureus* (Sarkar et al., 2003).

11.6 Animal feeds

Growth in the production of antibiotics and their success in livestock performance have resulted in their intensive use. However, investigations have led to increased concern about the prevalence of microbial resistance among human pathogens due to the continuous use of antibiotics as performance enhancers in commercial animals. Thus, the trend of finding available alternatives has increased (Gheisar and Kim, 2018).

Using phytogenic plants in poultry and swine nutrition has recently gained increasing interest. They can be added to the animal feeds to improve their productivity by enhancing diet properties, promoting animal production performance, and improving products derived from animals. For example, it was found that broilers fed with a diet containing cloves had higher final weight and weight gain than those provided with a control diet (Olayemi et al., 2020). Another study reported that broilers were given 400 ppm of CVO, resulting in significantly better dry matter, crude protein, and crude fat digestibility (Dalkiliç and Güler, 2009). The addition of CVO significantly improved duck-laying performance but not affected all observed essential amino acids. The different results from several studies might be due to varying compositions of basal feed, level of feed intake, hygienic standards, and environmental temperature (Widodo et al., 2019).

Liderfeed is a product from Taragona, Spain, intended to feed chickens for fattening to increase performance. It contains 5% eugenol as the active ingredient with other excipients such as glyceryl polyethyleneglycol ricinoleate, silica amorphous, and Eudragit L 30 D-55. In 2017, the Panel on Additives and Products or Substances used in Animal Feed (FEEDAP) concluded that Liderfeed could improve the growth of chickens for fattening at the supplementation level of 100 mg/kg complete feed (FEEDAP, 2015).

Moreover, it was reported that cold-pressed clove oil could enhance growth performance, improve health status, and reduce intestinal pathogens in Japanese quails (Hussien et al., 2019).

Using phytogenic plants in pigs is less because of its concerning feed preference. It was reported that pigs are more sensitive to flavor than poultry and said they are less tolerant to moderate volatile oils than poultry (Roura et al., 2008). A report showed that clove supplementation with a low concentration of Colistin significantly improved the growth performances and shape of weaned pigs and decreased diarrhea chances. Hematological parameters, total bilirubin, and lipid profiles were not different among treatments in weaned and growing pigs. The pigs fed with a diet supplemented with clove tended to have better health than the control group (Sritharet and Trongvanichnam, 2012). In piglets, eugenol was found to be absorbed almost entirely in the gastrointestinal tract within 2 h after oral administration (Michiels et al., 2008).

In ruminants, the impact of clove supplementation was studied on feed efficiency and animal performance. It was shown that the inclusion of CVO at 3.5–7 g/animal/day in the diet of the beef bull improved animal growth performance up to 11%. The animal performance was attributed to the increased final body weight gain, average daily gain, dry matter intake, and crude protein intake (Ornaghi et al., 2017). Moreover, it was recently reported that CVO is a vital feed additive to improve milk yield and feed effectiveness in ruminants since it reduces ruminal methane yield. Eugenol is responsible for enhancing animal productivity by reducing methane-producing microflora and inhibiting protein degradation in the rumen (Saeed et al., 2021).

11.7 Fish anesthesia

Anesthesia reduces metabolic rate and oxygen consumption, allowing fish to be transported in higher densities more efficiently. Anesthesia has been used to handle and transport live fish for years in aquaculture studies to reduce stress and relax fish. Currently, tricaine methanesulfonate (MS-222) is widely used to anesthetize fish, including food fish. However, according to the FDA, it is considered mildly toxic as an edible anesthetic, requiring that fish treated with MS-222 undergo an obligatory 21-day withdrawal period (FDA/CVM, 2009).

CVO has been used as fish anesthesia in many places. CVO is effective as fish anesthesia in Far Eastern catfish *Silurus asotus*. Based on an optimal anesthetic time of approximately 1 min, the preferred concentrations of the anesthetics were 500 ppm for CVO and 600 ppm for MS-222. Both the anesthetic time and the recovery time were longer for the large fish than for the small fish (Park, 2019). Another study reported that CVO could be used for sergeant major fish (*Abudefduf vaigiensis*) anesthesia. A concentration of 30ppm showed the best result with a mean time affected of 51s, the average timeout was 190s, and the recovery time was 132s (Amris et al., 2020). In the ornamental Siamese fighting fish business, stress is one of the significant causes of fish injury leading to difficulty in fish transportation. This problem could be solved using 10mg/mL CVO as an anesthetic treatment (Pattanasiri et al., 2017). Eugenol at 65mg/L was shown to induce anesthesia in juvenile and subadult tambaqui fish within the desired time (Chaieb et al., 2007).

A report showed that CVO and eugenol were relatively more efficient anesthesia than synthetics, giving faster induction and recovery times. Compared to MS-222, CVO is an inexpensive material that no report about its residues was found (Purbosari et al., 2019). However, the drawback of using CVO impacts the flesh taste of some fish species, and because CVO possesses high volatility and water insolubility, high concentration is needed; this cause may limit its practical use. Reducing the use of CVO concentration for fish anesthesia will benefit either toxicity or the taste of fish. Self-micro-emulsifying and nano-emulsifying drug delivery systems were developed to overcome these problems (Alghaith et al., 2021; Kheawfu et al., 2021; Yostawonkul et al., 2019).

11.8 Mosquito control

CVO was evaluated to determine mortality rates and morphological abnormalities when used against third and fourth larval instars of *Aedes aegypti* and *Anopheles dirus*. The larval mortalities were 88%–100% at the end of the larval development. For *Aedes aegypti*, CVO caused various morphological abnormalities, such as deformed larvae, deformed pupae, and dead normal pupae, leading to larval mortality. In *Anopheles dirus*, morphological abnormalities were rare, and CVO was effective in causing mortality among all larval stages (Soonwera and Phasomkusolsil, 2016). However, the utilization of CVO in mosquito control has limited practicality because of its high volatility and water insolubility. New CVO preparations such as emulsion can be developed and tested for their efficacy.

The QSAR of eugenol and insecticidal activity were evaluated in *Aedes aegypti* larvae. The larvicidal activity of eugenol derivatives varied between 62.3 and 1614.9μg/mL. Oxidation of eugenol allylic bond to primary alcohol and removal of the phenolic proton resulted in decreased potency. A reasonable explanation for this result may be an increased number of hydroxyl groups preventing substance penetration in the larva cuticle from reaching its targets. However, oxidation of the same double bond in 1-benzoate-2-methoxy-4-(2-propen-1-yl)-phenol resulted in increased potency (Barbosa et al., 2012).

12. Conclusion

CVO is the mixture of volatiles from cloves that can be currently used for several purposes, mostly as flavoring agents, dental formularies, and a source of eugenol. CVO properties must meet the established standard from ISO or FCC, or SNI. Even the compositions of CVO are varied because of clove varieties, ages, organ maturities, and environmental factors; it is found that the main compositions are eugenol, eugenyl acetate, and β-caryophyllene. The exception occurs only in Indonesia's forest cloves containing only 20% eugenol; thus, it is interesting to investigate its chemical composition, which may inspire its novel utilities. CVO is generally safe if used under appropriate amounts in food processing, medicines, animal feeds, and biopesticides. However, specific concentration needs to explore for any future applications.

Acknowledgment

The author thanks Prof. Mohamed Fawzy Ramadan for providing helpful information and Ms. Urasa Charoonratana for materials.

References

Adeniyi, I.M., Adejoba, O.R., Akinlabi, F.M., Alao, O.J., 2016. Vegetable oils as clearing agents. Achiev. Life Sci. 10, 1–4.

Agus, H.H., 2021. Toxicology: oxidative stress and dietary antioxidants. In: Patel, V., Preedy, V. (Eds.), Terpene Toxicity and Oxidative Stress. Academic Press, Cambridge, pp. 33–42.

Alfaro, T., Solis, N., Villalobos, M., Vargas, N., DeGracia, B., 2019. Fatal foodborne bacteremia due to *Chromobacterium violaceum* and *Corynebacterium jeikeium*. Appl. Microbiol. Open Access 5 (1), 1000159.

Alfikri, F.N., Pujiarti, R., Wibisono, M.G., Hardiyanto, E.B., 2020. Yield, quality, and antioxidant activity of clove (*Syzygium aromaticum* L.) bud oil at the different phenological stages in young and mature trees. Scientifica 2020, 1–8.

Alghaith, A.F., Alshehri, S., Alhakamy, N.A., Hosny, K.M., 2021. Development, optimization and characterization of nanoemulsion loaded with clove oil-naftifine antifungal for the management of tinea. Drug Deliv. 28 (1), 343–356.

Alias, R.B., 2013. An Inexpensive & Safe Method for Preparation of Caryophyllene Oxide (Master dissertation). Universiti of Malaysia Pahang, Pahang, Malaysia. Retrieved fro http://malcat.uum.edu.my/kip/Record/ump.4644/UserComments.

Alighiri, D., Eden, W.T., Supardi, K.I., Masturi, Purwinarko, A., 2017. Potential development essential oil production of Central Java, Indonesia. J. Phys. Conf. Ser. 824, 1–5.

Alma, M.H., Ertas, M., Nitz, S., Kollmannsberger, H., 2007. Chemical composition and content of essential oil from the bud of cultivated Turkish clove (*Syzygium aromaticum* L.). Bioresources 2 (2), 265–269.

Al-Mariri, A., Safi, M., 2014. In vitro antibacterial activity of several plant extracts and oils against some Gram-negative bacteria. Iran. J. Med. Sci. 39 (1), 36–43.

Al-Sharif, I., Remmal, A., Aboussekhra, A., 2013. Eugenol triggers apoptosis in breast cancer cells through E2F1/survivin down-regulation. BMC Cancer 13, 600.

Al-Trad, B., Alkhateeb, H., Alsmadi, W., Al-Zoubi, M., 2019. Eugenol ameliorates insulin resistance, oxidative stress and inflammation in high fat-diet/streptozotocin induced diabetic rat. Life Sci. 216, 183–188.

Ame, S., El Boshra, I.M., El Califa, E.A.A., 2007. Nutritive value of clove (*Syzygium aromaticum*) and detection of antimicrobial effect of its bud oil. Res. J. Microbiol. 2 (3), 266–271.

Amelia, B., Saepudin, E., Cahyana, A.H., Rahayu, D.U., Sulistyoningrum, A.S., Haib, J., 2017. GC-MS analysis of clove (*Syzygium aromaticum*) bud essential oil from Java and Manado. AIP Conf. Proc. 1862 (1), 030082-1–030082-8.

Amris, A.A.U., Rahim, S.W., Yaqin, K., 2020. Effectiveness of clove oil as anesthesia of Sergeant Major Abudefduf vaigiensis (Quoy & Gaimard, 1825). Akuatikisle: J. Aqua. Coast. & Isle 4 (1), 21–28.

Apriati, D., Anggrahini, S., Supriyadi, 2018. Rendemen dan Karakteristik Minyak Atsiri Bunga Cengkeh (Sizygium Aromaticum) Yang Diberi Perlakuan Pemeraman. Retrieved fro http://etd.repository.ugm.ac.id/penelitian/detail/158629.

Argyropoulos, D., Mullera, J., 2011. Effect of convective drying on quality of lemon balm (*Melissa Officinalis* L.). Procedia Food Sci. 1, 1932–1939.

Arrijani, K., Kapahang, A., 2017. Characteristics of clove leaf essential oil (*Eugenia aromatica* o.k) in various range of elevation. J. Medi. Plants Stud. 5 (5), 27–32.

Azadi, M., Alemi, F., Sadeghi, S., Mohammadi, M., Rahimi, N.A., Mirzaie, S., et al., 2020. An integrative in silico mathematical modelling study of the anti-cancer effect of clove extract (*Syzygium aromaticum*) combined with in vitro metabolomics study using ^{1}HNMR spectroscopy. Iran. J. Biotechnol. 18 (3), e2336.

Badan Standardisasi Nasional Standar Standar Indonesia, 2011. Susu segar Bagian 1: Sapi. SNI 3141.1:2011. Jakarta.

Barbosa, J.D.F., Silva, V.B., Alves, P.B., Gumina, G., Santos, R.L.C., Sousa, D.P., Cavalcanti, S.C.H., 2012. Structure–activity relationships of eugenol derivatives against Aedes aegypti (Diptera: Culicidae) larvae. Pest Manag. Sci. 68 (11), 1478–1483.

Barnes, J., Anderson, L.A., Phillipson, J.D., 2007. Clove. In: Herbal Medicines, third ed. Pharmaceutical Press, Grayslake, pp. 166–167.

Baskaran, S.A., Kollanoor-Johny, A., Nair, M.S., Venkitanarayanan, K., 2016. Efficacy of plant-derived antimicrobials in controlling enterohemorrhagic *Escherichia coli* virulence in vitro. J. Food Prot. 79, 1965–1970.

Bhuiyan, M.N.I., Begum, J., Nandi, N.C., Akter, F., 2010. Constituents of the essential oil from leaves and buds of clove (*Syzigium caryophyllatum* (L.) Alston). African J. Plant Sci. 4 (11), 451–454.

Blascheck, W., Ebel, S., Hackenthal, E., Holzgrabe, U., Keller, K., Reichling, J., Schulz, V., 2008. *Syzygium Caryophylli* Aetheroleum (Nelkenöl). Springer Verlag, Heidelberg.

Blumenthal, M., 1998. Herbs and phytomedicines in the European Community. In: Blumenthal, M. (Ed.), The Complete German Commission E Monographs. Therapeutic Guide to Herbal Medicines. American Botanical Council, Austin, Texas.

Boughendjioua, H., 2018. Essential oil composition of *Syzygium aromaticum* L. Int. Res. J. Pharm. Med. Sci. 1 (3), 26–28.

Brat, P., Yahia, A., Chillet, M., Bugaud, C., Bakry, F., Reynes, M., et al., 2004. Influence of cultivar, growth altitude and maturity stage on banana volatile compound composition. Fruits 59 (2), 75–82.

Burall, L.S., Harro, J.M., Li, X., Lockatell, C.V., Himpsl, S.D., Hebel, J.R., et al., 2004. Proteus mirabilis genes that contribute to pathogenesis of urinary tract infection: identification of 25 signature-tagged mutants attenuated at least 100-fold. Infect. Immun. 72, 2922–2938.

Carvalho, R.P.R., de Almeida Lima, G.D., Machado-Neves, M., 2021. Effect of eugenol treatment in hyperglycemic murine models: a meta-analysis. Pharmacol. Res. 165, 105315.

Casida, J.E., Durkin, K.A., 2013. Neuroactive insecticides: targets, selectivity, resistance, and secondary effects. Annu. Rev. Entomol. 58, 99–117.

Chaieb, K., Hajlaoui, H., Zmantar, T., Kahla-Nakbi, A.B., Rouabhia, M., Mahdouani, K., Bakhrouf, A., 2007. The chemical composition and biological activity of clove ssential oil, *Eugenia caryophyllata* (*Syzigium aromaticum* L. Myrtaceae): a short review. Phytother. Res. 21, 501–506.

Chapman, M.J., Ginsberg, H.N., Amarenco, P., Andreotti, F., Boren, J., Catapano, A.L., et al., 2011. Triglyceride-rich lipoproteins and high-density lipoprotein cholesterol in patients at high risk of cardiovascular disease: evidence and guidance for management. Eur. Heart J. 32, 1345–1361.

Charoonratana, T., Pongnuruk, K., Puttarak, P., 2021. Establishment of quality standards for raw material of *Cyanthillium cinereum* (L.) H. Rob. Nat. Prod. J. 11 (5), 734–747.

Chen, C., Xu, Z., Ma, Y., Liu, J., Zhang, Q., Tang, Z., et al., 2017. Properties, vapour-phase antimicrobial and antioxidant activities of active poly (vinyl alcohol) packaging films incorporated with clove oil. Food Control 88, 105–112.

Clark, J.M., Symington, S., 2012. Advances in the mode of action of pyrethroids (pp. 49-72). In: Matsuo, N., Mori, T. (Eds.), Pyrethroids. Springer, Berlin, Heidelberg.

Committee on Codex Specifications, Food and Nutrition Board, Division of Biological Sciences, Assembly of Life Sciences, National Research Council, 1996. Food Chemicals Codex. National Academy Press, Washington, DC.

Correa, Y.D.C.G., Faroni, L.R.A., Haddi, K., Oliveira, E.E., Pereira, E.J.G., 2015. Locomotory and physiological responses induced by clove and cinnamon essential oils in the maize weevil *Sitophilus zeamais*. Pestic. Biochem. Physiol. 125, 31–37.

Da Conceicao Machado, K., Islam, M.T., Ali, E.S., Rouf, R., Uddin, S.J., Dev, S., et al., 2018. A systematic review on the neuroprotective perspectives of beta-caryophyllene. Phytother. Res. 32 (12), 2376–2388.

Dalkiliç, B., Güler, T., 2009. The effects of clove extract supplementation on performance and digestibility of nutrients in broilers. Vet. Hekim. Der. Derg. 23, 161–166.

Dan Syafaruddin, E.R., 2011. Keragaan pohon cengkeh terpilih tipe zanzibar dan siputih palabuhanratu. Bull. RISTRI 2 (3), 405–410.

Danthu, P., Simanjuntak, R., Fawbush, F., Leong Pock Tsy, J.M., Razafimamonjison, G., Abdillahi, M.M., et al., 2020. The clove tree and its products (clove bud, clove oil, eugenol): prosperous today but what of tomorrow's restrictions? Fruits 75 (5), 224–242.

Dayton, T.M., Diefenbach, K.A., Fuller, L.M., Valtos, J., Niederhoffer, E.C., 1996. Production of superoxide dismutases from *Proteus mirabilis* and *Proteus vulgaris*. Biometals 9, 131–137.

De Lima, R.K., Cardoso, M.G., Andrade, M.A., Nascimento, E.A., de Morais, S.A., Nelson, D.L., 2010. Composition of the essential oil from the leaves of tree domestic varieties and one wild variety of the guava plant (*Psidium guajava* L., Myrtaceae). Braz. J. Pharmacogn. 20 (1), 41–44.

DeGraff, W.G., 1983. Mutagenicity evaluation of B161 (clove leaf oil) in the ames salmonella/microsome plate test. Unpublished Report, Study No. 6866, Litton Bionetics, Kensington, MD.

De Melo, T.M.P., da Silva, E.M., da Silva, A.G., da Costa Viera, G.H., Lopes, B.G., 2019. *Syzygium aromaticum* essential oil to the control of *Tenuipalpus heveae*. J. Agric. Sci. 11 (8), 295–301.

De Oliveira, T.L., das Gracas Cardoso, M., de Araujo Soares, R., Ramos, E.M., Piccoli, R.H., Tebaldi, V.M., 2013. Inhibitory activity of *Syzygium aromaticum* and *Cymbopogon citratus* (DC.) Stapf. essential oils against *Listeria monocytogenes* inoculated in bovine ground meat. Braz. J. Microbiol. 44 (2), 357–365.

Devi, K.P., Sakthivel, R., Nisha, S.A., Suganthy, N., Pandian, S.K., 2013. Eugenol alters the integrity of cell membrane and acts against the nosocomial pathogen *Proteus mirabilis*. Arch. Pharm. Res. 36, 282–292.

Ebadi, M.T., Azizi, M., Sefidkon, F., Ahmadi, N., 2015. Influence of different drying methods on drying period, essential oil content and composition of *Lippia citriodora* Kunth. J. Appl. Res. Med. Aromat. Plants 2 (4), 182–187.

EFSA FEEDAP Panel (EFSA Panel on Additives and Products or Substances used in Animal Feed), 2015. Scientific opinion on the safety and efficacy of Liderfeed® (eugenol) for chickens for fattening. EFSA J. 13 (11), 4273.

Elwakeel, H.A., Moneim, H.A., Farid, M., Gohar, A.A., 2007. Clove oil crea a new effective treatment for chronic anal fissure. Color. Dis. 9, 549–552.

Enan, E., 2001. Insecticidal activity of essential oils: octopaminergic sites of action. Comp. Biochem. Physiol. C 130, 325–337.

European Food Safety Authority, 2011. Scientific opinion on the safety and efficacy of allyhydroxybenzenes (chemical group 18) when used as flavourings for all animal species. EFSA J. 9, 1–56.

European Scientific Cooperative on Phytotherapy, 2003. E/S/C/O/P Monographs: The Scientific Foundation for Herbal Medicinal Products. European Scientific Cooperative on Phytotherapy, Exeter, UK.

Farag, R.S., Daw, Z.Y., Abo Raya, S.H.J., 1989. Influences of some spice essential oils on *Aspergillus parasiticus* growth and production of aflatoxins in synthetic medium. J. Food Sci. 54, 74–76.

Farrel, K.T., 1990. Spices, Condiments and Seasoning, second ed. Van Nostrand, New York.

FDA, 2011. Exemptions for pesticides of a character not requiring FIFRA regulation, 40 C.F.R. 152.25.

FDA, 2020a. Food Additives, 21 C.F.R. 170.3(o)(12).

FDA, 2020b. Labeling of drug preparations containing significant proportions of wintergreen oil, 21 C.F.R. 201.303.

FDA, 2020c. Labeling of drug preparations containing salicylates, 21 C.F.R. 201.314.

FDA., U.S, 1978. In: N.T.I. Service (Ed.), Scientific Literature Review (SLR) C8. Scientific Literature Review of Eugenol and Related Substances. PB283-501/AS.

FDA/CVM, 2009. Concerns Related to the Use of Clove Oil as an Anesthetic for Fish. FDA Guidance for Industry No. 150. Center for Veterinary Medicine, Food and Drug Administration, USA.

Ferlay, J., Soerjomataram, I., Dikshit, R., Eser, S., Mathers, C., Rebelo, M., et al., 2015. Cancer incidence and mortality worldwide: sources, methods and major patterns in GLOBOCAN 2012. Int. J. Cancer 136, E359–E386.

Flint, P.W., Haughey, B.H., Lund, V.J., Niparko, J.K., Richardson, M.A., Robbins, K.T., et al., 2010. Cummings Otolaryngology Head and Neck Surgery. Mosby, Philadelphia, PA, pp. 1944–1949.

Frandsen, C.S., Dejgaard, T.F., Madsbad, S., 2016. Non-insulin drugs to treat hyperglycaemia in type 1 diabetes mellitus. Lancet Diabetes Endocrinol. 4, 766–780.

Gheisar, M.M., Kim, H., 2018. Phytobiotics in poultry and swine nutrition—a review. Ital. J. Anim. Sci. 17 (1), 92–99.

Ghoneem, K.M., Saber, W.I.A., El-Awady, A.A., Rashad, Y.M., Al-Askar, A.A., 2016. Clove essential oil for controlling white mold disease, sprout suppressor and quality maintainer for preservation of Jerusalem artichoke tubers. Egypt. J. Biol. Pest Control 26 (3), 601–608.

Golmakani, M., Zare, M., Razzaghi, S., 2017. Eugenol enrichment of clove bud essential oil using different microwave-assisted distillation methods. Food Sci. Technol. Res. 23 (3), 385–394.

Gooderham, N.J., Cohen, S.M., Eisenbrand, G., Fukushima, S., Guengerich, P., Hecht, S.S., et al., 2020. FEMA GRAS assessment of natural flavor complexes: clove, cinnamon leaf and West Indian bay leaf-derived flavoring ingredients. Food Chem. Toxicol. 145, 111585.

Gopalakrishnan, N., Narayanan, C.S., 1988. Composition of clove leaf oil during leaf growth. Indian Perfum. 32 (2), 130–132.

Gopalakrishnan, N., 1994. Studies on the storage quality of CO_2-extracted cardamom and clove bud oils. J. Agric. Food Chem. 42, 796–798.

Greenberg, M., Dodds, M., Tian, M., 2008. Naturally occurring phenolic antibacterial compounds show effectiveness against oral bacteria by a quantitative structure-activity relationship study. J. Agric. Food Chem. 56, 11151.

Gruenwald, J., Brendler, T., Jaenicke, C. (Eds.), 2004. PDR for Herbal Medicines, third ed. Thomson PDR, Montvale, NJ.

Gu, Y., Yang, L., Huo, H.Q., Zhou, L.Z., Wen, R.S., Zhu, Y.J., 2019. Essential oil yield in *Eucalyptus citriodora* in relation to foliage age after pollarding and tree age. J. Trop. For. Sci. 31 (4), 452–458.

Hajlaoui, H., Snoussi, M., Noumi, E., Zanetti, S., Ksouri, R., Bakhrouf, A., 2010. Chemical composition, antioxidant and antibacterial activities of the essential oils of five Tunisian aromatic plants. Ital. J. Food Sci. 3 (22), 323–332.

Handayani, P.A., Sofiarani, A.W., Kurniawan, A.B., 2020. Response surface optimization of essential oils production from clove leaf waste by microwave-assisted hydro distillation. In: Proceeding of the 8th Engineering International Conference 2019. IOP Publishing, Bristol.

Hariyadi, S., Mahulette, A.S., Yahya, S., Wachjar, A., 2020a. Agro-morphologies and physicochemical properties of flower bud, stem and leaf oils in two clove varieties (*Syzygium aromaticum* L Merr. and Perry.) Originated from Ambon Island. CMU J. Nat. Sci. 19 (3), 516–530.

Hariyadi, S., Mahulette, A.S., Yahya, S., Wachjar, A., 2020b. Morphological characters and essential oil constituents extracted of two clove varieties (*Syzygium aromaticum* (L.) MERR. & L.M. Perry.) from Ambon island, Indonesia. Plant Arch. 20 (S1), 2208–2214.

Hasson, R.H., Al-Zubaidi, H.H., 2015. The acaricidal effect of clove leaf plant extract *Eugenia caryophyllus* against *Argas* spp. soft ticks in Iraq. International. J. Adv. Res. 3 (9), 257–262.

Hastuti, L.T., Saepudin, E., Cahyana, A.H., Rahayu, D.U.C., Murni, V.W., Haib, J., 2017. The influence of sun drying process and prolonged storage on composition of essential oil from clove buds (*Syzygium aromaticum*). AIP Conf. Proc. 1862 (1), 030092-1–030092-6.

Hatami, T., Meireles, M.A.A., Zahedi, G., 2010. Mathematical modeling and genetic algorithm optimization of clove oil extraction with supercritical carbon dioxide. J. Supercrit. Fluids 51, 331–338.

He, M., Du, M., Fan, M., Bian, Z., 2007. In vitro activity of eugenol against *Candida albicans* biofilms. Mycopathologia 163 (3), 137–143.

Herlina, Aziz, S.A., Kurniawati, A., Faridah, D.N., 2017. Changes of thymoquinone, thymol, and malondialdehyde content of black cumin (*Nigella sativa* L.) in response to Indonesia tropical altitude variation. HAYATI J. Biosci. 24, 156–161.

Higashi, Y., Fujii, Y., 2010. HPLC-UV analysis of eugenol in clove and cinnamon oils after pre-column derivatization with 4-fluoro-7-nitro-2,1,3-benzoxadiazole. J. Liq. Chromatogr. Relat. Technol. 34 (1), 18–25.

Homa, K., Barney, W.P., Ward, D.L., Wyenandt, C.A., Simon, J.E., 2014. Evaluation of fungicides for the control of *Peronospora belbahrii* on sweet basil in New Jersey. Plant Dis. 98, 1561–1566.

Hu, Q., Zhou, M., Wei, S., 2018. Progress on the antimicrobial activity research of clove oil and eugenol in the food antisepsis field. J. Food Sci. 83 (6), 1476–1483.

Hu, Y., Zeng, Z., Wang, B., Guo, S., 2017. Trans-caryophyllene inhibits amyloid β (Aβ) oligomer-induced neuroinflammation in BV-2 microglial cells. Int. Immunopharmacol. 51, 91–98.

Hua, H., Xing, F., Selvaraj, J.N., Wang, Y., Zhao, Y., Zhou, L., et al., 2014. Inhibitory effect of essential oils on *Aspergillus ochraceus* growth and ochratoxin A production. PLoS One 9, e108285.

Huang, X., Feng, Y., Huang, Y., Li, H., 2013. Chemical composition, antioxidant and the possible use as skin-care ingredient of clove oil (*Syzygium aromaticum* (L.) Merr. & Perry) and citronella oil (*Cymbopogon goeringii*) from China. J. Essent. Oil Res. 25 (4), 315–323.

Hussain, A., Priyani, A., Sadrieh, L., Brahmbhatt, K., Ahmed, M., Sharma, C., 2012. Concurrent sulforaphane and eugenol induces differential effects on human cervical cancer cells. Integr. Cancer Ther. 11, 154–165.

Hussien, M.M.A., Abd El-Hack, M.E., Mahgoub, S.A., Saadeldin, I.M., Swelum, A.A., 2019. Effects of clove (*Syzygium aromaticum*) oil on quail growth, carcass traits, blood components, meat quality, and intestinal microbiota. Poult. Sci. 98, 319–329.

Isa, A.M., Nazarifah, I., 1986. Quality of clove bud and clove bud oil. MARDI Res. Bull. 14 (2), 148–153.

Ishidate, M., Takizawa, Y., Sakabe, Y., Ishizaki, M., Watanbe, M., Tachi, M., et al., 1988. Mutagenicity test of food additives (part 9). Toxicol. Forum 11, 663–669.

Ivanovic, J., Dimitrijevic-Brankovic, S., Misic, D., Ristic, M., Zizovic, I., 2013. Evaluation and improvement of antioxidant and antibacterial activities of supercritical extracts from clove buds. J. Funct. Foods 5, 416–423.

Ivanovic, J., Zizovic, I., Ristic, M., Stamenic, M., Skala, D., 2011. The analysis of simultaneous clove/oregano and clove/thyme supercritical extraction. J. Supercrit. Fluids 55, 983–991.

Jaganathan, S.K., Mazumdar, A., Mondhe, D., Mandal, M., 2011. Apoptotic effect of eugenol in human colon cancer cell lines. Cell Biol. Int. 35, 607–615.

Jaganathan, S.K., Supriyanto, E., 2012. Antiproliferative and molecular mechanism of eugenol-induced apoptosis in cancer cells. Molecules 17, 6290–6304.

Jahanshiri, Z., Shams-Ghahfarokhi, M., Allameh, A., Razzaghi-Abyaneh, M., 2015. Inhibitory effect of eugenol on aflatoxin B1 production in *Aspergillus parasiticus* by downregulating the expression of major genes in the toxin biosynthetic pathway. World J. Microbiol. Biotechnol. 31, 1071–1078.

Jairoce, C.F., Teixeira, C.M., Nunes, C.F., Nunes, A.M., Pereira, C.M., Garcia, F.R., 2016. Insecticide activity of clove essential oil on bean weevil and maize weevil. Rev. Bras. Eng. Agric. Ambient. 20 (1), 72–77.

Janes, S.E.J., Price, C.S.G., Thomas, D., 2005. Essential oil poisoning: N-acetylcysteine for eugenol-induced hepatic failure and analysis of a national database. Eur. J. Pediatr. 164, 520–522.

Javed, H., Azimullah, S., Haque, M.E., Ojha, S.K., 2016. Cannabinoid type 2 (CB2) receptors activation protects against oxidative stress and neuroinflammation associated dopaminergic neurodegeneration in rotenone model of Parkinson's disease. Front. Neurosci. 321, 1–14.

Jayalekshmi, H., Omanakuttan, A., Pandurangan, N., Vargis, V.S., Maneesh, M., Nair, G.B., et al., 2016. Clove bud oil reduces kynurenine and inhibits pqs A gene expression in *P. aeruginosa*. Appl. Microbiol. Biotechnol. 100, 3681–3692.

Jayanti, H.D., 2011. Pengendalian Mutu Proses Pembuatan Minyak Atsiri Daun Cengkeh (Clove Leaf Oil) (Thesis). Sebelas Maret University, Jawa Tengah, Indonesia. Retrieved fro https://digilib.uns.ac.id/dokumen/detail/19819/Pengendalian-Mutu-Proses-Pembuatan-Minyak-Atsiri-Daun-Cengkeh-Clove-Leaf-Oil.

Jayawardee, A.L., Balachandran, S., 1979. Effects of storage on essential oils. In: Proceeding of 8th Annual Sessions Institue of Chemistry, Colombo.

JECFA, 1982. Toxicological Evaluation of Certain Food Additives and Contaminants. WHO Food Additive Series No. 17.

Jesudasan, J.S., Abdul Wahab, P.A., Muthu Sekhar, M.R., 2015. Effectiveness of 0.2% chlorhexidine gel and aeugenol-based paste on postoperative alveolar osteitis in patients having third molars extracted: a randomised controlled clinical trial. Br. J. Oral Maxillofac. Surg. 53 (9), 826–830.

Jirovetz, L., Buchbauer, G., Stoilova, I., Stoyanova, A., Krastanov, A., Schmidt, E., 2006. Chemical composition and antioxidant properties of clove leaf essential oil. J. Agric. Food Chem. 54, 6303–6307.

Joint FAO/WHO Expert Committee on Food Additives (2001: Rome, Italy), World Health Organization & Food and Agriculture Organization of the United Nations, 2002. Evaluation of Certain Food Additives and Contaminants: Fifty-Seventh Report of the JOINT FAO/WHO Expert Committee on Food Additives. World Health Organization.

Joint FAO/WHO Expert Committee on Food Additives (2004: Geneva, Switzerland), World Health Organization & Food and Agriculture Organization of the United Nations, 2004. Evaluation of Certain Food Additives and Flavoring Agents: Sixty-Third Report of the Joint FAO/WHO Expert Committee on Food Additives. World Health Organization.

Joint FAO/WHO Expert Committee on Food Additives (2005: Geneva, Switzerland), World Health Organization & Food and Agriculture Organization of the United Nations, 2005. Evaluation of Certain Food Additives and Flavoring Agents: Sixty-Fifth Report of the Joint FAO/WHO Expert Committee on Food Additives. World Health Organization.

Joint FAO/WHO Expert Committee on Food Additives (2009: Geneva, Switzerland), World Health Organization & Food and Agriculture Organization of the United Nations, 2009. Evaluation of Certain Food Additives: Sixty-Ninth Report of the Joint FAO/WHO Expert Committee on Food Additives. World Health Organization.

Joint FAO/WHO Expert Committee on Food Additives (2014: Geneva, Switzerland), World Health Organization & Food and Agriculture Organization of the United Nations, 2014. Evaluation of Certain Veterinary Drug Residues in Food: Seventy-Eighth Report of the Joint FAO/WHO Expert Committee on Food Additives. World Health Organization.

Jumbo, L.O.V., Faroni, L.R., Oliveira, E.E., Pimentel, M.A., Silva, G.N., 2014. Potential use of clove and cinnamon essential oils to control the bean weevil, *Acanthoscelides obtectus* say, in small storage units. Ind. Crop. Prod. 56, 27–34.

Juntila, J.R., Niemela, S.I., Hirn, J., 1988. Minimum growth temperature of *Listeria monocytogenes* and non-hemolytic Listeria. J. Appl. Bacteriol. 65, 321–327.

Kapadiya, S., Desai, M.A., 2017. Isolation of essential oil from buds of *Syzygium aromaticum* using hydrodistillation: multi-response optimization and predictive modelling. Int. J. Adv. Res. Sci. Eng. 6 (1), 405–418.

Karagoz, S.G., Yilmazer, M., Ozkan, G., Carbonell-Barrachina, A.A., Kiralan, M., Ramadan, M.F., 2017. Effect of cultivar and harvest time on C_6 and C_5 volatile compounds of Turkish olive oils. Euro. Food Res. Technol. 243, 1193–1200.

Khan, M.S., Ahmad, I., 2012. Biofilm inhibition by *Cymbopogon citratus* and *Syzygium aromaticum* essential oils in the strains of *Candida albicans*. J. Ethnopharmacol. 140 (2), 416–423.

Khan, K., Qadir, S.S., Khattak, K.N., Anderson, R.A., 2006. Cloves improve glucose, cholesterol and triglycerides of people with type 2 diabetes mellitus. FASEB J. 20 (5), A990.

Khan, M.S., Zahin, M., Hasan, S., Husain, F.M., Ahmad, I., 2009. Inhibition of quorum sensing regulated bacterial functions by plant essential oils with special reference to clove oil. Lett. Appl. Microbiol. 49 (3), 354–360.

Kheawfu, K., Pikulkaew, S., Rades, T., Mullertz, A., Jorgensen, L.V.G., Okonogi, S., 2021. Design and optimization of self-nanoemulsifying drug delivery systems of clove oil for efficacy enhancement in fish anesthesia. J. Drug Deliv. Sci. Technol. 61, 102241.

Kim, S.S., Oh, O.J., Min, H.Y., Park, E.J., Kim, Y., Park, H.J., et al., 2003. Eugenol suppresses cyclooxygenase-2 expression in lipopolysaccharide stimulated mouse macrophage RAW264.7 cells. Life Sci. 73, 337–348.

Kouidhi, B., Zmantar, T., Bakhrouf, A., 2010. Anticariogenic and cytotoxic activity of clove essential oil (*Eugenia caryophyllata*) against a large number of oral pathogens. Ann. Microbiol. 60, 599–604.

Koul, O., Walia, S., Dhaliwal, S., 2008. Essential oils as green pesticides: potential and constraints. Biopestici. Int. 4 (1), 63–84.

Kramer, R.E., 1985. Antioxidants in clove. J. Am. Oil Chem. Soc. 62, 111–113.

Kumar, P., Febriyanti, R., Sofyan, F., Luftimas, D.E., Abdulah, R., 2014. Anticancer potential of *Syzygium aromaticum* L. in MCF-7 human breast cancer cell lines. Pharm. Res. 6, 350–354.

Kumarswamy, A., 2016. Multimodal management of dental pain with focus on alternative medicine: a novel herbal dental gel. Contemp. Clin. Dent. 7 (2), 131–139.

Kurniasari, F., 2016. Evaluation of hedonic test and irritation test of lotion of clove leaf essential oil (*Eugenia aromatic* L.). Indonesian J. Pharm. 13 (2), 163–170.

Lane, B.W., Ellenhorn, M.J., Hulbert, T.V., McCarron, M., 1991. Clove oil ingestion in an infant. Hum. Exp. Toxicol. 10 (4), 291–294.

Lange, H., Smart, C., Seaman, A., 2020. Cucurbit powdery mildew (*Podosphaera xanthii*) control with Organic Materials Review Institute (OMRI)-listed products. NYS IPM Type: Vegetables IPM Report. New York State Integrated Pest Management Program, New York.

Lapin, B.A., Yakovleva, L.Y., 1957. The use of essential oils in histologic practice. Bull. Exp. Biol. Med. 44, 1019–1021.

Lee, K.G., Shibamoto, T., 2001. Inhibition of malonaldehyde formation from blood plasma oxidation by aroma extracts and aroma components isolated from clove and eucalyptus. Food Chem. Toxicol. 39, 1199–1204.

Lekjing, S., Chinarak, K., 2018. Effect of clove essential oil on film properties, antioxidant and antimicrobial activities of chitosan film. KKU Res. J. 19 (4), 172–189.

Lerin, L.A., Catani, M., Oliveira, D., Massi, A., Bortolini, O., Cavazzini, A., et al., 2015. Continuous ion-exchange resin catalysed esterification of eugenol for the optimized production of eugenyl acetate using a packed bed microreactor. RSC Adv. 94, 76898–76903.

Leung, A.Y., 1980. Encyclopedia of Common Natural Ingredients Used in Food, Drugs and Cosmetics. Wiley, New York, pp. 131–132.

Liang, D., Xing, F., Selvaraj, J.N., Liu, X., Wang, L., Hua, H., et al., 2015. Inhibitory effect of cinnamaldehyde, citral, and eugenol on aflatoxin biosynthetic gene expression and aflatoxin b1 biosynthesis in *Aspergillus flavus*. J. Food Sci. 80, M2917–M2924.

Lou, J., Cao, G., Li, R., Liu, J., Dong, Z., Xu, L., 2016. β-Caryophyllene attenuates focal cerebral ischemia-reperfusion injury by Nrf2/HO-1 pathway in rats. Neurochem. Res. 41, 1291–1304.

Lutfi, M., Andajani, K., Ilhamuddin, Utami, H.N., Affifah, F.N., 2020. Appropriate technology application of traditional clove oil production: effort to upgrade quality. Adv. Food Sci. Sustain. Agric. Agroind. Eng. 3 (2), 75–80.

Mactavish, H.S., Menary, R.C., 1997. The effect of flower maturity and harvest timing on floral extract from *Boronia megastigma* (Nees). Ann. Bot. 80, 299–303.

Mahakittikun, V., Soonthornchareonnon, N., Foongladda, S., Boitano, J.J., Wangapai, T., Ninsanit, P., 2013. A preliminary study of the acaricidal activity of clove oil, *Eugenia caryophyllus*. Asian Pac. J. Allergy Immunol. 32 (1), 46–52.

Mahulette, A.S., Hariyadi, S., Yahya, S., Wachjar, A., 2020. Physico-chemical properties of clove oil from three forest clove accession groups in Maluku. In: Proceeding of the 1st International Conference on Sustainable Plantation.

Majeed, H., Antoniou, J., Fang, Z., 2014. Apoptotic effects of eugenol-loaded nanoemulsions in human colon and liver cancer cell lines. Asian Pac. J. Cancer Prev. 15, 9159–9164.

Manrique, Y., Gibis, M., Schmidt, H., Weiss, J., 2017. Influence of application sequence and timing of eugenol and lauric arginate (LAE) on survival of spoilage organisms. Food Microbiol. 64, 210–218.

Mansourian, A., Boojarpour, N., Ashnagar, S., Momen Beitollahi, J., Shamshiri, A.R., 2014. The comparative study of antifungal activity of *Syzygium aromaticum*, *Punica granatum* and nystatin on *Candida albicans*; an in vitro study. J. Med. Mycol. 24 (4), e163–e168.

Marwati, T., Rusli, M.S., Noor, E., 2017. The effect of adsorption agent on color and heavy metal of clove leaf oil as biopesticide material. In: Proceeding of Orgatrop 2017: International Conference on Organic Agriculture in the Tropics State-of-the-Art, Challenges and Opportunities. IOP Publishing, Bristol.

Mazzarrino, G., Paparella, A., Chaves-López, Faberi, A., Sergi, M., Sigismondi, C., Serio, A., 2015. Salmonella enterica and Listeria monocytogenes inactivation dynamics after treatment with selected essential oils. Food Control 50, 794–803.

Mehdizadeh, L., Pirbalouti, A.G., Moghaddam, M., 2017. Storage stability of essential oil of cumin (*Cuminum Cyminum* L.) as a function of temperature. Int. J. Food Prop. 20, 1742–1750.

Mehra, L.K., MacLean, D.D., Shewfelt, R.L., Smith, K.C., Scherm, H., 2013. Effect of postharvest biofumigation on fungal decay, sensory quality, and antioxidant levels of blueberry fruit. Postharvest Biol. Technol. 85, 109–115.

Menon, K.Y., Garg, S.R., 2001. Inhibitory effect of clove oil on *Listeria monocytogenes* in meat and cheese. Food Microbiol. 18, 647–650.

Michiels, J., Missotten, J., Dierick, N., Fremaut, D., Maene, P., De Smet, S., 2008. In vitro degradation and in vivo passage kinetics of carvacrol, thymol, eugenol and trans-cinnamaldehyde along the gastrointestinal tract of piglets. J. Sci. Food Agric. 88 (13), 2371–2381.

Miladi, H., Zmantar, T., Kouidhi, B., Chaabouni, Y., Mahdouani, K., Bakhrouf, A., et al., 2017. Use of carvacrol, thymol, and eugenol for biofilm eradication and resistance modifying susceptibility of *Salmonella enterica* serovar *typhimurium* strains to nalidixic acid. Microb. Pathog. 104, 56–63.

Mishra, R., Gupta, A.K., Kumar, A., Lal, R.K., Saikia, D., Chanotiya, C.S., 2018. Genetic diversity, essential oil composition, and in vitro antioxidant and antimicrobial activity of *Curcuma longa* L. germplasm collections. J. Appl. Res. Med. Aromat. Plants 10, 75–84.

Morán, A., Gutiérrez, S., Martínez-Blanco, H., Ferrero, M.A., Monteagudo-Mera, A., Rodíguez-Aparicio, L.B., 2014. Non-toxic plant metabolites regulate Staphylococcus viability and biofilm formation: a natural therapeutic strategy useful in the treatment and prevention of skin infections. Biofouling 30, 1175–1182.

Murni, V.W., Saepudin, E., Cahyana, D.U.C., Rahayu, L., Haib, J., 2017. Effect of oven drying and storage on essential oil composition of clove (*Syzygium aromaticum*) from Toli-Toli. AIP Conf. Proc. 1862 (1), 030084-1–030084-6.

Musta, R., Nurliana, L., 2019. Kinetic studies effectiveness of clove leaf (*Syzygium aromaticum*) oil as antifungal of *Candida albicans*. Indones. J. Chem. Res. 6 (2), 107–114.

Musthafa, K.S., Voravuthikunchai, S.P., 2016. Eugenyl acetate inhibits growth and virulence factors of drug-resistant *Acinetobacter baumannii*. Flavour Fragance J. 31 (6), 448–454.

Musthafa, K.S., Hmoteh, J., Thamjarungwong, B., Voravuthikunchai, S.P., 2016. Antifungal potential of eugenyl acetate against clinical isolates of Candida species. Microb. Pathog. 99, 19–29.

Mutombo, A.B., Tozin, R., Kanyiki, H., Geertruyden, J.V., Jacquemyn, Y., 2019. Impact of antiviral AV2 in the topical treatment of HPV-associated lesions of the cervix: results of a phase III randomized placebo-controlled trial. Contemp. Clin. Trials Commun. 15, 100377.

Nair, D.V., Nannapaneni, R., Kiess, A., Schilling, W., Sharma, C.S., 2014. Reduction of Salmonella on turkey breast cutlets by plant-derived compounds. Foodborne Pathog. Dis. 11, 981–987.

Nathan, D.M., 2015. Diabetes: advances in diagnosis and treatment. J. Am. Med. Assoc. 314, 1052–1062.

National Toxicology Program, 2010. NTP Technical Report on the Toxicology and Carcinogenesis Studies of Isoeugenol (CAS No. 97-54-1) in F344/N Rats and B6C3F1 Mice (Gavage Studies). National Toxicology Program, U.S. Dept. of Health and Human Services, National Institutes of Health, Research Triangle Park, NC.

Niinimäki, A., 1984. Delayed-type allergy to spices. Contact Dermatitis 11, 34–40.

Nirmala, M.J., Durai, L., Gopakumar, V., Nagarajan, R., 2019. Anticancer and antibacterial effects of a clove bud essential oil-based nanoscale emulsion system. Int. J. Nanomedicine 14, 6439–6450.

Nurdjannah, N., Bermawie, N., 2012. Cloves. Woodhead Publishing, Cambridge.

Nurdjannah, N., Supriyadi, D., Ferdinanti, E., 2001. Antibacterial activity of gargle solution containing clove oils. In: Proceedings of the International Seminar on Natural Products Chemistry and Utilization of Natural Resources.

Nurdjannah, N., Rusli, S., 1988. Some factors affecting quality of clove. Bul. Penelit. Tanam. Rempah. Obat. 6 (2), 55–60.

Nurdjannah, N., Hardjo, S., Mirna., 1993. Distillation method influences the yield and quality of clove leaf oil. Ind. Crops Res. J. 3 (2), 18–26.

Nurhadianty, V., Cahyani, C., Nirwana, W.O.C., Dewi, L.K., Abdillah, G., Pratama, A.R., 2017. Peningkatan yield minyak daun cengkeh (Syzygium aromaticum) dengan fermentasi selulotik menggunakan Trichoderma harzianum. J. Rekayasa Bahan Alam dan Energi Berkelanjutan 1 (1), 36–41.

OECD, 2010. Explanatory background document to the OECD test guideline on in vitro skin irritation testing. OECD series on testing and assessment No. 137. In: Environment Directorate, Joint Meeting of the Chemicals Committee and the Working Party on Chemicals, Pesticides and Biotechnology. ENV/JM/MONO(2010)36. OECD, Paris.

Ojha, S., Javed, H., Azimullah, S., Haque, M.E., 2016. β-Caryophyllene, a phytocannabinoid attenuates oxidative stress, neuroinflammation, glial activation, and salvages dopaminergic neurons in a rat model of Parkinson disease. Mol. Cell. Biochem. 418, 59–70.

Omara, S.M., Al-Ghamdi, K.M., Mahmoud, M.A., Sharawi, S.E., 2015. Repellency and fumigant toxicity of clove and sesame oils against American cockroach (*Periplaneta americana* (L.)). Afr. J. Biotechnol. 12 (9), 963–970.

Olayemi, W.A., Williams, G.A., Olatidoye, O.P., Omofunmilola, E.O., 2020. Influence of dietary inclusion of phytobiotics on growth performance, carcass and organ weight of broiler chickens. J. Agric. Food Sci. 18 (2), 26–38.

Opdyke, D.L.J., 1975. Monographs on fragrance raw materials: clove bud oil. Food Cosmet. Toxicol. 13, 761–762.

Orav, A., Stulova, I., Kailas, T., Muurisepp, M., 2004. Effect of storage on the essential oil composition of *Piper nigrum* L. fruits of different ripening states. J. Agric. Food Chem. 52, 2582–2586.

Ornaghi, M.G., Passetti, R.A., Torrecilhas, J.A., Mottin, C., Vital, A.C.P., Guerrero, A., et al., 2017. Essential oils in the diet of young bulls: effect on animal performance, digestibility, temperament, feeding behavior and carcass characteristics. Anim. Feed Sci. Technol. 234, 274–283.

Osguthorpe, D., Nielsen, D.R., 2006. Otitis externa: review and clinical update. Am. Fam. Physician 74, 1510–1516.

Oyedemi, S.O., Okoh, A.I., Mabinya, L.V., Pirochenva, G., Afolayan, A.Y., 2009. The proposed mechanism of bactericidal action of eugenol, α-terpineol and γ-terpinene against *Listeria monocytogenes*, *Streptococcus pyogenes*, *Proteus vulgaris* and *Escherichia coli*. Afr. J. Biotechnol. 8 (7), 1280–1286.

Ozdemir, N., Ozgen, Y., Kiralan, M., Bayrak, A., Aslan, N., Ramadan, M.F., 2017. Effect of different drying methods on the essential oil yield, composition and antioxidant activity of *Origanum vulgare* L. and *Origanum onites* L. J. Food Meas. Charact. 12, 820–825.

Palmer, A.S., Stewart, J., Fyle, L., 1998. Antimicrobial properties of plant essential oils and essences against five important food-borne pathogens. J. Appl. Microbiol. 26, 118–122.

Panahi, Y., Akhavan, A., Sahebkar, A., Hosseini, S.M., Taghizadeh, M., Akbari, H., et al., 2014. Investigation of the effectiveness of *Syzygium aromaticum*, *Lavandula angustifolia* and *Geranium robertianum* essential oils in the treatment of acute external otitis: a comparative trial with ciprofloxacin. J. Microbiol. Immunol. Infect. 47, 211–216.

Park, I., 2019. The anesthetic effects of clove oil and MS-222 on far eastern catfish, *Silurus asotus*. Dev. Reprod. 23 (2), 183–191.

Park, A., Sim, Y., Lee, J., Kim, S., Kang, Y., Jung, J., et al., 2011. The analgesic effects and mechanisms of orally administered eugenol. Arch. Pharm. Res. 34 (3), 501–507.

Parkin, D.M., Bray, F., Ferlay, J., Jemal, A., 2014. Cancer in Africa 2012. Cancer Epidemiol. Biomark. Prev. 23, 953–966.

Pasay, C., Mounsey, K., Stevenson, G., Davis, R., Arlian, L., Morgan, M., et al., 2010. Acaricidal activity of eugenol based compounds against scabies mite. PLoS One 5 (8), e12079.

Pattanasiri, T., Taparhudee, W., Suppakul, P., 2017. Acute toxicity and anaesthetic effect of clove oil and eugenol on Siamese fighting fish, *Betta splendens*. Aquac. Int. 25, 163–175.

Pencarian Koleksi Perpustakaan Universitas Islam Indonesia, 2012. Penentuan sifat fisika dan kimia minyak cengkeh dan eugenol hasil isolasi kuncup bunga cengkeh (*Eugenia caryophyllata* Thunb) tipe zanzibar dari tolitoli sulawesi tengah. Retrieved from https://simpus.uii.ac.id/search_adv/?n=002222&l=100&b=I&j=SK.

Pérez-Conesa, D., Cao, J., Chen, L., McLandsborough, L., Weiss, J., 2011. Inactivation of *Listeria monocytogenes* and *Escherichia coli* O157:H7 biofilms by micelle-encapsulated eugenol and carvacrol. J. Food Prot. 74, 55–62.

Pillai, P., Ramaswamy, K., 2012. Effect of naturally occurring antimicrobials and chemical preservatives on the growth of *Aspergillus parasiticus*. J. Food Sci. Technol. 49, 228–233.

Pino, J.A., Marbot, R., Agüero, J., Fuentes, V., 2001. Essential oil from buds and leaves of clove (*Syzygium aromaticum* (L.) Merr. et Perry) grown in Cuba. J. Essent. Oil Res. 13 (4), 278–279.

Phothisuwan, S., Matan, N., Preechatiwong, W., 2018. Effects of clove oil released in active packages on shelf life extension of egg during storage under refrigerator. J. Sci. Technol. Ubon Ratchathani University 20 (2), 224–237.

Prado, J.M., Prado, G.H.C., Meireles, M.A.A., 2011. Scale-up study of supercritical fluid extraction process for clove and sugarcane residue. J. Supercrit. Fluids 56, 231–237.

Prasad, M.M., Seenayya, G., 2000. Effect of spices on the growth of red *halophylic cocci* isolated from salt cured fish and solar salt. Food Res. Int. 33, 793–798.

Priestley, C.M., Williamson, E.M., Wafford, K.A., Sattelle, D.B., 2003. Thymol, a constituent of thyme essential oil, is a positive allosteric modulator of human GABA receptors and a homo-oligomeric GABA receptor from *Drosophila melanogaster*. Br. J. Pharmacol. 140, 1363–1372.

Purbosari, N., Warsiki, E., Syamsu, K., Santoso, J., 2019. Natural versus synthetic anesthetic for transport of live fish: a review. Aquacult. Fish. 4, 129–133.

Purkait, S., Bhattacharya, A., Bag, A., Chattopadhyay, R.R., 2020. Synergistic antibacterial, antifungal and antioxidant efficacy of cinnamon and clove essential oils in combination. Arch. Microbiol. 202, 1439–1448.

Purseglove, J.W., Brown, E., Green, C., Robbins, S., 1981. Spices. vol. 1 Cambridge University Press, Cambridge.

Raina, V.K., Srivastava, S.K., Aggarwal, K.K., Syamasundar, K.V., Kumar, S., 2001. Essential oil composition of *Syzygium aromaticum* leaf from Little Andaman, India. Flavour Fragr. J. 16, 334–336.

Rajkowska, K., Kunicka-Styczńska, A., 2018. Typing and virulence factors of food-borne *Candida* spp. isolates. Int. J. Food Microbiol. 279, 57–63.

Rajkowska, K., Kunicka-Styczńska, A., Kunicka-Styczynska, 2019. Effect of clove and thyme essential oils on *Candida* biofilm formation and the oil distribution in yeast cells. Molecules 24, 1954.

Ramadan, M.F., Asker, M.M.S., Tadros, M., 2013. Lipid profile, antiradical power and antimicrobial properties of *Syzygium aromaticum* oil. Int. J. Fats Oils 64 (5), 509–520.

Rana, I.S., Rana, A.S., Rajak, R.C., 2011. Evaluation of antifungal activity in essential oil of the *Syzygium aromaticum* (L.) by extraction, purification and analysis of its main component eugenol. Braz. J. Microbiol. 42 (4), 1269–1277.

Ranasinghe, L., Jayawardena, B., Abeywickrama, K., 2002. Fungicidal activity of essential oils of *Cinnamomum zeylanicum* (L.) and *Syzygium aromaticum* (L.) Merr et L.M. Perry against crown rot and anthracnose pathogens isolated from banana. Lett. Appl. Microbiol. 35, 208–211.

Rathinam, P., Vijay Kumar, H.S., Viswanathan, P., 2017. Eugenol exhibits anti-virulence properties by competitively binding to quorum sensing receptors. Biofouling 33, 624–639.

Ratri, P.J., Ayurini, M., Khumaini, K., Rohbiya, A., 2020. Clove oil extraction by steam distillation and utilization of clove buds waste as potential candidate for eco-friendly packaging. J. Bahan Alam Terbarukan 9 (1), 47–54.

Razafimamonjison, G., Jahiel, M., Duclos, T., Ramanoelina, P., Fawbush, F., Danthu, P., 2014. Bud, leaf and stem essential oil composition of *Syzygium aromaticum* from Madagascar, Indonesia and Zanzibar. Int. J. Basic Appl. Sci. 3 (3), 224–233.

Razafimamonjison, G., Boulanger, R., Jahiel, M., Rakotoarison, M., Sandratriniaina, R., Rasoarahona, J., Ramanoelina, P., Fawbush, F., Lebrun, M., Danthu, P., 2016a. Pour l'optimisation de la qualite des produits du giroflier de Madagascar (clous et hulies essentielles): Étude des facteurs de leur variabilité. In: Duchafour, H., Razfimbelo-Andriamifidy, T., Rakotoarisoa, J., Ramamonjisoa, B., Rakotondravao (Eds.), Recherches Interdisciplinaires pour le Développement Durable et la Biodiversité Des Espaces Ruraux Malgaches. FSP PARRUR, Antananarivo, Madagascar, pp. 57–106.

Razafimamonjison, G., Boulanger, R., Jahiel, M., Ramanoelina, P., Fawbush, F., Lebruu, M., et al., 2016b. Variations in yield and composition of leaf essential oil from *Syzygium aromaticum* at various phases of development. Int. J. Basic Appl. Sci. 5 (1), 90–94.

Razafimamonjison, G., Jahiel, M., Ramanoelina, P., Fawbush, F., Danthu, P., 2013. Effects of phenological stages on yield and composition of essential oil of *Syzygium aromaticum* buds from Madagascar. Int. J. Basic Appl. Sci. 2 (4), 312–318.

Reineccius, G., 1994. Natural Flavouring Materials. Sourcebook of Flavours, second ed. Chapman & Hall, New York.

Reverchon, E., Marrone, C., 1997. Supercritical extraction of clove bud essential oil: isolation and mathematical modeling. Chem. Eng. Sci. 52 (20), 3421–3428.

Reynolds, J.E.F., 1989. Martindale: The Extra Pharmacopoeia, twenty-ninth ed. Pharmaceutical Press, London.

Rietjens, I.M.C.M., Cohen, S.M., Fukushima, S., Gooderham, N.J., Hecht, S., Marnett, L.J., et al., 2014. Impact of structural and metabolic variations on the toxicity and carcinogenicity of hydroxy and alkoxy-substituted allyl- and propenylbenzenes. Chem. Res. Toxicol. 27, 1092–1103.

Roura, E., Humphrey, B., Tedo Gipharraguerre, I., 2008. Unfolding the codes of short-term feed appetence in farm and companion animals. A comparative oronasal nutrient-sensing biology review. Can. J. Anim. Sci. 88, 535–558.

Saeed, M., Khan, M.S., Alagawany, M., Farag, M.R., Alqaisi, O., Aqib, A.I., et al., 2021. Clove (*Syzygium aromaticum*) and its phytochemicals in ruminant feed: an updated review. Rend. Lincei Sci. Fis. Nat.

Santin, J.R., Lemos, M., Klein-Júnior, L.C., Machado, I.D., Costa, P., de Oliveira, A.P., et al., 2011. Gastroprotective activity of essential oil of the *Syzygium aromaticum* and its major component eugenol in different animal models. Naunyn-Schmied. Arch. Pharmacol. 383, 149–158.

Sarkar, R.K., De, P., Chauhan, P.D., 2003. Bacteria-resist finish on cotton fabrics using natural herbal extracts. Indian J. Fibre Text. Res. 28, 322–331.

Saraphanchotiwitthaya, A., Khorana, N., Sripalakit, P., 2019. Comparative anti-inflammatory activity of eugenol and eugenyl acetate on the murine immune response in vitro. Songklanakarin J. Sci. Technol. 41 (3), 641–648.

Scalzo, R.L., Fibiani, M., Pietromarchi, P., Mandala, C., Torre, A.L., 2012. Effect of different fungicide treatments on grape, must and wine quality. Commun. Agric. Appl. Biol. Sci. 77 (3), 151–161.

Scholze, A., Jankowski, J., Pedraza-Chaverri, J., Evenepoel, P., 2016. Oxidative stress in chronic kidney disease. Oxidative Med. Cell. Longev. 2016, 1–2.

Shiferaw, Y., Kassahun, A., Tedla, A., Feleke, G., Abebe, A.A., 2019. Investigation of essential oil composition variation with age of *Eucalyptus globulus* growing in Ethiopia. Nat. Prod. Chem. Res. 7 (2), 360.

Singh, P., Jayaramaiah, R.H., Agawane, S.B., 2016. Potential dual role of eugenol in inhibiting advanced glycation end products in diabetes: proteomic and mechanistic insights. Sci. Rep. 7 (6), 18798.

Singh, A., Singh, R.K., Bhunia, A.K., Singh, N., 2003. Efficacy of plant essential oils as antimicrobial agents against Listeria monocytogenes in hotdogs. Lebensm. Wiss. Technol. 36, 787–794.

Smith, R.L., Waddell, W.J., Cohen, S.M., Feron, V.J., Marnett, L.J., Portoghese, et al., 2009. GRAS 24: the 24th publication by the FEMA expert panel presents safety and usage data on 236 new generally recognized as safe flavoring ingredients. Food Technol. 63 (6), 46–105.

Soderlund, D., 2012. Molecular mechanisms of pyrethroid insecticide neurotoxicity: recent advances. Arch. Toxicol. 86, 165–181.

Sohilait, H.J., Kainama, H., Nindatu, M., 2018. Essential oil components of *Eugenia caryophylata* leaf from West Amboina Island. In: Proceeding of The 3rd International Seminar on Chemistry. AIP Publishing, College Park.

Soonwera, M., Phasomkusolsil, S., 2016. Effect of *Cymbopogon citratus* (lemongrass) and *Syzygium aromaticum* (clove) oils on the morphology and mortality of *Aedes aegypti* and Anopheles dirus larvae. Parasitol. Res. 115 (4), 1691–1703.

Sosa, V., Zunino, P., 2009. Effect of Ibicella lutea on uropathogenic *Proteus mirabilis* growth, virulence and biofilm formation. J. Infect. Dev. Ctries. 3, 762–770.

Sparagano, O., Khallaayoune, K., Duvallet, G., Nayak, S., George, D., 2013. Comparing terpenes from plant essential oils as pesticides for the poultry red mite (*Dermanyssus gallinae*). Transbound. Emerg. Dis. 60 (s2), 150–153.

Srivastava, A.K., Srivastava, S.K., Syamsundar, K.V., 2005. Bud and leaf essential oil composition of *Syzygium aromaticum* from India and Madagascar. Flavour Fragance J. 20, 51–53.

Sulistyoningrum, A.S., Saepudin, E., Cahyana, A.H., Rahayu, D.U.C., Amelia, B., Haib, J., 2017. Chemical profiling of clove bud oil (*Syzygium aromaticum*) from Toli-Toli and Bali by GC-MS analysis. In: Proceeding of International Symposium on Current Progress in Mathematics and Sciences 2016. AIP Publishing, College Park.

Sritharet, N., Trongvanichnam, K., 2012. Effects of clove oil supplemented in diet on growth performances, fecal condition and blood parameters of weaned-growing swine. Thai Sci. Technol. J. 1, 57–68.

Suci, D., Mangundap, P., Mangundap, F., Lestari, W.I., 2013. Pemurnian eugenol minyak daun cengkeh dengan menggunakan proses adsorpsi. J. Teknik Kimia 7 (2).

Sun, G., Worley, S.D., 2005. Chemistry of durable and regenarable biocidal textiles. J. Chem. Educ. 82 (1), 60–64.

Tahir, M., Chuzaemi, H.S., Widodo, E., 2020. Chemical compounds and antioxidant contents of cloves leaves essential oil. Agric. Sci. J. 7 (1), 37–44.

Taskinen, M.R., 2003. Diabetic dyslipidaemia: from basic research to clinical practice. Diabetologia 46, 733–749.

The United States Pharmacopeia, 2006. Clove oil. In: United States Pharmacopeial Convention. National Formulary, Rockville, MD, p. 3315.

Tischer, J.S., Possan, H., Luiz, J., Malagutti, N.B., Martello, R., Valerio, A., et al., 2019. Synthesis of eugenyl acetate through heterogeneous catalysis. J. Essent. Oil Res. 31 (4), 312–318.

Toda, S., Ohnishi, M., Kimura, M., Toda, T., 1994. Inhibitory effects of eugenol and related compounds on lipid peroxidation induced by reactive oxygen. Planta Med. 60, 282.

Tombe, M., Kobayashi, K., Ma'mun, T., Sukamto, D., 1992. Eugenol dan daun tanaman cengkeh untuk pengendalian penyakit tanaman industri. Bul. Penelit. Tanam. Rempah. Obat. 8.

Torre, A.L., Caradonia, F., Matere, A., Battaglia, V., 2015. Using plant essential oils to control *Fusarium* wilt in tomato plants. Eur. J. Plant Pathol. 144, 487–496.

Tripathi, Y.C., Hazarika, P., 2014. Impact of harvesting cycle, maturity stage, drying and storage on essential oil content of patchouli leaves grown in northeast region of India. J. Essent. Oil Bear. Plants 17 (6), 1389–1396.

Tunça, M.T., Koca, I., 2019. Ohmic heating assisted hydrodistillation of clove essential oil. Ind. Crop. Prod. 141, 111763.

Turek, C., Stintzing, F.C., 2012. Impact of different storage conditions on the quality of selected essential oils. Food Res. Int. 46, 341–353.

Turkez, H., Togar, B., Tatar, A., Geyikoglu, F., Hacimuftuoglu, A., 2014. Cytotoxic and cytogenetic effects of α-copaene on rat neuron and N2a neuroblastoma cell lines. Biologia 69 (7), 936–942.

Upadhyay, A., Upadhyaya, I., Kollanoor-Johny, A., Venkitanarayanan, K., 2013. Antibiofilm effect of plant derived antimicrobials on *Listeria monocytogenes*. Food Microbiol. 36, 79–89.

Usai, M., Marchetti, M., Foddai, M., Caro, A.D., Desogus, R., Sanna, I., Piga, A., 2011. Influence of different stabilizing operations and storage time on the composition of essential oil of thyme (*Thymus officinalis* L.) and rosemary (*Rosmarinus officinalis* L.). LWT - Food Sci. Technol. 44, 244–249.

Van Balen, F.A., Smit, W.M., Zuithoff, N.P., Verheij, T.J., 2005. Clinical efficacy of three common treatments in acute otitis externa in primary care: randomised controlled trial. BMJ 327, 1201–1205.

Vidhya, N., Devaraj, S.N., 2011. Induction of apoptosis by eugenol in human breast cancer cells. Indian J. Exp. Biol. 49, 871–878.

Vieira, T.F., Araujo, M.F., Fernandes, M.J.G., Pereira, D.M., Fortes, A.G., Elisabete, M.S., et al., 2020. In silico identification of protein targets associated to the insecticide activity of eugenol derivatives. In: Proceeding of the 24th International Electronic Conference on Synthetic Organic Chemistry. MDPI, Basel.

Vladu, A.F., Marin, S., Neacsu, I.A., Trusca, R.D., Albukaya, M.G., Kaya, D.A., et al., 2020. Spongious fillers based on collagen-hydroxyapatite-eugenol acetate with therapeutic potential in bone cancer. Farmacia 68 (2), 313–321.

Voda, K., Boh, B., Vrtacnik, M., 2004. A quantitative structure–antifungal activity relationship study of oxygenated aromatic essential oil compounds using data structuring and PLS regression analysis. J. Mol. Model. 10, 76–84.

Waman, A.A., 2020. Essential oil composition of clove and nutmeg from Andaman and Nicobar Islands, India. Indian J. Biochem. Biophys. 57, 95–100.

Wei, A., Shibamoto, T., 2010. Antioxidant/lipoxygenase inhibitory activities and chemical compositions of selected essential oils. J. Agric. Food Chem. 58, 7218–7225.

Wenqiang, G., Shufen, L., Ruixiang, Y., Shaokun, T., Can, Q., 2007. Comparison of essential oils of clove buds extracted with supercritical carbon dioxide and other three traditional extraction methods. Food Chem. 101, 1558–1564.

Widayat, Cahyono, B., Hadiyanto, Ngadiwiyana, 2014. Improvement of clove oil quality by using adsorption-distillation process. Res. J. Appl. Sci. Eng. Technol. 7 (18), 3867–3871.

Widodo, E., Natsir, M.H., Djunaidi, I.H., Pitono, R., 2019. Effect of sardine oil and tomato powder with or without addition of clove oil on laying duck performances and digestibility of nutrients. Int. J. Eng. Adv. Technol. 8 (35), 361–363.

Wijaya, C., Jayuska, A., Alimuddin, A.H., 2015. Peningkatan rendemen minyak atsiri daun cengkeh (*Syzygium aromaticum*) dengan metode delignifikasi dan fermentasi. J. Kimia Khatulistiwa 4 (4), 15–20.

World Health Organization, 2002. WHO Monographs on Selected Medicinal Plants. World Health Organization, Geneva.

Xijun, W., Haijun, L., 2014. Method for Preparing Eugenol From Clove Oil Plant. China Patent. CN103819317A.

Yang, Y., Wei, M., Hong, S., 2014. Ultrasound-assisted extraction and quantitation of oils from *Syzygium aromaticum* flower bud (clove) with supercritical carbon dioxide. J. Chromatogr. A 1323, 18–27.

Yazdani, F., Mafi, M., Farhadi, F., Tabar-Heidar, K., Aghapoor, K., Mohsenzadeh, F., Darabi, H.R., 2005. Supercritical CO_2 extraction of essential oil from clove bud: effect of operation conditions on the selective isolation of eugenol and eugenyl acetate. Z. Naturforsch. 60b, 1197–1201.

Yostawonkul, J., Kitiyodom, S., Kaewmalun, S., Suktham, K., Nittayasut, N., Khongkow, M., et al., 2019. Bifunctional clove oil nanoparticles for anesthesia and anti-bacterial activity in Nile tilapia (*Oreochromis niloticus*). Aquaculture 503, 589–595.

Yun, W., Jilin, D., 2012. Method for Separating and Preparing Caryophyllene Oxide, Beta-Farnesene and Caryophyllene. China Patent, CN102603458B.

Zhang, C., Li, C., Abdel-Samie, M.A., Cui, H., Lin, L., 2020. Unraveling the inhibitory mechanism of clove essential oil against *Listeria monocytogenes* biofilm and applying it to vegetable surfaces. LWT - Food Sci. Technol. 134, 110210.

Zimmet, P.Z., Magliano, D.J., Herman, W.H., Shaw, J.E., 2014. Diabetes: a 21st-century challenge. Lancet Diabetes Endocrinol. 2, 56–64.

Further reading

EFSA FEEDAP Panel (EFSA Panel on Additives and Products or Substances used in Animal Feed), Rychen, G., Aquilina, G., Azimonti, G., Bampidis, V., Bastos, M.L., Bories, G., et al., 2017. Scientific Opinion on the safety and efficacy of an essential oil from *Origanum vulgare* subsp. *hirtum* (Link) letsw. var. *Vulkan* when used as a sensory additive in feed for all animal species. EFSA J. 15 (12), 5095.

FDA, 2021. Clove and its derivatives, 21 C.F.R. 184. 1257(a).

Santos, M.R.A., Innecco, R., 2004. Adubação orgânica e altura de corte da erva-cidreira brasileira. Hortic. Bras. 22 (2), 182–185.

CHAPTER

Secondary metabolites of clove (*Syzygium aromaticum*)

10

Niranjan Das[a], Akash Dey[b], Sudip Kumar Mandal[b], Debanjan Chatterjee[c], Rajan Logesh[d], and Hari Prasad Devkota[e]

Department of Chemistry, Iswar Chandra Vidyasagar College, Belonia, Tripura, India[a] Department of Pharmaceutical Chemistry, Dr. B.C. Roy College of Pharmacy and Allied Health Sciences, Durgapur, West Bengal, India[b] Department of Natural Products, National Institute of Pharmaceutical Education and Research, Ahmedabad, Gandhinagar, Gujarat, India[c] TIFAC-CORE in Herbal Drugs, Department of Pharmacognosy and Phytopharmacy, JSS College of Pharmacy (JSS Academy of Higher Education and Research), Rockland's, Ooty, Tamil Nadu, India[d] Graduate School of Pharmaceutical Sciences, Kumamoto University, Kumamoto, Japan[e]

1. Introduction

Syzygium aromaticum (L.) Merr. & L.M.Perry, also called clove tree, is an aromatic evergreen tree that grows to be around 9–12 m high and belongs to the Myrtaceae family (Cortés-Rojas et al., 2014; Begum et al., 2014). The aromatic flower buds of *S. aromaticum* are called cloves (Bao et al., 2012). The plant is native to the islands of Indonesia. Still, it has recently been farmed in different places worldwide, including Brazil, Haiti, India, Kenya, Madagascar, Malaysia, Mauritius, Mexico, Seychelles, Sri Lanka, and Tanzania (Cortés-Rojas et al., 2014). Clove has been traditionally used as a spice. It has a wide range of therapeutic potentials, including local anesthetic, carminative, stimulant, antimalarial, antimicrobial, antibiotic, antifungal, dentistry, anticancer, antioxidant, dyspepsia, stomach problems, nausea, vomiting, antiseptic, and analgesic efficacies (Cortés-Rojas et al., 2014; Begum et al., 2014; Kiran et al., 2019; Devkota and Adhikari-Devkota, 2020). Nevertheless, clove has acquired much interest among the other spices only due to its strong antimicrobial and antioxidant activities (Shan et al., 2005). One of the major essential oil constituents in clove, namely eugenol, has been used as a powerful local anesthetic in dentistry (Begum et al., 2014). The clove oil is generally used in root canal therapy in clinical dentistry in temporary fillings. It exhibits strong antimicrobial activity against oral bacteria related to dental caries and periodontal disease (Cai and Wu, 1996). It is reported that Indonesia is the world's largest clove producer around the world. The most recent statistics on clove production by Food and Agriculture Organization (FAO) reported that Indonesia produces 70% of the world's cloves and is mostly cultivated by smallholder farmers. In addition, Madagascar is the next largest producer and produced around 16% of the world's cloves (World Bank, 2017). Two countries, namely Madagascar and Zanzibar (Tanzania), are Africa's largest clove suppliers and producing 20,000–27,000 tonnes per year and at the same time Comoros, Kenya, and Togo, as well as Sri Lanka, Malaysia, China, and Grenada, producing 5000–7000 tonnes per year (Nurdjannah and Bermawie, 2012). The geographical distribution of cloves with their average annual productions is shown in Fig. 1. The major

Clove (*Syzygium aromaticum*). https://doi.org/10.1016/B978-0-323-85177-0.00020-3

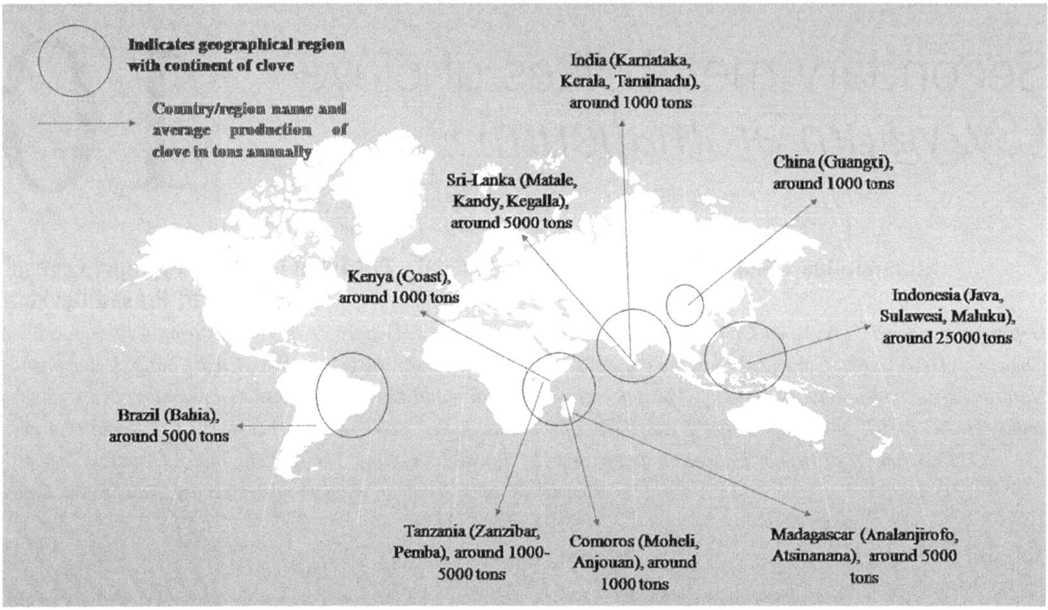

FIG. 1

Geographical distribution of clove is depicted above, and is found that it is cultivated in some parts of India, China, Kenya, and Comoros (average annual production 1000 tons), in Tanzania (average annual production 1000–5000 tons), in Sri Lanka, Madagascar, and Brazil (average annual production 5000 tons), and in Indonesia (25,000 tons) (Danthu et al., 2020).

countries involved in the production and exportation of cloves around the world with estimated average values in tons from the year 2012 to 2018 are shown in Table 1.

2. Economic value of clove

Clove is mainly cultivated by the small farmers of Indonesia, Madagascar, Tanzania, Sri Lanka, and Comoros. The major clove production countries worldwide are Indonesia and Madagascar, followed by Tanzania, Sri Lanka, Brazil, Comoros, Kenya, China, and India (Danthu et al., 2020). For the past few years, the annual global demand for cloves has seen little development. There are significant direct trade exchanges of cloves between producers and consumers, especially between Brazil and the countries of North and South America and between Madagascar and Sri Lanka and Asia and Europe. The traditional cigarette industry in Indonesia is the largest industrial consumer of cloves. According to Mintel Global Group, clove cultivation started in Java around the 1870s and now accounts for up to 90% of Indonesia's output (Hanusz, 2000; World Bank, 2017). A significant percentage of cloves grown in Zanzibar and Comoros are exported to Zanzibar and the rest of Africa's coast to help make up for domestic shortages. There are nearly half a billion adult male smokers in Indonesia, with 91%

Table 1 The major countries involved in the production and exportation of clove around the world.

Country	Production of clove (in tons)	Exportation of clove (in tons)
Madagascar	19,405	19,194
Indonesia	122,454	10,748
Singapore	–	8642
Tanzania	6829	≈5020 (for the year 2014–18)
Comoros	2328	4327
Sri Lanka	5517	3834
Brazil	≈5000	3080
United Arab Emirates	–	2336
Europe (including Netherlands)	–	≈1646 (686)
India	1220	964
Kenya	1973	372 (for the year 2012–17)
China	1229	105

These are estimated average figures in tons, covering the years from the year 2012 to 2018 (Danthu et al., 2020).

preferring kretek, reflecting the entire 54 million-strong population (Palipudi et al., 2015). Cloves are used more in the agro-food sector than in any other industry. It is available in two forms: whole and crushed. Clove is used in various gastronomic preparations, such as savory and food and alcoholic foods, on every continent except Japan (Danthu et al., 2020).

3. Phytochemistry

The phytochemical analysis of *S. aromaticum* has reported various classes of chemical compounds such as sesquiterpenes, monoterpenes, hydrocarbons, and phenolic compounds (Batiha et al., 2020). Clove is a major source of phenolic compounds such as flavonoids, hydroxybenzoic acids, hydroxyphenylpropenes, hydroxycinnamic acids, and eugenol (Cortés-Rojas et al., 2014; Batiha et al., 2020). The significant bioactive constituents that have been reported from the different parts of *S. aromaticum* included eugenol, *β*-caryophyllene, eugenitin, eugenin, carvacrol, biflorin, vanillin, ellagic acid, gallic acid, kaempferol, rhamnetin, myricetin, quercetin, crategolic acid (maslinic acid), oleanolic acid, campesterol, stigmasterol, and bicornin (Fig. 2) (Batiha et al., 2020). The clove is a rich source of gallic acid derivatives such as hydrolyzable tannins, obtained in large quantities in the fresh plant (Cortés-Rojas et al., 2014; Bao et al., 2012; Batiha et al., 2020). An aqueous acetone extract of dried clove reported the presence of 18 hydrolyzable tannins, including a new *C*-glucosidic ellagitannin named aromatinin A (Bao et al., 2012). The hydrolyzable tannins included syzyginin A, bicornin, platycaryanin A, alunusnin A, rugosin C, 1,2,3-tri-*O*-galloyl-*β*-D-glucose, 1,2,3,6-tetra-*O*-galloyl-*β*-D-glucose, tellimagrandin II, casuarictin, heterophylliin D, rugosin D, rugosin F, euprostin A, 1,2-di-*O*-galloyl-3-*O*-digalloyl-4,6-*O*-(*S*)-hexahydroxydiphenoy-*β*-D-glucose, alienanin B, squarrosanin A, and casuarinin (Bao et al.,

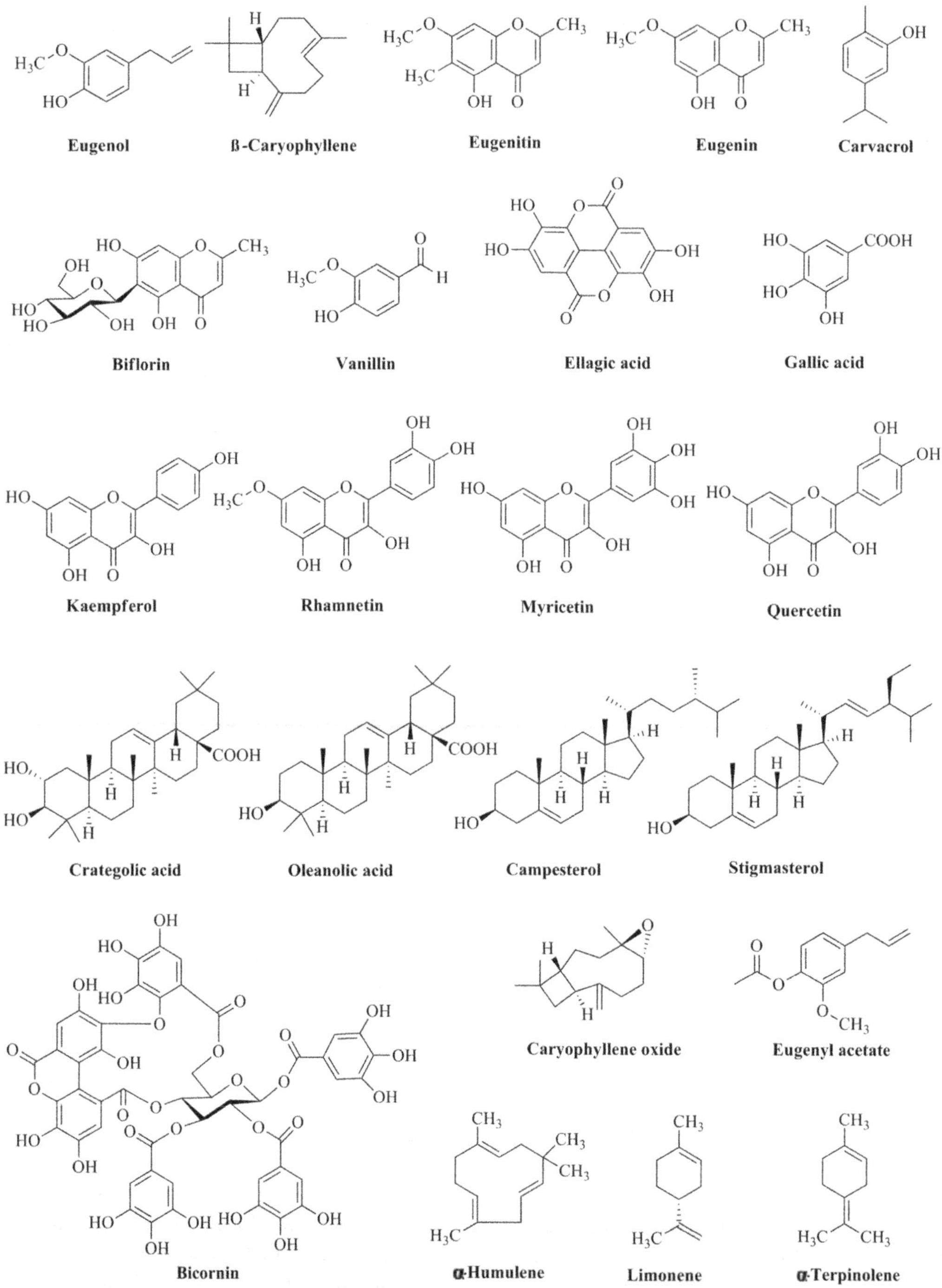

FIG. 2

Chemical structures of major bioactive compounds along with major essential oil constituents reported from the different parts of clove (*S. aromaticum*).

2012). Flavonoids such as quercetin and kaempferol and phenolic acids such as ferulic, caffeic, ellagic, and salicylic acids are also present in clove (Cortés-Rojas et al., 2014). Bioassay-guided chromatographic fractionation of crude MeOH extract of dried buds of *S. aromaticum* reported the presence of seven antibacterial phenolic constituents, namely 7-dihydroxy-2-methylchromone 8-C-*β*-D-glucopyranoside, biflorin, kaempferol, rhamnocitrin, myricetin, gallic acid, ellagic acid, and along with one antibacterial triterpenoid constituent, oleanolic acid (Cai and Wu, 1996).

The methanol extract of *S. aromaticum* contains several alkaloids, carbohydrates, flavonoids, ketones, terpenoids, glycosides, tannins, aldehydes, and steroids (Jimoh et al., 2017). The gas chromatography coupled with flame ionization detector (GC-FID) analysis of *S. aromaticum* reported 46 bioactive phenolic compounds. The significant compounds included gallic acid, syringic acid, protocatechuic acid, caffeic acid, eugenin, eugenitin, *p*-hydroxybenzoic acid, salicylic acid, kaempferol, quercetin, rhamnetin, phenylacetic acid, myricetin, and isorhamnetin (Fig. 3) (Jimoh et al., 2017). A tetracyclic triterpenoid compound such as limonin and an aromatic aldehyde, ferulic aldehyde, and eugenol are also isolated from the dichloromethane extract of buds of *S. aromaticum*

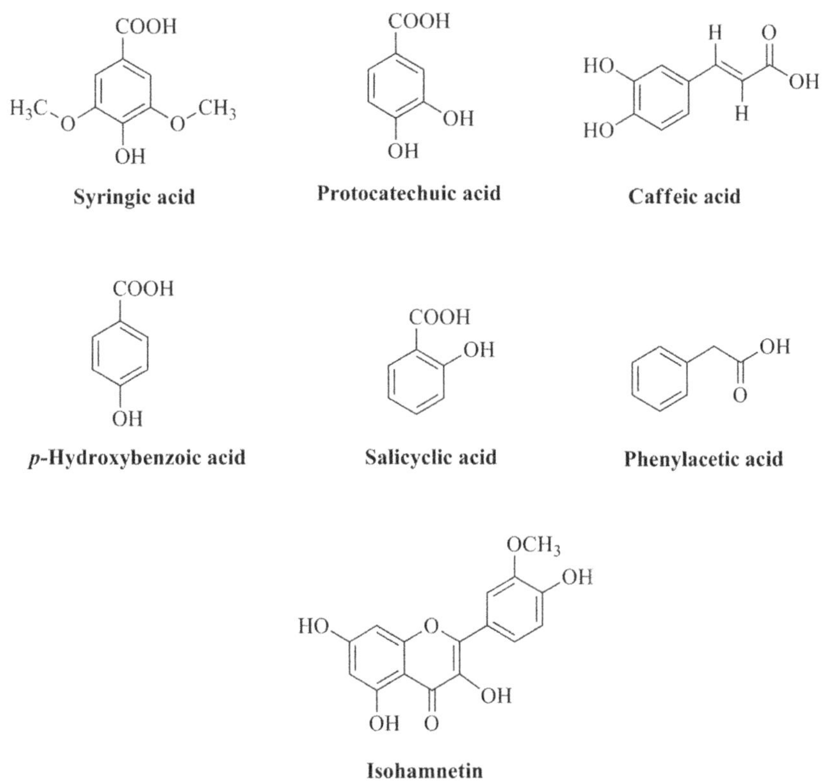

FIG. 3

Chemical structure of other phenolic compounds reported from different parts of clove (*S. aromaticum*).

(Nassar et al., 2007). Tamarixetin 3-*O*-*β*-D-glucopyranoside, ombuin 3-*O*-*β*-D-glucopyranoside, and quercetin are the flavonoids reported from the ethanol extract of buds of *S. aromaticum* (Nassar et al., 2007). The phytochemical analysis of *S. aromaticum* also reported the presence of several bioactive compounds such as oleanolic acid lactone, *β*-sitosterol, nigricin, flavaellagic acid, 2*α*-hydroxyolanolic acid, 3*β*-hydroxy-11-oxo-olean-12-en-28-oic acid, and *β*-sitosterol 3-*O*-*β*-D-glucopyranoside (Begum et al., 2014). Two new secondary metabolites such as a ferulic acid derivative, sabrinic acid, and a benzophenone derivative, 4,4′-diacetoxybenzophenone, along with few existing significant compounds, namely kaempferol-3,5-dimethyl ether, 4-methyl benzoic acid, asiatic acid, arjunolic acid, 5-hydroxy-4′-methoxy-6,7-methylenedioxy isoflavone, and 5,4′-dimethoxy-6,7-methylenedioxy isoflavone, have been isolated from *S. aromaticum* (Kiran et al., 2019; Ali et al., 2018).

The essential oil of clove buds has excellent pharmacological and biological potentials and is extensively used in the medicine, fragrance, and flavoring industries (Xu et al., 2016). The phytochemical analysis of dichloromethane extract of clove buds oil showed terpenoids, steroids, tannins, glycosides, sterols, carbohydrates, and phenolic compounds (Kumar et al., 2010). The essential oils constituents in clove buds reported the existence of monoterpenes (89.1%), oxygenated monoterpenes (8.06%), sesquiterpenes (10.8%), and linear components (0.03%) (Fankem et al., 2017). It is reported that monoterpenes are the major constituents as essential oils in the clove buds (Fankem et al., 2017; Mohamed et al., 2018). The flower buds of clove comprise 18% essential oil, eugenol, eugenol acetate, and *β*-caryophyllene (Jirovetz et al., 2006). The most significant constituents in clove oil are eugenyl acetate, eugenol, and *β*-caryophyllene (Alma et al., 2007; Marya et al., 2012; Batiha et al., 2020). The major essential oil constituents present in *S. aromaticum* are eugenol, eugenyl acetate, *β*-caryophyllene, limonene, and *α*-terpinolene (Fig. 2) (Jimoh et al., 2017). The gas chromatography coupled with mass spectrometry (GC-MS) analysis of the *n*-hexane extract of the buds of *S. aromaticum* has reported the presence of 16 volatile compounds, and out of that, eugenol (71.5%) and eugenol acetate (8.99%) are the major compounds (Nassar et al., 2007). The GC and GC-MS analysis of essential oils of clove buds reported the existence of 22 constituents and out of that eugenol (76.2%), *β*-caryophyllene (11.5%), caryophyllene oxide (4.29%), and eugenyl acetate (1.76%) (Fig. 2) were found to be the major components followed by *α*-caryophyllene (0.64%), *β*-Selinene (0.25%), *α*-Selinene (0.16%), eucalyptol (0.14%), 4-allylanisole (0.13%), (−)-*β*-cadinene (0.12%), anethol (0.11%), chavicol (0.09%), jasmone (0.07%), methyl salicylate (0.06%), *α*-copaene (0.05%), globulol (0.04%), *α*-Pinene (0.03%), ledol (0.03%), 2-pinene (0.02%), cedrene (0.02%), *α*-muurolene (0.01%), and valencene (0.01) (Xu et al., 2016). The GC-MS analysis of hydro-distillation fraction of clove essential oil revealed 36 compounds, including eugenol, *β*-caryophyllene, eugenyl acetate, ethyl hexanoate, 2-heptanone, *α*-humulene, calacorene, humulenol, and calamen-ene as major constituents (Batiha et al., 2020). The essential oil of the clove leaf contains a characteristic pleasant odor and light-yellow color. The existence of 23 constituents has been reported by the analysis of GC and GC-MS of essential oils of clove leaf. Among them, eugenol (76.8%), *β*-caryophyllene (17.4%), *α*-humulene (2.1%), and eugenyl acetate (1.2%) are the major constituents (Fig. 2) (Jirovetz et al., 2006). A list of reported various phytochemicals from the different parts of clove (*S. aromaticum*) is given in Table 2.

Table 2 List of phytochemicals reported from the flower buds of clove (*S. aromaticum*).

Phytochemical	References
Phenolic compounds	
2-Phenylethyl-*β*-*O*-glucoside	Jimoh et al. (2017)
3-*O*-Caffeoylquinic	Jimoh et al. (2017)
Apigenin	Jimoh et al. (2017)
Caffeic acid	Jimoh et al. (2017)
Carvacrol	Jimoh et al. (2017)
Catechin	Jimoh et al. (2017)
Chlorogenic acid	Jimoh et al. (2017)
Cinnamic acid	Jimoh et al. (2017)
Coumarin	Jimoh et al. (2017)
Curcumin	Jimoh et al. (2017)
Daidzein	Jimoh et al. (2017)
Delphinidin	Jimoh et al. (2017)
Epicatechin	Jimoh et al. (2017)
Epigallocatechin	Jimoh et al. (2017)
Eugenin	Jimoh et al. (2017)
Eugenitin	Jimoh et al. (2017)
Ferulic acid	Jimoh et al. (2017)
Gallic acid	Jimoh et al. (2017)
Genistein	Jimoh et al. (2017)
Gentisic acid	Jimoh et al. (2017)
Gingerol	Jimoh et al. (2017)
Glycitein	Jimoh et al. (2017)
Guaiacol	Jimoh et al. (2017)
Isohamnetin	Jimoh et al. (2017)
Kaempferol	Jimoh et al. (2017)
Luteolin	Jimoh et al. (2017)
Malvidin	Jimoh et al. (2017)
Myricetin	Jimoh et al. (2017)
Naringenin	Jimoh et al. (2017)
Naringenin chalcone	Jimoh et al. (2017)
O-Coumaric acid	Jimoh et al. (2017)
p-Coumaric acid	Jimoh et al. (2017)
Pentunidin	Jimoh et al. (2017)
Phenylacetic acid	Jimoh et al. (2017)
Phloroglucinol	Jimoh et al. (2017)
p-Hydroxybenzoic acid	Jimoh et al. (2017)
Piperic acid	Jimoh et al. (2017)
Protocatechuic acid	Jimoh et al. (2017)

Continued

Table 2 List of phytochemicals reported from the flower buds of clove (*S. aromaticum*)—cont'd

Phytochemical	References
Quercetin	Jimoh et al. (2017)
Rhamnetin	Jimoh et al. (2017)
Rosmarinio acid	Jimoh et al. (2017)
Rutin	Jimoh et al. (2017)
Salicyclic acid	Jimoh et al. (2017)
Sinapinic acid	Jimoh et al. (2017)
Syringic acid	Jimoh et al. (2017)
Vanillic acid	Jimoh et al. (2017)
Monoterpenes	
1,8-Cineole	Fankem et al. (2017), Jimoh et al. (2017)
Borneol	Jimoh et al. (2017)
Borneol acetate	Jimoh et al. (2017)
Camphenol	Fankem et al. (2017)
Carveol	Fankem et al. (2017)
Dihydro-eugenol	Fankem et al. (2017)
Eugenol	Fankem et al. (2017), Jimoh et al. (2017), Chaieb et al. (2007)
Eugenyl acetate	Fankem et al. (2017), Chaieb et al. (2007)
Fenchone	Fankem et al. (2017)
Geranial	Jimoh et al. (2017)
Geraniol	Fankem et al. (2017)
Isoeugenol	Fankem et al. (2017)
Limonene	Jimoh et al. (2017)
Linalool	Jimoh et al. (2017)
Linalyl acetate	Jimoh et al. (2017)
Myrcene	Jimoh et al. (2017)
Neral	Jimoh et al. (2017)
p-Cymene	Fankem et al. (2017), Nassar et al. (2007)
Pinene-2-ol	Jimoh et al. (2017)
Sabinene	Jimoh et al. (2017)
Terpinen-4-ol	Jimoh et al. (2017)
Terpinolene/α-terpinolene	Fankem et al. (2017), Jimoh et al. (2017)
α-Pinene	Jimoh et al. (2017)
α-Terpineol	Jimoh et al. (2017)
α-Thujene	Jimoh et al. (2017)
β-Pinene	Jimoh et al. (2017)
Υ-Terpinene	Jimoh et al. (2017)
Sesquiterpenes	
Bicyclogermacrene	Jimoh et al. (2017)
Calacorene	Chaieb et al. (2007)

Table 2 List of phytochemicals reported from the flower buds of clove (*S. aromaticum*)—cont'd	
Phytochemical	**References**
Caryophyllene oxide	Jimoh et al. (2017), Nassar et al. (2007)
Cubenol	Fankem et al. (2017)
Elemol	Fankem et al. (2017)
epi-(*E*)-Caryophyllene	Fankem et al. (2017)
Germacrene D	Jimoh et al. (2017), Fankem et al. (2017)
Guaiol	Fankem et al. (2017), Nassar et al. (2007)
Nootkatin	Nassar et al. (2007)
Spathulenol	Fankem et al. (2017)
trans-Isolongifolanone	Nassar et al. (2007)
α-Bergamotene	Jimoh et al. (2017)
α-Copaene	Jimoh et al. (2017)
α-Humulene/α-caryophyllene	Jimoh et al. (2017), Chaieb et al. (2007)
β-Bisabolene	Fankem et al. (2017), Jimoh et al. (2017)
β-Bisabolol	Fankem et al. (2017)
β-Caryophyllene	Fankem et al. (2017), Chaieb et al. (2007)
β-Elemene	Jimoh et al. (2017)
β-Selinene	Fankem et al. (2017)
γ-Eudesmol	Fankem et al. (2017)
γ-Gurjunene	Fankem et al. (2017)
Δ-Cadinene	Fankem et al. (2017)
δ-Elemene	Fankem et al. (2017)
Linear compounds	
2-Heptanone	Chaieb et al. (2007)
3(*E*)-Hexenol	Fankem et al. (2017)
4-Heptanol	Fankem et al. (2017)
5-Hexene-2-one	Nassar et al. (2007)
9,17-Octadeca-dienal	Nassar et al. (2007)
Alloocimene	Jimoh et al. (2017)
cis-Ocimene	Jimoh et al. (2017)
Ethyl hexanoate	Chaieb et al. (2007)
Hexadecanoic acid	Nassar et al. (2007)
Isoartemisia	Jimoh et al. (2017)
Octadecanoic acid butyl ester	Nassar et al. (2007)
Aromatic compounds	
Acetyleugenol	Jimoh et al. (2017)
Benzene-1-butylheptyl	Nassar et al. (2007)
Benzyl alcohol	Jimoh et al. (2017)
Benzyl benzoate	Jimoh et al. (2017)
Calamenene	Chaieb et al. (2007)

Continued

Table 2 List of phytochemicals reported from the flower buds of clove (*S. aromaticum*)—cont'd

Phytochemical	References
Dodecatrienoic acid-3,7,11-trimethylethyl ester	Nassar et al. (2007)
Elemicin	Jimoh et al. (2017)
Ethyl cinnamate	Jimoh et al. (2017)
Phenanthrene	Jimoh et al. (2017)
Dehydrodiisoeugenol	Nassar et al. (2007)
Thymol	Nassar et al. (2007)
Thymol methyl ether	Jimoh et al. (2017)
Vitamin E acetate	Nassar et al. (2007)

4. Biological activities

The different parts of the *S. aromaticum* (clove) exhibited a wide range of biological activities, including antibacterial, antioxidant, antifungal, anti-inflammatory, anticancer, nematicidal, anesthetic, herbicidal, insecticidal activities. The significant biological applications of clove are described below.

4.1 Antibacterial activity

Clove essential oil has been shown to have antibacterial activity against *Staphylococcus aureus* in pasteurized milk and *Listeria monocytogenes* in pasteurized milk (Cava et al., 2007; Matan, 2012). Clove oil displayed high antimicrobial resistance to *Penicillium* sp., *Aspergillus flavus*, and *Staphylococcus aureus* found on dried fish (*Decapterus maruadsi*) (Matan, 2012; Zengin and Baysal, 2015). Clove oil has antimicrobial activity against three gram-positive bacteria such as *Listeria innocua*, *Carnobacterium divergens*, *Staphylococcus aureus*, and four gram-negative bacteria, *Salmonella typhimurium*, *Escherichia coli*, *Serratia liquefaciens*, and *Shewanella putrefactions,* using the broth microdilution method (Kaur and Kaushal, 2019). Clove essential oil prevented the growth of all bacteria, except *Listeria* and *Shewanella*, which were reported to be oil resistant. Clove oil was also found to be effective against gram-negative bacteria (*Penicillium aeruginosa*, *Escherichia coli*, *Staphylococcus chloeraesius*, and *Yersinia enterocolitica*) and gram-positive bacteria (*Penicillium aeruginosa*, *Escherichia coli*, *Staphylococcus chloeraesius*, and *Yersinia enter*, *Bacillus cereus*, *Listeria monocytogenes*, *Staphylococcus aureus*, and *Enterococcus faecalis*) (Mytle et al., 2006).

When *Listeria monocytogenes* was treated with 1% and 2% clove oil, Mytle et al. (2006) found that the growth rate of the bacteria was reduced. The presence of eugenol, 2-heptanone, methyl salicylate, kaempferol, gallic acid, isoeugenol, and oleanolic acid in clove oil was reported by Fu et al. (2007), Yang et al. (2003), and Chaieb et al. (2007). These chemicals metabolized proteins that reacted with cell membranes, causing them to lose their permeability. Using an agar diffusion test, Warnke et al. (2009) found that clove bud oil had antimicrobial activity against six *Staphylococcus* strains, including methicillin-resistant *Staphylococcus aureus* (MRSA) three *Candida* strains, and four *Streptococcus* strains.

Clove oil exhibited significant antimicrobial activity, with an inhibition zone diameter of 12–20 mm. Using the paper disk diffusion method, Duraipandiyan et al. (2006) investigated the antibacterial activity of clove oil against *Enterococcus facalis*, *Bacillus subtilis*, *Ervinia* sp., *Staphylococcus epidermidis*, and *Proteus vulgaris*. In addition, clove oil inhibited bacterial colonization of *Klebsiella pneumoniae* in the lungs of mice, according to Saini et al. (2009).

4.2 Antioxidant activity

The presence of phenolic compounds such as eugenol, thymol, and eugenol acetate in clove oil resulted in high antioxidant activity (Yadav and Bhatnagar, 2007; Dai et al., 2013; Nam and Kim, 2013). The antioxidant function of eugenol in clove oil was comparable to synthetic antioxidants, pyrogallol, and BHA (Dorman et al., 2000). When treated with 15 μg/mL clove oil, Gulcin et al. (2012) reported a 97% inhibition of lipid peroxidation of linoleic acid emulsion. Under the same conditions, standard antioxidants like Trolox, butylated hydroxy anisole (BHA), tocopherol, and butylated hydroxy toluene (BHT) showed 95.6%, 95.4%, 84.6%, and 99.7% inhibition, respectively. According to Abdel-Wahhab and Aly (2005), phenolic compounds in clove essential oil inhibited the CYP_{450} enzyme, resulting in the formation of epoxide aflatoxin B1 and increased the ability of liver microsomes to catalyze aflatoxin-glutathione conjugation. According to a recent report, clove essential oil has 10 times the antiradical activity of BHT and seven times the antiradical activity of a mixture of cocoa butter and clove essential oil (Fankem et al., 2017).

4.3 Antifungal activity

Clove oil and eugenol have been shown to have antifungal efficacy against filamentous fungi, yeast as human pathogenic fungi (Gayoso et al., 2005), and food-borne fungi (Hammer et al., 1999; Pinto et al., 2009). Clove oil destroyed *Candida albicans* by causing lesions in the plasma membrane due to the presence of carvacrol, according to Pina-Vaz et al. (2004). Clove oil's antifungal potential was investigated using the disk diffusion method against Candida species: *Candida gullermondi*, *Candida albicans*, *Candida glabrata*, and *Candida parapsilosis*. All of which were isolated from bloodstream infections. According to the findings, clove oil had the strongest antifungal activity against *Candida tropicalis*, *Candida albicans*, and *Candida guilliermondii* (Kumar et al., 2012).

Chami et al. (2005) claimed that clove oil's antifungal activity was due to the existence of phenolic components (eugenol and carvacrol), while Pinto et al. (2009) claimed that clove oil's inhibitory function was due to a decrease in the amount of ergosterol, a particular component of fungal cell membranes. By using the agar dilution process, Viuda-Martos et al. (2007) demonstrated the antifungal ability of clove oil against the food spoilage fungi *Aspergillus flavus* and *Aspergillus niger*. Clove oil was found to be more effective at inhibiting *Aspergillus niger* than *Aspergillus flavus*. Bansod and Rai (2008) used three methods to demonstrate the antifungal activity of various essential oils, including clove oil, against *Aspergillus niger* and *Aspergillus fumigates* disk diffusion broth dilution and agar dilution. Clove oil had a zone of inhibition ranging from 10 to 15 mm for *Aspergillus fumigates* and 8 to 14 mm for *Aspergillus niger* in the disk diffusion process, with concentrations ranging from 12.5 to 100 mg. For both the fungi *Aspergillus fumigates* and *Aspergillus niger*, the minimum inhibitory concentration (MIC) of clove oil obtained by the agar dilution method was 0.12% (v/v).

Similarly, clove oil's MIC and minimum cidal concentration (MCC) for *Aspergillus fumigates* by broth microdilution method were 0.06% (v/v) and 0.12% (v/v) and values for *Aspergillus niger* (% v/v) were 0.12 and 0.06, respectively. Tullio et al. (2007) used a vapor contact and broth microdilution assay to show that clove essential oil has antifungal activity against filamentous fungi. Clove essential oil had a greater inhibitory effect in the vapor process than in the broth microdilution assay. Using the oxford cup process, Estrada-Cano et al. (2017) demonstrated that encapsulated clove oil has a good antifungal inhibitory effect against *Fusarium oxysporum*. The results revealed that naked clove oil had the strongest inhibitory effect on *Fusarium oxysporum* at the onset of the study. Micro capsulated clove oil produced maximum efficiency after 8h. The antifungal effect of clove oil microcapsule on meat products was investigated by Wang et al. (2018). According to the findings, clove oil microcapsules had antiseptic effects on meat products at concentrations greater than 0.07%. Boiling meat for half an hour improved the productivity of clove oil microcapsules to 0.08% (Wang et al., 2018).

4.4 Anti-inflammatory activity

Al-Ameedi et al. (2017) used a formalin test with 24 mice divided into four classes to assess the anti-inflammatory activity of alcoholic *S. aromaticum* extract (SAE). T1 and T2 groups were fed 100 and 200mg/kg (SAE), respectively, while T3 and T4 were fed 0.3mg/kg meloxicam and distilled water. In animals exposed to different concentrations of SAE, the findings showed a substantial increase in analgesia time and a decrease in licking number. At a concentration of 0.011%, Han and Parker (2017) found that clove essential oil had significant antiproliferative effects on human dermal fibroblasts. Furthermore, it was particularly resistant to the expression of many pro-inflammatory biomarkers such as interferon-inducible T-cell-chemoattractant (I-TAC), vascular cell adhesion molecule-1 (VCAM-1), and collagen III at both the gene and protein levels.

Using an enzyme-linked immunosorbent assay, Bachiega et al. (2012) stated that clove and eugenol had an anti-inflammatory effect on cytokine development in vitro (ELISA). Clove (100μg/mL) inhibited the development of IL-10, IL-1, and IL-6, while eugenol (50–100μg/mL) inhibited the production of IL-10 and IL-6. Using ELISA, Koh et al. (2013) demonstrated that eugenol has anti-inflammatory properties in IL-stimulated pulp cells and gingival fibroblasts. Eugenol was found to have anti-inflammatory properties in pulp cells but not in gingival fibroblasts (HGFs).

Using an ELISA test, Rodrigues et al. (2009) discovered that clove oil and eugenol inhibited macrophage synthesis of IL-1 and IL-6. Clove oil has anti-inflammatory and antipyretic properties, according to Nikoui et al. (2017). Clove oil treatment reduced band neutrophils, neutrophils, and white blood cells but no effect on hematocrit or red blood cells (Nikoui et al., 2017).

4.5 Anticancer activity

Kumar et al. (2014) used the MTT and brine shrimp lethality test (BSLT) assay to investigate the anticancer capacity of various concentrations of water, ethanol extract, and essential oil of clove in vitro against MCF-7 human breast cancer cells. Essential oil had a strong cytotoxic effect in both MTT and BSLT, with an LD_{50} of 37μg/mL in BSLT after 24h. The IC_{50} values for the MTT assay after 24 and 48h were 36.4 and 17.6μg/mL, respectively. Clove essential oil contains phenylpropanoids and terpenoids that have antitumor activity in cell lines and tumors in animals, according to Lesgards et al. (2014).

The protein caspases cause cancer cells to commit apoptosis (cell death). In addition, the formation of free radicals in cancer cells, the overexpression and control of liver detoxification enzymes, and the inhibition of angiogenesis were all observed. Using the MTT assay, Dwivedi et al. (2011) investigated the anticancer capacity of clove oil, its ethanol, and water extract against DU-145 prostate cancer, HeLa cervical cancer, TE-13 esophageal cancer, MDA-MB-231 (ER – ve), and MCF-7 (ER + ve) breast cancer, and normal human peripheral blood lymphocytes. Clove oil at 300 μL/mL caused the most cytotoxic activity and cell death (80%) in TF-13 cells within 24 h, while the same dose caused moderate cell death in DU-145 cells but no cytotoxicity in human peripheral blood mononuclear cells.

Using western blotting analysis during benzopyrene (BP)-induced lung carcinogenesis in strain A mice, Banerjee et al. (2006) demonstrated the chemopreventive capacity of aqueous clove infusion for lung cancer. Clove infusion decreased the number of proliferating cells while increasing the number of apoptotic cells. Via an MTT colorimetric assay with IC_{50} values ranging from 15.7 to 200 μg/mL, Kouidhi et al. (2010) investigated the cytotoxic and anticarcinogenic effect of clove essential oil on normal cells (MRC-5) and cancer cell lines (A549, raw 269.7, HT29, and Hep2).

4.6 Nematicidal activity

Using a greenhouse assay, Taniwiryono et al. (2009) investigated the nematicidal activity of *S. aromaticum* against *Meloidogyne incognita* root-knot nematode. They found that clove was successful in killing nematodes. Still, the EC_{50} value was 5–10 times lower than the EC_{50} value of synthetic pesticides chlorpyrifos, deltamethrin. At a concentration of 0.2 mL/L, Bala and Sukul (1987) found that eugenol, a major constituent of clove, effectively decreased root-knot nematode infection and stimulated plant growth in *Hibiscus esculentus*. Meyer et al. (2008) found the nematicidal suppression of *Meloidogyne incognita* (southern root-knot) nematode and its phytotoxicity on *Cucumis melo* (muskmelon), *Solanum lycopersicum* (tomato), *Cumumis sativus* (cucumber), and *Capsicum annuum* (pepper) seedlings in a greenhouse at concentrations within 0.1%–0.3%. According to the findings, clove oil decreased the nematode population and was not phytotoxic to the tested vegetable crops.

4.7 Anesthetic effect

Under semi-arid conditions, Diyaware et al. (2017) demonstrated the anesthetic effect of clove seed extract on *Clarias gariepinus*. The study found that mortality rates were higher in fish anesthetized with clove seed extract concentrations of 150 and 125 mg/L. According to Fujimoto et al. (2018), clove oil reduces muscle contraction capacity. It induces anesthesia in three Amazonian fish species: *Pterophyllum scalare* (angelfish), *Heros severus* (banded cichlid), and *Parachheirodon axelrodi* (cardinal tetra).

The anesthetic effect of clove oil-loaded nanoemulsion (CLN) on *Oreochromis niloticus* was investigated by Kheawfu et al. (2017). Clove oil, at a concentration of 90 μL/L, induced deep anesthesia (less than 3 min) in cardinal tetra and angelfish, while clove oil, at a concentration of 60 μL/L, induced anesthesia in banded cichlids. The best CLN contained 20% (w/w) clove oil and 15% (w/w) polysorbate 20. In Nile tilapia, the CLN had a stronger anesthetic effect than the clove ethanolic solution.

Kheawfu et al. (2017) looked at how clove oil and eugenol affected the muscle contraction of silkworms (*Bombyx mori*). In addition to D-glutamic acid, which caused complete loss of equilibrium and muscle tone in fish at stage 4 at 20 and 40 mM, the results showed that eugenol and clove oil at 60 ppm induced anesthesia in stage 4 fish. The anesthetic effect of eugenol on *Oreochromis niloticus* larvae and juveniles was also recorded by Ribeiro et al. (2015). The result revealed that a larva weighing 11.6 g was anesthetized using a concentration of 150–175 mg/L (Ribeiro et al., 2015).

4.8 Herbicidal activity

The herbicidal action of clove oil on cucumber seedlings in light and dark conditions was studied by Park et al. (2011). The clove oil treatment increased superoxide dismutase (SOD) activity while decreasing catalase activity, whereas the paraquat treatment reduced SOD and catalase activity. The findings revealed that clove oil has a different mechanism of herbicidal action than paraquat (Park et al., 2011).

Evans et al. (2009) investigated the herbicidal properties of clove oil and vinegar against *Abutilon theopharst* (velvetleaf) and *Amaranthus retroflexus* (redroot pigweed). Redroot pigweed was found to be easier to handle for both items than velvetleaf. Tworkoski (2002) investigated the phytotoxicity of 25 essential oils on detached dandelion leaves, including clove and cinnamon oil. Clove and cinnamon essential oils were applied to Johnson grass, common lambsquarters, and common ragweed in a greenhouse using dandelion leaf disk assay and whole plant assay. According to the findings, shoot death occurred between 1 h and 1 day after the application of gasoline.

Bainard et al. (2006) investigated the herbicidal function of clove oil and its primary constituent eugenol on *Amaranthus retroflexus* (redroot pigweed) and *Chenopodium album* (common lambsquarters), as well as the impact on leaf cell membrane integrity and seedling growth on *Amaranthus retroflexus* (redroot pigweed) and *Chenopodium album* (common lambsquarters). According to the findings, clove oil and its main constituent eugenol reduced cell membrane integrity and inhibited seedling development.

4.9 Insecticidal activity

Direct contact and residual contact bioassays with various essential oil-based products to combat bed bugs (*Climex lectuarius*) were used. In both the residual and direct contact assays, clove oil (0.3%) combined with peppermint oil (1%) and sodium laruyl sulfate (1.3%) caused more than 90% mortality of bed bug nymphs. Akhtar et al. (2008) documented insecticidal activity of clove oil against two noctuid caterpillars (*Psedudoletia unipuncta* and *Trichoplusia ni*). Clove oil was found to be more effective as an oral insecticide than as a touch insecticide. Compared to other essential oils, noctuid caterpillars (*Psedudoletia unipuncta* and *Trichoplusia ni*) were more receptive to clove oil. For two animals, the EC_{50} values were 6900 and 400 ppm, respectively. Clove oil has insecticidal activity against *Sitophilus zeamais* (maize weevil) and *Acanthoscelides obtectus*, according to Jairoce et al. (2016). The LC_{50} for *Sitophilus zeamais* was 10.15 μL/g and for *Acanthoscelides obtectus* was 9.45 μL/g, resulting in 100% mortality in both species. The toxic effect of terpenoids present in clove essential oil against granary weevils was stated by Plata-Rueda et al. (2018). Eugenol was found to be more toxic than α-humulene, α-pinene, and caryophyllene oxide in *Sitophilus granaries*. The insects' respiratory and mobility rates were also reduced (Plata-Rueda et al., 2018).

5. Conclusions and future prospective

Clove is a widely used commodity that is used as a spice and available in the market. In the future, we can clearly view the outspread of clove more and more throughout the world by its geoeconomic value. It will be more popularly used as a condiment throughout the entire world. Its position in the global market appears secure, and farmers of Indonesia, Madagascar, Zanzibar, and Comoros benefit from its development. The fast-developing pharmaceutical fields are bound to impact the designing of clove extract formulations. However, the effort is needed to initiate investigations that could bring the findings hidden on the research benches to effective clinical use.

References

Abdel-Wahhab, M.A., Aly, S.E., 2005. Antioxidant property of Nigella sativa (black cumin) and Syzygium aromaticum (clove) in rats during aflatoxicosis. J. Appl. Toxicol. Int. J. 25 (3), 218–223.

Akhtar, Y., Yeoung, Y.R., Isman, M.B., 2008. Comparative bioactivity of selected extracts from Meliaceae and some commercial botanical insecticides against two noctuid caterpillars, Trichoplusia ni and Pseudaletia unipuncta. Phytochem. Rev. 7 (1), 77–88.

Al-Ameedi, A.I., Faris, J.K., Rabee, A.H., Naji, H.H., Obayes, A.J., Obaid, W.F., 2017. Analgesic and anti-inflammatory effects of hydro alcoholic extract of (Syzygium aromaticum) in Albino mice. Kufa J. Vet. Med. Sci. 8, 56–62.

Ali, S.N., Begum, S., Siddiqui, B.S., 2018. Chemical constituents of Syzygium aromaticum. Chem. Nat. Compd. 54 (6), 1192–1193.

Alma, H.M., Ertas, M., Nitz, S., Kollmannsberger, H., 2007. Research on essential oil content and chemical composition of Turkish clove (Syzygium aromaticum L.). BioResources 2 (2), 265–269.

Bachiega, T.F., de Sousa, J.P.B., Bastos, J.K., Sforcin, J.M., 2012. Clove and eugenol in noncytotoxic concentrations exert immunomodulatory/anti-inflammatory action on cytokine production by murine macrophages. J. Pharm. Pharmacol. 64 (4), 610–616.

Bainard, L.D., Isman, M.B., Upadhyaya, M.K., 2006. Phytotoxicity of clove oil and its primary constituent eugenol and the role of leaf epicuticular wax in the susceptibility to these essential oils. Weed Sci. 54 (5), 833–837.

Bala, S.K., Sukul, N.C., 1987. Systemic nematicidal effect of eugenol. Nematropica 17, 219–222.

Banerjee, S., Panda, C.K., Das, S., 2006. Clove (Syzygium aromaticum L.), a potential chemopreventive agent for lung cancer. Carcinogenesis 27 (8), 1645–1654.

Bansod, S., Rai, M., 2008. Antifungal activity of essential oils from Indian medicinal plants against human pathogenic Aspergillus fumigatus and A. niger. World J. Med. Sci. 3 (2), 81–88.

Bao, L.M., Eerdunbayaer, Nozaki, A., Takahashi, E., Okamoto, K., Ito, H., Hatano, T., 2012. Hydrolysable tannins isolated from Syzygium aromaticu structure of a new c-glucosidic ellagitannin and spectral features of tannins with a tergalloyl group. Heterocycles 85 (2), 365–381. https://doi.org/10.3987/COM-11-12392.

Batiha, G.E.S., Alkazmi, L.M., Wasef, L.G., Beshbishy, A.M., Nadwa, E.H., Rashwan, E.K., 2020. Syzygium aromaticum L. (Myrtaceae): traditional uses, bioactive chemical constituents, pharmacological and toxicological activities. Biomol. Ther. 10 (2), 202. https://doi.org/10.3390/biom10020202.

Begum, S., Siddiqui, B.S., Khatoon, R., Aftab, F., 2014. Phytochemical studies on Syzygium aromaticum Linn. J. Chem. Soc. Pak. 36 (3), 512.

Cai, L., Wu, C.D., 1996. Compounds from Syzygium aromaticum possessing growth inhibitory activity against oral pathogens. J. Nat. Prod. 59 (10), 987–990.

Cava, R., Nowak, E., Taboada, A., Marin-Iniesta, F., 2007. Antimicrobial activity of clove and cinnamon essential oils against Listeria monocytogenes in pasteurized milk. J. Food Prot. 70 (12), 2757–2763.

Chaieb, K., Hajlaoui, H., Zmantar, T., Kahla-Nakbi, A.B., Rouabhia, M., Mahdouani, K., Bakhrouf, A., 2007. The chemical composition and biological activity of clove essential oil, Eugenia caryophyllata (Syzigium aromaticum L. Myrtaceae): a short review. Phytother. Res. 21 (6), 501–506.

Chami, N., Bennis, S., Chami, F., Aboussekhra, A., Remmal, A., 2005. Study of anticandidal activity of carvacrol and eugenol in vitro and in vivo. Oral Microbiol. Immunol. 20 (2), 106–111.

Cortés-Rojas, D.F., de Souza, C.R.F., Oliveira, W.P., 2014. Clove (Syzygium aromaticum): a precious spice. Asian Pac. J. Trop. Biomed. 4 (2), 90–96.

Dai, J.P., Zhao, X.F., Zeng, J., Wan, Q.Y., Yang, J.C., Li, W.Z., Chen, X.X., Wang, G.F., Li, K.S., 2013. Drug screening for autophagy inhibitors based on the dissociation of Beclin1-Bcl2 complex using BiFC technique and mechanism of eugenol on anti-influenza A virus activity. PLoS One 84 (4). https://doi.org/10.1371/journal.pone.0061026, e61026.

Danthu, P., Simanjuntak, R., Fawbush, F., Tsy, L.P., Razafimamonjison, J.M., Abdillahi, G., Jahiel, M.M., Penot, E., 2020. The clove tree and its products (clove bud, clove oil, eugenol): prosperous today but what of tomorrow's restrictions? Fruits 75 (5), 224–242. https://doi.org/10.17660/th2020/75.5.5.

Devkota, H.P., Adhikari-Devkota, A., 2020. Cold pressed clove (Syzygium aromaticum) oil. In: Cold Pressed Oils. Academic Press, pp. 273–276.

Diyaware, M.Y., Suleiman, S.A.B., Akinwande, A.A., Aliyu, M., 2017. Anesthetic effects of clove (Eugenia aromaticum) seed extract on Clarias gariepinus (Burchell, 1822) fingerlings under semi-arid conditions. J. Agric. Sci. (Belgr.) 62 (4), 411–421.

Dorman, H.D., Figueiredo, A.C., Barroso, J.G., Deans, S.G., 2000. In vitro evaluation of antioxidant activity of essential oils and their components. Flavour Fragance J. 15 (1), 12–16.

Duraipandiyan, V., Ayyanar, M., Ignacimuthu, S., 2006. Antimicrobial activity of some ethnomedicinal plants used by Paliyar tribe from Tamil Nadu, India. BMC Complement. Altern. Med. 6 (1), 1–7.

Dwivedi, V., Shrivastava, R., Hussain, S., Ganguly, C., Bharadwaj, M., 2011. Comparative anticancer potential of clove (Syzygium aromaticum)—an Indian spice—against cancer cell lines of various anatomical origin. Asian Pac. J. Cancer Prev. 12 (8), 1989–1993.

Estrada-Cano, C., Castro, M.A.A., Muñoz-Castellanos, L.N.A.O.A., García-Triana, N.A.O.A., Hernández-Ochoa, L., 2017. Antifungal activity of microcapsulated clove (Eugenia caryophyllata) and Mexican oregano (Lippia berlandieri) essential oils against Fusarium oxysporum. J. Microb. Biochem. Technol. 9 (1). https://doi.org/10.4172/1948-5948.1000342.

Evans, G.J., Bellinder, R.R., Goffinet, M.C., 2009. Herbicidal effects of vinegar and a clove oil product on redroot pigweed (Amaranthus retroflexus) and velvet leaf (Abutilon theophrasti). Weed Technol. 23 (2), 292–299.

Fankem, P.M., Kwanga, S.N., Sameza, M.L., Tchoumbougnang, F., Tchabong, R., Ngouné, L.T., Dongmo, P.M.J., 2017. Antioxidant and antifungal activities of cocoa butter (Theobroma cacao), essential oil of Syzygium aromaticum and a combination of both extracts against three dermatophytes. Am. Sci. Res. J. Eng. Technol. Sci. 37 (1), 255–272.

Fu, Y., Zu, Y., Chen, L., Shi, X., Wang, Z., Sun, S., Efferth, T., 2007. Antimicrobial activity of clove and rosemary essential oils alone and in combination. Phytother. Res. 21 (10), 989–994.

Fujimoto, R.Y., Pereira, D.J.M., Silva, J.C.S., de Oliveira, L.C.A., Inoue, L.A.K.A., Hamoy, M., de Mello, V.J., Torres, M.F., Barbas, L.A.L., 2018. Clove oil induces anaesthesia and blunts muscle contraction power in three Amazon fish species. Fish Physiol. Biochem. 44 (1), 245–256. https://doi.org/10.1007/s10695-017-0430-8.

Gayoso, C.W., Lima, E.O., Oliveira, V.T., Pereira, F.O., Souza, E.L., Lima, I.O., Navarro, D.F., 2005. Sensitivity of fungi isolated from onychomycosis to Eugenia cariophyllata essential oil and eugenol. Fitoterapia 76 (2), 247–249.

Gulcin, I., Elmastaş, M., Aboul-Enein, H.Y., 2012. Antioxidant activity of clove oil—a powerful antioxidant source. Arab. J. Chem. 5 (4), 489–499.

Hammer, K.A., Carson, C.F., Riley, T.V., 1999. Antimicrobial activity of essential oils and other plant extracts. J. Appl. Microbiol. 86 (6), 985–990.

Han, X., Parker, T.L., 2017. Anti-inflammatory activity of clove (Eugenia caryophyllata) essential oil in human dermal fibroblasts. Pharm. Biol. 55 (1), 1619–1622.

Hanusz, M., 2000. Kretek—The Culture and Heritage of Indonesia's Clove Cigarette. EquinoxPubl, Jakarta, Indonesia, p. 203.

Jairoce, C.F., Teixeira, C.M., Nunes, C.F., Nunes, A.M., Pereira, C.M., Garcia, F.R., 2016. Insecticide activity of clove essential oil on bean weevil and maize weevil. Rev. Bras. Eng. Agríc. Ambient. 20 (1), 72–77.

Jimoh, S.O., Arowolo, L.A., Alabi, K.A., 2017. Phytochemical screening and antimicrobial evaluation of Syzygium aromaticum extract and essential oil. Int. J. Curr. Microbiol. Appl. Sci. 6 (7), 4557–4567.

Jirovetz, L., Buchbauer, G., Stoilova, I., Stoyanova, A., Krastanov, A., Schmidt, E., 2006. Chemical composition and antioxidant properties of clove leaf essential oil. J. Agric. Food Chem. 54 (17), 6303–6307.

Kaur, K., Kaushal, S., 2019. Phytochemistry and pharmacological aspects of Syzygium aromaticu a review. J. Pharmacogn. Phytochem. 8, 398–406.

Kheawfu, K., Pikulkaew, S., Chaisri, W., Okonogi, S., 2017. Nanoemulsion: a suitable nanodelivery system of clove oil for anesthetizing Nile tilapia. Drug Discov. Ther. 11 (4), 181–185.

Kiran, Z., Begum, S., Sara, S., Bano, Z., Siddiqui, B.S., 2019. Novel ferulic acid and benzophenone derivatives from the flower buds of Syzygium aromaticum. Nat. Prod. Res. https://doi.org/10.1080/14786419.2019.1698570.

Koh, T., Murakami, Y., Tanaka, S., Machino, M., Sakagami, H., 2013. Re-evaluation of anti-inflammatory potential of eugenol in IL-1β-stimulated gingival fibroblast and pulp cells. In Vivo 27 (2), 269–273.

Kouidhi, B., Zmantar, T., Bakhrouf, A., 2010. Anticariogenic and cytotoxic activity of clove essential oil (Eugenia caryophyllata) against a large number of oral pathogens. Ann. Microbiol. 60 (4), 599–604.

Kumar, U., Kumar, B., Bhandari, A., Kumar, Y., 2010. Phytochemical investigation and comparison of antimicrobial screening of clove and cardamom. Int. J. Pharm. Sci. Res. 1 (12), 138–147.

Kumar, A., Thakur, S., Thakur, V.C., Kumar, A., Patil, S., Vohra, M.P., 2012. Antifungal activity of some natural essential oils against Candida species isolated from blood stream infection. J. Krishna Inst. Med. Sci. Univ. 1 (1), 61–66.

Kumar, P.S., Febriyanti, R.M., Sofyan, F.F., Luftimas, D.E., Abdulah, R., 2014. Anticancer potential of Syzygium aromaticum L. in MCF-7 human breast cancer cell lines. Pharm. Res. 6 (4), 350.

Lesgards, J.F., Baldovini, N., Vidal, N., Pietri, S., 2014. Anticancer activities of essential oils constituents and synergy with conventional therapies: a review. Phytother. Res. 28 (10), 1423–1446.

Marya, C.M., Satija, G., Nagpal, R., Kapoor, R., Ahmad, A., 2012. In vitro inhibitory effect of clove essential oil and its two active principles on tooth decalcification by apple juice. Int. J. Dent. 2012. https://doi.org/10.1155/2012/759618.

Matan, N., 2012. Antimicrobial activity of edible film incorporated with essential oils to preserve dried fish (Decapterus maruadsi). Int. Food Res. J. 19 (4), 1733–1738.

Meyer, S.L., Lakshman, D.K., Zasada, I.A., Vinyard, B.T., Chitwood, D.J., 2008. Phytotoxicity of clove oil to vegetable crop seedlings and nematotoxicity to root-knot nematodes. HortTechnology 18 (4), 631–638.

Mohamed, M.S., Abdallah, A.A., Mahran, M.H., Shalaby, A.M., 2018. Potential alternative treatment of ocular bacterial infections by oil derived from Syzygium aromaticum flower (clove). Curr. Eye Res. 43 (7), 873–881.

Mytle, N., Anderson, G.L., Doyle, M.P., Smith, M.A., 2006. Antimicrobial activity of clove (Syzgium aromaticum) oil in inhibiting Listeria monocytogenes on chicken frankfurters. Food Control 17 (2), 102–107.

Nam, H., Kim, M.M., 2013. Eugenol with antioxidant activity inhibits MMP-9 related to metastasis in human fibrosarcoma cells. Food Chem. Toxicol. 55, 106–112.

Nassar, M.I., Gaara, A.H., El-Ghorab, A.H., Farrag, A., Shen, H., Huq, E., Mabry, T.J., 2007. Chemical constituents of clove (Syzygium aromaticum, Fam. Myrtaceae) and their antioxidant activity. Rev. Latinoam. Quim. 35 (3), 47.

Nikoui, V., Ostadhadi, S., Bakhtiarian, A., Abbasi-Goujani, E., Habibian-Dehkordi, S., Rezaei-Roshan, M., Foroohandeh, M., Giorgi, M., 2017. The anti-inflammatory and antipyretic effects of clove oil in healthy dogs after surgery. PharmaNutrition 5 (2), 52–57. https://doi.org/10.1016/j.phanu.2017.02.003.

Nurdjannah, N., Bermawie, N., 2012. 11—Cloves. In: Peter, K.V. (Ed.), Handbook of Herbs and Spices, second ed. Woodhead Publishing, Cambridge, UK, pp. 197–215.

Palipudi, K., Mbulo, L., Kosen, S., Tjandra, A., Kadarmanto, Qureshi, F., Andes, L., Sinha, D., Asma, S., 2015. A cross sectional study of kretek smoking in Indonesia as a major risk to public health. Asian Pac. J. Cancer Prev. 16, 6883–6888. https://doi.org/10.7314/APJCP.2015.16.16.6883.

Park, K.W., Choi, S.H., Ahn, J.Y., Sohn, Y.G., Kim, C.G., Lee, J.J., 2011. Herbicidal action of clove oil on cucumber seedlings. Weed Biol. Manage. 11 (4), 235–240.

Pina-Vaz, C., Gonçalves Rodrigues, A., Pinto, E., Costa-de-Oliveira, S., Tavares, C., Salgueiro, L., Cavaleiro, C., Gonçalves, M.J., Martinez-de-Oliveira, J., 2004. Antifungal activity of Thymus oils and their major compounds. J. Eur. Acad. Dermatol. Venereol. 18 (1), 73–78. https://doi.org/10.1111/j.1468-3083.2004.00886.x.

Pinto, E., Vale-Silva, L., Cavaleiro, C., Salgueiro, L., 2009. Antifungal activity of the clove essential oil from Syzygium aromaticum on Candida, Aspergillus and dermatophyte species. J. Med. Microbiol. 58 (11), 1454–1462.

Plata-Rueda, A., Campos, J.M., da Silva Rolim, G., Martínez, L.C., Dos Santos, M.H., Fernandes, F.L., Serrão, J. E., Zanuncio, J.C., 2018. Terpenoid constituents of cinnamon and clove essential oils cause toxic effects and behavior repellency response on granary weevil, Sitophilus granarius. Ecotoxicol. Environ. Saf. 156, 263–270. https://doi.org/10.1016/j.ecoenv.2018.03.033.

Ribeiro, P.A., Miranda-Filho, K.C., Melo, D.C.D., Luz, R.K., 2015. Efficiency of eugenol as anesthetic for the early life stages of Nile tilapia (Oreochromis niloticus). An. Acad. Bras. Cienc. 87 (1), 529–535.

Rodrigues, T.G., Fernandes Jr., A., Sousa, J.P.B., Bastos, J.K., Sforcin, J.M., 2009. In vitro and in vivo effects of clove on pro-inflammatory cytokines production by macrophages. Nat. Prod. Res. 23 (4), 319–326.

Saini, A., Sharma, S., Chhibber, S., 2009. Induction of resistance to respiratory tract infection with Klebsiella pneumoniae in mice fed on a diet supplemented with tulsi (Ocimum sanctum) and clove (Syzgium aromaticum) oils. J. Microbiol. Immunol. Infect. 42 (2), 107–113.

Shan, B., Cai, Y.Z., Sun, M., Corke, H., 2005. Antioxidant capacity of 26 spice extracts and characterization of their phenolic constituents. J. Agric. Food Chem. 53 (20), 7749–7759. https://doi.org/10.1021/jf051513y.

Taniwiryono, D., Berg, H., Riksen, J.A.G., Rietjens, I.M.C.M., Djiwantia, S.R., Kammenga, J.E., Murk, A.J., 2009. Nematicidal activity of plant extracts against the root-knot nematode, Meloidogyne incognita. Open Nat. Prod. J. 2, 77–85.

Tullio, V., Nostro, A., Mandras, N., Dugo, P., Banche, G., Cannatelli, M.A., Cuffini, A.M., Alonzo, V., Carlone, N. A., 2007. Antifungal activity of essential oils against filamentous fungi determined by broth microdilution and vapour contact methods. J. Appl. Microbiol. 102 (6), 1544–1550. https://doi.org/10.1111/j.1365-2672.2006.03191.x.

Tworkoski, T., 2002. Herbicide effects of essential oils. Weed Sci. 50 (4), 425–431.

Viuda-Martos, M., Ruiz-Navajas, Y., Fernández-López, J., Pérez-Álvarez, J.A., 2007. Antifungal activities of thyme, clove and oregano essential oils. J. Food Saf. 27 (1), 91–101.

Wang, Y.F., Jia, J.X., Tian, Y.Q., Shu, X., Ren, X.J., Guan, Y., Yan, Z.Y., 2018. Antifungal effects of clove oil microcapsule on meat products. LWT 89, 604–609.

Warnke, P.H., Becker, S.T., Podschun, R., Sivananthan, S., Springer, I.N., Russo, P.A.J., Wiltfang, J., Fickenscher, H., Sherry, E., 2009. The battle against multi-resistant strains: renaissance of antimicrobial essential oils as a promising force to fight hospital-acquired infections. J. Craniomaxillofac. Surg. 37 (7), 392–397. https://doi.org/10.1016/j.jcms.2009.03.017.

World Bank, 2017. The Economics of Clove Farming in Indonesia. http://documents.worldbank.org/curated/en/166181507538499946/pdf/120318-REVISED-WP-WBGIndoCloveFarmingweb.pdf. (Accessed 9 May 2021).

Xu, J.G., Liu, T., Hu, Q.P., Cao, X.M., 2016. Chemical composition, antibacterial properties and mechanism of action of essential oil from clove buds against Staphylococcus aureus. Molecules 21 (9), 1–13. https://doi.org/10.3390/molecules21091194 (1194).

Yadav, A.S., Bhatnagar, D., 2007. Free radical scavenging activity, metal chelation and antioxidant power of some of the Indian spices. Biofactors 31 (3–4), 219–227.

Yang, Y.C., Lee, S.H., Lee, W.J., Choi, D.H., Ahn, Y.J., 2003. Ovicidal and adulticidal effects of Eugenia caryophyllata bud and leaf oil compounds on Pediculus capitis. J. Agric. Food Chem. 51 (17), 4884–4888. https://doi.org/10.1021/jf034225f.

Zengin, H., Baysal, A.H., 2015. Antioxidant and antimicrobial activities of thyme and clove essential oils and application in minced beef. J. Food Process. Preserv. 39 (6), 1261–1271. https://doi.org/10.1111/jfpp.12344.

CHAPTER

Health-promoting activities of clove (*Syzygium aromaticum*) 11

Hari Prasad Devkota[a], Rajan Logesh[b], and Niranjan Das[c]

Graduate School of Pharmaceutical Sciences, Kumamoto University, Kumamoto, Japan[a] TIFAC-CORE in Herbal Drugs, Department of Pharmacognosy and Phytopharmacy, JSS College of Pharmacy (JSS Academy of Higher Education and Research), Rockland's, Ooty, Tamil Nadu, India[b] Department of Chemistry, Iswar Chandra Vidyasagar College, Belonia, Tripura, India[c]

1. Introduction

Syzygium aromaticum (L.) Merr. & L.M.Perry. (Synonym: *E. caryophyllata* Thunb.) (Fig. 1) belonging to the family Myrtaceae is native to some islands of the Malay Archipelago, especially the Moluccas, and is cultivated in various countries such as Indonesia, Mauritius, Sri Lanka, and India (Sharma et al., 2000). Dried flower buds (Fig. 2), commonly known as clove in English, "Devapuspa" in Sanskrit, "Lwang" in Nepali, and "Lavanga" in Hindi, are used as a spice all over the world for their strong aromatic smell (Cortés-Rojas et al., 2014; Devkota and Adhikari-Devkota, 2020; Khanal et al., 2021). It is widely used in various food recipes such as curry, biryani, and many others including spicy tea. Clove is well known for its aromatic, stimulant, and carminative properties and it is used for the treatment of gastric irritation, dyspepsia, nausea, vomiting, and flatulence in Nepal (Rajbhandari et al., 1995) and also included in many Ayurveda formulations (Ministry for Health and Family Welfare, 2010). Phenylpropanoids (such as eugenol), tannins, sesquiterpenoids (such as β-caryophyllene), and triterpenoids are reported as the main chemical constituents of clove flower buds (Fig. 2).

The essential oil obtained from cloves, known as clove oil, is one of the most widely used products. It has a long history of being used to relieve toothache as an antiseptic and analgesic agent in dental care (Chaieb et al., 2007). It also has potent antispasmodic, carminative, and rubifacient activities (Rajbhandari et al., 1995). Eugenol is the main constituent of clove oil and its content is reported to vary from about 45% to almost 90% (Chaieb et al., 2007). The Japanese Pharmacopeia describes clove oil as the volatile oil (colorless or light yellow liquid with characteristic aroma and burning taste) distilled with steam from the flower buds or leaves of *S. aromaticum* with not less than 80% eugenol (The Ministry of Health, Labor and Welfare of Japan, 2016). Similarly, the European Pharmacopeia mentions that the Caryophylli Flos consists of the whole flower buds of *S. aromaticum* containing not less than 150 mL/kg of essential oil (European Medicines Agency, 2011). Other chemical constituents of clove oil are eugenyl acetate, β-caryophyllene, cinnamaldehyde, carvacol, calacorene, thymol, and 2-heptanone, among others (Chaieb et al., 2007). Clove oil possesses strong anesthetic, antimicrobial, antitumor, and insecticidal properties (Chaieb et al., 2007; Devkota and Adhikari-Devkota, 2020; Issac et al., 2015; Li et al., 2020; Tennyson et al., 2013; Zhu et al., 2001). Cold-pressed clove oil is also

Clove (*Syzygium aromaticum*). https://doi.org/10.1016/B978-0-323-85177-0.00009-4

FIG. 1

Photographs of the plant (A), flowers (B), and flower buds (C) and (D).

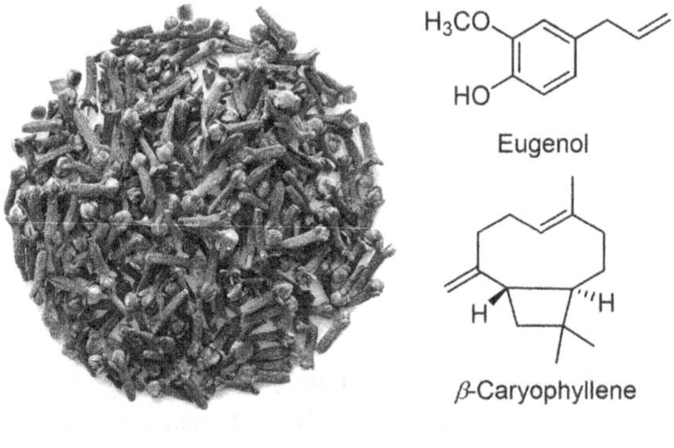

FIG. 2

Photograph of clove flower buds and structures of main compounds.

gaining interest in recent years due to its potent antioxidant and other pharmacological activities (Devkota and Adhikari-Devkota, 2020) The main aim of this chapter is to collect and analyze the scientific information about the health-promoting activities of clove.

2. Health-promoting and other pharmacological activities of clove extracts

The extracts, oil, and compounds obtained from the clove flower buds have shown various health-promoting activities. Eugenol is one of the most widely studied compounds for its biological activities (Chaieb et al., 2007; Pramod et al., 2010). Some of these activities are discussed in detail in the following sections.

2.1 Antimicrobial activities

Clove and its products are well known as potent antimicrobial agents. Nuñez and Aquino reported the antibacterial activity of clove essential oil against *Escherichia coli*, *Staphylococcus aureus*, and *Pseudomonas aeruginosa* (Nuñez and Aquino, 2012). Clove oil is also reported to have potent bactericidal activity against multidrug-resistant (MDR) *Streptococcus suis* (Wongsawan et al., 2019). Ács et al. reported the potent antibacterial activity of several essential oils along with clove oil against respiratory tract pathogens such as *Streptococcus pneumoniae*, *S. mutans*, *S. pyogenes*, *Haemophilus influenzae*, *H. parainfluenzae*, and *Moraxella catarrhalis* using liquid medium and vapor phase. Clove oil showed potent antibacterial activity against S. *pneumonia* and *S. pyogens* in a liquid medium (Ács et al., 2018). Clove oil also showed potent antibacterial activity against extended-spectrum beta-lactamase-producing *E. coli* (Mejía-Argueta et al., 2020). Clove oil is also studied as a potent antibacterial agent against foodborne pathogenic species such as *Cronobacter sakazakii* and *Salmonella enteritidis*. It was reported as a suitable agent for incorporation in marketed milk, yogurt, and kefir at various ranges for the production of bio-functional dairy products and also controlling the growth of the pathogenic bacteria in these products (Chon et al., 2020). Similarly, cold-pressed clove oil is also reported as a potent antibacterial agent against various Gram-positive and Gram-negative bacteria (Assiri and Hassanien, 2013; Ramadan et al., 2013).

2.2 Antioxidant activities

Clove extracts and clove essential oils are reported as potent antioxidants. Methanolic extract of clove flower buds had shown potent 1,1,-diphenyl-2-picrylhydrazyl ($DPPH^{\bullet}$) free radical inhibitory activity (Devkota et al., 2006). Gülçin et al. evaluated the antioxidant activity of hot water and ethanol extracts of clove buds using various in vitro methods such as reductive potential, free radical scavenging, superoxide anion radical scavenging, and metal chelating activities and reported that these extracts showed potent activities (Gülçin et al., 2004). The total phenolic compounds estimation revealed that the water and ethanol extracts had 179 and 264 μg gallic acid equivalent in 1 mL of the extracts, respectively. Gülçin et al. also evaluated the antioxidant activity of the clove oil using similar methods and reported that the oil also possessed potent antioxidant activities (Gülçin et al., 2012). The water-based emulsion nanoparticles of clove oil and eugenol are also reported to retain the antioxidant activities along with having the potential to be used in water-based formulations of various products such as disinfectants, flavoring of foods, etc. (Hamed et al., 2012). Cold-pressed clove oil is also reported as having potent antioxidant activity. For example, Ramadan et al. (2013) evaluated the free radical scavenging activity of cold-pressed clove oil and virgin clove oil. Authors reported that the cold-pressed oil showed strong radical scavenging activity, which was suggested to be due to the higher content of total phenolics in cold-pressed clove oil.

2.3 Anti-inflammatory activities

Ibrahim et al. studied the anti-inflammatory activities of cold-pressed clove oil in carrageenan-induced edema model rats and the oil administration reduced the rat paw edema (Ibrahim et al., 2017). Han and Parker evaluated the anti-inflammatory activity of clove essential oil using human dermal fibroblasts. Clove essential oil showed antiproliferative activity and also inhibited the production of pro-inflammatory biomarkers (Han and Parker, 2017). Similarly, Bachiega et al. also reported the anti-inflammatory and immunomodulatory activities of methanol extract of clove buds and eugenol using murine macrophages. Clove extract inhibited the lipopolysaccharide (LPS)-induced production of interleukin (IL)-1β, IL-6, and IL-10, whereas eugenol inhibited the production of IL-6 and IL-10 (Bachiega et al., 2012). Rodrigues et al. reported the inflammatory cytokines production inhibitory activity of water-soluble fraction of hydroalcoholic extract and essential oil of clove in macrophages obtained from BALB/c mice (Rodrigues et al., 2009).

2.4 Analgesic activities

Several studies have reported the analgesic activity of clove extracts and clove essential oil. Asl et al. evaluated the analgesic activity of water and ethanol extract of the dried clove buds in BALB/c male mice using the hot plate test method and reported that the water extract had more potent activity as compared to ethanol extract (Asl et al., 2013). Taher et al. reported the analgesic activity of clove oil using the hot plate test and acetic-acid-induced writhing test (Taher et al., 2015). The major compound, eugenol is also widely reported as a potent analgesic compound (Chaieb et al., 2007; Pramod et al., 2010). Analgesic activity was examined using acetic-acid-induced abdominal constrictions and the hotplate test.

2.5 Hepatoprotective activities

Various studies have reported the hepatoprotective activity of clove formulations. Administration of clove bud powder (CBP) to high-fat diet and low-dose streptozotocin-induced type 2 diabetic rats reduced the level of liver enzymes such as alanine aminotransferase, alkaline phosphatase, and aspartate aminotransferase (Adefegha et al., 2014). El-Hadary and Ramadan evaluated the hepatoprotective activity of cold-pressed clove oil against carbon tetrachloride-induced hepatotoxicity in healthy male albino rats. The administration of clove oil attenuated the increased levels of liver enzymes such as alanine transaminase, alkaline phosphatase, and aspartate transaminase. The levels of serum total bilirubin (TB) and direct bilirubin (B) were also decreased (El-Hadary and Ramadan Hassanien, 2016).

2.6 Antidiabetic activities

Administration of clove bud powder (CBP) to high-fat diet and low-dose streptozotocin-induced type 2 diabetic rats resulted in the reduced blood glucose level and reduced α-glucosidase activity as compared to control group (Adefegha et al., 2014). CBP administration also reduced the lipids concentrations except for high-density lipoprotein cholesterol and increased the levels of antioxidant enzymes. Kuroda et al. evaluated the hypoglycemic activity of ethanolic extract of clove buds in genetically diabetic KK-A^y mice. Administration of clove extract lowered the blood glucose levels as compared to the control group without clove extract administration. The authors further isolated eight compounds from the extract namely eugenol, dehydrofieugenol, dehydrofieugenol B, oleanolic acid, arjunolic acid, corosolic acid, asiatic acid, and betulinic acid. These compounds were evaluated for their peroxisome proliferator-activated receptor-gamma (PPAR)-γ ligand-binding activity and preadipocyte differentiation activity in 3T3-L1 adipocytes. Dehydrofieugenol and dehydrofieugenol B were reported as having potent activities in both of these assays and were suggested as major active compounds in clove extract (Kuroda et al., 2011).

Mohan et al. reported the hypoglycemic activity of polyphenol-rich hydroethanolic extract of clove buds in healthy and prediabetic volunteers after administering 250 mg of extract daily for 30 days. Authors also evaluated the cell-based activities and cell-free enzyme inhibitory activities of the extract, where it enhanced the glucose uptake in L6 monocytes and inhibited the glucose production in HeG2 cells along with inhibiting the activities of α-glucosidase and α-amylase in cell-free enzymatic assays (Mohan et al., 2019).

3. Conclusions

Clove is widely used as a spice all over the world for its strong aromatic smell and as a crude drug in various traditional medicine formulations. It is used as a popular remedy for toothache for its potent antiseptic and analgesic activities. Various health-promoting and other pharmacological activities are reported for the extracts, oils, and compounds such as antimicrobial, antioxidant, hepatoprotective, analgesic, and antidiabetic activities. Compared to in vitro and animal studies, there are relatively less clinical studies to evaluate the efficacy of clove, its extracts, and compounds. Further studies should focus on these aspects along with the safety studies and studies related to bioactivity-guided isolation and identification of active compounds.

References

Ács, K., Balázs, V.L., Kocsis, B., Bencsik, T., Böszörményi, A., Horváth, G., 2018. Antibacterial activity evaluation of selected essential oils in liquid and vapor phase on respiratory tract pathogens. BMC Complement. Altern. Med. 18 (1), 1–9. https://doi.org/10.1186/S12906-018-2291-9.

Adefegha, S.A., Oboh, G., Adefegha, O.M., Boligon, A.A., Athayde, M.L., 2014. Antihyperglycemic, hypolipidemic, hepatoprotective and antioxidative effects of dietary clove (*Szyzgium aromaticum*) bud powder in a high-fat diet/streptozotocin-induced diabetes rat model. J. Sci. Food Agric. 94 (13), 2726–2737. https://doi.org/10.1002/JSFA.6617.

Asl, M.K., Nazariborun, A., Hosseini, M., 2013. Analgesic effect of the aqueous and ethanolic extracts of clove. Avicenna J. Phytomed. 3 (2), 186.

Assiri, A.M.A., Hassanien, M.F.R., 2013. Bioactive lipids, radical scavenging potential, and antimicrobial properties of cold pressed clove (*Syzygium aromaticum*) oil. J. Med. Food 16, 1046–1056. https://doi.org/10.1089/jmf.2012.0288.

Bachiega, T.F., de Sousa, J.P., Bastos, J.K., Sforcin, J.M., 2012. Clove and eugenol in noncytotoxic concentrations exert immunomodulatory/anti-inflammatory action on cytokine production by murine macrophages. J. Pharm. Pharmacol. 64 (4), 610–616. https://doi.org/10.1111/J.2042-7158.2011.01440.X.

Chaieb, K., Hajlaoui, H., Zmantar, T., Kahla-Nakbi, A.B., Rouabhia, M., Mahdouani, K., Bakhrouf, A., 2007. The chemical composition and biological activity of clove essential oil, *Eugenia caryophyllata* (*Syzigium aromaticum* L. Myrtaceae): a short review. Phytother. Res. 21, 501–506. https://doi.org/10.1002/ptr.2124.

Chon, J.-W., Seo, K.-H., Bae, D., Kim, B., Jeong, D., Song, K.-Y., 2020. Antibacterial activity of clove oil against foodborne pathogenic bacteria and sensory attributes in clove oil-enriched dairy products: a preliminary study. J. Dairy Sci. Biotechnol. 38 (4), 197–206. https://doi.org/10.22424/JDSB.2020.38.4.197.

Cortés-Rojas, D.F., de Souza, C.R.F., Oliveira, W.P., 2014. Clove (*Syzygium aromaticum*): a precious spice. Asian Pac. J. Trop. Biomed. 4 (2), 90–96. https://doi.org/10.1016/S2221-1691(14)60215-X.

Devkota, H.P., Adhikari-Devkota, A., 2020. Cold pressed clove (*Syzygium aromaticum*) oil. In: Cold Pressed Oils. Academic Press, pp. 273–276, https://doi.org/10.1016/B978-0-12-818188-1.00023-2.

Devkota, H.P., Anjana, A., Poudel, S., Sushila, G.C., Takano, A., Basnet, P., 2006. Antioxidative activity of common natural medicines in Nepal. J. Nepal Pharm. Assoc. 26, 39–46.

El-Hadary, A.E., Ramadan Hassanien, M.F., 2016. Hepatoprotective effect of cold-pressed *Syzygium aromaticum* oil against carbon tetrachloride (CCl_4)-induced hepatotoxicity in rats. Pharm. Biol. 54 (8), 1364–1372. https://doi.org/10.3109/13880209.2015.1078381.

European Medicines Agency, 2011. Assessment report on *Syzygium aromaticum* (L.) Merill et L.M. Perry, flos and *Syzygium aromaticum* (L.) Merill et L.M. Perry, floris aetheroleum.

Gülçin, I., Elmastaş, M., Aboul-Enein, H.Y., 2012. Antioxidant activity of clove oil—a powerful antioxidant source. Arab. J. Chem. 5 (4), 489–499. https://doi.org/10.1016/J.ARABJC.2010.09.016.

Gülçin, İ., Şat, I.G., Beydemir, Ş., Elmastaş, M., Küfrevioğlu, Ö.I., 2004. Comparison of antioxidant activity of clove (*Eugenia caryophylata* Thunb) buds and lavender (*Lavandula stoechas* L.). Food Chem. 87 (3), 393–400. https://doi.org/10.1016/J.FOODCHEM.2003.12.008.

Hamed, S.F., Sadek, Z., Edris, A., 2012. Antioxidant and antimicrobial activities of clove bud essential oil and eugenol nanoparticles in alcohol-free microemulsion. J. Oleo Sci. 61 (11), 641–648. https://doi.org/10.5650/JOS.61.641.

Han, X., Parker, T.L., 2017. Anti-inflammatory activity of clove (*Eugenia caryophyllata*) essential oil in human dermal fibroblasts. Pharm. Biol. 55 (1), 1619–1622. https://doi.org/10.1080/13880209.2017.1314513.

Ibrahim, F.M., Attia, H.N., Maklad, Y.A.A., Ahmed, K.A., Ramadan, M.F., 2017. Biochemical characterization, anti-inflammatory properties and ulcerogenic traits of some cold-pressed oils in experimental animals. Pharm. Biol. 55, 740–748. https://doi.org/10.1080/13880209.2016.1275705.

Issac, A., Gopakumar, G., Kuttan, R., Maliakel, B., Krishnakumar, I.M., 2015. Safety and anti-ulcerogenic activity of a novel polyphenol-rich extract of clove buds (*Syzygium aromaticum* L). Food Funct. 6 (3), 842–852. https://doi.org/10.1039/c4fo00711e.

Khanal, A., Devkota, H.P., Kaundinnyayana, S., Gyawali, P., Ananda, R., Adhikari, R., 2021. Culinary herbs and spices in Nepal: a review of their traditional uses, chemical constituents, and pharmacological activities. Ethnobot. Res. Appl. 21, 40.

Kuroda, M., Mimaki, Y., Ohtomo, T., Yamada, J., Nishiyama, T., Mae, T., et al., 2011. Hypoglycemic effects of clove (*Syzygium aromaticum* flower buds) on genetically diabetic KK-Ay mice and identification of the active ingredients. J. Nat. Med. 66 (2), 394–399. https://doi.org/10.1007/S11418-011-0593-Z.

Li, H., Sureda, A., Devkota, H.P., Pittalà, V., Barreca, D., Silva, A.S., et al., 2020. Curcumin, the golden spice in treating cardiovascular diseases. Biotechnol. Adv. 38. https://doi.org/10.1016/j.biotechadv.2019.01.010.

Mejía-Argueta, E.L., Santillán-Benítez, J.G., Canales-Martinez, M.M., Mendoza-Medellín, A., 2020. Antimicrobial activity of *Syzygium aromaticum* L. essential oil on extended-spectrum beta-lactamases-producing Escherichia coli. Bull. Natl. Res. Cent. 44 (1), 1–7. https://doi.org/10.1186/S42269-020-00458-X.

Ministry for Health and Family Welfare, 2010. The Ayurvedic Pharmacopoeia of India. Government of India, Ministry for Health and Family Welfare, pp. 110–111.

Mohan, R., Jose, S., Mulakkal, J., Karpinsky-Semper, D., Swick, A.G., Krishnakumar, I.M., 2019. Water-soluble polyphenol-rich clove extract lowers pre- and post-prandial blood glucose levels in healthy and prediabetic volunteers: an open label pilot study. BMC Complement. Altern. Med. 19 (1), 1–9. https://doi.org/10.1186/S12906-019-2507-7.

Nuñez, L., Aquino, M.D., 2012. Microbicide activity of clove essential oil (*Eugenia caryophyllata*). Braz. J. Microbiol. 43 (4), 1255–1260. https://doi.org/10.1590/S1517-83822012000400003.

Pramod, K., Ansari, S.H., Ali, J., 2010. Eugenol: a natural compound with versatile pharmacological actions. Nat. Prod. Commun. 5 (12), 1999–2006. https://doi.org/10.1177/1934578X1000501236.

Rajbhandari, T., Joshi, M., Shrestha, T., Joshi, S., Acharya, B., 1995. Medicinal Plants of Nepal for Ayurvedic Drugs. Department of Plant Resources, Kathmandu.

Ramadan, M.F., Asker, M.M.S., Tadros, M., 2013. Lipid profile, antiradical power and antimicrobial properties of *Syzygium aromaticum* oil. Grasas Aceites 64, 509–520. https://doi.org/10.3989/gya.011713.

Rodrigues, T.G., Fernandes, A., Sousa, J.P.B., Bastos, J.K., Sforcin, J.M., 2009. In vitro and in vivo effects of clove on pro-inflammatory cytokines production by macrophages. Nat. Prod. Res. 23 (4), 319–326. https://doi.org/10.1080/14786410802242679.

Sharma, P., Yelne, M., Dennis, T., 2000. Database on Medicinal Plants Used in Ayurveda. Central Council for Research in Ayurveda and Siddha, New Delhi.

Taher, Y.A., Samud, A.M., El-Taher, F.E., Ben-Hussin, G., Elmezogi, J.S., Al-Mehdawi, B.F., Salem, H.A., 2015. Experimental evaluation of anti-inflammatory, antinociceptive and antipyretic activities of clove oil in mice. Libyan J. Med. 10 (1), 28685. https://doi.org/10.3402/LJM.V10.28685.

Tennyson, S., Samraj, D.A., Jeyasundar, D., Chalieu, K., 2013. Larvicidal efficacy of plant oils against the dengue vector *Aedes aegypti* (L.) (Diptera: Culicidae). Middle East J. Sci. Res. 13 (1), 64–68. https://doi.org/10.5829/idosi.mejsr.2013.13.1.64107.

The Ministry of Health, Labor and Welfare of Japan, 2016. In: The Japanese Pharmacopoeia, seventeenth ed. The Ministry of Health, Labor and Welfare of Japan, Tokyo.

Wongsawan, K., Chaisri, W., Tangtrongsup, S., Mektrirat, R., 2019. Bactericidal effect of clove oil against multidrug-resistant *Streptococcus suis* isolated from human patients and slaughtered pigs. Pathogens 9 (1), 14. https://doi.org/10.3390/PATHOGENS9010014.

Zhu, B.C.R., Henderson, G., Chen, F., Fei, H., Laine, R.A., 2001. Evaluation of vetiver oil and seven insect-active essential oils against the formosan subterranean termite. J. Chem. Ecol. 27 (8), 1617–1625. https://doi.org/10.1023/A:1010410325174.

CHAPTER

Nanocomposites using clove (*Syzygium aromaticum*) chemical constituents 12

Nusrat M. Begum[a], Abdulrahman M. Elbagory[b], Ateeq Rahman[c], Ahmed A. Hussein[b], Natascha Cheikhyoussef[d], and Ahmad Cheikhyoussef[e]

Department of Environmental Health, Water and Health Research Centre, University of Johannesburg, Doornfontein, South Africa[a] Chemistry Department, Cape Peninsula University of Technology, Cape Town, Western Cape, South Africa[b] Department of Physics, Chemistry & Material Science, Faculty of Agriculture, Engineering & Natural Sciences, University of Namibia, Windhoek, Namibia[c] Ministry of Higher Education, Technology and Innovation, Windhoek, Namibia[d] Science and Technology Division, Multidisciplinary Research Unit, Centre for Research Services, University of Namibia, Windhoek, Namibia[e]

1. Introduction

Green chemistry has the opportunity to ensure that the world of chemistry produces more effective and efficient materials for various applications and contributes to protecting public health and the environment. Furthermore, the technology optimizes all necessary parameters to minimize if not eliminate environmental pollution either through the production of benign by-products or employing safe reactants and solvents, and minimizing packaging waste.

Nanocomposites are a heterogeneous combination of two or more materials to yield new materials with new and unique properties. The unique characteristics of the newly formed nanocomposites can be controlled and optimized by using an effective mixture of original components. As a result, the applications of nanocomposite materials are expanding rapidly in many fields (Shirdar et al., 2019; Nguyen-Tri et al., 2018).

The Food and Agriculture Organization (FAO) of the United Nations estimated that 30% of the produced food is lost or wasted each year because of improper storage. Generally, food loss occurs at the production, handling, storage, and processing levels, while food waste indicates the losses at distribution, market, and final consumer levels (Ishangulyyev et al., 2019; Jiang et al., 2019). The global food packaging market was estimated at over USD 300 billion in 2019 and had a predicted increase rate of 5.2% yearly. The paper-based segment represents 31.9% of the total value and dominates as a biodegradable material. On the other hand, plastic food packaging, which is mostly nonbiodegradable, is dominating the market due to superior properties and lower prices; however, it raises environmental concerns (Wohner et al., 2020) and encourages the search for new safe and renewable natural-based packaging materials. Therefore, it becomes necessary to pay considerable attention to the development of biodegradable materials prepared from natural polymers. Thus, various natural

Clove (*Syzygium aromaticum*). https://doi.org/10.1016/B978-0-323-85177-0.00008-2

polymers are under study to develop improved products. In addition, many random research works have been documented in the literature on the development of biodegradable packaging film using renewable polymers (Shah et al., 2015; Tunma, 2018).

The market is shifting toward more sustainable packaging solutions like bio-based polymers (Wohner et al., 2020). The development of antimicrobial packages using natural agents such as clove oil is a promising path for actively controlling the bacterial and fungal proliferation that leads to food spoilage (Kerry et al., 2006; O'Callaghan and Kerry, 2014; Rai et al., 2019; Vilas et al., 2020). This type of package increases the latency phase and reduces the growth of microorganisms, enhancing the food quality and safety, ensuring a longer shelf life (Jayasena and Jo, 2013; Szabo et al., 2020; Motelica et al., 2020).

The clove essential oil (CEO) is an aromatic oil containing mainly eugenol (71.6%) and eugenol acetate (9.0%), while others have reported higher concentrations with the presence of β-caryophyllene as well. The various phytochemical compounds present exhibit a broad spectrum of pharmacological activities such as anticancer, antimutagenic, anti-inflammatory, antiallergic, antimicrobial, and antioxidant properties (Barboza et al., 2018). Such multi-pharmacological activities of the CEO give merits to be part of different nanocomposite packaging formulations that could enhance the food quality and can be employed in numerous applications.

2. Nanocomposites

The food industry containment has evolved to attract its customers, especially in packaging. The packaging technology has advanced to eliminate breakage and inhibit entry of moisture, oxygen, UV light, and movement of food solvent. Some natural films that have been used previously contain biodegradable polymers, waxes, polysaccharides, proteins, lipids, and/or a mixture. The research development in such lining can lead to edible films and coating to serve the current lifestyle of consumers. The design ensures that the inner lining further protects the food and reduces its oxidation process to warrant its safety. Thus, increasing shelf life and retaining the integrity of food's original taste. Food storage is crucial, and containers have improved over the years leading to various packaging technology.

Incorporating bioactive packaging that contains antimicrobial and antioxidant agents is beneficial compared to preservatives in the food industry. Preservatives can affect the flavor of the food; hence, natural alternatives would be desirable and safer. This is because they would be considered harmless given their natural origin and subsequently preserve the food. This has led to active packaging that provides passive protection that retains the food quality and further diminishes or hinders microorganisms' growth. The debatable controversy of butylated hydroxytoluene as a preservative due to its possible risks in cancer and asthma enables the utilization of CEO more attractive.

Antimicrobial in CEO activity equips the packaging material to inhibit the growth of microbial spoilage pathogens such as *Escherichia coli*, *Salmonella*, *Listeria monocytogenes*, *Staphylococcus aureus*, and *Bacillus cereus* (Liu et al., 2017). The phenolic compounds are multifunctional. They react with phospholipids of the microbial cell membrane that affects its porosity which denatures its proteins. β-Caryophyllene is the dominating nonvolatile compound in CEO possessing antioxidants property. It prevents the generation of free radicals produced during oxidation while the food deteriorates. It is, therefore, an ideal additive in food packaging as it can sustain the unique taste of food and inhibit the growth of food pathogens.

Polymers in food packaging have been designed to provide excellent barriers as they can be engineered to possess improved oxygen permeability and mechanical properties. This, however, has its drawback if affected during transportation which is overcome using CEO. However, phytochemical compounds that make CEO are volatile; hence, if added to packaging material, it is unstable to ensure that it retains its bioactivity. The volatility will diminish the antimicrobial and antioxidant activities over time, impeding the packaging material's efficiency and triggering food spoilage before the product reaches the final consumer. This drawback can be rectified if the EO is packaged within the packing inner lining in such a way that its volatility is reduced if not eliminated. This is achieved using nanotechnology to strengthen the existing properties of polymer by equipping with bioactive EO via the development of nanocomposites.

Polylactic acid (PLA) is a special renewable, biodegradable polyesters polymer with attractive thermoplastic, mechanical, and biological properties that has earned its approval by the US Food and Drug Administration in biomedical and packaging materials (Scaffaro et al., 2018). In the aforementioned applications, the development of polymeric systems with antimicrobial property reduces microbial growth. Treatment of such promising polymers with CEO is an attractive approach as it has previously been demonstrated to produce plasticized nanocomposite (Munteanu et al., 2018).

Functionalized electrospun PLA film was coated with a 100-nm-thick cationic chitosan layer encapsulated with CEO via coaxial electrospinning. The high phenolic content from the immobilized CEO resulted in synergistic and higher antibacterial activity against *E. coli*, *L. monocytogenes*, and *Salmonella typhymurium* (Lu et al., 2021). The study showed that EO added mechanical property in addition to bactericidal effects compared to the unloaded films. Thus, chitosan displays antimicrobial activity but lacks mechanical property. On the other hand, PLA lacks any mechanical activity, but EO introduces antimicrobial property and pronged shelf life and mechanical property upon loading.

A comparative study utilizing chitosan and polyvinyl alcohol (PVA) blended films was prepared via solvent casting and phase inversion (Antunes et al., 2021). The two hydrophilic polymers formed a potent blend due to hydrogen bonding from the abundant —OH groups. This, in turn, effectively encapsulated the CEO with further phenolic interaction, which displayed enhanced thermal property. In addition, the synergistic inhibitory effect of microbial growth against Gram-positive *Staphylococcus aureus* and Gram-negative *Pseudomonas aeruginosa* was observed via the agar diffusion test. The nano-biocomposites also displayed higher antimicrobial activity compared to the isolated polymers.

Spherical and stable chitosan nanoparticles between 40 and 100 nm were synthesized by a two-step process (Hasheminejad et al., 2019). First, to minimize the evaporation and oxidation, nanoencapsulation of CEO of this biopolymer was achieved via an oil-in-water emulsion and ionic gelation of emulsion droplets technique. Second, a controlled release with prolonged activity displayed enhanced antifungal activity against *Aspergillus niger*. Similarly, chitosan nanoparticles and clove oil nano-biocomposites displayed superior antifungal activity against postharvest pathogenic fungus *Pythium aphanidermatum* (Radhakrishnan et al., 2021).

Another reinforcement of CEO nanoencapsulation with a potato-starch material was recently employed via 3D nanonetwork (Fang et al., 2020). The resultant nanocomposite possessed improved stability against thermal treatment and prolonged antimicrobial activity against Gram-negative bacteria (*E. coli*, ATCC 35150) and Gram-positive bacteria (*Bacillus subtilis*, ATCC 6633, and *Staphylococcus aureus*, ATCC 25923). This makes CEO a promising biomolecule in bacterial and fungal infection in medical, food, and agriculture industries.

Eugenol is the principal constituent of CEO and is widely used in food, pharmaceutical, cosmetics, and active packaging applications, owing to its effective antimicrobial and antioxidant properties (Baskaran et al., 2010; Devi et al., 2010). However, most plant-derived bioactive compounds, including eugenol, are highly volatile, unstable, and sensitive to oxygen, light, and heat during processing, utilization, and storage (Choi et al., 2009; Garg and Singh, 2011; Gaysinsky et al., 2005). Therefore, eugenol is a good candidate as an antioxidant for various thermal processing applications, including bioactive plastics for food packaging. The thermal stability of eugenol has been improved by encapsulation into chitosan nanoparticles through a two-step method, i.e., the formation of an oil-in-water emulsion and the ionic gelation of emulsion droplets. The extrusion of eugenol-loaded chitosan nanoparticles and thermoplastic flour, a model plastic, at a temperature up to 155°C confirmed the improved thermal stability of encapsulated eugenol compared with nakedness eugenol (Woranuch and Yoksan, 2013). Furthermore, the emulsion-diffusion has been used as an effective method to improve the oxidation stability of eugenol; the compound was encapsulated by inclusion with β-cyclodextrin to be protected from light oxidation during storage time due to their complete wrapping of eugenol by polycaprolactone layer (Choi et al., 2009).

Lee et al. (2018) developed chitosan-based films containing CEO as an active agent by incorporating halloysite nanotubes (HNTs) to stabilize the oil droplets and reported their application as food packaging materials. Furthermore, they suggested that such films containing natural ingredients have the potential to be applied as active packaging solutions in food packaging systems due to their antioxidant and antifungal properties alongside their enhanced barrier property against water vapor (Lee et al., 2018). Chitosan nanoparticles loaded with CEO are very well-known for their antioxidant and antimicrobial activities; therefore, they have been applied in the food industry for food safety purposes (Hadidi et al., 2020). In addition, a recent study on the eugenol microemulsion reinforced with silver nanocomposite electrospun mats was developed for wound dressing applications (Sethuram et al., 2021).

3. Conclusions

Postharvest loss caused by microorganisms is considered a significant problem in the food industry. The recent packaging sector trends focus mainly on developing active packaging materials with inherent antimicrobial and improved barrier properties. The antimicrobial activity of the CEO is well known and documented, and it contributes to the final aims of the packaging materials as a protective layer against the microbial attack. Eugenol and CEO have been integrated into different polymers such as polylactic acid, PVA, potato-starch based, and chitosan in composite to produce effective and safe packaging materials; as the environmental impacts increase, the biodegradable natural components in the food industry become an urgent task. The efficiency of CEO nanocomposite was found to have the utmost potential as antimicrobial packaging material to ensure food safety by enhancing the shelf life and preventing postharvest loss. However, more research needs to be conducted to reveal the potential of the CEO as an effective component in food packaging due to the availability, safety, and intense antibacterial activity, and its major component, eugenol.

References

Antunes, J.C., Tavares, T.D., Teixeira, M.A., Teixeira, M.O., Homem, N.C., Amorim, M., Felgueiras, H.P., 2021. Eugenol-containing essential oils loaded onto chitosan/polyvinyl alcohol blended films and their ability to eradicate *Staphylococcus aureus* or *Pseudomonas aeruginosa* from infected microenvironments. Pharmaceutics 13 (2), 195.

Barboza, J.N., da Silva Maia Bezerra Filho, C., Silva, R.O., Medeiros, J., de Sousa, D.P., 2018. An overview on the anti-inflammatory potential and antioxidant profile of eugenol. Oxidative Med. Cell. Longev. 2018, 3957262.

Baskaran, Y., Periyasamy, V., Venkatraman, A.C., 2010. Investigation of antioxidant, anti-inflammatory and DNA-protective properties of eugenol in thioacetamide-induced liver injury in rats. Toxicology 268, 204–212.

Choi, M.J., Soottitantawat, A., Nuchuchua, O., Min, S.G., Ruktanonchai, U., 2009. Physical and light oxidative properties of eugenol encapsulated by molecular inclusion and emulsion-diffusion method. Food Res. Int. 42, 148–156.

Devi, K.P., Nisha, S.A., Sakthivel, R., Pandian, S.K., 2010. Eugenol (an essential oil of clove) acts as an antibacterial agent against *Salmonella typhi* by disrupting the cellular membrane. J. Ethnopharmacol. 130, 107–115.

Fang, Y., Fu, J., Liu, P., Cu, B., 2020. Morphology and characteristics of 3D nanonetwork porous starch-based nanomaterial via a simple sacrifice template approach for clove essential oil encapsulation. Ind. Crop. Prod. 143, 111939.

Garg, A., Singh, S., 2011. Enhancement in antifungal activity of eugenol in immunosuppressed rats through lipid nanocarriers. Colloids Surf. B: Biointerfaces 87, 280–288.

Gaysinsky, S., Davidson, P.M., Bruce, B.D., Weiss, J., 2005. Stability and antimicrobial efficiency of eugenol encapsulated in surfactant micelles as affected by temperature and pH. J. Food Prot. 68, 1359–1366.

Hadidi, M., Pouramin, S., Adinepour, F., Haghani, S., Jafari, S.M., 2020. Chitosan nanoparticles loaded with clove essential oil: characterization, antioxidant and antibacterial activities. Carbohydr. Polym. 236, 116075.

Hasheminejad, N., Khodaiyan, F., Safari, M., 2019. Improving the antifungal activity of clove essential oil encapsulated by chitosan nanoparticles. Food Chem. 275, 113–122.

Ishangulyyev, S., Kim, S., Lee, S.H., 2019. Understanding food loss and waste—why are we losing and wasting food? Foods 8, 297.

Jayasena, D.D., Jo, C., 2013. Essential oils as potential antimicrobial agents in meat and meat products: a review. Trends Food Sci. Technol. 34, 96–108.

Jiang, B., Na, J.X., Wang, L.L., Li, D.M., Liu, C.H., Feng, Z.B., 2019. Reutilization of food waste: one-step extraction, purification, and characterization of ovalbumin from salted egg white by aqueous two-phase flotation. Foods 8, 286.

Kerry, J.P., O'Grady, M.N., Hogan, S.A., 2006. Past, current and potential utilisation of active and intelligent packaging systems for meat and muscle-based products: a review. Meat Sci. 74, 113–130.

Lee, M.H., Kim, S.Y., Park, H.J., 2018. Effect of halloysite nanoclay on the physical, mechanical, and antioxidant properties of chitosan films incorporated with clove essential oil. Food Hydrocoll. 84, 58–67.

Liu, Q., Meng, X., Li, Y., Zhao, C.-N., Tang, G.-Y., Li, H.-B., 2017. Antibacterial and antifungal activities of spices. Int. J. Mol. Sci. 18 (6), 1283.

Lu, W., Cui, R., Zhu, B., Qin, Y., Cheng, G., Li, L., Yuan, M., 2021. Influence of clove essential oil immobilized in mesoporous silica nanoparticles on the functional properties of poly (lactic acid) biocomposite food packaging film. J. Mater. Res. Technol. 11, 1152–1161.

Motelica, L., Ficai, D., Oprea, O.C., Ficai, A., Andronescu, E., 2020. Smart food packaging designed by nanotechnological and drug delivery approaches. Coatings 10, 806.

Munteanu, B.S., Sacarescu, L., Vasiliu, A.-L., Hitruc, G.E., Pricope, G.M., Sivertsvik, M., Rosnes, J.T., Vasile, C., 2018. Antioxidant/antibacterial electrospun nanocoatings applied onto PLA films. Materials 11 (10), 1973.

Nguyen-Tri, P., Nguyen, T.A., Carriere, P., Ngo Xuan, C., 2018. Nanocomposite coatings: preparation, characterization, properties, and applications. Int. J. Corros. 2018. https://doi.org/10.1155/2018/4749501.

O'Callaghan, K.A.M., Kerry, J.P., 2014. Assessment of the antimicrobial activity of potentially active substances (nanoparticled and non-nanoparticled) against cheese-derived micro-organisms. Int. J. Dairy Technol. 67, 483–489.

Radhakrishnan, E., Jose, A., Saranya, A., Chacko, C., Jacob, D.M., 2021. Polyvinyl Alcohol Based Nanocomposite Films Engineered With Clove Oil and Chitosan Nanoparticles to Combat the Postharvest Fungal Diseases. Research Square.

Rai, M., Ingle, A.P., Gupta, I., Pandit, R., Paralikar, P., Gade, A., Chaud, M.V., dos Santos, C.A., 2019. Smart nanopackaging for the enhancement of food shelf life. Environ. Chem. Lett. 17, 277–290.

Scaffaro, R., Lopresti, F., Marino, A., Nostro, A., 2018. Antimicrobial additives for poly (lactic acid) materials and their applications: current state and perspectives. Appl. Microbiol. Biotechnol. 102 (18), 7739–7756.

Sethuram, L., Thomas, J., Mukherjee, A., Chandrasekaran, N., 2021. Eugenol micro-emulsion reinforced with silver nanocomposite electrospun mats for wound dressing strategies. Mater. Adv. 2, 2971–2988.

Shah, U., Gani, A., Ashwar, B.A., Shah, A., Ahmad, M., Gani, A., Wani, I.A., Masoodi, F.A., 2015. A review of the recent advances in starch as active and nanocomposite packaging films. Cogent Food Agric. 1, 1–9.

Shirdar, M.R., Farajpour, N., Shahbazian-Yassar, R., Shokuhfar, T., 2019. Nanocomposite materials in orthopedic applications. Front. Chem. Sci. Eng. 13 (1), 1–13.

Szabo, K., Teleky, B.E., Mitrea, L., Calinoiu, L.F., Martau, G.A., Simon, E., Varvara, R.A., Vodnar, D.C., 2020. Active packaging-poly (vinyl alcohol) films enriched with tomato by-products extract. Coatings 10, 141.

Tunma, S., 2018. Starch-based nanocomposites in active packaging for the extended shelf life of fresh fruits. Walailak J. Sci. Technol. 15, 273–281.

Vilas, C., Mauricio-Iglesias, M., Garcia, M.R., 2020. Model-based design of smart active packaging systems with antimicrobial activity. Food Packag. Shelf Life 24, 100446.

Wohner, B., Gabriel, V.H., Krenn, B., Krauter, V., Tacker, M., 2020. Environmental and economic assessment of food-packaging systems with a focus on food waste. Case study on tomato ketchup. Sci. Total Environ. 738, 139846.

Woranuch, S., Yoksan, R., 2013. Eugenol-loaded chitosan nanoparticles: I. Thermal stability improvement of eugenol through encapsulation. Carbohydr. Polym. 96 (2), 578–585.

Further reading

Shruthy, R., Jancy, S., Preetha, R., 2021. Cellulose nanoparticles synthesised from potato peel for the development of active packaging film for enhancement of shelf life of raw prawns (*Penaeus monodon*) during frozen storage. J. Food Sci. Technol. 56 (8), 3991–3999.

CHAPTER

Food applications of clove (*Syzygium aromaticum*) buds

13

Rajan Logesh[a], Niranjan Das[b], and Hari Prasad Devkota[c]

TIFAC-CORE in Herbal Drugs, Department of Pharmacognosy and Phytopharmacy, JSS College of Pharmacy (JSS Academy of Higher Education and Research), Rockland's, Ooty, Tamil Nadu, India[a] Department of Chemistry, Iswar Chandra Vidyasagar College, Belonia, Tripura, India[b] Graduate School of Pharmaceutical Sciences, Kumamoto University, Kumamoto, Japan[c]

1. Introduction

Clove (*Syzygium aromaticum* (L.) Merr. & L.M. Perry., Family: Myrtaceae) (Fig. 1), one of the most valuable spices, is being used for centuries in food recipes and also for its medicinal properties (Cortés-Rojas et al., 2014; Devkota and Adhikari-Devkota, 2020). The clove tree, also called a miracle tree of nature, is believed to be originated from the Moluccas Islands (also known as Spice Island). The plant part that is used is the unopened flower bud (Fig. 2) and it is most commonly used as a culinary spice in biryani, pickles, salads, garam masala, etc. The clove bud possesses an intense fragrance and burning taste and has a deep brown color, a powerful fragrant odor that is warm, pungent, and strongly sweet, and slightly astringent (Milind and Deepa, 2011). This chapter summarizes the food and traditional medicinal applications of clove flower buds.

2. Food applications (culinary uses)

Clove actually denotes the symbol of dignity and is one of the most valuable spices worldwide. Dried cloves are the key ingredient in Indian tea. Dried cloves are an important part of spices in South Indian cuisine, and they are also used in combination with many other spices. Clove is widely used to prepare tea, curry, and other foods such as cookies, chewing gum, spiced fruits, drinks, cakes, pickles, etc. It is also used widely as a spice in other South Asian countries and other parts of the world (Cai and Wu, 1996; Hussain et al., 2017).

3. Traditional medicinal uses

Clove is widely used in various traditional medicine systems around the world. Known as Devapuspa in Sanskrit, Lavanga in Hindi, and Lwang in the Nepali language, it is used as a crude drug in Ayurveda and other traditional medicine systems in South Asia (Ministry for Health and Family Welfare, 2010;

Clove (*Syzygium aromaticum*). https://doi.org/10.1016/B978-0-323-85177-0.00006-9

FIG. 1

Photographs of clove tree (A), flowers (B), and flower buds (C, D).

Devkota and Adhikari-Devkota, 2020). Having aromatic, carminative, and stimulant properties, it is used to treat dyspepsia, flatulence, nausea, and vomiting (Rajbhandari et al., 1995). Cloves are useful in treating various digestive problems such as indigestion, diarrhea, and gastric irritability (Hochenegg, 2010). Clove also promotes the discharge of mucous in the respiratory tract and is useful for cold and cough. Cloves are good expectorants that promote the discharge of mucous and secretions in the respiratory passage. It is reported that the sniffing of clove buds with a strong aroma reduces drowsiness, headache, and irritability (Hussain et al., 2017). Clove is also reported to enhance memory retention (Trongtokit et al., 2005). In Persian medicine, clove oil is used as an anticonvulsant agent and for the treatment of epilepsy (Pourgholami et al., 1999). Clove oil is also used to treat stress, parasitic infestations, cough, headache, and blood impurities (Rimpler, 1999; Hänsel and Sticher, 2007). Clove and clove oil are also used as aphrodisiac agents (Jurkowska, 2005; Brud and Konopacka-Brut, 2009).

Due to strong antibacterial properties, clove is used in various mouthwashes, dental creams, throat sprays, and toothpaste to kill pathogens. It is also used to relieve sore gums (Cai and Wu, 1996; Hussain et al., 2017). Clove has strong anesthetic properties, and clove oil is widely practiced to treat toothache (Bhowmik et al., 2012). Clove plays important therapeutic role as antibacterial, antifungal, antiseptic, stress reliever, and anti-inflammatory agent (Wichtl, 1989; Schönfelder and Schönfelder, 2004; Körfers and Sun, 2009).

FIG. 2

Photographs of fresh (A) and dried (B) flower buds of clove.

Nowadays, clove oil, an important natural antibacterial drug, is used in many fields, including dentistry, pharmaceuticals, and aromatherapy.

4. Other uses

Cloves are also widely used for animals to treat tooth pain in dogs and cats. Peppermint tea with a sprinkle of cloves and ginger is used to treat vomiting in dogs (Cai and Wu, 1996; Hussain et al., 2017).

Clove oil is also an effective mosquito and ant repellent (Trongtokit et al., 2005; Jabłońska-Trypuć and Fabiszewski, 2008; Heinrich and Barnes, 2004; Bhowmik et al., 2012). It is also used to flavor many other products such as perfumes and soaps (Góra and Lis, 2000). Clove and its oil are also used widely as air fresheners (Heinrich and Barnes, 2004; Bhowmik et al., 2012).

References

Bhowmik, D., Kumar, K.S., Yadav, A., Srivastava, S., Paswan, S., Dutta, A.S., 2012. Recent trends in Indian traditional herbs *Syzygium aromaticum* and its health benefits. J. Pharmacogn. Phytother. 1 (1), 13–22.

Brud, S., Konopacka-Brut, I., 2009. Podstawy Perfumerii. Historia, pochodzenie i zastosowanie substancji zapachowych. Oficyna Wydawnicza MA, Łódź.

Cai, L., Wu, C.D., 1996. Compounds from *Syzygium aromaticum* possessing growth inhibitory activity against oral pathogens. J. Nat. Prod. 59 (10), 987–990.

Cortés-Rojas, D.F., de Souza, C.R., Oliveira, W.P., 2014. Clove (*Syzygium aromaticum*): a precious spice. Asian Pac. J. Trop. Biomed. 4 (2), 90–96.

Devkota, H.P., Adhikari-Devkota, A., 2020. Cold pressed clove (*Syzygium aromaticum*) oil. In: Cold Pressed Oils. Academic Press, pp. 273–276.

Góra, J., Lis, A., 2000. Najcenniejsze olejki eteryczne. Wydawnictwo Uniwersytetu Mikołaja Kopernika, Toruń.

Hänsel, R., Sticher, O., 2007. Anhang: Nelkenöl and Eugenol in der konservierende Zahnheilkunde. Pharmakognosie–Phytopharmacie, 8. Auflage, Ed 25.8.6. Springer Medizin Verlag, Heidelberg, Germany, pp. 1133–1134.

Heinrich, M., Barnes, J., 2004. Clove, *Syzygium aromaticum* (L.) Merr. & L.M. Perry. In: Fundamentals of Pharmacognosy and Phytotherapy. Churchill Livingstone, Edinburgh-London, Great Britain, pp. 275–276.

Hochenegg, B., 2010. Evaluation of the Traditional and Well-Established Use of *Tormentillae rhizoma*, *Caryophylli flos* and *Caryophylli aetheroleum* (Thesis). Univ. Wien, Germany.

Hussain, S., Rahman, R., Mushtaq, A., Belaskri, A.E., 2017. Clove: a review of a precious species with multiple uses. Int. J. Chem. Biochem. Sci. 11, 129–133.

Jabłońska-Trypuć, Fabiszewski, R., 2008. Sensoryka i podstawy perfumerii. Pharm Polska, Wrocław.

Jurkowska, S., 2005. Surowce kosmetyczne, Wyższa Szkoła Fizykoterapii z siedzibą we Wrocławiu.

Körfers, A., Sun, Y., 2009. Caryophylli flos– Gewürznelken– Ding X. Traditionelle Chinesische Medizin– Arzneidrogen und Therapie. Wissenschaftliche Verlagsgesellschaft mbH Stuttgart Germany, pp. 397–398.

Milind, P., Deepa, K., 2011. Clove: a champion spice. Int. J. Res. Ayurveda Pharm. 2 (1), 47–54.

Ministry for Health and Family Welfare, 2010. The Ayurvedic Pharmacopoeia of India. Government of India, Ministry for Health and Family Welfare, pp. 110–111.

Pourgholami, M.H., Kamalinejad, M., Javadi, M., Majzoob, S., Sayyah, M., 1999. Evaluation of the anticonvulsant activity of the essential oil of *Eugenia caryophylata* in male mice. J. Ethnopharmacol. 64, 167–171.

Rajbhandari, T., Joshi, M., Shrestha, T., Joshi, S., Acharya, B., 1995. Medicinal Plants of Nepal for Ayurvedic Drugs. Department of Plant Resources, Kathmandu.

Rimpler, H., 1999. Caryophylli floris aetheroleum. Biogene Arzneistoff, 2. Auflage, Edn 498. Deutscher Apotheker Verlag Stuttgart, Germany.

Schönfelder, I., Schönfelder, P., 2004. *Syzygium aromaticum* (L.) MERR & PERRY (*Eugenia caryophyllata* THUNB *Jambosa caryophyllus* (SPRENG.) (NIEDENZU)). Das neue Handbuch der Heilpflanzen. Kosmos-Verlag, Stuttgart, Germany, pp. 431–432.

Trongtokit, Y., Rongsriyam, Y., Komalamisra, N., Apiwathnasorn, C., 2005. Comparative repellency of 38 essential oils against mosquito bites. Phytother. Res. 19 (4), 303–309.

Wichtl, M., 1989. Gewürznelken. Teedrogen, 2. Auflage. Wissenschaftliche Verlagsgesellschaft mbH Stuttgart Germany, pp. 193–194.

Further reading

Brud, S., Konopacka-Brud, I., 1997. Aromatoterapia w Farmacji i Medycynie. Wiadomości Zielarskie 1, 12–13.

Cimanga, K., Kambu, K., Tona, L., Apers, S., De Bruyne, S., Hermans, N., Totté, J., Pieters, L., Vlietinck, A.J., 2002. Correlation between chemical composition and antibacterial activity of essential oils of some aromatic medicinal plants growing in the Democratic Republic of Congo. J. Ethnopharmacol. 79, 213–220.

Gülçin, I., Güngör, S.I., Beydemir, S., Elnastaş, M., Küfrevioğlu, Ö.I., 2004. Comparison of antioxidant activity of clove (*Eugenia caryophylata* Thunb) buds and lavender (*Lavandula stoechas* L.). Food Chem. 87, 394–400.

Kluge, H., 2008. Wielka księga Hildegardy z Bingen. Jedność, Kielce.

Podlewski, J.K., Chalibogowska-Podlewska, A., 2010. Leki współczesnej terapii. Medical Tribune Polska, Warszawa.

CHAPTER 14

Clove (*Syzygium aromaticum*) phenolics: Extraction, compositions, and biological activities

Ahmad Cheikhyoussef[a], Natascha Cheikhyoussef[b], Ateeq Rahman[c], and Ahmed A. Hussein[d]

Science and Technology Division, Multidisciplinary Research Unit, Centre for Research Services, University of Namibia, Windhoek, Namibia[a] Ministry of Higher Education, Technology and Innovation, Windhoek, Namibia[b] Department of Physics, Chemistry & Material Science, Faculty of Agriculture, Engineering & Natural Sciences, University of Namibia, Windhoek, Namibia[c] Chemistry Department, Cape Peninsula University of Technology, Cape Town, Western Cape, South Africa[d]

1. Introduction

Clove herb is a tropical evergreen tree of *Syzygium aromaticum* L. Merr and Perry (syn. *Eugenia caryophyllata, Caryophyllus aromaticus*, L.). The herbal part is the unopened flower buds (Singletary, 2014). Clove belongs to the Myrtaceae family, which is native to India, Pakistan, and Indonesia (Chaieb et al., 2007; Maheswari et al., 2020). There are four key varieties of the clove: Ambon, Siputih, Zanzibar, and Sikotok, with the Zanzibar variety superiority to produce a higher amount of flowering buds than others considered the broadest cultivated variety (Alfikri et al., 2020). The clove tree has a vegetative cycle that can reach more than 100 years, with a height of 10–20 m; also, it has smooth oval leaves and reddish flowers with several terminal clusters positioned at the end of the branches (Cortés-Rojas et al., 2014). Clove fruit has an elongated shape with a distinctively strong aroma (Cortés-Rojas et al., 2014).

Clove buds are among the richest sources of clove plant parts in dietary phenolics with several uses in traditional medicine (Issac et al., 2015). Phenolic compounds, also known as polyphenols, are ubiquitous molecules in the vegetable kingdom and occur in a rich, diversified range of structures with a well-known essential role for plants growth and reproduction and specific function to protect injured plants against pathogens (Chiorcea-Paquim et al., 2020). Phenolic compounds are one of the largest secondary metabolites groups containing one or more hydroxyl groups binding to aromatic compounds (Rampadarath et al., 2014). Major classes of phenolic compounds are flavonoids, phenolic acids (hydroxycinnamic and hydroxybenzoic acids), and tannins (Chew and Nyam, 2019). Hydroxycinnamic acid derivatives include caffeic, ferulic, and *p*-coumaric acids with C3-C6 structure (Ramadan, 2019), meanwhile hydroxybenzoic acids derivatives such as gallic, syringic, and vanillic acids with C1-C6 skeleton (Boualem et al., 2017).

Phenolic compounds possess a wide range of biological activities, including anti-allergic, anticarcinogenic, anti-inflammatory, antimutagenic, modulating enzymatic activities, pharmacological

Clove (*Syzygium aromaticum*). https://doi.org/10.1016/B978-0-323-85177-0.00036-7

actions against oxidative damage diseases, including age-related diseases, coronary heart disease, and strokes (Chiorcea-Paquim et al., 2020). The reported biomolecular mechanisms for their action include angiogenesis inhibition, binding of electrophiles, apoptosis rate increment, H-donation, inhibition of cellular proliferation, reactive oxygen species (ROS) inactivation, protective enzymes induction, lipid peroxidation inhibition, and inhibition of DNA oxidation (Patil and Masand, 2018; Chiorcea-Paquim et al., 2020). This chapter provides an overview of the clove (*Syzygium aromaticum*) phenolics composition, structure, and biological activities, as well as their reported mode/mechanism of action.

2. Clove phenolics extraction, structure, and composition

2.1 Clove phenolics extraction

Generally, the extraction process for the phenolic compounds is mainly determined by solubility phenomena (Bodoira and Maestri, 2020). Phenolics compounds naturally occur in all plants with over 8000 known structures. They are non-nutritive bioactive compounds secreted through the shikimate pathway or the acetate pathway, in which structural characteristics are featured with at least one aromatic ring with one or several hydroxyl groups attached (Crozier et al., 2009; Dai and Mumper, 2010). To avoid lower degradation and losing bioactivity of bioactive compounds from herbs and spices and to ensure that they do not contain any microbial and chemical contaminants and extend their shelf life, the choice and the selection of a suitable extraction method should be carefully handled (El-Sayed and Youssef, 2019; Das and Sharangi, 2020).

Menon and Narayanan (1992) extracted bound volatiles from the dried clove buds and fresh green clove leaves using Svendsen's method (β-glucosidase and α-amyloglucosidase for enzyme hydrolysis). They reported the presence of eugenol, isoeugenol, farnesol, and nerolidol as the main aglycones in the extracts. Hydrodistillation, supercritical carbon dioxide, and superheated water extraction techniques have all been applied to the extraction of clove buds and maceration, which remain the most popular extraction method for clove buds an industrial scale (Alexandru et al., 2013). Alexandru et al. (2013) reported that the use of flow ultrasound-assisted extraction (UAE) operated at high flow rates yielded the highest phenolic compound extraction (215 mg GAE/L) and that because as the sonication flow rate increases, there is a corresponding increase in the total phenolic content of the clove crude extracts.

Singletary (2014) reported that the composition of clove extracts is not well documented, and the method used for clove extraction can vary considerably with various procedures being used such as hot water extraction, extraction with 80% ethanol, and steam distillation. Therefore, the chemical composition of clove extracts will be varied widely in their content of flavonoids, phenolics, and hydrolyzable tannins (Singletary, 2014; Tan and Karim, 2018). For example, El-Maati et al. (2016) reported that extracting phenolics and flavonoids from cloves with water resulted in higher antioxidant and antimicrobial properties than other used solvents. However, a better yield for clove *n*-hexane extract was reported with higher antimicrobial activity and antioxidant activities in terms of total phenol content (TPC) of 54 mg GAE/g, total flavonoid content (TFC) of 15.5 mg GAE/g, DPPH· of 0.29 mg/mL, and FRAP of 0.69 mg/mL in comparison to other clove extracts from other solvation systems including alcohol, petroleum ether and water (Ishaq et al., 2019).

According to Alara et al. (2021), there is a need to establish a standardized process for the recovery of phenolic compounds from plant materials through the consideration of the following factors:

(a) sample type and the targeted compounds, e.g., total phenolics, a specific class of phenolics, a specific phenolic, etc. (b) the object of the analysis (quantification or structure elucidation purpose), and lastly, (c) the availability of the technique to be used. Omran et al. (2017) used an extraction system consisting of various ratios of ethanol and methanol by using EtOH:MeOH:HCl:H_2O to obtain the best phenolics and flavonoids extraction, and H_2O:HCl at ratio 20:1 (v/v), that reported the best water ratio for the flavonoids and the phenolic extraction (Xu, 2012; Toma's-Barbera'n and Ferreres, 2012). They further reported that phenolic extraction occurred in acidic solvents as reported to be the best for recovering the bounded phenolic compounds from tissues by Toma's-Barbera'n and Ferreres (2012). Frohlich et al. (2019) utilized supercritical extraction with CO_2 and Soxhlet with *n*-hexane to prepare the clove leaf extract and reported that the best result for supercritical extraction was obtained with 220 bars at 40°C, giving the highest yield for eugenol with 29.8%.

Mohammed et al. (2020) compared the use of two different extraction methods (maceration and sonication) to extract bioactive components of clove buds ethanol extract. They detected 48 compounds for maceration and 43 compounds for sonication using gas chromatography triple quad time of flight (GC-Q-TOF). They recommended that sonication with GC-Q-TOF is a valuable and suitable combined technique for quick clove bioactive compounds. Furthermore, they emphasized that sonication has significant advantages over maceration in the qualitative analysis at the analytical level due to fast and efficient timing (30 min) compared to 24 h extraction time required for maceration. However, ultrasound-assisted equipment is much more costly compared to maceration cost (Mohammed et al., 2020).

2.2 Clove phenolics structure and composition

2.2.1 Phenolic acids and flavonoids

Phenolics acids, flavonoids, and tannins are the common classes of plant phenolics; meanwhile, stilbenes and lignans are the less common phenolics (Dai and Mumper, 2010). The first class, phenolic acids, can be then divided into two subclasses: (1) derivatives of benzoic acid such as gallic acid, and (2) derivatives of cinnamic acid such as coumaric, caffeic, and ferulic acid (Dai and Mumper, 2010). Meanwhile, the second common class of phenolics is flavonoids, classified under the benzopyran derivatives, and they are the most frequent polyphenols found in plants (~4000 natural flavonoids have been discovered). Mainly, they are composed of the flavan unit, containing 15 carbon atoms arranged in 3 rings which 3 kinds of substituents can link (C6-C3-C6), hydroxyl (A), methoxyl (B), and glucosides (C) (Dai and Mumper, 2010; Chiorcea-Paquim et al., 2020). According to the oxidation state of the central C ring, flavonoids are divided into six subgroups: flavones, flavonols, flavanols, flavanones, isoflavones, and anthocyanins (Dai and Mumper, 2010).

Cai and Wu (1996) reported the presence of gallic acid (Fig. 1), ellagic acid, kaempferol, rhamnocitrin, and myricetin in the methanolic extract of *S. aromaticum* using several spectroscopic techniques. The phenolic acids, especially gallic acid, are believed to be the main compound present in higher concentrations, alongside other gallic acid derivates such as hydrolyzable tannins are also present in higher concentrations compared to other compounds (Shan et al., 2005). In addition, caffeic, ferulic, ellagic, and salicylic acids can also be found among the other phenolic acids in clove (Shan et al., 2005; Mohammed et al., 2016).

Atawodi et al. (2011) quantified several phenolic compounds in the methanol extract of *S. aromaticum* using a combination of LC/MSD and UV dual-array detector following HPLC separation using a

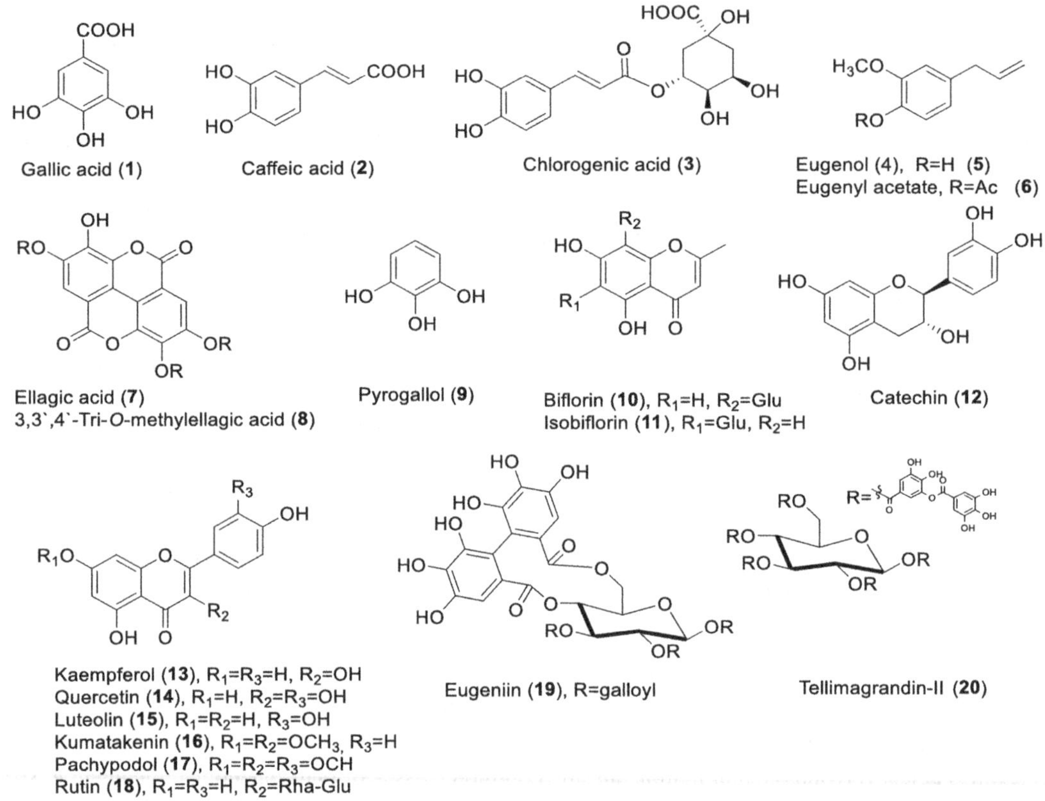

FIG. 1

Main phenolic compound in clove (*S. aromaticum*) extracts.

C-18 reverse phase column; these compounds include ellagic acid (9.1 mg/kg), ellagic acid derivative (3.5 mg/kg), gallic acid (28.6 mg/kg), quercetin glucoside (3.5 mg/kg), and some other unidentified phenolic compounds. Clove has been reported as the primary source of phenolic molecules (Fig. 1) like hydroxybenzoic acids, hydroxyphenyl propens, hydroxycinnamic acids, flavonoids, and eugenol (Neveu et al., 2010; Mohammed et al., 2016; Batiha et al., 2020; Saeed et al., 2021). Several phenolic acids and flavonoids in clove buds extracts have been quantified (Adefegha et al., 2016), such as quercitrin (30.2 mg/g), quercetin (29.7 mg/g), kaempferol (20.4 mg/g), chlorogenic acid (20.1 mg/g), gallic acid (14.0 mg/g), caffeic acid (13.9 mg/g), luteolin (9.28 mg/g), rutin (9.11 mg/g), catechin (1.85 mg/g) and ellagic acid (1.61 mg/g) (Fig. 1).

Flavonoids such as kaempferol, quercetin, and their glycosylated derivatives were also found in cloves in lower concentrations (Jirovetz et al., 2006). Clove essential oil can be found in the clove flower buds (up to 18%), which is composed of eugenol (89%) and 5%–15% of eugenol acetate and β-caryophyllene (Jirovetz et al., 2006).

Woo et al. (2017) isolated kumatakenin (3,7-dimethoxy kaempferol) from the EtOAc fraction of cloves with potential anticancer activity against ovarian cancer cells. They isolated 19 flavonoids from the EtOAc fraction, of which 6 are flavonoids (kumatakenin, pachypodol, quercetin, luteolin, rhamnazin-3-*O*-β-D-glucuronide-6″-methyl ester, and rhamnazin-3-*O*-β-D-glucoside) shown some cytotoxicity and growth inhibitory activity in 2 ovarian cancer cells: A2780 and SKOV3 (Ryu et al., 2016a). Fathoni et al. (2017) reported several nonvolatile compounds using LC-MS/MS in cloves from Manado, Indonesia, which is responsible for the cloves' flavor, more specifically to their taste and known to possess two main flavor perceptions including astringent, and bitter. They reported the presence of asquinic acid, biflorin, digalloylglucose, dimethoxyluteolin, ellagic acid, gallic acid, hydroxy gallic acid, isobiflorin, isorhamnetin, kaempferol, luteolin, monogalloylglucose, naringenin, quercetin, and rhamnetin (Fathoni et al., 2017). Biflorin (5,7-dihydroxy-2-methylchromone-6-*C*-β-D-glucopyranoside) and isobiflorin (5,7-dihydroxy-2-methylchromone-8-*C*-β-D-glucopyranoside), are two major chromone *C*-glucosides, they have been being isolated from an ethanolic extract of *E. caryophyllata* (Zhang and Chen, 1997). Han and Paik (2010) isolated two acetophenone glucosides (2,6-dihydroxy-4-methoxyacetophenone 3-*C*-β-D-glucoside and 2,6-dihydroxy-4-methoxyacetophenone 3-*C*-β-D-(6′-*O*-galloyl)glucoside), eugenin, 8-*C*-glucosylnoreugenin, and gallic acid from the EtOAc soluble fraction of the MeOH extract of clove buds.

Acetophenones, chromones, and phenylpropanoids belong to the plant's phenolic compounds (Alu'datt et al., 2017). The flavonol glucuronides were isolated for the first time from the flower buds of *S. aromaticum* (Ryu et al., 2016a) using a repeated chromatography of the EtOAc-soluble fraction from the 70% EtOH extract.

Two acetophenones; **compound 1:** 2,4,6-trihydroxy-3-methyl acetophenone-2-*O*-β-D-glucoside (Fig. 2A) and **compound 2:** 2,4,6-trihydroxy acetophenone-3-*C*-β-D-glucoside (Fig. 2B), four chromone glycosides (Fig. 3); biflorin, 6-*C*-β-D-(6′-*O*-galloyl) glucosylnoreugenin (Fig. 3A), isobiflorin, and 8-*C*-β-D-(6′-*O*-galloyl) glucosylnoreugenin (Fig. 3B), six phenylpropanoids (Fig. 4): eugenol, eugenyl acetate (Fig. 4A), *trans*-coniferyl aldehyde (Fig. 4B), 3-(4-hydroxy-3-methoxyphenyl) propane-1,2-diol, 1-*O*-methyl-guaiacylglycerol (Fig. 4C), and epoxiconiferyl alcohol (Fig. 4D) have been isolated from the leaves and flower buds of *S. aromaticum* (Ryu et al., 2016b).

Eugenol is the main bioactive compound and present in around 9.314.7% and ranging from 9381 to 14,650 mg per 100 g of fresh plant material (Neveu et al., 2010; Cortés-Rojas et al., 2014; Hussein et al., 2019). Volatile and non-volatile phenolic compounds, such as biflorin, isobiflorin, and acetyl eugenol, in addition to gallic acid, have been reported in the flower buds extracts of *S. aromaticum* (Ladurner et al., 2017). Hemalatha et al. (2016) conducted a qualitative analysis on three crude extracts of *E. caryophyllus* and confirmed the presence of phenols, flavonoids, and tannins; meanwhile, the

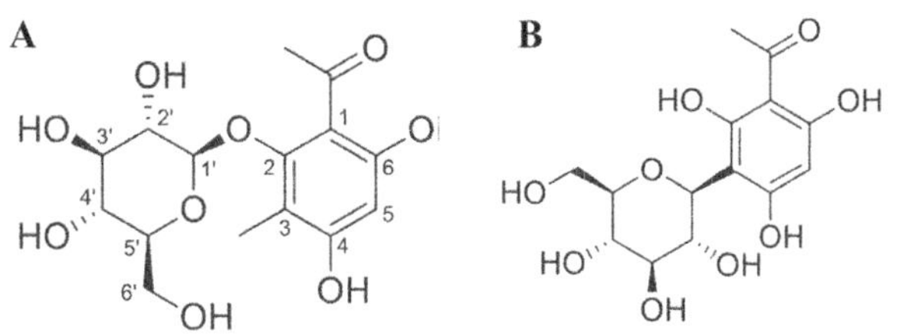

FIG. 2

Acetophenones from *S. aromaticum*; 2,4,6-trihydroxy-3-methyl acetophenone-2-*O*-β-D-glucoside (A) and 2,4,6-trihydroxy acetophenone-3-*C*-β-D-glucoside (B).

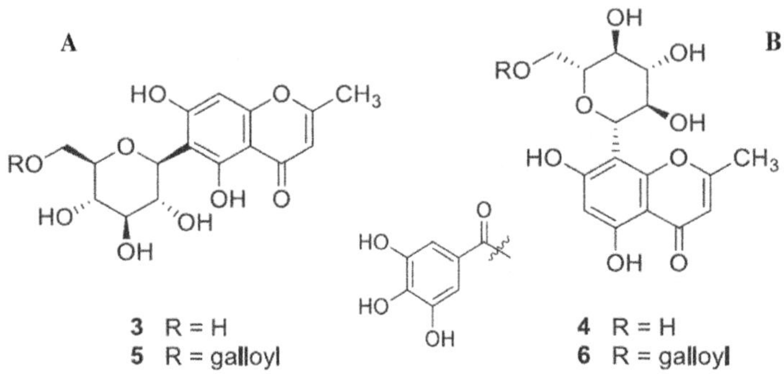

FIG. 3

Chromone glycosides from *S. aromaticum*; biflorin and 6-*C*-β-D-(6′-*O*-galloyl) glucosylnoreugenin (A), isobiflorin and 8-*C*-β-D-(6′-*O*-galloyl) glucosylnoreugenin (B).

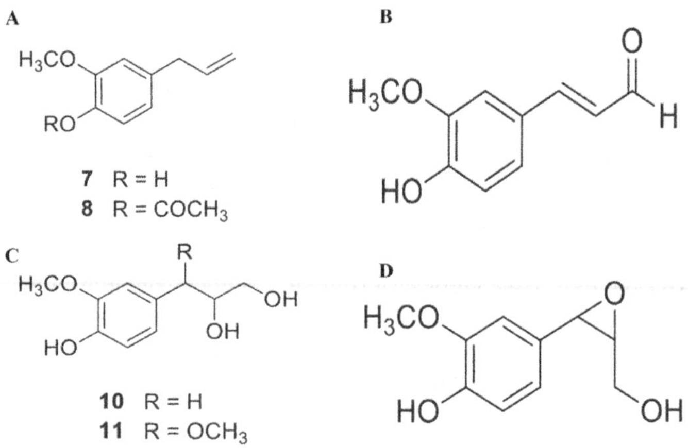

FIG. 4

Phenylpropanoids from *S. aromaticum*; eugenol, eugenyl acetate (A), *trans*-coniferyl aldehyde (B), 3-(4-hydroxy-3-methoxyphenyl) propane-1,2-diol,1-*O*-methyl-guaiacylglycerol (C), and epoxy coniferyl alcohol (D).

quantitative of the methanol extract in terms of the total phenolics content (TPC) and total flavonoids content (TFC) yielded 596 and 62.6 mg/g of the extract respectively. Total phenolic content (TPC) of Clove (*Syzygium aromaticum*) extract of 16,047 mg/100 g Fresh Weight (FW) was reported as per the phenol-Explorer Database (Neveu et al., 2010; Pérez-Jiménez et al., 2010a,b).

2.2.2 Tannins

The third common class is tannins, which are another major group of phenolics and subsequently divided into two subclasses: (1) hydrolyzable tannins (blends of simple phenols, e.g., ellagic and gallic acids, with a carbohydrate) and (2) condensed tannins (two or more monomers of flavan-3-ol units, linked together mainly by bonds between the A rings of the flavanol units and the pyran rings of other flavanols) (Dai and Mumper, 2010; Durazzo et al., 2019; Albuquerque et al., 2021). Fresh clove also contains several tannin compounds with concentrations ranging between 10% and 13% and has the same chemical composition as gallotannic acid and ellagitannin, which belongs to hydrolyzable tannins in addition to the presence of eugenin (Milind and Deepa, 2011). Joshi et al. (2011) revealed the presence of tannins and glycosides in the ethanol extracts of *E. caryophyllata* (clove). Monogalloylglucose, reported by Fathoni et al. (2017), belongs to hydrolyzable tannin, which is a water-soluble polyphenolic compound (Zimmer, 1999; Meyers et al., 2006) that might contribute to the odorless and astringent taste of cloves (Meyers et al., 2006).

Eugeniin, an ellagitannin compound, was isolated from the bud of *S. aromatica* (Nonaka et al., 1980; Takechi and Tanaka, 1981). A tannin, 3,3′,4′-tri-*O*-methylellagic acid (Fig. 1) was isolated by a mixed solvents system (*n*-hexane, EtOAc, and BuOH) soluble fractions from the flower buds of *S. aromaticum* and obtained a concentration of 3.1 mg was purified by recrystallization using physical (UV, IR) and spectroscopic data (^{1}H NMR, ^{13}C NMR, 2D NMR, and MS) measurements (Ryu et al., 2016b).

Tellimagrandin-II (Fig. 1), a hydrolyzable tannin, was isolated from the dried flower bud of *S. aromaticum* L. and has shown promising antiviral activity (Umashankar et al., 2021). Tellimagrandin-II contains hexahydrodiphenoyl (HHDP), and galloyl groups (Salminen, 2018) was also reported as an important bioactive plant phenolic (Moilanen et al., 2013). Umashankar et al. (2021) reported Tellimagrandin-II as the first polyphenolic compound (ellagitannin) formed from 1,2,3,4,6-pentagalloyl-glucose, and is an isomer of punicafolin or nupharin A, also known as cornustannin 2 or eugeniin.

2.2.3 Volatile phenolics

Twenty-two compounds have been identified in the extracts of clove buds by gas chromatography and gas chromatography/mass spectrometry (GC/MS) with major aroma compounds of eugenol (24.37 mg/g) and eugenyl acetate (2.354 mg/g). Most notably, the oxidation of hexanal was inhibited by eugenol (99%), eugenyl acetate (99%), and benzyl alcohol (82%) for a period of 30 days at 500 μg/mL (Lee and Shibamoto, 2001). Furthermore, the GC-MS analysis by Hemalatha et al. (2016) for the methanol extracts of *E. caryophyllus* confirms the presence of eugenol (54.8%), phenol,2-methoxy-4-(2-propenyl) (19.4%), pyrogallol (5. 48%), and caryophyllene (4.48%), as well as eugenyl acetate, which contribute to clove, extracts antioxidant and antibacterial activities.

Gowri and Manimegalai (2019) screened different organic solvents for the powdered *S. aromaticum* extracted and detected flavonoids and phenols in the ethanol extract of *S. aromaticum*. They further analyzed the ethanol extract using GC-MS and revealed the presence of 14 compounds with 64.4% of phenol, 2-methoxy-3-(2-propenyl), 14.9% eugenol, and 2.93% caryophyllene.

GC-MS analysis of *S. aromaticum* extracts, of which the ethyl acetate extract revealed that eugenol (58.8%) was the most abundant active component, followed by eugenyl acetate (23.8%), and other compounds (Yassin et al., 2020). Meanwhile, the main bioactive component of the diethyl ether extract was eugenol (66.4%), followed by eugenyl acetate (31.7%). On the other hand, the methanol extract for *S. aromaticum* yielded eugenol (55.5%), eugenyl acetate (19.8%), and other VOCs such as trans-caryophyllene (15.7%), α-humulene (2.46%), and caryophyllene oxide (1.10%), musk ketone (2.78%), α-cubebene (0.79%), chavicol (0.78%), β-cadinene (0.59%), and α-farnesene (0.37%) (Yassin et al., 2020).

Clove samples from the three different development stages of clove buds (phase I: green; phase II: pink; and phase III: red) were adsorbed onto polydimethylsiloxane-coated devices, and then gas chromatography-mass spectrometry was used to analyze volatile organic compounds (VOCs) (Kasai et al., 2016). They reported the presence of the following VOCs: eugenol, β-caryophyllene, α-humulene, α-farnesene, and eugenol acetate were the predominant constituents in most phases; meanwhile, the buds ripened additionally contain methyl salicylate, α-copaene, β-ocimene, 2-nonanone, ethyl hexanoate, and acetophenone. However, when the clove leaves were analyzed using the same method, the following VOCs were identified: β-caryophyllene, eugenol, α-humulene, 3-hexenol, 3-hexenyl acetate, and hexyl acetate (Kasai et al., 2016). Jimoh et al. (2017) also revealed the presence of tannins, alkaloids, terpenoids, carbohydrates, glycosides, ketones, aldehydes, and 46 phenolic compounds in methanol extract of clove with different phenolic compounds being identified using Gas Chromatography-Flame Ionization Detector (GC-FID). In addition, they reported the presence of caffeic acid, eugenin, eugenitin, gallic acid, isorhamnetin, kaempferol, myricetin, phenylacetic acid, protocatechuic acid, *p*-hydroxybenzoic acid, quercetin, rhamnetin, salicylic acid, and syringic acid.

Another study done by Ereifej et al. (2016) reported that Total Phenolic Compounds (TPC) concentrations had been affected significantly by the extractant (solvents) used and temperature employed, whereby high phenolic compound level in cloves acetone extract (781 mg GAE/100 g), ethanolic extract (548 mg GAE/100 g) and 305 mg GAE/100 g for methanol extract at 60°C. They further added that the level of phenolic compounds extracted by methanol, ethanol, and acetone at 20°C and 40°C varied significantly from when it was conducted at 60°C, and the level of phenolics had increased when the temperature was elevated from 20°C to 40°C and 60°C. Omran et al. (2017) reported a TPC for cloves seeds of 3.86 (mg GAE/gDW) and total flavonoids content (TFC) of 2.73 (mg QE/gDW) with antioxidant activity of 107 μg/mL. Saleem et al. (2019) utilized bioassay-guided isolation, which led to the identification of previously known isobiflorin, biflorin (Zhang and Chen, 1997), and eugeniin (Kurokawa et al., 1998; Nonaka et al., 1980; Takechi and Tanaka, 1981) from the flower buds fractions of *S. aromaticum*. Eugenol is a phenolic compound with a molecular formula of $C_{10}H_{12}O_2$, and is also named 4-allyl-2-methoxyphenol with a simple structure with three active sites: hydroxyl, allylic and aromatic groups. Eugenol possesses interesting chemistry with the potential impact on the synthesis of novel drugs and compounds useful for human medication (Abdou et al., 2021). A flavor constituent, 1-phenylethyl acetate (FEMA: 2684, CAS 93-92-5), was extracted from clove buds (Gassenmeier et al., 2017) and identified using headspace solid-phase microextraction-gas chromatography/mass spectrometry (HS-SPME-GC/MS), HPLC, and NMR Spectroscopy (Bitterling et al., 2020). Gassenmeier et al. (2017) reported that 1-phenylethyl acetate occurs naturally in clove buds, and probably, it is the product of enzymatic biosynthesis and not an artifact of subsequent processing. El Ghallab et al. (2020) reported that ethanol and hexane extracts of *S.* aromaticum buds possess a total phenolic content (351 and 348 mg of gallic acid equivalent/g, respectively) and moderate total flavonoid content in the aqueous extract of 21.9 mg of quercetin equivalent/g. They further performed a thin layer chromatography analysis and confirmed that these two extracts contain eugenol, which mainly contributes to the scavenging activity and tends to be more active in a mixture with terpenes, like in essential oil (El Ghallab et al., 2020).

3. Clove phenolics biological activities

Herbs and spices are becoming of high interest and have received increased attention from the food industry to prevent/delay spoilage of foods, and they are considered important sources of many effective antioxidants (Suhaj, 2006; Yashin et al., 2017; Jiang, 2019). Phenolic compounds bioactivities and health benefits depend on the absolute amounts of intake and their digestive fate when passing through the human digestive tract (Feng et al., 2021). Generally, bioactivity can be assessed by using in vivo or in vitro bioaccessibility and bioavailability (Cardoso et al., 2015; Bento-Silva et al., 2020), of which bioaccessibility refers to the accessibility of phenolics ready for absorption after ingestion; meanwhile, bioavailability refers to the actual amounts that can undergo absorption/assimilation into intestinal epithelium cells and are involved in the hepatic metabolism which then, it can be detected in plasma (Cardoso et al., 2015). The presence of free or conjugated phenolic compounds in food matrices determines their bioavailability which will be affected during food processing (Bento-Silva et al., 2020). Generally, phenolics have low bioavailability, with an estimated rate of absorption ranging from 0.3% to 43% (Caleja et al., 2017). Phenolic compounds are responsible for the organoleptic characteristics (taste, color, etc.) of plant food and the bitterness of fruits due to their interaction with salivary glycoprotein (Alara et al., 2021).

Cai and Wu (1996) reported that flavones (kaempferol and myricetin), isolated from the methanolic extract of *S. aromaticum*, possess potent growth-inhibitory activity against periodontal pathogens of *P. gingivalis* and *P. intermedia*. Furthermore, clove phenolic compounds possess a wide range of biological activities, such as antioxidant activities, anti-glycation, antinociceptive and antimicrobial properties that enhance bioactivities (Ryu et al., 2016a; Idowu et al., 2021). According to Omran et al. (2017), there are three mechanisms to explain the antioxidant properties of phenolic compounds, these are scavenging the radical species such as reactive oxygen species (ROS) and reactive nitrogen species (RNS) by suppressing ROS/RNS formation through the inhibition of specific enzymes or trace metals chelation that is involved in free radical production, and last, through the up-regulation or protection of the antioxidant defense system. When the methanolic extract of *S. aromaticum* was tested with the xanthine oxidase and the 2-deoxyguanosine assay models, they displayed very potent antioxidant potential ($IC_{50} = 2\,\mu L$) and radical scavenging activity ($IC_{50} = 55\,\mu L$), respectively (Atawodi et al., 2011).

High quantities of phenolic compounds have been reported from cloves and spices extracts after they exhibited high antioxidant and antimicrobial activity compared to many fruits, vegetables, and other spices (Cortés-Rojas et al., 2014). Saeed et al. (2021) reported that clove possesses high antioxidants because of its high levels of phenolic and other bioactive compounds. Furthermore, they reported that the phenylpropanoids in the clove bud possess antibacterial activity against several bacterial species via the interaction with their cell membrane that may affect the membrane structure's conformation, responsible for its fluidity and expansion (Tajodini et al., 2014). The disruption of membrane stability will impact the ionic leakage across the cellular membrane and reduce the ionic gradient across the bacterial cell membrane, which may slow down bacterial growth (Saeed et al., 2021).

Clove extracts (CE) obtained from total buds of clove have been broadly investigated to have outstanding antioxidant activities in meat and meat products (Shan et al., 2005). In addition, the principal phenolic components of CE are eugenol and eugenyl acetate, which reveal highly possible antioxidant characteristics due to their high total phenolic content of 293 mg GAE/g extract (El-Maati et al., 2016). According to Radha Krishnan et al. (2014), eugenol and caryophyllene are the main ingredients of clove extract and the most active constituents responsible for clove antioxidant activity. In addition, eugeniin was isolated from herbal extract (*S. aromaticum*) that had exhibited anti-HSV activity in mice (Kurokawa et al., 1998, 2001).

A phenylpropanoid, eugenol (4-allyl-2-methoxyphenol), plays a crucial role in dental and oral hygiene preparations (Pavithra, 2014; Batiha et al., 2020) with a wide range of biological activities as antifungal, antimicrobial, anti-inflammatory, antioxidant, analgesic, anticancer, and antiparasitic (Abdou et al., 2021). According to the US, Food and Drug Administration (FDA) and the World Health Organization (WHO), clove buds, clove oil, eugenol, and oleoresins are considered generally recognized as safe (GRAS) and can be used as food additives (US-FDA, 1978; Nisar et al., 2021). Eugenol's antimicrobial activity against both Gram-Positive and Gram-negative bacteria was attributed to the presence of the hydroxyl group (Tajodini et al., 2014). In addition, Eugenol was reported to inhibit the *Salmonella typhi* growth by disrupting the cellular membrane (Devi et al., 2010) and to increase apoptotic cell death via the activation of caspase-3 and p53 in colon cancer cells (Anita et al., 2012), as well as it can induce apoptosis through the ERα pathway in MCF7 breast cancer cells (Jaganathan et al., 2011). Another animal model reported that doses of eugenol ranging from 200 and 400 mg/kg have anti-inflammatory activity whereby they can reduce the volume of pleural exudates without changing the total count of blood leukocytes (Daniel et al., 2009).

Yassin et al. (2020) reported that eugenol exhibited antifungal potency against *C. albicans*, *C. glabrata*, and *C. tropicalis* demonstrating inhibition zone diameters of 15.6 mm. Another study reported by Pavesi et al. (2018) indicated the potency of eugenol at a concentration of 57 μg/disk against *C. albicans* with an inhibition zone diameter of 12.1 mm. Possible mechanisms could be attributed to this antifungal potency of *S. aromaticum* extracts due to the high content of eugenol (55.5%–66.4%), which inhibits the biosynthesis of ergosterol a component of the microbial cell membranes (Yassin et al., 2020). The ergosterol biosynthesis inhibition may disrupt microbial cell membrane permeability, causing cell death (de Oliveira Pereira et al., 2013).

Eugenol has shown antithrombotic effects as a platelet inhibitor, preventing blood clots (García-Mediavilla et al., 2007). Furthermore, at the molecular level via in vitro, eugenol has been shown to inhibit arachidonic acid-induced platelet aggregation, prostaglandin biosynthesis, and the formation of thromboxane B2 (Rasheed et al., 1984). However, when combined with acetyl eugenol, it produces stronger effects than acetylsalicylic acid in inhibiting the platelet aggregation induced by arachidonic acid, adrenaline, and collagen, presenting the potential of an anti-aggregation activity that can be compared to indomethacin (Srivastava, 1993).

Eugenol was reported to be responsible for the regulation of the cellular inflammatory pathways of nuclear factor kappa-light-chain-enhancer of activated B cells (NF-κB), extracellular-signal-regulated kinase/mitogen-activated protein kinase (ERK/MAPK), and pro-inflammatory interleukins production (Bahramsoltani and Rahimi, 2020). Recently, Saeed et al. (2021) reported that eugenol could enhance animal productivity via its reduced impact on the methane-producing microflora and protein degradation inhibition in the rumen. In addition, another mechanism indicated that the antifungal activity of eugenyl acetate could be to germ tube formation, which inhibition, prevention of the formation of candidal biofilms, and enhances phagocytic activity of macrophages against *C. albicans* species (Musthafa et al., 2016).

Issac et al. (2015) standardized phenolics-rich extracts as a water-soluble free-flowing powder (Clovinol) and suggested their suitability for functional food applications despite their pungency and aroma characteristics. Clovinol possesses significant antioxidant and anti-inflammatory effects as measured by cellular antioxidant levels and the ability to inhibit carrageenan-induced paw swelling in mice. They further reported its significant anti-ulcerogenic activity (>97% inhibition of ethanol-induced stomach ulcers in Wistar rats when orally administered at 100 mg/kg b.w.) and up-regulation of in vivo antioxidants, such as superoxide dismutase (SOD), glutathione (GSH), and catalase (CAT) (Issac et al., 2015). In addition, Clovinol could reduce the extent of lipid peroxidation among ulcer-induced rats, indicating its usefulness in ameliorating oxidative stress and improving gastrointestinal health, especially upon chronic alcohol consumption (Issac et al., 2015). Johannah et al. (2015) reported that nonvolatile polyphenols in clove buds are bioactive and can be used as dietary supplements. However, in-vitro model studies utilizing clovinol (41.2% polyphenols) did not produce any significant changes in body weight, food and water consumption, hematological and/or biochemical parameters when it was evaluated for its toxicity in Wistar rats, and mutagenicity in *Salmonella typhimurium* strains (Vijayasteltar et al., 2016). Furthermore, clovinol demonstrated radical scavenging activities and detoxification potential by enhancing the cellular antioxidant defense enzymes (glutathione, glutathione peroxidase, superoxide dismutase, and catalase) when it was tested in both animals and humans (Issac et al., 2015; Johannah et al., 2015).

Ryu et al. (2016b) reported that chromone *C*-glycosides with the galloyl group; 6-*C*-β-D-(6′-*O*-galloyl) glucosylnoreugeniesis showed weak cytotoxicity (IC_{50} = 66.7 μM) and the same for 8-*C*-β-D-(6′-*O*-galloyl)glucosylnoreugenin (IC_{50} = 87.5 μM), meanwhile, biflorin and isobiflorin were not effective (IC_{50} > 100 μM). Therefore, they proposed that galloyl groups may increase the cytotoxicity of these two chromone *C*-glycosides. Furthermore, they also observed that 6-*C*-β-D-(6′-*O*-galloyl) glucosylnoreugenin was more active than 8-*C*-β-D-(6′-*O*-galloyl) glucosylnoreugenin due to the different position of *C*-glucose- 6′-*O*-galloyl group (*C*-6 and *C*-8, respectively). Additionally, for the phenylpropanoids only *trans*-coniferyl aldehyde showed weak cytotoxicity with IC_{50} = 78.4 μM. The tannin compound of 3,3′,4′-tri-*O*-methylellagic acid reported weak cytotoxicity with IC_{50} = 87.6 μM against human ovarian cancer cells (A2780) using MTT assays (Ryu et al., 2016b). The above categories of compounds, acetophenone glucosides, chromone *C*-glucosides, and phenylpropanoids have been reported to show neuroprotective (Han and Paik, 2010), antioxidant (Han and Paik, 2012), and antimutagenic activity (Miyazawa and Hisama, 2003), respectively.

Biflorin and isobiflorin (Fig. 1) were isolated by Lee et al. (2016) from the flower buds of *S. aromaticum*. They reported their anti-inflammatory activities on lipopolysaccharide (LPS)-induced production of nitric oxide (NO) and prostaglandin E2 (PGE2) in RAW 264.7 macrophages whereby biflorin exhibited an IC_{50} = 51.7 and 37.1 μM, respectively; meanwhile, isobiflorin exhibited an IC_{50} > 60 and 46.0 μM respectively. Furthermore, they confirmed that biflorin could inhibit the LPS-induced production and mRNA expression of tumor necrosis factor-α (TNF-α) and interleukin (IL)-6. Older studies by Lee et al. (1995) and Hong et al. (2002) reported that clove extracts could inhibit LPS-induced NO, PGE2, IL-6, IL-1β, and IL-8.

According to Hemalatha et al. (2016), eugenol can reduce two or more DPPH· radicals by forming dimers. Pyrogallol detected in the methanol extracts of *E. caryophyllus* possesses antibacterial activities due to three OH groups attached to the phenyl ring, responsible for the toxicity against microorganisms. Kumatakenin, a flavonoid that was isolated from the flower buds of *S. aromaticum*, shows anticancer activities with a reported mode of action through the apoptosis of ovarian cancer cells (A2780 and SKOV3) using propidium iodide and Annexin V-FITC staining assay and inhibition of the alternative activation of tumor-associated macrophages (TAMs) (Woo et al., 2017). According to Woo et al. (2017), the inhibition of the alternative activation of TAMs is achieved via the reduction of the expressions of MCP-1 and RANTES, which are major determinants of macrophage recruitment, as well as inhibition of the M2 markers expression and cancer-promoting factors (IL-10, MMP-2/-9, and VEGF) in macrophages stimulated by the ovarian cancer cells.

Saleem et al. (2019) isolated eugeniin, isobiflorin, and biflorin from the flower buds of *S. aromaticum* and reported their antiviral activity against dengue virus protease with $IC_{50}=94.7$ nM of eugeniin against the proteases of DENV serotype-2 and $IC_{50}=7.5$ μM of eugeniin against the proteases of DENV serotype-3. They further reported that the mechanism of these compounds' antiviral activity is a competitive type of inhibition, which was further substantiated by computational docking and saturation transfer difference (STD)-NMR spectroscopy. At the molecular level, when these compounds bind at the protease's active site, extensive interactions are mediated by a network of hydrogen bonds and hydrophobic contacts (Saleem et al., 2019). A potent scavenging activity for ethanol extract ($EC_{50}=0.41$ mg/mL) and hexane extract ($EC_{50}=0.37$ mg/mL) of *S. aromaticum* buds (El Ghallab et al., 2020). Tellimagrandin-II from *S. aromaticum* L. was reported to have a binding energy of −8.2 kcal/mol and binding free energy of −32.0 kcal/mol; it was found to be highly potential due to its higher binding affinity and significant binding free energy (MM-PBSA), along with favorable ADMET properties and stable intermolecular interactions with hotspots including the ASN343 glycosylation site (Umashankar et al., 2021). Tellimagrandin-II belongs to ellagitannins or hydrolyzable tannins, with a well-known antiviral activity; this hydrolyzable tannins class is also well-known with several antiviral agents (Yoshida et al., 2000). Vicidomini et al. (2021) reported that clove buds contain β-caryophyllene, kaempferol, and rhamnetin, which belong to flavonoids that are believed to contribute to clove anti-inflammatory properties (Rho et al., 2011; Jnawali et al., 2014; Novo Belchor et al., 2017).

Recent computational studies (Joshi et al., 2020; Pandey et al., 2021) recommended the use of kaempferol, a flavonoid extracted from cloves, as potent anti-COVID-19 drug whereby the in silico results showed that it binds the substrate-binding pocket of SARS-CoV-2 main protease with high affinity leading to the interaction with Cys145 and His41; the active site residues through hydrophobic interactions and hydrogen bonding (Rehman et al., 2021). Furthermore, bicornin (9.2 kcal/mol) and biflorin (8.5 kcal/mol) for M^{pro} have shown potential inhibitory activity via molecular docking studies with high reported affinities of clove compounds (Rehman et al., 2021).

4. Conclusions

Clove plant parts extracts are rich in phytochemicals, such as alkaloids, glycosides, phenolic compounds, and terpenoids. Among these important phyto-compounds are phenolic compounds ranging from simple aromatic acids, phenolic acids like gallic acid, and its derivatives to complex tannins with diverse structures and functions. Eugenol is the major phenolic in cloves with different established biological activities and food, pharmaceutical, and medical applications.

References

Abdou, A., Elmakssoudi, A., El Amrani, A., et al., 2021. Recent advances in chemical reactivity and biological activities of eugenol derivatives. Med. Chem. Res. 30, 1011–1030. https://doi.org/10.1007/s00044-021-02712-x.

Adefegha, S.A., Oboh, G., Oyeleye, S.I., Osunmo, K., 2016. Alteration of starch hydrolyzing enzyme inhibitory properties, antioxidant activities, and phenolic profile of clove buds (*Syzygium aromaticum* L.) by cooking duration. Food Sci. Nutr. 4 (2), 250–260. https://doi.org/10.1002/fsn3.284.

Alara, O.R., Abdurahman, N.H., Ukaegbu, C.I., 2021. Extraction of phenolic compounds: a review. Curr. Res. Food Sci. 4, 200–214. https://doi.org/10.1016/j.crfs.2021.03.011.

Albuquerque, B.R., Heleno, S.A., Oliveira, M., Barros, L., Ferreira, I., 2021. Phenolic compounds: current industrial applications, limitations and future challenges. Food Funct. 12 (1), 14–29. https://doi.org/10.1039/d0fo02324h.

Alexandru, L., Cravotto, G., Giordana, L., Binello, A., Chemat, F., 2013. Ultrasound-assisted extraction of clove buds using batch-and flow-reactors: a comparative study on a pilot scale. Innovative Food Sci. Emerg. Technol. 20, 167–172. https://doi.org/10.1016/j.ifset.2013.07.011.

Alfikri, F.N., Pujiarti, R., Wibisono, M.G., Hardiyanto, E.B., 2020. Yield, quality, and antioxidant activity of clove (*Syzygium aromaticum* L.) bud oil at the different Phenological stages in young and mature trees. Scientifica 2020, 9701701. https://doi.org/10.1155/2020/9701701.

Alu'datt, M.H., Rababah, T., Alhamad, M.N., Al-Mahasneh, M.A., Almajwal, A., Gammoh, S., Ereifej, K., Johargy, A., Alli, I., 2017. A review of phenolic compounds in oil-bearing plants: distribution, identification and occurrence of phenolic compounds. Food Chem. 218, 99–106. https://doi.org/10.1016/j.foodchem.2016.09.057.

Anita, Y., Radifar, M., Kardono, L.B., Hanafi, M., Istyastono, E.P., 2012. Structure-based design of eugenol analogs as potential estrogen receptor antagonists. Bioinformation 8 (19), 901–906. https://doi.org/10.6026/97320630008901.

Atawodi, S.E., Atawodi, J.C., Pfundstein, B., Spiegelhalder, B., Bartsch, H., Owen, R., 2011. Assessment of the polyphenol components and in vitro antioxidant properties of *Syzygium aromaticum* (L.) Merr. & Perry. Elec. J. Env. Agricult. Food Chem. Title 10 (3), 1970–1978.

Bahramsoltani, R., Rahimi, R., 2020. An evaluation of traditional Persian medicine for the management of SARS-CoV-2. Front. Pharmacol. 11. https://doi.org/10.3389/fphar.2020.571434, 571434.

Batiha, G.E., Alkazmi, L.M., Wasef, L.G., Beshbishy, A.M., Nadwa, E.H., Rashwan, E.K., 2020. *Syzygium aromaticum* L. (Myrtaceae): traditional uses, bioactive chemical constituents, pharmacological and toxicological activities. Biomolecules 10 (2), 202. https://doi.org/10.3390/biom10020202.

Bento-Silva, A., Koistinen, V.M., Mena, P., Bronze, M.R., Hanhineva, K., Sahlstrøm, S., Kitrytė, V., Moco, S., Aura, A.M., 2020. Factors affecting intake, metabolism and health benefits of phenolic acids: do we understand individual variability? Eur. J. Nutr. 59 (4), 1275–1293. https://doi.org/10.1007/s00394-019-01987-6.

Bitterling, H., Schäfer, U., Krammer, G., Meier, L., Brückner, S.I., Hartmann, B., Ongouta, J., Carle, R., Steingass, C.B., 2020. Investigations into the natural occurrence of 1-phenylethyl acetate (styrallyl acetate). J. Agric. Food Chem. 68 (32), 8613–8620. https://doi.org/10.1021/acs.jafc.0c02758.

Bodoira, R., Maestri, D., 2020. Phenolic compounds from nuts: extraction, chemical profiles, and bioactivity. J. Agric. Food Chem. 68 (4), 927–942. https://doi.org/10.1021/acs.jafc.9b07160.

Boualem, M., Mokhtar, M., Saiah, F., Benourad, F., Bouhadiba, R., Berkani, A., 2017. Identification of *Mentha piperita* L. and *Ricinus communis* L. polyphenols by HPLC-DAD-ESI-MS and evaluation of their insecticidal properties against *Aphis spiraecola* P. *south Asian*. J. Exp. Biol. 7 (1), 28–34.

Cai, L., Wu, C.D., 1996. Compounds from *Syzygium aromaticum* possessing growth inhibitory activity against oral pathogens. J. Nat. Prod. 59 (10), 987–990. https://doi.org/10.1021/np960451q.

Caleja, C., Ribeiro, A., Barreiro, M.F., Ferreira, I., 2017. Phenolic compounds as nutraceuticals or functional food ingredients. Curr. Pharm. Des. 23 (19), 2787–2806. https://doi.org/10.2174/1381612822666161227153906.

Cardoso, C., Afonso, C., Lourenço, H., Costa, S., Nunes, M.L., 2015. Bioaccessibility assessment methodologies and their consequences for the risk-benefit evaluation of food. Trends Food Sci. Technol. 41 (1), 5–23. https://doi.org/10.1016/j.tifs.2014.08.008.

Chaieb, K., Hajlaoui, H., Zmantar, T., Kahla-Nakbi, A.B., Rouabhia, M., Mahdouani, K., Bakhrouf, A., 2007. The chemical composition and biological activity of clove essential oil, *Eugenia caryophyllata* (*Syzigium aromaticum* L. Myrtaceae): a short review. Phytother. Res.: PTR 21 (6), 501–506. https://doi.org/10.1002/ptr.2124.

Chew, S.C., Nyam, K.L., 2019. Kenaf (*Hibiscus cannabinus* L.) seed oil. In: Ramadan, M. (Ed.), Fruit Oils: Chemistry and Functionality. Springer, Cham, https://doi.org/10.1007/978-3-030-12473-1_23.

Chiorcea-Paquim, A.M., Enache, T.A., De Souza Gil, E., Oliveira-Brett, A.M., 2020. Natural phenolic antioxidants electrochemistry: towards a new food science methodology. Compr. Rev. Food Sci. Food Saf. 19 (4), 1680–1726. https://doi.org/10.1111/1541-4337.12566.

Cortés-Rojas, D.F., de Souza, C.R., Oliveira, W.P., 2014. Clove (*Syzygium aromaticum*): a precious spice. Asian Pac. J. Trop. Biomed. 4 (2), 90–96. https://doi.org/10.1016/S2221-1691(14)60215-X.

Crozier, A., Jaganath, I.B., Clifford, M.N., 2009. Dietary phenolics: chemistry, bioavailability and effects on health. Nat. Prod. Rep. 26 (8), 1001–1043. https://doi.org/10.1039/b802662a.

Dai, J., Mumper, R.J., 2010. Plant phenolics: extraction, analysis and their antioxidant and anticancer properties. Molecules 15 (10), 7313–7352. https://doi.org/10.3390/molecules15107313.

Daniel, A.N., Sartoretto, S.M., Schmidt, G., Caparroz-Assef, S.M., Bersani-Amado, C.A., Cuman, R.K.N., 2009. Anti-inflammatory and antinociceptive activities A of eugenol essential oil in experimental animal models. Rev. Bras. Farm. 19, 212–217. https://doi.org/10.1590/S0102-695X2009000200006.

Das, S., Sharangi, A.B., 2020. Nanotechnology: A potential tool in exploring herbal benefits. In: Thangadurai, D., Sangeetha, J., Prasad, R. (Eds.), Functional Bionanomaterials. Nanotechnology in the Life Sciences. Springer, Cham, https://doi.org/10.1007/978-3-030-41464-1_2.

de Oliveira Pereira, F., Mendes, J.M., de Oliveira Lima, E., 2013. Investigation on mechanism of antifungal activity of eugenol against *Trichophyton rubrum*. Med. Mycol. 51 (5), 507–513. https://doi.org/10.3109/13693786.2012.742966.

Devi, K.P., Nisha, S.A., Sakthivel, R., Pandian, S.K., 2010. Eugenol (an essential oil of clove) acts as an antibacterial agent against *salmonella typhi* by disrupting the cellular membrane. J. Ethnopharmacol. 130 (1), 107–115. https://doi.org/10.1016/j.jep.2010.04.025.

Durazzo, A., Lucarini, M., Souto, E.B., Cicala, C., Caiazzo, E., Izzo, A.A., Novellino, E., Santini, A., 2019. Polyphenols: a concise overview on the chemistry, occurrence, and human health. Phytother. Res.: PTR 33 (9), 2221–2243. https://doi.org/10.1002/ptr.6419.

El Ghallab, Y., Al Jahid, A., Jamal Eddine, J., Haj Said, A.A., Zarayby, L., Derfoufi, S., 2020. *Syzygium aromaticum* L.: phytochemical investigation and comparison of the scavenging activity of essential oil, extracts and eugenol. Orient Pharm Exp Med 20, 153–158. https://doi.org/10.1007/s13596-019-00416-7.

El-Maati, M.F.A., Mahgoub, S.A., Labib, S.M., Al-Gaby, A.M.A., Ramadan, M.F., 2016. Phenolic extracts of clove (*Syzygium aromaticum*) with novel antioxidant and antibacterial activities. Eur. J. Intern. Med. 8, 494–504. https://doi.org/10.1016/j.eujim.2016.02.006.

El-Sayed, S.M., Youssef, A.M., 2019. Potential application of herbs and spices and their effects in functional dairy products. Heliyon 5 (6). https://doi.org/10.1016/j.heliyon.2019.e01989, e01989.

Ereifej, K., Feng, H., Rababah, T., Tashtoush, S., Al-U'datt, M., Gammoh, S., Al-Rabadi, G., 2016. Effect of Extractant and temperature on phenolic compounds and antioxidant activity of selected spices. Food Nutr. Sci. 7, 362–370. https://doi.org/10.4236/fns.2016.75038.

Fathoni, A., Saepudin, E., Cahyana, A.H., Rahayu, D.U.C., Haib, J., 2017. Identification of nonvolatile compounds in clove (*Syzygium aromaticum*) from Manado. AIP Conf. Proc. 1862, 030079 (2017) https://doi.org/10.1063/1.4991183.

Feng, S., Yi, J., Li, X., Wu, X., Zhao, Y., Ma, Y., Bi, J., 2021. Systematic review of phenolic compounds in apple fruits: compositions, distribution, absorption, metabolism, and processing stability. J. Agric. Food Chem. 69 (1), 7–27. https://doi.org/10.1021/acs.jafc.0c05481.

Frohlich, P.C., Santos, K.A., Palú, F., Cardozo-Filho, L., da Silva, C., da Silva, E.A., 2019. Evaluation of the effects of temperature and pressure on the extraction of eugenol from clove (*Syzygium aromaticum)* leaves using supercritical CO2. J. Supercrit. Fluids 143, 313–320. https://doi.org/10.1016/j.supflu.2018.09.009.

García-Mediavilla, V., Crespo, I., Collado, P.S., Esteller, A., Sánchez-Campos, S., Tuñón, M.J., González-Gallego, J., 2007. The anti-inflammatory flavones quercetin and kaempferol cause inhibition of inducible nitric oxide synthase, cyclooxygenase-2 and reactive C-protein, and down-regulation of the nuclear factor kappaB pathway in Chang liver cells. Eur. J. Pharmacol. 557 (2–3), 221–229. https://doi.org/10.1016/j.ejphar.2006.11.014.

Gassenmeier, K., Schwager, H., Houben, E., Clery, R., 2017. Unequivocal identification of 1-Phenylethyl acetate in clove buds (*Syzygium aromaticum* (L.) Merr. & L.M.Perry) and clove essential oil. Foods 6 (7), 46. https://doi.org/10.3390/foods6070046.

Gowri, G., Manimegalai, K., 2019. Phytochemical screening and GC-MS analysis of ethanol extract of *Syzygium aromaticum* L. J. Pharmacogn. Phytother. 8 (5), 746–750.

Han, A.R., Paik, Y.S., 2010. Identification and PEP inhibitory activity of acetophenone glucosides from the clove buds (*Syzygium aromaticum*). J. Korean Soc. Appl. Biol. Chem. 53, 847–851. https://doi.org/10.3839/jksabc.2010.129.

Han, A.R., Paik, Y.S., 2012. Antioxidant and prolyl endopeptidase inhibitory capacities of chromone C-glucosides from the clove buds (*Syzygium aromaticum*). J. Appl. Biol. Chem. 53 (3), 195–198. https://doi.org/10.3839/jabc.2012.031.

Hemalatha, R., Nivetha, P., Mohanapriya, C., Sharmila, G., Muthukumaran, C., Gopinath, M., 2016. Phytochemical composition, GC-MS analysis, in vitro antioxidant and antibacterial potential of clove flower bud (*Eugenia caryophyllus*) methanolic extract. J. Food Sci. Technol. 53 (2), 1189–1198. https://doi.org/10.1007/s13197-015-2108-5.

Hong, C.H., Hur, S.K., Oh, O.J., Kim, S.S., Nam, K.A., Lee, S.K., 2002. Evaluation of natural products on inhibition of inducible cyclooxygenase (COX-2) and nitric oxide synthase (iNOS) in cultured mouse macrophage cells. J. Ethnopharmacol. 83 (1–2), 153–159. https://doi.org/10.1016/s0378-8741(02)00205-2.

Hussein, M., Abd El-Hack, M.E., Mahgoub, S.A., Saadeldin, I.M., Swelum, A.A., 2019. Effects of clove (*Syzygium aromaticum*) oil on quail growth, carcass traits, blood components, meat quality, and intestinal microbiota. Poult. Sci. 98 (1), 319–329. https://doi.org/10.3382/ps/pey348.

Idowu, S., Adekoya, A.E., Igiehon, O.O., Idowu, A.T., 2021. Clove (*Syzygium aromaticum*) spices: a review on their bioactivities, current use, and potential application in dairy products. J. Food Meas. Charact. 15. https://doi.org/10.1007/s11694-021-00915-9.

Ishaq, A., Syed, Q.A., Khan, M.I., Zia, M.A., 2019. Characterizing and optimising antioxidant and antimicrobial properties of clove extracts against food-borne pathogenic bacteria. Int. Food Res. J. 26, 1165–1172.

Issac, A., Gopakumar, G., Kuttan, R., Maliakel, B., Krishnakumar, I.M., 2015. Safety and anti-ulcerogenic activity of a novel polyphenol-rich extract of clove buds (*Syzygium aromaticum* L). Food Funct. 6 (3), 842–852. https://doi.org/10.1039/c4fo00711e.

Jaganathan, S.K., Mazumdar, A., Mondhe, D., Mandal, M., 2011. Apoptotic effect of eugenol in human colon cancer cell lines. Cell Biol. Int. 35 (6), 607–615. https://doi.org/10.1042/CBI20100118.

Jiang, T.A., 2019. Health benefits of culinary herbs and spices. J. AOAC Int. 102 (2), 395–411. https://doi.org/10.5740/jaoacint.18-0418.

Jimoh, S.O., Arowolo, L.A., Alabi, K.A., 2017. Phytochemical screening and antimicrobial evalution of *Syzygium aromaticum* extract and essential oil. Int. J. Curr. Microbiol. App. Sci. 6, 4557–4567.

Jirovetz, L., Buchbauer, G., Stoilova, I., Stoyanova, A., Krastanov, A., Schmidt, E., 2006. Chemical composition and antioxidant properties of clove leaf essential oil. J. Agric. Food Chem. 54 (17), 6303–6307. https://doi.org/10.1021/jf060608c.

Jnawali, H.N., Lee, E., Jeong, K.W., Shin, A., Heo, Y.S., Kim, Y., 2014. Anti-inflammatory activity of rhamnetin and a model of its binding to c-Jun NH2-terminal kinase 1 and p38 MAPK. J. Nat. Prod. 77 (2), 258–263. https://doi.org/10.1021/np400803n.

Johannah, N., Renny, R., Gopakumar, G., Balu, M., Sureshkumar, D., Krishnakumar, I.M., 2015. Beyond the flavour: a de-flavoured polyphenol rich extract of clove buds (*Syzygium aromaticum* L) as a novel dietary antioxidant ingredient. Food Funct. 6 (10), 3373–3382. https://doi.org/10.1039/c5fo00682a.

Joshi, T., Joshi, T., Sharma, P., Mathpal, S., Pundir, H., Bhatt, V., Chandra, S., 2020. In silico screening of natural compounds against COVID-19 by targeting Mpro and ACE2 using molecular docking. Eur. Rev. Med. Pharmacol. Sci. 24 (8), 4529–4536. https://doi.org/10.26355/eurrev_202004_21036.

Joshi, B., Sah, G., Basnet, B., Bhatt, M., Sharma, D., Subedi, K., Pandey, J., Malla, R., 2011. Phytochemical extraction and antimicrobial properties of different medicinal plants: *Ocimum sanctum* (Tulsi), *Eugenia caryophyllata* (clove), *Achyranthes bidentate* (Datiwan) and *Azadirachta indica* (neem). J. Microbiol. Antimicrob. 3 (1), 1–7. https://doi.org/10.5897/JMA.9000046.

Kasai, H., Shirao, M., Ikegami-Kawai, M., 2016. Analysis of volatile compounds of clove (*Syzygium aromaticum*) buds as influenced by growth phase and investigation of antioxidant activity of clove extracts. Flavour Fragr. J. 31, 178–184. https://doi.org/10.1002/ffj.3299.

Kurokawa, M., Hozumi, T., Basnet, P., Nakano, M., Kadota, S., Namba, T., Kawana, T., Shiraki, K., 1998. Purification and characterization of eugeniin as an anti-herpesvirus compound from *Geum japonicum* and *Syzygium aromaticum*. J. Pharmacol. Exp. Ther. 284 (2), 728–735.

Kurokawa, M., Hozumi, T., Tsurita, M., Kadota, S., Namba, T., Shiraki, K., 2001. Biological characterization of eugeniin as an anti-herpes simplex virus type 1 compound in vitro and in vivo. J. Pharmacol. Exp. Ther. 297 (1), 372–379.

Ladurner, A., Zehl, M., Grienke, U., Hofstadler, C., Faur, N., Pereira, F.C., Berry, D., Dirsch, V.M., Rollinger, J. M., 2017. Allspice and clove as source of triterpene acids activating the G protein-coupled bile acid receptor TGR5. Front. Pharmacol. 8, 468. https://doi.org/10.3389/fphar.2017.00468.

Lee, K.G., Shibamoto, T., 2001. Antioxidant property of aroma extract isolated from clove buds [*Syzygium aromaticum* (L.) Merr. Et Perry]. Food Chem. 74 (4), 443–448. https://doi.org/10.1016/S0308-8146(01)00161-3.

Lee, G.I., Ha, J.Y., Min, K.R., Nakagawa, H., Tsurufuji, S., Chang, I.M., Kim, Y., 1995. Inhibitory effects of oriental herbal medicines on IL-8 induction in lipopolysaccharide-activated rat macrophages. Planta Med. 61 (1), 26–30. https://doi.org/10.1055/s-2006-957992.

Lee, H.H., Shin, J.S., Lee, W.S., Ryu, B., Jang, D.S., Lee, K.T., 2016. Biflorin, isolated from the flower buds of *Syzygium aromaticum* L., suppresses LPS-induced inflammatory mediators via STAT1 inactivation in macrophages and protects mice from endotoxin shock. J. Nat. Prod. 79 (4), 711–720. https://doi.org/10.1021/acs.jnatprod.5b00609.

Maheswari, P., Ponnusamy, S., Harish, S., Muthamizhchelvan, C., Hayakawa, Y., 2020. Syntheses and characterization of *Syzygium aromaticum*, *Elettaria cardamomum* and *Cinnamomum verum* modified TiO2 and their biological applications. Mater. Sci. Semicond. Process. 105, 104724.

Menon, A.N., Narayanan, C.S., 1992. Glycosidically bound volatiles of clove *Syzygium aromaticum* (L.) merr. et perry (Myrtaceae). Flavour Fragr. J. 7, 155–157. https://doi.org/10.1002/ffj.2730070311.

Meyers, K.J., Swiecki, T.J., Mitchell, A.E., 2006. Understanding the native Californian diet: identification of condensed and hydrolyzable tannins in tanoak acorns (*Lithocarpus densiflorus*). J. Agric. Food Chem. 54 (20), 7686–7691. https://doi.org/10.1021/jf061264t.

Milind, P., Deepa, K., 2011. Clove: a champion spice. Int. J. Res. Ayurveda Pharm. 2 (1), 47–54.

Miyazawa, M., Hisama, M., 2003. Antimutagenic activity of phenylpropanoids from clove (*Syzygium aromaticum*). J. Agric. Food Chem. 51 (22), 6413–6422. https://doi.org/10.1021/jf030247q.

Mohammed, K.A., Abdulkadhim, H.M., Noori, S.I., 2016. Chemical composition and anti-bacterial effects of clove (*Syzygium aromaticum*) flowers. Int. J. Curr. Microbiol. App. Sci. 5 (2), 483–489. https://doi.org/10.20546/ijcmas.2016.502.054.

Mohammed, K.A.A., Wasfy, A.A.H.F., Bazalou, M.S., 2020. Comparison of bioactive components of clove buds as extracted by two different methods and analyzed by gas chromatography triple quad time-flight technology. Int. J. Pharm. Qual. Assur. 11 (4), 439–451.

Moilanen, J., Sinkkonen, J., Salminen, J.-P., 2013. Characterization of bioactive plant ellagitannins by chromatographic, spectroscopic and mass spectrometric methods. Chemoecology 23, 165–179. https://doi.org/10.1007/s00049-013-0132-3.

Musthafa, K.S., Hmoteh, J., Thamjarungwong, B., Voravuthikunchai, S.P., 2016. Antifungal potential of eugenyl acetate against clinical isolates of *Candida* species. Microb. Pathog. 99, 19–29. https://doi.org/10.1016/j.micpath.2016.07.012.

Neveu, V., Perez-Jiménez, J., Vos, F., Crespy, V., du Chaffaut, L., Mennen, L., Knox, C., Eisner, R., Cruz, J., Wishart, D., Scalbert, A., 2010. Phenol-explorer: an online comprehensive database on polyphenol contents in foods. Database: J. Biological Databases Curation 2010, bap024. https://doi.org/10.1093/database/bap024.

Nisar, M.F., Khadim, M., Rafiq, M., Chen, J., Yang, Y., Wan, C.C., 2021. Pharmacological properties and health benefits of eugenol: a comprehensive review. Oxidative Med. Cell. Longev. 2021, 2497354. https://doi.org/10.1155/2021/2497354.

Nonaka, G.-I., Harada, M., Nishioka, I., 1980. Eugeniin, a new ellagitannin from cloves. Chem. Pharm. Bull. 28, 685–687.

Novo Belchor, M., Hessel Gaeta, H., Fabri Bittencourt Rodrigues, C., Ramos da Cruz Costa, C., de Oliveira Toyama, D., Domingues Passero, L.F., Dalastra Laurenti, M., Hikari Toyama, M., 2017. Evaluation of Rhamnetin as an inhibitor of the pharmacological effect of secretory phospholipase A2. Molecules 22 (9), 1441. https://doi.org/10.3390/molecules22091441.

Omran, R., Taee, Z.M., Jassani, M.J., Hashim, H.O., 2017. Extraction of phenolic compounds as antioxidants from some plants and their cytotoxic activity against breast cancer cell line. Asian J. Pharm. Clin. Res. 10 (7), 349–355. https://doi.org/10.22159/ajpcr.2017.v10i7.18673.

Pandey, P., Singhal, D., Khan, F., Arif, M., 2021. An in silico screening on *Piper nigrum*, *Syzygium aromaticum* and *Zingiber officinale* roscoe derived compounds against SARS-CoV-2: a drug repurposing approach. Biointerface Res. Appl. Chem. 11 (4), 11122–11134. https://doi.org/10.33263/BRIAC114.1112211134.

Patil, V.M., Masand, N., 2018. Anticancer potential of flavonoids: chemistry, biological activities, and future perspectives. Stud. Nat. Prod. Chem. 59, 401–430. https://doi.org/10.1016/B978-0-444-64179-3.00012-8.

Pavesi, C., Banks, L.A., Hudaib, T., 2018. Antifungal and antibacterial activities of eugenol and non-polar extract of *Syzygium aromaticum* L. J. Pharm. Sci. Res. 10 (2), 337–339.

Pavithra, B., 2014. Eugenol-a review. J. Pharm. Sci. Res. 6 (3), 153–154.

Pérez-Jiménez, J., Neveu, V., Vos, F., Scalbert, A., 2010a. Identification of the 100 richest dietary sources of polyphenols: an application of the phenol-explorer database. Eur. J. Clin. Nutr. 64 (Suppl 3), S112–S120. https://doi.org/10.1038/ejcn.2010.221.

Pérez-Jiménez, J., Neveu, V., Vos, F., Scalbert, A., 2010b. Systematic analysis of the content of 502 polyphenols in 452 foods and beverages: an application of the phenol-explorer database. J. Agric. Food Chem. 58 (8), 4959–4969. https://doi.org/10.1021/jf100128b.

Radha Krishnan, K., Babuskin, S., Azhagu Saravana Babu, P., Sasikala, M., Sabina, K., Archana, G., Sivarajan, M., Sukumar, M., 2014. Antimicrobial and antioxidant effects of spice extracts on the shelf life extension of raw chicken meat. Int. J. Food Microbiol. 171, 32–40. https://doi.org/10.1016/j.ijfoodmicro.2013.11.011.

Ramadan, M.F., 2019. Fruit Oils: Chemistry and Functionality. Springer, Switzerland, https://doi.org/10.1007/978-3-030-12473-1.

Rampadarath, S., Puchooa, D., Ranghoo-Sanmukhiya, V.M., 2014. A comparison of polyphenolic content, antioxidant activity and insecticidal properties of *Jatropha* species and wild *Ricinus communis* L. found in Mauritius. Asian Pac J Trop Med 7 (Suppl. 1), S384–S390. https://doi.org/10.1016/S1995-7645(14)60263-7.

Rasheed, A., Laekeman, G., Totté, J., Vlietinck, A.J., Herman, A.G., 1984. Eugenol and prostaglandin biosynthesis. N. Engl. J. Med. 310 (1), 50–51. https://doi.org/10.1056/nejm198401053100115.

Rehman, M.T., AlAjmi, M.F., Hussain, A., 2021. Natural compounds as inhibitors of SARS-CoV-2 Main protease (3CLpro): a molecular docking and simulation approach to combat COVID-19. Curr. Pharm. Des. 27 (33), 3577–3589. https://doi.org/10.2174/1381612826999201116195851.

Rho, H.S., Ghimeray, A.K., Yoo, D.S., Ahn, S.M., Kwon, S.S., Lee, K.H., Cho, D.H., Cho, J.Y., 2011. Kaempferol and kaempferol rhamnosides with depigmenting and anti-inflammatory properties. Molecules 16 (4), 3338–3344. https://doi.org/10.3390/molecules16043338.

Ryu, B., Kim, H.M., Lee, J.S., Lee, C.K., Sezirahiga, J., Woo, J.H., Choi, J.H., Jang, D.S., 2016a. New Flavonol glucuronides from the flower buds of *Syzygium aromaticum* (clove). J. Agric. Food Chem. 64 (15), 3048–3053. https://doi.org/10.1021/acs.jafc.6b00337.

Ryu, B., Kim, H.M., Woo, J.H., Choi, J.H., Jang, D.S., 2016b. A new acetophenone glycoside from the flower buds of *Syzygium aromaticum* (cloves). Fitoterapia 115, 46–51. https://doi.org/10.1016/j.fitote.2016.09.021.

Saeed, M., Khan, M.S., Alagawany, M., et al., 2021. Clove (*Syzygium aromaticum*) and its phytochemicals in ruminant feed: an updated review. Rendiconti Lincei. Scienze Fisiche e Naturali 32, 273–285. https://doi.org/10.1007/s12210-021-00985-3.

Saleem, H.N., Batool, F., Mansoor, H.J., Shahzad-Ul-Hussan, S., Saeed, M., 2019. Inhibition of dengue virus protease by eugeniin, isobiflorin, and biflorin isolated from the flower buds of *Syzygium aromaticum* (cloves). ACS Omega 4, 1525–1533. https://doi.org/10.1021/acsomega.8b02861.

Salminen, J.P., 2018. Two-dimensional tannin fingerprints by liquid chromatography tandem mass spectrometry offer a new dimension to plant tannin analyses and help to visualize the tannin diversity in plants. J. Agric. Food Chem. 66 (35), 9162–9171. https://doi.org/10.1021/acs.jafc.8b02115.

Shan, B., Cai, Y.Z., Sun, M., Corke, H., 2005. Antioxidant capacity of 26 spice extracts and characterization of their phenolic constituents. J. Agric. Food Chem. 53 (20), 7749–7759. https://doi.org/10.1021/jf051513y.

Singletary, K., 2014. Clove; overview of potential health benefits. Nutr. Today 49 (4), 207–224.

Srivastava, K.C., 1993. Antiplatelet principles from a food spice clove (Syzygium aromaticum L). Prostaglandins Leukot. Essent. Fat. Acids 48 (5), 363–372. https://doi.org/10.1016/0952-3278(93)90116-e.

Suhaj, M., 2006. Spice antioxidants isolation and their antiradical activity: a review. J. Food Compos. Anal. 19 (6–7), 531–537.

Tajodini, M., Moghbeli, P., Saeedi, H.R., Effati, M., 2014. The effect of medicinal plants as a feed additive in ruminant nutrition. Iran. J. Appl. Anim. Sci. 4 (4), 681–686.

Takechi, M., Tanaka, Y., 1981. Purification and characterization of antiviral substance from the bud of *Syzygium aromatica*. Planta Med. 42, 69–74.

Tan, S.-H., Karim, M.R., 2018. Effects of three drying treatments on the polyphenol content, antioxidant and antimicrobial properties of *Syzygium aromaticum* extract. Chiang Mai J. Sci. 45 (2), 937–948.

Toma's-Barbera'n, F.A., Ferreres, F., 2012. Analytical methods of flavonols and flavones. In: Xu, Z., Howard, L. R. (Eds.), Analysis of Antioxidant-Rich Phytochamicals, first ed. Wiley-Blackwell, Wiley & Sons, Ltd., UK, pp. 207–246.

Umashankar, V., Deshpande, S.H., Hegde, H.V., Singh, I., Chattopadhyay, D., 2021. Phytochemical moieties from Indian traditional medicine for targeting dual hotspots on SARS-CoV-2 spike protein: an integrative in-silico approach. Front. Med. 8. https://doi.org/10.3389/fmed.2021.672629, 672629.

US-FDA, 1978. Scientific Literature Review of Eugenol and Related Substances in Flavor Usage Flavor and Extract Manufacturers Association of the United States. US-FDA, Washington DC.

Vicidomini, C., Roviello, V., Roviello, G.N., 2021. Molecular basis of the Therapeutical potential of clove (*Syzygium aromaticum* L.) and clues to its anti-COVID-19 utility. Molecules 26 (7), 1880. https://doi.org/10.3390/molecules26071880.

Vijayasteltar, L., Nair, G.G., Maliakel, B., Kuttan, R., I M, K., 2016. Safety assessment of a standardized polyphenolic extract of clove buds: subchronic toxicity and mutagenicity studies. Toxicol. Rep. 3, 439–449. https://doi.org/10.1016/j.toxrep.2016.04.001.

Woo, J.H., Ahn, J.H., Jang, D.S., Lee, K.T., Choi, J.H., 2017. Effect of Kumatakenin isolated from cloves on the apoptosis of Cancer cells and the alternative activation of tumor-associated macrophages. J. Agric. Food Chem. 65 (36), 7893–7899. https://doi.org/10.1021/acs.jafc.7b01543.

Xu, Z., 2012. Analysis method of phenolic acids. In: Xu, Z., Howard, L.R. (Eds.), Analysis of Antioxidant-Rich Phytochamicals, first ed. Wiley-Blackwell, Wiley & Sons, Ltd., UK, pp. 69–104.

Yashin, A., Yashin, Y., Xia, X., Nemzer, B., 2017. Antioxidant activity of spices and their impact on human health: a review. Antioxidants 6 (3), 70. https://doi.org/10.3390/antiox6030070.

Yassin, M.T., Mostafa, A.A., Al-Askar, A.A., 2020. In vitro anticandidal potency of *Syzygium aromaticum* (clove) extracts against vaginal candidiasis. BMC Complementary Med. Ther. 20 (1), 25. https://doi.org/10.1186/s12906-020-2818-8.

Yoshida, T., Hatano, T., Ito, H., Okuda, T., 2000. Chemical and biological perspectives of ellagitannin oligomers from medicinal plants. Stud. Nat. Prod. Chem. 23, 395–453. https://doi.org/10.1016/S1572-5995(00)80134-9.

Zhang, Y., Chen, Y., 1997. Isobiflorin, a chromone C-glucoside from cloves (*Eugenia caryophyllata*). Phytochemistry 45 (2), 401–403.

Zimmer, M., 1999. The fate and effects of ingested Hydrolyzable tannins in *Porcellio scaber*. J. Chem. Ecol. 25, 611–628. https://doi.org/10.1023/A:1020962105931.

CHAPTER

Clove (*Syzygium aromaticum*) in poultry feed

15

Jet Saartje Mandey

Department of Animal Nutrition, Animal Husbandry Faculty, University of Sam Ratulangi, Manado, North Sulawesi, Indonesia

1. Introduction

Aromatic plants and essential oils are becoming more important due to their antimicrobial effects and their stimulating effect on animal digestive systems. Aromatic plants have been used traditionally in the therapy of some diseases for a long time in the world. Essential oils in aromatic plants are used extensively in the medicine, food, and cosmetic industries. The species, herbs, vegetables, and some of the plants have a role as growth promoters in broiler diets to improve the efficiency of the growth conversion and reduce the cost of feed. Adding herbal plants as growth promoters in broiler diets improves their feed conversion ratio, mortality rate, and body weight gain.

All plant organs synthesize essential oils, i.e., buds, flowers, leaves, stems, twigs, seeds, fruits, roots, wood, or bark, stored in secretory cells, cavities, canals, epidermic cells, or glandular trichomes (Bakkali et al., 2008). Essential oils, also known as volatile oils, are complex mixtures of volatile constituents biosynthesized by plants, including two biosynthetically related groups (Pichersky et al., 2006). These main groups include terpenes and terpenoids and aromatic and aliphatic constituents, all characterized by low molecular weight. A blend of encapsulated essential oils and organic acids supplementation improved growth performance and gut health in necrotic enteritis due to *Clostridium perfringens* infected broiler chickens by strengthening the intestinal barrier function, positively modulating the gut microbiota community, and regulating intestinal immune responses (Pham et al., 2020).

Clove (*Eugenia caryophyllus*, syn. *Syzygium aromaticum*) is considered as a spice. It contains 10% volatile oil, most of it is eugenol, a substance with an anesthetic effect. Also, contains vitamins B and C, and the last vitamin is involved in stress hormones synthesis, thus significantly reducing body temperature by enhancing heat dissipation through blood vessels surrounding the body to maintain a relatively constant temperature. In addition, the clove contains phenol compounds that act as antibacterial agents. Clove oils and aqueous extract also have been used in food products, perfumery, antiseptic and digestion stimulant, anesthetics, anticarcinogenic, antiparasitic, and antioxidants.

Antibiotics have been used in poultry nutrition to improve growth performance, prevent some specific pathogenic microorganisms, and increase some beneficial microorganisms in intestinal microflora over the years. However, antibiotics used as growth promoters in animal feeds have been banned recently due to the potential development of antibiotic-resistant human pathogenic bacteria. Nowadays, the possibility of using new natural alternative additives instead of antibiotics in animal diets

Clove (*Syzygium aromaticum*). https://doi.org/10.1016/B978-0-323-85177-0.00015-X

is being researched. Most research and studies conducted on the clove and clove oil were in the laboratory animals but poultry rear.

Clove extract is commonly used in the food industry because of its unique aroma and natural safety. In addition, the essential oil from clove also exhibited strong antibacterial properties. Furthermore, antiseptic, appetite and digestion stimulant, anti-inflammatory, and antioxidant activities of clove and its ingredients have been reported. However, clove essential oil has been studied by very few as a performance enhancer of poultry.

In this chapter, the focus will be on clove as an alternative phyto-additive in the poultry diet. Therefore, clove research in poultry will only be touched upon within this context. The themes to be discussed include the role of clove as antibacterial, antioxidant, and the mechanism in improving growth performance through digestion stimulants. The discussion cannot be exhaustive but will cover those preventing some specific pathogenic microorganisms and increasing some beneficial microorganisms in intestinal microflora. Regarding the role of clove as alternative phyto-additive, these mechanisms are beginning to be understood.

2. Clove as alternative of phyto-additive

Clove is a medium-size tree (8–12 m) from the Myrtaceae family native to the Maluku Islands in East Indonesia (Kamatou et al., 2012). Cloves are considered one of the most versatile spices. They contain many biologically active compounds, such as eugenol, eugenol acetate, and β-caryophyllene (Jimoh et al., 2017). Eugenol is the most biologically active compound in cloves and makes up 70%–80% of clove oil (Xu et al., 2016). The main components in the clove leaf oil were eugenol (74.3%), eucalyptol (5.8%), caryophyllene (3.85%), and α-cadinol (2.43%), while the main components in bud oil were eugenol (49.7%), caryophyllene (18.9%), benzene,1-ethyl-3-nitro (11.1%), and benzoic acid, 3-(1-methylethyl) (8.9%) (Bhuiyan et al., 2010).

Clove has attracted attention due to its potent antioxidant and antimicrobial activities standing out among the other spices (Gülçin et al., 2012). In addition, it contains 10% volatile oils, most of it is eugenol, a substance with an anesthetic effect (Mukhtar, 2011). Also, contains vitamins B and C, and the last vitamin involved (Kamatou et al., 2012) in stress hormones synthesis thus has a major role in reducing body temperature by enhancing heat dissipation through blood vessels that are surrounding the body to maintain a relatively constant temperature (Hussein et al., 2019). In addition, the clove contains phenolic compounds that act as antibacterial agents. Thus, clove oils and aqueous extract have been used in food products, perfumery, antiseptic and digestion stimulant, anesthetics, anticarcinogenic, antiparasitic, and antioxidant (Gülçin et al., 2012).

It is known that the clove possesses growth-promoting properties such as antimicrobial appetite and digestion stimulating properties (Chaieb et al., 2007). In addition, *S. aromaticum* ethanol extract showed no mortality or signs of toxicity in rats at the doses used (250 and 500 mg/kg extract). Furthermore, the study demonstrated the potential anti-inflammatory effect of the extract against rat paw edema (Saeed et al., 2017). Cloves contain vitamin A (retinol), β-carotene, and vitamins K, B6, B1, and C (Dorman and Deans, 2000). Most importantly, cloves contain a phenolic compound (eugenol), which is its main bioactive ingredient (72%–90%) (Kaur and Kaushal, 2019).

The active ingredients in cloves increase the productive performance of broilers through stimulation of digestion, and have an antibiotic effect against organisms in the digestive tract. This material causes greater efficiency in the utilization of feed and leads to improvement in growth performance. Many studies have reported that clove was rich in trace minerals, essential for protein and carbohydrate metabolism. Therefore, reduced fatty acid synthesis and cholesterol could improve broiler performance (Al-Mufarrej et al., 2019b). It has been reported that the clove stimulates bile salt secretion and digestive enzyme activities of intestinal mucosa and pancreas (Mahrous et al., 2017). The results released a significant decrease in cholesterol levels because the main components of cloves and essential oils inhibited the activity of 3-hydroxy-3-methylglutaryl coenzyme (HMG-COA) hepatic reductase and caused hypoxia (Yang et al., 2018; Gole et al., 2020).

The liver is the primary site of nutrient metabolism in the body (Al-Mufarrej et al., 2019a). Cloves are one of the most abundant sources of dietary polyphenols with many traditional medicinal uses (Chaieb et al., 2007). Cloves contain a eugenol compound, which mitigates thioacetamide-induced liver damage (Kamatou et al., 2012). However, it has been found that clove essential oil and clove extracts have hepatoprotective effects by reducing liver toxicity (Gole et al., 2020). Nwaopara et al. (2007) found that excessive consumption of cloves could affect the liver hepatocytes in rabbits. Abdel-Rahman and El-Megeid (2006) found that the addition of 500 mg/kg of clove powder in the diet of mice improved the histology of hepatic cells. However, another study reported that cloves did not show any effect on the enzyme activity in the liver of rabbits (Kazi et al., 2017).

Mandey et al. (2019) reported that phytochemical screening on clove was positive in flavonoid, tannin, saponin, and triterpenoid. The quantitative analysis found a total phenolics of 80.0% (w/w) and β-carotene 1.43% on clove. Cloves contain dry matter 82.5%, ash 7.23%, crude protein 7.53%, crude fat 4.48%, crude fiber 14.1%, Ca 2.28%, P 0.08%, and a gross energy of 4652.04 kcal. Many fatty acids, such as lauric, palmitic, linolenic, linoleic, oleic, stearic, and myristic acids, have antibacterial and antifungal properties. Mandey et al. (2020) reported that dominant fatty acids in the flower of clove were linoleic acid (C18:2n6c), palmitic acid (C16:0), *cis*-13,16-docosadienoic acid (C22:2), stearic acid (C18:0), oleic acid (C18:1n9c), linolenic acid (C18:3n3), γ-linolenic acid (C18:3n6), and the total fatty acid was 9.59% (w/w), whereas fat content of clove was 4.19% (Table 1 and Fig. 1). Also, clove demonstrated the highest zone of inhibition at 10% concentration on *S. aureus* (8.64 mm) and the lowest inhibition on *E. coli* with a diameter of 6 mm. The clove generally proved effective against Gram-positive bacteria used in this study even at the highest concentration, while other Gram-negative organisms showed resistance at the lowest concentration (Table 2).

Clove (*Syzygium aromaticum* (L.) Merr. and Perry) is used in folk medicinal and traditional plants to treat various diseases. This plant is widely used and has commercial value because of the content of secondary metabolite compounds. Differences in the compound content of the plant have contributed to the taxonomic position of the plant. The highest compound content is 80.1% eugenol found in var. siputik followed by 28.8% β-caryophyllene in var. sikotok, α-humulene 3.05% in var. zanzibar, and 5.08% caryophyllene oxide in var. raja (Wael et al., 2018). The potential of phytochemicals in spices was studied and, if consumed in high amounts, could offer antioxidative properties and regulate key digestive enzymes, which may lead to prevention or decreased progression of diseases such as cancer, diabetes, and cardiovascular diseases (Chimbetete et al., 2019).

Coronavirus (c) is spreading overall quickly, and its control is troublesome because there is no compelling immunization or medications accessible in the medical sector. Moreover, this contagious disease has been associated with numerous respiratory issues. Therefore, there is an urgent need to elucidate plant-derived compounds that exhibit inhibitory potential against corona virus targets and enhance immunity. Research findings strongly recommended that phyto-compounds derived from

Table 1 Fatty acids of clove (Mandey et al., 2020).

Fat content	4.19
Fatty acids of clove (%, w/w)	
Caprilic acid C8:0	0.06
Capric acid C10:0	0.02
Lauric acid C12:0	0.11
Tridecanoic acid C13:0	0.11
Myristic acid C14:0	0.20
Pentadecanoic acid C15:0	0.02
Palmitic acid C16:0	1.63
Palmitoleic acid C16:1	0.06
Heptadecanoic acid C17:0	0.06
cis-10-Heptadecanoic acid C17:1	0.04
Stearic acid C18:0	0.65
Elaidic acid C18:1n9t	–
Oleic acid C18:1n9c	0.62
Linoleic acid C18:2n6c	2.72
Arachidic acid C20:0	0.31
γ-Linolenic acid C18:3n6	0.39
cis-11-Eicosenoic acid C20:1	0.07
Linolenic acid C18:3n3	0.52
Heneicosanoic acid C21:0	0.02
cis-11,14-Eicosedienoic acid C20:2	–
Behenic acid C22:0	0.21
Arachidonic acid C20:4n6	–
cis-13,16-Docosadienoic acid C22:2	1.28
Tricosanoic acid C23:0	–
Lignoceric acid C24:0	0.51
Fatty acid total	9.59

black pepper, clove, and ginger could be very useful in battling the COVID-19 pandemic era (Pandey et al., 2021). Srivastava et al. (2020) stated that their research would help researchers and industries to identify and scrutinize potential medicinal plants that can meet their interests for various applications, including the development of herbal/Ayurvedic antiviral drugs, designing antimicrobial/antiviral materials, development of immunopotentiator, preparation of potential sanitizer systems, etc. Nanda (2020) presents the perspectives of using phytochemicals to mitigate SARS-CoV-2 and the possible evolution of these phytochemicals into phytomedicines.

3. Clove in poultry diet

Many studies have evaluated the effects of clove powder on performance, immune response, blood parameters, and lymphoid organs in broiler chicken (Mahrous et al., 2017; Wasman and Mustafa,

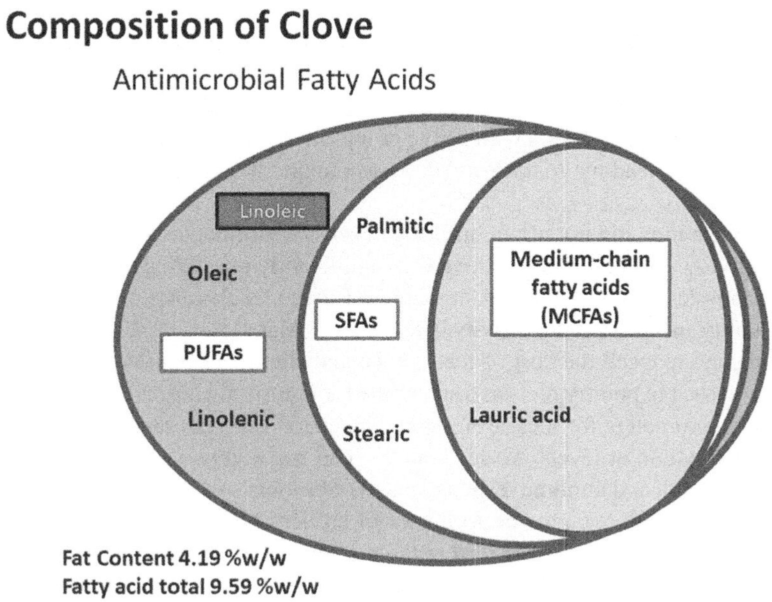

FIG. 1

Fatty acids of clove.

Table 2 Diameter zone of inhibition of clove for bacteria (Mandey et al., 2020).

		Zone of inhibition (mm)	
Sample	**Concentration (%)**	***E. coli***	***S. aureus***
Clove	2.5	6.00	6.00
	5	6.00	6.00
	10	6.00	8.64
Chloramphenicol (C+)	0.1	7.69	10.52

2020). For example, Petrovic et al. (2012) studied the effect of diet supplementation with 1% clove buds powder combined with either 0.2% lemon balm extract or 0.2% agrimony extract on the body weight, total feed intake, feed conversion ratio, and carcass yield showed no statistically significant differences in the growth performance indices such as body weight, total feed intake, feed conversion ratio, and carcass yield between the groups of chickens. However, the addition of pomegranate peels powder at 5% and clove powder at 2% into the poultry diet-induced significant increase in chick's weight compared with control (diet contaminated with mycotoxin non-treated with plant powders). Results revealed that feeding chicks on a diet amended with pomegranate peels and clove powders induces a significant increase in chicks' weight compared with chicks fed on a contaminated non-treated diet. The increase in chick's weight may be attributed to the ability of certain compounds in the medicinal plant powders to combine with the toxin forming complex nonabsorbable by the chick cells and get out of the body. It is also possible that some compounds degraded the mycotoxins in the diet in the powder through interaction with active groups in the mycotoxin, causing breaks in the active ring of the mycotoxin leading to destroy the mycotoxin or convert it to nontoxic compounds (Hussein, 2015).

Clove supplementation did not affect broiler chickens' performance, while it enhances serum immunoglobulins, INF-γ, and IL-10. Also, clove was significantly improving the antioxidant potential of broiler chicken muscles in a dose-dependent manner. Clove (1 g clove/kg diet) gave the best effect in increasing immunity and antioxidant activity in the broiler (Mahrous et al., 2017). Furthermore, the use of clove and oregano reduced the cost of feed intake and significantly affected meat cost. Thus, this supplement can be used in poultry diets as a substitute for industrial antioxidants under heat stress conditions and growth promoters for increasing feed conversion ratio (Borazjanizadeh et al., 2011).

Clove supplementation at levels greater than 2% can negatively affect performance without improving the liver's health and immune responsiveness of broilers. An increased level of cloves (greater than 2%) had a negative effect on the performance of broiler chickens, which may be attributed to an adverse effect on the palatability of the feed without affecting the health of the liver and immune responsiveness (Al-Mufarrej et al., 2019b). However, clove extract has positive effects on performance and the digestion process. In addition, it is a natural and safe feed additive, so that 400 ppm supplementation of clove extract to diets is considered an alternative natural growth promoter for poultry instead of antibiotics (Dalkiliç and Güler, 2009).

Eight treatments: negative control and unvaccinated; positive control and vaccinated; and treatments supplemented with different levels of clove powder (10, 20, 30, 40, 50, and 60 g clove powder/kg of diet, respectively) and all vaccinated showed that higher levels of clove powder (30–60 g) resulted in a gradual decrease in the live bodyweight of the birds at 21 and 35 days old. Relative liver weight increased in 50 and 60 g clove powder at 35 days, but not at 21 days. A shorter small intestine was recorded in birds with 50 and 60 g clove powder, while the proportional length of small intestinal parts did not show differences among treatments at 21 and 35 days old. The 30–60 g clove powder supplementation resulted in a gradual decrease in intestinal weight at 21 days but was not evident at 35 days. Villus height, villus surface area, and villus crypt depth gradually decreased from 10 to 60 g clove powder at 35 days old. Results indicated that a 10 g clove powder per kg diet could improve the growth rate without affecting the health of the liver and intestines of broilers (Al-Mufarrej et al., 2019a).

Agostini et al. (2012) reported that clove powder (0.1–2.5 g/kg diet) positively affected growth performance, feed efficiency, and changes in the intestine epithelium of broiler chicken. On the other hand, the inclusion levels of 1.0 and 1.5 g of clove powder/kg diet did not affect growth performance parameters but improved the health of broiler chickens (Mahrous et al., 2017). The use of cloves at levels higher than these levels may have minimal effect on growth rate through a reduction in feed palatability or may affect the intestinal epithelium and liver cells that alter digestion, absorption, and metabolism. However, the clove is a safe plant when consumed in concentrations lower than 1.5 g/kg (Cortés-Rojas et al., 2014). However, this does not constitute a biological constant because of factors such as species, age, and diet (Adam et al., 2013). Therefore, it is important to establish appropriate inclusion levels of cloves and their essential oil as growth promoters (Lee et al., 2020).

Clove significantly increased serum protein, globulin, IgG, IgM, INF-γ, IL-10, muscle GSH levels, T.SOD, and GST activities. At the same time, the total serum cholesterol, TAG, and muscle MDA levels were significantly decreased. Moreover, BWG, FCR, and PER were improved in a based diet containing vitamin E, and a basal diet containing 0.5 g clove/kg diet, compared to control and other clove supplemented groups. Although, higher doses of clove did not affect growth performance parameters with the enhancement of immunity and antioxidant activities in broiler chickens that concomitantly provide a healthy broiler's meat with less MDA favorable to human consumption (Mahrous et al., 2017).

Research with treatments: 1%–6% clove supplementation levels, a positive control (vaccinated), and negative control (unvaccinated) found that levels of dietary clove between 2% and 6% supplementation resulted in a gradual decrease in weight gain and feed intake with an increase in feed conversion ratio at 2 and 3 weeks, whereas those that received 4%–6% supplementation had reduced weight gain and the 3%–6% supplemented animals consumed the least amount of feed at 4 and 5 weeks. High levels of clove supplementation (4%, 5%, and 6%) resulted in reduced total serum protein and albumin and the greatest activity of aspartate aminotransferase at 35 days. Antibodies to viruses against which the birds had been vaccinated were not affected by different levels of clove supplementation compared with the positive control. However, they were increased in birds in the negative control group without vaccination. Relative weights of lymphoid organs were not affected by any treatments at 21 and 35 days. These results indicate that clove supplementation at levels greater than 2% can negatively affect performance without improving the health of the liver and immune responsiveness of broilers (Al-Mufarrej et al., 2019b).

Five inclusion levels of clove were tested: 0, 100, 200, 1000, and 2500 mg/kg diet. In the first experiment, 100 and 200 mg/kg levels showed the lower feed-to-gain ratio. Enterobacteriaceae counts were not affected by clove supplementation at 21 days of age, but the number of Lactobacillus was increased compared to the control diet (9.62 vs 9.18 log cfu/g digesta). In addition, the inclusion of cloves 100–200 mg/kg increased the intraepithelial lymphocyte count (1.55 vs 2.22 lymphocytes/100 enterocytes) and the lamina propria cell density (5.98 vs 6.84 cells/1000 mm^2) of broiler intestinal epithelium, 21 days old, and improve feed efficiency (Agostini et al., 2012).

The chickens' final BW was significantly different between the treatments, which decreased linearly with increasing levels of clove seed inclusion. Broiler chickens fortified with clove seeds did not significantly affect the chickens' carcass characteristics and body composition. However, an increment in weights of carcass components (liver, heart, and gizzard) was observed in low clove seed levels. In addition, water-holding capacity, cooking loss percentages, and tenderness of the meat were improved owing to clove seed inclusion (1% or 2%) compared with the control group (Suliman et al., 2020).

Bacterial counts significantly decreased with the increasing concentration of eugenol in feed. The population of *Escherichia coli* and *Salmonella* sp. at 2% eugenol and *Lactobacillus* sp. in 1.75% eugenol decreased to zero, meaning all of these bacteria died. The treatment that gave the highest nutrient utilization was the eugenol 15 g/kg. It is then concluded that 15 g/kg clove oil in broiler feed showed the optimum nutrient utilization (Tahir et al., 2019).

Administration of clove powder and thyme oil improved the symptoms induced by *C. albicans* and increased PCV%, ALT, and AST levels, decreasing erythrocyte count and total leukocyte count. Clove powder increased creatinine levels, while thyme oil decreased them. The tested plants induced marked histopathological alteration in crop, lung, intestine, and gizzard of treated birds. Clove powder and thyme oil can be used as alternative medicine to treat candidiasis in chickens (Abdel Fatah et al., 2020).

The inclusion of 0.50% clove powder/kg diet can be used as a growth promoter in Japanese quail (Tariq et al., 2015). However, dietary supplementation with clove leaf and candlenut seed meals up to 3% level could not decrease feed conversion and increase the egg index of laying hens (Ambawati et al., 2020). Furthermore, the addition of clove, cinnamon, and oil to the quail rations under heat stress led to lower relative density of heat shock proteins HSP40, HSP70, HSP90, concentrations of corticosterone hormone, MDA, and oxidative stimulator enzyme CK in blood plasma, and the ratio of heterophil-lymphocyte (H/L); however, the concentration of antioxidant enzymes AOA, GPx, SOD, and CAT in blood plasma was significantly higher in all additive treatments compared to the control group. So, treatment with the oil additives resulted in higher improvement than the powder additives (Mustafa and Wasman, 2020).

Our recent research about broiler chickens fed a diet containing 1% of clove and treated with carrot in drinking water showed significantly different blood lipid profiles (Table 3) and showed higher meat essential amino acids (Table 4).

4. Essential oil of clove

Essential or volatile oils are aromatic oily liquids extracted by distillation from plant parts, such as flowers, buds, seeds, leaves, twigs, bark, wood, fruits, and roots. Essential oils contain complex and highly variable constituents consisting of terpenoid groups (monoterpenes and sesquiterpenes), aromatic compounds (aldehydes, alcohols, phenols, methoxy derivatives, etc.), and terpenoids (isoprenoids) (Bakkali et al., 2008). Various essential oils have many properties in common, e.g., they can be vaporized with steam, are lipophilic, liquid at 18°C, optically active, and well soluble in ethanol, propylene glycol, or lipids (Gopi, 2014).

Essential oils are concentrated hydrophobic liquids containing the volatile aroma compounds of plants. They are also known as volatile or ethereal oils since they are volatile in steam. They differ in both chemical and physical properties from the so-called fixed oils. The essential oils are not simple compounds but various compounds (mainly terpenes and terpene derivatives).

Table 3 Blood lipid profiles of broiler chickens fed diet with clove and carrot.

	Treatment				
Blood lipid profile	**R0**	**R1**	**R2**	**R3**	***P*-value**
Triglyceride	62.50	41.80	47.20	45.50	.19
Total cholesterol	136.5	158.0	131.5	150.0	.006
HDL-cholesterol	96.00	114.0	107.5	108.5	.000
LDL-cholesterol	25.00	33.00	18.50	30.00	.000
SGOT	206.0	210.5	195.5	185.5	.008
Blood glucose	219.0	223.0	210.0	207.0	.000

R0 = 0 mL/L carrot + 1% clove; R1 = 10 mL/L carrot + 1% clove; R2 = 20 mL/L carrot + 1% clove; R3 = 30 mL/L carrot + 1% clove.

Table 4 Amino acids of broiler meat fed clove and carrot.

	Treatment				Standard (ppm)		
Amino acids	**R0**	**R1**	**R2**	**R3**	**50**	**100**	**250**
Aspartic acid, %	12.74	13.08	13.15	11.68	9.15	9.14	8.31
Glutamic acid, %	15.55	15.42	15.15	15.10	6.11	6.12	5.90
Serine, %	4.22	4.13	4.25	4.26	6.36	6.42	6.29
Glycine, %	5.45	5.42	5.51	6.40	7.36	7.67	7.66
Alanine, %	7.88	7.74	7.77	7.95	7.78	7.87	8.14
Valine, %	8.16	8.36	8.16	7.70	11.12	10.86	10.90
Isoleucine, %	7.84	7.74	7.48	7.21	10.03	9.84	9.92
Leucine, %	10.87	10.88	10.79	10.83	8.18	8.07	8.21
Lycine, %	5.70	5.36	5.53	6.26	2.31	2.53	2.71
Methionine, %	2.47	2.49	2.59	2.97	6.98	6.99	7.17

R0 = 0 mL/L carrot + 1% clove; R1 = 10 mL/L carrot + 1% clove; R2 = 20 mL/L carrot + 1% clove; R3 = 30 mL/L carrot + 1% clove.

Previous studies indicated that the main chemical compositions of clove essential oil are eugenol, β-caryophyllene, and eugenol acetate (Teixeira et al., 2014). Clove at the flowering stage produced the best essential oil ingredient and the most efficient source of natural antioxidants with slight differences in quality between the young and mature trees (Alfikri et al., 2020). These compounds have the main essential effects of clove as an anti-inflammatory, analgesic, antimicrobial, antiparasitic, antioxidant, antimutagenic, anticonvulsant, and anticarcinogenic agent. Furthermore, the study indicated that nanoemulsion of clove essential oil has good antimicrobial and antioxidant properties under in vitro conditions (Shahbazi, 2019). Furthermore, the study indicated that hake protein films, prepared with by-products, incorporated with clove, garlic, and origanum essential oils, show interesting properties (water vapor barrier and antibacterial and antioxidant activities) to be good candidates to be used in biodegradable food packaging systems (Teixeira et al., 2014).

The encapsulation process of clove essential oil by emulsion extrusion is an efficient technique to maintain the oil in the capsule (Kopp et al., 2019). The clove-based silver nanoparticles were efficient in controlling the growth of *Escherichia coli* and *Staphylococcus aureus* bacteria, and the minimum inhibitory concentration ranged from 60 to 100 μL/mL. Clove essential oil is feasible as an alternative for the synthesis of silver nanoparticles with a simple, inexpensive, and eco-friendly method (Maciel et al., 2019).

Analysis of the chemical composition of clove essential oil identified eugenol as its major compound. Clove essential oil and eugenol inhibit the growth of Gram-positive bacteria *S. aureus* and have antioxidant potential. Inhibitory activity against the epimastigotes and intracellular amastigotes of *T. cruzi*, associated with low cytotoxicity, demonstrated the selectivity of clove essential oil against the parasite (Teles et al., 2021). The selected essential oils can be developed as natural preservatives to prevent fungal contamination in the food industry. In vivo inhibitory activity of selected EOs on naturally infected bread demonstrated that cinnamon and clove EOs could be used as natural antifungal agents (Hu et al., 2019).

Clove oil nanoparticles can be produced using polyethylene glycol as encapsulated by the melt dispersion method. All of the findings presented here also showed the stability and potential use of clove oil nanoparticles to manage storage pests. The Rusty grain beetle, *Cryptolestes ferruginous* (Steph.), is one of the destructive pests of stored grain and has wide dissemination. This insect pest is widespread in temperate and tropical regions and in the wet and dry zones (Ikawati et al., 2020).

5. Essential oil in diet of poultry

Several essential oil components exhibit antimicrobial action, some more strongly than others. Phenols, alcohols, ketones, and aldehydes are mainly associated with antibacterial actions, although the exact mechanism has not been fully understood (Nazzaro et al., 2013). Another interesting issue concerns the antioxidant properties of essential oils. Applied in poultry production, they can reduce lipid peroxidation in the muscles of meat chickens. Thigh muscles are susceptible to oxidation due to a higher content of polyunsaturated fatty acids, oxidation of which produces peroxides, lipids, oxysterols, or malondialdehyde (Tongnuanchan and Benjakul, 2014). In addition, some essential oils positively influence the avian immune system since they promote the production of immunoglobulins, enhance lymphocytic activity, and boost interferon-γ release (Awaad et al., 2016; Gopi, 2014; Krishan and Narang, 2014; Gharechopogh et al., 2021).

Essential oils contain compounds known to possess potent anti-inflammatory properties, mainly terpenoids and flavonoids, which suppress the metabolism of inflammatory prostaglandins (Krishan and Narang, 2014). In addition, essential oils positively affect the avian digestive system since they help restore the microbiota balance and increase nutrient absorption, which may chiefly be attributed to terpenoid compounds (Wang et al., 2021). Therefore, it is very important in feed conversion (Mathlouthi et al., 2012; Wang et al., 2021).

Spice plants have been popular and widely used throughout the entire history of humanity. These uses are based on important biological activities, such as antimicrobial activity, antioxidant activity, anti-inflammatory activity, anti-plasmodial activity, cytotoxic activity, cytotoxicity against human cancer cells, cytotoxicity against human epithelial cells, etc. (Kamatou et al., 2012). In addition, the essential oil extracted from aromatic plants has been shown to possess antifungal, anti-coccidial, antibacterial, and antioxidant (Botsoglou et al., 2004) activities.

Essential oils enhance digestive secretions and nutrient absorption, reduce the pathogenic stress in the gut, exert antioxidant properties, and enhance the animal's immune status (Zeng et al., 2015). The application of medicinal plants as alternative feed additives is of essential value to the performance and the antioxidant status of birds (El-Far et al., 2016). The active principles of essential oils act as a digestibility enhancer, balancing the gut microbial ecosystem and stimulating the secretion of endogenous digestive enzymes, thus improving poultry growth performance (Barakat et al., 2016). Julianus Sohilait (2015) investigated clove essential oils' chemical composition and stated that clove essential oil mainly had about 81.1%–84.4% eugenol, 11.6%–5.02% eugenyl acetate, and 3.45%–4.60% β-caryophyllene in the oil from the bud.

Clove is an evergreen tree that is native to Indonesia and is mainly grown for its aromatic flowers. Clove essential oil (CEO) has been used in traditional medicine mainly as a pain reliever in dental care and the treatment of burns and wounds. The main component in CEO is eugenol, a phenolic compound that contributes to the oil's warm, spicy scent and exhibits many pharmacological properties (Batiha et al., 2020). The antioxidant, anti-inflammatory, antimicrobial (Pavesi et al., 2018), and analgesic (Ghofran et al., 2017) effects of eugenol have been previously documented. However, many studies investigating the impact of CEO and its constituents on immune responses have resulted in conflicting findings. For example, Carrasco et al. (2009) examined the effect of CEO containing >98% eugenol on humoral and cell-mediated immune responses in mice in vivo.

Essential oil combined with organic acids administration in a feed displayed synergistic modulation effects on immunity and gut health which is very important for exploring new alternatives for antibiotic growth promoters in broiler production. Administration of a combination of clove essential oil and organic acids exerts a positive effect on gut health and enhances immunity with a concomitant enhancement in the economics of broiler production (Gole et al., 2020). However, Mukhtar (2011) showed that the effect of clove oil on the broiler performance had no significant difference for chicks' performance.

Dietary supplementation with 2 g/kg of clove or cinnamon powders and 1 mL/kg clove and cinnamon oils can improve broiler performance without causing any evident adverse effects. Moreover, there was an improvement in small intestine-ileum Lactobacillus count, goblet cells numbers, villus height, and crypt depth. The proposed supplementation also decreased the harmful bacteria (*Enterococcus*, *E. coli*) and fungi. The studied characteristics of the birds were improved with cinnamon and clove essential oils better than their powders (Wasman and Mustafa, 2020). In vitro studies have demonstrated the antibacterial activity of essential oils (Burt, 2004). The inhibitory impacts of clove on microbes are attributed to the fact that this herb contains essential oils composed of components such as eugenol (Asha et al., 2014). Herbal essential oils are also used to help beneficial microbes colonize inside the gastrointestinal tract (Yang et al., 2018). Such oils are also antifungal (Shin and Lim, 2004), stimulate digestion, develop intestine morphology, and promote enzymatic activities (Mustafa et al., 2017).

The influence of varying concentrations of essential oil was determined in vitro using the agar dilution test. Concentrations as low as 0.25% and 0.12% of thyme and clove oils, respectively, could completely inhibit the growth of *Clostridium perfringens*. An in vivo study was conducted in commercial broiler chickens, where two groups of 10-day-old broilers were fed a balanced ration with the addition of thyme (0.25%) or clove (0.125%). Birds were then challenged with *Clostridium perfringens* (1×108 CFU). Starting 3 days after infection (DPI), birds were fed with diet containing thyme and clove oil 0.25% and 0.125% compared to the control group who used antibiotics for 5 days. The results showed that EOs improved broiler chickens' growth performance, especially when used as a preventive feed additive continuously. In addition, the thyme essential oil showed better anticlostridial activity in terms of reduced *Clostridium perfringens* counts in the intestine and alleviation of intestinal damage caused by infection (Eid et al., 2018).

The plant extracts contain essential oils, which are volatile, natural, complex compounds characterized by a strong odor formed by aromatic plants as secondary metabolites (Bakkali et al., 2008). In nature, essential oils play an essential role in protecting the plants as antibacterials, antivirals, antifungals, insecticides, and also against herbivores by reducing their appetite for such plants. The chemical profile of the essential oil products differs in the number of molecules and the stereochemical types of molecules extracted according to the type of extraction method. In addition, these plant oils (e.g., thyme and clove oils) have antiseptic effects as bactericidal, virucidal, and fungicidal, and medicinal properties (Masotti et al., 2003).

The control diet, diet with probiotic protexin, and diets supplemented with 150, 300, and 450 ppm clove essential oil were fed to the chicken. Clove essential oil at 450 ppm increased feed intake, body weight gain, and improved feed conversion ratio in the finisher period. Different levels of clove essential oil and probiotics had no significant effect on dressing percentage, abdominal fat, and internal organs percentage (liver, heart, and gizzard). The results revealed a significant decrease in total cholesterol concentration in diets supplemented with probiotics and 450 ppm clove essential oil at 21 days of age. Total cholesterol concentration, low-density lipoprotein (LDL), and cholesterol/high-density lipoprotein (HDL) ratio at 42 days of age were significantly improved in the probiotic protexin group compared to those of other groups. Dietary supplementation with 450 ppm clove essential oil improved feed intake, body weight gain, feed conversion ratio, and decreased total cholesterol concentration. Probiotics could decrease serum cholesterol, low-density lipoprotein (LDL), HDL-cholesterol, and improve HDL/LDL at day 42 broiler chickens (Mehr et al., 2014).

It has been reported that clove oil is very rich in manganese, a trace mineral necessary for protein and carbohydrate metabolism, so it can improve the performance of broilers (Mukhtar, 2011). Chicks fed on a diet with 600 mg/kg clove oil recorded more in feed intake and body weight than both control and antibiotic groups, but this increase was not significant for feed conversion ratio (FCR) and body weight gain, chicks fed on diet with antibiotic recorded the best value numerically. In addition, chicks fed on diets containing clove oil recorded the lowest mortality rate compared to the control group (Mukhtar, 2011).

The groups that received 100, 300, and 500 mg/kg clove essential oil showed better growth performance than the control treatment. Among the different treatments, the antibiotic treatment resulted in a higher feed intake and weight gain than the other treatments, although the group receiving 300 mg/kg clove essential oil had an improved feed conversion ratio compared to the other groups in the entire period. Supplementation of 100, 300, and 500 mg/kg clove essential oil did not affect serum lipids at 42 days of age. Adding 100, 300, and 500 mg/kg clove essential oil to broiler diets improved gut microflora (as measured by changes in *Escherichia coli* and *Lactobacillus*). However, increased villus height in the clove essential oil treatments compared to the control suggested that clove essential oil can provide a suitable condition for improving performance in broiler chickens (Mohammadi et al., 2014).

Nowadays, utilizing these herbs is being paid much attention, especially in the industries of quails (Wasman and Mustafa, 2020) and laying hens (Çabuk et al., 2006). Dietary clove essential oil prevented cross-contamination in eggs in an experimental infection of commercial layers (Ordóñez et al., 2008). Dietary supplementation with antimicrobial cold-pressed clove oil (1.5 mL/kg diet) could enhance growth performance, improve health status, and reduce intestinal pathogens in Japanese quails (Hussein et al., 2019).

Poultry raw meat is a perishable and nutritious food and is considered a prime source of protein. However, its high moisture content favors bacterial growth and leads to deteriorative changes. Bacterial contamination and lipid oxidation reduce the quality and stability of chicken meat and ultimately result in social health concerns. Treatment containing 1% chitosan solution +2% starch solution +1.5 mL clove oil performed best due to their synergistic barrier properties against aerobic microbes and increased the shelf stability of raw meat compared to uncoated meat samples (Mukhtar et al., 2018).

The combinations of cloves and lemon basil EOs to preserve chicken meat showed a better microbial growth inhibitory activity and preservation potential than those of the single-use. This meat preservation effect might be related to the presence of high fractions of oxygenated compounds, mainly eugenol, Z-citral, and E-citral, in both clove and lemon basil EOs (Hartanti et al., 2020).

6. Clove as phyto-additive in drinking water

Essential oils also find application in drinking water (Khosravinia, 2015; Galal et al., 2015). Adding aqueous extract of clove flower at different concentrations led to improved productive and physiological effects of broiler chicken (Al-Tabari et al., 2018). The supplementation of clove in the diet or drinking water of birds leads to impairment of their performance (Petrovic et al., 2012). Supplementation with the tulsi, black pepper, and clove extract, 1 mL/L drinking water causes a significant increase in live body weight and improvement in weekly weight gain and feed efficiency compared to that of the control group of broiler. Polyherbal supplementation in the broiler rations may be helpful for the safe, economic, and efficient production of broiler, and this formulation could be used as an alternative to commercial growth promoters (Nath et al., 2013).

Treatment without supplementing clove to diet and drinking water compared to treatments with 0.4% and 0.8% of clove powder to diet; treatment with 0.8% of clove oil to diet; and treatment with 0.4% of aqueous clove extract to drinking water have a slight improvement on some productive and physiological parameters of broiler exposed to heat stress (Salman and Ibrahim, 2012).

7. Phytochemicals of clove for intestinal health

A literature review indicated that most of the in vivo studies evaluating the effect of adding phytochemicals to broiler diets positively affected growth performance parameters, such as weight gain, feed intake, and feed conversion. Most of the phytochemicals (extracts, e.g., thyme, oregano, agronomy, clove, lemon, balm, red pepper, black cumin seed, and its essential oils) were feed additives (Valenzuela-Grijalva et al., 2017).

Herbal and its applications have been used in the prevention and treatment of diseases. Several plants and their extracts have affected T cells and cytokine and antibody production at the cellular and molecular levels (Burt, 2004). Cinnamon bark oil and clove bud oil increased villi height in the duodenum, jejunum, and ileum of broilers. Antibody titers against Newcastle disease virus were also increased compared to those of control groups. Essential oils have been identified as alternatives to antibiotic growth promoters in broilers. This review highlights the cellular effects induced by EOs, as well as the molecular impacts of EOs on cytokines, immunoglobulins, or regulatory pathways (Sandner et al., 2020).

Two well-accepted mechanisms for improving weight gain and feed conversion ratio are the stimulation of digestive enzyme secretion and the stabilization of the ecosystem of gut microflora (O'Bryan et al., 2015). The positive effects of essential oils on digestive enzyme secretion from the pancreas and intestinal mucosal have been reported in many broiler studies (Jang et al., 2002). These effects were confirmed by the increased digestibility of nutrients but did not improve growth performance (Amad et al., 2011). Various essential oils (thymol, carvacrol, cinnamaldehyde, eugenol, coriander, star anise, ginger, garlic, rosemary, turmeric, basil, caraway, lemon, and sage) have been used individually or mixed to improve animal health and performance (Gadde et al., 2017).

Lillehoj et al. (2018), in review, when considering using phytochemicals as antibiotic alternatives, reported the need to consider dose and variations in active compounds in plants and plant-derived products in animal target; target organs/tissues affected by phytochemicals; and safety of phytochemical residues in humans. In the use of different concentrations of spray dryer microencapsulated eucalyptol and eugenol in broiler chickens' diet as alternative growth promoters for avilamycin, eugenol demonstrated potent antioxidant activity, and both eugenol and eucalyptol showed antimicrobial activity. Thus, research suggests that the microencapsulated phytotherapeutic agents are a potential alternative to avilamycin as a growth promoter in broilers during the broilers' growing phase (Scherer et al., 2014).

The broad spectrum of eugenol's protective effects can be observed in various foodborne pathogens. For example, eugenol worked against the common nalidixic acid-resistant *S. enteritidis* found in contaminated eggs (Olasupo et al., 2003; Sharma et al., 2017), and it was found to inhibit the growth and colonization in the chicken reproductive tract (De Buck et al., 2004). Furthermore, supplementation of essential oil blend containing cinnamon oil, clove oil, and ajwain oil in equal proportion at 500 mg/kg to a diet deficient in digestible amino acids significantly increased villi height duodenum, antibody titers against ND vaccine, and serum catalase activity in broiler chickens. In addition, essential oils could improve gut morphology and immunity in broiler chickens fed on amino acids deficient diet (Parveen et al., 2020).

The treatment of organic acids + essential oils and the AGP led to increased growth performance, nutrient digestibility, and intestinal health of challenged broilers. The organic acids present in this blend are fumaric, sorbic, malic, and citric acids and the essential oils are thymol, vanillin, and eugenol. A marked difference occurred in favor of the OA + EO, suggesting that this blend may be used to improve intestinal health and broiler growth performance in AGP-free programs (Stefanello et al., 2020).

Campylobacter is a major cause of bacterial diarrhea in humans, and poultry is a major reservoir and source of the transmission of campylobacteriosis to humans (Kaakoush et al., 2015). In-feed supplementation of two generally recognized as safe (GRAS) compounds, eugenol (extract of clove) and trans-cinnamaldehyde (an extract of cinnamon) of select doses of eugenol or trans-cinnamaldehyde had inconsistent effects when used to prophylactically reduce Campylobacter colonization in 14 days broiler chicks (Arsi et al., 2019). Administration of clove powder (2%) and thyme oil (200 mL/kg) in diet and fluconazole (5 mg/kg) improved the symptoms induced by *C. albicans* and increased PCV%, ALT, and AST level, with a decrease in erythrocyte count and total leukocyte count. Clove powder increased creatinine levels, while thyme oil decreased them. In addition, the tested plants induced marked histopathological alteration in crop, lung, intestine, and gizzard of treated birds. Clove powder and thyme oil can be used as alternative medicine to treat candidiasis in chickens (Abdel Fatah et al., 2020). Provision of fermented clove leaf extract of 1%–2% through drinking water can increase lactic acid bacteria, reduce pathogenic bacteria, and increase feed digestibility, thereby increasing the growth of native chickens aged 2–10 weeks (Bulu et al., 2020). The roles of cloves in poultry were summarized in Fig. 2.

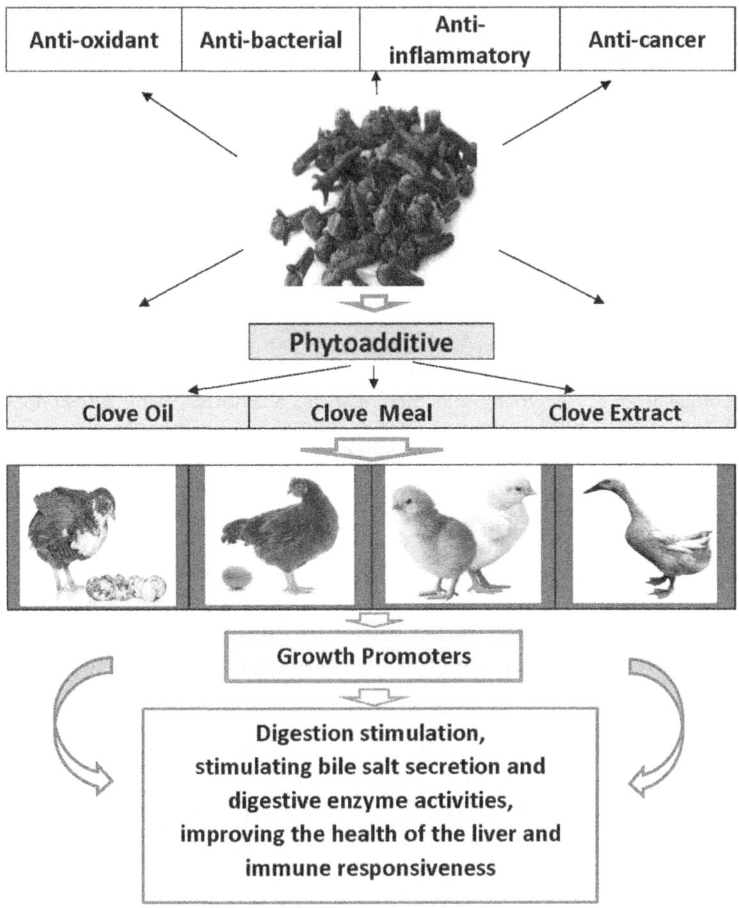

FIG. 2

The role of clove in poultry.

8. Conclusion

The application of herbal plants as alternative feed additives is of essential value to the performance and the antioxidant status of birds. Many studies have evaluated the effects of clove powder on performance, immune response, blood parameters, and lymphoid organs in broiler chickens, quails, and laying hens. Clove may be used in poultry diets as alternative growth promoter for improvement of feed conversion ratio. Essential oils of clove enhance digestive secretions and nutrient absorption, reduce the pathogenic stress in the gut, exert antioxidant properties, and enhance the animal's immune status. Dietary clove essential oil prevented cross-contamination in eggs in an experimental infection of commercial layers. Dietary supplementation with antimicrobial cold-pressed clove oil could enhance growth performance, improve health status, and reduce intestinal pathogens in Japanese quails.

References

Abdel Fatah, G., et al., 2020. The efficacy of clove and thyme against experimentally induced candidiasis in broilers. Mansoura Vet. Med. J. 21 (2), 25–31. https://doi.org/10.35943/mvmj.2020.21.2.0205.

Abdel-Rahman, M.K., El-Megeid, A.A., 2006. Hepatoprotective effect of soapworts (*Saponaria officinalis*), pomegranate peel (*Punica granatum* L.) and cloves (*Syzygium aromaticum* Linn.) on mice with CCl_4 hepatic intoxication. World J. Chem. 1 (1), 41–46.

Adam, S.I.Y., Mohamed, S.B., Abdelgadir, W.S., 2013. Effects of the aqueous extract of clove (*Syzygium aromaticum*) on Wistar rats. Br. J. Pharmacol. Toxicol. 4 (6), 262–266. https://doi.org/10.19026/bjpt.4.5410.

Agostini, P.S., et al., 2012. Role of in-feed clove supplementation on growth performance, intestinal microbiology, and morphology in broiler chicken. Livest. Sci. 147 (1–3), 113–118. https://doi.org/10.1016/j.livsci.2012.04.010.

Alfikri, F.N., et al., 2020. Yield, quality, and antioxidant activity of clove (*Syzygium aromaticum* L.) bud oil at the different phenological stages in young and mature trees. Scientifica 2020. https://doi.org/10.1155/2020/9701701.

Al-Mufarrej, S.I., Al-Baadani, H.H., Fazea, E.H., 2019a. Effect of level of inclusion of clove (*Syzygium aromaticum*) powder in the diet on growth and histological changes in the intestines and livers of broiler chickens. S. Afr. J. Anim. Sci. 49 (1), 166–175. https://doi.org/10.4314/sajas.v49i1.19.

Al-Mufarrej, S.I., et al., 2019b. Effects of clove powder supplementation on performance, blood biochemistry, and immune responses in broiler chickens. S. Afr. J. Anim. Sci. 49 (5), 835–844. https://doi.org/10.4314/SAJAS.V49I5.6.

Al-Tabari, A.S., Al-Zuhairi, Z.A., Abdulrazzaq, M., 2018. 15- Study the effect of adding aqueous extract of clove (*Eugenia caryophyllus*) to drinking water in productivity and physiological efficiency of broiler chicken. Bas. J. Vet. Res. 17 (1), 165–175. https://doi.org/10.33762/bvetr.2018.144949.

Amad, A.A., et al., 2011. Effects of a phytogenic feed additive on growth performance and ileal nutrient digestibility in broiler chickens. Poult. Sci. 90 (12), 2811–2816. https://doi.org/10.3382/ps.2011-01515. Poultry Science Association Inc.

Ambawati, L., Gading, B.M.W.T., Purwanto, H., 2020. Penambahan Tepung Daun Cengkih (*Syzygium aromaticum*) Dan Biji Kemiri. pp. 452–458.

Arsi, K., et al., 2019. Effect of dietary supplementation of essential oils, eugenol or trans-cinnamaldehyde, on enteric colonization of campylobacter in broiler chickens. Int. J. Adv. Sci. Eng. Technol. (2), 30–32.

Asha, K., Sunil, B., George, G., 2014. Effect of clove on the bacterial quality and shelf life of chicken meat. J. Meat Sci. Technol. 2 (2), 37–39.

Awaad, M.H.H., Abdel-Alim, G.A., Sayed, K.S., Ahmed, K.A., Nada, A.A., Metwalii, A.S.Z., Alkhala, A., 2016. Immunostimulant effects of essential oils of peppermint and eucalyptus in chickens. NeuroMethods 119 (May), 137–153. https://doi.org/10.1007/978-1-4939-5611-1_5.

Bakkali, F., et al., 2008. Biological effects of essential oils- a review. Food Chem. Toxicol., 446–475. https://doi.org/10.1016/j.fct.2007.09.106. Pergamon.

Barakat, D.M., et al., 2016. Anise (*Pimpinella anisum*) enhances the growth performance, immunity and antioxidant activities in broilers. Int. J. Pharm. Sci. Rev. Res. 37 (2), 134–140.

Batiha, G.E.S., et al., 2020. *Syzygium aromaticum* L. (myrtaceae): traditional uses, bioactive chemical constituents, pharmacological and toxicological activities. Biomolecules 10 (2). https://doi.org/10.3390/biom10020202.

Bhuiyan, N.I., et al., 2010. Constituents of the essential oil from leaves and buds of clove (*Syzygium caryophyllatum* (L.) Alston). African J. Plant Sci. 4 (11), 451–454. Available at: http://www.researchgate.net/publication/228673419_Constituents_of_the_essential_oil_from_leaves_and_buds_of_clove_(Syzigium_caryophyllatum_(L.)_Alston.

Borazjanizadeh, M., et al., 2011. The effect of clove and oregano on economic value of broiler chickens diet under hot weather of Khuzestan. J. Anim. Vet. Adv., 169–173. https://doi.org/10.3923/javaa.2011.169.173.

Botsoglou, N.A., et al., 2004. The effect of a mixture of herbal essential oils or α-tocopheryl acetate on performance parameters and oxidation of body lipid in broilers. S. Afr. J. Anim. Sci. 34 (1), 52–61. https://doi.org/10.4314/sajas.v34i1.4039.

Bulu, S., Mahardika, I.G., Nuriyasa, I.M., 2020. Giving the extract of clove leaves (*Syzygium aromaticum* L.) fermented through drinking water against growth of kampong chicken. Int. J. Life Sci. 4 (3), 50–58. https://doi.org/10.29332/ijls.v4n3.538.

Burt, S., 2004. Essential oils: their antibacterial properties and potential applications in foods- a review. Int. J. Food Microbiol., 223–253. https://doi.org/10.1016/j.ijfoodmicro.2004.03.022. Elsevier.

Çabuk, M., et al., 2006. Effect of a herbal essential oil mixture on growth and internal organ weight of broilers from young and old breeder flocks. S. Afr. J. Anim. Sci. 36 (2), 135–141. https://doi.org/10.4314/sajas.v36i2.3996.

Carrasco, F.R., et al., 2009. Immunomodulatory activity of *Zingiber officinale* Roscoe, *Salvia officinalis* L. and *Syzygium aromaticum* L. essential oils: evidence for humor- and cell-mediated responses. J. Pharm. Pharmacol. 61 (7), 961–967. https://doi.org/10.1211/jpp/61.07.0017.

Chaieb, K., et al., 2007. The chemical composition and biological activity of clove essential oil, Eugenia caryophyllata (*Syzigium aromaticum* L. Myrtaceae): a short review. Phytother. Res. https://doi.org/10.1002/ptr.2124.

Chimbetete, N., et al., 2019. Phytochemical content, radical scavenging ability & enzyme inhibiting activities of selected spices (cinnamon, cardamom and cloves). Food Nutr. Sci. 10 (03), 266–275. https://doi.org/10.4236/fns.2019.103020.

Cortés-Rojas, D.F., de Souza, C.R.F., Oliveira, W.P., 2014. Clove (*Syzygium aromaticum*): a precious spice. Asian Pac. J. Trop. Biomed. 4 (2), 90–96. https://doi.org/10.1016/S2221-1691(14)60215-X.

Dalkiliç, B., Güler, T., 2009. The effects of clove extract supplementation on performance and digestibility of nutrients in broilers. F. Ü. Sağ. Bil. Vet. Derg. 23 (3), 161–166. Available at: http://www.fusabil.org.

De Buck, J., et al., 2004. Colonization of the chicken reproductive tract and egg contamination by Salmonella. J. Appl. Microbiol. 97 (2), 233–245. https://doi.org/10.1111/j.1365-2672.2004.02294.x.

Dorman, H.J.D., Deans, S.G., 2000. Antimicrobial agents from plants: antibacterial activity of plant volatile oils. J. Appl. Microbiol. 88 (2), 308–316. https://doi.org/10.1046/j.1365-2672.2000.00969.x.

Eid, N.M., et al., 2018. Anticlostridial activity of the thyme and clove essential oils against experimentally induced necrotic enteritis in commercial broiler chickens. Vet. Sci. Res. Rev. 4 (1), 25–34.

El-Far, A.H., Ahmed, H.A., Shaheen, H.M., 2016. Dietary supplementation of Phoenix dactylifera seeds enhances performance, immune response, and antioxidant status in broilers. Oxid. Med. Cell. Longev. 2016. https://doi.org/10.1155/2016/5454963.

Gadde, U., et al., 2017. Alternatives to antibiotics for maximizing growth performance and feed efficiency in poultry: a review. Anim. Health Res. Rev. 18 (1), 26–45. https://doi.org/10.1017/S1466252316000207.

Galal, A.A.A.E.-G., et al., 2015. Positive impact of oregano essential oil on growth performance, humeral immune responses and chicken interferon alpha signalling pathway in broilers. Adv. Anim. Vet. Sci. 4 (1), 57–65. https://doi.org/10.14737/journal.aavs/2016/4.1.57.65.

Gharechopogh, A.M., et al., 2021. Performance, immune responses, and blood biochemistry of broiler chickens fed with plant growth compound. Trop. Anim. Sci. J. 44 (1), 62–70. https://doi.org/10.5398/tasj.2021.44.1.62.

Ghofran, O., Safari, T., Shahraki, M., 2017. Effects of eugenol on pain response to the formalin test and plasma antioxidant activity in high fructose drinking water in male rats. Int. J. Prev. Med. 8. https://doi.org/10.4103/ijpvm.IJPVM.

Gole, M., et al., 2020. The impact of feeding clove essential oils and organic acids on immunity, gut health and economics of broiler production. J. Pharmacogn. Phytochem. 9 (3), 1417–1422.

Gopi, M., 2014. Essential oils as a feed additive in poultry nutrition. Adv. Anim. Vet. Sci. 2 (1), 1–7. https://doi.org/10.14737/journal.aavs/2014.2.1.1.7.

Gülçin, I., Elmastaş, M., Aboul-Enein, H.Y., 2012. Antioxidant activity of clove oil - A powerful antioxidant source. Arab. J. Chem. 5 (4), 489–499. https://doi.org/10.1016/j.arabjc.2010.09.016.

Hartanti, D., Septiyaningrum, N.A., Hamad, A., 2020. Combination of clove and lemon basil essential oils for preservation of chicken meat. J. Food Qual. Hazards Control 7 (2), 84–93. https://doi.org/10.18502/JFQHC.7.2.2888.

Hu, F., et al., 2019. Comparison of antifungal activity of essential oils from different plants against three fungi. Food Chem. Toxicol. 134, 110821. https://doi.org/10.1016/j.fct.2019.110821. Elsevier.

Hussein, H.Z., 2015. Activity of pomegranate peels and clove powders in detoxification of aflatoxin B1 and Ochratoxin A from contaminated poultry diet. J. Plant Pathol. Microbiol. 06 (01). https://doi.org/10.4172/2157-7471.1000249.

Hussein, M.M.A., et al., 2019. Effects of clove (*Syzygium aromaticum*) oil on quail growth, carcass traits, blood components, meat quality, and intestinal microbiota. Poult. Sci. 98 (1), 319–329. https://doi.org/10.3382/ps/pey348. Poultry Science Association Inc.

Ikawati, S., et al., 2020. Thermostability, photostability, and toxicity of clove oil nanoparticles against *cryptolestes ferrugineus* (Stephens) (coleoptera: Laemophloeidae). Biodiversitas 21 (10), 4764–4771. https://doi.org/10.13057/biodiv/d211041.

Jang, I.S., et al., 2002. Influence of Essential Oils on Performance and Digestive Functions. pp. 394–400.

Jimoh, S.O., Arowolo, L.A., Alabi, K.A., 2017. Phytochemical screening and antimicrobial evaluation of *Syzygium aromaticum* extract and essential oil. Int. J. Curr. Microbiol. Appl. Sci. 6 (7), 4557–4567. https://doi.org/10.20546/ijcmas.2017.607.476.

Julianus Sohilait, H., 2015. Chemical composition of the essential oils in *Eugenia caryophylata* Thunb from Amboina Island. Sci. J. Chem. 3 (6), 95. https://doi.org/10.11648/j.sjc.20150306.13.

Kaakoush, N.O., et al., 2015. Global epidemiology of campylobacter infection. Clin. Microbiol. Rev. 28 (3), 687–720. https://doi.org/10.1128/CMR.00006-15.

Kamatou, G.P., Vermaak, I., Viljoen, A.M., 2012. Eugenol- from the remote Maluku Islands to the international market place: a review of a remarkable and versatile molecule. Molecules 17 (6), 6953–6981. https://doi.org/10.3390/molecules17066953.

Kaur, K., Kaushal, S., 2019. Phytochemistry and pharmacological aspects of *Syzygium aromaticum*: a review. J. Pharmacogn. Phytochem. 8 (1), 398–406.

Kazi, S., et al., 2017. Reversal of paracetamol induced hepatotoxicity in animals model: clove, an alternative medicine beyond the spices. J. Liaquat Univ. Med. Health Sci. 16 (3), 154–157. https://doi.org/10.22442/jlumhs.171630525.

Khosravinia, H., 2015. Litter quality and external carcass defects in broiler chicken influenced by supplementation of drinking water with savory (*Satureja Khuzistanica*) essential oils. Glob. J. Anim. Sci. Res. 3 (1), 247–252.

Kopp, V.V., et al., 2019. Microencapsulation of clove essential oil with gelatin and alginate. In: 35th IULTCS Congress 2019: 'Benign by Design' Leather- The Future Through Science and Technology, pp. 1–8.

Krishan, G., Narang, A., 2014. Use of essential oils in poultry nutrition: a new approach. J. Adv. Vet. Anim. Res. 1 (4), 156–162. https://doi.org/10.5455/javar.2014.a36.

Lee, J.W., et al., 2020. Dietary encapsulated essential oils improve production performance of coccidiosis-vaccine-challenged broiler chickens. Animals 10 (3). https://doi.org/10.3390/ani10030481.

Lillehoj, H., et al., 2018. Phytochemicals as antibiotic alternatives to promote growth and enhance host health. Vet. Res. 49 (1), 1–18. https://doi.org/10.1186/s13567-018-0562-6. BioMed Central.

Maciel, M.V.d.O.B., et al., 2019. *Syzygium aromaticum* L. (clove) essential oil as a reducing agent for the green synthesis of silver nanoparticles. Open J. Appl. Sci. 09 (02), 45–54. https://doi.org/10.4236/ojapps.2019.92005.

Mahrous, H., et al., 2017. Effects of different levels of clove bud (*Syzygium Aromaticum*) dietary supplementation on immunity, antioxidant status, and performance in broiler chickens. Alex. J. Vet. Sci. 54 (2), 29. https://doi.org/10.5455/ajvs.272231.

Mandey, J.S., et al., 2019. Nutrient and bioactive potentials of clove and carrot as natural feed additive/water additive candidates for organic broiler chicken. IOP Conf. Ser. Earth Environ. Sci. https://doi.org/10.1088/1755-1315/387/1/012126. Institute of Physics Publishing.

Mandey, J.S., Wolayan, F.R., Pontoh, C.J., Kowel, J.H., 2020. Fatty acid composition and antibacterial activity of clove (*Syzygium aromaticum*) and carrot (*Daucus carota*) as candidate of additive for broiler chickens. Sci. Papers Ser. D Anim. Sci. LXIII (1).

Masotti, V., et al., 2003. Seasonal and phenological variations of the essential oil from the narrow endemic species Artemisia molinieri and its biological activities. J. Agric. Food Chem. 51 (24), 7115–7121. https://doi.org/10.1021/jf034621y.

Mathlouthi, N., et al., 2012. Use of rosemary, oregano, and a commercial blend of essential oils in broiler chickens: in vitro antimicrobial activities and effects on growth performance. J. Anim. Sci. 90 (3), 813–823. https://doi.org/10.2527/jas.2010-3646.

Mehr, M.A., Hassanabadi, A., Moghaddam, H.N., 2014. Supplementation of clove essential oils and probiotic to the broiler's diet on performance, carcass traits and blood components. Iran. J. Appl. Anim. Sci. 4, 117–122.

Mohammadi, Z., Ghazanfari, S., Adib Moradi, M., 2014. Einfluss eines Zusatzes von essentiellen ölen der nelke zum futter auf die mikroflora und die morphologie im darm sowie auf blutparameter und leistung von broilern. Eur. Poult. Sci. 78, 1–11. https://doi.org/10.1399/eps.2014.51.

Mukhtar, M.A., 2011. The effect of dietary clove oil on broiler performance. Aust. J. Basic Appl. Sci. 5 (7), 49–51.

Mukhtar, S., Zahoor, T., Randhawa, M.A., Iqbal, R., Shabbir, A., Liaqat, A., Ahsan, S., 2018. Synergistic effect of chitosan and clove oil on raw poultry meat. J. Food Process. Technol. 09 (06), 6–10. https://doi.org/10.4172/2157-7110.1000737.

Mustafa, M.A., Wasman, P.H., 2020. The impact of powders and oil additives of cinnamon and clove in quails diet as antistressor and antioxidant during hot months. Iraqi J. Agric. Sci. 51 (3), 760–766. https://doi.org/10.36103/ijas.v51i3.1031.

Mustafa, M.A., Sabir, P.S., Mustafa, N.A., 2017. Effect of functional feed additives on egg production, hatchability and hematological traits of Japanese quails during summer condition. Iraqi J. Agric. Sci. 48 (Special issue), 80–85. https://doi.org/10.36103/ijas.v48ispecial.248.

Nanda, S., 2020. From phytochemicals to phytomedicines : potential roles of plant-based biomolecules in the covid era. Biosci. Biotechnol. Res. Commun. 13 (12), 39–43.

Nath, D.D., Rahman, M.M., Akter, F., Mostofa, M., 2013. Effects of Tulsi, black pepper and cloves extract as A growth promoter in broiler. Bangladesh Vet. 29 (2), 69–77. https://doi.org/10.3329/bvet.v29i2.14345.

Nazzaro, F., et al., 2013. Effect of essential oils on pathogenic bacteria. Pharmaceuticals 6 (12), 1451–1474. https://doi.org/10.3390/ph6121451.

Nwaopara, A.O., et al., 2007. The combined effects of excessive consumption of ginger, clove, red pepper and black pepper on the histology of the liver. Pak. J. Nutr. 6 (6), 524–527. https://doi.org/10.3923/pjn.2007.524.527.

O'Bryan, C.A., et al., 2015. Potential of plant essential oils and their components in animal agriculture—in vitro studies on antibacterial mode of action. Front. Vet. Sci. 2 (SEP), 1–8. https://doi.org/10.3389/fvets.2015.00035.

Olasupo, N.A., et al., 2003. Activity of natural antimicrobial compounds against *Escherichia coli* and *Salmonella enterica* serovar Typhimurium. Lett. Appl. Microbiol. 37 (6), 448–451. https://doi.org/10.1046/j.1472-765X.2003.01427.x.

Ordóñez, G., Llopis, N., Peñalver, P., 2008. Efficacy of eugenol against a *Salmonella enterica* serovar enteritidis experimental infection in commercial layers in production. J. Appl. Poultry Res. 17 (3), 376–382. https://doi.org/10.3382/japr.2007-00109.

Pandey, P., et al., 2021. An in silico screening on *Piper nigrum*, *Syzygium aromaticum* and *Zingiber officinale* roscoe derived compounds against sars-cov-2: a drug repurposing approach. Biointerface Res. Appl. Chem. 11 (4), 11122–11134. https://doi.org/10.33263/BRIAC114.1112211134.

Parveen, S., et al., 2020. Effect of essential oil blend on intestinal morphology, gut microbiota and immune response of broiler chickens. Indian J. Anim. Health 59 (2-Spl), 228–236. https://doi.org/10.36062/ijah.59.2spl.2020.228-236.

Pavesi, C., Banks, L.A., Hudaib, T., 2018. Antifungal and antibacterial activities of eugenol and non-polar extract of *Syzygium aromaticum* L. J. Pharm. Sci. Res. 10 (2), 337–339.

Petrovic, V., et al., 2012. The effect of supplementation of clove and agrimony or clove and lemon balm on growth performance, antioxidant status and selected indices of lipid profile of broiler chickens. J. Anim. Physiol. Anim. Nutr. 96 (6), 970–977. https://doi.org/10.1111/j.1439-0396.2011.01207.x.

Pham, V.H., et al., 2020. Dietary encapsulated essential oils and organic acids mixture improves gut health in broiler chickens challenged with necrotic enteritis. J. Anim. Sci. Biotechnol. 11 (1), 1–18. https://doi.org/10.1186/s40104-019-0421-y.

Pichersky, E., Noel, J.P., Dudareva, N., 2006. Biosynthesis of plant volatiles: nature's diversity and ingenuity. Science 311 (5762), 808–811. https://doi.org/10.1126/science.1118510.

Saeed, T.A., et al., 2017. Safety assessment and potential anti-inflammatory effect of ethanolic extract of *Syzygium aromaticum* in albino rats. Adv. Biosci. Biotechnol. 08 (11), 411–420. https://doi.org/10.4236/abb.2017.811030.

Salman, K.A.A., Ibrahim, D., 2012. Test the activity of supplementation clove (*Eugenia caryophyllus*) powder, oil and aqueous extract to diet and drinking water on performance of broiler chickens exposed to heat stress. Poult. Sci. 11 (10), 635–640.

Sandner, G., Heckmann, M., Weghuber, J., 2020. Immunomodulatory activities of selected essential oils. Biomol. Ther. 10 (8), 1–16. https://doi.org/10.3390/biom10081139.

Scherer, R., et al., 2014. Microencapsulated eucalyptol and eugenol as growth promoters in broilers. Rev. Bras. Pesqui. Aliment. 5 (1), 26. https://doi.org/10.14685/rebrapa.v5i1.157.

Shahbazi, Y., 2019. Antioxidant, antibacterial, and antifungal properties of nanoemulsion of clove essential oil. Nanomed. Res. J. 4 (4), 204–208. https://doi.org/10.22034/NMRJ.2019.04.001.

Sharma, A., et al., 2017. Antifungal activities of selected essential oils against Fusarium oxysporum f. sp. lycopersici 1322, with emphasis on *Syzygium aromaticum* essential oil. J. Biosci. Bioeng. 123 (3), 308–313. https://doi.org/10.1016/j.jbiosc.2016.09.011. Elsevier Ltd.

Shin, S., Lim, S., 2004. Antifungal effects of herbal essential oils alone and in combination with ketoconazole against Trichophyton spp. J. Appl. Microbiol. 97 (6), 1289–1296. https://doi.org/10.1111/j.1365-2672.2004.02417.x.

Srivastava, A., et al., 2020. Role of medicinal plants of traditional use in recuperating devastating role of medicinal plants of traditional use in recuperating devastating COVID-19 situation. Med. Aromat. Plants 9 (5), 1–16. https://doi.org/10.35248/2167-0412.20.9.359. Copyright.

Stefanello, C., et al., 2020. Protected blend of organic acids and essential oils improves growth performance, nutrient digestibility, and intestinal health of broiler chickens undergoing an intestinal challenge. Front. Vet. Sci. 6 (January), 1–10. https://doi.org/10.3389/fvets.2019.00491.

Suliman, G.M., et al., 2020. The effects of clove seeds (*Syzygium aromaticum*) dietary administration on carcass characteristics, meat quality and sensory attributes of broiler chickens. Poult. Sci. https://doi.org/10.1016/j.psj.2020.12.009. Elsevier Inc.

Tahir, M., et al., 2019. The performance of broilers given eugenol of clove leaf essential oil as a feed additive. Russ. J. Agric. Socio-Econ. Sci. 95 (11), 200–205. https://doi.org/10.18551/rjoas.2019-11.28.

Tariq, H., et al., 2015. Effect of Aloe vera and clove powder supplementation on carcass characteristics, composition and serum enzymes of Japanese quails. Vet. World 8 (5), 664–668. https://doi.org/10.14202/vetworld.2015.664-668.

Teixeira, B., et al., 2014. Characterization of fish protein films incorporated with essential oils of clove, garlic and origanum: physical, antioxidant and antibacterial properties. LWT- Food Sci. Technol. 59 (1), 533–539. https://doi.org/10.1016/j.lwt.2014.04.024.

Teles, A.M., et al., 2021. GC-MS characterization of antibacterial, antioxidant, and antitrypanosomal activity of *Syzygium aromaticum* essential oil and eugenol. Evid. Based Complement. Alternat. Med. 2021, 1–12. https://doi.org/10.1155/2021/6663255.

Tongnuanchan, P., Benjakul, S., 2014. Essential oils: extraction, bioactivities, and their uses for food preservation. J. Food Sci. 79 (7), 1231–1249. https://doi.org/10.1111/1750-3841.12492.

Valenzuela-Grijalva, N.V., et al., 2017. Dietary inclusion effects of phytochemicals as growth promoters in animal production. J. Anim. Sci. Technol. 59 (1), 1–17. https://doi.org/10.1186/s40781-017-0133-9.

Wael, S., et al., 2018. Secondary metabolites production in clove (*Syzygium aromaticum*): chemical compounds. J. Biol. Sci. 18 (7), 399–406. https://doi.org/10.3923/jbs.2018.399.406.

Wang, J., et al., 2021. Effect of a phytogenic feed additive on growth performance, nutrient digestion, and immune response in broiler-fed diets with two different levels of crude protein. Animals 11 (3), 1–11. https://doi.org/10.3390/ani11030775.

Wasman, P.H., Mustafa, M.A.G., 2020. The dietary impact of clove and cinnamon powders and oil supplementations on the performance, ileum morphology, and intestine bacterial population of quails. Plant Arch. 20 (1), 1503–1509.

Xu, J.G., et al., 2016. Chemical composition, antibacterial properties and mechanism of action of essential oil from clove buds against staphylococcus aureus. Molecules 21 (9), 1–13. https://doi.org/10.3390/molecules21091194.

Yang, X., et al., 2018. Impact of essential oils and organic acids on the growth performance, digestive functions and immunity of broiler chickens. Anim. Nutr. 4 (4), 388–393. https://doi.org/10.1016/j.aninu.2018.04.005. Elsevier Ltd.

Zeng, Z., et al., 2015. Essential oil and aromatic plants as feed additives in non-ruminant nutrition: a review. J. Anim. Sci. Biotechnol. 6 (1). https://doi.org/10.1186/s40104-015-0004-5.

CHAPTER

Clove (*Syzygium aromaticum*) and its derivatives in fish feed

16

Muhammad Saeed[a], Muhammad Sajjad Khan[a], Umair Younas[a], Faisal Siddique[a], Kamal Niaz[a], Zahid Farooq[a], Mayada R. Farag[b], Shaaban S. Elnesr[c], Mahmoud Alagawany[c], and Mohamed Fawzy Ramadan[d]

Cholistan University of Veterinary and Animal Sciences Bahawalpur, Bahawalpur, Pakistan[a] Forensic Medicine and Toxicology Department, Veterinary Medicine Faculty, Zagazig University, Zagazig, Egypt[b] Poultry Production Department, Faculty of Agriculture, Fayoum University, Fayoum, Egypt[c] Department of Clinical Nutrition, Faculty of Applied Medical Sciences, Umm Al-Qura University, Makkah, Saudi Arabia[d]

Abbreviations

CBE clove bud extract
CEO clove essential oil
DPPH 2,2-diphenyl-1-picrylhydrazyl
ECBE *Eugenia caryophyllata* bud extract
FAO Food and Agriculture Organization
ORAC oxygen radical absorption capacity
RBA respiratory burst activity
SOD superoxide dismutase

1. Introduction

The clove (*Syzygium aromaticum*) is a useful, valuable, and essential household spice belonging to the family Myrtaceae that is used as a food preservative, and its medicinal benefits have been known for centuries. Clove has a historical background and it is believed to have originated in the 1st century BC (Hussain et al., 2017). Its unique odor and sweet taste are commonly used as a spice worldwide (Milind and Deepa, 2011; Saeed et al., 2021).

Clove and its derivatives are used as herbal medicine for the treatment of various diseases worldwide (Saeed et al., 2021). The hepatoprotective role of clove essential oil (CEO) was observed at the dose rate of 0.1 and 0.2 g/kg in paracetamol-intoxicated hepatic damage assay, and it was noted that CEO restored the concentration of alanine aminotransferase and aspartate aminotransferase (Thuwaini et al., 2016). Some researchers (Hussain et al., 2017) reported the anesthetic action of clove oil derivative (eugenol). However, toxicity may appear due to overdose, and the safe limits of clove oil used in the dose are considered as <1.5 g/kg. The WHO recommends the acceptable dose of 2.5 mg/kg/day for humans (Doleželová et al., 2011). Due to having higher levels of phenolics and bioactive compounds,

Clove (*Syzygium aromaticum*). https://doi.org/10.1016/B978-0-323-85177-0.00038-0

the clove is considered the best antioxidant-rich plant present in nature. The clove has an oxygen radical absorption capacity (ORAC) of over 10 million, which is much higher than the other antioxidants present in nature (Bhowmik et al., 2012). The main antioxidant agents in the clove are the phenolic and flavonoid compounds (Hussain et al., 2017), whereas there is a positive correlation between phenolics and antioxidant potential.

In another study, CEO exhibited protective impact on experimental *Lactococcus garvieae* infection in tilapia and the potential to substitute the antibiotics for disease control. In this experiment, four herbs (clove, *Syzygium aromaticum*; cinnamon, *Cinnamomum verum*; holy basil, *Ocimum sanctum*, and ginger, *Zingiber officinale*) were tested for their antimicrobial effect and their mechanism of action against *Lactococcus garvieae* (pathogenic fish bacteria causing lactococcosis). Clove oil showed the most potent inhibitory effect than other herbs and showed a bactericidal mode of action against the pathogenic bacteria (Rattanachaikunsopon and Phumkhachorn, 2009).

Dietary supplements containing herbs or medicinal plants have a significant impact on the production of fish growth. Therefore, it is essential to protect fish from bacterial infections. Antibiotics are commonly used in aquaculture to prevent and control bacterial diseases, but they can also cause resistant bacterial strains. Therefore, people are interested in developing feed additives instead of traditional cost-effective alternatives against bacterial diseases to maintain environmentally friendly aquaculture (Adeshina et al., 2019). The main objectives of the following sections are to give an overview of the use of clove and its derivatives in fish feeding. The chemistry and antimicrobial properties and their effect on fish immunity, performance, and other health aspects will be precious for physiologists, scientists, nutritionists, veterinarians, pharmacists, and fish breeders.

1.1 Pharmacological properties of clove and its derivatives

Natural substances have been used for a long time for the treatment of bacterial infections. The clove and its derivatives have been the focus of aquaculture studies owing to their traits (e.g., antimicrobial, antioxidant, antiviral, antiparasitic, antifungal, and immunomodulatory effects) (Fig. 1). Pharmacologically, the clove has been tested toward different pathogenic parasites and microorganisms, including pathogenic bacteria, *Babesia*, *Theileria* parasites, *Plasmodium*, *Herpes simplex*, and hepatitis C viruses (Batiha et al., 2020). The antiviral and cleansing properties of cloves purify the body, augmenting the disease resistance (Bhowmik et al., 2012). Clove oil can prevent toxicity associated with exposure to environmental pollution (Bhowmik et al., 2012).

Some investigations have shown the efficacy of cloves against various strains of bacteria. The CEO and eugenol have antibacterial potential against Gram-negative bacteria and Gram-positive bacteria (Palmer et al., 1998; Mak et al., 2019). In aquaculture, this compound has been used as an anesthetic and bactericidal agent for different fish species (Brum et al., 2017). The essential oil of clove has also been reported to possess antibacterial effect against a broad spectrum of fish pathogens, indicating its impact to substitute commercial antibiotics as antimicrobial agents for use in aquaculture (Lee et al., 2009). Pathirana et al. (2019) stated that the antibacterial effect of CEO and eugenol was effective against pathogenic fish bacteria; these components might be advantageous to treat bacterial infections in aquaculture. Yostawonkul et al. (2019) illustrated that clove oil nanoparticles displayed good antibacterial effect against two important tilapia pathogens, namely *Flavobacterium columnare* and *Streptococcus agalactiae*. The inhibitory effect of clove is due to the presence of bioactive constituents, mainly eugenol, β-caryophyllene, and eugenyl acetate, and other active components (Lee et al., 2009).

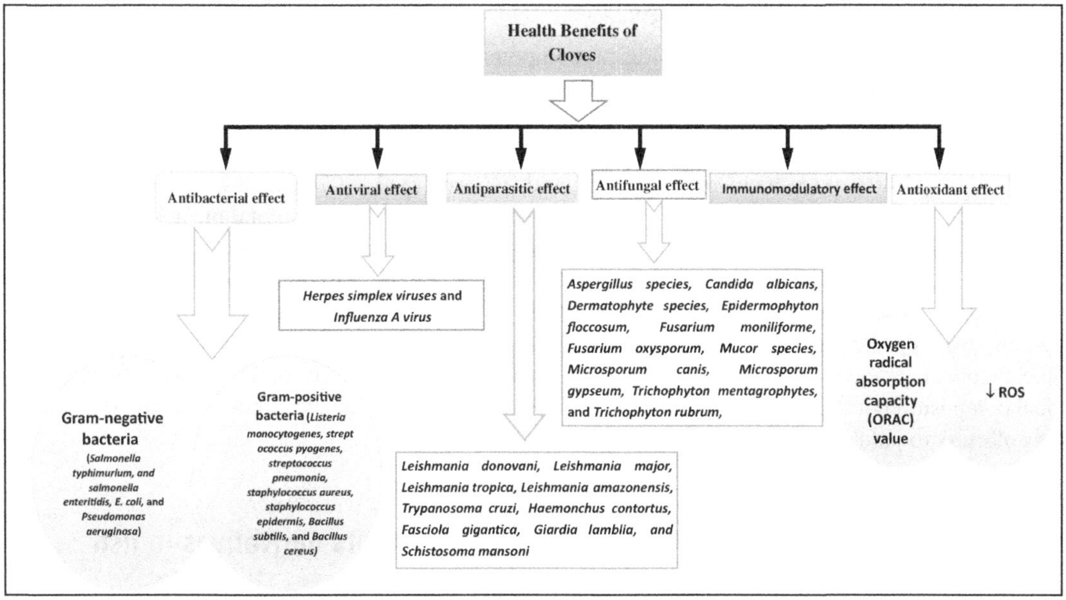

FIG. 1

Biological effects of clove.

The mechanism of antibacterial activity of these compounds is considered by challenging the cell membrane permeability, as these compounds could denature proteins and react with cell membrane phospholipids, increasing the permeability of the cell membrane that results in the disruption of the cytoplasmic membrane (Mittal et al., 2014).

Clove is among plant sources that protect polyunsaturated fatty acids (PUFA) against oxidation owing to its high antioxidative compounds (e.g., flavonoids and phenolics) that may exert antioxidative influence by various mechanisms such as the inhibition of oxidizing enzymes, and scavenging of free radicals, metal chelation, oxygen scavenging, and singlet oxygen quenching (Dudonne et al., 2009). Abdelkhalek et al. (2020) revealed that clove oil has a potent antioxidant role by inducing antioxidant enzyme activities [superoxide dismutase (SOD) and glutathione peroxidase] and antagonizing lipid peroxidation in Nile tilapia. Silva et al. (2021) clarified that the addition of clove leaf essential oil in zebrafish diets increased the antioxidant capacity through increasing values of DPPH·(2,2-diphenyl-1-picrylhydrazyl).

The clove and its derivatives are found active against different fungal species such as *Aspergillus* species, *Candida albicans*, *Dermatophyte* species, *Epidermophyton floccosum*, *Fusarium moniliforme*, *Microsporum canis*, *Fusarium oxysporum*, *Mucor* species, *Microsporum gypseum*, *Trichophyton mentagrophytes*, and *Trichophyton rubrum* (Fig. 1). They act as fungicide agents by increasing the cell permeability and altering the cell morphology of hyphae (Park et al., 2007; Pinto et al., 2009; Rana et al., 2011). Besides, eugenol has been reported fatal for the growth and multiplication of many parasites, namely *Fasciola gigantica*, *Haemonchus contortus*, *Giardia lamblia*, and *Schistosoma mansoni* (Machado et al., 2011; El-Kady et al., 2019).

1.2 Chemical structure of cloves

There are several distinct kinds of biological compounds found in cloves. Cloves include terpenoids, sugars, glycosides, phenolic compounds, and alkaloids (Kaur and Kaushal, 2019). Clove is the primary source of phenolic derivatives such as hydroxybenzoic acids, flavonoids, hydroxyphenyl propens, hydroxycinnamic acids, and eugenol. The key bioactive compound, i.e., present in the amount of 9.38–14.6mg/100g of fresh clove plant, is eugenol (Hussain et al., 2017). Besides, 10%–13% tannins are found in the fresh clove plant, having a similar chemical structure to gallotannic acid, eugenin, and ellagitannin (Milind and Deepa, 2011). Flavonoids, specifically quercetin and kaempferol, and phenolic acids like caffeic, gallic, ferulic, ellagic, and salicylic acids are also present in clove (Cortés-Rojas et al., 2014). There are considerable concentrations of sitosterol, stigmasterol, campesterol, and sesquiterpenes reported in the clove (Bhowmik et al., 2012). Studies have been performed to find out the chemical composition of the clove and its derivatives. However, most of the work on the biochemical composition of CEO was done. The clove oil is extracted from the clove bud, leaf, and stem which is colorless to pale yellow.

1.3 Health benefits and beneficial uses of clove and its derivatives in fish

1.3.1 Cloves effect on fish immunity

Clove and its derivatives are used as herbal medicine for the treatment of various diseases worldwide. The immunomodulatory effects of the active ingredients of cloves on the immune system need to be identified. Two types of immune systems are found in fish including adaptive and innate immune systems (Stratev et al., 2018). It consists of a broad range of nutrients and antioxidants that have been beneficial to the growth, boosting the immune system of fish. It can also be used in traditional medicine in Australia and some Asian countries for dental disorders, respiratory diseases, headaches, and sore throats. The antimicrobial, antiviral, antifungal, cytotoxic, anesthetic, and anti-inflammatory properties of cloves in fish have been reported (Abdel-Tawwab et al., 2018). The principal components of cloves are eugenol, cariophyllene, thymol, eugenyl acetate, and tanene. Although the exact mechanisms remain unknown, these components have been shown to modulate many of the impacts mentioned earlier and specific immune responses (Van Hai, 2015).

Immunomodulation is a crucial management method for disease safety in aquaculture (Abdel Rahman et al., 2019). The bacterial challenge test was used to confirm that clove bud powder has an immunomodulatory effect. Recent research investigated Nile tilapia's immune response to the pathogenic bacteria *Proteus mirabilis* using 3% clove bud powder in feed. The Immune indicator levels such as lysozyme activity, catalase, SOD, and respiratory bursts have been dramatically increased (Abdel Rahman et al., 2020).

Adeshina et al. (2019) reported that *Eugenia caryophyllata* bud extract (ECBE) usage as a feed supplement could improve fish immunity against the challenge infection with pathogenic *Aeromonas hydrophila*. Different markers have been used to classify the effect of immune response to ECBE, such as SOD, lysozyme, catalase, and respiratory burst activity (RBA). These chemicals were released by immune cells such as neutrophils and lymphocytes when contact with various pathogens was established. Antioxidants and immunomodulatory effects were related to eugenol. Study findings have shown the immunomodulating effects of ECBE on African catfish when administered with feed at a concentration of 15g/kg. In addition, that amount will increase the resistance of fish against pathogenic *Aeromonas hydrophila*. In earlier experiments using other herbal components on rainbow trout, white fish, Nile tilapia, and African catfish, similar types of findings were reported (Talpur, 2014; Saeidi Asl et al., 2017; Abdel-Tawwab et al., 2018).

The immune regulatory effect of edible clove essential oil has been studied in Nile tilapia against *Streptococcus suis* infection. Clove oil contains up to 90% eugenol and has been approved by the FDA as a food additive. It also has antibacterial activity against several bacteria, such as *Escherichia coli*, *Staphylococcus*, *Salmonella*, and *Listeria monocytogenes*. It has been shown that clove oil enhances humoral immunity and inhibits cell-mediated immunity. Fish feed supplemented with clove oil increased respiratory burst, lysozyme, and phagocytosis stimulation. Phagocytes and respiratory burst play an essential role in improving the innate immunity of fish and combating against streptococcal infection. The activity of lysozymes can be increased using clove oil as feed supplements, known to be part of the humoral immune response in aquatic organisms (Abdelkhalek et al., 2020).

Stress negatively impacts a fish's immune system by stimulating the hypothalamus and pituitary nervous system. The stress mechanism is allowed during the handling and transportation of fish from pond to pond (Bulfon et al., 2015). To mitigate these effects, due to their different properties, such as anesthetics, antioxidants, and antimicrobial agents, plant essential oils have become the focus of aquaculture study. It has been shown that biochemical and endocrine changes can be minimized, thereby improving the health of fish. Essential oils including CEO are applied to fish as a feed additive that serves as an immunostimulant (Souza et al., 2019). Immunostimulant markers such as leukocyte peroxidase, lysozyme, protease, and alkaline phosphatase have been elevated in fish feeding following the administration of CEO. Furthermore, the use of clove oil as a feed additive has been reported to increase or stimulate humoral and mucosal immunity in fish (Soltanian et al., 2018).

1.3.2 Clove effect on fish performance

The Food and Agriculture Organization (FAO) has stressed the importance of improving aquaculture development to meet the overwhelming task of feeding a global population predicted to hit 9.6 billion people by 2050. Currently, fish population growth has increased worldwide, primarily due to massive innovations in aquaculture development. As a result, aquaculture currently provides over half of the fish eaten by humans and is expected to grow to 62% by 2030 to satisfy rising demand (FAO, 2014).

In the last 25 years, there have been significant changes in the diet of aquaculture, which has led to increased fish growth in the fish farm and increased global aquaculture production. Although aquaculture has a long history, the lack of understanding of the specific nutritional needs of fish has hampered the development of extreme aquaculture, in which fish need to be fed nutritious food. The fish feed is based on experience rather than a realistic formula. This changed with the development of a nutritious, semi-pure fish diet. Using this diet, you can measure the growth status of the fish and estimate the various levels and dietary requirements, thus supplementing an essential nutrient (Small et al., 2016).

Although different types of technologies have been investigated to improve fish growth and efficiency, productivity largely depends on genetic, dietary, and environmental interactions. Dietary supplements containing herbs or medicinal plants have a significant impact on the production of fish growth, among others. Therefore, it is crucial to protect fish from bacterial infections in commercial agriculture. Antibiotics are commonly used in aquaculture to prevent and control bacterial diseases, but they can also cause resistant bacterial strains. Therefore, people are very interested in developing feed additives instead of traditional cost-effective alternatives against bacterial diseases to maintain environmentally friendly aquaculture (Adeshina et al., 2019). Medicinal herbs are rich in terms of nutrients and antioxidants that significantly affect fish farming (Awad and Awaad, 2017). They can also be used as natural feed supplements and chemotherapeutics. Since antibiotics are dispirited, numerous studies have been performed to investigate the impact of medicinal plants on fish growth (Hoseinifar et al., 2017). Clove (*Eugenia caryophyllata*; *Syzygium aromaticum*, Myrtaceae) bud is a promising herbal medicine considered to have almost no side effects or resistance problems. It has been efficiently used for the prevention of many bacterial and fungal diseases. Ethanol extract from clove buds has been shown to possess antioxidants and microbicidal activities. Therefore, it can be proposed that the use of clove bud extract (CBE) as a feed supplement may be beneficial in assessing the growth and health of fish (Batiha et al., 2020).

To protect cells from harm, defense mechanisms have been developed in fish, such as the production of antioxidant enzymes, e.g., catalase, glutathione reductase, SOD, glutathione *S*-transferase, and glutathione peroxidase. If the production rate of free radicals is greater than the defense system's removal capability, an exogenous antioxidant that may come from the diet is required. In this respect, clove oil has the maximum antioxidant potential among commonly marketed essential oils (Silva et al., 2021). Eugenol is a critical component of clove oil, of which several antioxidant properties are linked. The lipid peroxidation inhibition rate of linoleic acid is 97% when administered in feed at 15 µg/L concentration. According to recent reports, 0.5% clove oil can be added to the diet in *Labeo rohita* to minimize SOD and lipid peroxidase levels. It has been recorded worldwide that fish growth has increased with the addition of clove oil (Asimi and Sahu, 2016).

1.3.3 Clove effect on semen quality

In today's culture, the prevalence of male infertility has increased due to multiple reasons, such as industrial contamination, anxiety, and habits. Male infertility has a psychological and emotional effect on a stable life because infertility is considered an incomplete male in the Indus Valley (Smith et al., 2009). Due to significant new technologies in male reproductive health, infertility therapy in underdeveloped countries is beyond the reach of the average human being because it is so complex and expensive. Natural medicinal products are as prehistoric as modern civilization, and even currently 70%–80%of the worldwide people mainly depend on herbal medicine to meet their basic medical needs (WHO, 2005).

Syzygium aromaticum flower buds, commonly called cloves, are traditional medicines popularly used in Australia, Southeast Asia, and the Indus valley. In Ayurvedic and Greek medicines, *Syzygium aromaticum* is known as aphrodisiacs and can treat sexually transmitted diseases (Tajuddin et al., 2003). According to reports, treatment with *Syzygium aromaticum* seeds may increase the frequency of normal male mice and rats' mounting. Even so, as far as researchers know, the effects of *Syzygium aromaticum* on the function and physiology of male sex organs have not been observed. For this purpose, the flower bud extract was used to evaluate its impact on the functional biology of male reproductive organs such as the seminal vesicle and epididymis through the use of mice as an animal model. The results showed that increased sperm motility and secretory activity of the seminal vesicles and epididymis were dose-dependent. Lower doses of 15 mg increased sperm motility and secretion activity compared to high doses of 60 mg. The sex drive was not hindered in males treated with CBE. There was also a marked decline in litter in females treated with high doses of CBE (Mishra and Singh, 2013).

Amin et al. (2011) studied the impacts of different levels of clove extract (0, 50, 200, 400, and 600 mg) on some hematological indices and sperm function and in Prussian carp (*Carassius gibelio*). They found that the use of clove extract in lower dosages than 400 ppm did not affect hematological parameters and sperm motility traits in Prussian carp.

Baghshahi et al. (2014) studied the effect of CBE as an antioxidant that impacts semen quality parameters following cryopreservation of ram sperm cells. For this purpose, tris, glycerol, and egg yolk have been used as extenders. Different concentrations such as 0–115 μg/mL of CBE have been added to these extenders. The results showed that high antioxidant activity of CBE was measured compared to vitamin E. Improved semen quality and characteristics were observed by the addition of 35–75 μg/mL of CBE to semen extenders. After cryopreservation, semen exposure consisting of egg yolk + SDS and 75 μg CBE, semen dynamics, and plasma membrane integrity were also higher. However, the high content of CBE adversely affects all semen quality parameters and significantly reduces sheep sperm motility and plasma membrane integrity. The result is that CBE has antioxidant capacity, which makes it useful for adding mini-supplements, and for best results the maximum CBE is 75 μg/mL. The combined effects of egg yolk and detergent have the ability to improve sperm motility after freezing-thawing method.

The latest study aimed to determine the antibacterial effect of CBE on semen when used as an antibiotic substitute and to study its effect on semen quality attributes following cryopreservation of bovine sperm. Replace antibiotics with CBE at concentrations of 0, 25, 50, and 75 μg/mL in extender solution. The bacterial pressure in fresh semen and frozen semen has been analyzed in all groups. The conclusions observed that the amount of CBE is 50 and 75 μg/mL, and its antimicrobial activity is analogous to that of antibiotics. Conversely, the microbiological quality of semen accompanied by an extract of 25 μg/mL is relatively better. Two significant parameters, such as motility and morphology, showed better results after the addition of CBE to the extender solution. It is observed that the parameters of motion and morphological parameters are better. Therefore, it could be concluded that the extract of clove bud has antibacterial potential, making it more suitable for addition to semen supplements. The optimum extract of the clove bud is 50 μg/mL for bovine semen (Virmani et al., 2020).

2. Conclusion

It can be inferred, based on the knowledge and analysis presented in this chapter, that clove itself is an essential nutritional and medicinal plant and has great potential to substitute various drug therapies. It is an excellent food preservative and an ideal option for pharmaceutical medicinal or herbal preparations because of the nature of strong biological compounds and antioxidant and antibacterial properties. The clove and its derivatives can improve the productive and reproductive performance of fish and increase aquaculture production. It is possible to improve the immunity and productivity of fish, but the amount of food must be maintained within a healthy range to prevent toxic problems. Why this plant has been used for centuries is not surprising. Several investigations are needed to shed more light on the impact of clove or their derivatives on fish production.

References

Abdel Rahman, A., El-Bouhy, Z., Wahbah, M., Ahmed, S., 2020. Effects of dietary turmeric and clove powder on growth and immune response of the Nile tilapia. Egypt. J. Aquat. Biol. Fish. 24 (5), 589–608.

Abdel Rahman, A.N., ElHady, M., Shalaby, S.I., 2019. Efficacy of the dehydrated lemon peels on the immunity, enzymatic antioxidant capacity and growth of Nile tilapia (*Oreochromis niloticus*) and African catfish (*Clarias gariepinus*). Aquaculture 505, 92–97.

Abdelkhalek, N.K., Risha, E., Mohamed, A., Salama, M.F., Dawood, M.A., 2020. Antibacterial and antioxidant activity of clove oil against *Streptococcus iniae* infection in Nile tilapia (*Oreochromis niloticus*) and its effect on hepatic hepcidin expression. Fish Shellfish Immunol. 104 (2020), 478–488.

Abdel-Tawwab, M., Adeshina, I., Jenyo-Oni, A., Ajani, E.K., Emikpe, B.O., 2018. Growth, physiological, antioxidants, and immune response of African catfish, *Clarias gariepinus* (B.), to dietary clove basil, *Ocimum gratissimum*, leaf extract and its susceptibility to *Listeria monocytogenes* infection. Fish Shellfish Immunol. 78, 346–354.

Adeshina, I., Jenyo-Oni, A., Emikpe, B.O., Ajani, E.K., Abdel-Tawwab, M., 2019. Stimulatory effect of dietary clove, *Eugenia caryophyllata*, bud extract on growth performance, nutrient utilization, antioxidant capacity, and tolerance of African catfish, *Clarias gariepinus* (B.), to *Aeromonas hydrophila* infection. J. World Aquacult. Soc. 50 (2), 390–405.

Amin, F., Kasiri, M., Talebi, A., Sudagar, M., 2011. Effects of clove extract as an anesthetic on sperm motility traits and some hematological parameters in Prussian carp *Carassius gibelio*. Adv. Environ. Biol. 5, 1406–1410.

Asimi, O.A., Sahu, N.P., 2016. Effect of antioxidant rich spices, clove and cardamom extracts on the metabolic enzyme activity of *Labeo rohita*. J. Fish. Livest. Prod. 4 (1), 1–6.

Awad, E., Awaad, A., 2017. Role of medicinal plants on growth performance and immune status in fish. Fish Shellfish Immunol. 67, 40–54.

Baghshahi, H., Riasi, A., Mahdavi, A.H., Shirazi, A., 2014. Antioxidant effects of clove bud (*Syzygium aromaticum*) extract used with different extenders on ram spermatozoa during cryopreservation. Cryobiology 69 (3), 482–487.

Batiha, G.E.S., Alkazmi, L.M., Wasef, L.G., Beshbishy, A.M., Nadwa, E.H., Rashwan, E.K., 2020. *Syzygium aromaticum* L. (Myrtaceae): traditional uses, bioactive chemical constituents, pharmacological and toxicological activities. Biomol. Ther. 10, 202.

Bhowmik, D., Kumar, K.S., Yadav, A., Srivastava, S., Paswan, S., Dutta, A.S., 2012. Recent trends in Indian traditional herbs *Syzygium aromaticum* and its health benefits. J. Pharmacogn. Phytochem. 1, 13–22.

Brum, A., Pereira, S.A., Owatari, M.S., Chagas, E.C., Chaves, F.C.M., Mouriño, J.L.P., Martins, M.L., 2017. Effect of dietary essential oils of clove basil and ginger on Nile tilapia (*Oreochromis niloticus*) following challenge with *Streptococcus agalactiae*. Aquaculture 468, 235–243.

Bulfon, C., Volpatti, D., Galeotti, M., 2015. Current research on the use of plant-derived products in farmed fish. Aquac. Res. 46, 513–551.

Cortés-Rojas, D.F., de Souza, C.R., Oliveira, W.P., 2014. Clove (*Syzygium aromaticum*): a precious spice. Asian Pac. J. Trop. Med. 4, 90–96.

Doleželová, P., Mácová, S., Plhalová, L., Pištěková, V., Svobodová, Z., 2011. The acute toxicity of clove oil to fish *Danio rerio* and *Poecilia reticulata*. Acta Vet. Brno 80 (3), 305–308.

Dudonne, S., Vitrac, X., Coutiere, P., Woillez, M., Mérillon, J.M., 2009. Comparative study of antioxidant properties and total phenolic content of 30 plant extracts of industrial interest using DPPH, ABTS, FRAP, SOD, and ORAC assays. J. Agric. Food Chem. 57 (5), 1768–1774.

El-Kady, A.M., Ahmad, A.A., Hassan, T.M., El-Deek, H.E.M., Fouad, S.S., Althagfan, S.S., 2019. Eugenol, a potential schistosomicidal agent with anti-inflammatory and antifibrotic effects against *Schistosoma mansoni*, induced liver pathology. Infect. Drug Resist. 12, 709–719.

FAO (Food and Agriculture Organization), 2014. The State of World Fisheries and Aquaculture 2014: Opportunities and Challenges. Food & Agriculture Org.

Hoseinifar, S.H., Dadar, M., Khalili, M., Cerezuela, R., Esteban, M.A., 2017. Effect of dietary supplementation of palm fruit extracts on the transcriptomes of growth, antioxidant enzyme, and immune-related genes in common carp (*Cyprinus carpio*) fingerlings. Aquac. Res. 48, 3684–3692.

Hussain, S., Rahman, R., Mushtaq, A., El Zerey-Belaskri, A., 2017. Clove: a review of a precious species with multiple uses. Int. J. Chem. Biochem. Sci. 11, 129–133.

Kaur, K., Kaushal, S., 2019. Phytochemistry and pharmacological aspects of *Syzygium aromaticum*: a review. J. Pharmacogn. Phytochem. 8, 398–406.

Lee, S., Najiah, M., Wendy, W., Nadirah, M., 2009. Chemical composition and antimicrobial activity of the essential oil of *Syzygium aromaticum* flower bud (clove) against fish systemic bacteria isolated from aquaculture sites. Front. Agric. China 3 (3), 332–336.

Machado, M., Dinis, A.M., Salgueiro, L., Custódio, J.B., Cavaleiro, C., Sousa, M.C., 2011. Anti-Giardia activity of *Syzygium aromaticum* essential oil and eugenol: effects on growth, viability, adherence, and ultrastructure. Exp. Parasitol. 127, 732–739.

Mak, K.K., Kamal, M., Ayuba, S., Sakirolla, R., Kang, Y.B., Mohandas, K., Pichika, M., 2019. A comprehensive review on eugenol's antimicrobial properties and industry applications: a transformation from ethnomedicine to industry. Pharmacogn. Rev. 13, 1–9.

Milind, P., Deepa, K., 2011. Clove: a champion spice. Int. J. Res. Ayurveda Pharm. 2, 47–54.

Mishra, R.K., Singh, S.K., 2013. Reproductive effects of lipid soluble components of *Syzygium aromaticum* flower bud in male mice. J. Ayurveda Integr. Med. 4 (2), 94.

Mittal, M., Gupta, N., Parashar, P., Mehra, V., Khatri, M., 2014. Phytochemical evaluation and pharmacological activity of *Syzygium aromaticum*: a comprehensive review. Int. J. Pharm. Pharm. Sci. 6, 67–72.

Palmer, A.S., Stewart, J., Fyle, L., 1998. Antimicrobial properties of plant essential oils and essences against five important food-borne pathogens. Appl. Microbiol. 26, 118–122.

Park, M.J., Gwak, K.S., Yang, I., Choi, W.S., Jo, H.J., Chang, J.W., Jeung, E.B., Choi, I.G., 2007. Antifungal activities of the essential oils in *Syzygium aromaticum* (L.) Merr. Et Perry and *Leptospermum petersonii* bailey and their constituents against various dermatophytes. J. Microbiol. 45, 460–465.

Pathirana, H.N.K.S., Wimalasena, S.H.M.P., De Silva, B.C.J., Hossain, S., Heo, G., 2019. Antibacterial activity of clove essential oil and eugenol against fish pathogenic bacteria isolated from cultured olive flounder (*Paralichthys olivaceus*). Slov. Vet. Res. 56, 31–38.

Pinto, E., Vale-Silva, L., Cavaleiro, C., Salgueiro, L., 2009. Antifungal activity of the clove essential oil from *Syzygium aromaticum* on Candida, Aspergillus and dermatophyte species. J. Med. Microbiol. 58 (11), 1454–1462.

Rana, I.S., Rana, A.S., Rajak, R.C., 2011. Evaluation of antifungal activity in essential oil of the *Syzygium aromaticum* (L.) by extraction, purification and analysis of its main component eugenol. Braz. J. Microbiol. 42, 1269–1277.

Rattanachaikunsopon, P., Phumkhachorn, P., 2009. Protective effect of clove oil-supplemented fish diets on experimental *Lactococcus garvieae* infection in Tilapia. Biosci. Biotechnol. Biochem. 73 (9), 2085–2089. https://doi.org/10.1271/bbb.90294.

Saeidi Asl, M.R., Adel, M., Caipang, C.M.A., Dawood, M.A.O., 2017. Immunological responses and disease resistance of rainbow trout (*Oncorhynchus mykiss*) juveniles following dietary administration of stinging nettle (*Urtica dioica*). Fish Shellfish Immunol. 71, 230–238.

Saeed, M., Khan, M.S., Alagawany, M., Farag, M.R., Alqaisi, O., Aqib, A.I., Qumar, M., Siddique, F., Ramadan, M.F., 2021. Clove (*Syzygium aromaticum*) and its phytochemicals in ruminant feed: an updated review. Rend. Lincei Sci. Fis. Nat. https://doi.org/10.1007/s12210-021-00985-3.

Silva, T.C.D., Silva, M., Carbonera, F., Visentainer, J.V., Utsunomiya, K.S., Gasparino, E., Pereira, R., 2021. Flaxseed oil and clove leaf essential oil in Zebrafish diet (*Danio rerio*). Acta Sci. Anim. Sci. 43. https://doi.org/10.4025/actascianimsci.v43i1.48126, e48126.

Small, B.C., Hardy, R.W., C.S., 2016. Tucker, enhancing fish performance in aquaculture. Anim. Front. 6, 42–49.

Smith, J.F., Walsh, T.J., Shindel, A.W., Turek, P.J., Wing, H., Pasch, L., et al., 2009. Sexual, marital, and social impact of a man's perceived infertility diagnosis. J. Sex. Med. 6, 2505–2515.

Soltanian, S., Hoseinifar, S.H., Gholamhosseini, A., 2018. Modulation of rainbow trout (*Oncorhynchus mykiss*) cutaneous mucosal immune responses following anesthesia: a comparative study on different anesthetic agents. Fish Shellfish Immunol. 80, 319–324.

Souza, C.F., Baldissera, M.D., Baldisserotto, B., Heinzmann, B.M., Martos-Sitcha, J.A., Mancera, J.M., 2019. Essential oils as stress-reducing agents for fish aquaculture: a review. Front. Physiol. 10, 785.

Stratev, D., Zhelyazkov, G., Noundou, X.S., Krause, R.W.M., 2018. Beneficial effects of medicinal plants in fish diseases. Aquac. Int. 26, 289–308.

Tajuddin, A.S., Latif, A., Qasmi, I.A., 2003. Aphrodisiac activity of 50% ethanolic extract of *Myristica fragrans* Houtt. (nutmeg) and *Syzygium aromaticum* (L) Merr. and Perry. (clove) in male mice: a comparative study. BMC Complement. Altern. Med. 3, 1–10.

Talpur, A.D., 2014. *Mentha piperita* (peppermint) as feed additive enhanced growth performance, survival, immune response and disease resistance of Asian seabass, *Lates calcarifer* (Bloch) against *Vibrio harveyi* infection. Aquaculture 420, 71–78.

Thuwaini, M., Abdul-Mounther, M., Kadhem, H., 2016. Hepatoprotective effects of the aqueous extract of clove (*Syzygium aromaticum*) against paracetamol induced hepatotoxicity and oxidative stress in rats. Euro. J. Pharm. Med. Res. 3 (8), 36–42.

Van Hai, N., 2015. The use of medicinal plants as immunostimulants in aquaculture: a review. Aquaculture 446, 88–96.

Virmani, M., Malik, R.K., Singh, M., Patil, C.S., Singh, P., Vijayalakshmy, K., 2020. Enrichment of semen extender with *Syzygium aromaticum* extract: it's impact on bull sperm preservation as an alternative to antibiotics. Ruminant Sci. 9 (1), 63–68.

WHO (World Health Organization), 2005. Geneva: Traditional Medicine Strategy 2002–2005.

Yostawonkul, J., Kitiyodom, S., Kaewmalun, S., Suktham, K., Nittayasut, N., Khongkow, M., Yata, T., 2019. Bifunctional clove oil nanoparticles for anesthesia and anti-bacterial activity in Nile tilapia (*Oreochromis niloticus*). Aquaculture 503, 589–595.

CHAPTER

Clove (*Syzygium aromaticum*) and eugenol toxicity

17

Zeynep Aksoylu Özbek and Pelin Günç Ergönül

Department of Food Engineering, Faculty of Engineering, Manisa Celal Bayar University, Yunusemre, Manisa, Turkey

Abbreviations

AChE	acetylcholinesterase
ADI	acceptable daily intake
AFC_{50}	median antifeedant concentration
ALAT	L-alanine aminotransferase
AOEL	acceptable operator exposure level
ASAT	L-aspartate aminotransferase
BHT	butylated hydroxyl toluene
CFR	Code of Federal Regulations
CoE	Council of Europe
EDI	estimated daily intake
EFSA	European Food Safety Authority
EPA	US Environmental Protection Agency
FAO	Food and Agriculture Organization
FCC	Food Chemical Codex
FDA	The United States Food and Drug Administration
FEMA	The Flavor and Extract Manufacturers Association of the United States
GC-MS	gas chromatography-mass spectrometry
GRAS	Generally Recognized as Safe
HPLC	high-performance liquid chromatography
IARC	The International Agency for Research on Cancer
IC_{50}	half maximal inhibitory concentration
IFRA	The International Fragrance Association
IOFI	The International Organization of Flavor Industries
JECFA	The Joint FAO/WHO Expert Committee on Food Additives;
LC_{50}	median lethal concentration
LD_{50}	median lethal dose
MLT_{50}	mean lethal time
NOAEL	no observed adverse effect level
NOP	National Organic Program of the United States Department of Agriculture
$SC\text{-}CO_2$	supercritical carbon dioxide;
WHO	World Health Organization

Clove (*Syzygium aromaticum*). https://doi.org/10.1016/B978-0-323-85177-0.00029-X

1. Introduction

According to several international associations such as the United States Food and Drug Administration (FDA), Codex Alimentarium, Food Chemical Codex (FCC), The Flavor and Extract Manufacturers Association of the United States (FEMA), The International Organization of Flavor Industries (IOFI), and the Council of Europe (CoE), three classes of toxic constituents were constituted: The first class categorizes as low toxicity compounds, at the fifth percentile no observed adverse effect level (NOAEL) of 3.0 mg/kg per day; the second class includes components less safe than the first class, which do not raise major toxicity intuitions, at the fifth percentile NOAEL of 0.91 mg/kg per day; and finally the third class comprises components with significant toxicity, which decreases the safety of essential oils, at the fifth percentile NOAEL of 0.15 mg/kg per day. Unidentified components are also classified under the third class (Dima and Dima, 2015).

The world's largest producer of cloves, which accounts for almost 75% of the global production, is Indonesia. Other leading clove-producing countries are Madagascar, United Republic of Tanzania, and Comoros (Food and Agriculture Organization of the United Nations, 2019). The main products of clove plant are whole clove buds (as spice), ground clove (as spice), essential (volatile) oil, and oleoresins (Nurdjannah and Bermawie, 2012). Moreover, the majority of cloves harvested (80%–90%) in Indonesia are used to produce kretek (clove) cigarettes (Suprihanti et al., 2018). This valuable spice has been used to treat indigestion, nausea, cough, cold, vomiting, infections, diarrhea, diabetes, memory loss, toothache, memory loss, arthritis, and erectile dysfunction in folk medicine for ages (Singletary, 2014). However, several issues have been raised regarding the toxic potential of clove and clove essential oil originating from some bioactive compounds.

From toxicological perspective, the harmful effects of any compound may occur in different ways. As shown in Fig. 1, these include cytotoxicity, cardiotoxicity, hepatotoxicity, neurotoxicity, immunotoxicity, mutagenicity, carcinogenicity, teratotoxicity, and reproductive toxicity. The type and formation mechanism of toxicity caused by a component is determined by in vivo and/or in vitro experiments. This chapter deals with the toxicity of clove and its dominant constituent, eugenol. Toxicity of clove and eugenol were discussed separately.

2. Bioactive compounds of clove and their toxic effects

The aromatic flower buds of *Syzygium aromaticum* are called "cloves" and this part is usually consumed as spice (Kuete, 2017). Although the main component of clove is eugenol, numerous phytochemicals including chavibetol, β-caryophyllene, caryophyllene oxide, α-copaene, quercetin-3-*O*-β-D-glucuronide, isorhamnetin-3-*O*-β-D-glucuronide, kaempferol-3-*O*-β-D-glucuronide-6″-methyl ester, quercetin-3-*O*-β-D-glucuronide-6″-methyl ester, isorhamnetin-3-*O*-β-D-glucuronide-6″-methyl ester, kaempferol-3-*O*-β-D-glucoside, quercetin-3-*O*-β-D-glucoside, isorhamnetin-3-*O*-β-D-glucoside, rhamnazin-3-*O*-β-D-glucoside, quercetin-7-*O*-β-D-glucoside, luteolin, quercetin, rhamnocitrin, kumatakenin, pachypodol, α-humulene (α-caryophyllene), lucenin-2, acetyl tributyl citrate (Citroflex A), 11,13-dihydroxy-tetradec-5-ynoic acid, methyl ester, and α-cubebene have been isolated and identified in clove so far (Bagavan et al., 2011; Boraie et al., 2014; Ryu et al., 2016; Suleiman et al., 2019; Šernaitė et al., 2020). Most of these biologically active substances provide a wide range of health benefits. However, at the same time, they may pose toxicological risks such as acute toxicity, sub-chronic

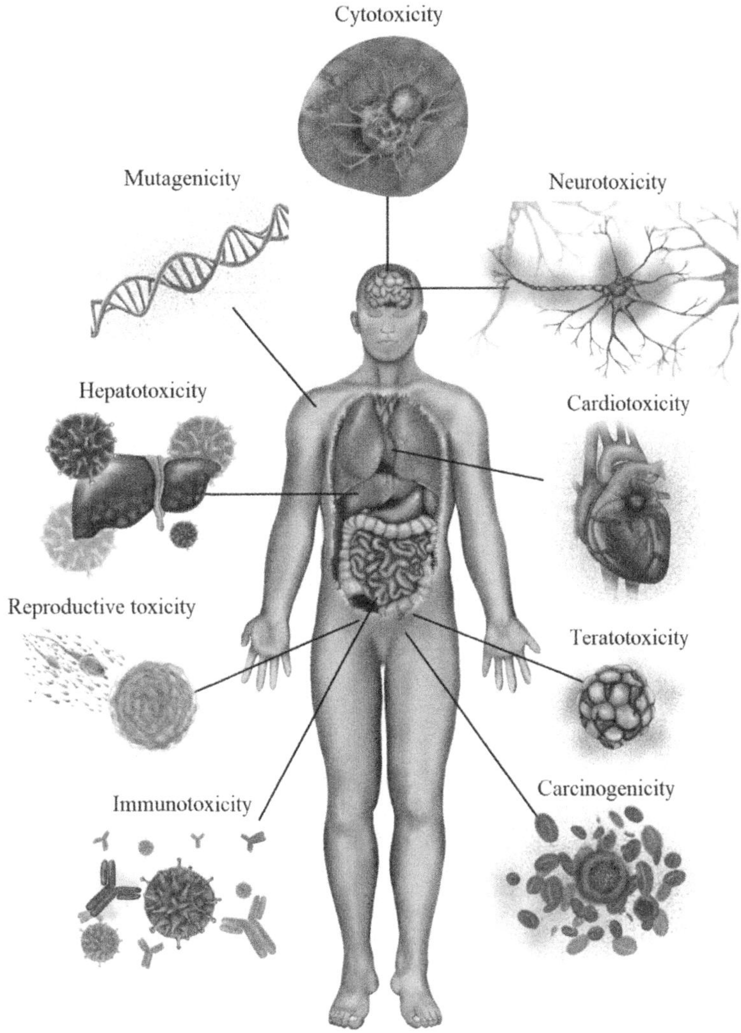

FIG. 1

Toxicity types occurred as a result of exposure to toxic substances.

toxicity, genotoxicity, cytotoxicity, or hepatotoxicity. Several studies have been carried out to understand the underlying mechanisms of toxic effects resulting from these phytochemicals. Table 1 summarizes some properties of four principal bioactives—except eugenol—(chavibetol, β-caryophyllene, caryophyllene oxide, and α-humulene) of cloves.

As can be understood from Table 1, clove includes some compounds that have toxic effects. Its toxicity toward different organisms such as fish, pests, or insects has been assessed until now. Fig. 2 illustrates the fish, snails, insects, and pests that clove has shown to be toxic.

Table 1 Molecular formula, biological functions, and toxicity of chavibetol, β-caryophyllene, caryophyllene oxide, and α-humulene.

Compound	Molecular formula	Biological functions	Toxicity	References
Chavibetol	$C_{10}H_{12}O_2$	Antioxidant, antibacterial, anti-hyperthyroid, and hepatoprotective effects	–	Rathee et al. (2006), Kurniawan et al. (2007), and Panda et al. (2019)
β-Caryophyllene	$C_{15}H_{24}$	Antimicrobial, antioxidant, anti-inflammatory, immunomodulatory, neuroprotective, cardioprotective, hepatoprotective, antidiabetic, anticancer, analgesic, anti-biofilm (against *Streptococcus mutans* biofilm), antiangiogenic, anxiolytic, antispasmodic, anticonvulsant, antidepressant, anti-alcohol (treatment of alcohol addiction disorder), antimutagenic, in vivo antigenotoxic, chemo-sensitizing, and anti-allodynic effects	– Contact ($LD_{50}=0.65\times10^{-1}$ μg/individual) and deterrent toxicity ($AFC_{50}=0.87\times10^{3}$ mg/L in 48 h, 1.63×10^{3} mg/L in 72 h) against *Aphis gossypii* – Contact ($LD_{50}=0.91$ μg/cm^2 for *Dermatophagoides farinae* and $LD_{50}=1.11$ μg/cm^2 for *Dermatophagoides pteronyssinus*) and fumigant toxicity ($LD_{50}=3.13$ μg/cm^2 for *Dermatophagoides farinae* and $LD_{50}=3.58$ μg/cm^2 for *Dermatophagoides pteronyssinus*) against house dust mites – Contact ($LD_{50}=138.51$ μg/adult and $LD_{90}=241.27$ μg/adult in 24 h; $LD_{50}=134.72$ μg/adult and $LD_{90}=238.69$ μg/adult in 48 h) and fumigant toxicity ($LC_{50}=128.15$ mg/L and $LC_{90}=205.67$ mg/L in the absence of wheat grains; $LC_{50}=172.60$ mg/L and $LC_{90}=244.78$ mg/L in the presence of wheat grains) against *Sitophilus granarius*	Opdyke (1979), Kubo et al. (1996), da Silva et al. (2008), Molina-Jasso et al. (2009), Liu et al. (2010), Abbas et al. (2013), Calleja et al. (2013), Chang et al. (2013), Al Mansouri et al. (2014), Álvarez-González et al. (2014), Assis et al. (2014), Basha and Sankaranarayanan (2014), Oh et al. (2014), Cho et al. (2015), Kim et al. (2015), Di Giacomo et al. (2016, 2017, 2018), Fidyt et al. (2016), Schmitt et al. (2016), Baldissera et al. (2017), Coté et al. (2017), Fontes et al. (2017), Segat et al. (2017), Cao et al. (2018a), da Machado et al. (2018), da Oliveira et al. (2018), Wang et al. (2018), Yoo and Jwa (2018), Chung et al. (2019), Yoo and Jwa (2019), Zhang et al. (2019), Bastaki et al. (2020), Pang et al. (2020), and Paventi et al. (2020)

			– Contact toxicity against *Tribolium castaneum* (LD_{50} = 36 μg/adult, LD_{90} = 66.10 μg/adult), *Liposcelis serricone* (LD_{50} = 51 μg/adult, LD_{90} = 111.60 μg/adult), and *Liposcelis bostrychophila* (LD_{50} = 50.80 μg/cm^2, LD_{90} = 121.90 μg/cm^2) adults – Contact toxicity against *Tribolium castaneum* (LD_{50} = 41.72 μg/adult) and *Liposcelis bostrychophila* (LD_{50} = 74.11 μg/cm^2) adults – Fumigant ($LC_{50} \geq 50$ mg/L air for *Tribolium castaneum, Lasioderma serricorne*, and $LC_{50} \geq 5$ mg/L air for *Liposcelis bostrychophila*) and contact toxicity (LD_{50} = 25.86 μg/adult for *Tribolium castaneum*, LD_{50} = 43.79 μg/adult for *Lasioderma serricorne*, LD_{50} = 52.52 μg/cm^2 for *Liposcelis bostrychophila*) – Fumigant toxicity against *Thrips palmi* adults ($LD_{50} \geq 0.23$ mg/cm^3 in 24 h) – Cytotoxicity against human cancer cell lines Caco-2 (IC_{50} = 1103.34 μM), CCRF/CEM (IC_{50} = 311.59 μM), and CEM/ADR5000 (IC_{50} = 368.48 μM) – No genotoxic effect in male mice up to 2000 mg/kg body weight	

Continued

Table 1 Molecular formula, biological functions, and toxicity of chavibetol, β-caryophyllene, caryophyllene oxide, and α-humulene—cont'd

Compound	Molecular formula	Biological functions	Toxicity	References
			– In vitro cytotoxicity against HeLa cells (IC_{50} = 3.86 μg/mL), BT-20 breast carcinoma cells (IC_{50} = 3.92 μg/mL), mouse B16 melanoma cells (IC_{50} = 3.88 μg/mL), and human HTB carcinoma cells (4.86 μg/mL) – In vitro cytotoxicity against human cervical carcinoma HeLa (IC_{50} = 21.80 μg/mL), human lung carcinoma A-549 (IC_{50} = 29.30 μg/mL), human colon adenocarcinoma HT-29 (IC_{50} = 40.90 μg/mL) cell lines, while lower in vitro cytotoxicity against monkey kidney Vero cells (IC_{50} = 88.30 μg/mL) and mice macrophages (IC_{50} = 107.30 μg/mL) – Cytotoxicity against human skin fibroblasts WS1 (IC_{50} = 59 μg/mL in 48 h), human lung carcinoma A-549 (IC_{50} = 55 μg/mL in 48 h), and human colon adenocarcinoma DLD-1 (IC_{50} = 97 μg/mL in 48 h) cell lines – Cytotoxicity against human lung adenocarcinoma A549 (IC_{50} = 47.05 μg/mL) and	

			NCI-H358 (IC_{50} = 54.78 μg/mL) cells – No oral sub-chronic toxicity (90 days) in male and female Wistar rats (700 mg/kg body weight/day) – No acute (14 days) and repeated-dose (28 days) toxicity in female Swiss mice up to 2000 mg/kg body weight – NOAEL of 222 mg/kg body weight/day for males and 263 mg/kg body weight/day for females – Oral acute $LD_{50} \geq 5$ g/kg for rats and acute dermal $LD_{50} \geq 5$ g/kg for rabbits	
Caryophyllene oxide	$C_{15}H_{24}O$	Antimicrobial, anticholinesterase, antioxidant, anticancer, anti-inflammatory, analgesic, antimutagenic, in vitro antifungal, anti-alcohol (treatment of alcohol addiction disorder), leishmanicidal, chemo-sensitizing, and trypanocidal effects	– Contact (LD_{50} = 0.44 μg/cm^2 for *Dermatophagoides farinae* and LD_{50} = 0.47 μg/cm^2 for *Dermatophagoides pteronyssinus*) and fumigant toxicity (LD_{50} = 1.36 μg/cm^2 for *Dermatophagoides farinae* and LD_{50} = 1.38 μg/cm^2 for *Dermatophagoides pteronyssinus*) against house dust mites – Contact toxicity against *Liposcelis bostrychophila* (LD_{50} = 27.20 μg/adult; μg/cm^2) – Contact toxicity against *Tribolium castaneum* (LD_{50} ≥50 μg/adult), *Liposcelis serricorne* (LD_{50} = 45.70 μg/adult, LD_{90} = 116.60 μg/	Yang et al. (2000), Sibanda et al. (2004), Setzer et al. (2007), Monzote et al. (2009), Kim et al. (2010, 2015), Chavan et al. (2010), Jun et al. (2011), Ogendo et al. (2011), Park et al. (2011), Di Sotto et al. (2013), Oh et al. (2014), Sain et al. (2014), Ambrož et al. (2015), Di Giacomo et al. (2016, 2017, 2018), Fidyt et al. (2016), Hanušová et al. (2017), Coté et al. (2017), Cao et al. (2018a,b), Martínez et al. (2018), Moreno et al. (2018), Plata-Rueda et al. (2018), Oppong-Damoah et al. (2019), Karakaya et al. (2020), Nararak et al. (2020), Pang et al. (2020), and Polanco-Hernández et al. (2021)

Continued

Table 1 Molecular formula, biological functions, and toxicity of chavibetol, β-caryophyllene, caryophyllene oxide, and α-humulene—cont'd

Compound	Molecular formula	Biological functions	Toxicity	References
			adult), and *Liposcelis bostrychophila* (LD_{50} = 46.40 μg/cm^2, LD_{90} = 96.90 μg/cm^2) adults – Contact toxicity against *Tenebrio molitor* (LC_{50} = 9.22 μg/insect and LD_{90} = 22.99 μg/insect for larva; LC_{50} = 14.64 μg/insect and LD_{90} = 30.96 μg/insect for pupa; LC_{50} = 25.46 μg/insect and LD_{90} = 54.59 μg/insect for adult in 48 h) – Fumigant ($LC_{50} \geq 50$ mg/L air for *Tribolium castaneum*, *Lasioderma serricorne*, and $LC_{50} \geq 5$ mg/L air for *Liposcelis bostrychophila*) and contact toxicity ($LD_{50} \geq 50$ μg/adult for *Tribolium castaneum*, LD_{50} = 37.56 μg/adult for *Lasioderma serricorne*, LD_{50} = 35.40 μg/cm^2 for *Liposcelis bostrychophila*) – Contact and fumigant toxicity against *Tribolium castaneum adults* (LD_{50} = 0.00018 mg/cm^3 in 24 h) – Contact toxicity against *Sitophilus granaries* (LC_{50} = 2.78 μL/mL, LC_{90} = 6.95 μL/mL in 24 h)	

			– Fumigant toxicity against *Thrips palmi* adults ($LD_{50} \geq 0.23\,mg/cm^3$ in 24 h) – Fumigant toxicity against *Sitophilus oryzae* (L.) ($LC_{50} = 65.84\,\mu L/L$ air in 24 h, $LC_{50} = 47.10\,\mu L/L$ air in 72 h, $LC_{50} = 31.50\,\mu L/L$ air in 120 h, $LC_{50} = 28.40\,\mu L/L$ air in 168 h), *Rhyzopertha dominica* (F.) ($LC_{50} = 32.48\,\mu L/L$ air in 24 h, $LC_{50} = 0.52\,\mu L/L$ air in 72 h, $LC_{50} = 0.20\,\mu L/L$ air in 120 h, $LC_{50} = 0.07\,\mu L/L$ air in 168 h), *Tribolium castaneum* ($LC_{50} = 58.22\,\mu L/L$ air in 24 h, $LC_{50} = 40.06\,\mu L/L$ air in 72 h, $LC_{50} = 28.30\,\mu L/L$ air in 120 h, $LC_{50} = 28.30\,\mu L/L$ air in 168 h), *Callosobruchus chinensis* (L.) ($LC_{50} = 29.56\,\mu L/L$ air in 24 h, $LC_{50} = 5.25\,\mu L/L$ air in 72 h), and *Oryzaephilus surinamensis* (L.) ($LC_{50} = 78\,\mu L/L$ air in 168 h) – No cytotoxicity against human lymphocyte cultures up to the concentration of 100 μM, while cytotoxic at concentrations higher than 0.9 mg/plate for *Salmonella typhimurium* TA98, *Salmonella typhimurium* TA100, and *Escherichia coli* WP2uvrA; no genotoxicity at chromosomal level	

Continued

Table 1 Molecular formula, biological functions, and toxicity of chavibetol, β-caryophyllene, caryophyllene oxide, and α-humulene—cont'd

Compound	Molecular formula	Biological functions	Toxicity	References
			– Cytotoxicity against human cancer cell lines Caco-2 (IC_{50} = 332.30 μM), CCRF/CEM (IC_{50} = 235.18 μM), and CEM/ADR5000 (IC_{50} = 297.98 μM) – Cytotoxicity against human breast cancer cell lines MDA-MB-231 (IC_{50} = 69 μg/mL in 72 h) and MCF7 (IC_{50} = 24 μg/mL in 72 h) – Cytotoxicity against HeLa human cervical adenocarcinoma cells (IC_{50} = 13.55 μM), HepG2 human leukemia cancer cells (IC_{50} = 3.95 μM), AGS human lung cancer cells (IC_{50} = 12.60 μM), SNU-1 human gastric cancer cell (IC_{50} = 16.79 μM), and SNU-16 human stomach cancer cells (IC_{50} = 27.39 μM) – Cytotoxicity against PC-3 human prostatic adenocarcinoma (LC_{50} = 150 μM), MDA-MB-231 human mammary adenocarcinoma (LC_{50} = 254 μM), Hs 578 T human mammary ductal carcinoma (LC_{50} = 351 μM), MCF7 human mammary adenocarcinoma	

			(LC_{50} = 178 μM), SK-MEL-28 human melanoma (LC_{50} = 324 μM), and 5637 human primary bladder carcinoma (LC_{50} = 147 μM) cell lines – Cytotoxicity against Caco-2 cancer cells (IC_{50} = 57.70 μg/mL in 72 h) – Cytotoxicity against Vero cells (IC_{50} = 113.10 μg/mL in 48 h) – Cytotoxicity against human skin fibroblasts WS1 (IC_{50} = 50 μg/mL in 48 h), human lung carcinoma A-549 (IC_{50} = 36 μg/mL in 48 h), and human colon adenocarcinoma DLD-1 (IC_{50} = 43 μg/mL in 48 h) cell lines – Cytotoxicity against peritoneal macrophages from BALB/c mice (IC_{50} = 2.10 μg/mL) – In vitro cytotoxicity against mouse normal fibroblast (BALB/c 3T3) cell lines (IC_{50} = 13.23–34.79 μg/mL) – No cytotoxicity up to 5 μg/mL and in vitro genotoxicity at chromosomal level against Chinese hamster ovary (CHO)-K1 cell line – No hepatotoxicity against primary culture of rat hepatocytes up to 50 μg/mL in 24 h	

Continued

Table 1 Molecular formula, biological functions, and toxicity of chavibetol, β-caryophyllene, caryophyllene oxide, and α-humulene—cont'd

Compound	Molecular formula	Biological functions	Toxicity	References
α-Humulene	$C_{15}H_{24}$	Anti-inflammatory, antiproliferative, anticancer, antibacterial, anti-biofilm (against *Bacteroides fragilis* biofilm), larvicidal, and insecticidal effects	– Contact toxicity against *Sitophilus granaries* (LC_{50} = 4.61 μL/mL, LC_{90} = 9.35 μL/mL in 24 h) – Contact toxicity against *Tribolium castaneum* (LD_{50} = 14.23 μg/adult) and *Liposcelis bostrychophila* (LD_{50} = 118.56 μg/cm^2) adults – Contact toxicity against *Sitophilus zeamais* (LD_{50} = 20 μg/mg, LD_{90} = 29 μg/mg in 7 days) and *Tribolium castaneum* (LD_{50} = 31 μg/mg, LD_{90} = 70 μg/mg in 7 days) – Contact toxicity against *Tenebrio molitor* (LC_{50} = 15.27 μg/insect and LD_{90} = 30.95 μg/insect for larva; LC_{50} = 21.41 μg/insect and LD_{90} = 40.45 μg/insect for pupa; LC_{50} = 31.85 μg/insect and LD_{90} = 56.76 μg/insect for adult in 48 h) – Contact (LD_{50} = 41.87 μg/adult and LD_{90} = 73.51 μg/adult in 24 h; LD_{50} = 26.83 μg/adult and LD_{90} = 49.49 μg/adult in 48 h) and fumigant toxicity (LC_{50} = 127.23 mg/L and LC_{90} = 188.49 mg/L in the	Fernandes et al. (2007), da Silva et al. (2008), Rogerio et al. (2009), Legault and Pichette (2010), El Hadri et al. (2010), Ogendo et al. (2011), Suthisut et al. (2011), Vinholes et al. (2014), Ambrož et al. (2015), Govindarajan and Benelli (2016), Coté et al. (2017), Benelli et al. (2018), Martínez et al. (2018), Plata-Rueda et al. (2018), Zhang et al. (2019), Chen et al. (2019), de Oliveira et al. (2019), Paventi et al. (2020), and Jang et al. (2020)

			absence of wheat grains; $LC_{50}=164.95$ mg/L and $LC_{90}=234.39$ mg/L in the presence of wheat grains) against *Sitophilus granariu* – Fumigant toxicity against *Sitophilus oryzae* (L.) ($LC_{50}=135.80$ μL/L air in 24 h, $LC_{50}=6.91$ μL/L air in 72 h, $LC_{50}=5.42$ μL/L air in 120 h, $LC_{50}=3.85$ μL/L air in 168 h), *Rhyzopertha dominica* (F.) ($LC_{50}=106.30$ μL/L air in 24 h, $LC_{50}=27.50$ μL/L air in 72 h, $LC_{50}=22.20$ μL/L air in 120 h, $LC_{50}=9.80$ μL/L air in 168 h), *Callosobruchus chinensis* (L.) ($LC_{50}=0.48$ μL/L air in 24 h, $LC_{50}=0.01$ μL/L air in 72 h), and *Oryzaephilus surinamensis* (L.) ($LC_{50}=25.70$ μL/L air in 24 h, $LC_{50}=7.22$ μL/L air in 72 h, $LC_{50}=4.82$ μL/L air in 120 h, $LC_{50}=4.20$ μL/L air in 168 h) – Fumigant toxicity against *Dorymyrmex thoracius* ($LC_{50}=3.75$ μL/L, $LC_{95}=8.68$ μL/L in 48 h) – Acute toxicity against early third-instar larvae of *Anopheles subpictus* ($LC_{50}=6.19$ μg/mL, $LC_{90}=12.03$ μg/mL in 24 h), *Aedes albopictus* ($LC_{50}=6.86$ μg/mL, $LC_{90}=12.98$ μg/mL in 24 h),	

Continued

Table 1 Molecular formula, biological functions, and toxicity of chavibetol, β-caryophyllene, caryophyllene oxide, and α-humulene—cont'd

Compound	Molecular formula	Biological functions	Toxicity	References
			and *Culex tritaeniorhynchus* ($LC_{50}=7.39$ μg/mL, $LC_{90}=13.68$ μg/mL in 24 h), while low acute toxicity against *Gambusia affinis* ($LC_{50}=1024.95$ μg/mL, $LC_{90}=2016.95$ μg/mL in 48 h) – Ingestion (acute) toxicity against third-instar larvae of *Helicoverpa armigera* ($LC_{50}=20.86$ μg/mL, $LC_{90}=41.72$ μg/mL in 24 h) – Cytotoxicity against Caco-2 cancer cells ($IC_{50}=24.40$ μg/mL in 72 h) – Cytotoxicity against human skin fibroblasts WS1 ($IC_{50}=24$ μg/mL in 48 h), human lung carcinoma A-549 ($IC_{50}=28$ μg/mL in 48 h), and human colon adenocarcinoma DLD-1 ($IC_{50}=43$ μg/mL in 48 h) cell lines – In vitro cytotoxicity against adenocarcinoma human cell line HCT-116 ($IC_{50}=64.20$ μg/mL), human breast cancer cell MCF-7 ($IC_{50}=81.90$ μg/mL), and murine macrophage cell line RAW264.7 ($IC_{50}=38.40$ μg/mL)	

			– In vitro cytotoxicity against human hepatocellular carcinoma Huh7 (IC_{50} = 15.09 μg/mL in 12 h), SMMC-7721 (IC_{50} = 17.31 μg/mL in 12 h), HepG2 (IC_{50} = 11.22 μg/mL in 12 h), and Hep3B (IC_{50} = 13.78 μg/mL in 12 h) cells, while lower in vitro cytotoxicity against normal hepatocytes L0–2 (IC_{50} = 115.69 μg/mL in 12 h) – In vitro cytotoxicity against human cervical carcinoma HeLa (IC_{50} = 35.20 μg/mL), human lung carcinoma A-549 (IC_{50} = 18.90 μg/mL), and human colon adenocarcinoma HT-29 (IC_{50} = 24.80 μg/mL) cell lines, while lower in vitro cytotoxicity against monkey kidney Vero cells (IC_{50} = 129.90 μg/mL) and mice macrophages (IC_{50} = 159 μg/mL) – No hepatotoxicity against primary culture of rat hepatocytes up to 50 μg/mL in 24 h	

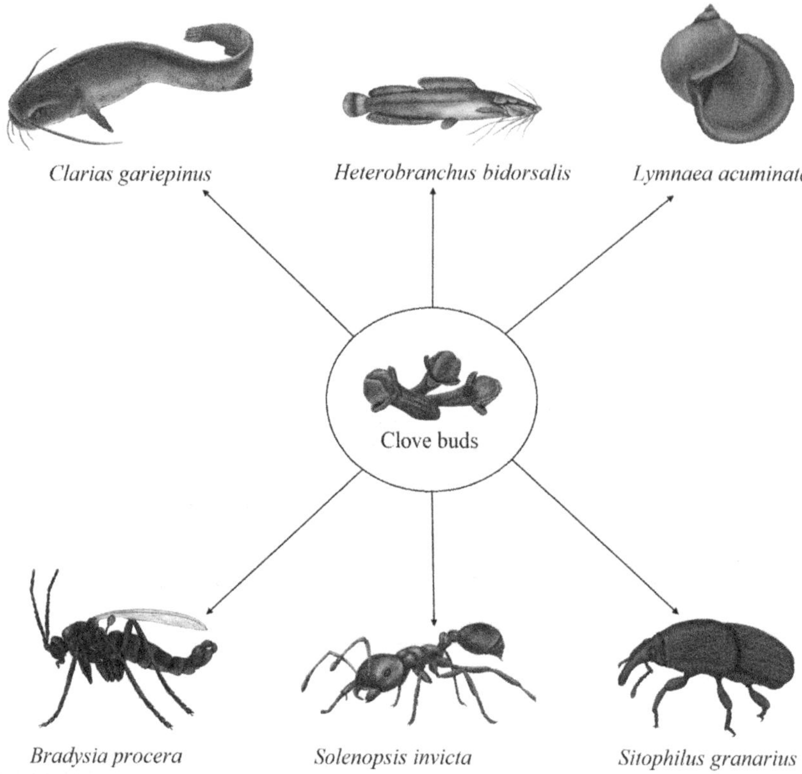

FIG. 2

Some organisms to which clove exhibits toxicity.

2.1 Toxicity of clove against fish

Toxicological impacts of clove buds' powder against *Clarias gariepinus* and *Heterobranchus bidorsalis* fingerlings (African catfish species) were investigated in Nigeria. In this study, 10 fish from every two species were placed in aquariums containing various concentrations of clove powder solution (50, 60, 70, 80, 100, and 120 mg/L). Then, the researchers recorded behavioral changes and cumulative mortality rates at 12th, 24th, 48th, 72nd, and 96th hours. Loss of balance, air gulping, erratic swimming, sudden fast movement, mucus secretion, hypersensitivity, and rolling movement were considered as stressful behaviors probably resulting from deteriorative effects of some components like tannins, alkaloids, or saponins on respiratory and nervous systems. LC_{50} values for acute toxicity of clove powder against *Clarias gariepinus* were reported to be 96.56, 78.29, 75.36, 58.57, and 54.33 mg/L in 12, 24, 48, 72, and 96 h, respectively. Accordingly, safe concentrations of clove powder for *Clarias gariepinus* decreased from 9.65 to 5.23 mg/L depending on the increase of exposure time (from 12 to 96 h), while its toxicity factor increased from 1 to 1.84. On the other side, lower LC_{50} values (88.27 mg/L in 12 h, 75.18 mg/L in 24 h, 68.46 mg/L in 48 h, 53.16 mg/L in 72 h, and 52.97 mg/L in 96 h) were reported for *Heterobranchus bidorsalis*. Similarly, lower concentrations of clove powder were regarded as safe

(8.83 mg/L for 12 h, 7.52 mg/L for 24 h, 6.85 mg/L for 48 h, 6.32 mg/L for 72 h, and 5.50 mg/L for 96 h). Mean lethal time (MLT_{50}) that was expected to cause death in 50% of *Clarias gariepinus* population was 101.55 h for clove powder solution at a concentration of 50 mg/L, whereas 8.38 h were required if the solution concentration increased to 120 mg/L. Longer exposure times were estimated to be lethal for 95% (MLT_{95}) of *Clarias gariepinus* population (122.51 h for 50 mg/L, 79.66 h for 60 mg/L, 60.66 h for 70 mg/L, 26.21 h for 80 mg/L, 15.89 h for 100 mg/L, and 12.17 h for 120 mg/L). With respect to *Heterobranchus bidorsalis*, higher MLT_{50} (ranging from 116.90 to 6.66 h) and MLT_{95} (ranging from 167.36 to 17.67 h) values were determined. The authors concluded that the susceptibility of *Clarias gariepinus* against clove powder was higher than that of *Heterobranchus bidorsalis* and clove powder solutions having concentrations below 60 mg/L can be used for anesthetic purposes in fish without mortality (Okey et al., 2018a). Also, the same research group reported that aqueous solutions of clove flower bud powder with concentrations ranging from 80 mg/L to 140 mg/L cause deep anesthesia between 3 and 10 min for *Clarias gariepinus* and *Heterobranchus bidorsalis* fingerlings, and so it may be used as an alternative anesthetic for these catfish species (Okey et al., 2018b).

2.2 Toxicity of clove against pests

Pesticides, which are chemicals with component-specific inherent toxicity, are commonly utilized to improve productivity in agricultural practices all over the world (Shammi et al., 2020). On the other hand, pesticides and their residues may pose serious health risks for both farmers and consumers (Damalas and Eleftherohorinos, 2011). Therefore, significant attention has been focused on seeking natural alternatives (i.e., essential oils, extracts, phenolics), namely biopesticides, to well-known and widely used pesticides (Damalas and Koutroubas, 2018). Attempts have been made to reveal and understand the toxic effects of numerous spices against different pests. For instance, Tripathi et al. (2009) evaluated the contact and fumigant toxicity potential of powders of large cardamom fruit, turmeric rhizome, camphor leaf, Indian cassia leaf, green cardamom fruit, clove bud, Ajowan fruit, and ginger rhizome against *Tribolium castaneum* (red flour beetle) and *Callosobruchus maculatus* F. (cowpea weevil, cowpea seed beetle). The contact toxicity of clove powder against *Callosobruchus maculatus* F. was lower (LD_{50} = 23.02 g/100 g food substrate) than large cardamom (LD_{50} = 17.80 g/100 g food substrate), turmeric (LD_{50} = 19.87 g/100 g food substrate), and ginger (LD_{50} = 20.05 g/100 g food substrate), whereas its fumigant toxicity was reported to be the highest (LC_{50} = 9.81 g/L). Regarding *Tribolium castaneum*, only ginger exhibited contact toxicity (LD_{50} = 20.94 g/100 g food substrate), while only large cardamom, green cardamom, and camphor powders were found to be fumigant toxic among the spice powders tested. Clove addition at a ratio of 1.5 g/50 g yielded a high repellency against *Callosobruchus maculatus* (92.50%) and complete repellency against *Tribolium castaneum* (100%). Moreover, the maximum dose of clove (5 g/100 g food substrate) prevented progeny production in both pests. This research proved the contact and fumigant toxicity of clove, large cardamom, and green cardamom powders against *Callosobruchus maculatus* and *Tribolium castaneum* (Tripathi et al., 2009). The superiority of clove powder over powders of black currant seeds, Africa breadfruit seeds, and Tazargade plant with regard to toxicity and anti-oviposition activity against *Callosobruchus maculatus* was also demonstrated by Oparaeke and Daria (2005). In addition to these pests, clove exhibited the highest insecticidal activity against *Callosobruchus chinensis* (pulse beetle) compared to red pepper, greater cardamom, and cinnamon (Aslam et al., 2002). In another study, insecticidal properties of powders of spices' various parts (Ajowan seed, bay leaves, cappers fruit, cinnamon bark and leaves, clove flower bud, coriander seed and leaves, cumin fruit, caraway fruit, fennel seed, fenugreek fruit, ginger rhizome, mace flower, nutmeg seed, pepper fruit, red chilly fruit, dried fruit of star anise, turmeric rhizome) at different dosages (1% and 5% (w/w) by wheat weight) against *Sitophilus oryzae* (rice weevil) were investigated. At 1% concentration, powder of clove flower bud was ranked the third most effective spice in terms of mortality rate (59.40%), after mace flower (100%), pepper fruit (100%), and nutmeg seed (74.70%). When the dosage of clove flower bud powder increased to 5%, it provided a mortality rate of 100% in 14 days. Hence, the authors categorized it as a highly effective spice (mortality rates between 85% and 100%) along with mace, pepper, nutmeg, cinnamon, and star anise and suggested their use in postharvest protection of grains (Devi and Devi, 2013). A similar observation about the contact toxicity and repellent activity of powder of clove flower bud against *Sitophilus oryzae* has been reported recently (Ismail and Sleem, 2021). Based on the toxic potential of clove, weight loss in sorghum grains caused by *Sitophilus oryzae* reduced considerably (Ileke et al., 2014). Efficiency of clove flower bud powder at different concentrations (1%, 3%, 5%, 7%, and 9% (w/w) by soil weight) against larvae of *Oryctes agamemnon* (rhinoceros beetle) was assessed to find out its biopesticide potential. As the proportion of the powder increased to 9% (w/w), larval mortality reached up to 68.50%. On the other side, the stability of the powder in the soil decreased over time. Both toxic effect against larvae and stability of the powder in soil were dose-dependent. For instance, at 1% concentration, the mortality rate of larvae was 2% on the first day, while it reduced to 0% on the third day. However, the mortality rate was reported to be 71.30% on the first day of application and decreased to 3.50% after 7 days and no toxicity was detected on the eighth day. The authors recommended further studies in order to determine the side effects of active constituents of clove on soil-active arthropods or other natural enemies and to define its optimum application dosage (Ibrahim and Alahmadi, 2015). Another study showed that utilization of clove powder, compared to hot pepper fruit and lavender leaves powders, provided the highest mortality rate, the lowest female fecundity, fertility, and viability of eggs of *Callosobruchus maculatus* in Morocco (Allali et al., 2018). However, Rasha et al. (2012) suggested the use of the combination of clove powder and katel-sous dust to achieve higher mortality rates in *Callosobruchus maculatus*, while mixtures of powders of *Vepris heterophylla* (stem bark or leaves) and clove seed exhibited contact toxicity and oviposition deterrence against *Callosobruchus maculatus* adults throughout storage of cowpea seeds (Noël et al., 2020). In addition to these studies, Khalaf Al-Joboory (2019) used clove powder with a concentration ranging from 2% to 6% (w/w, by grain weight) against flour beetle *Tribolium castaneum*. After a 48-h treatment, expulsion rates for adults were reported to be 4%, 5.83%, and 8.17% for clove powder concentrations of 2%, 4%, and 6%, respectively. Furthermore, addition of clove powder on ground wheat grains caused increased mortality rates (2.85%–4.35%) for adults of *Tribolium castaneum*. The authors attributed this result to the presence of toxic compounds in the spices (clove, black pepper, cinnamon, coriander) that may destroy digestion system or nervous system of insects in different ways (Khalaf Al-Joboory, 2019). Additionally, the repellency, and toxic effects of clove powder against *Solenopsis invicta* Buren (red imported fire ant) workers was explained by lipophilic compounds—mainly eugenol, eugenol acetate, and β-caryophyllene—which immigrate through cuticular lipids of the ants, and then gradually penetrate into hemocoel and finally nervous system (Kafle and Shih, 2013).

The potency of clove buds against snails has also been tested. Snails are regarded as significant pests of various vegetables and crops in terms of ecology and economy (Kumar, 2020). In this context, the toxicity of flower bud powder of clove against juveniles and adults of land snails (*Monacha cartusiana*) was examined in Egypt (Ismail and AbdEl-Kader, 2011). The findings of the study showed that the mortality rate depended on both concentration of powder and exposure time. For instance, at 10% concentration, the mortality rate was 20% for adults on the 1st day, and then reached 70% on the 21st day under laboratory conditions. Likewise, the powder with a concentration of 40% provided a mortality rate of 37.50% for adults on the 1st day and increased to 45%, 70%, 93.50%, and 100% on the 3rd, 7th, 14th, and 21st days, respectively. Nevertheless, these rates considerably reduced under field conditions (with a maximum mortality rate of 62.40% on the 21st day) (Ismail and AbdEl-Kader, 2011). Accordingly, the toxicity of clove flower-bud powder against *Lymnaea acuminata* (LC_{50} = 172.75 mg/L in 24 h, LC_{50} = 130.73 mg/L in 48 h, LC_{50} = 73.08 mg/L in 72 h, LC_{50} = 51.98 mg/L in 96 h) was reported by Kumar and Singh (2006) who explained the time-dependent effect of clove by (a) increasing amount of toxic compound in snail body depending on increasing exposure time, or (b) conversion of active compounds into more toxic forms in treatment medium or in snail body through activities of enzymes. Based on these observations, it can be generally concluded that clove has molluscicidal activity.

In order to understand the most effective compound of clove in terms of toxicity, Hong et al. (2018) carried out a research dealing with the toxicity assessment of each compound, which were previously identified in methanol extract and hydrodistillate of clove using convenient instrumental techniques (high-performance liquid chromatography (HPLC), gas chromatography-mass spectrometry (GC-MS)), against third-instar larvae and eggs of *Bradysia procera*. The ascending order of toxicity to third-instar larvae of *Bradysia procera* of the individual compounds tested was **methyl salicylate** (LC_{50} = 5.26 μg/cm^2, LC_{90} = 21.15 μg/cm^2) > **2-nonanone** (LC_{50} = 8.77 μg/cm^2, LC_{90} = 31.92 μg/cm^2) > **2-heptyl acetate** (LC_{50} = 10.200 μg/cm^2, LC_{90} = 38.97 μg/cm^2) > **2-heptanone** (LC_{50} = 10.40 μg/cm^2, LC_{90} = 35.97 μg/cm^2) > **natural eugenol** (LC_{50} = 14.09 μg/cm^2, LC_{90} = 32.57 μg/cm^2) > **pure eugenol** (LC_{50} = 13.58 μg/cm^2, LC_{90} = 37.02 μg/cm^2) > **eugenyl acetate** (LC_{50} = 15.40 μg/cm^2, LC_{90} = 33.04 μg/cm^2) > ***p*-allylphenol** (LC_{50} = 138.15 μg/cm^2, LC_{90} = 304.62 μg/cm^2) > **α-humulene** (LC_{50} = 163.36 μg/cm^2, LC_{90} = 547.87 μg/cm^2) > **caryophyllene oxide** (LC_{50} = 172.77 μg/cm^2, LC_{90} = 669.55 μg/cm^2) > **α-pinene** (LC_{50} = 178.62 μg/cm^2, LC_{90} = 669.91 μg/cm^2) > **α-copaene** (LC_{50} = 242.14 μg/cm^2, LC_{90} = 468.05 μg/cm^2). As the LC_{50} values determined for β-caryophyllene, *p*-cymene, limonene, and linalool were higher than 300 μg/cm^2, they were not considered to be toxic. On the other side, mortality caused by these compounds depended on the type of application. Methyl salicylate and eugenol provided the mortality rate (100%) in third-instar larvae of *Bradysia procera* when applied in closed containers. However, in open containers, mortality rates reduced to 2% and 0% for methyl salicylate and eugenol, respectively. While all compounds tested for larvicidal potential (methyl salicylate, 2-heptanone, 2-nonanone, eugenol, 2-heptyl acetate, and eugenyl acetate) led to considerable mortality rates (93%–100%) in closed containers, only eugenyl acetate (14%) and 2-heptyl acetate (9%) were reported to be promising agents in open-container treatment. Regarding potential of egg-hatching inhibition, methyl salicylate, 2-heptanone, 2-nonanone, eugenol, and eugenyl acetate were efficient compounds. Furthermore, α-pinene (IC_{50} = 26.90 mM) and α-copaene (IC_{50} = 27.70 mM) acted as strong in vitro acetylcholinesterase (AChE) inhibitors, whereas other compounds had also more or less inhibitory potential (IC_{50} = 49.30–168.70 mM). The authors concluded that each compound of clove has unique activity on eggs and larvae of *Bradysia procera*, and therefore clove buds-derived preparations can be considered as safe and accessible alternatives to common pesticides (Hong et al., 2018). Based on the findings of different research, the toxicity of clove constituents against *Solenopsis invicta* was sorted in descending order as eugenol < β-caryophyllene < eugenol acetate (Kafle and Shih, 2013). Another clove constituents-related study revealed that eugenol (LC_{50} = 2.77 μL/mL, LC_{90} = 5.73 μL/mL) had the highest contact toxicity against *Sitophilus granaries*, and followed by caryophyllene oxide ($LC_{50}=2.78\,\mu L/mL$, $LC_{90}=6.95\,\mu L/mL$), α-pinene ($LC_{50}=4.24\,\mu L/mL$, $LC_{90}=8.21\,\mu L/mL$), and α-humulene ($LC_{50}=4.61\,\mu L/mL$, $LC_{90}=9.35\,\mu L/mL$). Additionally, the authors declared that toxicity of these terpenoids may be due to their neurotoxic effects including alteration of behavior locomotor and muscle contractions and modulation of nerve-muscle interaction which result in rapid mortality (Plata-Rueda et al., 2018).

3. Eugenol

Eugenol is naturally found in numerous plant essential oils such as cloves (*Szygium aromaticum, Eugenia aromatica*, or *Eugenia caryophyllus*), cinnamon (*Cinnamomum* spp.), basil (*Ocimum* spp.), allspice (*Pimenta dioica*), bay laurel (*Laurus nobilis*), turmeric (*Curcuma longa*), and other plants (Moore et al., 2007). It is an allyl chain substituted guaiacol that is slightly soluble in water. Primarily, it was isolated from clove in 1826 and clove essential oil still remains the main commercial source of eugenol. Eugenol contents of clove bud oil, clove stem oil, and clove leaf oil range between 60% and 90%, 90% and 95%, and 82% and 88%, respectively (Baker and Grant, 2018). Mean eugenol contents of essential oils extracted from various parts of clove and the chemical structure of eugenol are shown in Fig. 3.

Eugenol ($C_{10}H_{12}O_2$; 4-allyl-2-methoxy phenol) is a phenylpropanoid, an allyl chain-substituted guaiacol as seen in Fig. 1. It is a volatile phenolic compound of clove essential oil obtained from *Eugenia caryophyllata* buds and leaves. This colorless to light yellowish-colored fluid is commonly found in essential oils of spices such as cinnamon, bay leaf, nutmeg, basil, and clove. Eugenol molecule is a weak acid and soluble in organic solvents. It also can be produced synthetically by the allylation of guaiacol with allylchlorid (Mohammadi Nejad et al., 2017). Eugenol has a therapeutic potential due to its ability to denature proteins and can react with cell membrane phospholipids by changing their

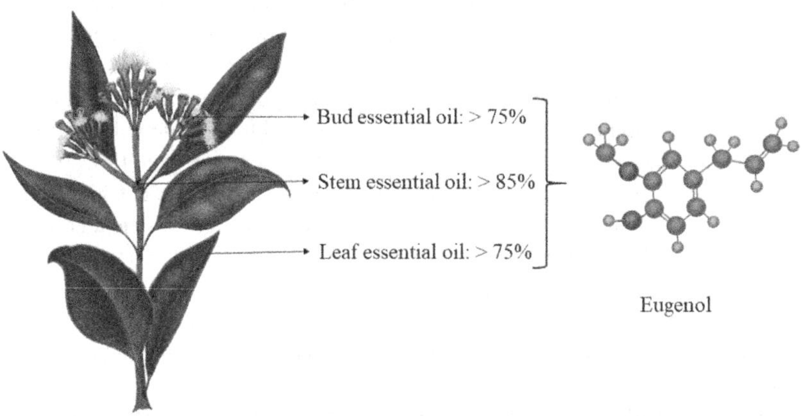

FIG. 3

Eugenol contents of essential oils extracted from different parts of clove and chemical structure of eugenol.

permeability. It has also great biological activities such as antifungal, antibacterial, antioxidant, and anti-inflammatory (Chatterjee and Bhattacharjee, 2013a).

Conventionally, eugenol is extracted from clove buds by steam distillation and solvent extraction which induce some problems such as thermal degradation, hydrolysis, water solubilization of constituents as well as environmental and health hazards (Guan et al., 2007). These issues restrict its applicability in food systems (Chatterjee and Bhattacharjee, 2013a). Eugenol is also extracted using synthetic solvents such as dichloromethane through hydrodistillation method. Alternatively, supercritical carbon dioxide extraction (SC-CO_2) may be applied to increase oil yield from clove buds, but this technique decreases eugenol content (Baker and Grant, 2018). Furthermore, microwave-assisted and ultrasound-assisted extractions are promising nonthermal techniques to obtain eugenol (Khalil et al., 2017). In this context, the effects of different extraction techniques (subcritical CO_2, SC-CO_2, liquid CO_2, steam distillation, Clevenger extraction, solvent extraction) on extraction yield and properties of eugenol extracted from *Ocimum sanctum* Linn. were investigated by Ghosh et al. (2013). In terms of eugenol concentration and phytochemical properties, SC-CO_2 yielded an extract with the highest eugenol and phenolic contents as well as the best reducing power, antimicrobial, anti-inflammatory, and antioxidant activities. Then, Chatterjee and Bhattacharjee (2013b) effectively applied both encapsulated and nonencapsulated forms of eugenol-rich clove extract to soybean oil as natural antioxidant alternative to butylated hydroxyl toluene (BHT).

4. Uses areas of eugenol

Eugenol is widely used in the fields of pharmaceuticals, cosmetics, dentistry, food, agriculture, and pesticides. Anti-insect activity of eugenol principally depends on the organic structure of its phenolic ring with specific hydroxyl functional group (Shahavi et al., 2019). Insecticide activity of eugenol against numerous pests including *Callosobruchus maculatus*, *Tribolium castaneum*, *Periplaneta americana* (L.), *Sitophilus granarius*, *Sitophilus zeamais*, *Prostephanus truncates*, *Cacopsylla chinensis*, *Musca domestica* L., *Blatta lateralis* (Walker), and *Ctenocephalides felis felis* (the most prevalent parasitic insect in cats and dogs) has been reported until now (Obeng-Ofori and Reichmuth, 1997; Ngoh et al., 1998; Huang et al., 2002; Liska et al., 2010; Tian et al., 2015; Gaire et al., 2017; da Silva et al., 2019a; González Armijos et al., 2019; Lambert et al., 2020). Additionally, it is meant to be used against several bacteria such as *Listeria monocytogenes*, *Staphylococcus aureus*, *Bacillus cereus*, *Escherichia coli*, *Chryseobacterium balustinum*, *Enterobacter* sp., *Pantoea agglomerans*-1, *Bacillus pumilus*, *Pseudomonas fluorescens*-1, *Pseudomonas fluorescens*-2, *Acinetobacter calcoaceticus*-1, *Acinetobacter calcoaceticus*-2, *Stenotrophomonas maltophilia*-1, *Stenotrophomonas maltophilia*-2, *Klebsiella pneumonia*, *Dickeya. zeae*, *Pectobacterium carotovorum* subsp. *Carotovorum* Pcc21, *Salmonella typhimurium*, *Vibrio parahaemolyticus*, *Leuconostoc carnosum*, *Enterococcus faecalis*, *Weissella viridescens*, *Lactobacillus plantarum*, *Leuconostoc brevis*, *Leuconostoc citreum*, *Mycobacterium tuberculosis*, *Staphylococcus epidermis*, *Streptococcus pyogenes*, *Streptococcus pneumoniae*, *Streptococcus mutans*, *Streptococcus sanguinis*, *Streptoccus sobrinus*, *Streptococcus criceti*, *Streptococcus anginosus*, *Streptococcus gordonii*, *Aggregatibacter actinomycetemcomitans*, *Fusobacterium nucleatum*, *Prevotella intermedia*, *Porphylomonas gingivalis*, *Streptococcus milleri*, *Streptococcus mitis*, *Prevotella buccae*, *Prevotella oris*, and *Prevotella intermedia* (Karapinar and Esen Aktuğ, 1987;

Didry et al., 1994; Leite et al., 2007; Tippayatum and Chonhenchob, 2007; Moon et al., 2011; Zheng et al., 2013; de Almeida et al., 2019; Takahashi et al., 2021).

Utilization of eugenol includes also household products such as fragrance in soaps and cosmetics, skin care products, flavoring substance for food, dental, and pharmaceutical items and it helps in the penetration of some drugs into skin (Mohammadi Nejad et al., 2017). Beyond these uses, eugenol plays an important role in agricultural practices and food industry due to its antibacterial, antifungal, and insecticide activities as well as antioxidant and anti-browning properties (Abbaszadeh et al., 2014; Chen et al., 2017; Talón et al., 2019; Ulanowska and Olas, 2021). Some examples of these applications are biopesticides, edible films, and active food packaging materials (i.e., bionanocomposites) (Walia et al., 2017; Garrido-Miranda et al., 2018; Navikaite-Snipaitiene et al., 2018; Wieczyńska and Cavoski, 2018; Zheng et al., 2019; Sahlan et al., 2021). Additionally, eugenol is effective in the treatment of skin diseases and disorders like contact dermatitis, inflammations (i.e., acne), and lesions (i.e., leishmaniasis) (Tsai et al., 2017; de Araújo Lopes et al., 2018; Moemenbellah-Fard et al., 2020). Nevertheless, it is regarded as a mild dermal irritant, whereas intake of eugenol orally may induce hepatotoxicity (Saraphanchotiwitthaya et al., 2019).

Analgesic effect of eugenol through different mechanisms has been proved, and therefore it is commonly used in dentistry (Lee et al., 2005; Park et al., 2006). Moreover, it has anti-inflammatory, antibacterial, and local anesthetic effects and also acts as therapeutic agent for dental caries and periodontal infections (Markowitz et al., 1992; Jadhav et al., 2004; Xu et al., 2013; Pramod et al., 2016). Nonetheless, cytotoxic effects of eugenol on normal function of stem cells found in human dental pulp tissue and human dental pulp fibroblasts of primary teeth have been observed (Anpo et al., 2011; Escobar-García et al., 2016). It may also induce allergic reactions in sensitized patients (Sarrami et al., 2002). All these issues restrict its usefulness in dentistry applications. Other limitations are its low solubility in water, chemical instability against light or high temperatures, and necessity of a high dose, which may cause toxicity, to achieve a therapeutic effect. The recommended limits vary depending on the type of food. The highest concentrations indicated 25 ppm for meat products and the lowest was 2.83 ppm for nonalcoholic beverages (Saraphanchotiwitthaya et al., 2019).

5. Toxicity of eugenol and its effects

As a food additive, eugenol is considered to be safe, but great concerns have been raised about its toxicity in recent years due to different applications and widespread use of clove oil (Mohammadi Nejad et al., 2017). On the other side, eugenol has been determined as a potential carcinogen by the Ames test. However, the National Toxicology Program and the International Agency for Research on Cancer have not presently categorized it as a carcinogen (Baker and Grant, 2018).

Higher concentrations of undiluted clove oil may result in some symptoms because of the presence of eugenol. So, an excessive dose of eugenol has considered to be toxic (Basch et al., 2008). Oral intake of eugenol in different doses during a 15-day period may cause some changes in blood chemistry that cause an increase in aspartate aminotransferase, alanine aminotransferase, and total bilirubin levels, but it seems that such effects are not dose-dependent (Li et al., 2012). Similarly, negative effects of overdose of eugenol on human fibroblasts and endothelial cells have been also reported (Prashar et al., 2006).

In humans, eugenol is principally metabolized through the formation of conjugates with sulfate and glucuronide by liver (Fischer et al., 1990). Overdose leads to saturation of these sulfate and glucuronide pathways, and conjugation of eugenol with glutathione. Incubation of isolated rat hepatocytes with eugenol causes intracellular glutathione depletion, leading to hepatocellular death (Thompson et al., 1991). In vitro studies with rabbit and human platelets have demonstrated that the antiplatelet aggregation activity of eugenol was due to the inhibition of cyclooxygenase-dependent formation of thromboxane A2 and B2 (Chen et al., 1996; Raghavendra and Naidu, 2009). In inhibiting arachidonic acid induced platelet aggregation, eugenol was found to be 29 times more effective than aspirin (Goulet et al., 2011).

The findings of a case study on a 2-year boy pointed out that ingestion of 5–10 mL of clove oil led to distressing and crying. After 3 h of ingestion, conditions deteriorated. The acidosis was marked and the patient was in deep coma. During 8 h, he suffered from a generalized seizure, and within 24 h, failure in liver function and also disseminated intravascular coagulopathy occurred. Within 1 week after ingestion and after wicked symptomatic treatment, the patient gained consciousness and recovered. Due to the hepatotoxic effects, results showed some similarities between eugenol and paracetamol poisoning (Hartnoll et al., 1993). The results of another study showed that eugenol was absorbed and metabolized rapidly after oral administration by male and female healthy volunteers. Within 24 h, it was completely excreted in the urine containing conjugates of eugenol. Less than 1% of administrated dose was excreted as non-metabolized in urine. Urine analysis also demonstrated that more than 90% of metabolic products are phenolic conjugates and 50% of them were eugenol-glucuronide and eugenol-sulfate (Sipes and Mattia, 2006).

The antioxidative, cytotoxic, and genotoxic effects of eugenol were tested as the ability to regulate resistance against the damaging effects of H_2O_2 on DNA of different strains of human cells as malignant HepG2 hepatoma cells, malignant Caco-2 colon cells, and nonmalignant human VH10 fibroblast. Results revealed that eugenol has remarkable antioxidative potential at all the concentrations tested. In regard to toxicity, it presented strong genotoxic effects (DNA damaging) on human VH10 fibroblast, medium genotoxic effects on Caco-2 colon cells, and non-DNA-damaging effects on HepG2 hepatoma cells (Slameňová et al., 2009). There are limited human studies on the pharmacokinetics and effects of eugenol. Animal studies have suggested that after inhalation of the smoke of clove cigarettes, minor amounts of eugenol could be absorbed from the lung. On the other side, some in vitro studies indicated that eugenol undergoes biotransformation in hepatocytes (Marin Municipal Water District, 2010).

Intravenous infusion of eugenol at 4 and 8 μL caused acute respiratory upset with hemorrhagic pulmonary edema in rats, associated with oxidative stress (Mohammadi Nejad et al., 2017). In an in vitro experiment on isolated rat hepatocytes, more than 85% of the cells showed hepatotoxicity and cell damage after 5 h of exposure to eugenol. Acetyl cysteine administration has been shown to have preventive effects on cell death (Soundran et al., 1994). In an in vivo model study, the incorporation of eugenol in mice diet 2 weeks before coupling and from zeroth day to fourth day of pregnancy, supported difficulty in implantation of the blastocyst in the uterus, raising the incidence of cell death (Domarácký et al., 2007). The half-life ($t1/2$) of eugenol was reported to be about 14 h in plasma and 18 h in blood in rats (Guénette et al., 2007). Moreover, LD_{50} of eugenol was 11 mg/kg in male F-344 rats and 17 mg/kg in male Syrian golden hamsters (Lavoie et al., 1986).

In a study, three gelatin capsules, each containing 50 mg of eugenol (total dose 150 mg; 1.70–2.90 mg/kg bw) were given to four healthy male and female volunteers for breakfast. After 3, 6, 12, and 24 h of consumption, urine was collected and venous blood was tested at 0, 15, 20, 25,

30, 40, 50, 60, 80, 100, and 120 min. In all samples tested, mainly the conjugated form of eugenol was detected. In a 3-h period, 71.30% of the 150 mg dose was reported in the urine as conjugated forms. The main metabolites of eugenol in the urine were glucuronic acid and sulfate conjugates of the phenolic hydroxyl group. Additionally, while more than 87% of the given dose was excreted after 6 h, 94% of the given dose was eliminated after 24 h in the urine (Sipes and Mattia, 2006).

The effects of oral administration of eugenol during preimplantation and organogenesis periods in Wistar rats were investigated by da Silva et al. (2019b). Treatment with eugenol lasted 10 days on organogenesis and 6 days on preimplantation. They declared that the disclosure time may not have been sufficient to affect female body weight gain, but the reduction in the ingestion of food and water could be the commencement of toxicity. There was also a decrease in absolute and relative placental masses at all doses (da Silva et al., 2019b). Biotransformation of several compounds like xenobiotics and different drugs through various enzymes which are essential for reduction, oxidation, or conjugation reactions takes place in placenta. This organ is also effective in metabolic reactions involving CYP450 complex enzymes (Al-Enazy et al., 2017). Eugenol may perform prooxidant activity in the presence of these enzymes, resulting in a metabolic intermediate with cytotoxicity called "quinone methide." This quinone methide metabolite of eugenol was also reported to be responsible for the hepatotoxic effect of eugenol in liver cells of rats (Thompson et al., 1998). In the presence of hydrogen peroxide, eugenol may be metabolized in vitro by placental peroxidases and this causes placental dysfunction (Zhang et al., 2000). Taking this into consideration, da Silva et al. (2019b) reported that the reduction in placental indexes related to the reduction in placental mass and constriction of blood vessels had indicated the crossing of eugenol through the transplacental barrier and its interference with maternal metabolism that exerts a toxic action on the fetus. Another indicator of eugenol toxicity is the occurrence of pulmonary edema, hepatic, renal, and adrenal alterations after the treatment of pregnant rats with eugenol during the preimplantation (da Silva et al., 2019b). In another research, the toxicity of eugenol was evaluated in female African clawed frogs at anesthetic levels. The damage in kidney and some morphological alteration and cell apoptosis in renal cells were observed. The results suggest that smaller frogs are more susceptible to the toxic effects of eugenol, and that the therapeutic index of eugenol in amphibians may be related to metabolic maturity (Goulet et al., 2011).

6. Acute toxicity of eugenol

The findings of studies dealing with acute toxicity of eugenol are summarized in Table 2. Acute toxicity of eugenol on diabetic rats was investigated by Öntürk and Özbek (2009). It has been formed 11 different groups, each containing 8 male mice and only 0.20 mL of 0.90% saline was applied to the control group. Treatment groups received 0.20, 0.40, 0.80, 1.60, and 3.20 mL/kg of eugenol, respectively. It was concluded that LD_{10}, LD_{50}, LD_{90}, and LD_{99} for eugenol were 0.32, 0.47, 0.67, and 0.90 mL/kg, respectively (Öntürk and Özbek, 2009). Another study carried out by Xie et al. (2015) tested acute toxicity of eugenol and its analogs (methyl eugenol, acetyleugenol, isoeugenol, methylisoeugenol, acetylisoeugenol) against termites (*Reticulitermes chinensis*) and white-rot fungi (*Trametes hirsuta, Schizphylhls commne, Pycnoporus sanguineus*). Eugenol and isoeugenol were found to be highly toxic against termites and fungi, whereas eugenol was more toxic than isoeugenol. Similarly, eugenol exhibited higher toxicity against *Reticulitermes chinensis* than isoeugenol. The findings showed that the presence of the double bond of the propenyl group in the compounds played a key role in toxicity. They

Table 2 An overview of the acute toxicity studies related to eugenol.

Study	Findings	References
Administration of eugenol through intragastric instillation by stomach tube—male and female albino rats	– First observable effect upon administration → weakness of the hind legs – 70% of rats → became prostrate and could not move, and 54 of these animals went into complete coma; only 5 of them survived coma – Time of death → 41% of all deaths in the first 6h; 33% in the next 18h; 22% during the second day following administration – Greatest damages in lungs, kidney, liver, stomach, and duodenum – LD_{50} for albino rats → 1.93g/kg body weight	Sober et al. (1950)
Administration of eugenol by stomach tube—adult male and female Osborne-Mendel or Sherman rats	– Acute oral LD_{50} for rats → 2680mg/kg – Comatose soon after treatment – Death time → maximum 2h after treatment	Taylor et al. (1964)
Administration of eugenol through intubation—adult male and female Osborne-Mendel rats, male and female guinea pigs, mice	– Acute oral LD_{50} for rats → 2680mg/kg body weight and coma soon after treatment; death time: approximately 1h – Acute oral LD_{50} for mice → 3000mg/kg body weight; severe depression immediately after treatment; death time: from few minutes to 2 days – Acute oral LD_{50} for guinea pigs → 2130mg/kg; depression; death time: from 4h to 3 days	Jenner et al. (1964)
Administration of eugenol by gavage to birds (red-winged blackbird, starling, coturnix)	– Acute oral LD_{50} for red-winged blackbirds → >316mg/kg – Acute oral LD_{50} for starling → >316mg/kg	Schafer et al. (1983)
Intraperitoneal or oral delivery of eugenol to male Swiss-Webster mice	– LD_{50} for mice through intraperitoneal delivery → 1109.60mg/kg body weight – No mortality through oral administration of eugenol up to maximum dose (14,794mg/kg body weight) – Administration of eugenol by both routes → induced anaphase mutations in polychromatic erythrocytes and formation of micronuclei – Eugenol → mutagenic capacity	Woolverton et al. (1986)
Intratracheal instillation of different concentrations of eugenol to male Fischer-344 rats (0.75, 1.50, 3, 6, 12, 24mg/kg body weight) and male Syrian golden hamsters (3, 6, 12, 24, 48, 96mg/kg)	– LD_{50} for rats → 11mg/kg – LD_{50} for hamsters → 17mg/kg – Rats received 12 and 24mg/kg eugenol → presence of red frothy fluid in the trachea; lungs were severely congested; development of acute emphysema and acute edema led to mortality	Lavoie et al. (1986)

Continued

Table 2 An overview of the acute toxicity studies related to eugenol—cont'd

Study	Findings	References
	– Rats received 12 and 24 mg/kg eugenol and survived for more than 24 h → thickening and collapse of alveoli (atelectasis), alveolar edema, acute bronchopneumonia, and compensatory peripheral emphysema – Hamsters received 48 and 96 mg/kg eugenol → congestion of lungs with petechial hemorrhage and several hamsters with acute emphysema – Hamsters received 96 mg/kg eugenol → hemorrhagic alveolitis with compensatory emphysema and focal pulmonary edema – Toxicity of eugenol through intratracheal instillation was at least 200 times higher than that of administration through gavage	
4 h Inhalation exposure to atmosphere containing different concentrations of eugenol (0.77, 1.37, 2.58 mg/L) as respirable droplets—5 female, 5 male Sprague-Dawley rats	– No rat died during a 14-day observation period – During exposure to high-dose eugenol atmosphere → increased salivation and restlessness, but on the first day of observation, all clinical signs disappeared – All rats exposed to eugenol-containing atmosphere → weight loss, reduction in food and water consumption overnight following exposure – Histological examination of the lungs after a 14-day observation period → normal – A few incidental and spontaneous lesions → unlikely to be related to inhalation of eugenol and no toxicological importance	Clark (1988)
Intraperitoneal delivery of eugenol to male CF1 mice	– LD_{50} for mice through intraperitoneal delivery → 700 mg/kg	Ellahueñe et al. (1994)
Intrathecal administration of eugenol to male Sprague-Dawley rats	– 40 mg/kg of eugenol for 30 days → no toxicity, blood hematology and biochemistry results were good, no lesions – Eugenol → alleviation of neuropathic pain, both allodynia and hyperalgesia	Lionnet et al. (2010)
Acute oral toxicity	– LD_{50} for rats → 1930 mg/kg body weight (R22: harmful if swallowed) – LD_{50} for dogs → < 2000 mg/kg body weight	European Food Safety Authority (2012)
Acute dermal toxicity	LD_{50} for rats → not available	
Skin irritation	Irritating (R38: irritating to skin)	
Eye irritation	Irritating (R36: irritating to eyes) → based on the data related to skin irritation	
Skin sensitization	Skin sensitizer (R43: may cause skin sensitization by skin contact)	

found that eugenol killed all termites after testing for 3 days, with an LC_{50} value of 12.10 μg/g. IC_{50} values of eugenol against *Trametes hirsuta*, *Schizphylhls commne*, and *Pycnoporus sanguineus* were determined as 83.60, 116.10, and 102.40 μg/mL, respectively (Xie et al., 2015). Pavela (2015) determined acute toxicity of 30 aromatic compounds and their mixtures against *Culex quinquefasciatus* Say (Diptera: Culicidae) larvae. It was indicated that the mixture (a mixing ratio of 1:3) of eugenol and isoeugenol caused 90.20% mortality. Additionally, nine substances including borneol, menthone, isoeugenol, limonene, camphor, 4-allylanisole, carvacrol, carvone, and trans-anethol exhibited synergistic effects on mortality of *Culex quinquefasciatus* Say larvae. Moreover, acute toxicity of eugenol against Siamese fighting fish (*Betta splendens*) (48-h LC_{50} = 29.95 mg/L), zebrafish (*Danio rerio*) (LC_{50} = 70.30 mg/L for those exposed to eugenol at the middle of the light phase and 104.90 mg/L for those exposed to eugenol at the middle of the dark phase), *Daphnia magna* (microcrustacean) (EC_{50} = 0.70 mg/L), *Aliivibrio fischeri* (EC_{50} = 2.97 mg/L), and *Daphnia pulex* (freshwater zooplankton) (LC_{50} = 42.50 mg/L) were also proved (Sánchez-Vázquez et al., 2011; Sutili et al., 2014; Gueretz et al., 2017; Pattanasiri et al., 2017).

Acute toxicity of eugenol was reported to be low for oral intake, with LD_{50} values ranging from 1.93 to 2.68 g/kg in rats (Sober et al., 1950; Taylor et al., 1964). On the other side, oral intake of eugenol at a dose of 150 mg per day by healthy and nonsmoking men for 7 days did not cause any change in clinical biochemical parameters (L-aspartate aminotransferase (ASAT), L-alanine aminotransferase (ALAT), γ-glutamyl transpeptidase, lactate dehydrogenase, alkaline phosphatase) and toxic effect (Rompelberg et al., 1996). However, Soundran et al. (1994) reported that intramuscularly administration of eugenol (20 or 30 μg/100 g body weight/day) to male albino rats for 10 days resulted in increased total and differential bilirubin concentrations and activities of alkaline phosphatase, transminases, lactate dehydrogenase as well as reduced activity of glytamyl transferase that indicate toxic effect (hepatotoxicity) of eugenol on liver.

Skin sensitization induced by eugenol was observed in mice and classified as irritant to skin based on the findings of standard human 4-h patch test (York et al., 1996; Bertrand et al., 1997). Results for the murine local lymph node assay and guinea pig maximization test were both positive which indicates the contact sensitizer effect and potential to cause allergic reactions of eugenol (Hilton et al., 1996). Therefore, as a fragrance allergen, the presence of eugenol in women's and men's perfumes was investigated. The mean amounts of eugenol were reported to be 140.44 and 69 mg/kg in women's and men's perfumes, respectively. The authors concluded that analyzed perfumes did not contain sufficient amounts of eugenol which may pose considerable dermal risk (Soo Lim et al., 2018). In addition to its skin sensitizing and irritating effect, eugenol is considered an eye irritant (EFSA Panel on Additives and Products or Substances used in Animal Feed (FEEDAP), 2011). However, it also has anticonvulsant and sedative effects, and thanks to these properties, eugenol has been widely utilized in dentistry (Sharma et al., 2012). Fatal overdoses rarely occurred and there is no known antidote effect (Baker and Grant, 2018).

7. Subacute and sub-chronic toxicity of eugenol

Findings of researches related to subacute and sub-chronic toxicity of eugenol are summarized in Table 3. Although some studies reported side effects on lung, liver, and nervous system upon accidental ingestion of eugenol, it is classified as category 3 with an oral LD_{50} value of higher than 1930 mg/kg for

Table 3 An overview of the subacute and sub-chronic toxicity studies of eugenol.

Study	Findings	Reference
34-Day repeated oral dose (increasing doses from 1400 to 4000 ppm/day) toxicity study in male and female Osborne-Mendel rats	– Liver enlargement slightly – Adrenal enlargement slightly accompanied with considerable yellowish discoloration – Moderately severe hyperplasia and hyperkeratosis of the stratified squamous epithelium associated with focal ulceration in the forestomach	Hagan et al. (1965)
34-Day repeated dose (gradually increasing doses from 1400 to 4000 ppm/day) toxicity study in male Osborne-Mendel rats—administration through stomach tube	– Only 8 rats survived until the end of experimental period (34 days) – Only 15 rats survived enough to receive the maximum dose of 4000 ppm – Coalescent areas covered by thick flaky white substance punctuated by minute ulcers were reported as macroscopic observations in mucosa of the forestomach – Moderately severe hyperkeratosis of the stratified squamous epithelium in the forestomach was reported as microscopic observation – Negligible degree of osteoporosis in bones	Hagan et al. (1967)
13-Week repeated oral dose (800, 1500, 3000, 6000, and 12,500 ppm) toxicity study in male and female rats	– No mortality – 800 ppm → weight loss by 1% in males and females compared to control groups – 1500 ppm → weight loss by 3% in males and 1% in females compared to control groups – 3000 ppm → weight loss by 3% in males and females compared to control groups – 6000 ppm → weight loss by 7% in males and 3% in females compared to control groups – 12,500 ppm → weight loss by 10% in males and 6% in females compared to control group	NTP (1983)
13-Week repeated oral dose (400, 800, 1500, 3000, and 6000 ppm/day) toxicity study in male and female mice	– No mortality – 400 ppm → weight gain by 6% in males and no difference in females compared to control groups – 800 ppm → weight gain by 3% in males and no difference in females compared to control groups – 1500 ppm → weight gain by 3% in males and weight loss by 4% in females compared to control groups	

Table 3 An overview of the subacute and sub-chronic toxicity studies of eugenol—cont'd

Study	Findings	Reference
	– 3000 ppm → no weight difference in males and weight loss by 4% in females compared to control groups – 6000 ppm → no weight difference in males and females compared to control groups	
28-Day repeated oral dose (20,000 ppm/day) toxicity study in F344 male rats	Lower body weight (10%–15%) compared to control group	Hirose et al. (1987)
14-Day repeated oral dose (7.34 and 70 mg/kg) toxicity in male and female Swiss albino mice	– Low-dose eugenol (7.34 mg/kg) administered mice → object recognition memory ↓ – High-dose eugenol (70 mg/kg) administered mice → object location memory dysfunction – Antidepressant activity of eugenol was dose-dependent – Eugenol administration → Mice aggression and anxiety/depression ↓, cognitive behavior ↑ – Food and water intake of mice were comparable to those of control group mice – No adverse effect of eugenol on immune system	Dhara and Tripathi (2020)
Short-term oral toxicity—NOAEL	– Dog: 100 mg/kg body weight/day – Rat: 600 mg/kg body weight/day – Mouse: 900 mg/kg body weight/day	European Food Safety Authority (2012)
Short-term dermal toxicity—NOAEL	No data available	
Short-term inhalation toxicity—NOAEL	No data available	

rodents by the U.S. Environmental Protection Agency (EPA) (Sellamuthu, 2014). Furthermore, EPA concluded that some human health cases resulting from the use of eugenol-containing pesticides could not be directly correlated with the presence of eugenol, as these pesticides also contained other active ingredients like capsaicin, phenethyl propionate, pyrethrins, and (r,z)-5-(1-decynyl) dihydro-2(3H)-furanone (Baker and Grant, 2018). However, exposure of operators and bystanders to eugenol in the form of biopesticide, as well as consumers, are also essential. Acceptable operator exposure level (AOEL) for eugenol with dermal and respiratory absorption rates of 75% and 100%, respectively, is 1 mg/kg body weight/day (Hernández-Moreno et al., 2013).

Eugenol is recognized as an allergen compound. For instance, in dental practices, Sarrami et al. (2002) reported two different cases related to the use of eugenol of which symptoms ranged from low-grade local reactions resembling typical contact allergic dermatitis to more severe conditions like necrosis and even anaphylactic reaction. Accordingly, another patient (an 8-year-old boy) experienced type 1 immediate hypersensitive reaction mediated by IgE antibodies upon exposure to eugenol during dental treatment (Tammannavar et al., 2013). It was shown that repeated exposure to dental preparations (i.e., periodontal dressings) including eugenol resulted in irritation of the contact area as well as formation of contact mucositis (Barkin et al., 1984). Similarly, burning mouth syndrome has been recently correlated with allergic contact stomatitis caused by eugenol as a result of repeated exposure to eugenol-containing mouthwash (Bui et al., 2019). Moreover, a study carried out on guinea pigs also showed eugenol-induced allergic contact dermatitis (Ohsumi et al., 1996). On the other hand, application of eugenol solution with a concentration of 2.70% (in diethyl phthalate: ethanol (2:98)) for 4 weeks led to the risk of allergic contact dermatitis (Svedman et al., 2012). In one study using patch test to understand contact allergenicity of 5% eugenol preparation, 2.60% of 155 patients with cosmetic dermatitis, 3.10% of 159 patients with eczema and dermatitis, and 2.10% of control patients with melasma, acne, or other noninflammatory skin disorders developed allergic reaction (Itoh, 1982). Likewise, Schnuch et al. (2007) reported that the frequency of sensitization induced by 1% eugenol was 0.40% and women were more sensitive to contact allergy of eugenol than men. While isoeugenol, which is widely used in various products such as soap, shampoo, perfumes, and cosmetics as a fragrance chemical, triggers eugenol allergy, eugenol also does in the same way (cross-react) (Buckley et al., 2000). Scholes et al. (1994) correlated the skin sensitization response to eugenol by epidermal cytochrome P4501A as it plays a key role in metabolization of eugenol and isoeugenol into less harmful moieties. The allergic effects of eugenol vary due to the sensitive populations (Baker and Grant, 2018). Therefore, The International Fragrance Association (IFRA) recommends the restriction of eugenol amount as a fragrance ingredient in some household items as summarized in Table 4.

Despite it is considered a contact allergen, there is no clear correlation between eugenol and respiratory allergy (Hilton et al., 1996). Eugenol has been associated with a reddening irritation (erythema), and in some cases ulcers and diffuse inflammation in some dental applications. Inhalation of eugenol through clove cigarettes is associated with pulmonary edema (Baker and Grant, 2018).

8. Chronic toxicity of eugenol

Chronic toxicity studies have been carried out to determine the potential target organs of toxicity; the reversibility of toxicities observed; the NOAEL; and the potential health risks regarding the anticipated clinical dose following long-term exposure (Colerangle, 2017). Oral, dermal, and inhalation are the main routes of administration in chronic toxicity studies (OECD, 2018). Chronic studies of 6 months duration in rodents and of 9 months duration in nonrodents are acceptable (ICH, 1999).

A detailed report including findings of a 2-year research related to carcinogenicity effects of eugenol on rats and mice was published in 1983 by the US Department of Health and Human Services. In this context, F344/N male rats, B6C3F$_1$ male and female mice received 3000 (low dose) or 6000 ppm (high dose) of eugenol, while 6000 (low dose) or 12,500 ppm (high dose) of eugenol were administered to F344/N female rats for 103 weeks. Female rats that received high dose of eugenol had lower body weights among groups. Alveolar/bronchiolar adenomas or carcinomas in the lung of low-dose eugenol received male rats with a higher incidence were observed, while any significant increase for high-dose male rats was not reported. On the other side, low-dose female rats experienced C-cell adenomas in the thyroid with a higher incidence rate (22%). Regarding uterus, the incidence of endometrial stromal polyps or sarcoma was higher in high-dose eugenol-treated female rats (32%). Surprisingly, mammary gland fibroadenoma occurred more frequently in control female rats. With regard to mice experiments, low dose received male mice had higher incidence rates for hepatocellular adenoma or carcinoma,

Table 4 Restriction limits for eugenol as a fragrance ingredient in various household products (IFRA, 2020).

IFRA product category	Products	Restriction limit of eugenol in the finished product (%)
1	Products applied to the lips	0.45
2	Products applied to the axillae	0.14
3	Products applied to the face/body using fingertips	1.40
4	Products related to fine fragrances	2.50
5A	Products applied to the body using the hands (palms), primarily leave-on (Body lotion)	0.64
5B	Products applied to the face using the hands (palms), primarily leave-on (face moisturizer)	0.64
5C	Products applied to the hands using the hands (palms), primarily leave-on (hand cream)	0.64
5D	Products applied to babies using the hands (palms), primarily leave-on (baby cream, oil, talc)	0.21
6	Products with oral and lip exposure	0.64
7A	Rinse-off products applied to the hair with some hand contact	1.40
7B	Leave-on products applied to the hair with some hand contact	1.40
8	Products with significant anogenital exposure (tampon)	0.21
9	Products with body and hand exposure, primarily rinse-off	4.90
10A	Household care products with mostly hand contact (excluding aerosol/spray products)	4.90
10B	Household care products with mostly hand contact (household aerosol/spray products)	18
11A	Products with intended skin contact but minimal transfer of fragrance to skin from inert substrate without UV exposure	0.21
11B	Products with intended skin contact but minimal transfer of fragrance to skin from inert substrate with potential UV exposure	0.21
12	Other air care products not intended for direct skin contact, minimal or insignificant transfer to skin	No restriction

while no significant difference was detected in female mice groups. The incidence of follicular cell adenoma (6%) was higher for high-dose eugenol received male mice. The study showed that oral administration of female rats and mice with high doses of eugenol led to reductions in body weights. In conclusion, eugenol did not exhibit carcinogenicity in male and female rats. However, there was equivocal evidence of carcinogenicity for mice (NTP, 1983). On the other side, Miller et al. (1983) reported that hepatic tumor (hepatomas) formation in female CD-1 mice was not induced by the administration of 0.5% of eugenol for 12 months.

The mutagenicity of eugenol was assessed using different test methods. The results of direct bacterial assay, microsomal mutagenesis assay, and host-mediated assay revealed that eugenol was not mutagenic for strains of *Salmonella typhimurium* (TA 1530, TA 1531, TA 1532, TA 1964) (Green and Savage, 1978). Similar observations were reported by Swanson et al. (1979), Sekizawa and Shibamoto (1982), and Azizan and Blevins (1995) who tested mutagenic activity of eugenol for *Salmonella typhimurium* strains (TA1535, TA100, TA98, TA1537, TA1538). Eugenol did not exhibit mutagenic activity for *Escherichia coli* WP2 *uvrA trp* (Sekizawa and Shibamoto, 1982). In vivo mutagenicity of eugenol in eukaryotic hosts of mice was reported by Woolverton et al. (1986). Then, its mutagenic effect on granuloma cells of female Sprague-Dawley rats was investigated. When eugenol was administered orally or directly injected into granuloma pouch, no in vivo mutagenic activity was detected (Maura et al., 1989). According to data on the widespread exposure and absence of human epidemiological given by The International Agency for Research on Cancer (IARC), no estimation could be made on the carcinogenicity of eugenol to humans—even if eugenol is a potential carcinogen. Therefore, eugenol was considered by IARC as "not classifiable" to its carcinogenicity to humans (Group 3) (IARC, 1985). Eugenol has been categorized as one of a number of naturally occurring pesticides that may be potential carcinogens by Ames et al. (1990). Tests carried out on the cultures of Chinese hamster cells showed significantly greater mutation and reduplication than the control, which indicated the genotoxicity and possible carcinogenicity of eugenol (Maralhas et al., 2006). In California, eugenol is not considered to be a chemical that is known to cause cancer. However, methyl eugenol was listed as a substance with "low carcinogenicity concern" by California in 2001 (Baker and Grant, 2018).

Eugenol exhibited cytotoxic effect on malignant cell line Caco-2 (colon carcinoma cells), non-malignant diploid human fibroblasts VH10, PC-3 prostate cancer cell line, U2OS cells of human osteogenic sarcoma, 153BR human fibroblasts, human normal dermal fibroblasts, HMEC-1 SV40-transformed human dermal microvascular endothelial cells, human foreskin fibroblasts, human hepatoma cell line HepG2, murine ES cell line E14, and Balb/c 3T3 cells, whereas it wasn't cytotoxic for malignant cell line HepG2 (hepatoma carcinoma cells) (Babich et al., 1993; Ho et al., 2006; Prashar et al., 2006; Slameňová et al., 2009; Chen et al., 2010; Sharma et al., 2017). Furthermore, Mohammadi Nejad (2014) reported that eugenol higher than 340 μM had in vitro cytotoxic effect in V79 cells and this cytotoxicity was dose-dependent. However, Babich et al. (1993) declared the levels of intracellular glutathione as an important parameter in terms of prevention of eugenol cytotoxicity. On the other side, Jeng et al. (1994) stated that clinical use of eugenol is generally safe for humans as its exposure time, dosage, and frequency are low, but high concentrations of eugenol, repeated or long-exposure time may be hazardous because of its cytotoxicity to oral mucosal fibroblasts which then may induce oral cancer.

Genotoxicity of eugenol has also been investigated so far. Ellahueñe et al. (1994) declared that eugenol exhibited cytotoxic potential in mice which was determined through bone marrow micronucleus assay. Contrarily, eugenol concentrations lower than IC_{50} values did not lead to any in vitro genotoxicity on human peripheral lymphocytes (Mohammadi Nejad, 2014). Nevertheless, DNA damage occurred in eugenol-treated dental pulp fibroblasts was confirmed by Escobar-García et al. (2016).

Regarding reproductive toxicity, eugenol impaired maternal reproductive capacity in preimplantation phase and causes embryo-fatal toxicity, and fetal malformations in organogenesis phase in Wistar rats (da Silva et al., 2019b). Similarly, Mohammed et al. (2016) declared that moderate and high concentrations of eugenol, which are taken by pregnant women in the early stages of gestation, are potentially teratogenic. On the other side, treatment of chick embryo with eugenol in early stages influenced the development of embryo negatively (El-Hak and Gerges, 2018). Furthermore, eugenol was classified as "strongly embryotoxic" by Chen et al. (2010).

9. Ecotoxicity of eugenol

As eugenol is considered a natural pesticide, its potential effects on environment should be also evaluated. In this context, ecotoxicity of eugenol is tested using indicator organisms, also known as bioindicators. Until now, it has been shown that eugenol is toxic against *Daphnia magna* (microcrustacean), *Aliivibrio fischeri* (bacterium), *Desmodesmus subspicatus* (algae), and *Daphnia pulex* (freshwater zooplankton) (Sutili et al., 2014; Gueretz et al., 2017). However, emulsification (nanoemulsion) was reported to be a promising approach to produce eugenol-based natural fungicide without ecotoxicity to *Folsomia candida* (springstail) (da Silva Gündel et al., 2019).

10. Regulatory requirements

Currently, there is no regulation about recommended intake amount of clove. As its toxicity is mostly due to eugenol, restrictions have been made for eugenol by regulatory authorities. Recommendations for daily eugenol intake differ. While The Joint FAO/WHO Expert Committee on Food Additives (JECFA) established an ADI of 2.50 mg/kg body weight for eugenol, an ADI value of 1 mg/kg body weight/day for eugenol was determined by European Food Safety Authority (EFSA) (WHO, 2006; European Food Safety Authority, 2012). For instance, a recent study showed that the highest estimated daily intake (EDI) of eugenol through consumption of instant herbal beverages available on Indonesian market was 46.90 μg/kg body weight/day (Suparmi et al., 2019). Furthermore, as eugenol is commonly utilized as fish anesthetic, researchers investigated the residue eugenol levels in market fish (seawater and freshwater fish) sold in China. They reported that residue eugenol contents of fish ranged from 3.11 to 36,090 μg/kg. The EDIs of eugenol (max <0.06 mg/kg body weight/day) through fish consumption showed that the risk of eugenol residue in fish fillets was insignificant in terms of human health (Ke et al., 2018). The estimated daily exposure amounts to eugenol are 1107 and 3364 μg/day in Europe and in the United States, respectively (Martins et al., 2018).

As regards to Code of Federal Regulations (CFR) Title 40 Part 180, no tolerance or exemption has been established for eugenol. Thus, it is only permitted for nonfood uses. As of 2017, there were 14 products registered with EPA that contained eugenol as an active ingredient (Baker and Grant, 2018). By the Food Chemicals Codex specification, food-grade eugenol is classified as "Generally Recognized as Safe (GRAS)," along with other clove derivatives. Its use is authorized for over-the-counter drugs for the analgesic and anesthetic uses, fever blister and cold sores, poison oak, poison ivy, and poison sumac treatments, and as an astringent. Zinc oxide mixed with eugenol is allowed to be used as dental paste. However, eugenol and other clove oil constituents are not allowed as animal drugs, and the FDA has presented drawbacks about the use of eugenol as a fish anesthetic (FDA/CVM, 2007). Synthetic eugenol is prevented for organic farming uses under the USDA's National Organic Program (NOP). Natural extracts of eugenol permitted by the NOP can be used as a botanical pesticide (Baker and Grant, 2018).

Eugenol has been tested for carcinogenicity in long-term studies in rats and mice, in addition to some special carcinogenicity studies such as skin painting and promotion studies. The findings suggested that it was not carcinogenic in a limited dietary study with groups of 30 CD-1 mice, in a limited gavage study in 55 male and 59 female CD-1 mice and after intraperitoneal injection in 52 male CD-1 mice. After reviewing the sub-chronic and chronic studies on eugenol, a NOAEL for noncarcinogenic effects of 300mg/kg body weight/day was determined (Scientific Panel on Food Additives, Flavourings, Processing Aids and Materials in Contact With Food, 2009).

11. Conclusion

Clove is a widely used spice all over the world. Its buds are principally used for food purposes along with limited anesthetic uses due to the presence of different bioactive compounds in its composition. On the other side, essential oil of clove is utilized as plant-based natural anesthetic in various seawater and freshwater fish species. Toxicity of clove against some fish, pests, and insects has been proven up to now. Furthermore, clove cigarettes (kretek) were also evaluated in terms of toxicological effects of clove. However, there is not enough study yet to confirm the toxicity of cloves for humans. Most of the biological effects of clove result from high contents of eugenol. Therefore, toxicity studies related to clove have already focused on eugenol. Regulatory bodies have set limits for eugenol, but there is no regulation for clove. Many scientists have demonstrated the acute, subacute, sub-chronic, cytotoxic, genotoxic, teratotoxic, embryotoxic, and ecotoxic effects of eugenol until now. In some cases, the toxicity of eugenol against some pests is desired as it can be utilized as biopesticide. However, residue levels of eugenol and their possible health and environment consequences should be criticized carefully. In conclusion, different properties and activities of eugenol are still undiscovered and should be further explored by more in vitro and in vivo long-term animal and human studies.

Acknowledgment

The authors gratefully thank Mrs. Süheyla Tuncer (MSc in Biology) for the illustrations entitled "Figs. 1–3" in this chapter.

References

Abbas, M.A., Taha, M.O., Zihlif, M.A., Disi, A.M., 2013. β-Caryophyllene causes regression of endometrial implants in a rat model of endometriosis without affecting fertility. Eur. J. Pharmacol. 702, 12–19. https://doi.org/10.1016/j.ejphar.2013.01.011.

Abbaszadeh, S., Sharifzadeh, A., Shokri, H., Khosravi, A.R., Abbaszadeh, A., 2014. Antifungal efficacy of thymol, carvacrol, eugenol and menthol as alternative agents to control the growth of food-relevant fungi. J. Mycol. Med. 24, e51–e56. https://doi.org/10.1016/j.mycmed.2014.01.063.

Al Mansouri, S., Ojha, S., Al Maamari, E., Al Ameri, M., Nurulain, S.M., Bahi, A., 2014. The cannabinoid receptor 2 agonist, β-caryophyllene, reduced voluntary alcohol intake and attenuated ethanol-induced place preference and sensitivity in mice. Pharmacol. Biochem. Behav. 124, 260–268. https://doi.org/10.1016/j.pbb.2014.06.025.

Al-Enazy, S., Ali, S., Albekairi, N., El-Tawil, M., Rytting, E., 2017. Placental control of drug delivery. Adv. Drug Deliv. Rev. https://doi.org/10.1016/j.addr.2016.08.002.

Allali, A., Rezouki, S., Eloutassi, N., Louaste, B., Bouchelta, Y., Fadli, M., 2018. Insecticidal activity of the powder from Capsicum frutescens (Solanaceae), Lavandula stoechas (Lamiaceae) and Syzygium aromaticum

(Myrtaceae) against Callosobruchus maculatus insect pest of chickpea seeds. Int. J. Appl. Environ. Sci. 1, 21–27.

Álvarez-González, I., Madrigal-Bujaidar, E., Castro-García, S., 2014. Antigenotoxic capacity of beta-caryophyllene in mouse, and evaluation of its antioxidant and GST induction activities. J. Toxicol. Sci. 39, 849–859. https://doi.org/10.2131/jts.39.849.

Ambrož, M., Boušová, I., Skarka, A., Hanušová, V., Králová, V., Matoušková, P., Szotáková, B., Skálová, L., 2015. The influence of sesquiterpenes from Myrica rubra on the antiproliferative and pro-oxidative effects of doxorubicin and its accumulation in cancer cells. Molecules 20, 15343–15358. https://doi.org/10.3390/molecules200815343.

Ames, B.N., Profet, M., Gold, L.S., 1990. Dietary pesticides (99.99% all natural). Proc. Natl. Acad. Sci. 87, 7777–7781. https://doi.org/10.1073/PNAS.87.19.7777.

Anpo, M., Shirayama, K., Tsutsui, T., 2011. Cytotoxic effect of eugenol on the expression of molecular markers related to the osteogenic differentiation of human dental pulp cells. Odontology 99, 188–192. https://doi.org/10.1007/s10266-011-0009-2.

Aslam, M., Khan, K.A., Bajwa, M.Z.H., 2002. Potency of some spices against Callosobruchus chinensis linnaeus. J. Biol. Sci. 2, 449–452. https://doi.org/10.3923/jbs.2002.449.452.

Assis, L.C., Straliotto, M.R., Engel, D., Hort, M.A., Dutra, R.C., de Bem, A.F., 2014. β-Caryophyllene protects the C6 glioma cells against glutamate-induced excitotoxicity through the Nrf2 pathway. Neuroscience 279, 220–231. https://doi.org/10.1016/j.neuroscience.2014.08.043.

Azizan, A., Blevins, R.D., 1995. Mutagenicity and antimutagenicity testing of six chemicals associated with the pungent properties of specific spices as revealed by the Ames Salmonella/microsomal assay. Arch. Environ. Contam. Toxicol. 28. https://doi.org/10.1007/BF00217624.

Babich, H., Stern, A., Borenfreund, E., 1993. Eugenol cytotoxicity evaluated with continuous cell lines. Toxicol. In Vitro 7, 105–109. https://doi.org/10.1016/0887-2333(93)90119-P.

Bagavan, A., Rahuman, A.A., Kamaraj, C., Elango, G., Zahir, A.A., Jayaseelan, C., Santhoshkumar, T., Marimuthu, S., 2011. Contact and fumigant toxicity of hexane flower bud extract of Syzygium aromaticum and its compounds against Pediculus humanus capitis (Phthiraptera: Pediculidae). Parasitol. Res. 109, 1329–1340. https://doi.org/10.1007/s00436-011-2425-1.

Baker, B.P., Grant, J.A., 2018. Eugenol Profile—Active Ingredient Eligible for Minimum Risk Pesticide Use, Integrated Pest Management Program. New York State IPM Program, Geneva, NY.

Baldissera, M.D., Souza, C.F., Grando, T.H., Stefani, L.M., Monteiro, S.G., 2017. β-Caryophyllene reduces atherogenic index and coronary risk index in hypercholesterolemic rats: the involvement of cardiac oxidative damage. Chem. Biol. Interact. 270, 9–14. https://doi.org/10.1016/j.cbi.2017.04.008.

Barkin, M.E., Boyd, J.P., Cohen, S., 1984. Acute allergic reaction to eugenol. Oral Surg. Oral Med. Oral Pathol. 57, 441–442. https://doi.org/10.1016/0030-4220(84)90166-X.

Basch, E., Gasparyan, A., Giese, N., Hashmi, S., Miranda, M., Sollars, D., Seamon, E., Tanguay-Colucci, S., Ulbricht, C., Varghese, M., Vora, M., Weissner, W., 2008. Natural standard professional monograph. J. Diet. Suppl. 5, 117–146. https://doi.org/10.1080/19390210802335391.

Basha, R.H., Sankaranarayanan, C., 2014. β-Caryophyllene, a natural sesquiterpene, modulates carbohydrate metabolism in streptozotocin-induced diabetic rats. Acta Histochem. 116, 1469–1479. https://doi.org/10.1016/j.acthis.2014.10.001.

Bastaki, M., Api, A.M., Aubanel, M., Bauter, M., Cachet, T., Demyttenaere, J.C.R., Diop, M.M., Harman, C.L., Hayashi, S., Krammer, G., Lu, V., Marone, P.A., Mendes, O., Renskers, K.J., Schnabel, J., Tsang, S.Y., Taylor, S.V., 2020. Dietary administration of β-caryophyllene and its epoxide to Sprague-Dawley rats for 90 days. Food Chem. Toxicol. 135, 110876. https://doi.org/10.1016/j.fct.2019.110876.

Benelli, G., Govindarajan, M., Rajeswary, M., Vaseeharan, B., Alyahya, S.A., Alharbi, N.S., Kadaikunnan, S., Khaled, J.M., Maggi, F., 2018. Insecticidal activity of camphene, zerumbone and α-humulene from

Cheilocostus speciosus rhizome essential oil against the Old-World bollworm, Helicoverpa armigera. Ecotoxicol. Environ. Saf. 148, 781–786. https://doi.org/10.1016/j.ecoenv.2017.11.044.

Bertrand, F., Basketter, D.A., Roberts, D.W., Lepoittevin, J.-P., 1997. Skin sensitization to eugenol and isoeugenol in mice: possible metabolic pathways involving ortho-quinone and quinone methide intermediates. Chem. Res. Toxicol. 10, 335–343. https://doi.org/10.1021/tx960087v.

Boraie, D.M., Al-Akhdar, H.H., Sadek, M.G., 2014. Insecticidal activity of commercial oil and petroleum ether extract of clove buds (Syzygium aromaticum) on some stored grain pests. Acarines 8, 83–90.

Buckley, D.A., Wakelin, S.H., Seed, P.T., Holloway, D., Rycroft, R.J.G., White, I.R., McFadden, J.P., 2000. The frequency of fragrance allergy in a patch-test population over a 17-year period. Br. J. Dermatol. 142, 279–283. https://doi.org/10.1046/j.1365-2133.2000.03298.x.

Bui, T.N.P.T., Mose, K.F., Andersen, F., 2019. Eugenol allergy mimicking burning mouth syndrome. Contact Dermatitis 80, 54–55. https://doi.org/10.1111/cod.13103.

Calleja, M.A., Vieites, J.M., Montero-Meterdez, T., Torres, M.I., Faus, M.J., Gil, A., Suárez, A., 2013. The antioxidant effect of β-caryophyllene protects rat liver from carbon tetrachloride-induced fibrosis by inhibiting hepatic stellate cell activation. Br. J. Nutr. 109, 394–401. https://doi.org/10.1017/S0007114512001298.

Cao, J.-Q., Guo, S.-S., Wang, Y., Pang, X., Geng, Z.-F., Du, S.-S., 2018a. Toxicity and repellency of essential oil from Evodia lenticellata Huang fruits and its major monoterpenes against three stored-product insects. Ecotoxicol. Environ. Saf. 160, 342–348. https://doi.org/10.1016/j.ecoenv.2018.05.054.

Cao, J.-Q., Guo, S.-S., Wang, Y., Pang, X., Geng, Z.-F., Du, S.-S., 2018b. Contact toxicity and repellency of the essential oils of Evodia lenticellata Huang and Evodia rutaecarpa (Juss.) Benth. Leaves against three stored product insects. J. Oleo Sci. 67, 1027–1034. https://doi.org/10.5650/jos.ess17251.

Chang, H.-J., Kim, J.-M., Lee, J.-C., Kim, W.-K., Chun, H.S., 2013. Protective effect of β-caryophyllene, a natural bicyclic sesquiterpene, against cerebral ischemic injury. J. Med. Food 16, 471–480. https://doi.org/10.1089/jmf.2012.2283.

Chatterjee, D., Bhattacharjee, P., 2013a. Supercritical carbon dioxide extraction of eugenol from clove buds: process optimization and packed bed characterization. Food Bioprocess Technol. 6, 2587–2599. https://doi.org/10.1007/s11947-012-0979-2.

Chatterjee, D., Bhattacharjee, P., 2013b. Comparative evaluation of the antioxidant efficacy of encapsulated and un-encapsulated eugenol-rich clove extracts in soybean oil: shelf-life and frying stability of soybean oil. J. Food Eng. 117, 545–550. https://doi.org/10.1016/j.jfoodeng.2012.11.016.

Chavan, M.J., Wakte, P.S., Shinde, D.B., 2010. Analgesic and anti-inflammatory activity of caryophyllene oxide from Annona squamosa L. bark. Phytomedicine 17, 149–151. https://doi.org/10.1016/j.phymed.2009.05.016.

Chen, S.J., Wang, M.H., Chen, I.J., 1996. Antiplatelet and calcium inhibitory properties of eugenol and sodium eugenol acetate. Gen. Pharmacol. 27, 629–633. https://doi.org/10.1016/0306-3623(95)02089-6.

Chen, R., Chen, J., Cheng, S., Qin, J., Li, W., Zhang, L., Jiao, H., Yu, X., Zhang, X., Lahn, B.T., Xiang, A.P., 2010. Assessment of embryotoxicity of compounds in cosmetics by the embryonic stem cell test. Toxicol. Mech. Methods 20, 112–118. https://doi.org/10.3109/15376510903585450.

Chen, X., Ren, L., Li, M., Qian, J., Fan, J., Du, B., 2017. Effects of clove essential oil and eugenol on quality and browning control of fresh-cut lettuce. Food Chem. 214, 432–439. https://doi.org/10.1016/j.foodchem.2016.07.101.

Chen, H., Yuan, J., Hao, J., Wen, Y., Lv, Y., Chen, L., Yang, X., 2019. α-Humulene inhibits hepatocellular carcinoma cell proliferation and induces apoptosis through the inhibition of Akt signaling. Food Chem. Toxicol. 134, 110830. https://doi.org/10.1016/j.fct.2019.110830.

Cho, H.-I., Hong, J.-M., Choi, J.-W., Choi, H.-S., Hwan Kwak, J., Lee, D.-U., Kook Lee, S., Lee, S.-M., 2015. β-Caryophyllene alleviates d-galactosamine and lipopolysaccharide-induced hepatic injury through suppression of the TLR4 and RAGE signaling pathways. Eur. J. Pharmacol. 764, 613–621. https://doi.org/10.1016/j.ejphar.2015.08.001.

Chung, K.-S., Hong, J.Y., Lee, J.-H., Lee, H.-J., Park, J.Y., Choi, J.-H., Park, H.-J., Hong, J., Lee, K.-T., 2019. β-Caryophyllene in the essential oil from chrysanthemum boreale induces g1 phase cell cycle arrest in human lung cancer cells. Molecules 24, 3754. https://doi.org/10.3390/molecules24203754.

Clark, G.C., 1988. Acute inhalation toxicity of eugenol in rats. Arch. Toxicol. 62, 381–386. https://doi.org/10.1007/BF00293627.

Colerangle, J.B., 2017. Preclinical development of nononcogenic drugs (small and large molecules). In: Faqi, A.S. (Ed.), A Comprehensive Guide to Toxicology in Nonclinical Drug Development. Academic Press, London, UK, pp. 659–683, https://doi.org/10.1016/B978-0-12-803620-4.00025-6.

Coté, H., Boucher, M.-A., Pichette, A., Legault, J., 2017. Anti-inflammatory, antioxidant, antibiotic, and cytotoxic activities of Tanacetum vulgare L. essential oil and its constituents. Medicines 4, 34. https://doi.org/10.3390/medicines4020034.

da Machado, K.C., Islam, M.T., Ali, E.S., Rouf, R., Uddin, S.J., Dev, S., Shilpi, J.A., Shill, M.C., Reza, H.M., Das, A.K., Shaw, S., Mubarak, M.S., Mishra, S.K., de Melo-Cavalcante, A.A.C., 2018. A systematic review on the neuroprotective perspectives of beta-caryophyllene. Phytother. Res. 32, 2376–2388. https://doi.org/10.1002/ptr.6199.

da Oliveira, G.L.S., Machado, K.C., Machado, K.C., da Silva, A.P.d.S.C.L., Feitosa, C.M., de Castro Almeida, F. R., 2018. Non-clinical toxicity of β-caryophyllene, a dietary cannabinoid: absence of adverse effects in female Swiss mice. Regul. Toxicol. Pharmacol. 92, 338–346. https://doi.org/10.1016/j.yrtph.2017.12.013.

da Silva Gündel, S., dos Reis, T.R., Copetti, P.M., Favarin, F.R., Sagrillo, M.R., da Silva, A.S., Segat, J.C., Baretta, D., Ourique, A.F., 2019. Evaluation of cytotoxicity, genotoxicity and ecotoxicity of nanoemulsions containing Mancozeb and Eugenol. Ecotoxicol. Environ. Saf. 169, 207–215. https://doi.org/10.1016/j.ecoenv.2018.11.023.

da Silva, S.L., Chaar, J.D.S., Figueiredo, P.D.M.S., Yano, T., 2008. Cytotoxic evaluation of essential oil from Casearia sylvestris Sw on human cancer cells and erythrocytes. Acta Amazon. 38, 107–112. https://doi.org/10.1590/S0044-59672008000100012.

da Silva, B.C., Melo, D.R., Franco, C.T., Maturano, R., Fabri, R.L., Daemon, E., 2019a. Evaluation of eugenol and (E)-cinnamaldehyde insecticidal activity against larvae and pupae of Musca domestica (Diptera: Muscidae). J. Med. Entomol. 57, 181–186. https://doi.org/10.1093/jme/tjz121.

da Silva, J.L., de Marinho, A.C.T.A., dos Prazeres, L.D.K.T., Da Silva, T.G., da Merlo, K.C., Costa, H.B.A., Wanderley, A.G., 2019b. Reproductive toxicity of eugenol in wistar rats. Rev. Eletrôn. Acervo Saúde 11, e1864. https://doi.org/10.25248/reas.e1864.2019.

Damalas, C.A., Eleftherohorinos, I.G., 2011. Pesticide exposure, safety issues, and risk assessment indicators. Int. J. Environ. Res. Public Health 8, 1402–1419. https://doi.org/10.3390/ijerph8051402.

Damalas, C.A., Koutroubas, S.D., 2018. Current status and recent developments in biopesticide use. Agriculture 8, 13. https://doi.org/10.3390/agriculture8010013.

de Almeida, A.L., Caleffi-Ferracioli, K.R., de L Scodro, R.B., Baldin, V.P., Montaholi, D.C., Spricigo, L.F., Nakamura-Vasconcelos, S.S., Hegeto, L.A., Sampiron, E.G., Costacurta, G.F., dos S Yamazaki, D.A., de F Gauze, G., Siqueira, V.L., Cardoso, R.F., 2019. Eugenol and derivatives activity against Mycobacterium tuberculosis, nontuberculous mycobacteria and other bacteria. Future Microbiol. 14, 331–344. https://doi.org/10.2217/fmb-2018-0333.

de Araújo Lopes, A., da Fonseca, F.N., Rocha, T.M., de Freitas, L.B., Araújo, E.V.O., Wong, D.V.T., Lima Júnior, R.C.P., Leal, L.K.A.M., 2018. Eugenol as a promising molecule for the treatment of dermatitis: antioxidant and anti-inflammatory activities and its nanoformulation. Oxidative Med. Cell. Longev. 2018, 1–13. https://doi.org/10.1155/2018/8194849.

de Oliveira, B.M.S., Melo, C.R., Santos, A.C.C., Nascimento, L.F.A., Nízio, D.A.C., Cristaldo, P.F., Blank, A.F., Bacci, L., 2019. Essential oils from Varronia curassavica (Cordiaceae) accessions and their compounds (E)-caryophyllene and α-humulene as an alternative to control Dorymyrmex thoracius (Formicidae: Dolichoderinae). Environ. Sci. Pollut. Res. 26, 6602–6612. https://doi.org/10.1007/s11356-018-4044-1.

Devi, K.C., Devi, S.S., 2013. Insecticidal and oviposition deterrent properties of some spices against coleopteran beetle, Sitophilus oryzae. J. Food Sci. Technol. 50, 600–604. https://doi.org/10.1007/s13197-011-0377-1.

Dhara, L., Tripathi, A., 2020. Sub-acute toxicological and behavioural effects of two candidate therapeutics, cinnamaldehyde and eugenol, for treatment of ESBL producing-quinolone resistant pathogenic Enterobacteriaceae. Clin. Exp. Pharmacol. Physiol. 47, 977–988. https://doi.org/10.1111/1440-1681.13276.

Di Giacomo, S., Mazzanti, G., Di Sotto, A., 2016. Mutagenicity of cigarette butt waste in the bacterial reverse mutation assay: the protective effects of β-caryophyllene and β-caryophyllene oxide. Environ. Toxicol. 31, 1319–1328. https://doi.org/10.1002/tox.22136.

Di Giacomo, S., Di Sotto, A., Mazzanti, G., Wink, M., 2017. Chemosensitizing properties of β-caryophyllene and β-caryophyllene oxide in combination with doxorubicin in human cancer cells. Anticancer Res. 37, 1191–1196. https://doi.org/10.21873/anticanres.11433.

Di Giacomo, S., Abete, L., Cocchiola, R., Mazzanti, G., Eufemi, M., Di Sotto, A., 2018. Caryophyllane sesquiterpenes inhibit DNA-damage by tobacco smoke in bacterial and mammalian cells. Food Chem. Toxicol. 111, 393–404. https://doi.org/10.1016/j.fct.2017.11.018.

Di Sotto, A., Maffei, F., Hrelia, P., Castelli, F., Sarpietro, M.G., Mazzanti, G., 2013. Genotoxicity assessment of β-caryophyllene oxide. Regul. Toxicol. Pharmacol. 66, 264–268. https://doi.org/10.1016/j.yrtph.2013.04.006.

Didry, N., Dubreuil, L., Pinkas, M., 1994. Activity of thymol, carvacrol, cinnamaldehyde and eugenol on oral bacteria. Pharm. Acta Helv. 69, 25–28. https://doi.org/10.1016/0031-6865(94)90027-2.

Dima, C., Dima, S., 2015. Essential oils in foods: extraction, stabilization, and toxicity. Curr. Opin. Food Sci. 5, 29–35. https://doi.org/10.1016/j.cofs.2015.07.003.

Domaracký, M., Rehák, P., Juhás, Š., Koppel, J., 2007. Effects of selected plant essential oils on the growth and development of mouse preimplantation embryos in vivo. Physiol. Res. 56, 97–104.

EFSA Panel on Additives and Products or Substances used in Animal Feed (FEEDAP), 2011. Scientific opinion on the safety and efficacy of allylhydroxybenzenes (chemical group 18) when used as flavourings for all animal species. EFSA J. https://doi.org/10.2903/j.efsa.2011.2440.

El Hadri, A., Del Río, M.Á.G., Sanz, J., Coloma, A.G., Idaomar, M., Ozonas, B.R., González, J.B., Reus, M.I.S., 2010. Cytotoxic activity of α-humulene and transcaryophyllene from Salvia officinalis in animal and human tumor cells. An. R. Acad. Nac. Farm. 76, 343–356.

El-Hak, H.N.G., Gerges, M.M., 2018. Evaluating the teratogenic effect of eugenol in the development of the chick embryos. J. Biol. Res. 1, 59–75.

Ellahueñe, M.F., Pérez-Alzola, L.P., Orellana-Valdebenito, M., Muñoz, C., Lafuente-Indo, N., 1994. Genotoxic evaluation of eugenol using the bone marrow micronucleus assay. Mutat. Res. 320, 175–180. https://doi.org/10.1016/0165-1218(94)90044-2.

Escobar-García, M., Rodríguez-Contreras, K., Ruiz-Rodríguez, S., Pierdant-Pérez, M., Cerda-Cristerna, B., Pozos-Guillén, A., 2016. Eugenol toxicity in human dental pulp fibroblasts of primary teeth. J. Clin. Pediatr. Dent. 40, 312–318. https://doi.org/10.17796/1053-4628-40.4.312.

European Food Safety Authority, 2012. Conclusion on the peer review of the pesticide risk assessment of the active substance plant oils/clove oil. EFSA J. 10, 2506. https://doi.org/10.2903/j.efsa.2012.2506.

FDA/CVM, 2007. Guidance for Industry—Concerns Related to the Use of Clove Oil as an Anesthetic for Fish. FDA Center for Veterinary Medicine, Rockville, Maryland.

Fernandes, E.S., Passos, G.F., Medeiros, R., da Cunha, F.M., Ferreira, J., Campos, M.M., Pianowski, L.F., Calixto, J.B., 2007. Anti-inflammatory effects of compounds alpha-humulene and (−)-trans-caryophyllene isolated from the essential oil of Cordia verbenacea. Eur. J. Pharmacol. 569, 228–236. https://doi.org/10.1016/j.ejphar.2007.04.059.

Fidyt, K., Fiedorowicz, A., Strządała, L., Szumny, A., 2016. β-Caryophyllene and β-caryophyllene oxide-natural compounds of anticancer and analgesic properties. Cancer Med. 5, 3007–3017. https://doi.org/10.1002/cam4.816.

Fischer, I.U., Von Unruh, G.E., Dengler, H.J., 1990. The metabolism of eugenol in man. Xenobiotica 20, 209–222. https://doi.org/10.3109/00498259009047156.

Fontes, L.B.A., dos Dias, D.S., Aarestrup, B.J.V., Aarestrup, F.M., Da Silva Filho, A.A., do Corrêa, J.O.A., 2017. β-Caryophyllene ameliorates the development of experimental autoimmune encephalomyelitis in C57BL/6 mice. Biomed. Pharmacother. 91, 257–264. https://doi.org/10.1016/j.biopha.2017.04.092.

Food and Agriculture Organization of the United Nations, 2019. FAOSTAT [WWW Document]. URL http://www.fao.org/faostat/en/?#data/QC/visualize.

Gaire, S., O'Connell, M., Holguin, F.O., Amatya, A., Bundy, S., Romero, A., 2017. Insecticidal properties of essential oils and some of their constituents on the turkestan cockroach (Blattodea: Blattidae). J. Econ. Entomol. 110, 584–592. https://doi.org/10.1093/jee/tox035.

Garrido-Miranda, K.A., Rivas, B.L., Pérez-Rivera, M.A., Sanfuentes, E.A., Peña-Farfal, C., 2018. Antioxidant and antifungal effects of eugenol incorporated in bionanocomposites of poly(3-hydroxybutyrate)-thermoplastic starch. LWT Food Sci. Technol. 98, 260–267. https://doi.org/10.1016/j.lwt.2018.08.046.

Ghosh, S., Chatterjee, D., Das, S., Bhattacharjee, P., 2013. Supercritical carbon dioxide extraction of eugenol-rich fraction from Ocimum sanctum Linn and a comparative evaluation with other extraction techniques: process optimization and phytochemical characterization. Ind. Crop. Prod. 47, 78–85. https://doi.org/10.1016/j.indcrop.2013.02.030.

González Armijos, M.J., Viteri Jumbo, L., Faroni, L.R., Oliveira, E.E., Flores, A.F., Heleno, F., Haddi, K., 2019. Fumigant toxicity of eugenol and its negative effects on biological development of Callosobruchus maculatus L. Rev. Ciênc. Agron. 36, 5–15. https://doi.org/10.22267/rcia.193601.94.

Goulet, F., Vachon, P., Hélie, P., 2011. Evaluation of the toxicity of eugenol at anesthetic doses in African clawed frogs (Xenopus laevis). Toxicol. Pathol. 39, 471–477. https://doi.org/10.1177/0192623311399785.

Govindarajan, M., Benelli, G., 2016. α-Humulene and β-elemene from Syzygium zeylanicum (Myrtaceae) essential oil: highly effective and eco-friendly larvicides against Anopheles subpictus, Aedes albopictus, and Culex tritaeniorhynchus (Diptera: Culicidae). Parasitol. Res. 115, 2771–2778. https://doi.org/10.1007/s00436-016-5025-2.

Green, N.R., Savage, J.R., 1978. Screening of safrole, eugenol, their ninhydrin positive metabolites and selected secondary amines for potential mutagenicity. Mutat. Res. Fundam. Mol. Mech. Mutagen. 57, 115–121. https://doi.org/10.1016/0027-5107(78)90257-9.

Guan, W., Li, S., Yan, R., Tang, S., Quan, C., 2007. Comparison of essential oils of clove buds extracted with supercritical carbon dioxide and other three traditional extraction methods. Food Chem. 101, 1558–1564. https://doi.org/10.1016/j.foodchem.2006.04.009.

Guénette, S.A., Ross, A., Marier, J.F., Beaudry, F., Vachon, P., 2007. Pharmacokinetics of eugenol and its effects on thermal hypersensitivity in rats. Eur. J. Pharmacol. 562, 60–67. https://doi.org/10.1016/j.ejphar.2007.01.044.

Gueretz, J.S., Somensi, C.A., Martins, M.L., de Souza, A.P., 2017. Evaluation of eugenol toxicity in bioassays with test-organisms. Cienc. Rural 47. https://doi.org/10.1590/0103-8478cr20170194.

Hagan, E.C., Jenner, P.M., Jones, W.I., Fitzhugh, O.G., Long, E.L., Brouwer, J.G., Webb, W.K., 1965. Toxic properties of compounds related to safrole. Toxicol. Appl. Pharmacol. 7, 18–24. https://doi.org/10.1016/0041-008X(65)90069-4.

Hagan, E.C., Hansen, W.H., Fitzhugh, O.G., Jenner, P.M., Jones, W.I., Taylor, J.M., Long, E.L., Nelson, A.A., Brouwer, J.B., 1967. Food flavourings and compounds of related structure. II. Subacute and chronic toxicity. Food Cosmet. Toxicol. 5, 141–157. https://doi.org/10.1016/S0015-6264(67)82961-4.

Hanušová, V., Caltová, K., Svobodová, H., Ambrož, M., Skarka, A., Murínová, N., Královа, V., Tomšík, P., Skálová, L., 2017. The effects of β-caryophyllene oxide and trans-nerolidol on the efficacy of doxorubicin in breast cancer cells and breast tumor-bearing mice. Biomed. Pharmacother. 95, 828–836. https://doi.org/10.1016/j.biopha.2017.09.008.

Hartnoll, G., Moore, D., Douek, D., 1993. Near fatal ingestion of oil of cloves. Arch. Dis. Child. 69, 392–393. https://doi.org/10.1136/adc.69.3.392.

Hernández-Moreno, D., Soffers, A.E.M.F., Wiratno, Falke, H.E., Rietjens, I.M.C.M., Murk, A.J., 2013. Consumer and farmer safety evaluation of application of botanical pesticides in black pepper crop protection. Food Chem. Toxicol. 56, 483–490. https://doi.org/10.1016/j.fct.2013.01.033.

Hilton, J., Dearman, R.J., Fielding, I., Basketter, D.A., Kimber, I., 1996. Evaluation of the sensitizing potential of eugenol and isoeugenol in mice and guinea pigs. J. Appl. Toxicol. 16, 459–464. https://doi.org/10.1002/(SICI)1099-1263(199609)16:5<459::AID-JAT366>3.0.CO;2-H.

Hirose, M., Masuda, A., Imaida, K., Kagawa, M., Tsuda, H., Ito, N., 1987. Induction of forestomach lesions in rats by oral administrations of naturally occurring antioxidants for 4 weeks. Jpn. J. Cancer Res. 78, 317–321. https://doi.org/10.20772/cancersci1985.78.4_317.

Ho, Y.-C., Huang, F.-M., Chang, Y.-C., 2006. Mechanisms of cytotoxicity of eugenol in human osteoblastic cells in vitro. Int. Endod. J. 39, 389–393. https://doi.org/10.1111/j.1365-2591.2006.01091.x.

Hong, T.K., Perumalsamy, H., Jang, K.H., Na, E.S., Ahn, Y.J., 2018. Ovicidal and larvicidal activity and possible mode of action of phenylpropanoids and ketone identified in Syzygium aromaticum bud against Bradysia procera. Pestic. Biochem. Physiol. 145, 29–38. https://doi.org/10.1016/j.pestbp.2018.01.003.

Huang, Y., Ho, S.-H., Lee, H.-C., Yap, Y.-L., 2002. Insecticidal properties of eugenol, isoeugenol and methyleugenol and their effects on nutrition of Sitophilus zeamais Motsch. (Coleoptera: Curculionidae) and Tribolium castaneum (Herbst) (Coleoptera: Tenebrionidae). J. Stored Prod. Res. 38, 403–412. https://doi.org/10.1016/S0022-474X(01)00042-X.

IARC, 1985. IARC Monographs on the Evaluation of the Carcinogenic Risk of Chemicals to Humans—Allyl Compounds, Aldehydes, Epoxides, and Peroxides. Lyon, France.

Ibrahim, R.A., Alahmadi, S.S., 2015. Effect of syzygium aromaticum cloves on larvae of the rhinoceros beetle, oryctes agamemnon (Coleoptera: Scarabaeidae). Afr. Entomol. 23, 458–466. https://doi.org/10.4001/003.023.0215.

ICH, 1999. ICH Safety Guideline—S4: Duration of Chronic Toxicity Testing in Animals (Rodent and Non Rodent Toxicity Testing), ICH Harmonised Tripartite Guideline. European Medicines Agency, London, UK.

IFRA, 2020. IFRA Standard (Amendment 49)—Eugenol. The International Fragrance Association, Geneva, Switzerland.

Ileke, K.D., Ogungbite, O.C., Olayinka-Olagunju, J.O., 2014. Powders and extracts of Syzygium aromaticum and Anacardium occidentale as entomocides against the infestation of Sitophilus oryzae (L.) [Coleoptera: Curculionidae] on stored sorghum grains. Afr. Crop. Sci. J. 22, 267–273.

Ismail, S., AbdEl-Kader, S., 2011. Clove: is it has a molluscicidal activity against land snails (Monacha cartusiana)? J. Plant Prot. Pathol. 2, 561–569. https://doi.org/10.21608/jppp.2011.86499.

Ismail, S.M., Sleem, F.M.A., 2021. Contact toxicity, repellent activity and residual effect of Syzygium aromaticum L. (Myrtaceae) and Piper nigrum L. (Pipeaceae) powders against Sitophilus oryzae L. (Coleoptera: Curculionidae). J. Entomol. Zool. Stud. 8, 2053–2058.

Itoh, M., 1982. Sensitization potency of some phenolic compounds. J. Dermatol. 9, 223–233. https://doi.org/10.1111/j.1346-8138.1982.tb02629.x.

Jadhav, B.K., Khandelwal, K.R., Ketkar, A.R., Pisal, S.S., 2004. Formulation and evaluation of mucoadhesive tablets containing eugenol for the treatment of periodontal diseases. Drug Dev. Ind. Pharm. 30, 195–203. https://doi.org/10.1081/DDC-120028715.

Jang, H.-I., Rhee, K.-J., Eom, Y.-B., 2020. Antibacterial and antibiofilm effects of α-humulene against Bacteroides fragilis. Can. J. Microbiol. 66, 389–399. https://doi.org/10.1139/cjm-2020-0004.

Jeng, J.H., Hahn, L.J., Lu, F.J., Wang, Y.J., 1994. Eugenol triggers different pathobiological effects on human oral mucosal fibroblasts 1. J. Dent. Res. 73, 1050–1055. https://doi.org/10.1177/00220345940730050601.

Jenner, P.M., Hagan, E.C., Taylor, J.M., Cook, E.L., Fitzhugh, O.G., 1964. Food flavourings and compounds of related structure I. Acute oral toxicity. Food Cosmet. Toxicol. 2, 327–343. https://doi.org/10.1016/S0015-6264(64)80192-9.

Jun, N.J., Mosaddik, A., Moon, J.Y., Jang, K.C., Lee, D.S., Ahn, K.S., Cho, S.K., 2011. Cytotoxic activity of β-caryophyllene oxide isolated from Jeju Guava (Psidium cattleianum Sabine) leaf. Rec. Nat. Prod. 5, 242–246.

Kafle, L., Shih, C.J., 2013. Toxicity and repellency of compounds from clove (Syzygium aromaticum) to red imported fire ants Solenopsis invicta (Hymenoptera: Formicidae). J. Econ. Entomol. 106, 131–135. https://doi.org/10.1603/EC12230.

Karakaya, S., Yilmaz, S.V., Özdemir, Ö., Koca, M., Pınar, N.M., Demirci, B., Yıldırım, K., Sytar, O., Turkez, H., Baser, K.H.C., 2020. A caryophyllene oxide and other potential anticholinesterase and anticancer agent in Salvia verticillata subsp. amasiaca (Freyn & Bornm.) Bornm. (Lamiaceae). J. Essent. Oil Res. 32, 512–525. https://doi.org/10.1080/10412905.2020.1813212.

Karapinar, M., Esen Aktuğ, Ş., 1987. Inhibition of foodborne pathogens by thymol, eugenol, menthol and anethole. Int. J. Food Microbiol. 4, 161–166. https://doi.org/10.1016/0168-1605(87)90023-7.

Ke, C., Liu, Q., Li, L., Chen, J., Zhao, C., Xu, J., Huang, K., Mengsong, M., Li, L., 2018. Residual levels and risk assessment of eugenol and its isomers in fish from China markets. Aquaculture 484, 338–342. https://doi.org/10.1016/j.aquaculture.2017.07.034.

Khalaf Al-Joboory, R., 2019. Evaluation of the effect of different concentrations of plant powders on the red (Coleoptera: Tenebrionidae) Herbest (flour beetles Tribolium castaneum). J. Phys. Conf. Ser. 1294. https://doi.org/10.1088/1742-6596/1294/6/062070.

Khalil, A.A., ur Rahman, U., Khan, M.R., Sahar, A., Mehmood, T., Khan, M., 2017. Essential oil eugenol: sources, extraction techniques and nutraceutical perspectives. RSC Adv. 7, 32669–32681. https://doi.org/10.1039/C7RA04803C.

Kim, S.-I., Yoon, J.-S., Jung, J.W., Hong, K.-B., Ahn, Y.-J., Kwon, H.W., 2010. Toxicity and repellency of origanum essential oil and its components against Tribolium castaneum (Coleoptera: Tenebrionidae) adults. J. Asia Pac. Entomol. 13, 369–373. https://doi.org/10.1016/j.aspen.2010.06.011.

Kim, K.-H., Yi, C.-G., Ahn, Y.-J., Kim, S.I., Lee, S.-G., Kim, J.-R., 2015. Fumigant toxicity of basil oil compounds and related compounds to Thrips palmi and Orius strigicollis. Pest Manag. Sci. 71, 1292–1296. https://doi.org/10.1002/ps.3925.

Kubo, I., Chaudhuri, S., Kubo, Y., Sanchez, Y., Ogura, T., Saito, T., Ishikawa, H., Haraguchi, H., 1996. Cytotoxic and antioxidative sesquiterpenoids from Heterotheca inuloides. Planta Med. 62, 427–430. https://doi.org/10.1055/s-2006-957932.

Kuete, V., 2017. Other health benefits of african medicinal spices and vegetables. In: Kuete, V. (Ed.), Medicinal Spices and Vegetables From Africa. Academic Press, London, UK, pp. 329–349, https://doi.org/10.1016/B978-0-12-809286-6.00013-3.

Kumar, R., 2020. Problem of snails and snugs in vegetable crops and their management. In: Vishwakarma, R., Kumar, R. (Eds.), Management of Insect Pests in Vegetable Crops—Concepts and Approaches. Apple Academic Press, Burlington, ON, Canada, pp. 279–307.

Kumar, P., Singh, D.K., 2006. Molluscicidal activity of Ferula asafoetida, Syzygium aromaticum and Carum carvi and their active components against the snail Lymnaea acuminata. Chemosphere 63, 1568–1574. https://doi.org/10.1016/j.chemosphere.2005.08.071.

Kurniawan, A., Andang, M.A., Hayati, A.T., 2007. The effect of Beetle leaves (Piper Betle Linn) for dental caries formation. Padjajaran J. Dent. 19. https://doi.org/10.24198/pjd.vol19no3.14162.

Lambert, M.M., Campos, D.R., Borges, D.A., de Avelar, B.R., Ferreira, T.P., Cid, Y.P., Boylan, F., Scott, F.B., de Almeida Chaves, D.S., Coumendouros, K., 2020. Activity of Syzygium aromaticum essential oil and its main constituent eugenol in the inhibition of the development of Ctenocephalides felis felis and the control of adults. Vet. Parasitol. 282, 109126. https://doi.org/10.1016/j.vetpar.2020.109126.

Lavoie, E.J., Adams, J.D., Reinhardt, J., Rivenson, A., Hoffmann, D., 1986. Toxicity studies on clove cigarette smoke and constituents of clove: determination of the LD50 of eugenol by intratracheal instillation in rats and hamsters. Arch. Toxicol. 59, 78–81. https://doi.org/10.1007/BF00286727.

Lee, M.H., Yeon, K.-Y., Park, C.-K., Li, H.-Y., Fang, Z., Kim, M.S., Choi, S.-Y., Lee, S.J., Lee, S., Park, K., Lee, J.-H., Kim, J.S., Oh, S.B., 2005. Eugenol inhibits calcium currents in dental afferent neurons. J. Dent. Res. 84, 848–851. https://doi.org/10.1177/154405910508400913.

Legault, J., Pichette, A., 2010. Potentiating effect of β-caryophyllene on anticancer activity of α-humulene, isocaryophyllene and paclitaxel. J. Pharm. Pharmacol. 59, 1643–1647. https://doi.org/10.1211/jpp.59.12.0005.

Leite, A.M., de Lima, E.O., de Souza, E.L., de Diniz, M.F.F.M., Trajano, V.N., de Medeiros, I.A., 2007. Inhibitory effect of beta-pinene, alpha-pinene and eugenol on the growth of potential infectious endocarditis causing Gram-positive bacteria. Rev. Bras. Ciênc. Farm. 43, 121–126. https://doi.org/10.1590/S1516-93322007000100015.

Li, J., Yu, Y., Yang, Y., Liu, X., Zhang, J., Li, B., Zhou, X., Niu, J., Wei, X., Liu, Z., 2012. A 15-day oral dose toxicity study of aspirin eugenol ester in Wistar rats. Food Chem. Toxicol. 50, 1980–1985. https://doi.org/10.1016/j.fct.2012.03.080.

Lionnet, L., Beaudry, F., Vachon, P., 2010. Intrathecal eugenol administration alleviates neuropathic pain in male Sprague-Dawley rats. Phytother. Res. 24, 1645–1653. https://doi.org/10.1002/ptr.3174.

Liska, A., Rozman, V., Kalinovic, I., Ivezic, M., Balicevic, R., 2010. Contact and fumigant activity of 1,8-cineole, eugenol and camphor against Tribolium castaneum (Herbst). Julius-Kühn-Archiv 425, 716–720. https://doi.org/10.5073/jka.2010.425.093.

Liu, Y.-Q., Xue, M., Zhang, Q.-C., Zhou, F.-Y., Wei, J.-Q., 2010. Toxicity of β-caryophyllene from Vitex negundo (Lamiales: Verbenaceae) to Aphis gossypii Glover (Homoptera: Aphididae) and its action mechanism. Acta Entomol. Sin. 53, 396–404.

Maralhas, A., Monteiro, A., Martins, C., Kranendonk, M., Laires, A., Rueff, J., Rodrigues, A.S., 2006. Genotoxicity and endoreduplication inducing activity of the food flavouring eugenol. Mutagenesis 21, 199–204. https://doi.org/10.1093/mutage/gel017.

Marin Municipal Water District, 2010. Clove Oil (Eugenol), Marin Municipal Water District Vegetation Management Plan—Herbicide Risk Assessment DRAFT-1/1/2010. Marin Municipal Water District, Corte Madera, CA.

Markowitz, K., Moynihan, M., Liu, M., Kim, S., 1992. Biologic properties of eugenol and zinc oxide-eugenol. Oral Surg. Oral Med. Oral Pathol. 73, 729–737. https://doi.org/10.1016/0030-4220(92)90020-Q.

Martínez, L.C., Plata-Rueda, A., Colares, H.C., Campos, J.M., Dos Santos, M.H., Fernandes, F.L., Serrão, J.E., Zanuncio, J.C., 2018. Toxic effects of two essential oils and their constituents on the mealworm beetle, Tenebrio molitor. Bull. Entomol. Res. 108, 716–725. https://doi.org/10.1017/S0007485317001262.

Martins, C., Rueff, J., Rodrigues, A.S., 2018. Genotoxic alkenylbenzene flavourings, a contribution to risk assessment. Food Chem. Toxicol. 118, 861–879. https://doi.org/10.1016/j.fct.2018.06.020.

Maura, A., Pino, A., Ricci, R., 1989. Negative evidence in vivo of DNA-damaging, mutagenic and chromosomal effects of eugenol. Mutat. Res. Lett. 227, 125–129. https://doi.org/10.1016/0165-7992(89)90008-0.

Miller, E.C., Swanson, A.B., Phillips, D.H., Liem, A., Miller, J.A., Fletcher, T.L., 1983. Structure-activity studies of the carcinogenicities in the mouse and rat of some naturally occurring and synthetic alkenylbenzene derivatives related to safrole and estragole. Cancer Res. 43, 1124–1134.

Moemenbellah-Fard, M.D., Abdollahi, A., Ghanbariasad, A., Osanloo, M., 2020. Antibacterial and leishmanicidal activities of Syzygium aromaticum essential oil versus its major ingredient, eugenol. Flavour Fragr. J. 35, 534–540. https://doi.org/10.1002/ffj.3595.

Mohammadi Nejad, S., 2014. Evaluation of Cytotoxicity and Genotoxicity of Eugenol in Lymphocytes. Hacettepe University.

Mohammadi Nejad, S., Özgüneş, H., Başaran, N., 2017. Öjenolün Farmakolojik Ve Toksikolojik Özellikleri. Turkish J. Pharm. Sci. 14, 201–206. https://doi.org/10.4274/tjps.62207.

Mohammed, O.J., McAlpine, R., Chiewhatpong, P., Latif, M.L., Pratten, M.K., 2016. Assessment of developmental cardiotoxic effects of some commonly used phytochemicals in mouse embryonic D3 stem cell differentiation and chick embryonic cardiomyocyte micromass culture models. Reprod. Toxicol. 64, 86–97. https://doi.org/10.1016/j.reprotox.2016.04.011.

Molina-Jasso, D., Álvarez-González, I., Madrigal-Bujaidar, E., 2009. Clastogenicity of beta-caryophyllene in mouse. Biol. Pharm. Bull. 32, 520–522. https://doi.org/10.1248/bpb.32.520.

Monzote, L., Stamberg, W., Staniek, K., Gille, L., 2009. Toxic effects of carvacrol, caryophyllene oxide, and ascaridole from essential oil of Chenopodium ambrosioides on mitochondria. Toxicol. Appl. Pharmacol. 240, 337–347. https://doi.org/10.1016/j.taap.2009.08.001.

Moon, S.-E., Kim, H.-Y., Cha, J.-D., 2011. Synergistic effect between clove oil and its major compounds and antibiotics against oral bacteria. Arch. Oral Biol. 56, 907–916. https://doi.org/10.1016/j.archoralbio.2011.02.005.

Moore, S.J., Lenglet, A., Hill, N., 2007. Plant-based insect repellents. In: Debboun, M., Frances, S.P., Strickman, D. (Eds.), Insect Repellents: Principles, Methods, and Uses. CRC Press, Boca Raton, FL, pp. 275–3030.

Moreno, É.M., Leal, S.M., Stashenko, E.E., García, L.T., 2018. Induction of programmed cell death in Trypanosoma cruzi by Lippia alba essential oils and their major and synergistic terpenes (citral, limonene and caryophyllene oxide). BMC Complement. Altern. Med. 18, 225. https://doi.org/10.1186/s12906-018-2293-7.

Nararak, J., Di Giorgio, C., Sukkanon, C., Mahiou-Leddet, V., Ollivier, E., Manguin, S., Chareonviriyaphap, T., 2020. Excito-repellency and biological safety of β-caryophyllene oxide against Aedes albopictus and Anopheles dirus (Diptera: Culicidae). Acta Trop. 210, 105556. https://doi.org/10.1016/j.actatropica.2020.105556.

Navikaite-Snipaitiene, V., Ivanauskas, L., Jakstas, V., Rüegg, N., Rutkaite, R., Wolfram, E., Yildirim, S., 2018. Development of antioxidant food packaging materials containing eugenol for extending display life of fresh beef. Meat Sci. 145, 9–15. https://doi.org/10.1016/j.meatsci.2018.05.015.

Ngoh, S.P., Choo, L.E.W., Pang, F.Y., Huang, Y., Kini, M.R., Ho, S.H., 1998. Insecticidal and repellent properties of nine volatile constituents of essential oils against the American cockroach, Periplaneta americana (L.). Pestic. Sci. 54, 261–268. https://doi.org/10.1002/(SICI)1096-9063(1998110)54:3<261::AID-PS794>3.0.CO;2-C.

Noël, M.S.G., Habiba, K., Illiassa, N., Elias, N.N., 2020. Bioefficacity of mixed powders of Vepris heterophylla and Syzygium aromaticum on the developing stages of Callosobruchus maculatus (Coleoptera: Bruchidae) pest control of Vigna unguiculata (Fabacae) in storage in the Far-North region of Cameroun. J. Entomol. Zool. Stud. 8, 1358–1364.

NTP, 1983. NTP Technical Report on the Carcinogenesis Studies of Eugenol (CAS No. 97-53-0) in F344/N Rats and B6C3F1 Mice (Feed Studies). North Carolina.

Nurdjannah, N., Bermawie, N., 2012. Cloves. In: Peter, K.V. (Ed.), Handbook of Herbs and Spices. Woodhead Publishing Limited, Cambridge, UK, pp. 197–215, https://doi.org/10.1533/9780857095671.197.

Obeng-Ofori, D., Reichmuth, C., 1997. Bioactivity of eugenol, a major component of essential oil of Ocimum suave (Wild.) against four species of stored-product Coleoptera. Int. J. Pest Manag. 43, 89–94. https://doi.org/10.1080/096708797229040.

OECD, 2018. Test No. 452: Chronic Toxicity Studies, OECD Guidelines for the Testing of Chemicals, Section 4. OECD, https://doi.org/10.1787/9789264071209-en.

Ogendo, J.O., Deng, A.L., Kostyukovsky, M., Ravid, U., Matasyoh, J.C., Omolo, E.O., Kariuki, S.T., Kamau, A. W., Bett, P.K., Shaaya, E., 2011. Biocontrol potential of selected plant essential oil constituents as fumigants of insect pests attacking stored food commodities. Baraton Interdiscip. Res. J. 1, 13–19.

Oh, M.-S., Yang, J.-Y., Kim, M.-G., Lee, H.-S., 2014. Acaricidal activities of β-caryophyllene oxide and structural analogues derived from Psidium cattleianum oil against house dust mites. Pest Manag. Sci. 70, 757–762. https://doi.org/10.1002/ps.3608.

Ohsumi, T., Higashi, S., Ozumi, K., Kuroki, K., 1996. Study on allergic contact dermatitis of eugenol in guinea pig. Oral Ther. Pharmacol. 15, 63–68. https://doi.org/10.11263/jsotp1982.15.63.

Okey, I.B., Gabriel, U.U., Deekae, S.N., 2018a. Comparative effects of the acute toxicity of clove (Eugenia Aromatica) powder to clarias gariepinus and heterobranchus bidorsalis fingerlings. Int. J. Innov. Stud. Aquat. Biol. Fish. 4, 32–39. https://doi.org/10.20431/2454-7670.0402005.

Okey, I.B., Keremah, R.I., Gabriel, U.U., 2018b. The efficacy of clove (Eugenia caryophyllata) powder as anaesthesia on African catfishes (Clarias gariepinus and Heterobranchus bidorsalis) fingerlings. J. Aquac. Mar. Biol. 7, 182–188. https://doi.org/10.15406/jamb.2018.07.00206.

Öntürk, H., Özbek, H., 2009. Trans-karyofillen ve öjenol'ün akut toksisitesi ve hipoglisemik etkinliğinin diyabetik fareler üzerinde araştırılması. Genel Tip Derg. 19, 17–23.

Oparaeke, A.M., Daria, V.S., 2005. Toxicity of some plant powders to Callosobruchus maculatus (Fab.) on stored cowpea. Niger. J. Entomol. 22, 76–83.

Opdyke, D.L.J., 1979. Monographs on Fragrance Raw Materials. Pergamon Press, Oxford, UK.

Oppong-Damoah, A., Blough, B.E., Makriyannis, A., Murnane, K.S., 2019. The sesquiterpene beta-caryophyllene oxide attenuates ethanol drinking and place conditioning in mice. Heliyon 5, e01915. https://doi.org/10.1016/j.heliyon.2019.e01915.

Panda, S., Sharma, R., Kar, A., 2019. Chavibetol corrects thyrotoxicosis through alterations in thyroid peroxidase. Naunyn Schmiedeberg's Arch. Pharmacol. 392, 541–550. https://doi.org/10.1007/s00210-018-01606-x.

Pang, X., Almaz, B., Qi, X.-J., Wang, Y., Feng, Y.-X., Geng, Z.-F., Xi, C., Du, S.-S., 2020. Bioactivity of essential oil from atalantia buxifolia leaves and its major sesquiterpenes against three stored-product insects. J. Essent. Oil Bear. Plants 23, 38–50. https://doi.org/10.1080/0972060X.2020.1729245.

Park, C.-K., Li, H.Y., Yeon, K.-Y., Jung, S.J., Choi, S.-Y., Lee, S.J., Lee, S., Park, K., Kim, J.S., Oh, S.B., 2006. Eugenol inhibits sodium currents in dental afferent neurons. J. Dent. Res. 85, 900–904. https://doi.org/10.1177/154405910608501005.

Park, K.-R., Nam, D., Yun, H.-M., Lee, S.-G., Jang, H.-J., Sethi, G., Cho, S.K., Ahn, K.S., 2011. β-Caryophyllene oxide inhibits growth and induces apoptosis through the suppression of PI3K/AKT/mTOR/S6K1 pathways and ROS-mediated MAPKs activation. Cancer Lett. 312, 178–188. https://doi.org/10.1016/j.canlet.2011.08.001.

Pattanasiri, T., Taparhudee, W., Suppakul, P., 2017. Acute toxicity and anaesthetic effect of clove oil and eugenol on Siamese fighting fish, Betta splendens. Aquac. Int. 25, 163–175. https://doi.org/10.1007/s10499-016-0020-2.

Pavela, R., 2015. Acute toxicity and synergistic and antagonistic effects of the aromatic compounds of some essential oils against Culex quinquefasciatus Say larvae. Parasitol. Res. 114, 3835–3853. https://doi.org/10.1007/s00436-015-4614-9.

Paventi, G., de Acutis, L., De Cristofaro, A., Pistillo, M., Germinara, G.S., Rotundo, G., 2020. Biological activity of Humulus lupulus (L.) essential oil and its main components against Sitophilus granarius (L.). Biomolecules 10, 1108. https://doi.org/10.3390/biom10081108.

Plata-Rueda, A., Campos, J.M., da Silva Rolim, G., Martínez, L.C., Dos Santos, M.H., Fernandes, F.L., Serrão, J. E., Zanuncio, J.C., 2018. Terpenoid constituents of cinnamon and clove essential oils cause toxic effects and behavior repellency response on granary weevil, Sitophilus granarius. Ecotoxicol. Environ. Saf. 156, 263–270. https://doi.org/10.1016/j.ecoenv.2018.03.033.

Polanco-Hernández, G.M., Giménez-Turba, A., Salamanca, E., Getti, G., Rai, R., Acosta-Viana, K.Y., Arana-Argáez, V.E., Torres-Romero, J.C., Fernández-Martín, K.G., Segura-Campos, M.R., Moo-Puc, R.E., Peña-Rodríguez, L.M., 2021. Leishmanicidal activity and immunomodulatory effect of a mixture of lupenone and β-caryophyllene oxide. Rev. Bras. https://doi.org/10.1007/s43450-021-00143-8.

Pramod, K., Aji Alex, M.R., Singh, M., Dang, S., Ansari, S.H., Ali, J., 2016. Eugenol nanocapsule for enhanced therapeutic activity against periodontal infections. J. Drug Target. 24, 24–33. https://doi.org/10.3109/1061186X.2015.1052071.

Prashar, A., Locke, I.C., Evans, C.S., 2006. Cytotoxicity of clove (Syzygium aromaticum) oil and its major components to human skin cells. Cell Prolif. 39, 241–248. https://doi.org/10.1111/j.1365-2184.2006.00384.x.

Raghavendra, R.H., Naidu, K.A., 2009. Spice active principles as the inhibitors of human platelet aggregation and thromboxane biosynthesis. Prostaglandins Leukot. Essent. Fat. Acids 81, 73–78. https://doi.org/10.1016/j.plefa.2009.04.009.

Rasha, F.S., Mohamed, A.H., Salwa, M.S.A., Ragaa, K.A.H., 2012. Effect of katel-sous dust and clove powder and their mixtures on the cowpea seed beetle, Callosobruchus maculatus (F.) (Coleoptera: Bruchidae). Egypt. Acad. J. Biol. Sci. F. Toxicol. Pest Control 4, 23–33.

Rathee, J.S., Patro, B.S., Mula, S., Gamre, S., Chattopadhyay, S., 2006. Antioxidant activity of piper betel leaf extract and its constituents. J. Agric. Food Chem. 54, 9046–9054. https://doi.org/10.1021/jf061679e.

Rogerio, A.P., Andrade, E.L., Leite, D.F.P., Figueiredo, C.P., Calixto, J.B., 2009. Preventive and therapeutic anti-inflammatory properties of the sesquiterpene α-humulene in experimental airways allergic inflammation. Br. J. Pharmacol. 158, 1074–1087. https://doi.org/10.1111/j.1476-5381.2009.00177.x.

Rompelberg, C., Vogels, J., de Vogel, N., Bruijntjes-Rozier, G., Stenhuis, W., Bogaards, J., Verhagen, H., 1996. Effect of short-term dietary administration of eugenol in humans. Hum. Exp. Toxicol. 15, 129–135. https://doi.org/10.1177/096032719601500205.

Ryu, B., Kim, H.M., Lee, J.S., Lee, C.K., Sezirahiga, J., Woo, J.H., Choi, J.H., Jang, D.S., 2016. New flavonol glucuronides from the flower buds of Syzygium aromaticum (Clove). J. Agric. Food Chem. 64, 3048–3053. https://doi.org/10.1021/acs.jafc.6b00337.

Sahlan, M., Tejamaya, M., Arbianti, R., Baruji, T., Adawiyah, R., Hermansyah, H., 2021. The effects of nano-casein encapsulation and productions of a controlled-release on eugenol containing bio-pesticide toxicity. Key Eng. Mater. 874, 115–127. https://doi.org/10.4028/www.scientific.net/KEM.874.115.

Sain, S., Naoghare, P., Devi, S., Daiwile, A., Krishnamurthi, K., Arrigo, P., Chakrabarti, T., 2014. Beta caryophyllene and caryophyllene oxide, isolated from aegle marmelos, as the potent anti-inflammatory agents against lymphoma and neuroblastoma cells. Antiinflamm. Antiallergy Agents Med. Chem. 13, 45–55. https://doi.org/10.2174/18715230113129990016.

Sánchez-Vázquez, F.J., Terry, M.I., Felizardo, V.O., Vera, L.M., 2011. Daily rhythms of toxicity and effectiveness of anesthetics (MS222 and Eugenol) in Zebrafish (Danio Rerio). Chronobiol. Int. 28, 109–117. https://doi.org/10.3109/07420528.2010.538105.

Saraphanchotiwitthaya, A., Khorana, N., Sripalakit, P., 2019. Comparative anti-inflammatory activity of eugenol and eugenyl acetate on the murine immune response in vitro. Songklanakarin J. Sci. Technol. 41, 641–648. https://doi.org/10.14456/sjst-psu.2019.68.

Sarrami, N., Pemberton, M.N., Thornhill, M.H., Theaker, E.D., 2002. Adverse reactions associated with the use of eugenol in dentistry. Br. Dent. J. 193, 257–259. https://doi.org/10.1038/sj.bdj.4801539.

Schafer, E.W., Bowles, W.A., Hurlbut, J., 1983. The acute oral toxicity, repellency, and hazard potential of 998 chemicals to one or more species of wild and domestic birds. Arch. Environ. Contam. Toxicol. 12, 355–382. https://doi.org/10.1007/BF01059413.

Schmitt, D., Levy, R., Carroll, B., 2016. Toxicological evaluation of β-caryophyllene oil. Int. J. Toxicol. 35, 558–567. https://doi.org/10.1177/1091581816655303.

Schnuch, A., Uter, W., Geier, J., Lessmann, H., Frosch, P.J., 2007. Sensitization to 26 fragrances to be labelled according to current European regulation. Contact Dermatitis 57, 1–10. https://doi.org/10.1111/j.1600-0536.2007.01088.x.

Scholes, E.W., Pendlington, R.U., Sharma, R.K., Basketter, D.A., 1994. Skin metabolism of contact allergens. Toxicol. In Vitro 8, 551–553. https://doi.org/10.1016/0887-2333(94)90014-0.

Scientific Panel on Food Additives, Flavourings, Processing Aids and Materials in Contact With Food, 2009. Scientific Opinion of the Panel on Food Additives, Flavourings, Processing Aids and Materials in Contact With Food on a Request From Commission on Consideration of Eugenol and Related Hydroxyallylbenzene Derivatives Evaluated by JECFA (65th Meeting) Struc.

Segat, G.C., Manjavachi, M.N., Matias, D.O., Passos, G.F., Freitas, C.S., Costa, R., Calixto, J.B., 2017. Antiallodynic effect of β-caryophyllene on paclitaxel-induced peripheral neuropathy in mice. Neuropharmacology 125, 207–219. https://doi.org/10.1016/j.neuropharm.2017.07.015.

Sekizawa, J., Shibamoto, T., 1982. Genotoxicity of safrole-related chemicals in microbial test systems. Mutat. Res. 101, 127–140. https://doi.org/10.1016/0165-1218(82)90003-9.

Sellamuthu, R., 2014. Eugenol. In: Wexler, P. (Ed.), Encyclopedia of Toxicology. vol. 2. Academic Press, London, UK, pp. 539–541, https://doi.org/10.1016/B978-0-12-386454-3.01125-8.

Šernaitė, L., Rasiukevičiūtė, N., Valiuškaitė, A., 2020. The extracts of cinnamon and clove as potential biofungicides against strawberry grey mould. Plants 9, 15–20. https://doi.org/10.3390/plants9050613.

Setzer, W.N., Stokes, S.L., Bansal, A., Haber, W.A., Caffrey, C.R., Hansell, E., McKerrow, J.H., 2007. Chemical composition and cruzain inhibitory activity of croton draco bark essential oil from monteverde, costa rica. Nat. Prod. Commun. 2. https://doi.org/10.1177/1934578X0700200613. 1934578X0700200.

Shahavi, M.H., Hosseini, M., Jahanshahi, M., Meyer, R.L., Darzi, G.N., 2019. Evaluation of critical parameters for preparation of stable clove oil nanoemulsion. Arab. J. Chem. 12, 3225–3230. https://doi.org/10.1016/j.arabjc.2015.08.024.

Shammi, M., Sultana, A., Hasan, N., Mostafizur Rahman, M., Saiful Islam, M., Bodrud-Doza, M., Khabir Uddin, M., 2020. Pesticide exposures towards health and environmental hazard in Bangladesh: a case study on farmers' perception. J. Saudi Soc. Agric. Sci. 19, 161–173. https://doi.org/10.1016/j.jssas.2018.08.005.

Sharma, M., Rauniar, G., Das, B., 2012. Experimental study of various central nervous system effects of eugenol in mice and rats. Health Renaiss. 10, 208–214. https://doi.org/10.3126/hren.v10i3.7137.

Sharma, U.K., Sharma, A.K., Gupta, A., Kumar, R., Pandey, A., Pandey, A.K., 2017. Pharmacological activities of cinnamaldehyde and eugenol: antioxidant, cytotoxic and anti-leishmanial studies. Cell. Mol. Biol. 63, 73–78. https://doi.org/10.14715/cmb/2017.63.6.15.

Sibanda, S., Chigwada, G., Poole, M., Gwebu, E.T., Noletto, J.A., Schmidt, J.M., Rea, A.I., Setzer, W.N., 2004. Composition and bioactivity of the leaf essential oil of Heteropyxis dehniae from Zimbabwe. J. Ethnopharmacol. 92, 107–111. https://doi.org/10.1016/j.jep.2004.02.010.

Singletary, K., 2014. Clove—overview of potential health benefits. Nutr. Today 49, 207–224. https://doi.org/10.1097/NT.0000000000000036.

Sipes, I.G., Mattia, A., 2006. Eugenol and related hydroxyallylbenzene derivates. In: Sixty-Fifth Meeting of the Joint FAO/WHO Expert Committee on Food Additive (JECFA) (Ed.), Safety Evaluation of Certain Food Additives. World Health Organization, Geneva, Switzerland, pp. 155–200.

Slameňová, D., Horváthová, E., Wsólová, L., Šramková, M., Navarová, J., 2009. Investigation of anti-oxidative, cytotoxic, DNA-damaging and DNA-protective effects of plant volatiles eugenol and borneol in human-derived HepG2, Caco-2 and VH10 cell lines. Mutat. Res. Genet. Toxicol. Environ. Mutagen. 677, 46–52. https://doi.org/10.1016/j.mrgentox.2009.05.016.

Sober, H.A., Hollander, F., Sober, E.K., 1950. Toxicity of eugenol: determination of LD50 on rats. Exp. Biol. Med. 73, 148–151. https://doi.org/10.3181/00379727-73-17608.

Soo Lim, D., Min Choi, S., Kim, K.-B., Yoon, K., Kacew, S., Sik Kim, H., Lee, B.-M., 2018. Determination of fragrance allergens and their dermal sensitization quantitative risk assessment (QRA) in 107 spray perfumes. J. Toxic. Environ. Health A 81, 1173–1185. https://doi.org/10.1080/15287394.2018.1543232.

Soundran, V., Namagiri, T., Manonayaki, S., Vanithakumari, G., 1994. Hepatotoxicity of eugenol. Anc. Sci. Life 13, 213–217.

Suleiman, W.B., El Bous, M.M., El Said, M., El Baz, H., 2019. In vitro evaluation of syzygium aromaticum l. Ethanol extract as biocontrol agent against postharvest tomato and potato diseases. Egypt. J. Bot. 59, 81–94. https://doi.org/10.21608/ejbo.2018.3838.1180.

Suparmi, S., Ginting, A.J., Mariyam, S., Wesseling, S., Rietjens, I.M.C.M., 2019. Levels of methyleugenol and eugenol in instant herbal beverages available on the Indonesian market and related risk assessment. Food Chem. Toxicol. 125, 467–478. https://doi.org/10.1016/j.fct.2019.02.001.

Suprihanti, A., Sinaga, B.M., Harianto, H., Kustiari, R., 2018. The impact of clove import policy on clove market and cigarette production in Indonesia. Adv. Eng. Res. 172, 274–279. https://doi.org/10.2991/fanres-18.2018.55.

Suthisut, D., Fields, P.G., Chandrapatya, A., 2011. Contact toxicity, feeding reduction, and repellency of essential oils from three plants from the ginger family (Zingiberaceae) and their major components against Sitophilus zeamais and Tribolium castaneum. J. Econ. Entomol. 104, 1445–1454. https://doi.org/10.1603/EC11050.

Sutili, F.J., Gressler, L.T., Baldisserotto, B., 2014. Anthelmintic activity of the phytochemical eugenol against the fish parasite Gyrodactylus sp. and acute toxicity in Daphnia pulex. Pan-Am. J. Aquat. Sci. 9, 223–227.

Svedman, C., Engfeldt, M., Api, A.M., Politano, V.T., Belsito, D.V., Isaksson, M., Bruze, M., 2012. A pilot study aimed at finding a suitable eugenol concentration for a leave-on product for use in a repeated open application test. Contact Dermatitis 66, 137–139. https://doi.org/10.1111/j.1600-0536.2011.02041.x.

Swanson, A.B., Chambliss, D.D., Blomquist, J.C., Miller, E.C., Miller, J.A., 1979. The mutagenicities of safrole, estragole, eugenol, trans-anethole, and some of their known or possible metabolites for Salmonella typhimurium mutants. Mutat. Res. Fundam. Mol. Mech. Mutagen. 60, 143–153. https://doi.org/10.1016/0027-5107(79)90178-7.

Takahashi, H., Nakamura, A., Fujino, N., Sawaguchi, Y., Sato, M., Kuda, T., Kimura, B., 2021. Evaluation of the antibacterial activity of allyl isothiocyanate, clove oil, eugenol and carvacrol against spoilage lactic acid bacteria. LWT Food Sci. Technol. 145, 111263. https://doi.org/10.1016/j.lwt.2021.111263.

Talón, E., Vargas, M., Chiralt, A., González-Martínez, C., 2019. Antioxidant starch-based films with encapsulated eugenol. Application to sunflower oil preservation. LWT Food Sci. Technol. 113, 108290. https://doi.org/10.1016/j.lwt.2019.108290.

Tammannavar, P., Pushpalatha, C., Jain, S., Sowmya, S.V., 2013. An unexpected positive hypersensitive reaction to eugenol. Case Rep. 2013. https://doi.org/10.1136/bcr-2013-009464. bcr2013009464–bcr2013009464.

Taylor, J.M., Jenner, P.M., Jones, W.I., 1964. A comparison of the toxicity of some allyl, propenyl, and propyl compounds in the rat. Toxicol. Appl. Pharmacol. 6, 378–387. https://doi.org/10.1016/S0041-008X(64)80002-8.

Thompson, D., Barhoumi, R., Burghardt, R., 1998. Comparative toxicity of eugenol and its quinone methide metabolite in cultured liver cells using kinetic fluorescence bioassays. Toxicol. Appl. Pharmacol. 149 (1), 55–63. https://doi.org/10.1006/taap.1997.8348.

Thompson, D.C., Constantin-Teodosiu, D., Moldéus, P., 1991. Metabolism and cytotoxicity of eugenol in isolated rat hepatocytes. Chem. Biol. Interact. 77, 137–147. https://doi.org/10.1016/0009-2797(91)90069-J.

Tian, B.-L., Liu, Q.-Z., Liu, Z.-L., Li, P., Wang, J.-W., 2015. Insecticidal potential of clove essential oil and its constituents on Cacopsylla chinensis (Hemiptera: Psyllidae) in laboratory and field. J. Econ. Entomol. 108, 957–961. https://doi.org/10.1093/jee/tov075.

Tippayatum, P., Chonhenchob, V., 2007. Antibacterial activities of thymol, eugenol and nisin against some food spoilage bacteria. Kasetsart J. (Nat. Sci.) 41, 319–323.

Tripathi, A.K., Singh, A.K., Upadhyay, S., 2009. Contact and fumigant toxicity of some common spices against the storage insects Callosobruchus maculatus (Coleoptera: Bruchidae) and Tribolium castaneum (Coleoptera: Tenebrionidae). Int. J. Trop. Insect Sci. 29, 151–157. https://doi.org/10.1017/S174275840999018X.

Tsai, T.-H., Huang, W.-C., Lien, T.-J., Huang, Y.-H., Chang, H., Yu, C.-H., Tsai, P.-J., 2017. Clove extract and eugenol suppress inflammatory responses elicited by Propionibacterium acnes in vitro and in vivo. Food Agric. Immunol. 28, 916–931. https://doi.org/10.1080/09540105.2017.1320357.

Ulanowska, M., Olas, B., 2021. Biological properties and prospects for the application of eugenol—a review. Int. J. Mol. Sci. 22, 3671. https://doi.org/10.3390/ijms22073671.

Vinholes, J., Gonçalves, P., Martel, F., Coimbra, M.A., Rocha, S.M., 2014. Assessment of the antioxidant and antiproliferative effects of sesquiterpenic compounds in in vitro Caco-2 cell models. Food Chem. 156, 204–211. https://doi.org/10.1016/j.foodchem.2014.01.106.

Walia, S., Saha, S., Tripathi, V., Sharma, K.K., 2017. Phytochemical biopesticides: some recent developments. Phytochem. Rev. 16, 989–1007. https://doi.org/10.1007/s11101-017-9512-6.

Wang, G., Ma, W., Du, J., 2018. β-Caryophyllene (BCP) ameliorates MPP+ induced cytotoxicity. Biomed. Pharmacother. 103, 1086–1091. https://doi.org/10.1016/j.biopha.2018.03.168.

WHO, 2006. Evaluation of Certain Food Additives: Sixty-Fifth Report of the Joint FAO/WHO Expert Committee on Food Additives. Geneva, Switzerland.

Wieczyńska, J., Cavoski, I., 2018. Antimicrobial, antioxidant and sensory features of eugenol, carvacrol and trans-anethole in active packaging for organic ready-to-eat iceberg lettuce. Food Chem. 259, 251–260. https://doi.org/10.1016/j.foodchem.2018.03.137.

Woolverton, C.J., Fotos, P.G., Mokas, M.J., Mermigas, M.E., 1986. Evaluation of eugenol for mutagenicity by the mouse micronucleus test. J. Oral Pathol. Med. 15, 450–453. https://doi.org/10.1111/j.1600-0714.1986.tb00656.x.

Xie, Y., Yang, Z., Cao, D., Rong, F., Ding, H., Zhang, D., 2015. Antitermitic and antifungal activities of eugenol and its congeners from the flower buds of Syzgium aromaticum (clove). Ind. Crop. Prod. 77, 780–786. https://doi.org/10.1016/j.indcrop.2015.09.044.

Xu, J.-S., Li, Y., Cao, X., Cui, Y., 2013. The effect of eugenol on the cariogenic properties of Streptococcus mutans and dental caries development in rats. Exp. Ther. Med. 5, 1667–1670. https://doi.org/10.3892/etm.2013.1066.

Yang, D., Michel, L., Chaumont, J.P., Millet-Clerc, J., 2000. Use of caryophyllene oxide as an antifungal agent in an in vitro experimental model of onychomycosis. Mycopathologia 148, 79–82. https://doi.org/10.1023/A:1007178924408.

Yoo, H.-J., Jwa, S.-K., 2018. Inhibitory effects of β-caryophyllene on Streptococcus mutans biofilm. Arch. Oral Biol. 88, 42–46. https://doi.org/10.1016/j.archoralbio.2018.01.009.

Yoo, H.-J., Jwa, S.-K., 2019. Efficacy of β-caryophyllene for periodontal disease related factors. Arch. Oral Biol. 100, 113–118. https://doi.org/10.1016/j.archoralbio.2019.02.015.

York, M., Griffiths, H.A., Whittle, E., Basketter, D.A., 1996. Evaluation of a human patch test for the identification and classification of skin irritation potential. Contact Dermatitis 34, 204–212. https://doi.org/10.1111/j.1600-0536.1996.tb02175.x.

Zhang, R., Kulkarni, K.A., Kulkarni, A.P., 2000. Oxidation of eugenol by purified human term placental peroxidase. Placenta 21, 234–240. https://doi.org/10.1053/plac.1999.0472.

Zhang, H.M., Guo, S.S., Fan, B., Du, S.S., Wang, Y.Y., Deng, Z.W., 2019. Evaluation of efficacy of the essential oil from Ostericum viridiflorum (Turcz.) Kitagawa in control of stored product insects. Environ. Sci. Pollut. Res. 26, 1406–1413. https://doi.org/10.1007/s11356-018-3728-x.

Zheng, L., Bae, Y.-M., Jung, K.-S., Heu, S., Lee, S.-Y., 2013. Antimicrobial activity of natural antimicrobial substances against spoilage bacteria isolated from fresh produce. Food Control 32, 665–672. https://doi.org/10.1016/j.foodcont.2013.01.009.

Zheng, K., Xiao, S., Li, W., Wang, W., Chen, H., Yang, F., Qin, C., 2019. Chitosan-acorn starch-eugenol edible film: physico-chemical, barrier, antimicrobial, antioxidant and structural properties. Int. J. Biol. Macromol. 135, 344–352. https://doi.org/10.1016/j.ijbiomac.2019.05.151.

CHAPTER

Anticancer effects of the active fraction from clove in vitro and in vivo

18

Gan Qiao[a] and Xiukun Lin[b]

Central Nervous System Drug Key Laboratory of Sichuan Province, School of Pharmacy, Southwest Medical University, Luzhou, China[a] Delisi Group, Weifang, Shandong, China[b]

1. Introduction

Cloves (*Syzygium aromaticum*), the dried bud of *Syzygium aromaticum* L., is a traditional Chinese medicinal herb (TCM) widely used in many Asian countries. Clove is believed to increase gastrointestinal function and alleviate pain in Chinese medicine, and the herb is used to treat nausea, gastric spasm, and sore throat in traditional Chinese medicine. Modern study reveals that clove displays broad biological activity, including antioxidant (El-Maati et al., 2016; Lee and Shibamoto, 2001), antibacterial (Nuñez and D'Aquino, 2012; Selles et al., 2020; Nzeako et al., 2006), anti-inflammatory (Banerjee et al., 2020; Nikoui et al., 2017), anti-diabetic (Salim et al., 2017; Topal, 2019), anti-neurodegeneration (Shekhar et al., 2018; Pant et al., 2016), antiviral activities (Mehmood et al., 2020), and anticancer activities (Neto, 2011; Tompkins and Thorburn, 2019; Venugopal et al., 2017; Wang et al., 2017; Yi et al., 2015).

Our studies confirmed that the active fractions of clove, including liposoluble and aqueous fractions, display potent anticancer effects toward different malignancies. The liposoluble fractions of the clove is able to inhibit the growth of tumor via PI3K/Akt/mTOR pathway while the aqueous fractions of the clove is capable of suppressing tumor growth through autophagy and apoptosis via AMPK/ULK Pathway (Li et al., 2019; Liu et al., 2018). Our studies provide evidence that clove possesses the potential to be developed as an adjuvant in the treatment of human tumors.

2. Anticancer effect of liposoluble fraction of clove

Clove was extracted with ethanol to prepare an ethanol extract of clove (EEC), and EEC was further purified with ethyl acetate to prepare a liposoluble fractions of clove (Active fractions of clove, AFC). We investigated the effects of AFC on different human colorectal cancer cells using CCK-8 assay. The results showed that AFC was able to inhibit the growth of several cancer cells, and displayed more sensitivity toward human colorectal cancer cells with IC_{50} values of 113.5 ± 7.83, 174.9 ± 9.52, 211.7 ± 7.29, 232.3 ± 15.42, and 280.9 ± 12.61, for cells HCT116, HT-29, SW620, HCT8, and LoVo,

Clove (*Syzygium aromaticum*). https://doi.org/10.1016/B978-0-323-85177-0.00021-5

respectively. Compared with other cells, HCT116 cell was the most sensitive to AFC. So, in subsequent experiments, we choose HCT116 cells to investigate the potential anticancer mechanisms of AFC.

The effects of AFC on cell viability in human colorectal cancer HCT-116 cells were analyzed by CCK-8 assay. The results showed that after treatment with AFC at 25, 50, 100, and 200 μg/mL, for 72 h, the cell viability of HCT-116 cells was 74.3%, 61.4%, 15.3%, and 4.34%; respectively (Fig. 1A). The data indicate that AFC could inhibit the cell growth of HCT-116 cells in a dose-dependent manner. Next, we evaluated the effect of AFC on the induction of apoptosis in HCT-116 cells by Hoechst 33258 staining assay. As shown in Fig. 1B, treatment with AFC resulted in obvious typical morphological features of apoptosis; chromatin condensation, nuclear shrinkage, and the formation of a few apoptotic bodies. We further investigated the quantity of apoptosis in HCT-116 cells by cytometric analysis using Annexin V-FITC/PI double staining. The result showed that the apoptotic rates of the cells were 8.35%, 17.6%, and 26.9% when treating the cells with AFC at 25, 50, and 100 μg/mL for 48 h, respectively (Fig. 1C and D). To study the underlying mechanism of AFC-induced apoptosis, we further detected the expression of cleavage of PARP, caspase-3, and caspase-9 by Western blotting. As shown in Fig. 1E and F, AFC clearly increased the level of cleaved PARP, pro-caspase 3, and pro-caspase 9 to their active forms. These data indicated that AFC inhibits cell viability by inducing apoptosis in HCT-116 cells.

To study the AFC-induced autophagy, we first determined the AFC-induced autophagic flow of LC3-puncta. HCT-116 cells were transfected with Ad-mCherry-GFP-LC3B and the fluorescence of GFP-LC3 puncta was observed under a fluorescence microscope. As shown in Fig. 2A, the fluorescence of GFP-LC3 puncta was frequently observed in HCT-116 cells treated with AFC, whereas the cells treated with medium (vehicle control) displayed a relatively homogeneous LC3 expression pattern. In addition, a large number of autophagic bodies and autophagy-lysosomes were observed in HCT-116 cells treated with AFC (Fig. 2B) under transmission electron microscopy.

We next examined the effect of AFC on the expressions of LC3 and Beclin-1. As shown in Fig. 2C–E, the expression levels of LC3-II and Beclin-1 in HCT-116 cells were significantly enhanced by AFC (25, 50, and 100 μg/mL) in a dose-and time-dependent manner. These data indicated that AFC treatment not only results in apoptosis but also induces autophagy in HCT-116 cells.

It has been documented that autophagy could facilitate cell survival in the adverse microenvironments and inhibition of autophagy triggers the increase of cytotoxicity and induction of apoptosis in cells (Degenhardt et al., 2006; Tompkins and Thorburn, 2019). We, therefore, investigated the effect of inhibition of autophagy on apoptosis induced by AFC in HCT-116 cells. Two autophagic inhibitors, 3-MA and BA are usually used to inhibit autophagic action. 3-MA is able to suppress class III phosphatidylinositol 3-kinase (PI3K); essential for the initiation of the early stages of autophagy (Wu et al., 2010), while BA, an inhibitor of the vacuolar-type ATPase, is capable of blocking the fusion of autophagosomes with lysosomes, preventing the autophagic degradation (Klionsky et al., 2008). As shown in Fig. 3A, the cell viability was significantly decreased in HCT-116 cells treated with the combination of AFC (100 μg/mL) and 3-MA (2 mM) or BA (1 nM); the viability of the cells decreased to 40.33 ± 2.52% and 54.34 ± 4.04%, respectively, compared to 73.25 ± 5.13% in the cells treated with AFC alone by CCK-8 assay. This result indicated that the inhibitory effect of AFC on HCT-116 cell growth was enhanced by autophagic inhibitors of 3-MA and BA. The effects of AFC on LC3-II expression in the presence of 3-MA or BA were studied by Western blotting. As shown in Fig. 3B, 3-MA (2 mM) decreased the expression of LC3 but BA (1 nM) resulted in a higher level of LC3 expression. The opposite effect of 3-MA and BA on LC3 is associated with blocking autophagy at different stages by the two inhibitors; 3-MA inhibited autophagosome formation from the beginning, whereas BA prevented the degradation of LC3 in autophagolysosomes and turn increased LC3 level. Additionally, 3-MA significantly decreased LC3 expression induced by AFC, while

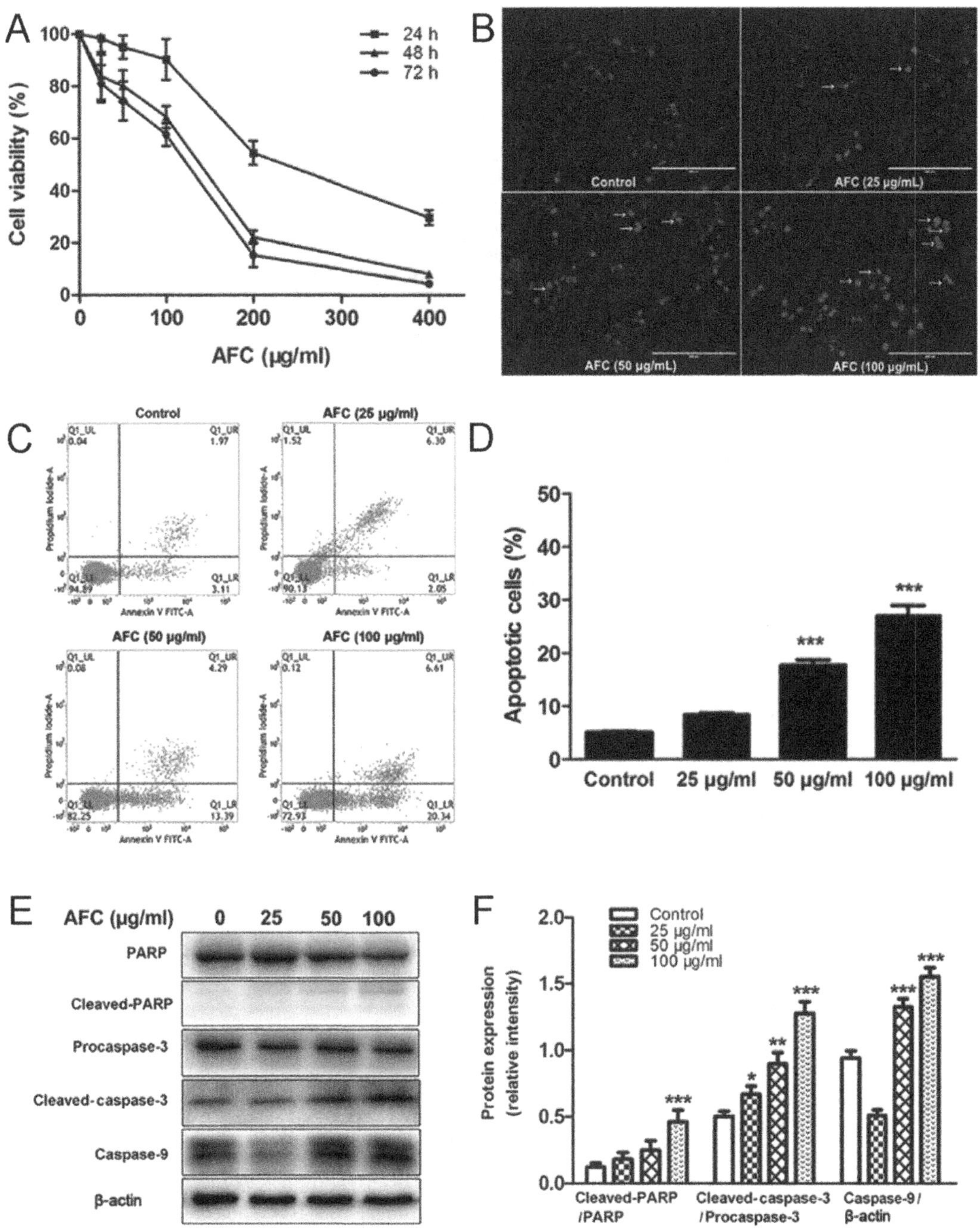

FIG. 1

AFC inhibited the growth and induced apoptosis in HCT-116 cells. (A) HCT-116 cells were treated with AFC at 25, 50, 100, 200, or 400 μg/mL for 24, 48, and 72 h, respectively, and the cell viability was analyzed by CCK-8 assay.

(Continued)

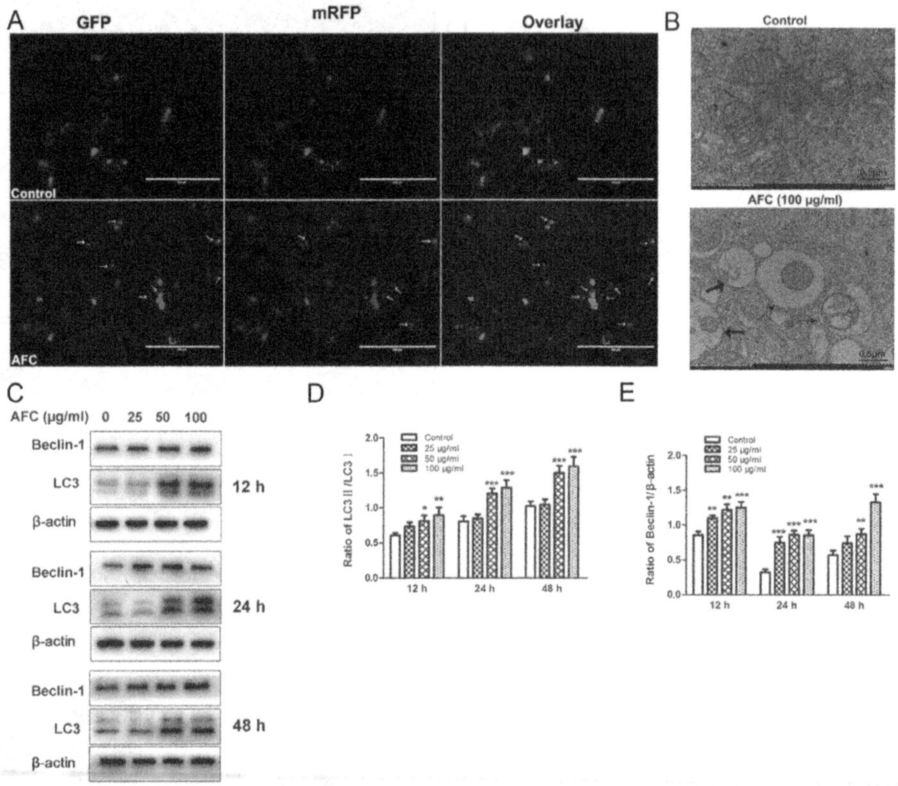

FIG. 2

AFC induced autophagy in HCT-116 cells. (A) Stable expression of GFP-LC3 was observed under a fluorescence microscope in HCT-116 cells after the cells were treated with 100 μg/mL AFC for 48 h. *Arrowheads* indicate autophagic cells. (B) Autophagy body *(thin arrows)* and autophagy-lysosome *(thick arrows)* were observed in HCT-116 cells under a transmission electron microscope (×400) after the cells were treated with 100 μg/mL AFC for 48 h. (C) Expressions of LC3 and Beclin-1 in HCT-116 cells by Western blotting analysis. (D and E) The intensities of bands were quantified by Image J software. Actin was used as a loading control. All data are expressed as mean ± SD from three independent experiments in triplicate. $^*P<.05$, $^{**}P<.01$ and $^{***}P<.001$ vs vehicle control.

FIG. 1, CONT'D (B) Hoechst 33258 staining was used to detect the morphological features of apoptosis including chromatin condensation and nuclear fragmentation in HCT-116 cells. *Arrowheads* indicate apoptotic cells. (C) HCT-116 cells were treated with 25, 50, or 100 μg/mL AFC for 48 h, and apoptosis was determined with Annexin V-FITC/PI double staining by flow cytometry. (D) The percentage of apoptosis cells was represented with a histogram from three independent experiments (mean ± SD). (E) HCT-116 cells were treated with AFC at 25, 50, or 100 μg/mL for 48 h, and the cleavage of PARP, caspase 3, and caspase 9 was detected by Western blotting analysis. Actin was used as a loading control. (F) The expressions of cleavage of PARP, caspase 3, and caspase 9 after the intensities of bands were quantified by Image J software. All data are expressed as mean ± SD from three independent experiments in triplicate. $^*P<.05$, $^{**}P<.01$ and $^{***}P<.001$ vs vehicle control.

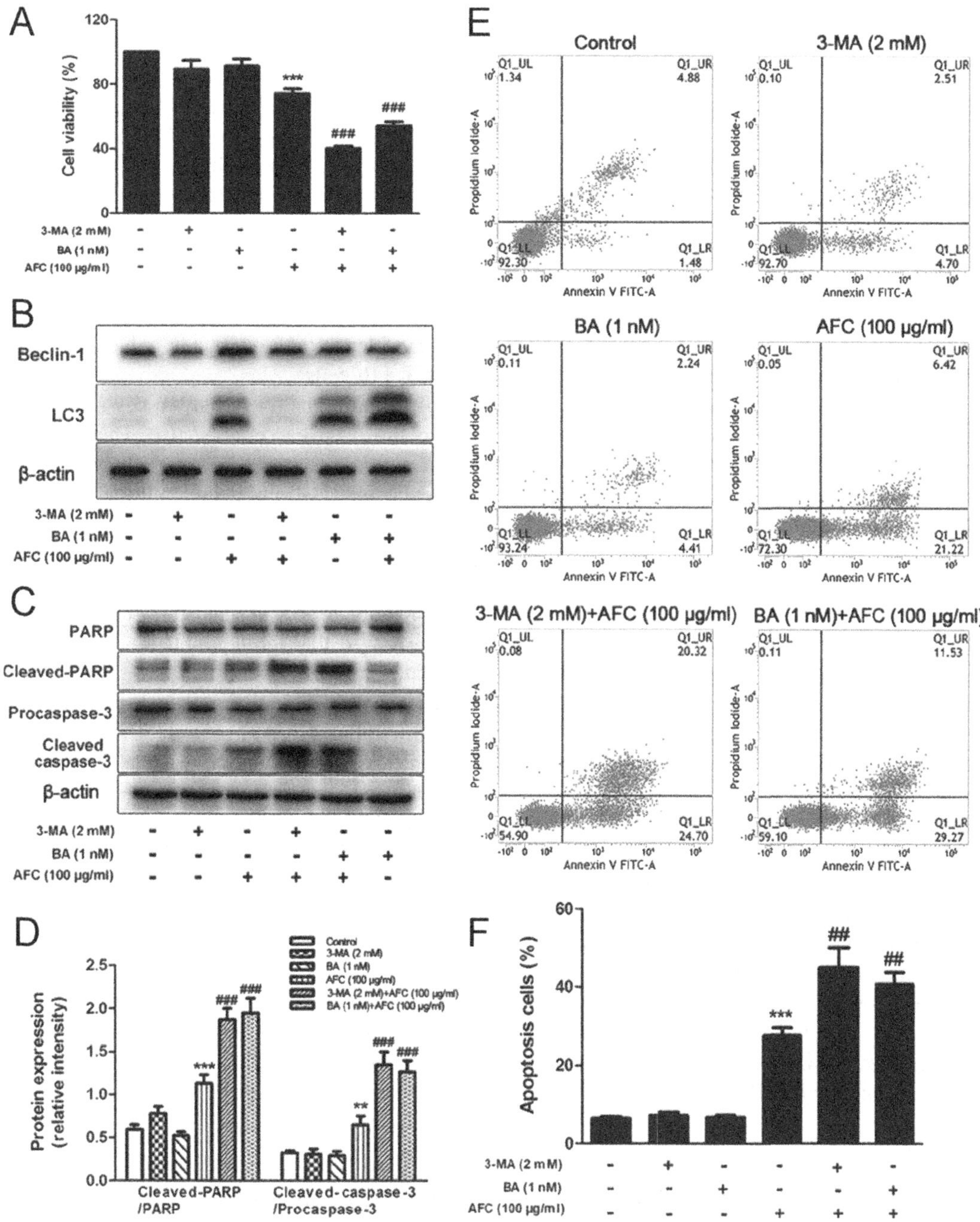

FIG. 3

Inhibition of autophagy enhances apoptosis induced by AFC in HCT-116 cells. (A) Effect of 3-MA and BA on cell viability induced by AFC in HCT-116 cells. (B) The expressions of LC3 and Beclin-1 proteins in HCT-116 cells by Western blotting analysis. (C) The expressions of cleaved caspase-3 and PARP by Western blotting analysis. (D) The intensities of bands were quantified by Image J software. (E) The apoptotic ratios in HCT-116 cells with Annexin V-FITC/PI double staining by flow cytometry analysis. (F) The percentage of apoptosis cells was represented with a histogram. Results are expressed as the means $\pm$ SD ($n = 3$) in triplicate. *$P < .05$, **$P < .01$ and ***$P < .001$ vs vehicle control; ##$P < .01$ and ###$P < .001$ vs AFC alone.

BA increased LC3-II level induced by AFC treatment (Fig. 3B). These data suggested that 3-MA and BA blocked autophagy induced by AFC at different stages (Hsin et al., 2012).

Furthermore, we determined the effect of AFC on the expression of apoptotic genes in the presence of 3-MA or BA. As shown in Fig. 3C and D, the expressions of cleaved-PARP and cleaved-caspase 3 were increased significantly ($P < .05$) with the combination of AFC and 3-MA or BA compared to either AFC, 3-MA, or BA alone in HCT-116 cells by Western blotting analysis. Similar effects were further confirmed by Annexin V-FITC/PI double staining. The results showed that combined treatment of AFC and 3-MA or BA more obviously augmented apoptosis than that of AFC, 3-MA, or BA alone in HCT-116 cells (Fig. 3E and F). These data provide more evidence that autophagy induced by AFC exerted a suppressive effect on apoptotic pathways in HCT-116 cells.

It has been well documented that the PI3K/Akt/mTOR signaling pathway plays a key role in regulating both apoptosis and autophagy through divergent pathways (Shao et al., 2016). Therefore, we studied the effect of AFC on the PI3K/Akt/mTOR signaling pathway in HCT-116 cells by Western blotting analysis. Treatment of AFC inhibited PI3K phosphorylation, decreased the levels of phospho-Akt, and down-regulated phospho-mTOR in a dose-dependent manner (Fig. 4A). The ratios of p-Akt/Akt and p-mTOR/mTOR were significantly decreased after AFC treatment by Image J software analysis (Fig. 4B). Next, we studied the effect of the combination of AFC (100 μg/mL) and LY294002 (10 μM), an inhibitor of PI3K, on the expressions of p-Akt/Akt and p-mTOR/mTOR in HCT-116 cells by Western blotting analysis. The results showed that the combination of AFC and LY294002 was more effective in decreasing the level of the ratios of p-Akt/Akt ($P < .001$) and p-mTOR/mTOR ($P < .05$) compared to that using AFC or LY294002 alone (Fig. 4C and D). We then studied the effects of AFC in the presence of IGF-I. As shown in Fig. 4E and F, IGF-I (50 ng/mL) treatment increased the phosphorylation of Akt and mTOR significantly. In contrast, when cells were pretreated with IGF-I (50 ng/mL), the effect of AFC was attenuated significantly. Moreover, significant differences in the expression of LC3-II/LC3-I, cleaved-PARP/PARP, and cleaved-caspase 3/procaspase-3 ratio were observed between AFC treatment alone and the combination of AFC and IGF-I (Fig. 4G and H). These data indicate that the effect of AFC on induction of apoptosis and autophagy is related to the inhibition of the PI3K/Akt/mTOR signaling pathway.

Our previous study has shown that there are two main components in the AFC; oleanolic acid (OA), eugenol, and maslinic acid (MA). Our study confirmed that oleanolic acid is able to inhibit tumor growth by switching pyruvate kinase type M isoforms; OA could decrease PKM2 abundance, and increase PKM1 expression, in both PC-3 and MCF-7 cancer cell lines, in a dose- and time-dependent manner (Fig. 5) (Liu et al., 2014b), while MA is able to induce the autophagy of pancreatic cancer by down-regulated HSPA-8 (Fig. 6) (Tian et al., 2018).

3. The anticancer effect of the aqueous fraction of clove

We also studied the anticancer effect of the aqueous fractions of cloves (AEC) and found that the aqueous extract of cloves (AEC) displayed more potent antitumor activity. AEC was prepared by extraction with H_2O and fractioned with macroporous adsorption resin (Li et al., 2019).

The cytotoxicity of AEC was analyzed using an MTS assay. As showing in Table 1, AEC was able to inhibit the growth of several cancer cells, with IC_{50} values of 330.9, 234, 324.5, 218.9, 470.3, 574, 156.5, 218.4, 369.2, 150.6, and 115.1 μg/mL for cells HepG2, Bel-7402, A431, HeLa, A549,

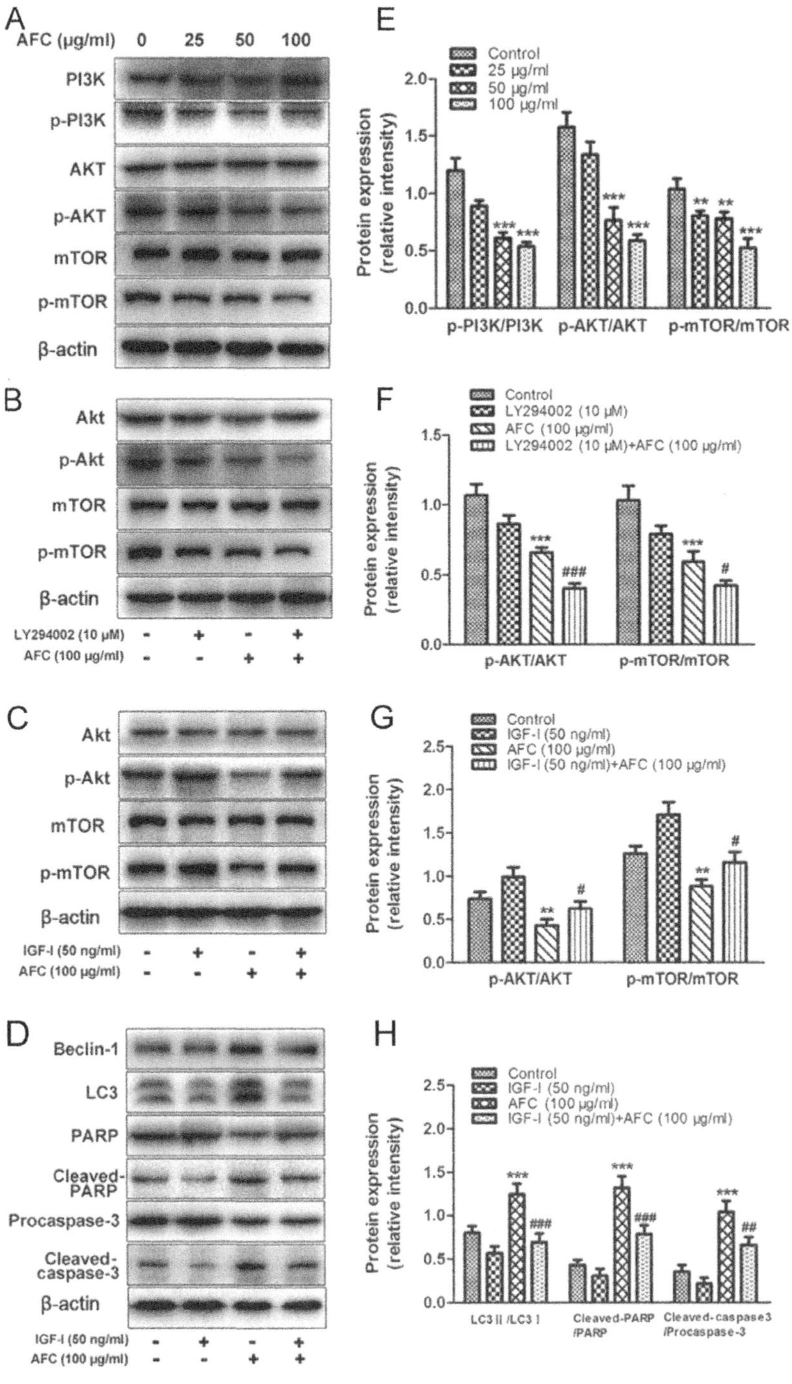

FIG. 4

AFC inhibited the activation of PI3K/Akt/mTOR signaling pathways. (A) The expressions of phospho-PI3K, phospho-Akt, and phospho-mTOR in HCT-116 cells by Western blotting analysis. (B) The intensities of bands

(Continued)

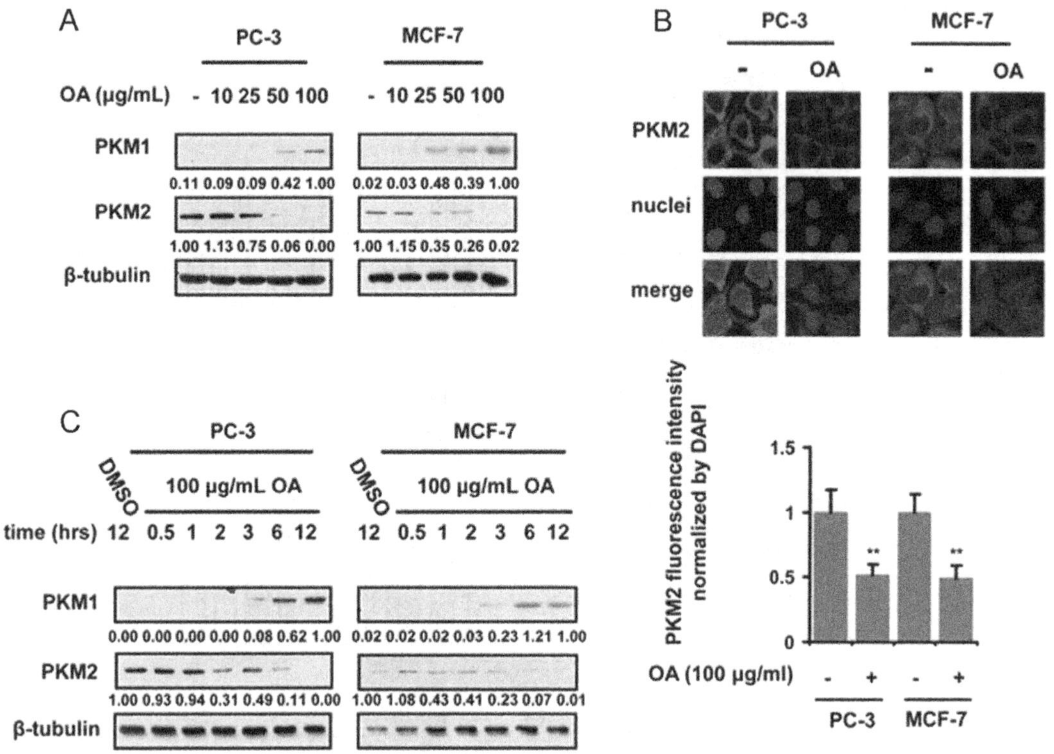

FIG. 5

OA treatment affected the expression profile of PKM isoforms in cancer cells in a dose-and time-dependent manner. PKM1 and PKM2 expression level was evaluated by immunoblotting assays in PC-3 and MCF-7 cells treated with 50mg/mL (A) or 100mg/mL OA for 0.5, 1, 2, 3, 6, 12h, respectively (B). Beta-tubulin was used as endogenous reference. (C) The PKM2 levels in PC-3 and MCF-7 cells were treated with 100mg/mL OA for 12h, respectively, as determined by immunofluorescent staining. The nuclei were stained with DAPI. The merged pictures were shown at the bottom of the panel. The fluorescence intensity was quantified with ImageJ software. The ratio of each group to control was presented here. The bars represented the average values of 5 randomly selected experiments.

FIG. 4, CONT'D

were quantified by densitometric analysis. (C) The expressions of phospho-Akt and phospho-mTOR in HCT-116 cells by Western blotting analysis. (D) The intensities of bands were quantified by Image J software. (E) The expressions of phospho-Akt and phospho-mTOR in HCT-116 cells. (F) The intensities of bands were quantified by Image J software. (G) The expressions of LC3, Beclin-1, caspase-3, and PARP proteins in HCT-116 cells by Western blotting analysis. (H) The intensities of bands were quantified by densitometric analysis. Actin was used as a loading control. All data are expressed as mean ± SD from three independent experiments in triplicate. $^{**}P<.01$ and $^{***}P<.001$ vs vehicle control; $^{\#}P<.05$, $^{\#\#}P<.01$ and $^{\#\#\#}P<.001$ vs AFC alone.

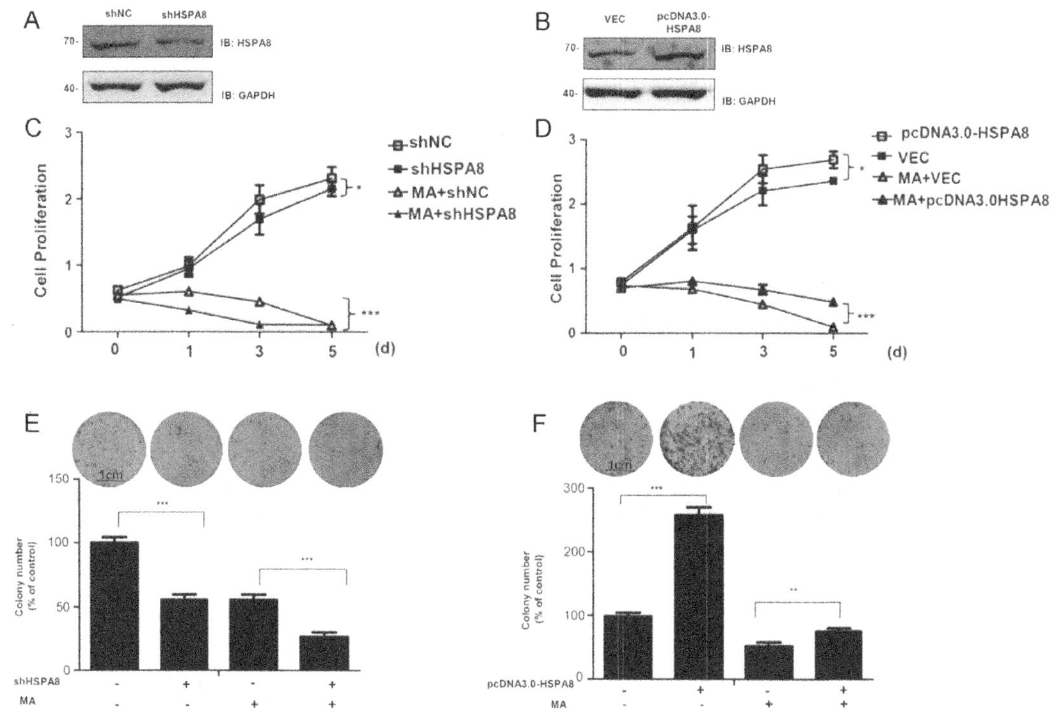

FIG. 6

MA inhibited the growth of Panc-28 cells by down-regulation of HSPA8. Panc-28 cells were transfected with HSPA8 shRNA plasmid or pcDNA3.1-HSPA8 plasmid. The expression of HSPA8 in cells transfected with HSPA8 shRNA (A) and pcDNA3.1-HSPA8 plasmid(B) were analyzed using Western blotting. The effect of MA (25 μM) on the growth of Panc-28 cells with knockdown overexpression (C) or overexpression of HSPA8 (D) was analyzed using an MTS assay. Effect of MA (25 μM) on colony formation in Panc-28 cells with knockdown (E) or overexpression (F) of HSPA8 ($n \geqq 3$ assays, *t*-test, $*P < .05$, $**P < .01$, $***P < .001$).

Table 1 Cytotoxic analysis of AEC on different cancer cells.

Cell line	IC_{50} (μg/mL)
HepG2	330 ± 19.7
Bel-7402	234 ± 30.2
A431	324 ± 18.7
HeLa	218.9 ± 23.4
A549	470.3 ± 13.5
MDA-MB-468	574.6 ± 30.2
HT-29	156.5 ± 11.2
HCT-116	218.3 ± 14.1
Mia-pac	369.2 ± 15.6
Panc-28	150.6 ± 16.1
ASPC-1	115.1 ± 8.3
L02	No effect

Cancer cells were untreated or treated with 50, 100, 150, 200, 250, 300, 350, and 400 μg/mL AEC for 48 h. The IC_{50} is determined by MTS assay, and the IC_{50} value was calculated using SPSS.

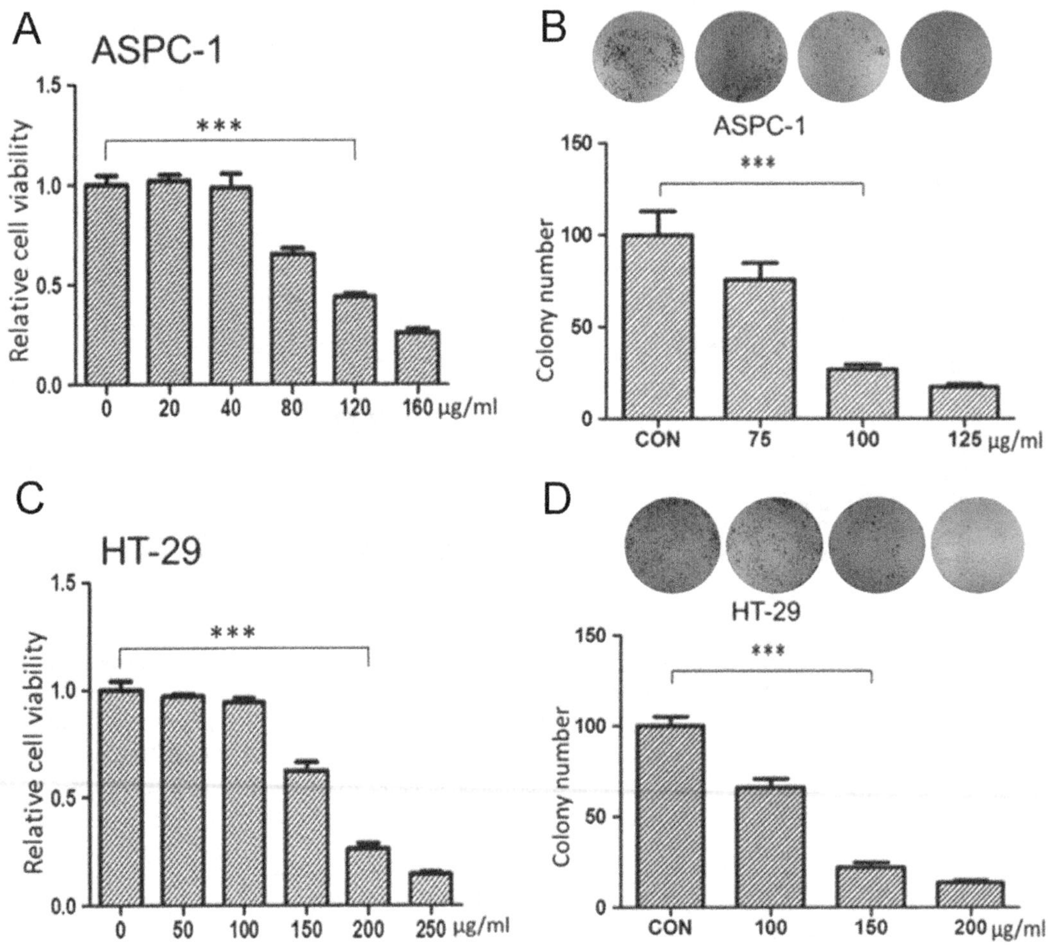

FIG. 7

AEC inhibited the cancer cell growth and colony formation in ASPC-1 and HT-29 cancer cells. ASPC-1 (A) and HT-29 cells (B) were treated with a certain the concentration of AEC for 48h, and the cell viability was determined by MTS assay. (C) and (D) presented the inhibitory effect of AEC on colony formation in ASPC-1 and HT-29 cells (*t*-test, $P < .01$, $^{***}P < .001$).

MDA-MB-468, HT-29, HCT-116, Mia-paca-2, Panc-28, and ASPC-1, respectively. However, AEC showed a very weak anti-proliferation effect on L02 cells, and the inhibition rate was only 0.44% when treating the cells with 350 μg/mL AEC for 48 h. Since the human pancreatic cancer cells, ASPC-1, and human colon cancer cells HT-29 were much more sensitive to AEC treatment than that of other cells, these cells were used for our further study.

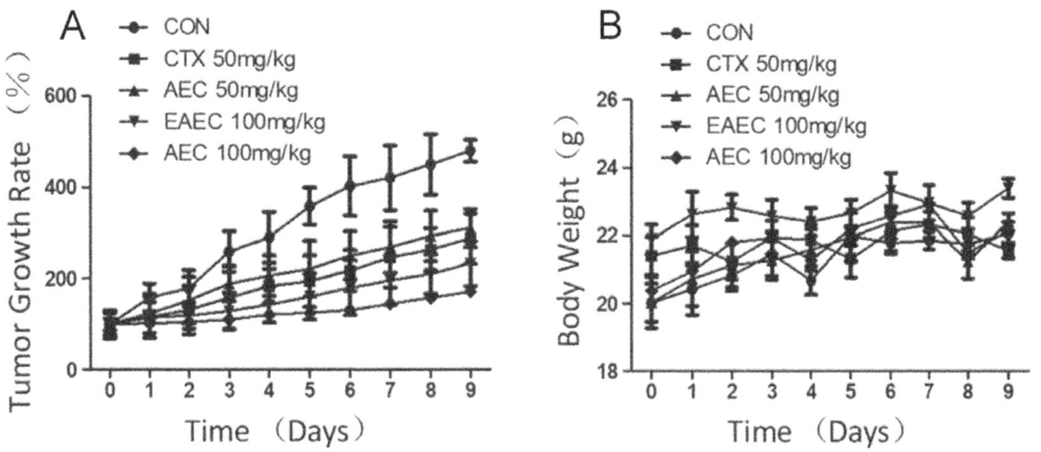

FIG. 8

AEC inhibited tumor growth in HT-29 tumor xenograft mice model. AFC-induced autophagy in HCT-116 cells. HT-29 cells (1 × 107, 500 μL) were injected into BALB/C null mice. When the tumors grew up to $100\,mm^3$. Mice were treated with CTX (50 mg/kg), AFC (100 mg/kg), AEC (50 mg/kg, 100 mg/kg). Tumor volume (A) and body weight (B) of the mice were measured every day. The tumor size was analyzed every day using a vernier caliper (*t*-test, $*P<.05$, $**P<.01$, $***P<.001$).

MTS analysis further confirmed that AEC inhibited cancer cell growth in a dose-dependent manner (Fig. 7A and B). The effect of AEC on colony-formation was also evaluated, and the results showed that AEC inhibited the colony-formation ability significantly in both HT-29 and ASPC-1 cells; in treatment of ASPC-1 cells with 100 μg/mL of AEC, the colony number was decreased to 40.2% as compared with that in the control group. Similar results were also found in HT-29 cells; the colony formation was decreased to 35.6% when treating the cells with 150 μg/mL of AEC (Fig. 7C and D). These results suggested that AEC is capable of inhibiting the growth and colony formation in cancer cells.

To study the antitumor effect of AEC in vivo, HT-29 xenograft BALB/c nude mice were treated P. O. with AEC. As shown in Fig. 8A, AEC was able to inhibit the tumor growth significantly; the inhibitory rate is 35.2% and 51.6%, respectively, when treating the mice with 50 mg/kg and 100 mg/kg of AEC. respectively. Meanwhile, the body weight of the mice was not affected significantly and no obvious difference in the body weight was observed among those treated with the AEC, AFC, vehicle, and CTX groups (Fig. 8B). These results suggested that AEC displayed a potent antitumor effect in vivo with low toxicity.

To study the underlying mechanism of clove, TEM was used to observe the formation of autophagosomes. Compared with control, the number of autophagosomes (indicated by the arrow) was increased significantly after AEC treatment in both ASPC-1 (Fig. 9A) and HT-29 cells (Fig. 9B). We then used flow cytometry to quantify the autophagy induced by AEC. As shown in Fig. 9C and

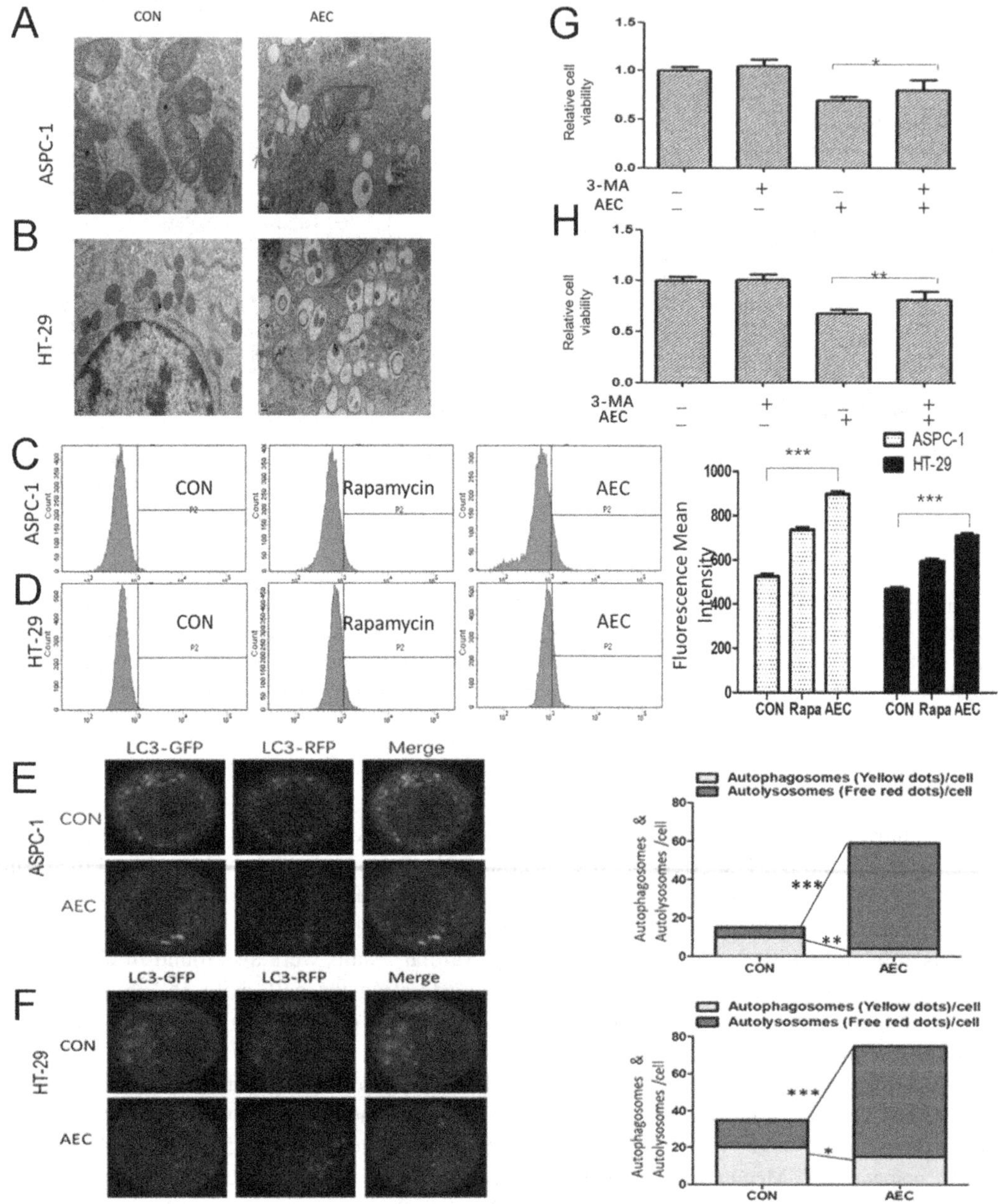

FIG. 9

AEC-induced autophagic cell death. AEC induced autophagic cell death in ASPC-1 and HT-29 cells. ASPC-1 (A) and HT-29 (B) cells were treated with or without AEC for 48 h. Autophagosomes were observed by transmission electron microscope. Quantification of autophagy was detected by flow cytometry for ASPC-1 (C) and HT-29 cancer cells (D). Cells were plated in confocal dishes and cultured for 24 h. Cells were then transfected with mRFP-GFP-LC3 adenovirus and incubated for another 24 h. The cells were treated with or without AEC for another 48 h, and autophagy flow was determined by confocal fluorescence microscopy in ASPC-1 (E) and HT-29 cancer cells (F). ASPC-1 (G) and HT-29 (H) cancer cells were treated with AEC of 125 and 150 μg/mL or the combination of AEC and 3-MA (10 μM) for 48 h. The relative viability was determined by MTS analysis (*t*-test, *$P < .05$, **$P < .01$, ***$P < .001$).

D, the number of autophagosomes was greatly increased in both ASPC-1 and HT-29 cancer cells treated with a certain concentration of AEC; the average relative value of fluorescence intensity was 897 in ASPC-1 cells treated with 125 μg/mL of AEC, compared with 526 in control cells and 736 in cells treated with positive drug Rapamycin (500 nM). Similar results were also found in HT-29 cells (Fig. 9D), with the average relative value of fluorescence intensity of 594 and 711 in HT-29 cells treated with Rapamycin (500 nM) and AEC (150 μg/mL), compared with 468 in control cells.

We next performed autophagic flow analysis to study the effect of AEC on autophagic flow. Cells were transfected with mRFP-GFP-LC3 adenovirus and then treated with AEC for 48 h. As shown in Fig. 9E, more red dots were observed in ASPC-1 cells treated with AEC than those in control cells. The same tendency also appeared in the HT-29 cells treated with AEC (Fig. 9F). The above results strongly confirmed that AEC is able to induce autophagy in both ASPC-1 and HT-29 cells.

To determine the properties of AEC-induced autophagy, we used 3-MA, a small molecule inhibitor of autophagy, to evaluate the effect of AEC on cancer cells. As shown in Fig. 9G, the cell viability was increased significantly in ASPC-1 cells treated with the combination of AEC and 3-MA, compared to those treated with AEC alone; the relative cell viability was 87.9% in cells treated with the combination of AEC and 3-MA, while the relative viability was decreased to 67.6% in cells treated with AEC alone. The same tendency was also detected in the HT-29 cells (Fig. 9H). This result provided evidence that AEC induces autophagic cell death in cancer cells.

It has been well established that activation of the AMPK pathway plays an important role in the autophagic process. To determine the effect of AEC on the AMPK/ULK-mediated pathway, a Western blotting assay was used to examine the related gene expression induced by AEC. As shown in Fig. 10, AEC treatment increased the expression of p-AMPK and p-ULK1 significantly and the ratio of LC3-II/LC3-I was also enhanced in both ASPC-1 (Fig. 10A) and HT-29 cells (Fig. 10B). In order to further verify that the AEC triggered autophagy, we determined the effect of AEC in the presence of compound C (an AMPK inhibitor) and AICAR (an AMPK activator). Western blotting analysis showed that the expression of p-AMPK, p-ULK1, and the ratio of LC3-II/LC3-I was significantly increased following the treatment with a combination of AEC and AICAR (0.1 mM), compared with those treated with AEC alone (Fig. 10C and D). In contrast, AEC-induced autophagy was attenuated in the presence of AMPK inhibitor compound C (Fig. 10E and F). Treatment with a combination of AEC and compound C (5.0 μM) diminished the expression of p-AMPK, p-ULK1 and the ratio of LC3-II/LC3-I significantly compared with that treated with AEC alone.

To confirm that the anticancer activity of AEC is associated with induction of AMPK-induced autophagy, we studied that AEC-induced inhibition of colony formation in the presence of compound C and AICAR. As showing in Fig. 11A, the relative colony number rate was 55.5% in cells treated with AEC (75 μg/mL) alone in ASPC-1 cells. In the presence of AICAR (0.1 mM), the relative colony number rate decreased to 37.3%. In contrast, the relative colony number rate increased to 66.3% in ASPC-1 cells treated with the combination of AEC (75 μg/mL) and compound C (5.0 μM). Similar results were also found in HT-29 cells (Fig. 11B); treatment with the combination of AEC (100 μg/mL)

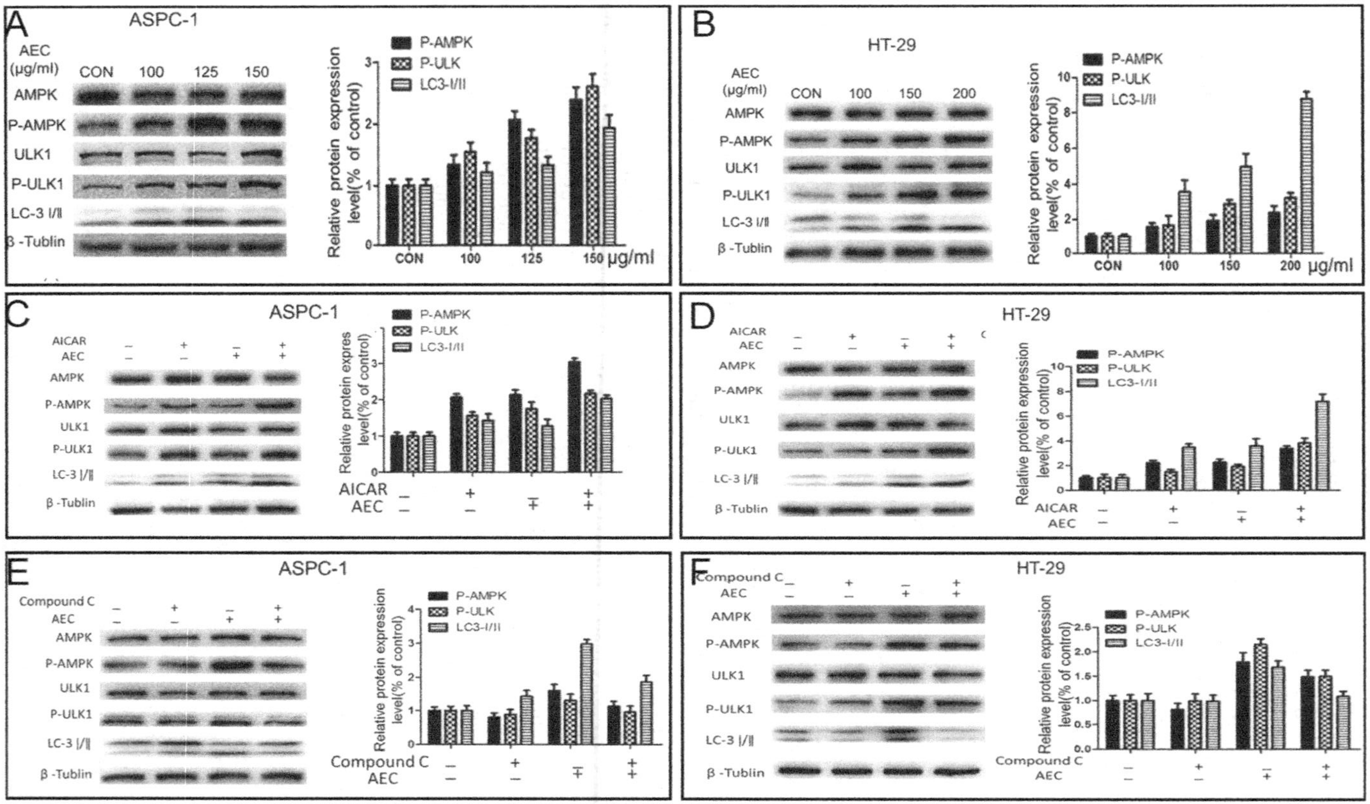

FIG. 10

MPK/ULK pathway played an important role in autophagy induced by AEC. (A) Stable expression of GFP-LC3 was observed under a fluorescence microscope in HCT-116 cells after the cells were treated with 100 μg/mL AFC for 48 h. Arrowheads indicate autophagic cells. (B) Autophagy body *(thin arrows)* and autophagy-lysosome *(thick arrows)* were observed in HCT-116 cells under a transmission electron microscope (×400) after the cells were treated with 100 μg/mL AFC for 48 h. (C) Expressions of LC3 and Beclin-1 in HCT-116 cells by Western blotting analysis. (D and E) The intensities of bands were quantified by Image J software. Actin was used as a loading control. All data were expressed as mean ± SD from three independent experiments in triplicate. $*P < .05$, $**P < .01$ and $***P < .001$ vs vehicle control.

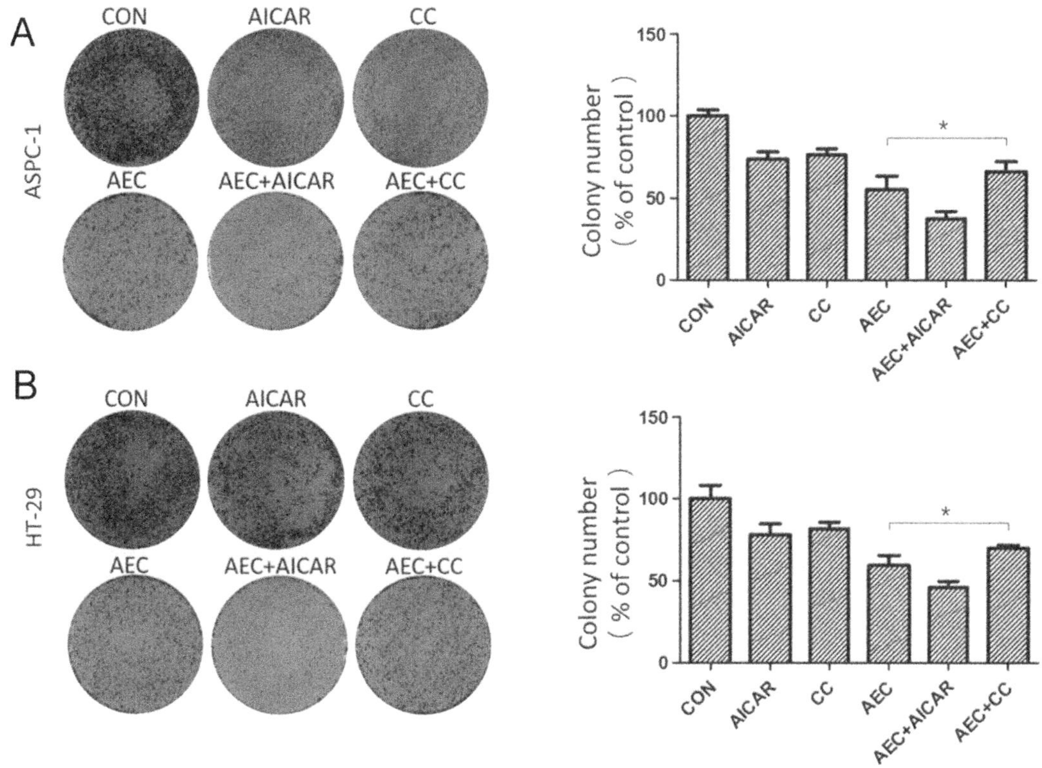

FIG. 11

Effects of AMPK function on colony formation, Cells (1 × 103) were seeded in 6-well plates. After incubation overnight, cells were treated without or with AICAR (0.1 mM), Compound C (5.0 μM), 75 μg/mL of AEC in ASPC-1, and 100 μg/mL of AEC in HT-29 cells. After incubation for 12 days, the colony number was observed under a light microscope. Data were expressed as means ± SD (*t*-test, **P* < .05, ***P* < .01, ****P* < .001).

and AICAR decreased the colony formation capacity; the relative colony number rate was about 45.9%, while the relative colony number rate was increased to 70.3% in cells treated with the combination AEC and compound C group, compared with 59.3% in cells treated with AEC alone. The results provide evidence that AEC induced autophagy through AMPK/ULK pathway in both ASPC-1 and HT-29 cells.

LC-MS analysis showed that there were 11 chemical components contained in AEC (Fig. 12A); including acid compound (compounds 1, 2), glucuronide (compounds 4, 5, 7, 8), and glycoside (compounds 6, 9). Fig. 12B showed the chemical structure of predicated components contained in AEC.

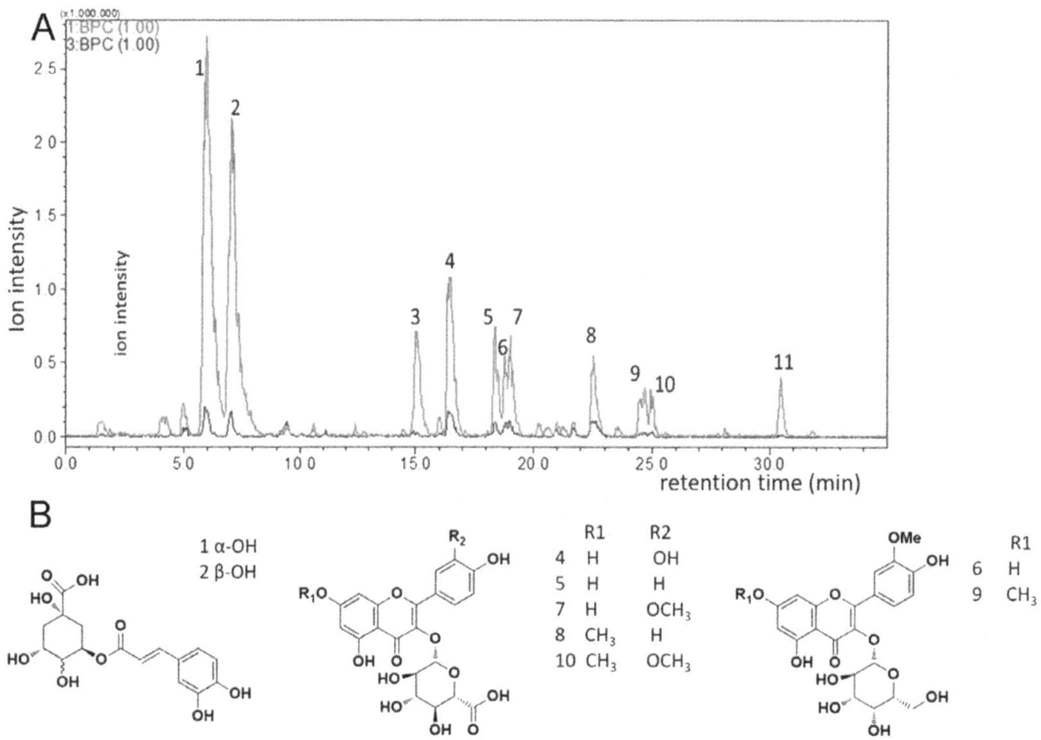

FIG. 12

Analysis of AEC components by LC-MS. The components of AEC were analyzed by LC-MS assay. A represented base peak chromatogram (BPC) of AEC. The chemical structures of nine identified components were presented in B.

4. Discussion

Our present study confirmed that both the liposoluble and aqueous fractions of clove displays a potent antitumor effect on several cancer cells. In China, TCM is widely used to alleviate the side effects of chemotherapy and improve the outcome of conventional treatment of cancer (Chan et al., 2014). Toad venom (venenumbufonis, also called Chan'su), derived from the dried skin secretions of giant toads (Bufogargarizans Cantor or Bufomelanostictus Suhneider) is used alone or in combination with other herbal ingredients for cancer treatment over centuries in China. An injectable formulation of toad venom called Cinobufacin is widely used in clinical therapy for various cancers in China (Qi et al., 2011). Clinical studies have shown that cinobufacin significantly increased the antitumor efficacy of docetaxel or cisplatin in combination therapy. Decreased toxicity and improved life quality were also observed in the clinical trial with cancer patients. Clove is traditionally believed to aid gastrointestinal function and alleviate pain in Chinese medicine, and historically the herb is used to treat nausea, gastric spasm, and sore throat (Pourgholami et al., 1999). Clove also displays chemo-preventive effect by inhibition of anti-proapoptotic gene expression, including bcl-2, VEGFA. and CD44 (Banerjee et al., 2006). In the present study, we found that both AFC and AEC are capable of inducing autophagy and inhibiting tumor growth. Considering the low toxicity and its potent antitumor effects, the active components of clove have the potential to be developed as novel adjuvants in the treatment of human tumors.

We have identified several active components; OA, eugenol, and malonic acid from the AFC (Liu et al., 2014a). OA has been widely studied for its anticancer activity and displays broad anticancer efficacy against various human cancer cells (Duan et al., 2019; Neto, 2011; Wang et al., 2017). Previous study in our laboratory has shown that OA is able to induce protective autophagy in multiple cancer cells (Neto, 2011). OA could decrease PKM2 abundance and increase PKM1 expression, in both PC-3 and MCF-7 cancer cell lines, in dose-dependent and time-dependent manner (Liu et al., 2014b). We also found that MA is able to inhibit the autophagy of pancreatic cancer by down-regulated HSPA-8 (Tian et al., 2018). Studies have shown that Eugenol can trigger apoptosis in breast cancer cells through E2F1/surviving down-regulation (Al-Sharif et al., 2013). The combination of myricetin and methyl Eugenol was able to enhance the anticancer activity of cisplatin against Hela cervical cancer cells (Yi et al., 2015). However, whether there is a synergistic effect of anticancer activity among OA, MA and Eugenol remains unclear. Additionally, it is possible that the AFC contains some other components which may display anticancer activity. More studies are needed to address the complicated mechanism of clove on its anticancer activity.

The aqueous fractions of clove (AEC) displays a more potent anticancer effect than that of the AFC. AEC is able to inhibit tumor growth both in vitro and in vivo. The result of LC-MS/MS analysis shows that AEC contains nine components; including chlorogenic acid, isocholorgenic acid, quercetin-3-*O*-β-D-glucuronide, kaempferol 3-*O*-β-glucuronide, isorhamnetin-3-*O*-β-D-glucoside, isorhamnetin-3-*O*-β-D-glucuronide, rhamnetin-3-*O*-β-D-glucuronide, rhamnazin-3-*O*-β-D-glucoside, and rhamnazin-3-*O*-β-D-glucuronide. Some of the components have been reported to possess significant anticancer effects. Chlorogenic acid can induce apoptosis in A549 human lung cancer cells by down-regulating the expression of bcl-2 gene and up-regulating Bax and capase-3 genes (Wang et al., 2017); isorhamnetin-3-*O*-β-D-glucoside showed an anticancer effect on MCF-7 human breast cancer cells and HCT-116 human colon cancer cells (Duan et al., 2019). Quercetin-3-*O*-β-D-glucuronide possesses strong antitumor effects through induction of a ROS-dependent apoptosis pathway in MCF-7 cells (Liu et al., 2014c). Study is in progress in our laboratory to address if any one of the components or the combination of the active components is responsible for the anticancer activity of AEC.

Taking together, our studies showed that both the liposoluble and aqueous fractions of clove exhibit potent anticancer effects and there is potential to develop these agents as an adjuvant for the treatment of human malignancy. The autophagic effect plays an important role in clove-induced growth inhibition of cancer cells. Some of the components of AFC or AEC in the clove are responsible for their anticancer effects. More studies are needed to address the synergistic effects among the compounds contained in the clove.

References

Al-Sharif, I., Remmal, A., Aboussekhra, A., 2013. Eugenol triggers apoptosis in breast cancer cells through E2F1/survivin down-regulation. BMC Cancer 13 (1), 1–10.

Banerjee, K., Madhyastha, H., Sandur, R., Manikandanath, N.T., Thiagarajan, N., Thiagarajan, P., 2020. Anti-inflammatory and wound healing potential of a clove oil emulsion. Colloids Surf. B Biointerfaces 193, 111102.

Banerjee, S., Panda, C.K., Das, S., 2006. Clove (*Syzygium aromaticum* L.), a potential chemopreventive agent for lung cancer. Carcinogenesis 27 (8), 1645–1654.

Chan, H.Y., Chui, Y.Y., Chan, C.W., Cheng, K.K., Shiu, A.T., So, W.K., Ho, S.S., Chan, M.M., 2014. Exploring the influence of Traditional Chinese Medicine on self-care among Chinese cancer patients. Eur. J. Oncol. Nurs. 18 (5), 445–451.

Degenhardt, K., Mathew, R., Beaudoin, B., Bray, K., Anderson, D., Chen, G., Mukherjee, C., Shi, Y., Gélinas, C., Fan, Y., 2006. Autophagy promotes tumor cell survival and restricts necrosis, inflammation, and tumorigenesis. Cancer Cell 10 (1), 51–64.

Duan, L., Yang, Z., Jiang, X., Zhang, J., Guo, X., 2019. Oleanolic acid inhibits cell proliferation migration and invasion and induces SW579 thyroid cancer cell line apoptosis by targeting forkhead transcription factor A. Anticancer Drugs 30 (8), 812–820.

El-Maati, M.F.A., Mahgoub, S.A., Labib, S.M., Al-Gaby, A.M.A., Ramadan, M.F., 2016. Phenolic extracts of clove (*Syzygium aromaticum*) with novel antioxidant and antibacterial activities. Eur. J. Integr. Med. 8 (4), 494–504. https://doi.org/10.1016/j.eujim.2016.02.006.

Hsin, I.-L., Sheu, G.-T., Jan, M.-S., Sun, H.-L., Wu, T.-C., Chiu, L.-Y., Lue, K.-H., Ko, J.-L., 2012. Inhibition of lysosome degradation on autophagosome formation and responses to GMI, an immunomodulatory protein from *Ganoderma microsporum*. Br. J. Pharmacol. 167 (6), 1287–1300.

Klionsky, D.J., Elazar, Z., Seglen, P.O., Rubinsztein, D.C., 2008. Does bafilomycin A1 block the fusion of autophagosomes with lysosomes? Autophagy 4 (7), 849–850.

Lee, K.-G., Shibamoto, T., 2001. Antioxidant property of aroma extract isolated from clove buds [*Syzygium aromaticum* (L.) Merr. et Perry]. Food Chem. 74 (4), 443–448. https://doi.org/10.1016/S0308-8146(01)00161-3.

Li, C., Xu, H., Chen, X., Chen, J., Li, X., Qiao, G., Tian, Y., Yuan, R., Su, S., Liu, X., 2019. Aqueous extract of clove inhibits tumor growth by inducing autophagy through AMPK/ULK pathway. Phytother. Res. 33 (7), 1794–1804.

Liu, H., Schmitz, J.C., Wei, J., Cao, S., Beumer, J.H., Strychor, S., Cheng, L., Liu, M., Wang, C., Wu, N., 2014a. Clove extract inhibits tumor growth and promotes cell cycle arrest and apoptosis. Oncol. Res. 21 (5), 247–259.

Liu, J., Wu, N., Ma, L., Liu, M., Liu, G., Zhang, Y., Lin, X., 2014b. Oleanolic acid suppresses aerobic glycolysis in cancer cells by switching pyruvate kinase type M isoforms. PLoS One 9 (3), e91606.

Liu, M., Zhao, G., Zhang, D., An, W., Lai, H., Li, X., Cao, S., Lin, X., 2018. Active fraction of clove induces apoptosis via PI3K/Akt/mTOR-mediated autophagy in human colorectal cancer HCT-116 cells. Int. J. Oncol. 53 (3), 1363–1373.

Liu, J., Zheng, L., Zhong, J., Wu, N., Liu, G., Lin, X., 2014c. Oleanolic acid induces protective autophagy in cancer cells through the JNK and mTOR pathways. Oncol. Rep. 32 (2), 567–572.

Mehmood, Y., Farooq, U., Yousaf, H., Riaz, H., Mahmood, R.K., Nawaz, A., Abid, Z., Gondal, M., Malik, N.S., Barkat, K., 2020. Antiviral activity of green silver nanoparticles produced using aqueous buds extract of Syzygium aromaticum. Pak. J. Pharm. Sci. 33, 839–845.

Neto, C.C., 2011. Ursolic acid and other pentacyclic triterpenoids: anticancer activities and occurrence in berries. In: Berries and Cancer Prevention. Springer, pp. 41–49.

Nikoui, V., Ostadhadi, S., Bakhtiarian, A., Abbasi-Goujani, E., Habibian-Dehkordi, S., Rezaei-Roshan, M., Foroohandeh, M., Giorgi, M., 2017. The anti-inflammatory and antipyretic effects of clove oil in healthy dogs after surgery. PharmaNutrition 5 (2), 52–57.

Nuñez, L., D'Aquino, M., 2012. Microbicide activity of clove essential oil (*Eugenia caryophyllata*). Braz. J. Microbiol. 43 (4), 1255–1260.

Nzeako, B.C., Al-Kharousi, Z.S.N., Al-Mahrooqui, Z., 2006. Antimicrobial activities of clove and thyme extracts. Sultan Qaboos Univ. Med. J. 6 (1), 33–39. https://pubmed.ncbi.nlm.nih.gov/21748125.

Pant, A., Prakash, P., Pandey, R., Kumar, R., 2016. *Syzygium aromaticum* (L.) elicits lifespan extension and attenuates age-related Aβ-induced proteotoxicity in *Caenorhabditis elegans*. Cogent Biol. 2 (1), 1218412.

Pourgholami, M.H., Kamalinejad, M., Javadi, M., Majzoob, S., Sayyah, M., 1999. Evaluation of the anticonvulsant activity of the essential oil of *Eugenia caryophyllata* in male mice. J. Ethnopharmacol. 64 (2), 167–171.

Qi, F., Li, A., Inagaki, Y., Kokudo, N., Tamura, S., Nakata, M., Tang, W., 2011. Antitumor activity of extracts and compounds from the skin of the toad Bufo bufo gargarizans Cantor. Int. Immunopharmacol. 11 (3), 342–349.

Salim, B., Said, G., Noureddine, M., Hocine, A., Angelika, B.A., 2017. A note study on antidiabetic effect of main molecules contained in clove using molecular modeling interactions with DPP-4 enzyme. Int. J. Comput. Theor. Chem. 5 (1), 9.

Selles, S.M.A., Kouidri, M., Belhamiti, B.T., Amrane, A.A., 2020. Chemical composition, in-vitro antibacterial and antioxidant activities of *Syzygium aromaticum* essential oil. J. Food Meas. Charact. 1.

Shao, X., Lai, D., Zhang, L., Xu, H., 2016. Induction of autophagy and apoptosis via PI3K/AKT/TOR pathways by azadirachtin a in *Spodoptera litura* cells. Sci. Rep. 6 (1), 1–12.

Shekhar, S., Yadav, Y., Singh, A.P., Pradhan, R., Desai, G.R., Dey, A.B., Dey, S., 2018. Neuroprotection by ethanolic extract of *Syzygium aromaticum* in Alzheimer's disease like pathology via maintaining oxidative balance through SIRT1 pathway. Exp. Gerontol. 110, 277–283.

Tian, Y., Xu, H., Farooq, A.A., Nie, B., Chen, X., Su, S., Yuan, R., Qiao, G., Li, C., Li, X., 2018. Maslinic acid induces autophagy by down-regulating HSPA8 in pancreatic cancer cells. Phytother. Res. 32 (7), 1320–1331.

Tompkins, K.D., Thorburn, A., 2019. Focus: death: regulation of apoptosis by autophagy to enhance cancer therapy. Yale J. Biol. Med. 92 (4), 707.

Topal, F., 2019. Anticholinergic and antidiabetic effects of isoeugenol from clove (*Eugenia caryophylata*) oil. Int. J. Food Prop. 22 (1), 583–592.

Venugopal, K., Rather, H.A., Rajagopal, K., Shanthi, M.P., Sheriff, K., Illiyas, M., Rather, R.A., Manikandan, E., Uvarajan, S., Bhaskar, M., 2017. Synthesis of silver nanoparticles (Ag NPs) for anticancer activities (MCF 7 breast and A549 lung cell lines) of the crude extract of *Syzygium aromaticum*. J. Photochem. Photobiol. B Biol. 167, 282–289.

Wang, R., Zheng, Q., Wang, W., Feng, L., Li, H., Huai, Q., 2017. Design and synthesis of new anticancer glycyrrhetinic acids and oleanolic acids. Biol. Pharm. Bull. 40 (5), 703–710.

Wu, Y.-T., Tan, H.-L., Shui, G., Bauvy, C., Huang, Q., Wenk, M.R., Ong, C.-N., Codogno, P., Shen, H.-M., 2010. Dual role of 3-methyladenine in modulation of autophagy via different temporal patterns of inhibition on class I and III phosphoinositide 3-kinase. J. Biol. Chem. 285 (14), 10850–10861.

Yi, J.-L., Shi, S., Shen, Y.-L., Wang, L., Chen, H.-Y., Zhu, J., Ding, Y., 2015. Myricetin and methyl eugenol combination enhances the anticancer activity, cell cycle arrest and apoptosis induction of cis-platin against HeLa cervical cancer cell lines. Int. J. Clin. Exp. Pathol. 8 (2), 1116.

CHAPTER

Clove and its constituents against urban pests: Examples from ants and cockroaches

19

Lekhnath Kafle[a] and Suwimol Chinkangsadarn[a,b]

Department of Tropical Agriculture and International Cooperation, National Pingtung University of Science and Technology, Pingtung, Taiwan, ROC[a] Faculty of Animal Sciences and Agricultural Technology, Silpakorn University, Cha-Am, Phetchaburi, Thailand[b]

1. Urban pests

Urban pests are organisms that infest the human habitat. They interfere with food and fiber, interrupt humans' daily lives, and cause health problems, inconvenience, damage, or economic loss to people or areas where they occupy. Urban pests can be structural pests (termites and wood-boring beetles), public health insects (mosquitoes, flies, flea, cockroaches, etc.), stored products pests (beetles, moths, etc.), fabric pests (carpet beetles, clothes moths, silverfish, etc.), annoying insects (fleas, bedbugs, bird lice, mites), and other regular pests (e.g., earwigs, thrips, ants, etc.) (Gerozisis et al., 2008).

Among urban pests, ants and cockroaches are categorized under household insect pests and are related to public health pests. In this chapter, the application of clove and its constituents to control ants and cockroaches is discussed.

2. Control methods of urban pests

Integrated Pest Management (IPM) is the key to control urban pests. Among many pest control methods of IPM, preventive measures are used the most to fight against urban pests, especially ants and cockroaches. Ants and cockroaches infest houses in search of water, food, and shelter. Therefore, limiting access to resources is the most efficient way to control ants and cockroaches. If preventive measures are nonfunctional or infestation is already observed, environmentally and animal-friendly control measures (mechanical/physical control or application of botanicals) are preferred.

Urban pests live near or sometimes live in the human habitat. Because of this, pest control strategies should be very sensitive. Control measures may start from preventive measures and then advance to mechanical/physical control. If unsuccessful, the next option would be to use chemical control. Synthetic toxic chemicals should be avoided as much as possible. Baiting is relatively a better control method for urban pests, including ants and cockroaches. Fumigation is also recommended for larger areas where baiting is not possible. However, fumigation leaves residues and pollutes structures and

Clove (*Syzygium aromaticum*). https://doi.org/10.1016/B978-0-323-85177-0.00026-4

materials. As such, environmental, human, and animal-friendly control methods are always in high demand. Botanicals are better alternatives to the conventional synthetic toxic pesticides against urban pests. Volatile compounds in several plant species contain secondary metabolites that are toxic and repel pests.

For years, many plant species have been used globally in pest control/management. Botanical pesticides are safe for animals, biodegradable and environmentally friendly, and do not affect nontarget species. Botanicals are readily available and affordable to purchase locally.

3. Application of clove and its constituents to control ants and cockroaches

Essential oils from clove (*Syzygium aromaticum*), clove powder, or its constituents, have repellent, toxic, and anesthetic properties against many urban pest insects, including mosquitos, head louse (*Pediculus capitis*), a red imported fire ant (RIFA) (*Solenopsis invicta*), German cockroaches (*Blattella germanica*), American cockroaches *(Periplaneta americana)*, Turkestan cockroaches (*Blatta lateralis*), and other pests (Chaieb et al., 2007; El Hag et al., 1999; Gaire et al., 2017; Kafle and Shih, 2013; Neupane et al., 2020; Omara et al., 2013; Trongtokit et al., 2005; Yang et al., 2003).

4. Mortality and repellency of clove products against ants

The RIFA is an invasive omnivorous pest native to Central America and is distributed in Caribbean islands, New Zealand, Australia, China, Korea, Japan, Taiwan, and many other countries (Ascunce et al., 2011; Chen et al., 2020; Kafle et al., 2010). The tropical fire ant (TFA) (*Solenopsis geminata*) is the most widely distributed species among fire ants and has colonized almost all the world's tropical areas. This is one of the most widespread, tropical, and subtropical ants. Wetterer (2011) identified southwestern Mexico as the most likely source for invasive populations.

The RIFA and TFA are general ants but are invasive in most parts of the world and known to be pests in urban and agroecosystems. Their sting or bite causes intense and immediate pain in humans and animals. They pose threats to local biodiversity, human and animal health, and institutional structures (Gotzek et al., 2015; Kafle et al., 2010; Perfecto and Vandermeer, 2011; Wetterer, 2011).

Usually, baits or chemical insecticides are used to control fire ants. Due to several negative issues associated with chemical applications, researchers are now focusing on investigating more natural, safe, and nonpolluting materials, such as plant extracts or plant bioactive compounds. Among different plant materials, clove products exhibited toxic and repellent effects against RIFA and TFA (Kafle and Shih, 2013; Nalini and Sasinathan, 2020).

There are limited natural products that have been identified to control or repel those invasive fire ants. Therefore, in this chapter, the application of clove and its constituents on the mortality and repellency of RIFA and TFA is discussed.

4.1 Mortality

The mortality of RIFA and TFA caused by clove powder, clove oil, and bioactive compounds of clove oil and clove powder (eugenol, eugenyl acetate, and β-caryophyllene) were investigated in contact toxicity bioassays (Kafle and Shih, 2013; Nalini and Sasinathan, 2020). Both studies used contact toxicity bioassays to investigate mortality. The clove products were spread on the surface of *Petri* dishes, ants were released inside, and mortalities were observed. Clove products used in their studies were clove powders, clove oil, eugenol, eugenyl acetate, and β-caryophyllene. Clove powder was found to be highly toxic to RIFA and TFA. The concentration of 0.34 $\mu g\,cm^{-2}$ clove powder killed 98% of RIFA in 6h. Clove powder concentration and RIFA mortality were positively correlated, while LT_{50} values were negatively correlated. When clove powder concentration increased to 1, 3, 9, or 12 $\mu g\,cm^{-2}$, 100% ant mortality in 6h was observed in all treatments. Likewise, when clove powder concentration increased from 0.34 to 3 $\mu g\,cm^{-2}$, the LT_{50} value was reduced by 2.5 folds (1.2h to 0.5h) (Kafle and Shih, 2013).

Similar trends of RIFA mortality and LT_{50} value were observed when TFA was treated with different clove powder concentrations. The 0.34 $\mu g\,cm^{-2}$ concentration caused a 93% TFA mortality in 6h. As clove powder concentration increased to 1, 3, 9, or 12 $\mu g\,cm^{-2}$, the mortality of TFA increased to 98%–100%. Moreover, when clove powder concentration was increased from 0.34 to 3 $\mu g\,cm^{-2}$, the LT_{50} value was reduced by 1.7 folds (1.8h to 1.1h) (Nalini and Sasinathan, 2020).

Similarly, clove oils were also toxic to both RIFA and TFA (Kafle and Shih, 2013; Nalini and Sasinathan, 2020). The 0.25 $\mu L\,cm^{-2}$ of clove oil concentration killed 24% of RIFA in 6h. The RIFA mortality reached 100% once clove oil concentration increased to 2 $\mu L\,cm^{-2}$ in 6h. When the clove oil concentration was increased from 0.25% to 2% $\mu L\,cm^{-2}$, the LT_{50} value was reduced by 42.8 folds (21.4h to 0.5h) (Kafle and Shih, 2013).

The TFA mortality and LT_{50} value had similar trends as RIFA when TFA was treated with different concentrations of clove oils. The 0.5% clove oil concentration had 82% ant mortality in 30min. When the concentration of clove oil increased up to 2.5%, 100% ant mortality was observed. When the concentration of clove oil increased from 0.5% to 2%, 100% ant mortality in 30min was observed (Nalini and Sasinathan, 2020).

Extensive studies on the application of bioactive compounds against ants still need to be further investigated. So far, there is only one report on this (Kafle and Shih, 2013). The study included extraction of compounds (eugenol, eugenyl acetate, and β-caryophyllene) from clove oil and was used to test its efficacy on RIFA mortality and repellency responses.

The RIFA mortality by eugenol had increased by 2.7 folds (37.5% to 100%) in 6h when the application rate of eugenol increased from 0.25 $\mu L\,cm^{-2}$ to 0.5 $\mu L\,cm^{-2}$. When the application rate of eugenol increased from 0.25 $\mu L\,cm^{-2}$ to 0.5 $\mu L\,cm^{-2}$, the LT_{50} value decreased by 13.7 folds (9.6h to 0.7h) (Kafle and Shih, 2013).

The RIFA mortality by eugenyl acetate had increased by 1.3 folds (64% to 82.5%) in 6h when the application rate of eugenyl acetate increased from 0.25 $\mu L\,cm^{-2}$ to 0.5 $\mu L\,cm^{-2}$. However, RIFA mortality increased to 99% when the application rate of eugenyl acetate increased to 2 $\mu L\,cm^{-2}$. As application rate increased from 0.25 $\mu L\,cm^{-2}$ to 0.5 $\mu L\,cm^{-2}$, the LT_{50} value decreased by 1.6 folds (1.9h to 1.2h) (Kafle and Shih, 2013). Moreover, RIFA mortality by β-caryophyllene had increased by 2.7 folds (96% to 100%) when the application rate of β-caryophyllene increased from 0.25 $\mu L\,cm^{-2}$ to

$0.5\,\mu L\,cm^{-2}$. When the application rate of β-caryophyllene rate increased from $0.25\,\mu L\,cm^{-2}$ to $0.5\,\mu L\,cm^{-2}$, the LT_{50} value decreased by 2.5 folds (1.9h to 1h) (Kafle and Shih, 2013).

When the LT_{50} value of clove oil, eugenol, eugenyl acetate, and β-caryophyllene was compared at $2\,\mu L\,cm^{-2}$, the LT_{50} value of clove oil was similar to eugenol, but was 1.4–2 folds shorter than the other two compounds evaluated (Kafle and Shih, 2013).

4.2 Repellency

To determine the repellency effect of clove products, different scholars used different bioassays. Kafle and Shih (2013) used the repellency bioassay without harborages, when Scocco et al. (2012) used bioassay with harborages. In both bioassays, ants could either stay in treated areas with test material or stay in non-treated areas. However, Chen (2009) used a two-choice digging bioassay which took advantage of the digging behavior of RIFA. The ants would either not dig or dig less in sand containing the eugenol as a repellent. In the same study, anti-digging behavior suppression was used as an indicator of repellency by eugenol (Chen, 2009).

Clove powder was observed to be a powerful repellent against RIFA, even if only a small amount was applied. When the application rate of clove powder increased from 0.34 to $1\,\mu g\,cm^{-2}$ in 3h, the mean repellency of ants ranged from 98.6% to 99.4% (Kafle and Shih, 2013).

When clove oil, with a concentration of $0.25\,\mu L\,cm^{-2}$, was used against RIFA for 3h, the mean repellency of ants reached 98.1% (Kafle and Shih, 2013). When three bioactive compounds (eugenol, eugenyl acetate, and β-caryophyllene) at $0.25\,\mu L\,cm^{-2}$ were tested on their repellency against RIFA, the mean repellency of ants was observed at 96.5%, 92.4%, and 96.1% for eugenol, eugenyl acetate, and β-caryophyllene, respectively, in 3h (Kafle and Shih, 2013).

Clove oil was evaluated (Scocco et al., 2012) using harborage bioassay for repellency at 0.1%, 1%, and 10% when applied as fresh oil and one-week-old oil against Argentine ant (*Linepithema humile*). They found that fresh clove oil repelled a significantly higher number of ants, even when the lowest amount (0.1%) of clove oil was applied. However, when the application rates increased to 1% or 10%, the repellency did not differ between the fresh and the one-week-old clove oils. Fresh 0.1%, 1%, and 10% clove oil had repelled 65%, 86%, and 91% of Argentine ants, respectively. Similarly, one-week-old 0.1%, 1%, and 10% clove oil had repelled 42%, 82%, and 84% of ants, respectively, in 2h (Scocco et al., 2012).

When clove oil was fresh, all concentrations repelled a higher number of ants. At 7-days aged clove oil at low concentration (0.1%) no longer repelled ants, but at 1% or 10%, it exhibited a repellency response against ants (Scocco et al., 2012). During the digging bioassay (Chen, 2009), it was found that the eugenol repelled a significantly higher number of RIFA when mixed with sand at 10 and $100\,mg\,kg^{-1}$; however, it failed to repel ants when eugenol was mixed with sand at $1.0\,mg\,kg^{-1}$.

The repellency and toxicity of clove products against ants are summarized in Table 1.

5. Mortality and repellency of clove products against cockroaches

Cockroaches are domiciliary insects globally distributed in human habitats (Schapheer et al., 2018), particularly the German cockroach (*Blattella germanica*) and the American cockroach (*Periplaneta americana*). They are obnoxious domestic pests of public health importance (Adenusi et al., 2018) that

Table 1 Repellency and toxicity of clove products against ants.

Items tested	Ant species	Bioassay	Reference
Clove powder Clove oil	Red imported fire ant (*Solenopsis invicta*) Tropical fire ant (*Solenopsis geminate*)	Repellency Surface contact toxicity	Kafle and Shih (2013), Nalini and Sasinathan (2020)
Eugenol Eugenyl acetate β-caryophyllene Clove powder	Red imported fire ant (*S. invicta*)	Repellency Surface contact toxicity	Kafle and Shih (2013)
Eugenol	Red imported fire ant (*S. invicta*)	Repellency (Two-choice digging bioassay)	Chen (2009)
Clove oil	Argentine ant (*Linepithema humile*)	Repellency (Harborage bioassay)	Scocco et al. (2012)

mechanically carry and transmit numerous pathogenic microorganisms, such as fungi, viruses, protozoa, parasitic worms (Etim et al., 2013; Ifeanyi and Olawumi, 2015), and bacteria with multidrug-resistant strains (*Klebsiella pneumoniae*, *Escherichia coli*, *Salmonella* spp., *Shigella* spp., and *Citrobacter* spp.) (Czajka et al., 2003; Moges et al., 2016). Moreover, cockroach feces and exuviate are sources of allergen that can provoke allergic reactions in sensitive people and lead to allergic rhinitis and asthma morbidity (Phillips et al., 2010; Pomés et al., 2019; Yeguerman et al., 2020). Consequently, several cockroach control practices (sprays, gel baits, and dusting powders) have been routinely applied to residents and food businesses to maintain acceptable hygiene standards (Sharawi et al., 2013).

Recent concerns with public health and environment safety, the investigation of essential oils to control cockroach population has been conducted as alternative agents to replace synthetic insecticides (Gaire et al., 2017; Lee et al., 2017; Phillips and Appel, 2010; Phillips et al., 2010; Sharififard et al., 2016; Sittichok et al., 2013; Thavara et al., 2007; Yeguerman et al., 2020) which impact human health, natural enemies, pollinators, and other nontarget species in both terrestrial and aquatic environments (Neupane et al., 2020; Yeguerman et al., 2020). In addition, repeated use of synthetic insecticides has led to the developing resistance of cockroaches due to their short generation time and high fecundity (Phillips and Appel, 2010). German cockroaches had reported resistance to numerous insecticides since the 1950s; for instance, acetamiprid, β-cyfluthrin, bendiocarb, bifenthrin, carbamates, cyfluthrin, fenvalerate, fipronil, indoxacarb, lambda-cyhalothrin, permethrin, phenothrin, propoxur, pyrethrin, and pyrethroids allethrin (Cochran, 1989; Fardisi et al., 2017; Pietri et al., 2018).

Clove has been widely studied for its repellency and insecticidal activities against domiciliary cockroaches since 1989 (Ngoh et al., 1998). It has been used in several forms, such as clove oil derived from dried clove flower bud (Gaire et al., 2017; Neupane et al., 2020; Omara et al., 2013; Sharawi et al., 2013; Sittichok et al., 2013), eugenol and eugenyl acetate that are two major clove oil constituents (Gaire et al., 2017; Neupane et al., 2020; Ngoh et al., 1998; Phillips and Appel, 2010; Phillips et al., 2010), and clove bud powder (Neupane et al., 2020). The main purposes of cockroach control can be divided into two parts: mortality (topical application, surface contact, and fumigation method) and repellency, which are different in functions, bioassays, and results.

5.1 Mortality

Three commonly employed bioassays for insecticide toxicity evaluation against cockroaches are topical toxicity, surface contact toxicity, and fumigant toxicity. For topical application, insecticide is directly applied to the body surface of an insect. On the contrary, insecticide is introduced into a glass container instead of on the insect for surface contact and fumigant application (Paramasivam and Selvi, 2017). Toxicity is usually expressed in terms of the median lethal dose (LD_{50}), concentration (LC_{50}), median lethal time (LT_{50}), and median knockdown dose (KD_{50}).

5.1.1 Topical application

In this bioassay, clove oil and its constituents dissolved in a relatively nontoxic and volatile solvent (e.g., acetone) are applied, with an operated micro applicator, between the metathoracic legs of each cockroach (Paramasivam and Selvi, 2017). An insecticide could reach the target site by passing through the cuticle, fat body, and other tissues.

Phillips et al. (2010) investigated the topical toxicity of eugenol on the different stages of German cockroaches. The results found that small nymphs have higher susceptibility than medium nymphs, adult males, large nymphs, gravid females, and adult females. The LD_{50} values ranged from 0.07 mg per cockroach to 0.29 mg per cockroach. Homogeneity of response (slope of the log-dose probit relationship) was similar for small nymphs, medium nymphs, and adult females (Phillips et al., 2010). Besides the common domiciliary cockroaches, Turkestan cockroach (*B. lateralis*), an invasive urban pest throughout the Southwestern United States, was evaluated in topical toxicity of clove oil and eugenol on fourth to fifth instar nymphs. The results showed that both test items had contact toxicity against Turkestan cockroach nymphs with LD_{50} value of 1.65 mg per nymph for clove oil and 1.56 mg per nymph for eugenol (Gaire et al., 2017).

5.1.2 Surface contact application

In this method, clove oil and its constituents are poured into acetone and applied to a glass container. Acetone was evaporated by rotating the container to equally spread the insecticide over the surface, leaving a residual film. Alternatively, clove oil and its constituents can be impregnated evenly on filter paper (or other materials) and allowed to dry before exposing the insects to the residual deposits (Neupane et al., 2020; Perry et al., 1998).

The contact toxicity test on American cockroaches using filter paper impregnation with eugenol demonstrated that adult female cockroaches were better killed than knockdown (LC_{50} higher than KD_{50}) at a low concentration of eugenol. However, at a high concentration of eugenol ($0.4\,mg\,cm^{-2}$), the killing (LC_{95}) and 95% of knockdown effect were not significantly different in cockroaches (Ngoh et al., 1998). Regarding surface contact toxicity of clove oil against American cockroaches, the first instar nymphs were more sensitive ($LC_{50}=0.0001\,\mu L\,cm^{-2}$) than the fourth instar ($LC_{50}=0.0077\,\mu L\,cm^{-2}$) after 24 h, while adults were the least sensitive with LC_{50} values at $1.5375\,\mu L\,cm^{-2}$ after 48 h. The LT_{50} values negatively correlated with concentrations in continuous exposure tests (Sharawi et al., 2013). Not only clove oil and its constituents are applied against cockroaches but Neupane et al. (2020) also tested clove powders, grounded from clove dried flower buds, as an insecticidal agent by dusting clove powders evenly on the inner bottom surface of the container. The toxicity of clove bud powders exhibited a good result by killing 92% of adult German cockroaches in 6 h at $30\,mg\,cm^{-2}$ clove powder concentration. For clove oil, eugenol, and eugenyl acetate, similar concentrations at

4.00 mL cm^{-2} provided a 95%, 85%, and 87% adult German cockroaches mortality in 24, 6, and 24 h, respectively (Neupane et al., 2020).

5.1.3 Fumigant application

For the fumigation toxicity test, either clove oil or eugenol were placed into a sealed container together with cockroaches; then mortality rates were recorded at different time intervals until the end of the exposure period (Paramasivam and Selvi, 2017; Perry et al., 1998). With this application, an insecticide could enter the tracheal system of the cockroach through spiracles to reach the target site, which differs from topical application and contact surface (Phillips and Appel, 2010).

Fumigant application on adult female American cockroaches using filter paper impregnated with eugenol exhibited low or negligible toxicity after being incubated for 24 h even at concentrations over 0.7 mg cm^{-2} (Ngoh et al., 1998). However, another study reported total mortality (100%) after 24 h fumigation in eugenol at the 10 μL/L air for first and fourth instar nymphs, and 20 μL/L air for adults of American cockroaches (Omara et al., 2013). For German cockroaches, the fumigation of eugenol's LC_{50} values ranged from 14.5 mg/L air for small nymphs to 1000 mg/L air for adult females (Phillips and Appel, 2010). An investigation in Turkestan cockroaches exhibited mortality (LC_{50}) from eugenol and clove oil fumigation at 251.20 mg/L air and 318.9 mg/L air, respectively (Gaire et al., 2017).

5.2 Repellent

Clove oil and its constituents are volatile and have been found to have repellent properties by providing a vapor barrier that prevents insects from flying to or landing on the surface (Gaspar et al., 2018; Nerio et al., 2010). These repellents, serving as runways for cockroaches, are helpful to protect delicate equipment or products in transport and storage from being disrupted by cockroaches (Ngoh et al., 1998).

Several studies have investigated the repellency percentage against different stages of cockroaches under laboratory conditions. The results varied in concentrations and forms of repellants (clove oil, eugenol, and eugenyl acetate), exposure time, cockroach species, and age. Younger nymphs are more sensitive and exhibited a higher response in negative chemotaxis to the lower concentration of repellant than older nymphs. Repellency percentages have been found to vary between 80% and 100%. Ngoh et al. (1998) investigated the repellency of eugenol against American cockroach nymphs. They found that eugenol exhibited a 50% repellency at a concentration of 77.1 μg cm^{-2}. In another study, the repellency percentage of clove oil was evaluated in different stages of the American cockroach (nymphs and adults). Complete repellency (100%) was recorded against first and fourth nymphs after 24 h at a concentration of 4% and 8%, respectively, while clove oil greatly repelled adult cockroaches at a 90% repellency after 48 h at a concentration of 10% (Omara et al., 2013). However, with a shorter exposure time (24 h), the same repellency percentage (90%) was reported while using clove oil concentration (10%) against American cockroaches (Sittichok et al., 2013). Adult German cockroaches treated with 2.0 mL cm^{-2} of clove oil, 1.0 mL cm^{-2} of eugenol, and 2.5 mL cm^{-2} of eugenyl acetate showed 80%, 85%, 85% repellency, respectively, within 30 min (Neupane et al., 2020).

The repellency and toxicity of clove products against cockroaches are summarized in Table 2.

Table 2 Repellency and toxicity of clove products against cockroaches.

Item tested	Cockroach species	Bioassay	Cockroach stage	Reference
Eugenol	American cockroach (*Periplaneta americana*)	- Repellent - Surface contact toxicity - Fumigant toxicity	- Nymphs (repellency activity) - Adult (female) (insecticidal test)	Ngoh et al. (1998)
Eugenol	German cockroach (*Blattella germanica*)	- Fumigant toxicity	- Nymphs (large, medium, and small) - Adult (female and male) - Gravid female	Phillips and Appel (2010)
Eugenol	German cockroach (*B. germanica*)	- Topical toxicity	- Nymphs (large, medium, and small) - Adult (female and male) - Gravid female	Phillips et al. (2010)
Clove oil	American cockroach (*P. americana*)	- Repellent - Fumigant toxicity	- Nymphs (1st and 4th instar) - Adult	Omara et al. (2013)
Clove oil	American cockroach (*P. americana*)	- Surface contact toxicity	- Nymphs (1st and 4th instar) - Adult	Sharawi et al. (2013)
Clove oil	American cockroach (*P. americana*)	- Repellent	- Adult	Sittichok et al. (2013)
Clove oil Eugenol	Turkestan cockroach (*Blatta lateralis*)	- Topical toxicity - Fumigant toxicity	- Nymphs (4th and 5th instar)	Gaire et al. (2017)
Clove oil Eugenol Eugenyl acetate Clove powder	German cockroach (*B. germanica*)	- Repellent - Surface contact toxicity	- Adult (female and male)	Neupane et al. (2020)

6. Mode of action of bioactive compounds of cloves

Clove contains 33.2% β-caryophyllene, 20.6% eugenol, and 31.1% eugenyl acetate in its chemical composition (Kafle and Shih, 2013). These lipophilic compounds are absorbed into cuticular lipids and then slowly enter the hemocoel and nervous system of insects. Once eugenol enters the insect's body, it acts as a neurotoxin inside the ants (Enan, 2001; Kafle and Shih, 2013).

Studies have reported that eugenyl acetate, eugenol, and β-caryophyllene can penetrate the chitin, become absorbed by cuticular lipids of the insects, and then slowly move into their hemocoel and nervous system. Another mode of toxicity of those bioactive compounds of clove is when absorbed into the tracheal system (Appel et al., 2004; Ntonifor et al., 2010), followed by their slow movement into the hemocoel and nervous system of insects. These bioactive compounds of clove are delayed natural toxicants in insects. After contact with these compounds, insects become paralyzed, twitched, or dead. Eugenol was also reported to be neurotoxic to carpenter ants (*Camponotus pennsylvanicus*).

The octopaminergic system in insects mediates the insecticidal activity of eugenol. Octopamine was found in high concentrations in insect tissues after the application of eugenol. Eugenol interferes with the octopaminergic nerves and becomes a neurotoxin (Enan, 2001; Isman, 2006). The eugenol also increases the permeability of cell membranes, disrupts the cytoplasmic membrane, and interacts with proteins (Devi et al., 2010; Hyldgaard et al., 2012; Neupane et al., 2020). The hydroxyl group present in eugenol binds proteins and interferes with their activities. It also inhibits ATPase, histidine decarboxylase, amylase, and protease enzymes (Gill and Holley, 2006; Hyldgaard et al., 2012; Thoroski et al., 1989; Wendakoon and Sakaguchi, 1995). Because eugenol increases cell membrane permeability, the cellular contents could be released. In bacteria, nonspecific permeabilization of the cytoplasmic membrane caused by eugenol increased the transport of potassium and ATP outside of the cells (Gill and Holley, 2006; Hemaiswarya and Doble, 2009; Hyldgaard et al., 2012; Neupane et al., 2020). These mechanisms of killing insects through bioactive compounds of clove also apply to ants and cockroaches.

7. Conclusion

Clove powder, clove oil, and bioactive compounds from the clove are efficient repellents and toxicants against ants and cockroaches. These botanical products are readily available, cheaper, and environment friendly. Therefore, these are better products to control urban insect pests.

References

Adenusi, A.A., Akinyemi, M.I., Akinsanya, D., 2018. Domiciliary cockroaches as carriers of human intestinal parasites in Lagos Metropolis, Southwest Nigeria: implications for public health. J. Arthropod. Borne Dis. 12 (2), 141.

Appel, A.G., Gehret, M.J., Tanley, M.J., 2004. Repellency and toxicity of mint oil granules to red imported fire ants (Hymenoptera: Formicidae). J. Econ. Entomol. 97 (2), 575–580.

Ascunce, M.S., Yang, C.C., Oakey, J., Calcaterra, L., Wu, W.J., Shih, C.J., Goudet, J., Ross, K.G., Shoemaker, D., 2011. Global invasion history of the fire ant *Solenopsis invicta*. Science 331 (6020), 1066–1068.

Chaieb, K., Hajlaoui, H., Zmantar, T., Kahla-Nakbi, A.B., Rouabhia, M., Mahdouani, K., Bakhrouf, A., 2007. The chemical composition and biological activity of clove essential oil, *Eugenia caryophyllata* (*Syzygium aromaticum* L. Myrtaceae): a short review. Phytother. Res. 21 (6), 501–506.

Chen, J., 2009. Repellency of an over-the-counter essential oil product in China against workers of red imported fire ants. J. Agric. Food Chem. 57 (2), 618–622.

Chen, S., Ding, F., Hao, M., Jiang, D., 2020. Mapping the potential global distribution of red imported fire ant (*Solenopsis invicta* Buren) based on a machine learning method. Sustainability 12 (23), 10182.

Cochran, D.G., 1989. Monitoring for insecticide resistance in field-collected strains of the German cockroach (Dictyoptera: Blattellidae). J. Econ. Entomol. 82 (2), 336–341.

Czajka, E., Pancer, K., Kochman, M., Gliniewicz, A., Sawicka, B., Rabczenko, D., Stypułkowska-Misiurewicz, H., 2003. Characteristics of bacteria isolated from body surface of German cockroaches caught in hospitals. Przegl. Epidemiol. 57 (4), 655–662.

Devi, K.P., Nisha, S.A., Sakthivel, R., Pandian, S.K., 2010. Eugenol (an essential oil of clove) acts as an antibacterial agent against *Salmonella typhi* by disrupting the cellular membrane. J. Ethnopharmacol. 130 (1), 107–115.

El Hag, E.A., El Nadi, A.H., Zaitoon, A.A., 1999. Toxic and growth retarding effects of three plant extracts on *Culex pipiens* larvae (Diptera: Culicidae). Phytother. Res. 13 (5), 388–392.

Enan, E., 2001. Insecticidal activity of essential oils: octopaminergic sites of action. Comp. Biochem. Physiol. Part C Toxicol. Pharmacol. 130 (3), 325–337.

Etim, S.E., Okon, O.E., Akpan, P.A., Ukpong, G.I., Oku, E.E., 2013. Prevalence of cockroaches (*Periplaneta americana*) in households in Calabar: public health implications. J. Public Health Epidemiol. 5 (3), 149–152.

Fardisi, M., Gondhalekar, A.D., Scharf, M.E., 2017. Development of diagnostic insecticide concentrations and assessment of insecticide susceptibility in German cockroach (Dictyoptera: Blattellidae) field strains collected from public housing. J. Econ. Entomol. 110 (3), 1210–1217.

Gaire, S., O'Connell, M., Holguin, F.O., Amatya, A., Bundy, S., Romero, A., 2017. Insecticidal properties of essential oils and some of their constituents on the Turkestan cockroach (Blattodea: Blattidae). J. Econ. Entomol. 110 (2), 584–592.

Gaspar, E.M., Renato, D., José, C.S., 2018. Volatile composition and antioxidant properties of clove products. Biomed. J. Sci. Tech. Res. 9 (4), 1–7.

Gerozisis, J., Hadlington, P.W., Staunton, I., 2008. Urban Pest Management in Australia. UNSW Press.

Gill, A.O., Holley, R.A., 2006. Disruption of *Escherichia coli*, *Listeria monocytogenes* and *Lactobacillus sakei* cellular membranes by plant oil aromatics. Int. J. Food Microbiol. 108 (1), 1–9.

Gotzek, D., Axen, H.J., Suarez, A.V., Helms Cahan, S., Shoemaker, D., 2015. Global invasion history of the tropical fire ant: a stowaway on the first global trade routes. Mol. Ecol. 24 (2), 374–388.

Hemaiswarya, S., Doble, M., 2009. Synergistic interaction of eugenol with antibiotics against Gram negative bacteria. Phytomedicine 16 (11), 997–1005.

Hyldgaard, M., Mygind, T., Meyer, R.L., 2012. Essential oils in food preservation: mode of action, synergies, and interactions with food matrix components. Front. Microbiol. 3 (12), 1–23.

Ifeanyi, O.T., Olawumi, O., 2015. Microbiology of cockroaches-a public health concern. Int. J. Sci. Res. 4 (4), 485–488.

Isman, M.B., 2006. Botanical insecticides, deterrents, and repellents in modern agriculture and an increasingly regulated world. Annu. Rev. Entomol. 51 (1), 45–66.

Kafle, L., Shih, C.J., 2013. Toxicity and repellency of compounds from clove (*Syzygium aromaticum*) to red imported fire ants *Solenopsis invicta* (Hymenoptera: Formicidae). J. Econ. Entomol. 106 (1), 131–135.

Kafle, L., Wu, W.J., Shih, C.J., 2010. A new fire ant (Hymenoptera: Formicidae) bait base carrier for moist conditions. Pest Manag. Sci. 66 (10), 1082–1088.

Lee, H.R., Kim, G.H., Choi, W.S., Park, I.K., 2017. Repellent activity of Apiaceae plant essential oils and their constituents against adult German cockroaches. J. Econ. Entomol. 110 (2), 552–557.

Moges, F., Eshetie, S., Endris, M., Huruy, K., Muluye, D., Feleke, T., Ayalew, G., Nagappan, R., 2016. Cockroaches as a source of high bacterial pathogens with multidrug resistant strains in Gondar town, Ethiopia. Biomed. Res. Int. 2016, 1–7.

Nalini, T., Sasinathan, S., 2020. Evaluation of toxicity of botanicals and essential oils against *Solenopsis geminata* (Fabricius) (Hymenoptera: Formicidae). Ecol. Environ. Conserv. 26, S210–S215.

Nerio, L.S., Olivero-Verbel, J., Stashenko, E., 2010. Repellent activity of essential oils: a review. Bioresour. Technol. 101 (1), 372–378.

Neupane, A.C., Sapakuka, S., Tao, P., Kafle, L., 2020. Repellancy and contact toxicity of clove bud oil and its constituents against German cockroaches, *Blattella germanica* (Dictyoptera: Blattellidae), under laboratory conditions. Int. J. Pest Manag. 66 (4), 289–297.

Ngoh, S.P., Choo, L.E., Pang, F.Y., Huang, Y., Kini, M.R., Ho, S.H., 1998. Insecticidal and repellent properties of nine volatile constituents of essential oils against the American cockroach, *Periplaneta americana* (L.). Pestic. Sci. 54 (3), 261–268.

Ntonifor, N.N., Oben, E.O., Konje, C.B., 2010. Use of selected plant-derived powders and their combinations to protect stored cowpea grains against damage by *Callosobruchus maculatus*. J. Agric. Biol. Sci. 5 (5), 13–21.

Omara, S.M., Al-Ghamdi, K.M., Mahmoud, M.A., Sharawi, S.E., 2013. Repellency and fumigant toxicity of clove and sesame oils against American cockroach *Periplaneta americana* (L.). Afr. J. Biotechnol. 12 (9), 963–970.

Paramasivam, M., Selvi, C., 2017. Laboratory bioassay methods to assess the insecticide toxicity against insect pests-a review. J. Entomol. Zool. Stud. 5 (3), 1441–1445.

Perfecto, I., Vandermeer, J., 2011. Discovery dominance tradeoff: the case of *Pheidole subarmata* and *Solenopsis geminata* (Hymenoptera: Formicidae) in neotropical pastures. Environ. Entomol. 40 (5), 999–1006.

Perry, A.S., Yamamoto, I., Ishaaya, I., Perry, R., 1998. Methods of testing chemicals on insects. In: Insecticides in Agriculture and Environment. Springer, pp. 12–13.

Phillips, A.K., Appel, A.G., 2010. Fumigant toxicity of essential oils to the German cockroach (Dictyoptera: Blattellidae). J. Econ. Entomol. 103 (3), 781–790.

Phillips, A.K., Appel, A.G., Sims, S.R., 2010. Topical toxicity of essential oils to the German cockroach (Dictyoptera: Blattellidae). J. Econ. Entomol. 103 (2), 448–459.

Pietri, J.E., Tiffany, C., Liang, D., 2018. Disruption of the microbiota affects physiological and evolutionary aspects of insecticide resistance in the German cockroach, an important urban pest. PLoS One 13 (12), e0207985.

Pomés, A., Glesner, J., Calatroni, A., Visness, C.M., Wood, R.A., O'Connor, G.T., Kattan, M., Bacharier, L.B., Wheatley, L.M., Gern, J.E., 2019. Cockroach allergen component analysis of children with or without asthma and rhinitis in an inner-city birth cohort. J. Allergy Clin. Immunol. 144 (4), 935–944.

Schapheer, C., Sandoval, G., Villagra, C.A., 2018. Pest cockroaches may overcome environmental restriction due to anthropization. J. Med. Entomol. 55 (5), 1357–1364.

Scocco, C.M., Suiter, D.R., Gardner, W.A., 2012. Repellency of five essential oils to *Linepithema humile* (Hymenoptera: Formicidae). J. Entomol. Sci. 47 (2), 150–159.

Sharawi, S.E., Omara, S.M., Al-Ghamdi, K.M., Abd-Alla, S.M., 2013. Surface contact toxicity of clove and rosemary oils against American cockroach, *Periplaneta americana* (L.). Afr. Entomol. 21 (2), 324–332.

Sharififard, M., Safdari, F., Siahpoush, A., Kassiri, H., 2016. Evaluation of some plant essential oils against the brown-banded cockroach, *Supella longipalpa* (Blattaria: Ectobiidae): a mechanical vector of human pathogens. J. Arthropod. Borne Dis. 10 (4), 528–537.

Sittichok, S., Phaysa, W., Soonwera, M., 2013. Repellency activity of essential oil on thai local plants against American cockroach (*Periplaneta americana* L.; Blattidae: Blattodea). J. Agric. Technol. 9 (6), 1613–1620.

Thavara, U., Tawatsin, A., Bhakdeenuan, P., Wongsinkongman, P., Boonruad, T., Bansiddhi, J., Chavalittumrong, P., Komalamisra, N., Siriyasatien, P., Mulla, M.S., 2007. Repellent activity of essential oils against cockroaches (Dictyoptera: Blattidae, Blattellidae, and Blaberidae) in Thailand. Southeast Asian J. Trop. Med. Public Health 38 (4), 663–673.

Thoroski, J., Blank, G., Biliaderis, C., 1989. Eugenol induced inhibition of extracellular enzyme production by *Bacillus subtilis*. J. Food Prot. 52 (6), 399–403.

Trongtokit, Y., Rongsriyam, Y., Komalamisra, N., Apiwathnasorn, C., 2005. Comparative repellency of 38 essential oils against mosquito bites. Phytother. Res. 19 (4), 303–309.

Wendakoon, C.N., Sakaguchi, M., 1995. Inhibition of amino acid decarboxylase activity of *Enterobacter aerogenes* by active components in spices. J. Food Prot. 58 (3), 280–283.

Wetterer, J.K., 2011. Worldwide spread of the tropical fire ant, *Solenopsis geminata* (Hymenoptera: Formicidae). Myrmecol. News 14 (1), 21–35.

Yang, Y.-C., Lee, S.H., Lee, W.J., Choi, D.H., Ahn, Y.J., 2003. Ovicidal and adulticidal effects of *Eugenia caryophyllata* bud and leaf oil compounds on *Pediculus capitis*. J. Agric. Food Chem. 51 (17), 4884–4888.

Yeguerman, C., Jesser, E., Massiris, M., Delrieux, C., Murray, A.P., González, J.W., 2020. Insecticidal application of essential oils loaded polymeric nanoparticles to control German cockroach: design, characterization and lethal/sublethal effects. Ecotoxicol. Environ. Saf. 189 (2020), 110047.

CHAPTER

20 Clove (*Syzygium aromaticum*) and clove extracts applications in the meat industry

Oscar Abel Sánchez-Velázquez[a], Alan Javier Hernández-Álvarez[b], and Guiomar Melgar-Lalanne[c]

Facultad de Ciencias Químico-Biológicas, Universidad Autónoma de Sinaloa, Culiacán, Sinaloa, Mexico[a] *School of Food Science and Nutrition, University of Leeds, Leeds, United Kingdom*[b] *Instituto de Ciencias Básicas, Universidad Veracruzana, Xalapa, Veracruz, Mexico*[c]

1. Introduction

Meat is a food rich in high-quality proteins, vitamins, and minerals. In general, it is also rich in lipids and deficient in carbohydrates. It has high water content when fresh, and its free water is also high (aw = 0.98). These characteristics make meat a food that is very susceptible to microbial contamination and oxidation, resulting in a very short shelf life. During meat spoilage, undesirable volatile compounds, color changes, superficial slime, and risk of borne diseases are produced (da Silva et al., 2021).

Due to its short life, meat has been preserved in many ways using thermal and nonthermal technologies (Fig. 1), including salt and spices. In recent times, consumers' habits have increased the development of minimally processed foods, mainly ready-to-eat products. These products require an extended shelf life without the use of artificial preservatives and antioxidants, which has increased the use of essential oils (EO) and natural extracts as preservatives, colorants, and flavorings (Jayasena and Jo, 2013).

Traditionally, the safety and quality of meat and its by-products have been determined through sensory, microbiological, and physicochemical techniques (Fig. 1). However, more sustainable and economical methods have been developed to control meat's final quality to improve organoleptic properties as well as consumer acceptance.

Essential oils are volatile liquids extracted from barks, seeds, flowers, peels, roots, leaves, wood, fruits, or even the whole plant and are commonly named for the plant from which they are obtained. Essential oils can be extracted by many different methods, such as hydrodistillation, steam distillation, hydro-diffusion, solvent extraction, and supercritical fluid extraction (Sharma et al., 2021; Melgar-Lalanne et al., 2017).

Clove (*S. aromaticum* (L.) Merr. & L.M.Perry, synonym: *Eugenia caryophyllus* Thunb.) is the aromatic dried flower bud of the homonymous tree in the family Myrtaceae (Myrtales), which includes other perfumed plants of economic importance such as guava (*Psidium guajava* L.), bay rum tree (*Pimenta racemosa* (Mill.) J. W. Moore), and eucalyptus (*Eucalyptus* spp. L'Her.). Clove is an evergreen tree that is 8–20 m tall and has straight stems with grayish bark, single-lanceolate and leathery

Clove (*Syzygium aromaticum*). https://doi.org/10.1016/B978-0-323-85177-0.00032-X

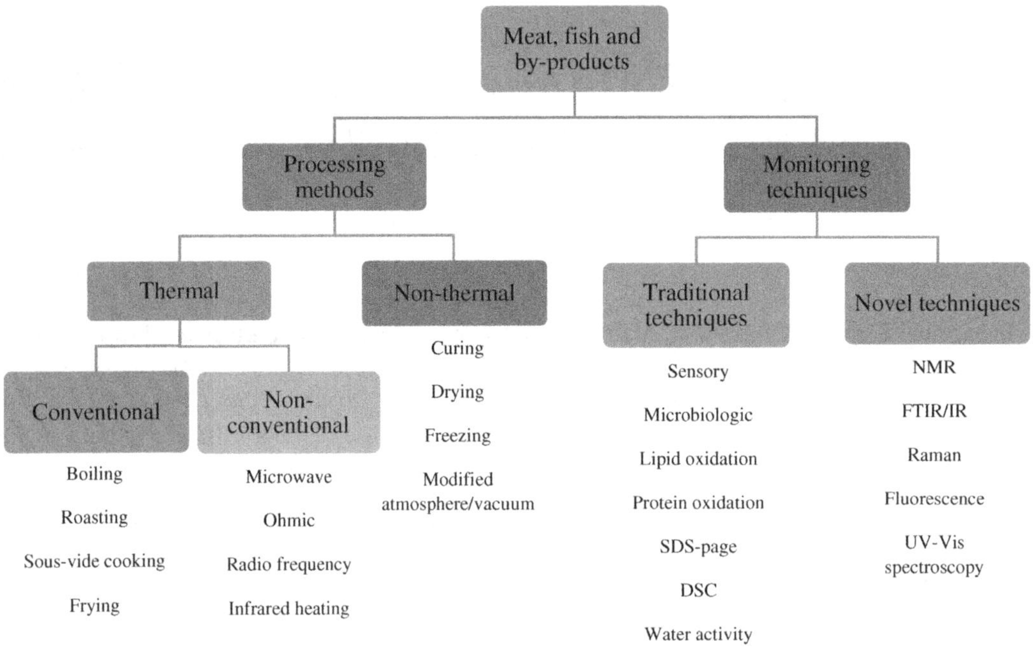

FIG. 1

Main processing methods and monitoring techniques applied to meat, fish, and by-products. *SDS-PAGE*, sodium dodecyl sulfate-polyacrylamide gel electrophoresis; *DSC*, differential scanning calorimetry.

long bright-green leaves, and sanguine flowers forming a terminal inflorescence arranged in branches (Danthu et al., 2014). These flower buds also have a long calyx ending in four spread-out sepals, four unopened petals with yellow pistils, and an ovary with abundant carpels (Pandey and Singh, 2011). During flower development, the flower buds gradually change color from pale green to bright reddish-brown and grow to 1.5–2.0cm long and 4mm wide, indicating the harvest time, usually four to six months after flowering starts. They are then carefully picked by hand (El-Saber Batiha et al., 2020). Once dry, cloves develop their characteristic strong odor and spicy, pungent, and aromatic taste (Pandey and Singh, 2011).

The clove tree is a strictly tropical plant, thriving in warm-humid climate regions of 20–30°C, ranging between sea level and 1500m ASL (Thangaselvabai et al., 2010). Clove trees flower between September and February (depending on altitude) and begin to yield after the 7th year, reach a full bearing stage at 15–20years old, and continue flowering until 70–80years old (Thangaselvabai et al., 2010). One of these trees may produce up to 34kg/year of clove buds. The Maluku Islands in Indonesia, also known as Moluccas Islands, are the origin place of the clove tree, and this country leads the clove production in the world, followed by Madagascar, Tanzania (Zanzibar Islands), Brazil, Sri Lanka, Comoros, Kenya, China, India, Malaysia, and Grenada, together reaching an annual net production of more than 200,000 tons[3] per year (Danthu et al., 2014). Archeological remains indicate the presence of cloves in Syria around 1729BCE, thousands of kilometers from its evolutionary place of origin. In China, clove is mentioned in documents of the Han dynasty, around 200BCE. It was a precious product that was a luxury good of the maritime commercial caravans of the Silk Road in the 4th century, which was its first introduction to the Western world. After 1511, with the discovery of the Moluccas by the Portuguese expeditions, clove was popularized in Europe and taken to the New World. Nowadays, clove is appreciated worldwide for its versatility in gastronomic and medicinal uses and as an antiseptic, room air freshener, aromatic, antiparasitic, and antimicrobial agent (Burt, 2004). Products derived from clove, such as EO, have been used since ancient times to avoid food decomposition (Seow et al., 2014).

This chapter aims to provide an overview of clove's antioxidant and antimicrobial properties and the effects of its extracts on meat and meat products' quality and shelf life.

2. Clove oil production

Clove oil has ancestral uses in aromatherapy and ethnomedicine. However, it is often used for meat preservation (Sánchez-González et al., 2011). In general, clove oils are obtained from clove buds using a single extraction method or a combination of methods (Aguilar-Gonzalez and López-Malo, 2013; Ratri et al., 2020). Table 1 contains a summary of some of the most popular explored techniques to better understand oil extraction from clove samples.

Traditionally, mechanical methods such as milling or grinding are popular techniques employed for oil extraction from clove leaves and buds (Kapadiya and Desai, 2017). This method does not require sophisticated equipment, preserves the chemical profile of the extract, and has low cost, but it requires a lot of starting raw material and is very time-consuming. The chemical extraction of clove oil is

Table 1 Advantages and limitations of clove bud oil extraction methods.

Characteristics	Mechanical extraction	Steam distillation extraction	Supercritical fluid extraction	Microwave extraction	Chemical extraction
Lots of raw starting material	X	X			X
Short time required		X	X	X	
Sophisticated equipment			X		
High cost		X	X	X	
Preparative extraction	X	X	X	X	X
Quantitative extraction		X	X	X	X
Pre-treatment of raw material		X	X	X	X
Exclusion of contaminants	X		X		
Increase in yield of extracted components		X	X	X	
High efficiency		X	X	X	

supported by the use of solvents of diverse isoelectric affinity (Danlami et al., 2014). Solvent carrying is a cheap method for oil extraction from powdered clove buds and is useful for quantitative extraction, but some contaminants derived from the interaction between the solvent and the phytocompounds in the clove samples could contaminate the end product.

Sophisticated techniques often used for oil extraction, such as steam distillation, microwave extraction, and supercritical fluid extraction, have some advantages over the abovementioned methods, including the short time of operation, increment in yield of extracted components, and high efficiency in extracting phytochemicals (Aguilar-Gonzalez and López-Malo, 2013). However, they usually require complex equipment or have a high cost. Choosing the most adequate extraction method depends on the purposes of the oil to be obtained.

3. Chemical composition and mechanisms of action of clove

Clove EO has been extensively popularized for its attributes including its organoleptic properties, wide availability, and high yield (Dev et al., 2021). Essential oil is the principal constituent in cloves, representing at least 70% to 92% of the chemical composition in clove buds, and >80% in clove leaves and stems (Kamatou et al., 2012; Wang et al., 2018; Pulikottil and Nath, 2015; Idowu et al., 2021; Srivastava et al., 2005). Clove EO presents flavor notes that are very appreciated in perfumery and aromatherapies (Sharmeen et al., 2021).

In addition to the sensory stimulation properties produced by this fragrant liquid, antiviral, antimicrobial, antifungal, anticancer, antiparasitic, anti-inflammatory, and antioxidant effects have also been reported by several studies (Kamatou et al., 2012; Wang et al., 2018; Pulikottil and Nath, 2015; Idowu et al., 2021). The oldest use of clove EO is in dentistry as a natural anesthetic agent to treat tooth pain since the 16th century (Dagli et al., 2015).

3.1 Eugenol

Eugenol is a guaiacol with an allyl chain-substituted (4-allyl-2-methoxy phenol) and a member of the allylbenzenes groups with the general formula $C_{10}H_{12}O_2$ (Fig. 2) (Kamatou et al., 2012). This phytochemical is found in numerous aromatic plants from the families Lamiaceae (basils, *Ocimum* spp.), Lauraceae (cinnamons, *Cinnamomum* spp. J.Presl), Myristicaceae (nutmeg, *Myristica fragrans* Houtt.), and others. However, among the family Myrtaceae and the rest of the plants, clove (*S. aromaticum*) is considered as the principal natural source of eugenol. It can be found in clove flower buds, leaves, and stems (Kamatou et al., 2012).

Eugenol is a phenylpropanoid that is slightly soluble in water and soluble in organic solvents, as well as weakly acidic. It is a colorless to pale yellow aromatic oily liquid present in several plant sources, such as cinnamon, nutmeg, bay leaf, and especially clove bud (80% to 90%) and leaf oils (82% to 88%). This allylbenzene was isolated for the first time in 1929 and industrialized from the 1940s. This liposoluble compound is highly versatile and is used as ingredient of several cosmetics and perfumery in restricted proportions, such as spice-flavored cigarettes, alcoholic beverages, desserts, candies, and baking (Kamatou et al., 2012). Its ester form has powerful spicy notes; thus, it is not common to find it in food and cosmetic products (Huang et al., 2013). Clove oil contains between 60% and 90% eugenol, which is the source of its anesthetic and antiseptic properties (Danthu

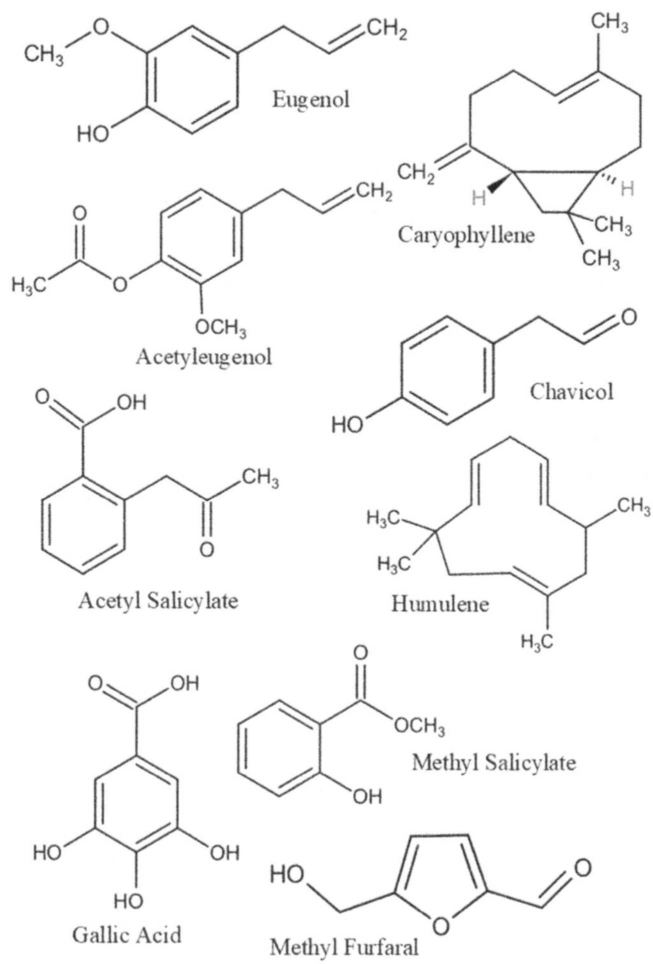

FIG. 2

Common phytochemicals found in clove extracts.

et al., 2014). In combination with zinc oxide, eugenol forms a polymeric adhesive for temporary filling, pulp capping agents, surgical dressing, and cavity liners (Lewis et al., 2011). Despite the synthetic eugenol produced by allylation of guaiacol with allyl chloride, its obtention from natural oil sources is still predominant in this industry (Yogalakshmi et al., 2010) Also, eugenol was used for the artificial synthesis of other phytochemicals in the past, such as vanillin, but nowadays, eugenol is used for its derivatives, such as isoeugenol and methyl eugenol (Barceloux, 2008; Li et al., 2004).

Eugenol has attracted the attention of researchers because of its chemopreventive and chemotherapeutic effects (Kamatou et al., 2012). Studies attribute these bioactivities to the presence of a phenolic group (Yogalakshmi et al., 2010; Kamatou et al., 2012). However, in recent years, many works have focused on the food preservation effects of eugenol and other clove extracts, with industry interest in retarding meat oxidation and extending the shelf life of meat products (de Oliveira Monteschio et al., 2017).

4. Quality control of meats with clove and clove eugenol

Muscle foods, including meat, fish, and their products, play essential roles in human nutrition due to their rich content of high-quality proteins, fats, vitamins, and minerals (Table 2). Nevertheless, it is well known that fresh meat and fish and their by-products are highly perishable. These products have a short shelf life because of their high free water content and nutrients for microbial growth. Meat deterioration can occur rapidly unless an acceptable processing method is used (Hassoun et al., 2020).

The use of EO as bio-preservatives in the food industry has gained increased interest in reducing or eliminating the pathogenic organisms present in food and boosting food functionality and quality. Moreover, some health risks are associated with consuming synthetic additives, increasing consumer awareness, and subsequent demand for safe food products containing ingredients from natural sources. These natural preservatives are bioactive substances derived from plants, animals, and beneficial microorganisms. Although plant-derived EO are generally recognized as safe (GRAS), the performance of these substances is under study, especially their efficacy, which depends on the vegetable source, the extraction method, the solvent, the concentration, the target pathogenic microorganism, and the final food where it is used (El-Saber Batiha et al., 2021).

Table 2 Nutrient content of different meats expressed per 100 g of edible portion.

Species and part	Energy (kcal/100 g)	Protein (g/100 g)	Fat (g/100 g)	Fiber (g/100 g)	Cholesterol (mg/100 g)
Mutton leg	196.56	15.12	15.12	–	65.52
Veal leg	85.32	15.72	2.45	–	56.09
Horse meat	109	21.5	2.5	–	75
Pork shoulder	13.2	16.89	7.05	–	50.02
Beef sirloin	112	20.1	3.5	–	59
Rabbit	123.24	16.59	6.32	–	51.35
Goose	140.63	5.78	13.04	–	32.8
Duck	199.04	8.64	18.30	–	48.64
Turkey breast	83	19.2	0.7	–	49
Turkey drumstick	100	16.6	3.7	–	81
Chicken breast	98	21.5	1.3	–	58
Chicken drumstick	125	17.8	6	–	84

Adapted from Orkusz, A. 2021. Edible insects versus meat—nutritional comparison: knowledge of their composition is the key to good health. Nutrients 13, 1207 https://doi.org/10.3390/nu13041207.

5. Antibacterial activity of clove and clove eugenol in meat and meat products

Clove powder, clove oil, and clove EO have shown high antimicrobial activity against some of the most commonly found pathogens in meat (Table 3).

Table 3 Antimicrobial activity of clove and clove EO in meat and meat by-products.

Meat/meat by-product	Microorganism target	Main findings	Reference
Refrigerated chicken breast	Entorobacteriaceae, lactic acid bacteria, *Pseudomonas* spp., psychrotrophic count, yeast, and molds	Sodium alginate coating containing 0.5% of clove and 0.5% of lemon verbena is the most effective combination to reduce the microbiological count.	Hosseini et al. (2021)
Refrigerated ray beef	*Listeria monocytogenes*	The maximum effect of phage + clove oil + UV light	Ishaq et al. (2021)
Meat samples (raw steak, raw mutton, raw mince, and raw pork)	*Eschericia coli* and *Enterococci* spp.	Water clove extracts (1:10; *w/v*) evaporated in a water bath was the best inhibitor.	Tshabalala et al. (2021)
Chicken patties	Total plate count, *S. aureus*, and coliforms	Chitosan enriched with 0.5% of clove EO. The shelf life was enhanced by 20 days.	Shukla et al. (2020)
Deboned chicken meat protein films	*B. subtilis*, *S. aureus*, and *L. monocytogenes*	Clove EO at 1.5% resulted in the high reduction of the microbial count.	Sarıcaoglu and Turhan (2020)
Raw and processed beef meat	*Bacillus* spp.	Clove powder (0.5 and 1%, w/w) and EO (0.25% to 0.5%, v/w) showed high antibacterial activity. Essential oils are stronger than the species powder.	Hetta et al. (2020)
Meat and fish products	*Escherichia coli* O157:H7	Clove water extract resulted in a high inhibition activity.	Abdelmotilib (2020)
Retail chicken meat	*Campylobacter jejuni*	Clove oil and cinnamon extracts showed antibacterial potential.	Samad et al. (2019)
Carcase traits of Japanese quails	Coliforms, *E. coli*, and *Salmonella* spp. in the ileal content	1.5 mL of clove oil/mg in the diet showed a reduction.	Hussein et al. (2019)
Fermented meat products	*E. coli*, *L. monocytogenes*, *Salmonella* spp., and *Staphylococcus aureus*	Cinnamon and clove food-grade commercial EO showed intermediate antimicrobial activity against all tested bacteria.	Laranjo et al. (2019)
Raw ground pork during chilled storage	Total viable count, Enterobacteriaceae count, and *Pseudomonas* spp.	Clove and cardamom ethanol extracts were the most potent antimicrobial extracts.	Muzolf-Panek et al. (2019)

6. Antioxidant activity of clove in meat and meat products

Endogenous and exogenous reactive oxygen species (ROS) and reactive nitrogen species (RNS) (as well as other free radicals) generate damage in macromolecules when the oxidant-antioxidant system is unbalanced in organisms (Di Meo et al., 2016). Oxidation of meat from biotic and abiotic origins causes a decrease in quality (Falowo et al., 2014). Market supply of livestock products is seriously compromised by biotic and abiotic agents (de Oliveira Monteschio et al., 2017). Artificial chemicals are usually incorporated into meat and other animal products to extend shelf life, avoid premature decomposition, remove bacteria, etc. (Burt, 2004; Ishaq et al., 2021). However, these added substances often have secondary effects on sensory properties and cooking characteristics and sometimes have negative consequences on consumers' health (long term) (Torrecilhas et al., 2021). In recent years, natural alternatives to the use of these food preservatives have been sought, such as antioxidant compounds from clove EO (Tanabe et al., 2002; Wojtunik-Kulesza et al., 2020).

Antioxidant compounds are incorporated into the diet to transfer them from food to muscle for moderating oxidative stress states and improving meat quality (Falowo et al., 2014; de Oliveira Monteschio et al., 2017). The dietary antioxidant compounds considered for meat quality and preservation include phytochemicals found in clove essential oil, including eugenol and vanillin (de Oliveira Monteschio et al., 2017), since beef samples from purebred Nellore heifers fed with diets supplied with encapsulated clove EO compounds have shown significantly higher antioxidant activity assessed by ABTS, DPPH, and FRAP than rosemary (*Rosmarinus officinalis*) EO or control (without EO).

The antioxidant activity of clove EO in animal farming and food preservation areas is shown in Table 4. Antioxidant activity of clove extracts or oils also has been observed on "white meat" products. An assay of Köytür Chicken products capped with MDCMP films containing clove EO obtained by hydrodistillation method with a Clevenger apparatus had statistically higher antioxidant activity evaluated by FRAP (31.57 ± 1.58 mmol Fe^{+2}/mL) and DPPH ($EC_{50} = 3.04 \pm 0.25$ μL) than thyme and rosemary EO (Sarıcaoglu and Turhan, 2020). Chicken legs and a lipid fraction from meat samples exposed to a clove bud extract, obtained by grinding 15 g buds with 225 mL of 50% aqueous ethanol and stirring in the dark, exhibited ABTS scavenging activity (2495.85 ± 31.41 μM/g), which was highly positively correlated with DPPH ($r = .997$ at $P = .000$, Student's t-test) and FRAP assays results ($r = .998$ at $P = .000$, Student's t-test). After nine days of exposure of raw chicken meat emulsion to ground clove powder, DPPH• inhibition ability was >48% than the other treatments, as well as 15% to 49% as ABTS$^{\bullet}$ inhibitory activity (Singh et al., 2014).

Chevon and chicken meat exposed to commercial clove EO showed in 0–21 days FRAP 4.9–6.3 mmol Trolox eq/kg chevon and 5.0–6.5 mmol Trolox eq/kg chicken meat, as well as ABTS radical scavenging activity of 65% to 73% and 62% to 70% in chevon and chicken meat, respectively, and DPPH radical scavenging activity of 65% to 68% in chevon and 62% to 68% in chicken meat (Sultana et al., 2017). The authors also found that clove EO had an antioxidant activity that was higher than butylated hydroxyanisole, tocopherol, and L-ascorbic acid, but similar to clove oleoresin and tertiary butyl hydroquinone. An ethanolic extract of clove buds also exhibited a higher DPPH radical scavenging activity ($EC_{50} = 0.21$ mg/mL) than that of rosemary ($EC_{50} = 0.49$ mg/mL) on raw chicken meat, but in the ferrous ion-chelating capacity, rosemary exhibited a higher ferrous ion-chelating ability ($EC_{50} = 0.41$ mg/mL) than cloves did (EC_{50} 0.60 mg/mL) (Zhang et al., 2017).

Fish protein films incorporated with clove EO obtained from distillation of clove buds presented radical DPPH inhibition >80% and reducing power >70 mg ascorbic acid/g, statistically higher

Table 4 Antioxidant activity of clove EO in animal farming and food preservation areas

Sample	Extraction	Antioxidant activity	Chemical properties	References
Mechanically deboned chicken meat protein	Hydrodistillation with a Clevenger apparatus	FRAP 31.57 mmol Fe^{+2}/mL; DPPH• EC_{50} 3.04 μL	α-pinene (20.65%), camphane (3.97%), 3-carene (11.62%), α-phellandrene (1.22%), cymol (26.29%), γ-terpinene (7.05%), terpinolene (1.45%), eugenol (17.02%), copaene (1.08%), caryo-phyllene (2.37%), eugenol acetate (1.24%)	Sarıcaoglu and Turhan (2020)
Fish protein films	Bud distillation	DPPH• ↓ 80%; reducing power ↓ 70 mg ascorbic acid/g	N.M.	Teixeira et al. (2014)
Chevon and chicken meat	Commercial clove oil	FRAP 5.0–6.5 mmol Trolox eq/kg chicken meat and 4.9–6.3 mmol Trolox eq/kg chevon, ABTS• of 62% to 70% and 65% to 73% in chicken meat and chevon, respectively; DPPH• 62% to 68% in chicken meat and 65% to 68% in chevon	N.M.	Sultana et al. (2017)
Beef sausages	Commercial clove oil	TBARS 0.263–760 mg malonaldehyde/kg fresh sausage formula or in combination with nitrite 0.212–0.562 mg MDA/kg fresh sausage formula in beef sausages of zero to three months without any additives and formulated with nitrite	N.M.	Sedki et al. (2020)
Chicken legs/ lipid fraction from meat samples	Ethanolic extract of dry clove buds	ABTS• 2495.85 μM/g (high positive correlation with DPPH and FRAP methods)	Protocatechuic acid (79.57 mg/kg dw), 2,5-dihydroxy-benzoic acid (1.33 mg/kg dw), 4-hyrdoxybenzoic acid (40.12 mg/kg dw), gallic acid (0.36 mg/kg dw), syringic acid (1.22 mg/kg dw), *t*-cinnamic acid (29.5 mg/kg dw), caffeic acid (52.9 mg/kg dw),	Muzolf-Panek et al. (2019)

Continued

Table 4 Antioxidant activity of clove EO in animal farming and food preservation areas—cont'd

Sample	Extraction	Antioxidant activity	Chemical properties	References
			p-coumaric acid (5.3 mg/kg dw), ferulic acid (2.1 mg/kg dw), chlorogenic acid (10.2 mg/kg dw), apigenin (0.07 mg/kg dw), luteolin (0.06 mg/kg dw), catechin (0.39 mg/kg dw), kaempferol (0.05 mg/kg dw)	
Beef	Commercial clove EO	Lipid peroxidation ↓ 0.4–0.8 mg MDA/kg from 1 to 15 days	Eugenol, phenolics	Ishaq et al. (2021)
Young crossbred (Brown Swiss × Nellore) bulls	Commercial clove EO	Lipid oxidation ↓ 20% to 30%	Eugenol (84.5 g/kg), carophylenne (13.3 g/kg), eugenyl acetate (1.3 g/kg)	Torrecilhas et al. (2021)
Chinese-style pork sausages	Ethanolic clove bud extracts	21 days of cold storage Chinese-style pork sausages treated with clove 1% and 2% extracts had MDA ↓ 50% lower in sausages with clove extract >1% than control (0% clove extract); 2% clove extract ↓ MDA at day 7 compared with day 0.	Eugenol	Zhang et al. (2017)
Raw chicken meat emulsion	Clove grinder powder	MDA after nine days, DPPH• ↑ 48% than the other treatments; as well as ↑ 15% to 49% as ABTS•	N.M.	Singh et al. (2014)

N.M., *not mentioned;* MDA, *malondialdehyde;* dw, *dry weight.*

($P < .05$) than garlic (*Allium sativum*) and oregano (*Origanum vulgare*) EO (Teixeira et al., 2014). Eugenol is mentioned (Nam and Kim, 2013) as a phenolic compound able to inhibit lipid peroxidation and scavenge DPPH free radicals. However, eugenol could also protect DNA from oxidation induced by hydroxyl radicals.

There are many nonmetallic free radicals and transition metals in the body with oxidative potential, of which iron is one of the most important lipid oxidation pro-oxidants. Iron in ferrous state promotes lipid oxidation by breaking down hydrogen and lipid peroxides into reactive free radicals via the Fenton reaction (Öztürk et al., 2007). Knowing the role of clove phytochemicals in this phenomenon and other lipid oxidative routes is essential for meat product preservation with good quality for long shelf life.

Malondialdehyde (MDA) is an organic compound synthesized when cell or tissue damage occurs by oxidative stress and other biological situations (Ayala et al., 2014). Malondialdehyde is commonly found in beef and meat as a lipid oxidation product, affecting meat quality (Macho-González et al., 2020; Ayala et al., 2014). Clove samples showed a lower efficacy in conjugated diene decrement. However, clove has a higher MDA formation inhibitory activity than nutmeg, allspice, bay leaf, black seed, caraway, and cardamom (Muzolf-Panek et al., 2019). The authors attributed this bioactivity to the high content and diversity of phenolic compound in the clove ethanolic extract, such as protocatechuic (79.57 mg/kg DW), 2,5-dihydroxy-benzoic (1.33 mg/kg DW), 4-hydroxybenzoic (40.12 mg/kg DW), gallic (0.36 mg/kg DW), syringic (1.22 mg/kg DW), *t*-cinnamic (29.5 mg/kg DW), caffeic (52.9 mg/kg DW), *p*-coumaric (5.3 mg/kg DW), ferulic (2.1 mg/kg DW), and chlorogenic (10.2 mg/kg DW) acids, as well as apigenin (0.07 mg/kg DW), luteolin (0.06 mg/kg DW), catechin (0.39 mg/kg DW), and kaempferol (0.05 mg/kg DW) (Muzolf-Panek et al., 2019).

In raw chicken meat emulsion exposed to clove, ground powder presented lower MDA synthesis in 1, 3, 5, 7, and 9 days than the control. Studies of other experimental treatments, such as garlic and ginger pastes (Singh et al., 2014; de Oliveira Monteschio et al., 2017), reported that lipid oxidation was influenced by diet and aging. After 14 days, diets containing encapsulated eugenol + thymol + vanillin blend and clove EO were significantly lower than control and rosemary EO diets.

However, refrigerated raw beef treated synergistically with a bacteriophage + UV light + clove EO had a significantly lower MDA produced after 15 days of aging than control and other experimental treatments (Ishaq et al., 2021). The authors attribute the inhibition of lipid peroxidation to eugenol. Pork meat treated with 1–5 μL of an extract of clove buds with potassium phosphate buffer (pH 7.4) displayed the lowest MDA formation (threefold reduction) in comparison with several herb/fruit extracts, including parsley, peppercorn, caraway, coriander, and tarragon, and had up to four times less MDA than pork meat without plant-based treatment (Tanabe et al., 2002).

In 0–3-month-old beef sausage without any additives and formulated with nitrite, clove EO alone and clove EO in combination with nitrite decreased MDA in 0.263–760 mg malonaldehyde/kg fresh sausage formula and 0.212–0.562 mg MDA/Kg fresh sausage formula, respectively. However, thyme EO samples showed better values (Sedki et al., 2020). Other sausages, Chinese-style pork sausages treated with clove 1% and 2% extracts, did not show statistical differences ($P > .05$) in MDA amount after 21 days of cold storage (Zhang et al., 2017). The MDA was <50% lower in sausages with clove extract >1% than the control (0% clove extract) after 21 days. The authors also observed a significant ($P > .05$) decreased in MDA to the 2% clove extract on day 7 compared with day 0. In this study, the protein oxidation was also measured by carbonyl groups synthesis, the values of which were statistically lower ($P > .05$) in >1% clove extract than in the control, but in all cases, carbonyl content on day 21 was >4.5-fold compared to day 0.

The compounds found in clove EO may maintain/improve meat quality during its shelf life and cold storage (de Oliveira Monteschio et al., 2017). In most of these studies, eugenol was the main active phenolic compound in clove extracts with oxidant protective effects.

7. Conclusion and perspectives

Meat and meat products are highly prone to microbial deterioration, which affects their shelf life and final quality. Clove, clove extracts, and clove EO are shown to possess excellent antimicrobial and antioxidant activities. However, some limitations, mainly related to sensorial characteristics, have been reported due to the intense flavor and taste of clove. In some cases, the interaction with some food ingredients could decrease its effectiveness. Moreover, the action of clove EO is lower when tested in meat and meat products than that of in vitro studies.

Studies should be conducted to maximize the antimicrobial and antioxidant activities of the clove extracts and EO with the lowest concentration to avoid undesirable sensorial properties in meat and meat products. Novel technologies such as nanoemulsions and nanoencapsulation can help to avoid these problems.

References

Abdelmotilib, N.M., 2020. Anti-*Escherichia coli* O157: H7 as natural preservative to control and prevent food contamination in meat and fish products. Pak. J. Biol. Sci. 23 (5), 674–684.

Aguilar-Gonzalez, A., López-Malo, A., 2013. Extractos y aceite esencial del clavo de olor (*Sizygium aromaticum*) y su potencial aplicación como agente antimicrobiano en alimentos. Temas Selectos Ing. Aliment. 7 (2), 35–41.

Ayala, A., Muñoz, M.F., Argüelles, S., 2014. Lipid peroxidation: production, metabolism, and signaling mechanisms of malondialdehyde and 4-hydroxy-2-nonenal. In: Ramana, K.V. (Ed.), Oxidative Medicine and Cellular Longevity 2014, p. 360438, https://doi.org/10.1155/2014/360438.

Barceloux, D.G., 2008. Medical Toxicology of Natural Substances: Foods, Fungi, Medicinal Herbs, Plants, and Venomous Animals. John Wiley & Sons.

Burt, S., 2004. Essential oils: their antibacterial properties and potential applications in foods—a review. Int. J. Food Microbiol. 94 (3), 223–253. https://doi.org/10.1016/j.ijfoodmicro.2004.03.022.

da Silva, B.D., Bernardes, P.C., Pinheiro, P.F., Fantuzzi, E., Roberto, C.D., 2021. Chemical composition, extraction sources and action mechanisms of essential oils: natural preservative and limitations of use in meat products. Meat Sci. 176. https://doi.org/10.1016/j.meatsci.2021.108463, 108463.

Dagli, N., Dagli, R., Mahmoud, R.S., Baroudi, K., 2015. Essential oils, their therapeutic properties, and implication in dentistry: a review. J. Int. Soc. Prev. Community Dent. 5 (5), 335–340. https://doi.org/10.4103/2231-0762.165933.

Danlami, J.M., Arsad, A., Ahmad Zaini, M.A., Sulaiman, H., 2014. A comparative study of various oil extraction techniques from plants. Rev. Chem. Eng. 30 (6).

Danthu, P., Penot, E., Mahafaka Ranoarisoa, K., Chrysostôme Rakotondravelo, J., Michel, I., Tiollier, M., Michels, T., Normand, F., Razafimamonjison, G., Fawbush, F., 2014. The clove tree of Madagascar: a success story with an unpredictable future. Bois For. Trop. 320 (2), 83–96.

de Oliveira Monteschio, J., de Souza, K.A., Vital, A.C.P., Guerrero, A., Valero, M.V., Kempinski, E.M.B.C., Barcelos, V.C., Nascimento, K.F., do Prado, I.N., 2017. Clove and rosemary essential oils and encapsuled active principles (eugenol, thymol and vanillin blend) on meat quality of feedlot-finished heifers. Meat Sci. 130, 50–57. https://doi.org/10.1016/j.meatsci.2017.04.002.

Dev, M., Ghosh, M., Bhattacharyya, D.K., 2021. Physico-chemical, antimicrobial, and organoleptic properties of roasted aromatic spice (clove bud) in baked product. Appl. Biochem. Biotechnol. https://doi.org/10.1007/s12010-021-03504-0.

Di Meo, S., Reed, T.T., Venditti, P., Victor, V.M., 2016. Role of ROS and RNS sources in physiological and pathological conditions. In: Franco, R. (Ed.), Oxidative Medicine and Cellular Longevity, p. 1245049, https://doi.org/10.1155/2016/1245049.

El-Saber Batiha, G., Alkazmi, L.M., Wasef, L.G., Beshbishy, A.M., Nadwa, E.H., Rashwan, E.K., 2020. *Syzygium aromaticum* L. (Myrtaceae): traditional uses, bioactive chemical constituents, pharmacological and toxicological activities. Biomolecules 10, 202. https://doi.org/10.3390/biom10020202.

El-Saber Batiha, G., Hussein, D.E., Algammal, A.M., George, T.T., Jeandet, P., Esmail Al-Snafi, A., Tiwari, A., et al., 2021. Application of natural antimicrobials in food preservation: recent views. Food Control 126. https://doi.org/10.1016/j.foodcont.2021.108066, 108066.

Falowo, A.B., Fayemi, P.O., Muchenje, V., 2014. Natural antioxidants against lipid–protein oxidative deterioration in meat and meat products: a review. Food Res. Int. 64, 171–181. https://doi.org/10.1016/j.foodres.2014.06.022.

Hassoun, A., Guðjónsdóttir, M., Prieto, M.A., Garcia-Oliveira, P., Simal-Gandara, J., Marini, F., Di Donato, F., D'Archivio, A.A., Biancolillo, A., 2020. Application of novel techniques for monitoring quality changes in meat and fish products during traditional processing processes: reconciling novelty and tradition. PRO. https://doi.org/10.3390/pr8080988.

Hetta, H., Kh Meshaal, A., Algammal, A.M., Yahia, R., Makharita, R.R., Marraiki, N., Ajmal Shah, M., Hassan, H.-A.M., El-Saber Batiha, G., 2020. In-vitro antimicrobial activity of essential oils and spices powder of some medicinal plants against Bacillus species isolated from raw and processed meat. Infect. Drug Resist. 13 (December), 4367–4378. https://doi.org/10.2147/IDR.S277295.

Hosseini, M., Jamshidi, A., Raeisi, M., Azizzadeh, M., 2021. Effect of sodium alginate coating containing clove (*Syzygium aromaticum*) and Lemon verbena (*Aloysia citriodora*) essential oils and different packaging treatments on shelf life extension of refrigerated chicken breast. J. Food Process. Preserv. 45 (3). https://doi.org/10.1111/jfpp.14946, e14946.

Huang, X.-w., Feng, Y.-c., Huang, Y., Li, H.-l., 2013. Chemical composition, antioxidant and the possible use as skin-care ingredient of clove oil (*Syzygium aromaticum* (L.) Merr. & Perry) and citronella oil (*Cymbopogon goeringii*) from China. J. Essent. Oil Res. 25 (4), 315–323. https://doi.org/10.1080/10412905.2013.775082.

Hussein, M.M.A., Abd El-Hack, M.E., Mahgoub, S.A., Saadeldin, I.M., Swelum, A.A., 2019. Effects of clove (*Syzygium aromaticum*) oil on quail growth, carcass traits, blood components, meat quality, and intestinal microbiota. Poult. Sci. 98 (1), 319–329. https://doi.org/10.3382/ps/pey348.

Idowu, S., Adekoya, A.E., Igiehon, O.O., Idowu, A.T., 2021. Clove (*Syzygium aromaticum*) spices: a review on their bioactivities, current use, and potential application in dairy products. J. Food Meas. Charact. 15, 3419–3435. https://doi.org/10.1007/s11694-021-00915-9.

Ishaq, A., Syed, Q.A., Ebner, P.D., Ubaid, H., ur Rahman., 2021. Multiple hurdle technology to improve microbial safety, quality and oxidative stability of refrigerated raw beef. LWT- Food Sci. Technol. 138. https://doi.org/10.1016/j.lwt.2020.110529, 110529.

Jayasena, D.D., Jo, C., 2013. Essential oils as potential antimicrobial agents in meat and meat products: a review. Trends Food Sci. Technol. 34 (2), 96–108. https://doi.org/10.1016/j.tifs.2013.09.002.

Kamatou, G.P., Vermaak, I., Viljoen, A.M., 2012. Eugenol—from the remote Maluku Islands to the international market place: a review of a remarkable and versatile molecule. Molecules. https://doi.org/10.3390/molecules17066953.

Kapadiya, S., Desai, M.A., 2017. Isolation of essential oil from buds of *Syzygium aromaticum* using hydrodistillation: multi-response optimization and predictive modelling. Int. J. Adv. Res. Sci. Eng. 6 (1), 405–418.

Laranjo, M., Fernández-León, A.M., Agulheiro-Santos, A.C., Potes, M.E., Elias, M., 2019. Essential oils of aromatic and medicinal plants play a role in food safety. J. Food Process. Preserv. https://doi.org/10.1111/jfpp.14278, e14278.

Lewis, N., Bohnert, H.J., Nguyen, H., 2011. Bioengineering and Molecular Biology of Plant Pathways. Elsevier.

Li, Y.-H., Sun, Z.-H., Zheng, P., 2004. Determination of vanillin, eugenol and Isoeugenol by RP-HPLC. Chromatographia 60 (11), 709–713. https://doi.org/10.1365/s10337-004-0440-4.

Macho-González, A., Garcimartín, A., López-Oliva, M.E., Bastida, S., Benedí, J., Ros, G., Nieto, G., Sánchez-Muniz, F.J., 2020. Can meat and meat-products induce oxidative stress? Antioxidants. https://doi.org/10.3390/antiox9070638.

Melgar-Lalanne, G., Alan Javier, A.J., Hernández-Álvarez, M.J.-F., Azuara, E., 2017. Oleoresins from Capsicum Spp.: extraction methods and bioactivity. Food Bioprocess Technol. 10 (1), 51–76. https://doi.org/10.1007/s11947-016-1793-z.

Muzolf-Panek, M., Kaczmarek, A., Tomaszewska-Gras, J., Cegielska-Radziejewska, R., Majcher, M., 2019. Oxidative and microbiological stability of raw ground pork during chilled storage as affected by plant extracts. Int. J. Food Prop. 22 (1), 111–129. https://doi.org/10.1080/10942912.2019.1579834.

Nam, H., Kim, M.-M., 2013. Eugenol with antioxidant activity inhibits MMP-9 related to metastasis in human fibrosarcoma cells. Food Chem. Toxicol. 55, 106–112. https://doi.org/10.1016/j.fct.2012.12.050.

Öztürk, M., Aydoğmuş-Öztürk, F., Duru, M.E., Topçu, G., 2007. Antioxidant activity of stem and root extracts of rhubarb (Rheum Ribes): an edible medicinal plant. Food Chem. 103 (2), 623–630. https://doi.org/10.1016/j.foodchem.2006.09.005.

Pandey, A., Singh, P., 2011. Antibacterial activity of *Syzygium aromaticum* (clove) with metal ion effect against food borne pathogens. Asian J. Plant Sci. Res. 1 (2), 69–80.

Pulikottil, S.J., Nath, S., 2015. Potential of clove of *Syzygium Aromaticum* in development of a therapeutic agent for periodontal disease: a review. S. Afr. Dent. J. 70.

Ratri, P.J., Ayurini, M., Khumaini, K., Rohbiya, A., 2020. Clove oil extraction by steam distillation and utilization of clove buds waste as potential candidate for eco-friendly packaging. J. Bahan Alam Terbarukan 9 (1), 47–54.

Samad, A., Abbas, F., Ahmed, Z., Akbar, A., Naeem, M., Sadiq, M.B., Ali, I., et al., 2019. Prevalence, antimicrobial susceptibility, and virulence of campylobacter Jejuni isolated from chicken meat. J. Food Saf. 39 (2). https://doi.org/10.1111/jfs.12600, e12600.

Sánchez-González, L., Vargas, M., González-Martínez, C., et al., 2011. Use of essential oils in bioactive edible coatings: a review. Food Eng. Rev. 3, 1–16.

Sarıcaoglu, F.T., Turhan, S., 2020. Physicochemical, antioxidant and antimicrobial properties of mechanically deboned chicken meat protein films enriched with various essential oils. Food Packag. Shelf Life 25. https://doi.org/10.1016/j.fpsl.2020.100527, 100527.

Sedki, A.G., El-Zainy, A.R., Rajab, B.T., 2020. Thyme and clove essential oils as antioxidants and antimicrobial in beef sausage. J. Food Nutr. Sci. 8 (5), 117–126.

Seow, Y.X., Yeo, C.R., Chung, H.L., Yuk, H.-G., 2014. Plant essential oils as active antimicrobial agents. Crit. Rev. Food Sci. Nutr. 54 (5), 625–644. https://doi.org/10.1080/10408398.2011.599504.

Sharma, S., Barkauskaite, S., Jaiswal, A.K., Jaiswal, S., 2021. Essential oils as additives in active food packaging. Food Chem. 343. https://doi.org/10.1016/j.foodchem.2020.128403, 128403.

Sharmeen, J.B., Mahomoodally, F.M., Zengin, G., Maggi, F., 2021. Essential oils as natural sources of fragrance compounds for cosmetics and cosmeceuticals. Molecules (Basel, Switzerland) 26 (3), 666. https://doi.org/10.3390/molecules26030666.

Shukla, V., Mendiratta, S.K., Zende, R.J., Agrawal, R.K., Jaiswal, R.K., 2020. Effects of chitosan coating enriched with *Syzygium aromaticum* essential oil on quality and shelf-life of chicken patties. J. Food Process. Preserv. 44 (11). https://doi.org/10.1111/jfpp.14870, e14870.

Singh, P., Sahoo, J., Chatli, M.K., Biswas, A.K., 2014. Shelf life evaluation of raw chicken meat emulsion incorporated with clove powder, ginger and garlic paste as natural preservatives at refrigerated storage (4 ± 1°C). Int. Food Res. J. 21 (4), 1363.

Srivastava, A.K., Srivastava, S.K., Syamsundar, K.V., 2005. Bud and Leaf essential oil composition of *Syzygium aromaticum* from India and Madagascar. Flavour Fragance J. 20 (1), 51–53. https://doi.org/10.1002/ffj.1364.

Sultana, K., Jayathilakan, K., Pandey, M.C., 2017. Evaluation of antioxidant activity, radical scavenging, and reducing power of clove oil and clove oleoresin in comparison with natural and synthetic antioxidants in Chevon (Capra Aegagrus Hircus) and chicken meat. Def. Life Sci. J. 3 (1), 51.

Tanabe, H., Yoshida, M., Tomita, N., 2002. Comparison of the antioxidant activities of 22 commonly used culinary herbs and spices on the lipid oxidation of pork meat. Anim. Sci. J. 73 (5), 389–393. https://doi.org/10.1046/j.1344-3941.2002.00054.x.

Teixeira, B., Marques, A., Pires, C., Ramos, C., Batista, I., Saraiva, J.A., Nunes, M.L., 2014. Characterization of fish protein films incorporated with essential oils of clove, garlic and origanum: physical, antioxidant and antibacterial properties. LWT Food Sci. Technol. 59 (1), 533–539. https://doi.org/10.1016/j.lwt.2014.04.024.

Thangaselvabai, T., Richard Kennedy, R., Prem Joshua, J., Jayasekar, M., 2010. Clove (*Syzigium aromaticum*)—the spicy flower bud of significance-a review. Agric. Rev. 31 (1).

Torrecilhas, J.A., Ornaghi, M.G., Passetti, R.A.C., Mottin, C., Guerrero, A., Ramos, T.R., Vital, A.C.P., Sañudo, C., Malheiros, E.B., Nunes, I., do Prado., 2021. Meat quality of young bulls finished in a feedlot and supplemented with clove or cinnamon essential oils. Meat Sci. 174. https://doi.org/10.1016/j.meatsci.2020.108412, 108412.

Tshabalala, R., Kabelinde, A., Christ-Donald, K.T., Ateba, C.N., Manganyi, M.C., 2021. Effect of clove (*Syzygium aromaticum*) spice as microbial inhibitor of resistant bacteria and organoleptic quality of meat. Saudi J. Biol. Sci. 28, 3855–3863. https://doi.org/10.1016/j.sjbs.2021.03.052.

Wang, Y.-F., Jia, J.-X., You-Qiu Tian, X., Shu, X.-J.R., Guan, Y., Yan, Z.-Y., 2018. Antifungal effects of clove oil microcapsule on meat products. LWT- Food Sci. Technol. 89, 604–609. https://doi.org/10.1016/j.lwt.2017.11.042.

Wojtunik-Kulesza, K., Oniszczuk, A., Oniszczuk, T., Combrzyński, M., Nowakowska, D., Matwijczuk, A., 2020. Influence of in vitro digestion on composition, bioaccessibility and antioxidant activity of food polyphenols—a non-systematic review. Nutrients 12, 1401. https://doi.org/10.3390/nu12051401.

Yogalakshmi, B., Viswanathan, P., Anuradha, C.V., 2010. Investigation of antioxidant, anti-inflammatory and DNA-protective properties of eugenol in thioacetamide-induced liver injury in rats. Toxicology 268 (3), 204–212. https://doi.org/10.1016/j.tox.2009.12.018.

Zhang, H., Peng, X., Li, X., Jingjuan, W., Guo, X., 2017. The application of clove extract protects Chinese-style sausages against oxidation and quality deterioration. Korean J. Food Sci. Anim. Resour. 37 (1), 114–122. https://doi.org/10.5851/kosfa.2017.37.1.114.

CHAPTER

21 Nanostructured pharmaceutical formulations for topical application of clove oil and eugenol

Talita Nascimento da Silva[a], Stephani Araujo Cardoso[b], and Thaís Nogueira Barradas[c]

Postgraduate Program in Pharmaceutical Sciences, Federal University of Rio de Janeiro, University City, Rio de Janeiro, Brazil[a] Postgraduate Program in Nanobiossystems, Federal University of Rio de Janeiro, University City, Rio de Janeiro, Brazil[b] Department of Pharmaceutical Sciences, Federal University of Juiz de Fora, Juiz de Fora, Brazil[c]

1. Introduction

Essential oils (EO) comprise naturally complex products derived from a plant's secondary metabolism to respond to several environmental stimuli like predators and pathogens, extreme temperatures, humidity, and injuries. Thus, EO is considered safe and biocompatible, providing comprehensive therapeutical benefits, which can be justified by their complex composition comprising fatty acids, terpenes, triterpenes, and other lipophilic components. Such complex matrixes feature interesting biological activities such as antimicrobial, repellent, analgesic, antiinflammatory activity, antineoplastic, wound healing, being a valuable source of therapeutic agents for pharmaceutical purposes (Donsì and Ferrari, 2016; Andreu et al., 2015; Bonferoni et al., 2017; de Matos et al., 2019). Moreover, terpenes are widely explored in developing topical pharmaceutical formulations due to their penetration-enhancing properties and capacity to increase the solubility of poorly soluble drugs.

Clove essential oil presents many benefits for pharmaceutical applications such as pain relief, antifungal, antimicrobial, antioxidant, and antiinflammatory activities (Shahavi et al., 2016; Jirovetz et al., 2006). Clove oil is mainly constituted by eugenol (major component), β-karyophylene, and benzyl alcohol in lesser proportions. The synergy among such components provides interesting antibacterial, antifungal, antiviral, antitumor, anesthetic, antioxidant, and insecticidal activity (Jirovetz et al., 2006; Chaieb et al., 2007a,b). Eugenol is a volatile monoterpene with a molecular weight of 164.20, a boiling point of 254°C, low water solubility (2460 mg/L), and freely miscible in ethanol and chloroform. It is safe for human consumption according to the Food and Drug Administration (FDA) and the European Commission (Yalkowsky et al., 2010; Sebaaly et al., 2015; Ahad et al., 2016).

Currently, the search for naturally originated bioactive molecules, especially natural antioxidants and antimicrobial agents, as a substitute for synthetic drugs is emerging. EO frequently features excellent antioxidant potential, reducing the formation of free radicals, often pointed to as responsible for cellular damage and risk factors for many chronic diseases. In this context, the research efforts to design and develop effective and stable pharmaceutical formulations have been rising. On the other hand, the development of topical pharmaceutical formulations containing EO is challenging due to their low

Clove (*Syzygium aromaticum*). https://doi.org/10.1016/B978-0-323-85177-0.00019-7

water solubility, uncontrollable volatility, and reduced stability in hydrophilic formulations. As EOs are highly lipophilic molecules, their aqueous pharmaceutical formulations could be limited. Thus, encapsulation strategies are technological tools that protect the encapsulated content while preserving their biological and pharmaceutical properties. Additionally, most EO is highly unstable under environmental conditions, as light exposure and oxidation. All those obstacles should be considered before developing topical pharmaceutical formulations (Badgujar et al., 2014; Herman and Herman, 2015; Moghimi et al., 2016).

The encapsulation of EO and terpenes into nanostructured materials is an exciting strategy that enables easier manufacturing, topical administration, and incorporation into a pharmaceutical conventional dosage form, like gels, lotions, or creams. Besides, nanoencapsulation of EO brings technological benefits due to their small-sized particle size, as increased water solubility, rate, and site-controlled delivery, protect from environmental degradation and prevent premature and uncontrolled evaporation (Dias et al., 2014; Sutradhar and Amin, 2013), which can improve the pharmacological effect.

Additionally, nanosized vehicles proved to be more efficient in overcoming several biological membranes, such as the *stratum corneum* (SC), eukaryotic cell, and microbial membranes. They provided high antimicrobial and antioxidant activity compared to free EO, as their encapsulation into nanosized particles can lead to a great disruptive activity of essential oils on cell membranes of microorganisms (McClements and Jafari, 2018). Several nanostructured systems for topical drug delivery have been designed with clove oil aiming to overcome the well-known obstacles of topical administration route, using technological approaches as mucoadhesive excipients that increase residence time on the mucosae and the SC and using permeation enhancers (Sajomsang et al., 2012; Jadhav et al., 2004). This chapter aims to show the current state-of-the-art research of novel nanostructured pharmaceutical formulations design containing clove oil and/or eugenol as the sole bioactive compounds or an excipient for topical administration routes.

2. Clove oil as an excipient in pharmaceutical formulations for topical drug delivery

Clove oil and many other essential oils have been widely explored by food and cosmetic industries due to their flavor and fragrance potential (Cui et al., 2015; Lasoń, 2020). Clove oil is obtained by distillation of the flowers *Eugenia aromatica* or *Eugenia caryophyllata*, and it has been shown safe for humans and is generally considered a safe compound by FDA (Mylonas et al., 2005). Eugenol as primary bioactive and other derivatives (α-caryophyllene, acetoeugenol, and humulene) present antimicrobial, antioxidant, antiinflammatory, and antipyretic properties (Banerjee et al., 2020). Due to these properties, clove oil and eugenol have been applied to pharmaceutical formulations. For example, clove oil has been used for topical application to relieve pain and promote healing (Moon et al., 2011).

The development of novel drug delivery systems requires a careful and rational selection of their excipients to meet some critical product attributes from the physicochemical and therapeutical point of view. Pramod et al. (2015) studied the compatibility of eugenol, the main constituent of clove oil (around 80%), with several common excipients of nanostructured colloidal carriers such as Tween 80, Carbopol 940, chitosan, sodium alginate, and polycaprolactone (PCL). Besides visual observations of possible incompatibilities, the authors conducted eugenol-excipient compatibility studies using physicochemical methods as differential scanning calorimetry (DSC), infrared spectroscopy (FTIR), and high-performance thin-layer chromatography (HPTLC). The results showed no significant visual changes, suggesting compatibility of eugenol with the selected excipients. DSC thermograms and FTIR results showed no significant changes in the peaks of eugenol in the presence of all excipients, confirming that eugenol is not subjected to any significant interactions with Tween 80 Carbopol 940, chitosan, sodium alginate, and PCL. HPTLC could not detect the presence of degradation products from eugenol, thereby corroborating the compatibility of eugenol with the proposed excipients.

2.1 Clove oil as a solvent for poorly soluble drugs

The search for reducing the use of organic solvents, such as DMSO and ethanol, for solubilizing poorly soluble drugs has been rising. In this context, there is a rapidly increasing need for "greener" effective substitutes for organic solvents that can be potentially harmful to both human use and the environment. EO and their terpenes are natural and renewable substances that considered safer and more suitable substitutes for conventional solvents. Terpenes are hydrocarbons constituted by C_5H_8 isoprene units, acyclic, bicyclic, or monocyclic compounds (Tanzi et al., 2012).

The primary chemical structure of terpenes and terpenoids is classified according to the number of isoprene units. For example, hemiterpenes are formed by one isoprene unit; monoterpenes comprise two isoprene units (C10), sesquiterpenes formed by three (C15), and diterpenes by four (C20) isoprene units. Terpenes volatility of terpenes and terpenoids decreases with an increased number of isoprene units. Additionally, smaller molecules are more susceptible to degradation, mainly by oxidation, isomerization, and temperature (Lasoń, 2020).

Clove oil promotes higher solubilization capacity for poorly soluble drugs (Barradas et al., 2017, 2018; De Campos et al., 2016; Senna et al., 2015). Different mechanisms can justify the increased drug solubility promoted by clove oil components, especially eugenol that has been reported to promote higher solubility of poorly soluble drugs, with greater efficiency for aromatic molecules.

One of the most critical steps to develop novel drug carriers is selecting the solvent and core materials to ensure maximum drug loading and encapsulation efficiency. Traditionally, such a task involves the necessity to use organic solvents like chloroform. However, such an approach is not environment friendly and highly toxic for human consumption (Marturano et al., 2018). Clove oil has been used as an alternative solvent for this purpose, as previously reported with clove oil-loaded nanocarriers for topical delivery of 8-methoxipsoralen (Barradas et al., 2017, 2018), minoxidil (Cardoso and Barradas, 2020), praziquantel (De Campos et al., 2016), zinc and chloroaluminum phthalocyanine (Senna et al., 2015), and azilsartan medoxomil (Madan et al., 2021). Besides the solvent properties, EO like clove oil provide beneficial characteristics, such as their antioxidant and antimicrobial properties (Marturano et al., 2018).

Senna et al. (2015) showed that clove oil was more effective in dissolving zinc phthalocyanine than DMSO. Polarized optical microscopy was used to assess the presence of undissolved microcrystals in suspension, and no residual drug was found for clove oil-based drug solution. Besides, the solvent volume required to solubilize 1 mg of zinc phthalocyanine was 4 mL for clove oil. The authors conducted further studies to access zinc phthalocyanine using UV-vis spectrophotometry. The results showed that the main characteristic bands of zinc phthalocyanine were observed in clove oil-based drug solutions, suggesting that the oil successfully solubilized the drug. Clove oil was effective substitute for toxic organic solvents as core material for the development of nanostructured drug carriers.

The increase in the solubility of poorly soluble drugs, especially as it was the case for zinc phthalocyanine (Senna et al., 2015), is essential for designing nanostructured pharmaceutical topical formulations, as it provides greater drug loading efficiency of such nanosized drug carriers. Such solubilization and loading capacity are essential for topical drug application, which often is limited by the low doses tolerated by topical administration sites. Even though the number of papers dedicated to thoroughly investigate and explore the solvent properties attributed to clove oil and eugenol is still limited and the application of clove oil as "greener," safer drug solvent remains under-explored.

2.2 Clove oil as a penetration enhancer

The transdermal drug delivery route is a conventional, painless and noninvasive approach for systemic treatment with several drugs that feature low oral bioavailability. Topical administration of pharmaceutical dosage forms also includes local drug delivery to treat epidermal or dermal disorders, which require drug skin penetration. In order to achieve this, drugs must overcome several obstacles imposed by SC, which is the most outer layer of the nonviable epidermis and constitutes a rate-limiting lipophilic barrier against both drug penetration and transepidermal water loss (Herman and Herman, 2015).

In this context, permeation enhancers are often applied as an excipient in formulations to enable and increase drug diffusion through SC layers by modifying the characteristics of the lipid intercellular matrix, which becomes less compact and more fluid. Such property is beneficial to developing pharmaceutical topical or skin formulations, as the SC constitutes the main barrier for drug skin penetration or transdermal absorption. In topical applications, EO as clove oil can potentially improve transdermal drug delivery by the interaction between the liquid crystals formed by highly ordered intercellular lipids in SC (Baptista-Silva et al., 2020). Essential oils are enhancers, absorption promoters, and accelerants for drug skin adsorption (Adorjan and Buchbauer, 2010). Moreover, EO mainly contain terpenes, which are recognized for their penetration-enhancing effects with low skin irritation at low concentrations (Sapra et al., 2008).

Both EO and terpenes facilitate topical administration of drugs through dermal, ocular, vaginal, nasal, and rectal routes. Moreover, FDA recognizes terpenes and EO as Generally Recognized as Safe (GRAS) compounds, being considered safer than conventional synthetic permeation enhancers (de Matos et al., 2019; Herman and Herman, 2015). Their role as penetration enhancers may favor drug transport across the skin by reducing barrier resistance with different mechanisms such as disruption of lipid structure, alteration of polarity, lipid extraction, phase separation, and fluidization from lipid bilayers (Marjukka Suhonen et al., 1999; Wang et al., 2005; Lopes and Garcia, 2015; Dragicevic and Maibach, 2015). Fig. 1 shows how EO as clove act facilitate drug penetration through skin layers and the possible mechanisms of increasing skin permeability with penetration enhancers.

In general, terpenes as eugenol are known and widely used in pharmaceutical formulations for their capacity to improve the skin penetration and transdermal permeation of lipophilic substances, being considered permeation enhancers, mostly due to their capacity of extracting SC intercellular lipids, reducing SC resistance to drug diffusion, without damaging viable cells (Aqil et al., 2007; Ahad et al., 2016). According to Herman and Herman (2015), terpenes can enable enhanced drug penetration through at least 4 main mechanisms: (i) disruption of the highly compact and organized intercellular lipid matrix in SC, (ii) interaction with intercellular proteins, inducing conformational changes in keratin molecular structure, (iii) increasing drug partitioning in the SC lipids due to the solvent properties of terpenes, and (iv) interaction with desmosomes between corneocytes. Several chemical structures

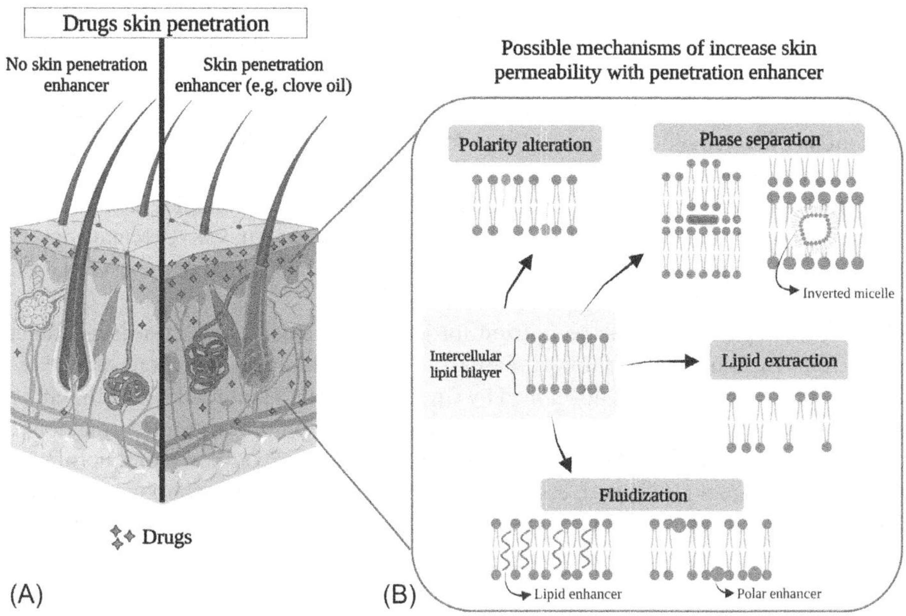

FIG. 1

Difference between drug transportation through the skin with and without penetration enhancers (A) and possible mechanisms of increased skin permeability with penetration enhancers (B).

Based on Lopes, L.B., Garcia, M.T.J., Bentley, M.V.L., 2015. Chemical penetration enhancers. Ther. Deliv. 6(9), 1053–1061. https://doi.org/10.4155/tde.15.61; Dragicevic, N., Maibach, H.I., 2015. Percutaneous Penetration Enhancers Chemical Methods in Penetration Enhancement: Modification of the Stratum Corneum. pp. 1–411. https://doi.org/10.1007/978-3-662-47039-8; Trommer, H., Neubert, R.H.H., 2006. Overcoming the Stratum corneum: the modulation of skin penetration. A review. Skin Pharmacol. Physiol. 19(2), 106–121. https://doi.org/10.1159/000091978; Williams, A.C., Barry, B.W., 2012. Penetration enhancers. Adv. Drug Deliv. Rev. 64(Suppl), 128–137. https://doi.org/10.1016/j.addr.2012.09.032. Figure created with BioRender.com.

and physicochemical properties of terpenes are responsible for their permeation-enhancing properties. For instance, liquid terpenes are often related to having superior enhancing characteristics than solid terpenes. Moreover, terpenes with higher log P values can be more effective enhancers since they feature an easier mixing with SC intercellular lipids than those with smaller log P. Besides, log P has been considered to be an essential property to predict the permeability enhancement property in QSAR models (Ahad et al., 2016). Eugenol presents a relatively high log P (2.27), classified as a lipophilic terpene, suitable for solubilizing poorly soluble drugs. Specifically, eugenol showed lower skin protein melting points, suggesting keratin denaturation and enabling intracellular permeation mechanism, as an additional mechanism besides the extraction of lipid bilayers often provided by terpenes (Ahad et al., 2016; Herman and Herman, 2015).

Moreover, hydrogen bonding between eugenol and SC ceramides can weaken lipid layers, contributing to the disruption of the lipid layers and creating new pathways for drug penetration (Herman and Herman, 2015). In this context, the presence of terpenes can improve the permeation of both lipophilic drugs (such as testosterone, ibuprofen, hydrocortisone) and hydrophilic drugs (such as propranolol) (Aqil et al., 2007; Dragicevic and Maibach, 2015). Since more than 40% of therapeutically active compounds are lipophilic and their oral delivery can be a challenge due to their poor variable bioavailability, topical delivery is an excellent alternative to improve water-insoluble drug pharmacokinetics parameters (Sengupta and Chatterjee, 2017).

Although topical formulations can be beneficial, as they avoid the first-pass metabolism, topical skin formulations often remain ineffective due to low percutaneous through SC. In order to overcome some, technological approaches like the use of surfactants, nanoparticles, and permeation enhancers can be used during the design of pharmaceutical formulations. As an example, Shen et al. (2007) developed a Carbopol gel containing clove oil to investigate in vitro and in vivo percutaneous absorption ibuprofen. In this study, the permeation ratio from the formulation containing the essential oil was 7.3 times superior to the control sample.

In this way, the topical route can be utilized for local treatments such as psoriasis, infection skin diseases and wound healing, and systemic diseases, such as hypertension. Another example of clove oil is applied to ibuprofen delivery was developed by Chen et al. (2015) that used five different EO (clove oil, angelica oil, cyperus oil, cinnamon oil, and chuanxiong oil) to improve transdermal ibuprofen delivery to treat dysmenorrhea. The formulation chosen was a hydrogel composed of ibuprofen (1% w/v), carbopol (1% w/v), ethanol (10% w/v), essential oil (3% w/v), and triethanolamine (1% w/v). The hydrogel formulations were tested in skin permeation assays using rat abdomen skin fragment and also in primary dysmenorrhoea mice in vivo model. The percutaneous permeation results showed that the highest permeation rates were obtained with clove oil (62.8%) and chuanxiong oil (67.2%). The in vivo model confirmed the superior results for both chuanxiong oil and clove oil, with enhanced transdermal delivery of ibuprofen. Hydrogel formulations containing chuanxiong oil provided a decrease in Ca^{2+} levels, responsible for modulating myometrial contractility and increasing nitric oxide (NO), which is negatively correlated to pain. Still, clove oil-containing formulations presented high NO levels in the uterus, being favorable for other applications. Even though Chuanxiong oil was more effective in decreasing uterus contraction, clove oil could modulate pain because NO's signs were high. Therefore, the formulation can be considered effective in releasing ibuprofen and could be applied to other organs.

El Khayat Noha et al. (2018) also incorporated eugenol as a permeation enhancer for transdermal delivery. The authors developed a microemulsion using pseudo-ternary phase diagrams constituted of Tween 40 (surfactant), propylene glycol (PG) (cosurfactant), eugenol, and water. The permeation results presented a suitable release rate of $143.2\,\mu g\,cm^{-2}\,h^{-1}$ of microemulsion containing a higher amount of eugenol (30%) in the presence of PG. This confirms a synergism between the essential oil and the surfactant as PG, which collaborate by fluidizing the interface microemulsion-skin, reducing the droplet size and surface tension (El Maghraby, 2008). Table 1 shows other examples of the application of clove oil as a permeation enhancer.

Besides skin penetration enhancing and drug-carrying properties, clove oil has been shown to increase drug follicle targeting for minoxidil, a poorly soluble drug widely used to treat alopecia areata, a chronic inflammatory disease characterized by hair loss. The authors produced clove oil-loaded nanoemulsions prepared by ultrasonication and studied the influence of 3 independent variables at two levels with three replicates of the center point. Optimized stable nanoemulsions showed monomodal size distributions showed a mean droplet size of 10 nm and provided controlled drug release two times more effective than the control sample. Drug skin penetration and follicular drug penetration from clove oil-loaded nanoemulsions were raised by more than 9 times, 26 times, respectively, compared with conventional ethanol-based solutions, suggesting that the proposed clove oil-based nanoemulsions

Table 1 Applying clove oil/eugenol as permeation enhance for topical and transdermal applications.

Application	Pharmaceutical active	Formulation	References
Hypertension	Omelsartan	Nanoemulsion	Ahad et al. (2016)
Psoriasis	Psoralen	Hydrogel-nanoemulsion	Barradas et al. (2017b)
Cutaneous candidiasis	Clove oil	Ethosomal gel	Shetty et al. (2019)
Drug delivery model	Felodipine	Niosomes	Eid et al. (2019)
Wound healing	Curcumin	Nanoemulsion	Ahmad et al. (2019)
Cosmetic formulation	Clove oil	Emulsion	Makuch et al. (2020)
Ulcerative colitis	Eugenol and cinnamaldehyde	Ethosomes	Zhang et al. (2020)
Alopecia	Minoxidil	Nanoemulsion	Cardoso and Barradas (2020)

constitute a promising approach for drug delivery of drugs in hair follicles and consequently for the treatment of several follicle-related diseases (Barradas et al., 2018).

3. Clove oil as a bioactive molecule in pharmaceutical formulations: Biological and pharmacological role

Clove oil composition includes mainly phenylpropanoids: eugenol, the sesquiterpene b-caryophyllene, and the monoterpene ester, eugenyl acetate, and α-humulene (Salvia-Trujillo et al., 2015; Goñi et al., 2016; Chaieb et al., 2007a,b). Due to its complex composition, clove oil can be used in several areas, such as the food, health, biomedical, and pharmaceutical industries due to its recognized therapeutical properties, such as pain relief and healing (Chaieb et al., 2007a,b), due to analgesic, anesthetic effects (Mylonas et al., 2005), and anticarcinogenic properties (Hadidi et al., 2020). Moreover, clove oil can be used as preservatives for pharmaceutical formulations, as it features antibacterial, antifungal, acaricidal, and antiviral activities (Kim et al., 2003; Chaieb et al., 2007a,b).

Besides, due to eugenol, clove oil can also present antimutagenic (Miyazawa and Hisama, 2003), antioxidant (Radünz et al., 2019; Jirovetz et al., 2006; de Meneses et al., 2017), antiinflammatory, and antinociceptive (Daniel et al., 2009; Chaieb et al., 2007a,b; Banerjee et al., 2020) properties. Table 2 summarizes clove oil these advantages for pharmaceutical application.

3.1 Antiinflammatory, antioxidant, and antinociceptive properties

It has been a trend to seek natural and safe antioxidants such as clove oil to reduce oxidative stress. This biological condition occurs when there is an imbalance between the production of reactive oxygen species and the system of reactive oxygen species. The accumulation of these species causes skin aging

Table 2 Clove oil advantages for pharmaceutical applications.

Property	References
Antibacterial	Salvia-Trujillo et al. (2015), Pathania et al. (2018), and Radünz et al. (2019)
Antifungal	Wan et al. (2019), Chaieb et al. (2007a), and Manganyi et al. (2015)
Insecticidal/insect repellent	Shapiro (2012), Jairoce et al. (2016), and Trongtokit et al. (2005)
Antioxidant	de Meneses et al. (2017), Jirovetz et al. (2006), and Gülçin (2011)
Anticancer	El-Darier et al. (2018), Nirmala et al. (2019), and Carrasco et al. (2008)
Wound healing	Banerjee et al. (2020) and Alam et al. (2017)
Antiinflammatory, antinociceptive, and antipyretic	Banerjee et al. (2020), Nikoui et al. (2017), and Daniel et al. (2009)

and many chronic diseases and cancer (Fukui et al., 2002). Another important property of clove oil is its antiinflammatory activity since it can inhibit the expression of the enzyme cyclooxygenase two and tumor necrosis factor signaling (Magalhães et al., 2010).

The antioxidant activity of clove oil is related to the presence of phenolic compounds, eugenol included, and may occur via different mechanisms, as free radical-scavenging or metal chelating capacity, specifically against Fe^{+3} ions, preventing the oxidative reaction cascade (Choudhary et al., 2009). Being the major phenolic component of clove oil, eugenol can be considered responsible for the antioxidant properties of EO. The structure-activity study of eugenol conducted by Gülçin (2011) showed that eugenol allows the donation of a hydrogen atom and subsequently fixes the phenoxyl radicals, which results in the formation of stable entities that do not engage oxidation cascade. Moreover, eugenol, being a phenolic compound with conjugated dienes groups in its carbon chain linked with the aromatic ring, enables the stabilization of phenoxyl radicals by resonance.

Chaieb et al. (2007a,b) tested the antioxidant properties of clove oil, and their results showed a significant radical scavenging activity. Such antioxidant capacity was observed by reducing the stable radical 2,2-diphenyl-l-1-picrylhydrazil ($DPPH^{\bullet}$) radical to the yellow-colored DPPH-H, reaching a 50% reduction with an IC_{50} of 0.2 μg mL, which was smaller than that observed for *tert*-butylated hydroxytoluene (BHT), used as a positive control.

Several studies showed that encapsulated clove oil could either maintain or increase antioxidant activity compared to free nonencapsulated oil. For example, Sebaaly et al. (2016) produced clove oil-loaded cyclodextrin-in-liposomes vesicles. Their $DPPH^{\bullet}$ scavenging results showed antioxidant activity ranging from 89.3% to 92.3% for clove EO encapsulated in cyclodextrin, comparable to free clove oil. Such behavior was justified by the fact that both the CD and the liposomes enable the interaction of the hydroxyl group from eugenol to interact with $DPPH^{\bullet}$ radicals.

de Meneses et al. (2017) produced clove oil-loaded poly(thioether-ester) nanoparticles. Their results showed that encapsulation led to an increase in $DPPH^{\bullet}$ scavenging activity that ranged from 66% to 79%, with a consequent improvement in antioxidant activity. This increase in the scavenging activity of $DPPH^{\bullet}$ radicals and inhibition of peroxidation showed that the eugenol antioxidant activity was preserved after the in situ polymerization method used to produce nanoparticles and can be considered an interesting approach to improve the antioxidant activity of clove oil.

Clove oil was encapsulated into chitosan nanoparticles for application as a natural antioxidant and antimicrobial agent for active packaging development. The antioxidant activity of encapsulated clove oil was significantly higher than plain clove oil. The same results were observed for in vitro antibacterial activity against *Listeria monocytogenes* and Staphylococcus *aureus* (Hadidi et al., 2020). Similar results had been observed earlier by Hasheminejad et al. (2019) with clove oil-loaded chitosan nanoparticles by emulsion-ionic gelation using tripolyphosphate (TPP) as a crosslinking agent. Their results showed that clove oil-loaded nanoparticles were more effective against *Aspergillus niger* when compared with free clove oil. Moreover, the in vitro release study showed that chitosan nanoparticles provided a controlled release of clove oil (Hasheminejad et al., 2019).

Natural products have been used as analgesics over human history in folk medicine (Abdel-Rahman, 2017; Yunes et al., 2005). Nowadays, such substances are still used in their native form, or they can be a template for new drugs produced from natural products, producing semisynthetic compounds. Indeed, several natural products have contributed to the development of modern analgesic drugs, such as opioids, morphine, and salicylate derivatives (Li and Vederas, 2009; Lenardão et al., 2015).

Terpenes and essential oils are known for their antinociceptive action and can be used as pain relievers in pharmaceutical products (Lenardão et al., 2015). Besides, the common adverse effects of nonsteroidal antiinflammatories drugs (NSAIDs) and opioids have constantly stimulated the search for safer yet effective analgesic substances.

The analgesic properties of both aqueous and ethanol extracts of clove have observed that aqueous and medium-polarity compounds, e.g., alkalenes, some short-chain fatty acids, and sterols, are not responsible for the analgesic effect of clove. On the other hand, the authors suggested that polar constituents of clove are related to the anticonvulsive activity observed for clove extracts (Asl et al., 2013). However, some reports that eugenol and clove EO feature analgesic properties (Sell and Carlini, 1976; Wang et al., 2015). Indeed, the analgesic effect of clove has been often related to the presence of eugenol, a natural capsaicin congener, which has been reported to be comparable with pentobarbital in animal models (Chaieb et al., 2007a,b; Sell and Carlini, 1976; Li et al., 2007).

Eugenol and some derivatives, namely: methyl eugenol, isoeugenol, and methyl isoeugenol, induce anesthesia in mice and rats, with methyl eugenol being pointed as the most effective (Sell and Carlini, 1976). Therefore, Wang et al. (2015) suggested that methyleugenol could be a more effective analgesic for the treatment of toothache than clove oil and eugenol, suggesting eugenol can be a template for the obtention of more effective semisynthetic molecules, which would enable the development of more potent pharmaceutical anesthetic formulations for dentistry purposes.

The analgesic and antiinflammatory activity of eugenol and its derivatives are related to its capacity to inhibit proinflammatory enzymes, as nitric oxide synthase, cyclooxygenase, and lipoxygenase, preventing the formation of prostaglandins, leukotrienes, and other inflammation intermediates (Cho et al., 2008). Besides, eugenol can regulate pain signaling through ion channels and the activity of pain receptors and the conduction of action potential, through the inhibition of voltage-activated Ca^{+2}, Na^{+}, and voltage-gated K^{+} channels, besides being considered a GABA receptor potentializator (Sell and Carlini, 1976; Wang et al., 2015; Li et al., 2007).

The use of eugenol and clove oil for analgesic purposes is often observed in dentistry (Wang et al., 2015). Lee et al. (2005) showed that eugenol inhibited high-voltage-activated calcium channels in dental primary afferent neurons, which could explain the analgesic effect of eugenol in dentistry (Lee et al., 2005). Later, Park et al. (2006) reported the eugenol capacity to inhibit voltage-gated sodium channel currents (I(Na)) in dental primary afferent neurons in animal models (Park et al., 2006), which might contribute to eugenol's analgesic effect. The same group also reported the anesthetic mechanism of action of eugenol trigeminal ganglion neurons of rats. Their results showed that eugenol produced a dose-dependent decrease of both thermal nociception and capsaicin-induced thermal hyperalgesia. Additionally, it blocked the potential of action conduction on the inferior alveolar nerve. Taken together, such data corroborate and encourage the use of eugenol as a local anesthetic for several pathological pain conditions, other than its traditional use in dentistry.

Daniel et al. (2009) studied the antiinflammatory and antinociceptive activities of eugenol in animal models in vivo by accessing the inflammatory exudates volume and leukocytes migration in rats. The antinociceptive activity was evaluated using the acetic acid-induced writhing and hot-plate tests in mice. Eugenol showed the capacity to reduce pleural exudates' volume, while the total blood leukocyte counts remained unchanged. Additionally, eugenol prevented the occurrence of carrageenan-induced edema after 2 to 4 h of the injection. Besides, eugenol significantly affected the response latency, which is blocked by meperidine in the hot-plate test. A remarkable antinociceptive effect in the test of acetic-acid-induced abdominal writhing was observed for eugenol at doses of 50, 75, and 100 mg/kg. Such data demonstrated and support the folk use of eugenol and clove oil for pain relief and antiinflammatories purposes.

3.2 Antimicrobial applications

Environmental conditions such as climate, soil, and harvesting time, and the production method and transportation conditions of essential oils directly influence their composition and, consequently, their antimicrobial activity (Szweda et al., 2015). Essential oils are recognized as safe substances by FDA, so their use in formulations has grown in recent years (Campelo et al., 2017). Clove oil and eugenol have been widely recognized for their excellent antimicrobial activity in studies, effectiveness against fungi, and a range of both gram-negative and gram-positive bacteria (Marchese et al., 2017). Thus, both clove oil and eugenol constitute clinically important bioactive natural compounds with broad-spectrum against human and food pathogens.

Currently, there has been an increase in tolerance to synthetic antifungals drugs, which represents a significant problem for the clinical area, leading to the need to develop new antifungal therapies (Shukla et al., 2016). Eugenol has shown to cause significant morphological disruption in the lipid envelope of *Saccharomyces cerevisiae* and *Candida albicans*, impacting membrane fluidity and permeability. Moreover, eugenol can significantly impair fungal pathogenicity, as it affects fungi adhesion to host cells and their transition to hyphae form, which reduces infectivity and capacity to colonize host tissues (Marchese et al., 2017). Eugenol is known to present antifungal activity against opportunistic pathogenic fungi like *Candida* albicans, *Cryptococcus neoformans, Saccharomyces cerevisiae*, and dermatophytes like *Microsporum* and *Trichophyton* spp.

Previous studies have shown that clove oil features antifungal activity against *Candida* spp. (Gandhi et al., 2020). However, as previously stated, the direct application of essential oils onto the skin is not recommended as they can be highly irritating. In order to circumvent this inconvenience, Siqueira and coworkers developed NE containing clove oil to assess the in vitro inhibitory potential of nanoformulations against *Candida* spp. It was observed that due to the small size of the droplets of NE, the interaction between nanodroplets and microorganisms was able to reduce the minimum inhibitory concentration of *Candida albicans* and *Candida glabrata* (de Oliveira et al., 2019).

Ahmad et al. (2015) showed the synergic and additive effects of different derivative eugenol with fluconazole against *Candida albicans*. Moreover, several eugenol semisynthetic derivatives were reported to show a significantly increased antifungal efficacy. The MIC of fluconazole-based formulations was reduced twofold to ninefold when combined with eugenol derivatives, which significantly impaired ergosterol biosynthesis in all *Candida albicans* strains. Moreover, results showed that both eugenol and its derivatives provided a downregulation of ergosterol biosynthesis pathway gene-ERG11, confirmed by docking studies. Such results clarify eugenol mechanistic antifungal activity and suggest that such compounds are potentially effective as antifungals. A free hydroxyl group in eugenol molecular structure justifies the antimicrobial activity commonly attributed to eugenol. The hydroxyl group enables binding to several bacterial proteins, impairing several metabolic important enzymatic-based functions, as is the case for *Bacillus subtitles*, *Enterobacteraerogenes*, *Escherichia coli*, and *Listeria monocytogenes*. In other bacteria species, such as *Pseudomonas fluorescens*, *Escherichia coli*, *Brochothrix thermosphacta*, and *Salmonella typhi*, for example, both eugenol and clove oil can cause a significant disruption of cytoplasmatic membrane fatty acids, with a consequent increase in nonspecific permeability of cell membrane, which affects K^+ ions and ATP transport for bacterial cells (Marchese et al., 2017).

The increase in antibiotic-resistant gram-positive or gram-negative bacteria constitutes a tremendous public health concern and has also triggered the interest in research with clove and oil is its antibacterial activity. Clove oil can induce the lyse the walls and cell membranes of pathogens such as Escherichia coli O157:H7, *L. monocytogenes,* Campylobacter jejuni, Salmonella enterica, Staphylococcus aureus, Lactobacillus sakei, and Helicobacter py*lor*i (Hemaiswarya and Doble, 2009; Hadidi et al., 2020; Cui et al., 2018).

Eugenol has shown to be effective against vancomycin-resistant Staphylococcus *aureus* due to the production of intracellular reactive oxygen species (ROS), which lead to DNA damage and cell death (Das et al., 2016). Besides, clove oil antimicrobial activity has been observed for multidrug-resistant bacteria, such as *Acinetobacter baumanni, Pseudomonas* aeruginosa, Staphylococcus aureus, and *Enterococcus faecalis*. The resistance to antibiotics brings the urgency in finding novel antimicrobial drugs to circumvent drug resistance. In this context, both clove oil and eugenol can be the basis for novel natural or semisynthetic drugs intended to treat antibiotic-resistant infections.

The rise of infections by Methicillin-resistant Staphylococcus aureus (MRSA) is a typical leading cause of hospital-acquired severe infections, and they have been commonly associated with higher morbidity and mortality; being a significant public health concern. Both eugenol and clove oil had antimicrobial activity against MRSA (Mandal, 2011; Kyaw et al., 2012; Qiu et al., 2010; Thosar et al., 2013). Additionally, eugenol has provided MIC values for methicillin-susceptible Staphylococcus *aureus* (MSSA), which was observed for both MRSA. Eugenol also impaired the formation of biofilm and the establishment of cell-to-cell connections. Such results were reported by Yadav et al. (2015) and showed that both biofilm-forming strains could be equally susceptible to eugenol, which can be an attractive bioactive molecule to treat biofilm-related infections and decrease antibiotics burden (Yadav et al., 2015).

Qiu et al. (2010) have shown that eugenol can affect the physiology of microorganisms, as MRSA through several distinct mechanisms of action that can interfere with different membrane functions and suppresses the secretion of virulence factors, such as toxins, enzymes. Such multifactorial activity seems beneficial since it can avoid bacterial resistance and biofilm formation (Qiu et al., 2010).

Eugenol and clove oil are well recognized for being topically used in odontology for mainly their analgesic and anesthetic properties, which are beneficially complemented by their antiinflammatory and antimicrobial properties. Moon et al. (2011) reported the antibacterial activity and MIC of eugenol against common oral cariogenic and periodontopathogenic bacteria species, such as *Streptococcus gordonii* and *Porphyromonas* gingivalis (MIC: 100 μg/mL), *Streptococcus mutans*, *Fusobacterium nucleatum*, and *Prevotella* intermedia (MIC: 100 μg/mL). Eugenol was also effective against other oral bacteria like *Streptococcus sobrinus*, *Streptococcus sanguinis*, *Strongyloses ratti*, *Streptococcus criceti*, *Streptococcus anginosus*, and Aggregatibacter actinomycetemcomitans, with MIC values ranging from 200 to 800 μg/mL. Moreover, a synergistic interaction among both clove oil and eugenol alone with ampicillin and gentamicin was observed, resulting in an increased rate of killing in units of CFU/mL. Such results show that either clove oil or isolated eggnog can be explored as natural antibacterial agents in odontology (Moon et al., 2011).

Several studies have reported that eugenol can be effective against *Heliobacter pylori*, a gram-negative bacterium challenging to treat and that causes chronic inflammation of gastric mucosa and, in some cases, lead to severe ulcers and cancer (Ali et al., 2005). Eugenol has been shown to inhibit the growth of *H. pylori*, including metronidazole-resistant strains. Ali et al. (2005) observed that high eugenol concentrations caused a significant change in *H. pylori* cell morphology after 4 h of exposure (Ali et al., 2005). Such findings point to both eugenol and clove oil as antibacterial agents for the development of local drug delivery systems for *H. pylori* infections.

3.3 Anticancer and antiproliferative properties

Several EO, including clove oil, can be applied in conjunction with chemotherapeutic drugs to mitigate the side effects of such drugs, helping to reduce patient fatality and mortality (Gautam et al., 2014). Many of the chemotherapy drugs traditionally used in the treatment of cancer are capable of inducing cell apoptosis. Therefore, natural antiproliferative compounds with the ability to induce apoptosis and low toxicity are exciting approaches for the synergic treatment of cancer and other antiproliferative molecules. Eugenol induces apoptosis in cancer cells, as HL-60, to generate reactive oxygen species and a mitochondria-dependent mechanism (Yoo et al., 2005). Jaganathan and coworkers also studied the antiproliferative effect of eugenol against colon cancer cells. In this study, eugenol was able to promote the apoptosis of these cells, making it a candidate compound for chemopreventive agent against this type of cancer (Jaganathan et al., 2011).

Eugenol can also inhibit the proliferation of human malignant melanoma cells by blocking the cell cycle in the S phase and inducing apoptosis. Another action of eugenol is to cause deregulation of the E2F family of transcription factors that are responsible for stimulating the proliferation of melanoma cells; E2F deregulation indicates that eugenol may be of great importance for inhibiting the growth of melanoma (Ghosh et al., 2005).

It is worth noting that β-caryophyllene, despite being in smaller concentrations in clove oil, can be responsible for some of the biological properties attributed to clove. Irrera and colleagues studied the in vitro action of β-caryophyllene in a glioblastoma model since it is a sort of cancer that is quite aggressive with intense cell proliferation and inflammation. β-Caryophyllene is a well-known agonist of the cannabinoid receptor 2 (CB2), and it was able to cause the inhibition of cell proliferation, induction of apoptosis, and reduction of inflammation through the modulation of the cannabinoid system (Irrera et al., 2020).

Additionally, both clove oil and eugenol have been reported to show anticancer properties against leukemia, lung, breast, skin tumors, melanoma, prostate gastric, and colorectal cancers by oncogene regulation and the caspase-dependent pathway (Gaber et al., 2020). Han and Parker showed the anticancer properties of clove oil, which resulted from its impact on cancer biology and the capacity of cell cycle control. In small concentrations (0.0011%), clove oil showed significant antiproliferative activity and inhibition of proinflammatory biochemical markers, as vascular cell adhesion molecule-1 (VCAM-1), interferon C-induced protein 10 (IP-10), interferon-inducible T-cell chemoattractant (I-TAC), and monokine induced by interferon-C (MIG). Besides, important remodeling proteins, as collagen-I, collagen-III, macrophage colony-stimulating factor (M-CSF), and tissue inhibitor of metalloproteinase 2 (TIMP-2), were suppressed, and global gene expression was modulated by clove oil. Such report provided crucial evidence that clove oil can affect signaling pathways important for proinflammatory and cancer signaling processes at gene expression levels in human dermal fibroblasts, supporting clove oil and eugenol anticancer therapeutical properties (Han and Parker, 2017).

Moreover, Chaieb et al. (2007a,b) reported that eugenol could trigger apoptosis in cancer cells, justifying its antiproliferative effect. Several eugenol semisynthetic derivatives have been pointed as potential anticancer chemotherapeutics due to their capacity to induce partial or total unprogrammed apoptosis in human cancer cells (Gaber et al., 2020; Kumar et al., 2014). Kumar et al. (2014) showed the anticancer potential of clove oil in MCF-7 strains of human cancer cells. The anticancer activity of clove oil showed to be superior to the ethanol extract. Aqueous extracts failed to produce antiproliferative activity in the brine shrimp lethality test and showed 0% of lethality in concentrations up to 125 μg/mL. Clove oil reached LD50 concentration at 37 μg/mL, whereas the ethanol extract reached LD_{50} at 103 μg/mL. The same pattern of anticancer effect was observed for clove oil, ethanol, and aqueous extracts by using MTT assays to study cytotoxic potential on MCF-7 cells. Clove essential oil showed the highest anticancer potential with IC_{50} concentrations of 36.4 and 17.6 μg/mL, after 24 and 48 h, respectively. Ethanol extracts showed an IC_{50} of 61.2 μg/mL after 24 h and 16.7 μg/mL after 48 h. However, aqueous extract of clove showed no cytotoxic activity on the MCF-7 cancer cells (Kumar et al., 2014).

Furthermore, eugenol is reported to block oncogenes, as NF-B and cyclin D1 and cancer multiplication genes in a p53-independent pathway. Besides, eugenol can upregulate the flexible cyclin-dependent kinase inhibitor protein, enabling an antiproliferative activity in xenograft human breast tumors. Moreover, eugenol used in cytotoxic concentrations reduced ATP concentrations and increased the glycolytic metabolites and polyamines in normal oral cells and oral squamous cell carcinoma, suggesting a stimulus in the unprogrammed cell death cancer cells (Al-Sharif et al., 2013).

Interestingly, eugenol has shown to be efficient against tumors and against metastasis associated with oxidative stress by blocking metalloproteinase-9 expression in HT1080 human fibrosarcoma cells, related to the already known eugenol antioxidant effect (Nam and Kim, 2013). Such data suggest that clove oil and eugenol can be used to assist and enhance cancer treatment as a synergic agent, as shown by Hemaiswarya and Doble that combined eugenol with 5-fluorouracil (5FU) in cervical cancer cells. Eugenol interacted synergistically with 5FU, enabling the reduction in the concentration of 5FU necessary for to anticancer effect. Their results indicated that both eugenol and 5FU act on different cell cycle targets and induce apoptosis in the cancer cells (Hemaiswarya and Doble, 2013).

4. Wound healing properties

They were rebuilding the functional skin after a wound is a challenge due to the complexity of the healing process. The repair of skin wounds is a complex biological process involving orchestrated cell signaling events and defined biochemical cascades. The incidence of chronic wounds is increasing and reaching epidemic proportions, which justifies the great interest in finding more efficient therapies. Due to the economic and assistance impacts of wound healing, it is understandable that increasingly, the so-called "natural" approaches to wound care are being studied and developed.

There is a general trend suggesting that essential oils have healing properties for wounds, but little understanding of their mechanisms and components might be responsible for these benefits. Nevertheless, the already known antimicrobial and antioxidant effects of essential oils from various plant species are well reported in the literature. Additionally, it is a common understanding that ROS and free radicals are directly involved in the inflammatory response, which often contributes to tissue necrosis. Therefore, as the bases of wound management are (i) the prevention of infection, (ii) treatment of infections and (iii) tissue repair, it is easy to assume that EO can directly contribute to the healing process (Kennewell et al., 2019).

Clove oil is also known to have antiinflammatory and wound healing potential effects on human dermal fibroblasts and murine macrophages. Eugenol is reported to affect the production of proinflammatory cytokines, i.e., IL-6 and IL-10 in preinflamed human skin keratinocytes cells (HaCat) in a biochemical cascade involving immunomodulatory/antiinflammatory events by inhibiting lipopolysaccharide (LPS) effects. The hypothesis is that eugenol provides the suppression of the nuclear factor-κB pathway. The effect of eugenol on IL-6 production efficiently prevented the effects of LPS when added either before or after LPS incubation. On the other hand, eugenol effects on IL-10 production showed antagonism to LPS action when added after LPS incubation (Bachiega et al., 2012). Moreover, both eugenol and clove oil already have known antioxidants, and antibiotic potential is also beneficial to wound healing.

Recently, Banerjee et al. (2020) produced topical formulations with clove oil by phase inversion method combined with ultrasonication, followed by an extensive physicochemical characterization. Cytotoxicity was studied in vitro using m55 skin fibroblast cells and showed that clove oil concentrations up to 15% maintained more than 85% of cell viability, unchanged structural and morphological characteristics in the cells, and with no significant production or accumulation of ROS. Besides, the in vitro wound scratch assay showed the wound healing and proproliferative potential of the formulations, with a migration pattern of fibroblast cells 4h after the addition of the formulation with 15% of clove oil. The wound healing potential was accessed in female Wistar rats and histopathological analysis by the incision model, which measured the extent of recovery of skin integrity and elasticity and tissue repair after an injury. Additionally, wound breaking strength (WBS) is accessed as the minimal force necessary to break open a recently healed wound. WBS can be related to the proper formation and repair of collagen fiber complexes, which is closely linked to the tissue healing phenomenon and the degree of skin integrity and elasticity. WBS was significantly lower (283.3 g) in the untreated animals in comparison with the group treated with clove oil-based formulations (405.4 g) 10 days after wound induction. Moreover, the group treated with clove oil formulations showed a similar healing pattern to those treated with neomycin (positive control). These results raise the hypothesis of different mechanisms of modulations, with healing mainly being related to hydration, able to maintain a hydration balance in the environment of wounded skin, which can enable the absorption of fluids and exudates, the retention of protease solutes onto the injured skin, preventing skin prevents drying. Moreover, hydrated wounds enable the skin penetration of bioactive molecules across the injured SC and reepithelization. The mean epithelialization period (days) for the animals treated with clove oil was 10.67 in comparison to those treated with neomycin (11.29) and the untreated animals (13.45), respectively. All these results showed that clove oil-loaded formulations were comparable to commercially available diclofenac gel and neomycin cream, used as control of antiinflammatory and wound healing effects, and they can be topically used as a potential substitute for synthetic drugs-loaded gels and antibiotic-based formulations (Banerjee et al., 2020).

Alam et al. (2017) produced clove oil-loaded NE to wound healing application in wound excision in vivo rat models. Their results showed that clove oil-loaded NE provided wound contraction, collagen content, and healing effects comparable to those observed in the group treated with gentamycin, used as positive control and superior to the nonencapsulated pure clove oil. Moreover, the time for complete reepithelization was significantly shorter for clove oil-loaded NEs and gentamycin treated groups than pure clove oil-treated rats and control groups. Histopathological analysis showed that formulations were safe and nontoxic with no sign of proinflammatory cells. From these data, it is possible to infer that clove oil was more effective when nano encapsulated in NE than its free form due to the enhancement in reepithelization and proproliferative effects of clove oil (Alam et al., 2017).

5. Nanostructured systems used in the encapsulation of clove oil

Both clove oil and eugenol are sensitive to light, heat, and oxidation and prone to rapid volatilization, which could lead to a loss of therapeutic efficacy as well as short storage stability (Guan et al., 2007; Sebaaly et al., 2015). Additionally, clove oil in its native liquid form features a further drawback in its lack of portability and commercial viability. In this context, nanoencapsulation of clove oil promotes protecting the core material from several environmental conditions (Thonggoom et al., 2016; Unalan et al., 2019; Radünz et al., 2019). Clove oil alone presents pharmacological properties and interesting excipient application; it can be considered an important constituent of pharmaceutical formulations.

For these reasons, to preserve its biomedical benefits, encapsulation of clove oil into nanostructures can control its release and decrease its toxicity (Jabbari et al., 2020). Several nanocarriers systems as nanoemulsions (Barradas et al., 2017b, 2018), liposomes (Sebaaly et al., 2015), nanoemulgels (Aman et al., 2020), polymer nanoparticles (Hadidi et al., 2020; de Meneses et al., 2017; Thonggoom et al., 2016), lipid-based nanoparticles (Fazly Bazzaz et al., 2018; Rodenak-Kladniew et al., 2019), and cyclodextrins (Sajomsang et al., 2012) have been employed as a strategy for improve pharmaceutical effects of clove and prevent instability, volatility and improve oil solubility (Cortés-Rojas and de Souza, 2014). Fig. 2 shows the nanostructures that have been used as a vehicle to clove oil in pharmaceutical formulations.

Such nanostructured carrier systems feature colloidal particles, often with dimensions between 1 and 100nm with suitable pharmaceutical properties like biocompatibility and biodegradability (Montenegro et al., 2016; Cevc, 2004). Hence, clove oil can constitute different nanostructures, nanoparticles, and other structures as nanogels and nanofibers. Fig. 2 shows the nanostructures used as a vehicle to clove oil in pharmaceutical formulations.

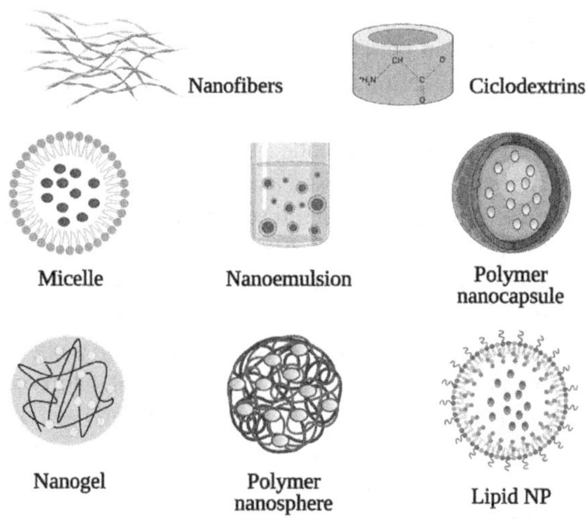

FIG. 2

Clove oil nanostructures.

Based on Gürten, B., Yenigül, E., Sezer, A.D., Malta, S., 2018. Complexation and enhancement of temozolomide solubility with cyclodextrins. Braz. J. Pharm. Sci. 54(2). https://doi.org/10.1590/s2175-97902018000217513; Zhang, H., Zhai, Y., Wang, J., Zhai, G., 2016. New progress and prospects: the application of nanogel in drug delivery. Mater. Sci. Eng. C Mater. Biol. Appl. https://doi.org/10.1016/j.msec.2015.11.041; Hossen, S., Khalid Hossain, M., Basher, M.K., Mia, M.N.H., Rah-man, M.T., Jalal Uddin, M., 2019. Smart nanocarrier-based drug delivery systems for cancer therapy and toxicity studies: a review. J. Adv. Res. 15, 1–18. https://doi.org/10.1016/j.jare.2018.06.005; Mitchell, M.J., Billingsley, M.M., Haley, R.M., Wechsler, M.E., Peppas, N.A., Langer, R., 2021. Engineering precision nanoparticles for drug delivery. Nat. Rev. Drug Discov. 20(2), 101–124. https://doi.org/10.1038/s41573-020-0090-8; de Meneses, A.C., Marques, E.B.P., Leimann, F.V., Gonçalves, O.H., Ineu, R.P., de Araújo, P.H.H., de Oliveira, D., Sayer, C., 2019. Encapsulation of clove oil in nanostructured lipid carriers from natural waxes: preparation, characterization and in vitro evaluation of the cholinesterase enzymes. Colloids Surf. A Physicochem. Eng. Asp. 583. https://doi.org/10.1016/j.colsurfa.2019.123879. Figure created with BioRender.com.

5.1 Nanoemulsions

Nanoemulsions (NEs) are kinetically stable liquid-in-liquid dispersions with droplet sizes in the order of 100 nm. NEs intended for pharmaceutical purposes are mainly oil-in-water (O/W), consisting of tiny oil droplets dispersed in an aqueous dispersing phase. On the other hand, water-in-oil (W/O) NE consists of small water droplets dispersed in an oily external phase (Barradas and de Holanda e Silva, 2020, 2021; Singh et al., 2017). Due to the droplet size, NEs have suitable properties like high surface area, optically transparent appearance, and tunable rheology (Gupta et al., 2016a,b). Also, different from thermodynamically stable microemulsions, NEs feature kinetic stability, and they can undergo for months or years without instability effects like creaming, coalescence, flocculation, and Ostwald ripening (Sonneville-Aubrun et al., 2004; Rao and McClements, 2012). Despite the kinetic advantages and comparing to microemulsions, NE is thermodynamically unstable (Barradas and de Holanda e Silva, 2021). Still, NEs are interesting drug carriers for pharmaceutical formulations, personal care, or topical dosage forms (Lu et al., 2018).

As NEs are thermodynamically unstable, they are not spontaneously obtained. Thus, an input of energy is required to produce them. The high-energy methods rely on a mechanical energy input from specific high shear devices such as high-pressure homogenizers or ultrasonicators (Leong et al., 2009).

This mechanical energy supplies a certain amount of high shear energy to assist the emulsification process, breaking the oil-water interface and reducing droplet size. The low-energy methods mainly depend on the intrinsic physicochemical properties of surfactants and the oily phase (Yukuyama et al., 2016; Peng et al., 2010). They are capable of producing fine oil droplets in surfactant-oil-water mixtures due to phase inversion approaches that can be achieved with specific surfactants that undergo a change in their curvature upon drastic changes in the dispersing/dispersed phase proportion or temperature (Rao and McClements, 2012; Mayer et al., 2013).

Surfactants as nonionic are molecules that combine both hydrophilic and lipophilic groups, and they are classified according to hydrophilic-lipophilic balance (HLB). Once the surfactants are added to the formulation, their molecule broad-chain length distribution and the HLB stabilize the nanosized droplets dispersion (Al-Sabagh, 2002; Lu et al., 2018). As examples of surfactants, Tween 80 and Span 20 are considered safe additives, and therefore, they are often applied to obtain clove oil NE.

Shahavi et al. (2019) studied the effects of manufacturing and formulation parameters, including the ultrasonic cycles and intensity, clove oil, and surfactant content in the production of clove oil-loaded NEs. A surfactant blend of Tween 80/Span 80 was used. As expected, the average droplet sizes decreased with an increased ultrasonication cycle, although pulsed ultrasonication seemed to be more efficient than the continuous method, with the advantage of saving energy. The optimized formulation and obtention conditions were defined at ultrasonication time of 300 s, 10% of clove oil, surfactants concentration of 5 wt%, hydrophilic-lipophilic balance number of 9, duty cycle 0.75%, and ultrasonic intensity of 208 W/cm^2. Clove oil-loaded NEs showed an average droplet of 43 nm, which remained unchanged for at least 30 days. The formulations showed suitable physical stability during a storage time of 6 months. A slight increase in the droplet size in some formulations during storage time after 30 days was observed, yet within the range of defined NE droplets size and with no sign of Ostwald ripening or phase separation. Such findings indicate that the ultrasonication emulsification method efficiently produces clove oil-loaded NE with low and stable droplet size.

The enhanced antimicrobial activity commonly attributed to NE regarding other nanostructured carriers relies on their nanosized flexible and soft droplets composed of significant amounts of surfactants, which enable them to fuse with the lipid envelope of microorganisms. Such enhanced interaction, combined with the effective surface area for contact, causes significant disturbances in microbial membrane morphology, leading to the leakage of intracellular components, protein denaturation, and cell death (Lu et al., 2018; Ramalingam et al., 2013).

De Siqueira et al. (2019) used Pluronic F-127 as a surfactant in different proportions with clove oil to develop NE against skin infections of *Candida albicans* and *Candida glabrata*. As Pluronic F-127 presents a relatively longer ethylene oxide chain length as a hydrophilic portion, this surfactant was added into the aqueous phase. To form NE, the oil phase was added to the aqueous containing different amounts of surfactant. After this, both phases were vortexed for a coarse homogenization. Then, an ultrasonic processor was applied as a high source of energy input with different amplitudes. The optimization of the parameters allowed to achieve NE formulations with droplet sizes less than 40 nm and polydispersion (PDI) 0.2 nm. The in vitro results confirmed that NE provided a better interaction of clove oil components and microorganisms than the free nonencapsulated oil. Thus, this clove oil NE using Pluronic presented uniform size and PDI through 420 days and improved EO activity, being better than free oil and preventing skin irritation during the treatment for candidiasis.

As for vehicles for lipophilic drugs, Aqil et al. (2016) developed a clove oil-loaded NE containing Tween 20 for skin delivery of Olmesartan. The goal of this study was maximizing the drug transdermal flux, high blood pressure treatment. First, drug solubility was evaluated in many different vegetable oils such as almond oil, olive oil, black seed oil, eucalyptus oil, sesame oil, babchi oil, and clove oil. In this solubility test, other surfactants (Tween 20, Tween 80, Span 20, Span 80) were combined with cosurfactants, as PEG and ethanol in different proportions. As a result, Olmesartan exhibited maximum solubility in clove oil with approximately 15 mg/mL, containing Tween 20 (40 mg/mL) as surfactant and PEG (15 mg/mL) cosurfactant. To be more suitable for transdermal skin application, NE was converted into a gel formulation. Such a strategy can increase drug residence time on the skin, which, combined with the small droplet size (53 nm), produced Olmesartan-loaded NE with a better pharmacokinetics parameter than the conventional oral dosage.

Other examples of clove oil NE for skin application were produced (Barradas et al., 2017b, 2018). Both works' main goal was to produce 8-methoxy psoralen (8-MOP) NE for topical treatment of psoriasis. In these works, clove oil was chosen as phase oil due to the higher drug solubility. Barradas et al. (2017) showed that clove oil presented drug solubility (± 4.42 mg/mL), which was higher than sweet fennel oil (±1.46 mg/mL), another well-known EO with important biological properties. Due to this, clove oil was chosen as an oil phase to be added into the aqueous phase containing Cremophor RH40 or Pluronic F68 as surfactants. A high-pressure homogenizer (HPH) was the high energy source applied to obtain NE in this work. To optimize adhesiveness and favor the topical treatment, a mucoadhesive polymer, chitosan, was added to the outer phase of O/W NE to produce nanocomposite hydrogels. As a result, the use of Pluronic F68 showed a suitable formulation that allowed excellent drug skin retention without significant transdermal absorption, being satisfactory for the topical photochemotherapy of psoriasis.

In more recent works, it is observed that clove oil was chosen as a vehicle for pharmaceutical drugs. As an example, Schuenck-Rodrigues et al. (2020) developed a clove oil nanoemulsion containing zinc phthalocyanine (ZnPc) for oral infections of methicillin-resistant Staphylococcus *aureus* (MRSA). In this work, the O/W nanoemulsion was prepared by high emulsification energy with ZnPc solution incorporated into the oil phase (clove oil), and the oil phase was added in an aqueous phase containing Pluronic F127 as surfactant. As for size results, the nanoemulsion presented a diameter of 30 nm approximately and was monodispersed (PDI $<$ 0.2) for 180 days. These results show suitable stability from nanoemulsion. Also, the formulation presented antimicrobial activity in vitro against *Enterococcus faecalis* and methicillin-resistant Staphylococcus *aureus* with minimum inhibitory concentration (MIC) of 1.09 and 0.065 μg/mL, respectively. Additionally, the formulation without the ZnPc still showed antibacterial effects due to clove oil activity, leading to synergy from both actives. Thus, clove oil exhibited a dual role in improving drug solubilization and antimicrobial activity in this work.

6. Lipid-based nanoparticles: Liposomes, SLN, and NLC

Liposomes are phospholipid-based spherical vesicles with an aqueous inner cavity. Due to their natural lipids content, liposomes are biodegradable, nontoxic, and biocompatible nanocarriers. Liposomes are a convenient approach to encapsulate EO, including clove oil and eugenol. Sebaaly et al. (2015) reported the encapsulation of clove oil into liposomes prepared by the ethanol injection method. The effects of natural hydrogenated and nonhydrogenated soybean phospholipid on liposome morphology and physicochemical properties were reported. Different eugenol and clove oil concentrations were also tested to access stability, particle size, polydispersity (PdI), and loading efficiency.

The results showed that liposomes exhibited a nanometric size of around 200 to 300 nm, with spherical-shaped oligolamellar vesicles, which protected eugenol from degradation induced by UV exposure and maintained the antioxidant activity of free eugenol. Liposomes constituted a suitable system for encapsulating volatile, unstable essential oil constituents and sustained clove oil stability for at least 2 months stored at 4°C.

Cui et al. (2015) produced clove oil-loaded liposomes with stimuli-responsive properties and enhanced selective antimicrobial activity for Staphylococcus *aureus*. Selective antimicrobial activity for Staphylococcus aureus by utilizing pore-forming toxins (PFTs) triggered clove oil release from liposome, which was not observed in the absence of PFT, secreted by Staphylococcus aureus. PFT induced the formation of pores into the liposome membranes, triggering clove oil release under these conditions. Moreover, the liposomes showed important bacteria-targeted drug delivery, increasing shelf-life of clove oil, as premature oil leakage was prevented. The dose of clove oil released showed to be self-regulated, following the severeness of Staphylococcus *aureus*-borne infections, since the higher bacterial population, the higher PFT released, and the higher clove oil concentration released. Such a strategy is a novel approach to deliver a variety of antimicrobials substances to treat various infections caused by bacteria or other organisms that secrete PFTs.

Lipid-based nanoparticles also comprise solid lipid nanoparticles (SLN) and nanostructured lipid carriers (NLC). Different from O/W nanoemulsions, both SLN and NLC feature an inner lipid core, which is totally (for SLN) or mainly (NLC) composed of solid lipids at room temperature (Garcês et al., 2018). Since the core lipid material is solid at room, the drug mobility and diffusivity in a solid lipid matrix are impaired, resulting in higher storage stability and an optimized controlled drug release (Schäfer-Korting et al., 2007). Besides, SLN and NLC provide several advantages such as direct and greater uptake of nanoparticles through biological membranes, such as the skin and mucosae, an increase of permeability through biological barriers, adhesiveness, increase in skin hydration through occlusion, and the possibility of encapsulating lipophilic and hydrophilic drugs (Puglia et al., 2008; Montenegro et al., 2016; Junyaprasert et al., 2009). Table 3 summarizes the advantages of both SLN and NLC.

Table 3 Lipid-based nanoparticle advantages for pharmaceutical applications.

SLN and NLC advantages	References
Biodegradability from the lipids Nontoxicity	Martins et al. (2007) and Montenegro et al. (2016)
The controlled drug release rate Low-cost production	Puglia et al. (2008) and Bose et al. (2017)
Higher drug content	Severino et al. (2012), Lim et al. (2012), and Paliwal et al. (2020)
Improved stability of pharmaceuticals	Pereira et al. (2018) and Garcês et al. (2018)
Water-based technology	Shirodkar et al. (2019)
Affordable feasibility in scale-up Avoidance of organic solvents in the production	Shirodkar et al. (2019), Martins et al. (2007), and Ganesan and Narayanasamy (2017)
Suitable for oral, intravenous, pulmonary, ocular, and transdermal routes	Martins et al. (2007), Severino et al. (2012), and Paliwal et al. (2020)

Regarding composition and production, SLN is formed by O/W emulsion with 0.1% (w/w) to 30% (w/w) of solid lipid dispersed in an aqueous phase, stabilized with preferably 0.5% (w/w) to 5% (w/w) of surfactant (Ganesan and Narayanasamy, 2017). Since SLN is composed of a solid inner phase, it can lead to drug expulsion if lipid crystallization occurs, leading to one of the most common limitations to high drug loading efficiency and shelf stability (Garcês et al., 2018; Ganesan and Narayanasamy, 2017; Shirodkar et al., 2019). To overcome SLN limitations, NLC was developed, and their composition includes an inner phase constituted by a blend of both solid and liquid lipids, leading to an amorphous solid phase (López-García and Ganem-Rondero, 2015). Regarding the methods to produce SLN and NLC, several methods can be conducted, as multiple emulsion, solvent emulsification/evaporation, spray drying method, double emulsion method, and ultrasonication, being the hot and cold homogenization the most commonly used (Pottoo et al., 2020).

A top-down approach for SLN and NLC production is often based on incorporating the drug in the melted lipid mixed with an aqueous surfactant solution, followed by immediate cooling with lipid recrystallization and nanoparticle formation (Teixeira et al., 2017).

Fazly Bazzaz et al. (2018) prepared SLN containing clove oil obtained from *Eugenia caryophyllata* flower to improve oil antimicrobial and antifungal activity. In this work, the essential oil was incorporated into different solid lipid phases formed by precirol, stearic acid (SA), and glyceryl monostearate (GMS); Tween 80 and Poloxamer 188 were used as surfactants in the external phase. To obtain nanoparticles, the high-shear homogenization and ultrasound methods were applied. Nanoparticle characterization results showed particle size decreased (around 200 nm) from the SA formulation in the presence of both surfactants when compared to GMS (around 700 nm). The formulations were tested against *Salmonella typhi*, *Pseudomonas aeruginosa*, Staphylococcus *aureus*, and *Candida albicans* to evaluate antimicrobial activity. Consequently, in vitro assays of SA-based nanoformulation showed an improved passive cellular absorption and antimicrobial activity from clove oil, mainly against *Candida albicans*. As nanoparticle size was lower than cells, essential oil absorption was improved, and antimicrobial activity increased with the nanoformulation.

More recently, Rodenak-Kladniew et al. (2019) developed eugenol-based SLN as a vehicle for Ofloxacin (Ofx) and bacterial infections of *Pseudomonas aeruginosa* and Staphylococcus *aureus*. The lipid matrix was formed by myristyl myristate (MM), eugenol, and 0.5% (w/v) of chitosan. Pluronic F68 was used as a polymeric surfactant. The homogenization/ultrasonication methods were used to obtain SLN. Drug Release assays showed that chitosan Ofx-loaded nanoparticles led to increased drug expulsion from the SLN lipid matrix. This occurred due to the higher solubility of Ofx in low pH values, which were required by chitosan incorporation. Eugenol was added to the lipid phase to overcome such drawbacks, which increased Ofx encapsulation efficiency (33.5%) and improved the drug release profile, which showed a slow drug release of 20% of the total drug amount during 21 h. The addition of eugenol into SLN led to the obtention of NLC, improving physicochemical and biological properties such as pharmaceutical formulations. Since NLC was revealed to be an intelligent approach to overcome limited drug-loading capacity and drug expulsion during storage from SLN, they have received increased attention as a drug delivery strategy, being preferentially chosen for incorporate lipophilic drugs (Fang et al., 2013).

Due to NLC advantages, de Meneses et al. (2019) developed NLC by hot homogenization/ultrasound emulsification to encapsulate clove oil to inhibit cholinesterase enzyme (AChE) Alzheimer's disease multiple system atrophy and other conditions. The essential oil was well incorporated in lipid phase matrix associated with beeswax or carnauba wax (solid lipids) and medium-chain triglycerides

(MCT) or crodamol (liquid lipids). The addition of liquid lipids disorganized the crystalline packing commonly observed for waxes, leading to a polymorphism confirmed by the transition temperatures in the differential scanning calorimetry (DSC). The DSC results presented the following melting temperatures: 72.7°C for isolated carnauba wax; 51.4°C, for isolated beeswax. The addition of liquid lipids shift melting temperature toward lower temperatures, i.e., around 54.1°C for liquid lipids and carnauba wax blend and 43.5°C for the liquid lipids and beeswax mixture. Finally, NLC presented activity against AChE during in vitro assays with *Drosophila melanogaster*, used as a biological model.

Both NLCs and SLNs can be widely applied as carriers of anticancer drugs, antimicrobial agents, gene vector, transdermal drugs, and bioimaging agents (Souto et al., 2020). Because of the low risk/benefit ratio and versatility, several SLNs therapeutic patents are emerging as FDA-approved formulations (Paliwal et al., 2020).

7. Polymeric nanoparticles

Although there are many advantages of SLN, they can often provide inadequate effects, such as drug burst release (Ganesan and Narayanasamy, 2017; Severino et al., 2012). Also, for skin as tissue target, polymeric nanoparticles tend to behave better than solid lipid nanoparticles, due to their higher drug accumulation and loading efficiency, with more controllable drug release rates as already seen by Abdel-Mottaleb and Lamprecht (2011). Generally, polymers can be adequately screened to provide biocompatibility and low toxicity for biomedical applications. Thus, either clove oil or eugenol can be encapsulated and delivered as the main bioactive molecules or compose the core of polymeric nanoparticles to deliver other drugs.

Polymer nanoparticles (PNPs) have attracted interest due to their wide range of tunable properties in which their structure can be modified/functionalized, leading to many advantages for biomedical applications (Mallakpour and Behranvand, 2016; El-Say and El-Sawy, 2017; Kenry et al., 2020; Hyeon and Rotello, 2012; Crucho and Barros, 2017; Banik et al., 2016). PNPs present many advantages for biomedical applications such as controlled release, the ability to combine both therapy and imaging (theranostics), allow stimuli response (pH and temperature) protection of drug molecules, decreased drug toxicity, polymeric degradation, and site-specific targeting with surface ligands (Cismaru and Popa, 2010; Morachis et al., 2012; Wong et al., 2020; Ahlin Grabnar and Kristl, 2011; Sarcan et al., 2018; Umerska et al., 2018; López-López et al., 2019; Muhamad and Selvakumaran, 2016; Crucho and Barros, 2017). Further, biosafety and biocompatibility are also relevant characteristics from the polymers applied into the matrix of PNPs used as drug delivery systems (Amoabediny et al., 2018). To form PNPs matrixes, different polymers can be applied either from natural or synthetic sources. Table 4 exemplifies polymers applied to PNPs production.

Considering both structure and morphology, PNPs can be classified as nanocapsules or nanospheres; nanocapsules are composed of a solid polymer shell surrounding a liquid or semisolid core. On the other hand, nanospheres comprise particles composed of a polymeric matrix blended with the encapsulated material, providing a homogeneous biphasic system (Cismaru and Popa, 2010; Mallakpour and Behranvand, 2016). These nanosystems can be prepared either by preformed polymers or by direct polymerization from the monomers (Rao and Geckeler, 2011). Solvent evaporation, emulsion, double emulsion, nanoprecipitation, salting-out, dialysis, and supercritical fluid technology are

Table 4 Natural and synthetic polymers applied to PNPs production.

Source	Polymer	Main polymer advantages for biomedical application	References
Natural	Chitosan	Biocompatibility, biodegradability, nontoxicity, reactivity, film/fiber-forming, the possibility of crosslinking, capacity for adsorption, cellular tight junction opening, mucoadhesion, and faster kinetics	Dash et al. (2011), Hadidi et al. (2020), and Anitha et al. (2014)
	Gelatin	Cell-adhesive structure, low cost, off-the-shelf availability, high biocompatibility, biodegradability, and low immunogenicity	Hoffman (2002) and Su and Wang (2015)
	Sodium alginate	Used as a drug delivery system for bone tissue engineering due to its nontoxic, biodegradable, biocompatible properties and its low price	Oliveira et al. (2014) and Senna et al. (2018)
	Albumin	Versatile nanocarrier for drug delivery, biocompatibility, lack of toxicity, lack of immunogenicity, inexpensive production, great storage, and in vivo stability	Chen and Liu (2016) and Valmikinathan et al. (2009)
Synthetic	Polylactides (PLA)	Low rigidity, manageability, processability, controlled biodegradation, thermal plasticity, suitable mechanical properties, eco-friendly, nontoxicity, and biocompatibility	Fattahi et al. (2019) and Tyler et al. (2016)
	Polyglycolides (PGA)	Biodegradability, biocompatibility, and nontoxicity are used in medical fields such as in drug delivery systems and dental and orthopedic applications	Bin Park et al. (2021), Budak et al. (2020), and Fattahi et al. (2019)
	Poly(ε-caprolactone) (PCL)	Compatibility with a wide range of drugs, slow degradation rate, increase aqueous drug solubility, easy processability, proper mechanical strength, and biodegradation	Umerska et al. (2018) and Malikmammadov et al. (2017)
	Poly(lactide *co*-glycolides) (PLGA)	Biodegradability, biocompatibility, and potential of PLGA nanoparticles for target specific, controlled delivery, mechanical strength, and more straightforward modification	Fattahi et al. (2019), López-López et al. (2019), and Mir et al. (2012)
	Poly(*N*-vinyl pyrrolidone)	The drug-controlled release improves the bioavailability of poorly water-soluble drugs, protects the active compound against external agents (pH, temperature, and oxygen), and masking unpleasant odors/flavors	Feldstein et al. (1996) and Franco and de Marco (2020)
	Polyglutamic acid	Biocompatibility, nonimmunogenicity, biodegradability, used as composites for vaccine adjuvants, cancer therapy, medical devices, and tissue regeneration	Bin Park et al. (2021)
	Poly(vinyl alcohol) (PVA)	Chemical versatility, biocompatibility, biodegradability, employed as wound dressings, wound management, drug delivery system, artificial organs, and contact lenses	Da Silva et al. (2018), Oh et al. (2004), Opanasopit et al. (2013), and Paradossi et al. (2003)
	Poly(ethylene glycol) (PEG)	Biocompatibility, biodegradability, decreased opsonization, act as a surfactant, gold standard for stealth polymers, high solubility in organic solvents, solubility in water, excipient for parenteral, topical, nasal, and ocular applications, the principle active for laxatives	Anjum et al. (2016), Feldstein et al. (1996), and Knop et al. (2010)

some of the most used methods to obtain PNPs applying the preformed polymers (Amoabediny et al., 2018; Crucho and Barros, 2017; Rao and Geckeler, 2011).

The physicochemical PNPs properties as size, porosity, polydispersion, and functionalization can be achieved according to the optimization from the preparation methods. This versatility also made PNPs suitable vehicles to deliver clove oil/eugenol and combine this essential oil with other drugs.

As an example for PNPs containing clove oil, Thonggoom et al. (2016) developed amphiphilic diblock copolymer nanoparticles of poly (ethylene glycol) (PEG) and PCL. The PNPs were obtained by the self-assembly particles formed with the nanoprecipitation technique. Both copolymer and clove oil were dissolved in tetrahydrofuran, and this solution was added dropwise in water under vigorous agitation. After solvent evaporation, nanoparticles with a diameter range from 59 to 150 nm and PdI less than 0.3 were formed, without noticeable aggregation. The monodispersity is relevant since the properties of the nanoparticles influence drug release efficacy. Also, encapsulation efficiency varied as the PCL block length increased due to the strong hydrophobic interactions and hydrogen bond from PCL and clove oil. Furthermore, the clove oil release rate was also controlled by PCL molecular weight due to the increased oil solubilization in the polymer phase and the slow polymer degradability, which led to the controlled release from polymeric particles.

Eugenol-loaded chitosan nanoparticles were developed via emulsion-ionic gelation and cross-linked with tripolyphosphate and showed an average size smaller than 100 nm. Encapsulation phenomena showed to increase by 8-folds thermal stability and 2.7-fold antioxidant activity of eugenol compared to free eugenol (Woranuch and Yoksan, 2013).

Besides thermal stability and controlled drug release rate, clove oil incorporation into PNPs can also improve biomedical activity from essential oils. Following this goal, de Meneses et al. (2017) incorporated clove oil into poly(thioether-ester) nanoparticles to improve natural clove oil antioxidant activity. The miniemulsion polymerization was the method of in situ nanoparticle production, initiated by di-anhydro-D-glucityl diundec-10-enoate (DGU) as a monomer and Lutensol AT 80 as a surfactant, both in the aqueous phase. Meanwhile, the lipophilic phase was compounded by different amounts of clove oil. After stirring both phases, butanedithiol was added as a reaction initiator, producing PNPs. Besides, the addition of clove oil implied physicochemical alterations in PNPs. It was observed that higher concentrations, 20 wt%, of oil reduced the melting temperature for PNPs by 15°C. Such observation suggested that clove oil can be incorporated into the polymer matrix with a plasticizer, leading to the reduction of polymer crystallinity. Since the eugenol phenol group reduces polymerization rate by acting as a radical scavenger or by a double bond with the allylic group, the increase of clove oil amount in formulation resulted in a reduced oil encapsulated polymeric binding chain during the reaction. Once oil encapsulation efficiency by bound decreased from 60% to 36%, the remaining oil was encapsulated by noncovalent bound with the polymeric matrix. The presence of clove oil in the PNP formulation confirmed the improvement of antioxidant effect from ~53% (without oil) to ~79% (with 6% of oil). This result confirmed that the bioactivity was preserved even during the polymerization process with the inhibition of peroxidation. Finally, the antioxidant activity from the oil was not interfered by the polymerization process.

Xu et al. (2019) produced PNPs containing eugenol by in situ polymerization, using eugenyl methacrylate (EMA) and dopamine methacrylamide (DMA) as monomers, which were combined to form an antibacterial and adhesive coating for periodontal implants. FTIR characterized the PNPs, and the results confirmed the synthesis of poly (DMA-*co*-EMA) with the signal at 1653 cm^{-1} related to the amidation reaction of DMA, the ester peak at 1737 cm^{-1}, and the disappearance of the phenolic group

($3550\,cm^{-1}$ peak), which confirmed the esterification from eugenol with methacryloyl chloride of DMA. Also, the PNPs were tested by immersion of titanium (Ti) plates in the solution, and after being rinsed, Ti plates were observed by scanning electronic microscopy (SEM) to confirm PNPs adhesion. The SEM results showed that particles remained on Ti plates, indicating great adhesion from PNPs even after the dopamine modification, providing suitable materials for eugenol release for mucobuccal application.

Recently, Uma Maheswari et al. (2021) incorporated eugenol and hesperidin, both cytotoxic drugs, into bovine serum albumin (BSA) conjugated to folic acid nanoparticles (BSA-CFNP), aiming at a more selective vectorization for breast cancer treatment. Based on previous work, the nanoparticles were prepared by the solution-gelation (sol-gel) method followed by freeze-drying. Drug solutions of eugenol and hesperidin were added dropwise into BSA-CFNP solution, ensuring drug encapsulation. The encapsulation efficacy showed high results for both drugs (62.9% for hesperidin and 84.5% for eugenol). Meanwhile, the release rate was improved by the acidification (pH 1.2) from the medium, which is suitable for cancer drug delivery. Also, in vitro assays with breast cancer cells showed that folic acid conjugation into PNPs surface increased the interaction and targeting with the cancer cells, contributing to a higher cytotoxic effect and more efficient anticancer treatment. Therefore, polymers being applied as the matrix of nanosystems can be optimized by changing the PNPs surface or using other nanomaterials such as nanofibers.

8. Electrospun nanofibers

Electrospinning methods also provide nanofibers that can be used to deliver actives such as clove oil. It involves the application of electrostatic force between polymer solutions or melt kept in a syringe connected to a high voltage source and a grounded counter metal (collector) kept at a suitable distance (Sundaray et al., 2004; Tucker et al., 2012; Wang et al., 2008; Cramariuc et al., 2013; Hu et al., 2014). As the syringe is pumped, the polymer solution is ejected, forming a droplet that receives the high voltage, and because of the electric field, the droplet produces a jet that travels into collector direction, evaporating along the way and reaching the collector solid fibers (Mohammadian and Eatemadi, 2017). The formation of the fibers can be controlled by processing parameters, including applied voltage, the flow rate of polymer solution, and the distance between the tip of the spinneret and the collector (Thakkar and Misra, 2017; Bhattarai et al., 2018).

The optimization from the electrospinning parameters also modifies nanofiber morphology, and it depends on several aspects: the polymer chemistry, volume of polymer solution, viscosity, the type of solvent, and process temperature (Megelski et al., 2002). As such parameters are altered, the fibers can present size in the nanometer range, providing a scaffold with a high surface area. The scaffold featuring a high surface area improves cell adhesion, migration, proliferation and maintains nutrients transport (Sridhar et al., 2015; Nascimento et al., 2021). Therefore, electrospinning is applied to produce functional scaffolds suitable for tissue engineering, being employed for many organs or tissues.

Following the benefits from electrospinning for tissue engineering, Unalan et al. (2019) produced PCL-gelatin (PCL-GEL) electrospun nanofibers containing clove oil for wound healing applications. Various nanofibers containing different amounts of clove oil were fabricated, and glacial acetic acid (GAA) was used as a solvent. The essential oil was added in different amounts (1.5%, 3%, and 6%, v/v) into PCL-GEL solution, then left overnight for homogenization. The flow rate was settled at 0.6 mL/h.

Meanwhile, the applied voltage was 19 kV, and the collector distance of 12 cm. As a result, the nanofibers microscopy showed smooth and uniform surfaces.

Meanwhile, the contact angle assay presented a wettability increase of 18%–27% from fibers with different amounts of clove oil (1.5%–6%). The wettability results were suitable since they affect cell adhesion on nanofiber surfaces and, consequently, cell proliferation during the wound healing process (Ballester-Beltrán et al., 2011). Besides, the nanofibers did not present significant cytotoxicity and produced antibacterial effects against Staphylococcus *aureus* and *Escherichia coli*, reducing bacterial viability at 6 and 24 h, respectively. Such results support that PCL-GEL nanofibers containing clove oil have the potential for wound healing application.

9. Hydrogel-based nanomaterials

Along with NE, lipid nanoparticles, polymeric nanoparticles, and nanofibers, hydrogel-based nanomaterials present many advantages for biomedical applications. Either natural or synthetic polymers can constitute hydrogels, and they have been considered attractive matrices for repairing and regenerating tissues for tissue engineering purposes. Based on this, Huerta et al. (2020) produced emulsion-loaded cellulose nanofiber-based hydrogels containing clove oil to provide a scaffold that presented cytocompatibility with human gingival fibroblast cells. The hydrogel was obtained by applying the high-intensity ultrasonicator into cellulose fibers extracted from canola straws. After this, high-pressure homogenizer and high-intensity ultrasonication methods were applied to produce the emulsion-filled hydrogels.

As a consequence of essential oil incorporation and increased hydrophobicity, hydrogels showed a decreased water retention and swelling capacity. Still, the presence of the clove oil up to 0.5 wt% maintained good cytocompatibility, with cell viability shown to be higher than 74%. Those results showed that clove oil-loaded hydrogel-based scaffolds could be considered as an alternative treatment for periodontal disease.

Unlike hydrogels, hybrid nanomaterials are composed of both inorganic and organic components, modulating their properties and retaining beneficial features from both (Taylor-Pashow et al., 2010). The hybrid nanomaterials feature the advantage of combining the favorable therapeutic efficacy characteristics and enhanced penetration attributed to nanocarriers with physical stability related to scaffolds (Nascimento et al., 2021). Based on this, hybrid materials are considered valuable materials for various medical/therapeutical devices such as drug delivery, scaffolds for tissue engineering, vascular stents, and dental implants (Ribeiro et al., 2017; Deng et al., 2019). Hence, clove oil is also vehiculated into these materials, improving the essential oil performance (Shahavi et al., 2016; de Meneses et al., 2017; Hadidi et al., 2020).

As an example of hybrid composites, nano zinc oxide (ZnO) nanostructures were used to modify polymer hydrogel matrices, producing novel materials with new properties. In this context, Ashjazadeh et al. (2019) investigated nanofibers-based hydrogels containing eugenol for tissue engineering application. The authors produced nanofibers-based hydrogels with wound healing potential, and these hybrid materials were evaluated in Wistar rats. Nanofibers constituted by PCL and ZnO were also tested in order to prove eugenol superior healing effectiveness. The wound dressings were changed every 24 h through 21 days of treatment. After the treatment, histological assessments were made by evaluating several neutrophils, macrophages, fibroblast cells, collagen volume, and new blood vessels. As a result, the eugenol-loaded hydrogels presented relevant superior biochemical markers for healing as collagen, leukocytes, and new blood vessels. In addition, histological graphs confirmed the best granulation tissue due to collagen production. Thus, comparing to the PCL-ZnO nanofibers-based hydrogel, eugenol-loaded wound dressings presented higher tissue healing aspects.

10. Other vesicle-derived carriers

Vesicular systems, such as ethosomes, have already been widely used as carriers for essential oils in developing topical pharmaceutical dosages. They entrap active molecules and delivery into or across the skin and act as reservoirs for sustained release of active compounds. Other than liposomes, innovative vesicular-derived carriers include niosomes, transferosomes, and ethosomes (Lohani and Verma, 2017).

Ethosomes are unique vesicular carriers, and they are distinguished from the other lipid nanocarriers by some characteristics as the vesicular bilayer fluidity and permeation enhancement (Ainbinder et al., 2010). As advantages, ethosomes can overcome some problems associated with topical drug delivery due to their drug delivery properties superior to liposomes and micelles (Faisal et al., 2018). Due to these advantages, Shetty et al. (2019) produced ethosomes containing clove oil to treat cutaneous candidiasis. In this work, the ethosomes were formulated by diluting the essential oil with ethanol and adding an aqueous solution of soy phosphatidylcholine. To increase mucoadhesivity, hydrogels composed of Carbopol 974 were added into ethosomes colloidal suspension. The drug release results showed that ethosomes alone presented appropriate drug release, with more than 90% being released due to increased clove oil solubility by ethosomes. Although ethosome-loaded hydrogels showed a lightly lesser drug release (86.3%), in vitro assays showed enhanced activity against fungus *Candida albicans* (21 mm of inhibition) being similar to nystatin (28 mm of inhibition), which is the standard drug. Therefore, it was observed the potential to apply clove oil for cutaneous infection of *Candida albicans*.

Niosomes are characterized by nonionic surfactant-base vesicles often combined with cholesterol to increase stability against core leakage. Eid et al. (2019) produced clove oil-loaded rigid niosomes with cholesterol using Span 60 as a surfactant for controlled release felodipine. Niosomes showed an average size of around 300 nm and low PdI. DSC results indicated that the incorporation of clove oil produced a fluidizing effect on rigid cholesterol-rich niosomes membranes. However, such plasticizer effect did not produce drug leakage, reflected by the formulations' zero-order drug release profile. However, the incorporation of clove oil in niosomes provided a significant increase in drug transdermal flux compared with oil-free niosomes, reflecting the penetration-enhancing properties of clove oil.

Transfersomes are elastic, highly flexible, and deformable vesicles, commonly obtained from liposomes, composed of phospholipids, cholesterol, and sodium cholate. Similarly, invasomes are soft innovative soft, and deformable vesicles composed of phospholipids, ethanol, and often terpenes as penetration enhancers. Both of them have been developed to overcome common drawbacks that restrict the application of liposomes to drug skin delivery (Nangare and Dugam, 2020; Lohani and Verma, 2017). Dwivedi et al. (2017) produced isotretinoin-loaded invasomes with 148 nm and high entrapment efficiency (>85%), aiming at the pilosebaceous follicular unit. Eugenol was used as a penetration enhancer and solvent for isotretinoin, which justified the high entrapment efficiency observed for these systems. Additionally, high concentrations of eugenol provided high vesicle deformability and cumulative isotretinoin permeation. Confocal laser scanning microscopy (CLSM) showed that invasomes reached the pilosebaceous follicular unit due to their deformable and flexible nature.

11. Inclusion complexes and cyclodextrins

Cyclodextrins (CDs) are starch-derived structures that feature a ring-like structure composed of glucopyranoses units. CDs provide a hydrophobic environment in their core, which enables the formation of inclusion complexes. CD can also provide nanostructured carriers for hydrophobic biomolecules. CD are cyclic water-soluble oligosaccharides obtained from starch hydrolysis. Their inner hydrophobic cavity constitutes an inclusion complex with lipophilic substances, as volatile essential oils and drugs. The most common CD are α-cyclodextrin (α-CD), β-cyclodextrin (β-CD) e γ-cyclodextrin (γ-CD) (Loftsson et al., 2004; Nostro et al., 2002). As a consequence of molecular inclusion, several properties are modified, such as vapor pressure, volatility, solubility, and lipophilicity. The guest molecule-CD bond is dynamic equilibrium, and binding strength depends on the interactions between guest molecules functional groups and those from CDs (Brewster and Loftsson, 2007; Loftsson et al., 2005).

Inclusion complexes based on β-CD and three derivatives: heptakis-2,6-di-β-CD (DM-β-CD), mono [2-*O*-(2-hydroxyethyl)]-β-CD (HE-β-CD), and mono [2-*O*-(2-hydroxypropyl)]-β-CD (HP-β-CD) have been developed, and their interaction capacity with eugenol was investigated by fluorescence and FTIR. Additionally, the thermodynamic parameters of the inclusion process (DG, DH, and DS) were determined. The results showed an increase in fluorescence intensity, which is a direct consequence of the increase in electronic density after eugenol complexation into the hydrophobic cavity of CDs. The IR spectra confirmed the inclusion of eugenol inside the cavity of CDs. Moreover, complexation capacity was strongly affected by pH, which impacts eugenol ionization constant and by the type of derivative. HP-β-CD provided the highest interaction capacity with eugenol, followed by DM-β-CD, HE-β-CD, and β-CD. Moreover, the thermodynamic results showed that the complexation is a spontaneous process triggered by an exothermic event related to the release of 'high-energy water' molecules from the inner CDs cavities during the complexation phenomenon. Additionally, the possible mechanism of eugenol-CDs complexation was attributed by Van der Waals forces and hydrophobic interactions, hydrogen bonding formation beside the release of high-energy water molecules from the cavities of CDs (Zhan et al., 2008; Sajomsang et al., 2012).

β-CD inclusion complexes containing different essential oils were developed by Hill et al. (2013) for the application of antimicrobial delivery systems. Cinnamon bark extract, *trans*-cinnamaldehyde, clove bud oil, eugenol, and a 2:1 trans-cinnamaldehyde: eugenol blend were encapsulated into CD the freeze-drying method. The authors showed that the inclusion into CD not only did not reduce the antimicrobial activity of the essential oils against *Salmonella enterica* serovar Typhimurium LT2 and Listeria innocua but was able to inhibit bacterial growth at lower concentrations than the nonencapsulated free oils. Such increased activity is due to the higher water solubility of essential oils enabled by their encapsulation into CD and the increased contact surface between pathogen cells and the essential oils.

Choi and coworkers produced inclusion complexes with eugenol using β-CD, 2-HP-β-CD with an average size of 320 nm and high encapsulation efficiency (90.9% and 89.1%, respectively). β-CD showed a higher encapsulation efficiency for eugenol compared to 2-HP-β-CD. The side chain of the hydroxypropyl group in the 2-HP-β-CD structure can impair the inclusion of eugenol molecules inside the 2-HP-β-CD cavity, which can explain such a result. However, the molecular inclusion method was compared with the encapsulation with PCL, which was demonstrated to be more suitable to maintain eugenol stability against UV light-induced oxidation of eugenol (Choi et al., 2009).

Sajomsang et al. (2012) produced novel mucoadhesive systems based on self-aggregation of β-CD grafted with chitosan and their inclusion complex with eugenol. Their mucoadhesive properties were accessed by surface plasmon resonance (SPR), and the results showed that mucoadhesion was mainly ruled by electrostatic interactions, followed by hydrogen between mucin and the β-CD/chitosan derivatives bonding. Their antimicrobial activities against *Candida albicans*, *Streptococcus oralis*, and *Streptococcus mutans* showed enhanced CFU reduction compared with free eugenol, primarily due to the increased eugenol solubility and the interactions between eugenol and microorganisms, favored by the mucoadhesive provided by chitosan. The antimicrobial activities of eugenol-loaded inclusion complexes were ranked as *Candida albicans* > *Streptococcus mutans* > *Streptococcus oralis*, suggesting that CD-based formulations can increase eugenol solubility and antibacterial properties also be tunable to achieve interesting pharmaceutical properties, as mucoadhesion (Sajomsang et al., 2012).

12. Conclusion

Nanostructured pharmaceutical carriers are unique strategies for delivering lipophilic components as clove oil and eugenol entrapped into such nanosized vehicles. They can provide more stable, effective, readily manufacturable, and acceptable formulations for topical application of essential oils and terpenes. Additionally, nanocarriers also protect the entrapped bioactive molecules against the effects of environmental conditions, as they encapsulate bioactive molecules in the core of nanosized capsules. They exhibit high surface area, stability, and tunable release properties, which can improve the biological effects of clove oil and eugenol, enabling the replacement and optimization of several treatments based on synthetic drugs and making them less toxic and more effective. Recently, a growing interest in using natural compounds and essential oils as substitutes for traditional synthetic drugs has been taking place since they have proved to show significant antimicrobial, antioxidant, anticancer, anesthetic, and antiinflammatory properties. However, much research is still needed before those nanotechnology-based products can be commercially available for the general public. On the other hand, there is a growing research effort with more papers being published in this field. From this chapter, it could be concluded that before nanostructured pharmaceutical topical dosage forms become widespread in the pharmaceutical field, several challenges must be overcome. This chapter presents the main aspects of nanostructured pharmaceutical formulations for topical delivery of clove oil and eugenol. Their application as bioactive molecules or as excipients in such formulations are reported, and the most common and important nanostructured vehicles are also reviewed.

References

Abdel-Mottaleb, M.M.A., Lamprecht, A., 2011. Standardized in vitro drug release test for colloidal drug carriers using modified USP dissolution apparatus I. Drug Dev. Ind. Pharm. 37 (2), 178–184. https://doi.org/10.3109/03639045.2010.502534.

Abdel-Rahman, R.F., 2017. Natural products as a source for novel analgesic compounds. In: Pain Relief—From Analgesics to Alternative Therapies. InTech, https://doi.org/10.5772/66770.

Adorjan, B., Buchbauer, G., 2010. Biological properties of essential oils: an updated review. Flavour Fragr. J. 25 (6), 407–426. https://doi.org/10.1002/ffj.2024.

Ahad, A., Aqil, M., Ali, A., 2016. The application of anethole, menthone, and eugenol in transdermal penetration of valsartan: enhancement and mechanistic investigation. Pharm. Biol. 54 (6), 1042–1051. https://doi.org/10.3109/13880209.2015.1100639.

Ahlin Grabnar, P., Kristl, J., 2011. The manufacturing techniques of drug-loaded polymeric nanoparticles from preformed polymers. J. Microencapsul. 28 (4), 323–335. https://doi.org/10.3109/02652048.2011.569763.

Ahmad, A., Wani, M.Y., Khan, A., Manzoor, N., Molepo, J., 2015. "Synergistic interactions of eugenol-tosylate and its congeners with fluconazole against Candida albicans." Edited by Chaminda Jayampath Seneviratne. PLoS One 10 (12), e0145053. https://doi.org/10.1371/journal.pone.0145053.

Ahmad, N., Ahmad, R., Al-Qudaihi, A., Alaseel, S.E., Fita, I.Z., Khalid, M.S., Pottoo, F.H., 2019. Preparation of a novel curcumin nanoemulsion by ultrasonication and its comparative effects in wound healing and the treatment of inflammation. RSC Adv. 9 (35), 20192–20206. https://doi.org/10.1039/C9RA03102B.

Ainbinder, D., Paolino, D., Fresta, M., Touitou, E., 2010. Drug delivery applications with ethosomes. J. Biomed. Nanotechnol. 6 (5), 558–568. https://doi.org/10.1166/jbn.2010.1152.

Alam, P., Ansari, M.J., Khalid Anwer, M., Raish, M., Kamal, Y.K.T., Shakeel, F., 2017. Wound healing effects of nanoemulsion containing clove essential oil. Artif. Cells Nanomed. Biotechnol. 45 (3), 591–597. https://doi.org/10.3109/21691401.2016.1163716.

Ali, S.M., Khan, A.A., Ahmed, I., et al., 2005. Antimicrobial activities of eugenol and cinnamaldehyde against the human gastric pathogen *Helicobacter pylori*. Ann. Clin. Microbiol. Antimicrob. 4 (20), 455–478.

Al-Sabagh, A.M., 2002. The relevance HLB of surfactants on the stability of asphalt emulsion. Colloids Surf. A Physicochem. Eng. Asp. https://doi.org/10.1016/S0927-7757(01)01115-3.

Al-Sharif, I., Remmal, A., Aboussekhra, A., 2013. Eugenol triggers apoptosis in breast cancer cells through E2F1/survivin down-regulation. BMC Cancer 13 (1), 600. https://doi.org/10.1186/1471-2407-13-600.

Aman, R.M., Hashim, I.I.A., Meshali, M.M., 2020. Novel clove essential oil nanoemulgel tailored by Taguchi's model and scaffold-based nanofibers: phytopharmaceuticals with promising potential as cyclooxygenase-2 inhibitors in external inflammation. Int. J. Nanomedicine 15, 2171–2195. https://doi.org/10.2147/IJN.S246601.

Amoabediny, G., Haghiralsadat, F., Naderinezhad, S., Helder, M.N., Kharanaghi, E.A., Arough, J.M., Zandieh-Doulabi, B., 2018. Overview of preparation methods of polymeric and lipid-based (niosome, solid lipid, liposome) nanoparticles: a comprehensive review. Int. J. Polym. Mater. Polym. Biomater. 67 (6), 383–400. https://doi.org/10.1080/00914037.2017.1332623.

Andreu, V., Mendoza, G., Arruebo, M., Irusta, S., 2015. Smart dressings based on nanostructured fibers containing natural origin antimicrobial, anti-inflammatory, and regenerative compounds. Materials 8 (8), 5154–5193. https://doi.org/10.3390/ma8085154.

Anitha, A., Sowmya, S., Sudheesh Kumar, P.T., S., D., Chennazhi, K.P., Ehrlich, H., Tsurkan, M., Jayakumar, R., 2014. Chitin and chitosan in selected biomedical applications. Prog. Polym. Sci. 39 (9).

Anjum, S., Arora, A., Alam, M.S., Gupta, B., 2016. Development of antimicrobial and scar preventive chitosan hydrogel wound dressingsAuthor links open overlay. Int. J. Pharm. 508 (1), 92–101. https://doi.org/10.1016/j.ijpharm.2016.05.013.

Aqil, M., Ahad, A., Sultana, Y., Ali, A., 2007. Status of terpenes as skin penetration enhancers. Drug Discov. Today 12, 1061–1067. https://doi.org/10.1016/j.drudis.2007.09.001.

Aqil, M., Kamran, M., Ahad, A., SarimImam, S., 2016. Development of clove oil based nanoemulsion of olmesartan for transdermal delivery: Box–Behnken design optimization and pharmacokinetic evaluation. J. Mol. Liquids 214, 238–248. https://doi.org/10.1016/j.molliq.2015.12.077.

Ashjazadeh, M.A., Jahandideh, A., Abedi, G., Akbarzadeh, A., Hesaraki, S., 2019. Histopathology and histomorphological study of wound healing using clove extract nanofibers (eugenol) compared to zinc oxide nanofibers on the skin of rats. Arch. Razi Inst. 74 (3), 267–277. https://doi.org/10.22092/ari.2018.120170.1184.

Asl, M.K., Nazariborun, A., Hosseini, M., 2013. Analgesic effect of the aqueous and ethanolic extracts of clove. Avicenna J. Phytomed. 3 (2), 186–192. http://www.ncbi.nlm.nih.gov/pubmed/25050273.

Bachiega, T.F., Barreto, J.P., de Sousa, J., Bastos, K., Sforcin, J.M., 2012. Clove and eugenol in noncytotoxic concentrations exert immunomodulatory/anti-inflammatory action on cytokine production by murine macrophages. J. Pharm. Pharmacol. 64 (4), 610–616. https://doi.org/10.1111/j.2042-7158.2011.01440.x.

Badgujar, S.B., Patel, V.V., Bandivdekar, A.H., 2014. Foeniculum vulgare mill: a review of its botany, phytochemistry, pharmacology, contemporary application, and toxicology. Biomed. Res. Int. https://doi.org/10.1155/2014/842674.

Ballester-Beltrán, J., Rico, P., Moratal, D., Wenlong, S., Mano, J.F., Salmerón-Sánchez, M., 2011. Role of superhydrophobicity in the biological activity of fibronectin at the cell–material interface. Soft Matter. (22).

Banerjee, K., Madhyastha, H., Rajendra, S.V., Manikandanath, N.T., Thiagarajan, N., Thiagarajan, P., 2020. Anti-inflammatory and wound healing potential of a clove oil emulsion. Colloids Surf. B: Biointerfaces 193, 111102. https://doi.org/10.1016/j.colsurfb.2020.111102.

Banik, B.L., Fattahi, P., Brown, J.L., 2016. Polymeric nanoparticles: the future of nanomedicine. Wiley Interdiscip. Rev. Nanomed. Nanobiotechnol. 8 (2), 271–299. https://doi.org/10.1002/wnan.1364.

Baptista-Silva, S., Borges, S., Ramos, O.L., Pintado, M., Sarmento, B., 2020. The progress of essential oils as potential therapeutic agents: a review. J. Essent. Oil Res. 32 (4), 279–295. https://doi.org/10.1080/10412905.2020.1746698.

Barradas, T.N., de Holanda e Silva, K.G., 2020. Nanoemulsions as optimized vehicles for essential oils. In: Sustainable Agriculture Reviews. vol. 44. Springer Nature Switzerland AG, pp. 115–167, https://doi.org/10.1007/978-3-030-41842-7_4.

Barradas, T.N., de Holanda e Silva, K.G., 2021. Nanoemulsions of essential oils to improve solubility, stability and permeability: a review. Environ. Chem. Lett. 19 (2), 1153–1171. https://doi.org/10.1007/s10311-020-01142-2.

Barradas, T.N., Senna, J.P., Cardoso, S.A., Nicoli, S., Padula, C., Santi, P., Francesca Rossi, K., de Holanda e Silva, G., Elias Mansur, C.R., 2017. Hydrogel-thickened nanoemulsions based on essential oils for topical delivery of psoralen: permeation and stability studies. Eur. J. Pharm. Biopharm. 116, 38–50. https://doi.org/10.1016/j.ejpb.2016.11.018.

Barradas, T.N., Senna, J.P., Cardoso, S.A., de Holanda e Silva, K.G., Mansur, C.R.E., 2018. Formulation characterization and in vitro drug release of hydrogel-thickened nanoemulsions for topical delivery of 8-methoxypsoralen. Mater. Sci. Eng. C 92, 245–253. https://doi.org/10.1016/j.msec.2018.06.049.

Bhattarai, R., Bachu, R., Boddu, S., Bhaduri, S., 2018. Biomedical applications of electrospun nanofibers: drug and nanoparticle delivery. Pharmaceutics 11 (1), 5. https://doi.org/10.3390/pharmaceutics11010005.

Bin Park, S., Sung, M.-H., Uyama, H., Han, D.K., 2021. Poly(glutamic acid): production, composites, and medical applications of the next-generation biopolymer. Prog. Polym. Sci. 113. https://doi.org/10.1016/j.progpolymsci.2020.101341.

Bonferoni, M.C., Rossi, S., Cornaglia, A.I., Mannucci, B., Grisoli, P., Vigani, B., Saporito, F., et al., 2017. Essential oil-loaded lipid nanoparticles for wound healing. Int. J. Nanomedicine 13, 175–186. https://doi.org/10.2147/IJN.S152529.

Bose, R.J.C., Ravikumar, R., Karuppagounder, V., Bennet, D., Rangasamy, S., Thandavarayan, R.A., 2017. Lipid–polymer hybrid nanoparticle-mediated therapeutics delivery: advances and challenges. Drug Discov. Today 22 (8), 1258–1265. https://doi.org/10.1016/j.drudis.2017.05.015.

Brewster, M.E., Loftsson, T., 2007. Cyclodextrins as pharmaceutical solubilizers. Adv. Drug Deliv. Rev. 59 (7), 645–666. https://doi.org/10.1016/j.addr.2007.05.012.

Budak, K., Sogut, O., Sezer, U.A., 2020. A review on synthesis and biomedical applications of polyglycolic acid. J. Polym.r Res. 27. https://doi.org/10.1007/s10965-020-02187-1.

Campelo, P.H., Junqueira, L.A., Vilela de Resende, J., Zacarias, R.D., Fernandes, R.V.d.B., Botrel, D.A., Borges, S.V., 2017. Stability of lime essential oil emulsion prepared using biopolymers and ultrasound treatment. Int. J. Food Prop. 20 (Sup 1), S564–S579. https://doi.org/10.1080/10942912.2017.1303707.

Cardoso, S.A., Barradas, T.N., 2020. Developing formulations for drug follicular targeting: nanoemulsions loaded with minoxidil and clove oil. J. Drug Delivery Sci. Technol. 59, 101908. https://doi.org/10.1016/j.jddst.2020.101908.

Carrasco, H., Espinoza, L., Cardile, V., Gallardo, C., Cardona, W., Lombardo, L., Catalán, K., Cuellar, M., Russo, A., 2008. Eugenol and its synthetic analogues inhibit cell growth of human cancer cells (part I). J. Braz. Chem. Soc. 19 (3), 543–548. https://doi.org/10.1590/S0103-50532008000300024.

Cevc, G., 2004. Lipid vesicles and other colloids as drug carriers on the skin. Adv. Drug Deliv. Rev. https://doi.org/10.1016/j.addr.2003.10.028.

Chaieb, K., Hajlaoui, H., Zmantar, T., Kahla-Nakbi, A.B., Rouabhia, M., Mahdouani, K., Bakhrouf, A., 2007a. The chemical composition and biological activity of clove essential oil, Eugenia caryophyllata (*Syzygium aromaticum* L. Myrtaceae): a short review. Phytother. Res. 21 (6), 501–506. https://doi.org/10.1002/ptr.2124.

Chaieb, K., Zmantar, T., Ksouri, R., Hajlaoui, H., Mahdouani, K., Abdelly, C., Bakhrouf, A., 2007b. Antioxidant properties of the essential oil of Eugenia caryophyllata and its antifungal activity against a large number of clinical candida species. Mycoses 50 (5), 403–406. https://doi.org/10.1111/j.1439-0507.2007.01391.x.

Chen, J., Jiang, Q.-D., Ye-Ming, W., Liu, P., Yao, J.-H., Qing, L., Zhang, H., Duan, J.-A., 2015. Potential of essential oils as penetration enhancers for transdermal administration of ibuprofen to treat dysmenorrhoea. Molecules 20 (10), 18219–18236. https://doi.org/10.3390/molecules201018219.

Chen, Q., Liu, Z., 2016. Albumin carriers for cancer theranostics: a conventional platform with new promise. Adv. Mater. 28 (47), 10557–10566.

Cho, J.S., Kim, T.H., Lim, J.-M., Song, J.-H., 2008. Effects of eugenol on Na+ currents in rat dorsal root ganglion neurons. Brain Res. 1243, 53–62. https://doi.org/10.1016/j.brainres.2008.09.030.

Choi, M.-J., Soottitantawat, A., Nuchuchua, O., Min, S.-G., Ruktanonchai, U., 2009. Physical and light oxidative properties of eugenol encapsulated by molecular inclusion and emulsion-diffusion method. Food Res. Int. 42 (1), 148–156. https://doi.org/10.1016/j.foodres.2008.09.011.

Choudhary, M.I.I., Jalil, S., Sarfraz Ahmad, S.A., Nawaz, K.M., Tareen, R.B.R.B., Azizuddin, S.J., Khan, K.M., 2009. Antiinflammatory and lipoxygenase inhibitory compounds from Vitex agnus-castus. Phytother. Res. 23 (9), 1336–1339. https://doi.org/10.1002/ptr.

Cismaru, L., Popa, M., 2010. Polymeric nanoparticles with biomedical applications. Rev. Roum. Chim. 55 (8), 433–442.

Cortés-Rojas, D.F., de Souza, C.R.F., Oliveira, W.P., 2014. Clove (Syzygium aromaticum): a precious spice. Asian Pac. J. Trop. Biomed. 4 (2), 90–96. https://doi.org/10.1016/S2221-1691(14)60215-X.

Cramariuc, B., Cramariuc, R., Scarlet, R., Manea, L.R., Lupu, I.G., Cramariuc, O., 2013. Fiber diameter in electrospinning process. J. Electrost. 71 (3), 189–198. https://doi.org/10.1016/j.elstat.2012.12.018.

Crucho, C.I.C., Barros, M.T., 2017. Polymeric nanoparticles: a study on the preparation variables and characterization methods. Mater. Sci. Eng. C. https://doi.org/10.1016/j.msec.2017.06.004.

Cui, H., Zhao, C., Lin, L., 2015. The specific antibacterial activity of liposome-encapsulated clove oil and its application in tofu. Food Control 56, 128–134. https://doi.org/10.1016/j.foodcont.2015.03.026.

Cui, H., Bai, M., Rashed, M.M.A., Lin, L., 2018. The antibacterial activity of clove oil/chitosan nanoparticles embedded gelatin nanofibers against *Escherichia coli* O157:H7 biofilms on cucumber. Int. J. Food Microbiol. 266 (February), 69–78. https://doi.org/10.1016/j.ijfoodmicro.2017.11.019.

Da Silva, T.N., Gonçalves, R.P., Rocha, C.L., Archanjo, B.S., Barbosa, C.A., Pierre, M.B.R., Reynaud, F., Picciani, P.H.S., 2018. Controlling burst effect with PLA/PVA coaxial electrospun scaffolds loaded with BMP-2 for bone guided regeneration. Mater. Sci. Eng. C 97. https://doi.org/10.1016/j.msec.2018.12.020.

Daniel, A.N., Sartoretto, S.M., Schmidt, G., Caparroz-Assef, S.M., Bersani-Amado, C.A., Roberto, K.N., Cuman., 2009. Anti-inflammatory and antinociceptive activities A of eugenol essential oil in experimental animal models. Rev. Bras 19 (1b), 212–217. https://doi.org/10.1590/S0102-695X2009000200006.

Das, B., Mandal, D., Dash, S.K., Chattopadhyay, S., Tripathy, S., Dolai, D.P., Dey, S.K., Roy, S., 2016. Eugenol provokes ROS-mediated membrane damage-associated antibacterial activity against clinically isolated multidrug-resistant *Staphylococcus aureus* strains. Infect. Dis. *9* (*January*). https://doi.org/10.4137/IDRT.S31741. IDRT.S31741.

Dash, M., Chiellini, F., Ottenbrite, E., Chiellini, E., 2011. Chitosan—a versatile semi-synthetic polymer in biomedical applications. Prog. Polym. Sci. 36 (8). https://doi.org/10.1016/j.progpolymsci.2011.02.001.

De Campos, V.E.B., Cerqueira-Coutinho, C.S., Mansur, C.R.E., Santos-Oliveira, R., Pyrrho, A.S., 2016. Biodistribution of praziquantel (PZQ) nanoemulsion (NE) in healthy wistar rats: evaluation of biological behavior. J. Bionanosci. 10 (6), 486–490. https://doi.org/10.1166/jbns.2016.1403.

de Matos, S.P., Teixeira, H.F., de Lima, Á.A.N., Veiga-Junior, V.F., Koester, L.S., 2019. Essential oils and isolated terpenes in nanosystems designed for topical administration: a review. Biomolecules 9 (4), 138. https://doi.org/10.3390/biom9040138.

de Meneses, A.C., dos Santos, P.C.M., Machado, T.O., Sayer, C., de Oliveira, D., de Araújo, P.H.H., 2017. Poly (thioether-ester) nanoparticles entrapping clove oil for antioxidant activity improvement. J. Polym. Res. 24 (11). https://doi.org/10.1007/s10965-017-1353-x.

de Meneses, A.C., Marques, E.B.P., Leimann, F.V., Gonçalves, O.H., Ineu, R.P., de Araújo, P.H.H., de Oliveira, D., Sayer, C., 2019. Encapsulation of clove oil in nanostructured lipid carriers from natural waxes: preparation, characterization and in vitro evaluation of the cholinesterase enzymes. Colloids Surf. A Physicochem. Eng. Asp. 583. https://doi.org/10.1016/j.colsurfa.2019.123879.

de Oliveira, B., de Siqueira, L., dos Santos Matos, A.P., da Silva Cardoso, V., Villanova, J.C.O., Guimarães, B.d.C. L.R., Santos, E.P.d., Vermelho, A.B., Santos-Oliveira, R., Junior, E.R., 2019. Clove oil nanoemulsion showed potent inhibitory effect against *Candida* spp. Nanotechnology 30 (42), 425101. https://doi.org/10.1088/1361-6528/ab30c1.

De Siqueira, L.B.O., Matos, A.P.D.S., Cardoso, V.S., Villanova, J.C.O., Guimarães, B.C.L.R., Dos Santos, E.P., Vermelho, A.B., Santos-Oliveira, R., Ricci Junior, E., 2019. Clove oil nanoemulsion showed potent inhibitory effect against *Candida* spp. Nanotechnology 30 (42). https://doi.org/10.1088/1361-6528/ab30c1.

Deng, L., Liu, Z., Li, L., 2019. Hybrid nanocomposites for imaging-guided synergistic theranostics. In: Nanomaterials for Drug Delivery and Therapy. Elsevier, https://doi.org/10.1016/B978-0-12-816505-8.00017-5.

Dias, D.D.O., Colombo, M., Kelmann, R.G., Kaiser, S., Lucca, L.G., Teixeira, H.F., Limberger, R.P., Veiga, V.F., Koester, L.S., 2014. Optimization of copaiba oil-based nanoemulsions obtained by different preparation methods. Ind. Crop. Prod. 59, 154–162. https://doi.org/10.1016/j.indcrop.2014.05.007.

Donsì, F., Ferrari, G., 2016. Essential oil nanoemulsions as antimicrobial agents in food. J. Biotechnol. https://doi.org/10.1016/j.jbiotec.2016.07.005.

Dragicevic, N., Maibach, H.I., 2015. Terpenes and Essential Oils as Skin Penetration Enhancers. Percutaneous Penetration Enhancers Chemical Methods In Penetration Enhancement: Modification Of The Stratum Corneum, I. II Springer, San Francisco, CA, USA, pp. 1–411.

Dwivedi, M., Sharma, V., Pathak, K., 2017. Pilosebaceous targeting by isotretenoin-loaded invasomal gel for the treatment of eosinophilic pustular folliculitis: optimization, efficacy and cellular analysis. Drug Dev. Ind. Pharm. 43 (2), 293–304. https://doi.org/10.1080/03639045.2016.1239628.

Eid, R.K., Essa, E.A., El Maghraby, G.M., 2019. Essential oils in niosomes for enhanced transdermal delivery of felodipine. Pharm. Dev. Technol. 24 (2), 157–165. https://doi.org/10.1080/10837450.2018.1441302.

El Khayat Noha, W., Donia, A.A., Mady, O.Y., El Maghraby, G.M., 2018. Optimization of eugenol microemulsion for transdermal delivery of indomethacin. J. Drug Delivery Sci. Technol. 48, 311–318. https://doi.org/10.1016/j.jddst.2018.10.006.

El Maghraby, G.M., 2008. Transdermal delivery of hydrocortisone from eucalyptus oil microemulsion: effects of cosurfactants. Int. J. Pharm. 355 (1–2), 285–292. https://doi.org/10.1016/j.ijpharm.2007.12.022.

El-Darier, S.M., El-Ahwany, A.M.D., Elkenany, E.T., Abdeldaim, A.A., 2018. An in vitro study on antimicrobial and anticancer potentiality of thyme and clove oils. Rend. Lincei Sci. Fis. Nat. 29 (1), 131–139. https://doi.org/10.1007/s12210-018-0672-0.

El-Say, K.M., El-Sawy, H.S., 2017. Polymeric nanoparticles: promising platform for drug delivery. Int. J. Pharm. 528 (1–2), 675–691. https://doi.org/10.1016/j.ijpharm.2017.06.052.

Faisal, W., Soliman, G.M., Hamdan, A.M., 2018. Enhanced skin deposition and delivery of voriconazole using ethosomal preparations. J. Liposome Res. 28 (1), 14–21. https://doi.org/10.1080/08982104.2016.1239636.

Fang, C.-L., Al-Suwayeh, S.A., Fang, J.-Y., 2013. Nanostructured lipid carriers (NLCs) for drug delivery and targeting. Recent Pat. Nanotechnol. 7 (1), 41–55. https://doi.org/10.2174/187221051130701004l.

Fattahi, F.-S., Khoddami, A., Avinc, O., 2019. Poly(lactic acid) (PLA) nanofibers for bone tissue engineering. J. Text. Polym. 7 (2), 47–64.

Fazly Bazzaz, B.S., Khameneh, B., Namazi, N., Iranshahi, M., Davoodi, D., Golmohammadzadeh, S., 2018. Solid lipid nanoparticles carrying Eugenia caryophyllata essential oil: the novel nanoparticulate systems with broad-spectrum antimicrobial activity. Lett. Appl. Microbiol. 66 (6), 506–513. https://doi.org/10.1111/lam.12886.

Feldstein, M.M., Tohmakhchi, V.N., Malkhazov, L.B., Vasiliev, A.E., Platé, N.A., 1996. Hydrophilic polymeric matrices for enhanced transdermal drug delivery. Int. J. Pharm. 131 (2), 229–242.

Franco, P., de Marco, I., 2020. The use of poly(N-vinyl pyrrolidone) in the delivery of drugs: a review. Polymers 12 (5). https://doi.org/10.3390/polym12051114.

Fukui, K., Omoi, N.-O., Hayasaka, T., Shinnkai, T., Suzuki, S., Abe, K., Urano, A.S., 2002. Cognitive impairment of rats caused by oxidative stress and aging, and its prevention by vitamin E. Ann. N. Y. Acad. Sci. 959 (1), 275–284. https://doi.org/10.1111/j.1749-6632.2002.tb02099.x.

Gaber, E.-S.B., Alkazmi, L.M., Wasef, L.G., Beshbishy, A.M., Nadwa, E.H., Rashwan, E.K., 2020. *Syzygium aromaticum* L. (Myrtaceae): traditional uses, bioactive chemical constituents, pharmacological and toxicological activities. Biomolecules 10 (2), 202. https://doi.org/10.3390/biom10020202.

Gandhi, J., Suthar, D., Patel, H., Shelat, P., Parejiya, P., 2020. Development and characterization of microemulsion based topical gel of essential oil of clove (*Syzygium aromaticum*) for superficial fungal infections. Adv. Tradit. Med. https://doi.org/10.1007/s13596-020-00462-6.

Ganesan, P., Narayanasamy, D., 2017. Lipid nanoparticles: different preparation techniques, characterization, hurdles, and strategies for the production of solid lipid nanoparticles and nanostructured lipid carriers for oral drug delivery. Sustain. Chem. Pharm. 6, 37–56. https://doi.org/10.1016/j.scp.2017.07.002.

Garcês, A., Amaral, M.H., Sousa Lobo, J.M., Silva, A.C., 2018. Formulations based on solid lipid nanoparticles (SLN) and nanostructured lipid carriers (NLC) for cutaneous use: a review. Eur. J. Pharm. Sci. 112, 159–167. https://doi.org/10.1016/j.ejps.2017.11.023.

Gautam, N., Mantha, A.K., Mittal, S., 2014. Essential oils and their constituents as anticancer agents: a mechanistic view. BioMed Res. Int. 20 (14). https://doi.org/10.1155/2014/154106.

Ghosh, R., Nadiminty, N., Fitzpatrick, J.E., Alworth, W.L., Slaga, T.J., Kumar, A.P., 2005. Eugenol causes melanoma growth suppression through inhibition of E2F1 transcriptional activity. J. Biol. Chem. 280 (7), 5812–5819. https://doi.org/10.1074/jbc.M411429200.

Goñi, M.G., Roura, S.I., Ponce, A.G., Moreira, M.R., 2016. Clove (*Syzygium aromaticum*) oils. In: Preedy, V.R. (Ed.), Essential Oils in Food Preservation, Flavor and Safety. Academic Press.

Guan, W., Li, S., Yan, R., Tang, S., Quan, C., 2007. Comparison of essential oils of clove buds extracted with supercritical carbon dioxide and other three traditional extraction methods. Food Chem. 101 (4), 1558–1564. https://doi.org/10.1016/j.foodchem.2006.04.009.

Gülçin, İ., 2011. Antioxidant activity of eugenol: a structure-activity relationship study. J. Med. Food. https://doi.org/10.1089/jmf.2010.0197.

Gupta, A., Burak Eral, H., Alan Hatton, T., Doyle, P.S., 2016a. Controlling and predicting droplet size of nanoemulsions: scaling relations with experimental validation. Soft Matter. https://doi.org/10.1039/C5SM02051D.

Gupta, A., Burak Eral, H., Alan Hatton, T., Doyle, P.S., 2016b. Nanoemulsions: formation, properties and applications. Soft Matter 12 (11). https://doi.org/10.1039/C5SM02958A.

Hadidi, M., Pouramin, S., Adinepour, F., Haghani, S., Jafari, S.M., 2020. Chitosan nanoparticles loaded with clove essential oil: characterization, antioxidant and antibacterial activities. Carbohydr. Polym. 236 (2019), 116075. https://doi.org/10.1016/j.carbpol.2020.116075.

Han, X., Parker, T.L., 2017. Anti-inflammatory activity of clove (*Eugenia caryophyllata*) essential oil in human dermal fibroblasts. Pharm. Biol. 55 (1), 1619–1622. https://doi.org/10.1080/13880209.2017.1314513.

Hasheminejad, N., Khodaiyan, F., Safari, M., 2019. Improving the antifungal activity of clove essential oil encapsulated by chitosan nanoparticles. Food Chem. 275, 113–122. https://doi.org/10.1016/j.foodchem.2018.09.085.

Hemaiswarya, S., Doble, M., 2009. Synergistic interaction of eugenol with antibiotics against gram negative bacteria. Phytomedicine 16 (11), 997–1005. https://doi.org/10.1016/j.phymed.2009.04.006.

Hemaiswarya, S., Doble, M., 2013. Combination of phenylpropanoids with 5-fluorouracil as anti-cancer agents against human cervical cancer (HeLa) cell line. Phytomedicine 20 (2), 151–158. https://doi.org/10.1016/j.phymed.2012.10.009.

Herman, A.A.P., Herman, A.A.P., 2015. Essential oils and their constituents as skin penetration enhancer for transdermal drug delivery: a review. J. Pharm. Pharmacol. https://doi.org/10.1111/jphp.12334.

Hill, L.E., Gomes, C., Matthew Taylor, T., 2013. Characterization of beta-cyclodextrin inclusion complexes containing essential oils (trans-cinnamaldehyde, eugenol, cinnamon bark, and clove bud extracts) for antimicrobial delivery applications. LWT Food Sci. Technol. 51 (1), 86–93. https://doi.org/10.1016/j.lwt.2012.11.011.

Hoffman, A.S., 2002. Hydrogels for biomedical applications. Adv. Drug Deliv. Rev. 54 (1), 3–12. https://doi.org/10.1016/s0169-409x(01)00239-3.

Hu, X., Liu, S., Zhou, G., Huang, Y., Xie, Z., Jing, X., 2014. Electrospinning of polymeric nanofibers for drug delivery applications. J. Control. Release 185 (1), 12–21. https://doi.org/10.1016/j.jconrel.2014.04.018.

Huerta, R.R., Silva, E.K., El-Bialy, T., Saldaña, M.D.A., 2020. Clove essential oil emulsion-filled cellulose nanofiber hydrogel produced by high-intensity ultrasound technology for tissue engineering applications. Ultrason. Sonochem. 64, 104845. https://doi.org/10.1016/j.ultsonch.2019.104845.

Hyeon, T., Rotello, V., 2012. Nanomedicine. Chemical Society Reviews, I. I Royal Society of Chemistry, London, pp. 2537–2539.

Irrera, N., D'Ascola, A., Pallio, G., Bitto, A., Mannino, F., Arcoraci, V., Rottura, M., et al., 2020. β-Caryophyllene inhibits cell proliferation through a direct modulation of CB2 receptors in glioblastoma cells. Cancer 12 (4), 1038. https://doi.org/10.3390/cancers12041038.

Jabbari, N., Eftekhari, Z., Roodbari, N.H., Parivar, K., 2020. Evaluation of encapsulated eugenol by chitosan nanoparticles on the aggressive model of rheumatoid arthritis. Int. Immunopharmacol. 85, 106554. https://doi.org/10.1016/j.intimp.2020.106554.

Jadhav, B.K., Khandelwal, K.R., Ketkar, A.R., Pisal, S.S., 2004. Formulation and evaluation of mucoadhesive tablets containing eugenol for the treatment of periodontal diseases. Drug Dev. Ind. Pharm. 30 (2), 195–203. https://doi.org/10.1081/DDC-120028715.

Jaganathan, S.K., Mazumdar, A., Mondhe, D., Mandal, M., 2011. Apoptotic effect of eugenol in human colon cancer cell lines. Cell Biol. Int. 35 (6), 607–615. https://doi.org/10.1042/CBI20100118.

Jairoce, C.F., Teixeira, C.M., Nunes, C.F.P., Nunes, A.M., Pereira, C.M.P., Garcia, F.R.M., 2016. Insecticide activity of clove essential oil on bean weevil and maize weevil. Rev. Bras. de Eng. Agricola e Ambient. 20 (1), 72–77. https://doi.org/10.1590/1807-1929/agriambi.v20n1p72-77.

Jirovetz, L., Buchbauer, G., Stoilova, I., Stoyanova, A., Krastanov, A., Schmidt, E., 2006. Chemical composition and antioxidant properties of clove leaf essential oil. J. Agric. Food Chem. 54 (17), 6303–6307. https://doi.org/10.1021/jf060608c.

Junyaprasert, V.B., Teeranachaideekul, V., Souto, E.B., Boonme, P., Müller, R.H., 2009. Q10-loaded NLC versus nanoemulsions: stability, rheology and in vitro skin permeation. Int. J. Pharm. https://doi.org/10.1016/j.ijpharm.2009.05.020.

Kennewell, T.L., Mashtoub, S., Howarth, G.S., Cowin, A.J., Kopecki, Z., 2019. Antimicrobial and healing-promoting properties of animal and plant oils for the treatment of infected wounds. Wound Pract. Res. 27 (4). https://doi.org/10.33235/wpr.27.4.175-183.

Kenry, T.Y., Manghnani, P.N., Middha, E., Pan, Y., Chen, H., Lim, C.T., Liu, B., 2020. Mechanistic understanding of the biological responses to polymeric nanoparticles. ACS Nano 14 (4), 4509–4522. https://doi.org/10.1021/acsnano.9b10195.

Kim, E.-H., Kim, H.-K., Ahn, Y.-J., 2003. Acaricidal activity of clove bud oil compounds against Dermatophagoides farinae and Dermatophagoides pteronyssinus (Acari: Pyroglyphidae). J. Agric. Food Chem. 51 (4), 885–889. https://doi.org/10.1021/jf0208278.

Knop, K., Hoogenboom, R., Fischer, D., Schubert, U.S., 2010. Poly(ethylene glycol) in drug delivery: pros and cons as well as potential alternatives. Angew. Chem. 49 (36). https://doi.org/10.1002/anie.200902672.

Kumar, P.S., Febriyanti, R.M., Sofyan, F.F., Luftimas, D.E., Abdulah, R., 2014. Anticancer potential of Syzygium aromaticum L. in MCF-7 human breast cancer cell lines. Pharm. Res. 6 (4), 350. https://doi.org/10.4103/0974-8490.138291.

Kyaw, B.M., Arora, S., Lim, C.S., 2012. Bactericidal antibiotic-phytochemical combinations against methicillin resistant Staphylococcus aureus. Braz. J. Microbiol. 43 (3), 938–945. https://doi.org/10.1590/S1517-83822012000300013.

Lasoń, E., 2020. Topical administration of terpenes encapsulated in nanostructured lipid-based systems. Molecules 25 (23), 5758. https://doi.org/10.3390/molecules25235758.

Lee, M.H., Yeon, K.-Y., Park, C.-K., Li, H.-Y., Fang, Z., Kim, M.S., Choi, S.-Y., et al., 2005. Eugenol inhibits calcium currents in dental afferent neurons. J. Dent. Res. 84 (9), 848–851. https://doi.org/10.1177/154405910508400913.

Lenardão, E.J., Savegnago, L., Jacob, R.G., Victoria, F.N., Martinez, D.M., 2015. Antinociceptive effect of essential oils and their constituents: an update review. J. Braz. Chem. Soc. https://doi.org/10.5935/0103-5053.20150332.

Leong, T.S.H., Wooster, T.J., Kentish, S.E., Ashokkumar, M., 2009. Minimising oil droplet size using ultrasonic emulsification. Ultrason. Sonochem. 16 (6), 721–727. https://doi.org/10.1016/j.ultsonch.2009.02.008.

Li, J.W.-H., Vederas, J.C., 2009. Drug discovery and natural products: end of an era or an endless frontier? Science 325 (5937), 161–165. https://doi.org/10.1126/science.1168243.

Li, H.Y., Park, C.-K., Jung, S.J., Choi, S.-Y., Lee, S.J., Park, K., Kim, J.S., Oh, S.B., 2007. Eugenol inhibits K + currents in trigeminal ganglion neurons. J. Dent. Res. 86 (9), 898–902. https://doi.org/10.1177/154405910708600918.

Loftsson, T., Másson, M., Brewster, M.E., 2004. Self-association of cyclodextrins and cyclodextrin complexes. J. Pharm. Sci. 93 (5), 1091–1099. https://doi.org/10.1002/jps.20047.

Lim, S.B., Banerjee, A., Önyüksel, H., 2012. Improvement of drug safety by the use of lipid-based nanocarriers. J. Controll. Release 163 (1), 34–45. https://doi.org/10.1016/j.jconrel.2012.06.002.

Loftsson, T., Hreinsdóttir, D., Másson, M., 2005. Evaluation of cyclodextrin solubilization of drugs. Int. J. Pharm. 302 (1–2), 18–28. https://doi.org/10.1016/j.ijpharm.2005.05.042.

Lohani, A., Verma, A., 2017. Vesicles: potential nano carriers for the delivery of skin cosmetics. J. Cosmet. Laser Ther. 19 (8), 485–493. https://doi.org/10.1080/14764172.2017.1358451.

Lopes, L.B., Garcia, M.T.J., Bentley, M.V.L., 2015. Chemical penetration enhancers. Ther. Deliv. 6 (9), 1053–1061. https://doi.org/10.4155/tde.15.61.

López-García, R., Ganem-Rondero, A., 2015. Solid lipid nanoparticles (SLN) and nanostructured lipid carriers (NLC): occlusive effect and penetration enhancement ability. J. Cosmet. Dematol. Sci. Appl. 05 (02), 62–72. https://doi.org/10.4236/jcdsa.2015.52008.

López-López, M., Fernández-Delgado, A., Moyá, M.L., Blanco-Arévalo, D., Carrera, C., de la Haba, R.R., Ventosa, A., Bernal, E., López-Cornejo, P., 2019. Optimized preparation of levofloxacin loaded polymeric nanoparticles. Pharmaceutics 11 (2), 1–13. https://doi.org/10.3390/pharmaceutics11020057.

Lu, W.C., Huang, D.W., Wang, C.C.R., Yeh, C.H., Tsai, J.C., Huang, Y.T., Li, P.H., 2018. Preparation, characterization, and antimicrobial activity of nanoemulsions incorporating citral essential oil. J. Food Drug Anal. 26 (1), 82–89. https://doi.org/10.1016/j.jfda.2016.12.018.

Madan, J.R., Patil, K., Awasthi, R., Dua, K., 2021. Formulation and evaluation of solid self-microemulsifying drug delivery system for Azilsartan medoxomil. Int. J. Polym. Mater. Polym. Biomater. 70 (2), 100–116. https://doi.org/10.1080/00914037.2019.1695206.

Magalhães, C.B., Riva, D.R., DePaula, L.J., Brando-Lima, A., Koatz, V.L.G., Leal-Cardoso, J.H., Zin, W.A., Faffe, D.S., 2010. In vivo anti-inflammatory action of eugenol on lipopolysaccharide-induced lung injury. J. Appl. Physiol. 108 (4), 845–851. https://doi.org/10.1152/japplphysiol.00560.2009.

Makuch, E., Nowak, A., Günther, A., Pełech, R., Kucharski, Ł., Duchnik, W., Klimowicz, A., 2020. Enhancement of the antioxidant and skin permeation properties of eugenol by the esterification of eugenol to new derivatives. AMB Express 10 (1), 187. https://doi.org/10.1186/s13568-020-01122-3.

Malikmammadov, E., Tanir, T.E., Kiziltay, A., Hasirci, V., Hasirci, N., 2017. PCL and PCL-based materials in biomedical applications. J. Biomater. Sci. Polym. Ed. 28 (7), 863–893. https://doi.org/10.1080/09205063.2017.1394711.

Mallakpour, S., Behranvand, V., 2016. Polymeric nanoparticles: recent development in synthesis and application. Express Polym Lett 10 (11), 895–913. https://doi.org/10.3144/expresspolymlett.2016.84.

Mandal, S., 2011. In vitro antibacterial activity of three Indian spices against methicillin-resistant Staphylococcus aureus. Oman Med. J. 26 (5), 319–323. https://doi.org/10.5001/omj.2011.80.

Manganyi, M.C., Regnier, T., Olivier, E.I., 2015. Antimicrobial activities of selected essential oils against *Fusarium oxysporum* isolates and their biofilms. S. Afr. J. Bot. 99 (July), 115–121. https://doi.org/10.1016/j.sajb.2015.03.192.

Marchese, A., Barbieri, R., Coppo, E., Orhan, I.E., Daglia, M., Nabavi, S.F., Izadi, M., Abdollahi, M., Nabavi, S. M., Ajami, M., 2017. Antimicrobial activity of eugenol and essential oils containing eugenol: a mechanistic viewpoint. Crit. Rev. Microbiol. 43 (6), 668–689. https://doi.org/10.1080/1040841X.2017.1295225.

Marjukka Suhonen, T., Bouwstra, J.A., Urtti, A., 1999. Chemical enhancement of percutaneous absorption in relation to stratum corneum structural alterations. J. Control. Release 59 (2), 149–161. https://doi.org/10.1016/S0168-3659(98)00187-4.

Martins, S., Sarmento, B., Ferreira, D.C., Souto, E.B., 2007. Lipid-based colloidal carriers for peptide and protein delivery- -liposomes versus lipid nanoparticles. Int. J. Nanomed. 2 (4), 595–607.

Marturano, V., Bizzarro, V., De Luise, A., Calarco, A., Ambrogi, V., Giamberini, M., Tylkowski, B., Cerruti, P., 2018. Essential oils as solvents and core materials for the preparation of photo-responsive polymer nanocapsules. Nano Res. 11 (5), 2783–2795. https://doi.org/10.1007/s12274-017-1908-5.

Mayer, S., Weiss, J., McClements, D.J., 2013. Vitamin E-enriched nanoemulsions formed by emulsion phase inversion: factors influencing droplet size and stability. J. Colloid Interface Sci. https://doi.org/10.1016/j.jcis.2013.04.016.

McClements, D.J., Jafari, S.M., 2018. General aspects of nanoemulsions and their formulation. In: Nanoemulsions. Elsevier, pp. 3–20, https://doi.org/10.1016/B978-0-12-811838-2.00001-1.

Megelski, S., Stephens, J.S., Bruce Chase, D., Rabolt, J.F., 2002. Micro- and nanostructured surface morphology on electrospun polymer fibers. Macromolecules 35 (22), 8456–8466. https://doi.org/10.1021/ma020444a.

Mir, M., Ahmed, N., Rehman, A.U., 2012. Recent applications of PLGA based nanostructures in drug delivery. Colloids Surf. B 159, 217–231. https://doi.org/10.1016/j.colsurfb.2017.07.038.

Miyazawa, M., Hisama, M., 2003. Antimutagenic activity of phenylpropanoids from clove (*Syzygium aromaticum*). J. Agric. Food Chem. 51 (22), 6413–6422. https://doi.org/10.1021/jf030247q.

Moghimi, R., Ghaderi, L., Rafati, H., Aliahmadi, A., McClements, D.J., 2016. Superior antibacterial activity of nanoemulsion of Thymus daenensis essential oil against E. coli. Food Chem. 194, 410–415. https://doi.org/10.1016/j.foodchem.2015.07.139.

Mohammadian, F., Eatemadi, A., 2017. Drug loading and delivery using nanofibers scaffolds. Artif. Cells Nanomed. Biotechnol. 45 (5), 881–888. https://doi.org/10.1080/21691401.2016.1185726.

Montenegro, L., Lai, F., Offerta, A., Sarpietro, M.G., Micicchè, L., Maccioni, A.M., Valenti, D., Fadda, A.M., 2016. From nanoemulsions to nanostructured lipid carriers: a relevant development in dermal delivery of drugs and cosmetics. J. Drug Delivery Sci. Technol. https://doi.org/10.1016/j.jddst.2015.10.003.

Moon, S.-E., Kim, H.-Y., Cha, J.-D., 2011. Synergistic effect between clove oil and its major compounds and antibiotics against oral bacteria. Arch. Oral Biol. 56 (9), 907–916. https://doi.org/10.1016/j.archoralbio.2011.02.005.

Morachis, J.M., Mahmoud, E.A., Almutairi, A., 2012. Physical and chemical strategies for therapeutic delivery by using polymeric nanoparticles. Pharmacol. Rev. 64 (3), 505–519. https://doi.org/10.1124/pr.111.005363.

Muhamad, I.I., Selvakumaran, S., 2016. Designing Polymeric Nanoparticles for Targeted Drug Delivery System Designing Polymeric Nanoparticles for Targeted Drug Delivery System. Nanomedicine, I. I One Central Press, Cheshire, UK, pp. 287–313.

Mylonas, C.C., Cardinaletti, G., Sigelaki, I., Polzonetti-Magni, A., 2005. Comparative efficacy of clove oil and 2-phenoxyethanol as anesthetics in the aquaculture of European sea bass (*Dicentrarchus labrax*) and gilthead sea bream (*Sparus aurata*) at different temperatures. Aquaculture 246 (1–4), 467–481. https://doi.org/10.1016/j.aquaculture.2005.02.046.

Nam, H., Kim, M.-M., 2013. Eugenol with antioxidant activity inhibits MMP-9 related to metastasis in human fibrosarcoma cells. Food Chem. Toxicol. 55 (May), 106–112. https://doi.org/10.1016/j.fct.2012.12.050.

Nangare, S., Dugam, S., 2020. Smart invasome synthesis, characterizations, pharmaceutical applications, and pharmacokinetic perspective: a review. Future J. Pharm. Sci. 6 (1), 123. https://doi.org/10.1186/s43094-020-00145-8.

Nascimento, T., de Souza Picciani, H., de Holanda e Silva, K.G., Barradas, T.N., 2021. Hybrid smart materials for topical drug delivery: application of scaffolds. In: Atta-ur-Rahman, Choudhary, M.I. (Eds.), Frontiers in Drug Design and Discovery. vol. 1. Bentham Science Publishers, pp. 168–237, https://doi.org/10.2174/9789811421563121100007.

Nikoui, V., Ostadhadi, S., Bakhtiarian, A., Abbasi-Goujani, E., Habibian-Dehkordi, S., Rezaei-Roshan, M., Foroohandeh, M., Giorgi, M., 2017. The anti-inflammatory and antipyretic effects of clove oil in healthy dogs after surgery. PharmaNutrition 5 (2), 52–57. https://doi.org/10.1016/j.phanu.2017.02.003.

Nirmala, M.J., Durai, L., Gopakumar, V., Nagarajan, R., 2019. Anticancer and antibacterial effects of a clove bud essential oil-based nanoscale emulsion system. Int. J. Nanomedicine 14, 6439–6450. https://doi.org/10.2147/IJN.S211047.

Nostro, P., Fratoni, L., Baglioni, P.L., 2002. Modification of a cellulosic fabric with β-cyclodextrin for textile finishing applications. J. Incl. Phenom. 44, 423–427.

Oh, K.S., Han, S.K., Choi, Y.W., Lee, J.H., Lee, J.Y., Yuk, S.H., 2004. Hydrogen-bonded polymer gel and its application as a temperature-sensitive drug delivery system. Biomaterials 25 (12), 2393–2398. https://doi.org/10.1016/j.biomaterials.2003.09.008.

Oliveira, E.F., Paula, H.C.B., Paula, R.C., 2014. Alginate/cashew gum nanoparticles for essential oil encapsulation. Colloids Surf. B 113, 146–151. https://doi.org/10.1016/j.colsurfb.2013.08.038.

Opanasopit, P., Sila-on, W., Rojanarata, T., Ngawhirunpat, T., 2013. Fabrication and properties of capsicum extract-loaded PVA and CA nanofiber patches. Pharm. Dev. Technol. 18 (5), 1140–1147. https://doi.org/10.3109/10837450.2012.727004.

Paliwal, R., Paliwal, S.R., Kenwat, R., Kurmi, B.D., Sahu, M.K., 2020. Solid lipid nanoparticles: a review on recent perspectives and patents. Expert Opin. Ther. Pat. 30 (3), 179–194. https://doi.org/10.1080/13543776.2020.1720649.

Paradossi, G., Cavalieri, F., Ester, C., Spagnoli, C., Cowman, M.K., 2003. Poly(vinyl alcohol) as versatile biomaterial for potential biomedical applications. J. Mater. Sci. Mater. Med. 14, 687–691.

Park, C.-K., Li, H.Y., Yeon, K.-Y., Jung, S.J., Choi, S.-Y., Lee, S.J., Lee, S., Park, K., Kim, J.S., Oh, S.B., 2006. Eugenol inhibits sodium currents in dental afferent neurons. J. Dent. Res. 85 (10), 900–904. https://doi.org/10.1177/154405910608501005.

Pathania, R., Khan, H., Kaushik, R., Khan, M.A., 2018. Essential oil nanoemulsions and their antimicrobial and food applications. Curr. Res. Nutr. Food Sci. https://doi.org/10.12944/crnfsj.6.3.05.

Peng, L.C., Liu, C.H., Kwan, C.C., Huang, K.F., 2010. Optimization of water-in-oil nanoemulsions by mixed surfactants. Colloids Surf. A Physicochem. Eng. Asp. https://doi.org/10.1016/j.colsurfa.2010.08.060.

Pereira, I., Zielińska, A., Ferreira, N.R., Souto, E.B., 2018. Optimization of linalool-loaded solid lipid nanoparticles using experimental factorial design and long-term stability studies with a new centrifugal sedimentation method. Int. J. Pharm. 549, 1–2. https://doi.org/10.1016/j.ijpharm.2018.07.068.

Pottoo, F.H., Shrestha Sharma, M., Noushad Javed, M., Barkat, A., Harshita, M., Alam, S., Naim, M.J., et al., 2020. Lipid-based nanoformulations in the treatment of neurological disorders. Drug Metab. Rev. 52 (1), 185–204. https://doi.org/10.1080/03602532.2020.1726942.

Pramod, K., Suneesh, C.V., Shanavas, S., Ansari, S.H., Ali, J., 2015. Unveiling the compatibility of eugenol with formulation excipients by systematic drug-excipient compatibility studies. J. Anal. Sci. Technol. 6 (1), 34. https://doi.org/10.1186/s40543-015-0073-2.

Puglia, C., Blasi, P., Rizza, L., Schoubben, A., Bonina, F., Rossi, C., Ricci, M., 2008. Lipid nanoparticles for prolonged topical delivery: an in vitro and in vivo investigation. Int. J. Pharm. https://doi.org/10.1016/j.ijpharm.2008.01.045.

Qiu, J., Feng, H., Jing, L., Xiang, H., Wang, D., Dong, J., Wang, J., Wang, X., Liu, J., Deng, X., 2010. Eugenol reduces the expression of virulence-related exoproteins in Staphylococcus aureus. Appl. Environ. Microbiol. 76 (17), 5846–5851. https://doi.org/10.1128/AEM.00704-10.

Radünz, M., Martins, M.L., da Trindade, T., Camargo, M., Radünz, A.L., Borges, C.D., Gandra, E.A., Helbig, E., 2019. Antimicrobial and antioxidant activity of unencapsulated and encapsulated clove (*Syzygium aromaticum*, L.) essential oil. Food Chem. 276, 180–186. https://doi.org/10.1016/j.foodchem.2018.09.173.

Ramalingam, K., Frohlich, N.C., Lee, V.A., 2013. Effect of nanoemulsion on dental unit waterline biofilm. J. Dent. Sci. 8 (3), 333–336. https://doi.org/10.1016/j.jds.2013.02.035.

Rao, J.P., Geckeler, K.E., 2011. Polymer nanoparticles: preparation techniques and size-control parameters. Prog. Polym. Sci. 36 (7), 887–913. https://doi.org/10.1016/j.progpolymsci.2011.01.001.

Rao, J., McClements, D.J., 2012. Food-grade microemulsions and nanoemulsions: role of oil phase composition on formation and stability. Food Hydrocoll. 29 (2), 326–334. https://doi.org/10.1016/j.foodhyd.2012.04.008.

Ribeiro, L.N.M., Alcântara, A.C.S., Rodrigues, G.H., Franz-montan, M., Nista, S.V.G., Castro, S.R., Couto, V.M., Guilherme, V.A., De Paula, E., 2017. Advances in Hybrid Polymer-Based Materials for Sustained Drug Release.

Rodenak-Kladniew, B., Scioli Montoto, S., Sbaraglini, M.L., Di Ianni, M., Ruiz, M.E., Talevi, A., Alvarez, V.A., Durán, N., Castro, G.R., Islan, G.A., 2019. Hybrid ofloxacin/eugenol co-loaded solid lipid nanoparticles with enhanced and targetable antimicrobial properties. Int. J. Pharm. 569 (October), 118575. https://doi.org/10.1016/j.ijpharm.2019.118575.

Sajomsang, W., Nuchuchua, O., Gonil, P., Saesoo, S., Sramala, I., Soottitantawat, A., Puttipipatkhachorn, S., Ruktanonchai, U.R., 2012. Water-soluble β-cyclodextrin grafted with chitosan and its inclusion complex as a mucoadhesive eugenol carrier. Carbohydr. Polym. 89 (2), 623–631. https://doi.org/10.1016/j.carbpol.2012.03.060.

Salvia-Trujillo, L., Rojas-Graü, A., Soliva-Fortuny, R., Martín-Belloso, O., 2015. Physicochemical characterization and antimicrobial activity of food-grade emulsions and nanoemulsions incorporating essential oils. Food Hydrocoll. https://doi.org/10.1016/j.foodhyd.2014.07.012.

Sapra, B., Jain, S., Tiwary, A.K., 2008. Percutaneous permeation enhancement by terpenes: mechanistic view. AAPS J. 10 (1), 120–132. https://doi.org/10.1208/s12248-008-9012-0.

Sarcan, E.T., Silindir-Gunay, M., Yekta Ozer, A., 2018. Theranostic polymeric nanoparticles for NIR imaging and photodynamic therapy. Int. J. Pharm. 551 (1–2), 329–338. https://doi.org/10.1016/j.ijpharm.2018.09.019.

Schäfer-Korting, M., Mehnert, W., Korting, H.C., 2007. Lipid nanoparticles for improved topical application of drugs for skin diseases. Adv. Drug Deliv. Rev. https://doi.org/10.1016/j.addr.2007.04.006.

Schuenck-Rodrigues, R.A., Betzler De Oliveira De Siqueira, L., dos Santos Matos, A.P., da Costa, S.P., da Silva Cardoso, V., Vermelho, A.B., Colombo, A.P.V., Oliveira, C.A., Santos-Oliveira, R., Ricci-Júnior, E., 2020. Development, characterization and photobiological activity of nanoemulsion containing zinc phthalocyanine for oral infections treatment. J. Photochem. Photobiol. B Biol. 211, 112010. https://doi.org/10.1016/j.jphotobiol.2020.112010.

Sebaaly, C., Jraij, A., Fessi, H., Charcosset, C., Greige-Gerges, H., 2015. Preparation and characterization of clove essential oil-loaded liposomes. Food Chem. 178, 52–62. https://doi.org/10.1016/j.foodchem.2015.01.067.

Sebaaly, C., Charcosset, C., Stainmesse, S., Fessi, H., Greige-Gerges, H., 2016. Clove essential oil-in-cyclodextrin-in-liposomes in the aqueous and lyophilized states: from laboratory to large scale using a membrane contactor. Carbohydr. Polym. 138, 75–85. https://doi.org/10.1016/j.carbpol.2015.11.053.

Sell, A.B., Carlini, E.A., 1976. Anesthetic action of methyleugenol and other eugenol derivatives. Pharmacology 14 (4), 367–377. https://doi.org/10.1159/000136617.

Sengupta, P., Chatterjee, B., 2017. Potential and future scope of nanoemulgel formulation for topical delivery of lipophilic drugs. Int. J. Pharm. 526 (1–2), 353–365. https://doi.org/10.1016/j.ijpharm.2017.04.068.

Senna, J.P., Barradas, T.N., Cardoso, S.A., Castiglione, T.C., Serpe, M.J, Silva, K.G.H., Mansur, C.R.E., 2018. Dual alginate-lipid nanocarriers as oral delivery systems for amphotericin B. Colloids Surf. B 166 (1), 187–194. https://doi.org/10.1016/j.colsurfb.2018.03.015.

Senna, J.P., Ricci-Junior, E., Mansur, C.R.E., 2015. Development and evaluation of nanoemulsions containing phthalocyanines for use in photodynamic cancer therapy. J. Nanosci. Nanotechnol. 15 (6), 4205–4214. https://doi.org/10.1166/jnn.2015.9609.

Severino, P., Andreani, T., Macedo, A.S., Fangueiro, J.F., Santana, M.H.A., Silva, A.M., Souto, E.B., 2012. Current state-of-art and new trends on lipid nanoparticles (SLN and NLC) for oral drug delivery. J. Drug Deliv. 2012 (November), 1–10. https://doi.org/10.1155/2012/750891.

Shahavi, M.H., Hosseini, M., Jahanshahi, M., Meyer, R.L., Darzi, G.N., 2016. Clove oil nanoemulsion as an effective antibacterial agent: Taguchi optimization method. Desalin. Water Treat. 57 (39), 18379–18390. https://doi.org/10.1080/19443994.2015.1092893.

Shahavi, M.H., Hosseini, M., Jahanshahi, M., Meyer, R.L., Darzi, G.N., 2019. Evaluation of critical parameters for preparation of stable clove oil nanoemulsion. Arab. J. Chem. 12 (8), 3225–3230. https://doi.org/10.1016/j.arabjc.2015.08.024.

Shapiro, R., 2012. Prevention of vector transmitted diseases with clove oil insect repellent. J. Pediatr. Nurs. 27 (4), 346–349. https://doi.org/10.1016/j.pedn.2011.03.011.

Shen, Q., Li, W., Li, W., 2007. The effect of clove oil on the transdermal delivery of ibuprofen in the rabbit by in vitro and in vivo methods. Drug Dev. Ind. Pharm. 33 (12), 1369–1374. https://doi.org/10.1080/03639040701399346.

Shetty, S., Jose, J., Kumar, L., Narayana, R., Charyulu., 2019. Novel ethosomal gel of clove oil for the treatment of cutaneous candidiasis. J. Cosmet. Dermatol. 18 (3), 862–869. https://doi.org/10.1111/jocd.12765.

Shirodkar, R.K., Kumar, L., Mutalik, S., Lewis, S., 2019. Solid lipid nanoparticles and nanostructured lipid carriers: emerging lipid based drug delivery systems. Pharm. Chem. J. 53 (5), 440–453. https://doi.org/10.1007/s11094-019-02017-9.

Shukla, P.K., Singh, P., Yadav, R.K., Pandey, S., Bhunia, S.S., 2016. Past, Present, and Future of Antifungal Drug Development. Communicable Diseases of the Developing World. 29 Springer, pp. 125–167.

Singh, Y., Meher, J.G., Raval, K., Khan, F.A., Chaurasia, M., Jain, N.K., Chourasia, M.K., 2017. Nanoemulsion: concepts, development and applications in drug delivery. J. Control. Release 252, 28–49. https://doi.org/10.1016/j.jconrel.2017.03.008.

Sonneville-Aubrun, O., Simonnet, J.T., L'Alloret, F., 2004. Nanoemulsions: a new vehicle for skincare products. Adv. Colloid Interf. Sci. https://doi.org/10.1016/j.cis.2003.10.026.

Souto, E.B., Baldim, I., Oliveira, W.P., Rao, R., Yadav, N., Gama, F.M., Mahant, S., 2020. SLN and NLC for topical, dermal, and transdermal drug delivery. Expert Opin. Drug Deliv. 17 (3), 357–377. https://doi.org/10.1080/17425247.2020.1727883.

Sridhar, R., Lakshminarayanan, R., Madhaiyan, K., Barathi, V.A., Limh, K.H.C., Ramakrishna, S., 2015. Electrosprayed nanoparticles and electrospun nanofibers based on natural materials: applications in tissue regeneration, drug delivery and pharmaceuticals. Chem. Soc. Rev. 44 (3), 790–814. https://doi.org/10.1039/c4cs00226a.

Su, K., Wang, C., 2015. Recent advances in the use of gelatin in biomedical research. Biotechnol. Lett. 37, 2139–2145. https://doi.org/10.1007/s10529-015-1907-0.

Sundaray, B., Subramanian, V., Natarajan, T.S., Xiang, R.Z., Chang, C.C., Fann, W.S., 2004. Electrospinning of continuous aligned polymer fibers. Appl. Phys. Lett. 84 (7), 1222–1224. https://doi.org/10.1063/1.1647685.

Sutradhar, K.B., Amin, L., 2013. Nanoemulsions: increasing possibilities in drug delivery. Eur. J. Nanomed. 5 (2), 97–110. https://doi.org/10.1515/ejnm-2013-0001.

Szweda, P., Gucwa, K., Kurzyk, E., Romanowska, E., Dzierżanowska-Fangrat, K., Jurek, A.Z., Kuś, P.M., Milewski, S., 2015. Essential oils, silver nanoparticles and propolis as alternative agents against fluconazole resistant *Candida albicans*, *Candida glabrata* and *Candida krusei* clinical isolates. Indian J. Microbiol. 55 (2), 175–183. https://doi.org/10.1007/s12088-014-0508-2.

Tanzi, D., Celine, M.A., Vian, C.G., Elmaataoui, M., Chemat, F., 2012. Terpenes as green solvents for extraction of oil from microalgae. Molecules 17 (7), 8196–8205. https://doi.org/10.3390/molecules17078196.

Taylor-Pashow, K.M.L., Rocca, J.D., Huxford, R.C., Lin, W., 2010. Hybrid nanomaterials for biomedical applications. Chem. Commun. 46 (32), 5832. https://doi.org/10.1039/c002073g.

Teixeira, M.C., Carbone, C., Souto, E.B., 2017. Beyond liposomes: recent advances on lipid based nanostructures for poorly soluble/poorly permeable drug delivery. Prog. Lipid Res. 68, 1–11. https://doi.org/10.1016/j.plipres.2017.07.001.

Thakkar, S., Misra, M., 2017. Electrospun polymeric nanofibers: new horizons in drug delivery. Eur. J. Pharm. Sci. 107, 148–167. https://doi.org/10.1016/j.ejps.2017.07.001.

Thonggoom, O., Punrattanasin, N., Srisawang, N., Promawan, N., Thonggoom, R., 2016. In vitro controlled release of clove essential oil in self-assembly of amphiphilic polyethylene glycol-block-polycaprolactone. J. Microencapsul. 33 (3), 239–248. https://doi.org/10.3109/02652048.2016.1156173.

Thosar, N., Basak, S., Bahadure, R.N., Rajurkar, M., 2013. Antimicrobial efficacy of five essential oils against oral pathogens: an in vitro study. Eur. J. Dent. 07 (S 01), S071–S077. https://doi.org/10.4103/1305-7456.119078.

Trongtokit, Y., Rongsriyam, Y., Komalamisra, N., Apiwathnasorn, C., 2005. Comparative repellency of 38 essential oils against mosquito bites. Phytother. Res. 19 (4), 303–309. https://doi.org/10.1002/ptr.1637.

Tucker, N., Stanger, J.J., Staiger, M.P., Razzaq, H., Hofman, K., 2012. The history of the science and technology of electrospinning from 1600 to 1995. J. Eng. Fibers Fabr. 7 (3), 63–73. https://doi.org/10.1177/155892501200702s10.

Tyler, B., Gullotti, D., Mangraviti, A., Utsuki, T., Brem, H., 2016. Polylactic acid (PLA) controlled delivery carriers for biomedical applications. Adv. Drug Deliv. Rev. 107, 163–175. https://doi.org/10.1016/j.addr.2016.06.018.

Uma Maheswari, P., Rhea Muthappa, K., Bindhya, P., Begum, K.M.M.S., 2021. Evaluation of folic acid functionalized BSA-CaFe2O4 nanohybrid carrier for the controlled delivery of natural cytotoxic drugs hesperidin and eugenol. J. Drug Delivery Sci. Technol. 61, 102105. https://doi.org/10.1016/j.jddst.2020.102105.

Umerska, A., Gaucher, C., Oyarzun-Ampuero, F., Fries-Raeth, I., Colin, F., Villamizar-Sarmiento, M.G., Maincent, P., Sapin-Minet, A., 2018. Polymeric nanoparticles for increasing oral bioavailability of curcumin. Antioxidants 7 (4), 1–18. https://doi.org/10.3390/antiox7040046.

Unalan, I., Endlein, S.J., Slavik, B., Buettner, A., Goldmann, W.H., Detsch, R., Boccaccini, A.R., 2019. Evaluation of electrospun poly(ε-caprolactone)/gelatin nanofiber mats containing clove essential oil for antibacterial wound dressing. Pharmaceutics 11 (11), 570. https://doi.org/10.3390/pharmaceutics11110570.

Valmikinathan, C.M., Defroda, S., Yu, X., 2009. Polycaprolactone and bovine serum albumin based nanofibers for controlled release of nerve growth factor. Biomacromolecules 10 (5), 1084–1089. https://doi.org/10.1021/bm8012499.

Wan, J., Zhong, S., Schwarz, P., Chen, B., Rao, J., 2019. Physical properties, antifungal and mycotoxin inhibitory activities of five essential oil nanoemulsions: impact of oil compositions and processing parameters. Food Chem. 291, 199–206. https://doi.org/10.1016/j.foodchem.2019.04.032.

Wang, Y., Thakur, R., Fan, Q., Michniak, B., 2005. Transdermal iontophoresis: combination strategies to improve transdermal iontophoretic drug delivery. Eur. J. Pharm. Biopharm. 60 (2), 179–191. https://doi.org/10.1016/j.ejpb.2004.12.008.

Wang, X., Zhang, K., Zhu, M., Hao, Y., Zhou, Z., Chen, Y., Hsiao, B.S., 2008. Continuous polymer nanofiber yarns prepared by self-bundling electrospinning method. Polymer 49 (11), 2755–2761. https://doi.org/10.1016/j.polymer.2008.04.015.

Wang, Z.-J., Tabakoff, B., Levinson, S.R., Heinbockel, T., 2015. Inhibition of Nav1.7 channels by methyl eugenol as a mechanism underlying its antinociceptive and anesthetic actions. Acta Pharmacol. Sin. 36 (7), 791–799. https://doi.org/10.1038/aps.2015.26.

Wong, K.H., Aiping, L., Chen, X., Yang, Z., 2020. Natural ingredient-based polymeric nanoparticles for cancer treatment. Molecules 25 (16), 1–24. https://doi.org/10.3390/molecules25163620.

Woranuch, S., Yoksan, R., 2013. Eugenol-loaded chitosan nanoparticles: I. Thermal stability improvement of eugenol through encapsulation. Carbohydr. Polym. 96 (2), 578–585. https://doi.org/10.1016/j.carbpol.2012.08.117.

Xu, H., Zhang, D., Li, J., 2019. Antibacterial nanoparticles with universal adhesion function based on dopamine and eugenol. J. Bioresour. Bioprod. 4 (3), 177–182. https://doi.org/10.12162/jbb.v4i3.006.

Yadav, M.K., Chae, S.-W., Im, G.J., Chung, J.-W., Song, J.-J., 2015. Eugenol: a phyto-compound effective against methicillin-resistant and methicillin-sensitive Staphylococcus aureus clinical strain biofilms. Edited by Christophe Beloin. PLoS One 10 (3), e0119564. https://doi.org/10.1371/journal.pone.0119564.

Yalkowsky, S.H., He, Y., Jain, P., 2010. Handbook of Aqueous Solubility Data, second ed. vol. 687 CRC Press, Boca Raton, FL.

Yoo, C.-B., Han, K.-T., Cho, K.-S., Ha, J., Park, H.-J., Nam, J.-H., Kil, U.-H., Lee, K.-T., 2005. Eugenol isolated from the essential oil of Eugenia caryophyllata induces a reactive oxygen species-mediated apoptosis in HL-60 human promyelocytic leukemia cells. Cancer Lett. 225 (1), 41–52. https://doi.org/10.1016/j.canlet.2004.11.018.

Yukuyama, M.N., Ghisleni, D.D.M., Pinto, T.J.A., Bou-Chacra, N.A., 2016. Nanoemulsion: process selection and application in cosmetics—a review. Int. J. Cosmet. Sci. https://doi.org/10.1111/ics.12260.

Yunes, R.A., Filho, V.C., Ferreira, J., Calixto, J.B., 2005. The Use of Natural Products as Sources of New Analgesic Drugs. Studies in Natural Products Chemistry. 30 Elsevier, pp. 191–212.

Zhan, H., Jiang, Z.-T., Wang, Y., Li, R., Dong, T.-S., 2008. Molecular microcapsules and inclusion interactions of eugenol with β-cyclodextrin and its derivatives. Eur. Food Res. Technol. 227 (5), 1507–1513. https://doi.org/10.1007/s00217-008-0873-3.

Zhang, Y., Zhang, H., Zhang, K., Li, Z., Guo, T., Wu, T., Hou, X., Feng, N., 2020. Co-hybridized composite nanovesicles for enhanced transdermal eugenol and cinnamaldehyde delivery and their potential efficacy in ulcerative colitis. Nanomed. Nanotechnol. Biol. Med. 28, 102212. https://doi.org/10.1016/j.nano.2020.102212.

CHAPTER

22

Extraction of bioactive compounds from clove (*Syzygium aromaticum*)

Muhammad Kamran Khan[a], Sadia Hassan[b], Muhammad Imran[a], and Muhammad Haseeb Ahmad[a]

Department of Food Science, Faculty of Life Sciences, Government College University, Faisalabad, Punjab, Pakistan[a]

Department of Nutritional Sciences, Faculty of Science and Technology, Government College Women University, Faisalabad, Pakistan[b]

1. Introduction

Cloves, *Syzygium aromaticum*, are the aromatic flower buds of a tree, belonging to the family Myrtaceae. The top producing countries include Indonesia with total production of 135 metric tonnes followed by Madagascar, Tanzania, Comoros, and Sri Lanka (FAO, 2019). Cloves are native to Indonesia and are commonly used as a spice particularly in Asian countries. Cloves are also known for their health benefits and constitute a major part of traditional medicines especially to treat dental issues. Their pharmacological activities are attributed to their rich phytochemical composition which includes monoterpenes, sesquiterpenes, alkaloids, and phenolic compounds (Table 1; Batiha et al., 2020). Eugenol (70%–85%), eugenyl acetate (15%), and β-caryophyllene (5.12%) are the characteristic compounds of the essential oil present in clove buds and various biological activities like anticancer, antioxidant, analgesic, antidepressant, antiseptic, anti-inflammatory, antispasmodic, antiviral, antifungal, and antibacterial activities are correlated to the presence of these compounds (Kaur and Kaushal, 2019). The major bioactive molecules present in the clove buds include phenolic molecules like flavonoids (quercetin and kaempferol), phenolic acids (hydroxybenzoic acids, hydroxyphenyl propens, hydroxycinamic acids), gallic acid derivatives, and eugenol. The activity of these compounds mainly depends on the method of extraction used, hence their concentrations in the extract.

2. Extraction of bioactive compounds

To detect and purify a range of bioactive compounds from different plant sources, the extraction process is considered an important phase. For the extraction of bioactive compounds including flavonoids and phenolics, the selection of suitable analytical technique is very important and it depends on the study purpose along with the nature of both analyte and sample. Any extraction technique must be focused on the goal to provide the maximum amount of targeted bioactive compound in the extract, as

Clove (*Syzygium aromaticum*). https://doi.org/10.1016/B978-0-323-85177-0.00010-0

Table 1 Different bioactive compounds isolated from *S. aromaticum* along with their chemical formula and names

Chemical compound	Chemical formula	Chemical structure	Chemical compound	Chemical formula	Chemical structure
Eugenol	$C_{10}H_{12}O_2$	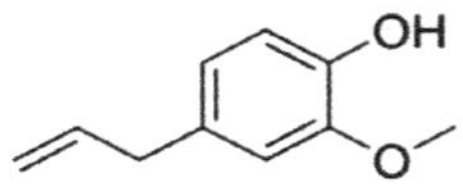	Gallic acid	$C_7H_6O_5$	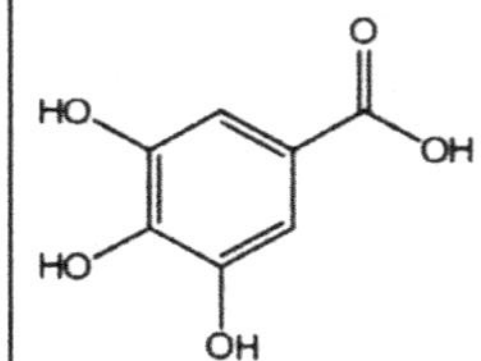
β-Caryophyllene	$C_{15}H_{24}$	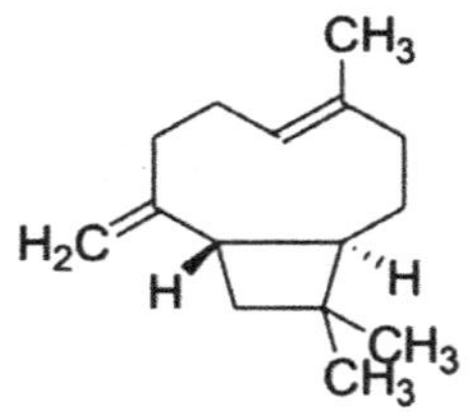	Biflorin	$C_{16}H_{18}O_9$	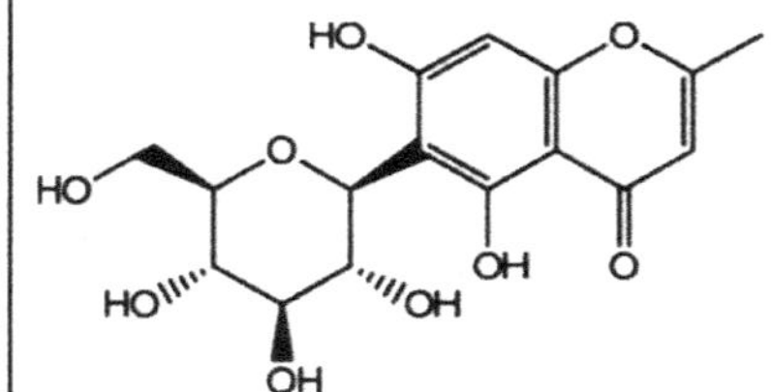
Vanillin	$C_8H_8O_3$	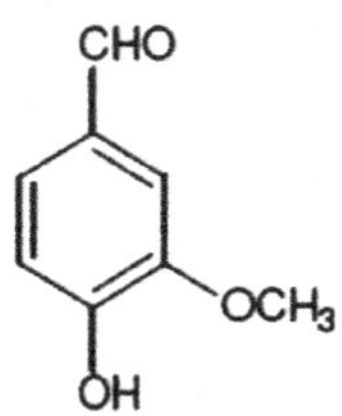	Myricetin	$C_{15}H_{10}O_8$	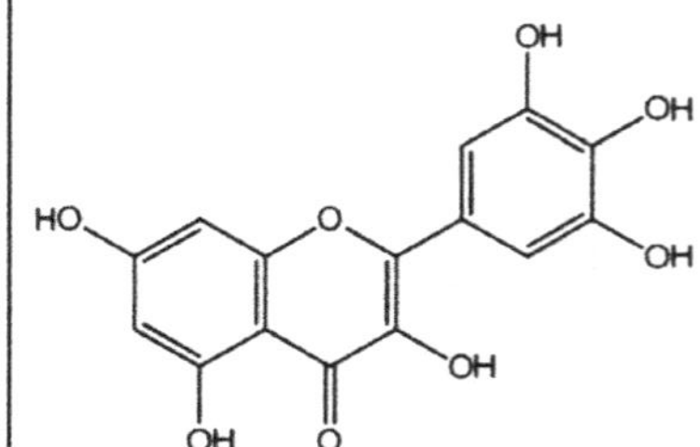
Crategolic acid	$C_{30}H_{48}O_4$	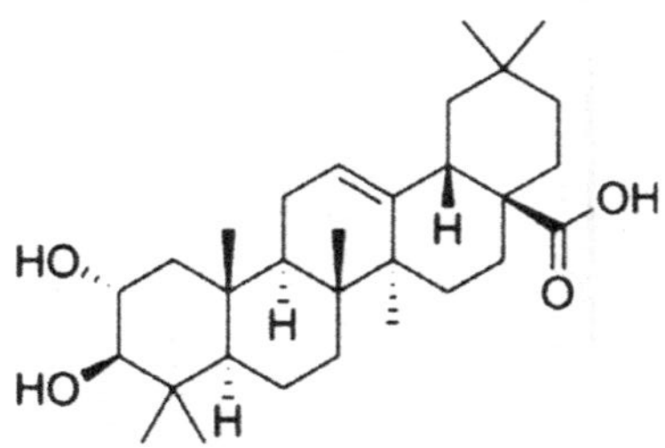	Campesterol	$C_{28}H_{48}O$	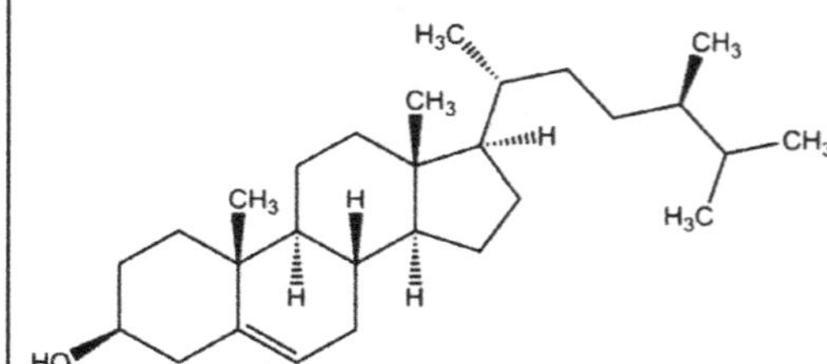

Kaempferol	$C_{15}H_{10}O_6$	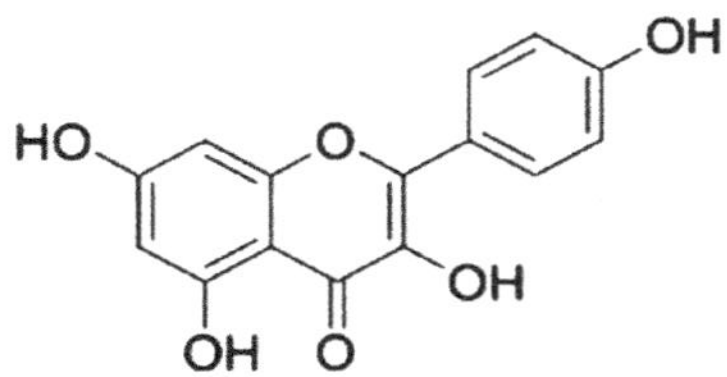	Stigmasterol	$C_{29}H_{48}O$	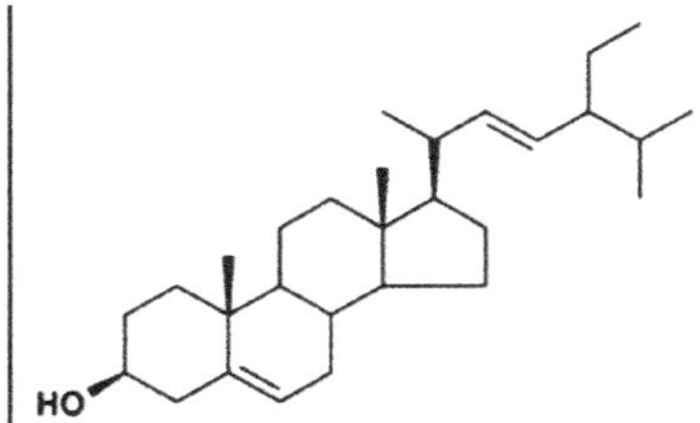
Rhamnetin	$C_{16}H_{12}O_7$	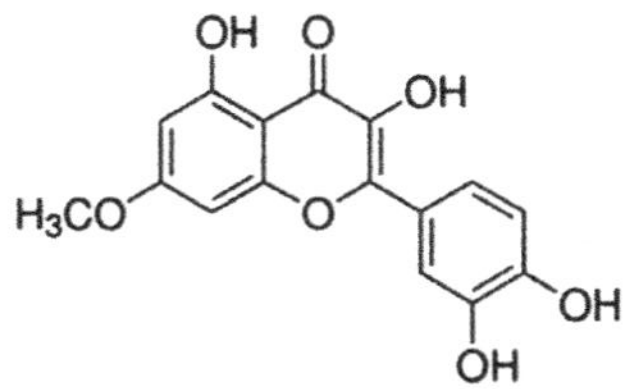	Oleanolic acid	$C_{30}H_{48}O_3$	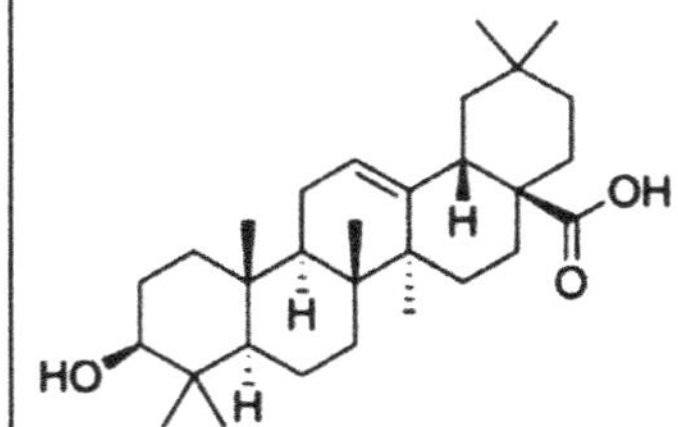
Eugenitin	$C_{12}H_{12}O_4$	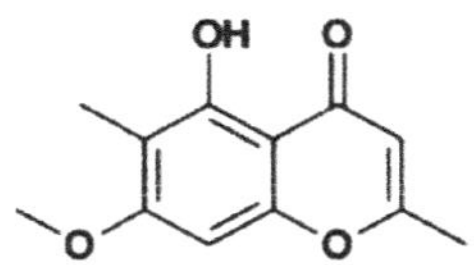	Bicornin	$C_{48}H_{32}O_{30}$	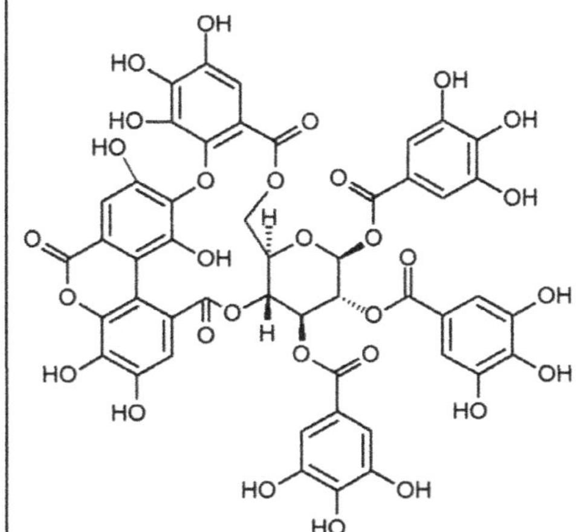

Continued

Table 1 Different bioactive compounds isolated from *S. aromaticum* along with their chemical formula and names—cont'd

Chemical compound	Chemical formula	Chemical structure	Chemical compound	Chemical formula	Chemical structure
Eugenin	$C_{11}H_{10}O_4$	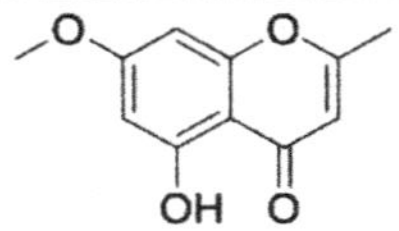	Quercetin	$C_{15}H_{10}O_7$	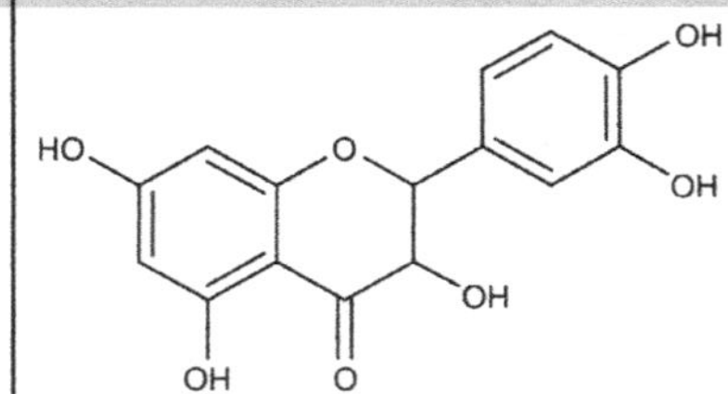
Ellagic acid	$C_{14}H_6O_8$	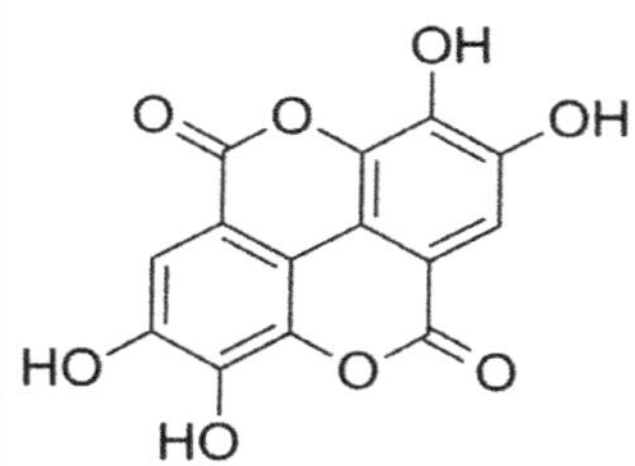	Carvacrol	$C_{10}H_{14}O$	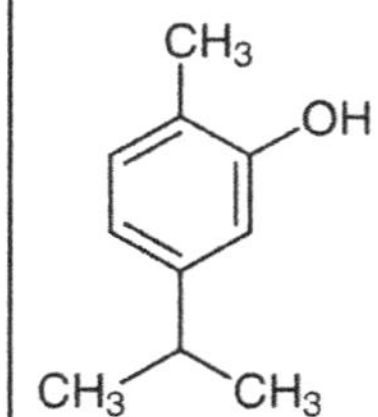

From: *Batiha, G.E.-S., Alkazmi, L.M., Wasef, L.G., Beshbishy, A.M., Nadwa, E.H., Rashwan, E.K., 2020. Syzygium aromaticum L. (Myrtaceae): traditional uses, bioactive chemical constituents, pharmacological and toxicological activities. Biomolecules, 10(2), https://doi.org/10.3390/biom10020202; Cortés-Rojas, D.F., de Souza, C.R.F., Oliveira, W.P., 2014. Clove (Syzygium aromaticum): a precious spice. Asian Pac. J. Trop. Biomed. 4(2), 90–96. https://doi.org/10.1016/S2221-1691(14)60215-X; Neveu, V., Perez-Jiménez, J., Vos, F., Crespy, V., du Chaffaut, L., Mennen, L., Knox, C., Eisner, R., Cruz, J., Wishart, D., Scalbert, A., 2010. Phenol-Explorer: an online comprehensive database on polyphenol contents in foods. Database 2010.*

well as a minimum level of contamination with different undesirable constituents (Barba et al., 2017). Therefore, different extraction processes are used for this purpose.

For the recovery of bioactive compounds from natural products, some of the major factors are the extraction method and extraction solvent. The most commonly used extraction solvents are acetone, alcohols, water, or their mixtures, and these solvents influence both the antioxidant activity of extracts and the yield of extraction. The diverse composition and structure of the plant materials along with the chemical properties of extraction solvents influence and differentiate the behavior of material-solvent systems. Therefore, in a single plant sample, a single solvent cannot extract all the antioxidants having different solubility and polarity (Adaramola and Onigbinde, 2016). Hence, the selection of appropriate solvent as well as extraction technique is an important feature to attain the maximum concentration of desired bioactive constituent from plant extracts.

The extraction of bioactive constituents from various sources mainly depends on the efficiency and productivity of the extraction processes (Azmir et al., 2013). Both conventional and novel extraction methods have been used to extract the primary and secondary plant metabolites from natural food/agricultural sources (Mehmood et al., 2019; Moreira et al., 2019; Sharayei et al., 2019). The oldest method for nonvolatile compounds was solid-liquid extraction, commonly known as maceration, in which a plant sample is brought into contact with a suitable solvent (Lefebvre et al., 2021). Various modern extraction methods have been developed with the use of "assisted" techniques (i.e., microwaves, pressure, and ultrasonic waves) to enhance the performance of extraction. These assisted technologies could reduce process duration, induce energy, and save solvents through different intensification mechanisms. These physical extraction methods could influence the extraction process probably by improving the extraction yields and inducing extraction selectivities as compared to the simple maceration method (Tirado et al., 2019). Conventional extraction methods using organic solvents have been in practice including maceration, hydrodistillation, and solid-liquid extraction (e.g., Soxhlet extraction) (Giacometti et al., 2018). However, the conventional techniques have several limitations like time-consuming, need high purity, and costly solvent requires a large volume of extraction solvents, and have low extraction selectivity (Aires, 2017). Based on the disadvantages of conventional methods, it is required to develop extraction techniques that are more eco-friendly, efficient, and sustainable. To overcome the limitations caused by conventional techniques, novel extraction methods have been offered (Putnik et al., 2018). Novel assisted extraction methods include microwave-assisted (Rodsamran and Sothornvit, 2019), ultrasound-assisted, enzyme-assisted (Marathe et al., 2019), pulsed electric field-assisted (Redondo et al., 2018), pressurized liquid (Pereira et al., 2019) and supercritical fluid (Pimentel-Moral et al., 2019) methods. In the food industry, all these methods are of great interest and through various studies, researchers have addressed that the application of novel extraction methods in combination can be an efficient approach for effective and fast extraction (Aires, 2017; Chemat et al., 2017).

2.1 Solvent extraction (maceration)

In medicinal plant research, maceration is the most widely used conventional extraction technique. This method involves the soaking of powder sample with extraction solvent in a container and keeping at room temperature with frequent agitation for a minimum of 3 days followed by filtration (Handa et al., 2008). The purpose of the soaking process is to release the soluble phytochemicals by softening and breaking the plant's cell wall. In this method, heat is transferred through conduction and convection

(Azwanida, 2015). Maceration is the simplest and easiest method. However, the main issue is an organic waste as a large solvent volume is used. Thus, proper waste management is needed. Alteration in extraction temperature and extraction solvents improves the efficiency of the extraction process, and these alterations can be used in the maceration technique (Yung et al., 2010).

Extraction of clove bud using 80% acetone extract resulted in the highest extraction yield with maximum values of TPC (200 mg GAE/g), TFC (501 mg QE/g), DPPH activity (73.6%), and reducing power (0.88). TPC and TFC were best extracted with aqueous semipolar organic solvent (acetone) as compared to most polar solvent (water) (Adaramola and Onigbinde, 2016). Ghallab et al. (2019) extracted clove buds by using absolute ethanol, ethyl acetate, *n*-hexane, and water having different polarities. Results exhibited that TPC contents were maximum for ethanol (351.8 mg GAE/g) and hexane (348 mg GAE/g) extract. Similarly, regarding scavenging activities, ethanol (0.41 mg/mL) and hexane (0.37 mg/mL) extracts also gave the highest values EC_{50}. Acetone extract of clove bud showed highest TFC (29.7 mg CE/g), while water extracts gave highest TPC (140.2 mg GAE/g), DPPH activity (IC50: 89.9–118.5 μg/mL), and antioxidant activity (84.17%) (Singh et al., 2016). In another study, ethyl acetate, ethanol (80%), and water were used for extraction, and results recommended water as the best solvent for extracting flavonoids (17.5 mg QE/g). In contrast, ethanol and water extract gave the highest value of TPC (ca. 230 mg GAE/g). The DPPH and ABTS activities varied from 25.3% to 91.4% and 49.4 to 99.4%, respectively. Ethanol exhibited better results than water for TPC and TFC extractions. It could be concluded that for clove extraction, solvents with high polarity are more effective as compared to those having low polarity (El-Maati et al., 2016). El-Refai et al. (2020) performed extraction from clove bud by using ethanol (95%) exhibited the maximum TPC (372.2 mg GAE/g), TFC (177.1 mg QE/g), and reducing power 1.031.

2.2 Soxhlet extraction

In Soxhlet extraction, a thimble containing a finely powdered sample is placed in the thimble chamber of the Soxhlet apparatus. In the bottom flask, the extraction solvent is heated, which vaporizes into a thimble and further condenses in the condenser, and finally, drips back. Once the liquid level reaches the siphon arm, it is again emptied into the bottom flask, and in the same manner, the process is continued. As compared to the maceration method, this technique works with a small amount of extraction solvent (Handa et al., 2008). On the other hand, this method also has a disadvantage like contact with harmful and flammable liquid organic solvents during extraction having potential toxic emissions. Solvents of high purity should be used in this extraction system which might be costly. As compared to other advanced extraction methods like supercritical fluid extraction, the Soxhlet method can induce pollution issues; therefore, it is not an environmentally friendly procedure (Naudé et al., 1998). Ideally, the sample should be dry and finely powdered as well as a sample-to-solvent ratio, agitation speed, and temperature should also be considered during extraction (Amid et al., 2010).

Singh et al. (2016) explored the effect of different extraction methods (centrifugation, mechanical shaking, Soxhlet extraction, and refluxing) on clove buds regarding phytochemicals and antioxidant activities using acetone, ethyl alcohol, and distilled water as extraction solvents. The extract obtained through Soxhlet extraction exhibited the highest value of TPC (127 mg GAE/g), TFC (26.7 mg CE/g), DPPH activity (range of IC50: 89.9–125 μg/mL), and antioxidant activity (85.8%). In another study, the essential oil was extracted from clove by using *n*-hexane as an extraction solvent. The analysis of essential oil showed its maximum yield of 54.7% as well as 5.72% eugenyl acetate was found in the extract (Yunusa et al., 2019). Quantitative analysis of clove extract obtained using 70% methanol gave the 7.70 g of eugenol in 100 g of sample along with the presence of tannin, saponins, flavonoids, terpenoids, and alkaloids in clove plant (Amin et al., 2013). Inam et al. (2014) also addressed the similar trend for the presence of eugenol in methanolic extract of clove bud powder. Clove samples were extracted with three solvents (petroleum ether, chloroform, and ethanol) using a Soxhlet method. Qualitative analysis of all extracts showed alkaloids, flavonoids, steroids, terpenoids, carbohydrates, protein, and phenolic compounds. The results exhibited the maximum percentage of phenol, 2-methoxy-3-(2-propenyl) (64.4%), followed by eugenol (14.9%) (Gowri and Manimegalai, 2019).

2.3 Ultrasound-assisted extraction

Ultrasound-assisted extraction is based on the application of ultrasound within the range of 20–2000 kHz (Handa et al., 2008). The cavitation effect induced by ultrasound waves is the main mechanical process that improves the permeability of the cell wall, enhancing the surface contact between sample and extraction solvents. Ultrasound waves induce the disruption of plant cell walls, thus changing the chemical and physical characteristics of plant materials. As a result, the whole process increases the mass transport of the extraction solvents into the sample plant cells, thus facilitating the release of chemical constituents (Dhanani et al., 2017). This technology is simple and low cost which can be used for phytochemicals extraction at both small and large scales. The main benefits of this extraction method are reduction in solvent consumption and extraction time (Kaufmann and Christen, 2002). A study was conducted to extract the essential oil from clove using *n*-hexane as an extraction medium through an ultrasound-assisted extraction method. This extraction was conducted with a power level of 100 W and 500 W at 20 min each. Quantitative analysis of samples showed that the yields of essential oil and eugenyl acetate were 71.5% and 9.24%, respectively (Yunusa et al., 2019).

Alexandru et al. (2013) assessed the efficiency of ultrasonic reactors, both batch and flow, regarding the extraction of dry clove bud samples. Extraction was conducted in a multi-horn reactor working in continuous flow mode (450, 900, 1350 mL/min). Power density for both flow and batch reactors was 700 and 18 W/L, respectively. While the specific energy input was 47.2 and 48.6 kJ/L for flow and batch reactors, respectively. The best results of maximum TPC and radical scavenging activity were found in underflow reactor operating at flow mode of 1350 mL/min.

Tekin et al. (2015) extracted the clove essential oils by using ethanol at three independent variables, including plant sample concentration (3% to 7%), extraction temperature (32–52°C), and extraction time (30 to 60 min). All extraction was carried out at a frequency of 53 kHz. Analysis of clove extract showed eugenol, α-caryophyllene, and 2-methoxy-4-(2-propenyl) phenol acetate. The maximum yield of clove extract was 28.2% obtained from 5% plant concentration at 60°C temperature, 45 min.

Wei et al. (2016b) extracted oil and α-humulene from clove by using ultrasound-assisted supercritical carbon dioxide (USC-CO_2) extraction technique at different pressures (9.0–25.0 MPa) and temperature (32–50°C) as compared to other methods like steam distillation and heat reflux extraction. Results exhibited the better extracted yield of oil and α-humulene from clove within a short extraction duration. Additionally, Wei et al. (2016a) stated that as compared to conventional extraction methods (steam distillation and heat reflux), the USC-CO_2 extraction method produced higher amounts of clove oil. The best conditions for maximum oil yield were 50°C temperature, 28.5 MPa pressure, 15 min extraction time, 0.355 mm sample particle size, a CO_2 flow rate of 0.35 g/min, 40 kHz ultrasonic frequency, power 185 W, and 75% ultrasound cycle.

2.4 Microwave-assisted extraction

Microwave-assisted extraction works using microwave energy lead to the partition of analytes from sample matrix to extraction solvent (Trusheva et al., 2007). In this method, microwave radiations interact with the sample and solvents cause their heating and heat will be transferred through the conduction process. Microwave radiations induce the molecular dipole rotation, which results in breakdown of the hydrogen bonding, enhances the movement of dissolved ions, and thus stimulates the extraction solvents penetration into sample material (Kaufmann and Christen, 2002). In the case of nonpolar solvents, energy is transferred only through dielectric absorption. Therefore, heating is not proper (Handa et al., 2008). Compared to conventional extraction techniques like maceration or Soxhlet extraction, this method works with less volume of extraction solvents and extraction time. This method has better reproducibility and analyte recoveries, but at the same time to prevent thermal degradation, it is essential to work under proper conditions (Kaufmann and Christen, 2002). On the other hand, this method is unsuitable for smaller molecules of phenolic compounds like phenolic acids, isoflavin, quercetin, and trans-resveratrol. Higher process conditions could cause oxidation of compounds leading to extreme yield reduction of these compounds. Similarly, at high temperature, anthocyanins and tannins could be possibly subjected to degradation (Trusheva et al., 2007). Kennouche et al. (2015) analyzed comparatively three methods by using distilled water to extract volatile oils from cloves, including microwave-assisted extraction, conventional hydrodistillation, and microwave steam distillation. According to results, microwave-assisted extraction (850 W for 130 min) and microwave steam distillation (500 W for 10 min) produced more valuable essential oils with high eugenol levels. Similarly, Gonzalez-Rivera et al. used traditional hydrodistillation and microwave hydrodistillation to extract essential oils from different herbs, including clove, fennel, lavender, rosemary, and sage. Results exhibited that eugenol and the essential oil extracted through microwave hydrodistillation gave more thermal stability. Moreover, as compared to the conventional extraction method, extraction time was quite less (González-Rivera et al., 2016). Haqqyana et al. (2020) used microwave hydrodistillation extraction of the essential oil from clove stem by using aquadest heated at different power levels of 300, 450, 600, and 800 W. Results concluded that an increase in the levels of irradiation power could influence the reaction rate leading to a higher extraction yield within the short time duration (Haqqyana et al., 2020).

2.5 Supercritical fluid extraction

A supercritical fluid is also known as dense gas and, at its critical point, shares the physical characteristics of both gas and liquid. Temperature and pressure are the two factors that put the substances into their critical regions. Supercritical fluids perform mostly like gas but also have the solvating feature like a liquid. The most common example of supercritical fluid is CO_2 that becomes SF at above 7380 kPa pressure and 31.1°C temperature. Increasing interest in supercritical CO_2 extraction is due to being an excellent extraction solvent for nonpolar compounds, as well as the easy availability of CO_2 at less cost with less toxicity level (Azwanida, 2015). Although for polar compounds, SC-CO_2 shows less solubility, the addition of ethanol and methanol in some amount permits the extraction of polar compounds. During this extraction method, analytes are produced in concentrated form because at room temperature CO_2 vaporizes. The strength of solvents used in this method can be easily changed by altering the pressure, temperature, and modifiers. All these practices, as a result, reduce the time for extraction. The main disadvantage of this technique is the higher initial cost of the equipment (Naudé et al., 1998).

Chatterjee and Bhattacharjee (2013) extracted the eugenol from clove buds using optimized parameters including 60°C temperature, 250 bar pressure, and 2 L/min CO_2 flow rate in 90 min extraction process. This process produced the eugenol amount of 129.8 mg/g clove powder. In another study, essential oils were extracted (14.4%) from cloves at 50°C temperature and 300 bar pressure. Quantitative analysis of essential oil demonstrated the presence of α-humulene, E-caryophyllene, eugenol, and eugenol acetate. Above 99% of volatile compounds were recognized in essential oil, including oxygenated monoterpenes (59.7%), phenolic compounds (19.7%), sesquiterpene hydrocarbons (19.7%), oxygenated sesquiterpenes (0.59%), and other compounds (0.055%) (de Oliveira et al., 2016). Gopalakrishnan et al. (1990) stated a higher extraction rate of clove oil through SC-CO_2 extraction as compared to liquid extraction. According to Reverchon and Marrone (1997), extracted yield of eugenol can be maximized by increasing the CO_2 flow rate in supercritical extraction. Moreover, Guan et al. (2007) comparatively studied the composition of clove bud oil extracted through SC-CO_2 and conventional methods (solvent extraction, steam distillation, molecular distillation), and the results showed a similar composition for the SC-CO_2 method as obtained by traditional techniques.

3. Applications of clove extracts

The scientific exploration of natural extracts for their nutritional and pharmacological benefits to humans and the harmful effects of synthetic additives have urged the industrial community to use natural extracts for food, cosmetic, and pharmaceutical products. The complexity of natural extracts usually provides more than one function to the final product. These natural extracts have been applied as natural preservatives (antimicrobial and antioxidant), colorants, flavoring and fragrance agents, supplements, essential oils, etc. The extracts from clove buds with a major contribution from its essential oil have a wide range of applications in pharmacology, cosmetic, and food products. For instance, the antimicrobial effect of clove oil was found better than clove ethanol extract and sodium propionate, a standard food preservative, against various bacterial and fungal strains (Hussain et al., 2017). The significant presence of eugenol in clove oil also provides good antioxidant potential, which proved the better shelf life of bakery products (Karunamay et al., 2019) and Chinese-style sausages (Zhang et al., 2017) in comparison to synthetic antioxidants. Moreover, eugenol-rich clove extract has also been employed as a flavoring agent in mayonnaise and showed better shelf life of the final product (Chatterjee and Bhattacharjee, 2013). Currently, clove oil-based composite film has also been employed in active food packaging and exhibited UV blocking as well as antioxidant properties (Sharma et al., 2020). Similarly, clove extract has shown various pharmacological properties, importantly analgesic, anticancer, and anti-inflammatory (Batiha et al., 2020; Hussain et al., 2017). It is often used in various traditional dentistry and novel medicinal pastes to relieve off teeth pain. Although extract of clove or its essential cannot be used directly on skin due to skin irritation or allergic reaction, however, clove essential oil is an important component of various oriental flavor perfume products (Nowak et al., 2012). Clove oil is also good as a blend in massage oils to relieve various muscular and rheumatic pains.

4. Conclusion

Clove extracts, particularly essential oil, have been widely investigated for their composition. Eugenol and its derivatives occupy the major part of clove extracts. The composition of the clove extract principally depends on the extraction technique, the solvent used, and the pretreatment of the clove. Although conventional techniques have been extensively applied to get clove extract, still investigations are required to get more yield using novel and environment-friendly technologies as pretreatment or as a major technique. Another area of clove extract that still needs researchers' attention is to find out its applications in the food and cosmetic industries and address challenges faced by them in using clove extract.

References

Adaramola, B., Onigbinde, A., 2016. Effect of extraction solvent on the phenolic content, flavonoid content and antioxidant capacity of clove bud. IOSR J. Pharm. Biol. Sci. 11 (3), 33–38.

Aires, A., 2017. Phenolics in foods: extraction, analysis and measurements. In: Soto-Hernandez, M., Palma-Tenango, M., Garcia-Mateos, M.d.R. (Eds.), Phenolic Compounds—Natural Sources, Importance and Applications. IntechOpen, pp. 61–88.

Alexandru, L., Cravotto, G., Giordana, L., Binello, A., Chemat, F., 2013. Ultrasound-assisted extraction of clove buds using batch- and flow-reactors: a comparative study on a pilot scale. Innov. Food Sci. Emerg. Technol. 20, 167–172. https://doi.org/10.1016/j.ifset.2013.07.011.

Amid, A., Salim, R.J.M., Adenan, M.I., 2010. The factors affecting the extraction condition for neuroprotective activity of Centella asiatica evaluated by metal chelating activity assay. J. Appl. Sci. 10 (10), 837–842. https://doi.org/10.3923/jas.2010.837.842.

Amin, M., Jassal, M., Tygi, S., 2013. Phytochemical screening and isolation of eugenol from *Syzygium aromaticum* by gas chromatography. Int. J. Res. Phytochem. Pharmacol. 3 (1), 74–77.

Azmir, J., Zaidul, I.S.M., Rahman, M.M., Sharif, K.M., Mohamed, A., Sahena, F., Jahurul, M.H.A., Ghafoor, K., Norulaini, N.A.N., Omar, A.K.M., 2013. Techniques for extraction of bioactive compounds from plant materials: a review. J. Food Eng. 117 (4), 426–436. https://doi.org/10.1016/j.jfoodeng.2013.01.014.

Azwanida, N., 2015. A review on the extraction methods use in medicinal plants, principle, strength and limitation. Med. Aromat. Plants 4, 2167.

Barba, F.J., Putnik, P., Bursać Kovačević, D., Poojary, M.M., Roohinejad, S., Lorenzo, J.M., Koubaa, M., 2017. Impact of conventional and non-conventional processing on prickly pear (Opuntia spp.) and their derived products: from preservation of beverages to valorization of by-products. Trends Food Sci. Technol. 67, 260–270. https://doi.org/10.1016/j.tifs.2017.07.012.

Batiha, G.E.S., Alkazmi, L.M., Wasef, L.G., Beshbishy, A.M., Nadwa, E.H., Rashwan, E.K., 2020. *Syzygium aromaticum* l. (Myrtaceae): traditional uses, bioactive chemical constituents, pharmacological and toxicological activities. Biomolecules 10 (2). https://doi.org/10.3390/biom10020202.

Chatterjee, D., Bhattacharjee, P., 2013. Supercritical carbon dioxide extraction of eugenol from clove buds. Food Bioprocess Technol. 6 (10), 2587–2599. https://doi.org/10.1007/s11947-012-0979-2.

Chemat, F., Rombaut, N., Sicaire, A.G., Meullemiestre, A., Fabiano-Tixier, A.S., Abert-Vian, M., 2017. Ultrasound assisted extraction of food and natural products. Mechanisms, techniques, combinations, protocols and applications. A review. Ultrason. Sonochem. 34, 540–560. https://doi.org/10.1016/j.ultsonch.2016.06.035.

de Oliveira, M.S., da Costa, W.A., Pereira, D.S., Botelho, J.R.S., de Alencar Menezes, T.O., de Aguiar Andrade, E. H., da Silva, S.H.M., da Silva Sousa Filho, A.P., de Carvalho, R.N., 2016. Chemical composition and phytotoxic activity of clove (Syzygium aromaticum) essential oil obtained with supercritical CO2. J. Supercrit. Fluids 118, 185–193. https://doi.org/10.1016/j.supflu.2016.08.010.

Dhanani, T., Shah, S., Gajbhiye, N.A., Kumar, S., 2017. Effect of extraction methods on yield, phytochemical constituents and antioxidant activity of *Withania somnifera*. Arab. J. Chem. 10, S1193–S1199. https://doi.org/10.1016/j.arabjc.2013.02.015.

El-Maati, M.F.A., Mahgoub, S.A., Labib, S.M., Al-Gaby, A.M.A., Ramadan, M.F., 2016. Phenolic extracts of clove (*Syzygium aromaticum*) with novel antioxidant and antibacterial activities. Eur. J. Integr. Med. 8 (4), 494–504. https://doi.org/10.1016/j.eujim.2016.02.006.

El-Refai, A.A., Sharaf, A.M., Azzaz, N.A.E., El-Dengawy, M.M., 2020. Antioxidants and antibacterial activities of bioactive compounds of clove (*Syzygium aromaticum*) and thyme (*Tymus vulgaris*) extracts. J. Food Dairy Sci., 265–269. https://doi.org/10.21608/jfds.2020.118367.

FAO, 2019. The State of Food and Agriculture 2019. FAO.

Ghallab, J., Eddine, J.J., Said, A.A.H., Zarayby, L., Derfoufi, S., 2019. *Syzygium aromaticum* L.: phytochemical investigation and comparison of the scavenging activity of essential oil, extracts and eugenol. Orient. Pharm. Exp. Med., 1–6.

Giacometti, J., Bursać Kovačević, D., Putnik, P., Gabrić, D., Bilušić, T., Krešić, G., Stulić, V., Barba, F.J., Chemat, F., Barbosa-Cánovas, G., Režek Jambrak, A., 2018. Extraction of bioactive compounds and essential oils from Mediterranean herbs by conventional and green innovative techniques: a review. Food Res. Int. 113, 245–262. https://doi.org/10.1016/j.foodres.2018.06.036.

González-Rivera, J., Duce, C., Falconieri, D., Ferrari, C., Ghezzi, L., Piras, A., Tine, M.R., 2016. Coaxial microwave assisted hydrodistillation of essential oils from five different herbs (lavender, rosemary, sage, fennel seeds and clove buds): chemical composition and thermal analysis. Innov. Food Sci. Emerg. Technol. 33, 308–318. https://doi.org/10.1016/j.ifset.2015.12.011.

Gopalakrishnan, N., Shanti, P.P.V., Narayanan, C.S., 1990. Composition of clove (*Syzygium aromaticum*) bud oil extracted using carbon dioxide. J. Sci. Food Agric. 50 (1), 111–117. https://doi.org/10.1002/jsfa.2740500113.

Gowri, G., Manimegalai, K., 2019. Phytochemical screening and GC-MS analysis of ethanol extract of *Syzygium aromaticum* L. J. Pharmacogn. Phytochem. 8 (5), 746–750.

Guan, W., Li, S., Yan, R., Tang, S., Quan, C., 2007. Comparison of essential oils of clove buds extracted with supercritical carbon dioxide and other three traditional extraction methods. Food Chem. 101 (4), 1558–1564. https://doi.org/10.1016/j.foodchem.2006.04.009.

Handa, S.S., Khanuja, S.P.S., Longo, G., Rakesh, D.D., 2008. Extraction Technologies for Medicinal and Aromatic Plants. Earth, Environmental and Marine Sciences and Technologies. United Nations Industrial Development Organization and the International Centre for Science and High Technology, Italy.

Haqqyana, H., Tania, V.F.W., Suyadi, A.M., Kusuma, H.S., Altway, A., Mahfud, M., 2020. Kinetic study in the extraction of essential oil from clove (*Syzgium aromaticum*) stem using microwave hydrodistillation. Mor. J. Chem. 8 (1), 64–71. http://revues.imist.ma/index.php?journal=morjchem.

Hussain, S., Rahman, R., Mushtaq, A., Zerey-Belaskri, A.E., 2017. Clove: a review of a precious species with multiple uses. Int. J. Chem. Biochem. Sci. 11, 129–133.

Inam, F., Deo, S., Narkhede, N., 2014. HPLC–UV method development and quantification of eugenol from methanolic extracts of some spices. Int. J. Chem. Phys. Sci. 3, 96–102.

Karunamay, S., Badhe, S.R., Shukla, V., Singh, N., Lali, K., Patil, S., 2019. Application of clove essential oil in food industry—a review. J. Food Res. Technol. 7 (4), 23–25.

Kaufmann, B., Christen, P., 2002. Recent extraction techniques for natural products: microwave-assisted extraction and pressurised solvent extraction. Phytochem. Anal. 13 (2), 105–113. https://doi.org/10.1002/pca.631.

Kaur, K., Kaushal, S., 2019. Phytochemistry and pharmacological aspects of *Syzygium aromaticum*: a review. J. Pharmacogn. Phytochem. 8 (1), 398–406.

Kennouche, A., Benkaci-Ali, F., Scholl, G., Eppe, G., 2015. Chemical composition and antimicrobial activity of the essential oil of *Eugenia caryophyllata* cloves extracted by conventional and microwave techniques. J. Biol. Act. Prod. Nat. 5 (1), 1–11. https://doi.org/10.1080/22311866.2014.961100.

Lefebvre, T., Destandau, E., Lesellier, E., 2021. Selective extraction of bioactive compounds from plants using recent extraction techniques: a review. J. Chromatogr. A 1635, 461770. https://doi.org/10.1016/j.chroma.2020.461770.

Marathe, S.J., Jadhav, S.B., Bankar, S.B., Kumari Dubey, K., Singhal, R.S., 2019. Improvements in the extraction of bioactive compounds by enzymes. Curr. Opin. Food Sci. 25, 62–72. https://doi.org/10.1016/j.cofs.2019.02.009.

Mehmood, A., Ishaq, M., Zhao, L., Yaqoob, S., Safdar, B., Nadeem, M., Munir, M., Wang, C., 2019. Impact of ultrasound and conventional extraction techniques on bioactive compounds and biological activities of blue butterfly pea flower (*Clitoria ternatea* L.). Ultrason. Sonochem. 51, 12–19. https://doi.org/10.1016/j.ultsonch.2018.10.013.

Moreira, S.A., Alexandre, E.M.C., Pintado, M., Saraiva, J.A., 2019. Effect of emergent non-thermal extraction technologies on bioactive individual compounds profile from different plant materials. Food Res. Int. 115, 177–190. https://doi.org/10.1016/j.foodres.2018.08.046.

Naudé, Y., De Beer, W.H.J., Jooste, S., Van Der Merwe, L., Van Rensburg, S.J., 1998. Comparison of supercritical fluid extraction and Soxhlet extraction for the determination of DDT, DDD and DDE in sediment. Water SA 24 (3), 205–214.

Nowak, K., Ogonowski, J., Jaworska, M., Grzesik, K., 2012. Clove oil—properties and applications. Chem. Aust. 66 (2), 145–152. http://chemikinternational.com/images/pdf_files/2012_02/2_2012_f.pdf.

Pereira, D.T.V., Tarone, A.G., Cazarin, C.B.B., Barbero, G.F., Martínez, J., 2019. Pressurized liquid extraction of bioactive compounds from grape marc. J. Food Eng. 240, 105–113. https://doi.org/10.1016/j.jfoodeng.2018.07.019.

Pimentel-Moral, S., Borrás-Linares, I., Lozano-Sánchez, J., Arráez-Román, D., Martínez-Férez, A., Segura-Carretero, A., 2019. Supercritical CO2 extraction of bioactive compounds from Hibiscus sabdariffa. J. Supercrit. Fluids 147, 213–221. https://doi.org/10.1016/j.supflu.2018.11.005.

Putnik, P., Lorenzo, J., Barba, F., Roohinejad, S., Režek Jambrak, A., Granato, D., Montesano, D., Bursać Kovačević, D., 2018. Novel food processing and extraction technologies of high-added value compounds from plant materials. Foods 7 (7), 106. https://doi.org/10.3390/foods7070106.

Redondo, D., Venturini, M.E., Luengo, E., Raso, J., Arias, E., 2018. Pulsed electric fields as a green technology for the extraction of bioactive compounds from thinned peach by-products. Innov. Food Sci. Emerg. Technol. 45, 335–343. https://doi.org/10.1016/j.ifset.2017.12.004.

Reverchon, E., Marrone, C., 1997. Supercritical extraction of clove bud essential oil: isolation and mathematical modeling. Chem. Eng. Sci. 52 (20), 3421–3428. https://doi.org/10.1016/S0009-2509(97)00172-3.

Rodsamran, P., Sothornvit, R., 2019. Extraction of phenolic compounds from lime peel waste using ultrasonic-assisted and microwave-assisted extractions. Food Biosci. 28, 66–73. https://doi.org/10.1016/j.fbio.2019.01.017.

Sharayei, P., Azarpazhooh, E., Zomorodi, S., Ramaswamy, H.S., 2019. Ultrasound assisted extraction of bioactive compounds from pomegranate (*Punica granatum* L.) peel. LWT 101, 342–350. https://doi.org/10.1016/j.lwt.2018.11.031.

Sharma, S., Barkauskaite, S., Duffy, B., Jaiswal, A.K., Jaiswal, S., 2020. Characterization and antimicrobial activity of biodegradable active packaging enriched with clove and thyme essential oil for food packaging application. Foods 9 (8), 1117. https://doi.org/10.3390/foods9081117.

Singh, I., Madan, V.K., Jangra, S.S., Singh, S., 2016. Effect of extraction techniques and solvents on various phytochemicals and antioxidant activity of clove (*Syzygium aromaticum* L.) buds. Asian J. Chem. 28 (4), 801–806. https://doi.org/10.14233/ajchem.2016.19522.

Tekin, K., Akalin, M.K., Şeker, M.G., 2015. Ultrasound bath-assisted extraction of essential oils from clove using central composite design. Ind. Crop. Prod. 77, 954–960. https://doi.org/10.1016/j.indcrop.2015.09.071.

Tirado, D.F., de le Fuente, E., Calvo, L., 2019. A selective extraction of hydroxytyrosol rich olive oil from alperujo. J. Food Eng. 263, 409–416. https://doi.org/10.1016/j.jfoodeng.2019.07.030.

Trusheva, B., Trunkova, D., Bankova, V., 2007. Different extraction methods of biologically active components from propolis; a preliminary study. Chem. Cent. J. 1 (1). https://doi.org/10.1186/1752-153X-1-13.

Wei, M.C., Lin, P.H., Hong, S.J., Chen, J.M., Yang, Y.C., 2016a. Development of a green alternative procedure for the simultaneous separation and quantification of clove oil and its major bioactive constituents. ACS Sustain. Chem. Eng. 4 (12), 6491–6499. https://doi.org/10.1021/acssuschemeng.6b01263.

Wei, M.C., Xiao, J., Yang, Y.C., 2016b. Extraction of α-humulene-enriched oil from clove using ultrasound-assisted supercritical carbon dioxide extraction and studies of its fictitious solubility. Food Chem. 210, 172–181. https://doi.org/10.1016/j.foodchem.2016.04.076.

Yung, O.H., Maskat, M.Y., Wan Mustapha, W.A., 2010. Kesan pengekstrakan terhadap kandungan polifenol, aktiviti antipengoksida dan pH ekstrak Pegaga (centella asiatica). Sains Malays. 39 (5), 747–752.

Yunusa, S., Yusuf, U., Haruna, I., 2019. Comparison of essential oils of clove buds extracted using Soxhlet and ultrasonic-assisted extraction methods. J. Exp. Res. 7 (1), 75–78.

Zhang, H., Peng, X., Li, X., Wu, J., Guo, X., 2017. The application of clove extract protects Chinese style sausages against oxidation and quality deterioration. Korean J. Food Sci. Anim. Resour. 37 (1), 114–122. https://doi.org/10.5851/kosfa.2017.37.1.114.

SECTION 2

CLOVE (*SYZYGIUM AROMATICUM*) FIXED OIL: CHEMISTRY, FUNCTIONALITY AND APPLICATIONS

CHAPTER

Composition and functionality of clove (*Syzygium aromaticum*) fixed oil

23

Mohamed Fawzy Ramadan

Department of Clinical Nutrition, Faculty of Applied Medical Sciences, Umm Al-Qura University, Makkah, Saudi Arabia

1. Introduction

Several spices and aromatic plants have been recognized to have medical traits and health-promoting effects, including antioxidant activity, antimicrobial, anti-inflammatory, digestive stimulant action, antimutagenic, hypolipidemic, and anticarcinogenic (Pizzale et al., 2002; Shan et al., 2005; Ramadan, 2013; Ramadan et al., 2013; Hassanien, 2014a,b; Abo Elmaati et al., 2016; El-Hadary and Ramadan Hassanien, 2016; Osman et al., 2019).

Clove (*Syzygium aromaticum* L., family Myrtaceae) is the flower bud, wherein *Syzygium aromaticum* tree is planted in tropical regions (Wenqiang et al., 2007; Abo Elmaati et al., 2016). *Syzygium aromaticum* is used in food processing because of its unique aroma and beneficial health traits. Clove fixed and essential oils are used in nonfood products such as antimicrobial agents against oral bacteria associated with dental caries and periodontal disease and using the oils as a protective agent against tissue injury in the lens and cardiac muscles (Cai and Wu, 1996; El-Hadary and Ramadan Hassanien, 2016; Osman et al., 2019). Treatment with clove could reduce lipid peroxidation and the blood sugar increases in streptozotocin-induced diabetic animals (Shukri et al., 2010).

Cloves are rich in bioactive phytochemicals such as sesquiterpenes, tannins, and triterpenoids. Clove oil has been considered a "Generally Regarded As Safe" substance by the FDA at levels not exceeding 1500 ppm in food products. For humans, WHO Expert Committee on Food Additives has established the acceptable daily human intake of clove oil at 2.5 mg/kg body weight (Kildeaa et al., 2004). The main aroma compound in buds, eugenol (4-allyl-2-methoxyphenol), exhibited functional traits (Lee and Shibamoto, 2002; Miyazawa and Hisama, 2003; El-Hadary and Ramadan Hassanien, 2016; Osman et al., 2019). As a food additive, eugenol was classified by the United States FDA to be a substance that is generally regarded as safe (Gulcin et al., 2012). In addition, *Syzygium aromaticum* volatile oil showed antioxidant traits in retarding oxidation of edible oils and fats (Ramadan, 2013; Ozcan and Arslan, 2011; El-Hadary and Ramadan Hassanien, 2016; Osman et al., 2019).

Several nontraditional fixed oils have been introduced to the international market, and their composition and functional characteristics have been studied. Clove fixed oil **(CFOil)** is rich in essential

Clove (*Syzygium aromaticum*). https://doi.org/10.1016/B978-0-323-85177-0.00035-5

fatty acids and bioactive phytochemicals (i.e., phenolic compounds and tocols). This chapter summarizes up-to-date knowledge on the composition and functional properties of CFOil.

2. Fatty acid profile of CFOil

Fatty acid profile of CFOil and some cold-pressed oils are given in Table 1 (Ramadan et al., 2013; Hassanien, 2014b). Nine fatty acids are detected in CFOil. Linoleic and oleic were the major fatty acids in CFOil (ca. 80%). CFOil contained a high amount of monounsaturated fatty acids (MUFA, ca. 39%) comparable to the blueberry, hemp, cranberry, onion, and milk thistle cold-pressed oils (Parker et al., 2003; Parry et al., 2005). Polyunsaturated fatty acids (PUFA) content in CFOil was 42%. Palmitic and stearic were the main saturated fatty acids (SFA, ca. 15%). CFOil contained ca. 18% SFA, which is lower than cardamom seed oil (31%) and comparable to 14% and 16% found in the milk thistle and pumpkin seed oils, respectively (Parry et al., 2006; Ramadan et al., 2013). A unique feature of CFOil was the high content of PUFA and MUFA. MUFA is reported to retain HDL-cholesterol and lower LDL-cholesterol (Ramadan, 2013). The fatty acid profile of CFOil evinces the oil as an excellent source of essential fatty acids. Besides, high amounts of MUFA and PUFA make CFOil a unique raw material for pharmaceuticals and nutraceuticals.

3. Lipid classes and subclasses of CFOil

In CFOil, the amount of neutral lipids (NL) was the greatest (ca. 95%), followed by glycolipids (GL, 0.7%) and phospholipids (PL, 0.4%). The neutral lipids (NL) classes included triacylglycerol (TG, ca. 96%), free fatty acids (FFA, 1.4%), diacylglycerol (DG), esterified sterols (STE), and monoacylglycerol (MG) in decreasing order. The GL classes found in CPCO were digalactosyldiglycerides (DGD), sulphoquinovosyldiacylglycerol (SQD), monogalactosyldiglycerides (MGD), cerebrosides (CER), sterylglycosides (SG), and esterified sterylglycosides (ESG). The PL classes in CFOil were PC, followed by PE, PI, and PS (Ramadan et al., 2013; Hassanien, 2014a,b).

Table 1 Fatty acid profile of CFOil and some cold-pressed seed oils.

	CFOil	Caraway	Cranberry	Hemp	Carrot	Black raspberry	Red raspberry
C 10:0	2.8	ND	ND	ND	ND	ND[a]	ND
C 12:0	0.1	ND	ND	ND	ND	ND	ND
C 14:0	0.1	0.2	ND	ND	ND	ND	ND
C 16:0	8.6	12.1	7.8	6.3	3.8	1.2	1.3
C 16:1	0.4	1.0	ND	ND	ND	ND	ND
C 18:0	6.6	3.4	1.9	2.7	0.5	ND	1.0
C 18:1	39.4	24.0	22.7	11.7	82.0	7.9	12.4
C 18:2	40.2	55.8	44.3	59.3	13.2	55.7	53.0
C 18:3	2.0	0.2	22.3	20.0	0.5	35.2	32.3

[a]*ND, not detected.*

4. Tocol profile of CFOil

In CFOil, levels of α-, β-, γ-, and δ-tocotrienols were 1.11, 0.055, 0.085, and 9.49 g/kg oil, respectively. On the other side, amounts of α-, β-, γ-, and δ-tocopherols in CFOil were 14.89, 0.056, 4.18, and 0.186 g/kg oil, respectively. All tocols isomers were present in CFOil, wherein α-tocopherol accounted for ca. 50% of tocols followed by δ-tocotrienol (32%) and γ-tocopherol (14%). Thus, levels of tocols in CFOil might contribute to the oxidative stability of CFOil (Ramadan et al., 2013; Hassanien, 2014a).

5. Antiradical potential of CFOil

Antiradical traits of the CFOil and olive oil were studied and compared using DPPH·and galvinoxyl free radicals. CFOil had more substantial antiradical potential than olive oil. After 1 h incubation with DPPH·, 70% of DPPH· were deactivated by CFOil, while olive oil inhibited 45%. In another test, CFOil inhibited 57% of galvinoxyl radicals, and olive oil inhibited 38%. CFOil is characterized by high levels of phenolic compounds (ca. 4.6 mg/g oil). Therefore, tocols and phenolics in edible oils significantly affect their antiradical and radical scavenging potential (Ramadan et al., 2013).

6. Antimicrobial activity of CFOil

The screening of the antimicrobial effect of CFOil using agar diffusion test revealed that CFOil inhibited the growth of microbial species, including Gram-negative bacteria, Gram-positive bacteria, and molds (yeasts and fungi). The highest inhibition zone obtained with CFOil against *Aspergillus flavus* and *Aspergillus niger* were 31 and 29 mm, respectively. In addition, CFOil exhibited broad-spectra activity against Gram-positive bacteria, Gram-negative bacteria, and molds (MIC between 0.6 and 1.0 mL/L). CFOil had a drastic impact on the biosynthesis of protein and total lipids in *Bacillus subtilis* cells. In addition, CFOil had a drastic impact on nucleic acids. The influence increased with increase of incubation period and the concentration of CFOil. In addition, CFOil affects the biosynthesis of proteins by inhibiting steps in the process of translation. CFOil could be used in applications as natural food preservatives to inhibit microbial growth (Ramadan et al., 2013).

7. Hepatoprotective effect of CFOil against CCl_4-induced toxicity in animals

Natural antioxidants and plant extracts are active in preventing the oxidative stress-related pathologies in liver due to interactions and synergisms (Yeh et al., 2013). Eugenol inhibited malonaldehyde formation from cod liver oil and the formation of hexanal (Gulcin et al., 2004). El-Hadary and Ramadan Hassanien (2016) studied the protective effect of CFOil against CCl_4-induced toxicity in the experimental animals. CFOil was administered (100 and 200 mg/kg) to rats with CCl_4. Treatment with

200 mg/kg CFOil resulted in decreased urea, creatinine, and uric acid levels. Levels of TAG, TL, TC, LDL-C, and VLDL-C were also decreased. CFOil reduced the activities of ALT, AST, and ALP as well as kidney function markers. Levels of hepatic malondialdehyde were reduced and glutathione levels were increased in CFOil-treated animals. Thus, CFOil possessed a protective impact against toxicity in the experimental animals, mediated possibly by CFOil antioxidant traits.

8. Application of CFOil in meat products

Osman et al. (2019) studied the antimicrobial and antioxidant impact of CFOil supplementation on minced beef properties during storage. The potential of CFOil supplementation (1%, 2%, and 4%, w/w) to minced beef on the chemical properties and the growth *Salmonella enteritidis* and *Listeria monocytogenes* inoculated artificially in meat samples during storage was tested. After 15 days of storage, *L. monocytogenes* was more resistant than *Salmonella enteritidis*. The shelf life of CFOil-supplemented meat was prolonged under cold conditions. In addition, oxidative stability of CFOil-supplemented meat was enhanced.

9. Application of CFOil in edible oil blends

Ramadan (2013) prepared blends of cold-pressed oils, including CFOil with sunflower oil (SFO). Oxidative stability and radical scavenging activity of SFO and blends stored under accelerated oxidative conditions were investigated. By increasing the proportion of CFOil in SFO, the linoleic acid amount decreased, while the tocol amounts increased. Inverse relations were noted between peroxide value (PV) and *p*-anisidine value (Av) and oxidative stability at the end of the storage experiment. Levels of conjugated dienes (CD) and conjugated trienes (CT) in SFO and SFO blends increased with an increase in time. The stabilities of CFOil blends were higher than SFO, likely due to changes in tocols and bioactive lipids in CFOil.

10. Conclusion

CFOil is a good source of essential fatty acids and lipid-soluble bioactives. Tocols and phenolics at the level determined in CFOil might be of nutritional importance as natural antioxidants and might react with and deactivate free radicals and prevent lipid oxidation. Therefore, CFOil could be nutritionally considered a nontraditional raw material for edible purposes, cosmetics, and pharmaceuticals.

References

Abo Elmaati, M.F., Mahgoub, S.A., Labib, S.M., Al-Gaby, A.M.A., Ramadan, M.F., 2016. Phenolic extracts of clove (*Syzygium aromaticum*) with novel antioxidant and antibacterial activities. Eur. J. Intern. Med. 8. https://doi.org/10.1016/j.eujim.2016.02.006.

Cai, L., Wu, C.D., 1996. Compounds from *Syzygium aromaticum* possessing growth inhibitory activity against oral pathogens. J. Nat. Prod. 59, 987–990.

El-Hadary, A.E., Ramadan Hassanien, M.F., 2016. Hepatoprotective effect of cold-pressed *Syzygium aromaticum* oil against carbon tetrachloride (CCl_4)-induced hepatotoxicity in rats. Pharm. Biol. 54 (8), 1364–1372.

Gulcin, I., Sat, I.G., Beydemir, S., Elmastas, M., Kufrevioglu, O.I., 2004. Comparison of antioxidant activity of clove (*Eugenia caryophylata* Thunb) buds and lavender (*Lavandula stoechas* L.). Food Chem. 87, 393–400.

Gulcin, I., Elmastas, M., Aboul-Enein, H.Y., 2012. Antioxidant activity of clove oil—a powerful antioxidant source. Arab. J. Chem. 5, 489–499.

Hassanien, M.F.R., 2014a. Composition and antiradical power of *Syzygium aromaticum* lipids. Chem. Nat. Compd. 50, 716–718. https://doi.org/10.1007/s10600-014-1060-3.

Hassanien, M.F.R., 2014b. Oil of clove (*Syzygium aromaticum*). Inform 25 (2), 120–122.

Kildeaa, M.A., Allanb, G.L., Kearney, R.E., 2004. Accumulation and clearance of the anaesthetics clove oil and AQUI-S from the edible tissue of silver perch (*Bidyanus bidyanus*). Aquaculture 232, 265–277.

Lee, K.G., Shibamoto, T., 2002. Antioxidant property of aroma extract isolated from clove buds [*Syzygium aromaticum* (L.) Merr. et Perry]. Food Chem. 74, 443–448.

Miyazawa, M., Hisama, M., 2003. Antimutagenic activity of phenylpropanoids from clove (*Syzygium aromaticum*). J. Agric. Food Chem. 51, 6413–6422.

Osman, A., Mahgoub, S., Mohamed, K., Hassanien, M.F.R., 2019. Antimicrobial and antioxidant infuence o *Syzygium aromaticum* oil supplementation on minced bee quality during cold storage. Arch. Leb. 70, 40–47. https://doi.org/10.2376/0003-925X-70-40.

Ozcan, M.M., Arslan, D., 2011. Antioxidant effect of essential oils of rosemary, clove and cinnamon on hazelnut and poppy oils. Food Chem. 129, 171–174.

Parker, T.D., Adams, D.A., Zhou, K., Harris, M., Yu, L., 2003. Fatty acid composition and oxidative stability of cold-pressed edible seed oils. J. Food Sci. 68, 1240–1243.

Parry, J., Su, L., Luther, M., Zhou, K., Yurawecz, M.P., Whittaker, P., Yu, L., 2005. Fatty acid composition and antioxidant properties of cold-pressed marionberry, boysenberry, red raspberry, and blueberry seed oils. J. Agric. Food Chem. 53, 566–573.

Parry, J., Su, L., Moore, J., Cheng, Z., Luther, M., Rao, J.N., Wang, J.-Y., Yu, L., 2006. Chemical compositions, antioxidant capacities, and antiproliferative activities of selected fruit seed flours. J. Agric. Food Chem. 54, 3773–3778.

Pizzale, L., Bortolomeazzi, R., Vichi, S., Uberegger, E., Conte, L.S., 2002. Antioxidant activity of sage (*Sal*Via *officinalis* and *S. fruticosa*) and oregano (*Origanum onites* and *O. indercedens*) extracts related to their phenolic compound content. J. Sci. Food Agric. 82, 1645–1651.

Ramadan, M.F., 2013. Healthy blends of high linoleic sunflower oil with selected cold pressed oils: functionality, stability and antioxidative characteristics. Ind. Crop. Prod. 43, 65–72.

Ramadan, M.F., Asker, M.M.S., Tadros, M., 2013. Lipid profile, antiradical power and antimicrobial properties of *Syzygium aromaticum* oil. Grasas Aceites 64, 509–520.

Shan, B., Cai, Y.Z., Sun, M., Corke, H., 2005. Antioxidant capacity of 26 spice extracts and characterization of their phenolic constituents. J. Agric. Food Chem. 53, 7749–7759.

Shukri, R., Mohamed, S., Mustapha, N.M., 2010. Cloves protect the heart, liver and lens of diabetic rats. Food Chem. 122, 1116–1121.

Wenqiang, G., Shufen, L., Ruixiang, Y., Shaokun, T., Can, Q., 2007. Comparison of essential oils of clove buds extracted with supercritical carbon dioxide and other three traditional extraction methods. Food Chem. 101, 1558–1564.

Yeh, Y.-H., Hsieh, Y.-L., Lee, Y.-T., 2013. Effects of yam peel extract against carbon tetrachloride-induced hepatotoxicity in rats. J. Agric. Food Chem. 61, 7387–7396.

CHAPTER

24 Food applications of clove (*Syzygium aromaticum*) fixed oil

Muhammad Imran[a] and Mohamed Fawzy Ramadan[b]

Department of Food Science, Faculty of Life Sciences, Government College University, Faisalabad, Punjab, Pakistan[a]

Department of Clinical Nutrition, Faculty of Applied Medical Sciences, Umm Al-Qura University, Makkah, Saudi Arabia[b]

1. Introduction

During the last decades, fixed oils have greatly influenced consumers due to their physicochemical characteristics and nutritional profile. Fixed oils are usually obtained by crushing oilseeds at room temperature without involving chemical reagents and high temperature. Seed fixed oils extracted through cold pressing have been recognized as safe and abundant in bioactive compounds, antioxidants, and flavors compared with seed fixed oils obtained after hot pressing, chemical solvents treatment, and hot temperature (200°C). There are several oilseeds and fruits from which naturally present fixed oils are extracted with total nutritive capacity and profile. Some of these sources are olive, peanut, canola, sesame seed, coconut, and sunflower seeds. Moreover, many innovative and conventional ingredients have been explored for natural preservation, value addition, shelf-life enhancement, and quality improvement of food products in food science and technology.

Clove (*Syzygium aromaticum* Linn) is an important spice known as botanically aromatic dried flower buds of the family Myrtaceae. Clove seeds or buds are processed to get fixed oil through a cold-pressing procedure. The quality of the extracted clove oil entirely depends on the buds' maturity stage, history, and origin. The extraction method, sample preparation, equipment, and solvent ratio also directly affect the clove oil yield. Clove contains fixed oil in the range of 8%–10% (saponifiable and nonvolatile) and 15%–20% essential oil (unsaponifiable and volatile). In terms of its nutraceutical and functional food applications, the use of clove oil in a natural liquid state has gained quite extensive attention in different applications. Many research and experimental studies have been conducted to summarize and comprehensively explain the health benefits of clove fixed oil. However, very limited information is available regarding the applications of clove fixed oil in different food products. This chapter focuses on providing information and an overview of clove fixed oil for potential food applications.

Clove (*Syzygium aromaticum*). https://doi.org/10.1016/B978-0-323-85177-0.00031-8

2. Applications of clove oil in edible oil blends

The preparation of oxidative-resistant blends of cold-pressed oils is increasing the popularity among the edible oil processing industries. The blends of cold-pressed oils (10% and 20%, w/w), including cumin oil, clove oil, black cumin oil, and coriander oil with high linoleic sunflower oil, were prepared (Ramadan, 2013). The prepared blends were stored under oxidative conditions (60°C) for 1 week and evaluated for radical scavenging activity and oxidative stability. The concentration of tocols level increased, and linoleic acid level decreased with increasing the level of cold-pressed oils in sunflower oil formulation. Furthermore, the prepared blends showed about 70% inhibition of DPPH radicals. The results also indicated that the oxidative stability of prepared oil blends was much better than sunflower oil (Ramadan, 2013).

The food products supplemented with clove oil have been demonstrated to possess better antimicrobial activity than the extracts and powders of clove spice. Clove oil can be effectively used as a colorant, preservative, and spice ingredient in different food recipes and products. Clove oil containing flavoring substances acts as a natural preservative, and it has been observed that the clove oil is not harmful when consumed in different types of refined food products. The mixture of clove oil with cinnamon was evaluated for inhibitory action against important four types of spoilage microorganisms usually present in intermediate moisture foods, and it was reported that the higher concentration levels of clove oil in combination with cinnamon oil were more active to inhibit the growth of *Aspergillus flavus* (Matan et al., 2006).

3. Applications of microencapsulated clove oil

Clove oil with strong flavor has worth in the sugar, beverage, and meat industries. Clove oil has been declared as GRAS (Generally Regarded As Safe) substance by the US-FDA (United States-Food and Drug Administration). Clove oil safe dose has been described below 1500 ppm in all food categories (Gülçin et al., 2012). The clove oil has been comprehensively explored against a wide range of pathogenic microorganisms and showed effective antimicrobial and antibacterial activities in different objective studies (Sebaaly et al., 2015). The antifungal preservation potential of clove oil microcapsules was investigated in different meat products such as cooked beef, cooked chicken, cooked fish, and cooked pork. This study reported that microcapsules of clove oil possessed a high inhibition effect and strong heat resistance when supplemented in different meat products. Moreover, the variation rates of mold spores and occurrence of disease indices were decreased. The results concluded that the clove oil microcapsules had practical applications as a preservative in meat products processed at high temperatures. Meanwhile, microencapsulation technology can improve the stability of clove oil (Wang et al., 2018).

4. Applications of clove oil in meat products

The food products with short storage life have more vulnerability to microorganism attack and deteriorate quickly. By applying clove oil, the delay in the growth of microorganisms can be achieved, leading to more food safety and prolonged storage quality. In a study by Menon and Garg (2001), the researchers examined the antilisteric activity of clove oil in minced meat (mutton) at both 30°C and 7°C. The application of clove oil at concentrations of 0.5% and 1% reduced the growth of studied microorganisms (*Listeria monocytogenes*) in the meat samples at both experimental temperatures. However, the concentration of 1% clove oil showed a more pronounced inhibitory effect as compared to a 0.5% concentration of clove oil during storage. Overall, this study suggested that clove oil has extensive commercial applications and can be utilized as a natural preservative in meat food items. The positive potential of cold-pressed clove oil supplemented in the range of 1%–4% to minced beef for the inhibition of microorganism growth and chemical traits during cold storage (4°C) for 15 days has been documented (Osman et al., 2019).

Different mixture ratios of clove oil with lactic acid and vitamin C were used for buffalo meat steaks, and these steaks were retained at 4°C up to the 12th day of storage with intervals of 3 days for studying the different parameters. The results indicated that treatments showed significant trends for studied parameters as clove oil combined with lactic acid and vitamin C significantly lowered the thiobarbituric acid reactive substances values, psychrotrophic counts, aerobic plate counts, and coliform counts. Meanwhile, these treatment combinations showed higher instrumental color CIE a^* and b^* values and improved sensoric acceptability in terms of color and odor compared with control treatment (distilled water application). Moreover, clove oil treatment with either lactic acid and/or vitamin C enhanced the storage life of buffalo meat steaks at 4°C (Naveena et al., 2006).

The survival and uncontrolled growth of *Listeria monocytogenes* strain in many types of ready-to-eat (RTE) poultry products at refrigeration temperature has been considered an excellent concern for the food production industry and discerning public health issues. However, it has been noted in a research study that the addition of clove oil at a concentration of 1% to 2% in chicken frankfurters (stored at 5°C and 15°C) inhibited the growth of all studied strains of *L. monocytogenes*. Moreover, based on the results, it was also explained that the sensory considerations should be observed using the clove oil in chicken frankfurters as an effective strategy to control *L. monocytogenes* (Mytle et al., 2006).

Çoban and Patir (2013) evaluated the antimicrobial and antioxidant impact of clove oil (*Eugenia caryophyllata*) on sliced smoked rainbow trout fillets during storage at 2°C. The study's findings documented that the peroxide value, thiobarbituric acid, and microbiological growth were very less in experimental sliced smoked rainbow trout fillet samples treated with clove oil ranging in concentration from 0.1% to 1% as compared to the control group (without clove oil treatment). In addition, the storage life of experimental samples treated with clove oil was noted to be extended up to 4 to 5 weeks compared to the control samples. Microbiological and chemical data concluded that the clove oil could be explored as a natural preservative to enhance the shelf life of smoked fish under chilled storage.

In a study, the aerobically packaged treatment containing chevon cutlets with clove oil (100 ppm) and with barley flour (4%) showed significantly lower values for various parameters like psychrophilic count, TBARS value, yeast, mold, and total plate count as compared to aerobically packaged control treatment samples (chevon cutlets without clove oil and without barley flour) stored at 4°C for 15 days (Singh et al., 2014).

In another study, the aerobically packaged chevon cutlets containing optimum level (6%) of sorghum flour and clove oil (100 ppm) were evaluated under refrigerated conditions (4°C) for storage quality up to 15 days. The analyzed parameters such as crude fiber, adhesiveness, chewiness, hardness, gumminess, springiness, cohesiveness, and product redness value increased significantly. At the same time, fat content, moisture percent, and overall acceptability significantly decreased when the level of incorporation of these substrates was increased in product formulation. During the storage, these product samples showed a significant increasing trend for psychrophilic count (cfu/g), total plate count (cfu/g), and thiobarbituric acid-reactive substance value (mg malonaldehyde/kg), whereas decreasing trend was noted for all the sensory parameters during storage days. Overall, the developed products can be stored without marked loss in quality up to 10 days under refrigerated conditions (Singh et al., 2015a).

The possibility of clove oil combined with processed oats was evaluated to enhance the shelf life of fiber-enriched designer chevon cutlets. In this experimental trial, clove oil (100 ppm) and processed oats (4%) were added to the formulation of chevon cutlets, and samples were packaged aerobically in low-density ethylene pouches at 4°C for consecutive 15 days. The psychrophilic, total plate count, and thiobarbituric acid reactive substances value were increased significantly, whereas the sensory attributes acceptance decreased significantly with the storage period. However, it was noted that the functional product could be refrigerated for 10 days without marked loss in the quality of the final product (Singh et al., 2015b).

Clove oil—supplemented broiler meat emulsion—packed in low-density polyethylene pouches and stored for 14 days under refrigerated (4°C) conditions showed gradual decrease in pH value, free fatty acid, and thiobarbituric acid reacting substances value increased with increasing days of storage. However, a higher sensory rating was received by the product samples enriched with clove oil (Para et al., 2017). The mixture combination of oleoresin and clove oil was evaluated for antioxidant potential in two species of meat (chevon and chicken) at refrigerated temperature (4°C). This study concluded that the clove oil and oleoresin possessed potential antioxidant impact for preserving cooked meat products by inhibiting lipid oxidation (Sultana et al., 2017).

5. Clove oil, baked product, and vegetable plant

Badei et al. (2002) documented that the clove oil supplemented in cookies suppressed the growth of microorganisms in the final product and it has also been observed that clove possessed the worth potential to retard the growth of Gram-positive and Gram-negative bacteria (Mannie, 1999). The combined treatment of clove oil (1, 2, and 4 mg/mL) and cold nitrogen plasma (400–600 W) showed a mildly negative effect on lettuce quality (Cui et al., 2016).

6. Applications of clove oil in edible coating

The edible coating has been recognized as an effective strategy on a commercial scale to improve the final product's storage quality and enhance the shelf life of food and meat products. The edible coating consisting of chitosan, gallic acid, gelatine, and clove oil improved the antimicrobial effects. Furthermore, this edible coating showed the best performance for preserving salmon fillet samples and prolonged the shelf life of experimental samples for at least 5 days at 4°C. Overall, the findings of this study suggested that such edible coatings developed using the practical commercial method have the potential for the preservation of fresh fish meat and products (Xiong et al., 2021).

Different food processing industries are developing bioactive films and/or edible coatings to maintain the quality and safety of food products in recent years. Bioactive edible coatings developed from chitosan, clove oil, and ethylenediaminetetraacetate (EDTA) were used to improve the microbial quality of meat. The study results documented that the addition of clove oil in bioactive edible coating promotes the inhibition rates of *Staphylococcus aureus* (96.4%) and *Escherichia coli* (99.1%). Moreover, the application of clove oil containing bioactive edible coating maintains the acceptable sensory characteristics and moderates the growth of total microbes in pork slices during storage for 7 days at 4°C (He et al., 2014).

Carboxymethyl cellulose-clove oil-polyvinyl alcohol films were developed using different mixing ratios of these substrates. Ground chicken meat samples (as a model food system) were packed in these developed films and were stored for 12 days. The results point out that these films reduced the total viable counts and increased the stability of packed meat samples compared to control chicken meat samples that were spoiled after 4 days during refrigeration. The study concluded that the clove oil containing films possessed the immense potential to be used as active packaging material for meat preservation (Muppalla et al., 2014).

The decrease in thiobarbituric acid reactive substances (36%) and peroxide value (28%) was noted in the smoked salmon packaged with the chicken feather protein/gelatin composite film containing 1.5% clove oil compared with the control during storage. The proposed mechanism was that the small fraction of clove oil in a free form was present in the film network. These oil contents did not interact with the protein matrix during the chicken feather protein/gelatin composite film formation. This clove oil-free fraction might migrate to the smoked salmon surface during storage and lead to prevent the oxidation of stored product (Song et al., 2014). Chitosan coatings with chitosan concentration (2%), with clove oil (1.5%), and without clove oil (control) were evaluated for determining the effects on shelf life stored at a refrigerated temperature (4°C) and the quality of cooked pork sausages over 25 days of storage. The rate of increase in L^* value, total viable count, peroxide value, psychrotrophic bacteria count, and the thiobarbituric acid reactive substances while the rate of decrease in a^* value, the b^* value, the sensory scores, and pH was slowest in cooked pork sausages treated with chitosan and clove oil. Moreover, the shelf life extended up to 20 days of cooked pork sausages treated with chitosan and clove oil under refrigerated storage (Lekjing, 2016).

In a research study, the bioactive edible coating solutions comprised of gelatin, chitosan, and clove oil with different concentrations were developed. The tuna fillet samples were dipped in developed edible coating solutions and stored under refrigerated conditions (4°C). The fillet samples without dip treatment were considered as control samples. The edible coating solution containing clove oil showed excellent preservative tuna fillets (Kumar et al., 2018). The antifungal activity of clove oil combined with chitosan against citrus green mold caused by *Penicillium digitatum* was tested in vitro and in vivo. It was found that clove oil combined with chitosan exhibited higher antifungal properties on *Penicillium digitatum* (Shao et al., 2015).

Antimicrobial-based edible coatings, also known as food grade preserving coatings, are widely used nowadays to maintain the quality and safety of different meat products. In a research trial, meat samples coated with ingredients mixture containing starch solution, chitosan solution, and clove oil showed a positive impact against aerobic microbes and sensory acceptance rating compared to uncoated meat samples (Mukhtar et al., 2018). The inhibition effect of clove oil could be attributed to the hydrophobic nature, which has a trend for partial partition in the lipid contents of the bacterial mitochondria and cell membrane. This connection usually produces disturbance in cell structure and renders it more permeable (Sikkema et al., 1994). It was also reported that the cold-pressed clove oil has a drastic effect on the biosynthesis of lipids and proteins in the cell of the tested *B. subtilis* microorganism (Ramadan et al., 2013).

The clove oil with its protentional nutritional worth has always expressed inhibitory effect against Gram-negative bacteria like *Salmonella*, *E. coli*, and *P. aeruginosa*, and Gram-positive bacteria such as *L. monocytogenes*, *Streptococcus*, and *Staphylococcus*, as well as *Aspergillus* as *A. parasiticus*, *A. flavus*, and *A. ochraceus*. The mechanism involved is expected to be virulence factor coding gene expression, biofilm synthesis, destruction of synthesized biofilm, and even bacterial migration. The scientists have also recommended setting up the defined application standards for clove oil and its bioactive components to minimize their side effects in different food formulations and their flavor profiles (Hu et al., 2018).

7. Conclusion

Food researchers and agricultural scientists are constantly searching for the optimum extraction and preservation of premium quality bioactive food components with maximum nutritional profile due to the rising global demand from discerning consumers and processing industries. Clove oil has been recognized as a raw substrate possessing health-promoting impacts such as antimicrobial and antioxidant traits. Thus, clove oil could be considered a natural preservative and/or potential ingredient for supplementing different food products.

References

Badei, A.Z., Faheid, S.M., El-Akel, A., Mahmoud, B.S., 2002. Application of some spices in flavoring and preservation of cookies: 2-antimicrobial and sensory properties of cardamom, cinnamon and clove. Dtsch. Lebensmitt. Rundsch. 98 (7), 261–265.

Çoban, Ö.E., Patir, B., 2013. Antimicrobial and antioxidant effects of clove oil on sliced smoked *Oncorhynchus mykiss*. J. Verbr. Lebensm. 8 (3), 195–199.

Cui, H., Ma, C., Lin, L., 2016. Synergetic antibacterial efficacy of cold nitrogen plasma and clove oil against *Escherichia coli* O157: H7 biofilms on lettuce. Food Control 66, 8–16.

Gülçin, İ., Elmastaş, M., Aboul-Enein, H.Y., 2012. Antioxidant activity of clove oil—a powerful antioxidant source. Arab. J. Chem. 5 (4), 489–499.

He, S., Yang, Q., Ren, X., Zi, J., Lu, S., Wang, S., Wang, Y., 2014. Antimicrobial efficiency of chitosan solutions and coatings incorporated with clove oil and/or ethylenediaminetetraacetate. J. Food Saf. 34 (4), 345–352.

Hu, Q., Zhou, M., Wei, S., 2018. Progress on the antimicrobial activity research of clove oil and eugenol in the food antisepsis field. J. Food Sci. 83 (6), 1476–1483.

Kumar, K.S., Chrisolite, B., Sugumar, G., Bindu, J., Venkateshwarlu, G., 2018. Shelf life extension of tuna fillets by gelatin and chitosan based edible coating incorporated with clove oil. Fish. Technol. 55, 104–113.

Lekjing, S., 2016. A chitosan-based coating with or without clove oil extends the shelf life of cooked pork sausages in refrigerated storage. Meat Sci. 111, 192–197.

Mannie, E., 1999. Natural antimicrobial ingredients for meat products. Meat Process. 12 (1), 80.

Matan, N., Rimkeeree, H., Mawson, A.J., Chompreeda, P., Haruthaithanasan, V., Parker, M., 2006. Antimicrobial activity of cinnamon and clove oils under modified atmosphere conditions. Int. J. Food Microbiol. 107 (2), 180–185.

Menon, K.V., Garg, S.R., 2001. Inhibitory effect of clove oil on *Listeria monocytogenes* in meat and cheese. Food Microbiol. 18 (6), 647–650.

Mukhtar, S., Zahoor, T., Randhawa, M.A., Iqbal, R., Shabbir, A., Liaqat, A., Ahsan, S., 2018. Synergistic effect of chitosan and clove oil on raw poultry meat. J. Food Process. Technol. 9 (6), 737–741.

Muppalla, S.R., Kanatt, S.R., Chawla, S.P., Sharma, A., 2014. Carboxymethyl cellulose-polyvinyl alcohol films with clove oil for active packaging of ground chicken meat. Food Packag. Shelf Life 2 (2), 51–58.

Mytle, N., Anderson, G.L., Doyle, M.P., Smith, M.A., 2006. Antimicrobial activity of clove (*Syzgium aromaticum*) oil in inhibiting *Listeria monocytogenes* on chicken frankfurters. Food Control 17 (2), 102–107.

Naveena, B.M., Muthukumar, M., Sen, A.R., Babji, Y., Murthy, T.R.K., 2006. Improvement of shelf-life of buffalo meat using lactic acid, clove oil and vitamin C during retail display. Meat Sci. 74 (2), 409–415.

Osman, A., Mahgoub, S.A., Mohamed, K., Wahdan, M., Ramadan, M.F., 2019. Antimicrobial and antioxidant influence o *Syzygium aromaticum* oil supplementation on minced bee quality during cold storage. Arch. Lebensm. 70, 40–47.

Para, P.A., Kumar, S., Raja, W.H., Ganguly, S., 2017. Effect of clove oil on some quality characteristics and sensory attributes of papaya pulp enriched enrobed chicken nuggets at refrigerated storage. Indian J. Poult. Sci. 52 (1), 96–103.

Ramadan, M.F., 2013. Healthy blends of high linoleic sunflower oil with selected cold pressed oils: functionality, stability and antioxidative characteristics. Ind. Crop. Prod. 43, 65–72.

Ramadan, M.F., Asker, M.M.S., Tadros, M., 2013. Lipid profile, antiradical power and antimicrobial properties of *Syzygium aromaticum* oil. Grasas Aceites 64 (5), 509–520.

Sebaaly, C., Jraij, A., Fessi, H., Charcosset, C., Greige-Gerges, H., 2015. Preparation and characterization of clove essential oil-loaded liposomes. Food Chem. 178, 52–62.

Shao, X., Cao, B., Xu, F., Xie, S., Yu, D., Wang, H., 2015. Effect of postharvest application of chitosan combined with clove oil against citrus green mold. Postharvest Biol. Technol. 99, 37–43.

Sikkema, J., de Bont, J.A.M., Poolman, B., 1994. Interaction of cyclic hydrocarbons with biological membranes. J. Biol. Chem. 269, 8022–8028.

Singh, P.K., Kumar, S., Bhat, Z.F., Kumar, P., 2014. Effect of clove oil on the storage quality of aerobically packaged fiber-enriched chevon cutlets. J. Meat Sci. 10 (1), 36–43.

Singh, P.K., Kumar, S., Bhat, Z.F., Kumar, P., 2015a. Effect of Sorghum bicolour and clove oil on the quality characteristics and storage quality of aerobically packaged chevon cutlets. Nutr. Food Sci. 45, 145–163. https://doi.org/10.1108/NFS-02-2014-0017.

Singh, P.K., Kumar, S., Bhat, Z.F., Kumar, P., Kumar, A., 2015b. Effect of processed oats and clove oil on the characteristics and storage quality of aerobically packaged chevon cutlets. Indian J. Small Rumin. 21 (1), 76–84.

Song, N.B., Lee, J.H., Al Mijan, M., Song, K.B., 2014. Development of a chicken feather protein film containing clove oil and its application in smoked salmon packaging. LWT - Food Sci. Technol. 57 (2), 453–460.

Sultana, K., Jayathilakan, K., Pandey, M.C., 2017. Evaluation of antioxidant activity, radical scavenging, and reducing power of clove oil and clove oleoresin in comparison with natural and synthetic antioxidants in Chevon (*Capra aegagrus* hircus) and chicken meat. Def. Life Sci. J. 3 (1), 51.

Wang, Y.F., Jia, J.X., Tian, Y.Q., Shu, X., Ren, X.J., Guan, Y., Yan, Z.Y., 2018. Antifungal effects of clove oil microcapsule on meat products. LWT - Food Sci. Technol. 89, 604–609.

Xiong, Y., Kamboj, M., Ajlouni, S., Fang, Z., 2021. Incorporation of salmon bone gelatine with chitosan, gallic acid and clove oil as edible coating for the cold storage of fresh salmon fillet. Food Control 125, 107994.

CHAPTER

25 Health-promoting activities of clove (*Syzygium aromaticum*) fixed oil

Zarina Mushtaq[a], Muhammad Imran[a], Muhammad Haseeb Ahmad[a], Ayesha Rasheed[a], Nosheen Asghar[a], Marwa Waheed[a], Muhammad Bilal Hussain[a], and Mohamed Fawzy Ramadan[b]

Institute of Home and Food Sciences, Faculty of Life Sciences, Government College University, Faisalabad, Punjab, Pakistan[a] Department of Clinical Nutrition, Faculty of Applied Medical Sciences, Umm Al-Qura University, Makkah, Saudi Arabia[b]

1. Introduction

The conventional therapeutic system composed of natural remedies still has a significant function in health curative purposes. Recently, therapeutic plants have been getting broad recognition because herbal flora have been used to produce herbal inventions, which showed minimum health issues and enhanced effectiveness compared to their artificial products (Abushouk et al., 2017). Recently, nearly 80% of people worldwide used conventional therapeutics for their fundamental health concerns (Ekor, 2014). These herbal plants possess antibacterial, antiviral, and antifungal effects. They can also be used in preservation and have microbicidal, antispasmodic, tranquilizing, analgesic, and general anesthetic potential (Batiha et al., 2018; Beshbishy et al., 2019). Various herbs have been given medicinal importance due to their bioactive components, including glycosides, saponins, flavonoids, steroids, tannins, alkaloids, and terpenes (Batiha et al., 2018). Natural cure systems have been acknowledged as vital resources for determining curative narrative components utilized to care for major maladies (Beshbishy et al., 2019).

Syzygium aromaticum, another name for clove, is the dehydrated bud of a flower related to the Myrtaceae family. It is native to the Maluku islands in Indonesia, although in recent years it has been cultivated in other places as well (Cortés-Rojas et al., 2014). The clove plant is based on leaves and buds (the marketable portion of the tree), and the bud of flower cultivation starts after 4 years on the farm. They are marketed for various curative purposes as well as in the perfume industry. Clove is a spice that can be significantly used as an additive in various food products (Cortés-Rojas et al., 2014; Chomchalow et al., 1996).

Clove has a great deal of concentration in various spices because of its strong microbicidal and antioxidant potential (Shan et al., 2005). The efficiency of clove in stopping various maladies is credited to the availability of many substances with antioxidant potential in rich amounts (Astuti et al., 2019). Clove essential oil (CEO) is conventionally used in healing tissues' curative purpose and as antispasmodic in tooth especially in the cure of tooth illnesses and tooth spasm. Additionally, it is used

Clove (*Syzygium aromaticum*). https://doi.org/10.1016/B978-0-323-85177-0.00023-9

in many trade uses and applied widely in scents and as a purification carrier in histological research (Sarrami et al., 2002). Cloves are used in Indian and Chinese conventional drugs as a warm and motivating ingredient (Batiha et al., 2018). Conventionally, cloves were utilized for hundreds of years to treat nausea as well as liver, bowel, and stomach discomfort. In the United States, clove was conventionally utilized to prevent foodborne toxins from various bacterial and protozoan illnesses (Bhowmik et al., 2012).

Various spicy, aromatic, and medicinal plants have been studied for their antioxidant potential (Chomchalow et al., 1996). The herbs and spices commonly used to flavor dishes are massive sources of phenolic compounds, which have shown good antioxidant activity. The influential role of clove in preventing various degenerative diseases is due to various chemical components in concentrations with antioxidant activity (Astuti et al., 2019). Chemical constituents with antioxidant activity found in high concentrations in plants determine their considerable role in preventing various degenerative diseases (Shan et al., 2005). The purpose of this work is to evaluate the chemical composition of clove fixed oil and its health-promoting effects from different parts of *Syzygium aromaticum* (clove).

2. Chemical composition of clove fixed oil

Nonvolatile components in clove fixed oil add flavor, mainly taste. At present, not much data are available about nonvolatile components in clove. However, the characterization of nonvolatile components is crucial to enhance clove's value. To identify nonvolatile components, compound extraction was done by maceration in ethanol. Nonvolatile compounds are characterized by liquid chromatography-tandem mass spectrometry (LC-MS/MS). LC-MS/MS was operated in negative mode with 0.1% formic acid in water and acetonitrile as the mobile phase. Nonvolatile components were characterized by fragment analysis and compared to references. Various compounds were identified and characterized such as monogalloylglucose, asquinic acid, digalloylglucose, gallic acid, isobiflorin, biflorin, ellagic acid, hydroxygallic acid, quercetin, luteolin, kaempferol, naringenin, dimethoxyluteolin, isorhamnetin, and rhamnetin (Fathoni et al., 2016).

Nonvolatile compounds of clove fixed oil were detected by Milind and Deepa (2011), including triterpenes, tannins, flavonoids, and sterols. The fixed oil of clove contains 10%–13% tannins, which have the same chemical composition as gallotannic acid. Therefore, ellagitannin and eugenin were extracted from cloves (Nonaka et al., 1980). Two ellagitannins—syzyginin A (1,2,3-tri-*O*-galloyl-4, 6-(*S*)-tergalloyl-β-D-glucoside) and syzyginin B—were isolated from the leaves. Cloves contain about 2% oleanolic acid and triterpene. In addition, eugenol glucoside gallate, a chromone *C*-glycoside, galloyl, and hexahydroxy diphenyl esters of 2,4,6-trihydroxy acetophenone-3-glucopyranoside were identified in clove leaves (Tanaka et al., 1993).

Sterols isolated from clove include sitosterol, stigmasterol, and campesterol (Brieskorn et al., 1975). Maslinic acid was also isolated from clove buds (Narayanan and Natu, 1974). From clove, 2α-hydroxyoleanolic acid was also isolated. From the ethanol extract of the seeds, apigenin 6-*C*-(β-D-xylopyranosyl-(1→2″)-β-D-galactopyranoside)-7-*O*-β-D-glucopyranoside and apigenin-6-*C*-(β-D-ylopyranosyl-(1→2″)-β-D-galactopyranoside)-7-*O*-β-D-(6-*O*-pcoumarylglucopyranoside) were isolated (Nassar, 2006). Also isolated were achromone C-glucoside from the ethanol extract of the seeds; apigenin 6-*C*-(β-D-xylopyranosyl-(1→2″)-, β-D-galactopyranoside)-7-*O*-β-D-glucopyranoside, and apigenin-6-*C*-(β-D-ylopyranosyl-(1→2″)-β-D-galactopyranoside)-7-*O*-β-D-(6-*O*-pcoumarylglucopyranoside) (Nassar, 2006).

The antioxidant activity of clove fixed oil was characterized with clove volatile oils extracted from the dry buds of *Syzygium aromaticum* with the reducing power assay, total flavonoid content (TFC), total phenolic content (TPC), and the scavenging activity on 1,1-diphenyl-2-picrylhydrazyl (DPPH$^{\bullet}$). It was observed that clove fixed oil had higher TFC and TPC than clove volatile oil by comparing with the standard (pyrogallol, quercetin). A DPPH$^{\bullet}$ radical scavenging activity assay assessed the ability of the clove extract to donate hydrogen or scavenge free radicals. As DPPH$^{\bullet}$ radical is a stable free radical, when reacting with an antioxidant compound that can donate hydrogen, it is reduced to diphenylpicryl hydrazine. When clove oil buds are mixed with the DPPH$^{\bullet}$, they are decolorized due to the hydrogen donating capability. It was concluded that the scavenging activity of clove bud fixed oil at all concentrations from 100 to 500 μg/mL is stronger than volatile oil. Milind and Deepa (2011) revealed that clove seeds contain 5.98% protein, 20% total fat, 61.2% carbohydrates, 34.2% fiber, and 5.88% ash. However, they also contain high potassium (1102 mg/100 g) and magnesium (268 mg/100 g) levels. Milind and Deepa (2011) also found total saturated fatty acids (5.38%), total monounsaturated fatty acids (1.47%), and total polyunsaturated acids (7.09%). Fahad Mohammed and Mohammed (2012) studied the composition of cress, mustard, black cumin, fenugreek, black pepper, and clove seeds. According to them, clove oil contains crude fat (16.6%), protein (6.9%) fiber (11.4%), and ash (5.9%). In addition, clove oil has been found to be a rich source of potassium, magnesium, calcium, zinc, manganese, copper, and iron (650, 97, 270, 0.7, 43, 0.8, and 36 mg/100 g, respectively). Total saturated fatty acids and total unsaturated fatty acids in clove oil were 12.1% and 87.8%. Myristic acid (1.29), palmitic acid (6.21%), palmitoleic acid (20.9%), oleic caid (13%), elaidic acid (6.2%), linoleic acid (44.7), α-Linolenic acid (2.9%), and arachidonic acid (4.68%) were reported. The differences in clove oil content and composition depend mainly on several factors such as pretreatments, variety, agro-ecological conditions, and extraction processes (Hastuti et al., 2017).

3. Extraction of clove oil

Different extraction procedures mainly show extremely varying results on the compositions of *Syzygium aromaticum* (Wei et al., 2016). Furthermore, various experiments with a similar extraction procedure can achieve different results on the constituents of clove oil. For example, when clove buds are extracted with steam distillation and detected with GC-MS, only four components were detected by are detected by Li et al. (2016): 3-allyl-2-methoxyphenol (69.7%), 3-phenyl-2-propen aldehyde (14.3%), caryophyllene (13.7%), and alpha-caryophyllene (2.17%). However, 20 kinds of components were identified by Kong et al. (2004), among which the relative content of eugenol was 68%. The content of isocaryophillene (16.6%), eugenol acetate (9.3%), α-caryophyllene (2.0%), and cadina-1(10),4-diene (1.2%) were higher. Thus, it can be concluded that varying constituents of *Syzygium aromaticum* are obtained from different producing regions and extraction methods.

Similarly, the characterization of nonvolatile components is essential to enhance clove's nutritional value. Compound extraction was carried out by maceration in ethanol. The fractionation process of the extract was carried out by utilizing gravity column chromatography on silica gel and Sephadex LH-20 as the stationary phase. The nonvolatile components of clove in clove fixed oil were characterized by LC-MS/MS. LC-MS/MS was operated in negative mode with 0.1% formic acid in water and acetonitrile as the mobile phase. Nonvolatile components were characterized by fragment analysis and compared to references (Fathoni et al., 2016).

4. Pharmacological activities of clove

4.1 Antioxidant/free radical scavenging activity

Clove essential oil has the highest antioxidant activity and is considered one of the best oils for food and supplementation. Because of these characteristics, it has been included in shelf-life extension formulations for various foods. Compared with the standard (quercetin, pyrogallol), fixed oil had the highest flavonoid and phenolic content over clove volatile oil. DPPH$^•$ scavenging activity assessed the ability of clove extract to scavenge free radicals or to donate hydrogen ions. DPPH$^•$ radical is suggested as a stable radical, but when it reacts with an antioxidant, it is reduced to diphenylpicryl hydrazine. When the clove oil buds mixed with DPPH$^•$, it was decolorized because of the hydrogen donating capacity. It was also observed that the scavenging ability of clove bud fixed oil ranged from 100 to 500 μg/mL stronger than volatile oil. According to the research studies of Attitalla et al. (2015), the extraction of both volatile and fixed oil from *Syzygium aromaticum* gives it the potential to be a natural antioxidant.

Eugenol and clove have a strong antioxidant effect, which can be compared to the activities of synthetic antioxidants such as pyrogallol and butylated hydroxyl anisole (BHA) (Dorman et al., 2000). At 15 μg/mL concentration, clove oil prevented 97% lipid peroxidation. The essential oil showed scavenging activity against the 2,2-diphenyl-1-picryl hydrazyl (DPPH$^•$) radical at concentrations lower than the concentrations of BHA, butylated hydroxytoluene (BHT), and eugenol (Gulcin et al., 2012). Likewise, in another study, it was observed that kidney functions, enhanced liver functions, and antioxidant level in rats treated with clove showed its protective role against H_2O_2-induced cell damage; this was because of the active compounds in the plant extract and essential oil (Abozid and Sayed, 2013).

4.2 Anticarcinogenic activity

Clove essential oil has been shown to have anticancer (Zheng et al., 1992) and antimutagenic effects due to its intense free radical scavenging activity (Miyazawa and Hisama, 2001). Because of clove fixed oil's antioxidant potential compared to clove essential oil, it can also serve as an antimutagenic and anticarcinogenic agent. A few studies recommended the chemopreventive role of clove essential oil, predominantly in the treatment of skin, digestive system, and lung cancers (Schonfelder and Schonfelder, 2004). Ethyl acetate extract of clove obstructs cancer or tumor growth and also encourages apoptosis and cell cycle arrest. Oleanolic acid is one of the components of the ethyl acetate extract of clove and was observed to be responsible for the anticancer effect. Its anticancer mechanism was the promotion of Go/G1 cell cycle arrest and apoptosis in a dose-dependent way (Liu et al., 2014).

4.3 Antibacterial activity

In a study by Attitalla et al. (2015), the fixed oil and volatile oil of *Syzygium aromaticum* were examined for antimicrobial properties against some pathogens such as *Pseudomonas aeruginosa*, *Escherichia coli*, *Acinetobacter* sp., *Klebsiella pneumonia*, and *Staphylococcus aureus* (Attitalla et al., 2015). The clove fixed oil and volatile oils isolated produced considerable inhibition against all these pathogens. Furthermore, the clove fixed oil extracted was more effective than the volatile oils extracted compared with standard drugs. Thus, this study concludes that both clove fixed and volatile oils isolated from *Syzygium aromaticum* are potential natural antioxidants.

Various studies have suggested the potential antibacterial effects of clove (Gupta et al., 2014; Fu et al., 2007), which is considered to be due to the presence of several constituents, mainly eugenol, eugenyl acetate, β-caryophyllene, 2-heptanone (Chaieb et al., 2007) acetyl-eugenol, α-humulene, methyl salicylate, isoeugenol, methyl-eugenol (Yang et al., 2003), phenyl propanoides, dehydrodieugenol, *trans*-confireryl aldehyde, biflorin, kaempferol, rhamnetin, myricetin, gallic acid, ellagic acid, and oleanolic acid (Cai and Wu, 1996). Most of these compounds are also present in clove fixed oil (Fathoni et al., 2016). These constituents denature proteins and react phospholipids within the cell membrane, changing their permeability. In addition, clove oil was effective against nontoxigenic strains of *E. coli* O157:H7 (Burst and Reinders, 2003). In another study, clove oil was observed to be effective against foodborne Gram-positive bacteria (*B. cereus*, *Staphylococcus aureus*, *Listeria monocytogenes*, and *Enterococcus faecalis*) and Gram-negative bacteria (*Yersinia enterocolitica*, *E. coli*, *P. aeruginosa*, and *Salmonella choleraesuis*) (Gupta et al., 2014). Demirpek et al. (2009) examined the ethanolic and aqueous extracts of clove bud oil and found it to inhibit the growth of methicillin-resistant clinical isolates at 1000 and 500mg/mL concentrations (Ali et al., 2005).

4.4 Antiinflammatory activity

Clove oil acts as an expectorant that clarifies respiratory passages and treats many upper respiratory tract infections, including eye styes, colds, sinus conditions, bronchitis, and asthma cough (Ozturk and Ozbek, 2005). Clove has been utilized as a traditional folk medicine to prevent musculoskeletal pain and nasal obstruction, suggesting its antiinflammatory properties and the property due to COX-2 inhibition (Kim et al., 2003). Clove oil comprises various flavonoids, including kaempferol, rhamnetin, and β-caryophyllene, contributing to its antioxidant and antiinflammatory properties (Ghelardini et al., 2001). Clove has an antiinflammatory effect compatible with etodolac at 0.025 and 0.1 mL/kg and with indomethacin at 0.05 and 0.2mL/kg doses (Daniel et al., 2009). Clove fixed oil has higher flavonoid content over clove volatile oil compared with the standard (pyrogallol, quercetin).

4.5 Antithrombotic activity

Clove oil effectively prevents human platelet aggregation induced by platelet-activating factor (PAF), arachidonic acid (AA), or collagen. Clove was found to be a more active inhibitor for aggregation induced by arachidonic acid and platelet-activating factor (6μM and IC_{50}: 4, respectively) than collagen (132μM: IC_{50}). A study on rabbits suggested that clove oil (50–100mg/kg) afforded 70% protection against AA (2.0mg/kg, i.v.) and 100% protection against PAF (11mg/kg, i.v.) induced thrombosis and shock due to pulmonary platelet thrombosis. It also inhibited thromboxane-A2 and 12-HETE production by human platelets (Kamatou et al., 2012; Saeed and Gilani, 1994).

4.6 Antidiabetic activities

In vivo studies were used to examine an increase in blood sugar and lipid peroxidation in diabetic rats, and this established the antioxidant enzyme levels with the nutrition administration of cloves (Koh et al., 2013). The clove fixed oil has been observed in various studies to have more antioxidant potential than clove essential oils.

The antidiabetic efficacy of *Syzygium aromaticum* extracts may be attributed to the existence of insulin-stimulating agents. Normal blood sugar has been enhanced in in vivo experiments in *Syzygium aromaticum* extract-treated mice (Kuroda et al., 2012; Sanae et al., 2014). Moreover, Jung et al. (2012) suggested that diet supplemented with *Syzygium aromaticum* extracts reduced serum insulin, leptin, and hepatic lipid levels. Similarly, the body weight of high-fat diet mice suggested its natural antiobesity property and its capability to reduce hepatic lipid accumulation. Thus, the antidiabetic property of *Syzygium aromaticum* extracts may be ascribed to the presence of insulin-stimulating agents (Kuroda et al., 2012). Furthermore, various studies suggested that normal blood sugar was improved in *Syzygium aromaticum* extract-treated mice (Kuroda et al., 2012; Sanae et al., 2014).

4.7 Other pharmacological activities

Clove oil is used in various medicines as an antiseptic to treat toothaches, oral diseases, allergy disorders, acne, scars, asthma, and rheumatoid arthritis (Kim et al., 2003; Wongsawan et al., 2019). To inhibit and control viral infections, several herbal medicines have been utilized to discover new pharmaceutical medicines (Astani Ekor et al., 2011). Moreover, other phytochemical components extracted from *Syzygium aromaticum,* including sanguinarine and benzo phenanthridine alkaloids, have been reported to protect against liver damage (Ali et al., 2014). Shyamala et al. (2003) revealed that clove consumption proved to recover ALT, urea, AST, and lipid levels in kidneys, serum, and liver compared to normal values in hyperlipidemic rats.

5. Toxicity doses

The safety of clove oil, clove buds, oleoresins, and eugenol as a food supplement has been confirmed by the US Food and Drug Administration (FDA). Moreover, recently there has been considerable attention toward their toxicity (Vijayasteltar et al., 2016). The acceptable daily quantity of clove in humans recommended by the World Health Organization (WHO) is 2.5 mg/kg body weight (Ogunwande et al., 2005). Johannah et al. (2015) suggested the considerable detoxification and cardiac health effects in humans by lowering lipid peroxidation and enhancing the endogenous redox enzyme levels.

6. Conclusion

In this chapter, all the phytochemical molecules isolated from *Syzygium aromaticum* fixed oil and their medicinal properties were discussed. Various compounds have been isolated from fixed oil, including asquinic acid, monogalloylglucose, gallic acid, digalloylglucose, isobiflorin, biflorin, ellagic acid, hydroxygallic acid, luteolin, quercetin, naringenin, kaempferol, isorhamnetin, dimethoxyluteolin, and rhamnetin. Pharmacologically, clove fixed oil elements possess antimicrobial, antioxidant, anticarcinogenic, antiinflammatory, antithrombotic, antidiabetic, and antibacterial effects.

References

Abozid, M.M., Sayed, E.L., 2013. Antioxidant and protective effect of clove extracts and clove essential oil on hydrogen peroxide treated rats. Int. J. ChemTech Res. 5 (4), 1477–1485.

Abushouk, A.I., Negida, A., Ahmed, H., Abdel-Daim, M.M., 2017. Neuroprotective mechanisms of plant extracts against MPTP induced neurotoxicity: future applications in Parkinson's disease. Biomed. Pharmacother. 85, 635–645.

Ali, S.M., Khan, A.A., Ahmed, I., Musaddiq, M., Ahmed, K.S., Polasa, H., 2005. Antimicrobial activities of eugenol and cinnamaldehyde against the human gastric pathogen *Helicobacter pylori*. Ann. Clin. Microbiol. Antimicrob. https://doi.org/10.1186/1476-0711-4-20.

Ali, S., Prasad, R., Mahmood, A., Routray, I., Shinkafi, T.S., Sahin, K., Kucuk, O., 2014. Eugenol-rich fraction of *Syzygium aromaticum* (clove) reverses biochemical and histopathological changes in liver cirrhosis and inhibits hepatic cell proliferation. J. Cancer Prev. 19 (4), 288–300.

Astani Ekor, A., Reichling, J., Schnitzler, P., 2011. Screening for antiviral activities of isolated compounds from essential oils. Evid. Based Complement. Alternat. Med. 2011. Article ID 253643. 8 pages.

Astuti, R.I., Listyowati, S., Wahyuni, W.T., 2019. Life span extension of model yeast Saccharomyces cerevisiae upon ethanol derived-clover bud extract treatment. IOP Conf. Ser.: Earth Environ. Sci. 299, 012059.

Attitalla, I.H., Elmhdwi, M.F., Dwivedi, S., 2015. Antibacterial screening of fixed and volatile oils extracted from *S. aromaticum* (clove). Int. J. Pharm. Life Sci. 6, 4740–4745.

Batiha, G.E.S., Beshbishy, A.A., Tayebwa, D.S., Shaheen, M.H., Yokoyama, N., Igarashi, I., 2018. Inhibitory effects of *Uncaria tomentosa* bark, *Myrtus communis* roots, *Origanum vulgare* leaves and *Cuminum cyminum* seeds extracts against the growth of Babesia and Theileria in vitro. Jpn. J. Vet. Parasitol. 17, 1–13.

Beshbishy, A.M., Batiha, G.E.S., Adeyemi, O.S., Yokoyama, N., Igarashi, I., 2019. Inhibitory effects of methanolic *Olea europaea* and acetonic *Acacia laeta* on the growth of Babesia and Theileria. Asian Pac J Trop Med 12, 425–434.

Bhowmik, D., Kumar, K.P.S., Yadav, A., Srivastava, S., Paswan, S., Dutta, A.S., 2012. Recent trends in Indian traditional herbs *Syzygium Aromaticum* and its health benefits. J. Pharmacol. Phytochem. 1, 13–23.

Brieskorn, C.H., Munzhuber, K., Unger, G., 1975. Crataegolsaure and steroid glukoside aus blutenknospen von *Syzygium aromaticum*. Phytochemistry 14, 2308–2309.

Burst, S.A., Reinders, R.D., 2003. Antibacterial activity of selected plant essential oils against *Escherichia coli* O157:H7. Lett. Appl. Microbiol. 3, 162–167.

Cai, L., Wu, C.D., 1996. Compounds from *Syzygium aromaticum* possessing growth inhibitory activity against oral pathogens. J. Nat. Prod. 59 (10), 987–990.

Chaieb, K., Zmantar, T., Ksouri, R., Hajlaoui, H., Mahdouani, K., Abdelly, C., Bakhrouf, A., 2007. Antioxidant properties of essential oil of *Eugenia caryophyllata* and its antifungal activity against a large number of clinical Candida species. J. Mycosis 50 (5), 403–406.

Chomchalow, N., 1996. Spice production in Asia-An overview; Proceedings of the Conference IBC's Asia Spice Markets 96 Conference; Singapore. 27–28 May 1996.

Cortés-Rojas, D.F., de Souza, C.R., Oliveira, W.P., 2014. Clove (*Syzygium aromaticum*): a precious spice. Asian Pac. J. Trop. Med. 4, 90–96.

Daniel, A.N., Sartoretto, S.M., Schmidt, G., Caparroz-Assef, S.M., Bersani-Amado, C.A., Cuman, R.K.N., 2009. Anti-inflammatory and antinociceptive activities of eugenol essential oil in experimental animal models. J. Rev. Bras. Farmacogn. 19, 212–217.

Demirpek, U., Olgun, A., Kısa, O., Guvenç, A., 2009. *In vitro* antibacterial activity of cloves (*Syzygium aromaticum*) against MRSA. J. Planta Med. 75, 59.

Dorman, H.J.D., Figueiredo, A.C., Barroso, J.G., Deans, S.G., 2000. *In vitro* evaluation of antioxidant activity of essential oils and their components. Flavour Fragr. J. 15, 12–16.

Ekor, M., 2014. The growing use of herbal medicines: issues relating to adverse reactions and challenges in monitoring safety. Front. Pharmacol. 4, 177.

Fahad Mohammed, A., Mohammed, S.A., 2012. Chemical composition and fatty acid content of some spices and herbs under Saudi Arabia conditions. Sci. World J. https://doi.org/10.1100/2012/859892. Article ID 859892. 5 pages.

Fathoni, A.E., Saepudin, A.H., Cahyana, D.U.C., Rahayu, Haib, J., 2016. Identification of nonvolatile compounds in clove (*Syzygium aromaticum*) from Manado. AIP Conf. Proc. 1862. https://doi.org/10.1063/1.4991183. 030079-1-030079-10.

Fu, Y., Zu, Y., Chen, L., Shi, X., Wang, Z., Sun, S., Efferth, T., 2007. Antimicrobial activity of clove and rosemary essential oils alone and in combination. J. Phytother. Res. 21 (10), 989–994.

Ghelardini, C., Galeotti, N., Di Cesare Mannelli, L., Mazzanti, G., Bartolini, A., 2001. Local anaesthetic activity of beta-caryophyllene. J. Farmacol. 56 (5–7), 387–389.

Gulcin, I., Elmastas, M., Hassan, Y., Enein, A., 2012. Antioxidant activity of clove oil—a powerful antioxidant source. Arab. J. Chem. 5, 489–499.

Gupta, N., Mittal, M., Parashar, P., Mehra, V., Khatri, M., 2014. Antibacterial potential of *Elletaria cardamomum*, *Syzygium aromaticum* and *Piper nigrum*, their synergistic effects and phytochemical determination. J. Pharm. Res. 8 (8), 1091–1097.

Hastuti, L.T., Saepudin, E., Cahyana, A.H., Rahayu, D.U.C., Murni, V.W., Haib, J., 2017. The influence of sun drying process and prolonged storage on composition of essential oil from clove buds (*Syzygium aromaticum*). AIP Conf. Proc. 2017 (1862), 030092.

Johannah, N., Renny, R., Gopakumar, G., Balu, M., Sureshkumar, D., Krishnakumar, I.M., 2015. Beyond the flavour: a de-flavored polyphenol-rich extract of clove buds (*Syzygium aromaticum* L) as a novel dietary antioxidant ingredient. Food Funct. 6, 3373–3382.

Jung, C.H., Ahn, J., Jeon, T.I., Kim, T.W., Ha, T.Y., 2012. *Syzygium aromaticum* ethanol extract reduces high-fat diet-induced obesity in mice through downregulation of adipogenic and lipogenic gene expression. Exp. Ther. Med. 4, 409–414.

Kamatou, G.P., Vermaak, I., Vilijoen, A.M., 2012. Eugenol-from the remote Maluku islands to the International market place: a review of a remarkable and versatile molecule. J. Mol. 17, 6953–6981.

Kim, E.H., Kim, H.K., Ahn, Y.J., 2003. Acaricidal activity of clove bud oil compounds against Dermatophagoides farinae and Dermatophagoides pteronyssinus (Acari: Pyroglyphidae). J. Agric. Food Chem. 51 (4), 885–889.

Koh, T., Murakami, Y., Tanaka, S., Machino, M., Sakagami, H., 2013. Re-evaluation of anti-inflammatory potential of eugenol in IL-1beta-stimulated gingival fibroblast and pulp cells. In Vivo 27, 269–273.

Kong, Q.L., Song, Y.Z., Zhang, L.L., Chen, L.Y., Li, Q.F., 2004. Natural antifungal compounds from *Syzygium aromaticum* (L.) Merr. et perry. Acta Agric. Shanghai 20, 68–72.

Kuroda, M., Mimaki, Y., Ohtomo, T., Yamada, J., Nishiyama, T., Mae, T., Kishida, H., Kawada, T., 2012. Hypoglycemic effects of clove (*Syzygium aromaticum* flower buds) on genetically diabetic KK-Ay mice and identification of the active ingredients. J. Nat. Med. 66, 394–399.

Li, S., Guo, C., Zhang, Y., Zhang, W., Wang, G., Chai, X., Wang, L., 2016. The detection of components in volatile oils of traditional Chinese medicine by gas chromatography-mass spectrometry. Heilongjiang Med. Pharm. 39, 8–9.

Liu, H., Schmitz, J.C., Wei, J., Cao, S., Beumer, J.H., Strychor, S., 2014. Clove extract inhibits tumor growth and promotes cell cycle arrest and apoptosis. J. Oncol. Res. 5, 247–259.

Milind, P., Deepa, K., 2011. Clove: a champion spice. Int. J. Res. Ayurveda Pharm. 2 (1), 47–54.

Miyazawa, M., Hisama, M., 2001. Suppression of chemical mutagen induced SOS response by alkylphenols from clove (*Syzygium aromaticum*) in *Salmonella typhymurium* TA153/pSK1002 umu test. J. Agric. Food Chem. 49, 4019–4025.

Narayanan, C.R., Natu, A.A., 1974. Triterpene acids of Indian clove buds. Phytochemistry 13 (9), 1999–2000.

Nassar, M.I., 2006. Flavonoid triglycosides from the seeds of *Syzygium aromaticum*. Carbohydr. Res. 341, 160–163.

Nonaka, G., Harada, M., Nishioka, I., 1980. Eugenin, a new ellagitannin from cloves. Chem. Pharm. Bull. 28, 685–687.

Ogunwande, I., Olawore, N., Ekundayo, O., Walker, T.M., Schmidt, J.M., Setzer, W.N., 2005. Studies on the essential oils composition, antibacterial and cytotoxicity of Eugenia uniflora L. Int. J. Aromather. 15, 147–152.

Ozturk, A., Ozbek, H., 2005. The anti-inflammatory activity of *Eugenia caryophyllata* essential oil. An animal model of anti-inflammatory activity. Eur. J. Gen. Med. 2 (4), 159–163.

Saeed, S.A., Gilani, A.H., 1994. Antithrombotic activity of clove oil. J. Pak. Med. Assoc. 44 (5), 112–115.

Sanae, F., Kamiyama, O., Ikeda-Obatake, K., Higashi, Y., Asano, N., Adachi, I., Kato, A., 2014. Effects of eugenol-reduced clove extract on glycogen phosphorylase b and the development of diabetes in db/db mice. Food Funct. 5, 214–219.

Sarrami, N., Pemberton, M., Thornhill, M., Theaker, E.D., 2002. Adverse reactions associated with the use of eugenol in dentistry. Br. Dent. J. 193, 253.

Schonfelder, I., Schonfelder, P., 2004. *Syzygium aromaticum* (L.) MERR & PERR *Eugenia caryophyllata* THUNB *Jambosa caryophyllus* (SPRENG.) (NIEDENZU). In: Das neue Handbuch der Heilpflanzen. 2004. Kosmos-Verlag, Stuttgart, Germany, pp. 431–432.

Shan, B., Cai, Y.Z., Sun, M., Corke, H., 2005. Antioxidant capacity of 26 spice extracts and characterization of their phenolic constituents. J. Agric. Food Chem. 53, 7749–7759.

Shyamala, M.P., Venukumar, M.R., Latha, M.S., 2003. Antioxidant potential of the *Syzygium aromaticum* (gaertn) Linn (cloves) in rats fed with high-fat diet. Indian J. Pharmacol. 35, 99–103.

Tanaka, T., Orii, Y., Nonaka, G., Nishioka, I., 1993. Tannins and related compounds. CXXIII. Chromone, acetophenone and phenyl propanoid glycosides and their galloyl and/or hexahydroxy phenoyl esters from leaves of *Syzygium aromaticum* Merr and Perry. Chem. Pharm. Bull. 28, 685–687.

Vijayasteltar, L., Nair, G.G., Maliakel, B., Kuttan, R., Krishnakumar, I.M., 2016. Safety assessment of a standardized polyphenolic extract of clove buds: subchronic toxicity and mutagenicity studies. Toxicol. Rep. 3, 439–449.

Wei, M.C., Xiao, J., Yang, Y.C., 2016. Extraction of α-Astuti mulene-enriched oil from clove using ultrasound-assisted supercritical carbon dioxide extraction and studies of its fictitious solubility. Food Chem. 210, 172–181.

Wongsawan, K., Chaisri, W., Tangtrongsup, S., Mektrirat, R., 2019. Bactericidal effect of clove oil against multidrug-resistant *Streptococcus suis* isolated from human patients and slaughtered pigs. Pathogens 9, E14.

Yang, Y.C., Lee, S.H., Lee, W.J., Choi, D.H., Ahn, Y.J., 2003. Ovicidal and adulticidal effects of *Eugenia cryophyllata* bud and leaf oil compounds on *Pediculus capitis*. J. Agric. Food Chem. 51 (17), 4884–4888.

Zheng, G.Q., Kenny, P.M., Lam, K.T., 1992. Sesquiterpenes from clove (*Eugenia carrophyllata*) as potential anticarcinogenic agents. J. Nat. Prod. 55, 999–1003.

CHAPTER

26

Clove oil as bio-additives in biodiesel-diesel fuel

Elaine Elaine and Kar Lin Nyam

Faculty of Applied Sciences, UCSI Heights, UCSI University, Kuala Lumpur, Malaysia

Abbreviations

BSFC	brake-specific fuel consumption
BTE	brake thermal efficiency
CI	compression ignition
CO	carbon monoxide
CO_2	carbon dioxide
HC	hydrocarbon
NO_x	nitrogen oxides
PM	particulate matters

1. Introduction

1.1 Clove oil

Clove (*Syzygium aromaticum*) is a native plant from the Maluku islands in the eastern part of Indonesia. It has a synonym scientific name *Eugenia caryophyllata,* part of the Myrtaceae family (Nowak et al., 2012). This aromatic flower bud contains eugenol as its main volatile constituent functional compound. Eugenol is responsible for the distinctive pleasant aroma of clove and other beneficial properties such as antioxidants, health-promoting properties, and spices. Therefore, it has been utilized in pharmaceutical, food, fragrance, agriculture, cosmetics, and other industrial sector applications (Nowak et al., 2012; Hatami et al., 2019). The economic trade of this valuable spice has been widely pursued, especially in Asian countries (Kamatou et al., 2012).

Clove essential oil can be obtained from its plant parts such as leaves, flowers, and stems. The dominant phenol is eugenol or 2-methoxy-4-(2-propenyl) (Khalil et al., 2017). Eugenol, the main constituent in clove oil, has high oxygen content in its molecular structure, contributing to desirable effects toward fuel properties. Besides eugenol, clove oil contained different phenolic acids: gallic, caffeic, ferulic, elagic, and salicylic acids. Flavonoids such as kaempferol, quercetin, and their glycosylated derivatives are also found in clove oil at a lower concentration. These compounds contribute to excellent antioxidant properties and other health-benefiting properties (Kamatou et al., 2012; Cortés-Rojas et al., 2014).

Clove (*Syzygium aromaticum*). https://doi.org/10.1016/B978-0-323-85177-0.00024-0

The versatile application of clove oil has been investigated by researchers in various sectors, including the fuel industry. It is commonly used to prevent lipid oxidation and is widely applied in the pharmaceutical and food industry, but limited information was investigated in the biofuel industry (Souza et al., 2018; Ramos et al., 2021). Several studies have been conducted to assess the benefits of using clove oil as fuel additives for the combustion process, which will be discussed in this chapter. Clove has its molecular formula of $C_7H_{12}ClN_3O_2$ with a molecular weight of 205.6 g/mol. In general, it has a boiling point around 250°C and a flashpoint around >121°C. The flashpoint indicates that the flammability temperature of clove oil is lower than pure diesel fuel. This is beneficial because it indicates that the fuel additives can be stored and transported at room temperature.

The physical state of clove oil at room temperature is liquid, and the density is 1.05 g/mL at 25°C. Due to its physical and chemical properties, few researchers had tried to investigate the suitability of clove oil as an alternative fuel for the engine (Mbarawa, 2008; Waweru et al., 2019). The cetane number of clove oil is not fulfilling the requirement as an alternative fuel, but it could be blended as additives (Mbarawa, 2008; Ramos et al., 2021). Its high source of antioxidants can contribute to good oxidation stability and emission reduction by improving engine performance and combustion process (Jeyakumar and Narayanasamy, 2019; Waweru et al., 2019). Besides, clove oil has a lower heating value than pure diesel fuel. The oxygen concentration in clove oil promotes the combustion process, which leads to lower exhaust emissions such as hydrocarbons (HC), particulate matter (PM), and carbon monoxide (CO) (Mbarawa, 2008).

1.2 Extraction of clove oil

There are conventional and modern extraction techniques that researchers are still using to obtain plant oils. Conventionally, clove oil can be obtained through solvent extraction, hydrodistillation, and molecular distillation extraction techniques. Hydrodistillation is the most widely used extraction technique to obtain essential oil. The clove was dried and pulverized into clove powder. Afterward, the clove oil was extracted through clove powder by steam distillation (Khalil et al., 2017; Waweru et al., 2019). Clove powder was weighed into a flat bottom flask and soaked in deionized water for 15 min. The steam distillation setup was assembled, and a high-temperature setting of 300°C was used to heat the mixture to boiling. Afterward, the cloudy suspensive was obtained and poured into a separatory funnel. An extraction solvent was added to dissolve the essential oil by several cycles of extraction. The layers containing bioactive compounds (typically top layers) were combined, cleaned, and dried over sodium sulfate (Waweru et al., 2019). In recent years, hydrodistillation has been the leading extraction technique that researchers have conducted to obtain clove oil for fuel experiments (Mbarawa, 2008; Jeyakumar and Narayanasamy, 2019; Waweru et al., 2019).

However, modern extraction techniques have been extensively investigated to obtain clove essential oils with desirable antioxidant properties. For instance, some of them are ultrasonic-assisted extraction and supercritical fluid extraction (Khalil et al., 2017). According to Nowak et al. (2012) and Hatami et al. (2019), the combination of supercritical fluid extraction assisted with cold pressing shows the good economic value and desirable bioactive yield of the clove oil. Besides producing high-quality essential oil, these modern extraction techniques are considered green processing technology that negatively disrupts the environment (Kapadiya and Desai, 2017; Khalil et al., 2017).

1.3 Biodiesel-diesel fuel

Fuel has been an essential source to support the engine's primary function, commonly used in transportation, household, manufacturing sectors, and others (Chauhan et al., 2016; Sriram, 2016; Soudagar et al., 2018). Conventionally, petroleum which consists of aliphatic hydrocarbons with different boiling temperatures from 130 to 370°C has been widely used as diesel fuel (Pirouzfar et al., 2012; Pirouzfar and Fayyazbakhsh, 2016). However, the usage of fossil fuels has emerging environmental issues such as the exploitation of unrenewable natural resources and air pollution produced when burning fossil fuels. Furthermore, these compounds present in the air contribute to climate change, leading to interactive hurricanes, heavy rains, floods, and even erosive enough to damage historical buildings and statues, especially in the urban areas (Vijayakumar et al., 2016; Fayyazbakhsh and Pirouzfar, 2017).

Biodiesel is the alternative fuel that has been studied to tackle environmental issues such as exhaust emission from the combustion process (Saluja et al., 2016; Vijayakumar et al., 2016). Biodiesel is the alkyl ester of fatty acids obtained through transesterification of plant or animal oils. Through the transesterification process, the complex organic compounds in the actual oil are broken down to ensure the quality of engine performance (Ramos et al., 2018; Gamayel et al., 2020a). One of the environment-friendly substitutions suggested is plant seed oils. It is sustainable, renewable, and cleaner to the environment, a suitable alternative fuel for compression ignition (CI) engines (Vijayakumar et al., 2016; Ramos et al., 2021). However, plant-based oils generally contain unsaturated fatty acids that contribute to biodiesel fuel's poor oxidative stability and might cause clogging in the engine (Dueso et al., 2018; Kusrini et al., 2018). The esters of unsaturated fatty acids are unstable to light, catalytic systems, and atmospheric oxygen (Ramos et al., 2021). Therefore, many automobile companies have not considered the destructive environmental impact of fossil fuel degradation when developing diesel engines and vehicles due to the good oxidative stability of fossil fuels (Tamilselvan et al., 2017; Soudagar et al., 2018).

Although natural plants contain lots of natural antioxidants, the transesterification process to biodiesel had reduced the amount significantly (Varatharajan and Pushparani, 2018). One of the significant issues associated with biodiesel usage is maintaining the fuel at specified standards for a more extended period. Biodiesel is more prone to oxidization than mineral diesel; the complete degradation occurs after four weeks (Xin et al., 2009). The final products of oxidation alter the physical and chemical properties of the fuel, resulting in the formation of insoluble gums that can plug fuel filters. This instability of biodiesel is a long-standing issue and it has not yet been satisfactorily resolved. Nevertheless, it could be a significant barrier to market expansion (Balaji and Cheralathan, 2015; Ramalingam et al., 2019).

In recent years, researchers have investigated the optimization strategies of using biodiesel fuel to reduce fossil fuel usage. Although biodiesel may considerably reduce PM, CO, and HC emissions, the nitrogen oxides (NO_x) emission increased (Subramanian et al., 2018; Varatharajan and Pushparani, 2018). Therefore, the blending of biodiesel and diesel fuel is introduced to reduce harmful exhaust emissions. The biodiesel-diesel fuel demonstrates a significant reduction of exhaust emission and good stability to improve the engine performance characteristics. According to the literature review, 20% of biodiesel incorporated with diesel blends gives better engine performance and reduces exhaust emission noticeably (Chauhan et al., 2016; Ramalingam et al., 2019).

1.4 Fuel additives

The usage of fuel additives is recommended to improve the stability and emission characteristics of the fuel combustion process. It might be a promising and cost-effective approach to improve the oxidative stability of biodiesel fuel (Varatharajan and Pushparani, 2018; Freitas et al., 2019). To impart desirable fuel combustion characteristics, different additives such as antioxidants, improvers for cetane number, and cold flow properties improvers can be added to fuel mixture (Dueso et al., 2018). Nevertheless, the research on additives for biodiesel remains in its infancy stage (Ali et al., 2016). Some antioxidants are added to biodiesel, such as tertiary butylhydroquinone (TBHQ), butylated hydroxytoluene (BHT), ascorbyl palmitate, beta-hydroxy acid (BHA), propyl gallate, tocopherols, and flavonoids (Xin et al., 2009; Souza et al., 2018). These synthetic antioxidants are convenient and effective against the instability of biodiesel due to their high purity (Vijayakumar et al., 2016; Freitas et al., 2019). However, synthetic antioxidants are expensive and toxic at higher concentrations (Fangsuwannarak et al., 2016; Souza et al., 2018; Waweru et al., 2019).

Bio-additives are getting attention due to the added advantages of being degradable and nontoxic (Abd Hadi, 2015; Saha et al., 2020). Natural plants contained many natural antioxidants favorable for antioxidant additives due to economic feasibility, easy extraction, and local availability (Waweru et al., 2019). Fangsuwannarak et al. (2016) and Setyaningsih and Siahaan (2018) reported that blended palm biodiesel with the incorporation of bio-additives showed a desirable result in the combustion process and exhaust emission. Therefore, it is advantageous for power recovery of biodiesel engines by improving the ignition and combustion performance of diesel-biodiesel. As a result, it might indirectly promote the economy and engine performance (Ali et al., 2016). This chapter reviews the effect of clove oil on biodiesel-diesel fuel mixture in terms of engine performance and exhaust emission characteristics during the combustion process. According to a few literature reviews, clove oil can be the oxygenate and antioxidant bio-additives for fuel.

2. The effect of clove oil on diesel-biodiesel fuel

2.1 General overview

Studies stated that the incorporation of clove oil into biodiesel-diesel fuel showed improved fuel properties, engine performance, and exhaust emissions (Dueso et al., 2018; Jeyakumar and Narayanasamy, 2019; Muhyi et al., 2019; Gamayel et al., 2020b). Based on DIRJEN MIGAS specification, clove oil was safe and fulfilled the physical specification for bio-additive fuel (Kadarohman et al., 2010). Researchers have confirmed that monomeric phenol compounds resulted in oxidative stability improvement in engine performance (Waweru et al., 2019; Saha et al., 2020).

The incorporation of biodiesel will lead to engine power loss due to its lower heating value than petroleum diesel. Many additives have been reviewed to improve engine performance ranging from synthetic to natural additives (Focke et al., 2012; Pirouzfar and Fayyazbakhsh, 2016; Varatharajan and Pushparani, 2018; Muhyi et al., 2019; Ramalingam et al., 2019). The biodiesel-diesel blend should has a higher viscosity, flash point, and density of fuel blend than conventional fuel (Chauhan et al., 2016). The increased viscosity and lower volatility in biodiesel promote improper vaporization of fuel and poor cold flow properties of fuel (Ramalingam et al., 2019). The poor cold flow properties cause the solidification of fuel during the starting of the engine. As a result, it deposited and clogged the engine combustion machine (Soudagar et al., 2018; Waweru et al., 2019).

2.2 Fuel properties

2.2.1 Antioxidant effect of clove oil

Clove is one of the richest sources of phenolic compounds such as eugenol, eugenol acetate, and gallic acid. The main constituents of clove essential oil include eugenol (76.7%), β-caryophyllene (17.4%), α-humulene (2.1), and eugenol acetate (1.2%) (Sriram, 2016; Hatami et al., 2019; Gamayel et al., 2020a). These compounds are inhibitory toward hydroxyl radicals and can also chelate irons (Waweru et al., 2019). Eugenol is the main bioactive compound of clove with a bulky structure with two oxygen atoms attached (Kamatou et al., 2012; Hatami et al., 2019; Gamayel et al., 2020b). It is highlighted as a radical scavenger and a commercial source of phenolics (Cortés-Rojas et al., 2014; Fayyazbakhsh and Pirouzfar, 2017; Jeyakumar and Narayanasamy, 2019).

The antioxidant action of eugenol can be explained by reducing properties whereby it breaks the chain of the oxidation reaction. Eugenol prevents the formation of free radicals during the degradation process by reacting with oxygen present in the midst (Khalil et al., 2017; Ramos et al., 2018). Souza et al. (2018) and Gulcin (2011) demonstrate that the antioxidant action of eugenol was more efficient than TBHQ in terms of relative variation of oxidation stability expressed in the form of the induction period. The effectiveness of antioxidant activity can be affected by the fuel mixture's volatility, heat sensitivity, and solubility. The nature of antioxidants might induce synergism or antagonism toward diesel-biodiesel fuel blend (Karavalakis et al., 2011; Waweru et al., 2019). Antioxidants can be classified as free radical terminators, metal ion chelators, and oxygen scavengers based on the action modes. Free-radical terminators are commonly called primary antioxidants, which convert high-energy lipid radicals into thermodynamically stable products. Meanwhile, metal ion chelators can catalyze lipid oxidation, and oxygen scavengers react with oxygen in closed systems (Karavalakis et al., 2011; Khalil et al., 2017). Clove represents a rich source of antioxidant activities, which showed strong hydrogen donating ability, metal chelating ability, and scavenging free radicals, hydrogen peroxides, and superoxide (Cortés-Rojas et al., 2014).

Eugenyl acetate is the form of oxygen enrichment from the eugenol compound, which has a bulkier structure. Therefore, it has higher oxygen content than eugenol, which is supposed to contribute to a better combustion process (Chauhan et al., 2016). However, the improvement of combustion performance is not linear with the increase of the oxygen supply, as shown by eugenyl acetate. Kadarohman et al. (2012) also reported that this is not the case because it also has a higher viscosity and lower volatility than eugenol, leading to poor atomization of fuel mixtures (Gamayel et al., 2020b).

Clove stems were tested as biodiesel with synthetic antioxidants and showed improvement in oxidation stability (Waweru et al., 2019; Ramos et al., 2021). These OH groups in clove oil help stabilize and inactivate the free radical by donating an electron hence damaging the oxidative chain reaction. Clove antioxidant continued performing well because of its high phenol content and its alkene part to donate hydrogen quickly. The addition of antioxidants to biodiesel fuel helped to decrease the kinematic viscosity, density, flash point, and calorific value (Focke et al., 2012; Saluja et al., 2016; Freitas et al., 2019).

2.2.2 Viscosity

In terms of viscosity, all types of biodiesel have a higher viscosity than fossil fuels. The poor atomization process of fuel mixture due to high viscosity leads to lower combustion efficiency (Gamayel et al., 2020a). Therefore, it increases brake-specific fuel consumption (BSFC) (Rahman et al., 2019). According to Jeyakumar and Narayanasamy (2019), clove oil showed a desirable effect on the physical fuel mixture. The higher the percentage of clove oil incorporated, the higher the heating temperature, which leads to reduced viscosity (Fayyazbakhsh and Pirouzfar, 2017; Jeyakumar and Narayanasamy, 2019). Eugenol in clove oil has a bulky structure which resulted in resonance delocalization of this aromatic compound. Therefore, the molecular interaction between eugenol and triglyceride makes regular back and forth movement of the molecule more active than before (Gamayel, 2016; Gamayel et al., 2020b).

Although it reduced the calorific value, it resulted in a drop in BSFC which exhibited that clove oil, eugenol, and eugenyl acetate had a great potential to reduce fuel consumption rate in a larger composition (Kadarohman et al., 2012; Muhyi et al., 2019). The addition of clove oil into biodiesel-diesel fuel blend showed BSFC reduction by 1.71% and 8.05% at 1000 and 2000 ppm, respectively (Jeyakumar and Narayanasamy, 2019).

2.2.3 High oxygen content

The hydroxyl group in molecules can contribute to high calorific values (Gamayel, 2016). Eugenol has a lesser hydroxyl group, therefore, eugenol has a lower heating value. Despite having a lower heating value, its physical properties of being volatile oil and lower viscosity contribute to a high quality of combustion in a faster and complete process (Kusrini et al., 2018; Rahman et al., 2019; Gamayel et al., 2020b). According to Waweru et al. (2019), eugenol in clove oil has fewer hydroxyl groups (OH) in phenol content than catechin structure in turpentine oil, but it showed better performance by increasing the reactivity of fuel combustion. Its bulky structure and high oxygen content lead to a good fuel combustion process (Kadarohman et al., 2010).

Eugenyl acetate is produced by the ester reaction of eugenol (Nowak et al., 2012). The bulky structure of the molecule can decrease the strength of the van der Waals bond in diesel fuel and the chain of carbon. As a result, oxygen content and molecular structure can increase the efficiency of fuel combustion (Kadarohman et al., 2010). With high oxygen content in fuel, it increases the total heat release, indicating combustion reactivity. A higher oxygen concentration improves engine performance and reduces several types of emissions except for NO_x (Ali et al., 2016; Rahman et al., 2019). The reduction of unwanted emissions is due to the cleaner combustion of biodiesel than pure diesel. It functions similarly as oxygenate bio-additive during fuel combustion (Abd Hadi, 2015).

2.3 Engine performance

In general, many standards can be followed to fulfill the specification of fuel. Table 1 showed the common standards that researchers have applied to evaluate the characteristics of biodiesel-diesel fuel. The common specifications that are followed by many countries are ASTM and EN standards (Marketing, 2007; Yang et al., 2013; Ramos et al., 2018). Mbarawa (2008) reported that the performance parameters of clove seed oil in blended fuels do not differ significantly from pure diesel fuel. During combustion, the eugenol may oxidize into eugenyl acetate, which contributes to oxygen enrichment that might improve diesel fuel performance. However, eugenyl acetate has a higher viscosity and lower volatility which might decrease cylinder peak pressure (Kadarohman et al., 2012).

The combustion behavior of the fuel blends was primarily due to the chemical kinetics of the mixture (Kadarohman et al., 2012; Sriram, 2016). Terpene compounds in clove oil were largely responsible for making a perfect solution between the bio-additive and pure fuel, leading to rapid combustion and shorter ignition delay in engines fuelled by clove oil fuel blend (Kadarohman et al., 2012).

Table 1 Biodiesel specification.

Parameters	ASTM 6751	EN14214	DIRJEN MIGAS
Kinematic viscosity (mm^2/s)	1.9–6.0	3.5–5.0	1.6–5.8
Density (kg/m^3)		860–900	
Cetane number	>47	>51	
Flash point (°C)	>93	>101	>60

Adapted from Jaaskelainen, H., 2009. Biodiesel Standards and Properties. ECOpoint [Online]. Available from: https://dieselnet.com/tech/fuel_biodiesel_std.php. (Accessed 19 December 2020) and Waweru, E.J., Pogrebnaya, T., Kivevele, T.T., 2019. Effect of antioxidants extracted from clove wastes and babul tree barks on the oxidation stability of biodiesel made from water hyacinth of Lake Victoria origin. Waste Biomass Valoriz. 1–10.

2.3.1 Brake thermal efficiency (BTE)

Brake Thermal Efficiency (BTE) was found to increase with biodiesel blends and decrease with diesel fuel blends (Dueso et al., 2018; Jeyakumar and Narayanasamy, 2019). Studies reported that the antioxidant additive might impair the incomplete and improper combustion due to the complex composition and chemical structure (Dueso et al., 2018). However, BTE increased with clove stem oil concentration because of the oxygen content, which increased fuel combustion efficiency (Sriram, 2016).

2.3.2 Brake-specific fuel consumption (BSFC)

The Brake-Specific Fuel Consumption (BSFC) is the ratio of the fuel consumed by the engine, and the brake power is produced. It measures the fuel efficiency of any prime mover that burns fuel and produces rotational or shaft power. Good engine performance should have a low number of BSFC. Biodiesel usage consumed a higher amount of fuel than conventional fossil fuel (Dueso et al., 2018).

Few studies have reported that bio-additives have a negative or minimum impact on brake power (Saha et al., 2020). Mbarawa (2008) stated that clove-diesel blended fuels have low heating value as compared to conventional fuel. This might cause slight power losses and an increased fuel consumption rate. However, Muhyi et al. (2019) reported that the maximum fuel consumption reduction of 7.7% was reached with 0.5% of clove oil. Similarly, Kadarohman et al. (2010) found that fuel incorporated with 6% of clove oil had reduced the fuel consumption rate to approximately 251.9 mL/h. It is 4.43$ lower than control diesel fuel. The mixture of clove and biodiesel-diesel fuel was comparable to diesel in terms of performance. The oxygen-containing compounds in clove oil lead to increase in cetane number and complete the fuel combustion process thus reducing the BSFC (Kadarohman et al., 2010). In general, oxygenate fuel additives have positive impact on BSFC.

The varied concentration of clove oil incorporated into fuel met the DIRJEN MIGAS specification (Table 1) in terms of fuel's physical and chemical characteristics. Therefore, clove oil is suitable for fuel bio-additives (Kadarohman et al., 2010). The combustion efficiency is essential to reduce fuel consumption. Eugenol and eugenyl acetate are the two main compounds that have noticeable oxygenated molecular structures. Therefore, both of them positively contribute to fuel performance and decrease exhaust emissions. The bulkier structure of eugenyl acetate lowered the BSFC better than eugenol.

2.3.3 Exhaust gas temperature (EGT)

Exhaust Gas Temperature (EGT) is an essential indicator for the heat release rate of fuel. It is defined as the in-cylinder temperature that exhausts gas, contributing to NO_x emission (Jeyakumar and Narayanasamy, 2019; Saha et al., 2020). In terms of exhaust gas temperature, biodiesel fuel showed lower values than conventional diesel fuel (Focke et al., 2012; Tamilselvan et al., 2017). The lower the EGT, the lower the exhaust emission produced.

Fuel additives that contribute to the high latent heat of vaporization lead to the cooling of the combustion chamber. The cooling effect took place because high latent heat helped absorb the temperature and reduce the EGT. Atmanli (2016) explained that oxygenate additives like clove oil lowered the EGT by increasing the latent heat of evaporation. However, this theory was refuted by Rahman et al. (2019) and Kadarohman et al. (2012) that oxygenate additives that have high cetane numbers promote the combustion process, which raises the EGT and increased the NO_x emission subsequently.

2.3.4 Ignition delay

Ignition delay is the duration between the injection start of fuel and the start of fuel combustion. It implies the engine efficiency, diesel emissions, and smoothness of engine operation (Soudagar et al., 2018). The ignition delays for the biodiesel mixtures are longer than that for diesel. This might be due to its higher viscosity and density than pure diesel. Longer ignition delay contributes to slow-burning characteristics of fuel which is undesirable for the combustion process. The ignition delay can be decreased by increasing the cetane number in the fuel mixture (Sriram, 2016). Cetane number is the measure of fuel ignition delay. A higher cetane number indicates a shorter ignition delay period which is preferable. Clove oil added into diesel fuel showed the highest cetane number burned quickly, resulting in a shorter ignition delay and complete combustion process. Ultimately, it enhanced the process of mixing between clove oil fuel blend and air, promoted rapid combustion, and lowered ignition delay (Gamayel et al., 2020a).

2.4 Exhaust emission

Exhaust emissions from engines consisted of various air pollutants such as unburnt HC, NO_x, and CO. The usage of bio-additives can reduce exhaust emissions up to 80% at specific operating conditions. In general, bio-additives can impart the maximum reduction of unburnt hydrocarbon, smoke emission, and exhaust gas temperature by more than 50%, 65%, and 14%, respectively, in comparison with the base fuels without additives (Saha et al., 2020). Based on Kadarohman et al. (2010), emissions such as unburned HC, CO, and smoke emissions were reduced by the blending of clove oil into diesel fuel. Jeyakumar and Narayanasamy (2019) also reported that the formulation of diesel fuel using eugenyl acetate achieved the lowest HC emission rate. The high oxygen in eugenyl acetate promoted complete combustion and cetane number (Jeyakumar and Narayanasamy, 2019).

However, Kadarohman et al. (2012) reported that a high cetane number contributes to the increase of exhaust gas temperature, which increases the NO_x formation rate. Despite that, many studies have proven that the high cetane number reduced the temperature and reduced the NO_x emission rate (Varatharajan and Cheralathan, 2012; Atmanli, 2016; Saha et al., 2020). After adding bio-additives, CO and HC increased with diesel blends and decreased with biodiesel blends, while NO_x emissions decreased with diesel and biodiesel blends (Yilmaz et al., 2017).

2.4.1 Nitrogen oxides (NO_x)

NO_x is the standard air pollutant that is colorless which is generated through the fuel combustion process. NO_x has been a particular concern when using biodiesel as a fuel for combustion. This compound may react with ozone and form acid rain in the presence of UV light and oxygen. During engine combustion, it produced three types of NO_x: thermal NO_x, fuel NO_x, and prompt NO_x (Dueso et al., 2018). The formation of NO_x is strongly dependent on temperature and oxygen content. The greater the quantity of oxygen present during the combustion process, the greater the NOx quantity (Kadarohman et al., 2012).

Thermal NO_x was highly affected by combustion temperature, and the molar concentration of N and O presented in the system. Fuel NO_x resulted from the oxidation of ionized nitrogen in the fuel (Rahman et al., 2019). Prompt NO_x was formed because nitrogen in the air combined with fuel-rich conditions and oxidized during the combustion process (Cox, 1999). Fuel NOx is obtained from the reaction between oxygen and nitrogen present in the fuel. Since biodiesel does not contain fuel-bound nitrogen, common NOx normally produced with biodiesel blend fuels were thermal NOx and prompt NOx (Jeyakumar and Narayanasamy, 2019). In this case, some researchers reported that the presence of oxygen improves combustion, which raises the temperature in the combustion chamber. Thus, it increased the formation rate of NO_x (Dueso et al., 2018). By controlling the temperature of the combustion chamber, the thermal NO_x can be controlled.

Although the application of biodiesel has reduced lots of unwanted emissions, the NO_x emission is still significant. Prompt NO_x is the primary concern of using biodiesel as fuel alternative. The kinetic action of prompt NO_x can be controlled by antioxidant additives (Kadarohman et al., 2012; Saha et al., 2020). With antioxidant bio-additives, it played a role in inhibiting the production of prompt NO_x. The highly reactive free radicals are delayed or inhibited by donating the electron or hydrogen atoms to free radical derivatives (Balaji and Cheralathan, 2015; Ali et al., 2016; Adam et al., 2017). The addition of clove oil by 2000 and 1000ppm decreases EGT at full-load conditions at 5.16% and 2.36%, respectively, since antioxidants act as a deterrent in the fuel reaction (Varatharajan and Cheralathan, 2012). By decreasing EGT, it reduced the NO_x emission indirectly. Balaji and Cheralathan (2015) and Subramanian et al. (2018) also reported that eugenol reduced the NO_x emission due to its phenolic and OH groups. NO_x emissions were increased remarkably when the engine was fuelled with the 50% clove stem oil-diesel blended fuel operation mode (Mbarawa, 2008).

The higher the fuel-burning efficiency, the higher energy is being released, contributing to higher emission of NO_x (Sriram, 2016). Several studies have demonstrated that oxygenate fuel might produce higher NO_x emissions because of the efficient combustion process. The produce of the NO_x is highly dependent on cylinder temperature, the oxygen content of the fuel, and seat time for the ignition reaction to take location (Fayyazbakhsh and Pirouzfar, 2017; Rahman et al., 2019). Despite that, clove oil may represent antioxidant and oxygenate additives that balance the physical and chemical characteristics of the fuel mixture. The addition of antioxidant additives to biodiesel-diesel fuel blend can decrease the NO_x emission by scavenging the free radicals present in the combustion chamber (Ramalingam et al., 2019).

2.4.2 Carbon monoxide (CO) and smoke

Carbon monoxide and smoke are considered air pollutants harmful to the human respiratory system (Varatharajan and Cheralathan, 2012; Jiaqiang et al., 2017). The incomplete combustion of carbon forms carbon monoxide and smoke. Incomplete combustion can be caused by oxygen content, temperature, and fuel properties such as cetane number and viscosity. The application of biodiesel had significantly reduced harmful emissions such as carbon monoxide and smoke through the air (Madiwale et al., 2017; Dueso et al., 2018). Besides, the application of additives also maintained the engine performance of biodiesel fuel and decreased exhaust emissions drastically. The application of clove in blended fuel significantly reduced the CO and smoke emissions during operation. Although the smoke emission of clove oil is relatively higher than other bio-additives, it is still lower than conventional petroleum fuel (Jiaqiang et al., 2017). The application of bio-additives contributes to a cleaner process of biodiesel-diesel combustion performance (Ali et al., 2016).

In general, all bio-additives can decrease CO emission at various speeds and variations in weight (Mokhtar et al., 2015; Dueso et al., 2018). However, the decrease in CO concentration indicates the increase in CO_2 emission. However, the production of CO_2 remains consistent as the high oxygen atoms that accelerate the combustion process shown in this study. However, it is considered neutral because plants readily absorb it for its growth. The carbon cycle is maintained in the balance as plants readily absorb the CO_2 for photosynthesis, whereby CO contributes to air pollutants that might induce human health complications (Balaji and Cheralathan, 2015).

In terms of smoke emission, it was produced from fuel decomposition when the local temperatures reached the smoke point. Bio-additives may have a maximum reduction of 65% at particular conditions due to the presence of high oxygen content (Saha et al., 2020). Mbarawa (2008) reported that the addition of clove stem oil to diesel fuel reduced the smoke emission remarkably due to high oxygen content. Oxygenate additives played a vital role in completing the combustion process.

2.4.3 Particulate matters and unburnt hydrocarbon (HC)

The complete fuel combustion process reduced the emission of particulate matter and unburnt hydrocarbon through the air (Kadarohman et al., 2012). The blending of biodiesel-diesel fuel with oxygen-containing compounds increases the cetane number and completes fuel combustion (Sriram, 2016; Vijayakumar et al., 2016; Dueso et al., 2018). The bulky structure of eugenol can decrease the strength of the van der Waals' bond in diesel fuel and the chain of carbon (Kadarohman et al., 2012). Therefore, oxygen content and the molecular structure can increase the efficiency of the fuel combustion process (Kadarohman et al., 2010).

Hydrocarbon (HC) emission is unburned fuel that escapes through the air with other volatile compounds. The level of HC emission increased and was attributed to low cetane number and high ignition delay (Sriram, 2016). Therefore, bio-additives have the optimum performance to decrease HC concentration up to 50% in exhaust emissions than base fuel without bio-additives (Mbarawa, 2008; Soudagar et al., 2018; Saha et al., 2020). The higher the concentration of bio-additives, the lower the concentration of hydrocarbons produced. This might be due to the increase in the fuel's oxygen content, which results in a cleaner combustion process (Mbarawa, 2008). However, Atmanli (2016) stated that high latent evaporation might cause a reduction in temperature, flame quenching, and cooling regions, leading to an increase in HC emissions. It caused higher evaporative cooling in the combustion chamber, which consequently leads to lower exhaust gas temperature. Noting with this, clove oil possesses high latent heat of evaporation, but the HC emission was lowered down with the addition of clove oil (Atmanli, 2016; Jeyakumar and Narayanasamy, 2019).

2.5 Cost

The usage of bio-additives improves the security of resources and economic status indirectly (Vijayakumar et al., 2016). The full utilization of clove biomass requires the initiative of the manufacturer site to manage the biomass properly.

2.6 Future trends

Currently, the processing of cloves is not widely emphasized for fuel additives purposes. Besides, the clove biodiesel-diesel blended fuel might not be competitive with pure diesel fuel. However, the environmental impact is valued monetarily as it still achieves the target requirement for fuel specification and reduces undesirable emissions (Mbarawa, 2008; Tamilselvan et al., 2017). Also, the fuel additives using plant extracts might need to consider the stability of bioactive compounds within extracts. Therefore, the application of nanoencapsulation might be feasible to improve the stability of the beneficial properties of natural plant extracts (Soudagar et al., 2018; Hatami et al., 2019). The reformulation of clove oil as bio-additives in biodiesel-diesel blends is required to achieve maximum engine performance considering exhaust emissions and sustainability.

3. Conclusion

In a nutshell, clove oil is a potential oxygenated bio-additive that can increase the cetane number and improve fuel properties. Research studies showed that it could reduce fuel consumption, CO, and unburned HC emissions and improve engine performance by contributing to an efficient fuel combustion process. Therefore, the incorporation of biodiesel-diesel fuel and clove oil bio-additives showed a promising combination to improve engine performance and tackle environmental issues produced from fuel combustion. Besides, further studies are required to tackle the high cloud point and NO_x of biodiesel-diesel fuel to increase fuel combustion efficiency and reduce pollutant emission. It is recommended to study the optimum formulation of biodiesel-diesel fuel and clove oil to reduce exhaust emission and improve engine performance.

References

Abd Hadi, S., 2015. Experimental Study of the Bio-Additives in Biodiesel Fuel on Performance and Emissions Characteristics of Diesel Engine. Universiti Tun Hussein Onn Malaysia.

Adam, I.K., Aziz, A.R.A., Heikal, M.R., Yusup, S., 2017. Performance and emission analysis of rubber seed methyl ester and antioxidant in a multicylinder diesel engine. Energy Fuel 31, 4424–4435.

Ali, O.M., Mamat, R., Abdullah, N.R., Abdullah, A.A., 2016. Commercial and synthesized additives for biodiesel fuel: a review. ARPN J. Eng. Appl. Sci. 11, 3650–3654.

Atmanli, A., 2016. Effects of a cetane improver on fuel properties and engine characteristics of a diesel engine fueled with the blends of diesel, hazelnut oil and higher carbon alcohol. Fuel 172, 209–217.

Balaji, G., Cheralathan, M., 2015. Experimental investigation of antioxidant effect on oxidation stability and emissions in a methyl ester of neem oil fueled DI diesel engine. Renew. Energy 74, 910–916.

Chauhan, B.S., Singh, R.K., Cho, H.M., Lim, H.C., 2016. Practice of diesel fuel blends using alternative fuels: a review. Renew. Sustain. Energy Rev. 59, 1358–1368.

Cortés-Rojas, D.F., De Souza, C.R.F., Oliveira, W.P., 2014. Clove (*Syzygium aromaticum*): a precious spice. Asian Pac. J. Trop. Biomed. 4, 90–96.

Cox, L., 1999. Nitrogen Oxides (NOx) Why and How they Are Controlled. Diane Publishing.

Dueso, C., Muñoz, M., Moreno, F., Arroyo, J., Gil-Lalaguna, N., Bautista, A., Gonzalo, A., Sánchez, J.L., 2018. Performance and emissions of a diesel engine using sunflower biodiesel with a renewable antioxidant additive from bio-oil. Fuel 234, 276–285.

Fangsuwannarak, K., Wanriko, P., Fangsuwannarak, T., 2016. Effect of bio-polymer additive on the fuel properties of palm biodiesel and on engine performance analysis and exhaust emission. Energy Procedia 100, 227–236.

Fayyazbakhsh, A., Pirouzfar, V., 2017. Comprehensive overview on diesel additives to reduce emissions, enhance fuel properties and improve engine performance. Renew. Sustain. Energy Rev. 74, 891–901.

Focke, W.W., Westhuizen, I.V.D., Grobler, A.B.L., Nshoane, K.T., Reddy, J.K., Luyt, A.S., 2012. The effect of synthetic antioxidants on the oxidative stability of biodiesel. Fuel 94, 227–233.

Freitas, J.P.A., França, F.R.M., Silva, M.S., Toms, R.J., Silva, G.F.D., 2019. Evaluation of antioxidant activity of natural extracts on the oxidative stability of cottonseed biodiesel. Braz. J. Chem. Eng. 36, 905–911.

Gamayel, A., 2016. Characteristics of the mixed crude jatropha oil-clove oil. Jurnal Teknik Mesin (JTM) 6, 63–67.

Gamayel, A., Mohammed, M., Al-Zubaidi, S., Yusuf, E., 2020a. Effect of clove oil in droplet combustion of crude jatropha oil. Int. J. Adv. Sci. Technol. 29, 1564–1571.

Gamayel, A., Mohammed, M., Al-Zubaidi, S., Yusuf, E., 2020b. A potential study on droplet combustion performance of vegetable oil-clove oil blend. Int. J. Adv. Sci. Technol. 29, 1555–1563.

Gulcin, I., 2011. Antioxidant activity of eugenol: a structure-activity relationship study. J. Med. Food 14, 975–985.

Hatami, T., Johner, J.C.F., Zabot, G.L., Meireles, M.A.A., 2019. Supercritical fluid extraction assisted by cold pressing from clove buds: extraction performance, volatile oil composition, and economic evaluation. J. Supercrit. Fluids 144, 39–47.

Jeyakumar, N., Narayanasamy, B., 2019. Clove as antioxidant additive in diesel-biodiesel fuel blends in diesel engines. Int. J. Green Energy 16, 284–292.

Jiaqiang, E., Pham, M., Zhao, D., Deng, Y., Le, D., Zuo, W., Zhu, H., Liu, T., Peng, Q., Zhang, Z., 2017. Effect of different technologies on combustion and emissions of the diesel engine fueled with biodiesel: a review. Renew. Sustain. Energy Rev. 80, 620–647.

Kadarohman, A., Hernani, H., Khoerunisa, F., Astuti, R.M., 2010. A potential study on clove oil, eugenol and eugenyl acetate as diesel fuel bio-additives and their performance on one cylinder engine. Transport 25, 66–76.

Kadarohman, A., Hernani, H., Rohman, I., Kusrini, R., Astuti, R.M., 2012. Combustion characteristics of diesel fuel on one cylinder diesel engine using clove oil, eugenol, and eugenyl acetate as fuel bio-additives. Fuel 98, 73–79.

Kamatou, G.P., Vermaak, I., Viljoen, A.M., 2012. Eugenol-from the remote Maluku Islands to the international market place: a review of a remarkable and versatile molecule. Molecules 17, 6953–6981.

Kapadiya, S., Desai, M.A., 2017. Isolation of essential oil from buds of *Syzygium aromaticum* using hydrodistillation: multi-response optimization and predictive modelling. Int. J. Adv. Res. Sci. Eng. 6, 406–418.

Karavalakis, G., Hilari, D., Givalou, L., Karonis, D., Stournas, S., 2011. Storage stability and ageing effect of biodiesel blends treated with different antioxidants. Energy 36, 369–374.

Khalil, A.A., Rahman, U.U., Khan, M.R., Sahar, A., Mehmood, T., Khan, M., 2017. Essential oil eugenol: sources, extraction techniques and nutraceutical perspectives. RSC Adv. 7, 32669–32681.

Kusrini, E., Suryantoro, M.T., Sugiarto, B., Setiapraja, H., Yubaidah, S., Mulyono, A.B., Attharik, M.I., Ariestiawan, M.R., Cohen, A., Juwono, F.H., Yatim, A., Setiawan, E.A., 2018. Biodiesel deposit characterization on hot chamber test rig with biodiesel fuels (B20) and bioaditif essential oil (atsiri). E3S Web Conf. 67, 02016.

Madiwale, S., Karthikeyan, A., Bhojwani, V., 2017. A comprehensive review of effect of biodiesel additives on properties, performance, and emission. IOP Conf. Ser. Mater. Sci. Eng., 012015.

Marketing, G., 2007. Diesel Fuels Technical Review. Chevron Corporation California.

Mbarawa, M., 2008. Performance, emission and economic assessment of clove stem oil-diesel blended fuels as alternative fuels for diesel engines. Renew. Energy 33, 871–882.

Mokhtar, A., Atan, N., Rahman, N., Khalid, A., 2015. Review of performance and emmissions characteristics of bio-additive fuel on SI engine fuelled by biopetrol. Appl. Mech. Mater. 773-774, 430–434.

Muhyi, A., Silitonga, D.J., Alfian, D.G.C., Supriyadi, D., Prahmana, R.A., 2019. Performance characterization of diesel engine generator set with the addition of clove oil as bio-additives for diesel fuel. IOP Conf. Ser. Earth Environ. Sci. 258, 012009.

Nowak, K., Ogonowski, J., Jaworska, M., Grzesik, K., 2012. Clove oil-properties and applications. Chemik 66, 145–152.

Pirouzfar, V., Fayyazbakhsh, A., 2016. Diesel Fuel Additives.

Pirouzfar, V., Zarringhalam Moghaddam, A., Mirza, B., 2012. Physicochemical properties and combustion performance of gas oil-fuel additives. J. Energy Resour. Technol. 134.

Rahman, S.M.A., Van, T.C., Hossain, F.M., Jafari, M., Dowell, A., Islam, M.A., Nabi, M.N., Marchese, A.J., Tryner, J., Rainey, T., Ristovski, Z.D., Brown, R.J., 2019. Fuel properties and emission characteristics of essential oil blends in a compression ignition engine. Fuel 238, 440–453.

Ramalingam, S., Ganesan, P., Murugasan, E., 2019. Effect of antioxidant additives on performance and emission behavior of biodiesel fueled DI diesel engine. Energy Sour. Part A 42, 1085–1096.

Ramos, T.C.P.M., De Souza, E.F., Pina, C.C., Cavalheiro, A.A., Fiorucci, A.R., Da Silva, M.S., 2018. Evaluation of natural antioxidants action in oxidative stability of commercial biodiesel. Orbital Electron J. Chem. 10. https://doi.org/10.17807/orbital.v10i1.1027.

Ramos, T.C.P.M., Santos, E.P.S., Ventura, M., Pina, J.C., Cavalheiro, A.A., Fiorucci, A.R., Silva, M.S., 2021. Eugenol and TBHQ antioxidant actions in commercial biodiesel obtained by soybean oil and animal fat. Fuel 286, 119374.

Saha, D., Sinha, A., Roy, B., 2020. A critical review of emission and performance characteristics of CI engine using bio-additives. Clean Technol. Environ. Policy 22, 1613–1638.

Saluja, R.K., Kumar, V., Sham, R., 2016. Stability of biodiesel—a review. Renew. Sustain. Energy Rev. 62, 866–881.

Setyaningsih, D., Siahaan, D.G., 2018. The influence of essential oil addition to oxidative stability of palmoil biodisel. IOP Conf. Ser. Earth Environ. Sci. 209, 012007.

Soudagar, M.E.M., Nik-Ghazali, N.-N., Abul Kalam, M., Badruddin, I.A., Banapurmath, N.R., Akram, N., 2018. The effect of nano-additives in diesel-biodiesel fuel blends: a comprehensive review on stability, engine performance and emission characteristics. Energy Convers. Manag. 178, 146–177.

Souza, E.F.D., Ramos, T.C.P.M., Santos, M.N., Ferraz, J.R., Silva, M.S.D., Batistote, M., Fiorucci, A.R., 2018. Effect of the natural antioxidant eugenol on quality preservation of commercial biodiesels produced with soybean oil or waste frying oil during storage at different temperatures. Ciência e Natura 40, 61.

Sriram, S., 2016. Performance and Emissions Testing of Mixtures of Diesel Extracts of Bio-Oil and Eugenol in Diesel.

Subramanian, T., Varuvel, E.G., Ganapathy, S., Vedharaj, S., Vallinayagam, R., 2018. Role of fuel additives on reduction of NOX emission from a diesel engine powered by camphor oil biofuel. Environ. Sci. Pollut. Res. Int. 25, 15368–15377.

Tamilselvan, P., Nallusamy, N., Rajkumar, S., 2017. A comprehensive review on performance, combustion and emission characteristics of biodiesel fuelled diesel engines. Renew. Sustain. Energy Rev. 79, 1134–1159.

Varatharajan, K., Cheralathan, M., 2012. Influence of fuel properties and composition on NOx emissions from biodiesel powered diesel engines: a review. Renew. Sustain. Energy Rev. 16, 3702–3710.

Varatharajan, K., Pushparani, D., 2018. Screening of antioxidant additives for biodiesel fuels. Renew. Sustain. Energy Rev. 82, 2017–2028.

Vijayakumar, C., Ramesh, M., Murugesan, A., Panneerselvam, N., Subramaniam, D., Bharathiraja, M., 2016. Biodiesel from plant seed oils as an alternate fuel for compression ignition engines—a review. Environ. Sci. Pollut. Res. Int. 23, 24711–24730.

Waweru, E.J., Pogrebnaya, T., Kivevele, T.T., 2019. Effect of antioxidants extracted from clove wastes and babul tree barks on the oxidation stability of biodiesel made from water hyacinth of Lake Victoria origin. Waste Biomass Valoriz., 1–10.

Xin, J., Imahara, H., Saka, S., 2009. Kinetics on the oxidation of biodiesel stabilized with antioxidant. Fuel 88, 282–286.

Yang, Z., Hollebone, B.P., Wang, Z., Yang, C., Landriault, M., 2013. Factors affecting oxidation stability of commercially available biodiesel products. Fuel Process. Technol. 106, 366–375.

Yilmaz, N., Atmanli, A., Trujillo, M., 2017. Influence of 1-pentanol additive on the performance of a diesel engine fueled with waste oil methyl ester and diesel fuel. Fuel 207, 461–469.

SECTION

CLOVE ESSENTIAL OIL: CHEMISTRY, FUNCTIONALITY AND APPLICATIONS

3

CHAPTER

Composition and functionality of clove (*Syzygium aromaticum*) essential oil

27

Sonia Kaushal and Rashmi

Department of Chemistry, Punjab Agricultural University, Ludhiana, Punjab, India

1. Introduction

The term "Spice" has been derived from the Latin word "Species," meaning asset of specific value. Since ancient times, spices have played a significant role in human civilization for enhancing the palatability and aroma of food (Jimoh et al., 2017). They are products of aromatic plants and used in folk and traditional medicines in Unan and Sino-Tibetan systems. They have been recognized as drug materials used to treat various human disorders in several countries like China, Europe, India, and United States. Clove (*Syzygium aromaticum*) is the most valuable spice that belongs to Myrtaceae native to the East Indonesian Maluku islands known as Spice Islands (Chaieb et al., 2007). Cloves are unopened flower buds that are dark brown with intense fragrance. The plant has various medicinal properties to control nausea, vomiting, gastrointestinal spasm, cough, diarrhea, dyspepsia, flatulence, and relieves pain (Ortes-Rojas et al., 2014). Besides this, they have diuretic, stomachic, odontalgic, and stimulant effects (Jagtap et al., 2019). United States Food and Drug Administration has listed clove oil as "Generally Regarded As Safe" and is widely used in medicine, flavoring and fragrance industries (Kildeaa et al., 2004).

2. Distribution

Clove (*Syzygium aromaticum*) is the second most important spice in world trade. Due to different harvesting seasons, they are grown throughout the year worldwide. The islands of Indonesia, Zanzibar, and Pemba are the main contributors to clove in the world. Indonesia alone produces 66% of the total world production of cloves. Around 1800, cloves were introduced to India by the East India Company and is the second contributor of cloves after Indonesia (Board, 2010). The plant resides for more than 100 years. They grow in regions with well-drained soil, high humidity above 70%, uniformly distributed rainfall above 150 cm, and atmospheric temperatures (25–35°C) (Danthu et al., 2014). In India, clove cultivation takes place in the black loamy red soils of Kerala and the hilly landscape of Western Ghats in Tamil Nadu and Karnataka (Byng, 2016).

Clove (*Syzygium aromaticum*). https://doi.org/10.1016/B978-0-323-85177-0.00037-9

3. Morphology

The name clove has been derived from the French word "Clou" (meaning nail) as the bud resembles small irregular nails. The clove plant is an evergreen, slender, and medium-sized tree with a conical and cylindrical appearance and grows up to 8–12 m in height (Alma et al., 2007). The branches are grayish, dense, and semi-erect. Leaves are simple, coriaceous, and glabrous with shiny and short red petioles. Leaves have many oil glands on the lower surface. Flowers are bisexual and red. Flowering starts after 7 years and continues up to 80 years or above. Initially, flower buds are pale yellow and turn green to bright red at maturity. Commercially, cloves are used as air-dried unopened flower buds, which are 2.5 cm in length and 1.25 cm wide. Fruit (mother of clove) is a berry, 2–2.5 cm long, red with one oblong seed of 1.5 cm height (Kamatou et al., 2012; Ortes-Rojas et al., 2014).

4. Chemical constituents of clove essential oil

Clove essential oil can be obtained from buds, leaves, and stems of the plant differing in their chemical composition, flavor, and color (Uddin et al., 2017).

4.1 Clove bud essential oil

The clove bud essential oil is denser than water and yellow. The oil has a fresh and fruity odor. Eugenol, present as a major compound in clove bud essential oil, is colorless or light-yellow oily liquid mainly responsible for the aroma of clove essential oil (Pruthi, 2001).

Various literature reports showed different terpenoids, mainly mono and sesquiterpenoids, in clove essential oil in variable amounts. This is because the nature and the amount of various secondary metabolites present in the essential oil are affected by genetic factors, cultivation techniques, postharvest processes, and agroclimatic conditions (Arslan et al., 2004). For instance, De Oliveiraa et al. (2016) reported that clove essential oil contained oxygenated monoterpenes (59.7%), phenolic compounds (19.8%), sesquiterpene hydrocarbons (19.7%), and oxygenated sesquiterpenes (0.59%). In contrast, Fankem et al. (2017) reported that clove essential oil consisted of oxygenated monoterpenes (89.1%), sesquiterpene hydrocarbons (10.6%), oxygenated sesquiterpenes (0.27%), monoterpene hydrocarbons (0.04%), and other components (0.03%) with eugenol as primary compound (87.6%). On the other hand, Kaur et al. (2019) reported that clove bud essential oil constituted of oxygenated monoterpenoids (69.7%), phenolic compounds (14.4%), sesquiterpene hydrocarbons (14.3%), oxygenated sesquiterpenoids (0.77%), and other compounds (0.77%).

Major compounds reported in clove bud essential oil included eugenol, *trans*-caryophyllene, eugenol acetate, caryophyllene oxide, and α-humulene belonging to the terpenoids group. Eugenol has been reported as the most abundant compound present in clove essential oil, and the second most abundant compound reported included β-caryophyllene. For instance, Moemenbellah-Fard et al. (2020) revealed the presence of eugenol (65.4%), *trans*-caryophyllene (12.6%), eugenol acetate (9.85%), caryophyllene oxide (3.00%), and α-humulene (1.73%) in clove essential oil. Pedro et al. (2020) revealed the presence of eugenol as the main component (87.4%), followed by β-caryophyllene (11.5%). Xu et al. (2016) reported the presence of 22 components in clove bud essential oil with eugenol (76.2%) as a major component followed by β-caryophyllene (11.5%), caryophyllene oxide (4.29%), and other minor components. Fu et al. (2007) investigated the presence of eugenol (68.5%), β-caryophyllene (19.0%), 2-methoxy-4-[2-propenyl] phenol acetate (10.1%) and α-caryophyllene (1.85%) in clove essential oil. Other studies also revealed the presence of eugenol (74.3%) followed by β-caryophyllene (15.9%) and eugenol acetate (5.8%) as major compounds in clove bud essential oil (Marya et al., 2012; Matta, 2010; Khan et al., 2009).

Eugenol acetate has also been reported as the second most abundant compound in clove bud essential oil. For example, Chaieb et al. (2007) identified 36 components in the bud oil with maximum concentration of eugenol (88.6%), eugenol acetate (5.62%), β-caryophyllene (1.39%), 2-heptanone (0.93%), ethyl hexanoate (0.66%), humulenol (0.27%), α-humulene (0.19%), calacorene (0.11%), and calamenene (0.10%). Sameza et al. (2015) investigated the presence of eugenol (79.4%), eugenol acetate (9.2%), and isocaryophyllene (7.0%) as major components in clove bud essential oil. Kaur et al. (2019) reported the presence of 21 compounds along with eugenol (69.7%) as major compound followed by eugenol acetate (14.4%), β-caryophyllene (12.2%), and α-humulene (1.50%), whereas Banerjee et al. (2020) identified the presence of eugenol (76.1%) and eugenol acetate (12.4%) as the major components. Chavibetol, also called meta eugenol, has also been reported as the second most abundant compound in clove bud essential oil. For example, Alma et al. (2007) investigated the presence of 18 compounds in the bud essential oil with eugenol (87%), chavibetol (19.7%), β-caryophyllene (13%), eugenol acetate (8.01%), trisiloxane1,1,1,5,5,5-hexa-methyl-3,3-bis [(trimethylsilyl)oxy] (1.7%), and other constituents.

4.2 Clove leaves essential oil

The clove leaves essential oil is light yellow and has a characteristic pleasant odor. The amount of eugenol in clove leaves essential oil accounted for 75%–95% compared to bud essential oil (65%–85%). Eugenol (91%) was also the most abundant compound in the clove leaves essential oil, followed by β-caryophyllene (8%) and α-humulene (1%) (de Meneses et al., 2019). Jirovetz et al. (2006) reported the presence of 23 compounds with eugenol (76.8%), β-caryophyllene (17.4%), α-humulene (2.1%), and eugenol acetate (1.2%) as major compounds. Raina et al. (2001) analyzed 16 compounds in leaf oil with eugenol (94.4%) as a major compound, followed by β-caryophyllene (2.9%). Srivastava et al. (2005) reported 22 compounds including eugenol, isoeugenol, α-humulene, β-caryophyllene, and γ-cadinene as major compounds. Structures of different compounds present in clove essential oil are shown in Table 1.

4.3 Clove stem oil

The constituents responsible for fresh and fruity odor are available in lesser quantity in clove stem essential oil. Therefore, they are not commercially used as bud oil. Patil and Dhale (2013) reported that eugenol was present in high concentration compared to bud oil.

5. The functionality of clove essential oil

Clove essential oil exhibits diverse biological functions in medical, agricultural, and veterinary sciences (Table 2). Some of the properties (Fig. 1) of clove essential oil are listed below:

Table 1 Structure of different compounds present in clove essential oil.

No.	Name	Structure
1	Eugenol	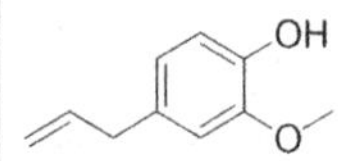
2	β-Caryophyllene	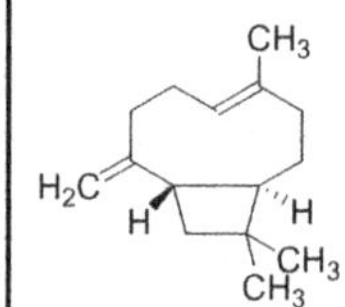
3	Eugenol acetate	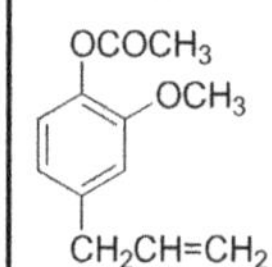
4	Caryophyllene oxide	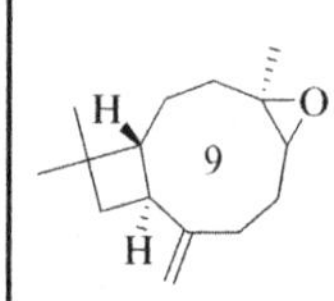
5	α-Humulene	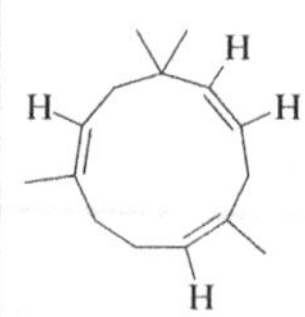
6	2-Methoxy-4-(2-propenylacetate phenol)	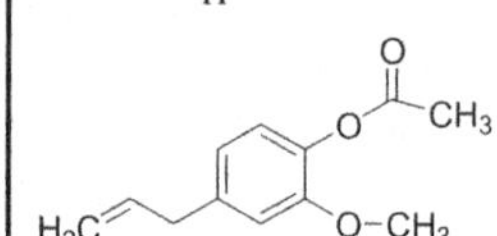
7	2-Heptanone	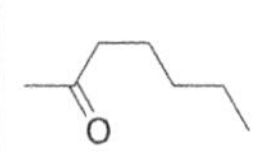
8	Ethyl hexanoate	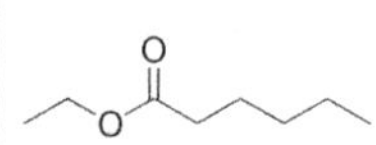
9	Humulenol	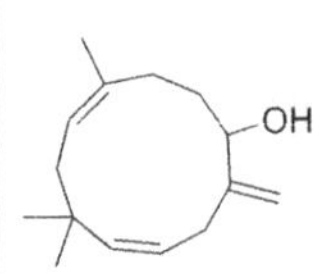
10	Calacorene	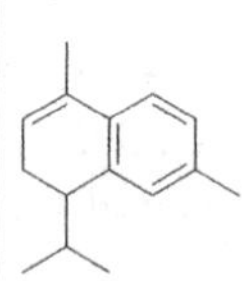

Table 1 Structure of different compounds present in clove essential oil—cont'd

No.	Name	Structure
11	Calamenene	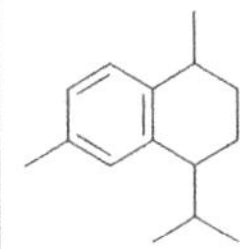
12	Isocaryophyllene	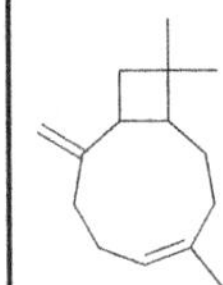
13	Chavibetol	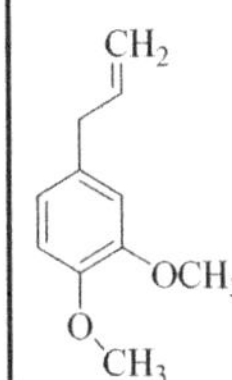
14	Trisiloxane1,1,1,5,5,5-hexa-methyl-3,3-bis [(trimethylsilyl)oxy]	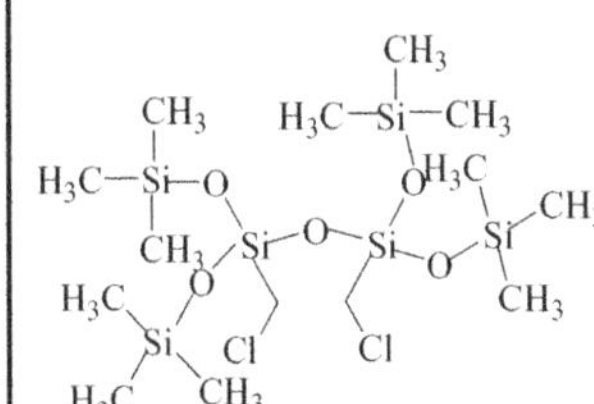
15	Isoeugenol	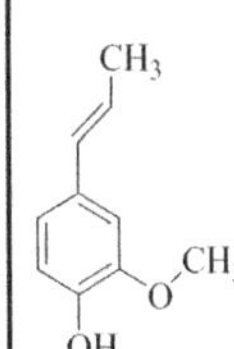
16	γ-Cadinene	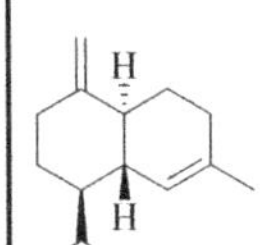
17	Cadinene	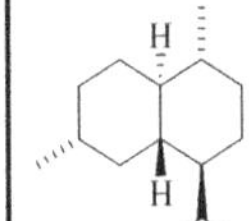

Table 2 Functionality of clove essential oil.

Activity	Concentration	Microorganism / strain/animal used	Tests used	Parameter noted	References
Antimicrobial activity	1%, 10%, 25%, 50%, and 100% (V/V)	*Edwardsiella tarda* (FP5060, ED47, Yoshida and ED45), *Photobacterium damselae* (FP4101), *Vibrio harveyi* (FP 8370), *Vibrio ichthyoenteri* (FP 4004), *Lactococcus garvieae* (FP5245), *Streptococcus iniae* (FP5228, S186, S530 and S131), and *Streptococcus parauberis* (FP5228, S124, S527 and S1466)	Disk diffusion assay	Gram-negative bacteria MIC values: 0.125%–0.5% (v/v) Gram-positive bacteria: 0.25%–0.5% (v/v)	Pathirana et al. (2019)
	10.9–0.34 mg/mL	*Escherichia coli F5, E. coli CS31A, Kluyvera* sp., *Klebsiella* sp., *Salmonella* sp., and *Serratia* sp.	Broth microdilution method	MIC values: 1.36–2.72 mg/mL MBC values: 5.45–10.9 mg/mL	Selles et al. (2020)
	100, 50, 25, 12.5, 6.125, 3.5, 1.56, and 0.78 mg/mL	*Bacillus subtilis* (NCTC 8236), *Staphylococcus aureus* (ATCC 25923) *Escherichia coli* (ATCC 25922), *Pseudomonas aeruginosa* (ATCC 27853)	Cup plate agar diffusion method and micro-dilution method	Gram positive bacteria MIC: 0.78 MBC: 3.125 mg/mL Gram negative bacteria MIC: 1.5 MBC: 6.25 mg/mL	Fagere and Al Magboul (2016)
	0.2% and 0.4%	*Escherichia coli Pseudomonas aeruginosa Staphylococcus aureus*	Bacterial death kinetics	Maximum inactivation rate *E. coli* (K=1.514) P. aeruginosa (K=1.334) S. aureus (K=0.995)	Nuñez and Aquino (2012)

Table 2 Functionality of clove essential oil—cont'd

Activity	Concentration	Microorganism / strain/animal used	Tests used	Parameter noted	References
	10%–0.039% (v/v)	*Acinetobacter baumannii*, *Pseudomonas aeruginosa*, *Staphylococcus aureus*, *Enterococcus faecalis*, *Staphylococcus aureus* ATCC2921, *Pseudomonas aeruginosa* ATCC27853	Agar well diffusion method	MIC values: 0.312%–1.25% (v/v)	Abdullah et al. (2015)
	0.156%–4.000% (v/v)	*Staphylococcus epidermidis*, *Staphylococcus aureus*, *Bacillus subtilis*, *Proteus vulgaris*, *Pseudomonas aeruginosa*, *Escherichia coli*	Agar disk diffusion method	MIC values: 0.062%–0.500% (v/v)	Fu et al. (2007)
	4500, 4000, 3000, 2000, 1500, 1000, and 500 ppm	*Listeria monocytogenes*	–	MIC and MBC: 3000 and 10,500 ppm	Cava et al. (2007)
Antioxidant activity	0.10–10 μg/mL	–	DPPH$^{\bullet}$ scavenging activity	IC_{50}: 0.2 μg/mL	Chaieb et al. (2007)
	–	–	DPPH$^{\bullet}$ scavenging activity, reducing power assay	IC_{50}: (4.82×10^{-2} μg/mL) EC_{50}: (3.47×10^{-2} μg/mL)	Selles et al. (2020)
	200–600 μg	–	DPPH$^{\bullet}$ scavenging assay, hydrogen peroxide assay, nitric oxide scavenging activity	16.19%, 22.50%, and 30.24% inhibition in DPPH$^{\bullet}$, hydrogen peroxide and nitric oxide scavenging assay	Binti Zaaba et al. (2016)

Continued

Table 2 Functionality of clove essential oil—cont'd

Activity	Concentration	Microorganism / strain/animal used	Tests used	Parameter noted	References
	15 and 45 µg/mL	–	DPPH• and ABTS radical scavenging activity, total antioxidant activity determination by ferric thiocyanate, total reducing ability determination by Fe^{3+}-Fe^{2+} transformation method, hydrogen peroxide and superoxide anion radical scavenging and ferrous ions (Fe^{2+}) chelating activities	83.6% and 100% DPPH• and ABTS scavenging activity, 97.3% lipid peroxidation of linoleic acid emulsion, 58.2% ferrous ions chelating activity, 22.9% hydrogen peroxide scavenging activity, 57.0% superoxide scavenging activity	Gulcin et al. (2012)
	15–45 µg/mL	–	DPPH• scavenging activity	IC_{50} values: 15.80–108.85 µg/mL	Alfikri et al. (2020)
	0.5, 1.0, and 3.0 mg/mL	–	Superoxide dismutase (SOD-3) or glutathione transferase (GST-4) expression	DAF-16 translocation level elevated by about 3.34 folds in comparison with control	Zhang et al. (2020)
Antifungal activity	10 mg	*Candida* species	Disk paper diffusion method	*Candida sake* showed maximum diameter of inhibition growth (21 mm)	Chaieb et al. (2007)
	0.02–20 µL/mL (v/v)	*Candida*, *Aspergillus*, and *dermatophyte* species	Germ tube inhibition assay	MIC values: 0.08–0.64 µg/mL	Pinto et al. (2009)
	–	*Aspergillus. niger* and *Candida albicans*	Agar disk diffusion method	Diameter of inhibition zones: 2.13–3.19 mm	Yasser (2019)
	0, 12.5, 50, 100, 200, 400, and 800 mg/mL	*Aspergillus niger*, *Alternaria alternata*, *Colletotrichum gloeosporioides*, *Lasiodiplodia theobromae*, *Phomopsis viticola*, and *Rhizopus stolonifer*	Inverted petriplate method	MIC values: 200, 200, 400, 800, 200, and 200 mg/mL, respectively	Sukatta et al. (2008)

Table 2 Functionality of clove essential oil—cont'd

Activity	Concentration	Microorganism / strain/animal used	Tests used	Parameter noted	References
	–	*Candida albicans*, *Epidermophyton floccosum*, *Microsporum audouinii*, *Trichophyton mentagrophytes*, and *Trichophyton rubrum*	Phytatray chamber assay	MIC values: 2.5%, 5%, 2.5%, 1%, and 1% (v/v) respectively	Chee and Lee (2007)
	12.5–100 mg	*Aspergillus niger* and *Aspergillus fumigates*	Disk diffusion method, broth dilution method, agar dilution method	Inhibition zones: 8–14 and 10–15 mm	Bansod and Rai (2008)
	100 μL	*Aspergillus nidulans*, *Alternaria alternate*, *Aspergillus flavus*, *Penicillium notatum*, and *Aspergillus fumigatus*	Agar well diffusion method	Diameter of inhibition of 18, 18, 10, 8, and 6 mm	Verma et al. (2015)
	100 and 1000 μL/L	*Collelotrichum gloeosporioides*	Vapor phase and contact phase	EC_{50} values: 20 and 80 μL/L MIC values: 100 and 300 μL/L	Wang et al. (2019)
	0.1, 0.2, 0.4, 0.5, 1, 2, and 5 μL/mL	*Fusarium oxysporum* f.sp. *radicis lycopercisi*, *F. commune*, and *F. redolens*	Agar diffusion plate method	complete inhibition at 1 μL/mL for *Fusarium oxysporum* f.sp. *radicis lycopersici* and *Fusarium redolens*, complete inhibition at 0.5 μL/mL for *F. commune*	Hamini-Kadar et al. (2014)
	150–500 ppm	*Rhizopus stolonifer* and *Fusarium solani*	Food poisoning technique	Total inhibition of the mycelia growth at 200 and 300 ppm, total inhibition of spores germination at 31.25 and 250 ppm, respectively	Sameza et al. (2015)

Continued

Table 2 Functionality of clove essential oil—cont'd

Activity	Concentration	Microorganism / strain/animal used	Tests used	Parameter noted	References
	5, 10, 25, 50, 75, 100, 200 µg/mL	*Fusarium moniliforme*, *Rhizocotonia solani*, and *Helminthosporium oryzae*	Poison food technique	ED_{50} values: 25.5, 28.0, 53.1 ppm	Kaur et al. (2019)
Anti-inflammatory activity	0.011%, 0.0037%, 0.0012%, and 0.00041% (v/v)	Human dermal fibroblasts	Enzyme-linked immunosorbent assay (ELISA)	Inhibit collagen-I, collagen-III, M-CSF, TIMP-2, VCAM-1, collagen III	Han and Parker (2017)
	10 mg/mL	Mice	Carrageenan-induced paw edema test	Inhibit mouse paw edema (50.6%)	Taher et al. (2015)
	1% (w/v)	Mice	Carrageenan-induced rat paw edema test	Inhibit cyclo-oxygenase (COX-2) enzyme and prostaglandins (25.6%)	Kandekar et al. (2016)
	25 mg/kg	Dogs	Measurement of edema, complete blood count, histopathology, and rectal temperature	Decrease in the band neutrophils, neutrophils and white blood cells	Nikoui et al. (2017)
	0.050, 0.025, 0.050, 0.100 and 0.200 mL/kg	Rats	Carrageenan paw edema test	Concentrations of 3 and 50 mg/kg reduced inflammation by 95.70% and 43.7%	Öztürk and Özbek (2005)
Acaricidal activity	0.1 and 1.0 µL	*Varroa destructor*	Glutathione-*S*-transferase (GST) assay kit	Decreasing water-soluble protein content, glutathione-*S*-transferase	Li et al. (2017)
	10 µL	House dust mites (*Dermatophagoide pteronyssinus*)	Direct contact	Efficacy of pipetting and spraying was 99% and 81%	Mahakittikun et al. (2014)
	–	*Dermatophagoides farinae*	Direct contact and vapor phase toxicity bioassays	LD_{50} 12.20 µg/cm^2	Li et al. (2009)

Table 2 Functionality of clove essential oil—cont'd

Activity	Concentration	Microorganism / strain/animal used	Tests used	Parameter noted	References
	50.9, 32.5, 25.5, 12.7, 6.4, 3.2, 1.6, 0.8, and 0.4 mg/cm^2	*Dermatophagoides farinae*	Impregnated fabric disk bioassay	94% mortality rate	Kim and Sharma (2011)
	2.5, 5.0, 50, and 100 mg/mL	*Rhipicephalus microplus*	Adult immersion test (AIT) and the larval immersion test (LIT)	Good efficacy at 50 mg/mL and 100% potency at 100 mg/mL	Ferreira et al. (2018)
	–	*Tyrophagus putrescentiae*	Impregnated fabric disk and fumigation methods	LD_{50} values: methyleugenol β-caryophyllene, isoeugenol, eugenol, α-humulene and acetyleugenol 1.18, 8.21, 11.7, 12.1, 12.9, and 28.7 mg/cm^2	Kim et al. (2003)
Leishmanicidal activity	–	*Leishmania major* and *Leishmania tropica*	–	IC_{50} value oil and eugenol against *Leishmania major* (180.2 and 233.5 μg/mL) and *Leishmania major* (654.7 and 517.1 μg/mL)	Moemenbellah-Fard et al. (2020)
	100 μg/mL	*Leishmania. donovani* promastigotes and intracellular amastigotes	Growth kinetics assay, growth reversibility assay	IC_{50} values: 21 and 15.2 μg/mL	Islamuddin et al. (2014)
Anesthetic activity	40, 80, 120, 160, 200, 240, and 280 mg/L	Olive flounder (*Paralichthys olivaceus*)	Hematocrit, hemoglobin levels, red blood cell count, osmolality, cortisol, glucose, lactic acid, Na^+, K^+, and Cl^- concentration, aspartate amino transferase and alanine amino transferase activities	Loss of motility, balance and consciousness	Hur et al. (2019)

Continued

Table 2 Functionality of clove essential oil—cont'd

Activity	Concentration	Microorganism / strain/animal used	Tests used	Parameter noted	References
	60 and 90 μL/L	*Parachheirodon axelrodi* (cardinal tetra), *Pterophyllum scalare* (angelfish), *Heros severus* (banded cichlid)	Electromyographic recordings	Deep anesthesia (<3 min)	Fujimoto et al. (2017)
Anticancer activity	For BSLT assay: 31.2, 62.5, 125, 250, 500 μg/mL and for MTT assay: 0.2, 0.4, 0.8, 1.6, 3.1, 6.3, 12.5, 25, and 50 μg/mL	MCF-7 human breast cancer cells	BSLT and MTT assay	LD_{50} values: 37 μg/mL (after 24 h) in BSLT assay, in MTT assay, IC_{50} values: 36.4 and 17.6 μg/mL after 24 and 48 h	Kumar et al. (2014)
	300 μL/mL	DU-145 prostate cancer, HeLa cervical cancer, TE-13 esophageal cancer, MDA-MB-231 (ER − ve) and MCF-7 (ER + ve) breast cancer, normal human peripheral blood lymphocytes	MTT assay	Maximum cytotoxic activity and cell deaths (80%) in TE-13 cells within 24 h	Dwivedi et al. (2011)
	20%–1.25%	Breast cancer cell line (MCF-7), leukemia cell line (K-562) and cervical cancer cell lines	MTT assay	Higher degree of cytotoxic activity on the K-562 cell line	Sujatha and Sirisha (2019)
	0.2, 0.4, 0.6, 0.8, and 1.0 μL/mL	Thyroid cancer cell line (HTh-7)	MTT assay, colony formation assay, and Annexin V-FITC assay	IC_{50} value: 0.7 μL/mL	Nirmala et al. (2019)
	–	Cancer cells (HT29 and Hep2, A549 and raw 269.7)	MTT colorimetric assay	IC_{50} values: 15.7–200 μg/mL	Kouidhi et al. (2010)
Nematicidal activity	31.12–500 ppm	*Meloidogyne graminicola* J2s	Juvenile mortality test	Mortality ranging from 70.0%–92.5% to 77.5%–97.5% at 31.1–500 ppm after 24 and 96 h	Ajith et al. (2020)

Table 2 Functionality of clove essential oil—cont'd

Activity	Concentration	Microorganism / strain/animal used	Tests used	Parameter noted	References
	1000 ppm	*Meloidogyne incognita*	Egg hatching and juvenile mortality test	62.5% and 54.2% egg hatching after 24, 4, and 96 h of exposure LD_{50} values of 11.8, 2.9, and 1.8 ppm after 4, 8, and 24 h in juvenile hatching test	Addabbo et al. (2020)
	0.1%–0.3%	*Meloidogyne incognita*	Juvenile mortality test	36%–37% reduction in nematode population	Meyer et al. (2008)
	–	*Bursaphelenchus xylophilus*	Immersion bioassay	LC_{50} value ranging from 0.57 to 0.88 mg/mL	Kong et al. (2006)
Herbicidal activity	2.5%	Lambsquarters (*Chenopodium album*)	Membrane integrity measurement	Phytotoxicity on cell membrane integrity	Stokłosa et al. (2012)
	5%–10% (V/V)	Common lambsquarters (*Chenopodium album*), common ragweed (*Ambrosia artemisiifolia*) and johnsongrass (*Sorghum halepense*)	Dandelion leaf disk assay and whole plant assay	Shoot death postpone from 1 h to 1 day	Tworkoski (2002)
	2.5%	Redroot pigweed (*Amaranthus retroflexus*) and common lambsquarters (*Chenopodium album*)	Membrane integrity measurement	Decline in leaf cell membrane integrity and seedling growth	Bainard et al. (2006)
	–	Cucumber (*Cucumis sativus*)	–	Increase superoxide dismutase (SOD) activity and decreased catalase activity	Park et al. (2011)

Continued

Table 2 Functionality of clove essential oil—cont'd

Activity	Concentration	Microorganism / strain/animal used	Tests used	Parameter noted	References
Larvicidal activity	1%, 5%, and 10%	House fly (*Musa domesticai* L.)	Oviposition deterrent bioassay	LC_{50} and LT_{50} values: 9.83% and 27.05 h	Soonwera (2015)
	5, 10, 20, 40, 60, 80, 100, and 120 mg/L	*Aedes aegypti*	Mortality test	LC_{50} value: 66.90 mg/L	Pandiyan et al. (2019)
	12.5, 25, 50, 75, 100, and 150 ppm	*Anopheles stephensi*	Larvicidal bioassay	LC_{50} and LC_{90} values: 57.49 and 93.14 ppm	Osanloo et al. (2018)

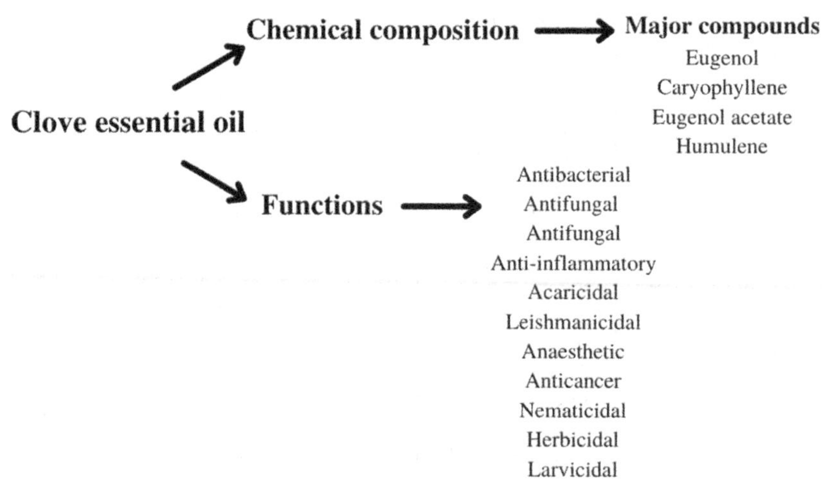

FIG. 1

Composition and functionality of clove essential oil.

5.1 Antibacterial activity

Various researchers have tested the antimicrobial potential of clove essential oil against various gram-positive and gram-negative bacteria. For example, clove essential oil exhibited antibacterial activity against nine gram-positive and gram-negative fish pathogenic bacteria isolated from the cultured olive flounder (*Paralichthys olivaceus*) with minimum inhibitory concentration (MIC) for gram-negative bacteria ranging from 0.125% to 0.5% (v/v) and for gram-positive strains from 0.25% to 0.5% (v/v) (Pathirana et al., 2019), against six enteropathogenic bacteria isolated from diarrhea of young calves with MIC and minimum bactericidal concentration (MBC) in the range of 1.36–2.72 and 5.45–10.9 mg/mL, respectively (Selles et al., 2020).

The oil possessed more antibacterial potential against gram-negative bacteria than gram-positive bacteria as the oil (0.2% and 0.4%) showed more inactivation rate against gram-negative bacteria: *Escherichia coli* (*E. coli*) ($K = 1.514$) and *Pseudomonas aeruginosa* (*P. aeruginosa*) ($K = 1.334$) than gram-positive bacteria: *Staphylococcus aureus* (*S. aureus*) ($K = 0.995$) (Nuñez and Aquino, 2012). This is further confirmed by the fact that MIC and MBC of clove essential oil against gram-positive bacteria: *Bacillus subtilis* (*B. subtilis*) (NCTC 8236) and *S. aureus* (ATCC 25923) have been reported to be less (0.78 and 3.12 mg/mL) as compared to gram-negative bacteria: *E. coli* (ATCC 25922) and *P. aeruginosa* (ATCC27853) (1.5 and 6.25 mg/mL) (Fagere and Al Magboul, 2016). Clove bud oil exhibited better antimicrobial activity than its major compound eugenol against *S. aureus, E. coli, P. aeruginosa,* and *Klebsiella pneumoniae* (*K. pneumoniae*) as IC_{50} values of clove essential oil (177.6, 300.5, 362.9, and 248.0 μg/mL) were less than that of eugenol (320.8, 692.9, 591.8, and 445.8 μg/mL), respectively (Moemenbellah-Fard et al., 2020).

MIC values of clove bud essential oil against multidrug-resistant strains, namely *Acinetobacter baumanni* (*A. baumanni*), *P. aeruginosa*, *S. aureus*, and *Enterococcus feacalis* (*E. feacalis*), and two standard strains, *S. aureus* ATCC29213 and *P. aeruginosa* ATCC27853, ranged from 0.31 to 1.25% (v/v) (Abdullah et al., 2015), whereas against *Staphylococcus epidermidis* (*S. epidermidis*), *S. aureus*, *B. subtilis*, *Proteus vulgaris* (*P. vulgaris*), *P. aeruginosa*, *E. coli* ranged from 0.06% to 0.50% (v/v) (Fu et al., 2007).

Listeria monocytogenes (*L. monocytogenes*) is an important foodborne pathogen contaminating many food products, posing a threat to public health. Through biofilms formation, it can easily adapt to the external environment and can tolerate low temperatures and high salt concentrations. Clove essential oil exhibited antimicrobial activity against *L. monocytogenes* by affecting cell membrane permeability, causing leakage of proteins, ATP and DNA. The treatment also reduced activities of β-galactosidase and alkaline phosphatase. The oil also affected the respiratory metabolism of *L. monocytogenes* by reducing activities of isocitrate dehydrogenase, citrate synthase, and α-ketoglutarate dehydrogenase in the Krebs cycle (Cuia et al., 2018). In another study, clove oil exhibited antimicrobial activity against *L. monocytogenes* in pasteurized milk with MIC and MBC of 3000 and 10,500 ppm, respectively (Cava et al., 2007).

5.2 Antioxidant activity

Various researchers have reported antioxidant potential of clove essential oil using different assays involving different parameters as with 2,2-diphenyl-1-picrylhydrazyl ($DPPH^{\bullet}$) scavenging assay having IC_{50} value of 0.2 μg/mL against synthetic antioxidant *tert*-butylated hydroxytoluene (BHT) with IC_{50} value of 11 μg/mL (Chaieb et al., 2007), IC_{50} (4.82×10^{-2} μg/mL) in $DPPH^{\bullet}$ assay along with reducing power assay with an EC_{50} value of 3.47×10^{-2} μg/mL (Selles et al., 2020), 16.2% inhibition in $DPPH^{\bullet}$ scavenging assay, 22.5% in hydrogen peroxide scavenging assay and 30.2% in nitric oxide scavenging assay (Binti Zaaba et al., 2016), inhibited 97.3% of lipid peroxidation of linoleic acid emulsion, exhibited 97.3% ferrous ions chelating activity, 22.9% hydrogen peroxide scavenging activity and 57.0% superoxide scavenging activity at 15 μg/mL, whereas at 45 μg/mL, it showed 83.6% $DPPH^{\bullet}$ scavenging activity and 100% (2,2′-azino-bis(3-ethylbenzothiazoline-6-sulfonic acid) (ABTS) scavenging activity. Standard antioxidants such as Trolox, butylated hydroxyanisole (BHA), α-tocopherol, and butylated hydroxytoluene (BHT) showed lesser inhibition at the same conditions (Gulcin et al., 2012). The key steps are responsible for the antioxidant properties of clove essential oil involved induction of expression of superoxide dismutase (SOD-3) or glutathione transferase (GST-4) (Zhang et al., 2020).

On comparing antioxidant activity of clove bud essential oil at the different phenological stages in young and mature trees, the highest antioxidant activity was noticed in the flowering stage of young trees, and IC_{50} values were ranged from 15.8 to 108 μg/mL (Alfikri et al., 2020).

5.3 Antifungal activity

Clove essential oil exhibited antifungal potential against 53 human pathogenic *Candida* species, which are the most common cause of opportunistic mycoses with a maximum diameter of inhibition of growth (21 mm) against *Candida sake* (*C. sake*) (Chaieb et al., 2007), against six fungi causing postharvest decay of grapes with MIC values in the range of 200–800 mg/mL (Sukatta et al., 2008), against *Aspergillus niger* (*A. niger*) and *Candida albicans* (*C. albicans*) with a diameter of inhibition zones in the range of 2.13–3.19 mm (Yasser, 2019). Clove essential oil also exhibited fungicidal potential against dermatophytic fungi (*C. albicans*, *Epidermophyton floccosum* (*E. floccosum*), *Microsporum audouinii* (*M. audouinii*), *Trichophyton mentagrophytes* (*T. mentagrophytes*), and *Trichophyton rubrum* (*T. rubrum*)) with MIC values of 2.5%, 5%, 2.5%, 1%, and 1% (v/v), respectively (Chee and Lee, 2007). According to Pinto et al. (2009), MIC values of clove essential oil against *Aspergillus*, *Candida*, and *dermatophyte* species ranged from 0.08 to 0.64 μg/mL.

Clove essential oil showed the diameter of inhibition zones in the range of 6–18 mm a 100 μL against different species of *Aspergillus* and *Penicillium notatum* (Verma et al., 2015), but according to Bansod and Rai (2008), the oil showed more values of inhibition zones against *Aspergillus fumigates* (*A. fumigates*) (10–15 mm) as compared to *A. niger* (8–14 mm) at a concentration ranging from 12.5 to 100 mg. However, the oil showed the same MIC of 0.12 (% v/v) against both the tested fungi. Clove oil also revealed antifungal potential against *Colletotrichum gloeosporioides* (*C. gleosporioides*), causing bitter rot in various vegetables in vitro and in vivo with EC_{50} (half maximal effective concentration) of 20 and 80 μL/L and MIC values of 100 and 300 μL/L in vapor and contact phase, respectively (Wang et al., 2019). Among three *Fusarium* sp., viz., *Fusarium oxysporum* f. sp. *radicis lycopercisi*, *F. commune*, and *F. redolens*, the oil exhibited maximum antifungal activity at 1 μL/mL against *F. oxysporum* f.sp. *radicis lycopercisi* and *F. redolens* and at 0.5 μL/mL against *F. commune* (Hamini-Kadar et al., 2014).

Clove bud essential oil exhibited antifungal potential against *Rhizopus stolonifer* (*R. stolonifer*) and *Fusarium solani* (*F. solani*), causing tuber rot in Yam (*Dioscorea rotundata* Poir.) which is the typical staple food in Africa. *Rhizopus* and *Fusarium* sp. are associated with post harvested yam tuber rot by penetrating the tuber through wounds affecting the inner tuber tissues. Clove bud essential oil caused complete mycelial growth inhibition of *R. stolonifer* and *F. solani* at 200 and 300 ppm, respectively, against standard (Ridomil) value of 1600 ppm for both pathogens. The treatment also caused complete inhibition of spores germination of *R. stolonifer* and *F. solani* at 31.25 and 250 ppm, respectively, against standard values of 200 and 1600 ppm, respectively (Sameza et al., 2015). On comparing the antifungal potential of clove oil against *Fusarium moniliforme* (*F. moniliforme*), *Rhizoctonia solani* (*R. solani*), and *Helminthosporium oryzae* (*H. oryzae*), it has been found that clove oil was most effective against *F. moniliforme* with ED_{50} value (25.5 ppm) followed by *H. oryzae* (28 ppm) and *R. solani* (53 ppm) (Kaur et al., 2019).

5.4 Anti-inflammatory activity

Clove oil (0.011%) exhibited antiproliferative effects on human dermal fibroblasts by enhancing the production of various pro-inflammatory biomarkers. The oil also inhibited tissue remodeling protein molecules and tissue inhibitors of metalloproteinase 2 (TIMP-2) (Han and Parker, 2017). Treatment with clove oil resulted in inhibition of mouse paw edema (50.6%) induced by carrageenan against 70.4% inhibition caused by indomethacin used as standard (Taher et al., 2015). The oil also exhibited anti-inflammatory potential in bio-adhesive gel, which inhibited cyclo-oxygenase (COX-2) enzyme and prostaglandins. The self-micro emulsifying drug delivery system (SMEDDS) gel exhibited 42.0% inhibition of edema by the end of 3h (Kandekar et al., 2016). Clove essential oil-treated dogs showed a decline in the band neutrophils, neutrophils, and white blood cells compared to control with no effect on hematocrit and red blood cells (Nikoui et al., 2017). Clove essential oil at concentrations of 3 and 50mg/kg reduced inflammation of right hind paw of rats by 95.7% and 43.7%, whereas concentrations ranging from 0.050, 0.025, 0.050, 0.100, and 0.200mL/kg reduced inflammation by 46.5%, 90.1%, 66.9%, and 82.8%, respectively (Öztürk and Özbek, 2005).

5.5 Acaricidal activity

Varroa destructor, an ectoparasitic mite of honey bees, poses a severe threat to honeybees colony health as it feeds on the hemolymph of honey bees leading to their attenuated lifespan and decreased immunity. Clove leaves essential oil exhibited acaricidal activity on *V. destructor* by affecting its metabolism by decreasing water-soluble protein content after exposure to 1.0μL/mL for 30min. The oil also increased superoxide dismutase and Ca^{2+}-Mg^{2+}-ATPase and decreased glutathione-*S*-transferase activity (Li et al., 2017). Clove oil (10μL) also exhibited acaricidal activity against house dust mites *Dermatophagoides pteronyssinus* (*D. pteronyssinus*) with 99% and 81% efficacy of pipetting and spraying respectively, whereas with placebo (ethyl alcohol) mortality was <5% (Mahakittikun et al., 2014). The oil (12.2μg/cm^2) also destroyed all the indoor dust mites *Dermatophagoides farina* (*D. farina*) in the contact toxicity bioassay after 24h of exposure showing an adequate acaricidal activity (Li et al., 2009). In another study, clove oil exhibited a 94% mortality rate against *D. farina* (Kim and Sharma, 2011).

Rhipicephalus microplus (*R. microplus*) (Ixodida: Ixodidae), a cattle tick, is an ectoparasite in cattle-breeding areas. Clove essential oil, its main constituent, eugenol, when applied in free or in formulation form against larvae and females of *R. microplus*, exhibited 100% mortality in larvae at initial applications of 2.5 and 5.0mg/mL, respectively. In the adult immersion tests, both eugenol and oil showed good efficacy at a 50mg/mL concentration and achieved 100% potency at 100mg/mL (Ferreira et al., 2018). Among clove bud oil components; acetyl eugenol, β-caryophyllene, eugenol, α-humulene, isoeugenol and methyl eugenol, methyl eugenol was most toxic to *Tyrophagus putrescentiae* (*T. putrescentiae*) (Acari: Acaridae) with LD_{50} value of 1.18mg/cm^2 followed by isoeugenol (8.21mg/cm^2), β-caryophyllene (11.7mg/cm^2), eugenol (12.1mg/cm^2), α-humulene (12.9mg/cm^2), and acetyl eugenol (28.7mg/cm^2). The activity of isoeugenol was comparable with benzyl benzoate with an LD_{50} value of 8.85mg/cm^2 (Kim et al., 2003).

5.6 Leishmanicidal activity

The leishmaniases constitute a group of skin diseases caused by a protozoan parasite from the Leishmania species. *Leishmania major* (*L. major*) and *Leishmania tropica* (*L. tropica*) are mainly responsible for human cutaneous leishmaniasis. Clove essential oil and eugenol showed leishmanicidal activity on *L. tropica* (IC_{50} of 180.2 and 233.5 μg/mL) and *L. major* (654 and 517 μg/mL), respectively (Moemenbellah-Fard et al., 2020). Clove bud essential oil also possessed remarkable activity against *Leishmania donovani* (*L. donovani*) against promastigotes and intracellular amastigotes with IC_{50} values of 21 and 15.2 μg/mL, respectively (Islamuddin et al., 2014).

5.7 Anesthetic activity

It is essential to give anesthesia to fishes as it minimizes stress during manual spawning, surgery, air exposure, vaccination, blood and tissue collection, antibiotics, and transportation in bulk. Synthetic anesthetics are associated with various side effects and also pose selective toxicity by changes in fish metabolism. This results in loss of motility, balance, and consciousness on increasing the anesthesia time and concentration. Clove oil was used as a natural anesthetic agent at 160 mg/L at 21°C for olive flounder (*Paralichthys olivaceus*), weighting 245 g (Hur et al., 2019). The oil also induced deep anesthesia (<3 min) in three Amazon fish species; *Parachheirodon axelrodi* (cardinal tetra) and *Pterophyllum scalare* (angelfish) at 60 μL/L and in *Heros severus* (banded cichlid) at 90 μL/L (Fujimoto et al., 2017).

5.8 Anticancer activity

Clove essential oil exhibited anticancer potential toward MCF-7 human breast cancer cells with an LD_{50} value of 37 μg/mL after 24 h through brine shrimp lethality test (BSLT) and 36.4 and 17.6 μg/mL after 24 and 48 h, respectively through 3-(4,5-dimethylthiazol-2-yl)-2,5-diphenyl tetrazolium bromide (MTT) assays (Kumar et al., 2014). Clove oil also possessed anticancer potential against DU-145 prostate cancer, HeLa cervical cancer, TE-13 esophageal cancer, MDA-MB-231 (ER − ve), and MCF-7 (ER + ve) breast cancer. Maximum cytotoxic activity and cell deaths (80%) were observed in TE-13 cells within 24 h at 300 μL/mL, whereas minimum cell deaths in DU-145 cells (Dwivedi et al., 2011). On comparing the anticancer activity of clove oil on breast cancer cell lines (MCF-7), leukemia cell lines (K-562), and cervical cancer cell lines, it was found that the oil represented a higher degree of cytotoxic activity on the K-562 cell line (Sujatha and Sirisha, 2019). Clove oil-based emulsion induced apoptosis and decreased cancer cell proliferation against thyroid cancer cell line (HTh-7) with IC_{50} value of 0.7 μL/mL (Nirmala et al., 2019). IC_{50} values of clove oil against cancer cells (HT29 and Hep2, A549 and raw 269.7) ranged from 15.7 to 200 μg/mL (Kouidhi et al., 2010).

5.9 Nematicidal activity

Clove oil exhibited nematicidal activity against *Meloidogyne graminicola* (*M. graminicola*) J2s with mortality of J2s ranging from 70.0% to 92.5% and 77.5% to 97.5% at 31.1–500 ppm after 24 and 96 h, respectively. Efficacy of the oil was compared with carbofuran at 62.5 ppm (76.0%–87.6%) and 31.2 ppm (70.6%–83.3%) after 24–96 h (Ajith et al., 2020). Clove oil also reduced the nematode population of *Meloidogyne incognita* (*M. incognita*) (36%–37%) and was non-phytotoxic to *Cucumis melo* (muskmelon), *Solanum lycopersicum* (tomato), *Cucumis sativus* (cucumber), and *Capsicum annuum* (pepper) at various concentrations ranging from 0.1% to 0.3% (Meyer et al., 2008). The oil was highly toxic to *M. incognita* J2 with LD_{50} values of 11.8, 2.9, and 1.8 ppm after 4, 8, and 24 h of exposure and less active on nematode eggs with percent hatching of 70.5, 62.5, and 54.2% after 24, 48, and 96 h of exposure at 1000 ppm (Addabbo et al., 2020). Clove oil also exhibited nematicidal activity against *Bursaphelenchus xylophilus* (causing pine wilt disease) with LC_{50} values ranging from 0.57 to 0.88 mg/mL (Kong et al., 2006).

5.10 Herbicidal activity

Clove oil (2.5%) induced phytotoxicity on cell membrane integrity in common lambsquarters (*Chenopodium album*) and broccoli at the nine and three-leaf stage due to its major constituent eugenol, while β-caryophyllene and α-humulene showed less or no role. Moreover, cell membrane damage by clove oil and eugenol sprays declined with enhancing light intensity (Stokłosa et al., 2012).

Clove oil (5%–10%) resulted in shoots death from 1 h to 1 day when applied on the shoots of common lambsquarters (*Chenopodium album*), common ragweed (*Ambrosia artemisiifolia*), and johnsongrass (*Sorghum halepense*) (Tworkoski, 2002). Clove oil and its primary constituent, eugenol, resulted in declined leaf cell membrane integrity and hinder seedling growth of *Amaranthus retroflexus* (redroot pigweed) and *Chenopodium album* (common lambsquarters) (Bainard et al., 2006). Clove oil and paraquat followed different herbicidal action mechanisms as clove oil increased superoxide dismutase (SOD) activity and decreased catalase activity on the cucumber (*Cucumis sativus*) seedling, while in paraquat treatment, SOD and catalase activity were decreased (Park et al., 2011).

5.11 Larvicidal activity

Clove essential oil showed higher larvicidal activity against house fly (*Musa domesticai* L.), having LC_{50} and LT_{50} values of 9.83% and 27.0 h, respectively, than cypermethrin having LC_{50} and LT_{50} values of 11.4% and 31.6 h (Soonwera, 2015). Clove essential oil in synergized combination star anise (*Illicium verum*) essential oil showed an LC_{50} value of 49.0 mg/L as compared to LC_{50} value of 66.9 mg/L shown by clove essential oil alone against *Aedes aegypti* (Pandiyan et al., 2019). Clove oil exhibited better larvicidal activity than eugenol against *Anopheles stephensi* having LC_{50} and LC_{90} values (57.4 and 93.1 ppm) lower than that of eugenol (86.9 and 128 ppm) (Osanloo et al., 2018).

References

Abdullah, B.H., Hatem, S.F., Jumaa, W., 2015. A comparative study of the antibacterial activity of clove and rosemary essential oils on multidrug resistant bacteria. UK J. Pharm. Biosci. 3, 18–22.

Addabbo, T.D., Argentieri, M.P., Laquale, S., Candido, V., Avato, P., 2020. Relationship between chemical composition and nematicidal activity of different essential oils. Plants 9, 1–19.

Ajith, M., Singh, P., Shakil, N., Kaushik, P., Rana, V., 2020. Chemical composition and nematicidal activity of essential oils and their major compounds against *Meloidogyne graminicola* (rice root-knot nematode). J. Essent. Oil Res. 32, 1–10.

Alfikri, F.N., Pujiarti, R., Wibisono, M.G., Hardiyanto, E.B., 2020. Yield, quality and antioxidant activity of clove (*Syzygium aromaticum* L.) bud oil at the different phenological stages in young and mature trees. Scientifica 2020, 1–8.

Alma, M., Murat, E., Siegfrie, N., Hubert, K., 2007. Chemical composition and content of essential oil from the bud of cultivated Turkish clove (*Syzygium aromaticum* L.). BioResources 2, 265–269.

Arslan, N., Gürbüz, B., Sarihan, E.O., 2004. Variation in essential oil content and composition in Turkish anise (*Pimpinella anisum* L.) populations. Turk. J. Agric. For. 28, 173–177.

Bainard, L.D., Isman, M.B., Upadhayaya, M.K., 2006. Phtyotoxicity of clove oil and its primary constituent eugenol and the role of leaf epicuticular wax in the susceptibitlity to these essential oils. Weed Sci. 54, 833–837.

Banerjee, K., Madhyastha, H., Rajendra, S.V., Manikandanath, N.T., Thiagarajan, N., Thiagarajan, P., 2020. Anti-inflammatory and wound healing potential of a clove oil emulsion. Colloids Surf. B: Biointerfaces 193, 1–9.

Bansod, S., Rai, M., 2008. Antifungal activity of essential oils from Indian medicinal plants against human pathogenic *Aspergillus fumigatus* and A. niger. World J. Med. Sci. 3, 81–88.

Binti Zaaba, N.A.A., Gayathri, G.R., Vishnu Priya, V.P.V., 2016. In vitro free radical scavenging activity of tea tree oil and clove oil. Asian J. Pharm. Clin. Res. 9, 179–183.

Board, N., 2010. Handbook on Spices. Asia Pacific Business Press Inc, Delhi, pp. 199–213.

Byng, 2016. Revision of Eugenia and Syzygium (Myrtaceae) from Comoro archipelago. Phytotaxa 252, 162–184.

Cava, R., Nowak, E., Taboada, A., Martin-Iniesta, F., 2007. Antimicrobial activity of clove and cinnamon essential oils against *Listeria monocytogenes* in pasteurized milk. J. Food Prot. 70, 2757–2763.

Chaieb, K., Hajlaoui, H., Zmantar, T., Nakbi, K.A.B., Rouabhia, M., Mahdouani, K., Bakhrouf, A., 2007. The chemical composition and biological activity of essential oil, *Eugenia cryophyllata* (*Syzygium aromaticum* L. Myrtaceae): a short review. Phytother. Res. 21, 501–506.

Chee, H.Y., Lee, M.H., 2007. Antifungal activity of clove essential oil and its volatile vapour against dermatophytic fungi. Mycobiology 35, 241–243.

Cuia, H., Zhanga, C., Lib, C., Lina, L., 2018. Antimicrobial mechanism of clove oil on *Listeria monocytogenes*. Food Control 94, 140–146.

Danthu, P., Penot, E., Ranoarisoa, K.M., Rakotondravelo, J.C., Michel, I., Tiollier, M., Michels, T., Normand, F., Razafimamonjison, G., Fawbush, F., Jahie, M., 2014. The clove tree of Madagascar: a success story with an unpredictable future. Bois For. Trop. 320, 83–96.

de Meneses, A.C., Sayer, C., Puton, B.M.S., Cansian, R.L., Araújo, P.H.H., de Oliveira, D., 2019. Production of clove oil nanoemulsion with rapid and enhanced antimicrobial activity against gram-positive and gram-negative bacteria. J. Food Process Eng. 42, 1–9.

De Oliveiraa, M.S., Da Costaa, W.A., Pereirac, D.S., Botelhob, J.R.S., Menezesf, T.O.A., Andraded, E.H.A., Da Silvae, S.H.M., da Silva, S.F.A.P., de Carvalho, J.R.N., 2016. Chemical composition and phytotoxic activity of clove (*Syzygium aromaticum*) essential oil obtained with supercritical CO_2. J. Supercrit. Fluids 118, 185–193.

Dwivedi, V., Shrivastava, R., Husain, S., Ganguly, C., Bharadwaj, M., 2011. Comparative anticancer potential of clove (*Syzygium aromaticum*)—an Indian species—against cancer cell lines of various anatomical origin. Asian Pac. J. Cancer Prev. 12, 1989–1993.

Fagere, Z.O., Al Magboul, A.Z., 2016. Antibacterial activity of clove oil against some microorganisms at Khartoum State. Adv. Med. Plant Res. 4, 122–128.

Fankem, P.M., Kwanga, S.N., Sameza, M.L., Tchoumbougnang, F., Tchabong, R., Ngoune, T., Dongmo, P.M.J., 2017. Antioxidant and antifungal activities of cocoa butter (*Theobroma cacao*), essential oil of *Syzygium aromaticum* and a combination of both extracts against three dermotophytes. Am. Acad. Sci. Res. J. Eng. Technol. Sci. 37, 255–272.

Ferreira, F.M., Delmonte, C.C., Novato, T.L.P., Monteriro, C.M.O., Daemon, E., Vilela, F.M.P., Amaral, M.P.H., 2018. Acaricidal activity of essential oil of *Syzygium aromaticum*, hyudrolated and eugenol formulated or free on larvae and engorged females of *Rhipicephalus microplus*. Med. Vet. Entomol. 32, 41–47.

Fu, Y., Zu, Y., Chen, L., Shi, X., Wang, Z., Sun, S., 2007. Antimicrobial activity of clove and rosemary essential oils alone and in combination. J. Phytother. Res. 59, 989–990.

Fujimoto, R.Y., Pereira, D.M., Silva, J.C.S., de Oliveria, L.C.A., Inoue, L.A.K.A., Hamoy, M., de Mello, V.J., Torres, M.F., Barbas, L.A.L., 2017. Clove oil induces anaesthesis and blunts muscle contraction power in three Amazon fish species. Fish Physiol. Biochem. 44, 245–256.

Gulcin, I., Elmastas, M., Adoul-Enein, H.Y., 2012. Antioxidant activity of clove oil—a powerful antioxidant source. Arab. J. Chem. 5, 489–499.

Hamini-Kadar, N., Hamdane, F., Boutoutaou, R., Kihal, M., Henni, J.E., 2014. Antifungal activity of clove (*Syzygium aromaticum* L.) essential oil against phytopathogenic fungi on tomato (*Solanum lycopersicum* L) in Algeria. J. Exp. Biol. Agric. Sci. 2, 447–454.

Han, X., Parker, T.L., 2017. Anti-inflammatory activity of clove (*Eugenia caryophyllata*) essential oil in human dermal fibroblasts. Pharm. Biol. 55, 1619–1622.

Hur, J.W., Gilb, H.W., Choic, S.H., Jungd, H.J., Kang, Y.J., 2019. Anesthetic efficacy of clove oil and the associated physiological responses in olive flounder (*Paralichthys olivaceus*). Aquac. Rep. 15, 1–6.

Islamuddin, M., Sahal, D., Afrin, F., 2014. Apoptosis-like death in *Leishmania donovani* promastigotes induced by eugenol-rich oil of *Syzygium aromaticum*. J. Med. Microbiol. 63, 74–85.

Jagtap, P.N., Mulani, S.P., Patil, R., 2019. Role of clove in medicine and dentistry—an interesting review article. Res. Rev. J. 4, 180–182.

Jimoh, S.O., Arowolo, L.A., Alabi, K.A., 2017. Phytochemical screening and antimicrobial evaluation of *Syzygium aromaticum* extract and essential oil. Int. J. Curr. Microbiol. Appl. Sci. 6, 4557–4567.

Jirovetz, L., Buchbauer, G., Stoilova, I., Stoyanova, A., Krastanov, A., Schmidt, E., 2006. Chemical composition and antioxidant properties of clove leaf essential oil. J. Agric. Food Chem. 54, 6303–6307.

Kamatou, G.P., Vermaak, I., Viljoen, A.M., 2012. Eugenol—from Maluka islands to the international market place: a review of a remarkable and versatile molecules. Molecules 17, 6953–6981.

Kandekar, S., Bhujbal, N., Deshpande, Y., 2016. Novel formulation for enhanced efficacy of *Lavang Taila* (clove oil) for anti-inflammatory activity. Int. J. Ayurveda Pharma Res. 4, 24–27.

Kaur, K., Kaushal, S., Rani, R., 2019. Chemical composition, antioxidant and antifungal potential of clove (*Syzygium aromaticum*) essential oil, its major compound and its derivatives. J. Essent. Oil Bear. Plants 22, 1195–1217.

Khan, M.S., Zahin, M., Hasan, S., Husain, F.M., 2009. Inhibition of quorum sensing regulated bacterial functions by plant essential oils with special reference to clove oil. Lett. Appl. Microbiol. 49, 354–360.

Kildeaa, M.A., Allanb, G.L., Kearney, R.E., 2004. Accumulation and clearance of the anaesthetics clove oil and AQUI-STM from the edible tissue of silver perch (*Bidyanus bidyanus*). Aquaculture 232, 265–277.

Kim, J.R., Sharma, S., 2011. Acaricidal activities of clove bud oil and red thyme oil using microencapsulated against HDMs. J. Microencapsul. 28, 82–91.

Kim, E.H., Kim, H.K., Choi, D.H., Ahn, Y.J., 2003. Acaricidal activity of clove oil compounds against *Tyrophagus putrescentiae* (Acari: Acaridae). Appl. Entomol. Zool. 38, 261–266.

Kong, J.-O., Lee, S.-M., Moon, Y.-S., Lee, S.-G., Ahn, Y.-J., 2006. Nematicidal activity of plant essential oils against *Bursaphelenchus xylophilus* (Nematoda: Aphelenchoididae). J. Asia Pac. Entomol. 9, 173–178.

Kouidhi, B., Zmantar, T., Bakhrouf, A., 2010. Anticarcinogenic and cytotoxic activity of clove essential oil (*Eugenia caryophyllata*) against a large number of oral pathogens. Ann. Microbiol. 60, 599–604.

Kumar, P.S., Febriyanti, R.M., Sofyan, F.F., Luftimas, D.E., Abdulah, R., 2014. Anticancer potential of *Syzygium aromaticum* L. in MCF-7 human breast cancer cell lines. Pharm. Res. 6, 350–354.

Li, J., Wu, H.Q., Liu, Z.G., 2009. Acaricidal activity of clove bud oil against *Dermatophagoides farina* (Acari: Pyroglyphidae). Zhongguo Ji Sheng Chong Xue Yu Ji Sheng Chong Bing Za Zhi 27, 492–493 (Chinese).

Li, L., Lin, Z.-G., Wang, S., Su, X.-L., Gong, H.-R., Li, H.-L., Hua, F.-L., Zheng, H.-Q., 2017. The effects of clove oil on the enzyme activity of Varroa destructor Anderson and Trueman (Arachnida: Acari: Varroidae). Saudi J. Biol. Sci. 24, 996–1000.

Mahakittikun, V., Soonthornchareonnon, N., Foongladda, S., Boitano, J., Wangapai, T., Ninsanit, P., 2014. A preliminary study of the acaricidal activity of clove oil, *Eugenia caryophyllus*. Asian Pac. J. Allergy Immunol. 32, 46–52.

Marya, C.M., Satija, G., Avinash, J., Nagpal, R., Ahmad, A., 2012. In vitro inhibitory of clove essential oil and its two active principles on tooth decalcification by apple juice. Int. J. Dent. 2012, 1–6.

Matta, F.B., 2010. Essential oils from six herbal plants for biocontrol of the maize weevil. Hortic. Sci. 45, 592–598.

Meyer, S.L.F., Lakshman, D.K., Zasade, I.A., Vinyard, B.T., Chitwood, D.J., 2008. Phytotoxicity of clove oil to vegetable crop seedling and nematotoxicitty to root-knot nematodes. HortTechnology 18, 631–638.

Moemenbellah-Fard, M.D., Abdollahi, A., Ghanbariasad, A., Osanloo, M., 2020. Antibacterial and leishmanicidal activities of *Syzygium aromaticum* essential oil versus its major ingredient, eugenol. Flavour Fragr. J. 35, 1–7.

Nikoui, V., Ostadhadi, S., Bakhtiarian, A., Abbasi-Goujani, E., Habibian-Dehkordi, S., Rezaei-Rooshan, M., Forroohandeh, M., Giorgi, M., 2017. The anti-inflammatory and antipyretic effects of clove oil in healthy dogs after surgery. PharmaNutrition 5, 52–57.

Nirmala, M.J., Durai, L., Gopakumar, V., Nagarajan, R., 2019. Anticancer and antibacterial effects of a clove bud essential oil-based nanoscale emulsion system. Int. J. Nanomedicine 14, 6439–6450.

Nuñez, L., Aquino, M.D., 2012. Microbicide activity of clove essential oil (*Eugenia caryophyllata*) of clove essential oil. Braz. J. Microbiol. 43, 1255–1260.

Ortes-Rojas, D.F., Femandes de Souza, C.R., Oliveria, W.P., 2014. Clove (*Syzygium aromaticum*): a precious spice. Asian Pac. J. Trop. Biomed. 4, 90–96.

Osanloo, M., Sedaghat, M.M., Esmaeili, E., Amani, A., 2018. Larvicidal activity of essential oil of *Syzygium aromaticum* (clove) in comparison with its major constituent, eugenol, against Anopheles stephensi. J. Arthropod. Borne Dis. 12, 361–369.

Öztürk, A., Özbek, H., 2005. The anti-inflammatory activity of *Eugenia Caryophyllata* essential oil: an animal model of anti-inflammatory activity. Eur. J. Gen. Med. 2, 159–163.

Pandiyan, G.N., Mathew, N., Munusamy, S., 2019. Larvicidal activity of selected essential oil in synergized combinations against *Aedes aegypti*. Ecotoxicol. Environ. Saf. 174, 549–556.

Park, K.W., Choi, S.H., Ahn, J.Y., Sohn, Y.G., Kim, C.G., Lee, L.L., 2011. Herbicidial action of the clove oil on cucumber seedlings. Weed Biol. Manage. 11, 235–240.

Pathirana, H.N.K.S., Wimalasena, S.H.M.P., De Silva, B.C.J., Hossain, S., Gang-Joon, H., 2019. Antibacterial activity of clove essential oil and eugenol against fish pathogenic bacteria isolated from cultured olive flounder (*Paralichthys olivaceus*). Slov. Vet. Res. 56, 31–38.

Patil, D.A., Dhale, D.A., 2013. Spices and Condiments Origin, History and Application. Daya Publishing House, New Delhi, India, pp. 76–80.

Pedro, T., Luis, J., Sarah, M., Khalid, H., Bruno, A., Tarcísio, M., Terezinha, L., Raimundo, W., Guy, S., Eugenio, O., 2020. Disentangling the ecotoxicological selectivity of clove essential oil against aphids and non-target ladybeetles. Sci. Total Environ. 718, 1–12.

Pinto, E., Vale-Silva, L., Cavaleiro, C., Salgueiro, L., 2009. Antifungal activity of the clove essential oil from *Syzygium aromaticum* on Candida, Aspergillus and dermatophyte species. J. Med. Microbiol. 58, 145–162.

Pruthi, J.S., 2001. Minor Spices and Condiments Crop Management and Post-Harvest Technology. Indian Council of Agricultural Research, New Delhi, India, pp. 308–331.

Raina, V.K., Srivastava, S.K., Aggarwal, K.K., Syamasundar, K.V., Kumar, S., 2001. Essential oil composition of *Syzygium aromaticum* leaf from Little Andaman, India. Flavour Fragr. J. 16, 334–336.

Sameza, M.L., Mabou, L.C.N., Tchameni, S.N., Bedine, M.A.B., Tchoumbougnang, F., Dongmo, P.M.J., Fekam, F.B., 2015. Evaluation of clove essential oil as a mycobiocide against *Rhizopus stolonifer* and Fusarium solani, tuber rot causing fungi in yam (*Dioscorea rotundata* Poir.). J. Phytopathol. 164, 1–8.

Selles, S.M.A., Kouidri, M., Belhamiti, B.T., Amrane, A.A., 2020. Chemical composition, in vitro antibacterial and antioxidant activities of *Syzygium aromaticum* essential oil. J. Food Meas. Charact. 14, 2352–2358.

Soonwera, M., 2015. Larvicidal and oviposition deterrent activities of essential oils against house fly (*Musca domestica* L.; Diptera: Muscidae). J. Agric. Technol. 11, 657–667.

Srivastava, A.K., Srivastava, S.K., Syamsundar, K.V., 2005. Bud and leaf essential oil composition of *Syzygium aromaticum* from India and Madagascar. Flavour Fragr. J. 20, 51–53.

Stokłosa, A., Matraszek, R., Isman, M.B., Upadhyaya, M.K., 2012. Phytotoxic activity of clove oil, its constituents, and its modification by light intensity in broccoli and common lambsquarters (*Chenopodium album*). Weed Sci. 60, 607–611.

Sujatha, K., Sirisha, K.B., 2019. Anti-cancer and anti-oxidant activity of essential oils of *Rosmarinus officinalis, Azadirachta indica, Syzygium aromaticum* and *Cymbopogon nardus*. J. Pharmacogn. Phytochem. 8, 4493–4498.

Sukatta, U., Haruthaithanasan, V., Chantarapanont, W., Dilokkunanant, U., Suppakul, P., 2008. Antifungal activity of clove and cinnamon oil and their synergistic against postharvest decay fungi of grape in vitro. Kasetsart J. (Nat. Sci.) 42, 169–174.

Taher, Y.A., Samud, A.M., El-Taher, F.E., Ben-Hussin, G., Elmezogi, J.S., Al-Mehdawi, B.F., Salem, H.A., 2015. Experimental evaluation of anti-inflammatory, antinociceptive and antipyretic activities of clove oil in mice. Libyan J. Med. 10, 1–7.

Tworkoski, T., 2002. Herbicidal effects of essential oils. Weed Sci. 50, 425–431.

Uddin, M.A., Shahinuzzaman, M., Rana, M.S., Yaakob, Z., 2017. Study of chemical composition and medicinal properties of volatile oil from clove buds (*Eugenia caryophyllus*). Int. J. Pharm. Sci. Res. 8, 895–899.

Verma, S., Karkun, A., Siddiqui, H.N., 2015. Comparative study of clove oil against bacteria and fungal species. Int. J. Clin. Biochem. Res. 2, 73–76.

Wang, D., Zhang, J., Jia, X., Xin, L., Zhai, H., 2019. Antifungal effects and potential mechanism of essential oils on *Collelotrichum gloeosporioides* in vitro and in vivo. Molecules 24, 1–12.

Xu, J.G., Liu, T., Hu, Q.P., Cao, X.M., 2016. Chemical composition, antibacterial properties and mechanism of action of essential oil from clove buds against *Staphylococcus aureus*. Molecules 21, 1–13.

Yasser, S., 2019. Antioxidant, antibacterial and antifungal properties of nanoemulsion of clove essential oil. Nanomed. Res. J. 4, 204–208.

Zhang, L., Gu, B., Wang, Y., 2020. Clove essential oil confers antioxidant activity and lifespan extension in *C. elegans* via the DAF-16/FOXO transcription factor. Comp. Biochem. Physiol. C 242, 1–8.

CHAPTER

Effect of extraction techniques on the yield, composition, and quality of clove (*Syzygium aromaticum*) essential oil

28

Jyotimala J. Jadhav, Girirajsinh C. Jadeja, and Meghal A. Desai
Department of Chemical Engineering, Sardar Vallabhbhai National Institute of Technology, Surat, Gujarat, India

1. Introduction

Cloves (*Syzygium aromaticum*) are aromatic, dried, unopened flower buds that belong to the Myrtaceae family. They are strong in aroma, sweet, and pungent in taste. Matured clove buds are 1.5–2.0 cm long in size and bright red in color. Clove cultivation is primarily done in Sri Lanka, India, Madagascar, Pakistan, and in the islands of Zanzibar, Pemba (Tanzania), and Indonesia (Razafimamonjison et al., 2014). Whole or ground cloves contain 15%–20% (w/w) volatile oil. Eugenol (70%–95%), eugenyl acetate (up to 17%), and β-caryophyllene (12%–15%) are the significant constituents found in the clove oil, whereas α-humulene and caryophyllene oxide exist in minor quantity (Nurdjannah and Bermawie, 2012). The concentration of the constituents in the oil may vary based on the source, i.e., stem, leaf, or clove bud, and the region and the season. Eugenol is a versatile molecule and it has been used as a spicy flavor in whisky, ice creams, candies, and bakery products (Kamatou et al., 2012). The clove has a wide range of pharmacological properties, including anticancer, antidiabetic, anti-inflammatory, antinociceptive, anti-infective (like antibacterial, antifungal, and antiprotozoal), antiviral, antioxidant, and antithrombotic properties (Mbaveng and Kuete, 2017). Therefore, it has found a wide range of applications in food industries as preservatives and flavoring agents, cosmetic industries, and pharmaceutical industries. Clove oil is used to control nausea, vomiting, cough, diarrhea, dyspepsia, flatulence, gastrointestinal spasm, stomach distension, and toothache. It also relieves pain, causes uterine contractions, and stimulates the nerves (Vitamins and Supplements A-Z List - C, 2019). Clove oil is also used as a fish anesthetic, insect repellant, pesticide, and fungicide in the drug and cosmetic industries (Sellamuthu, 2014). Importantly, the clove oil has exhibited anticancer activity and is also a potent solution against vector-borne diseases. Driven by the potential uses of the clove oil and its primary compound, various methods are being used to extract the valuable essential oil from buds, leaves, and stems of a clove plant. This chapter attempts to provide complete details about the extraction techniques employed to date for isolating essential oil from the clove. In the present chapter, conventional and advanced methods used for extracting clove oil will be briefly

Clove (*Syzygium aromaticum*). https://doi.org/10.1016/B978-0-323-85177-0.00027-6

discussed, focusing on underlying principles, mechanisms, benefits, and constraints. Critical parameters of a given technique will be given due importance and discussed as they affect the yield, composition, and biological potential of the oil. The chapter will also compare novel techniques in terms of processing conditions, responses, energy, and environmental impact.

2. Methods for extraction of essential oil

There are more than 90 types of essential oils. Every single type carries its distinctive aroma and health-promoting benefits depending on the raw material from which it is being extracted. Peppermint, lavender, sandalwood, bergamot, rose, tea tree, and clove are a few popular health claimed essential oils in the pharmaceutical and cosmetic industries (What Are Essential Oils?, 2020). The quality and the quantity of the essential oil differ based on multiple factors, viz., variety and origin of the plant material, raw material quality, method of extraction, pretreatment, and posttreatments to the extraction, and variation in the factors associated with the extraction method (Nurdjannah and Bermawie, 2012). The selection of the extraction method, therefore, becomes a vital step. Conventionally, the essential oil has been extracted using hydrodistillation (Kapadiya and Desai, 2017), steam distillation (Alma et al., 2007), and organic solvent extraction. They are associated with drawbacks like hydrolysis, water solubilization of certain constituents, thermal degradation, recovery of the oil, and solvent contamination in solvent extraction (Wenqiang et al., 2007). Conventional methods also require longer extraction time, excessive energy, and a large amount of raw material and solvent. Also, they produce a greater environmental impact which renders their applications relatively unsustainable. Therefore, researchers have explored advanced and more efficient green extraction methods to extract essential oil for the last few decades. These methods include supercritical fluid extraction (Prado et al., 2011; Scopel et al., 2014), microwave-assisted extraction (Kapadiya et al., 2018; Kennouche et al., 2015), ultrasound-assisted extraction (Alexandru et al., 2013; Tekin et al., 2015) and combinations of two or more extraction techniques like ultrasound-assisted supercritical carbon dioxide extraction (USC-CO_2) (Yang et al., 2014), cold pressing assisted supercritical fluid extraction (SFEAP) (Hatami et al., 2019), and ohmic heating assisted hydrodistillation (OAHD) (Tunc and Koca, 2019) to obtain an improved quality and quantity of the essential oil. Advanced extraction methods have distinctive and greener features like reduction in solvent consumption, reduced number of unit operations, minimized extraction time and energy demands, negligible environmental impact, speedy return on investment, and improved production rates. Advanced extraction methods have greater benefits, applications, and demands than traditional methods and can tender the sustainable solution. Table 1 summarizes the work done to date for isolating clove oil using various extraction techniques.

2.1 Classical methods

2.1.1 Steam distillation

Early in the 20th century, clove cultivation was started in the Mediterranean Sea region of Turkey. To study the detailed chemical composition of Turkish clove buds, Alma et al. employed steam distillation. Steam distillation is a conventional extraction technique in which superheated steam is used as an extracting agent to vaporize the high boiling point volatile compounds present in the plant material at a

Table 1 Various methods available for the extraction of clove essential oil.

No.	Method used	Operating conditions and their range	Optimum operating conditions	Results and remarks	References
1	Steam distillation	Equipment—industrial type distiller Time—3 h	–	Clove bud oil composition Eugenol—87.00% Eugenyl acetate—8.01% β-Caryophyllene—3.56%	Alma et al. (2007)
2	Hydrodistillation	Equipment—clevenger type apparatus Time—4 h	–	Total constituents in leaf oil—38 Eugenol—74.30% Eucalyptol—5.80% Caryophyllene—3.85% α-Cadinol—2.43% Total constituents in clove bud oil—31 Eugenol—49.70% Caryophyllene—18.90% 1-Ethyl-3-nitrobenzene—11.10% Benzoic acid, 3-(1-methylethyl)—8.90%	Bhuiyan et al. (2010)
3	Hydrodistillation	Equipment—clevenger type apparatus Time—4 h	–	Indian bud oil constituents—28 Madagascan bud oil constituents—35 Madagascan leaf oil constituents—22 Major constituents—eugenol and β-caryophyllene	Srivastava et al. (2005)
4	Hydrodistillation	Solid loading—10–50 g Volume of water—240–400 mL Particle size—whole buds to 565 μm Extraction time—5–45 min Power—360 W	Solid loading—20 g Volume of water—300 mL Size of clove buds—565 μm Extraction time—45 min	Yield of clove oil—11.35%, w/w Yield of eugenol –10.30%, w/w Zone of inhibition—21 mm	Kapadiya and Desai (2017)
5	Supercritical CO_2 extraction	CO_2 flow rate—0.6–1.2 kg/h External mass transfer resistance (k_e)—2.5×10^{-6}–1.6×10^{-5} m/s Extraction bed height—1/4 L-L where L = 15 cm Pressure—90 bar Temperature—50°C	CO_2 flow rate—1.2 kg/h $k_e = 1.6 \times 10^{-5}$ m/s Extraction bed height—1/4 L	Essential oil yield—20.80%	Reverchon and Marrone (1997)

Continued

Table 1 Various methods available for the extraction of clove essential oil—cont'd

No.	Method used	Operating conditions and their range	Optimum operating conditions	Results and remarks	References
6	Supercritical fluid extraction assisted by cold pressing	Torques—40 and 80 N m Pressure—150 bar Temperature—40°C Time—20 min	Torque—40 N m	Yield of the oil—22.19/100 g buds Eugenol—57.70% Eugenyl acetate—12.60% β-Caryophyllene—8.30% α-Humulene—0.90% Cost of manufacturing—US$13.42/kg volatile oil	Hatami et al. (2019)
7	Supercritical fluid extraction	Temperature—30–50°C Pressure—10–30 MPa Particle size index—0.7944, 0.6430, 0.5223	Temperature—50°C Pressure—10 MPa Particle size index—0.6430	Total constituents—23 Yield of the oil—19.56% Eugenol—58.77%	Wenqiang et al. (2007)
8	Supercritical fluid extraction	Temperature/pressure (T/P) conditions 9000 kPa/313.15 K, 10000 kPa/313.15 K, 9000 kPa/323.15 K, 10,000 kPa/323.15 K	Optimum conditions for Yield—9000 kPa/313.15 K Amount of eugenol—10,000 kPa/313.15 K Bacterial growth—9000 kPa/ 313.15 K	Essential oil yield—14.17% Eugenol—69.12% Minimum inhibitory concentration (MIC) *E. coli* and *E. faecalis*—1.0 mg/mL *S. aureus*—4.0 mg/mL	Scopel et al. (2014)
9	Ultrasound-assisted extraction	Continuous flow modes—450, 900, 1350 mL/min Clove origins—Indonesia, Madagascar, China	Optimum flow mode—1350 mL/min	Yield of clove bud crude extract—36 to 43%, w/w Inhibition—45% to 58% inhibition Total phenolic content range—191 to 215 mg GAE/L extract	Alexandru et al. (2013)
10	Ultrasound-assisted extraction	Temperature—32–52°C Extraction time—30–60 min Plant concentration—3–7% Ultrasound bath frequency—53 kHz	Temperature—30°C Extraction time—31 min Plant concentration—7%	Clove extract yield Predicted—22.86% Actual—22.43% The zone of inhibition (mm)—13 mm against *E. coli* 19 mm against *S. aureus*	Tekin et al. (2015)
11	Ultrasound-assisted extraction with supercritical carbon dioxide extraction (USC-CO_2)	Mean particle size—0.925, 0.725, 0.550, 0.355, and $\leq$0.21 mm Temperature—30–50°C Pressure—10–20 MPa Static extraction time (with or without ultrasound)—5–30 min	Particle size—0.355 mm Temperature—40°C Pressure—15 MPa Static time—15 min Dynamic time—100 min CO_2 flow rate—1.4 g/min	Yield for USC-CO_2—22.04%, w/w Eugenol—59.18% Caryophyllene—15.35% Eugenyl acetate—18.60% α-Humulene—1.93%	Yang et al. (2014)

Table 1 Various methods available for the extraction of clove essential oil—cont'd

No.	Method used	Operating conditions and their range	Optimum operating conditions	Results and remarks	References
		Dynamic time (with or without ultrasound)—20–140 min CO_2 flow rate—0.4–1.8 g/min Ultrasonic frequency—40 kHz Power—185 W Ultrasound cycle—75%			
12	Ultrasound-assisted extraction with supercritical carbon dioxide extraction (USC-CO_2)	Temperatures—32–50°C Pressures—9.0–28.5 MPa CO_2 flow rate—0.08–0.74 g/min Static extraction time—15 min Ultrasonic frequency—40 kHz Power—185 W Ultrasound cycle—75%	Temperature—50°C Pressure—28.5 MPa Particle size—0.355 mm CO_2 flow rate—0.35 g/min	Yield—23.19% Total heat of the solution—9.83 kJ/mol Heat of the solvation—19.03 kJ/mol Heat of the vaporization—9.69 kJ/mol	Wei et al. (2016a)
13	Ultrasound-assisted supercritical carbon dioxide extraction	Temperature—32–50°C Pressure—9.0–25.0 MPa Particle size—0.355 mm CO_2 flow rate—1.4 g/min Extraction time—115 min Ultrasonic frequency—40 kHz Power—185 W Ultrasound cycle—75%	Temperature—44°C Pressure—18.5 MPa	Yield—23.31% α-Humulene yield—2.75%	Wei et al. (2016b)
14	Microwave-assisted extraction (MAE)	Amount of plant material—10–40 g Volume of water—200–500 mL Power—300–850 W Extraction time—15–30 min	Amount of plant material—30 g Volume of water—200 mL Power—600 W Extraction time—30 min	Yield of oil—13.1%, w/w Yield of eugenol—11.93%, w/w Zone of inhibition (ZOI)—24.80 mm	Kapadiya et al. (2018)

Continued

Table 1 Various methods available for the extraction of clove essential oil—cont'd

No.	Method used	Operating conditions and their range	Optimum operating conditions	Results and remarks	References
15	Ohmic heating assisted hydrodistillation	Mass of clove bud—20–40 g Process time—30–120 min Voltage gradient—7.5–12.5 V/cm	Mass of clove bud—40 g Process time—97.96 min Voltage gradient—12.50 V/cm	Clove oil yield—13.18% Energy consumption—0.214 kW h/g	Tunc and Koca (2019)

temperature near 100°C. In the study performed using steam distillation, the Turkish clove buds were found in the comparable composition of eugenol, eugenyl acetate, and β-caryophyllene to those with naturally grown clove trees in their native places. The analysis was performed using GC and GC-MS, and the chemical composition of Turkish clove buds was obtained to be 87.00% eugenol, 8.01% eugenyl acetate, and 3.56% β-caryophyllene (Alma et al., 2007).

2.1.2 Hydrodistillation

Hydrodistillation is another conventional technique that uses water as an extracting solvent. In hydrodistillation, the mixture of water and plant material is allowed to boil in the presence of an external heating medium. The volatile compounds present in the plant material get vaporized along with water vapors which are further separated after condensation and decantation. Bhuiyan et al. incorporated hydrodistillation and performed analysis using GC-MS to obtain the composition of fresh leaves and dry buds qualitatively and quantitatively. The leaves and buds used in this study were grown in Bangladesh. In total, 31 components were detected in clove bud oil having 49.70% eugenol, 18.90% caryophyllene, 11.10% 1-ethyl-3-nitrobenzene, and 8.90% 3-(1-methylethyl)-benzoic acid, whereas leaf oil contains a total of 38 components having 74.30% eugenol, 5.80% eucalyptol, 3.85% caryophyllene, and 2.43% α-cadinol as the main constituents (Bhuiyan et al., 2010). Similarly, Srivastava et al. have performed hydrodistillation for isolation of essential oil from clove buds and leaves of clove trees grown in India and Madagascar. This study identified a total of 28 and 35 constituents from bud oil, representing 99.90% of oil from Indian and Madagascan origins, respectively. In contrast, a total of 22 constituents from leaf oil representing 99.90% of the oil were identified. In both the cases, eugenol and β-caryophyllene were observed as the main constituents (Srivastava et al., 2005).

Kapadiya and Desai have explored hydrodistillation for essential oil extraction from clove buds. The work was mainly focused on multi-response optimization of different process parameters by the Taguchi method and grey relational analysis. To observe the effect of various process parameters and the interaction effect of the parameters on the response, parametric study and analysis of variance (ANOVA) were performed, respectively. The process parameters were optimum at 20 g solid loading, 300 mL water, 45 min extraction time, and 565-μm size of clove buds at a constant power supply of 360 W. At these optimum conditions, clove oil yield and eugenol content were found to be 11.35%, w/w and 10.30%, w/w, respectively. In contrast, the zone of inhibition for *Bacillus subtilis* microbes was found to be 21 mm. Details of operating conditions are provided in Table 1. Model prediction was made using an artificial neural network (ANN) which provided better accuracy in the case of 4-8-3 topology and tangent sigmoidal transfer function as the hidden layer (Kapadiya and Desai, 2017).

2.2 Advanced/innovative methods

2.2.1 Supercritical fluid extraction

Supercritical fluid extraction (SFE) is an advanced extraction technique for extracting essential oil from the matrix. The fluid at its supercritical conditions, i.e., temperature and pressure above the critical point, is used as an extracting solvent. Supercritical fluids are extremely compressed gases with combined properties of liquids and gases like diffusivity, surface tension, and viscosity approaching the gaseous system, while density tending toward the liquid system. Gas-like properties improve the flow characteristics, and the fluid can easily solubilize the solute of interest from the solid matrices while liquid-like properties improve the solubilization capacity. Because of which the mass transfer is greatly improved, and complete recovery of the product and the solvent is possible by varying the pressure. However, a higher pressure causes economic and safety concerns. In SFE, the most commonly used supercritical fluid is carbon dioxide (CO_2) because of its reasonable cost, near room temperature critical value, and nontoxic nature. It has been exhaustively employed for isolating essential oil from various plant matrices. To date, many researchers studied supercritical fluid extraction for clove buds and explored the parametric effect of various operating parameters like temperature, pressure, CO_2 flow rate, and particle size.

In 1997, Reverchon and Marrone carried out the isolation of essential oil from clove buds using supercritical CO_2 extraction with a fractional separation unit where two separators were operated in series. At optimum values of temperature and pressure, the operating parameters, viz., extraction bed height and CO_2 flow rate were studied to determine their impact on the response parameters, viz., extraction yield and concentration at the exit of the extraction bed. Mass balance integration was used for model validation where a single adjustable parameter model was obtained as a best-fit model (Reverchon and Marrone, 1997).

Wenqiang et al. have compared the composition of essential oil obtained using SFE (extracting solvent -CO_2) with the composition of essential oil obtained using conventional methods like hydrodistillation, steam distillation, and solvent extraction. In this study, the effect of various parameters (pressure, temperature, and particle size) on extraction yield and eugenol content in the extract was inspected utilizing a three-level orthogonal array. Along with a total of 23 components, 19.56% yield and 58.77% eugenol content were found at 10 MPa pressure and 50°C temperature, whereas particle size and temperature were observed to be the most influencing parameters on oil yield and eugenol content, respectively. The comparison between SFE and traditional methods has concluded that the composition of the oil was found to be similar, whereas the relative concentration of the constituents may vary based on the method of extraction (Wenqiang et al., 2007). SFE was the most favorable technique for attaining the good quality of the clove oil.

Hatami et al. applied mathematical modeling and optimization to extract essential oil from clove buds using supercritical carbon dioxide. Optimization of operating parameters such as temperature and pressure was performed using a genetic algorithm. The authors also tried to establish a scale-up methodology using constant velocity and constant residence time conditions to reproduce the small-scale experimental data to the large-scale SFE processes. Based on the genetic algorithm, the optimum operating conditions obtained for the highest amount of clove oil were 10 MPa pressure and 304.2 K temperature (Hatami et al., 2010). Moreover, the authors also studied the impact of cold pressing on supercritical fluid extraction (SFE) of clove buds at 150 bar pressure, 40°C temperature, and two torques, 40 and 80 N m, respectively (Hatami et al., 2019). At 40 N m, the response was found to be greater and provided 22.19 g/100 g of volatile oil yield with 57.70% eugenol, 12.60% eugenyl acetate, 8.30% β-caryophyllene, and 0.90% α-humulene. This article also presented an economic evaluation of supercritical fluid extraction assisted by cold pressing (SFEAP) over SFE in terms of manufacturing cost, which covered mainly the scaling-up cost of the plant, procuring cost for raw materials, capital sponsoring, solvent-to-feed ratio, methods used for extraction, etc.

Prado et al. developed the simple scale-up criteria for the SFE process of Brazilian cultivated raw products, clove, and sugar cane. A constant solvent-to-feed mass ratio was used to scale-up the raw materials' overall extraction curves (OEC) by 15-folds from 290 mL to 5.15 L extraction vessel size. The OECs were almost found similar at a larger scale operation, except the yield was slightly higher (about 20% for clove) at the pilot scale (Prado et al., 2011).

Scopel et al. have studied the extraction kinetics to predict operating conditions for supercritical carbon dioxide extraction. It was the first time that the thermodynamic approach was implemented for mathematical modeling of solubility calculation for clove essential oil in supercritical CO_2. The authors predicted the solubility of oil using Peng-Robinson EoS coupled with three different mixing rules, van der Waals 1 (vdW1), van der Waals 2 (vdW2), and Mathias-Klotz-Prausnitz (MKP). The MKP mixing rule coupled with Peng-Robinson EoS has provided the best equilibrium for clove essential oil with reduced computation time and a smaller number of parameters with better accuracy of the model. Also, the minimum inhibitory concentration (MIC) of the extracts was determined by the microdilution method for *Staphylococcus aureus*, *Enterococcus faecalis*, and *Escherichia coli*. It was noted that all the extracts inhibited bacterial growth (Scopel et al., 2014). The results are summarized in Table 1.

Zabot et al. have conducted scientific research in terms of the influence of various process parameters on the scale-up of supercritical technology. Supercritical CO_2 was used to assess the influence of the extractor's bed geometry (the ratio of the height to the diameter of the bed (HB/DB)) on the composition of oil and overall extraction curves (OEC), where more emphasis was given on the mass transfer rate, the oil composition, and the temperature profiles. The study was conducted on two-bed geometries, viz., E-1 = 7.1 and E-2 = 2.7 while keeping volume and solvent-to-feed mass ratio at a constant value. Construction cost and the composition of the targeted compounds were other criteria for selecting bed geometries. Around 17.00% more α-humulene and 9.00% more eugenol were extracted in E-1 and E-2, respectively (Zabot et al., 2014). The systematic evaluation concluded that the equal S/F ratio and extraction time instead of equal interstitial solvent velocity were appropriate scale-up criteria. Too small particles have reduced the mass transfer since they compacted the bed leading to pressure drop and flow constraints. In terms of yield, bed E-2 has provided better results, whereas the influence of axial dispersion was observed more in bed E-1. Further, the relation between the mean particle diameter (d_p) and the porosity of the bed (ϵ) has been studied based on the variation between raw material and particle mean diameter to reduce back mixing, bypassing, nonuniform porosity distribution, viscosity gradient, and importantly dead space.

2.2.2 Ultrasound-assisted extraction

Ultrasound-assisted extraction is another intensified extraction method having distinctive features. In ultrasound-assisted extraction, the propagation of sonic waves in liquid media leads to the cavitation phenomenon. Breakdown of cavitation bubbles on or near to the product surface results in a microjet, which further provides either particle breakdown or surface cracking or erosion, whereas collapse of cavitation bubbles into the liquid medium results in micromixing or macro turbulences in the medium (Chemat et al., 2017). Thus, ultrasound-assisted extraction increases the rate of mass transfer by improving the interactions in solids and liquids. Few researchers have explored the ultrasound-assisted extraction either alone or in combination with supercritical fluid extraction for better yield while focusing on parameters like temperature, time, solid loading, and particle size.

Alexandru et al. aimed to investigate the efficiency of the batch and flow modes of ultrasonic-assisted extraction and compare the results obtained with the conventional maceration technique. In this study, the experiments were performed at three dissimilar flow rates 450, 900, and 1350 mL/min and with clove buds of Indian, Chinese, and Madagascan origin. Crude extract yield, phenolic content, and DPPH radical scavenging activity were attained as the responses where ultrasound-assisted extractor working at 1350 mL/min flow has shown prominent results for the response parameters (Alexandru et al., 2013). The process was also intensified by developing flow reactors having large-scale multiple transducers functioning at an elevated power density.

Tekin et al. performed ultrasound-assisted extraction of clove buds using an ultrasound bath with 53 kHz frequency. The focus was mainly given on optimization of process parameters, viz., temperature (32–52°C), extraction time (30–60 min), and plant concentration (3%–7%) using central composite design (CCD) of response surface methodology. This is an advanced optimization technique that can save the number of experiments and thus the amount of material, reagent, and the time. The extraction temperature was the most influencing parameter on the recovery of the oil with the lowest p-value, and the highest t-value was obtained with it. The extract was found with a higher eugenol content while 2-methoxy-4-(2-propenyl) phenol acetate and α-caryophyllene were in minor quantity (Tekin et al., 2015). The author also studied the antibacterial study and concluded that clove extract could be utilized as a bacteriostatic and bactericidal agent in the pharmaceutical industry.

Yang et al. have evaluated ultrasound-assisted extraction with supercritical carbon dioxide extraction (USC-CO_2) to determine yield and biologically active constituents like eugenol, α-humulene, β-caryophyllene, and eugenyl acetate. Further, the parametric study was collectively performed to validate the operating conditions and kinetic study. The parameters considered were CO_2 flow rate, pressure, temperature, particle size of the sample, and static and dynamic time required for the extraction. The results obtained using USC-CO_2 were compared with other methods like supercritical carbon dioxide (SC-CO_2) extraction and heat flux extraction. A total 13.50% increase in the yield of essential oil comparatively at lower operating conditions was observed in SC-CO_2 whereas it was around 20.80% in the case of heat flux extraction (Yang et al., 2014). The USC-CO_2 has the potential to be considered as a green alternative over the conventional methods since it can provide greater extraction efficiency and product quality at relatively mild operating conditions.

Wei et al. have explored ultrasound-assisted extraction with supercritical carbon dioxide extraction (USC-CO_2) for effective isolation and quantification of clove oil and its important biologically active components. The article predicted the solubility of clove oil and fictitious solubility of eugenol and correlated the data with different density-based models, viz., Bartle, Kumar-Johnston, and Chrastil (Wei et al., 2016a). The USC-CO_2 technique was useful in several ways as it could save time, consumption of raw materials, solvent, and energy without chemical waste. Also, it has provided a greater yield of essential oil and bioactive constituents. In addition, Wei et al. used supercritical CO_2 extraction with ultrasound (USC-CO_2) or without ultrasound (SC-CO_2) for effective isolation of clove oil and α-humulene. The yield was calculated by varying temperature (32–50°C) and pressure conditions (9.0–25.0 MPa), and the results were compared with the steam distillation and heat reflux extraction. Among various alternatives explored, USC-CO_2 has provided a better yield of oil and α-humulene at reduced extraction time by employing less severe operating conditions (Wei et al., 2016b). In this study, preliminary solubility data was obtained by performing experiments and correlated with the density-based models.

2.2.3 Microwave-assisted extraction

Microwave-assisted extraction, an advanced extraction technique, uses microwave energy for volumetric heating of the moisture present inside the plant material. Electromagnetic waves in the form of microwaves penetrate through the plant material and get absorbed by the water molecules or other polar molecules present. The electromagnetic energy is then converted into heat energy due to dipole rotation, or ionic conduction, or both. It causes a rise in temperature inside the cell, which further generates pressure inside the cell resulting in cell rupture and movement of volatile components from inside the cell to the surrounding medium (Desai et al., 2010). Kapadiya et al. have incorporated microwave-assisted extraction (MAE) for isolating essential oil from clove buds. This study also considered the grey-based Taguchi method for optimizing process parameters. The optimum conditions were found at 30 g clove buds in 200 mL water, 30 min extraction time, and 600 W power. Under these conditions, yield and eugenol content were obtained as 13.11% (w/w) and 11.93% (w/w), respectively, whereas inhibition zone (ZOI) was 24.80 mm. In addition, the utilization of waste generated either as a raw material or as a by-product was also explored. A comparison between MAE and traditional hydrodistillation has shown that MAE could offer superior oil quality, decrease in the extraction time, reduction in the energy consumption as well as the rejection of CO_2 to the environment (Kapadiya et al., 2018).

2.2.4 Ohmic heating assisted hydrodistillation

Ohmic heating associated with conventional hydrodistillation is an intensified technique used for the extraction of essential oil. In ohmic heating, electric resistance associated with the plant material itself generates heat when an electric current is passed through it. Tunc and Koca studied ohmic-assisted hydrodistillation (OAHD) for the extraction of clove bud essential oil. Response surface methodology was utilized to optimize process parameters in the range of 20–40 g mass of clove buds, 30–120 min process time, and 7.5–12.5 V/cm voltage gradient. The optimum operating conditions were observed to be 40 g, 97.96 min, and 12.50 V/cm for the mass of clove buds, process time, and voltage gradient, respectively. Under the best processing conditions, the response parameters were observed as 13.18% oil yield and 0.214 kW h/g energy consumption (Tunc and Koca, 2019). Further, the author compared the results of OAHD with conventional hydrodistillation carried out using distilled water and salted water since the electric conductivity for salted water is more than that of distilled water. Yields of CEO obtained by OAHD, HD-D (distilled water), and HD-S (salted water) were 13.18%, 8.23%, and 7.88%, respectively, whereas eugenol yield was found to be the maximum for OAHD (6.95%) than HD-D (5.16%) and HD-S (5.13%). No significant change in the energy consumption was observed with the variation in the method of extraction.

3. Comparative studies

It becomes imperative to assess the proposed technique concerning the existing techniques for its potential to be recognized as a sustainable alternative. Table 2 summarizes the work done to date, covering the comparative studies in yield, composition, biological activity, processing conditions, energy requirement, and environmental impact. While extracting the essential oil using supercritical CO_2, a

Table 2 Comparison of essential oil extraction methods.

No.	Purpose of the study	Extraction technique	Compared parameters and their values	References
1.	Comparison of essential oils of clove buds extracted with supercritical carbon dioxide with traditional extraction methods, viz., steam distillation, hydrodistillation, and solvent extraction	Supercritical fluid extraction	Yield—19.60% Eugenol + eugenol acetate—58.80 +19.60% Extraction period—2 h Color and texture—Pale yellow oil Organic solvent used—No	Wenqiang et al. (2007)
		Steam distillation	Yield—10.10% Eugenol + eugenol acetate—61.20 +10.20% Extraction period—8–10 h Color and texture—pale yellow oil Organic solvent used—no	
		Hydrodistillation	Yield—11.50% Eugenol plus eugenol acetate—50.30+3.20% Extraction period—4–6 h Color and texture—brown yellow oil Organic solvent used—no	
		Solvent extraction	Yield—41.80% Eugenol plus eugenol acetate—30.80+9.30% Extraction period—6 h	

Continued

Table 2 Comparison of essential oil extraction methods—cont'd

No.	Purpose of the study	Extraction technique	Compared parameters and their values	References
			Color and texture—brown ointment Organic solvent used—Yes	
2.	Comparison of extraction yields obtained by heat-reflux extraction, steam distillation, supercritical carbon dioxide, and ultrasound-assisted supercritical carbon dioxide methods	Heat-reflux extraction	Clove oil yield—17.20%, w/w α-Humulene—1.71%, w/w	Wei et al. (2016b)
		Steam distillation	Clove oil yield—20.04%, w/w α-Humulene—2.35%, w/w	
		Supercritical carbon dioxide	Clove oil yield—20.97%, w/w α-Humulene—1.89%, w/w	
		Ultrasound-assisted supercritical carbon dioxide	Clove oil yield—23.31%, w/w α-Humulene—2.75%, w/w	
3.	Comparison of microwave-assisted extraction with hydrodistillation in terms of processing conditions, responses, energy, and environmental impact	Microwave-assisted extraction using multi-response optimization	Yield of essential oil—13.11%, w/w Yield of eugenol—11.93%, w/w Eugenol in oil—90.10%, w/w ZOI—24.80 mm Power consumption—1.00 kW h per 100 g plant material CO_2 emission—800 g per 100 g plant material	Kapadiya et al. (2018)
		Hydrodistillation	Yield of essential oil—12.45%, w/w Yield of eugenol—10.96%, w/w Eugenol in oil—88.03%, w/w ZOI (mm)—21.40 mm Power consumption—6.00 kW h per 100 g	

Table 2 Comparison of essential oil extraction methods—cont'd

No.	Purpose of the study	Extraction technique	Compared parameters and their values	References
			plant material CO_2 emission—4800 g per 100 g plant material treated	
4.	Comparison of different hydrodistillation methods, viz., Ohmic heating assisted hydrodistillation (OAHD), hydrodistillation with distilled water (HD-D), hydrodistillation with salted water (HD-S)	Ohmic heating assisted hydrodistillation (OAHD)	Yield—13.18% Extraction rate—0.05 g/min Energy consumption—0.214 kWh/g CEO CO_2 emission—171.19 g/mL CEO FRAP—4.00 mmol Fe(II)/mL CEO $1/EC_{50}$—172.28 mL/μL CEO	Tunc and Koca (2019)
		Hydrodistillation with distilled water (HD-D)	Yield—8.23% Extraction rate—0.034 g/min Energy consumption—0.188 kWh/g CEO CO_2 emission—150.48 g/mL CEO FRAP—3.34 mmol Fe(II)/mL CEO $1/EC_{50}$—177.62 mL/μL CEO	
		Hydrodistillation with salted water (HD-S)	Yield—7.88% Extraction rate—0.032 g/min Energy consumption—0.215 kWh/g CEO CO_2 emission—172.12 g/mL CEO FRAP—3.29 mmol Fe(II)/mL CEO $1/EC_{50}$—168.66 mL/μL CEO	

better quality of the oil was obtained with a higher amount of eugenol and eugenol acetate (78.40%) compared to oil obtained by steam distillation, hydrodistillation, and solvent extraction (Wenqiang et al., 2007). Though a higher yield was obtained using solvent extraction, its quality was compromised as it contained the minimum amount of eugenol and its acetate (~40%), which might be due to the extraction of undesirable compounds along with the oil in the presence of an organic solvent. This could lead to the requirement of rigorous post-processing steps to obtain a better quality oil, thereby adding the cost to the operation. Also, all the operations except SFE were time-consuming and utilized organic solvent during the treatment. SFE could lead to efficient extraction of clove oil compared to conventional techniques. However, the comparison did not focus on energy requirements during operation and the carbon footprint. Wei et al. have combined sonication with SFE to improve the recovery of α-humulene and obtained the highest yield (23.31%) with a higher fraction of α-humulene (2.75%) compared to SFE, steam distillation, and heat-reflux extraction (Wei et al., 2016b). A symbiotic effect between sonication and SFE was observed in this study and paved a new path in the extraction.

In isolating clove oil using microwave radiation, the superior quality of the oil with the highest yield was obtained when compared with hydrodistillation. The oil exhibited a better antibacterial activity (~16% more) than that isolated by hydrodistillation. The power consumption was reduced to 1/6th of the requirement by hydrodistillation, while CO_2 release was shrunken to 800 g (Kapadiya et al., 2018). Thus, MAE provided a green and sustainable route for isolating the essential oil from clove buds.

Tunc and Koca have employed ohmic heating to improve the processing conditions of hydrodistillation. An increase in the yield (13.80% vs 8.23%) was obtained with a better extraction rate and similar power consumption and environmental impact. The efficacy of the oil in terms of radical scavenging activity was unaffected by the change in the method of heating (Tunc and Koca, 2019). The proposed method could tender the low-cost alternative for isolating clove oil.

4. Conclusion

This chapter explored both conventional and advanced extraction methods utilized to date for isolating essential oil from clove buds. The yield of essential oil, eugenol content, and other constituent's composition was given prime importance. It was observed that compared to conventional methods like steam distillation and hydrodistillation, advanced extraction methods or a combination of two or more extraction methods had provided better yield and content of major constituents at relatively lower operating parametric range. Therefore, advanced extraction techniques can provide better quality and quantity of essential oil at a minimized extraction time, better processing conditions, reduced energy demands, shrunken environmental impacts like CO_2 emission to the atmosphere, avoidance of organic solvents, etc. Due to such multiple benefits and greener aspects, advanced extraction techniques can be considered as an effective extraction methods over conventional methods and can be explored in detail at a higher scale operation.

References

Alexandru, L., Cravotto, G., Giordana, L., Binello, A., Chemat, F., 2013. Ultrasound-assisted extraction of clove buds using batch- and flow-reactors: a comparative study on a pilot scale. Innov. Food Sci. Emerg. Technol. 20, 167–172. https://doi.org/10.1016/j.ifset.2013.07.011.

Alma, M.H., Ertas, M., Nitz, S., Kollmannsberger, H., 2007. Chemical composition and content of essential oil from the bud of cultivated Turkish clove (*Syzygium aromaticum* L.). Bioresources 2 (2), 265–269.

Bhuiyan, M.N.I., Begum, J., Nandi, N.C., Akter, F., 2010. Constituents of the essential oil from leaves and buds of clove (Syzigium caryophyllatum (L.) Alston). African J. Plant Sci. 4 (11), 451–454.

Chemat, F., Rombaut, N., Sicaire, A.-G., Meullemiestre, A., Fabiano-Tixier, A.-S., Abert-Vian, M., 2017. Ultrasound assisted extraction of food and natural products. Mechanisms, techniques, combinations, protocols and applications. A review. Ultrason. Sonochem. 34, 540–560. https://doi.org/10.1016/j.ultsonch.2016.06.035.

Desai, M., Parikh, J., Parikh, P.A., 2010. Extraction of natural products using microwaves as a heat source. Sep. Purif. Rev. 39 (1–2), 1–32. https://doi.org/10.1080/15422111003662320.

Hatami, T., Meireles, M.A.A., Zahedi, G., 2010. Mathematical modeling and genetic algorithm optimization of clove oil extraction with supercritical carbon dioxide. J. Supercrit. Fluids 51, 331–338. https://doi.org/10.1016/j.supflu.2009.10.001.

Hatami, T., Johner, J.C.F., Zabot, G.L., Meireles, M.A.A., 2019. Supercritical fluid extraction assisted by cold pressing from clove buds: extraction performance, volatile oil composition, and economic evaluation. J. Supercrit. Fluids 144, 39–47. https://doi.org/10.1016/j.supflu.2018.10.003.

Kamatou, G.P., Vermaak, I., Viljoen, A.M., 2012. Eugenol—from the remote Maluku islands to the international market place: a review of a remarkable and versatile molecule. Molecules 17 (6), 6953–6981. https://doi.org/10.3390/molecules17066953.

Kapadiya, S., Desai, M.A., 2017. Isolation of essential oil from buds of *Syzygium aromaticum* using hydrodistillation: multi-response optimization and predictive modelling. Int. J. Adv. Res. Sci. Eng. 6 (1), 405–418.

Kapadiya, S.M., Parikh, J., Desai, M.A., 2018. A greener approach towards isolating clove oil from buds of *Syzygium aromaticum* using microwave radiation. Ind. Crop. Prod. 112, 626–632. https://doi.org/10.1016/j.indcrop.2017.12.060.

Kennouche, A., Benkaci-Ali, F., Scholl, G., Eppe, G., 2015. Chemical composition and antimicrobial activity of the essential oil of *Eugenia caryophyllata* cloves extracted by conventional and microwave techniques. J. Biol. Act. Prod. Nat. 5 (1), 1–11. https://doi.org/10.1080/22311866.2014.961100.

Mbaveng, A.T., Kuete, V., 2017. Syzygium aromaticum. In: Kuete, V. (Ed.), Medicinal Spices and Vegetables From Africa. Academic Press, pp. 611–625, https://doi.org/10.1016/b978-0-12-809286-6.00029-7.

Nurdjannah, N., Bermawie, N., 2012. Cloves. In: Peter, K.V. (Ed.), Handbook of Herbs and Spices, second ed. Woodhead Publishing Ltd, pp. 197–215.

Prado, J.M., Prado, G.H.C., Meireles, M.A.A., 2011. Scale-up study of supercritical fluid extraction process for clove and sugarcane residue. J. Supercrit. Fluids 56 (3), 231–237. https://doi.org/10.1016/j.supflu.2010.10.036.

Razafimamonjison, G., Jahiel, M., Duclos, T., Ramanoelina, P., Fawbush, F., Danthu, P., 2014. Bud, leaf and stem essential oil composition of Syzygium aromaticum from Madagascar, Indonesia and Zanzibar. Int. J. Basic Appl. Sci. 3 (3), 224–233. https://doi.org/10.14419/ijbas.v3i3.2473.

Reverchon, E., Marrone, C., 1997. Supercritical extraction of clove bud essential oil: isolation and mathematical modeling. Chem. Eng. Sci. 52 (20), 3421–3428. https://doi.org/10.1016/S0009-2509(97)00172-3.

Scopel, R., Falcao, M.A., Lucas, A.M., Almeida, R.N., Gandolfi, P.H.K., Cassel, E., Vargas, R.M.F., 2014. Supercritical fluid extraction from *Syzygium aromaticum* buds: phase equilibrium, mathematical modeling and antimicrobial activity. J. Supercrit. Fluids 92, 223–230. https://doi.org/10.1016/j.supflu.2014.06.003.

Sellamuthu, R., 2014. Eugenol. In: Wexler, P. (Editor-in-Chief), Encyclopedia of Toxicology, third ed. Academic Press/Elsevier, pp. 539–541, https://doi.org/10.1016/B978-0-12-386454-3.01125-8.

Srivastava, A.K., Srivastava, S.K., Syamsundar, K.V., 2005. Bud and leaf essential oil composition of *Syzygium aromaticum* from India and Madagascar. Flavour Fragr. J. 20 (1), 51–53. https://doi.org/10.1002/ffj.1364.

Tekin, K., Akalın, M.K., Seker, M.G., 2015. Ultrasound bath-assisted extraction of essential oils from clove using central composite design. Ind. Crop. Prod. 77, 954–960. https://doi.org/10.1016/j.indcrop.2015.09.071.

Tunc, M.T., Koca, I., 2019. Ohmic heating assisted hydrodistillation of clove essential oil. Ind. Crop. Prod. 141, 111763. https://doi.org/10.1016/j.indcrop.2019.111763.

Vitamins & Supplements A-Z list—C, 2019. https://www.rxlist.com/supplements/alpha_c.htm.

Wei, M.-C., Lin, P.-H., Hong, S.-J., Chen, J.-M., Yang, Y.-C., 2016a. Development of a green alternative procedure for simultaneous separation and quantification of clove oil and its major bioactive constituents. ACS Sustain. Chem. Eng. 4 (12), 6491–6499. https://doi.org/10.1021/acssuschemeng.6b01263.

Wei, M.-C., Xiao, J., Yang, Y.-C., 2016b. Extraction of α-humulene-enriched oil from clove using ultrasound-assisted supercritical carbon dioxide extraction and studies of its fictitious solubility. Food Chem. 210, 172–181. https://doi.org/10.1016/j.foodchem.2016.04.076.

Wenqiang, G., Shufen, L., Ruixiang, Y., Shaokun, T., Can, Q., 2007. Comparison of essential oils of clove buds extracted with supercritical carbon dioxide and other three traditional extraction methods. Food Chem. 101, 1558–1564. https://doi.org/10.1016/j.foodchem.2006.04.009.

Anon., 2020. What Are Essential Oils? https://pharmeasy.in/blog/health-benefits-of-essential-oils/.

Yang, Y.-C., Wei, M.-C., Hong, S.-J., 2014. Ultrasound-assisted extraction and quantitation of oils from *Syzygium aromaticum* flower bud (clove) with supercritical carbon dioxide. J. Chromatogr. A 1323, 18–27. https://doi.org/10.1016/j.chroma.2013.10.098.

Zabot, G.L., Moraes, M.N., Petenate, A.J., Meireles, M.A.A., 2014. Influence of the bed geometry on the kinetics of the extraction of clove bud oil with supercritical CO_2. J. Supercrit. Fluids 93, 56–66. https://doi.org/10.1016/j.supflu.2013.10.001.

CHAPTER

Health-promoting activities of clove (*Syzygium aromaticum*) essential oil

29

Eman T. El Kenany[a,b]

Botany and Microbiology Department, Faculty of Science, Alexandria University, Alexandria, Egypt[a] *Oral Biology Department, Faculty of Dentistry, Pharos University in Alexandria, Alexandria, Egypt*[b]

1. Introduction

For the anxiety of human health, scientists hope to reduce the dense dependence on synthetic medications and turn to phytotherapy, which has many advantages. Medicinal plants may offer a new source of anticancer, antiviral, and antibacterial agents to many important human pathogens that have developed resistance to antimicrobials (Barbour et al., 2004). Many herbal medications are thought to have therapeutic abilities, including treating some infectious diseases and cancer (Jin et al., 2006). Nowadays, modern medicine uses the therapeutic properties of essential oils and their constituents (Hassine et al., 2021).

Clove essential oil (CEO) contains a high amount of phenolic compounds with several biological activities, including antibacterial, antifungal, antiprotozoal, insecticidal, antioxidant, and many other properties (Guan et al., 2007; Golmakani et al., 2017; Immacolata et al., 2019; Hatami et al., 2019). Clove flower buds contain up to 18 main compounds, consisting of eugenol, eugenol acetate, β-caryophyllene, α-humulene, and other constituents (Sugihartini et al., 2019). Eugenol is the major component of the oil (Haro-González et al., 2021).

2. Health-promoting biological activities

2.1 Antibacterial activities

According to Holetz et al. (2002), compounds with minimum inhibition concentration (MIC) values below 100 μg/mL have good antibacterial activity, from 100 to 500 μg/mL have moderate activity, and from 500 to 1000 μg/mL have weak activity, and those above 1000 μg/mL are inactive. Thus, pathogenic bacteria are greatly affected by CEO. The inhibition mechanism of the CEO is mostly due to the interaction of —OH groups, located at the meta and ortho sites in the CEO's main chemical structure, with the cytoplasmic membrane of microbial cells (El-Darier et al., 2018; Behbahani et al., 2019; Marmouzi et al., 2019; Shahbazi, 2019). In addition, the lipophilic properties of CEO permit its penetration through the cell membrane, and the interaction of CEO with polysaccharides, fatty acids, and

Clove (*Syzygium aromaticum*). https://doi.org/10.1016/B978-0-323-85177-0.00033-1

phospholipids causes loss of cellular membrane protection, leading to the outflow of cellular contents, inconsistency in proton pump activity, and finally to cell death (Morsy, 2017; El-Darier et al., 2018; Behbahani et al., 2019; Shahbazi, 2019). The CEO can affect both Gram-positive and Gram-negative bacteria (Table 1), with greater inhibition to Gram-positive bacteria, due to its diffusible mucopeptide layer that makes them susceptible to antimicrobial agents. Meanwhile, the cell membrane of Gram-negative bacteria contains a complex layer of lipopolysaccharide which significantly reduces the rate of diffusion of lipophilic antibacterial compounds (Behbahani et al., 2019). Also, food-related pathogens have shown more sensitivity to CEO than probiotics (Shahbazi, 2019). Table 1 illustrates the recent relevant antibacterial effects of CEO and/or eugenol.

Table 1 Antibacterial activity of CEO and eugenol.

Substance (notice)	Bacterial species	MIC (%, V/V)	MBC (%, V/V)	IZ (mm)	References
CEO and eugenol (oral bacteria)	*Streptococcus* species	0.1			Saeki et al. (1989)
	Actinomyces species	0.1			
	Propionibacterium acnes	0.1			
	Actinobacillus species	0.01			
	Bacteroides species	0.01			
	Capnocytophaga gingivalis	0.01			
	Eikenella corrodens	0.01			
CEO	*Escherichia coli*	0.04	0.1		Smith-Palmer et al. (1998)
	Staphylococcus aureus	0.04	0.04		
	Listeria monocytogenes	0.03	0.04		
	Salmonella enteridis	0.04	0.075		
	Campylobacter jejuni	0.05	1		
Eugenol	*Klebsiella pneumoniae*	1.89			Nascimento et al. (2000)
	Shigella ssp.	0.94			
	Proteus ssp.	0.94			
CEO and eugenol	*Escherichia coli*	0.050	0.100		Rhayour et al. (2003)
	Bacillus subtilis	0.033	0.050		
CEO	*Staphylococcus aureus*	< 0.61			Prabuseenivasan et al. (2006)
	Bacillus subtilis	< 0.30			
	Klebsiella pneumoniae	< 0.61			
	Proteus vulgaris	< 0.30			
	Pseudomonas aeruginosa	< 0.15			
	Escherichia coli	< 0.15			
CEO (bacteria from respiratory infections)	*Streptococcus pyogenes*	1.25			Fabio et al. (2007)
	Streptococcus agalactiae	1.25			
	Streptococcus pneumoniae	1.25			
	Staphylococcus aureus	1.25			
	Haemophilus influenzae	1.25			
	Klebsiella pneumoniae	5.00			
	Stenotrophomonas maltophilia	1.25			

Table 1 Antibacterial activity of CEO and eugenol—cont'd

Substance (notice)	Bacterial species	MIC (%, V/V)	MBC (%, V/V)	IZ (mm)	References
CEO	*Staphylococcus epidermidis*	0.250	0.250		Fu et al. (2007)
	Staphylococcus aureus	0.125	0.250		
	Bacillus subtilis	0.125	0.250		
	Escherichia coli	0.125	0.125		
	Proteus vulgaris	0.125	0.250.		
	Pseudomonas aeruginosa	0.500	0.500		
CEO	*Propionibacterium acnes*	0.31	0.31	24.0	Fu et al. (2009)
CEO (foodborne pathogens)	*Escherichia coli*	1.00	2.00		Sofia et al. (2007)
	Bacillus cereus	1.00	2.00		
	Staphylococcus aureus	1.00	1.00		
Eugenol	*Acinetobacter calcoacetica*			15.4 ± 0.3	Dorman and Deans (2000)
	Aeromonas hydrophila			17.0 ± 0.4	
	Alcaligenes faecalis			12.3 ± 0.5	
	Bacillus subtilis			21.8 ± 0.4	
	Beneckea natriegens			20.8 ± 1.8	
	Brevibacterium linens			12.7 ± 0.1	
	Brocothrix thermosphacta			14.1 ± 0.2	
	Citrobacter freundii			9.1 ± 0.3	
	Clostridium sporogenes			9.7 ± 0.1	
	Enterococcus faecalis			10.0 ± 0.1	
	Enterobacter aerogenes			9.9 ± 0.1	
	Erwinia carotovora			10.0 ± 0.5	
	Escherichia coli			13.3 ± 0.2	
	Flavobacterium suaveolens			7.0 ± 0.6	
	Klebsiella pneumoniae			10.9 ± 0.3	
	Lactobacillus plantarum			21.5 ± 0.6	
	Leuconostoc cremoris			–	
	Micrococcus luteus			11.7 ± 0.7	
	Moraxella sp.			10.1 ± 0.6	
	Proteus vulgaris			8.3 ± 0.3	
	Pseudomonas aeruginosa			15.5 ± 0.6	
	Salmonella pullorum			12.9 ± 0.1	
	Serratia marcescens			22.9 ± 0.8	
	Staphylococcus aureus			11.5 ± 0.5	
	Yersinia enterocolitica			11.6 ± 0.4	
CEO (oral bacteria)	*Streptococcus mutans* ATCC 25175	0.2	0.8		Moon et al. (2011)
	S. sanguinis ATCC 10556	0.4	0.8		
	S. sobrinus ATCC 27607	0.2	0.8		

Continued

Table 1 Antibacterial activity of CEO and eugenol—cont'd

Substance (notice)	Bacterial species	MIC (%, V/V)	MBC (%, V/V)	IZ (mm)	References
	S. ratttb KCTC 3294	0.8	1.6		
	S. criceti KCTC 3292	0.4	1.6		
	S. anginosus ATCC 31412	0.8	1.6		
	S. gordonii ATCC 10558	0.1	0.2		
	Actinobacillus actinomycetemcomitans ATCC 43717	0.8	1.6		
	Fusobacterium nucleatum ATCC 51190	0.1	0.2		
	Porphyromonas intermedia ATCC 49049	0.1	0.2		
	P. gingivalis ATCC 3327	0.1	0.1		
Eugenol (oral bacteria)	*Streptococcus mutans* ATCC 25175	0.1	0.2		Moon et al. (2011)
	S. sanguinis ATCC 10556	0.4	0.8		
	S. sobrinus ATCC 27607	0.2	0.4		
	S. ratttb KCTC 3294	0.8	1.6		
	S. criceti KCTC 3292	0.4	0.8		
	S. anginosus ATCC 31412	0.4	0.8		
	S. gordonii ATCC 10558	0.1	0.1		
	Actinobacillus actinomycetemcomitans ATCC 43717	0.4	0.8		
	Fusobacterium nucleatum ATCC 51190	0.1	0.2		
	Porphyromonas intermedia ATCC 49049	0.1	0.2		
	P. gingivalis ATCC 3327	0.1	0.1		
β-Caryophyll-ene (oral bacteria)	*Streptococcus mutans* ATCC 25175	1.6	3.2		Moon et al. (2011)
	S. sanguinis ATCC 10556	1.6	3.2		
	S. sobrinus ATCC 27607	12.8	12.8		
	S. ratttb KCTC 3294	6.4	12.8		
	S. criceti KCTC 3292	3.2	12.8		
	S. anginosus ATCC 31412	6.4	12.8<		
	S. gordonii ATCC 10558	1.6	3.2		
	Actinobacillus actinomycetemcomitans ATCC 43717	6.4	12.8		
	Fusobacterium nucleatum ATCC 51190	3.2	12.8		

Table 1 Antibacterial activity of CEO and eugenol—cont'd

Substance (notice)	Bacterial species	MIC (%, V/V)	MBC (%, V/V)	IZ (mm)	References
	Porphyromonas intermedia ATCC 49049	1.6	3.2		
	P. gingivalis ATCC 3327	0.8	1.6		
Eugenol (nosocomial pathogen)	*Proteus mirabilis*	0.125	0.25		Pandima et al. (2013)
CEO (nanoemulsions)	*Escherichia coli* ATCC 25,922			27.27 ± 0.6	Zhang et al. (2017)
	Bacillus subtilis ATCC 6633			16.5 ± 1.32	
	Staphylococcus aureus ATCC 12600			24.57 ± 0.4	
	Salmonella typhimurium CMCC50093			30.0 ± 1.0	
CEO	*Escherichia coli*	10		17.0	El-Darier et al. (2018)
	Agrobacterium tumefaciens	10		15.0	
	Proteus sp.	10		10.0	
	Erwinia sp.	40		26.0	
	Klebsiella sp.	10		11.0	
	Pseudomonas aeruginosa	10		14.0	
CEO (oral bacteria)	*Streptococcus sanguinis*	0.02			Yanti et al. (2019)
CEO	*Escherichia coli*			8.85	Hosseini et al. (2019)
	Salmonella thyphimurium			8.45	
	Pseudomonas aeruginosa			7.60	
	Shigella dysenteriae			9.87	
	Staphylococcus aureus			9.51	
	Listeria monocytogenes			9.04	
	Bacillus cereus			10.09	
CEO (fish pathogenic bacteria)	*Vibrio harveyi* (FP8370)	0.125	0.25		Pathirana et al. (2019)
	V. ichthyoenteri (FP4004)	0.125	0.25		
	Photobacterium damselae (FP4101)	0.125	0.25		
	Edwardsiella tarda (FP5060)	0.25	0.5		
	E. tarda (ED47)	0.5	0.5		
	E. tarda (Yoshida)	0.125	0.25		
	E. tarda (ED45)	0.5	0.5		
	Lactococcus garvieae (FP5245)	0.5	1		
	Streptococcus iniae (FP3287)	0.5	1		

Continued

Table 1 Antibacterial activity of CEO and eugenol—cont'd

Substance (notice)	Bacterial species	MIC (%, V/V)	MBC (%, V/V)	IZ (mm)	References
	S. iniae (S186)	0.25	0.25		
	S. iniae (S530)	0.25	0.5		
	S. iniae (S131)	0.25	0.25		
	Streptococcus parauberis (FP5228)	0.5	0.5		
	S. parauberis (S124)	0.5	0.5		
	S. parauberis (S527)	0.5	0.5		
	S. parauberis (S1466)	0.25	0.5		
Eugenol (fish pathogenic bacteria)	*Vibrio harveyi* (FP8370)	0.0312	0.125		Pathirana et al. (2019)
	V. ichthyoenteri (FP4004)	0.0312	0.125		
	Photobacterium damselae (FP4101)	0.125	0.25		
	Edwardsiella tarda (FP5060)	0.0625	0.25		
	E. tarda (ED47)	0.125	0.5		
	E. tarda (Yoshida)	0.125	0.25		
	E. tarda (ED45)	0.125	0.5		
	Lactococcus garvieae (FP5245)	1	1		
	Streptococcus iniae (FP3287)	0.5	1		
	S. iniae (S186)	0.5	0.5		
	S. iniae (S530)	0.5	0.5		
	S. iniae (S131)	0.125	0.5		
	Streptococcus parauberis (FP5228)	1	1		
	S. parauberis (S124)	0.5	0.5		
	S. parauberis (S527)	1	1		
	S. parauberis (S1466)	0.5	0.5		
CEO (foodborne pathogenic bacteria)	*Cronobacter sakazakii* *Salmonella Enteritidis*	Positive inhibition			Chon et al. (2020)
CEO (from grounded buds)	*Bacillus cereus*			27.3	Hassine et al. (2021)
	Staphylococcus aureus			29.5	
	Listeria monocytogenes			28.0	
	Escherichia coli			28.6	
	Salmonella enteritidis			25.8	
CEO (from ungrounded buds)	*Bacillus cereus*			22.3	Hassine et al. (2021)
	Staphylococcus aureus			21.6	
	Listeria monocytogenes			17.2	
	Escherichia coli			26.0	
	Salmonella Enteritidis			21.4	

2.2 Antifungal activities

The common clinically used antifungals have various disadvantages in terms of toxicity, efficacy, and cost, and their frequent use has led to the emergence of resistant strains (Rana et al., 2011). Moreover, in recent years, community pressure to decrease the use of synthetic fungicides in agriculture has increased. Meanwhile, worries have been elevated about environmental influence and the possible health hazards associated with using these compounds (Abad et al., 2007).

Fungi were more susceptible than bacteria to CEO (MIC range: 0.062%–0.125%, v/v) (Marchese et al., 2017). So, it can be included in unified disease management programs at a lesser cost than conventional agrochemicals (Fuentes et al., 2019). Meanwhile, eugenol could be considered as strong antifungal agent and could be proposed as therapeutic agent for oral candidiasis (Chami et al., 2004). Table 2 shows the recent relevant antifungal effects of CEO and/or eugenol.

Table 2 Antifungal activity of CEO and eugenol.

Substance	Fungal species	Reference
CEO	*Fusarium verticillioides*	Velluti et al. (2002)
CEO	*Fusarium chlamdosporum* *Helminthosporium oryzae* *Rhizoctonia bataticola*	Beg and Ahmad (2002)
CEO	*Fusarium proliferatum*	Velluti et al. (2003)
CEO	*Tricophyton rubrum* *Tricophyton mentagrophytesvar* *Epidermophyton flocosum*	Gayoso et al. (2005)
CEO	*Microsporum canis* (KCTC 6591) *Trichophyton mentagrophytes* (KCTC 6077) *Trichophyton rubrum* (KCCM 60443) *Epidermophyton floccosum* (KCCM 11667) *Microsporum gypseum*	Park et al. (2007)
Eugenol	*Microsporum gypseum*	Lee et al. (2007)
CEO	*Aspergillus* sp.	Viuda-Martos et al. (2007)
CEO	*Candida albicans* ATCC 10231 *C. krusei* ATCC 6258 *C. tropicalis* ATCC 13803 *C. parapsilosis* ATCC 90018 *C. albicans* D5 *C. albicans* D1 *C. krusei* D39 *C. tropicalis* D42 *C. glabrata* D10R *Epidermophyton floccosum* FF9	Pinto et al. (2009)

Continued

Table 2 Antifungal activity of CEO and eugenol—cont'd

Substance	Fungal species	Reference
CEO	*Trichophyton rubrum* FF5 *T. mentagrophytes* FF7 *Microsporum canis* FF1 *M. gypseum* FF3 *Aspergillus flavus* F44 *A. fumigatus ATCC* 46645 *A. niger ATCC* 16404 *Fusarium moniliforme*, NCIM 1100 *F. oxysporum* MTCC 284 *Aspergillus* sp. *Mucor* sp. *Trichophyton rubrum* *Microsporum gypseum*	Rana et al. (2011)
CEO	*Candida* species	Kumar et al. (2012)
CEO	*Aspergillus parasiticus*	Pillai and Ramaswamy (2012)
CEO	*Saccharomyces cerevisiae*	Darvishi et al. (2013)
CEO	*Aspergillus ochratus*	Hua et al. (2014)
CEO	*Penicillium roqueforti*	Šimović et al. (2014)
CEO	*Botrytis cinerea*	Aguilar-González et al. (2015)
CEO	*Aspergillus* sp.	Císarová et al. (2016)
CEO	*Fusarium oxysporum*	Sharma et al. (2017)
CEO	*Phytophthora cinnamomi*	Fuentes et al. (2019)
CEO	*Aspergillus* sp. *Helminthosporium oryzae* *Rhizoctonia solani*	Kaur et al. (2019)
CEO	*Aspergillus ochraceus*	Moghadam et al. (2019)
CEO	*Aspergillus niger* *Aspergillus ochraceus* *Aspergillus oryzae*	Hu et al. (2019)
CEO	*Aspergillus niger*	Jahani et al. (2020)

2.3 Other biological activities and applications

The Food and Drug Administration (FDA) classified CEO as generally recognized as safe (GRAS); accordingly, it is used in perfumes, cosmetics, sanitary, medicines, and foods (Wei et al., 2016; Golmakani et al., 2017). Meanwhile, the World Health Organization (WHO) has established the acceptable daily uptake of cloves in humans at 2.5 mg/kg body weight (Ogunwande et al., 2005). Table 3 demonstrates other biological activities and applications, for human health, of CEO and/or eugenol.

Table 3 Biological activities and applications of CEO and eugenol.

Biological activities and/or applications	References
Antiviral	Astani et al. (2011), Pandey et al. (2020), Joshi et al. (2020), and Rehman et al. (2021)
Analgesic	Khalilzadeh et al. (2016) and Correia et al. (2017)
Anesthetic	Rapp (2007), Bhowmik et al. (2012), Khalilzadeh et al. (2016), Fujimoto et al. (2018), and da Silva et al. (2021)
Anticancer	Zheng et al. (1992), Miyazawa and Hisama (2001), Yoo et al. (2005), Kouidhi et al. (2010), Uju and Obioma (2011), Banerjee et al. (2006), Aisha et al. (2011), Aisha et al. (2012), Rajkowska et al. (2016), Han and Parker (2017), El-Darier et al. (2018), Behbahani et al. (2019), and Nirmala et al. (2019)
Anti-inflammatory	Drashan and Doreswamy (2004), García-Mediavilla et al. (2007), Yang et al. (2011), Han and Parker (2017), Sugihartini et al. (2019), Marmouzi et al. (2019), and Banerjee et al. (2020)
Antidiarrheal	Uchôa Lopes et al. (2020)
Antiangiogenic	Zheng et al. (1992) and Aisha et al. (2012)
Antimutagenic	Miyazawa et al. (2001)
Anticoagulant	García-Mediavilla et al. (2007), Yang et al. (2011), and Uchôa Lopes et al. (2020)
Antioxidant	Ogata et al. (2000), Lee and Shibamoto (2002), Jirovetz et al. (2006), Ramadan et al. (2015), Bakour et al. (2018), Behbahani et al. (2019), and Marmouzi et al. (2019)
Antinociceptive	Beltrán-Villalobos et al. (2017)
Antipyretic	Uchôa Lopes et al. (2020)
Hemolytic	Uchôa Lopes et al. (2020)
Wound healing	Alam et al. (2017)
Bone preserving	Karmakar et al. (2012)
Oral and dental diseases	Chami et al. (2004), Chami et al. (2005), Gupta et al. (2011), Thosar et al. (2013), Labib and Aldawsari (2015), and Yanti et al. (2019)
Insecticide	Reuss et al. (2020), and Hassine et al. (2021)
Antiprotozoal and/or trypanocidal	Ueda-Nakamura et al. (2006), Santoro et al. (2007), Machado et al. (2011), and El-Kady et al. (2019)

References

Abad, M.J., Ansuategui, M., Bermejo, P., 2007. Active antifungal substances from natural sources. Arkivoc 7, 116–145.

Aguilar-González, A.E., Palou, E., López-Malo, A., 2015. Antifungal activity of essential oils of clove (*Syzygium aromaticum*) and/or mustard (*Brassica nigra*) in vapor phase against gray mold (*Botrytis cinerea*) in strawberries. Innov. Food Sci. Emerg. Technol. 32, 181–185.

Aisha, A., Nassar, Z., Siddiqui, M., Abu-Salah, K., Alrokayan, S., Ismail, Z., Abdul Majid, A., 2011. Evaluation of antiangiogenic, cytotoxic and antioxidant effects of *Syzygium aromaticum* L. extracts. Asian J. Biol. Sci. 4, 282–290.

Aisha, A.F., Abu-Salah, K.M., Alrokayan, S.A., Siddiqui, M.J., Ismail, Z., Majid, A.M.S.A., 2012. *Syzygium aromaticum* extracts as good source of betulinic acid and potential anti-breast cancer. Rev. Bras. Farm. 22, 335–343.

Alam, P., Ansari, M.J., Anwer, M.K., Raish, M., Kamal, Y.K.T., Shakeel, F., 2017. Wound healing effects of nanoemulsion containing clove essential oil. Artif. Cells Nanomed. Biotechnol. 45, 591–597. https://doi.org/10.3109/21691401.2016.1163716.

Astani, A., Reichling, J., Schnitzler, P., 2011. Screening for antiviral activities of isolated compounds from essential oils. Evid. Based Complement. Alternat. Med. 2011, 253643. https://doi.org/10.1093/ecam/nep187.

Bakour, M., Soulo, N., Hammas, N., Fatemi, H., Aboulghazi, A., Taroq, A., Abdellaoui, A., Al-Waili, N., Lyussi, B., 2018. The antioxidant content and protective effect of argan oil and *Syzygium aromaticum* essential oil in hydrogen peroxide-induced biochemical and histological changes. Int. J. Mol. Sci. 19, 610. https://doi.org/10.3390/ijms19020610.

Banerjee, S., Panda, C.K., Das, S., 2006. Clove (*Syzygium aromaticum* L.), a potential chemopreventive agent for lung cancer. Carcinogenesis 27, 1645–1654.

Banerjee, K., Madhyastha, H., Sandur, V.R., Thiagarajan, P., 2020. Anti-inflammatory and wound healing potential of a clove oil emulsion. Colloids Surf. B Biointerfaces 193, 111102. https://doi.org/10.1016/j.colsurfb.2020.111102.

Barbour, E.K., Al Sharif, M., Sagherian, V.K., Habre, A.N., Talhouk, R.S., Talhouk, S.N., 2004. Screening of selected indigenous plants of Lebanon for antimicrobial activity. J. Ethnopharmacol. 93, 1–7.

Beg, A.Z., Ahmad, I., 2002. In vitro fungitoxicity of the essential oil of *Syzygium aromaticum*. World J. Microbiol. Biotechnol. 18, 313–315.

Behbahani, B.A., Noshad, M., Falah, F., 2019. Study of chemical structure, antimicrobial, cytotoxic, and mechanism of action of *Syzygium aromaticum* essential oil on foodborne pathogens. Potr. Slovak J. Food Sci. 13, 875–883. https://doi.org/10.5219/1226.

Beltrán-Villalobos, K.L., Déciga-Campos, M., Aguilar-Mariscal, H., González-Trujano, M.E., Martínez-Salazar, M.F., de los Ángeles Ramírez-Cisneros, M., Rios, M.Y., López-Muñoz, F.J., 2017. Synergistic antinociceptive interaction of *Syzygium aromaticum* or *Rosmarinus officinalis* coadministered with ketorolac in rats. Biomed. Pharmacother. 94, 858–864. https://doi.org/10.1016/j.biopha.2017.07.166.

Bhowmik, D., Kumar, K.S., Yadav, A., Srivastava, S., Paswan, S., Dutta, A.S., 2012. Recent trends in Indian traditional herbs *Syzygium aromaticum* and its health benefits. J. Pharmacogn. Phytochem. 1, 13–22.

Chami, N., Chami, F., Bennis, S., Trouillas, J., Remmal, A., 2004. Antifungal treatment with carvacrol and eugenol of oral candidiasis in immunosuppressed rats. Braz. J. Infect. Dis. 8, 217–226.

Chami, N., Bennis, S., Chami, F., et al., 2005. Study of anticandidal activity of carvacrol and eugenol in vitro and in vivo. Oral Microbiol. Immunol. 20, 106–111.

Chon, J., Seo, K., Bae, D., Kim, B., Dongkwang Jeong, D., Kwang-Young Song, K., 2020. Antibacterial activity of clove oil against foodborne pathogenic bacteria and sensory attributes in clove oil-enriched dairy products: a preliminary study. J. Dairy Sci. Biotechnol. 38 (4), 197–206.

Císarová, M., Tančinová, D., Medo, J., Kačániová, M., 2016. The in vitro effect of selected essential oils on the growth and mycotoxin production of *Aspergillus* species. J. Environ. Sci. Health Part B 51, 668–674.

Correia, A.M., Pedrazzani, A.S., Mendonça, R.C., Massucatto, A., Ozório, R.A., Tsuzuki, M.Y., 2017. Basil, tea tree and clove essential oils as analgesics and anaesthetics in *Amphiprion clarkii* (Bennett, 1830). Braz. J. Biol. 78, 436–442. https://doi.org/10.1590/1519-6984.166695.

da Silva, D.R., Arvigo, A.L., Giaquinto, P.C., Delicio, H.C., Barcellos, L.J.G., Barreto, R.E., 2021. Effects of clove oil on behavioral reactivity and motivation in Nile tilapia. Aquaculture 532, 736045. https://doi.org/10.1016/j..2020.736045.

Darshan, S., Doreswamy, R., 2004. Patented antiinflammatory plant drug development from traditional medicine. Phytother. Res. 18, 343–357.

Darvishi, H., Najafi, G., Hosainpour, A., Khodaei, J., Aazdbakht, M., 2013. Far-infrared drying characteristics of mushroom slices. Chem. Prod. Process Model. 8 (2), 107–117.

Dorman, H.J., Deans, S.G., 2000. Antimicrobial agents from plants: antibacterial activity of plant volatile oils. J. Appl. Microbiol. 88 (2), 308–316. https://doi.org/10.1046/j.1365-2672.2000.00969.x. 10736000.

El-Darier, S.M., El-Ahwany, A.M.D., Elkenany, E.T., Abdeldaim, A.A., 2018. An in vitro study on antimicrobial and anticancer potentiality of thyme and clove oils. Rend. Lincei Sci. Fis. Nat. 29, 131–139. https://doi.org/10.1007/s12210-018-0672-0.

El-Kady, A.M., Ahmad, A.A., Hassan, T.M., El-Deek, H.E.M., Fouad, S.S., Althagfan, S.S., 2019. Eugenol, a potential schistosomicidal agent with anti-inflammatory and antifibrotic effects against *Schistosoma mansoni*, induced liver pathology. Infect. Drug Resist. 12, 709–719.

Fabio, A., Cermelli, C., Fabio, G., Nicoletti, P., Quaglio, P., 2007. Screening of the antibacterial effects of a variety of essential oils on microorganisms responsible for respiratory infections. Phytother. Res. 21, 374–377. https://doi.org/10.1002/ptr.1968.

Fu, Y., Zu, Y., Chen, L., Shi, X., Wang, Z., Sun, S., Efferth, T., 2007. Antimicrobial activity of clove and rosemary essential oils alone and in combination. Phytother. Res. 21, 989–994. https://doi.org/10.1002/ptr.2179.

Fu, Y., Chen, L., Zu, Y., Liu, Z., Liu, X., Liu, Y., Yao, L., Efferth, T., 2009. The antibacterial activity of clove essential oil against *Propionibacterium acnes* and its mechanism of action. Arch. Dermatol. 145 (1), 86–88. https://doi.org/10.1001/archdermatol.2008.549.

Fuentes, Y.M.O., Pérez, A.H., Ortiz, J.C.D., Bautista, O.H., Chavez, E.C., Uribe, L.A., Vargas, L.M., 2019. In vitro organic control of *Phytophthora cinnamomi* with essential oils of oregano and clove. Rev. Mexicana Cienc. Agríc. 10 (4), 961–968.

Fujimoto, R.Y., Pereira, D.M., Silva, J.C.S., de Oliveira, L.C.A., Inoue, L.A.K.A., Hamoy, M., de Mello, V.J., Torres, M.F., Barbas, L.A.L., 2018. Clove oil induces anaesthesia and blunts muscle contraction power in three Amazon fish species. Fish Physiol. Biochem. 44, 245–256. https://doi.org/10.1007/s10695-017-0430-8.

García-Mediavilla, V., Crespo, I., Collado, P.S., Esteller, A., Sánchez-Campos, S., Tuñón, M.J., González-Gallego, J., 2007. The anti-inflammatory flavones quercetin and kaempferol cause inhibition of inducible nitric oxide synthase, cyclooxygenase-2 and reactive C-protein, and down-regulation of the nuclear factor kappaB pathway in Chang Liver cells. Eur. J. Pharmacol. 557, 221–229.

Gayoso, C.W., Lima, E.O., Oliveria, V.T., Pereira, F.O., Souza, E.L., Lima, I.O., Navarro, D.F., 2005. Sensitivity of fungi isolated from onychomycosis to *Eugenia caryopyllata* essential oil and eugenol. Fitoterapia 76, 247–249.

Golmakani, M.T., Zare, M., Razzaghi, S., 2017. Eugenol enrichment of clove bud essential oil using different microwave-assisted distillation methods. Food Sci. Technol. Res. 23, 385–394. https://doi.org/10.3136/fstr.23.385.

Guan, W., Li, S., Yan, R., Tang, S., Quan, C., 2007. Comparison of essential oils of clove buds extracted with supercritical carbon dioxide and other three traditional extraction methods. Food Chem. 101, 1558–1564. https://doi.org/10.1016/j.foodchem.2006.04.009.

Gupta, C., Kumari, A., Garg, A.P., Catanzaro, R., Marotta, F., 2011. Comparative study of cinnamon oil and clove oil on some oral microbiota. Acta Biomed. 82, 197–199.

Han, X., Parker, T.L., 2017. Anti-inflammatory activity of clove (*Eugenia caryophyllata*) essential oil in human dermal fibroblasts. Pharm. Biol. 55, 1619–1622. https://doi.org/10.1080/13880209.2017.1314513.

Haro-González, J.N., Castillo-Herrera, G.A., Martínez-Velázquez, M., Espinosa-Andrews, H., 2021. Clove essential oil (*Syzygium aromaticum* L. Myrtaceae): extraction, chemical composition, food applications and essential bioactivities for human health. Molecules 26, 6387. https://doi.org/10.3390/molecules26216387.

Hassine, D.B., El Euch, S.K., Rahmani, R., Ghazouani, N., Kane, R., Abderrabba, M., Bouajila, J., 2021. Clove buds essential oil: the impact of grinding on the chemical composition and its biological activities involved in consumer's health security. BioMed Res. Int. 2021, 9940591. https://doi.org/10.1155/2021/9940591.

Hatami, T., Johner, J.C.F., Zabot, G.L., Meireles, M.A.A., 2019. Supercritical fluid extraction assisted by cold pressing from clove buds: extraction performance, volatile oil composition, and economic evaluation. J. Supercrit. Fluids 144, 39–47. https://doi.org/10.1016/j.supflu.2018.10.003.

Holetz, F.B., Pessini, G.L., Sanches, N.R., Cortez, D.A., Nakamura, C.V., Filho, B.P., 2002. Screening of some plants used in the Brazilian folk medicine for the treatment of infectious diseases. Mem. Inst. Oswaldo Cruz 97 (7), 1027–1031.

Hosseini, M., Jamshidi, A., Raeisi, M., Azizzadeh, M., 2019. The antibacterial and antioxidant effects of clove (*Syzygium aromaticum*) and lemon verbena (*Aloysia citriodora*) essential oils. J. Hum. Environ. Health Promot. 5 (2), 86–93.

Hu, F., Tu, X.F., Thakur, K., Hu, F., Li, X.L., Zhang, Y.S., Zhang, J.G., Wei, Z.J., 2019. Comparison of antifungal activity of essential oils from different plants against three fungi. Food Chem. Toxicol. 134, 110821.

Hua, H., Xing, F., Selvaraj, J.N., Wang, Y., Zhao, Y., Zhou, L., Liu, X., Liu, Y., 2014. Inhibitory effect of essential oils on *Aspergillus ochraceus* growth and ochratoxin A production. PLoS ONE 9 (9), e108285. https://doi.org/10.1371/journal.pone.0108285.

Immacolata, A., Sabia, C., de Niederhäusern, S., et al., 2019. In vitro evaluation of the amoebicidal activity of rosemary (*Rosmarinus officinalis* L.) and cloves (*Syzygium aromaticum* L. Merr. & Perry) essential oils against *Acanthamoeba polyphaga* trophozoites. Nat. Prod. Res. 33 (4), 606–611.

Jahani, M., Pira, M., Aminifard, M.H., 2020. Antifungal effects of essential oils against *Aspergillus niger* in vitro and in vivo on pomegranate (*Punica granatum*) fruits. Sci. Hortic. 264, 109188.

Jin, Z., Charlock, T.P., Rutledge, K., Stamnes, K., Wang, Y., 2006. Analytical solution of radiative transfer in the coupled atmosphere system with a rough surface. Appl. Opt. 45, 7443–7455.

Jirovetz, L., Buchbauer, G., Stoilova, I., Stoyanova, A., Krastanov, A., Schmidt, E., 2006. Chemical composition and antioxidant properties of clove leaf essential oil. J. Agric. Food Chem. 54, 6303–6307.

Joshi, T., Joshi, T., Sharma, P., Mathpal, S., Pundir, H., Bhatt, V., Chandra, S., 2020. In silico screening of natural compounds against COVID-19 by targeting Mpro and ACE2 using molecular docking. Eur. Rev. Med. Pharmacol. Sci. 24, 4529–4536.

Karmakar, S., Choudhury, M., Das, A.S., Maiti, A., Majumdar, S., Mitra, C., 2012. Clove (*Syzygium aromaticum* Linn) extract rich in eugenol and eugenol derivatives shows bone-preserving efficacy. Nat. Prod. Res. 26, 500–509.

Kaur, K., Kaushal, S., Rani, R., 2019. Chemical composition, antioxidant and antifungal potential of clove (*Syzygium aromaticum*) essential oil, its major compound and its derivatives. J. Essent. Oil Bear. Plants 22, 1195–1217. https://doi.org/10.1080/0972060X.2019.1688689.

Khalilzadeh, E., Hazrati, R., Saiah, G., 2016. Effects of topical and systemic administration of *Eugenia caryophyllata* buds essential oil on corneal anesthesia and analgesia. Res. Pharm. Sci. 11, 293. https://doi.org/10.4103/1735-5362.189297.

Kouidhi, B., Zmantar, T., Bakhrouf, A., 2010. Anticariogenic and cytotoxic activity of clove essential oil (*Eugenia caryophyllata*) against a large number of oral pathogens. Ann. Microbiol. 60, 599–604.

Kumar, A., Thakur, S., Thakur, V.C., Kumar, A., Patil, S., Vohra, M.P., 2012. Antifungal activity of some natural essential oils against *Candida* species isolated from blood stream infection. J. Krishna Inst. Med. Sci. Univ. 1, 61–66.

Labib, G.S., Aldawsari, H., 2015. Innovation of natural essential oil-loaded Orabase for local treatment of oral candidiasis. Drug Des. Devel. Ther. 9, 3349.

Lee, K.G., Shibamoto, T., 2002. Antioxidant property of aroma extract isolated from clove buds (*Syzygium aromaticum* (L.) Merr. Et Perry). Food Chem. 74, 443–448.

Lee, S.J., Han, J.I., Lee, G.S., Park, M.J., Choi, I.G., Na, K.J., Jeung, E.B., 2007. Antifungal effect of eugenol and nerolidol against *Microsporum gypseum* in a guinea pig model. Biol. Pharm. Bull. 30, 184–188.

Machado, M., Dinis, A.M., Salgueiro, L., Custódio, J.B., Cavaleiro, C., Sousa, M.C., 2011. Anti-Giardia activity of *Syzygium aromaticum* essential oil and eugenol: effects on growth, viability, adherence and ultrastructure. Exp. Parasitol. 127, 732–739.

Marchese, A., Barbieri, R., Coppo, E., Oran, I.E., Daglia, M., Nabavi, S.F., Izadi, M., Abdollahi, M., Nabavi, S.M., Ajami, M., 2017. Antimicrobial activity of eugenol and essential oils containing eugenol: a mechanistic viewpoint. Crit. Rev. Microbiol. 43 (6), 668–689. https://doi.org/10.1080/1040841X.2017.1295225.

Marmouzi, I., Karym, E.M., Alami, R., El Jemli, M., Kharbach, M., Mamouch, F., Attar, A., Faridi, B., Cherrah, Y., Faouzi, M.E.A., 2019. Modulatory effect of *Syzygium aromaticum* and *Pelargonium graveolens* on oxidative and sodium nitroprusside stress and inflammation. Orient. Pharm. Exp. Med. 19, 201–210. https://doi.org/10.1007/s13596-018-0335-9.

Miyazawa, M., Hisama, M., 2001. Suppression of chemical mutagen induced SOS response by alkylphenols from clove (*Syzygium aromaticum*) in the *Salmonella typhimurium* TA1535/pSK1002 umu test. J. Agric. Food Chem. 49, 4019–4025.

Miyazawa, M., Sakano, K., Nakamura, S., Kosaka, H., 2001. Antimutagenic activity of isoflavone from *Pueraria lobata*. J. Agric. Food Chem. 49 (1), 336–341.

Moghadam, Z.A., Hosseini, H., Hadian, Z., Asgari, B., Mirmoghtadaie, L., Mohammadi, A., Shamloo, E., Javadi, N.H.S., 2019. Evaluation of the antifungal activity of cinnamon, clove, thymes, zataria multiflora, cumin and caraway essential oils against Ochratoxigenic *Aspergillus ochraceus*. J. Pharm. Res. Int. 26 (1), 1–16.

Moon, S.E., Kim, H.Y., Cha, J.D., 2011. Synergistic effect between clove oil and its major compounds and antibiotics against oral bacteria. Arch. Oral Biol. 56, 907–916.

Morsy, N.F.S., 2017. Chemical structure, quality indices and bioactivity of essential oil constituents. In: Active Ingredients From Aromatic and Medicinal Plants. InTech, London, UK, pp. 175–206, https://doi.org/10.5772/66231.

Nascimento, G.G.F., Localetti, J., Freitas, P.C., Silva, G.L., 2000. Antibacterial activity of plant extracts and phytochemicals on antibiotic resistant bacteria. Braz. J. Microbiol. 31, 247–256. https://doi.org/10.1590/S1517-83822000000400003.

Nirmala, M.J., Durai, L., Gopakumar, V., Nagarajan, R., 2019. Anticancer and antibacterial effects of a clove bud essential oil-based nanoscale emulsion system. Int. J. Nanomedicine 14, 6439–6450. https://doi.org/10.2147/IJN.S211047.

Ogata, M., Hoshi, M., Urano, S., Endo, T., 2000. Antioxidant activity of eugenol and related monomeric and dimeric compounds. Chem. Pharm. Bull. 48, 1467–1469.

Ogunwande, I., Olawore, N., Ekundayo, O., Walker, T., Schmidt, J., Setzer, W., 2005. Studies on the essential oils composition, antibacterial and cytotoxicity of *Eugenia uniflora* L. Int. J. Aromather. 15, 147–152.

Pandey, P., Singhal, D., Khan, F., Arif, M., 2020. An in silico screening on Piper nigrum, *Syzygium aromaticum* and Zingiber officinale roscoe derived compounds against SARS-CoV-2: a drug repurposing approach. Biointerface Res. Appl. Chem. 11, 11122–11134.

Pandima, K.D., Sakthivel, R., ArifNisha, S., Suganthy, N., Pandian, K.S., 2013. Eugenol alters the integrity of cell membrane and acts against the nosocomial pathogen Proteus mirabilis. Arch. Pharm. Res. 36, 282–292.

Park, M.J., Gwak, K.S., Yang, I., Choi, W.S., Jo, H., Chang, J.W., Jeung, E.B., Choi, I.G., 2007. Antifungal activities of the essential oils in *Syzygium aromaticum* (L.) Merr. Et Perry and *Leptospermum petersonii* Bailey and their constituents against various dermatophytes. J. Microbiol. 45 (5), 460–465. 17978807.

Pathirana, H.N.K.S., Wimalasena, S.H., De Silva, B.C., Hossain, S., Gang-Joon, H., 2019. Antibacterial activity of clove essential oil and eugenol against fish pathogenic bacteria isolated from cultured olive flounder (*Paralichthys olivaceus*). Slov. Vet. Res. 56 (1), 31–38.

Pillai, P., Ramaswamy, K., 2012. Effect of naturally occurring antimicrobials and chemical preservatives on the growth of *Aspergillus parasiticus*. J. Food Sci. Technol. 49, 228–233. https://doi.org/10.1007/s13197-011-0275-6.

Pinto, E., Vale-Silva, L., Cavaleiro, C., Salgueiro, L., 2009. Antifungal activity of the clove essential oil from *Syzygium aromaticum* on Candida, Aspergillus and dermatophyte species. J. Med. Microbiol. 58 (11), 1454–1462. https://doi.org/10.1099/jmm.0.010538-0. Epub 2009 Jul 9 19589904.

Prabuseenivasan, S., Jayakumar, M., Ignacimuthu, S., 2006. In vitro antibacterial activity of some plant essential oils. BMC Complement. Med. Ther. 6, 39. https://doi.org/10.1186/1472-6882-6-39.

Rajkowska, K., Nowak, A., Kunicka-Styczyńska, A., Siadura, A., 2016. Biological effects of various chemically characterized essential oils: investigation of the mode of action against *Candida albicans* and HeLa cells. RSC Adv. 6, 97199–97207. https://doi.org/10.1039/C6RA21108A.

Ramadan, M.M., Ali, M.M., Ghanem, K.Z., El-Ghorabe, A.H., 2015. Essential oils from Egyptian aromatic plants as antioxidant and novel anticancer agents in human cancer cell lines. Grasas Aceites 66, e080. https://doi.org/10.3989/gya.0955142.

Rana, I.S., Rana, A.S., Rajak, R.C., 2011. Evaluation of antifungal activity in essential oil of the *Syzygium aromaticum* (L.) by extraction, purification and analysis of its main component eugenol. Braz. J. Microbiol. 42 (4), 1269–1277.

Rapp, C., 2007. Clove oil as effective as topical anesthetic. HerbalGram 74, 26.

Rehman, M., AlAjmi, M.F., Hussain, A., 2021. Natural compounds as inhibitors of SARS-CoV-2 main protease (3CLpro): a molecular docking and simulation approach to combat COVID-19. Curr. Pharm. Des. 27 (33). https://doi.org/10.2174/138161282 6999201116195851.

Reuss, F., Kreß, A., Braun, M., Magdeburg, A., Pfenninger, M., Müller, R., Mehring, M., 2020. Knowledge on exotic mosquitoes in Germany, and public acceptance and effectiveness of Bti and two self-prepared insecticides against *Aedes japonicas japonicus*. Sci. Rep. 10, 18901. https://doi.org/10.1038/s41598-020-75780-5.

Rhayour, K., Bouchikhi, T., Tantaoui-Elarak, A., et al., 2003. The mechanism of bactericidal action of oregano and clove essential oils and of their phenolic major components on *Escherichia coli* and *Bacillus subtilis*. J. Essent. Oil Res. 15, 356–362.

Saeki, Y., Ito, Y., Shibata, M., Sato, Y., Okuda, K., Takazoe, I., 1989. Antimicrobial action of natural substances on oral bacteria. Bull. Tokyo Dent. Coll. 30, 129–135.

Santoro, G.F., Cardoso, M.G., Guimarães, L.G., Mendonça, L.Z., Soares, M.J., 2007. *Trypanosoma cruzi*: activity of essential oils from *Achillea millefolium* L., *Syzygium aromaticum* L. and *Ocimum basilicum* L. on epimastigotes and trypomastigotes. Exp. Parasitol. 116, 283–290.

Shahbazi, Y., 2019. Antioxidant, antibacterial, and antifungal properties of nanoemulsion of clove essential oil. Nanomed. Res. J. 4, 204–208. https://doi.org/10.22034/nmrj.2019.04.001.

Sharma, A., Rajendran, S., Srivastava, A., Sharma, S., Kundu, B., 2017. Antifungal activities of selected essential oils against *fusarium oxysporum* f. sp. *lycopersici* 1322, with emphasis on *Syzygium aromaticum* essential oil. J. Biosci. Bioenerg. 123, 308–313.

Šimović, M., Frane Delaš, F., Gradvol, V., Dragana Kocevski, D., Pavlović, H., 2014. Antifungal effect of eugenol and carvacrol against foodborne pathogens *Aspergillus carbonarius* and *Penicillium roqueforti* in improving safety of fresh-cut watermelon. J. Intercult. Ethnopharmacol. 3 (3), 91–96.

Smith-Palmer, A., Stewart, J., Fyfe, L., 1998. Antimicrobial properties of plant essential oils and essences against five important food-borne pathogens. Lett. Appl. Microbiol. 26, 118–122. https://doi.org/10.1046/j.1472-765X.1998.00303.x.

Sofia, P.K., Prasad, R., Vijay, V.K., Srivastava, A.K., 2007. Evaluation of antibacterial activity of Indian spices against common foodborne pathogens. Int. J. Food Sci. Technol. 42 (8), 910–915.

Sugihartini, N., Prabandari, R., Yuwono, T., Rahmawati, D.R., 2019. The anti-inflammatory activity of essential oil of clove (*Syzygium aromaticum*) in absorption base ointment with addition of oleic acid and propylene glycol as enhancer. Int. J. Appl. Pharm., 106–109. https://doi.org/10.22159/ijap.2019.v11s5.T0081.

Thosar, N., Basak, S., Bahadure, R.N., Rajurkar, M., 2013. Antimicrobial efficacy of five essential oils against oral pathogens: an in vitro study. Eur. J. Dent. 7, 71.

Uchôa Lopes, C.M., Saturnino de Oliveira, J.R., Holanda, V.N., Rodrigues, A.Y.F., Martins da Fonseca, C.S., Galvão Rodrigues, F.F., Camilo, C.J., Lima, V.L.d.M., Coutinho, H.D.M., Kowalski, R., et al., 2020. GC-MS analysis and hemolytic, antipyretic and antidiarrheal potential of *Syzygium aromaticum* (clove) essential oil. Separations 7, 35. https://doi.org/10.3390/separations7020035.

Ueda-Nakamura, T., Mendonça-Filho, R.R., Morgado-Díaz, J.A., Korehisa Maza, P., Prado Dias Filho, B., Aparício Garcia Cortez, D., Alviano, D.S., Rosa Mdo, S., Lopes, A.H., Alviano, C.S., et al., 2006. Antileishmanial activity of eugenol-rich essential oil from *Ocimum gratissimum*. Parasitol. Int. 55, 99–105.

Uju, D.E., Obioma, N.P., 2011. Anticariogenic potentials of clove, tobacco and bitter kola. Asian Pac. J. Trop. Med. 4, 814–818.

Velluti, A., Sanchis, V., Ramos, A.J., Egido, J., Marın, S., 2003. Inhibitory effect of cinnamon, clove, lemongrass, oregano and palmarose essential oils on growth and fumonisin B1 production by *Fusarium proliferatum* in maize grain. Int. J. Food Microbiol. 89, 145–154.

Velluti, A., Sanchis, V., Ramos, A.J., Marin, S., 2002. Control of *Fusarium verticillioides* growth and fumonisin B1 production in maize grain by the addition of cinnamon, clove, lemongrass, oregano and palmarose essential oils. J. Food Sci. 217, 332–337.

Viuda-Martos, M., Ruiz-Navajas, Y., Fernandes-Lopez, J., Perez-Alvarez, J.A., 2007. Antifungal activity of thyme, clove and oregano essential oils. J. Food Saf. 27, 91–101.

Wei, M.C., Xiao, J., Yang, Y.C., 2016. Extraction of α-humulene-enriched oil from clove using ultrasound-assisted supercritical carbon dioxide extraction and studies of its fictitious solubility. Food Chem. 210, 172–181. https://doi.org/10.1016/j.foodchem.2016.04.076.

Yang, Y.Y., Lee, M.J., Lee, H.S., Park, W.H., 2011. Screening of antioxidative, antiplatelet aggregation and antithrombotic effects of clove extracts. J. Physiol. Pathol. Korean Med. 25, 471–481.

Yanti, Y., Juniardi, S., Lay, B.W., 2019. *Syzygium aromaticum* essential oil prevents halitosis caused by oral bacteria *Streptococcus sanguinis*. Food Res. 3 (6), 814–820.

Yoo, C.B., Han, K.T., Cho, K.S., Ha, J., Park, H.J., Nam, J.H., Kil, U.H., Lee, K.T., 2005. Eugenol isolated from the essential oil of *Eugenia caryophyllata* induces a reactive oxygen species-mediated apoptosis in HL-60 human promyelocytic leukemia cells. Cancer Lett. 225, 41–52.

Zhang, S., Min Zhang, M., Fang, Z., Yaping Liu, Y., 2017. Preparation and characterization of blended cloves/cinnamon essential oil nanoemulsions. LWT - Food Sci. Technol. 75, 316–322.

Zheng, G.Q., Kenney, P.M., Lam, L.K., 1992. Sesquiterpenes from clove (*Eugenia caryophyllata*) as potential anticarcinogenic agents. J. Nat. Prod. 55, 999–1003.

Further reading

Aboubakr, H.A., Nauertz, A., Luong, N.T., Agrawal, S., El-Sohaimy, S.A.A., Youssef, M.M., Goyal, S.M., 2016. In vitro antiviral activity of clove and ginger aqueous extracts against feline calicivirus, a surrogate for human norovirus. J. Food Prot. 79 (6), 1001–1012. https://doi.org/10.4315/0362-028X.JFP-15-593.

Bachiega, T.F., de Sousa, J.P., Bastos, J.K., Sforcin, J.M., 2012. Clove and eugenol in noncytotoxic concentrations exert immunomodulatory/anti-inflammatory action on cytokine production by murine macrophages. J. Pharm. Pharmacol. 64, 610–616.

Bahramsoltani, R., Rahimi, R., 2020. An evaluation of traditional Persian medicine for the management of SARS-CoV-2. Front. Pharmacol. 11, 571434. https://doi.org/10.3389/fphar.2020.571434.

Chaves, D.S., Coumendouros, K., 2020. Activity of *Syzygium aromaticum* essential oil and its main constituent eugenol in the inhibition of the development of *Ctenocephalides felis* and the control of adults. Vet. Parasitol. 282, 109126. https://doi.org/10.1016/j.vetpar.2020.109126.

Hasheminejad, N., Khodaiyan, F., Safari, M., 2019. Improving the antifungal activity of clove essential oil encapsulated by chitosan nanoparticles. Food Chem. 275, 113–122.

Kanyinda, J.N.M., 2020. Coronavirus (COVID-19): a protocol for prevention and treatment. Eur. J. Med. Health Sci. 2, 1–4.

Keene, J.L., Noakes, D.L.G., Moccia, R.D., Soto, C.G., 2019. The efficacy of clove oil as an anaesthetic for rainbow trout, *Oncorhynchus mykiss* (Walbaum). Aquacult. Res. 29, 89–101.

Lambert, R.J.W., Skandamis, P.N., Coote, P.J., Nychas, G.J.E., 2001. A study of the minimum inhibitory concentration and mode of action of oregano essential oil, thymol and carvacrol. J. Appl. Microbiol. 91, 453–462.

Radünz, M., da Trindade, M.L.M., Camargo, T.M., Radünz, A.L., Borges, C.D., Gandra, E.A., et al., 2019. Antimicrobial and antioxidant activity of unencapsulated and encapsulated clove (*Syzygium aromaticum* L.) essential oil. Food Chem. 15, 180–186.

Saini, A., Sharma, S., Chhibber, S., 2009. Induction of resistance to respiratory tract infection with *Klebsiella pneumoniae* in mice fed on a diet supplemented with tulsi (*Ocimum sanctum*) and clove (*Syzgium aromaticum*) oils. J. Microbiol. Immunol. Infect. 42, 107–113.

Tunç, M.T., Koca, İ., 2019. Ohmic heating assisted hydrodistillation of clove essential oil. Ind. Crops Prod. 141, 111763. https://doi.org/10.1016/j.indcrop.2019.111763.

Xing, Y., Li, X., Xu, Q., Yun, J., Lu, Y., 2010. Antifungal activities of cinnamon oil against *Rhizopus nigricans*, *Aspergillus flavus* and *Penicillium expansum* in vitro and in vivo fruit test. Int. J. Food Sci. Technol. 45, 1837–1842.

CHAPTER

Encapsulation of clove (*Syzygium aromaticum*) essential oil

30

Onur Özdikicierler[a] and Pelin Günç Ergönül[b]

Faculty of Engineering, Food Engineering Department, Ege University, İzmir, Turkey[a] Department of Food Engineering, Faculty of Engineering, Manisa Celal Bayar University, Yunusemre, Manisa, Turkey[b]

1. Introduction

Food products may be contaminated both with pathogen bacteria and molds at different stages of the production flow, such as during processing, transportation, and storage. In addition, some fungal genera (e.g., *Fusarium*, *Penicillium*, *Aspergillus*, and *Alternaria*) may produce secondary metabolites, called mycotoxins, which may lead to carcinogenic, mutagenic, teratogenic, and immune suppressive effects on humans. Therefore, much recent research is focused on cost-effective and natural ways to inhibit microbial contamination and mycotoxin production (Mutlu-Ingok et al., 2020). Within this context, bioactive constituents of essential oils offer a great potential to reduce and control these hazardous microbial food contamination risks. Moreover, most essential oils can limit free radical formation, which causes cellular damage and chronicle diseases such as cancer (Barradas and de Holanda e Silva, 2020).

Essential oils (EOs) consist of bioactive components such as phenolic compounds that generally inhibit microbial growth owing to their inhibitory effect on microorganisms by influencing cellular metabolism. Although EOs are considered safe (GRAS), their use is limited by organoleptic criteria. In addition, EOs are highly volatile and are susceptible to chemical degradation and modification in the presence of ambient oxygen and light. For these reasons, the applications of encapsulation in nanoscale delivery systems have become widespread and revealed to improve the EOs bioactivity through the activation of passive cell absorption mechanisms (Rajaei et al., 2017).

Cloves (*Syzygium aromaticum*) contain a higher amount of essential oil (3.86 mL oil/100 g dry weight) when compared with other aromatic plants such as rosemary (*Rosmarinus officinalis* L.), thyme (*Thymus vulgaris* L.), dill (*Anethum graveolens* L.), basil (*Ocimum basilicum* L.), and ginger (*Zingiber officinale* Rosc) (Fadel et al., 2020). Like all essential oils, clove oil is rich in bioactive constituents, mainly eugenol and thymol. These bioactive constituents possess many biological activities such as antibacterial, antifungal, and antioxidant properties and are vulnerable to environmental factors (e.g., moisture, heat, oxygen, light, pH) (Wang et al., 2018a). Therefore, antiviral, antimicrobial, antifungal, and antioxidant properties of clove oil decrease in a storage period inevitably because of the degradation of these components. Fortification by encapsulation of clove oil is mainly aimed at preserving its physical stability and bioactive properties during a period and consequently renders possible the preservation of food products' nutritional value (Sebaaly et al., 2015).

Clove (*Syzygium aromaticum*). https://doi.org/10.1016/B978-0-323-85177-0.00028-8

The concept of encapsulation consists of two different terms: microencapsulation and nanoencapsulation. The microencapsulation procedure is basically coating tiny liquid droplets using a polymeric material to provide environmental protection. In general, most of the microencapsulated products have a diameter between 1 and 1000 μm. Nanoencapsulation is a similar technology referring to packing at a nanoscale range between 10 and 1000 nm. In nanoencapsulation, the encapsulated material is generally expressed as a nanocapsule or nanosphere. Since delivery of the material to a biological system is highly dependent on particle size, nanoencapsulated materials have more potential with enhanced bioavailability (Suganya and Anuradha, 2017).

Antimicrobial essential oils provide a great opportunity for the development of special additives. They could be incorporated into the packaging system since they are generally nontoxic and have significant food preservation properties. Some essential oils could be protected by nanocapsule coatings from the degradation effects of the processing equipment and might be used to produce active antimicrobial or high-performance antioxidant packaging (Zanetti et al., 2018).

This chapter outlines the general encapsulation methods of clove oil. Moreover, the rheological properties of encapsulated clove oil were discussed. Finally, biologic activities of encapsulated clove essential oil and potential uses of clove oil capsules in food matrices were also explained.

2. Clove oil encapsulation methods

2.1 Encapsulation using spray drying or lyophilization (freeze-drying)

Spray drying and lyophilization are well-known encapsulation methods applied for the encapsulation of oils. The principle of encapsulation by spray drying is mainly based on emulsification of the core material with a carrier material solution (wall material), followed by atomization and spraying the mixture into a hot air stream inside a drying chamber. The rapid evaporation of the solvent causes entrapment of the core material inside dried wall material particles. This method, known as a low-cost method, is suitable for a wide choice of wall materials, provides good encapsulation efficiency and good stability for the encapsulated products, and is suitable for scale-up studies. An inevitable degradation for sensitive materials may be expected because of the hot air stream applied to atomized feed. However, contrary to general expectations during spray drying operation, the temperature of the product will not exceed air output temperature; therefore, even heat-sensitive materials may be processed with negligible damage. The size of the particles may vary in a wide range (Suganya and Anuradha, 2017). All spray dryers use different systems to disperse the liquid or slurry into a controlled drop size spray. The most common spraying systems are rotating disks and nozzles, where both atomizing systems work with high pressured air. The particle size of the capsules obtained after spray drying is expected to be between 10 and 400 μm (Aksoylu and Günç Ergönül, 2017).

The principle of lyophilization or freeze-drying is generally based on entrapment by lyophilization of an emulsion containing the material mixed with the solution of carrier agent. This method is suitable for thermosensitive substances that are unstable in aqueous solutions. However, this process takes a significant time to complete and is an expensive procedure in terms of operation, storage, and transportation of the encapsulated product (Aksoylu and Günç Ergönül, 2017). The efficiency of encapsulation is mechanically dependent on the formation of solid wall material (maybe referred to as encapsulation agent, carrier material, or coating material) capsule around oil droplets in both spray drying and lyophilization (Suganya and Anuradha, 2017). The most used wall materials were maltodextrin (Cortés-Rojas et al., 2014; Nurliana et al., 2020; Talón et al., 2019) and whey protein (Talón et al., 2019).

Encapsulation of pure eugenol was studied by Talón et al. (2019), where potentials of whey protein, purified oleic acid, and soy lecithin mixed with maltodextrin as wall material on eugenol retention, antimicrobial and rheological properties of powders were compared. Atomizer was used to spray the feed into the drying chamber at a flow rate of 20 mL/min. Inlet temperature was 180°C, while outlet temperature was approximately around 80°C. The atomization was actualized with pressured air at 1.9 bar. Encapsulation efficiency of 95%–98% was achieved for both whey protein and lecithin, which implied that maltodextrin had a big role in the encapsulation (Talón et al., 2019).

Clove oil was encapsulated by spray drying and freeze-drying methods. The effect of both techniques on eugenol and eugenyl acetate retention was discussed comparatively by Cortés-Rojas et al. (2014). Three different emulsification methods were used before encapsulation processes; mechanical, ultrasound, and high-pressured homogenization. While inlet temperature was 90°C, outlet temperature was measured as 66.5°C during spray drying of clove oil. Therefore, the temperature of the encapsulated clove oil has risen to 66.5°C during the process. On the other hand, lyophilization started by freezing at −20°C and drying occurred at −80°C. Since eugenol and eugenyl acetate were heat-sensitive, freeze-drying was found more effective than spray drying for eugenol and eugenyl acetate retention after encapsulation (Cortés-Rojas et al., 2014).

Nurliana et al. (2020) used spray drying for encapsulation of clove oil oleoresin (essential oil) extracted by hydrodistillation (steam-water distillation) system from clove leaves. Clove essential oil was mixed with maltodextrin as a wall material at varying ratios (1:4 to 1:12, as essential oil: maltodextrin). The spray dryer was equipped with a nozzle atomizer. Inlet temperature was 110°C and outlet temperature was 68°C while feed flow rate was 300 mL/min (Nurliana et al., 2020).

An aqueous solution of β-cyclodextrin and soluble starch, as wall materials, was used in a previous study to encapsulate clove oil using spray drying. Clove oil and ethanol mixture was dripped into the aqueous solution and the mixture was emulsified with the help of sucrose ester as an emulsifier on a stirrer for 1.5 h at 60°C. This solution was fed into a spray dryer at a rate of 40 mL/min and at inlet temperature of 170°C (Wang et al., 2018b).

Encapsulation efficiency is one of the most important indicators for encapsulated clove oil obtained not only with spray drying but also for every method. This parameter indicates the mass rate of the clove oil entrapped within the capsules over theoretical clove oil in the weighted particles. The mass of the encapsulated clove oil is calculated by subtracting the mass of the free surface clove oil from theoretical clove oil in the weighted sample. Therefore, a mild extraction that determines free clove oil should be carried out. In general, small amounts of hexane, methanol, and petroleum ether were used within a short time to wash free oil from particles (Talón et al., 2019; Wang et al., 2018b).

2.2 Encapsulation by emulsification

Emulsions are mixtures of one immiscible liquid dispersed in another immiscible liquid as droplets generally through the use of a surfactant. At the same time, oil-in-water emulsions are defined as an emulsion of an aqueous phase that contains an oily dispersed phase. If the dispersed phase is an aqueous solution and the dominant phase is an oil or lipid phase, the emulsion is a water-in-oil emulsion. In many cases, an emulsifier agent or surfactant was used to reduce the interfacial tension between dispersed and continuous phases and provide stable emulsions. Emulsions are being divided into three groups according to their droplet size of dispersed phase and preparation techniques as macroemulsions, microemulsions, and nanoemulsions (Barradas and de Holanda e Silva, 2020; Ferreira and Nunes, 2019). Emulsification technology for encapsulation of clove oil was described by many researchers.

Radünz et al. (2019) defined an encapsulation method by emulsifying clove oil with 2% aqueous sodium alginate solution using glycerol monostearate and polyoxyethylene sorbitan monolaurate (Tween 20) as emulsifiers. Then, 1% of clove oil was added to calcium chloride solution and homogenized with sodium alginate and emulsifiers using Ultra-Turrax at 12000 rpm for 10 min. Then the solution was filtered and washed. The final product was dried at 30°C to obtain encapsulated clove oil particles (Radünz et al., 2019).

The dripping method is based on the droplet extrusion principle. The final form of capsules may be wet but can be dried using lyophilization. In general, particle sizes can be uniform, but encapsulated material has a higher chance to diffuse through a biopolymer network (Yang et al., 2020). A modified version of droplet extrusion/dripping was combined with the emulsification method to encapsulate clove oil by Hasheminejad et al. (2019). Briefly, droplet formation was achieved in the oil-in-water emulsion, prepared using chitosan and tripolyphosphate solution. Then droplets are solidified using the ionic gelation method. The emulsion was first obtained using a mechanical stirring at 700 rpm for 10 min at 25°C. Then tripolyphosphate solution was added to the emulsion of chitosan solution and clove oil. Particles were obtained with centrifugation and lyophilization. During the procedure, pH was mindfully controlled at every stage of encapsulation (Hasheminejad et al., 2019).

Ruiz-Gonzalez et al. (2019) studied encapsulation of clove oil by nano-emulsification method. Nanoemulsions were produced using 5% clove oil inside 10% inulin as a stabilizing agent and 3% polysorbate 80 as an emulsifier. Nanoemulsions were prepared using an ultrasonic processor at 84 μm amplitude for 15 min (Ruiz-Gonzalez et al., 2019). Researchers obtained the emulsion phase in a liquid form, unlike other researchers at dry powdered forms.

Ultrasonic emulsification to encapsulate clove oil was also performed by Shahavi et al. (2015). First, a full hydrated aqueous biopolymer solution was obtained by keeping a vortex-blended Arabic gum and whey protein mixture for 24 h at room temperature. Then, the emulsifier (Tween 80) and clove oil mixture was added to biopolymer solution emulsified using an ultrasonic homogenizer for 2 min at 70 W and 20 kHz. During emulsification, the temperature was kept under 30°C using an ice bath. Emulsions were kept at 4°C refrigeration conditions. No further drying was applied. The encapsulation efficiency and other indices were studied using an optimization approach (Shahavi et al., 2015).

Wang et al. (2018a,b) encapsulated clove oil by emulsification using a mechanical stirrer. First, 0.15 g chitosan was mixed with acidified water and the mixture was homogenized using an ultrasonic homogenizer. Meanwhile, clove oil was mixed with an emulsifier (Tween 80) in water. Microcapsules of clove oil were obtained by dripping of clove oil mixture inside chitosan solution under agitation using a mechanical stirrer. Obtained microcapsules were stored at refrigeration temperatures but stabilized for 24 h at 25°C before further trials (Wang et al., 2018a).

Encapsulation efficiency, particle size, surface charge measurements, scanning electron microscopy observation, thermokinetic assessment with Differential Scanning Calorimetry (DSC), and color assessment are generally applied to determine the encapsulation performance of an emulsification method (Hasheminejad et al., 2019; Radünz et al., 2019; Ruiz-Gonzalez et al., 2019; Shahavi et al., 2015; Yang et al., 2020).

2.3 Encapsulation by entrapment in liposomes

Liposomes are spherical vesicles with one lipid bilayer formed in aqueous solutions. Liposomes have many advantages of being biodegradable, nontoxic, non-immunogenic, and biocompatible particles. However, the production cost of liposomes is high, has low solubility, and lower prevention against degradation reactions when compared with other encapsulation methods (Anwekar et al., 2011; Sebaaly et al., 2015). Phospholipids are generally used for the development of liposome structures. A simple scheme of liposomes of phospholipids in an aqueous solution was given in Fig. 1.

Phosphatidylcholine or soy lecithin together with cholesterols was used to obtain liposomes to encapsulate clove oil. A multistaged yet fragile method was used for encapsulation of clove oil in liposomes. The main principle of encapsulation is based on dissolving clove oil, phospholipid mixture, and cholesterol with solvent, suspension of the liquid phase using phosphate buffer solution, homogenization, and filtration of liposome phase using membranes (Cui et al., 2015; Sebaaly et al., 2015). Two similar techniques were defined to obtain liposomes to encapsulate clove oil. Cui et al. (2015) dissolved soy lecithin: cholesterol mixture (5:1) in trichloromethane and added clove oil into this mixture. Then evaporated the mixture in a rotary evaporator. Drying was completed in a vacuum oven under 30°C. The lipid film was suspended in a 20mL phosphate buffer solution which contains 0.02g polyvinylpyrrolidone (PVP). After grinding, to obtain liposomes, the upper fluid phase was discarded by centrifugation and membrane filtration (Cui et al., 2015).

Clove oil liposomes were obtained using the ethanol injection method in a previous study. According to this method, phospholipid, cholesterol, and clove oil were dissolved in ethanol by stirring. The organic solution was injected into ultrapure water using a syringe pump under magnetic stirring. The injection was applied at a temperature above the transition temperature of the phospholipid used. Liposomes formed as soon as ethanolic solution was contacted with an aqueous solution. The liposome

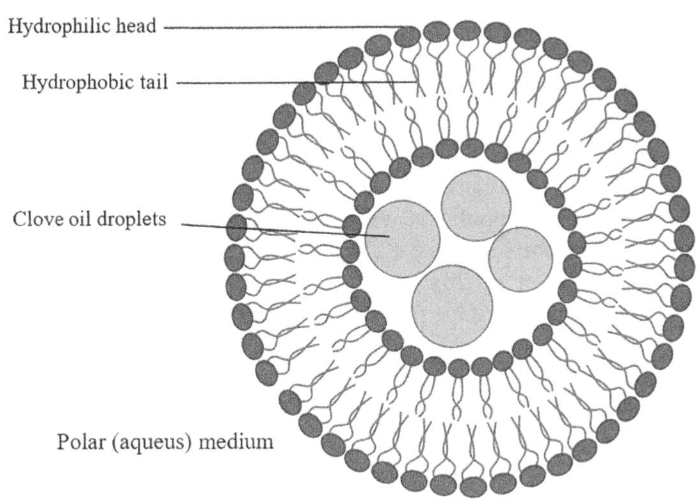

FIG. 1

Structure of a phospholipid liposome.

phase was stirred for 15 min at room temperature. Liposomes were obtained after evaporation of the ethanolic phase under reduced pressure at 40°C (Sebaaly et al., 2015).

Liposomes can be characterized by different methods. Particle size distribution and polydispersity index (PDI), turbidity, morphological characterization under the electron microscope, and zeta potential are the main rheological tests applied for clove oil liposomes in previous studies (Cui et al., 2015; Sebaaly et al., 2015).

3. Biologic activities of encapsulated clove essential oil

Among the clove (*Eugenia caryophyllata* Thunb) species, clove bud oil has biological activities, such as antibacterial, antifungal, insecticidal, and antioxidant properties. It is used traditionally as a flavoring agent and as an antimicrobial material in food (Passone et al., 2012). The high levels of eugenol contained in clove oil give it strong biological and antimicrobial activities (Nonsee et al., 2011). Also, it acts as a bactericide against some of the most important foodborne pathogens, including *Staphylococcus aureus*, *Escherichia coli*, *Listeria monocytogenes*, and *Salmonella typhimurium* due to the presence of eugenol and other phenolic compounds, and has anti-free radical and metal chelating activities (Radünz et al., 2019).

Antimicrobials are generally used to prohibit the bacterial spoilage and growth of pathogenic microorganisms. However, synthetic food preservatives have been discussed about possible toxic and carcinogenic effects. Therefore, innovation in microbial inhibition continues to look for a safer and effective novel natural solution from plant sources (Purkait et al., 2020). Studies revealed that clove oil has significant antimicrobial activity against a wide range of pathogenic microorganisms. Moreover, encapsulation improves its stability and water solubility.

Encapsulated pure eugenol using maltodextrin-based wall materials containing whey protein or lecithin provided a significant antimicrobial effect in a previous study. In general, powders with different formulations were more effective against *Escherichia coli* than *Listeria innocua*. The antibacterial effect of the powders against *Escherichia coli* and *Listeria innocua* was also coherent with the eugenol content of powders; however, powders with purified oleic acid provided an additional positive effect against *Listeria innocua* (Talón et al., 2019).

The liposome-encapsulated clove oil also showed good antimicrobial activity on tofu cheese by limiting the total microbial population compared with control samples during storage. By the end of 120 h of storage, the microbial population was approximately 5 log CFU/mL for liposome encapsulated clove oil applied tofu samples, while control samples had 10 log CFU/mL total population in the same period of storage (Cui et al., 2015).

Nonsee et al. (2011) investigated edible film's antimicrobial activity by incorporating encapsulated clove (*Eugenia caryophyllata* Thunb.) oil as a natural antibacterial agent. Antimicrobial edible films were prepared by incorporating encapsulated clove oil from 1, 2, 3 folds of minimal bactericidal concentration into 1.5% of hydroxypropyl methylcellulose (HPMC) film solution (Nonsee et al., 2011). The results showed that increasing encapsulated clove oil in the edible HPMC films from 1, 2, 3 folds of MBC yielded a higher inhibition zone from 18.50–26.75, 28.75–39.00, and 16.67–31.00 mm of *Escherichia coli* O175:H7, *Staphylococcus aureus*, and *Listeria monocytogenes*, respectively. Gómez-Estaca et al. (2010) found the strong antimicrobial activity of gelatin-chitosan-based films in which clove essential oil was loaded. The clove-loaded edible films caused inhibition of six studied microorganisms such as *Pseudomonas fluorescens*, *Shewanella putrefaciens*, *Photobacterium phosphoreum*, *Listeria innocua*, *Escherichia coli*, and *Lactobacillus acidophilus* (Gómez-Estaca et al., 2010).

In vitro, antioxidant and antimicrobial activities of the encapsulated clove essential oil using sodium alginate/emulsifiers system were evaluated by Radünz et al. (2019). They prepared three sets of particles as sodium alginate and clove essential oil (AO), sodium alginate, glycerol monostearate, clove essential oil (AMO), and sodium alginate polyoxyethylene sorbitan monolaurate, and clove essential oil (ATO). They found that AMO particles had a bacterial killing of *Salmonella typhimurium* and *Staphylococcus aureus* at a concentration of 23.66 mg/mL. AO particles supported the killing of *Staphylococcus aureus* at a concentration of 79.6 mg/mL, while no minimum bactericidal concentration was observed for the ATO particles. It has been shown in different studies that the bactericidal effect of essential oils depends on the wall material. In this study, no particles displayed a bactericidal effect for *Listeria monocytogenes*. As regards antioxidant activity, AO and AMO particles showed higher antioxidant activity for all tested radicals (DPPH and Hydroxyl) than ATO particles. DPPH scavenging activity was lower than the results reported in the literature due to the different wall material, particle characteristics such as size and load, and the method of analysis employed. Researchers found that poly(thioether-ester) nanoparticles via miniemulsion polymerization, using dianhydro-D-glucityl diundec-10-enoate and a dithiol monomer with different concentrations of clove oil were a powerful DPPH antioxidant scavenging activity than free clove essential oil (de Meneses et al., 2017; Radünz et al., 2019; Sebaaly et al., 2016).

Rajaei et al. (2017) investigated the antioxidant and antibacterial activities of clove essential oil encapsulated by chitosan (CS)-Myristic acid (MA) nanogel. The results showed that the antioxidant activity of clove essential oil was increased by encapsulation. Additionally, the antibacterial activity of free clove essential oil, CS-MA nanogel, and CS-MA encapsulated oils was evaluated against *Salmonella enteric* Ser. Enteritidis. The findings indicated that the minimum inhibitory concentration (MIC) and minimum bacteriostatic concentration (MBC) values of free CEOs, CS-MA nanogel, and encapsulated CEOs were 100–200, 10–20, and 5–10 mg/L, respectively. The results showed that major bioactive compounds such as monoterpenes, eugenol, cinnamic aldehyde, and thymol in clove essential oil are responsible for strong antimicrobial activity (Rajaei et al., 2017).

Gul et al. (2018) also investigated the antibacterial activity and the antioxidant activity of hazelnut meal proteins and clove essential oil nanoemulsion films applying the US treatment at different times and amplitudes. They indicated that US treatment of emulsions at different amplitudes and times significantly increased the antibacterial activity of films due to increasing acoustic energy supplied to the emulsion system. Additionally, the antioxidant capacity was increased due to the presence of phenolic compounds of both hazelnut protein and clove essential oil (Gul et al., 2018). Furthermore, the antibacterial activity of essential oils was well preserved when they were incorporated into the films (Froiio et al., 2019). Sebaaly et al. (2016) obtained DPPH antioxidant scavenging ranging from 89.3% to 92.3% for clove essential oil encapsulated in cyclodextrin. This result was probably allowing the greater release of the compounds present in the essential oil.

Clove essential oil and its main component, eugenol, with effect to fungal pathogens such as *Botrytis cinerea* and *Penicillium expansum* and also *Aspergillus* spp., such as *Aspergillus niger* and *Aspergillus flavus*, have been marked as natural fungicides (Passone et al., 2012). Clove essential oil was encapsulated with chitosan nanoparticles using emulsion-ionic gelation technique to progress its antifungal efficacy by Hasheminejad et al. (2019). In vitro studies lasted 56 days, demonstrating that clove essential oil-chitosan nanoparticles films had a controlled release for 56 days. They also determined that nanoencapsulated clove essential oil showed greater performance than chitosan nanoparticles and nonencapsulated oil against *Aspergillus niger* separated from spoiled pomegranate. Eventually, Hasheminejad et al. (2019) suggested the usability of clove essential oil-chitosan nanoparticles films as a natural antifungal agent in the food industry and agriculture.

Sagiri et al. (2016) indicated that microencapsulated clove, thyme, and cinnamon oils inside alginate microspheres showed high growth inhibition of *Aspergillus niger* and *Fusarium verticillioides*. After 8 days of storage, the oil encapsulated alginate microspheres demonstrated 30%–50% of the antifungal activity. In the case of free oils, the antifungal activity was lost after 2 days of incubation with fungal-infected grains (Sagiri et al., 2016).

4. Potential uses of encapsulated clove essential oil in food matrices

The antimicrobial activity of essential oils has been demonstrated and evaluated in vitro and food matrices by several researchers (Aureli et al., 1992; Cava et al., 2007; Friedman et al., 2002; Lis-Balchin and Deans, 1997; Smith-Palmer et al., 1998). Whereas there is limited study on the uses of encapsulated clove essential oil in foods. Types of essential oil and foods are two important factors that should be considered to apply essential oils as an antimicrobial agent. For example, some essential oils such as eugenol, oregano, coriander, thyme, and clove oils have more antimicrobial activity than other essential oils, in meat industry (Bahrami et al., 2020).

Clove essential oil encapsulated with liposome has shown some noticeable advantages when applied in tofu production for inhibition of *Staphylococcus aureus*, contributing to improving tofu stability and shelf life based on the use of bacteria-targeted antibiotics (Cui et al., 2015). The oil-framed emulsion was used for encapsulation of clove and lavender essential oils and vanillin into chitosan and alginate beads to control the growth of *Botrytis cinerea* in grapes (Sangsuwan and Sutthasupa, 2019). Using chitosan and alginate beads containing clove essential oil were maintained the freshness, firmness, flavor, odor, and overall acceptability of grapes.

The clove bud and oregano essential oils nanoemulsion, which was prepared by low-speed mixing and ultrasonication, immobilized in methylcellulose/polyethylene glycol films to decrease film rigidity and raise toughness, were used for coating bread slices to improve their antimicrobial properties and reduce the counts of yeasts and molds during the 15 days of storage (Otoni et al., 2014). The bread slices were placed into methylcellulose film/metalized polypropylene bags and stored at 25°C for 15 days. Counts of yeasts and molds, and microorganisms *(Aspergillus niger* or *Penicillium* sp.) that cause to spoil bakery products in a short time were analyzed after 0, 5, 10, and 15 days of storage. They have not recorded inhibition halos of *Aspergillus niger* and *Penicillium* sp. below 20 mg/mL of essential oil concentrations. Regarding counts of yeasts and molds, there was an increase in control bread slices packed without preservation, and bread slices contained commercial antifungals such as sorbic acid, calcium propionate, ethanol, and grain alcohol, after 5 days. At 15 days of storage, the counts of yeasts and molds in bread slices packed with antimicrobial films were found lower when compared with the control and commercial treatments. This showed that clove bud essential oil is effective in the decrease of microbial development in food matrices than other conventional techniques (Otoni et al., 2014).

Studies in recent years have been demonstrated that consumer demand for fresh-cut vegetables was raised due to their health benefits (Park et al., 2019). Fresh-cut vegetables are naturally exposed to microbial contamination with increasing consumption (Park et al., 2018). They could be contaminated by *Salmonella Typhimurium* and *Listeria monocytogenes* (Ma et al., 2017). So, it is necessary to make suitable washing treatments and postharvest technologies to decrease the microbiological risks and to extend their shelf life (Chun and Song, 2013). Hence, essential oils could be used as an alternative to sanitizers because of their vigorous antimicrobial activities. Essential clove bud oil is one of them. Regardless, it has limited uses in the washing process due to its potent odor (Park et al., 2019). Therefore, Park et al. (2019) improved an emulsion system containing clove bud essential oil/benzethonium chloride to inactivate *Salmonella typhimurium* and *Listeria monocytogenes* inoculated on fresh-cut pak choi. Firstly, fresh-cut pak choi was washed with emulsion and stored in a modified atmosphere for 14 days. During storage, the removal of pathogens was examined. While water-washing decreased the populations of *Salmonella typhimurium* and *Listeria monocytogenes* by 0.29 and 0.37 log CFU/g, washing with clove bud essential oil/benzethonium chloride emulsion reduced by 1.97 and 2.00 log CFU/g, respectively, during 14 days of storage. Thus, Park et al. (2019) has been proved that the washing process with clove bud essential oil/benzethonium chloride emulsion/modified atmosphere storage was an available alternative method to increase the microbiological safety and shelf life of the fresh-cut pack (Park et al., 2019).

Cui et al. (2018) used gelatin nanofibers to encapsulate clove essential oil to reduce the *Escherichia coli* biofilms population in cucumbers. Cucumbers were treated with clove oil-loaded chitosan nanoparticles incorporated in gelatin electrospun nanofiber. After 6 mg/mL and 9 mg/mL gelatin/clove oil-loaded chitosan nanoparticles treatment at 12°C for 4 days, in the population of *Escherichia coli* O157:H7 biofilm, 4.28 and 4.97 $\log_{10}$ reductions were observed, respectively (Cui et al., 2018). Mixtures of some bioactive compounds of essential oils such as thymol, carvacrol, trans-cinnamaldehyde, eugenol, and vanillin were used as the decontaminant on fresh lettuces against *Escherichia coli* O157:H7. The results showed that a combination of thymol/eugenol, carvacrol/eugenol, and *trans*-cinnamaldehyde/eugenol had better antimicrobial activity as synergistic effects when compared with the individual compounds (Zhu et al., 2021).

Rajaei et al. (2017) evaluated the potential effects of clove essential oil, CS-MA nanogel, and clove essential oil encapsulated by CS-MA nanogel to eliminate *Salmonella enterica Ser*. Enteritidis, in a real food system, i.e., beef during 12 days of refrigerated storage at 4°C. They found that during the storage period, *Salmonella enteric Ser*. Enteritidis population inoculated on the control sample increased from 4 to 6.8 log CFU/g. On day 0, the free and encapsulated CEOs and the CS-MA nanogel coatings caused an immediate reduction of about 3 log CFU/g in *Salmonella enterica Ser*. Enteritidis population (Rajaei et al., 2017). As a result, the CEOs encapsulated by the CS-MA nanogel had more efficiency in controlling the investigated pathogen than the free CEOs. This could be explained by the fact that the encapsulation of bioactive compounds may theoretically enhance the antimicrobial activity of these compounds by increasing their dispersion in an aqueous environment and stimulating passive adsorption mechanisms in cells (Donsì et al., 2011). In situ antimicrobial effect of encapsulated clove essential oil on burger-like meat products was examined. For this, they formulated a meat product as follows: 57% lean beef, 10% swine fat, 15% ice, 15% ice water, and 3% soy protein and then treated it with *Staphylococcus aureus*. Unfortunately, clove essential oil particles allowed *Staphylococcus aureus* growth. According to a study, essential clove oil integrated into whey protein edible films showed a great inhibition of microbial growth in poultry samples (mesophilic, *Enterobacteriaceae*, and *Pseudomonas*) at concentrations of 30 g/kg (Ponce et al., 2008).

Zhang et al. (2017) aimed in their study to utilize cloves/cinnamon essential oil nanoemulsions as natural preservatives in mushroom sauce to improve flavor. The major flavor compounds of the mushroom sauces are sulfur-containing compounds, oxygen-containing heterocycles, nitrogen-containing heterocycles, phenols, aldehydes, alcohols, acids, esters, and hydrocarbons. They are too similar compounds to the added nanoemulsion. As a result, they revealed that flavor compounds such as eugenol, 2-propenal, cinnamic aldehyde, methyl salicylate, ethanol, and benzyl alcohol were significantly increased after the addition of nanoemulsion. It may also have the potential as a natural antimicrobial agent (Zhang et al., 2017).

Clove essential oil comprised the novel cooling storage-ice glazing system to protect the fish from microbial contamination. Moreover, sea bass samples were plunged in clove essential oil solutions at different concentrations and stored at -1 ± 0.2°C for 25 days. The presence of clove essential oil increased the shelf-life and they were kept fresh for a longer time by holding up the oxidation and microbiological spoilage (He et al., 2018). Da Rocha et al. (2018) evaluated the effects of protein hydrolysate and clove essential oil incorporated on agar films applied to flounder fillets during chilled storage at 5°C for 15 days. They determined that clove essential oil films were less transparent than protein hydrolyzate and control films. In contrast, the existence of protein hydrolyzate increased the water solubility and water vapor permeability of the films. Additionally, fillets coated with films, especially clove oil films, had lower volatility and pH values and retarded the growth of H2S-producing microorganisms such as *Staphylococcus aureus, Yersinia enterocolitica, Aeromonas hydrophila, Debaryomyces hansenii*, and *Listeria innocua* from the third day. As a result, they suggested the usability of clove essential oil as a natural bio-preservative to increase the shelf-life of flounder fillets (Da Rocha et al., 2018).

Gómez-Estaca et al. (2010) used gelatin-chitosan-based edible films incorporated with clove essential oil in fish preservation. Before applying to the fish, their antimicrobial activity was tested against some microorganisms such as *Pseudomonas fluorescens, Shewanella putrefaciens, Photobacterium phosphoreum, Listeria innocua, Escherichia coli*, and *Lactobacillus acidophilus*. As a result, they found that the clove-containing films were inhibited in all these microorganisms. Then codfish fillets were covered with clove-containing films, vacuum packed, and left to storage at 2°C for 10 days. Researchers were determined that the growth of gram-negative bacteria (especially in *Enterobacteria*) was significantly decreased while lactic acid bacteria was not changed during chilled storage. However, there was no information on the clove essential oil effect on *Shewanella putrefaciens*, which is important for fish spoilage (Gómez-Estaca et al., 2010).

The encapsulated essential oils are also used in pharmacy for medicinal purposes because of their medical potential. For instance, encapsulated sage, tea tree, and clove essential oils are useful as skin antimicrobial agents, eucalyptus oil as a coughing agent, peppermint as a decongestant, and tea tree oil as an antiseptic agent (Fathi et al., 2021).

5. Conclusion

Clove (*Syzygium aromaticum*) oil exhibits significant biological activities such as antibacterial, antifungal, and antioxidant properties because of its phenolic compounds such as eugenol, thymol, and gallic acid. However, these compounds are vulnerable to external factors; hence technological solutions for better preservation offer significant opportunities for application to use bioavailability and antimicrobial effect of clove oil. In this aspect, various encapsulation techniques have been studied widely to preserve clove oil bioactive components. In general, spray drying, lyophilization, and emulsification are the main methods used in studies for clove oil encapsulation. In addition, nanoencapsulated clove oil particles have more efficient effects when bioavailability and the use of encapsulated clove oil in food matrices were considered. Nevertheless, studies revealed that, regardless of the capsule size range, clove oil bioactive compounds exhibit significant antimicrobial effects against molds, especially in vitro studies. Further studies may focus on developing more stable encapsulated clove oil particles that have a long shelf life and are suitable for various application purposes.

References

Aksoylu, Z., Günç Ergönül, P., 2017. A review on encapsulation of oils. Celal Bayar Üniv. Fen Bilim. Derg. 13 (2), 293–309. https://doi.org/10.18466/cbayarfbe.313358.

Anwekar, H., Patel, S., Singhai, A.K., 2011. Liposome—as drug carriers. Int. J. Pharm. Life Sci. 2 (7), 945–951.

Aureli, P., Costantinf, A., Zolea, S., 1992. Antimicrobial activity of some plant essential oils against *Listeria monocytogenes*. J. Food Prot. 55 (5), 344–348. https://doi.org/10.4315/0362-028X-55.5.344.

Bahrami, A., Delshadi, R., Assadpour, E., Jafari, S.M., Williams, L., 2020. Antimicrobial-loaded nanocarriers for food packaging applications. Adv. Colloid Interface Sci. 278, 102140. https://doi.org/10.1016/j.cis.2020.102140.

Barradas, T.N., de Holanda e Silva, K.G., 2020. Nanoemulsions of essential oils to improve solubility, stability and permeability: a review. Environ. Chem. Lett., 0123456789. https://doi.org/10.1007/s10311-020-01142-2.

Cava, R., Nowak, E., Taboada, A., Marin-Iniesta, F., 2007. Antimicrobial activity of clove and cinnamon essential oils against *Listeria monocytogenes* in pasteurized milk. J. Food Prot. 70 (12), 2757–2763. https://doi.org/10.4315/0362-028X-70.12.2757.

Chun, H.H., Song, K.B., 2013. The combined effects of aqueous chlorine dioxide, fumaric acid, and ultraviolet-C with modified atmosphere packaging enriched in CO2 for inactivating preexisting microorganisms and *Escherichia coli* O157: H7 and *Salmonella typhimurium* inoculated on buckwheat sprouts. Postharvest Biol. Technol. 86, 118–124. https://doi.org/10.1016/j.postharvbio.2013.06.031.

Cortés-Rojas, D.F., Souza, C.R.F., Oliveira, W.P., 2014. Encapsulation of eugenol rich clove extract in solid lipid carriers. J. Food Eng. 127, 34–42. https://doi.org/10.1016/j.jfoodeng.2013.11.027.

Cui, H., Zhao, C., Lin, L., 2015. The specific antibacterial activity of liposome-encapsulated clove oil and its application in tofu. Food Control 56, 128–134. https://doi.org/10.1016/j.foodcont.2015.03.026.

Cui, H., Bai, M., Rashed, M.M.A., Lin, L., 2018. The antibacterial activity of clove oil/chitosan nanoparticles embedded gelatin nanofibers against *Escherichia coli* O157:H7 biofilms on cucumber. Int. J. Food Microbiol. 266, 69–78. https://doi.org/10.1016/j.ijfoodmicro.2017.11.019.

Da Rocha, M., Aleman, A., Romani, V.P., Lopez-Caballero, M.E., Gomez-Guillen, M.C., Montero, P., Prentice, C., 2018. Effects of agar films incorporated with fish protein hydrolysate or clove essential oil on flounder (*Paralichthys orbignyanus*) fillets shelf-life. Food Hydrocolloids 81, 351–363. https://doi.org/10.1016/j.foodhyd.2018.03.017.

de Meneses, A.C., dos Santos, P.C.M., Machado, T.O., Sayer, C., de Oliveira, D., Hermes de Araújo, P.H., 2017. Poly(thioether-ester) nanoparticles entrapping clove oil for antioxidant activity improvement. J. Polym. Res. 24 (11). https://doi.org/10.1007/s10965-017-1353-x.

Donsì, F., Annunziata, M., Sessa, M., Ferrari, G., 2011. Nanoencapsulation of essential oils to enhance their antimicrobial activity in foods. LWT - Food Sci. Technol. 44 (9), 1908–1914. https://doi.org/10.1016/j.lwt.2011.03.003.

Fadel, H.H.M., El-Ghorab, A.H., Hussein, A.M.S., El-Massry, K.F., Lotfy, S.N., Sayed Ahmed, M.Y., Soliman, T. N., 2020. Correlation between chemical composition and radical scavenging activity of 10 commercial essential oils: impact of microencapsulation on functional properties of essential oils. Arab. J. Chem. 13 (8), 6815–6827. https://doi.org/10.1016/j.arabjc.2020.06.034.

Fathi, M., Vinceković, M., Jurić, S., Viskić, M., Režek Jambrak, A., Donsì, F., 2021. Food-grade colloidal systems for the delivery of essential oils. Food Rev. Intl. 37 (1), 1–45. https://doi.org/10.1080/87559129.2019.1687514.

Ferreira, C.D., Nunes, I.L., 2019. Oil nanoencapsulation: development, application, and incorporation into the food market. Nanoscale Res. Lett. 14 (9), 1–13. https://doi.org/10.1186/s11671-018-2829-2.

Friedman, M., Henika, P.R., Mandrell, R.E., 2002. Bactericidal activities of plant essential oils and some of their isolated constituents against *Campylobacter jejuni*, *Escherichia coli*, *Listeria monocytogenes*, and *Salmonella enterica*. J. Food Prot. 65 (10), 1545–1560. https://doi.org/10.4315/0362-028X-65.10.1545.

Froiio, F., Mosaddik, A., Morshed, M.T., Paolino, D., Fessi, H., Elaissari, A., 2019. Edible polymers for essential oils encapsulation: application in food preservation. Ind. Eng. Chem. Res. 58 (46), 20932–20945. https://doi.org/10.1021/acs.iecr.9b02418.

Gómez-Estaca, J., López de Lacey, A., López-Caballero, M.E., Gómez-Guillén, M.C., Montero, P., 2010. Biodegradable gelatin-chitosan films incorporated with essential oils as antimicrobial agents for fish preservation. Food Microbiol. 27 (7), 889–896. https://doi.org/10.1016/j.fm.2010.05.012.

Gul, O., Saricaoglu, F.T., Besir, A., Atalar, I., Yazici, F., 2018. Effect of ultrasound treatment on the properties of nano-emulsion films obtained from hazelnut meal protein and clove essential oil. Ultrason. Sonochem. 41 (August 2017), 466–474. https://doi.org/10.1016/j.ultsonch.2017.10.011.

Hasheminejad, N., Khodaiyan, F., Safari, M., 2019. Improving the antifungal activity of clove essential oil encapsulated by chitosan nanoparticles. Food Chem. 275 (September 2018), 113–122. https://doi.org/10.1016/j.foodchem.2018.09.085.

He, Q., He, J., Xiao, K., 2018. A novel superchilling storage-ice glazing (SS-IG) approach using anti-oxidative and antimicrobial essential oil (EO) for freshness-keeping of sea bass (*Dicentrarchus labrax*). Aquaculture. https://doi.org/10.1016/j.aquaculture.2018.10.022.

Lis-Balchin, M., Deans, S.G., 1997. Bioactivity of selected plant essential oils against *Listeria monocytogenes*. J. Appl. Microbiol. 82 (6), 759–762. https://doi.org/10.1046/j.1365-2672.1997.00153.x.

Ma, L., Zhang, M., Bhandari, B., Gao, Z., 2017. Recent developments in novel shelf life extension technologies of fresh-cut fruits and vegetables. Trends Food Sci. Technol. 64, 23–38. https://doi.org/10.1016/j.tifs.2017.03.005.

Mutlu-Ingok, A., Devecioglu, D., Dikmetas, D.N., Karbancioglu-Guler, F., Capanoglu, E., 2020. Antibacterial, antifungal, antimycotoxigenic, and antioxidant activities of essential oils: an updated review. Molecules 25 (20). https://doi.org/10.3390/molecules25204711.

Nonsee, K., Supitchaya, C., Thawien, W., 2011. Antimicrobial activity and the properties of edible hydroxypropyl methylcellulose based films incorporated with encapsulated clove (*Eugenia caryophyllata* Thunb.) oil. Int. Food Res. J. 18 (4), 1531–1541.

Nurliana, L., Kurniawati, D., Kadir, L.A., Dewi, F., Musta, R., Nurjana., 2020. Effectiveness of maltodextrin as a coating in microencapsulation of clove leaves oil (*Syzygium aromaticum*) for antibacterial applications. IOP Conf. Ser. Earth Environ. Sci. 465 (1). https://doi.org/10.1088/1755-1315/465/1/012040.

Otoni, C.G., Pontes, S.F.O., Medeiros, E.A.A., Soares, N.D.F.F., 2014. Edible films from methylcellulose and nanoemulsions of clove bud (*Syzygium aromaticum*) and oregano (*Origanum vulgare*) essential oils as shelf life extenders for sliced bread. J. Agric. Food Chem. 62 (22), 5214–5219. https://doi.org/10.1021/jf501055f.

Park, J.B., Kang, J.H., Song, K.B., 2018. Combined treatment of cinnamon bark oil emulsion washing and ultraviolet-C irradiation improves microbial safety of fresh-cut red chard. LWT 93, 109–115. https://doi.org/10.1016/j.lwt.2018.03.035.

Park, J.-B., Kang, J.-H., Song, K.B., 2019. Clove bud essential oil emulsion containing benzethonium chloride inactivates *Salmonella* Typhimurium and *Listeria monocytogenes* on fresh-cut pak choi during modified atmosphere storage. Food Control. https://doi.org/10.1016/j.foodcont.2019.01.001.

Passone, M.A., Girardi, N.S., Etcheverry, M., 2012. Evaluation of the control ability of five essential oils against Aspergillus section Nigri growth and ochratoxin A accumulation in peanut meal extract agar conditioned at different water activities levels. Int. J. Food Microbiol. 159 (3), 198–206. https://doi.org/10.1016/j.ijfoodmicro.2012.08.019.

Ponce, A.G., Roura, S.I., del Valle, C.E., Moreira, M.R., 2008. Antimicrobial and antioxidant activities of edible coatings enriched with natural plant extracts: in vitro and in vivo studies. Postharvest Biol. Technol. 49 (2), 294–300. https://doi.org/10.1016/j.postharvbio.2008.02.013.

Purkait, S., Bhattacharya, A., Bag, A., Chattopadhyay, R.R., 2020. Synergistic antibacterial, antifungal and antioxidant efficacy of cinnamon and clove essential oils in combination. Arch. Microbiol. 202 (6), 1439–1448. https://doi.org/10.1007/s00203-020-01858-3.

Radünz, M., da Trindade, M.L.M., Camargo, T.M., Radünz, A.L., Borges, C.D., Gandra, E.A., Helbig, E., 2019. Antimicrobial and antioxidant activity of unencapsulated and encapsulated clove (*Syzygium aromaticum*, L.) essential oil. Food Chem. 276 (September 2018), 180–186. https://doi.org/10.1016/j.foodchem.2018.09.173.

Rajaei, A., Hadian, M., Mohsenifar, A., Rahmani-Cherati, T., Tabatabaei, M., 2017. A coating based on clove essential oils encapsulated by chitosan-myristic acid nanogel efficiently enhanced the shelf-life of beef cutlets. Food Packag. Shelf Life 14 (March), 137–145. https://doi.org/10.1016/j.fpsl.2017.10.005.

Ruiz-Gonzalez, N., Lopez-Malo, A., Palou, E., Ramirez-Corona, N., Jimenez-Munguia, M.T., 2019. Antimicrobial activity and physicochemical characterization of oregano, thyme and clove leave essential oils, nonencapsulated and nanoencapsulated, using emulsification. Appl. Food Biotechnol. 6 (4), 237–246. https://doi.org/10.22037/afb.v6i4.25541.

Sagiri, S.S., Anis, A., Pal, K., 2016. Review on encapsulation of vegetable oils: strategies, preparation methods, and applications. Polym.-Plast. Technol. Eng. 55 (3), 291–311. https://doi.org/10.1080/03602559.2015.1050521.

Sangsuwan, J., Sutthasupa, S., 2019. Effect of chitosan and alginate beads incorporated with lavender, clove essential oils, and vanillin against *Botrytis cinerea* and their application in fresh table grapes packaging system. Packag. Technol. Sci. 32 (12), 595–605. https://doi.org/10.1002/pts.2476.

Sebaaly, C., Jraij, A., Fessi, H., Charcosset, C., Greige-Gerges, H., 2015. Preparation and characterization of clove essential oil-loaded liposomes. Food Chem. 178, 52–62. https://doi.org/10.1016/j.foodchem.2015.01.067.

Sebaaly, C., Charcosset, C., Stainmesse, S., Fessi, H., Greige-Gerges, H., 2016. Clove essential oil-in-cyclodextrin-in-liposomes in the aqueous and lyophilized states: from laboratory to large scale using a membrane contactor. Carbohydr. Polym. 138, 75–85. https://doi.org/10.1016/j.carbpol.2015.11.053.

Shahavi, M.H., Hosseini, M., Jahanshahi, M., Darzi, G.N., 2015. Optimization of encapsulated clove oil particle size with biodegradable shell using design expert methodology. Pak. J. Biotechnol. 12 (2), 149–160.

Smith-Palmer, A., Stewart, J., Fyfe, L., 1998. Antimicrobial properties of plant essential oils and essences against five important food-borne pathogens. Lett. Appl. Microbiol. 26 (2), 118–122. https://doi.org/10.1046/j.1472-765X.1998.00303.x.

Suganya, V., Anuradha, V., 2017. Microencapsulation and nanoencapsulation: a review. Int. J. Pharm. Clin. Res. 9 (3), 233–239. https://doi.org/10.25258/ijpcr.v9i3.8324.

Talón, E., Lampi, A.M., Vargas, M., Chiralt, A., Jouppila, K., González-Martínez, C., 2019. Encapsulation of eugenol by spray-drying using whey protein isolate or lecithin: release kinetics, antioxidant and antimicrobial properties. Food Chem. 295 (May), 588–598. https://doi.org/10.1016/j.foodchem.2019.05.115.

Wang, Y., Bian, W., Ren, X., Song, X., He, S., 2018a. Microencapsulation of clove essential oil improves its antifungal activity against *Penicillium digitatum* in vitro and green mould on Navel oranges. J. Hortic Sci. Biotech. 93 (2), 159–166. https://doi.org/10.1080/14620316.2017.1345332.

Wang, Y.F., Jia, J.X., Tian, Y.Q., Shu, X., Ren, X.J., Guan, Y., Yan, Z.Y., 2018b. Antifungal effects of clove oil microcapsule on meat products. LWT- Food Sci. Technol. 89 (August 2017), 604–609. https://doi.org/10.1016/j.lwt.2017.11.042.

Yang, M., Liang, Z., Wang, L., Qi, M., Luo, Z., Li, L., 2020. Microencapsulation delivery system in food industry—challenge and the way forward. Adv. Polym. Technol. 2020, 1–14. https://doi.org/10.1155/2020/7531810.

Zanetti, M., Carniel, T.K., Dalcanton, F., dos Anjos, R.S., Gracher Riella, H., de Araújo, P.H.H., de Oliveira, D., Antônio Fiori, M., 2018. Use of encapsulated natural compounds as antimicrobial additives in food packaging: a brief review. Trends Food Sci. Technol. 81, 51–60. https://doi.org/10.1016/j.tifs.2018.09.003.

Zhang, S., Zhang, M., Fang, Z., Liu, Y., 2017. Preparation and characterization of blended cloves/cinnamon essential oil nanoemulsions. LWT Food Sci. Technol. 75, 316–322. https://doi.org/10.1016/j.lwt.2016.08.046.

Zhu, Y., Li, C., Cui, H., Lin, L., 2021. Encapsulation strategies to enhance the antibacterial properties of essential oils in food system. Food Control 123 (October 2020), 107856. https://doi.org/10.1016/j.foodcont.2020.107856.

CHAPTER

Supercritical carbon dioxide extraction of clove essential oil: Optimization and characterization 31

Moncef Chouaibi

Higher School of Food Industries of Tunisia, Food Engineering Department, University of Carthage, Tunis, Tunisia

1. Introduction

The essential oil, also called aromatic oil or volatile oil, is obtained from the flowers, stems, leaves, roots, and fruits of plants through several extraction techniques. The essential oil has a complex chemical composition and mainly includes hydrocarbons (i.e., terpenes and sesquiterpenes) and oxygenated compounds hydrocarbons (i.e., phenols, epoxy compounds, ethers, esters, acids, alcohols, and ketones) (Yousefi et al., 2019). Due to these chemical groups, the essential oil has unique physicochemical characteristics, including durable fragrance, antimicrobial and antioxidant activities. Hence, essential oil has been widely employed in the cosmetic, perfume, flavor, food, and pharmaceuticals industries (Yousefi et al., 2019; Dima and Dima, 2015).

Essential oils are concentrated liquids of complex mixtures of volatile compounds and can be extracted from several plant organs. Essential oils are a good source of several bioactive compounds, which possess antioxidative and antimicrobial properties (Khaw et al., 2017). Some essential oils have been used as medicine. Furthermore, the uses of essential oils have received increasing attention as the natural additives for the shelf-life extension of food products due to the risk of using synthetic preservatives. Essential oils could be incorporated into packaging, in which they can provide multifunctions termed "active or smart packaging." Those essential oils can modify the matrix of packaging materials, thereby rendering improved properties.

Clove contains many effective components and has very intense aromas, so it has been used in *Syzygium aromaticum* traditional Chinese food as a condiment for more than 2000 years. Furthermore, in traditional Chinese medicine theory, clove improves digestion and inhibits microorganism growth, which modern pharmacological studies have proved. Therefore, the application of clove and its most important medicinal composition, eugenol ($C_{10}H_{12}O_2$, CAS No.: 97-53-0), in the food industry as an antiseptic has attracted many researchers (Hu et al., 2018).

Different extraction methods have a certain effect on the quality of essential oils. For example, patchouli oil has frequently been extracted by hydrodistillation (HD) (Tongnuanchan and Benjakul, 2014) or solvent extraction. However, the method of HD requires high temperature, which has many disadvantages such as high cost, thermal degradation of heat-sensitive compounds, long extraction

Clove (*Syzygium aromaticum*). https://doi.org/10.1016/B978-0-323-85177-0.00040-9

time, and low yield. Moreover, the solvent extraction method can cause environmental pollution and damage these sensitive compounds (Tongnuanchan and Benjakul, 2014).

Supercritical carbon dioxide (SC-CO_2) extraction is a new technology that may be a more efficient alternative to conventional solvent extraction, hydrodistillation, or mechanical pressing for essential oils and other bioactive components. It offers several advantages; it does not employ explosive or toxic solvents, leaves no toxic residue, and achieves high purity at a relatively low cost with better retention of aromatic compounds (Herrero et al., 2006; Pourmortazavi and Hajimirsadeghi, 2007; Wang et al., 2008).

Current methods for the extraction of essential oil are hydrodistillation, steam distillation, and solvent extraction. However, hydro- and steam distillation has several disadvantages, such as incomplete extraction of essential oils from plant materials, high operating temperatures with the consequent breakdown of thermally labile components, promotion of hydration reactions of chemical constituents, and require a post-extraction process to remove water. On the other hand, solvent extraction overcomes the drawbacks of distillation but has the major disadvantage of solvent residue in the extracts.

Indeed, SC-CO_2 extraction has been used to isolate oils and other bioactive compounds from plants and algae, including essential oils (Wei et al., 2009; Yamini et al., 2008), edible *Hibiscus cannabinus* seed oil, and other seed oils (Chan and Ismail, 2009; Liu et al., 2009), oleoresins (Fernandez-Ronco et al., 2010), therapeutic oils (Maxia et al., 2009), anthocyanins (Ku and Mun, 2008), and astaxanthin (Thana et al., 2008). A few studies have explored the possible use of SC-CO_2 extraction for isolating essential oils from *Cyperus rotundus* (Tam et al., 2007), but the conditions have not yet been optimized. In general, several factors, including temperature, pressure, extraction time, CO_2 flow rate, and concentration of solvents, can affect the efficiency of SC-CO_2 extraction.

Extraction of essential oil constituents at high pressure using SC-CO_2 has greatly increased interest, especially in food, cosmetic, and pharmaceutical industries, as an alternative to conventional methods (Assis and Lanças, 1999). Thermal degradation and decomposition of thermolabile components are avoided in SC-CO_2 extraction due to the operation at reduced temperature. In addition, SC-CO_2 extraction allows a constant modification in dissolution power and selectivity because of changes in the solvent density (Gamiz-Gracia and Luque de Castro, 2000).

Several important factors determine the efficiency of SCE, including the pressure, temperature, and extraction time. To study the effect of multiple variables on output, the response surface method (RSM) is an effective technique (Chouaibi et al., 2020a,b). The RSM is a collection of mathematical and statistical techniques useful for the modeling and analyzing problems in which several variables influence the response of interest, and the objective is to optimize this response (Chouaibi et al., 2020a,b). The RSM has been successfully employed to model and optimize supercritical CO_2 extraction of bioethanol production (Chouaibi et al., 2020a), protein isolate (Mechmeche et al., 2017; Chouaibi et al., 2018), and seed oils (Chouaibi et al., 2020b).

2. Supercritical SC-CO_2 apparatus

A 50mL stainless steel extraction column loaded with approximately 1kg of clove milled roots was connected to the system shown schematically in Fig. 1. The CO_2 pump (ISCO Model 260D Syringe pump) was cooled to 4°C, and the pressurized CO_2 was delivered to the extraction vessel through a heating coil. The extraction column and heating coil were immersed in a water tank, the temperature of which was controlled by a circulating heater (Thermoline). The outlet of the extraction column was

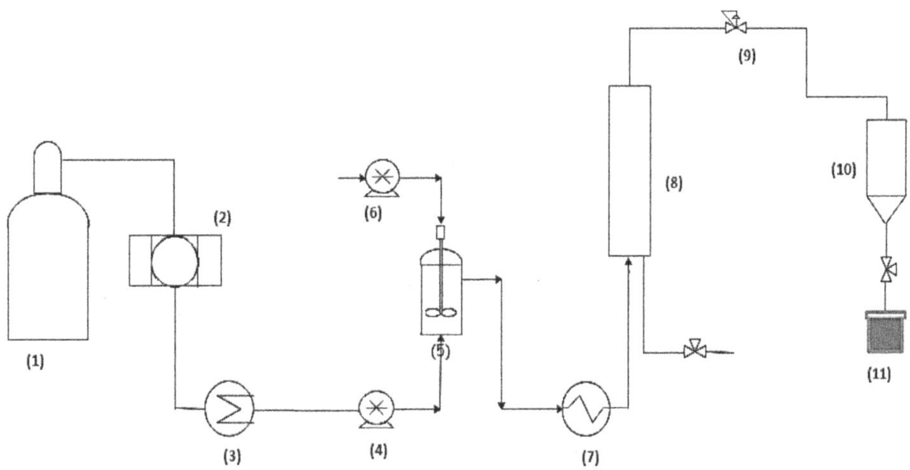

FIG. 1

Supercritical CO_2 apparatus used for essential oil extraction from Tunisian clove: (1) liquid CO_2 cylinder; (2) microfilter; (3) refrigerant; (4) pump; (5) mixer; (6) pump; (7) preheater; (8) extractor; (9) pressure regulator; (10) cyclone separator; and (11) beaker.

connected to an on-off valve that was placed upstream of a micro-metering valve. The extraction experiments were commenced when the system reached a predetermined pressure and temperature. There were two stages of extraction: static and dynamic. The static stage was 30 min for all experiments, and the dynamic stage varied from 33 to 117 min. The flow rate of CO_2 was kept at 4 mL/min measured at operating pressure and 4°C for all experiments. Supercritical CO_2 was expanded across the micro-metering valve, and the essential oil was collected in a glass tube. The glass tube contained 3 g of glass beads and was refrigerated between −20°C and −5°C in a cooling bath. The addition of glass beads aimed to increase the contact of CO_2 and essential oil with a cold surface to facilitate the separation of the oil and reduce the drag effect of CO_2. The extract was collected during the dynamic stage. After the experiment, the extract condensed along the tubing and valves was collected by washing with hexane. The hexane wash was then mixed with the extract collected in the glass tube. Finally, the extraction yield was calculated. The final volume of the extracts was diluted to 5 mL with ethyl ether and subjected to GC-MS analysis. The samples were then weighed and the essential oil yield was assessed according to the following expression (1):

$$\text{Yield}\,(\%) = \frac{\text{Weight of essential oil (g)}}{\text{Weight of clove}} \times 100 \tag{1}$$

3. Optimization of essential oil yield

Response optimization was conducted to predict the optimum levels of independent variables leading to the desired response objectives (Chouaibi et al., 2012, 2020a,b). In order to find the optimal extraction, numerical and graphical optimization procedures were applied. Chouaibi et al. (2012) indicated that a 3D response surface and contour plots were highly suggested for the graphical optimization.

Hence, the optimum conditions were found via superimposing all figures. In order to check the exact optimum points of factors resulting in the optimized conditions, a numerical procedure was used as well. The numerical optimization results demonstrated that the overall optimum area was predicted to be obtained by the SC-CO_2 extraction at the combined level of 195.6 bar and 43.8°C for pressure and temperature, respectively, with the desirability of 0.999% by response surface plots and response surface methodology. The corresponding predicted response value for yield was 1.15%, based on the final reduced model.

The analysis of variance (ANOVA) results are used to assess the significance of the predicted model terms. Based on these criteria, the variance analysis (ANOVA) revealed that the selected model is significant at $P < .0001$, thus indicating that it is valid. The effects of two operating conditions of clove essential extraction using SC-CO_2, namely extraction pressure and temperature on the extraction of yield essential oil, were investigated using the RSM procedure.

Analyses by multiple regression were implemented using experimental oil yields to identify the constants and coefficients of variables and investigate the linear effect of variables and their quadratic and interaction effects based on the Student's "t" test and P-values. In general, the larger the magnitude of t and the smaller the P-value, the more significant is the corresponding coefficient term as discussed by Chouaibi et al. (2012). Values of $P < .05$ indicate significant model terms. The estimated coefficients of linear, quadratic, and interaction factors, as well as the constants and coefficients are given by Eq. (2) as follows:

$$Y = 0.82 + 0.23X_1 + 0.010X_2 - 0.12X_1X_2 - 0.08X_1^2 - 0.10X_2^2 \tag{2}$$

In addition, the quadratic models of the responses (Y) predicted by ignoring the nonsignificant terms and expressed in terms of coded variables are given by the following expressions (Eq. 3):

$$Y = 0.82 + 0.23X_1 + 0.010X_2 - 0.12X_1X_2 - 0.10X_2^2 \tag{3}$$

where Y denotes the dependent variables (essential oil yield), and X_1 and X_2 signify the independent variables (pressure and temperature, respectively).

Furthermore, all the linear terms of the independent variables are significant at $P < .05$. The statistical data established that the extraction pressure is the major factor in the essential oil yield (%) of Tunisian cloves. However, the quadratic term of the extraction temperature revealed a negative and significant effect on the essential oil yield.

Among the two independent variables tested, only pressure ($P < .0010$) possesses a highly significant linear effect on the essential oil yield, whereas the effect of extraction temperature ($P = .03$) on essential oil yield was minimal. Therefore, extraction pressure also revealed a negative quadratic effect on essential yield ($P = .001$). However, extraction pressure did not have a significant quadratic effect; both temperature quadratic effect ($P = .021$) and interaction effect with pressure ($P = .01$) were important. The response equation fitted the experimental data with a regression coefficient ($R^2 = 0.98$), indicating that 98% of the variability in extraction yield of essential oil can be explained by the model presented in Eq. (2). Furthermore, the goodness of fit of the model to the experimental data has a correlation coefficient of 97.3%.

The increase in oil yield with pressure became larger as temperature increased. For example, as pressure increased from 75 to 225 bar, the oil yield increased from 0.64% to 1.14% at 40°C and from 0.35% to 1.05% at 50°C. Such increments in yields are due to the interaction between pressure and temperature.

3.1 Effect of extraction pressure

At low pressures, increments in temperature resulted in decreased essential oil yield; however, the oil yield increased with extraction temperature at higher pressures. For example, yield decreased from 0.64% to 0.35% when temperature increased from 40°C to 50°C at 100 bar, however at 220 bar, the same temperature increment corresponded to a change from 0.89% to 1.20%. A comparable trend was found in a study for the optimization of *Citrullus colocynthis* oil yield using the RSM and artificial neural network (ANN) techniques as described by Chouaibi et al. (2020b). This may be elucidated by the increase of the SC-CO_2 density emanating from the rise in the solubility of compounds. The distance between the molecules of the solvent and the lipophilic solutes decreases with the increase of the density of SC-CO_2. Thus, the interaction between oil and SC-CO_2 rises to enhance the oil solubility in SC-CO_2, hence increasing the mass transfer rate (Chouaibi et al., 2020b).

3.2 Effect of extraction temperature on the essential oil yield

Extraction temperature was also an essential factor. The effect of temperature on the overall extracted clove essential oil yield was conducted at pressures of 75 and 225 bar. The results are shown in Fig. 2. The overall results indicated that the yield increases as temperatures decrease at higher value pressures.

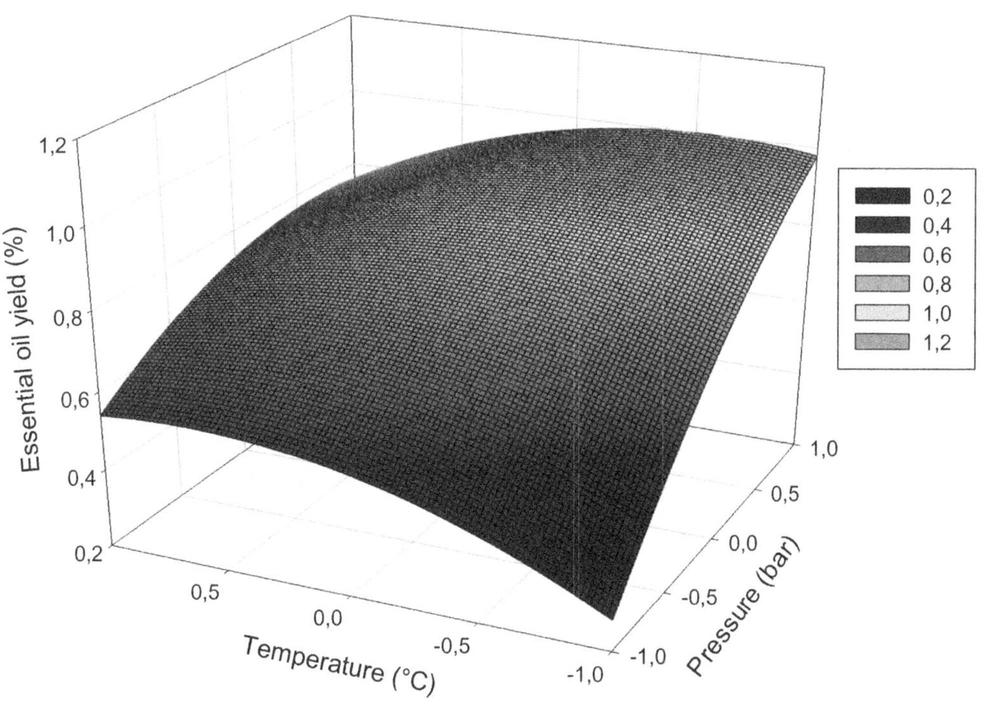

FIG. 2

Response surface (3D) and contour plots for essential oil yield (Y) of Tunisian clove as a function of independent variables (pressure and extraction temperature).

For example, at lower pressure of 75 bar, the difference in overall oil yield is 0.25% when temperature decreases from 60°C to 40°C. At a pressure of 225 bar, the yield at 40°C and 60°C was almost the same at 0.20% and 1.15%, respectively, while a drastic drop to 0.80% is observed when the temperature increases to 60°C.

At the higher pressure of 200 bar, the highest yield of clove seed oil, 1.15%, is achieved at 40°C. On the other hand, only 0.77% of oil yield was obtained at 60°C with a pressure of 225 bar, but the yield is higher than the extraction at a pressure of 100 bar. The higher temperature might lead to a saturated vapor pressure condition, oil volatility, and mass transfer elevation, which allows the solute to dissolve in SC-CO_2 easily. However, the increasing temperature at constant pressure can cause the SC-CO_2 density to decrease (Zhao et al., 2014; Rivero et al., 2015). Hence, the solubility of solute could have a positive or negative effect depending on whether the solute vapor pressure or the solvent density is predominant. The trend of the observations in this study can be explained by the latter, where the reduction of SC-CO_2 density overcomes the vapor pressure and further decreases the interaction of CO_2 with seed matrix and hence reduces the dissolution of solute in the solvent. A similar tendency of extraction was observed in the study by Zermane et al. (2012), in which the recovery of Algerian Rosemary oil using SC-CO_2 was observed to decrease significantly with increasing temperature at pressures of 10–22 MPa. Therefore, the yield is higher at the very early stage at the temperature of 40°C. However, the yield increases rapidly as the extraction time progresses. This phenomenon is called "crossover effect" where higher temperature produces low yield while lower temperatures produce high yield (Khaw et al., 2017). This effect has been observed in the extraction of ginger oil, where the increase of solute vapor pressure and reduction of solvent density were competing to become the main element in the extraction process (Roy et al., 1996). On the other hand, increasing the extraction temperature increases the oil vapor pressure, improving the oil extraction yield from the seeds of *Carthamus tinctorius*, as explained by Bala et al. (2016).

The contour plots represent elliptical shapes and hyperbolic branches that confirm that the interactions (pressure and extraction temperature) significantly affect the essential oil extraction yield from Tunisian cloves (Fig. 3). These findings agree well with those noticed in the variance analysis table (ANOVA).

3.3 Optimal conditions of essential oil extraction

To validate the predictability of the established model, the optimized parameters were tested in an experiment under the following conditions: extraction pressure of 195.6 bar and extraction temperature of 43.8°C (determined by Matlab version 19). There was no significant difference ($P > .05$) between experimental and predicted values (1.15% and 1.14%, respectively), suggesting that the response surface equation had a high capability to assess the parameters for clove essential oil.

4. Chemical characteristics

To extract natural volatile compounds from Tunisian cloves without thermal degradation and residue of organic solvents, supercritical fluid extraction (SFE) using carbon dioxide was employed to prepare essential oil of cloves in the current work. According to the optimal conditions, two important parameters, pressure and temperature, were selected as independent variables in the SC-CO_2. The

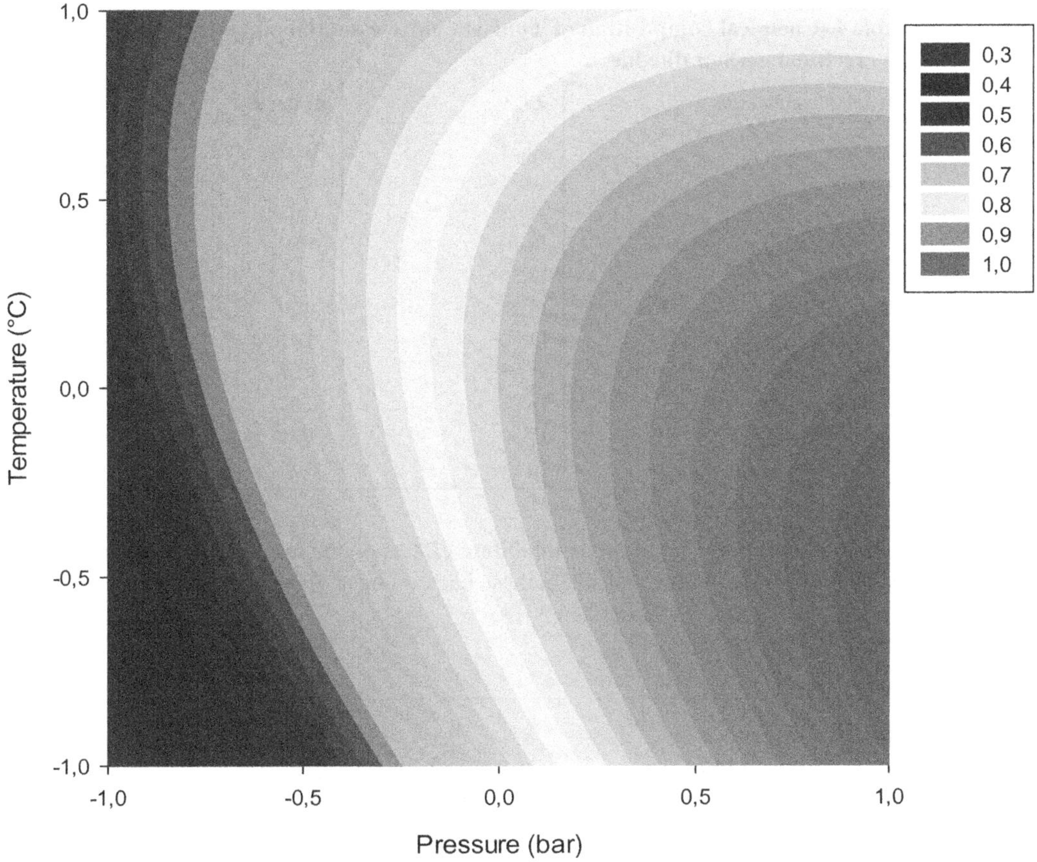

FIG. 3

Contour plots for essential oil yield (Y) of Tunisian cloves as a function of independent variables (pressure and extraction temperature).

composition of the essential oils varies depending on the geographical position, the plant's origin and species, harvest time, extraction methods, etc. Additionally, the composition of the essential oils and consequently their biological and therapeutic activities depend on the combination and ratio of their numerous different compounds. Therefore, GC-MS was employed to assess the essential oils' quality and quantity of chemical compounds. The clove essential oil possesses the simplest composition, such as eugenol, eugenol acetate, and β/α-caryophyllene, accompanied by several other sesquiterpenes listed in Table 1. Besides, the compounds identified by GC-MS showed 99.7% of the essential oil of clove extracted by SC-CO_2. These results are in great agreement with those found by Cai et al. (2015) and Wei et al. (2016).

Numerous reports show that clove oil and its main active component, eugenol have beneficial effects on common food source Gram-negative bacteria such as *Escherichia coli*, *Salmonella*, *Pseudomonas aeruginosa*, and so on, and Gram-positive bacteria such as *Staphylococcus*, *Streptococcus*,

Table 1 Chemical composition of Tunisian clove essential oil extracted by supercritical carbon dioxide.

Chemical compound	RT	Area (%)
p-Cymene	6.22	0.06
Chavicol	10.63	0.05
Thymol	11.43	0.03
Eugenol	18.50	85.80
α-Cubebene	13.24	0.01
β-Caryophyllene	14.54	4.50
α-Caryophyllene	14.80	0.91
Aromadendrene	14.93	0.01
α-Famesene	15.22	0.02
Eugenol acetate	15.80	7.91
Caryophyllene oxide	16.89	0.40

Listeria, and so on. Furthermore, based on its inhibiting effects on the migration, adhesion, virulence factors expression, biofilm formation, and so on, these bacteria have good application prospects in the food antisepsis field, as reported by Hu et al. (2018).

5. Conclusion

In contrast to the numerous researches concerning clove essential oils, the study of clove essential oil extracted by SC-CO_2 is relatively scarce. SC-CO_2 is used to extract the clove essential oil without thermal degradation and residue of organic reagents, which resulted in the extraction of benefic natural volatile compounds with high biological activities. The current study demonstrated that clove essential oil extracted by SC-CO_2 is rich in bioactive compounds like eugenol with high content of 85.7%. Furthermore, the essential oil of clove using SFE contained many volatile compounds and showed considerable biological activity. The information will contribute to developing clove essentials as raw materials in the healthcare and food flavor industries.

References

Assis, L.M., Lanças, F.M., 1999. High-resolution gas chromatography and high-resolution gas chromatography/mass spectrometry study of the volatile fraction obtained from high-inertinite Brazilian coal by supercritical fluid extraction. J. Microcolumn Sep. 11, 501–512.

Bala, M., Madhu, B., Tyagi, S.K., Gupta, R.K., 2016. Optimization of supercritical CO_2 extraction of safflower seed oil using response surface methodology. Asian J. Chem. 28, 1579–1583.

Cai, J., Lu, J., Li, Q., Guo, S., Dai, Y., 2015. Analysis on volatile components of *Caryophylli flos* from different habitats. Plant Sci. J. 33, 251–258. https://doi.org/10.11913/PSJ.2095-0837.2015.20251.

Chan, K.W., Ismail, M., 2009. Supercritical carbon dioxide fluid extraction of *Hibiscus cannabinus* L. seed oil: a potential solvent-free and high antioxidative edible oil. Food Chem. 114, 970–975.

Chouaibi, M., Rezig, L., Ben Daoued, K., Mahfoudhi, N., Bouhafa, H., Hamdi, S., 2012. Extraction of polysaccharide from *Zizyphus lotus* fruits. Int. J. Food Eng. 8, 1556–3758.

Chouaibi, M., Boussaid, A., Donsì, F., Ferrari, G., Hamdi, S., 2018. Optimization of the extraction process by response surface methodology of protein isolate from defatted jujube (*Zizyphus lotus* L.) seeds. Int. J. Pept. Res. Ther. 25 (4), 1509–1521.

Chouaibi, M., Ben Daoued, K., Riganne, K., Rouissi, T., Ferrari, G., 2020a. Production of bioethanol from pumpkin peel wastes: comparison between response surface methodology (RSM) and artificial neural networks (ANN). Ind. Crop. Prod. 155, 112822.

Chouaibi, M., Rigane, K., Ferrari, G., 2020b. Extraction of *Citrullus colocynthis* L. seed oil by supercritical carbon dioxide process using response surface methodology (RSM) and artificial neural network (ANN) approaches. Ind. Crop. Prod. 158, 113002.

Dima, S., Dima, C., 2015. Essential oils in foods: extraction, stabilization, and toxicity. Curr. Opin. Food Sci. 5, 29–35.

Fernandez-Ronco, M.P., Ortega-Noblejas, C., Gracia, I., De Lucas, A., Garcia, M.T., Rodriguez, J.F., 2010. Supercritical fluid fractionation of liquid oleoresin capsicum: statistical analysis and solubility parameters. J. Supercrit. Fluids 54, 22–29.

Gamiz-Gracia, L., Luque de Castro, M., 2000. Continuous subcritical water extraction of medicinal plant essential oil: comparison with conventional techniques. Talanta 51, 1179–1185.

Herrero, M., Cifuentes, A., Ibanez, E., 2006. Sub- and supercritical fluid extraction of functional ingredients from different natural sources: plants, food-byproducts, algae and microalgae: a review. Food Chem. 98, 136–148.

Hu, Q., Zhou, M., Wei, S., 2018. Progress on the antimicrobial activity research of clove oil and eugenol in the food antisepsis field. J. Food Sci. 83 (6), 1476–1483.

Khaw, K.Y., Parat, M.O., Shaw, P.N., Falconer, J.R., 2017. Solvent supercritical fluid technologies to extract bioactive compounds from natural sources: a review. Molecules 22, 1–22.

Ku, C.S., Mun, S.P., 2008. Optimization of the extraction of anthocyanin from Bokbunja (*Rubus coreanus Miq.*) marc produced during traditional wine processing and characterization of the extracts. Bioresour. Technol. 99, 8325–8330.

Liu, S., Yang, F., Zhang, C., Ji, H., Hong, P., Deng, C., 2009. Optimization of process parameters for supercritical carbon dioxide extraction of *Passiflora* seed oil by response surface methodology. J. Supercrit. Fluids 48, 9–14.

Maxia, A., Marongiu, B., Piras, A., Porcedda, S., Tuveri, E., Goncalves, M.J., Cavaleiro, C., Salgueiro, L., 2009. Chemical characterization and biological activity of essential oils from *Daucus carota* L. subsp. carota growing wild on the Mediterranean coast and on the Atlantic coast. Fitoterapia 80, 57–61. https://doi.org/10.1016/j.fitote.2008.09.008.

Mechmeche, M., Kachouri, F., Chouaibi, M., Ksontini, H., Setti, S., Hamdi, M., 2017. Optimization of extraction parameters of protein isolate from tomato seed using response surface methodology. Food Anal. Methods 10, 809–819.

Pourmortazavi, S.M., Hajimirsadeghi, S.S., 2007. Supercritical fluid extraction in plant essential and volatile oil analysis. J. Chromatogr. A 1163, 2–24.

Rivero, E., Costa, F.M., Watanabe, E.O., Hori, C.E., 2015. Study of soybean oil extraction with supercritical CO_2: effect of temperature on the yield and fatty acids composition. Cong. Bras. Eng. 19, 1–6.

Roy, B.C., Goto, M., Hirose, T., 1996. Extraction of ginger oil with supercritical carbon dioxide: experiments and modeling. Ind. Eng. Chem. Res. 35, 607–612.

Tam, C.U., Yang, F.Q., Zhang, Q.W., Guan, J., Li, S.P., 2007. Optimization and comparison of three methods for extraction of volatile compounds from *Cyperus rotundus* evaluated by gas chromatography-mass spectrometry. J. Pharm. Biomed. Anal. 44, 444–449.

Thana, P., Machmudah, S., Goto, M., Sasaki, M., Pavasant, P., Shotipruk, A., 2008. Response surface methodology to supercritical carbon dioxide extraction of astaxanthin from *Haematococcus pluvialis*. Bioresour. Technol. 99, 3110–3115.

Tongnuanchan, P., Benjakul, S., 2014. Essential oils: extraction, bioactivities, and their uses for food preservation. J. Food Sci. 79, 1231–1249.

Wang, L., Weller, C.L., Schlegel, V.L., Carr, T.P., Cuppett, S.L., 2008. Supercritical CO_2 extraction of lipids from grain sorghum dried distillers grains with solubles. Bioresour. Technol. 99, 1373–1382.

Wei, Z.J., Liao, A.M., Zhang, H.X., Liu, J., Jiang, S.T., 2009. Optimization of supercritical carbon dioxide extraction of silkworm pupal oil applying the response surface methodology. Bioresour. Technol. 100, 4214–4219.

Wei, M.C., Xiao, J., Yang, Y.C., 2016. Extraction of α-humulene-enriched oil from clove using ultrasound-assisted supercritical carbon dioxide extraction and studies of its fictitious solubility. Food Chem. 210, 172–181. https://doi.org/10.1016/j.foodchem.2016.04.076.

Yamini, Y., Khajeh, M., Ghasemi, E., Mirza, M., Javidnia, K., 2008. Comparison of essential oil compositions of *Salvia mirzayanii* obtained by supercritical carbon dioxide extraction and hydrodistillation methods. Food Chem. 108, 341–346.

Yousefi, M., Rahimi-Nasrabadi, M., Pourmortazavi, S.M., Wysokowski, M., Jesionowski, T., Ehrlich, H., Mirsadeghi, S., 2019. Supercritical fluid extraction of essential oils. TrAC Trends Anal. Chem. 118, 182–193.

Zermane, A., Meniai, A.H., Larkeche, O., Barth, D., 2012. Extraction and modeling of Algerian rosemary essential oil using supercritical CO_2: effect of pressure and temperature. Energy Procedia 18, 1038–1046.

Zhao, X.H., Zeng, J., Gao, H., Wang, Y., 2014. Optimization and composition of volatile oil from *Polygonatum odoratum* (Mill Druce) using supercritical fluid extraction. Trop. J. Pharm. Res. 13, 779–786.

Further reading

Belwal, T., Dhyani, P., Bhatt, I.D., Rawal, R.S., Pande, V., 2016. Optimization extraction conditions for improving phenolic content and antioxidant activity in *Berberis asiatica* fruits using response surface methodology (RSM). Food Chem. 207, 115–124.

Ennaifer, M., Bouzaine, T., Chouaibi, M., Hamdi, M., 2018. *Pelargonium graveolens* aqueous decoction: a new water-soluble polysaccharide and antioxidant rich-extract. Biomed. Res. Int. https://doi.org/10.1155/2018/2691513.

CHAPTER

Clove essential oil nanoemulsion: Optimization using artificial neural network 32

Moncef Chouaibi

Higher School of Food Industries of Tunisia, Food Engineering Department, University of Carthage, Tunis, Tunisia

1. Introduction

Natural antimicrobials have attracted considerable attention in recent years due to the increased consumer awareness of food quality and safety (Donsi et al., 2011). As a result, new approaches have been adopted in food preservation, including adding natural compounds with inhibitory or bactericidal activities, like essential oils (EO), as safe preservatives (Selim, 2011; Donsi et al., 2011). Essential oils are compounds derived from plant material that can be used as natural food preservatives (Holley and Patel, 2005; Nielsen et al., 2017). In addition, the growing consumer demand to use antimicrobial agents from natural sources justifies the incorporation of EO into food (Burt, 2004). For example, the EO of black pepper (*Piper nigrum* L.) and cinnamon (*Cinnamomum zeylanicum*) have been shown to have an antimicrobial effect against pathogenic and spoilage bacteria (Hamidpour et al., 2015).

Oil-in-water nanoemulsions, which consist of small lipid droplets ($d < 200$ nm) dispersed in water, are particularly suitable for encapsulating hydrophobic bioactive agents (Anton and Vandamme, 2011; Mason et al., 2006; McClements, 2012; Tadros et al., 2004). Indeed, they are gaining increasing popularity because of their ease of preparation and handling, high physical stability, and good oral bioavailability (Walker et al., 2015). As a result, the food, supplement, and pharmaceutical industries are actively developing nanoemulsion-based delivery systems to encapsulate, protect, and control the release of various types of hydrophobic bioactive agents (Acosta, 2009; Gleeson et al., 2016; McClements et al., 2007; Waraho et al., 2011).

Antimicrobial nanoemulsions could be fabricated using both low-energy and high-energy approaches. Recently, many researchers have focused on the encapsulation of carvacrol within nanoemulsions produced by low-energy methods because of the relatively low cost and simplicity of their preparation and the fact that no specialized homogenization equipment is required (Chang et al., 2013; Landry et al., 2014, 2015). Chang et al. (2013) reported that carvacrol could be encapsulated within oil-in-water nanoemulsions produced using a spontaneous emulsification method that simply involved titrating an essential oil/ripening inhibitor/surfactant mixture in the water. The ripening inhibitor had to be included in the formulation to prevent rapid droplet growth due to Ostwald ripening, i.e., diffusion of oil molecules from small droplets to large droplets.

Clove (*Syzygium aromaticum*). https://doi.org/10.1016/B978-0-323-85177-0.00002-1

Nevertheless, their incorporation in food products presents several limitations due to their low solubility and intense aroma at high concentrations (Sanchez-Gonzalez et al., 2011). The emulsification of EO is currently used for their dispersion into food products, but their functionality and long-term stability largely depend on the oil droplet size and distribution (Tadros et al., 2004). Nanoemulsions could be used as carriers of lipophilic bioactive compounds for their incorporation into food products. Nanoemulsions consist of at least one immiscible liquid dispersed in another with a surfactant (nonionic or polymeric) in the form of small droplets, with an average droplet size between 20 and 200 nm (Burguera and Burguera, 2012; Solans et al., 2005; Wulff-Perez et al., 2009). Nanoemulsions has several advantages than conventional emulsions (Qian and McClements, 2011; Tadros et al., 2004). First, they are optically transparent, so they might be good candidates to be incorporated in clear drinks or beverages (Qian and McClements, 2011). Second, they are kinetically stable colloidal systems (Solans et al., 2005). Third, they present a high-active surface area, with higher functionality (Qian and McClements, 2011). Although there are several methods to form nanoemulsions, yet high-energy methods are more commonly used. They require specialized mechanical devices such as high-pressure homogenizers and ultrasounds capable of generating intense mechanical disruptive forces inducing the breakup of oil droplets (Mason et al., 2006; Tadros et al., 2004).

However, nanoemulsions are susceptible to destabilization phenomena, so an optimal formulation is crucial for their long-term stability (Henry et al., 2009; Mirhosseini et al., 2008). Stability may be defined as resistance to physical changes. Moreover, emulsions are destabilized via two mechanisms: coalescence and Ostwald ripening (Henry et al., 2009; Nazarzadeh et al., 2013; Solans et al., 2005). When two oil droplets contact due to the weak stearic repulsion, coalescence occurs, and they unify in a sole larger droplet. Ostwald ripening occurs due to the diffusion of oil molecules from small to large droplets through the continuous phase in relatively water-soluble oils (such as EO), increasing the oil droplet size (Nazarzadeh et al., 2013; Rao and McClements, 2012). On the other hand, food hydrocolloids such as polysaccharides and proteins have stabilized emulsions in several studies (Ye and Singh, 2006). For example, pectin is a naturally sourced polysaccharide commonly used in the food and pharmaceutical industries as a gelling and thickening agent (Liu et al., 2007).

2. Optimization techniques

The large-scale application of O/W nanoemulsion needs study of optimization for different variables (mainly composition, speed of agitation, and duration of formulation process) to achieve the desired properties of formulated nanoemulsions, including minimum droplet size or minimum polydispersity, low dynamic viscosity, stability, and the specific utility effect of variables on properties of nanoemulsions seems to be the area of future research. Factors such as composition and preparation technique have been described as optimizing techniques for controlling droplet size, polydispersity index, dynamic viscosity, and turbidity. To obtain the scale-up values of variables affecting the preparation of essential oil nanoemulsion, the application of numerical approaches has been explored. Numerical optimization is discussed for more independent variables affecting the stability of EO nanoemulsions. Box-Behnken design (BBD) by Expert Design Software has been used to assess the influence of three independent variables or factors of the ultrasonic process for clove oil nanoemulsions. In this study, the three independent variables, responses, and factor levels are listed in Table 1. The ultrasonication temperature, Tween 80 concentration, sonication time ranged from 5°C and 50°C, from 1% to 2%, and from 2 to 6 min, respectively.

Table 1 Independent variables and their factor levels of clove nanoemulsion formulation using response surface methodology (RSM).

		Factor levels		
Independent variable	**Symbol**	**−1**	**0**	**1**
Temperature (°C)	X_1	5	27.50	50
Tween 80 (%)	X_2	1	1.5	2
Sonication time (min)	X_3	2	4	6

There are many techniques and methods used for the optimization of chemical and food processes.

2.1 Response surface methodology

Response surface methodology (RSM) is an effective and powerful statistical tool for optimizing the extraction process while reducing the experimental number required (Domingos et al., 2008; Chouaibi et al., 2012a,b). Using quadratic models, RSM is a statistical and mathematical method using a second-degree polynomial model to study the relationships between one or more dependent variables and several independent variables. The response surface methodology involves three steps: (1) central composite rotatable design (CCRD), (2) response surface modeling through regression analysis based on ANOVA table, and (3) process factor optimization using the RSM models. RSM has been successfully used to optimize SC-CO_2 extraction of oils, bioethanol production, protein isolate extraction, polysaccharide extraction, lactic acid bacteria production (Chouaibi et al., 2012a,b, 2018, 2020a,b; Mechmeche et al., 2017; Ennaifer et al., 2018). Table 2 presents the experimental design by RSM of clove oil encapsulation by emulsification process.

The univariate model is uncomplicated, but this method is time-consuming and lacks interactions between factors in the optimization. If interaction effect between factors exists, it is necessary to measure the interaction effects occurring between factors using a multivariate model. The optimum conditions are obtained by using complex experimental designs, including three-level full factorial design, Central Composite Designs (CCD), Box-Behnken design (BBD), or Doehlert matrix (DM).

The three independent variables were selected in this study based on the literature as detailed by Pongsumpum et al. (2020). The results revealed that the most significant independent variable for all studied responses was Tween 80, followed by sonication time and sonication temperature. All quadratic terms of independent variables were significant on diameter size, polydispersity index, and dynamic viscosity ($P < .05$). Also, there is significant interaction between factors (sonication temperature, Tween 80 concentration, and sonication time). For diameter size, the interaction between sonication time and the temperature was found. Also, the significant interaction of Tween 80 concentration and sonication temperature was observed in the polydispersity index (PDI). However, there is no interaction between variables for the viscosity of clove nanoemulsion.

The effect of sonication temperature and Tween 80 concentration on mean droplet size is depicted in Fig. 1A. It is clear that, at low sonication temperature, the increasing of Tween 80 concentration increased linearly the mean droplet size of clove EO nanoemulsion. However, the increase of sonication temperature increases the mean drople size of clove nanaoemulsion quadratically. These findings are in great agreement with these obtained by Pongsumpum et al. (2020). They concluded that

Table 2 Experimental design by RSM of clove oil encapsulation by emulsification process.

Run	Independent variable			Experimental response		
	X_1	X_2	X_3	Diameter (nm)	PDI	Dynamic viscosity (mPa s)
1	0	0	0	96 0.4	0.34	0.67
2	0	0	0	95.22	0.32	0.82
3	0	0	0	95.83	0.36	0.71
4	0	0	0	96.81	0.34	0.73
5	0	0	0	95.43	0.37	0.81
6	−1	0	−1	86.89	0.31	1.065
7	0	1	1	62.94	0.2	0.97
8	0	1	−1	86.76	0.34	1.06
9	0	−1	1	106.58	0.40	0.87
10	1	1	0	77.43	0.29	0.98
11	1	0	−1	107.69	0.4	0.825
12	−1	0	1	69.74	0.24	0.75
13	0	−1	−1	114.94	0.46	0.88
14	1	0	1	76.06	0.27	0.815
15	−1	1	0	49.19	0.18	0.885
16	−1	−1	0	108.12	0.43	0.875
17	1	−1	0	99.47	0.41	0.665

X_1, *sonication temperature;* X_2, *Tween 80 concentration;* X_3, *sonication time;* X_1, X_2, *and* X_3 *are coded variables;* PDI, *polydispersity index.*

surfactants consist of hydrophobic and hydrophilic structures that link the gap between oil and aqueous phases by decreasing interfacial tension at the oil-water interface, resulting in decreased free energy for the formation of nanoemulsion.

In addition, the effect of sonication time and temperature on mean droplet size was investigated in Fig. 1B. The obtained data revealed that the increase of temperature and sonication time increased quadratically, increasing mean droplet size. This was probably because a longer ultrasonic duration forces the emulsion droplets into smaller sizes. In addition, the droplet size augmented with increasing temperature but decreased when the temperature was larger than 30°C. This data agrees with the work of Alzorqi et al. (2016) and Pongsumpum et al. (2020), who indicated that higher ultrasonic time gave reduced droplet sizes of nanoemulsion.

The effect of sonication temperature and Tween 80 concentration on polydispersity index (PDI) is summarized in Fig. 1C. Form these results, the polydispersity index increased with increasing the Tween 80 concentration and sonication temperature. These results concordance with these observed by Pongsumpum et al. (2020) and Chouaibi et al. (2019a,b).

2.2 Artificial neural network

An Artificial neural network (ANN) is a computational model comprising many simple-processing elements called nodes or neurons organized in layers. Each neuron connects to other neurons using communication links. The neurons have an associated weight that represents the learned information through the neural network. These weights are used for solving problems. ANN is a computational

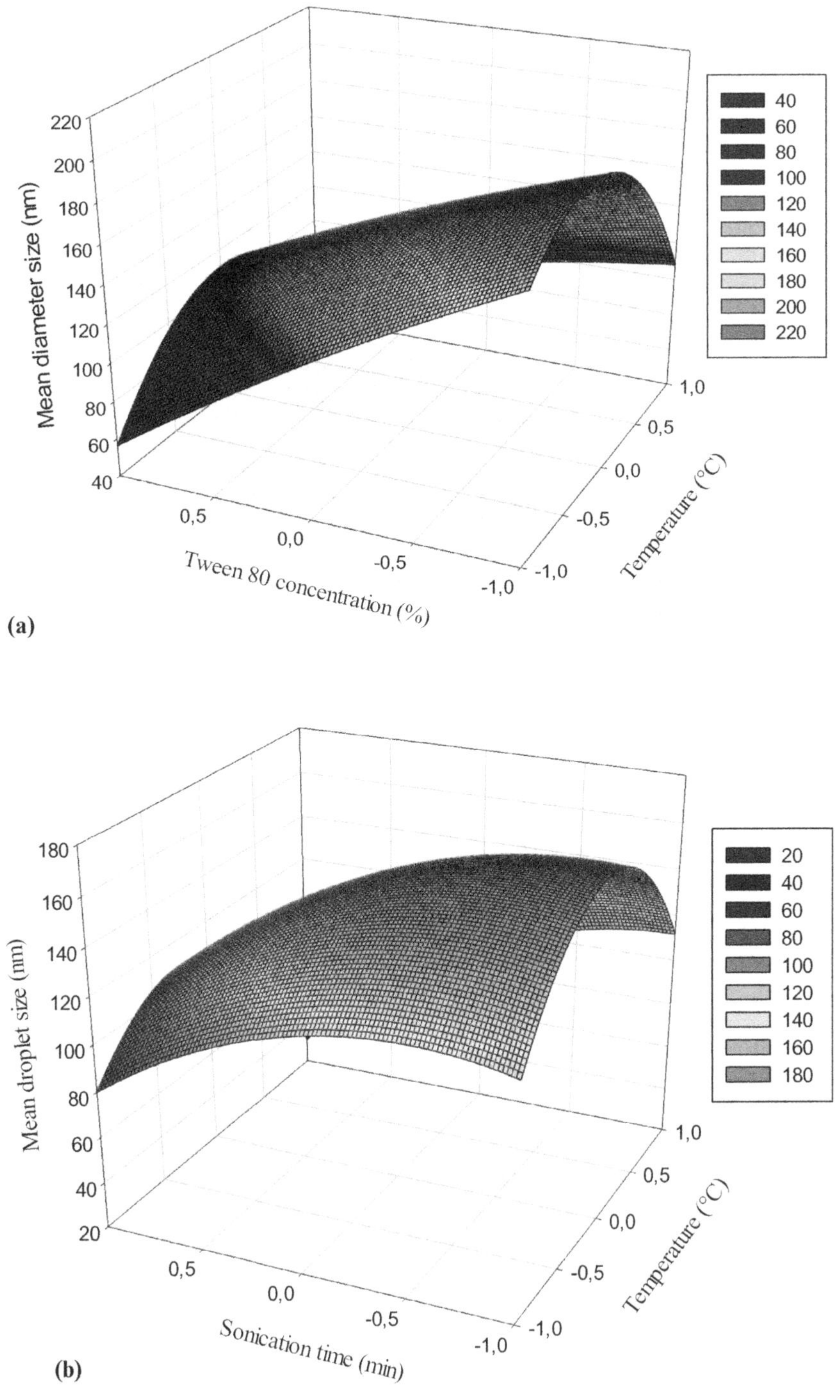

FIG. 1

Response surface (3D) for EO nanoemulsion of Tunisian clove as a function of independent variables (Tween 80 concentration, sonication time, and sonication temperature).

(Continued)

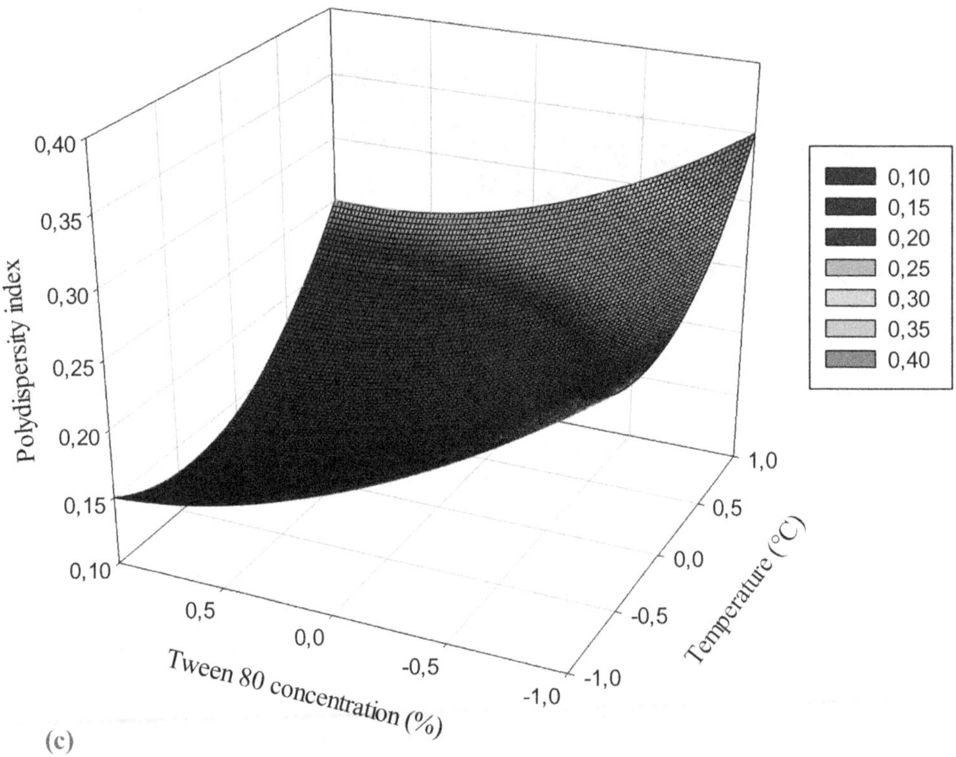

FIG. 1, CONT'D

and mathematical modeling method inspired by the biological neural networks in the human brain, emanating from their predictive reliability and estimation capabilities.

Several studies applied response surface methodology (RSM) for the optimization of bioactive compound extraction. However, some problems have not been solved, and therefore some researchers show interest toward ANN to improve the extraction yield. Also, ANN was the best approach to optimize the polyphenol and bioethanol production because of the high sensitivity of the mathematical models to small variations in the process variables (Ayoola et al., 2020; Chouaibi et al., 2020a). Furthermore, an overview of the literature revealed that the ANN method is more adequate and accurate than RSM in terms of estimation capabilities, precision, and predictability.

The obtained data shows that the best architecture and topology are selected based on the lowest value in root mean square error (RSME) and the highest in regression coefficient (R^2). Based on this criterion, the topological scheme (3-6-1) of the ultrasonic emulsification is composed of 3 input factors which are sonication temperature, Tween 80 concentration, and sonication time, one hidden layer composed of 6 neurons, 2 biases, and a single output (diameter of droplets, polydispersity index (PDI), and dynamic viscosity) as listed in Fig. 2.

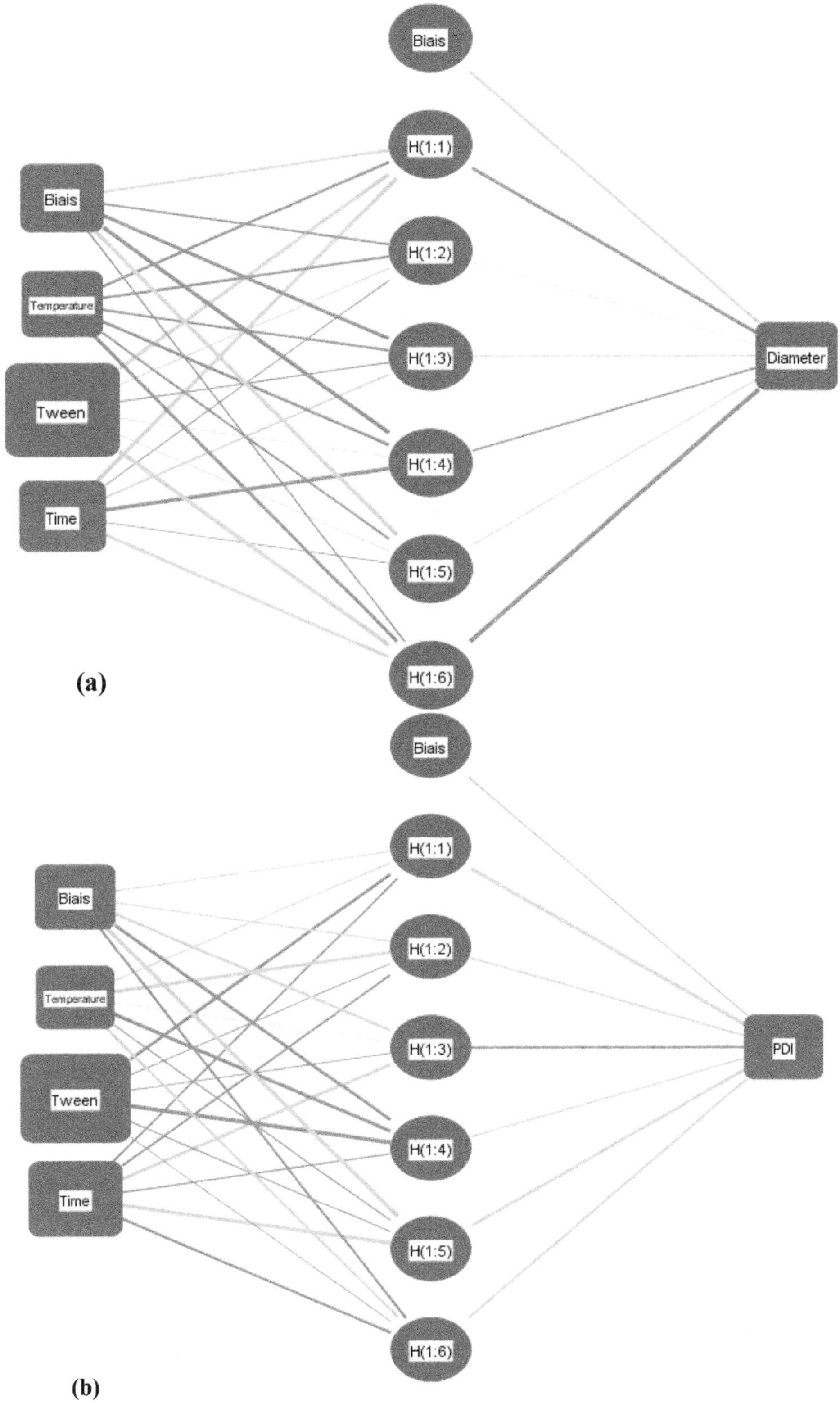

FIG. 2

See figure legend on next page.

(Continued)

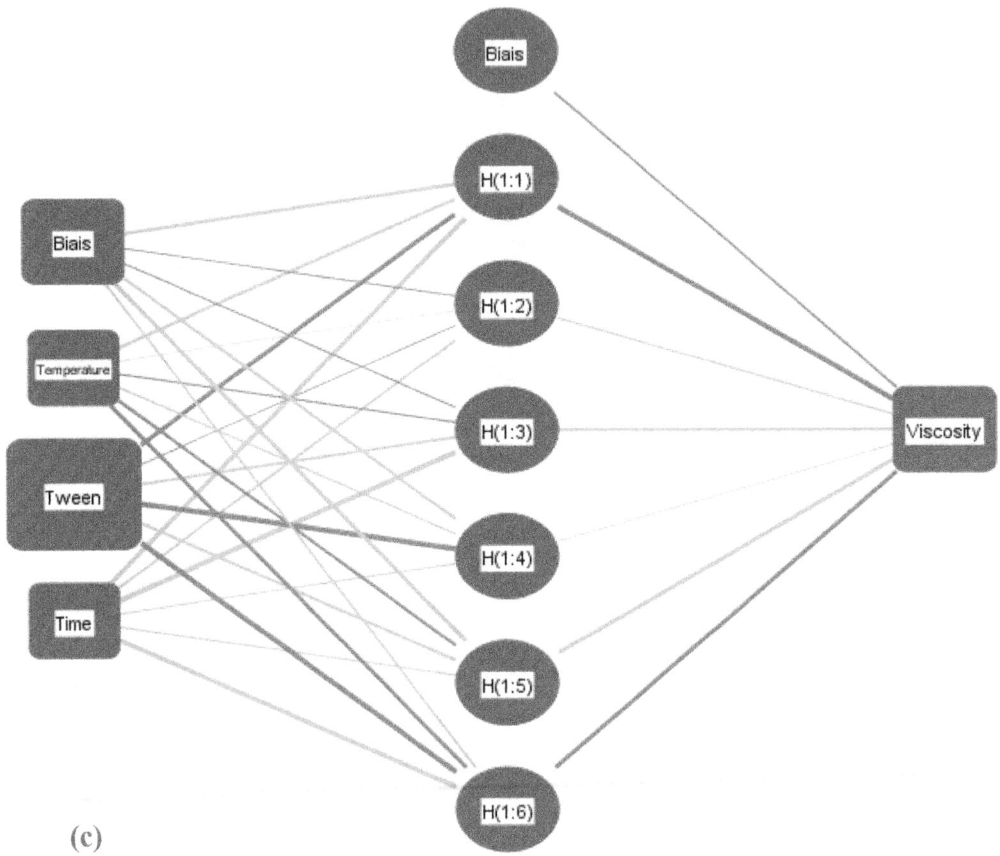

FIG. 2, CONT'D

ANN architecture for the formulation of clove oil nanoemulsion constituted of temperature, Tween 80 concentration, and sonication time, based on the diameter of droplets (A), polydispersity index (PDI) (B), and dynamic viscosity (C).

2.3 Adaptive neuro fuzzy inference system

Nowadays, the rapid development of computer-processing technologies and the creation of related software make it possible for us to benefit from the advantages of artificial intelligence. One example of which is the application of artificial neural networks (ANNs) that copy human brain functions, to some extent, in solving problems in systems and process modeling. Adaptive neuro fuzzy inference system (ANFIS) is a soft computation method applied to solve complex and nonlinear problems. ANFIS can be able to produce and predict one output. A significant advantage of the ANFIS model over statistical procedures is that they require no assumptions on a model. Instead, ANFIS is training with empirical data to yield a suitable topology; therefore, it is becoming a very popular prognostication methodology and is known to be less time-consuming and efficient in modeling complicated systems than mathematical and statistical models (Kaveh et al., 2018).

ANFIS is applied in many agricultural works for the modeling of food and chemical processing. However, no reports were found using ANN and ANFIS for essential oil encapsulation by O/W nanoemulsion. The fuzzy section established the relationship between input and output results. Neural networks assess the parameters of membership functions in the fuzzy part. Different input variables used for various predictions were sonication temperature, Tween 80 concentration, and sonication time. In the ANFIS models suggested for predicting droplet size, polydispersity index, and dynamic viscosity. The approach applied for ANFIS optimization is determined by MATLAB software (R2018b).

3. Techniques of essential oil nanoemulsions

Over the last few years, nanoemulsions have a significant impact in various fields such as pharmaceutics, cosmetics, agriculture, and food industries. Based on the application of nanoemulsions over the last decade, a higher percentage of involvement of nanoemulsions is found in pharmaceutics and cosmetics, followed by the food industry and other miscellaneous industries (Table 3). Nanoemulsions is a technique of encapsulation of some bioactive compounds and EO. They act as a delivery system for appropriate dosage of pesticide. Commercial application in pesticide delivery is yet to be explored due to the poor stability of direct nanoemulsions. The use of nanoemulsions formulation (water/non-ionic surfactant/methyl decanoate) for pesticide production by phase-inversion composition has been reported for commercial use (Pascual-Villalobos et al., 2017). Some authors have reported significant benefits for using oral nanoemulsion formulations compared to regular oral formulations (Table 3). To date, pharmaceutical applications have majorly benefited from the preparation of nanoemulsions due to essential characteristics exhibited by nanoemulsions formulated drugs in the body. It mainly includes

Table 3 Techniques, properties, and applications of encapsulated EO.

Essential oils	Techniques used	Properties	Applications	References
Finger citron	Spontaneous emulsification method	Antibacterial activities (*E. coli*, *B. subtilis*, and *S. aureus*)	Food and beverage products	Li et al. (2018)
Oregano	Homogenization (high-speed homogenizer (HSP))	Antioxidant and antibacterial activities	Fresh meat preservation	Xiong et al. (2020)
Basil	Ultrasonication method	Antibacterial activities	Fruit juices	Frankylyne et al. (2019)
Cinnamon	Ultrasonication and agitation methods	Antifungal activities	Fruit and legume preservation	Pongsumpum et al. (2020)
Thyme	Rotor-stator and high-pressure homogenization (HPH)	Antibacterial activities	Food preservation	Benjemaa et al. (2018)

Continued

Table 3 Techniques, properties, and applications of encapsulated EO—cont'd

Essential oils	Techniques used	Properties	Applications	References
Pinus koraiensis	Agitation process	Antitumor activity	Pharmaceutical application	Zhang et al. (2020)
Salvia mirzayanii	Ultrasonic process (400 W, 24 kHz)	Antioxidant and antibacterial activities	Food packaging	Hasheminya and Dehghannya (2021)
Lemongrass	Low energy emulsification technique	Antimicrobial activity	Disinfection of documentary heritage based on parchment	Saada et al. (2020)
Ginger	Low energy emulsification technique	Anticancer activities	Biomedicine and Pharmacotherapy	Al-Otaibi et al. (2018)
Origanum majorana L.	High-speed homogenization (HSH)	Antifungal, aflatoxin B1 inhibitory and antioxidant activity	Novel food preservative	Chaudhari et al. (2020)
Clove	High-pressure homogenization (HPH)	Antifungal activities (mycotoxin inhibitory activity)	Malting process	Wan et al. (2020)
Eucalyptus	Ultrasonication process	Antimicrobial activities	Food preservation	Sugumar et al. (2013)
Zataria multiflora Boiss and *Bunium persicum*	High-speed homogenization (HSH)/ ultrasonication	Antimicrobial activities	Coatings	Keykhosravy et al. (2020)
Coriander	Microfluidization	Repellent activity	Agricultural products	Pascual-Villalobos et al. (2017)
Tarragon	High-speed homogenization (HSH)	Larvicidal activity	Industrial larvicides	Osanlooa et al. (2017)
Ageratum conyzoides, *Achillea fragrantissima*, and *Tagetes minuta*	High-speed homogenization (HSH)/high-pressure homogenization (HPH)	Insecticidal activity	Against stored-grain insects/toxicological evaluation	Nenaah et al. (2015)
Laurel	Ultrasonic emulsification process	Antioxidant and antibacterial activities	Fish product preservation	Ozogul et al. (2017)

Table 3 Techniques, properties, and applications of encapsulated EO—cont'd

Essential oils	Techniques used	Properties	Applications	References
Rosemary	Ultrasonic homogenization	Antimicrobial activities	Preservation of fresh vegetables	Dávila-Rodríguez et al. (2019)
Lime	Low-energy method	Antimicrobial activities	Incorporation into water-based food products and beverages as flavoring and antimicrobial agents	Liew et al. (2020)
Myristica fragrans	Low-energy method	Antioxidant and antifungal activities	Preservative against fungal and aflatoxin contamination of food commodities	Das et al. (2020)
Sage	Ultrasonic homogenization	Antimicrobial activities	Antimicrobial agent	Yazgan (2020)
Fennel (*Foeniculum vulgare*)	Ultrasonic homogenization	Bioherbicidal properties	Agricultural and cereal products	Kaur et al. (2021)
Tea	Ultrasonic homogenization	Antimicrobial activities	Regulatory toxicology and pharmacology	Wei et al. (2021)
Spearmint	High-speed homogenization (HSH)	Anticancer activities	Medicine applications	Tubtimsri et al. (2021)
Myrtle	High-speed homogenization (HSH)/high-pressure homogenization (HPH)	Antibacterial activity	Food industries	Falleh et al. (2021)
Tunisian clove	Couette emulsification	Antifungal activities	Food and cosmetic application	Present study

reduced drug resistance, dose, and enhanced anticancer activity (Table 3). An O/W lipid nanoemulsion of diclofenac using a high-pressure homogenizer method and droplet size of 200 nm has been studied to treat arthritic conditions. Oral delivery of lipophilic drugs like antibiotics, hormones, steroids, cytotoxics, diuretics, and antifungals too benefited from the use of nanoemulsion. Remarkable applications of EO nanoemulsions have been used to manufacture beverages, i.e., fortified waters and soft drinks, by low-energy methods. Improvement in the shelf life of food has been exhibited with the use of nanoemulsions.

Nanoemulsions are expected to become the area of research and development gradually. However, many challenges still need to be overcome in terms of extensive energy use. Furthermore, due to cost implications, scaling up nanoemulsion production at laboratory bench scale to the commercial market in the various application areas need more attention. To overcome the issues, emulsion formulation by low-energy methods becomes a suitable option. Table 3 describes the type of EO, encapsulation techniques, and industrial applications of all encapsulated EO by nano emulsification process.

4. Comparison between RSM, ANN, and ANFIS

The root means square error values (RMSE) for all data obtained by the ANFIS model of nanoemulsification process were 0.006, 0.03, and 0.05 for mean droplet size, polydispersity index, and dynamic viscosity, respectively. Also, the regression coefficient (R^2) of mean droplet size, polydispersity index, and dynamic viscosity were 0.9989, 0.999, and 0.9979, respectively. Besides, the highest of R^2 and the lowest of RMSE values were observed in all responses predicted by ANFIS-models. Thus, the closeness of data error validates the accuracy of the ANFIS- model.

Moreover, ANFIS-model-predicted values were plotted against the experimental values of both response variables, such as mean droplet size, polydispersity index, and dynamic viscosity (Table 4). This study compares RSM, ANFIS and ANN models regarding their predictive performance and estimation capabilities. The predicted values of the target responses provided by ANFIS models are depicted in the figures (Table 4). A statistical assessment of the ANFIS model-predicted values of the three target responses (mean droplet size, polydispersity index, and dynamic viscosity) was carried out by drawing comparative resemblance plots. The findings showed that the ANFIS model was more accurate, with better precision and estimation capabilities in terms of experimental data fitting of all target responses than the ANN and RSM models. More deviation between the predicted and actual values was observed in the RSM and ANN model than in the ANFIS-model. Indeed, the ANFIS-model demonstrated higher R^2 and lower RMSE (compared to values of RSM, ANN, and ANFIS), which indicate the superiority of the ANFIS-model in terms of predictive and estimation capabilities. Besides, all experimental points were located closer to the straight line, signifying that the ANFIS-models can successfully predict the experimental design for both dependent variables (mean droplet size, polydispersity index, and dynamic viscosity) with reliable precision and validity. This ANFIS-model is needed to formulate EO nanoemulsions as it reduces time, cost, and effort associated with experimental techniques.

5. Optimal conditions of stable O/W nanoemulsion

The stability of O/W nanoemulsions is crucial for their applications in various food, pharmaceutical, cosmetical products. The studies by Mahfoudhi et al. (2014), Chouaibi et al. (2018, 2019a,b), and Snoussi et al. (2020) reported that the droplet size and polydispersity index are significant variables

Table 4 Statistical parameters (regression coefficient and root mean square error) of ANFIS-model.

	Regression coefficient (R^2)			RMSE		
Dependent variable	**ANFIS**	**RSM**	**ANN**	**ANFIS**	**RSM**	**ANN**
Mean droplet size	0.9989	0.90	0.983	0.006	0.6	0.02
Polydispersity index	0.999	0.88	0.990	0.03	0.23	0.10
Dynamic viscosity	0.9979	0.94	0.991	0.05	0.65	0.20

RMSE, *root mean square error;* RSM, *response surface methodology;* ANN, *artificial neural networks;* ANFIS, *adaptive neuro fuzzy inference system.*

to indicate emulsion stability. To validate the predictability of the applied ANFIS-model, the optimized factors (input variables) were established in an experiment under the following optimal conditions: the sonication time of 280 s, temperature 6.5°C, and Tween 80 of 2% (w/w). Thus, the droplet size, PDI, and dynamic viscosity were 60.4 nm, 0.13, and 4.70 mPa s, respectively. Interestingly, there were no significant variations ($P > .05$) between the ANFIS predicted and experimental responses (mean droplet size, polydispersity index, and dynamic viscosity), indicating that adaptive neuro fuzzy inference system (ANFIS) models had a high capability to assess the optimal conditions for mean droplet size, polydispersity index, and dynamic viscosity of clove nanoemulsion.

6. Conclusion

The ultrasonic process produced clove nanoemulsions with smaller droplet size, lower PDI, and higher viscosity as compared to the ultrasonic bath. The optimum condition for preparing the CEO nanoemulsions using ultrasonic probe was determined using the RSM based on BBD, ANN, and ANFIS. In particular, the BBD was employed to fit the second-degree polynomial model to produce the CEO nanoemulsions. The results suggested that sonication time, temperature, and Tween 80 concentration affected droplet size, PDI, and dynamic viscosity of the formulated CEO nanoemulsions. Based on the numerical optimization models, the sonication time of 280 s, temperature of 6.5°C, and Tween 80 of 2% (w/w) were suitable to give the optimum clove EO nanoemulsion with small droplet size, low PDI, and high dynamic viscosity. The obtained nanoemulsion had good stability at 4°C, 30°C, and 45°C for 100 days. The predictive power of obtained models was analyzed using statistical parameters such as regression coefficient and root means square error. The results indicated a high correlation between the obtained data from the ANFIS models and experimental results. As a result, all the developed ANFIS models in this study have higher predictive power than those RSM and ANN. Compared to coarse emulsion, the formulated clove nanonanoemulsion has a higher potential to be incorporated in food industries as a natural antimicrobial agent due to its improved activities.

References

Acosta, E., 2009. Bioavailability of nanoparticles in nutrient and nutraceutical delivery. Curr. Opin. Colloid Interface Sci. 14 (1), 3–15.

Al-Otaibi, W., Alkhatib, M.H., Noor Wali, A., 2018. Cytotoxicity and apoptosis enhancement in breast and cervical cancer cells upon coadministration of mitomycin C and essential oils in nanoemulsion formulations. Biomed. Pharmacother. 106, 946–955.

Alzorqi, I., Ketabchi, M.R., Sudheer, S., Manickam, S., 2016. Optimization of ultrasound induced emulsification on the formulation of palm-olein based nanoemulsions for the incorporation of antioxidant β-D-glucan polysaccharides. Ultrason. Sonochem. 31, 71–84. https://doi.org/10.1016/j.ultsonch.2015.12.004.

Anton, N., Vandamme, T.F., 2011. Nano-emulsions and micro-emulsions: clarifications of the critical differences. Pharm. Res. 28 (5), 978–985.

Ayoola, A.A., Hymore, F.K., Omonhinmin, C.A., Babalola, P.O., Fayomi, O.S.I., Olawole, O.C., Babalola, A., 2020. Response surface methodology and artificial neural network analysis of crude palm kernel oil biodiesel production. Chem. Data Collect., 100478. https://doi.org/10.1016/j.cdc.2020.100478.

Benjemaa, M., Neves, M.A., Falleh, H., Isoda, H., Ksouri, R., Nakajima, M., 2018. Nanoencapsulation of *Thymus capitatus* essential oil: formulation process, physical stability characterization and antibacterial efficiency activity. Ind. Crop. Prod. 113, 414–421.

Burguera, J.L., Burguera, M., 2012. Analytical applications of emulsions and microemulsions. Talanta 96, 11–20.

Burt, S., 2004. Essential oils: their antibacterial properties and potential applications in foods—a review. Int. J. Food Microbiol. 94, 223–253.

Chang, Y., McLandsborough, L., McClements, D.J., 2013. Physicochemical properties and antimicrobial efficacy of carvacrol nanoemulsions formed by spontaneous emulsification. J. Agric. Food Chem. 61 (37), 8906–8913.

Chaudhari, A.K., Singh, V.K., Deepika, S.D., Prasad, J., Dwivedy, A.K., Dubey, N.K., 2020. Improvement of *in vitro* and *in situ* antifungal, AFB1 inhibitory and antioxidant activity of *Origanum majorana* L. essential oil through nanoemulsion and recommending as novel food preservative. Food Chem. Toxicol. https://doi.org/10.1016/j.fct.2020.111536.

Chouaibi, M., Rezig, L., Ben Daoued, K., Mahfoudhi, N., Bouhafa, H., Hamdi, S., 2012a. Extraction of polysaccharide from *Zizyphus Lotus* fruits. Int. J. Food Eng. 8, 1–8.

Chouaibi, M., Rezig, L., Ben Daoued, K., Mahfoudhi, N., Bouhafa, H., Hamdi, S., 2012b. Extraction of polysaccharide from *Zizyphus Lotus* fruits. Int. J. Food Eng. 8, 1556–3758.

Chouaibi, M., Boussaid, A., Donsì, F., Ferrari, G., Hamdi, S., 2018. Optimization of the extraction process by response surface methodology of protein isolate from defatted jujube (*Zizyphus lotus* L.) seeds. Int. J. Pept. Res. Ther. 25 (4), 1509–1521.

Chouaibi, M., Mejri, J., Rezig, L., Abdelli, K., Hamdi, S., 2019a. Experimental study of quercetin microencapsulation using water-in-oil-in-water ($W_1/O/W_2$) double emulsion. J. Mol. Liq. 273, 183–191.

Chouaibi, M., Rezig, L., Lakoud, A., Boussaid, A., Hassouna, M., Ferrari, G., et al., 2019b. Exploring potential new galactomannan source of *Retama reatam* seeds for food, cosmetic and pharmaceuticals: characterization and physical, emulsifying and antidiabetic properties. Int. J. Biol. Macromol. 124, 1167–1176.

Chouaibi, M., Ben Daoued, K., Riganne, K., Rouissi, T., Ferrari, G., 2020a. Production of bioethanol from pumpkin peel wastes: comparison between response surface methodology (RSM) and artificial neural networks (ANN). Ind. Crop. Prod. 155, 112822.

Chouaibi, M., Rigane, K., Ferrari, G., 2020b. Extraction of *Citrullus colocynthis* L. seed oil by supercritical carbon dioxide process using response surface methodology (RSM) and artificial neural network (ANN) approaches. Ind. Crop. Prod. 158, 113002.

Das, S., Singh, V.K., Dwivedy, A.K., Chaudhari, A.K., Dubey, N.K., 2020. *Myristica fragrans* essential oil nanoemulsion as novel green preservative against fungal and aflatoxin contamination of food commodities with emphasis on biochemical mode of action and molecular docking of major components. LWT - Food Sci. Technol. https://doi.org/10.1016/j.lwt.2020.109495.

Dávila-Rodríguez, M., López-Malo, A., Palou, E., Ramírez-Corona, N., Jiménez-Munguía, T., 2019. Antimicrobial activity of nanoemulsions of cinnamon, rosemary, and oregano essential oils on fresh celery. LWT - Food Sci. Technol. https://doi.org/10.1016/j.lwt.2019.06.014.

Domingos, A.K., Saad, E.B., Wilhelm, H.M., Ramos, L.P., 2008. Optimization of the ethanolysis of *Raphanus sativus* (L. var.) crude oil applying the response surface methodology. Bioresour. Technol. 99, 1837–1845.

Donsì, F., Annunziata, M., Sessa, M., Ferrari, G., 2011. Nanoencapsulation of essential oils to enhance their antimicrobial activity in foods. Food Sci. Technol. 44 (9), 1908–1914.

Ennaifer, M., Bouzaine, T., Chouaibi, M., Hamdi, M., 2018. *Pelargonium graveolens* aqueous decoction: a new water-soluble polysaccharide and antioxidant rich-extract. Biomed. Res. Int. https://doi.org/10.1155/2018/2691513.

Frankylyne, J.S., Iyer, S., Ebenazer, A., Mukherjee, A., Chandrasekaran, N., 2019. Essential oil nanoemulsions: antibacterial activity in contaminated fruit juices. Int. J. Food Sci. Technol. 54, 2802–2810.

Gleeson, J.P., Ryan, S.M., Brayden, D.J., 2016. Oral delivery strategies for nutraceuticals: delivery vehicles and absorption enhancers. Trends Food Sci. Technol. 53, 90–101.

Hamidpour, R., Hamidpour, M., Hamidpour, S., Shahlari, M., 2015. Cinnamon from the selection of traditional applications to its novel effects on the inhibition of angiogenesis in cancer cells and prevention of Alzheimer's disease, and a series of functions such as antioxidant, anticholesterol, antidiabetes, antibacterial, antifungal, nematicidal, acaracidal, and repellent activities. J. Tradit. Complement. Med. 5, 66–70.

Hasheminya, S., Dehghannya, J., 2021. Development and characterization of novel edible films based on *Cordia dichotoma* gum incorporated with *Salvia mirzayanii* essential oil nanoemulsion. Carbohydr. Polym. 257, 117606. https://doi.org/10.1016/j.carbpol.2020.117606.

Henry, J.V.L., Fryer, P.J., Frith, W.J., Norton, I.T., 2009. Emulsification mechanism and storage instabilities of hydrocarbon-in-water sub-micron emulsions stabilized with tweens (20 and 80), Brij 96v and sucrose monoesters. J. Colloid Interface Sci. 338 (1), 201–206.

Holley, R.A., Patel, D., 2005. Improvement in shelf life and safety of perishable foods by plant essential oils and smoke antimicrobials. Food Microbiol. 22 (4), 273–292.

Kaur, P., Gupta, S., Kaur, K., Kaur, N., Kumar, R., Bhullar, M.S., 2021. Nanoemulsion of *Foeniculum vulgare* essential oil: a propitious striver against weeds of *Triticum aestivum*. Ind. Crop. Prod. 168 (2021), 113601. https://doi.org/10.1016/j.indcrop.2021.113601.

Kaveh, M., Sharabiani, V.R., Chayjan, R.A., Taghinezhad, E., Abbaspour-Gilandeh, Y., Golpour, I., 2018. ANFIS and ANNs model for prediction of moisture diffusivity and specific energy consumption potato, garlic and cantaloupe drying under convective hot air dryer. Inf. Process. Agric. 5, 372–387.

Keykhosravy, K., Khanzadi, S., Hashemi, M., Azizzadeh, M., 2020. Chitosan-loaded nanoemulsion containing *Zataria multiflora Boiss* and *Bunium persicum Boiss* essential oils as edible coatings: its impact on microbial quality of Turkey meat and fate of inoculated pathogens. Int. J. Biol. Macromol. https://doi.org/10.1016/j.ijbiomac.2020.02.092.

Landry, K.S., Chang, Y., McClements, D.J., McLandsborough, L., 2014. Effectiveness of a novel spontaneous carvacrol nanoemulsion against *salmonella enterica Enteritidis* and *Escherichia coli* O157:H7 on contaminated mung bean and alfalfa seeds. Int. J. Food Microbiol. 187, 15–21.

Landry, K.S., Micheli, S., McClements, D.J., McLandsborough, L., 2015. Effectiveness of a spontaneous carvacrol nanoemulsion against *salmonella enterica Enteritidis* and *Escherichia coli* O157:H7 on contaminated broccoli and radish seeds. Food Microbiol. 51, 10–17.

Li, Z-h., Cai, M., Liu, Y-s., Sun, P-l., 2018. Development of finger citron (*Citrus medica* L. var. sarcodactylis) essential oil loaded nanoemulsion and its antimicrobial activity. Food Control 94, 317–323.

Liew, S.N., Utra, U., Alias, A.K., Tan, T.B., Tan, C.P., Yussof, N.S., 2020. Physical, morphological and antibacterial properties of lime essential oil nanoemulsions prepared via spontaneous emulsification method. LWT - Food Sci. Technol. https://doi.org/10.1016/j.lwt.2020.109388.

Liu, L., Fishman, M.L., Hicks, K.B., 2007. Pectin in controlled drug delivery—a review. Cellulose 14 (1), 15–24.

Mahfoudhi, N., Sessa, M., Chouaibi, M., Ferrari, G., Donsì, F., Hami, S., 2014. Assessment of emulsifying ability of almond gum in comparison with gum arabic using response surface methodology. Food Hydrocoll. 37, 49–59.

Mason, T.G., Wilking, J.N., Meleson, K., Chang, C.B., Graves, S.M., 2006. Nanoemulsions: formation, structure, and physical properties. J. Phys. Condens. Matter 18, R635–R666.

McClements, D.J., 2012. Nanoemulsions versus microemulsions: terminology, differences, and similarities. Soft Matter 8 (6), 1719–1729.

McClements, D.J., Decker, E.A., Weiss, J., 2007. Emulsion-based delivery systems for lipophilic bioactive components. J. Food Sci. 72 (8), R109–R124.

Mechmeche, M., Kachouri, F., Chouaibi, M., Ksontini, H., Setti, S., Hamdi, M., 2017. Optimization of extraction parameters of protein isolate from tomato seed using response surface methodology. Food Anal. Methods 10, 809–819.

Mirhosseini, H., Tan, C.P., Aghlara, A., Hamid, N.S.A., Yusof, S., Chern, B.H., 2008. Influence of pectin and CMC on physical stability, turbidity loss rate, cloudiness and flavor release of orange beverage emulsion during storage. Carbohydr. Polym. 73 (1), 83–91.

Nazarzadeh, E., Anthonypillai, T., Sajjadi, S., 2013. On the growth mechanisms of nanoemulsions. J. Colloid Interface Sci. 397, 154–162.

Nenaah, N.E., Ibrahim, S.I.A., Al-Assiuty, B.A., 2015. Chemical composition, insecticidal activity and persistence of three *Asteraceae* essential oils and their nanoemulsions against *Callosobruchus maculatus* (F.). J. Stored Prod. Res. 61, 9–16.

Nielsen, C.K., Kjems, J., Mygind, T., Snabe, T., Schwarz, K., Serfert, Y., Meyer, R.L., 2017. Antimicrobial effect of emulsion-encapsulated isoeugenol against biofilms of food pathogens and spoilage bacteria. Int. J. Food Microbiol. 242, 7–12.

Osanlooa, M., Amania, A., Sereshtic, H., Abaid, M.R., Esmaeilia, F., Sedaghat, M.M., 2017. Preparation and optimization nanoemulsion of tarragon (*Artemisia dracunculus*) essential oil as effective herbal larvicide against *Anopheles stephensi*. Ind. Crop. Prod. 109, 214–219.

Ozogul, Y., Yuvka, I., Ucar, Y., Durmus, M., Kösker, A.R., Öz, M., Ozogul, F., 2017. Evaluation of effects of nanoemulsion based on herb essential oils (rosemary, laurel, thyme, and sage) on sensory, chemical and microbiological quality of rainbow trout (*Oncorhynchus mykiss*) fillets during ice storage. LWT - Food Sci. Technol. https://doi.org/10.1016/j.lwt.2016.10.009.

Pongsumpum, P., Iwaamoto, S., Siripatrawan, U., 2020. Response surface methodology for optimisation of cimmamon essential oil nanoemulsion with improved stability and antifungal activity. Ultrason. Sonochem. 60, 104604.

Qian, C., McClements, D.J., 2011. Formation of nanoemulsions stabilized by model food-grade emulsifiers using high-pressure homogenization: factors affecting particle size. Food Hydrocoll. 25 (5), 1000–1008.

Rao, J., McClements, D.J., 2012. Food-grade microemulsions and nanoemulsions: role of oil phase composition on formation and stability. Food Hydrocoll. 29 (2), 326–334.

Saada, N.S., Abdel-Maksoud, G., Abd El-Aziz, M.S., A.M.Youssef, A.M., 2020. Evaluation and utilization of lemongrass oil nanoemulsion for disinfection of documentary heritage based on parchment. Biocatal. Agric. Biotechnol. 29, 101839. https://doi.org/10.1016/j.bcab.2020.101839.

Sanchez-Gonzalez, L., Vargas, M., Gonzalez-Martínez, C., Chiralt, A., Chafer, M., 2011. Use of essential oils in bioactive edible coatings. Food Eng. Rev. 3, 1–16.

Selim, S., 2011. Antimicrobial activity of essential oils against vancomycin-resistant *enterococci (vre)* and *Escherichia coli* O157:H7 in feta soft cheese and minced beef meat. Braz. J. Microbiol. 42, 187–196.

Snoussi, A., Chouaibi, M., Bouzouita, N., Hamdi, S., 2020. Microencapsulation of catechin using water-in-oil-in-water (W1/O/W2) double emulsions: study of release kinetics, rheological, and thermodynamic properties. J. Mol. Liq. https://doi.org/10.1016/j.molliq.2020.113304.

Solans, C., Izquierdo, P., Nolla, J., Azemar, N., Garcia-Celma, M., 2005. Nanoemulsions. Curr. Opin. Colloid Interface Sci. 10 (3–4), 102–110.

Sugumar, S., Nirmala, J., Ghosh, V., Anjali, H., Mukherjee, A., Chandrasekaran, N., 2013. Biobased nanoemulsion formulation, characterization and antibacterial activity against food-borne pathogens. J. Basic Microbiol. 53, 677–685.

Tadros, T., Izquierdo, P., Esquena, J., Solans, C., 2004. Formation and stability of nano-emulsions. Adv. Colloid Interf. Sci. 108–109, 303–318.

Tubtimsri, S., Limmatvapirat, C., Limsirichaikul, S., Akkaramongkolporn, P., Piriyaprasarth, S., Patomchaiviwat, V., Limmatvapirat, S., 2021. Incorporation of fixed oils into spearmint oil-loaded nanoemulsions and their influence on characteristic and cytotoxic properties against human oral cancer cells. J. Drug Deliv. Sci. Technol. 63, 102443. https://doi.org/10.1016/j.jddst.2021.102443.

Walker, R., Decker, E.A., McClements, D.J., 2015. Development of food-grade nanoemulsions and emulsions for delivery of omega-3 fatty acids: opportunities and obstacles in the food industry. Food Funct. 6 (1), 42–55.

Wan, J., Jin, Z., Zhong, S., Schwarz, P., Chen, B., Rao, J., 2020. Clove oil-in-water nanoemulsion mitigates growth of *fusarium graminearum* and *Trichothecene mycotoxin* production during the malting of *fusarium* infected barley. Food Chem. https://doi.org/10.1016/j.foodchem.2019.126120.

Waraho, T., McClements, D.J., Decker, E.A., 2011. Mechanisms of lipid oxidation in food dispersions. Trends Food Sci. Technol. 22 (1), 3–13.

Wei, S., Zhao, X., Yu, J., Yin, S.J., Liu, J., Bo, R.N., Li, J.G., 2021. Characterization of tea tree oil nanoemulsion and its acute and subchronic toxicity. Regul. Toxicol. Pharmacol. 124, 104999. https://doi.org/10.1016/j.yrtph.2021.104999.

Wulff-Perez, M., Torcello-Gomez, A., Galvez-Ruíz, M.J., Martín-Rodríguez, A., 2009. Stability of emulsions for parenteral feeding: preparation and characterization of o/w nanoemulsions with natural oils and Pluronic f 68 as surfactant. Food Hydrocoll. 23 (4), 1096–1102.

Xiong, Y., Li, S., Warner, R.D., Fang, Z., 2020. Effect of oregano oil and resveratrol nanoemulsion loaded pectin edible coating on the preservation of pork loin in modified atmosphere packaging. Food Control 114, 107226.

Yazgan, H., 2020. Investigation of antimicrobial properties of sage essential oil and its nanoemulsion as antimicrobial agent. LWT: Food Sci. Technol. 130, 109669. https://doi.org/10.1016/j.lwt.2020.109669.

Ye, A., Singh, H., 2006. Heat stability of oil-in-water emulsions formed with intact or hydrolyzed whey proteins: influence of polysaccharides. Food Hydrocoll. 20, 269–276.

Zhang, Y., Xin, C., Cheng, C., Wang, Z., 2020. Antitumor activity of nanoemulsion based on essential oil of *Pinus koraiensis* pinecones in MGC-803 tumor-bearing nude mice. Arab. J. Chem. 13, 8226–8238.

Falleh, H., Ben Jemaa, M., Neves, M.A., Isoda, H., Nakajima, M., Ksouri, R. (2021). Peppermint and Myrtle nanoemulsions: formulation, stability, and antimicrobial activity. LWT: Food Sci. Technol., 152, 112377. doi:10.1016/j.lwt.2021.112377.

Pascual-Villalobos, M.J. Cantó-Tejero, M., Vallejo, R., Guirao, P., Rodríguez-Rojo, S., Cocero, M.J. (2017). Use of nanoemulsions of plant essential oils as aphid repellents. Ind. Crops Prod., 110, 45–57.

CHAPTER

Preparation of oil-in-water (O/W) clove essential oil nanoemulsion: Characterization and stability

33

Moncef Chouaibi

Higher School of Food Industries of Tunisia, Food Engineering Department, University of Carthage, Tunis, Tunisia

1. Introduction

Regardless of the many benefits, incorporating essential oils (EOs) generates problems in food industry. Most of them arise from their volatility and hydrophobic characteristics. EOs have an excessive interest as substitutes to synthetic food preservatives due to their antibacterial and antioxidant activities. Nevertheless, they are hydrophobic and facilitate vaporization. Commonly, EOs have high efficiency against food-borne pathogen and spoilage microorganisms in nutrient media, but this effect in food is only achieved with higher levels than those in media, producing some unpleasant sensorial effects (Almadiy et al., 2016; Topuz et al., 2016). Moreover, Burt (2004) indicated that when EOs are incorporated in food, hydrophobic interactions between them and food compositions like lipids, minerals, carbohydrates, and protein reduce their sanitizing effect and/or shorten the duration of effect on microorganisms. Consequently, the fabrication of nanoemulsions presents an alternative method to avoid these undesirable defects. In emulsion-based systems, EOs are encapsulated into oil-in-water (O/W) emulsions (radius >100 nm) or nanoemulsions (radius <100 nm). Based on the physicochemical mechanism of droplet disruption, nanoemulsion preparation methods are characterized as either high energy or low energy (Biasi-Garbin et al., 2015). Two techniques are available to formulate for nanoemulsion according to low-energy techniques: phase inversion and spontaneous emulsification methods. The noticeable advantage of the low-energy method is producing small droplet sizes than the high-energy method, as described by McClements and Rao (2011) and McClements (2012).

Nanoemulsions have shown potential advantages over conventional emulsions due to their large active surface area but are also susceptible to destabilization. The O/W nanoemulsions are principally appropriate as delivery systems because lipid-soluble compounds can be dissolved within oil droplets that are then dispersed into aqueous-based food products (Frede et al., 2014; Salvia-Trujillo et al., 2015). Numerous challenges have been made to preserve the bioactive compounds of EOs, and one of them is by fabricating encapsulated EOs. Encapsulation by nanoemulsions is the biphasic dispersion of two immiscible liquids, which are either O/W or water-in-oil and stabilized by an amphiphilic surfactant (Singh et al., 2017). In addition, nanoemulsions may lead to a higher carotenoid bioavailability since of the rapid rate of lipid digestion and mixed micelle formation associated with the high surface area of small lipid droplets (Boon et al., 2010; Khalil et al., 2012; Frede et al., 2014; Saberi et al., 2014;

Clove (*Syzygium aromaticum*). https://doi.org/10.1016/B978-0-323-85177-0.00016-1

Davidov-Pardo et al., 2016). Newly, our laboratory formulated β-carotene–loaded nanoemulsion-based delivery systems using vegetable oil as a lipid phase, Arabic and almond gums as an emulsifier, high-pressure homogenization (HPH) emulsification process (Mahfoudhi et al., 2012, 2014). However, these O/W nanoemulsions were unstable to particle aggregation at pH values around the isoelectric point of hydrocolloids. However, the chemical stability of the carotenoid could be enhanced by encapsulation in the O/W nanoemulsions (Davidov-Pardo et al., 2016).

Encapsulation by nanoemulsions are supposed to be long-term kinetically stable (Devarajan and Ravichandran, 2011); these systems are still delicate to environmental conditions such as heat, air, and light, pH variations that could destabilize the encapsulated bioactive compounds (McClements, 1999). The emulsification evaporation method has been applied to fabricate β-carotene nanoemulsions (Tan and Nakajima, 2005a,b). The authors showed that the mean droplet diameters varied from 85 to 140 nm and that during storage at 4°C, the β-carotene demonstrated a significant reduction. Yuan et al. (2008a) prepared O/W nanoemulsions of β-carotene by HPH, and the droplet sizes were within the range of 130–170 nm. However, during storage at 4°C and 25°C, up to 25% of β-carotene in the nanoemulsions was lost after 4 weeks, with a slightly greater loss occurring at 25°C as found by Yuan et al. (2008b). Also, astaxanthin nanoemulsions were investigated by Affandi et al. (2011) using HPH technique under different emulsifying conditions, where droplet size was ranged from 120 to 180 nm and astaxanthin stability was achieved during 3 months at 4°C and 25°C.

EOs are complex mixtures of volatile organic compounds extracted from different plant parts such as leaves, barks, stems, roots, flowers, and fruits (Burt, 2004; Calo et al., 2015). The EOs often exhibit antimicrobial and antioxidant activities, and the growing demand for natural preservatives has led to the consideration and application of EOs as possible alternatives for synthetic preservatives (Jayasena and Jo, 2013).

Clove oil nanoemulsions had the greatest antifungal and mycotoxin inhibitory activities among other EOs. Still, our findings have not shown if they share the same capacity in the actual storage process. Mainly, concerning nanoemulsion preparation, several food-grade emulsifiers are available for the preparation of nanoemulsions, including natural (e.g., saponins, dairy proteins, polysaccharides, and exudate gums) and industrial emulsifiers (e.g., Tween 80). However, the effect of emulsifier type on its antifungal, antioxidant, and mycotoxin inhibitory activities during the preservation process has not been examined.

2. Clove oil nanoemulsions

Characteristics and formulation of nanoemulsions prepared with clove EO as dispersed phase is given in Table 1.

3. Emulsification process

O/W nanoemulsions constituted of small and spherical oil droplets ($50 < d < 200$ nm) dispersed in a continuous aqueous phase, with oil droplets being surrounded by a protective coating of emulsifier molecules (Acosta, 2009). Nanoemulsions can be prepared through an energy-intensive comminution process, such as high-pressure homogenization, which produces intense, disruptive forces that break up the bulk oil phase so that forming fine oil droplets dispersed in the water phase (Chouaibi et al., 2012; Mahfoudhi et al., 2012, 2014; Leong et al., 2009).

Table 1 Characteristics and formulation of nanoemulsions prepared using clove essential oil as dispersed phase.

Formulation	Composition	Characteristics	Stability	References
Nanoemulsion (Ultra-Turrax homogenization Higher)	Clove oil (5%), NaCas (5%), and pectin (0.1%)	Higher antimicrobial activity than the non-nanoemulsion counterparts	10 days at refrigerated temperature	Zhang et al. (2017)
Nanoemulsion (ultrasonic process)	CEO (0.5 wt%) and Tween 80 (0.75 wt%) into distilled water	Improvement antibacterial activities	Not determined	Shen et al. (2021)
High-pressure homogenization (HPH)	0.5 wt% Tween 80, 2.5 wt% of EO, 2.5 wt% of MCT, and 94.5 wt% of water)	Improvement antifungal activities	30 days	Wan et al. (2019)
Low-energy method	10% clove oil and 60% polysorbate 20 and 30% ethanol	High miscibility in aqueous phase	90 days at 4°C	Kheawfu et al. (2018)
High-speed homogenization (HSH)	Clove oil phase (O) (1%–15%, v/v) NaCas (1%–15%, w/v) and pectin (0.1%–1%, w/v)	High antimicrobial bioactive substances	20 days	Sharma et al. (2017)
Nanoemulsion (Couette cell emulsification) +ultrasonic process	CO 4%, sodium caseinate 27%, MCT 9%, galactomannan 2%, and water 58%	High antioxidant and antimicrobial activities than clove essential oil	20 days at 4°C	Present study

The obtained data revealed that the optimal conditions for the formulation of the Tunisian clove nanoemulsions were oil to sodium caseinate ratio of 1:9. Therefore, this nanoemulsion constituted clove EO 4%, sodium caseinate 27%, MCT 9%, galactomannan 2%, and water 60%. With this formulation, clear and transparent Tunisian clove EO nanoemulsions were obtained. Hence, a ternary phase diagram was performed to obtain the optimum conditions based on mean particle size, polydispersity index (PDI), dynamic viscosity, and zeta potential to formulate small droplets at minimum energy input. The details of the experimental runs and the corresponding responses are described in Table 2.

Table 2 Composition of nanoemulsion formulations based on clove essential oil used in the current study.

Symbol	Composition
Control	CO 4%, MCT 9%, and water 87%
SC-nanoemulsion	CO 4%, sodium caseinate 27%, MCT 9%, and water 60%
G-nanoemulsion	CO 4%, MCT 9%, galactomannan 2%, and water 85%
SCG-nanoemulsion	CO 4%, sodium caseinate 27%, MCT 9%, galactomannan 2%, and water 58%

CO, *clove essential oil;* G, *galactomannan;* SC, *sodium caseinate;* SCG, *sodium caseinate-galactomannan.*

The encapsulation stability was evaluated during 20-day storage in the cold room at 4°C, ambient 25°C, and high temperatures of 40°C and compared with the control samples where the bacteria was inserted in the outer aqueous phase.

4. Nanoemulsion morphology

Fig. 1 revealed the morphology of clove nanoemulsions. All photographs were taken within 5 h after formulation to avoid macroscopic instability. It is clear that all droplets were spherical, and their size showed similar evolution (Fig. 1). Fig. 1 shows that the emulsion containing only SC was nonhomogeneous, indicating the poor stabilization of the clove nanoemulsions. However, when stabilized with galactomannan and sodium caseinate, the droplets were well dispersed and exhibited similar droplet diameters, agreeing with observations.

After adding galactomannan, the size of the droplets increased slowly: large oil droplet clusters were formed in the nanoemulsions demonstrating polydisperse distributions after 20 storage days. Thick droplet boundaries could be distinguished, implying that interfacial layers formed by the complexes were sufficient to resist droplet breakage during preparation and short storage time. After further adding galactomannan and sodium caseinate, the trend in oil droplet diameter reversed; it became smaller and homogeneous, demonstrating the high efficiency of the complexes in producing clove nanoemulsions. Droplets with sizes on the order of 100 nm could be observed, confirming the formation of a bimodal system. Overall, incorporating galactomannan and sodium caseinate extensively enhanced the emulsifying ability and led to stable clove nanoemulsions with tunable droplet size.

The microphotographs showed that the nanoemulsion containing SC-G was the more unstable and formed a stable network. Furthermore, the micrographs presented aggregates with size dependent on total ion concentration. Therefore, the structure may also be responsible for the stability of the caseinate

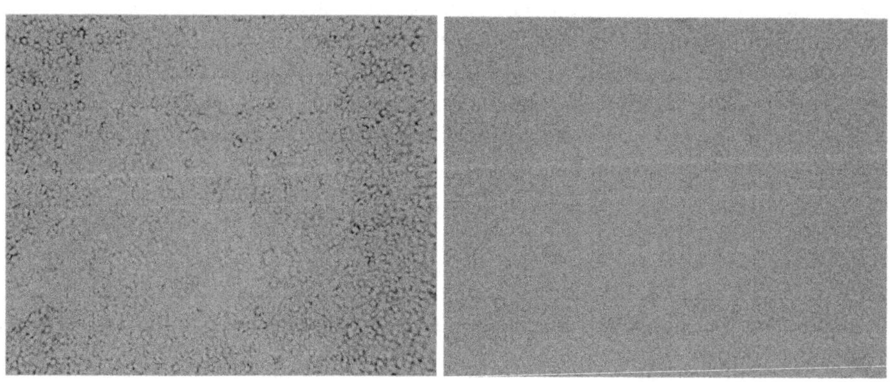

(a) (b)

FIG. 1

Micrograph of nanoemulsion made (A) clove essential oil 4%, sodium caseinate 27%, MCT 9%, galactomannan 2%, and water 58%, and (B) clove essential oil 4%, sodium caseinate 27%, MCT 9%, and water 60% stored at 5°C and 5 h after preparations.

in the nanoemulsion. Scanning electron micrographs showed spherical nanoemulsions with slight adherence and confirmed the morphology through discrete spherical droplets with defined core and coating material boundaries.

5. Stability studies of clove nanoemulsion

From an industrial point of view, it is crucial that the formulated food products have a sufficient physicochemical stability and can be achieved through an economical production process. Therefore, numerous researches have shown that a smaller particle size increases the antioxidant activity and functionality of bioactive compounds like EOs, polyphenols, and anthocyanins, as reported by Doost et al. (2019). Therefore, the droplet size distribution, mean droplet diameter (Z-averages), zeta potential, dynamic viscosity, and PDI are very important indicators that assess the quality, stability, uniformity, and dispersibility of the O/W nanoemulsions (Mason et al., 2006; Mahfoudhi et al., 2012; Chouaibi et al., 2012; Rezig et al., 2013; Mahfoudhi et al., 2014; Chouaibi et al., 2019a,b; Snoussi et al., 2020).

5.1 Effect storage period on droplet size

The prepared clove EO nanoemulsions must have sufficient physicochemical stability under storage conditions. An essential factor in assessing a nanoemulsion's physicochemical stability is the mean droplet size distributed in the continuous phase. To this end, the mean droplet diameter, the visual appearance, and microstructure of the O/W nanoemulsions were monitored during 20 days of storage at 4°C, 25°C, and 40°C.

The nanoemulsions were stored at 4°C, 25°C, and 40°C, and the particle size variations over the storage period are listed in Fig. 2. Notably, the nanoemulsions stabilized by sodium caseinate and galactomannan displayed good storage stability. No obvious phase separation or creaming was found for all the prepared samples. In addition, the mean droplet size determination was in concordance with the visual observations of all clove nanoemulsions. At the same time, the mean droplet diameter size of clove EO nanoemulsion increased insignificantly slightly at refrigerated storage temperature. At 4°C, the change in droplet size is relatively smaller, and the droplet size was almost within 10 nm during 20-day storage. However, at the higher temperature of 40°C, the droplet size increased a bit more to around 20 nm for the samples without galactomannan. This increase could be attributed to the theory of "Ostwald ripening," which is the growth of oil droplets at the cost of smaller oil droplets. Ostwald ripening is a common problem for the instability of EO emulsions, as indicated by Luo et al. (2019). Compared with the samples without galactomannan, the nanoemulsion stabilized by caseinate not only had a smaller initial droplet size but also had a smaller increase of droplet size during storage at a refrigerated temperature (4°C). For instance, the droplet size of clove nanoemulsions with high sodium caseinate increased by 120 and 204.1 nm at 4°C and 40°C, respectively, which has a significant difference.

Additionally, the droplet size distribution of the clove nanoemulsions mixed with medium-chain triglycerides (MCT) in the lipid presented also a noticeable shift toward bigger droplets after 20 days. This indicates that the transfer of the oil molecules from smaller droplets through the intervening medium occurred. Nonetheless, with galactomannan incorporation, the droplet size was almost the same

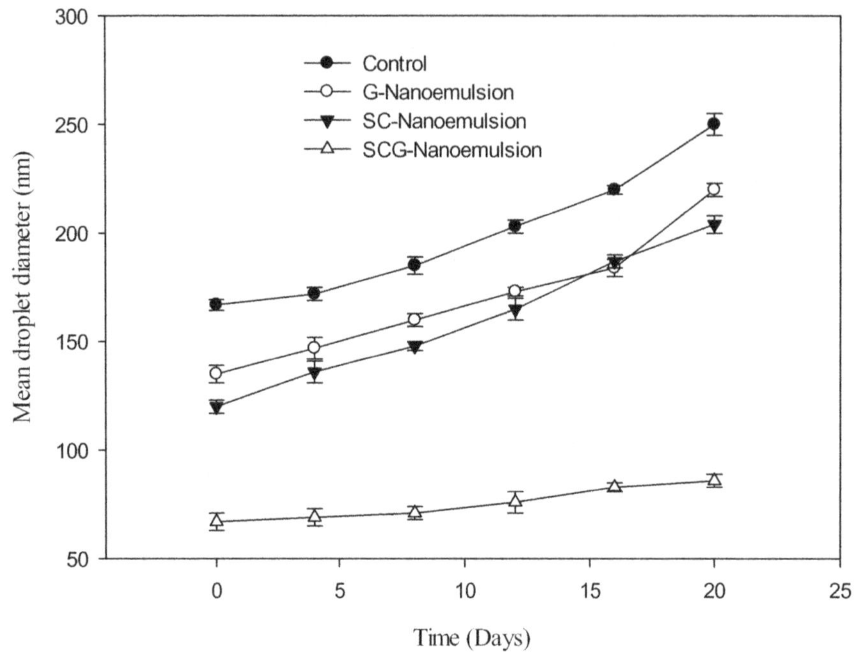

FIG. 2

Mean droplet diameter of the prepared clove nanoemulsions as a function of storage time at 4°C.

under different temperatures, indicating that the addition of galactomannan facilitated stabilizing the Tunisian clove nanoemulsion during storage under different storage temperatures.

The droplet size was significantly augmented during the storage period between the formulated O/W nanoemulsions ($P < .05$). Therefore, there are many reasons for the droplet size increase during storage, probably explained by the occurrence of flocculation or coalescence among the droplets that data in a quick increase in droplet size with storage period (Zhang et al., 2020). In addition, the increase may be due to the change in the pH values of the storage nanoemulsions. This is in concordance with Chen et al. (2019), who mentioned that when the pH values ranged from 2.0 to 6.0, the mean droplet diameter reduced from 9.17 to 0.31 μm, while when the pH varied from 7.0 to 10.0, the mean droplet diameter of prepared emulsions was less than 0.33 μm. Likewise, the emulsifier amount might be one reason why most of the individual droplets formed during the emulsification process are not retained in the final emulsion, resulting in larger droplet size and lower stability during storage (Dickinson, 2009). Thus, at a higher amount of surfactants, the size of the droplets during the emulsification would be lower, and there would be a minor increase in the size of the droplets during storage (Pimentel-Moral et al., 2018).

Furthermore, the size of droplets could be affected by the system viscosity; the high viscosity eliminates the occurrence of flocculation or coalescence of the dispersed phase and leads to high stability for emulsions during storage. However, the droplet size in the stored clove nanoemulsions at 4°C was more stable than that stored at 25°C and 40°C when compared to the degree of storage. These findings are in remarkable agreement with those found by Chouaibi et al. (2019a,b).

5.2 Effect storage period on PDI

The nanoemulsions' PDI are critical factors for estimating the emulsification effect and emulsion stability. The obtained results also showed that the PDI of the clove EO nanoemulsions stored at different temperatures ranging from 4°C to 40°C increased insignificantly over storage time. However, the PDI increased when stored at elevated temperatures longer than 12 days. This is probably because high temperature increases the coalescence rate. While nanoemulsions with a PDI value <0.30 are supposed to have a narrow distribution, as indicated by Chouaibi et al. (2019b). Hence, the obtained data suggested that clove essential nanoemulsions have good stability under the tested temperatures and storage period. These results are in remarkable agreement with those found by Pongsumpum et al. (2020).

These results could be explained by the action mechanism of emulsifiers, i.e., emulsifier surface excess principle. According to this, the formation of large droplets is inhibited by the formation of the emulsifier layer onto the surface of the droplets, reducing the interfacial tension of the oil and water interphase as higher ultrasonic power leads to greater surface excess that results in emulsions with smaller droplet size. In this study, the prepared clove nanoemulsion without galactomannan showed PDI values <0.30, but these values increased rapidly with storage time ($P < .05$). Shao et al. (2018) mentioned that the droplet disruption after sonication generates new smaller droplets with an increased overall surface area, approved this. An adequate amount of emulsifier thus covers these newly formed interfaces and prevents their coalescence. However, differences between emulsions' particle size after ultrasound treatments may be related either to the type of emulsion, surfactant, oil, or possible surfactants' molecular interactions at the interface of the emulsion (Velderrain-Rodríguez et al., 2019).

5.3 Effect storage period on ζ-potential

To better understand the stabilization mechanism of clove nanoemulsion, their ζ potential was determined. The nanoemulsions had relatively high negative zeta potential. According to Snoussi et al. (2021), the nanoemulsion was stable when zeta-potential was superior to −30 and +30 mV. The strong negative surface charge density of the Tunisian clove essential nanoemulsion droplets was thought to be due to carboxylic acid groups within the chemical structure of galactomannan. These highly negative charges provide a powerful electrostatic repulsion force between the formed droplets, leading to a stable colloidal dispersion, as Tippel et al. (2016) mentioned. The ζ-potential of the emulsion system is often determined by the type of EMs, which can exert a defining influence on the emulsion's pH, conductivity, and ionic strength (Wu et al., 2011). The desired ζ-potential (generally, > -30 mV) could provide electrostatic repulsion to reduce the clove EO aggregation during the preparing process (Li et al., 2015). Hence, there is no significant difference in ζ-potentials among all samples. The addition of galactomannan in nanoemulsions reduced the zeta potential insignificantly in the prepared samples ($P > .05$).

Comparable findings were found with Ni et al. (2020), who observed that after ultrasonic treatments, the zeta potential value of the resulting nanoparticle suspensions decreased sharply to approximately −42 mV. On the other hand, the increase of ultrasonic powers was led to a significant difference in ζ-potential values. Costa et al. (2018) also showed a related result, who indicated that ultrasonic treatments caused an increase of zeta potential values from −24.3 to −55.5 mV. Hereafter, the sonication forces on the emulsions' contents may play a role in ζ-potential variations found on the prepared emulsions. This is in line with Silva and Sato (2019), who attributed the reason for the low value of Zeta in his work of sonochemical reactions. The above data suggest that the size and distribution of clove EO droplets present the major difference among the four embedding emulsions.

5.4 Effect storage period on dynamic viscosity

Shear viscosity of the nanoemulsions is a vital test related to fundamental fluid dynamics problems and has numerous technological implications (Masalova et al., 2018). The variations in the apparent viscosity of the prepared clove nanoemulsions as a function of shear rates at day 0 and during storage demonstrated that all clove nanoemulsions are non-Newtonian fluids. On day 0, a noticeable shear-thinning behavior was found for all prepared nanoemulsions, and the apparent viscosity of clove nanoemulsions decreased with increasing shear rate. In addition, under the same fixed shear rate, a greater apparent viscosity was generally observed for the galactomannan added in nanoemulsions compared to control and these with sodium caseinate only. After 20 days at 25°C, there was a variation in the shear results, as it was noticed that the viscosity decreased less in control, followed by the caseinate nanoemulsion, compared to the rest of the galactomannan nanoemulsions. The clove nanoemulsions kept at 4°C also had deferent apparent viscosity high at galactomannan nanoemulsions, compared to the remaining emulsions, and the lowest values were found in control, with decreased apparent viscosity in the emulsions with increasing shear rate as well. Also, the shear viscosity findings indicated that formulated clove nanoemulsions manifested as pseudo-plastic fluids with shear-thinning behavior, which might be derived from the disintegration of the clusters of droplets during the shear process (Luo et al., 2019). Many researches have assessed that droplets were close enough to interact with each other and to form a network of aggregated droplets in concentrated emulsions. The network of aggregated droplets was deformed and consequently disrupted when the applied shear rate was increased sufficiently to reduce the viscosity (Luo et al., 2019). In this study, all emulsions at 4°C had higher viscosity after 28 days than the emulsions kept at 25°C, meaning that the high temperature affected the viscosity behavior with time. This agrees with Ilia and Nour (2010) and Chouaibi et al. (2012), who used several crude oils to study the effect of viscosity and droplet diameter on nanoemulsions, and they indicated that the viscosity reduces by increasing heating temperature.

In order to assess the rheological parameters, there are many rheological models listed in Table 3. The most important models used to describe the flow behavior of nanoemulsions are Ostwald, Herschel-Bulkley, Casson, and Carreau models.

5.5 Effect of storage time on antibacterial activity

The antibacterial activity of clove EO nanoemulsion after 20 days of storage was also examined. The comparison between the freshly-prepared clove EO nanoemulsion inhibition zone after 20 days of storage against *E. coli*, *Salmonella* spp., and *S. aureus* was investigated. From the result, all of the cloves nanoemulsions demonstrated a significant reduction in the diameter of the inhibition zone after 20 days of storage at 4°C, indicating a decrease in the antibacterial activity of clove nanoemulsion. This is because the components in clove EO are volatile and may be easily oxidized or hydrolyzed depending on their respective structure (Scott, 2005). A study conducted by Nguyen et al. (2009) revealed that there was a decrease in α-terpinene, limonene, β-phellandrene, β-caryophyllene, and citral together with an increase in *p*-cymene, limonene oxide, α-terpineol, and geranic acid upon the storage of cold-pressed lime oil. This indicated that some bioactive compounds in the clove EOs were degraded and lost during storage. Although the antibacterial activity of SCG-nanoemulsion was higher compared to that of

Table 3 Different models used to assess the rheological parameters of O/W nanoemulsions.

Bioactive compound	Model used	Expression	References
β-Carotene	Ostwald model	$\sigma = k\dot{\gamma}^n$	Mahfoudhi et al. (2014)
Catechin	Casson model	$\sigma = \sigma_c^{0.5} + \mu_C\dot{\gamma}^{0.5}$	Snoussi et al. (2020)
Quercetin	Hershel-Bulkley model	$\sigma = \sigma_c + k\dot{\gamma}^n$	Chouaibi et al. (2019a)
Peppermint essential oil	Hershel-Bulkley model	$\sigma = \sigma_c + k\dot{\gamma}^n$	Xie et al. (2021)
Rosemary essential oil	Ostwald model	$\sigma = k\dot{\gamma}^n$	Martin-Piñero et al. (2020)
Extracted green tea	Carreau model	$\mu = \frac{\mu_0}{\left(1 + \left(\frac{\dot{\gamma}}{\dot{\gamma}_c}\right)^2\right)^N}$	Lante and Friso (2013)
Clove essential oil	Casson model	$\sigma = \sigma_c^{0.5} + \mu_C\dot{\gamma}^{0.5}$	Present study

Where σ describes shear stress (Pa), σ_c describes yield stress (Pa), $\dot{\gamma}$ is shear rate (1/s), k describes consistency index (Pa s^n), μ_C is the plastic viscosity, and n describes flow behavior index.

control and the clove EO, it was found that SCG-nanoemulsion showed the highest percentage of reduction in terms of antibacterial activity after storage. Control nanoemulsion with the largest mean particle size and PDI value was less stable than SCG-nanoemulsions. The particles of control and G-nanoemulsions might have coalesced to form larger particles. The bioactive compounds such as thymol, eugenol, and many more might have degraded during storage, causing a higher percentage of average reduction in antibacterial performance after 20 days of storage (Salvia-Trujillo et al., 2014; Calo et al., 2015). It was suggested that SCG-nanoemulsions with significantly smaller particle sizes than control and SC-, and G-nanoemulsions were more stable upon storage and able to maintain their performance in terms of antibacterial activity.

6. Conclusion

In the current work, Tunisian clove oil was nanoencapsulated with a food-grade emulsifier (NaCas) using Couette emulsification process to address the critical issues like sensitivity to oxygen, light, temperature, or undesirable sensory attributes for its utilization in food, pharmaceutical, and cosmetic applications. The obtained results demonstrated that the mean droplet size, PDI, and zeta potential of the stable O/W nanoemulsion were lowest and highest, respectively. In addition, the nanoemulsion was stable during all the processing conditions encountered during food processing except at pH of 3.0–5.0. The formulation remained stable against coalescence and phase separation for 20 days (storage period). Therefore, it can be stated that clove nanoemulsions obtained by Couette cell and ultrasonic processes are more stable than the coarse emulsion prepared by a conventional high-speed blender. Therefore, galactomannan addition can encapsulate hydrophobic compounds, improve the stability of nanoemulsions, sustain the continuous release of active compounds, and increase the antimicrobial activity of nanoemulsions. In the future, the application of clove EO nanoemulsions in food packaging and matrix should be studied and evaluated to deepen the research. This study offers leadership for formulating emulsions with large dispersed volume fractions and high antioxidant activities, as required in many food, cosmetic, and pharmaceutical applications.

Acknowledgments

The authors wish to thank the Carthage University of Tunisia for financial support.

References

Acosta, E., 2009. Bioavailability of nanoparticles in nutrient and nutraceutical delivery. Curr. Opin. Colloid Interface Sci. 14 (1), 3–15.

Affandi, M.M.M., Julianto, T., Majeed, A., 2011. Development and stability evaluation of astaxanthin nanoemulsion. Asian J. Pharm. Clin. Res. 4, 142–148.

Almadiy, A.A., Nenaah, G.E., Al Assiuty, B.A., Moussa, E.A., Mira, N.M., 2016. Chemical composition and antibacterial activity of essential oils and major fractions of four Achillea species and their nanoemulsions against foodborne bacteria. LWT Food Sci. Technol. 69, 529–537.

Biasi-Garbin, R.P., Saori Otaguiri, E., Morey, A.T., et al., 2015. Effect of eugenol against Streptococcus agalactiae and synergistic interaction with biologically produced silver nanoparticles. Evid. Based Complement. Alternat. Med. 2015, 861497.

Boon, C.S., McClements, D.J., Weiss, J., Decker, E.A., 2010. Factors influencing the chemical stability of carotenoids in foods. Crit. Rev. Food Sci. Nutr. 50, 515–532.

Burt, S., 2004. Essential oils: their antibacterial properties and potential applications in foods—a review. Int. J. Food Microbiol. 94, 223–253.

Calo, J.R., Crandall, P.G., O'Bryan, C.A., Ricke, S.C., 2015. Essential oils as antimicrobials in food systems—a review. Food Control 54, 111–119. https://doi.org/10.1016/j.foodcont.2014.12.040.

Chen, X., Han, Y.u., Meng, H., Li, W., Li, Q., Luo, Y., Zhang, Y., 2019. Characteristics of the emulsion stabilized by polysaccharide conjugates alkali-extracted from green tea residue and its protective effect on catechins. Ind. Crop. Prod. 140, 111611. https://doi.org/10.1016/j.indcrop.2019.111611.

Chouaibi, M., Mahfoudhi, N., Rezig, L., Donsì, F., Ferrari, G., Hamdi, S., 2012. A comparative study on physicochemical, rheological, and surface tension properties of Tunisian jujube (*Zizyphus lotus* L.) seed and vegetable oils. Int. J. Food Eng. 2, 1–8.

Chouaibi, M., Mejri, J., Rezig, L., Abdelli, K., Hamdi, S., 2019a. Experimental study of quercetin microencapsulation using water-in-oil-in-water ($W_1/O/W_2$) double emulsion. J. Mol. Liq. 273, 183–191.

Chouaibi, M., Rezig, L., Lakoud, A., Boussaid, A., Hassouna, M., Ferrari, G., et al., 2019b. Exploring potential new galactomannan source of *Retama reatam* seeds for food, cosmetic and pharmaceuticals: characterization and physical, emulsifying and antidiabetic properties. Int. J. Biol. Macromol. 124, 1167–1176.

Costa, A.L.R., Gomes, A., Tibolla, H., Menegalli, F.C., Cunha, R.L., 2018. Cellulose nanofibers from banana peels as a Pickering emulsifier: high-energy emulsification processes. Carbohydr. Polym. 194, 122–131.

Davidov-Pardo, G., Gumus, C.E., McClements, D.J., 2016. Lutein-enriched emulsion-based delivery systems: influence of pH and temperature on physical and chemical stability. Food Chem. 196, 821–827. https://doi.org/10.1016/j.foodchem.2015.10.018.

Devarajan, V., Ravichandran, V., 2011. Nanoemulsions: as modified drug delivery tool. Int. J. Compr. Pharm. 4, 1–6.

Dickinson, E., 2009. Hydrocolloids as emulsifiers and emulsion stabilizers. Food Hydrocoll. 23 (6), 1473–1482.

Doost, A.S., Camp, J.V., Dewettinck, K., Van der Meeren, P., 2019. Production of thymol nanoemulsions stabilized using Quillaja Saponin as a biosurfactant/antioxidant activity enhancement. Food Chem. 293, 134–143.

Frede, K., Henze, A., Khalil, M., Baldermann, S., Schweigert, F.J., Rawel, H., 2014. Stability and cellular uptake of lutein-loaded emulsions. J. Funct. Foods 8, 118–127.

Ilia, A.A., Nour, A., 2010. Affect of Viscosity and Droplet Diameter on Water-in-Oil (W/O) Emulsions: An Experimental Study. vol. 62 World Academy of Science, Engineering and Technology, Malaysia.

Jayasena, D.D., Jo, C., 2013. Essential oils as potential antimicrobial agents in meat and meat products: a review. Trends Food Sci. Technol. 34 (2), 96–108. https://doi.org/10.1016/j.tifs.2013.09.002.

Khalil, M., Raila, J., Ali, M., et al., 2012. Stability and bioavailability of lutein ester supplements from tagetes flower prepared under food processing conditions. J. Funct. Foods 4, 602–610.

Kheawfu, K., Pikulkaew, S., Rades, T., Müllertz, A., Okonogi, S., 2018. Development and characterization of clove oil nanoemulsions and self micro emulsifying drug delivery systems. J. Drug Deliv. Sci. Technol. https://doi.org/10.1016/j.jddst.2018.05.028.

Lante, A., Friso, D., 2013. Oxidative stability and rheological properties of nanoemulsions with ultrasonic extracted green tea infusion. Food Res. Int. 54, 269–276.

Leong, T.S.H., Wooster, T.J., Kentish, S.E., Ashokkumar, M., 2009. Minimizing oil droplet size using ultrasonic emulsification. Ultrason. Sonochem. 16 (6), 721–727.

Li, M., Yaragudi, N., Afolabi, A., Dave, R., Bilgili, E., 2015. Sub-100 nm drug particle suspensions prepared via wet milling with low bead contamination through novel process intensification. Chem. Eng. Sci. 130, 207–220.

Luo, S.-Z., Hu, X.-F., Pan, L.-H., Zheng, Z., Zhao, Y.-Y., Cao, L.-L., et al., 2019. Preparation of camellia oil-based W/O emulsions stabilized by tea polyphenol palmitate: structuring camellia oil as a potential solid fat replacer. Food Chem. 276, 209–217.

Mahfoudhi, N., Chouaibi, M., Donsi, F., Ferrari, G., Hamdi, S., 2012. Chemical composition and functional properties of gum exudates from the trunk of the Almond (*Prunus dulcis*). Food Sci. Technol. Int. 18 (3), 241–250.

Mahfoudhi, N., Sessa, M., Chouaibi, M., Ferrari, G., Dondi, F., Hamdi, S., 2014. Assessment of emulsifying ability of almond gum in comparison with gum Arabic using response surface methodology. Food Hydrocoll. 37, 49–59.

Martin-Piñero, M.J., Muñoz, J., Alfaro-Rodriguez, M.C., 2020. Improvement of the rheological properties of rosemary oil nanoemulsions prepared by microfluidization and vacuum evaporation. J. Ind. Eng. Chem. 91, 340–346.

Masalova, I., Kapiamba, F., Tshilumbu, N., Malkin, A.Y., 2018. Shear stability of highly concentrated emulsions. Colloid J. 80 (1), 54–58.

Mason, T.G., Wilking, J.N., Meleson, K., Chang, C.B., Graves, S.M., 2006. Nanoemulsions: formation, structure, and physical properties. J. Phys. Condens. Matter 18, R635–R666.

McClements, D.J., 1999. Food Emulsions: Principles, Practices, and Techniques. CRC Press, Taylor & Francis Group, Boca Raton, FL, pp. 300–342.

McClements, D.J., 2012. Nanoemulsions versus microemulsions: terminology, differences, and similarities. Soft Matter 8, 1719–1729.

McClements, D.J., Rao, J., 2011. Food-grade nanoemulsions: formulation, fabrication, properties, performance, biological fate, and potential toxicity. Crit. Rev. Food Sci. Nutr. 51, 285–330.

Nguyen, H., Campi, E., Jackson, W., Patti, A., 2009. Effect of oxidative deterioration on flavour and aroma components of lemon oil. Food Chem. 112, 388–393.

Ni, Y., Li, J., Fan, L., 2020. Effects of ultrasonic conditions on the interfacial property and emulsifying property of cellulose nanoparticles from ginkgo seed shells. Ultrason. Sonochem. 70, 105335.

Pimentel-Moral, S., Ochando-Pulido, J.M., Segura-Carretero, A., Martinez-Ferez, A., 2018. Stabilization of W/O/W multiple emulsion loaded with Hibiscus sabdariffa extract through protein-polysaccharide complexes. LWT 90, 389–395.

Pongsumpum, P., Iwaamoto, S., Siripatrawan, U., 2020. Response surface methodology for optimisation of cimmamon essential oil nanoemulsion with improved stability and antifungal activity. Ultrason. Sonochem. 60, 104604.

Rezig, L., Chibani, F., Chouaibi, M., Dalgalarrondo, M., Hessini, K., Guéguen, J., Hamdi, S., 2013. Pumpkin (*Cucurbita maxima*) seed proteins: sequential extraction processing and fraction characterization. J. Agric. Food Chem. 61 (32), 7715–7721.

Saberi, A.H., Fang, Y., McClements, D.J., 2014. Effect of salts on formation and stability of vitamin E-enriched mini emulsions produced by spontaneous emulsification. J. Agric. Food Chem. 62, 11246–11253.

Salvia-Trujillo, L., Rojas-Graü, A., Soliva-Fortuny, R., Martín-Belloso, O., 2014. Physicochemical characterization and antimicrobial activity of food grade emulsions and nanoemulsions incorporating essential oils. Food Hydrocoll 43, 1–10.

Salvia-Trujillo, L., Sun, Q., Um, B.H., Park, Y., McClements, D.J., 2015. In vitro and in vivo study of fucoxanthin bioavailability from nanoemulsion-based delivery systems: impact of lipid carrier type. J. Funct. Foods 17, 293–304.

Scott, R., 2005. Essential oils. In: Encyclopedia of Analytical Science, second ed. Elsevier, pp. 554–561.

Shao, Y., Wu, C., Wu, T., Li, Y., Chen, S., Yuan, C., Hu, Y., 2018. Eugenol-chitosan nanoemulsions by ultrasound-mediated emulsification: formulation, characterization and antimicrobial activity. Carbohydr. Polym. 193, 144–152.

Sharma, M., Mann, B., Sharma, R., Bajaj, R., Prabin Sarkar, A.S., Pothuraju, R., 2017. Sodium caseinate stabilized clove oil nanoemulsion: physicochemical properties. J. Food Eng. https://doi.org/10.1016/j.jfoodeng.2017.05.006.

Shen, Y., Ni, Z.-J., Thakur, K., Zhang, J.-G., Hua, F., Wei, Z.-J., 2021. Preparation and characterization of clove essential oil loaded nanoemulsion and Pickering emulsion activated pullulan-gelatin based edible film. Int. J. Biol. Macromol. 181, 528–539.

Silva, K.C.G., Sato, A.C.K., 2019. Sonication technique to produce emulsions: the impact of ultrasonic power and gelatin concentration. Ultrason. Sonochem. 52, 286–293.

Singh, Y., Meher, J.G., Raval, K., Khan, F.A., Chaurasia, M., Jain, N.K., Chourasia, M.K., 2017. Nanoemulsion: concepts, development and applications in drug delivery. J. Control. Release 252, 28–49.

Snoussi, A., Chouaibi, M., Bouzouita, N., Hamdi, S., 2020. Microencapsulation of catechin using water-in-oil-in-water (W1/O/W2) double emulsions: study of release kinetics, rheological, and thermodynamic properties. J. Mol. Liq. https://doi.org/10.1016/j.molliq.2020.113304.

Snoussi, A., Chouaibi, M., Haj Koubaier, H., Bouzouita, N., 2021. Encapsulation of Tunisian thyme essential oil in O/W nanoemulsions: application for meat preservation. Meat Sci. 188, 108785. https://doi.org/10.1016/j.meatsci.2022.108785.

Tan, C.P., Nakajima, M., 2005a. Effect of polyglycerol esters of fatty acids on physicochemical properties and stability of b-carotene nanodispersions prepared by emulsification/evaporation method. J. Sci. Food Agric. 85, 121–126.

Tan, C.P., Nakajima, M., 2005b. Beta-carotene nanodispersions: preparation, characterization and stability evaluation. Food Chem. 92, 661–671.

Tippel, J., Lehmann, M., von Klitizing, R., Drusch, S., 2016. Interfacial properties of Quillaja saponins and its use for micellisation of lutein esters. Food Chem. 212, 35–42.

Topuz, O.K., Ozvural, E.B., Zhao, Q., Huang, Q., Chikindas, M., Gölükçü, M., 2016. Physical and antimicrobial properties of anise oil loaded nanoemulsions on the survival of foodborne pathogens. Food Chem. 203, 117–123.

Velderrain-Rodríguez, G.R., Acevedo-Fani, A., Gonzalez-Aguilar, G.A., Martín-Belloso, O., 2019. Encapsulation and stability of a phenolic-rich extract from mango peel within water-in-oil-in-water emulsions. J. Funct. Foods 56, 65–73.

Wan, J., Zhong, S., Schwarz, P., Chen, B., Rao, J., 2019. Physical properties, antifungal and mycotoxin inhibitory activities of five essential oil nanoemulsions: impact of oil compositions and processing parameters. Food Chem. 291, 199–206.

Wu, L., Zhang, J., Watanabe, W., 2011. Physical and chemical stability of drug nanoparticles. Adv. Drug Deliv. Rev. 63, 456–469.

Xie, H., Zhang, Y., Cao, M., Liu, C., Mao, Y., Ren, G., Wu, Z., Fang, S., Tian, S., Wu, D., 2021. Fabrication of PGFE/CN-stabilized β-carotene-loaded peppermint oil nanoemulsions: storage stability, rheological behavior and intelligent sensory analyses. LWT Food Sci. Technol. 138, 110688. https://doi.org/10.1016/j.lwt.2020.110688.

Yuan, Y., Gao, Y., Mao, L., Zhao, J., 2008a. Optimisation of conditions for the preparation of b-carotene nanoemulsions using response surface methodology. Food Chem. 107, 1300–1306.

Yuan, Y., Gao, Y., Zhao, J., Mao, L., 2008b. Characterization and stability evaluation of β-carotene nanoemulsions prepared by high pressure homogenization under various emulsifying conditions. Food Res. Int. 41, 61–68.

Zhang, K., Mao, Z., Huang, Y., Xu, Y., Huang, C., Guo, Y., Ren, X., Liu, C., 2020. Ultrasonic assisted water-in-oil emulsions encapsulating macro-molecular polysaccharide chitosan: influence of molecular properties, emulsion viscosity and their stability. Ultrason. Sonochem. 64, 105018. https://doi.org/10.1016/j.ultsonch.2020.105018.

Zhang, S., Zhang, M., Fang, Z., Liu, Y., 2017. Preparation and characterization of blended cloves/cinnamon essential oil nanoemulsions. LWT-Food Sci. Technol. 75, 316–322.

CHAPTER

Storage quality of clove (*Syzygium aromaticum*) oil

34

Muhammad Imran[a], Muhammad Kamran Khan[a], Muhammad Nadeem[b], Muhammad Haseeb Ahmad[a], Rabia Shabir Ahmad[a], and Muhammad Abdul Rahim[a]

Department of Food Science, Faculty of Life Sciences, Government College University, Faisalabad, Punjab, Pakistan[a]

Department of Dairy Technology, University of Veterinary and Animal Sciences, Lahore, Punjab, Pakistan[b]

1. Introduction

Clove (*Syzygium aromaticum*) is a dried reddish-brown flower bud from an evergreen clove tree. It belongs to the family Myrtaceae producing high-quality oil which is extracted from leaves and flower buds. Clove contains the essential oil with excellent yield ranging from 10% to 20%. The extraction of clove bud essential oil (CBEO) uses conventional distillation and technically advanced supercritical fluid extraction techniques. Eugenol, eugenyl acetate, and caryophyllene are the main components of CBEO with antioxidant, anti-inflammatory, antiseptic, anesthetic, and antimicrobial properties (Singletary, 2014; Das et al., 2020). It is most commonly used in the pharmaceutical, nutraceuticals, and functional food industries as a flavoring and antioxidant agent. It is an excellent source of minerals, vitamins, and essential fatty acids. It is used as a natural medicine for toothaches and other tooth pains, for dry and puffy skin, for spot bunions and plantar warts, improves the blood circulation in the human body, used as a painkiller, used to reduce stomach pain, used as a fragment and antioxidants in soaps, cosmetics, and perfumes. CBEO is suitable for patients with nausea, diarrhea, hernia, and phlegm cough (Idowu et al., 2021). The CBEO composition of fresh extract contains lower eugenol and higher acetyl eugenol as major components. After 4 to 5 months of storage, the CBEO composition is slightly changed due to the increase in eugenol while decreasing acetyl eugenol main components. Furthermore, the chemical process such as isomerization, evaporation, and oxidation can occur during storage, leading to the appearance or disappearance of some components (Hastuti et al., 2017).

2. Extraction of clove bud essential oil

Clove bud essential oil (CBEO) and eugenol are extracted from cloves using the various traditional extraction methods at optimum conditions. The quality of the oil depends on the maturity of clove buds and the extraction process. Golmakani et al. (2017) conducted a comparative study on microwave-assisted and traditional methods to extract the CBEO and eugenol from the best raw material. In this study, a significant difference was found between the microwave-assisted and traditional methods. The yield of CBEO extraction by traditional methods was 11.5% in steam distillation and 12.9% in the

Clove (*Syzygium aromaticum*). https://doi.org/10.1016/B978-0-323-85177-0.00034-3

hydrodistillation process using the Clevenger apparatus. Moreover, the CBEO yield of microwave-assisted methods was estimated at 12.7% in microwave-assisted steam distillation and 13.9% in microwave-assisted hydrodistillation. The results show that the hydrodistillation process recovered more CBEO values of yield than the steam distillation process. In another similar study, the CBEO was extracted from cloves using the microwave-assisted extraction process. In this study, the gray-based Taguchi method was used to obtain maximum yield and eugenol content. The results indicated that the maximum yield of CBEO was achieved at 30 g of plant material, 200 mL of water, 600 W of power, and 30 min of extraction time. Furthermore, the ethanol refluxing method was adopted to separate the eugenol content from cloves (Kapadiya et al., 2018).

Supercritical fluid extraction is an innovative extraction method commonly used in oil extraction. CO_2 is used as a solvent in this extraction process. It is relatively fast, cheap, and easier than oil produced from clove buds, leaves, and shoots. The amount and particle size of eugenol content in CBEO is one of the key factors affecting production. In this method, the maximum yield (19.5%) of CBEO was obtained at high temperature (50°C) and low pressure (10 MPa), while the maximum content of eugenol was 58.7%. Other conventional methods (steam distillation, hydrodistillation, and Soxhlet extraction) in which the higher yield was obtained by Soxhlet extraction (41.8%) and the maximum eugenol contents was 30.8%. The major components and composition of CBEO in various extraction methods were estimated using gas chromatography-mass spectrometry (GCMS). The CBEO pale-yellow color and texture appeared in the supercritical fluid extraction and steam distillation methods. The results show that the supercritical fluid extraction was considered the optimal process for obtaining CBEO with high quality (Guan et al., 2007). In the review study of Khalil et al. (2017), the CBEO and eugenol contents were extracted from clove buds using various extensively employed methods such as solvent extraction, hydro- and steam distillation, microwave-based extraction, supercritical fluid extraction, and ultrasound-assisted extraction. Supercritical fluid extraction was mostly preferred to extract the oil from plant and animal sources due to its relatively rapid, low in viscosities, obtain good quality of oil, and high diffusivities associated with supercritical fluid. In another study, the steam hydrodistillation method was used to extract the CBEO from wet and dry clove buds. The extraction procedure was performed in 3 to 6 h. Optimization of CBEO was obtained using the FTIR and GCMS results. In this study, the clove buds waste was processed with tapioca (20%) and chitosan (1%) in acid solution to prepare the cardboard; further, it was used as a packaging material in many food products (Ratri et al., 2020).

In another comparative research work, supercritical fluid extraction with and without ultrasound assistance was used to extract the CBEO from cloves at optimum levels. The heat reflux extraction and steam distillation methods were used as traditional methods. The results reveal that the high quality, short extraction time, and better yield of supercritical fluid extraction combined with ultrasound assistance of CBEO were noted than traditional methods (Wei et al., 2016b). In similar research, the ultrasound-assisted supercritical carbon dioxide (USC-CO_2) method combined with GCMS analysis was used to extract the CBEO and its major bioactive constituents from clove buds. The optimal conditions that obtained the highest amount of CBEO were pressure 10 MPa and temperature 31°C. The quality and color of CBEO were better than other traditional methods. The USC-CO_2 method was obtained 20.8% more than extraction yield compared with other heat reflux extraction methods. It was documented as a simple operation process and suitable for industrial production (Hatami et al., 2010; Yang et al., 2014; Wei et al., 2016a).

3. Chemistry of clove bud essential oil

The clove bud essential oil (CBEO) contained a total of 23 identified components. Eugenol, eugenyl acetate, and caryophyllene are the major constituents of CBEO. The color and quality of CBEO depend on the number of these major compounds. Eugenol is a type of phenol formally derived from the allyl chain-substituted guaiacol; it is a member of the allylbenzene class of chemical compounds. It provides a pale-yellow color liquid and is slightly bitter. The binomial name for eugenol is 4-allyl-2-methoxy phenol (Jirovetz et al., 2006; Giuliani, 2014). Eugenyl acetate is found naturally in CBEO in combination with eugenol. Its chemical structure is larger than the eugenol structure, and it also contains more oxygen components. It is produced by a chemical reaction that involves the acetylation of acetic anhydride with eugenol. Thus, it is an important aromatic and phytochemical component of CBEO. It provides CBEO a yellowish color and aroma and is found in crystalline solid (dos Santos et al., 2009). The contents of caryophyllene is a main plant volatile compound found in CBEO as a mixture of organic compounds. It is found naturally in the form of a cyclobutane ring. It is an isocaryophyllene component that gives flavor to CBEO (Chaieb et al., 2007).

4. Storage stability

The CBEO is a colorless or pale-yellow oil in a state of pure compound. The colorless or pale-yellow color appears in CBEO due to the presence of simple phenol solid. When CBEO is usually kept for 4 to 5 months, then it has a dark color due to oxidization if it reacts with air. The solubility of water depends on the number of hydroxyl groups. The solubility of polar organic solvents in water is improved by increasing the hydroxyl groups. Therefore, the phenol content is more soluble in the sodium hydroxide solution than in water (Alfikri et al., 2020).

The clove bud essential oil (CBEO) is extracted from mature buds, leaves, and shoots. It is colorless or pale yellow and darkens during storage. The darkness of CBEO occurred due to maximum caryophyllene contents and minimum eugenyl acetate contents. The major problem of CBEO yield is low in quality due to the high amount of eugenol content and dark color of oil during storage. Therefore, innovative extraction is required to improve the yield and quality of CBEO (Nowak et al., 2012). In another research work, the steam distillation process was used to extract the CBEO from fresh and dried clove buds. The extracted oil was stored in an aluminum polyethylene bag at a laboratory room temperature for 4 months. The results showed that the value of caryophyllene was slightly reduced during storage, while the eugenyl acetate content reduced during storage and eugenol content improved (Hastuti et al., 2017).

5. Storage quality of food products supplemented with clove bud essential oil

Clove bud essential oil (CBEO) contains various biological activities. Therefore, it is used in many food products for multiple purposes. The CBEO was used as a coating material to improve cooked pork sausages' quality and shelf life in research work. The sample was stored at a refrigerated temperature for 25 days. After storage, the peroxide value and sensory evaluation of the sample were better than the control sample. Furthermore, the CBEO was used as an antilisteric to prevent bacterial growth during various storage intervals in meat and cheese samples. The results show that the concentration of CBEO 0.5% to 1% significantly reduced the bacterial growth compared with control samples after 5 days at 30°C. A similar difference was recorded in meat and cheese samples at 30°C during storage (Menon and Garg, 2001; Lekjing, 2016).

The inhibitory effect of CBEO was used to retard the growth of *Listeria monocytogenes* in chicken frankfurters at low temperatures. All strains of *Listeria monocytogenes* were treated with 1% and 2% CBEO. The treated samples were stored at 5°C and 15°C for 7 to 14 days. The results showed that all treated samples of bacterial growth were inhibited under both storage conditions compared with the control sample (Mytle et al., 2006). Furthermore, the CBEO was also used as a natural antioxidant to retard the oxidation process, provided excellent flavor, and increased fish fillet quality during cooking. The results also showed that the CBEO reduced the lipid peroxidation, which improved the quality and flavor of fish fillet during grilling (Oskoueian et al., 2013).

The CBEO was used to develop the designer chevon cutlets, and its effect on the storage quality of the product was also assessed. The treated samples were determined for different physicochemical, microbial, and sensory attributes. The product was successfully stored for 15 days under storage conditions in the presence of 0.01% CBEO without any quality loss (Singh et al., 2015). The CBEO and cinnamon oil were used to protect the grains from insects. The results indicated that the oils significantly reduced insects' activity over time (Jumbo et al., 2014). A study carried out by Xiong et al. (2021) used various coating materials to improve the shelf life of fresh salmon fillets during storage conditions. In this research work, the combination of various bioactive ingredients significantly enhanced the antimicrobial and antioxidant effects on fresh salmon fillets under refrigeration conditions. The gelatine and chitosan with clove oil improved the shelf life of fresh salmon fillets for at least 5 days. The biological components of this excellent performance can be used on the commercial level to improve the shelf life of fish fillets.

The inhibitory effect of CBEO, which was applied to the surface of dried persimmon fruits, was measured on fungal species at 20°C for 4 weeks. *Rhizopus oryzae* and *Aspergillus niger* survived, and grew on control samples at 20°C, but growth was inhibited under combined treatments of UV-C light and CBEO. The addition of CBEO to dried persimmon fruits prevented mold growth and extended the shelf life by at least 28 days (Vurmaz and Gündüz, 2020). CBEO is a natural source of secondary metabolites with antioxidant properties. The stability of β-carotene is very low during storage and when exposed to light. The CBEO was used as a natural antioxidant agent to protect the emulsion of food-grade soy oil from photodegradation of β-carotene. The treated samples with CBEO did not affect the color of all three concentrations of the emulsion during dark storage (Zhou et al., 2021). Tajik et al. (2014) designed a research work to evaluate the inhibition mechanism of CBEO and grape seed extract on spoilage populations and apply it to buffalo patties to reduce the risk of contamination during storage at 8°C for 9 days. The CBEO treated samples showed a significant effect on the pathogen growth. On the other hand, the grape seed extract did not show an inhibitory effect on the spoilage microorganism.

The antioxidant properties of CBEO, carboxymethyl cellulose with CBEO, and their combination for improving the shelf life of pistachio during the storage period of 25 to 50 days at 25°C were investigated in this study. The results revealed that all treated samples retained the firmness of pistachio compared to the control sample. The amount of phenolic compounds, total chlorophyll of kernel, and kernel carotenoids were not maintained in the treated samples during storage (Mirdehghan, 2021). Taghavi et al. (2021) carried out a comparative study, the five essential oils are used to protect the strawberry from fungal growth and maintain fruit quality during storage. According to the results, these five essential oils improved the shelf life of strawberries and prevented the fungal growth during storage at 4°C and 95% humidity for 4 weeks.

The edible coating has been used to improve the product quality and extend the shelf life of different banana varieties. The CBEO and polyvinyl polychloride were used as an alternative for preserving the banana from fungi during the storage period. Cassava starch film containing CBEO and polyvinyl polychloride and their combination were used to inhibit biodegradation in fruits during the postharvest period. Edible coating of CBEO showed a limited reduction in fungal growth and was not effective against the yeast. Nevertheless, the results concluded that minimum weight loss and satisfactory quality and shelf life of bananas were improved during the storage period (de Figueiredo Sousa et al., 2019). In a research work of Boukaew et al. (2017), four essential oils were used to remove the fungus from infected maize seeds during the storage for 14 days at room temperature. According to the results, the high concentrations of clove and vatica oil compared to the control samples significantly affected the values of infected maize seeds.

The antioxidant properties of CBEO are well known; therefore, it was used in the boiler meat emulsion to improve the quality of chicken nuggets at refrigerated storage. The CBEO significantly affected the physicochemical properties of an experimental food product. During the storage period, the amount of chemical composition in the product increased, and the pH value decreased. The overall acceptability was determined for the CBEO-treated product, and it was packaged in pouches in refrigerated condition (Para et al., 2017). The addition of CBEO in the boneless chicken meat showed a significant effect on the physical, chemical, and microbial properties of beef sucuk. Thyme and CBEO were used as coating materials. The prepared product was stored under refrigeration temperature for 0, 15, 30, and 45 days. The CBEO-coated sample improved the storage quality of heat-treated sucuk at 4°C (Saricaoglu and Turhan, 2019).

6. Biological activities

The use of medicinal plants has been increased due to the effectiveness, no harmful side effects on health, high quality and low cost, and lack of access to modern drugs. For example, CBEO is a herbal medicine with major biological activities. Therefore, CBEO is very important in medicinal plants because it contains major phytochemical compounds and their derivatives. It is also used as a food flavoring agent in various food and cosmetic industries. In addition, it is used as a pharmaceutical and nutraceutical medicine due to its antifungal, antibacterial, antiseptic, antioxidant, anticarcinogenic, and antibiotic properties (Mittal et al., 2014).

1. The antibacterial activity of CBEO indicated that the major components could potentially denature the proteins and react with cell membrane phospholipids and change the permeability of the cell membrane. Therefore, it is very effective against harmful bacteria (Nuñez and D'Aquino, 2012).
2. Antifungal activity of eugenol is shown against phytopathogenic fungi due to lysis of the spores and micelles like saprophytic fungus, rhodina, *Cryptococcus neoformans*, canker diseases, and *Cryptococcus albicans* (Rana et al., 2011).

3. Eugenol is the main component of CBEO. It has strong antioxidant properties compared to synthetic antioxidants such as butylated hydroxyl anisole and phenolic antioxidants (Gülçin et al., 2012).
4. The CBEO contains the nonsteroidal anti-inflammatory drug, which is more effective against swelling of inflammation and other painful conditions (Han and Parker, 2017).
5. The CBEO also contains the penicillin-related antibiotic, which helps treat bacterial infections caused by *Staphylococcus aureus* (Chaieb et al., 2007).
6. The CBEO has anticarcinogenic activity due to the natural presence of oleanic acid. Therefore, it has potential therapeutic effects on various diseases and their symptoms (Kouidhi et al., 2010).

7. Multidisciplinary applications

Clove bud essential oil (CBEO) and its components have traditionally been used for a variety of purposes as follows:

1. CBEO for dry skin is great natural medicine. It can increase the blood flow to the skin and make it look brighter.
2. It is suitable for your teeth and gum due to its strong germicidal properties (Agrawal et al., 2014; Kaur and Chandrul, 2017).
3. It is very useful against unpleasant fungal infections commonly found in the female reproductive tract, nails, middle ear, and inner ear (Milind and Deepa, 2011; Cui et al., 2018).
4. It can be used to prevent bacterial infections (Santa Packyanathan and Prakasam, 2017).
5. Great for relieving respiratory conditions such as cough and asthma.
6. It can increase the blood flow through the area and help reduce the pain.
7. It is used to treat indigestion and stomach-related problems (Bhowmik et al., 2012).
8. It helps boost the immune system by reducing toxin levels in your blood.
9. It is used to reverse the damage caused by free radicals, including cell death and cancer (Rani and Jena, 2021).
10. It promotes hair growth and makes it stronger (Shahtalebi et al., 2016).
11. It can be used to increase sexual health.
12. It is perfect for heart health due to acetyl eugenol, a potent platelet inhibitor (Singletary, 2014).

8. Conclusions

Clove (*Syzygium aromaticum* L., family Myrtaceae) and clove bud essential oil can be used in different food products. Clove bud essential oil has gained great intention of food processing and biopharma industries to develop stable products using health-promoting traits for discerning consumers. However, more research areas should be explored in science and technological fields to use clove oil as a natural ingredient in pharmaceutical, nutraceutical, and/or food products.

References

Agrawal, M., Agrawal, S., Rastogi, R., Singh, P., Adyanthaya, B.R., Gupta, H.L., 2014. A review on uses of clove in oral and general health. Indian J. Res. Pharm. Biotechnol. 2 (4), 1321.

Alfikri, F.N., Pujiarti, R., Wibisono, M.G., Hardiyanto, E.B., 2020. Yield, quality, and antioxidant activity of clove (*Syzygium aromaticum* L.) bud oil at the different phenological stages in young and mature trees. Scientifica 2020, 1–8.

Bhowmik, D., Kumar, K.S., Yadav, A., Srivastava, S., Paswan, S., Dutta, A.S., 2012. Recent trends in Indian traditional herbs *Syzygium aromaticum* and its health benefits. J. Pharmacogn. Phytochem. 1 (1), 13–22.

Boukaew, S., Prasertsan, P., Sattayasamitsathit, S., 2017. Evaluation of antifungal activity of essential oils against aflatoxigenic *Aspergillus flavus* and their allelopathic activity from fumigation to protect maize seeds during storage. Ind. Crop. Prod. 97, 558–566.

Chaieb, K., Hajlaoui, H., Zmantar, T., Kahla-Nakbi, A.B., Rouabhia, M., Mahdouani, K., Bakhrouf, A., 2007. The chemical composition and biological activity of clove essential oil, Eugenia caryophyllata (*Syzigium aromaticum* L. Myrtaceae): a short review. Phytother. Res. 21 (6), 501–506.

Cui, H., Zhang, C., Li, C., Lin, L., 2018. Antimicrobial mechanism of clove oil on *Listeria monocytogenes*. Food Control 94, 140–146.

Das, M., Roy, S., Guha, C., Saha, A.K., Singh, M., 2020. In vitro evaluation of antioxidant and antibacterial properties of supercritical CO_2 extracted essential oil from clove bud (*Syzygium aromaticum*). J. Plant Biochem. Biotechnol., 1–5.

de Figueiredo Sousa, H.A., de Oliveira Filho, J.G., Egea, M.B., da Silva, E.R., Macagnan, D., Pires, M., Peixoto, J., 2019. Active film incorporated with clove essential oil on storage of banana varieties. Nutr. Food Sci. 39 (9), 655–661.

dos Santos, A.L., Chierice, G.O., Alexander, K., Riga, A., 2009. Crystal structure determination for eugenyl acetate. J. Chem. Crystallogr. 39 (9), 655–661.

Giuliani, F., 2014. The composition, structure, sources, and applications of eugenol. ESSAI 12 (1), 19.

Golmakani, M.T., Zare, M., Razzaghi, S., 2017. Eugenol enrichment of clove bud essential oil using different microwave-assisted distillation methods. Food Sci. Technol. Res. 23 (3), 385–394.

Guan, W., Li, S., Yan, R., Tang, S., Quan, C., 2007. Comparison of essential oils of clove buds extracted with supercritical carbon dioxide and other three traditional extraction methods. Food Chem. 101 (4), 1558–1564.

Gülçin, İ., Elmastaş, M., Aboul-Enein, H.Y., 2012. Antioxidant activity of clove oil-A powerful antioxidant source. Arab. J. Chem. 5 (4), 489–499.

Han, X., Parker, T.L., 2017. Anti-inflammatory activity of clove (*Eugenia caryophyllata*) essential oil in human dermal fibroblasts. Pharm. Biol. 55 (1), 1619–1622.

Hastuti, L.T., Saepudin, E., Cahyana, A.H., Rahayu, D.U.C., Murni, V.W., Haib, J., 2017. The influence of sun drying process and prolonged storage on composition of essential oil from clove buds (*Syzygium aromaticum*). AIP Conf. Proc. 1862 (1), 030092.

Hatami, T., Meireles, M.A.A., Zahedi, G., 2010. Mathematical modeling and genetic algorithm optimization of clove oil extraction with supercritical carbon dioxide. J. Supercrit. Fluids 51 (3), 331–338.

Idowu, S., Adekoya, A.E., Igiehon, O.O., Idowu, A.T., 2021. Clove (*Syzygium aromaticum*) spices: a review on their bioactivities, current use, and potential application in dairy products. J. Food Meas. Charact. 15 (4), 3419–3435.

Jirovetz, L., Buchbauer, G., Stoilova, I., Stoyanova, A., Krastanov, A., Schmidt, E., 2006. Chemical composition and antioxidant properties of clove leaf essential oil. J. Agric. Food Chem. 54 (17), 6303–6307.

Jumbo, L.O.V., Faroni, L.R., Oliveira, E.E., Pimentel, M.A., Silva, G.N., 2014. Potential use of clove and cinnamon essential oils to control the bean weevil, *Acanthoscelides obtectus* say, in small storage units. Ind. Crop. Prod. 56, 27–34.

Kapadiya, S.M., Parikh, J., Desai, M.A., 2018. A greener approach towards isolating clove oil from buds of *Syzygium aromaticum* using microwave radiation. Ind. Crop. Prod. 112, 626–632.

Kaur, D., Chandrul, K.K., 2017. *Syzygium aromaticum* L. (clove): a vital herbal drug used in periodontal disease. Indian J. Pharm. Biol. Res. 5 (02), 45–51.

Khalil, A.A., ur Rahman, U., Khan, M.R., Sahar, A., Mehmood, T., Khan, M., 2017. Essential oil eugenol: sources, extraction techniques and nutraceutical perspectives. RSC Adv. 7 (52), 32669–32681.

Kouidhi, B., Zmantar, T., Bakhrouf, A., 2010. Anticariogenic and cytotoxic activity of clove essential oil (*Eugenia caryophyllata*) against a large number of oral pathogens. Ann. Microbiol. 60 (4), 599–604.

Lekjing, S., 2016. A chitosan-based coating with or without clove oil extends the shelf life of cooked pork sausages in refrigerated storage. Meat Sci. 111, 192–197.

Menon, K.V., Garg, S.R., 2001. Inhibitory effect of clove oil on *listeria monocytogenes* in meat and cheese. Food Microbiol. 18 (6), 647–650.

Milind, P., Deepa, K., 2011. Clove: a champion spice. Int. J. Res. Ayurveda Pharm. 2 (1), 47–54.

Mirdehghan, H., 2021. Investigation of the effect of combined treatment of clove plant essential oil (*Syzygium aromaticum*) with carboxymethylcellulose on storage of fresh pistachio. Food Sci. Technol. 18 (111), 279–294.

Mittal, M., Gupta, N., Parashar, P., Mehra, V., Khatri, M., 2014. Phytochemical evaluation and pharmacological activity of *Syzygium aromaticum*: a comprehensive review. Int J Pharm Pharm Sci 6 (8), 67–72.

Mytle, N., Anderson, G.L., Doyle, M.P., Smith, M.A., 2006. Antimicrobial activity of clove (*Syzgium aromaticum*) oil in inhibiting *listeria monocytogenes* on chicken frankfurters. Food Control 17 (2), 102–107.

Nowak, K., Ogonowski, J., Jaworska, M., Grzesik, K., 2012. Clove oil-properties and applications. Chemik 66 (2), 145–152.

Nuñez, L., D'Aquino, M., 2012. Microbicide activity of clove essential oil (*Eugenia caryophyllata*). Braz. J. Microbiol. 43 (4), 1255–1260.

Oskoueian, E., Maroufyan, E., Goh, Y.M., Ramezani-Fard, E., Ebrahimi, M., 2013. Clove essential oil improves lipid peroxidation and antioxidant activity in tilapia fish fillet cooked by grilling and microwaving. Int. Sch. Sci. Res. Innov. 7 (12), 845–847.

Para, P.A., Kumar, S., Raja, W.H., Ganguly, S., 2017. Effect of clove oil on some quality characteristics and sensory attributes of papaya pulp enriched enrobed chicken nuggets at refrigerated storage. Indian J. Poult. Sci. 52 (1), 96–103.

Rana, I.S., Rana, A.S., Rajak, R.C., 2011. Evaluation of antifungal activity in essential oil of the *Syzygium aromaticum* (L.) by extraction, purification and analysis of its main component eugenol. Braz. J. Microbiol. 42 (4), 1269–1277.

Rani, R., Jena, M.K., 2021. Clove (*Syzygium aromaticum*): beneficial effects on human health: a review. Plant Arch. 21 (1), 1967–1972.

Ratri, P.J., Ayurini, M., Khumaini, K., Rohbiya, A., 2020. Clove oil extraction by steam distillation and utilization of clove buds waste as potential candidate for eco-friendly packaging. J. Bahan Alam Terbarukan 9 (1), 47–54.

Santa Packyanathan, J., Prakasam, G., 2017. Antibacterial effect of clove oil against clinical strains of *Escherichia coli*. J. Pharm. Sci. Res. 9 (7), 1203.

Saricaoglu, F.T., Turhan, S., 2019. Performance of mechanically deboned chicken meat protein coatings containing thyme or clove essential oil for storage quality improvement of beef sucuks. Meat Sci. 158, 107912.

Shahtalebi, M.A., Sadat-Hosseini, A., Safaeian, L., 2016. Preparation and evaluation of clove oil in emu oil self-emulsion for hair conditioning and hair loss prevention. J. HerbMed Pharmacol. 5 (2), 72–77.

Singh, P.K., Kumar, S., Bhat, Z.F., Kumar, P., 2015. Effect of Sorghum bicolour and clove oil on the quality characteristics and storage quality of aerobically packaged chevon cutlets. Nutr. Food Sci. 45 (1), 145–163.

Singletary, K., 2014. Clove: overview of potential health benefits. Nutr. Today 49 (4), 207–224.

Taghavi, T., Bahamdan, G., Bell, M., Patel, H., Taylor, E., Rafie, R., 2021, May. Comparison of five essential oils on postharvest strawberry fruit quality. In: IX International Strawberry Symposium 1309, pp. 941–946.

Tajik, H., Farhangfar, A., Moradi, M., Razavi Rohani, S.M., 2014. Effectiveness of clove essential oil and grape seed extract combination on microbial and lipid oxidation characteristics of raw buffalo patty during storage at abuse refrigeration temperature. J. Food Process. Preserv. 38 (1), 31–38.

Vurmaz, A.K., Gündüz, G.T., 2020. Inhibition of mold growth on the surface of dried persimmons using combined treatments of UV-C light and clove oil. Innov. Food Sci. Emerg. Technol. 61, 102336.

Wei, M.C., Lin, P.H., Hong, S.J., Chen, J.M., Yang, Y.C., 2016a. Development of a green alternative procedure for the simultaneous separation and quantification of clove oil and its major bioactive constituents. ACS Sustain. Chem. Eng. 4 (12), 6491–6499.

Wei, M.C., Xiao, J., Yang, Y.C., 2016b. Extraction of α-humulene-enriched oil from clove using ultrasound-assisted supercritical carbon dioxide extraction and studies of its fictitious solubility. Food Chem. 210, 172–181.

Xiong, Y., Kamboj, M., Ajlouni, S., Fang, Z., 2021. Incorporation of salmon bone gelatine with chitosan, gallic acid and clove oil as edible coating for the cold storage of fresh salmon fillet. Food Control 125, 107994.

Yang, Y.C., Wei, M.C., Hong, S.J., 2014. Ultrasound-assisted extraction and quantitation of oils from *Syzygium aromaticum* flower bud (clove) with supercritical carbon dioxide. J. Chromatogr. A 1323, 18–27.

Zhou, Y.M., Chang, H.T., Zhang, J.P., Skibsted, L.H., 2021. Clove oil protects β-carotene in oil-in-water emulsion against photodegradation. Appl. Sci. 11 (6), 2667.

SECTION 4

CLOVE EXTRACTS: CHEMISTRY, FUNCTIONALITY AND APPLICATIONS

CHAPTER

Composition and functionality of clove (*Syzygium aromaticum*) extracts

35

Kevser Karaman[a] and Mahmut Kaplan[b]

Faculty of Agriculture, Department of Agricultural Biotechnology, Erciyes University, Kayseri, Turkey[a] Faculty of Agriculture, Department of Field Crops, Erciyes University, Kayseri, Turkey[b]

Abbreviations

AAE	ascorbic acid equivalent
ABTS	2,2′-azinobis-(3-ethylbenzothiazoline-6-sulfonic acid)
AChE	acetylcholinesterase
BChE	butyrylcholinesterase
BHA	butylated hydroxyanisole
BHT	butylated hydroxytoluene
BW	body weight
CE	clove extract
DAD	diode array detector
DPPH	2,2-diphenyl-1-picrylhydrazyl
ESI-TOF-MS	electrospray ionization-time of flight-mass spectrometry
FRAP	ferric reducing antioxidant power
FTIR	Fourier transform infrared spectroscopy
GAE	gallic acid equivalent
GC-MS	gas chromatography-mass spectrometry
GLC	gas-liquid chromatography
GRAS	generally recognized as safe
HPLC	high-performance liquid chromatography
IC50	inhibitory concentration for 50%
LC-MS	liquid chromatography-mass spectrometry
LC-MS-IT-TOF	liquid chromatography-mass spectrometry-ion trap-time of flight
MAE	microwave-assisted extraction
MIC	minimum inhibitory concentration
ND	not determined
PEF	pulsed electric field
PVP	polyvinylpyrrolidone
QE	quercetin equivalent
SFE	supercritical fluid extraction
TAE	tannic acid equivalent
TBHQ	tert-butylhydroquinone
TPC	total phenolic content

Clove (*Syzygium aromaticum*). https://doi.org/10.1016/B978-0-323-85177-0.00013-6

UAE	ultrasound-assisted extraction
UHPLC	ultra high-performance liquid chromatography
UV	ultraviolet
XRD	X-ray diffraction

1. Clove (*Syzygium aromaticum*)

Syzygium aromaticum is a tree in the family of *Myrtaceae*, and it is known that it is native to the Maluku Islands in Indonesia and is also called as clove. *Syzygium aromaticum* tree is a tropical and evergreen which grows up to 20m in height and 30cm in trunk diameter. The tree has leaves and buds and the buds are the commercial parts of the plants (Batiha et al., 2020; Cortés-Rojas et al., 2014a). The clove is constituted 15%–20% essential oil by weight, and the major compound of the oil is eugenol (70%–95%). The chemical composition of fully developed clove buds contains 12.8% water, 16.4% oil, 4.7% ash, and 8.5% fiber (Nurdjannah and Bermawie, 2012). Clove is the aromatic flower bud of the tree, and it has wide usage as a spice for food flavoring or preservation. It is also one of the most important medicinal plants because of the rich bioactive constituents, and so it shows important pharmacological activities (Batiha et al., 2020). The clove sold as a spice or condiment is the dried and unexpanded flower of *Syzygium aromaticum*, and it has wide usage in the food and pharmaceutical industries. *Syzygium aromaticum* draws attention due to its biofunctional properties, and it has also been widely used in medicinal treatments such as dental care (El-Maati et al., 2016). Especially, the essential oil of the clove buds showed a significant effect on dental care, and it has a strong antimicrobial activity against many pathogenic bacteria (Cai and Wu, 1996; Ramadan et al., 2013).

2. Extraction of clove (*Syzygium aromaticum*)

The term extraction means that the separation of medicinally active parts of plants by using selective solvents (Handa et al., 2008). The main aim of the extraction process is to separate the soluble active compounds from the insoluble cellular residues. To determine the biological activity of the medicinal plants, extraction is the first and essential step before starting the processing. There are different extraction methods to obtain the soluble active compounds, such as maceration or Soxhlet extraction techniques which are commonly used traditionally and also different modern extraction methods are also developed, namely microwave-assisted or ultrasound-assisted extraction and supercritical fluid extraction, which provide an increased yield at lower cost (Azmir et al., 2013; Azwanida, 2015). Clove (*Syzygium aromaticum*) is a popular medicinal plant used by people worldwide because of its rich bioactive compounds. It was reported that the bioactive performance of clove buds is better than those of many fruits, vegetables, and some spices because its extract is composed of many biofunctional compounds, and also it is rich in essential oil showing better antioxidant and antimicrobial activities (Cortés-Rojas et al., 2014b; Miyazawa and Hisama, 2003). Clove buds are rich in essential oil, and also phenolic-rich extracts are produced from the clove buds by using different extraction techniques. Jirovetz et al. (2006) reported that the clove buds contained 18% essential oil, which is composed of eugenol, eugenyl acetate, and β-caryophyllene, and the differences in the clove essential oil levels and also the oil composition of the clove buds change depending on several factors such as variety, agroeconomical situations, and also the extraction process (Hastuti et al., 2017). Different scientific researches are reporting the extraction of clove buds and their bioactivity. El-Maati et al. (2016) investigated the biological characteristics of the solvent extracts produced by different solvents showing different polarity, namely ethanol (80%), ethyl acetate, and water, and they reported that the ethanol and water showed the superior phenolic extraction performance than that of ethyl acetate and also the water extracts showed the highest flavonoid content compared to ethanol and ethyl acetate. The extract yields were 11.7, 7.5, and 3.9 g extract/100 g for ethanol, water, and ethyl acetate, respectively. The significant differences in the yield values were explained by the differences in the polarity of the compounds in the plant structure. In addition to that, significant differences were observed in the total phenolic content (TPC) of the extracts and it was reported that the total phenolic content of water and ethanolic extracts (ca. 230 mg GAE/g extract) was significantly higher than that of ethyl acetate extract (58.8 mg GAE/g extract). For the flavonoid content, water was characterized as the best solvent for the isolation of flavonoids, with the total flavonoid level as 17.5 mg QE/g extract followed by ethanol (12 mg QE/g extract) and ethyl acetate (4.70 mg QE/g extract). In the same study, it was monitored that the water and ethanolic extracts of the clove buds performed a comparable antioxidant activity to the tert-butylhydroquinone (TBHQ), which is a popular and strong synthetic antioxidant substance (El-Maati et al., 2016). Köprü et al. (2020) carried out an optimization study to determine the best processing conditions of solvent extraction system using response surface methodology for the fabrication of clove extract, and they selected the liquid/solid ratio and ethanol concentration. The samples' total phenolic content and flavonoid content ranged between 13.8 and 23.8 mg GAE/mL and 3.64 and 7.83 mg QE/mL, respectively. Finally, they revealed that the optimum conditions for the highest phenolic concentration and antioxidant activity were observed for the clove extract produced by 53% ethanol and 4 mL/g sample. In another study, different solvents were used to extract clove buds under different dry heating techniques such as microwave and roasting, and it was reported that the heating provided a significant increase in bioactivity. Also, it was concluded that the highest amount of phenols and flavonoids were extracted in 80% acetone (4053–4064 mg TAE equivalent/100 g) and 80% methanol (11,271–11,370 mg QE/100 g), respectively (Nikousaleh and Prakash, 2016). Hemalatha et al. (2016) investigated the bioactive characteristics of clove bud extracts and reported that the highest extract yield was determined in methanolic extracts compared to acetone and chloroform. In the same study, as correlated with the yield, methanolic extracts showed better radical scavenging activity than other solvents. In another similar research, extract yield and total phenolic contents of the aqueous and ethanolic extracts of clove buds were reported as 3.23 and 3.20 g/20 g plant and 179 and 264 µg GAE/mg sample (Gülçin et al., 2004).

In the extraction process of food ingredients, due to some disadvantages of the maceration or solvent extraction such as time-consuming and the presence of organic solvent residue, which can be toxic and a source of pollution to the environment in addition to the variation of the yields according to the combination of the temperature, time, and solvent conditions, many modern extraction approaches have been improved. So, these techniques present high-efficiency, non-toxic characteristics, and use lower solvent levels. Some of these extraction techniques are ultrasound-assisted extraction (UAE), supercritical fluid extraction (SFE), microwave-assisted extraction (MAE), pulsed electric field (PEF), etc. Based on the processing approach of these extraction processes, it can be said that these methods show better extraction efficiencies than that of the conventional ones, with less time required and higher antioxidant activities and also these are more ecological and economical in the long term (Arias et al., 2009; Santos et al., 2010; Ferrentino et al., 2018; Rodríguez-Rojo et al., 2012; Molina et al., 2020).

Ultrasound-assisted extraction (UAE) is one of the modern and popular extraction systems used in extract production from medicinal plants. As is known, ultrasound can be defined as a special sound that is not in the spectrum of the human hearing capacity. In this technique, ultrasound waves pass through a medium by creating compression and expansion. During the ultrasound application, cavitation results in occurrence, growth, and collapse of bubbles, and high-level energy is released. Ultrasound energy can provide facilitation of organic and inorganic compounds leaching from the plant matrix (Herrera and Luque De Castro, 2005; Azmir et al., 2013). UAE provides an enhancement of the solvent penetration in the cells by the cavitation bubbles and the disruption of tissue surface structures (Vilkhu et al., 2008). UAE is a green technique and it provides acceleration for the extraction process and decreases energy consumption. This technique presents some advantages in some topics such as productivity, yield, and processing time and extracts quality (Chemat and Khan, 2011). Alexandru et al. (2013) performed interesting research about the application of ultrasound for the extraction of clove buds and used batch and flow reactors equipped with ultrasound processors for the extraction. They investigated the effect of flow and ultrasound application effect. They revealed that the highest extract yields were recorded for the flow ultrasound-assisted extraction system at higher flow rates than maceration, showing the lowest extraction yields. TPC of the samples was in the range of 191–215 mg GAE/L extract and the highest levels of the TPC were measured for the extract produced by flow ultrasound-assisted extraction system at higher flow rates, and a correlation was observed with the flow rate and total phenolic content of the crude extracts (Alexandru et al., 2013).

The other popular modern extraction process is supercritical fluid extraction (SFE) which can be used in plant extract production. This extraction technique is applied in environmental, pharmaceutical, and polymer applications and food analysis after this method has attracted wide scientific interest (Zougagh et al., 2004). As is well known, every substance can show three basic states such as solid, liquid, and gas. The term supercritical means a distinctive state, and this state can be obtained only if the matter is exposed to temperature and pressure beyond its critical point, which means that the characteristic temperature and pressure above which distinctive gas and liquid phases do not exist (Inczedy et al., 1998). So, the supercritical fluid shows a gas-like characteristic in the diffusion, viscosity, or surface tension properties, and these characteristic properties provide a superior extraction performance with high yield values from the plant tissues in a short time (Sihvonen et al., 1999). Ivanovic et al. (2013) investigated the effect of SFE of clove buds on the biofunctional performance and revealed that the extracts obtained by supercritical fluid showed the highest eugenol concentration (64%) and also the total phenolic contents (530 mg GAE/g extract). And these extracts also performed a superior antiradical activity comparable to synthetic antioxidant substances. So they revealed that supercritical fluid extraction is an efficient method to produce clove bud extract showing bioactive performance (Frohlich et al., 2019). The extraction of clove leaf by supercritical fluid extraction revealed a higher eugenol content compared to Soxhlet extraction approach, so higher antioxidant performance was observed for the extract produced by supercritical fluid extraction. At the end of the study, it was stated that supercritical fluid extraction proved to be an efficient extraction method compared to Soxhlet extraction because it works at lower temperatures and requires less time and does not require an additional step to remove the solvent.

3. Composition of clove extracts

The composition of clove extracts was determined with various instruments such as GC-MS, LC-MS, HPLC, GLC, etc. The common compound was reported as eugenol, among the identified components (Table 1). As can be seen in Table 1, eugenol was detected in all studies except from Li et al. (2019) and Hassan et al. (2021) who used water as a solvent to produce clove extract. Since Talón et al. (2019) stated that eugenol has limited water solubility. The results in Table 1 showed that the use of sole water as a solvent in the extraction process could not extract the eugenol from the clove.

Eugenol (4-allyl-2-methoxyphenol) is a phenolic compound from the class of phenylpropanoids and can also be found in soybean (*Glycine* max (L.) Merr.), beans, coffee, cinnamon (*Cinnamomum verum* J. Presl), basil (*Ocimum basilicum* L.), "canelinha" (*Croton zehntneri* Pax et Hoffm), banana, and bay laurel (*Laurus nobilis* L.) although it is known as the main component of the clove (Barboza et al., 2018).

Ulanowska and Olas (2021) reported that eugenol was isolated first in 1929 as a volatile compound from the clove, and commercial production began in the United States in 1940. Nowadays, it can be produced chemically or biotechnologically. In the biotechnological methods, a wide range of microorganisms, such as *Corynebacterium* spp., *Streptomyces* spp., and *Escherichia coli* were used (Ulanowska and Olas, 2021). Eugenol can also be produced from refining process of steam distilled clove bud/leaf oil, cinnamon leaf oil, or basil (Kamatou et al., 2012).

Its structure is an oily liquid that is clear, pale yellow, weakly acidic, slightly soluble in water, and soluble in organic solvents (Pramod et al., 2010; Kamatou et al., 2012). Eugenol constituted between 9381 and 14,650mg per 100g of clove (Pathak et al., 2004; Shan et al., 2005). It is used in perfumeries for spicy, clove-like, and oriental-type fragrances, and in food as a flavoring agent. Two other isomers are also used in perfumes, toiletries, detergents, insecticides, and foods (Bendre et al., 2016).

Antimicrobial, anti-inflammatory, analgesic, antioxidant, anticancer, antiseptic, anthelmintic, anti-inflammatory, and analgesic activities of eugenol were studied by different researchers (Kamatou et al., 2012; Marchese et al., 2017). Gülçin (2011) reported that eugenol inhibited 96.7% lipid peroxidation of a linoleic acid emulsion at a 15 μg/mL concentration and was more successful than BHA, BHT, α-tocopherol, and Trolox at the same concentration. And the author also informed that the principle of antioxidant activity is based on the availability of electrons to neutralize any free radical and is related to the number and the nature of the hydroxylation pattern on the aromatic ring. Eugenol owes its high antioxidant activity to the fact that it is a polyphenol containing an aromatic ring.

Marchese et al. (2017) informed that the antimicrobial activity of eugenol can be attributed to the presence of a free hydroxyl group in the molecule. In addition, the antimicrobial activity of eugenol was explained in different ways, such as alteration of membrane fatty acids, disruption of the cytoplasmic membrane, production of intracellular reactive oxygen species, and inhibition of enzymes.

Eugenol has been classified as GRAS by the US Food and Drug Administration (Expert Committee on Food Additives, 1982). However, it is necessary to determine the toxicity of eugenol due to the vast range of different applications and extensive use (Mohammadi Nejad et al., 2017). At high concentrations, eugenol has adverse effects, including inflammatory, ulcer formation, and allergic reactions (Gülçin, 2011; Kamatou et al., 2012), possibly due to the formation of phenoxyl radicals and, subsequently, of quinine intermediates, via its pro-oxidant activity (Gülçin, 2011).

Table 1 Major detected compounds in the clove extracts

Plant parts	Major compound	Solvent/extraction type	Equipment	References
Bud	Eugenol Phenol, 2-methoxy-4-(2-propenyl) 1,2,3-Benzenetriol Caryophyllene Asarone 2′,3′,4′-Trimethoxyacetophenone	Methanol	GC-MS	Hemalatha et al. (2016)
Bud	Eugenol	Ethanol Acetonitrile Acetone Dichloromethane Methanol	HPLC-DAD	Yun et al. (2010)
Leave	Chavicol Eugenol *n*-Pentacosane Hexacosanal Lupeol Octacosanol Tocopherol	Supercritical CO_2	GC-MS	Frohlich et al. (2019)
Whole plant	Eugenol Acetyl eugenol Caryophyllene Humulene	Ethanol	GC-MS	Ding et al. (2017)
Bud	Eugenol Eugenyl acetate Caryophyllene oxide	*n*-Hexane	GC-MS	Nassar et al. (2007)
Bud	Eugenol *trans*-Caryophyllene Eugenyl acetate	Supercritical CO_2	GC-MS	Ivanovic et al. (2011)
Bud	Eugenol Chlorogenic acid Monocaffeoylquinic acid Quercetin glucuronide Gallic acid	Ethanol	LC-MS-DAD	Kubatka et al. (2017)
Bud	Chlorogenic acid Isochlorgenic acid Quercetin-3-*o*-β-D-glucuronide Kaempferol-3-*o*-β-glucuronide Isorhamnetin-3-*o*-β-D-glucoside Isorhamnetin-3-*o*-β-D-glucuronide Rhamnetin-3-*o*-β-D-glucuronide Rhamnazin-3-*o*-β-D-glucoside Rhamnazin-3-*o*-β-D-glucuronide	Aqueous	LC-MS-IT-TOF	Li et al. (2019)

Table 1 Major detected compounds in the clove extracts—cont'd

Plant parts	Major compound	Solvent/extraction type	Equipment	References
ND	Eugenol Oleanolic acid	Ethyl acetate	HPLC	Liu et al. (2014)
ND	Eugenol β-Caryophyllene	Oil	GLC	Dwivedil et al. (2011)
ND	Flavonoids Saponin Phenol Alkaloids Hemagglutinin Tannins	ND	Standard methods	Uju and Obioma (2011)
Bud	Ellagic acid Benzoic acid Rutin Quercetin Myricetin	Aqueous	HPLC	Hassan et al. (2021)
Bud	Eugenol Eugenol acetate *trans*-Caryophyllene α-Humulene	Subcritical CO_2	GC-MS	Šernaitė et al. (2020)
Bud	Eugenol	Ethanol	GC/MS	Rodrigues et al. (2009)
Bud	Eugenol	Methanol	GC/MS	Bachiega et al. (2012)
Bud	Eugenol	Ethanol	HPLC	Tsai et al. (2017)
Bud	Eugenol Eugenol acetate *trans*-Caryophyllene α-Humulene Caryophyllene oxide	Ethyl acetate Diethyl ether Methanol *n*-Hexane	GC/MS	Yassin et al. (2020)
Bud	Quercetin Luteolin Apigenin Gallic acid Caffeic acid Chlorogenic acid Ellagic acid Oleanolic acid Eugenol	Ethanol	UPLC-ESI-TOF-MS	Issac et al. (2015)

Caryophyllene is one of the most detected compounds in clove extracts (Table 1). β-Caryophyllene is a sesquiterpene present in different natural products, e.g., cloves, cinnamon, and copaiba balsam (Sköld et al., 2006). β-Caryophyllene shows significant anticancer activities, affecting the growth and proliferation of numerous cancer cells and also enhancing the efficacy of some chemotherapeutics (Fidyt et al., 2016). In several studies, anti-inflammatory, antibiotic, antioxidant, and local anesthetic activities of β-caryophyllene were evaluated (Legault and Pichette, 2007). In addition, it possesses cardioprotective, hepatoprotective, gastroprotective, and nephroprotective activities as well as has a protective role in some nervous system related disorders, including pain, anxiety, spasm, convulsion, depression, alcoholism, and Alzheimer's disease (Machado et al., 2018). Bahi et al. (2014) also reported that β-caryophyllene, at the doses of 50 mg/kg (orally), displayed CB2R-mediated antidepressant activity in Swiss albino mice.

Oleanolic acid was also identified by Liu et al. (2014) and Issac et al. (2015) and they informed that oleanolic acid was responsible for the antitumor activity of ethyl acetate clove extract.

4. Functionality of clove extracts

4.1 Antimicrobial effect

The antimicrobial effect of a material refers to the killing or inhibiting ability of the microbes causing some diseases in alive organisms or spoilage in materials such as food, feed, etc. There are various antimicrobial agents used to prevent the growth of bacteria, fungi, or viruses. But among people, there is considerable doubt about the risk of using synthetic additives for human health (Gyawali and Ibrahim, 2014). Due to the negative images of the synthetic additives reported by the researchers, the plant extracts are started to be used as an antimicrobial agent, and many studies are performed to reveal the natural preservative effect of plant extracts (Fernández-López et al., 2005; Clarke et al., 2017; Gonelimali et al., 2018; Suppakul et al., 2016). Talib and Mahasneh (2010) reported that there are various phytochemicals such as flavonoids, alkaloids, tannins, and terpenoids performing antimicrobial efficiency.

Some studies about the antimicrobial activity of clove extracts could be seen in Table 2. Hemalatha et al. (2016) investigated the four levels (25, 50, 75, and 100 μL from 1 mg/mL extract) of methanolic clove extracts against Gram-negative and Gram-positive bacteria growth and stated that minimum zone diameter of 17 mm for *Vibrio cholerae* and *Klebsiella pneumonia* with a maximum zone diameter of 26 mm were observed. Similar concentrations were also used (50–100 μg/mL) in a study of El-Maati et al. (2016). They informed that an inhibition zone diameter as 10–17 mm with *Staphylococcus aureus, Escherichia coli, Serratia marcescens, Listeria monocytogenes,* and *Salmonella enteritidis* was occurred with using ethanol extracts of clove. Dua et al. (2014) reported the inhibition zone values for standard antibiotics to be 18–26 mm for chloramphenicol (25 μg/mL) and 36–44 mm for ampicillin (25 μg/mL). In a different study, clove extracts showed moderate antibacterial activities against selected *Staphylococcus* including methicillin-resistant *Staphylococcus aureus* and *Enterococcus* bacterial strains, while the addition of 50% of oregano extract to clove extract enhanced antibacterial activity against all tested strains (Ivanovic et al., 2013).

In a study, the anticandidal potency of clove extracts was evaluated by Yassin et al. (2020) who stated that ethyl acetate extract (10 mg/disc) displayed the highest anti-yeast activity against *Candida*

Table 2 Studies reporting the antimicrobial effects of clove extracts.

	Bacteria/fungi	Extraction type	MIC range (μg/mL)/ inhibition (zone mm)	References
Bud	*Salmonella enterica* serovar Typhimurium LT2 *Listeria innocua*	Nd	>1000 μg/mL 2000 μg/mL	Hill et al. (2013)
Bud	*Bacillus subtilis* *Staphyloccocus aureus* *Klebsiella pneumoniae* *Vibrio cholerae*	Methanol	24.3 mm (100 μL/ 1 mg/mL extract) 25 mm (100 μL/1 mg/mL extract) 26 mm (100 μL/1 mg/mL extract) 24 mm (100 μL/1 mg/mL extract)	Hemalatha et al. (2016)
Bud	*Staphyloccocus aureus* *Vibrio cholerae* *Pseudomonas aeruginosa*	Ethanol	0.025 μg/mL	Mehrotra and Srivastava (2010)
Bud	Marine bacteria	Aqueous	ND	Lakhan et al. (2020)
ND	*Staphyloccocus aureus* *Escherichia coli*	Aqueous	20 mm/100 μL 19 mm/100 μL	Nassan et al. (2015)
Bud	Methicillin-resistant *Staphylococcus aureus* Methicillin-resistant *Staphylococcus haemolyticus* *Staphylococcus aureus* ATCC 11632 *Enterococcus* sp.	SFE	320–640 μg/mL	Ivanovic et al. (2013)
Flower	*Staphylococcus aureus* ATCC 25923 *Pseudomonas aeruginosa* ATCC 27853 *Escherichia coli* PCM 2209 *Candida albicans*	Water	12.8 mm 0 mm 8.2 mm 0 mm	Parlinska-Wojtan et al. (2018)
Bud	*Candida albicans* *Candida tropicalis* *Candida glabrata*	Ethyl acetate	0.5–4 mg/disc 0.25–4 mg/disc 0.5–4 mg/disc	Yassin et al. (2020)
Bud	*Staphylococcus aureus* ATCC 6538 *Listeria monocytogenes* Scott A *Salmonella enteritidis* PT4 *Serratia marcescens* *Escherichia coli* ATCC 8739	Ethyl acetate Ethanol Water	100 μg/mL	El-Maati et al. (2016)
Bud	*Escherichia coli* *Pseudomonas aeruginosa* *Staphylococcus aureus* *Bacillus pumilus*	Methanol	24 mm (25 mg/mL) 30 mm (25 mg/mL) 28 mm (25 mg/mL) 28 mm (25 mg/mL)	Dua et al. (2014)

albicans, *Candida glabrata*, and *Candida tropicalis* with inhibition zone diameters of 20.9, 14.9, and 30.7 mm, respectively. Additionally, eugenol activity was lower than these inhibition zone results. In contrast to these results, Parlinska-Wojtan et al. (2018) monitored no inhibition zones against *Pseudomonas aeruginosa* and *Candida albicans*, while 12.8 and 8.2 mm of inhibition zones were obtained for *Staphylococcus aureus* and *Escherichia coli*, respectively. It could be concluded that the reason for antimicrobial activity differences among studies might be differences in extraction solvents.

4.2 Antioxidant and antiradical effect

Antioxidant and antiradical effects of a substance or compound refer to its scavenging ability of reactive oxygen species or free radicals. Antioxidant compounds provide an increment in the shelf life and retardation in lipid oxidation which is a major factor in the deterioration of food and pharmaceutical products during processing and storage. In food structures, antioxidants play a crucial role to prevent or retard the lipid peroxidation and the formation of secondary lipid peroxidation products to protect flavor, color, and texture of the food product during storage. It is very well known that the many phenolic compounds (in addition to tocopherols) present in the plant tissues are potential antioxidants and flavonoids, tannins and lignin precursors may work as a scavenging compound for the reactive oxygen species (Blokhina et al., 2003). Clove is one of the most popular medicinal plants and many scientific researches demonstrated that the clove extracts performed a remarkable antioxidant performance. El-Maati et al. (2016) investigated the antioxidant and antiradical activities of different solvent extracts (methanol, water, and ethyl acetate) and revealed that the ethanol and water extracts performed a comparable antioxidant activity to the synthetic antioxidant (TBHQ). In the same study, DPPH and $ABTS^{\bullet+}$ radical scavenging activities ranged between 25.3% and 91.4% and 49.4% and 99.4%, respectively. In another research, an optimization study was performed to determine the suitable solvent level and concentration for the clove bud extraction and it was reported that the $ABTS^{\bullet+}$ radicals scavenging activity and ferric reducing antioxidant activity (FRAP) values ranged from 44.78 to 174.7 μg Trolox/g sample and 62.7 to 87.7 mg AAE/L, respectively (Köprü et al., 2020). Ivanovic et al. (2013) studied the possible application of supercritical fluid extraction of clove buds and reported that the inhibition concentration level of clove extracts for the DPPH radical was 20.6 mg/mL and FRAP was 1040.7 μmol Fe^{2+}/g extract. Misharina and Alinkina (2017) carried out an investigation about the antiradical properties of the clove bud extracts and reported that the antiradical efficacy of clove bud extract was equal to the antiradical efficacy of ionol which is a synthetic antioxidant. In another study, antioxidant activity of water and ethanolic extracts obtained from clove was characterized. It was reported that the inhibition levels of lipid peroxidation of linoleic acid emulsion were 97.9% and 98.2% for water and ethanol extracts, respectively. In the same study, inhibition levels of lipid peroxidation of linoleic acid emulsion were reported as 96.5%, 99.2%, and 61.1% for the 60lg/mL of standard antioxidants such as BHA, BHT, and α-tocopherol (Gülçin et al., 2004). Johannah et al. (2015) reported that the hydroalcoholic extract of clove buds performed significantly higher antioxidant and antiradical activities compared to ascorbic acid which is accepted as a powerful antioxidant vitamin. They revealed that the superoxide radical scavenging ability with an IC_{50} value was12.5 μg/mL compared to vitamin C in which of IC_{50} was found to be 1900 μg/mL under the same conditions. Also, hydroxyl radical quenching activity and DPPH radical scavenging capabilities of hydroalcoholic extract of clove buds were found to be better than that of vitamin C with IC_{50} values of 76 μg/mL and 1550 μg/mL and 6 μg/mL and 14 μg/mL, respectively. In a similar study, radical scavenging performances of different solvent extracts of clove buds were compared and it was monitored that the radical inhibition levels of acetone, chloroform, and methanolic extracts (20 μg/mL) of clove buds were 75.2%, 65.6%, and 81.8%, respectively while the radical inhibition level of α-tocopherol was 17.2 at the same concentration (Hemalatha et al., 2016). Suantawee et al. (2014) reported that the aqueous clove extract exhibited antioxidant properties including DPPH radical scavenging activity ($IC_{50}=0.29$ mg/mL) and Trolox equivalent antioxidant capacity (4.69 μmol Trolox equivalents/mg dried extract).

4.3 Anticancer effect

Cancer is one of the most widespread diseases occurred by the uncontrollable growth of some of the body cells and increases by spreading to the other parts of the body. Recently, cancer cases are rapidly increasing worldwide, and the researches on cancer-preventing methods are continuing potently. Especially, there are many and important studies about the potential positive health effects of medicinal plants on some cancer types. Kubatka et al. (2017) investigated the anticancer effects of cloves in the mammary carcinoma model in vivo and in vitro. Clove extract was prepared from powder diluted in 40% ethanol, and in vitro antiproliferative effect on MCF-7 cells was determined. It was reported that significant pro-apoptotic, antiproliferative, anti-angiogenic, and antioxidant effects of clove were observed in the mammary carcinoma model. Besides, the authors recommended the regular consumption of cloves for the risk reduction of breast cancer according to in vivo results.

In a different study performed by Li et al. (2019), the anticancer effect of the aqueous extract of clove was evaluated. 3-(4,5-Dimethylthiazol-2-yl)-5-(3-carboxymethoxyphenyl)-2-(4-sulfophenyl)-2*H*-tetra-zolium (MTS) assay was established for cell viability and aqueous extract of clove was able to inhibit cancer cell growth in vitro on different cancer cell lines and the IC_{50} value was around 150 μg/mL for human pancreatic ASPC-1 and human colon HT- 29 cancer cells. The same research team investigated the ethyl acetate extract of cloves and the potential bioactive components responsible for its antitumor activity in a different study (Liu et al., 2014). Ethyl acetate extract of clove and oleanolic acid—a compound of extract—exhibited cytotoxicity against several human cancer cell lines. The authors concluded that clove extract might be used in the treatment of colorectal cancer. Khan et al. (2018) investigated the synergistic effect of fluorescent magnetic submicronic polymer nanoparticles (FMSP-nanoparticles) in combination with crude clove extracts on human breast cancer cells (MCF-7) to test the combination effect on the cancer cell death and revealed that the combination of FMSP nanoparticles along with clove extracts has a profound effect in reducing the cancer cell survivability and cell proliferation.

Water, ethanol, and oil extracts were screened for antiproliferative activity against HeLa (cervical cancer), MCF-7 (ER + ve) and MDA-MB-231 (ER − ve) breast cancer, DU-145 prostate cancer, and TE-13 esophageal cancer cell lines in a study of Dwivedil et al. (2011). The percentage of cell death was estimated by MTT [3-(4,5-dimethylthiazol-2-yl)-2,5-diphenyltetrazolium bromide] assay. It was reported that among extracts, oil extract had maximal cytotoxic activity (80% cell death at 300 μL/mL dose). The reason for the power of oil extract was attributed to eugenol by authors.

Thenmozhi (2020) coated the clove extract with polyvinylpyrrolidone (PVP) + iron oxide nanoparticles and showed apoptosis in MCF-7 breast cancer cell lines treated with nanoparticles.

4.4 Anti-inflammatory effect

Rodrigues et al. (2009) determined the anti-inflammatory effect of clove extract in vivo. The water-soluble part of a hydroalcoholic extract of clove was treated in mice at a dose of 200 mg/kg. After treatment, macrophages were collected as a source of IL-1β and IL-6, IL-1β, and IL-6, and production tests were performed to determine the inflammatory response. IL-1β and IL-6 production by peritoneal macrophages were significantly inhibited after treatment with the water-soluble part of hydroalcoholic extract of clove. The authors indicated that eugenol was responsible for the inhibition of IL-1β and IL-6 production by macrophages. Similarly, Bachiega et al. (2012) investigated the immunomodulatory/anti-inflammatory effect of methanolic (70%) clove extract on cytokine production (interleukin (IL)-1β, IL-6, and IL-10) in vitro. A 5, 10, 25, 50, or 100 μg/well concentrations of clove extracts were used, and clove extracts at a dose of 5, 10, and 25 μg/well significantly stimulated IL-1β production, whereas the concentration of 100 μg/well inhibited significantly IL-6 and IL-1β release. Additionally, clove extract did not inhibit IL-10 production while the concentration of 100 μg/well inhibited after the cells were incubated with LPS (lipopolysaccharide) (5 mg/mL) for 2 h. In contrast to these studies, IL-1β production with the treatment of clove extract did not significantly alter compared to the positive control (lipopolysaccharide-stimulated cells) in a different study by Rusmana et al. (2015).

In another study, the effects of clove extract on *Propionibacterium acnes*-induced inflammatory responses were studied in vitro. Treatments of ethanolic clove extract significantly suppressed *P. acnes*-stimulated NF-κB-mediated tumor necrosis factor-α, interleukin (IL)-1β, and IL-8 productions. Besides, in vivo extract studies showed positive results and the size of *P. acnes*-induced microabscess, ear swellings, and ear thickness was reduced by clove extract in mice (Tsai et al., 2017).

4.5 Antidiabetic effect

Diabetes (Diabetes mellitus) is defined as a metabolic disorder occurred because of some defects in insulin action, secretion, or both in the body. Because of diabetes, some problems in protein, lipid, and carbohydrate metabolism cause long-term complications in the body (Vareda et al., 2014). Diabetic patients show some important clinical cases such as polyuria, polyphagia, polydipsia, and ketosis, and diabetic patients are faced with some complications like notably retinopathy, neuropathy, nephropathy, and cardiovascular diseases that are known to be the major causes of morbidity and mortality (Oyedem et al., 2011). It is reported that the better treatment of diabetes by using current antidiabetic drugs looks like a challenging event because of side effects and ease of access. So, controlling the diabetes using medicinal plants in phytotherapy is being strengthened because the treatment is easily accessible, less costly, very effective, and with fewer side effects (Tafesse et al., 2017). Clove buds are tested in terms of antidiabetic activities in many scientific reports. Prasad et al. (2005) tested several different herb, spice, and medicinal plant extracts for insulin-like effects and a water-soluble extract of clove was determined that repressed hepatic PEPCK and G6Pase gene expression like insulin. In a different study, the effects of 50% aqueous ethanol extract of clove and also eugenol-reduced extract on glycogen phosphorylase b-inhibiting activity (the enzyme responsible for converting liver glycogen to glucose) were investigated in vivo(Singh et al., 2009). The authors informed that aqueous ethanol extract of cloves and the components, especially ellagitannins, displayed inhibitory activity toward glycogen phosphorylase b (IC_{50} value of 0.86 mg/mL for 50% aqueous:ethanol extract) and also inhibited glucagon-stimulated glucose production in primary rat hepatocytes; however, eugenol did not show any significant inhibitory activity toward glycogen phosphorylase b. Besides, eugenol-reduced clove extract supplementation improved hyperglycemia and hyperlipidemia in db/db mice (Sanae et al., 2014). In similar studies, hypoglycemic effects of clove extracts were investigated (Kuroda et al., 2012; Tu et al., 2014). As a human study, Mohan et al. (2019) conducted an open-label pilot study to test the effect of polyphenolic clove extract on glucose metabolism. According to the results, clove extract worked to control glucose homeostasis decreasing the hepatic glucose production.

4.6 Anti-stress effect

With a different perspective, Singh et al. (2009) evaluated the anti-stress effect of clove hydroalcoholic extract in vivo. A 70% v/v ethanol extract of clove was used, and rats were divided into four groups. Cold restraint stress-induced gastric ulcers, sound stress-induced biochemical changes, and anoxic stress-induced convulsion parameters were used for the determination of the anti-stress effect. The high dose of clove extract successfully prevented the development of gastric ulcers in the cold restraint stress-induced gastric ulcer model. It caused the decrement of the levels of biochemical markers of cell damage. An increased latency of anoxic stress-induced convulsions was also observed with clove extract treatment.

On the other hand, Issac et al. (2015) used ethanol-induced stomach ulcer model for the determination of anti-ulcerogenic activities of clove extract in Wistar rats. The high anti-ulcerogenic activity of clove extracts has been attributed to their high antioxidant, anti-inflammatory, and mucous-producing effects.

4.7 Effect on testicular function

Syzygium aromaticum has been used as indigenous medicine to treat male sexual disorders in Asian countries, and some studies were established to approve these mechanisms. Baghshahi et al. (2014) investigated the effects of clove ethanol extracts on semen quality parameters after cryopreservation of ram spermatozoa. Progressive motility and movement characteristics of the sperm were positively affected by adding 35 and/or 75 μg/mL of clove bud extract to semen extenders; however, higher levels of clove bud extract reduced the motility, movement parameters, viability, and plasma membrane integrity of ovine sperm. The authors informed the maximum dose of clove extract to be 75 μg/mL.

Hexane extracts of clove buds in three doses (15 mg, 30 mg, and 60 mg/kg BW) were evaluated on the testicular function of mice in a different study by Mishra and Singh (2008). The authors reported that the lower dose (15 mg) increased testosterone production, while higher doses (30 and 60 mg) caused reduction in testosterone production in spermatogenesis of mice.

4.8 Food practices

Chatterjee and Bhattacharjee (2013) used encapsulated eugenol-rich clove extract as a natural antioxidant in soybean oil. They stated that there was no significant difference between antioxidant activities of unencapsulated and encapsulated clove extracts in soybean oil through 30-days storage. On the other hand, the authors recommended the encapsulation process since unencapsulated clove extract could have pro-oxidative activity in soybean oil. The same authors investigated the antioxidant activity of clove extract in mayonnaise; differently, the eugenol-lean fraction (98% lower eugenol content than a eugenol-rich fraction) was used in the study. It was determined that clove extract improved color and other physical properties of mayonnaise and provided higher antioxidative properties compared to mustard-formulated mayonnaise and the market sample (Chatterjee and Bhattacharjee, 2015).

Similarly, Ashrafuzzaman Zahid et al. (2019) assessed the antioxidative activity of clove extract in fresh beef patties. The authors compared the antioxidative effect of clove extract with BHT and ascorbic acid and stated that the adding of BHT, ascorbic acid, and clove extract prevented lipid oxidation and provided the preservation of the color properties in beef patties. The authors concluded that clove extract could be used as a natural antioxidant instead of synthetic ones. The same authors tested the clove extract in cooked beef patties by comparing it with BHT and ascorbic acid and informed that among the three antioxidants, application of clove extract performed more efficient antioxidant activities (Zahid et al., 2020).

In a different study, clove extract was tested in fresh mutton for its antimicrobial activity during storage. Through 4 days of storage period, biochemical and microbiological analyses were performed, and it was determined that the clove extract prevented some spoilage and pathogenic bacteria growth as well as the formation of spoilage biomarkers such as cadaverine and free fatty acids (Kumudavally et al., 2011).

Clove extracts were not only used to prevent lipid oxidation but also protein oxidation. Chen et al. (2016) tested the effectiveness of clove extracts in the inhibition of oxidation in porcine myofibrillar protein. The addition of clove extract significantly inhibited carbonyl formation, enhanced solubility, and thermal stability, and improved the gel formation ability of myofibrillar proteins. Ashrafuzzaman Zahid et al. (2019) reported that the antioxidant effects of BHT, ascorbic acid, and clove extract on protein oxidation were less significant than the effects on lipid oxidation in fresh beef patties.

Bensid et al. (2014) evaluated the effect of ice-containing clove (0.02%, w/v) extracts on the quality parameters of anchovy. The clove extract was found to be less effective as an antibacterial agent for the preservation of anchovies because of its low concentration. According to sensory panel results, the shelf life of anchovy was 9 days for the control group and 12 days for the treated groups.

4.9 Encapsulation applications of clove extract

Encapsulation is a process that has two key factors: coating or entrapment material and one or a mixture of compounds (Tavares et al., 2021). It is a useful technology to preserve bioactive compounds from environmental factors such as heat, oxygen, moisture, air, and light (Comunian et al., 2021).

In some cases, clove extracts were used as an encapsulated form to preserve active ingredients and thus to enhance their bioactive properties. Parlinska-Wojtan et al. (2018) investigated the encapsulation process with Ag nanoparticles which were synthesized using clove extract with the aim to enhance their antibacterial properties. The results of the study showed that the antibacterial activity of AgNPs/CE was much more efficient against pathogenic bacteria than that of pure clove extracts.

Lakhan et al. (2020) also synthesized Ag nanoparticles with clove aqueous extract and characterized them using different methods. XRD, FTIR, and UV-Vis spectroscopy analyses confirmed the encapsulation process. The synthesized nanoparticles showed a higher antibacterial and antidiatom activity in the marine environment, compared to non-capsulated clove extract. Similarly, Ajitha et al. (2019) found that the silver nanoparticles synthesized with clove extract had chemo-catalytic activity against the degradation of three pollutant organic dyes, namely 4-nitrophenol, methylene blue, and rhodamine B. Cortés-Rojas et al. (2014b) tried a different encapsulation method using two different solid lipids, two surfactants, and three drying carriers, and spray drying. As a result, the spray-dried formulation composed of Compritol as solid lipid, Poloxamer 188 as a surfactant, and Maltodextrin DE10 as drying carrier was more successful in terms of higher retention of eugenol, eugenyl acetate, and higher product recovery.

Hill et al. (2013) investigated the β-cyclodextrin inclusion complexes for encapsulation of clove bud extract and determined their antibacterial activity against *Salmonella enterica serovar* Typhimurium LT2 and *Listeria innocua*. Clove bud extracts β-cyclodextrin inclusion complexes showed higher antimicrobial activity compared to *trans*-cinnamaldehyde and eugenol β-cyclodextrin inclusion complexes.

4.10 Other applications

Uju and Obioma (2011) investigated the antibacterial effects of *n*-hexane, hot water, and ethanol extracts of clove, bitter kola, and tobacco on *Streptococcus mutans* as a cavity-causing bacterium. According to the results, clove was more successful in inhibiting *S. mutans*, while *n*-hexane extract had the best antimicrobial activity. Besides, the authors recommended clove as an ingredient for toothpaste preparation or mouthwash formulation.

Dalai et al. (2014) determined the anti-cholinesterase activity of the methanol extract of clove. AChE inhibition (IC_{50}) value of methanol extract of clove was found to be 61.5 μg/mL while BChE inhibition was 103 μg/mL proved that clove extract showed better inhibition of AChE than BChE. On the other hand, as a component of clove extract, eugenol was more successful than extract in terms of anti-cholinesterase activity.

Clove and cinnamon extracts were tested for strawberry infection caused by *Botrytis cinerea* both on Petri plates and detached strawberry leaves in vitro. The authors reported that 12 mL/L concentration of clove extract was successful for inhibiting the gray mold infection on leaves while the minimal fungicidal concentration of the clove extract was 1400 μL/L as less effective compared to cinnamon extract (Šernaitė et al., 2020). As an agricultural application, potassium silicate and clove water extract with different rates of nitrogen were evaluated in terms of growth, productivity, and *Alternaria alternata* fungal infection resistance of hot pepper in a different study. Spraying treatment to hot pepper plants with clove water extract significantly decreased the infection severity and it was recommended the using clove water extract with potassium silicate and nitrogen in hot pepper growing, by authors (Hassan et al., 2021).

The clove bud extract was tested in navel oranges as an immersion method, and post-harvest quality and enzyme activities were determined. Clove extract treatment significantly reduced the fruit's decay rate and weight loss compared with those of the control group, while the authors recommended using the clove extract with a coating material such as chitosan and sodium alginate (Zeng et al., 2012).

Hoque et al. (2011) evaluated the gelatin-based film performance by incorporating different herb extracts including clove extract. The incorporation of herb extracts had a profound impact on the water vapor permeability of films.

Balamurugan et al. (2016) carried out the study related to the efficacy of clove extract as a low-cost and effective sedative agent for use in the transportation of clownfish. The authors concluded that the clove extract could be used as an alternative anesthetic agent to clove oil and other chemicals and possessed advantageous properties such as being easily available, relatively inexpensive, eco-friendly, and safer. As a different study in aquaculture, Jin and Cho (2011) stated that water extracts of clove exhibited hypolipidemic activity in a hypercholesterolemic zebrafish model.

References

Ajitha, B., Reddy, Y.A.K., Lee, Y., Kim, M.J., Ahn, C.W., 2019. Biomimetic synthesis of silver nanoparticles using *Syzygium aromaticum* (clove) extract: catalytic and antimicrobial effects. Appl. Organomet. Chem. 33 (5). https://doi.org/10.1002/aoc.4867.

Alexandru, L., Cravotto, G., Giordana, L., Binello, A., Chemat, F., 2013. Ultrasound-assisted extraction of clove buds using batch- and flow-reactors: a comparative study on a pilot scale. Innov. Food Sci. Emerg. Technol. 20, 167–172. https://doi.org/10.1016/j.ifset.2013.07.011.

Arias, M., Penichet, I., Ysambertt, F., Bauza, R., Zougagh, M., Ríos, A., 2009. Fast supercritical fluid extraction of low- and high-density polyethylene additives: comparison with conventional reflux and automatic Soxhlet extraction. J. Supercrit. Fluids 50 (1), 22–28. https://doi.org/10.1016/j.supflu.2009.04.012.

Ashrafuzzaman Zahid, M., Seo, J.K., Parvin, R., Ko, J., Yang, H.S., 2019. Comparison of butylated hydroxytoluene, ascorbic acid, and clove extract as antioxidants in fresh beef patties at refrigerated storage. Food Sci. Anim. Resour. 39 (5), 768–779. https://doi.org/10.5851/kosfa.2019.e67.

Azmir, J., Zaidul, I.S.M., Rahman, M.M., Sharif, K.M., Mohamed, A., Sahena, F., Jahurul, M.H.A., Ghafoor, K., Norulaini, N.A.N., Omar, A.K.M., 2013. Techniques for extraction of bioactive compounds from plant materials: a review. J. Food Eng. 117 (4), 426–436. https://doi.org/10.1016/j.jfoodeng.2013.01.014.

Azwanida, N., 2015. A review on the extraction methods use in medicinal plants, principle, strength and limitation. Med. Aromat. Plants 4 (3), 1000196.

Bachiega, T.F., De Sousa, J.P.B., Bastos, J.K., Sforcin, J.M., 2012. Clove and eugenol in noncytotoxic concentrations exert immunomodulatory/anti-inflammatory action on cytokine production by murine macrophages. J. Pharm. Pharmacol. 64 (4), 610–616. https://doi.org/10.1111/j.2042-7158.2011.01440.x.

Baghshahi, H., Riasi, A., Mahdavi, A.H., Shirazi, A., 2014. Antioxidant effects of clove bud (*Syzygium aromaticum*) extract used with different extenders on ram spermatozoa during cryopreservation. Cryobiology 69 (3), 482–487. https://doi.org/10.1016/j.cryobiol.2014.10.009.

Bahi, A., Al Mansouri, S., Al Memari, E., Al Ameri, M., Nurulain, S.M., Ojha, S., 2014. β-Caryophyllene, a CB2 receptor agonist produces multiple behavioral changes relevant to anxiety and depression in mice. Physiol. Behav. 135, 119–124. https://doi.org/10.1016/j.physbeh.2014.06.003.

Balamurugan, J., Ajith Kumar, T.T., Prakash, S., Meenakumari, B., Balasundaram, C., Harikrishnan, R., 2016. Clove extract: a potential source for stress free transport of fish. Aquaculture 454, 171–175. https://doi.org/10.1016/j.aquaculture.2015.12.020.

Barboza, J.N., da Silva Maia Bezerra Filho, C., Silva, R.O., Medeiros, J.V.R., de Sousa, D.P., 2018. An overview on the anti-inflammatory potential and antioxidant profile of eugenol. Oxidative Med. Cell. Longev. 2018. https://doi.org/10.1155/2018/3957262.

Batiha, G.E.S., Alkazmi, L.M., Wasef, L.G., Beshbishy, A.M., Nadwa, E.H., Rashwan, E.K., 2020. *Syzygium aromaticum* L.(Myrtaceae): traditional uses, bioactive chemical constituents, pharmacological and toxicological activities. Biomol. Ther. 10 (2), 202.

Bendre, R.S., Rajput, J.D., Bagul, S.D., Karandikar, P.S., 2016. Outlooks on medicinal properties of eugenol and its synthetic derivatives. Nat. Prod. Chem. Res. 4 (3), 1–6. https://doi.org/10.4172/2329-6836.1000212.

Bensid, A., Ucar, Y., Bendeddouche, B., Özogul, F., 2014. Effect of the icing with thyme, oregano and clove extracts on quality parameters of gutted and beheaded anchovy (Engraulis encrasicholus) during chilled storage. Food Chem. 145, 681–686. https://doi.org/10.1016/j.foodchem.2013.08.106.

Blokhina, O., Virolainen, E., Fagerstedt, K.V., 2003. Antioxidants, oxidative damage and oxygen deprivation stress: a review. Ann. Bot. 91, 179–194. https://doi.org/10.1093/aob/mcf118.

Cai, L., Wu, C.D., 1996. Compounds from *Syzygium aromaticum* possessing growth inhibitory activity against oral pathogens. J. Nat. Prod. 59 (10), 987–990. https://doi.org/10.1021/np960451q.

Chatterjee, D., Bhattacharjee, P., 2013. Comparative evaluation of the antioxidant efficacy of encapsulated and un-encapsulated eugenol-rich clove extracts in soybean oil: shelf-life and frying stability of soybean oil. J. Food Eng. 117 (4), 545–550. https://doi.org/10.1016/j.jfoodeng.2012.11.016.

Chatterjee, D., Bhattacharjee, P., 2015. Use of eugenol-lean clove extract as a flavoring agent and natural antioxidant in mayonnaise: product characterization and storage study. J. Food Sci. Technol. 52 (8), 4945–4954. https://doi.org/10.1007/s13197-014-1573-6.

Chemat, F., Khan, M.K., 2011. Applications of ultrasound in food technology: processing, preservation and extraction. Ultrason. Sonochem., 813–835. https://doi.org/10.1016/j.ultsonch.2010.11.023.

Chen, H., Diao, J., Li, Y., Chen, Q., Kong, B., 2016. The effectiveness of clove extracts in the inhibition of hydroxyl radical oxidation-induced structural and rheological changes in porcine myofibrillar protein. Meat Sci. 111, 60–66. https://doi.org/10.1016/j.meatsci.2015.08.017.

Clarke, D., Tyuftin, A.A., Cruz-Romero, M.C., Bolton, D., Fanning, S., Pankaj, S.K., Bueno-Ferrer, C., Cullen, P. J., Kerry, J.P., 2017. Surface attachment of active antimicrobial coatings onto conventional plastic-based laminates and performance assessment of these materials on the storage life of vacuum packaged beef sub-primals. Food Microbiol. 62, 196–201. https://doi.org/10.1016/j.fm.2016.10.022.

Comunian, T.A., Silva, M.P., Souza, C.J.F., 2021. The use of food by-products as a novel for functional foods: their use as ingredients and for the encapsulation process. Trends Food Sci. Technol. 108, 269–280. https://doi.org/10.1016/j.tifs.2021.01.003.

Cortés-Rojas, D.F., de Souza, C.R.F., Oliveira, W.P., 2014a. Clove (*Syzygium aromaticum*): a precious spice. Asian Pac. J. Trop. Biomed. 4 (2), 90–96. https://doi.org/10.1016/S2221-1691(14)60215-X.

Cortés-Rojas, D.F., Souza, C.R.F., Oliveira, W.P., 2014b. Encapsulation of eugenol rich clove extract in solid lipid carriers. J. Food Eng. 127, 34–42. https://doi.org/10.1016/j.jfoodeng.2013.11.027.

Dalai, M.K., Bhadra, S., Chaudhary, S.K., Bandyopadhyay, A., Mukherjee, P.K., 2014. Anti-cholinesterase activity of the standardized extract of *Syzygium aromaticum* L. Pharmacogn. Mag. 10 (2), S276. https://doi.org/10.4103/0973-1296.133275.

Ding, Y., Gu, Z., Wang, Y., Wang, S., Chen, H., Zhang, H., Chen, W., Chen, Y.Q., 2017. Clove extract functions as a natural fatty acid synthesis inhibitor and prevents obesity in a mouse model. Food Funct. 8 (8), 2847–2856. https://doi.org/10.1039/C7FO00096K.

Dua, A., Garg, G., Nagar, S., Mahajan, R., 2014. Methanol extract of clove (Syzygium aromaticum Linn.) damages cells and inhibits growth of enteropathogens. J. Innov. Biol. 1 (4), 200–205.

Dwivedil, V., Shrivastaval, R., Hussain, S., Ganguly, C., Bharadwaj, M., 2011. Comparative anticancer potential of clove (*Syzygium aromaticum*) – an Indian spice—against cancer cell lines of various anatomical origin. Asian Pac. J. Cancer Prev. 12 (8), 1989–1993. http://www.apocp.org/cancer_download/Volume12_No8/1989-93%20C.7.4%20Bharadwaj.pdf.

El-Maati, M.F.A., Mahgoub, S.A., Labib, S.M., Al-Gaby, A.M.A., Ramadan, M.F., 2016. Phenolic extracts of clove (*Syzygium aromaticum*) with novel antioxidant and antibacterial activities. Eur. J. Integr. Med. 8 (4), 494–504. https://doi.org/10.1016/j.eujim.2016.02.006.

Expert Committee on Food Additives, 1982. Evaluation of certain food additives and contaminants. In: Anonymus 2020. vol. 683.

Fernández-López, J., Zhi, N., Aleson-Carbonell, L., Pérez-Alvarez, J.A., Kuri, V., 2005. Antioxidant and antibacterial activities of natural extracts: application in beef meatballs. Meat Sci. 69 (3), 371–380. https://doi.org/10.1016/j.meatsci.2004.08.004.

Ferrentino, G., Morozova, K., Mosibo, O.K., Ramezani, M., Scampicchio, M., 2018. Biorecovery of antioxidants from apple pomace by supercritical fluid extraction. J. Clean. Prod. 186, 253–261. https://doi.org/10.1016/j.jclepro.2018.03.165.

Fidyt, K., Fiedorowicz, A., Strządała, L., Szumny, A., 2016. β-Caryophyllene and β-caryophyllene oxide—natural compounds of anticancer and analgesic properties. Cancer Med. 5 (10), 3007–3017. https://doi.org/10.1002/cam4.816.

Frohlich, P.C., Santos, K.A., Palú, F., Cardozo-Filho, L., da Silva, C., da Silva, E.A., 2019. Evaluation of the effects of temperature and pressure on the extraction of eugenol from clove (*Syzygium aromaticum*) leaves using supercritical CO_2. J. Supercrit. Fluids 143, 313–320. https://doi.org/10.1016/j.supflu.2018.09.009.

Gonelimali, F.D., Lin, J., Miao, W., Xuan, J., Charles, F., Chen, M., Hatab, S.R., 2018. Antimicrobial properties and mechanism of action of some plant extracts against food pathogens and spoilage microorganisms. Front. Microbiol. 9. https://doi.org/10.3389/fmicb.2018.01639.

Gülçin, I., 2011. Antioxidant activity of eugenol: a structure-activity relationship study. J. Med. Food 14 (9), 975–985. https://doi.org/10.1089/jmf.2010.0197.

Gülçin, I., Şat, I.G., Beydemir, S., Elmastaş, M., Küfrevioğlu, O.I., 2004. Comparison of antioxidant activity of clove (Eugenia caryophylata Thunb) buds and lavender (*Lavandula stoechas* L.). Food Chem. 87 (3), 393–400. https://doi.org/10.1016/j.foodchem.2003.12.008.

Gyawali, R., Ibrahim, S.A., 2014. Natural products as antimicrobial agents. Food Control 46, 412–429. https://doi.org/10.1016/j.foodcont.2014.05.047.

Handa, S., Khanuja, Longo, G., Rakesh, D., 2008. Extraction Technologies for Medicinal and Aromatic Plants, first ed. United Nations Industrial Development Organization and the International Centre for Science and High Technology, Italy. No. 66.

Hassan, S.M., El-Bebany, A.F., Salem, M.Z.M., Komeil, D.A., 2021. Productivity and post-harvest fungal resistance of hot pepper as affected by potassium silicate, clove extract foliar spray and nitrogen application. Plants 10 (4). https://doi.org/10.3390/plants10040662.

Hastuti, L.T., Saepudin, E., Cahyana, A.H., Rahayu, D.U.C., Murni, V.W., Haib, J., 2017. The influence of sun drying process and prolonged storage on composition of essential oil from clove buds (*Syzygium aromaticum*). AIP Conf. Proc. 1862, 030092.

Hemalatha, R., Nivetha, P., Mohanapriya, C., Sharmila, G., Muthukumaran, C., Gopinath, M., 2016. Phytochemical composition, GC-MS analysis, in vitro antioxidant and antibacterial potential of clove flower bud (*Eugenia caryophyllus*) methanolic extract. J. Food Sci. Technol. 53 (2), 1189–1198. https://doi.org/10.1007/s13197-015-2108-5.

Herrera, M.C., Luque De Castro, M.D., 2005. Ultrasound-assisted extraction of phenolic compounds from strawberries prior to liquid chromatographic separation and photodiode array ultraviolet detection. J. Chromatogr. A 1100 (1), 1–7. https://doi.org/10.1016/j.chroma.2005.09.021.

Hill, L.E., Gomes, C., Taylor, T.M., 2013. Characterization of beta-cyclodextrin inclusion complexes containing essential oils (trans-cinnamaldehyde, eugenol, cinnamon bark, and clove bud extracts) for antimicrobial delivery applications. LWT Food Sci. Technol. 51 (1), 86–93. https://doi.org/10.1016/j.lwt.2012.11.011.

Hoque, M.S., Benjakul, S., Prodpran, T., 2011. Properties of film from cuttlefish (*Sepia pharaonis*) skin gelatin incorporated with cinnamon, clove and star anise extracts. Food Hydrocoll. 25 (5), 1085–1097. https://doi.org/10.1016/j.foodhyd.2010.10.005.

Inczedy, J., Lengyel, T., Ure, A.M., 1998. Supercritical fluid chromatography and extraction. In: Compendium of Analytical Nomenclature (Definitive Rules 1997). Blackwell Science Oxford, UK.

Issac, A., Gopakumar, G., Kuttan, R., Maliakel, B., Krishnakumar, I.M., 2015. Safety and anti-ulcerogenic activity of a novel polyphenol-rich extract of clove buds (*Syzygium aromaticum* L). Food Funct. 6 (3), 842–852. https://doi.org/10.1039/c4fo00711e.

Ivanovic, J., Dimitrijevic-Brankovic, S., Misic, D., Ristic, M., Zizovic, I., 2013. Evaluation and improvement of antioxidant and antibacterial activities of supercritical extracts from clove buds. J. Funct. Foods 5 (1), 416–423. https://doi.org/10.1016/j.jff.2012.11.014.

Ivanovic, J., Zizovic, I., Ristic, M., Stamenic, M., Skala, D., 2011. The analysis of simultaneous clove/oregano and clove/thyme supercritical extraction. J. Supercrit. Fluids 55 (3), 983–991. https://doi.org/10.1016/j.supflu.2010.09.012.

Jin, S., Cho, K.H., 2011. Water extracts of cinnamon and clove exhibits potent inhibition of protein glycation and anti-atherosclerotic activity in vitro and in vivo hypolipidemic activity in zebrafish. Food Chem. Toxicol. 49 (7), 1521–1529. https://doi.org/10.1016/j.fct.2011.03.043.

Jirovetz, L., Buchbauer, G., Stoilova, I., Stoyanova, A., Krastanov, A., Schmidt, E., 2006. Chemical composition and antioxidant properties of clove leaf essential oil. J. Agric. Food Chem. 54 (17), 6303–6307. https://doi.org/10.1021/jf060608c.

Johannah, N.M., Renny, R.M., Gopakumar, G., Maliakel, B., Sureshkumar, D., Krishnakumar, I.M., 2015. Beyond the flavour: a de-flavoured polyphenol rich extract of clove buds (*Syzygium aromaticum* L) as a novel dietary antioxidant ingredient. Food Funct. 6 (10), 3373–3382. https://doi.org/10.1039/c5fo00682a.

Kamatou, G.P., Vermaak, I., Viljoen, A.M., 2012. Eugenol – from the remote Maluku Islands to the international market place: a review of a remarkable and versatile molecule. Molecules 17 (6), 6953–6981. https://doi.org/10.3390/molecules17066953.

Khan, F.A., Akhtar, S., Almohazey, D., Alomari, M., Almofty, S.A., 2018. Extracts of clove (*Syzygium aromaticum*) potentiate FMSP-nanoparticles induced cell death in MCF-7 cells. Int. J. Biomater. 2018. https://doi.org/10.1155/2018/8479439.

Köprü, S., Say, R., Karaman, K., Yilmaz, M.M., Kaplan, M., 2020. Optimization of processing parameters for the preparation of clove (*Syzygium aromaticum*) hydroalcoholic extract: a response surface methodology approach to characterize the biofunctional performance. J. Appl. Res. Med. Aromat. Plants 16. https://doi.org/10.1016/j.jarmap.2019.100236.

Kubatka, P., Uramova, S., Kello, M., Kajo, K., Kruzliak, P., Mojzis, J., Vybohova, D., Adamkov, M., Jasek, K., Lasabova, Z., Zubor, P., Fialova, S., Dokupilova, S., Solar, P., Pec, M., Adamicova, K., Danko, J., Adamek, M., Busselberg, D., 2017. Antineoplastic effects of clove buds (*Syzygium aromaticum* L.) in the model of breast carcinoma. J. Cell. Mol. Med. 21 (11), 2837–2851. https://doi.org/10.1111/jcmm.13197.

Kumudavally, K.V., Tabassum, A., Radhakrishna, K., Bawa, A.S., 2011. Effect of ethanolic extract of clove on the keeping quality of fresh mutton during storage at ambient temperature (25 ± 2 °C). J. Food Sci. Technol. 48 (4), 466–471. https://doi.org/10.1007/s13197-010-0181-3.

Kuroda, M., Mimaki, Y., Ohtomo, T., Yamada, J., Nishiyama, T., Mae, T., Kishida, H., Kawada, T., 2012. Hypoglycemic effects of clove (*Syzygium aromaticum* flower buds) on genetically diabetic KK-A y mice and identification of the active ingredients. J. Nat. Med. 66 (2), 394–399. https://doi.org/10.1007/s11418-011-0593-z.

Lakhan, M.N., Chen, R., Shar, A.H., Chand, K., Shah, A.H., Ahmed, M., Ali, I., Ahmed, R., Liu, J., Takahashi, K., Wang, J., 2020. Eco-friendly green synthesis of clove buds extract functionalized silver nanoparticles and evaluation of antibacterial and antidiatom activity. J. Microbiol. Methods 173. https://doi.org/10.1016/j.mimet.2020.105934.

Legault, J., Pichette, A., 2007. Potentiating effect of β-caryophyllene on anticancer activity of α-humulene, isocaryophyllene and paclitaxel. J. Pharm. Pharmacol. 59 (12), 1643–1647. https://doi.org/10.1211/jpp.59.12.0005.

Li, C., Xu, H., Chen, X., Chen, J., Li, X., Qiao, G., Tian, Y., Yuan, R., Su, S., Liu, X., Lin, X., 2019. Aqueous extract of clove inhibits tumor growth by inducing autophagy through AMPK/ULK pathway. Phytother. Res. 33 (7), 1794–1804. https://doi.org/10.1002/ptr.6367.

Liu, H., Schmitz, J.C., Wei, J., Cao, S., Beumer, J.H., Strychor, S., Cheng, L., Liu, M., Wang, C., Wu, N., Zhao, X., Zhang, Y., Liao, J., Chu, E., Lin, X., 2014. Clove extract inhibits tumor growth and promotes cell cycle arrest and apoptosis. Oncol. Res. 21 (5), 247–259. https://doi.org/10.3727/096504014X13946388748910.

Machado, K.D.C., Islam, M.T., Ali, E.S., Rouf, R., Uddin, S.J., Dev, S., Shilpi, J.A., Shill, M.C., Reza, H.M., Das, A.K., Shaw, S., Mubarak, M.S., Mishra, S.K., Melo-Cavalcante, A.A.D.C., 2018. A systematic review on the neuroprotective perspectives of beta-caryophyllene. Phytother. Res. 32 (12), 2376–2388. https://doi.org/10.1002/ptr.6199.

Marchese, A., Barbieri, R., Coppo, E., Orhan, I.E., Daglia, M., Nabavi, S.F., Izadi, M., Abdollahi, M., Nabavi, S. M., Ajami, M., 2017. Antimicrobial activity of eugenol and essential oils containing eugenol: a mechanistic viewpoint. Crit. Rev. Microbiol. 43 (6), 668–689. https://doi.org/10.1080/1040841X.2017.1295225.

Mehrotra, S., Srivastava, A.K., 2010. Comparative antimicrobial activities of Neem, Amla, Aloe, Assam Tea and Clove extracts against *Vibrio cholerae*, *Staphylococcus aureus* and *Pseudomonas aeruginosa*. J. Med. Plant Res. 4 (22), 2393–2398. https://doi.org/10.5897/JMPR10.243.

Misharina, T.A., Alinkina, E.S., 2017. Antiradical properties of essential oils and extracts from spices. In: Chemistry and Technology of Plant Substances: Chemical and Biochemical Aspects. Apple Academic Press, pp. 291–308, https://doi.org/10.1201/9781315207469.

Mishra, R.K., Singh, S.K., 2008. Safety assessment of *Syzygium aromaticum* flower bud (clove) extract with respect to testicular function in mice. Food Chem. Toxicol. 46 (10), 3333–3338. https://doi.org/10.1016/j.fct.2008.08.006.

Miyazawa, M., Hisama, M., 2003. Antimutagenic activity of phenylpropanoids from clove (*Syzygium aromaticum*). J. Agric. Food Chem. 51 (22), 6413–6422. https://doi.org/10.1021/jf030247q.

Mohammadi Nejad, S., Özgüneş, H., Başaran, N., 2017. Öjenolün farmakolojik ve toksikolojik özellikleri. Turkish J. Pharm. Sci. 14 (2), 201–206. https://doi.org/10.4274/tjps.62207.

Mohan, R., Jose, S., Mulakkal, J., Karpinsky-Semper, D., Swick, A.G., Krishnakumar, I.M., 2019. Water-soluble polyphenol-rich clove extract lowers pre- and post-prandial blood glucose levels in healthy and prediabetic volunteers: an open label pilot study. BMC Complement. Altern. Med. 19 (1). https://doi.org/10.1186/s12906-019-2507-7.

Molina, G.A., González-Fuentes, F., Loske, A.M., Fernández, F., Estevez, M., 2020. Shock wave-assisted extraction of phenolic acids and flavonoids from *Eysenhardtia polystachya* heartwood: a novel method and its comparison with conventional methodologies. Ultrason. Sonochem. 61. https://doi.org/10.1016/j.ultsonch.2019.104809.

Nassan, M.A., Mohamed, E.H., Abdelhafez, S., Ismail, T.A., 2015. Effect of clove and cinnamon extracts on experimental model of acute hematogenous pyelonephritis in albino rats: Immunopathological and antimicrobial study. Int. J. Immunopathol. Pharmacol. 28 (1), 60–68. https://doi.org/10.1177/0394632015572075.

Nassar, M.I., Gaara, A.H., El-Ghorab, A.H., Farrag, A., Shen, H., Huq, E., Mabry, T.J., 2007. Chemical constituents of clove (*Syzygium aromaticum*, Fam. Myrtaceae) and their antioxidant activity. Rev. Latinoam. Quim. 35 (3), 47.

Nikousaleh, A., Prakash, J., 2016. Antioxidant components and properties of dry heat treated clove in different extraction solvents. J. Food Sci. Technol. 53 (4), 1993–2000. https://doi.org/10.1007/s13197-015-2113-8.

Nurdjannah, N., Bermawie, N., 2012. Cloves. In: Handbook of Herbs and Spices, second ed. vol. 1. Elsevier Inc, pp. 197–215, https://doi.org/10.1533/9780857095671.197.

Oyedem, S.O., Yakubu, M.T., Afolayan, A.J., 2011. Antidiabetic activities of aqueous leaves extract of *Leonotis leonurus* in streptozotocin induced diabetic rats. J. Med. Plant Res. 5 (1), 119–125. http://www.academicjournals.org/JMPR/PDF/pdf 2011/4Jan/Oyedemi%20et%20al.pdf.

Parlinska-Wojtan, M., Depciuch, J., Fryc, B., Kus-Liskiewicz, M., 2018. Green synthesis and antibacterial effects of aqueous colloidal solutions of silver nanoparticles using clove eugenol. Appl. Organomet. Chem. 32 (4). https://doi.org/10.1002/aoc.4276, e4276.

Pathak, S.B., Niranjan, K., Padh, H., Rajani, M., 2004. TLC densitometric method for the quantification of eugenol and gallic acid in clove. Chromatographia 60 (3–4), 241–244. https://doi.org/10.1365/s10337-004-0373-y.

Pramod, K., Ansari, S.H., Ali, J., 2010. Eugenol: a natural compound with versatile pharmacological actions. Nat. Prod. Commun. https://doi.org/10.1177/1934578x1000501236, 1934578X1000501.

Prasad, R.C., Herzog, B., Boone, B., Sims, L., Waltner-Law, M., 2005. An extract of Syzygium aromaticum represses genes encoding hepatic gluconeogenic enzymes. J. Ethnopharmacol. 96 (1–2), 295–301. https://doi.org/10.1016/j.jep.2004.09.024.

Ramadan, M.F., Asker, M.M.S., Tadros, M., 2013. Lipid profile, antiradical power and antimicrobial properties of *Syzygium aromaticum* oil. Grasas Aceites 64 (5), 509–520. https://doi.org/10.3989/gya.011713.

Rodrigues, T.G., Fernandes, A., Sousa, J.P.B., Bastos, J.K., Sforcin, J.M., 2009. In vitro and in vivo effects of clove on pro-inflammatory cytokines production by macrophages. Nat. Prod. Res. 23 (4), 319–326. https://doi.org/10.1080/14786410802242679.

Rodríguez-Rojo, S., Visentin, A., Maestri, D., Cocero, M.J., 2012. Assisted extraction of rosemary antioxidants with green solvents. J. Food Eng. 109 (1), 98–103. https://doi.org/10.1016/j.jfoodeng.2011.09.029.

Rusmana, D., Elisabeth, M., Widowati, W., Fauziah, N., Maesaroh, M., 2015. Inhibition of inflammatory agent production by ethanol extract and eugenol of *Syzygium aromaticum* (L.) flower bud (clove) in LPS-stimulated raw 264.7 cells. Res. J. Med. Plant 9 (6), 264–274. https://doi.org/10.3923/rjmp.2015.264.274.

Sanae, F., Kamiyama, O., Ikeda-Obatake, K., Higashi, Y., Asano, N., Adachi, I., Kato, A., 2014. Effects of eugenol-reduced clove extract on glycogen phosphorylase b and the development of diabetes in db/db mice. Food Funct. 5 (2), 214–219. https://doi.org/10.1039/c3fo60514k.

Santos, D.T., Veggi, P.C., Meireles, M.A.A., 2010. Extraction of antioxidant compounds from Jabuticaba (*Myrciaria cauliflora*) skins: yield, composition and economical evaluation. J. Food Eng. 101 (1), 23–31. https://doi.org/10.1016/j.jfoodeng.2010.06.005.

Šernaitė, L., Rasiukevičiūtė, N., Valiuškaitė, A., 2020. The extracts of cinnamon and clove as potential biofungicides against strawberry grey mould. Plan. Theory 9 (5). https://doi.org/10.3390/plants9050613.

Shan, B., Cai, Y.Z., Sun, M., Corke, H., 2005. Antioxidant capacity of 26 spice extracts and characterization of their phenolic constituents. J. Agric. Food Chem. 53 (20), 7749–7759. https://doi.org/10.1021/jf051513y.

Sihvonen, M., Järvenpää, E., Hietaniemi, V., Huopalahti, R., 1999. Advances in supercritical carbon dioxide technologies. Trends Food Sci. Technol. 10 (6–7), 217–222. https://doi.org/10.1016/S0924-2244(99)00049-7.

Singh, A.K., Dhamanigi, S.S., Asad, M., 2009. Anti-stress activity of hydro-alcoholic extract of *Eugenia caryophyllus* buds (clove). Indian J. Pharmacol. 41 (1), 28–31. https://doi.org/10.4103/0253-7613.48889.

Sköld, M., Karlberg, A.T., Matura, M., Börje, A., 2006. The fragrance chemical β-caryophyllene—air oxidation and skin sensitization. Food Chem. Toxicol. 44 (4), 538–545. https://doi.org/10.1016/j.fct.2005.08.028.

Suantawee, T., Wesarachanon, K., Anantsuphasak, K., Daenphetploy, T., Thien-Ngern, S., Thilavech, T., Pasukamonset, P., Ngamukote, S., Adisakwattana, S., 2014. Protein glycation inhibitory activity and antioxidant capacity of clove extract. J. Food Sci. Technol. 52 (6), 3843–3850. https://doi.org/10.1007/s13197-014-1452-1.

Suppakul, P., Thanathammathorn, T., Samerasut, O., Khankaew, S., 2016. Shelf life extension of \fios de ovos\, an intermediate-moisture egg-based dessert, by active and modified atmosphere packaging. Food Control 70, 58–63. https://doi.org/10.1016/j.foodcont.2016.05.036.

Tafesse, T.B., Hymete, A., Mekonnen, Y., Tadesse, M., 2017. Antidiabetic activity and phytochemical screening of extracts of the leaves of Ajuga remota Benth on alloxan-induced diabetic mice. BMC Complement. Altern. Med. 17 (1). https://doi.org/10.1186/s12906-017-1757-5.

Talib, W.H., Mahasneh, A.M., 2010. Antimicrobial, cytotoxicity and phytochemical screening of Jordanian plants used in traditional medicine. Molecules 15 (3), 1811–1824. https://doi.org/10.3390/molecules15031811.

Talón, E., Vargas, M., Chiralt, A., González-Martínez, C., 2019. Antioxidant starch-based films with encapsulated eugenol. Application to sunflower oil preservation. LWT 113. https://doi.org/10.1016/j.lwt.2019.108290, 108290.

Tavares, L., Santos, L., Zapata Noreña, C.P., 2021. Bioactive compounds of garlic: a comprehensive review of encapsulation technologies, characterization of the encapsulated garlic compounds and their industrial applicability. Trends Food Sci. Technol. 114, 232–244. https://doi.org/10.1016/j.tifs.2021.05.019.

Thenmozhi, T., 2020. Functionalization of iron oxide nanoparticles with clove extract to induce apoptosis in MCF-7 breast cancer cells. 3 Biotech 10 (2). https://doi.org/10.1007/s13205-020-2088-7.

Tsai, T.H., Huang, W.C., Lien, T.J., Huang, Y.H., Chang, H., Yu, C.H., Tsai, P.J., 2017. Clove extract and eugenol suppress inflammatory responses elicited by *Propionibacterium acnes* in vitro and in vivo. Food Agric. Immunol. 28 (5), 916–931. https://doi.org/10.1080/09540105.2017.1320357.

Tu, Z., Moss-Pierce, T., Ford, P., Jiang, T.A., 2014. Syzygium aromaticum L. (clove) extract regulates energy metabolism in myocytes. J. Med. Food 17 (9), 1003–1010. https://doi.org/10.1089/jmf.2013.0175.

Uju, D.E., Obioma, N.P., 2011. Anticariogenic potentials of clove, tobacco and bitter kola. Asian Pac. J. Trop. Med. 4 (10), 814–818. https://doi.org/10.1016/S1995-7645(11)60200-9.

Ulanowska, M., Olas, B., 2021. Biological properties and prospects for the application of eugenol—a review. Int. J. Mol. Sci. 22 (7), 3671. https://doi.org/10.3390/ijms22073671.

Vareda, P.M.P., Saldanha, L.L., Camaforte, N.A.D.P., Violato, N.M., Dokkedal, A.L., Bosqueiro, J.R., 2014. Myrcia bella leaf extract presents hypoglycemic activity via PI3k/Akt insulin signaling pathway. Evid. Based Complement. Alternat. Med. 2014. https://doi.org/10.1155/2014/543606.

Vilkhu, K., Mawson, R., Simons, L., Bates, D., 2008. Applications and opportunities for ultrasound assisted extraction in the food industry – a review. Innov. Food Sci. Emerg. Technol. 9 (2), 161–169. https://doi.org/10.1016/j.ifset.2007.04.014.

Yassin, M.T., Mostafa, A.A.-F., Al-Askar, A.A., 2020. In vitro anticandidal potency of Syzygium aromaticum (clove) extracts against vaginal candidiasis. BMC Complement. Med. Ther. 20 (1). https://doi.org/10.1186/s12906-020-2818-8.

Yun, S.M., Lee, M.H., Lee, K.J., Ku, H.O., Son, S.W., Joo, Y.S., 2010. Quantitative analysis of eugenol in clove extract by a validated HPLC method. J. AOAC Int. 93 (6), 1806–1810. https://doi.org/10.1093/jaoac/93.6.1806.

Zahid, M.A., Choi, J.Y., Seo, J.K., Parvin, R., Ko, J., Yang, H.S., 2020. Effects of clove extract on oxidative stability and sensory attributes in cooked beef patties at refrigerated storage. Meat Sci. 161. https://doi.org/10.1016/j.meatsci.2019.107972.

Zeng, R., Zhang, A., Chen, J., Fu, Y., 2012. Postharvest quality and physiological responses of clove bud extract dip on "Newhall" navel orange. Sci. Hortic. 138, 253–258. https://doi.org/10.1016/j.scienta.2012.02.036.

Zougagh, M., Valcárcel, M., Ríos, A., 2004. Supercritical fluid extraction: a critical review of its analytical usefulness. TrAC Trends Anal. Chem. 23 (5), 399–405. https://doi.org/10.1016/S0165-9936(04)00524-2.

CHAPTER

Food applications of clove (*Syzygium aromaticum*) extracts

36

Mustafa Kiralan[a] and Onur Ketenoglu[b]

Faculty of Engineering, Department of Food Engineering, Balikesir University, Balikesir, Turkey[a]

Faculty of Agriculture, Department of Food Engineering, Eskisehir Osmangazi University, Eskisehir, Turkey[b]

1. Introduction

Culinary spices and herbs could be recognized as an essential part of the human diet and used in food products to enhance aroma, flavor, and color (Yashin et al., 2017; Jin et al., 2018). Besides flavoring properties, they also possess such therapeutic properties as antioxidative, antidiabetic, and antimicrobial activities (Yashin et al., 2017; El-Sayed and Youssef, 2019). These valuable spices and herbs have been generally used in food applications in different forms such as whole or ground, extracts, or encapsulated (Embuscado, 2015).

Bioactive compounds have many functionalities that provide an increased production diversity of nutraceuticals, functional food, and food additives (Joana Gil-Chávez et al., 2013). The diversity and quantity of compounds in the extracts from spices and herbs are affected by various factors such as extraction techniques, time and temperature of extraction, solvent type, and solvent-to-sample ratio (Da Silva et al., 2016; Aznar et al., 2020; Sánchez-Camargo et al., 2020).

Clove (*Syzygium aromaticum*) belongs to the Myrtaceae family and is one of the most valuable spices in the world trade used as a food preservative and for various medicinal purposes in many countries (Cortés-Rojas et al., 2014). Clove is indigenous to North Maluku Islands in Indonesia (Kaur and Kaushal, 2019). The proximate composition of ground cloves is exhibited in Table 1. According to USDA data, ground clove mainly contains carbohydrates (65.53 g), followed by dietary fiber (33.9 g) per 100 g.

Aside from major components, clove contains minor compounds that exhibit biological effects on human health. One of these chemical groups is phenolic compounds. Among these compounds, eugenol, eugenol acetate, and gallic acid exhibited great potential for such applications as pharmaceutical, cosmetic, and food (Cortés-Rojas et al., 2014). Chaieb et al. (2007) determined eugenol (Fig. 1) as the main compound in clove essential oil, comprising 89% of total volatile compounds, followed by eugenol acetate (Fig. 1). In addition to classical extraction systems, modern extraction systems are utilized to obtain higher amounts of eugenol. Frohlich et al. (2019) determined a high concentration of eugenol in the extract obtained from clove leaves by supercritical carbon dioxide (SC-CO_2) extraction at 40°C and 220 bar. The predominant compound in the clove extract was eugenol (29.7%), followed by chavicol (13.3%). Besides, vitamin E was also detected at 9.30%.

Clove (*Syzygium aromaticum*). https://doi.org/10.1016/B978-0-323-85177-0.00030-6

Table 1 The proximate composition of clove (g/100 g).

Compound	Amount
Water	9.87
Protein	5.97
Total lipid	13.0
Ash	5.63
Carbohydrate	65.5
Fiber, total dietary	33.9
Sugars	2.38

Source: USDA. 2020. USDA National Nutrient Database. Available at https://fdc.nal.usda.gov/fdc-app.html#/food-details/171321/nutrients (Accessed 9 July 2020).

Clove extracts show such biological activities as antioxidant and antimicrobial. These biological effects of extracts are attributed in particular to eugenol, one of the phenolic compounds. Eugenol exhibits antioxidant activity by donating a hydrogen atom and stabilizing the phenoxyl radicals to inhibit the oxidation (Gülçin, 2012). Due to its biological activities, studies have gained importance in using clove as a preservative in various foodstuffs. In this section, we will comprehensively discuss the applications of cloves in the food industry.

2. Antioxidant and antimicrobial properties of clove extracts

The extraction method, process parameters, and solvent type are essential factors in the extracts' yield and composition. Table 2 summarizes the findings of the reviewed studies.

Hemalatha et al. (2016) evaluated the antioxidant and antimicrobial properties of extracts using different solvents. According to the 1,1-diphenyl-2-picrylhydrazyl (DPPH•) assay results, all the extracts showed antioxidant activity in a 20–200 μg/mL concentration range. The major radical scavenging activity was observed in the methanol extract with an inhibition rate of 93%, followed by acetone and chloroform with 85 and 81%, respectively. Besides, antimicrobial activity was measured in this

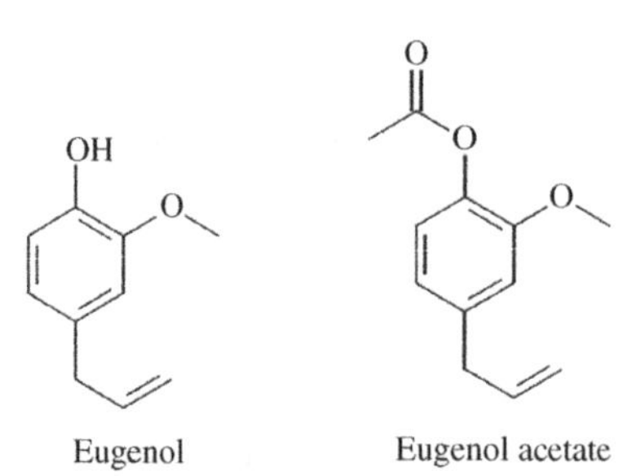

FIG. 1

Chemical structure of eugenol and eugenol acetate.

Table 2 Antioxidant activity of extracts from clove.

Type of extract	Used part	Analytical method	Dose	Findings	Reference
Acetone Chloroform Methanol	Flower buds	DPPH assay	20–200 μg/mL	- 75.2–85.3% inhibition for acetone extract - 65.5–81.4% inhibition for chloroform extract - 81.8–92.2% inhibition for methanol extract	Hemalatha et al. (2016)
Ethyl acetate Ethanol (80%) Water	Buds	- DPPH - ABTS - β-Carotene/linoleic acid bleaching - FRAP	- 100 μL of each extract (10 mg extract/10 mL solvent) - 10 μL of each clove extract (10 mg extract in 10 mL solvent) - 0.5 mL of extract (10 mg extract in 10 mL solvent) - 100 μL of each extracts (10 mg extract in10 mL solvent)	- In all assays, water and ethanol extracts showed higher antioxidant activity compared with ethyl acetate in addition to a comparable antioxidant activity to the synthetic antioxidant tert-butyl hydroquinone (TBHQ)	El-Maati et al. (2016)
Water Ethanol	Buds	- Reductive potential - DPPH - Superoxide anion radical scavenging Metal chelating activities	20–60 μg/mL	- All extracts showed intense antioxidant activity and exhibited higher values compared with α-tocopherol	Gülçin et al. (2004)
Water	Buds	- DPPH	5–200 μg/mL	- The highest inhibition rate was 73.5%, with the lowest IC_{50} (6.80 μg/mL)	Abdel-Wahab et al. (2020)
Water Methanol (80%) Acetone (80%)	Buds	- DPPH - Reducing power	10–100 μg/mL	- The highest radical inhibition rate (73.6%) and highest reducing power (0.88) found in the extracts using 80% acetone	Adaramola and Onigbinde (2016)

Continued

Table 2 Antioxidant activity of extracts from clove—cont'd

Type of extract	Used part	Analytical method	Dose	Findings	Reference
Hot water	Unspecified	- DPPH - Superoxide Anion Radical Scavenging Activity - Hydroxyl Radical Scavenging Activity	- 1 mg/mL for DPPH• experiment - 0.5 mg/mL for Superoxide Anion Radical Scavenging Activity	- Clove extract showed radical scavenging ability in the DPPH• test (84.2%) - Clove extract showed lower activity in Superoxide Anion and Hydroxyl radical scavenging activities	Kim et al. (2011)

study using the agar well diffusion method. Results revealed that methanol extract efficiently inhibited *Klebsiella pneumonia* than other tested bacterial strains.

Gülçin et al. (2004) examined the antioxidant activity of water and ethanol extracts from clove buds using many antioxidant assays, including reductive potential, free radical scavenging, superoxide anion radical scavenging, and metal chelating activities. Ethanol and water extracts showed potent inhibition (93%–98%) at a 20–60 μg/mL concentration range. Reductive potential, superoxide radical, and radical (DPPH) scavenging activity of water and ethanol extracts showed higher antioxidant activities than α-tocopherol.

Lee et al. (2004) evaluated the antimicrobial activity of clove extracts using different solvents against many food spoilage microorganisms, including *Bacillus subtilis*, *Staphylococcus aureus*, *Escherichia coli*, *Salmonella typhimurium*, and *Pseudomonas aeruginosa*. The study revealed that methanol extract was more effective than water extract, which indicated the inhibition effect against *E. coli*, *S. typhimurium*, and *S. aureus*, but there was no observed effect against *B. subtilis*. The methanol extract was found more effective for Gram (+) bacteria than Gram (−) bacteria.

El-Maati et al. (2016) evaluated the antioxidant activity of different extracts (ethyl acetate, ethanol (80%), and water) using various assays such as DPPH, 2,2′azino bis-(3-ethylbenzothiazoline-6-sulfonic acid (ABTS), β-carotene/linoleic acid bleaching, and ferric reducing antioxidant power (FRAP). Antioxidant tests exhibited that ethanol and water extracts from clove had higher activity than ethyl acetate extract. Besides, ethanol and water extracts exhibited comparable activity to the synthetic antioxidant, namely tert-butyl hydroquinone (TBHQ). In this study, antibacterial activities of extracts against *S. aureus*, *Listeria monocytogenes*, *Salmonella enteritidis*, *Serratia marcescens*, and *E. coli* were also evaluated. The extracts showed antibacterial activity against *S. aureus* and *E. coli* at concentrations ranging from 50 to 100 μg/mL.

Abdel-Wahab et al. (2020) revealed that a significant antimicrobial activity of clove extract was observed against Gram-negative bacteria (*E. coli*), Gram-positive bacteria (*Streptococcus pyogenes*), and yeast (*Candida albicans*) with minimum inhibition concentration (MIC, 12.5 mg/mL).

3. Food applications of clove extracts

3.1 Meat and meat products

Shan et al. (2009) evaluated possible uses of five spice and herb (cinnamon stick, oregano, clove, pomegranate peel, and grape seed) extracts in raw pork as a natural preservative. During storage at room temperature for nine days, raw pork treated with clove contained the smallest populations of three pathogen bacteria (*L. monocytogenes*, *S. aureus*, and *Salmonella enterica*) compared to other treated samples. During the same period, the sample treated with clove extract showed lower TBARS values than the control and antioxidant added samples.

The antimicrobial activity of ethanolic extract from clove was evaluated on fresh mutton during storage at 25°C by Kumudavally et al. (2011). Samples treated with clove extract had higher shelf lives of four days than the control sample (one day). In samples with clove extract, yeast, and molds, *E. coli* and *S. aureus* were not observed till the end of the fifth day of storage at 25°C. During storage, fewer losses were observed in the water holding capacity of samples treated with clove extract than the control sample. At the end of four days, the clove-treated samples exhibited slightly different hardness values than the control sample. Besides, cadaverine, which is recognized as an indicator of spoilage, occurred in one-day storage in the control sample, while this compound was absent up to four days in the treated mutton. Biogenic amine index and free fatty acids of treated samples with clove extract were lower than that of control samples during storage.

Chan et al. (2012) used a deodorized aqueous extract of clove bud in chicken meatballs at 200 ppm to reduce lipid oxidation. Peroxide value (PV) and 2-thiobarbituric acid reactive substances (TBARS) were used to track the lipid oxidation during storage at 8°C for 12 days. Besides, the color and sensory properties of samples were determined during storage. The samples supplemented with antioxidants exhibited a slight increase in PV and TBARS until the end of the storage period, while a sharp increase in these values was observed in the samples without any addition. During storage, the redness of meatballs was improved with clove bud extracts compared to other antioxidant-treated samples. Besides, the sensory analysis showed that the meat sample treated with clove bud extract did not exert any adverse effects.

Zhang et al. (2017) evaluated the effects of ethanol extract from clove at different concentrations (0.25%–2%) on lipid and protein oxidation of Chinese-style sausage stored for 21 days at 4°C. In lipid oxidation, TBARS values of the clove extract-treated group significantly decreased during storage compared to the control sample. As for protein oxidation, the carbonyl compounds content of sausages treated with clove showed lower values than the control sample. The addition of 2% of CE showed a significant inhibitory effect on lipid and protein oxidation. However, the sausages supplemented with clove extracts reduced thiol groups in sausages. According to color analysis, a decrease in L^* values and an increase in a^* values were observed in the sausages with clove extract compared to the control. Texture analysis showed that textural deterioration was retarded in the sausages with clove extract compared to the control sample. Moreover, no adverse effect on the sensory properties was observed in the sausages with clove extract.

Zahid et al. (2020) studied the effects of clove extract on the oxidative stability and sensory attributes in cooked beef patties at refrigerated storage by comparing BHT and ascorbic acid. A remarkable decrease in TBARS value of patties treated with antioxidants was observed during storage than the control sample. At the end of 10 days of storage, the TBARS value of the clove extract-treated patties was lower than BHT and ascorbic acid-treated samples. As for protein oxidation, carbonyl content was used to track protein oxidation in patties samples. At the end of storage, the carbonyl content in the samples mixed with clove extract and ascorbic acid showed lower values than the control and BHT treated samples. Clove extracts improved sensory attributes when compared to the control at refrigerated storage for 10 days.

Chen et al. (2016) investigated the effects of clove extracts in the inhibition of hydroxyl radical oxidation-induced structural and rheological changes in porcine myofibrillar protein (MP). Clove extracts reduced carbonyl formation in muscles after oxidation for 5 h compared to the control sample. The antioxidant activity of clove extract on protein oxidation was enhanced with the addition of clove extract. Besides, the loss in Ca-ATPase activities of MP decreased with the addition of clove extract. The incorporation of clove extract enhanced protein solubility and also thermal stability of MP. Regarding rheological properties, clove extract improved the gel formation ability (storage modulus, loss modulus) of MP.

3.2 Fish products

Bensid et al. (2014) examined the effects of ice-containing clove (0.02%, w/v) extracts on some quality parameters of anchovy (*Engraulis encrasicholus*). Sensory analysis revealed that anchovy stored in ice prepared with clove extracts had a shelf life of 12 days, while the batch stored in traditional ice had a shelf life of 9 days. At the end of 12 days of storage, indicators of lipid oxidation, PV, and TBA values were lower when compared to the control sample. Similarly, the lower values in total volatile basic nitrogen content were observed in anchovy samples during chilling storage under icing with clove extract conditions. This value could be used to determine fish deterioration.

Shi et al. (2014) examined the effects of clove extract on the lipid and protein oxidation of silver carp fillets during refrigerated storage over 18 days. During storage, the PV of the samples treated with clove extract was found lower than the control sample. Similarly, TBA values of the sample with clove extract were lower compared to the control sample during storage. The results from the salt-soluble protein and total sulfhydryl group indicated that extracts had less effect on protein oxidation.

Sardines (*Rastrineobola argentea*) are known to be a good source of health-promoting *omega*-3 fatty acids. Open sun drying is a traditional and widespread method to process sardines. However, these dried products could encounter lipid oxidation, causing the off-flavor in these valuable products. For this reason, Chaula et al. (2019) used different concentrations of clove water extract (5, 10, and 20 g/L) in sardines to reduce lipid oxidation. The total phenolic content of these extracts ranged between 18.1 and 28.7 μg GAE/mL. Clove extracts exhibited a higher inhibition of DPPH• (93%–95%). On the contrary, iron chelating activity was found at lower values (14.7%–22.2%). The peroxide value was significantly lower in sardines treated with clove (ranging from 8.29 to 11.5 meq O_2/kg oil) than that of the control sample (18.8 meq O_2/kg oil). Besides, volatiles from secondary oxidation products were present in lower amounts in clove-treated sardines when compared to the control sample. A remarkable decrease was observed in 1-penten-3-ol, 4-heptanal, and t, t-2, and 4-heptadienal, known as the decomposition products of EPA and DHA.

Slavin et al. (2016) studied the effects of two oven-drying conditions and four doses of clove water extract on the lipid oxidation and sensory attributes of omena fish (*R. argentea*). PV decreased as the clove concentration increased for the same oven-drying condition. Like PV, the TBARS value of samples with clove water extract was lower than that of the control sample. Samples with clove extract at 20 g/L concentration had the lowest values for PV and TBARS. Total phenolic content of water extracts changed between 4.91 and 23.9 mmol gallic acid equivalent (GAE)/L extract. Results of antiradical tests (ABTS and ORAC) showed that 20 g/L of clove water extract had significantly higher activity than those of the other extracts.

Asimi and Sahu (2016) examined the anti-stress effect of clove extract in *Labeo rohita*. The results revealed a significant lowering in superoxide dismutase (SOD) and catalase (CAT) activity level in liver, gill, and serum of *L. rohita* fed with clove extract at 0.5% in comparison to the control. SOD and CAT are the antioxidant defense systems in metabolism. The presence of these enzymes at a lower amount indicates the minimum production of free radicals.

3.3 Other food products

Shende et al. (2014) used clove oleoresin and steam-distilled clove extract to improve ghee's oxidation stability, which is anhydrous milk fat, having a pleasing and appetizing aroma. During storage at 80°C, the PV, diene, and TBARS values in ghee treated with clove extracts (0.5%) were significantly low than that of the control sample without any addition. In the Rancimat apparatus at 120°C, ghee's induction period, including clove extracts (21.0 h-oleoresin, 27.1 h-steam distilled), showed higher values than the control ghee sample (10.85 h). During the deep frying experiment at 180°C, ghee-containing clove extracts were more stable than the control sample. All oxidation tests showed that steam distilled clove extract exhibited stronger resistance against oxidation compared to clove oleoresin.

Rege and Momin (2018) examined the effects of clove oleoresin (0.5%, and 1.5%, w/v) and a eugenol-rich fraction (ERF) from clove oleoresin (2%, w/v) on the oxidation of refined sunflower oil at 60°C for 30 days. Clove oleoresin showed pro-oxidant activity, whereas ERF showed antioxidant activity. This difference is explained by the authors as follows: during the purification process from oleoresin, nonpolar compounds were removed from eugenol, and this form exhibited intense antioxidant activity.

Huang et al. (2019) studied the effects of clove extract at a concentration of 300 mg/kg on sensorial and quality attributes of the meat-based filler in frozen pork dumplings stored at −18°C. TBARS and protein carbonyl content in frozen dumpling fillers with clove extract exhibited lower values than control samples during storage at −18°C. Clove extract retarded lipid and protein oxidation in frozen dumpling fillers. During storage, the treated dumpling samples showed higher breaking strength and lower cooking loss than the control. Differential scanning calorimetry analysis proved that the treated samples with clove extract had higher thermal stability during storage, which indicated that clove extracts were effective against MP denaturation. Regarding rheological properties, the dumpling fillers supplemented with clove extract had higher G' values than the control, indicating that extracts could delay the deterioration in texture qualities. Sensory analysis showed that enriched dumplings have positive sensory scores in juiciness, texture, and overall acceptability and lower rancidity scores.

Chatterjee and Bhattacharjee (2013) evaluated the antioxidant efficacy of encapsulated clove extract, unencapsulated clove extract, and commercial antioxidant BHT in soybean oil. During the 30-day storage of frying oils, control samples with no added antioxidants had higher values in IC_{50} compared to all samples treated with antioxidants.

Chatterjee et al. (2012) used clove extract obtained by supercritical carbon dioxide extraction to prolong the shelf life of cookies during 30 days of storage. Oxidative rancidity was evaluated using electronic nose technology. Pentanal, hexanal, heptanal, octanal, nonanal, n-butanol, and pentanol, which are recognized as rancidity markers, were not present in the cookies containing clove extracts, while these compounds occurred in the control samples during the storage period. Besides, cookies with clove extract were rich in eugenol.

Harlina et al. (2018) evaluated the effects of clove extracts on the antioxidant activity, lipid oxidation, fatty acid composition, and volatile compounds of duck eggs over the salting period. Radical scavenging ability and reducing the power of salted duck eggs with 0.5% (w/v) clove extract was significantly higher than the control samples at 28 days of salting for egg white and egg yolk samples. Besides, such oxidation indicators as TBARS, conjugated diene, and *p*-anisidine values were lower in salted eggs supplemented with 0.5% (w/v) clove extract compared to the control sample. Eugenol was found as bioactive volatile in duck eggs enriched with clove extracts.

Boonpawa et al. (2019) investigated the antifungal activity of aqueous and ethanol extract of clove bud against *Aspergillus flavus* during the storage of dried chili. The ethanol extract of clove bud showed 2–5 times higher antifungal activity against *A. flavus* than the aqueous counterpart. At a concentration of 1000 ppm, ethanol extract of clove bud inhibited the fungal mycelial growth of *A. flavus* by 83%.

References

Abdel-Wahab, M., El-Sohaimy, S.A., Ibrahim, H.A., El-Makarem, H.S.A., 2020. Evaluation the efficacy of clove, sage and kiwifruit peels extracts as natural preservatives for fish fingers. Ann. Agric. Sci. 65 (1), 98–106. https://doi.org/10.1016/j.aoas.2020.06.002.

Adaramola, B., Onigbinde, A., 2016. Effect of extraction solvent on the phenolic content, flavonoid content and antioxidant capacity of clove bud. IOSR J. Pharm. Biol. Sci. 11 (3), 33–38.

Asimi, O.A., Sahu, N.P., 2016. Effect of antioxidant rich spices, clove and cardamom extracts on the metabolic enzyme activity of *Labeo rohita*. J. Fish. Livest. Prod. 4 (1). https://doi.org/10.4172/2332-2608.1000157, 100157.

Aznar, R., Rai, K., D., 2020. Purification and isolation techniques for enrichment of bioactive phytochemicals from herbs and spices. In: Hossain, M.B., Brunton, N.P., Rai, D.K. (Eds.), Herbs, Spices and Medicinal Plants: Processing, Health Benefits and Safety. Wiley-Blackwell, pp. 177–206, https://doi.org/10.1002/9781119036685.ch7.

Bensid, A., Ucar, Y., Bendeddouche, B., Özogul, F., 2014. Effect of the icing with thyme, oregano and clove extracts on quality parameters of gutted and beheaded anchovy (*Engraulis encrasicholus*) during chilled storage. Food Chem. 145, 681–686. https://doi.org/10.1016/j.foodchem.2013.08.106.

Boonpawa, R., Sriyoha, F., Kamnon, G., 2019. Antifungal activity of clove and betel extract against *Aspergillus flavus* during dried chili storage. Asia Pac. J. Public Health 5 (2), 17–22.

Chaieb, K., Hajlaoui, H., Zmantar, T., Kahla-Nakbi, A.B., Rouabhia, M., Mahdouani, K., Bakhrouf, A., 2007. The chemical composition and biological activity of clove essential oil, *Eugenia caryophyllata* (*Syzygium aromaticum* L. Myrtaceae): a short review. Phytother. Res. 21, 501–506. https://doi.org/10.1002/ptr.2124.

Chan, K.W., Khong, N.M.H., Iqbal, S., Ch'ng, S.E., Babji, A.S., 2012. Preparation of clove buds deodorized aqueous extract (CDAE) and evaluation of its potential to improve oxidative stability of chicken meatballs in comparison to synthetic and natural food antioxidants. J. Food Qual. 35, 190–199. https://doi.org/10.1111/j.1745-4557.2012.00445.x.

Chatterjee, D., Bhattacharjee, P., 2013. Comparative evaluation of the antioxidant efficacy of encapsulated and un-encapsulated eugenol-rich clove extracts in soybean oil: shelf-life and frying stability of soybean oil. J. Food Eng. 117 (4), 545–550. https://doi.org/10.1016/j.jfoodeng.2012.11.016.

Chatterjee, D., Bhattacharjee, P., Lechat, H., Ayouni, F., Vabre, V., 2012. Assessment of shelf-life of cookies formulated with clove extracts using electronic nose estimation of rancidity in cookies. In: 2012 Sixth International Conference on Sensing Technology (ICST), Kolkata, pp. 404–409, https://doi.org/10.1109/ICSensT.2012.6461709.

Chaula, D., Laswai, H., Chove, B., Dalsgaard, A., Mdegela, R., Jacobsen, C., Hyldig, G., 2019. Effect of clove (*Syzygium aromaticum*) and seaweed (*Kappaphycus alvarezii*) water extracts pretreatment on lipid oxidation in sun-dried sardines (*Rastrineobola argentea*) from Lake Victoria, Tanzania. Food Sci. Nutr. 7 (4), 1406–1416. https://doi.org/10.1002/fsn3.975.

Chen, H., Diao, J., Li, Y., Chen, Q., Kong, B., 2016. The effectiveness of clove extracts in the inhibition of hydroxyl radical oxidation-induced structural and rheological changes in porcine myofibrillar protein. Meat Sci. 111, 60–66. https://doi.org/10.1016/j.meatsci.2015.08.017.

Cortés-Rojas, D.F., de Souza, C.R.F., Oliveira, W.P., 2014. Clove (*Syzygium aromaticum*): a precious spice. Asian Pac. J. Trop. Biomed. 4 (2), 90–96. https://doi.org/10.1016/S2221-1691(14)60215-X.

Da Silva, R.P.F.F., Rocha-Santos, T.A.P., Duarte, A.C., 2016. Supercritical fluid extraction of bioactive compounds. TrAC Trends Anal. Chem. 76, 40–51. https://doi.org/10.1016/j.trac.2015.11.013.

El-Maati, M.F.A., Mahgoub, S.A., Labib, S.M., Al-Gaby, A.M.A., Ramadan, M.F., 2016. Phenolic extracts of clove (*Syzygium aromaticum*) with novel antioxidant and antibacterial activities. Eur. J. Integr. Med. 8 (4), 494–504. https://doi.org/10.1016/j.eujim.2016.02.006.

El-Sayed, S.M., Youssef, A.M., 2019. Potential application of herbs and spices and their effects in functional dairy products. Heliyon 5 (6). https://doi.org/10.1016/j.heliyon.2019.e01989, e01989.

Embuscado, M.E., 2015. Spices and herbs: natural sources of antioxidants-a mini review. J. Funct. Foods 18, 811–819. https://doi.org/10.1016/j.jff.2015.03.005.

Frohlich, P.C., Santos, K.A., Palu, F., Cardozo-Filho, L., da Silva, C., da Silva, E.A., 2019. Evaluation of the effects of temperature and pressure on the extraction of eugenol from clove (*Syzygium aromaticum*) leaves using supercritical CO_2. J. Supercrit. Fluids 143, 313–320. https://doi.org/10.1016/j.supflu.2018.09.009.

Gülçin, İ., 2012. Antioxidant activity of food constituents: an overview. Arch. Toxicol. 86, 345–391. https://doi.org/10.1007/s00204-011-0774-2.

Gülçin, İ., Şat, İ.G., Beydemir, Ş., Elmastaş, M., Küfrevioğlu, Ö.İ., 2004. Comparison of antioxidant activity of clove (*Eugenia caryophylata* Thunb) buds and lavender (*Lavandula stoechas* L.). Food Chem. 87 (3), 393–400. https://doi.org/10.1016/j.foodchem.2003.12.008.

Harlina, P.W., Ma, M., Shahzad, R., Gouda, M.M., Qiu, N., 2018. Effect of clove extract on lipid oxidation, antioxidant activity, volatile compounds and fatty acid composition of salted duck eggs. J. Food Sci. Technol. 55, 4719–4734. https://doi.org/10.1007/s13197-018-3367-8.

Hemalatha, R., Nivetha, P., Mohanapriya, C., Sharmila, G., Muthukumaran, C., Gopinath, M., 2016. Phytochemical composition, GC-MS analysis, in vitro antioxidant and antibacterial potential of clove flower bud (*Eugenia caryophyllus*) methanolic extract. J. Food Sci. Technol. 53, 1189–1198. https://doi.org/10.1007/s13197-015-2108-5.

Huang, L., Ding, B., Zhang, H., Kong, B., Xiong, Y.L., 2019. Textural and sensorial quality protection in frozen dumplings through the inhibition of lipid and protein oxidation with clove and rosemary extracts. J. Sci. Food Agric. 99 (10), 4739–4747. https://doi.org/10.1002/jsfa.9716.

Jin, W., Mujumdar, A.S., Zhang, M., Shi, W., 2018. Novel drying techniques for spices and herbs: a review. Food Eng. Rev. 10, 34–45. https://doi.org/10.1007/s12393-017-9165-7.

Joana Gil-Chávez, G., Villa, J.A., Fernando Ayala-Zavala, J., Basilio Heredia, J., Sepulveda, D., Yahia, E.M., González-Aguilar, G.A., 2013. Technologies for extraction and production of bioactive compounds to be used as nutraceuticals and food ingredients: an overview. Compr. Rev. Food Sci. Food Saf. 12, 5–23. https://doi.org/10.1111/1541-4337.12005.

Kaur, K., Kaushal, S., 2019. Phytochemistry and pharmacological aspects of *Syzygium aromaticum*: a review. J. Pharmacogn. Phytochem. 8, 398–406.

Kim, I.-S., Yang, M.-R., Lee, O.-H., Kang, S.-N., 2011. Antioxidant activities of hot water extracts from various spices. Int. J. Mol. Sci. 12 (6), 4120–4131. https://doi.org/10.3390/ijms12064120.

Kumudavally, K.V., Tabassum, A., Radhakrishna, K., Bawa, A.S., 2011. Effect of ethanolic extract of clove on the keeping quality of fresh mutton during storage at ambient temperature ($25 \pm 2°C$). J. Food Sci. Technol. 48, 466–471. https://doi.org/10.1007/s13197-010-0181-3.

Lee, O.-H., Jung, S.-H., Son, J.-Y., 2004. Antimicrobial activity of clove extract by extraction solvents. J. Korean Soc. Food Sci. Nutr. 33 (3), 494–499.

Rege, S.A., Momin, S.A., 2018. Synergistic antioxidant activity of clove oleoresin with Capsicum oleoresin and Kalonji seeds extract in sunflower oil. J. Nutr. Disord. Ther. 2 (1), 001–006.

Sánchez-Camargo, A.P., Montero, L., Mendiola, J.A., Herrero, M., Ibáñez, E., 2020. Novel extraction techniques for bioactive compounds from herbs and spices. In: Hossain, M.B., Brunton, N.P., Rai, D.K. (Eds.), Herbs, Spices and Medicinal Plants: Processing, Health Benefits and Safety. Wiley-Blackwell, pp. 95–128, https://doi.org/10.1002/9781119036685.ch5.

Shan, B., Cai, Y.-Z., Brooks, J.D., Corke, H., 2009. Antibacterial and antioxidant effects of five spice and herb extracts as natural preservatives of raw pork. J. Sci. Food Agric. 89, 1879–1885. https://doi.org/10.1002/jsfa.3667.

Shende, S., Patel, S., Arora, S., Sharma, V., 2014. Oxidative stability of ghee incorporated with clove extracts and BHA at elevated temperatures. Int. J. Food Prop. 17 (7), 1599–1611. https://doi.org/10.1080/10942912.2012.752382.

Shi, C., Cui, J., Yin, X., Luo, Y., Zhou, Z., 2014. Grape seed and clove bud extracts as natural antioxidants in silver carp (*Hypophthalmichthys molitrix*) fillets during chilled storage: effect on lipid and protein oxidation. Food Control 40, 134–139. https://doi.org/10.1016/j.foodcont.2013.12.001.

Slavin, M., Dong, M., Gewa, C., 2016. Effect of clove extract pretreatment and drying conditions on lipid oxidation and sensory discrimination of dried omena (*Rastrineobola argentea*) fish. Int. J. Food Sci. 51, 2376–2385. https://doi.org/10.1111/ijfs.13218.

Yashin, A., Yashin, Y., Xia, X., Nemzer, B., 2017. Antioxidant activity of spices and their impact on human health: a review. Antioxidants 6 (3), 70. https://doi.org/10.3390/antiox6030070.

Zahid, M.A., Choi, J.Y., Seo, J.-K., Parvin, R., Ko, J., Yang, H.-S., 2020. Effects of clove extract on oxidative stability and sensory attributes in cooked beef patties at refrigerated storage. Meat Sci. 161. https://doi.org/10.1016/j.meatsci.2019.107972, 107972.

Zhang, H., Peng, X., Li, X., Wu, J., Guo, X., 2017. The application of clove extract protects Chinese-style sausages against oxidation and quality deterioration. Korean J. Food Sci. Anim. Resour. 37 (1), 114–122. https://doi.org/10.5851/kosfa.2017.37.1.114.

CHAPTER 37

Health-promoting activities of clove (*Syzygium aromaticum*) extracts

A. Tomas[a], A. Maroyi[b], Natascha Cheikhyoussef[c], Ahmed A. Hussein[d], and Ahmad Cheikhyoussef[e]

Department of Biochemistry, Microbiology, and Biotechnology, Faculty of Agriculture, Engineering & Natural Sciences, University of Namibia, Windhoek, Namibia[a] Medicinal Plants and Economic Development (MPED) Research Centre, Department of Botany, University of Fort Hare, Alice, South Africa[b] Ministry of Higher Education, Technology and Innovation, Windhoek, Namibia[c] Chemistry Department, Cape Peninsula University of Technology, Cape Town, Western Cape, South Africa[d] Science and Technology Division, Multidisciplinary Research Unit, Centre for Research Services, University of Namibia, Windhoek, Namibia[e]

1. Introduction

Clove (*Syzygium aromaticum* or *Eugenia caryophyllus*) is a tropical plant in the Myrtaceae family with aromatic flower buds (Ryu et al., 2016). It can be classified as a large-leaved evergreen species, with a height between 8 and 12 m tall and crimson flowers grouped in terminal clusters (El-Shouny et al., 2019). The clove flower buds are initially pale hue; however, they transit to a bright red color when ready to be harvested at 1.5–2 cm long (Singh et al., 2018). The *S. aromaticum* dried flower buds are among the most valuable spices with a long history of applications in pharmaceutical, cosmetic, food, and agricultural industries (Cortés-Rojas et al., 2014). Because of these, cloves are now commercially farmed in different places worldwide, although they are known to be native to Maluku Islands in eastern Indonesia (Lambert et al., 2020). The world's top producers of cloves are Indonesia, India, Malaysia, Sri Lanka, West Indies, Madagascar, Tanzania, and Comoros (Idowu et al., 2021).

The ethnobotanical knowledge of clove provides a bird's eye view to understand a wide range of beneficial health-promoting activities, active biological properties, and unique aromatic structures (Cortés-Rojas et al., 2014). For instance, in Indian and Chinese traditional medicine, clove was documented to treat diverse complications like arthritis, digestive disorders, erectile dysfunction, diabetes, coughs and colds, diarrhea, toothaches, vomiting, memory loss, and nausea (Batiha et al., 2020). Over the last few years, numerous research studies have attempted to provide scientific evidence of the biological activities of clove as a potential drug candidate for these ailments (Cortés-Rojas et al., 2014; Ajiboye et al., 2016; Radünz et al., 2019; Batiha et al., 2020). Some of these studies were focused on the effects of clove toward suppression of microbial growth, improvement of neurological problems, and diabetes symptoms (Singletary, 2014). On the other hand, other preclinical studies supported the contribution of clove to counteract inflammation, inhibit oxidative stress, and oppose processes associated with cardiovascular disease (El-Maati et al., 2016; Mittal et al., 2014). Interestingly, various

Clove (*Syzygium aromaticum*). https://doi.org/10.1016/B978-0-323-85177-0.00018-5

constituents of *S. aromaticum* were analyzed to determine the relationship between pharmacological activity and a diverse array of secondary metabolites toward a wide range of beneficial health-promoting activities, active biological properties, and unique aromatic structures (Mittal et al., 2014; Oliveira et al., 2016). This chapter gives insights into the effects of clove extracts on health promotion and their potential applications.

2. Ethnobotanical knowledge and traditional uses

For centuries, clove and clove essential oil were used in Asian traditional medicine to treat multiple illnesses, including microbial diseases (Singletary, 2014). Some of these illnesses include arthritis, asthma, bronchitis, bruises, burns, cholera, colds, colic, coughs, diarrhea, digestive disorders, earaches, gum diseases, headaches, hypertension, impotence, inflammation of the pharynx, intestinal worms, nausea, malaria, rheumatism, scabies, toothache, tuberculosis, ulcers, vomiting, and wounds (Batiha et al., 2020; Singh et al., 2018).

In a culinary setting, *S. aromaticum* was documented as a flavoring addition to food and alcoholic beverages (El-Shouny et al., 2019). This was due to its ability to preserve property that enables it to inhibit food-borne pathogens like bacterial, fungal, viruses, worms, and protozoan contamination (Suliman et al., 2021). Histological data also revealed that clove was used to make perfumes, oral products, soaps, and detergents in various industrial applications (Ajiboye et al., 2016; Lambert et al., 2020; Mansourian et al., 2014). Furthermore, cloves were added to tobacco, known locally as kretek in Indonesia (El-Shouny et al., 2019). Other uses of clove include insect-repellent and growth-promoter agents. In addition, clove buds have been used as an antiepileptic remedy (Avicenna, 1988). They are traditionally used as spices in almost every household in India, with several therapeutic properties, including asthma, dental, digestive, and other respiratory disorders (Shah et al., 2012). Vicidomini et al. (2021) reported on the traditional therapeutic use of clove in respiratory disorders and its biological activity against several viruses, alongside its well-known anti-inflammatory, immunostimulatory, and antithrombotic properties, signifying its potential to fight COVID-19 disease. Two ethnobotanical studies that have been reported recently have shown the potential uses of cloves in the prevention and treatment of the COVID-19 pandemic in India (Singh et al., 2021) and Morocco (Chaachouay et al., 2021) in combination with other medicinal plants such as basil black pepper, cinnamon, garlic, ginger, and neem.

3. Biological activities

Literature has reported the antibacterial, antifungal, antibiotic, anticarcinogenic, antiseptic, anti-inflammatory, antioxidant, antiviral, antithrombotic, and antiparasitic properties of *S. aromaticum* (Ajiboye et al., 2016; Batiha et al., 2020; El-Maati et al., 2016; Lambert et al., 2020; Radünz et al., 2019). Active compounds in clove, like eugenol [2-methoxy-4-(2-propenyl)phenol] and other phenolic compounds, were believed to be the main contributors to its biological actions (Singh et al., 2018). In addition, cloves are well known for their antiemetic activity to relieve nausea and vomiting and their carminative properties (Singh et al., 2021). For this reason, clove has been considered safe for use in food and medicinal applications for a wide range of beneficial health-promoting activities.

3.1 Antioxidant activity

The clove extracts were documented with high antioxidant activity, both in vitro and in vivo (Cortés-Rojas et al., 2014; Radünz et al., 2019; Singletary, 2014). Clove represents one of the potent antioxidants and radical scavenger foods (Dudonné et al., 2009; Shan et al., 2005), with the ability to inhibit lipid peroxidation and malonaldehyde formation and to scavenge superoxide anions, thus assisting in promoting beneficial health activities (El-Maati et al., 2016). Generally, spice plants are known to have higher phenolic contents, and this is not an exception with clove extracts, whereby there is a strong correlation between the phenolic contents and their antioxidant activities (Kumar et al., 2014; Feng et al., 2021). Several studies attributed the antioxidant properties in clove to its secondary metabolites profile, like flavonoids, hydroxybenzoic acids, hydroxycinnamic acids, gallic acids, hydroxyphenyl propenes, flavonol glycosides, volatile phenolic oils (eugenol, acetyl eugenol, isoeugenol), and tannins (Mittal et al., 2014; Embuscado, 2015; Oliveira et al., 2016). These metabolites had effective reductive potential, free radical scavenging, superoxide anion radical scavenging, and metal chelating activities (Gülçin, 2011) whereby clove buds' aqueous and ethanol extracts inhibited 95% of both superoxide and 2,2- diphenyl-1-picrylhydrazyl (DPPH•) radicals activity (Gülçin et al., 2012).

Because of these, the antioxidant activity of clove can promote beneficial health activities by decreasing oxidative stress, reverting memory, and learning deficits caused by scopolamine in the short and long term (Halder et al., 2011; Mehta et al., 2010). Clove extracts were found to possess high antioxidant activity (DPPH• = 1353%), comparable to synthetic antioxidants against DPPH• radicals and the formation of peroxides (Ivanovic et al., 2013). Hemalatha et al. (2016) reported that the methanolic extract of *E. caryophyllus* showed a maximum antioxidant capacity of 93.2%; meanwhile, the antioxidant activity for the acetone extract was 85.4% and 81.4% for the chloroform extract. The ethanol clove leaf extract's effect on the oxidative stress response has been reported recently in *Schizosaccharomyces pombe*, which directly depends on ROS scavenging activity (Fauzya et al., 2019).

3.2 Antibacterial activity

The oldest study on clove antibacterial activity was reported by Bachmann (1916). Clove extracts were reported to be a potent antimicrobial agent of a variety of bacteria associated with foodborne diseases and human infections that include *Bacillus cereus*, *Escherichia coli*, *Helicobacter pylori*, *Listeria monocytogenes*, *Neisseria gonorrhoeae*, *Porphyromonas gingivalis*, *Prevotella intermedia*, *Propionibacterium acnes*, *Pseudomonas aeruginosa*, *Staphylococcus aureus*, *Staphylococcus epidermidis*, *Streptococcus mutans*, *S. pyogenes*, *Shigella* species, *Salmonella enterica serovar typhi*, and *Vibrio cholera* (Ajiboye et al., 2016; Radünz et al., 2019; El-Shouny et al., 2019). Literature provided evidence that clove extracts were bacteriostatic and bactericidal against most of these pathogens (El-Maati et al., 2016). Cai and Wu (1996) reported that the crude methanol extract of *S. aromaticum* exhibited a superior growth-inhibitory activity against Gram-negative oral pathogens, especially accompanying dental caries and periodontal diseases, e.g., *Porphyromonas gingivalis* and *Prevotella intermedia*. The antibacterial activity of *E. caryophyllus* extracts has been reported against several pathogenic bacteria such as *Klebsiella pneumonia*, *Bacillus subtilis*, *Staphylococcus aureus*, and *Vibrio chlorea* (Feres et al., 2005; Hemalatha et al., 2016).

According to Singletary (2014), clove extracts mechanisms of antibacterial action inhibit quorum sensing (QS) and biofilm formation (BF), disrupt cellular membranes, and reduce the expression of virulence-related *exo*-proteins. Based on these, clove is used to preserve food and treat human infections caused by bacteria, therefore improving health-promoting activities in our daily lives. The bud chloroform, hexane, and methanol extracts of *S. aromaticum* have been evaluated for their anti-QS property by estimating the reduction in quorum-controlled violacein pigment in biomarker strains of *Chromobacterium violaceum* CV026, *E. coli* pSB1075, and *Pseudomonas aeruginosa* PA01, suggesting clove buds extracts potential for anti-infective drugs development (Krishnan et al., 2012). Furthermore, the anti-virulence potential of partially purified *S. aromaticum* bud extracts against two clinically isolated *Pseudomonas aeruginosa* (RRLP1 and RRLP2) from patients with catheter-associated urinary tract infection has been reported by Rathinam and Viswanathan (2018). They have used bioactivity-guided partial purification, size exclusion chromatography, and GC-MS analysis to confirm that eugenol is the major bioactive compound with well-known medicinal properties (Nisar et al., 2021). QS is a well-known cell communication, and regulatory system bacteria use to signal their communications to the external environment. It influences antibiotic synthesis, bacterial growth and proliferation, biofilm formation, environmental adaptation, and virulence factor(s) production (Yang et al., 2020).

3.3 Antiviral activity

Crude extracts of clove were documented in vitro effective against human viruses like herpes simplex virus and hepatitis C virus strains (Singletary, 2014; Batiha et al., 2020). According to Cortés-Rojas et al. (2014), *S. aromaticum* attacks viral DNA synthesis by inhibiting the viral DNA polymerase. Therefore, the clove can enhance health-promoting activities by treating viral diseases such as sores around the mouth, lips, genitals, or rectum, hepatitis C, and some cancers such as liver cancer and lymphomas in humans. Furthermore, Batiha et al. (2020) reported a synergistic action between acyclovir and *S. aromaticum* extract with an important impact on the inhibition of viral replication.

Eugeniin was isolated from *S. aromaticum* has been identified as antiherpes simplex virus compound at 5 µg/mL concentration (Kurokawa et al., 1998; Singh et al., 2021), and its inhibitory action on the viral DNA synthesis as a selective inhibitor of the HSV-1 DNA polymerase and another well-known eugenol for its action on viral replication and reducing infection (Kurokawa et al., 1998).

A Feline Calicivirus (FCV) was used as a surrogate for human norovirus to test the aqueous extract of clove to test its antiviral activity. Clove extract scored similar antiviral activity to pure eugenol even though at a lower level (Aboubakr et al., 2016). Mehmood et al. (2020) synthesized silver nanoparticles using the aqueous extract of *S. aromaticum* flower buds, and they reported their novelty and efficacy against Newcastle Viral Disease (NDV) in vitro and embryonated eggs. Vicidomini et al. (2021) reported the anti-COVID-19 molecular basis and the therapeutical potential of *S. aromaticum*. Mputu Kanyinda (2020) reported a protocol that can help in the prevention and treatment of the early stages of COVID-19 disease using cloves, whereby a decoction is prepared via boiling cloves in water with other plant materials (*Eucalyptus globulus, Cymbopogon citratus*, and *Zingiber officinale*) for 15 min. The presented protocol further explained that it could be used either via inhalation of the released active volatile compounds for 5 min or taken as a drinkable decoction.

3.4 Anti-inflammatory activity

S. aromaticum extracts contain a diverse array of phenylpropanoids, like eugenol, which have a variety of pharmacologic activities, including anti-inflammatory action (Barboza et al., 2018). A study by Daniel Apparecido et al. (2009) has demonstrated that eugenol from clove has antioxidant and anti-inflammatory capacities; thus, it was more effective in reducing inflammation.

Literature (Barboza et al., 2018; Daniel Apparecido et al., 2009; Han and Parker, 2017; Sugihartini et al., 2019) reported that this could be achieved either:

a) Modulates the macrophage functions and regulates negatively the inflammation,
b) Decreased the in vitro and in vivo leukocytes migration in response to chemotactic factors by the modulation of rolling and adherence to perivascular tissue,
c) Modulates the bone remodeling,
d) Inhibition of enzymes and oxidative processes, for instance, inhibition of pro-inflammatory cytokines in pre-inflamed human dermal fibroblast cells.

Because of these, a diverse array of secondary metabolites in clove has the potential to boost health-promoting benefits and consequently improve the quality of life. Hypersensitivity was treated with the aqueous extract of *S. aromaticum* flower bud (Kim et al., 1998). In animal model (rate), the aqueous extract inhibited compound 48/80-induced systemic anaphylaxis with an $IC_{50}=31.2$ mg/kg, i.p., as well as it inhibited the local immunoglobulin E (IgE)-mediated passive cutaneous anaphylactic reaction with an $IC_{50}=17.7$ mg/kg, i.v. and $IC_{50}=19.8$ mg/kg, p.o. They emphasized that aqueous clove extract inhibits immediate hypersensitivity by inhibiting histamine release from mast cells in vivo and in vitro (Kim et al., 1998). Lipopolysaccharide (LPS)-induced lung inflammation was relieved through treating with both eugenol and whole clove aqueous extract (WCAE) via reducing tumor necrosis factor-alpha (TNF-α) and inhibiting the signaling of nuclear factor kappa-light-chain-enhancer of activated B cells (NF-$_{k}$B), alongside alveolar damage improvement (Magalhães et al., 2010; Chniguir et al., 2019).

3.5 Antifungal activity

Clove extracts (methanol, ethyl acetate, *n*-hexane, and diethyl ether) had effective inhibitory activity against fungi causing diseases such as vaginal candidiasis from *Candida* species (*C. albicans*, *C. glabrata*, and *C. tropicalis*) and *Trichophyton* species, both in vivo and in vitro (Khan et al., 2009; Singletary, 2014; Gonelimali et al., 2018; Yassin et al., 2020). For example, Yassin et al. (2020) reported that the minimum inhibitory concentration (MIC) of the *S. aromaticum* ethyl acetate extract was 250 μg/disc against *C. tropicalis* and 500 μg/disc against *C. albicans* and *C. glabrata*, while the minimum fungicidal concentration (MFC) was 0.5 mg/disc against *C. tropicalis* and 1 mg/disc against *C. albicans* and *C. glabrata*.

Various mechanisms of antifungal action have been reported, such as arresting the cell cycle, preventing biofilm formation, suppressing virulence factors, and disrupting the architecture of cell envelopes (Cortés-Rojas et al., 2014; Radünz et al., 2019). These actions were likely due to eugenol, the primary antifungal agent in clove (Mittal et al., 2014). In light of this, secondary metabolites in cloves promote health benefits by inhibiting the growth of fungal pathogens that cause human diseases and food poisoning. For example, the *S. aromaticum* and *Elettaria cardamomum* extracts demonstrated a protective effect on black tea by inhibiting the growth of *Xerophilic aflatoxigenic* fungi and advocated adding these extracts as preservation and flavoring agents in marketed black tea (Al-Sohaibani et al., 2011). Furthermore, a recent study on the aqueous extract of clove demonstrated protective effects on an animal model of pyelonephritis (Nassan et al., 2015), which is a recent kidney inflammation in COVID-19 patients that had been reported (Su et al., 2020).

3.6 Anticancer activity

Several studies reported the anticancer activities of clove extracts (Banerjee et al., 2006; Arung et al., 2011; Dwivedi et al., 2011; Kumar et al., 2014; Liu et al., 2014). In a mouse model of benzo[*a*]pyrene (BP)-induced lung cancer, an infusion prepared from clove has been shown to inhibit the incidence of hyperplasia and dysplasia (Banerjee et al., 2006). In addition, the ethanolic and ethyl acetate extracts of *S. aromaticum* were effective in inhibiting BEL-7402 liver cancer cells, HeLa cervical cancer cell line, ovarian cancer (SKOV-3 cell line), pancreatic cancer (PANC-1 cell line) (Liu et al., 2014), esophageal cancer (TE-13 cell line), and prostate cancer (DU-145 cell line) (Dwivedi et al., 2011). Secondary metabolites, like β-caryophyllene and tannin compounds such as eugeniin and tellimagrandin, were identified from cloves to possess anticancer actions (Arung et al., 2011). Aisha et al. (2012) investigated the anti-breast cancer effect of *S. aromaticum* extracts (bark, leaves, and stem) on the MCF-7 and MDA-MB-231 and prepared betulinic acid fraction by crystallization from clove leaves. They further reported that the stem extract induced apoptosis in MCF-7 cells by activating caspase 3/7 (Aisha et al., 2012).

Liu et al. (2014) reported the anticancer activity of ethyl acetate extract of cloves against colon cancer (HT-29) cells xenograft model with significant cell growth inhibition. The reported mechanism includes arresting the cell cycle, inducing apoptosis, increasing proteins expression, and decreasing cell cycle-regulated proteins. However, the same extract has shown a toxic effect against the non-tumorigenic colon 841 cells (Liu et al., 2014).

Vijayasteltar et al. (2016) have confirmed the safety of clove bud extracts in Wistar rats, whereby they reported the absence of toxicological changes in behavioral observations, body weights, organ weights, ophthalmic examinations, feed consumption, hematology, urinalysis, and clinical biochemistry parameters compared to the untreated group of animals confirming the safe use of clove bud extracts as dietary supplements. Fadipe et al. (2016) tested an isolated oleanoleic acid and its derivative, 3-*O*-Acetyl-Oleanolic acid (OAA), from the flower buds of *S. aromaticum* against both human embryonic kidney (HEK293) and hepatocellular carcinoma cell lines (HepG2) using the MTT assay with both compounds, exhibited LC_{50} of ˃300 μg/mL suggesting that these compounds may not be toxic at low to moderate concentrations and their potential to be used in the treatment of tuberculosis and malaria. In a preliminary in vitro screening against the human ovarian cancer cells (A2780) using MTT assays, Ryu et al. (2016) reported that the ethanol extract (70%) of the flower buds of *S. aromaticum* displayed a significant inhibitory activity with an IC_{50} value of 22.6 μM. Kubatka et al. (2017) reported the anti-neoplastic effects of clove buds in the model of breast carcinoma and have shown that the ethanol extract was able to decrease the MCF-7 cell survival through dose- and time-dependent mechanism via the induction of S phase cell cycle arrest and apoptosis (Liu et al., 2014). Other cell lines of MCF-7 and MDA-MB-231 were used to evaluate the anticancer activities of clove extracts (ethanol, oil, and water) and exhibited cytotoxic effects toward MCF-7 cells. However, only water and oil extracts were toxic toward MDA-MB-231 cells, whereby this effect was attributed to the induction of cancer cell death via apoptosis (Dwivedi et al., 2011). Oyedeji et al. (2014) reported two semisynthetic compounds, 3-acetoleanolic hydrazide and 3-acetoxyoleanolic hydrazine, derived from ethyl acetate fraction of *S. aromaticum* and have shown cytotoxic activity against MCF-7 cells. Ali et al. (2019) reported an in vitro inhibition of cell migration and cellular apoptosis caused by four chloroform extract fractions of *S. aromaticum* using human A549 and H1299 lung cancer cell lines. They reported inhibition of wound closure/cell migration of A549 and H1299 and apoptosis induction of H1299.

The anticancer activities of clove extracts were reviewed by Chua et al. (2019). The World Health Organization (WHO) stated 2.5 mg/kg body weight as the acceptable daily amount of clove in humans (Idowu et al., 2021). Yassin et al. (2020) tested the toxicity of different clove extracts (diethyl ether, ethyl acetate, methanol, and *n*-hexane) against the HUH7 cell line using the 3-(4,5-dimethylthiazol)-2,5-diphenyl tetrazolium bromide (MTT) assay. They reported that diethyl ether clove extracts showed the lowest toxic effect against the HUH7 cell line ($IC_{50}=62.4\,\mu g/mL$), while higher toxicity was obtained with the methanol extract ($IC_{50}=24.1\,\mu g/mL$); meanwhile, the *n*-hexane ($IC_{50}=42.1\,\mu g/mL$) and ethyl acetate ($IC_{50}=33.6\,\mu g/mL$) extracts of clove exhibited moderate toxicity (Yassin et al., 2020). In conclusion, the water extract of *S. aromaticum* at lower doses may be safe; meanwhile, ethanol extract and essential oil, even at minimal concentrations, possess toxicity (Chua et al., 2019).

3.7 Antimycobacterial and antiplasmodial activities

Fadipe et al. (2016) isolated oleanoleic acid from the flower buds of *Syzygium aromaticum*. Then, they synthesized 3-*O*-acetyl-oleanolic acid from oleanoleic acid and reported the antiplasmodial and antimycobacterial activities of both (oleanoleic acid and its derivative) against *Plasmodium falciparum* (Chloroquine Sensitive Strain) NF54 and *Mycobacterium tuberculosis* H37Rv, respectively, with IC_{50} of 4.3 µg/mL against *P. falciparum* and IC_{50} of 0.042 µg/mL against *M. tuberculosis* for OA (Fadipe et al., 2016).

3.8 Antidiabetic activity

Recently, the antidiabetic potential of *Syzygium* sp. has been reviewed and reported by Zulcafli et al. (2020). The polyphenol-rich extract of *S. aromaticum* in a dose-dependent manner showed better inhibition of α-glucosidase than α-amylase. It exhibited potential antioxidant activities and inhibition of Fe(2+)-induced lipid peroxidation in rat pancreas in vitro (Adefegha and Oboh, 2012). The EtOH extract clove flower buds were reported to have a significant suppression effect on increased blood glucose levels in type 2 diabetic KK-A(y) mice (Kuroda et al., 2012). Two main compounds of dehydrodieugenol and dehydrodieugenol B have shown potential peroxisome proliferator-activated receptor (PPAR)-γ ligand-binding activities in a GAL4-PPAR-γ chimera assay; meanwhile, the major compound in the EtOH extract, oleanolic acid, reported a moderate activity (Kuroda et al., 2012). Oleanolic acid and maslinic acid from the *S. aromaticum* extract were reported to improve glucose homeostasis and suppress postprandial hyperglycemia in diabetic rats through the inhibition of carbohydrate hydrolysis, glucose transporters reduction, and glycogen synthesis in the gastrointestinal tract (Khathi et al., 2013a,b). Another study by Sompong et al. (2016) reported beneficial effects (antihyperglycemic and antihyperlipidemic) of *S. aromaticum* extracts and its triterpene constituents and their ability to reduce the formation of cholesterol micelles in vitro. The aqueous extracts of clove and other culinary/medicinal plants have shown an insulin-like biological activity in vitro model (Broadhurst et al., 2000). Shukri et al. (2010) reported on the effects of *E. aromaticum* on some target organs of Streptozotocin (STZ or N-ethylnitrosocar-bamoyl-D-glucosamine)-induced oxidative stress in diabetic rats and confirmed that cloves could protect the heart, liver, and lens of diabetic rats. *S. aromaticum* extract was reported to regulate energy metabolism in C2C12 myocytes and has been shown to perform comparable insulin effect through significant reduction in blood glucose levels in animal models (Tu et al., 2014).

The in vitro and in vivo effects of the crude extracts and chemical components of cloves that attributed to their antidiabetic, lipid-lowering, and anti-obesity activities has been reported by Habtemariam (2019). Ladurner et al. (2017) reported that the flower buds extracts of *S. aromaticum* activated the bile acid receptor, Takeda G protein-coupled receptor 5(TGR5) at 100 mg/mL with a stronger effect than the positive control lithocholic acid (LCA) at 10 mM. In hepatocytes and hepatoma cells, *S. aromaticum* extracts act like insulin through their ability to reduce phosphoenolpyruvate carboxykinase (PEPCK) and glucose 6-phosphatase (G6Pase) gene expression (Prasad et al., 2005). Clove extracts can perform a comparable effect to insulin by significantly reducing blood glucose levels in animal models (C2C12 myocytes) and demonstrating that it could significantly increase glucose consumption (Tu et al., 2014). Clove extracts increase the phosphorylation of AMP-activated protein kinase (AMPK), as well as its substrate, acetyl-CoA carboxylase (ACC) as well as transcriptionally regulates sirtuin 1 (SIRT1) and PPARγ coactivator 1α (PGC1α) genes that are involved in metabolism. Furthermore, clove extracts in a dose-dependent can increase muscle glycolysis and mitochondrial function (Tu et al., 2014).

3.9 Anticonvulsant activity

Pharmacological studies demonstrated the anticonvulsant and anti-stress properties of eugenol (Dallmeier and Carlini, 1981). Kamkar Asl et al. (2013) reported the analgesic effects of aqueous and ethanol extracts of clove buds in mice whereby, the administration of three doses (50, 100, and 200 mg/kg) of the ethanol extract increased the maximal percent effect (MPE) in the animal groups and suggested a similar opioid system in the analgesic effect for clove extracts. Traditionally, clove was served as an analgesic agent for treating dental pain (Diaz and Sembrano, 1985).

3.10 Hepatoprotective activity

An in vivo study revealed the significant role of eugenol in protecting male rats during exposure to aflatoxins-induced hepato, nephrotoxicity, and oxidative stress (Abdel-Wahhab and Aly, 2005). In an animal model study using the liver in a hypercholesterolemia rabbit, the methanol extract of clove leaves has been reported to stabilize enzymes that are essential for oxidative stress response (Cu-SOD and Zn-SOD); at the same time, it reduces the malondialdehyde (MDA) which is known as one of the final products of polyunsaturated fatty acids peroxidation in cells (Gaweł et al., 2004; Munisa et al., 2015).

Hepatoprotective activity of eugenol-rich fraction of *S. aromaticum* was reported by Ali et al. (2014), which can protect the liver against injury. They reported its action via *S. aromaticum* eugenol-rich fraction's ability to reverse the biochemical and histopathological changes in liver cirrhosis induced by thioacetamide (0.03%) and inhibit hepatic cell proliferation. Ding et al. (2017) reported a FASN inhibitor-alcohol extract of clove (AEC) using a fast microplate method whereby the AEC supplementation could significantly reduce body weight and abdominal adipose tissue weight, lowered lipid accumulation in the liver and epididymal adipose tissue when it was compared with the HFD control group. In addition, they confirmed that the AEC could regulate the content of total triglyceride (TG) and low-density lipoprotein cholesterol (LDL-C) in the serum lipid profiles. Rahman Alizadeh et al. (2017) investigated and reported that dietary administration of clove bud powder to hens given a low n-6 to n-3 ratio diet could modulate serum antioxidant indices and enzyme alterations in birds and improve the heightened susceptibility of hepatocytes to lipid peroxidation. Abdelrahman et al. (2018) reported that clove in combination with honey has antioxidant activity with the ability to reverse liver fibrosis via cytokines reduction (TGF-β, TNF-α, and EGF) and modulation of liver function enzymes including ALT, AST, and GGT activities (58.5, 56; 119, 121, and 6.1; 6.3) in comparison to the CCl_4 control group (97, 191, and 10.1).

3.11 Skin protective effect

Hwang et al. (2018) reported on the *S. aromaticum* role in resisting skin photoaging whereby a 50% ethanol extract of *S. aromaticum* and eugenol was tested for antiaging effects of UVB-irradiated normal human dermal fibroblasts (NHDFs) and hairless mice. The ethanolic extract of *S. aromaticum* significantly improved procollagen type I and elastin levels through TGF/Smad signaling. They further reported the histopathological effect of *S. aromaticum* extract and its ability to reduce wrinkles and increase filament aggregating protein (filaggrin), which is known to repair the skin barrier function and improve the skin's hydration.

3.12 Neuroprotection effect

In Ayurvedic medicine, clove oil has been used as a cognitive enhancer and nerves stimulant. The treatment of eugenol in the animal model led to improved cognitive impairment under oxidative stress conditions and overall improvement of the nervous system (Nunomura et al., 2001). The ethanol extract of *S. aromaticum* has been reported to possess a neuroprotective activity in Alzheimer's disease model via maintaining oxidative balance through Silence information regulator 1 (SIRT1) pathway whereby *S. aromaticum* extract was able to scavenge ROS and elevate the antioxidant enzymes percentage, alongside with the activation and elevation of the level of SIRT1 and downregulation of γ-secretase level (Shekhar et al., 2018a). SIRT1 is a member of the sirtuin family that targets histones and many nonhistone proteins and takes part in several physiological functions, including antioxidative effects and mitochondria protection (Khan et al., 2015), autophagy activation (Shekhar et al., 2018b), and neuroinflammation inhibition (Kauppinen et al., 2013) and more recently, their upregulation can slow down the progression of Alzheimer's disease (Corpas et al., 2017). Shekhar et al. (2018a) reported that *S. aromaticum* possesses a maximum level of oxygen radical absorbance capacity (ORAC) as it was reported by the National Institute on Aging in the National Institutes of Health (NIH) as per the published data by the United States Department of Agriculture (USDA). Another neuro-function of *S. aromaticum* is the stimulation of nerves due to vata pacification (Halder et al., 2011). They further emphasized that *S. aromaticum* can be considered an anticholinesterase (ACh) agent for the management of cognitive ailments such as Alzheimer's disease (AD) to make up for ACh deficiency during AD (Shekhar et al., 2018a). The anticholinesterase activity of the methanol extract of clove, eugenol, and clove oil has been reported with eugenol possessing higher inhibition of acetylcholinesterase

(AChE) and butyrylcholinesterase (BChE) than clove extract and its oil (Dalai et al., 2014). *S. aromaticum* can enhance catalase's catalytic activity in a dose-dependent manner and maintain the oxidative balance in cells under oxidative and proteotoxic stress by maintaining the level of ROS and oxidative enzymes that may lead to down-streaming the effect of SIRT1 pathway (Shekhar et al., 2018a) which promotes neuroprotection via the prevention of oxidative stress (Braidy et al., 2012).

3.13 Immunomodulation effects

Active compounds such as demethyldiisoeugenol that are responsible for clove extracts have been shown to modulate some of the immune responses such as anti-inflammatory effects (Ko et al., 1995), suppression of T cell cellular immunity, and enhancement of humoral immune responses with modulation of cytokine pattern and Th2 responses and humoral immunity cytokines acceleration (Dibazar et al., 2014). Dibazar et al. (2014) further indicated that researchers who studied clove benefits in traditional medicine settings proposed that its effects could be centralized on the immune system and its components. Previous studies on eugenol and its structural analog, isoeugenol by Park et al. (2007), reported the inhibition of IL-2 gene expression (typical T cell cytokine). Isoeugenol and its analogs (eugenol and allylbenzene) clove ingredients were reported to inhibit nitric oxide production and induce nitric oxide synthetase expression in a dose-dependent manner LPS-stimulated RAW 264.7 murine macrophages (Choi et al., 2007). The water-soluble part of the hydroalcoholic extract of clove demonstrated pro-inflammatory cytokines (IL-1beta and IL-6) production by macrophages of BALB/c mice, with eugenol being the major component of the clove extract responsible for the cytokine inhibition (Rodrigues et al., 2009).

Grespan et al. (2012) reported that eugenol in the mice model could reduce IFN- secretion in accordance with its ability to decrease the secretion of IFN- from T cells. The same compound (eugenol) enhanced the IL-10 cytokine production (Grespan et al., 2012). Bachiega et al. (2012) investigated the effect of clove water extract administration over a short term to BALB/c mice on T helper 1 (Th1, IFN-, and IL-2) and Th2 (IL-4 and IL-10) cytokine production with no influence on the Th1/Th2 cytokine balance been observed. They further confirmed that clove immunomodulatory/anti-inflammatory effects could be attributed to the LPS inhibition alongside another possible mechanism of action for eugenol involving suppressing the nuclear factor-κB pathway (Bachiega et al., 2012). Dibazar et al. (2015) reported the immunomodulatory effects of clove constituents isolated from the aqueous and ethanol extracts on the macrophages via in vitro evaluations. They demonstrated that clove extracts do not have a specific cytotoxic activity that impacts macrophage cytokine activity in a dose-dependent pattern.

4. Other health benefits

The aphrodisiac activity of 50% ethanol extracts of *S. aromaticum* and nutmeg in male mice yielded an enhanced sexual behavior in male Swiss mice with no adverse effect (Tajuddin et al., 2003). Hemphill (2006) reported on the culinary uses of clove, including meat flavoring, especially ham, stewed fruits, pickles, curries, pies, salads, and some spicy alcoholic beverages. They also have been added as food antioxidants to mayonnaise (Chatterjee and Bhattacharjee, 2015) and meat sausages (Zhang et al., 2017). Clove is used as a spice, and its extracts have a remarkable aroma, flavor, and vast beneficial health characteristics; therefore, they have been utilized in the food industries (Ramadan et al., 2013;

Idowu et al., 2021). Radha Krishnan et al. (2014) reported that the inclusion of clove extracts could lead to efficient oxidation-reduction, sensory properties enhancement, and prolongation of meat products' shelf life at refrigeration conditions during storage. Takahashi et al. (2004) reported on the ability of cloves to mask spoiled food smells via the interference in the odor maps in the olfactory bulb of the forebrain. Scientific literature has supported the health benefits of clove toward reducing microbial growth, improving nutrition values, diabetes symptoms, and neurological problems (Cortés-Rojas et al., 2014; Singletary, 2014). Studies have documented the contribution of *S. aromaticum* to counteract inflammation, oppose processes associated with cardiovascular disease, and inhibit oxidative stress (Batiha et al., 2020; Gülçin, 2011). Secondary metabolites in clove, such as eugenol, are likely to contribute to these actions and other health-promoting activities (Oliveira et al., 2016).

The pediculocidal activity using the *S. aromaticum* flower bud hexane extract against *Pediculus humanus capitis* was examined by direct contact and fumigant toxicity (closed- and open-container methods) bioassay (Bagavan et al., 2011). They reported that chavibetol (5-allyl-2-methoxyphenol) is the major compound (58.8%) of clove flower bud hexane extract. The flower buds of *S. aromaticum* are relieving flatulence, stimulant, natural anthelmintic, and antimalarial and used for dyspepsia, gastric trouble, nausea, and vomiting, as well as it is used to relieve toothache (Alqareer et al., 2006; Tanko et al., 2008; Gowri and Manimegalai, 2019). Chaula et al. (2019) demonstrated that clove extracts gave higher retention of *omega*-3 fatty acids and lower concentrations of secondary lipid oxidation products in comparison to the seaweed; *Kappaphycus alvarezii* showing its feasibility of pretreating whole, *omega*-3-rich small sardines (*Rastrineobola argentea*) with natural clove antioxidants to prevent lipid oxidation during preservation of the small sardines via sun drying. Saeed et al. (2021) reported that clove extracts were used as feed additives to manipulate rumen fermentation through its main bioactive compound, eugenol, via the propionate production increment, acetate and methane production reduction, altering the proteolysis pattern and peptidolysis, and the deamination of the amino acid in the rumen. Positive effects were also reported regarding cloves to broiler chicken diets on the performance, carcass, and production traits of broilers (Boyraz and Ozcan, 2006; Salman and Ibrahim, 2012; Suliman et al., 2021).

5. Conclusions

Clove is considered a viable plant with several agricultural, cosmetic, food, and pharmaceutical applications. Regardless of its pleasant aroma and nutritional values, literature has identified the anticancer, antidiabetic, anti-inflammatory, antinociceptive, antibacterial, antifungal, antiprotozoal, antioxidant, and antithrombotic properties of *S. aromaticum*. Clove extracts possess potent antibacterial and antioxidant activities; therefore, they have great potential to be used as natural food preservatives. In addition, clove is one of the richest sources of phenolic compounds such as eugenol, gallic acid, and tannins, with great potential for pharmaceutical and cosmetic applications. Due to its numerous pharmacological activities, clove can be considered a potential drug candidate for health-promoting activities.

References

Abdelrahman, M.T., Maina, E.N., Elshemy, H.A., 2018. Clove (*Syzygium aromaticum*) and honey extracts significantly reduce inflammatory cytokines and liver function enzymes in experimental rats fed on carbon tetrachloride (CCl_4). J. Radiat. Res. Appl. Sci. 11 (4), 416–422. https://doi.org/10.1016/j.jrras.2018.08.003.

Abdel-Wahhab, M.A., Aly, S.E., 2005. Antioxidant property of *Nigella sativa* (black cumin) and *Syzygium aromaticum* (clove) in rats during aflatoxicosis. J. Appl. Toxicol. 25 (3), 218–223. https://doi.org/10.1002/jat.1057.

Aboubakr, H.A., Nauertz, A., Luong, N.T., Agrawal, S., El-Sohaimy, S.A., Youssef, M.M., Goyal, S.M., 2016. In vitro antiviral activity of clove and ginger aqueous extracts against feline calicivirus, a surrogate for human norovirus. J. Food Prot. 79 (6), 1001–1012. https://doi.org/10.4315/0362-028X.JFP-15-593.

Adefegha, S.A., Oboh, G., 2012. In vitro inhibition activity of polyphenol-rich extracts from *Syzygium aromaticum* (L.) Merr. & Perry (Clove) buds against carbohydrate hydrolyzing enzymes linked to type 2 diabetes and Fe(2+)-induced lipid peroxidation in rat pancreas. Asian Pac. J. Trop. Biomed. 2 (10), 774–781. https://doi.org/10.1016/S2221-1691(12)60228-7.

Aisha, A.F.A., Abu-Salah, K.M., Alrokayan, S.A., Siddiqui, M.J., Ismail, Z., Majid, A.M.S.A., 2012. *Syzygium aromaticum* extracts as good source of betulinic acid and potential anti-breast cancer. Rev. Bras. Farm. 22 (2), 335–343. https://doi.org/10.1590/S0102-695X2011005000185.

Ajiboye, T.O., Mohammed, A.O., Bello, S.A., Yusuf, I.I., Ibitoye, O.B., Muritala, H.F., Onajobi, I.B., 2016. Antibacterial activity of *Syzygium aromaticum* seed: studies on oxidative stress biomarkers and membrane permeability. Microb. Pathog. 95, 208–215. https://doi.org/10.1016/j.micpath.2016.03.011.

Ali, I., Naqshbandi, M.F., Husain, M., 2019. Cell migration and apoptosis in human lung cancer cells by clove (*Syzygium aromaticum*) dried flower buds extract. J. Taibah Univ. Sci. 13, 1163–1174. https://doi.org/10.1080/16583655.2019.1691480.

Ali, S., Prasad, R., Mahmood, A., Routray, I., Shinkafi, T.S., Sahin, K., Kucuk, O., 2014. Eugenol-rich fraction of *Syzygium aromaticum* (clove) reverses biochemical and histopathological changes in liver cirrhosis and inhibits hepatic cell proliferation. J. Cancer Prev. 19 (4), 288–300. https://doi.org/10.15430/JCP.2014.19.4.288.

Alqareer, A., Alyahya, A., Andersson, L., 2006. The effect of clove and benzocaine versus placebo as topical anesthetics. J. Dent. 34 (10), 747–750. https://doi.org/10.1016/j.jdent.2006.01.009.

Al-Sohaibani, S., Murugan, K., Lakshimi, G., Anandraj, K., 2011. Xerophilic aflatoxigenic black tea fungi and their inhibition by *Elettaria cardamomum* and *Syzygium aromaticum* extracts. Saudi J. Biol. Sci. 18 (4), 387–394. https://doi.org/10.1016/j.sjbs.2011.06.005.

Arung, E.T., Matsubara, E., Kusuma, I.W., Sukaton, E., Shimizu, K., Kondo, R., 2011. Inhibitory components from the buds of clove (*Syzygium aromaticum*) on melanin formation in B16 melanoma cells. Fitoterapia 82 (2), 198–202. https://doi.org/10.1016/j.fitote.2010.09.008.

Avicenna, A., 1988. Ghanoon dar Teb. Soroosh, Tehran, pp. 244–251.

Bachiega, T.F., de Sousa, J.P., Bastos, J.K., Sforcin, J.M., 2012. Clove and eugenol in noncytotoxic concentrations exert immunomodulatory/anti-inflammatory action on cytokine production by murine macrophages. J. Pharm. Pharmacol. 64 (4), 610–616. https://doi.org/10.1111/j.2042-7158.2011.01440.x.

Bachmann, F.M., 1916. Inhibiting action of certain spices on the growth of microorganisms. J. Ind. Eng. Chem. 8, 620–623.

Bagavan, A., Rahuman, A.A., Kamaraj, C., Elango, G., Zahir, A.A., Jayaseelan, C., Santhoshkumar, T., Marimuthu, S., 2011. Contact and fumigant toxicity of hexane flower bud extract of *Syzygium aromaticum* and its compounds against *Pediculus humanus capitis* (Phthiraptera: Pediculidae). Parasitol. Res. 109 (5), 1329–1340. https://doi.org/10.1007/s00436-011-2425-1.

Banerjee, S., Panda, C.K., Das, S., 2006. Clove (*Syzygium aromaticum* L.), a potential chemopreventive agent for lung cancer. Carcinogenesis 27 (8), 1645–1654. https://doi.org/10.1093/carcin/bgi372.

Barboza, J.N., da Silva Maia Bezerra Filho, C., Silva, R.O., Medeiros, J., de Sousa, D.P., 2018. An overview on the anti-inflammatory potential and antioxidant profile of eugenol. Oxid. Med. Cell. Longev. 2018, 3957262. https://doi.org/10.1155/2018/3957262.

Batiha, G.E., Alkazmi, L.M., Wasef, L.G., Beshbishy, A.M., Nadwa, E.H., Rashwan, E.K., 2020. *Syzygium aromaticum* L. (Myrtaceae): traditional uses, bioactive chemical constituents, pharmacological and toxicological activities. Biomolecules 10 (2), 202. https://doi.org/10.3390/biom10020202.

Boyraz, N., Ozcan, M., 2006. Inhibition of phytopathogenic fungi by essential oil, hydrosol, ground material and extract of summer savory (*Satureja hortensis* L.) growing wild in Turkey. Int. J. Food Microbiol. 107 (3), 238–242. https://doi.org/10.1016/j.ijfoodmicro.2005.10.002.

Braidy, N., Jayasena, T., Poljak, A., Sachdev, P.S., 2012. Sirtuins in cognitive ageing and Alzheimer's disease. Curr. Opin. Psychiatry 25 (3), 226–230. https://doi.org/10.1097/YCO.0b013e32835112c1.

Broadhurst, C.L., Polansky, M.M., Anderson, R.A., 2000. Insulin-like biological activity of culinary and medicinal plant aqueous extracts in vitro. J. Agric. Food Chem. 48 (3), 849–852. https://doi.org/10.1021/jf9904517.

Cai, L., Wu, C.D., 1996. Compounds from *Syzygium aromaticum* possessing growth inhibitory activity against oral pathogens. J. Nat. Prod. 59 (10), 987–990. https://doi.org/10.1021/np960451q.

Chaachouay, N., Douira, A., Zidane, L., 2021. COVID-19, prevention and treatment with herbal medicine in the herbal markets of Salé prefecture, North-Western Morocco. Eur. J. Intern. Med. 42. https://doi.org/10.1016/j.eujim.2021.101285, 101285.

Chatterjee, D., Bhattacharjee, P., 2015. Use of eugenol-lean clove extract as a flavoring agent and natural antioxidant in mayonnaise: product characterization and storage study. J. Food Sci. Technol. 52 (8), 4945–4954. https://doi.org/10.1007/s13197-014-1573-6.

Chaula, D., Laswai, H., Chove, B., Dalsgaard, A., Mdegela, R., Jacobsen, C., Hyldig, G., 2019. Effect of clove (*Syzygium aromaticum*) and seaweed (*Kappaphycus alvarezii*) water extracts pretreatment on lipid oxidation in sun-dried sardines (*Rastrineobola argentea*) from Lake Victoria, Tanzania. Food Sci. Nutr. 7 (4), 1406–1416. https://doi.org/10.1002/fsn3.975.

Chniguir, A., Zioud, F., Marzaioli, V., El-Benna, J., Bachoual, R., 2019. *Syzygium aromaticum* aqueous extract inhibits human neutrophils myeloperoxidase and protects mice from LPS-induced lung inflammation. Pharm. Biol. 57 (1), 56–64. https://doi.org/10.1080/13880209.2018.1557697.

Choi, C.Y., Park, K.R., Lee, J.H., Jeon, Y.J., Liu, K.H., Oh, S., Kim, D.E., Yea, S.S., 2007. Isoeugenol suppression of inducible nitric oxide synthase expression is mediated by down-regulation of NF-kappaB, ERK1/2, and p38 kinase. Eur. J. Pharmacol. 576 (1–3), 151–159. https://doi.org/10.1016/j.ejphar.2007.07.034.

Chua, L.K., Lim, C.L., Ling, A., Chye, S.M., Koh, R.Y., 2019. Anticancer potential of Syzygium species: a review. Plant Foods Hum. Nutr. 74 (1), 18–27. https://doi.org/10.1007/s11130-018-0704-z.

Corpas, R., Revilla, S., Ursulet, S., Castro-Freire, M., Kaliman, P., Petegnief, V., Giménez-Llort, L., Sarkis, C., Pallàs, M., Sanfeliu, C., 2017. SIRT1 overexpression in mouse Hippocampus induces cognitive enhancement through proteostatic and neurotrophic mechanisms. Mol. Neurobiol. 54 (7), 5604–5619. https://doi.org/10.1007/s12035-016-0087-9.

Cortés-Rojas, D.F., de Souza, C.R., Oliveira, W.P., 2014. Clove (*Syzygium aromaticum*): a precious spice. Asian Pac. J. Trop. Biomed. 4 (2), 90–96. https://doi.org/10.1016/S2221-1691(14)60215-X.

Dalai, M.K., Bhadra, S., Chaudhary, S.K., Bandyopadhyay, A., Mukherjee, P.K., 2014. Anti-cholinesterase activity of the standardized extract of *Syzygium aromaticum* L. Pharmacogn. Mag. 10 (Suppl 2), S276–S282. https://doi.org/10.4103/0973-1296.133275.

Dallmeier, K., Carlini, E.A., 1981. Anesthetic, hypothermic, myorelaxant and anticonvulsant effects of synthetic eugenol derivatives and natural analogues. Pharmacology 22 (2), 113–127. https://doi.org/10.1159/000137479.

Daniel Apparecido, N., Sartoretto Simone, M., Schmidt, G., Caparroz-Assef Silvana, M., Bersani-Amado Ciomar, A., Cuman Roberto Kenji, N., 2009. Anti-inflammatory and antinociceptive activities A of eugenol essential oil in experimental animal models. Rev. Bras 19 (1b), 212–217. https://doi.org/10.1590/S0102-695X2009000200006.

Diaz, M.R., Sembrano, J.M., 1985. A comparative study of the efficacy of garlic and eugenol as palliative agents against dental pain of pulpal origin. J. Philipp. Dent. Assoc. 35 (1), 3–10.

Dibazar, S.P., Fateh, S., Daneshmandi, S., 2014. Clove (*Syzygium aromaticum*) ingredients affect lymphocyte subtypes expansion and cytokine profile responses: an in vitro evaluation. J. Food Drug Anal. 22 (4), 448–454. https://doi.org/10.1016/j.jfda.2014.04.005.

Dibazar, S.P., Fateh, S., Daneshmandi, S., 2015. Immunomodulatory effects of clove (*Syzygium aromaticum*) constituents on macrophages: in vitro evaluations of aqueous and ethanolic components. J. Immunotoxicol. 12 (2), 124–131. https://doi.org/10.3109/1547691X.2014.912698.

Ding, Y., Gu, Z., Wang, Y., Wang, S., Chen, H., Zhang, H., Chen, W., Chen, Y.Q., 2017. Clove extract functions as a natural fatty acid synthesis inhibitor and prevents obesity in a mouse model. Food Funct. 8 (8), 2847–2856. https://doi.org/10.1039/c7fo00096k.

Dudonné, S., Vitrac, X., Coutière, P., Woillez, M., Mérillon, J.M., 2009. Comparative study of antioxidant properties and total phenolic content of 30 plant extracts of industrial interest using DPPH, ABTS, FRAP, SOD, and ORAC assays. J. Agric. Food Chem. 57 (5), 1768–1774. https://doi.org/10.1021/jf803011r.

Dwivedi, V., Shrivastava, R., Hussain, S., Ganguly, C., Bharadwaj, M., 2011. Comparative anticancer potential of clove (*Syzygium aromaticum*)-an Indian spice-against cancer cell lines of various anatomical origin. Asian Pac. J. Cancer Prev. 12 (8), 1989–1993.

El-Maati, M.F.A., Mahgoub, S.A., Labib, S.M., Al-Gaby, A.M.A., Ramadan, M.F., 2016. Phenolic extracts of clove (*Syzygium aromaticum*) with novel antioxidant and antibacterial activities. Eur. J. Integr. Med. 8 (4), 494–504. https://doi.org/10.1016/j.eujim.2016.02.006.

El-Shouny, W.A., Ali, S.S., Hegazy, H.M., Abd Elnabi, M.K., Ali, A., Sun, J., 2019. *Syzygium aromaticum* L.: traditional herbal medicine against *cag*A and *vac*A toxin genes-producing drug resistant *Helicobacter pylori*. J. Tradit. Complement. Med. 10 (4), 366–377. https://doi.org/10.1016/j.jtcme.2019.05.002.

Embuscado, M.E., 2015. Spices and herbs: natural sources of antioxidants – a mini review. J. Funct. Foods 18, 811–819. https://doi.org/10.1016/j.jff.2015.03.005.

Fadipe, V.O., Opoku, A.R., Shintre, S.A., Singh, P., Mongalo, N.I., 2016. Antimycobacterial, antiplasmodial and toxicological effect of oleanolic acid and its derivative from *Syzygium aromaticum* Linn (Myrtaceae). S. Afr. J. Bot. 103, 314. https://doi.org/10.1016/j.sajb.2016.02.043.

Fauzya, A.F., Astuti, R.I., Mubarik, N.R., 2019. Effect of ethanol-derived clove leaf extract on the oxidative stress response in yeast *Schizosaccharomyces pombe*. Int. J. Microbiol. 2019, 2145378. https://doi.org/10.1155/2019/2145378.

Feng, S., Yi, J., Li, X., Wu, X., Zhao, Y., Ma, Y., Bi, J., 2021. Systematic review of phenolic compounds in apple fruits: compositions, distribution, absorption, metabolism, and processing stability. J. Agric. Food Chem. 69 (1), 7–27. https://doi.org/10.1021/acs.jafc.0c05481.

Feres, M., Figueiredo, L.C., Barreto, I.M., Coelho, M.H., Araujo, M.W., Cortelli, S.C., 2005. In vitro antimicrobial activity of plant extracts and propolis in saliva samples of healthy and periodontally-involved subjects. J. Int. Acad. Periodontol. 7 (3), 90–96.

Gaweł, S., Wardas, M., Niedworok, E., Wardas, P., 2004. Dialdehyd malonowy (MDA) jako wskaźnik procesów peroksydacji lipidów w organizmie [Malondialdehyde (MDA) as a lipid peroxidation marker]. Wiad. Lek. 57 (9–10), 453–455.

Gonelimali, F.D., Lin, J., Miao, W., Xuan, J., Charles, F., Chen, M., Hatab, S.R., 2018. Antimicrobial properties and mechanism of action of some plant extracts against food pathogens and spoilage microorganisms. Front. Microbiol. 9, 1639. https://doi.org/10.3389/fmicb.2018.01639.

Gowri, G., Manimegalai, K., 2019. Phytochemical screening and GC-MS analysis of ethanol extract of *Syzygium aromaticum* L. J. Pharmacogn. Phytochem. 8 (5), 746–750.

Grespan, R., Paludo, M., Lemos, H., Barbosa, C.P., Bersani-Amado, C.A., Dalalio, M.M., Cuman, R.K., 2012. Anti-arthritic effect of eugenol on collagen-induced arthritis experimental model. Biol. Pharm. Bull. 35 (10), 1818–1820. https://doi.org/10.1248/bpb.b12-00128.

Gülçin, İ., 2011. Antioxidant activity of eugenol: a structure-activity relationship study. J. Med. Food 14 (9), 975–985. https://doi.org/10.1089/jmf.2010.0197.

Gülçin, İ., Elmastaş, M., Aboul-Enein, H.Y., 2012. Antioxidant activity of clove oil-a powerful antioxidant source. Arab. J. Chem. 5 (4), 489–499. https://doi.org/10.1016/j.arabjc.2010.09.016.

Habtemariam, S., 2019. Cloves' chemical and pharmacological basis (*Syzygium aromaticum* (L.) Merr. & L.M. Perry) as potential therapy for type 2 diabetes and associated diseases. In: Habtemariam, S. (Ed.), Medicinal Foods as Potential Therapies for Type-2 Diabetes and Associated Diseases. 2019. Academic Press, ISBN: 9780081029220, pp. 551–578, https://doi.org/10.1016/B978-0-08-102922-0.00016-X.

Halder, S., Mehta, A.K., Kar, R., Mustafa, M., Mediratta, P.K., Sharma, K.K., 2011. Clove oil reverses learning and memory deficits in scopolamine-treated mice. Planta Med. 77 (8), 830–834. https://doi.org/10.1055/s-0030-1250605.

Han, X., Parker, T.L., 2017. Anti-inflammatory activity of clove (*Eugenia caryophyllata*) essential oil in human dermal fibroblasts. Pharm. Biol. 55 (1), 1619–1622. https://doi.org/10.1080/13880209.2017.1314513.

Hemalatha, R., Nivetha, P., Mohanapriya, C., Sharmila, G., Muthukumaran, C., Gopinath, M., 2016. Phytochemical composition, GC-MS analysis, in vitro antioxidant and antibacterial potential of clove flower bud (*Eugenia caryophyllus*) methanolic extract. J. Food Sci. Technol. 53 (2), 1189–1198. https://doi.org/10.1007/s13197-015-2108-5.

Hemphill, I., 2006. The Spice and Herb Bible, second ed. Robert Rose, Inc, Toronto, ON, Canada, p. 215Y220.

Hwang, E., Lin, P., Ngo, H., Yi, T.H., 2018. Clove attenuates UVB-induced photodamage and repairs skin barrier function in hairless mice. Food Funct. 9 (9), 4936–4947. https://doi.org/10.1039/c8fo00843d.

Idowu, S., Adekoya, A.E., Igiehon, O.O., Idowu, A.T., 2021. Clove (*Syzygium aromaticum*) spices: a review on their bioactivities, current use, and potential application in dairy products. J. Food Meas. Charact. 15 (4), 3419–3435. https://doi.org/10.1007/s11694-021-00915-9.

Ivanovic, J., Dimitrijevic-Brankovic, S., Misic, D., Ristic, M., Zizovic, I., 2013. Evaluation and improvement of antioxidant and antibacterial activities of supercritical extracts from clove buds. J. Funct. Foods 5 (1), 416–423. https://doi.org/10.1016/j.jff.2012.11.014.

Kamkar Asl, M., Nazariborun, A., Hosseini, M., 2013. Analgesic effect of the aqueous and ethanolic extracts of clove. Avicenna J. Phytomed. 3 (2), 186–192.

Kauppinen, A., Suuronen, T., Ojala, J., Kaarniranta, K., Salminen, A., 2013. Antagonistic crosstalk between NF-κ B and SIRT1 in the regulation of inflammation and metabolic disorders. Cell. Signal. 25 (10), 1939–1948. https://doi.org/10.1016/j.cellsig.2013.06.007.

Khan, R., Islam, B., Akram, M., Shakil, S., Ahmad, A., Ali, S.M., Siddiqui, M., Khan, A.U., 2009. Antimicrobial activity of five herbal extracts against multi drug resistant (MDR) strains of bacteria and fungus of clinical origin. Molecules 14 (2), 586–597. https://doi.org/10.3390/molecules14020586.

Khan, S.A., Sathyanarayan, A., Mashek, M.T., Ong, K.T., Wollaston-Hayden, E.E., Mashek, D.G., 2015. ATGL-catalyzed lipolysis regulates SIRT1 to control PGC-1α/PPAR-α signaling. Diabetes 64 (2), 418–426. https://doi.org/10.2337/db14-0325.

Khathi, A., Masola, B., Musabayane, C.T., 2013b. Effects of *Syzygium aromaticum*-derived oleanolic acid on glucose transport and glycogen synthesis in the rat small intestine. J. Diabetes 5 (1), 80–87. https://doi.org/10.1111/j.1753-0407.2012.00230.x.

Khathi, A., Serumula, M.R., Myburg, R.B., Van Heerden, F.R., Musabayane, C.T., 2013. Effects of *Syzygium aromaticum*-derived triterpenes on postprandial blood glucose in streptozotocin-induced diabetic rats following carbohydrate challenge. PLoS One 8 (11). https://doi.org/10.1371/journal.pone.0081632, e81632.

Kim, H.M., Lee, E.H., Hong, S.H., Song, H.J., Shin, M.K., Kim, S.H., Shin, T.Y., 1998. Effect of *Syzygium aromaticum* extract on immediate hypersensitivity in rats. J. Ethnopharmacol. 60 (2), 125–131. https://doi.org/10.1016/s0378-8741(97)00143-8.

Ko, F.N., Liao, C.H., Kuo, Y.H., Lin, Y.L., 1995. Antioxidant properties of demethyldiisoeugenol. Biochim. Biophys. Acta 1258 (2), 145–152. https://doi.org/10.1016/0005-2760(95)00111-o.

Krishnan, T., Yin, W.F., Chan, K.G., 2012. Inhibition of quorum sensing-controlled virulence factor production in *Pseudomonas aeruginosa* PAO1 by Ayurveda spice clove (*Syzygium aromaticum*) bud extract. Sensors 12 (4), 4016–4030. https://doi.org/10.3390/s120404016.

Kubatka, P., Uramova, S., Kello, M., Kajo, K., Kruzliak, P., Mojzis, J., Vybohova, D., Adamkov, M., Jasek, K., Lasabova, Z., Zubor, P., Fialova, S., Dokupilova, S., Solar, P., Pec, M., Adamicova, K., Danko, J., Adamek, M., Busselberg, D., 2017. Antineoplastic effects of clove buds (*Syzygium aromaticum* L.) in the model of breast carcinoma. J. Cell. Mol. Med. 21 (11), 2837–2851. https://doi.org/10.1111/jcmm.13197.

Kumar, S., Sandhir, R., Ojha, S., 2014. Evaluation of antioxidant activity and total phenol in different varieties of *Lantana camara* leaves. BMC. Res. Notes 7, 560. https://doi.org/10.1186/1756-0500-7-560.

Kuroda, M., Mimaki, Y., Ohtomo, T., Yamada, J., Nishiyama, T., Mae, T., Kishida, H., Kawada, T., 2012. Hypoglycemic effects of clove (*Syzygium aromaticum* flower buds) on genetically diabetic KK-Ay mice and identification of the active ingredients. J. Nat. Med. 66 (2), 394–399. https://doi.org/10.1007/s11418-011-0593-z.

Kurokawa, M., Hozumi, T., Basnet, P., Nakano, M., Kadota, S., Namba, T., Kawana, T., Shiraki, K., 1998. Purification and characterization of eugeniin as an anti-herpesvirus compound from *Geum japonicum* and *Syzygium aromaticum*. J. Pharmacol. Exp. Ther. 284 (2), 728–735.

Ladurner, A., Zehl, M., Grienke, U., Hofstadler, C., Faur, N., Pereira, F.C., Berry, D., Dirsch, V.M., Rollinger, J. M., 2017. Allspice and clove as source of triterpene acids activating the G protein-coupled bile acid receptor TGR5. Front. Pharmacol. 8, 468. https://doi.org/10.3389/fphar.2017.00468.

Lambert, M.M., Campos, D.R., Borges, D.A., de Avelar, B.R., Ferreira, T.P., Cid, Y.P., Boylan, F., Scott, F.B., de Almeida Chaves, D.S., Coumendouros, K., 2020. Activity of *Syzygium aromaticum* essential oil and its main constituent eugenol in the inhibition of the development of Ctenocephalides felis felis and the control of adults. Vet. Parasitol. 282. https://doi.org/10.1016/j.vetpar.2020.109126, 109126.

Liu, H., Schmitz, J.C., Wei, J., Cao, S., Beumer, J.H., Strychor, S., Cheng, L., Liu, M., Wang, C., Wu, N., Zhao, X., Zhang, Y., Liao, J., Chu, E., Lin, X., 2014. Clove extract inhibits tumor growth and promotes cell cycle arrest and apoptosis. Oncol. Res. 21 (5), 247–259. https://doi.org/10.3727/096504014X13946388748910.

Magalhães, C.B., Riva, D.R., DePaula, L.J., Brando-Lima, A., Koatz, V.L., Leal-Cardoso, J.H., Zin, W.A., Faffe, D.S., 2010. In vivo anti-inflammatory action of eugenol on lipopolysaccharide-induced lung injury. J. Appl. Physiol. (1985) 108 (4), 845–851. https://doi.org/10.1152/japplphysiol.00560.2009.

Mansourian, A., Boojarpour, N., Ashnagar, S., Momen Beitollahi, J., Shamshiri, A.R., 2014. The comparative study of antifungal activity of *Syzygium aromaticum*, *Punica granatum* and nystatin on *Candida albicans*; an in vitro study. J. Mycol. Med. 24 (4), e163–e168. https://doi.org/10.1016/j.mycmed.2014.07.001.

Mehmood, Y., Farooq, U., Yousaf, H., Riaz, H., Mahmood, R.K., Nawaz, A., Abid, Z., Gondal, M., Malik, N.S., Barkat, K., Khalid, I., 2020. Antiviral activity of green silver nanoparticles produced using aqueous buds extract of *Syzygium aromaticum*. Pak. J. Pharm. Sci. 33 (2), 839–845.

Mehta, K.D., Garg, G.R., Mehta, A.K., Arora, T., Sharma, A.K., Khanna, N., Tripathi, A.K., Sharma, K.K., 2010. Reversal of propoxur-induced impairment of memory and oxidative stress by 4′-chlorodiazepam in rats. Naunyn Schmiedebergs Arch. Pharmacol. 381 (1), 1–10. https://doi.org/10.1007/s00210-009-0475-z.

Mittal, M., Gupta, N., Parashar, P., Mehra, V., Khatri, M., 2014. Phytochemical evaluation and pharmacological activity of *Syzygium aromaticum*: a comprehensive review. Int J Pharm Pharm Sci 6 (8), 67–72.

Mputu Kanyinda, J.-N., 2020. Coronavirus (COVID-19): A protocol for prevention and treatment (Covalyse®). Eur. J. Med. Health Sci. 2 (3), 1–4. https://doi.org/10.24018/ejmed.2020.2.4.340.

Munisa, A., Manalu, W., Wresdiyati, T., Kusumorini, N., 2015. The effect of clove leaf methanol extract on the profiles of superoxide dismutase and malondialdehyde in the liver of rabbits under hypercholesterolemia condition. Transl. Biomed. 6 (2), 1–5. https://doi.org/10.21767/2172-0479.100012.

Nassan, M.A., Mohamed, E.H., Abdelhafez, S., Ismail, T.A., 2015. Effect of clove and cinnamon extracts on experimental model of acute hematogenous pyelonephritis in albino rats: Immunopathological and antimicrobial study. Int. J. Immunopathol. Pharmacol. 28 (1), 60–68. https://doi.org/10.1177/0394632015572075.

Nisar, M.F., Khadim, M., Rafiq, M., Chen, J., Yang, Y., Wan, C.C., 2021. Pharmacological properties and health benefits of eugenol: a comprehensive review. Oxid. Med. Cell. Longev. 2021, 2497354. https://doi.org/10.1155/2021/2497354.

Nunomura, A., Perry, G., Aliev, G., Hirai, K., Takeda, A., Balraj, E.K., Jones, P.K., Ghanbari, H., Wataya, T., Shimohama, S., Chiba, S., Atwood, C.S., Petersen, R.B., Smith, M.A., 2001. Oxidative damage is the earliest event in Alzheimer disease. J. Neuropathol. Exp. Neurol. 60 (8), 759–767. https://doi.org/10.1093/jnen/60.8.759.

Oliveira, M.S., Costa, W.A., Pereira, D.S., Botelho, J.R.S., Menezes, T.O.A., Andrade, E.H.A., Silva, S.H.M., Filho, A.P.S.S., Carvalho Junior, R.N., 2016. Chemical composition and phytotoxic activity of clove (*Syzygium aromaticum*) essential oil obtained with supercritical CO_2. J. Supercrit. Fluids 118, 185–193.

Oyedeji, O.O., Shode, F.O., Oyedeji, A.O., Songca, S.P., Gwebu, E.T., Hill, G.M., Setzer, W.N., 2014. Semisynthesis of nitrogen derivatives of oleanolic acid and effect on breast carcinoma MCF-7 cells. Anticancer Res 34 (8), 4135–4139.

Park, K.R., Lee, J.H., Choi, C., Liu, K.H., Seog, D.H., Kim, Y.H., Kim, D.E., Yun, C.H., Yea, S.S., 2007. Suppression of interleukin-2 gene expression by isoeugenol is mediated through down-regulation of NF-AT and NF-kappaB. Int. Immunopharmacol. 7 (9), 1251–1258. https://doi.org/10.1016/j.intimp.2007.05.015.

Prasad, R.C., Herzog, B., Boone, B., Sims, L., Waltner-Law, M., 2005. An extract of *Syzygium aromaticum* represses genes encoding hepatic gluconeogenic enzymes. J. Ethnopharmacol. 96 (1–2), 295–301. https://doi.org/10.1016/j.jep.2004.09.024.

Radha Krishnan, K., Babuskin, S., Azhagu Saravana Babu, P., Sasikala, M., Sabina, K., Archana, G., Sivarajan, M., Sukumar, M., 2014. Antimicrobial and antioxidant effects of spice extracts on the shelf life extension of raw chicken meat. Int. J. Food Microbiol. 171, 32–40. https://doi.org/10.1016/j.ijfoodmicro.2013.11.011.

Radünz, M., da Trindade, M., Camargo, T.M., Radünz, A.L., Borges, C.D., Gandra, E.A., Helbig, E., 2019. Antimicrobial and antioxidant activity of unencapsulated and encapsulated clove (*Syzygium aromaticum*, L.) essential oil. Food Chem. 276, 180–186. https://doi.org/10.1016/j.foodchem.2018.09.173.

Rahman Alizadeh, M., Mahdavi, A.H., Rahmani, H.R., Jahanian, E., 2017. Clove bud (*Syzygium aromaticum*) improved blood and hepatic antioxidant indices in laying hens receiving low n-6 to n-3 ratios. J. Anim. Physiol. Anim. Nutr. 101 (5), 881–892. https://doi.org/10.1111/jpn.12502.

Ramadan, M.F., Asker, M.M.S., Tadros, M., 2013. Lipid profile, antiradical power, and antimicrobial properties of *Syzygium aromaticum* oil. Grasas Aceites 64, 509.

Rathinam, P., Viswanathan, P., 2018. Anti-virulence potential of eugenol-rich fraction of *Syzygium aromaticum* against multidrug resistant uropathogens isolated from catheterized patients. Avicenna J. Phytomed. 8 (5), 416–431.

Rodrigues, T.G., Fernandes Jr., A., Sousa, J.P., Bastos, J.K., Sforcin, J.M., 2009. In vitro and in vivo effects of clove on pro-inflammatory cytokines production by macrophages. Nat. Prod. Res. 23 (4), 319–326. https://doi.org/10.1080/14786410802242679.

Ryu, B., Kim, H.M., Woo, J.H., Choi, J.H., Jang, D.S., 2016. A new acetophenone glycoside from the flower buds of *Syzygium aromaticum* (cloves). Fitoterapia 115, 46–51. https://doi.org/10.1016/j.fitote.2016.09.021.

Saeed, M., Khan, M.S., Alagawany, M., et al., 2021. Clove (*Syzygium aromaticum*) and its phytochemicals in ruminant feed: an updated review. Rend. Fis. Acc. Lincei 32, 273–285. https://doi.org/10.1007/s12210-021-00985-3.

Salman, K.A.A.-A., Ibrahim, D.K., 2012. Test the activity of supplementation clove (*Eugenia caryophyllus*) powder, oil, and aqueous extract to diet and drinking water on performance of broiler chickens exposed to heat stress. Int. J. Poult. Sci. 11 (10), 635–640.

Shah, B., Sheth, F., Parabia, M., 2012. Folk herbal knowledge on the management of respiratory disorders prevailing in ethnic society of Valsad district, Gujarat. Indian J. Nat. Prod. Resour. 3, 438–447.

Shan, B., Cai, Y.Z., Sun, M., Corke, H., 2005. Antioxidant capacity of 26 spice extracts and characterization of their phenolic constituents. J. Agric. Food Chem. 53 (20), 7749–7759. https://doi.org/10.1021/jf051513y.

Shekhar, S., Yadav, Y., Singh, A.P., Pradhan, R., Desai, G.R., Dey, A.B., Dey, S., 2018a. Neuroprotection by ethanolic extract of *Syzygium aromaticum* in Alzheimer's disease like pathology via maintaining oxidative balance through SIRT1 pathway. Exp. Gerontol. 110, 277–283. https://doi.org/10.1016/j.exger.2018.06.026.

Shekhar, S., Yadav, Y., Singh, A.P., Pradhan, R., Desai, G.R., Dey, A.B., Dey, S., 2018b. Neuroprotection by ethanolic extract of *Syzygium aromaticum* in Alzheimer's disease like pathology via maintaining oxidative balance through SIRT1 pathway. Exp. Gerontol. 110, 277–283. https://doi.org/10.1016/j.exger.2018.06.026.

Shukri, R., Mohamed, S., Mustapha, N.M., 2010. Cloves protect the heart, liver and lens of diabetic rats. Food Chem. 122, 1116–1121. https://doi.org/10.1016/j.foodchem.2010.03.094.

Singh, N.A., Kumar, P., Jyoti, Kumar, N., 2021. Spices and herbs: potential antiviral preventives and immunity boosters during COVID-19. Phytother. Res. https://doi.org/10.1002/ptr.7019. doi:10.1002/ptr.7019.

Singh, V., Pahuja, C., Ali, M., Sultana, S., 2018. Analysis and bioactivities of essential oil of the flower buds of *Syzygium aromaticum* (L.) Merr. et. LM Perry. J. Med. Plant Stud. 6 (6), 79–83.

Singletary, K., 2014. Clove: overview of potential health benefits. Nutr. Today 49 (4), 207–224.

Sompong, W., Muangngam, N., Kongpatpharnich, A., Manacharoenlarp, C., Amorworasin, C., Suantawee, T., Thilavech, T., Adisakwattana, S., 2016. The inhibitory activity of herbal medicines on the keys enzymes and steps related to carbohydrate and lipid digestion. BMC Complement. Altern. Med. 16 (1), 439. https://doi.org/10.1186/s12906-016-1424-2.

Su, H., Yang, M., Wan, C., Yi, L.X., Tang, F., Zhu, H.Y., Yi, F., Yang, H.C., Fogo, A.B., Nie, X., Zhang, C., 2020. Renal histopathological analysis of 26 postmortem findings of patients with COVID-19 in China. Kidney Int. 98 (1), 219–227. https://doi.org/10.1016/j.kint.2020.04.003.

Sugihartini, N., Prabandari, R., Yuwono, T., Rahmawati, D.R., 2019. The anti-inflammatory activity of essential oil of clove (*Syzygium aromaticum*) in absorption base ointment with addition of oleic acid and propylene glycol as enhancer. Int. J. Appl. Pharm 11 (5), 16–109. https://doi.org/10.22159/ijap.2019.v11s5.T0081.

Suliman, G.M., Alowaimer, A.N., Al-Mufarrej, S.I., Hussein, E.O.S., Fazea, E.H., Naiel, M.A.E., Alhotan, R.A., Swelum, A.A., 2021. The effects of clove seed (*Syzygium aromaticum*) dietary administration on carcass characteristics, meat quality, and sensory attributes of broiler chickens. Poult. Sci. 100 (3). https://doi.org/10.1016/j.psj.2020.12.009, 100904.

Tajuddin, Ahmad, S., Latif, A., Qasmi, I.A., 2003. Aphrodisiac activity of 50% ethanolic extracts of *Myristica fragrans* Houtt. (nutmeg) and *Syzygium aromaticum* (L) Merr. & Perry. (clove) in male mice: a comparative study. BMC Complement. Altern. Med. 3, 6. https://doi.org/10.1186/1472-6882-3-6.

Takahashi, Y.K., Nagayama, S., Mori, K., 2004. Detection and masking of spoiled food smells by odor maps in the olfactory bulb. J. Neurosci. 24 (40), 8690–8694. https://doi.org/10.1523/JNEUROSCI.2510-04.2004.

Tanko, Y., Mohammed, A., Okasha, M.A., Umar, A.H., Magaji, R.A., 2008. Anti-nociceptive and anti-inflammatory activities of ethanol extract of *Syzygium aromaticum* flower bud in Wistar rats and mice. Afr. J. Tradit. Complement. Altern. Med. 5 (2), 209–212. https://doi.org/10.4314/ajtcam.v5i2.31275.

Tu, Z., Moss-Pierce, T., Ford, P., Jiang, T.A., 2014. *Syzygium aromaticum* L. (clove) extract regulates energy metabolism in myocytes. J. Med. Food 17 (9), 1003–1010. https://doi.org/10.1089/jmf.2013.0175.

Vicidomini, C., Roviello, V., Roviello, G.N., 2021. Molecular basis of the therapeutical potential of clove (*Syzygium aromaticum* L.) and clues to its anti-COVID-19 utility. Molecules 26 (7), 1880. https://doi.org/10.3390/molecules26071880.

Vijayasteltar, L., Nair, G.G., Maliakel, B., Kuttan, R., Krishnakumar, I.M., 2016. Safety assessment of a standardized polyphenolic extract of clove buds: subchronic toxicity and mutagenicity studies. Toxicol. Rep. 3, 439–449. https://doi.org/10.1016/j.toxrep.2016.04.001.

Yang, M., Meng, F., Gu, W., Li, F., Tao, Y., Zhang, Z., Zhang, F., Yang, X., Li, J., Yu, J., 2020. Effects of natural products on bacterial communication and network-quorum sensing. Biomed. Res. Int. https://doi.org/10.1155/2020/8638103, 8638103.

Yassin, M.T., Mostafa, A.A., Al-Askar, A.A., 2020. In vitro anticandidal potency of *Syzygium aromaticum* (clove) extracts against vaginal candidiasis. BMC Complement. Med. Ther. 20 (1), 25. https://doi.org/10.1186/s12906-020-2818-8.

Zhang, H., Peng, X., Li, X., Wu, J., Guo, X., 2017. The application of clove extract protects Chinese-style sausages against oxidation and quality deterioration. Korean J. Food Sci. Anim. Resour. 37 (1), 114–122. https://doi.org/10.5851/kosfa.2017.37.1.114.

Zulcafli, A.S., Lim, C., Ling, A.P., Chye, S., Koh, R., 2020. Antidiabetic potential of *Syzygium* sp.: an overview. Yale J. Biol. Med. 93 (2), 307–325.

CHAPTER

38

Clove (*Syzygium aromaticum*)-mediated metallic nanoparticles: Synthesis, characterization, and possible pharmacological and industrial applications

Abdulrahman M. Elbagory[a], Ateeq Rahman[b], Natascha Cheikhyoussef[c], Ahmad Cheikhyoussef[d], Nusrat M. Begum[e], and Ahmed A. Hussein[a]

Chemistry Department, Cape Peninsula University of Technology, Cape Town, Western Cape, South Africa[a] Department of Chemistry and Biochemistry, Faculty of Science, University of Namibia, Windhoek, Namibia[b] Ministry of Higher Education, Technology and Innovation, Windhoek, Namibia[c] Science and Technology Division, Multidisciplinary Research Unit, Centre for Research Services, University of Namibia, Windhoek, Namibia[d] Department of Environmental Health, Water and Health Research Centre, University of Johannesburg, Doornfontein, South Africa[e]

Abbreviations

AFM atomic force microscopy
AgNPs silver nanoparticles
AlNPs alumina nanoparticles
AuNPs gold nanoparticles
CB clove bud
CuNPs copper nanoparticles
CuONPs copper oxide nanoparticles
DLS dynamic light scattering
FeNPs iron nanoparticles
FTIR Fourier transform infrared spectroscopy
GC-MS gas chromatography/mass spectroscopy
h hour
IC_{50} 50% inhibitory concentration
MgONPs magnesium oxide nanoparticles
MIC minimum inhibitory concentration
min minutes
MnONPs manganese oxide nanoparticles
MNPs metal nanoparticles
MoNPs molybdenum nanoparticles
NiNPs nickel nanoparticles

Clove (*Syzygium aromaticum*). https://doi.org/10.1016/B978-0-323-85177-0.00011-2

NPs	nanoparticles
NR	not reported
PdNPs	palladium nanoparticles
sec	seconds
SEM	scanning electron microscopy
TEM	transmission electron microscopy
TiO_2NPs	titanium oxide nanoparticles
UV-vis	ultra violet-visible
ZnONPs	zinc oxide nanoparticles
ZnSNPs	zinc sulfide nanoparticles

1. Introduction

Nanotechnology offers practical applications to different fields, stretching from chemical techniques to environmental and medicinal innovations. This is mainly because nanomaterials have completely different physicochemical from individual atoms, compounds, and bulk materials (Etheridge et al., 2013). Nowadays, NPs are being explored for a plethora of industrial applications such as electronics, energy, water purification, agriculture, and cosmetics (Kargozar and Mozafari, 2018). Different types of NPs have been studied extensively for biomedical applications like drug delivery, medical imaging, diagnosis, and therapeutics (Ramos et al., 2017).

Several metal precursors have been used to synthesize MNPs in mono and multimetallic forms. MNPs possess unique optoelectrical properties due to their localized surface plasmon resonance, which can be utilized in several applications (Khan et al., 2019). Applying these MNPs in the different fields is dependent on the type of metal(s) used. To exemplify, the metal silver has been historically known for its antibacterial activity (Clement and Jarrett, 1994), and consequently, AgNPs are mainly studied for their antibacterial potential (Le Ouay and Stellacci, 2015). Plant phytochemicals have enabled researchers to green synthesize and stabilize MNPs without the need for applying expensive, toxic, and environmentally harmful materials in the conventional chemical synthesis methods (Gour and Jain, 2019). Besides contributing to the synthesis, these phytochemicals surround the MNPs rendering them more biologically active (Afreen et al., 2020).

The widely used Indonesian spice *Syzigium aromaticum* (L.) (clove) belongs to the family Myrtaceae (Batiha et al., 2019). The aromatic, nail-shaped CB is a clove's dried, unopened flower bud (Adaramola and Onigbinde, 2016). The traditional use of cloves is known, which is also supported by the scientific reports that investigated the clove's extracts, essential oil, and isolated compounds for their antiseptic (Meeker and Linke, 1988), antibacterial (Ajiboye et al., 2016), antifungal (Kaur et al., 2019), antiviral (Cortés-Rojas et al., 2014), and anticancer activities (Banerjee, 2005). Clove essential oil has enormous applications in the food, sanitary, biomedical, pharmaceutical, active packaging, and cosmetics industries due to its biological properties, including antioxidant, antimicrobial, antiseptic, analgesic, and anticancer activities (Nowak et al., 2012). A recent review by Batiha et al. (2019) summarizes the reports describing the traditional use, chemical constituents, and the pharmacological actions of different clove components.

2. Role of clove phytochemicals in the synthesis

The green chemical reduction of the metal salts by plants involves many phytochemicals found in their extracts. These reducing phytochemicals can reduce the metal ions to their pure atoms (e.g., Ag^{3+} to Ag^0). The electron donor functional groups can achieve this to complete the reduction step. Afterward, the agglomeration of these reduced metal atoms occurs, leading to the development of oligomeric clusters followed by the formation of the MNPs (Iravani et al., 2014). Depending on the chemical nature of the plant phytochemicals, the stabilization of the formed MNPs can be done without adding external stabilizers. However, when using the plant extract in the synthesis of the MNPs, it is not easy to single out a particular phytocomponent that is responsible for the reduction and stabilization of the MNPs because of the presence of several and different phytochemicals in the plant extract (Ovais et al., 2018). Therefore, only speculated mechanisms were reported to explain the mechanism of the MNPs synthesis using plant extracts (Dauthal and Mukhopadhyay, 2016). Till now, phytochemicals such as phenolics (e.g., flavonoids, phenolic acid, and terpenoids), organic acids, and proteins are deemed accountable for the synthesis of the MNPs (Ahmad et al., 2019; Mallikarjuna et al., 2014; Ovais et al., 2018).

Several studies also have suggested that different phytochemicals of different functional groups (e.g., R—CH, C—O, C=O, N—H, C—N, and O—H) from *S. aromaticum* (Table 1) are responsible for the synthesis of MNPs (Bagyalakshmi et al., 2019; A. Jain et al., 2018; Kanneganti et al., 2014; Kumar et al., 2017; Lakhan et al., 2020; Maheswari et al., 2020; Parlinska-Wojtan et al., 2018; Rajesh et al., 2018; Sathishkumar et al., 2018; Shanthi et al., 2017; Wang and Bai, 2020). These studies based these conclusions on analyzing the FTIR spectra of the CB or clove leave extracts and the produced MNPs.

The chemistry of *S. aromaticum* has been investigated, and this spice is known to include flavonoids, tannins, phenolic acids, and sesquiterpenes. GC-MS analysis has shown that more than 75% of clove essential comprises terpenoid eugenol (Jirovetz et al., 2006). The functional groups in the terpenoid structures are believed to have a significant role in synthesizing and stabilizing MNPs (Ur Mashwani et al., 2016). Several of the studies reviewed herein also suggested that eugenol is mainly responsible for synthesizing MNPs from CB. The following mechanism was suggested for the MNPs synthesis from eugenol: the presence of strong electron-withdrawing methoxy and allyl groups in the *ortho* and *para* positions to the O—H group facilitates the release of proton and renders eugenol in anionic form. Afterward, resonating anionic structures of eugenol are formed (Fig. 1), and two electrons are released in the process (Kaur et al., 2013; Parlinska-Wojtan et al., 2018). This is also justified by the presence of transmittance of C—O and C=O peaks and the disappearance of the O—H in the FTIR spectra of the generated MNPs from CB (Parlinska-Wojtan et al., 2018).

Eugenol is practically insoluble in water (Yuwono et al., 2002), and hence the use of water to extract eugenol and essential oils, in general, has shown less yield compared to the use of other organic solvents and distillation methods. Clove oil yield was low as 0.44%, with eugenol content less than 5% when macerating the buds with water for isolation for five days (Shafira et al., 2020). The study showed that the highest eugenol content could be obtained using the Soxhlet apparatus (Shafira et al., 2020). Another study also detected lower clove oil yield (around 15%) when using water than other solvent systems such as 80% acetone and 80% methanol (Adaramola and Onigbinde, 2016). Eugenol was not even detected in the water extract of *S. aromaticum* in one study using thin-layer chromatography

Table 1 Summary of the different types of MNPs from *S. aromaticum*.

Metal nanoparticles	Size (nm)	Shape	Applications	Plant part used	General comments on the synthesis	Reference
AgNPs	30–60	Spherical	NR	Buds	Synthesis was done using microwave	Deshpande et al. (2012)
	20–149	Spherical	NR	Buds	Stirring at room temperature for 24 h	Vijayaraghavan et al. (2012)
	90	Quasi-spherical	Substrate for detection of explosives	Buds	Tested variable volumes of clove extract and silver salt	Sil et al. (2014)
	59 to 98	Polydispersed	Wound healing and antimicrobial	Buds	Stirring at room temperature for 24 h	Varalakshmi et al. (2015)
	12 to 85	Polydispersed	Antimicrobial	Buds	Stirring in the dark for overnight	Parlinska-Wojtan et al. (2018)
	5.0 to 40	Spherical	Anticancer		Stirring at different temperatures for 10 min	Kumar et al. (2017)
	23	Mostly spherical	Antimicrobial/ dye degradation	Buds	Different volumes and concentrations	Ajitha et al. (2019)
	50	Polydispersed	Antimicrobial	Buds	Stirring at 27°C for 1 h	Kaur et al. (2013)
	25 to 50	Cubic	Antimicrobial	Eugenol from buds	Synthesis from eugenol	Tekin et al. (2019)
	31 to 72	Polydispersed	Antimicrobial	Clove essential oil	Synthesis from clove essential oil under different pH levels	de Maciel et al. (2019)
	>200	Spherical	Antibacterial	Clove essential oil	Microemulsion technique	Gao et al. (2017)
	9.42	Spherical	Antibacterial/ antidiatom	Buds	Stirring in dark for 24 h	Lakhan et al. (2020)
AuNPs	100 to 300	Mainly triangular when using 1:1 clove extract to gold salt	NR	Buds	Optimization by using different volumes of the CB extract	Singh et al. (2010)

Table 1 Summary of the different types of MNPs from *S. aromaticum*—cont'd

Metal nanoparticles	Size (nm)	Shape	Applications	Plant part used	General comments on the synthesis	Reference
	5.0 to 100	Irregular	NR	Buds	Stirring the CB extract with the metal salt for 1 h	Raghunandan et al. (2010)
	NR	Irregular	Anticancer	Buds	Same methodology as reported in Raghunandan et al. (2010)	Raghunandan et al. (2011)
	NR	Square	NR	Buds	Mixing the CB extract into preheated (at 50°C) gold salt solution	Sharma et al. (2017)
FeNPs	50 to 100	Spherical	NR	Buds	Stirring the aqueous extract with $FeCl_3$ at 50–60°C, followed by stabilization with 1% chitosan and 1% PVA	Pattanayak et al. (2013)
	NR	Spherical	NR	Buds	Mixing Fe(I) and Fe(II) sulfate solutions with the CB aqueous extract at room temperature	Sujin Jeba Kumar et al. (2020)
	13.5	Spherical	Water treatment	Buds	Reaction at pH4, precipitation of clove-FeNPs occurred after 40 min	Mondal and Purkait (2018)
	52	Irregular	Anticancer	Buds	Stirring the aqueous extract with $FeCl_3$ at 70°C for 5 min, followed by incubation at room temperature	Bagyalakshmi et al. (2019)
PdNPs	10 ± 1.42	Spherical	Catalysis	Buds	Mixing Palladium Chloride with a premixed solution of *S. aromaticum* and acacia water extracts at pH 9 and 55°C	Saxena et al. (2018)

Continued

Table 1 Summary of the different types of MNPs from *S. aromaticum*—cont'd

Metal nanoparticles	Size (nm)	Shape	Applications	Plant part used	General comments on the synthesis	Reference
	4.49	Spherical	Catalysis	Leaves	Wet impregnation of clove leaves and palladium chloride followed by stirring for 3h at 50°C	Wang and Bai (2020)
	20±5	Spherical	Anticancer	Buds	Stirring with Palladium Chloride solution at 70°C for 30min	Shanthi et al. (2017)
ZnSNPs	165.9	Spherical	Antimicrobial/ photocatalysis	NR	*S. aromaticum* was used to cap the ZnSNPs synthesized using zinc sulfate and thiourea	Sathishkumar et al. (2018)
	46	Spherical	Antimicrobial	Buds	Stirring with zinc sulfate for 2h	Sathishkumar et al. (2017)
ZnONPs	35.69	Triangular/ hexagonal	Antifungal	Buds	Synthesis was done using combustion method with zinc nitrate	Lakshmeesha et al. (2019)
	NR	NR	Antibacterial	Buds	Stirring methanol CB extract with the metal precursor for 1h followed by thermal annealing	Osaili et al. (2019)
	50	Hexagonal	Photocatalysis/ Antimicrobial/ Antioxidant	Buds	Used ethanol CB extract	Anvarinezhad et al. (2020)
CuNPs	15	Spherical	Antimicrobial	Buds	Stirring with copper acetate for 15min at 30°C	Rajesh et al. (2018)
	NR	Spherical	Antimicrobial	Buds	Stirring the CB aqueous extract with copper nitrate at pH 14	Aziz and Jassim (2018)

Table 1 Summary of the different types of MNPs from *S. aromaticum*—cont'd

Metal nanoparticles	Size (nm)	Shape	Applications	Plant part used	General comments on the synthesis	Reference
CuONPs	NR	NR	Antibacterial	Buds	Stirring methanol CB extract with the metal precursor for 1 h. The NPs were recovered by annealing	Osaili et al. (2019)
MgONPs	2.0 to 4.0	Spherical	Colorimetric detection of Fe^{+3}	Buds	Stirring with magnesium acetate for 30 min at room temperature	Jain et al. (2018)
	~7.0	Mostly spherical	pH sensing	Buds	A similar procedure was reported (Jain et al., 2018) and modified with L-lysin	Jain et al. (2019)
MnONPs	4.0	Spherical	Electrochemical sensing of P-nitrophenol	Buds	Optimization by varying the concentration and volume of the metal salt and CB extract as well as the temperature	Kumar et al. (2017)
AlNPs	3.0 to 5.0	Spherical	NR	Buds	Microwave-assisted synthesis	Hasanpoor et al. (2017)
TiO_2NPs	9.5	Spherical	Anticancer/ Antibacterial	Buds	Modification of chemically synthesized TiO_2NPs with CB aqueous extract via hydrothermal method	Maheswari et al. (2020)
MoNPs	19.47	Spherical	NR	Buds	Stirring the extract with the metal salt for 15 min at 60°C until complete evaporation, followed by calcination at 700°C	Kanneganti et al. (2014)

Continued

Table 1 Summary of the different types of MNPs from *S. aromaticum*—cont'd

Metal nanoparticles	Size (nm)	Shape	Applications	Plant part used	General comments on the synthesis	Reference
NiNPs	40	NR	Antimicrobial	Buds	Stirring the extract with the metal salt for 15 min in a 1:1 ratio	Pavan et al. (2016)
Ag-Au-PdNPs	11.61 ± 6.5		Catalysis/antimicrobial	Buds	One-step chemical reduction of the metal precursors using CB in combination with *Aegle marmelos* leaves extract	Rao and Paria (2015)
Ag-AuNPs	Variable according to the ratio of the metal salts	Polydispersed	Catalysis	Buds	Chemical reduction with different ratios of the metal salts at room temperature for 15 min	Sharma et al. (2020)

(El Ghallab et al., 2019). Therefore, presenting any assumption regarding the role of eugenol in the MNPs synthesis should be made with caution, specifically when using the water extract. Indeed, most of the reviewed papers herein described the synthesis of the MNPs from CB water extract. On the other hand, only a few described the synthesis from CB essential oil (de Maciel et al., 2019), isolated eugenol (Tekin et al., 2019), or CB extract using other organic solvents or techniques, e.g., steam distillation (Varalakshmi et al., 2015), hexane (Lakshmeesha et al., 2019), 30% ethanol (Anvarinezhad et al., 2020), or 80% methanol (Osaili et al., 2019).

2.1 AgNPs

AgNPs have unique physicochemical properties and are employed in different biological applications. The use of colloid silver, silver salts, and other silver products has been documented since ancient times (Alexander, 2009). Silver is mainly known for its antimicrobial uses, and consequently, many studies aimed to evaluate the effect of AgNPs against microbes (Dube et al., 2020; Majoumouo et al., 2019; Yin et al., 2020). Other studies also reported the wound healing (Konop et al., 2016), anti-inflammatory (Abdelhafez et al., 2020), and anticancer activities of AgNPs (Barua et al., 2017).

The aqueous phyto-constituents of CB produced AgNPs of the broad particle size range (20–149 nm) at room temperature using silver nitrate solution for 24 h (Vijayaraghavan et al., 2012). To accelerate the nanoparticles synthesis, Deshpande et al. used microwave radiation to synthesize

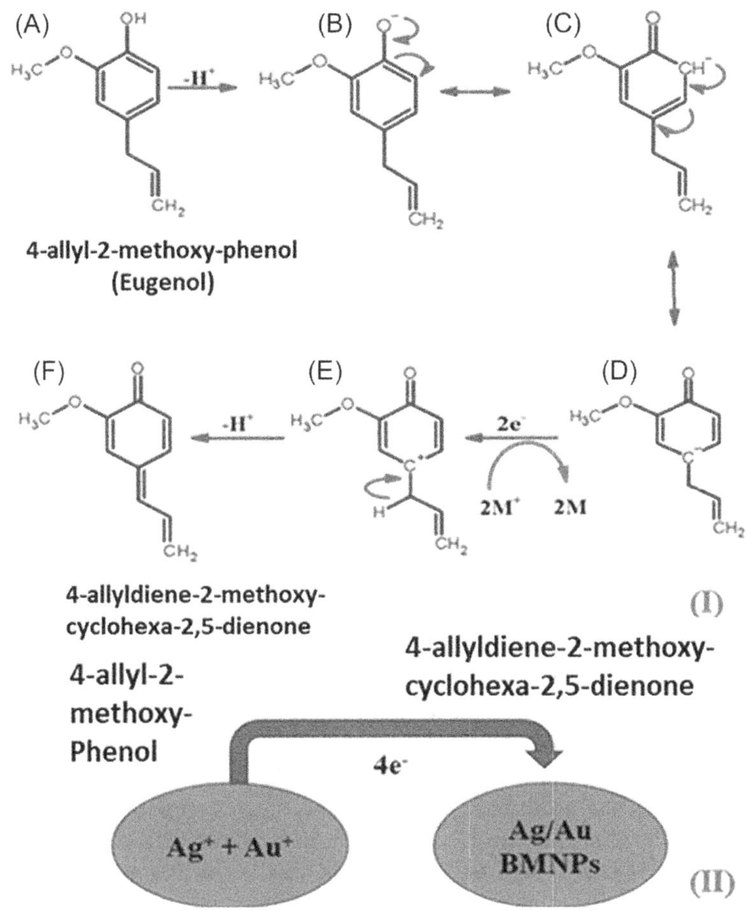

FIG. 1

Suggested mechanism of MNPs synthesis by eugenol (Sharma et al., 2020).

clove-AgNPs in a remarkably short time (120 s) (Deshpande et al., 2012). From the FTIR results, the authors suggested that the water-soluble polyphenolic (flavonoids) content of the CB was responsible for the reduction and the stabilization of the clove-AgNPs (Deshpande et al., 2012). The same methodology was adapted by Parveen et al. to reproduce clove-AgNPs with similar characteristics (Parveen et al., 2018). In this study, the clove-AgNPs, formulated in a cream form, were further evaluated for their wound healing properties. The cream containing 5% clove-AgNPs exhibited faster wound healing (28 days) compared to the 3% clove-AgNPs cream (46 days) (Parveen et al., 2018). The 5% clove-AgNPs cream wound healing mechanism was attributed to the increased collagen deposition, fibroblast production, and the reduction of macrophages (Parveen et al., 2018). Another study also reported the fabrication of clove-AgNPs based sponge dressing with antimicrobial activity, which showed almost 97% wound healing in rats compared to only 58% in control-treated rats (Varalakshmi et al., 2015).

Clove-AgNPs exhibited significant antimicrobial activity when tested with various bacteria and fungi. Parlinska-Wojtan et al. (2018) found that 40 X dilution of clove-AgNPs (original concentration was not reported) had bacteriostatic activity against *Escherichia coli*, *Staphylococcus aureus*, and *Candida albicans*, whereas the aqueous clove extract did not have any bacteriostatic action. Another interesting study was conducted by Ajitha et al. (2019), in which the authors tested different volumes of *S. aromaticum* aqueous extract and different concentrations of silver nitrate solution. The transmission electron microscopy (TEM) analysis data revealed an inverse proportional relationship between the concentration of silver nitrate and the size of the clove-AgNPs; the reverse was true concerning the volume of the *S. aromaticum* aqueous extract (Ajitha et al., 2019). Zone inhibitions produced by the clove-AgNPs indicated strong antimicrobial activity, with the largest zones observed against *Pseudomonas* spp. and *Bacillus* spp. (Ajitha et al., 2019). By observing the changes in UV-vis spectra, the authors also assessed the catalytic activity of phyto-mediated AgNPs, which reduced the absorbance of three pollutant organic dyes, viz., 4-nitrophenol, methylene blue, and rhodamine B (Ajitha et al., 2019). Another study showed that clove-AgNPs have more antibacterial activity against *E. coli*, *S. aureus*, and *Salmonella typhi* strains than the CB aqueous extract (Sameen et al., 2014). A study by Kaur et al. (2013) showed that clove-AgNPs have bactericidal effects against the growth of *Bacillus subtilis*. Tekin et al. (2019) adapted the same methodology from Kaur et al. (2013) to synthesize AgNPs from isolated eugenol (eugenol-AgNPs) from the methanol extract of *S. aromaticum*. The zone inhibitions of the eugenol-AgNPs showed antibacterial activity against *E. coli* and *S. aureus* but less antifungal activity against *C. albicans* (Tekin et al., 2019). de Maciel et al. (2019) synthesized AgNPs from the essential oil of *S. aromaticum*. Their UV-vis, DLS, and TEM analysis data showed that the synthesis could be optimized by varying the pH conditions (de Maciel et al., 2019). For instance, the AgNPs synthesized at pH 8.0 gave spherical NPs an average size of 35.5 nm, while the synthesis at pH 10 formulated NPs of different shapes with an average diameter (37.9 nm) according to their TEM micrographs (de Maciel et al., 2019). The pH variable AgNPs also showed different antimicrobial activity as determined by the broth microdilution assay. For example, the AgNPs at pH 8.0 gave MIC values of 60 and 80 μL/mL against *E. coli* and *S. aureus*, respectively. On the other hand, the AgNPs at pH 9.0 had an MIC of 100 μL/mL against both bacteria (de Maciel et al., 2019). The AgNPs produced from clove oil via the microemulsion method inhibited the growth of *E. coli* and *S. aureus*. This activity was decreased upon removal of clove oil from the AgNPs by ultrasonic cleaning, which shows the role of clove oil in enhancing the antibacterial activity of AgNPs (Gao et al., 2017). The study by Lakhan et al. (2020) showed that clove-AgNPs inhibited the growth of marine community bacteria using the disk diffusion method. The clove-AgNPs inhibited the growth of the marine algae *Nitzschia closterium*. The number of the algae was significantly decreased after the treatment with clove-AgNPs in a dose-dependent manner. This inhibition was found consistent for seven days. On the other hand, the CB extract did not show similar antidiatom activity (Fig. 2). The authors concluded that clove-AgNPs could be utilized to prevent microorganisms responsible for the formation of micro-fouling in artificial marine surfaces (Lakhan et al., 2020).

The anti-inflammatory effects of clove-AgNPs were also reported. Varghese et al. compared the anti-inflammatory activities of aqueous clove extract and its AgNPs. The results showed that clove-AgNPs induced higher scavenging activity and protein denaturation inhibition in a dose-dependent mode, reducing anti-inflammatory responses (Varghese et al., 2017). Moreover, clove-AgNPs down-regulated the in vitro expression of interleukin-1 beta by 80% compared to 60% achieved by the aqueous clove extract. Unfortunately, the authors did not report the physical characteristics of the clove-AgNPs, such as their size or morphology.

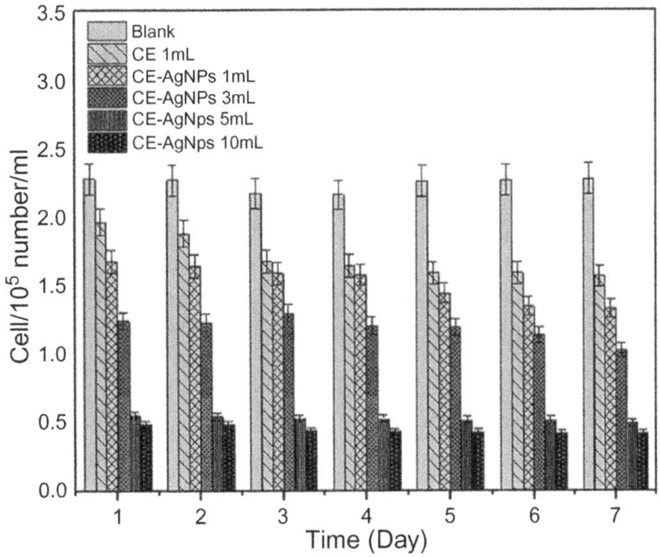

FIG. 2

Effect of clove-AgNPs (denoted CE-AgNPs) and CB aqueous extract (denoted CE) treatment on the *N. closterium* diatom cell number (Lakhan et al., 2020).

The anticancer actions of clove-AgNPs were also studied. Venugopal et al. showed that clove-AgNPs were able to induce cell death by increasing the treatment concentrations. The calculated IC_{50} value of clove-AgNPs was 60 and 50 μg/mL against breast and lung cancer cell lines, respectively. On the other hand, the aqueous extract of *S. aromaticum* exhibited less anticancer effect with an IC_{50} value of 70 μg/mL against both the cell lines (Venugopal et al., 2017). Furthermore, the frames obtained from the fluorescent microscope of the cells stained with acridine orange, ethidium bromide, and DAPI revealed that clove-AgNPs induced cell death via apoptosis (Venugopal et al., 2017).

2.2 AuNPs

AuNPs have attracted the attention of biomedical researchers globally not only because of their unique optical and electrical properties and ease of synthesis, but also for two main reasons: first, AuNPs are proven biocompatible agents due to their chemical and physical stability (Hu et al., 2020) and secondly, S- and N-containing AuNPs are easy to be functionalized with biologically active molecules attributed to the ability of AuNPs to form stable chemical bonds with S- and N-containing organic molecules (Bai et al., 2020).

One of the first reports about the synthesis of AuNPs from clove (clove-AuNPs) was done by Singh et al. (2010). The resultant clove-AuNPs varied in size distribution and morphology by changing the ratio between the CB aqueous extract and the gold salt (Singh et al., 2010). The authors used L-cysteine to functionalize the clove-AuNPs, useful for future biological conjugation (Singh et al., 2010). In the same year, Deshpande et al. mixed solutions of macerated aqueous extract of *S. aromaticum* and gold salt to synthesize clove-AuNPs in 1 h. Different methods were utilized, such as UV-vis, TEM, SEM, and AFM analyses, to characterize the clove-AuNPs, and the formation of the NPs has attributed to the water-soluble flavonoids as shown by FTIR results (Raghunandan et al., 2010). Later, the same methodology was used with modification to synthesize clove-AuNPs in 4h (Hameed et al., 2014). The synthesized clove-AuNPs inhibited the urease and carbonic anhydrase enzymes, which can be useful in treating many diseases such as glaucoma, epilepsy, and stones formation prevention (Hameed et al., 2014). Also, to explore the anticancer potential of clove-based AuNPs, different cancer cell lines were treated with biofabricated AuNPs from CB (Raghunandan et al., 2011). The results from the XTT assay showed that clove-AuNPs had an IC_{50} as low as 20 µg/mL on HeLa (human cervix), HEK-293 (human kidney), and HT-29 (human colorectal) cell lines, and 30 µg/mL against K-562 (leukemia). The authors suggested that the antiproliferation activity of clove-AuNPs was due to their irregular morphology and the free-radical scavenging as indicated by the nitric oxide assay (Raghunandan et al., 2011).

2.3 FeNPs

The number of studies reporting the green synthesis of FeNPs is fewer compared to the reports pertaining to the green synthesis of Au- and AgNPs. It is mainly because reduced iron is easily oxidized in solution due to its particle instability and high aggregate tendency attributed to its low surface charge. Therefore, it is more difficult to obtain FeNPs (Sujin Jeba Kumar et al., 2020). Additionally, due to the superparamagnetic properties, modifiable surfaces, and low toxicity (Chang et al., 2011), FeNPs have been highly considered for cancer treatment, mainly as a drug carrier (Kanagesan et al., 2013). FeNPs are also useful in catalysis because of their uniform distribution in solutions, high reactivity, and ease of separation owing to their magnetic properties (Sujin Jeba Kumar et al., 2020).

A study conducted by Sebastian et al. evaluated the ability of 24 plant water extracts collected from peninsular India (Sebastian et al., 2018). FeNPs were only synthesized from the water extract of eight plants, including the leaves water extract of *S. aromaticum*, which immediately produced the FeNPs (clove-FeNPs) (Sebastian et al., 2018). Interestingly, the amount of phenolics in the reaction medium was reduced after the synthesis by more than 20%, citing the role of phenolic contents of *S. aromaticum* in the formation of clove-FeNPs (Sebastian et al., 2018). In another study, CB aqueous extract was used to produce multiphase FeNPs instantaneously (Sujin Jeba Kumar et al., 2020). Other plants were also tested but did not give positive results. However, the XPS analysis confirmed the presence of both Fe^{2+} and Fe^{3+} to determine the phases of the FeNPs (Fig. 3). Furthermore, the authors established the superparamagnetic behavior of the produced clove-FeNPs, which render these NPs useful as drug carriers or for hyperthermia treatment of cancer (Sujin Jeba Kumar et al., 2020).

Nowadays, there is increased usage of magnetic nanoparticles in catalysis owing to their high dispersibility, increased reactivity, and ease of separation. This is facilitated by their magnetic properties, nanosize, and having both homogenous and heterogeneous catalysis properties. Therefore, Clove-FeNPs were evaluated for their water detoxification properties. As a result, the clove-FeNPs loaded on a polymeric membrane were found to reduce nitrobenzene to aniline and adsorb fluorides, which can be utilized to remove harmful nitro-containing compounds and fluorides from water (Mondal and Purkait, 2018).

One study by Thenmozhi et al. (2019) successfully combined the anticancer properties of both the FeNPs and *S. aromaticum* by loading the methanolic extract on PVP-stabilized FeNPs (Thenmozhi et al., 2019). The MTT results showed that the biofunctionalized FeNPs inhibited the growth of the MCF-7 cell line (breast cancer) (Thenmozhi et al., 2019). A follow-up study on the same NPs

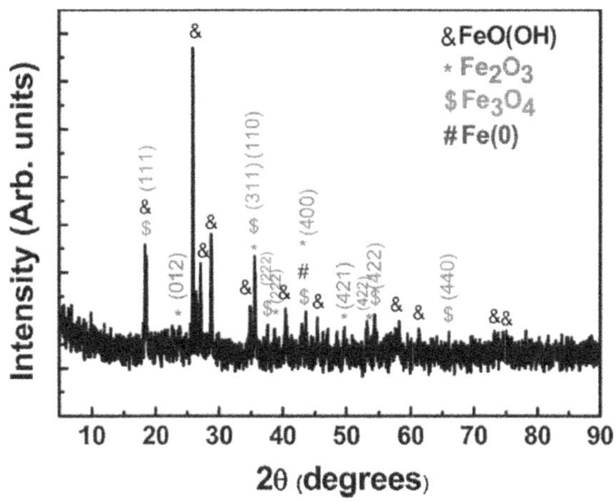

FIG. 3

XRD pattern of the multiphase clove-FeNPs (Sujin Jeba Kumar et al., 2020).

confirmed, using flow cytometry assays, that these NPs induce apoptosis on MCF-7 by triggering caspase-3, caspase-8, and caspase-9 (Thenmozhi, 2020). Furthermore, clove-FeNPs enhanced bee venom's antioxidant and anticancer activities against lung cancer (Bagyalakshmi et al., 2019). The authors described the biological effect of mixing clove-FeNPs with bee venom at different ratios. Based on the DPPH assay, it was found that the ratio 1:1 possessed higher antioxidant effects with IC_{50} of 56 μg/mL compared to 66 μg/mL when testing the bee venom alone. Furthermore, the same mixture ratio induced more than 50% cell death at a 250 μg/mL concentration after 4 h of incubation with lung cancer cell line (A549) (Bagyalakshmi et al., 2019). The limitation in this study was the absence of a comparative study to show the anticancer effect of bee venom and clove-FeNPs individually in their results.

2.4 PdNPs

Like other metal nanoparticles, PdNPs have been studied because of their imitable optical, electronic, and chemical properties, useful in several industrial and biomedical applications. However, unlike the NPs from other noble metals, PdNPs are more considered for applications such as catalysis, sensors, and fuel cells (Saxena et al., 2018). Pd metal is highly considered in catalysis as it is efficient in regulating various chemical processes such as oxidation, carbon-carbon, and carbon-heteroatom bond formation, hydrogenation, and carbonylation dehalogenations (Tsuji, 2004). For example, Pd is used as a catalyst in Suzuki coupling reaction, which, due to its mild conditions and tolerance for functional groups, is essential in synthesizing building molecules for several pharmaceutical compounds (Martin and Buchwald, 2008). The catalytic activity of PdNPs is further enhanced because of their high surface-to-volume ratio and unique SPR (Wang and Bai, 2020). Furthermore, PdNPs can be easily immobilized on different solids, which allows their recycling from the reaction medium upon catalysis (Monguchi et al., 2017).

Saxena et al. used a combination of the CB extract and the plant surfactant (acacia) to produce PdNPs (Clove-PdNPs) in a controlled pH medium (Saxena et al., 2018). The fabricated clove-PdNPs were successfully immobilized on glass and wafer silicon slides using the evaporation-induced self-assembly technique. The authors compared the chemical and electrical catalytic ability of both the immobilized and free clove-PdNPs. The heterogeneous catalytic activity of the immobilized clove-PdNPs was observed by reducing nitrobenzene in the presence of sodium borohydride, which was found to be higher than the free PdNPs. Clove-PdNPs immobilized on a fluorine-doped tin oxide glass surface (anode) exhibited higher current density and lowered charge-transfer resistance when compared to the free clove-PdNPs (Saxena et al., 2018). These results indicate the potential use of clove-PdNPs in fuel cells.

Furthermore, Wang and Bai explored the catalytic efficiency of clove-PdNPs synthesized from clove leaves powder for Suzuki coupling reaction (Wang and Bai, 2020). Using micro-infrared spectroscopy, the synthesis of clove-PdNPs was attributed to a hydroxyl functional group found in the phenolic, polysaccharides, and flavonoids content of clove leaves. The clove-PdNPs were also found to increase the yield of 4-bromonitrobenzene and phenylboronic acid (products in Suzuki coupling reaction) by 98%. This percentage was decreased to 91% after seven reaction cycles. However, the percentage remained the same after removing the catalyst (clove-PdNPs), which shows that clove-PdNPs are excellent heterogeneous catalysts for Suzuki coupling reaction (Fig. 4) (Wang and Bai, 2020).

The potential use of clove-PdNPs in cancer treatment was also investigated. Shanthi et al. (2017) reported a selective dose-dependent inhibition of HeLa cells by clove-PdNPs ($IC_{50}=15\,\mu g/mL$ after 48 h of incubation). This selective cytotoxicity was evident by the higher IC_{50} value against normal cells (H8 cell line) ($IC_{50}=45\,\mu g/mL$ after 48 h incubation). The authors utilized confocal microscopy to show the internalization of the clove-PdNPs inside the nucleus at different time intervals, with 80% of the NPs taken up by the cells after 48 h. The study concluded that the clove-PdNPs internalization in HeLa cells was accomplished via endocytosis (Shanthi et al., 2017).

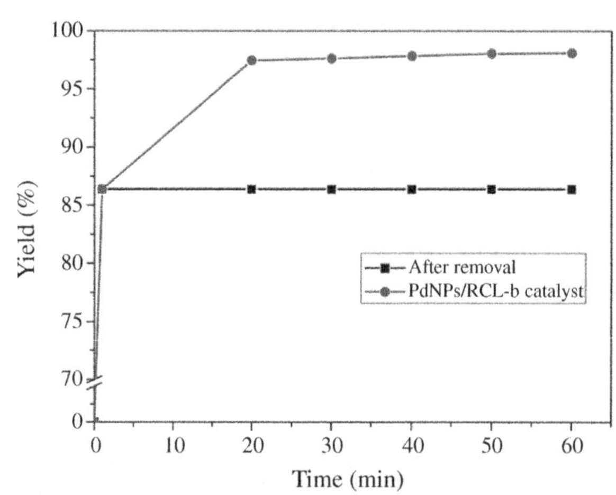

FIG. 4

Yield of Suzuki reaction using clove-PdNPs (denoted: PdNPs/RCL-b) at different time intervals (Wang and Bai, 2020).

2.5 ZnSNPs/ZnONPs

Zn is a nontoxic semiconductor metal of II–VI group with promising applications due to its direct gap energy (~3.63 eV), high refractive index, and dielectric constant (Liu et al., 2006; Rahchamani et al., 2015; Wang et al., 2010). ZnSNPs were therefore considered in several applications in optoelectronics, solar cells, sensors, photocatalysis, and as an antimicrobial agent (Azizi-Lalabadi et al., 2019; Khalafi et al., 2019; Lee et al., 2020; Mishra et al., 2013; Sarkar et al., 2016).

A study used the methanol extract of *S. aromaticum* to cap the ZnSNPs synthesized chemically using thiourea as a reducing agent (Sathishkumar et al., 2018). The bandgap of ZnSNPs from thiourea was 3.85 eV as calculated from the UV-vis spectrum; however, the addition of the *S. aromaticum* was found to increase the bandgap of the ZnSNPs to 4.12 eV. Additionally, the ZnSNPs incorporated with *S. aromaticum* methanol extract increased the photocatalytic degradation of methylene blue and methyl orange dyes compared to the photocatalytic effect of ZnSNPs from thiourea only (pure ZnSNPs). Also, the *S. aromaticum* mediated ZnSNPs exhibited a potent antimicrobial effect, which was significantly higher than the antimicrobial activity of pure ZnSNPs (Sathishkumar et al., 2018). A similar antimicrobial activity was also detected for the ZnSNPs synthesized entirely from CB (clove-ZnSNPs) (Sathishkumar et al., 2017). The disc diffusion assay showed that the highest inhibition zone detected for clove-ZnSNPs was 25 mm against *C. albicans* compared to 16 mm from the positive control (Sathishkumar et al., 2017). Another study by Lakshmeesha et al. (2019) synthesized ZnONPs from CB hexane extract (clove-ZnONPs) via the combustion method. The clove-ZnONPs were found to have antifungal activity by inhibiting the growth of *Fusarium graminearum* (plant pathogen). The authors also studied the possible antifungal mechanism attributed to ROS formation, the increase in lipid peroxidation, and the reduction of ergosterol levels (Lakshmeesha et al., 2019). The ethanol CB extract was also used to produce ZnONPs with bactericidal activity against *E. coli* and *S. aureus* (Anvarinezhad et al., 2020). Interestingly, the degree of bactericidal activity was changed by changing synthesis parameters, specifically the CB extract volume and zinc salt concentration (Anvarinezhad et al., 2020). Osaili et al. (2019) synthesized ZnONPs and CuONPs either via thermal annealing or by using the methanol CB extract. The study compared the antibacterial activity of the different synthesized NPs (individually and in combination) against different foodborne pathogens, *E. coli* and *Listeria monocytogenes*. It was found that the methanol CB extract enhanced the antibacterial activity of the NPs against *E. coli* but not L. *monocytogenes*. The best antibacterial activity could be attained when a mixture of clove-ZnONPs and clove-CuONPs was used (MBC = 300 μg/mL) rather than using the individual NPs (MBC = 350 and above 400 μg/mL for clove-ZnONPs and clove-CuONPs, respectively) (Osaili et al., 2019).

2.6 Other mono-MNPs

Several researchers have explored the use of CuNPs in several applications such as electrical, catalytic, optical antibacterial, and antifungal applications (Bogdanović et al., 2014; Jain et al., 2015; Mohindroo et al., 2016; Pham et al., 2019). The abundance and low cost of Cu are advantageous compared to other expensive metals such as Au, Ag, and Pt (Gawande et al., 2016). Cu has been known as an effective antimicrobial agent with minimum toxicity on the body. However, the exact mechanism in which Cu exerts its antimicrobial activity remains largely speculative. Recently, it was suggested that the release of Cu ions from the Cu-based nanomaterials inside the bacterial cells drives the antimicrobial activity of Cu (Bastos et al., 2020). CuNPs were synthesized from CB (clove-CuNPs) after mixing the aqueous extract with copper acetate solution for 15 min (Rajesh et al., 2018). The biosynthesized clove-CuNPs showed antibacterial and antifungal activities with an 8.0 and 6.0 mm zone inhibition against *Bacillus* spp. and *Penicillium* spp., respectively (Rajesh et al., 2018).

Jain et al. reported the biosynthesis of amorphous MgONPs from CB aqueous extract (clove-MgONPs). The synthesis was done at room temperature. Moreover, the authors optimized the synthesis by changing reaction parameters such as reaction time, the concentration, and the volume of magnesium acetate and the CB aqueous extract (Jain et al., 2018). The clove-MgONPs produced were utilized for colorimetric detection of Fe^{+3}. Although Fe^{+3} is the most abundant transition metal and is involved in several metabolic processes (Aisen et al., 1999), still Fe^{+3} biological imbalance can lead to several ailments, e.g., anemia, diabetes, liver, kidney, and heart diseases (Que et al., 2008). The results from this study detected a shift in the maximal absorbance of the clove-MgONPs from 347 nm to 358 nm and a color change from pale yellow to black after the addition of Fe^{+3}, indicating the ability of the NPs to sense Fe^{+3} in solution. These observations were not detected upon the addition of other metals to the clove-MgONPs, which also indicates the selectivity of sensing toward Fe^{+3} (Jain et al., 2018). Jain et al. (2019) formulated simple, reliable, nontoxic, and affordable nano-based pH paper strips (ranging between 2.0 and 13.0) based on L-lysine-modified MgONPs that were synthesized by the green approach using the CB aqueous extract.

CB extract was also used to synthesize MnONPs (clove-MnONPs) (Kumar et al., 2017). The reaction parameters were changed to optimize the synthesis. The optimization was monitored by observing the UV-vis spectra of the produced clove-MnONPs using different volumes, concentrations, and ratios of the CB extract and the metal salt. MnONPs are widely used in contaminant sensing. Therefore, the authors used cyclic voltammetry, differential pulse voltammetry, and amperometry to show that clove-MnONPs can specifically detect P-nitrophenol (Kumar et al., 2017).

TiO_2NPs are another type of metallic nanoparticles reported from CB. TiO_2NPs have been shown to have antibacterial and anticancer activities (Abdel-Ghany et al., 2020; Abdolmajid et al., 2019; Behnam et al., 2018; Santhoshkumar et al., 2014). Maheswari et al. used *S. aromaticum,* individually and mixed with other plants, to investigate the possibility of enhancing the antibacterial and anticancer activities of the chemically synthesized TiO_2NPs from isopropyl alcohol (Maheswari et al., 2020). The study utilized the XRD and UV-vis techniques to confirm the incorporation of the plant extracts into the TiO_2NPs by comparing the spectra of the free and plant-modified TiO_2NPs. The study has found that the clove modified TiO_2NPs inhibited the growth of the KB oral cancer cell line more than the free TiO_2NPs. Furthermore, the clove-TiO_2NPs gave the highest inhibition zones against several bacterial strains compared to the free and the other plant-modified TiO_2NPs (Maheswari et al., 2020).

2.7 Multi-MNPs

MNPs can be categorized into mono- and multimetallic (bi- or tri-metallic) NPs according to the number of metals involved in the synthesis. Bi- and tri-MNPs can be formulated as alloy or core-shell structures. The synthesis of the multi-MNPs in solution is usually accomplished by the successive reduction of the metal precursors (Venkatesan and Santhanalakshmi, 2010). Due to the different metals' synergistic effects, multi-MNPs present enhanced physicochemical properties and greater efficiency in several industrial and nanomedicine applications, especially in catalysis (Nasrollahzadeh et al., 2020; Sharma et al., 2019).

A study by Rao and Paria (2015) was successful in synthesizing Ag-Au-Pd-NPs in alloy using a mixture of the CB and *Aegle marmelos* leaves aqueous extracts (ratio 1:1) in combination with the natural surfactant *Acacia* (Rao and Paria, 2015). The quick synthesis was done by the successive reduction of Ag, Au, and Pd salts at 50°C in a controlled pH medium. The study utilized the catalytic activity of the formed tri-MNPs in glucose oxidation to gluconic acid (important compound in food and pharmaceutical industries) with a conversion rate of 80% in just 190 min. The authors attributed this high catalytic activity to the electron charge transfer between the three metals, which leads to higher electron density at the catalytic site. Further, the study also showed that the Ag-Au-Pd-NPs have antibacterial activity against *E. coli*. This activity was determined by disc diffusion assay, and the morphological changes in the bacteria were also observed using SEM. The authors concluded that the biofabricated Ag-Au-Pd-NPs could produce gluconic acid with antimicrobial protection of the reaction medium during the manufacturing process (Rao and Paria, 2015). A recent study also explored the catalytic activity of bimetallic Ag-Au-NPs alloy from CB aqueous extract (Sharma et al., 2020). The authors described the effect of varying the metal salts mixture ratios on the physicochemical characteristics of the formulated bi-MNPs.

Moreover, the study also synthesized the respective monometallic NPs (Clove-AgNPs and clove-AuNPs) and compared the antioxidant of all the NPs using three different assays (ABTS, DPPH, and hydroxyl radical scavenging assays), which showed that the bi-MNPs have greater antioxidant activity compared to the individual NPs synthesized from CB. This antioxidant activity was attributed to the phytochemicals of the CB extract capping the NPs. Furthermore, three organic dyes (methylene blue, methyl orange, and P-nitrophenol) were used to study the catalytic activity of the NPs. The results indicated that the catalytic activity of the bi-MNPs is higher than the mono-MNPs owing to the higher number of Au and Ag interfaces in the bi-MNPs available for the redox process (Sharma et al., 2020).

3. Conclusions

The clove's extracts and bio-components have been shown to have a rich source of phytochemicals that can synthesize MNPs using cost-effective and safe methods. Most of the methodologies reported for the MNPs from clove followed typical chemical reduction fundamentals. The data have shown that more studies are needed to determine the clove's chemical component responsible for reducing and stabilizing the MNPs. The clove-mediated MNPs have exhibited interesting pharmacological activities that can be utilized in biomedical applications in the future, for example, as antibacterial and for wound healing. Additionally, clove-MNPs can play a role in industrial applications like water treatment and fuel cells.

References

Abdel-Ghany, S.E., El-Sayed, E., Ashraf, N., Mokhtar, N., Alqosaibi, A., Cevik, E., et al., 2020. Titanium oxide nanoparticles improve the chemotherapeutic action of erlotinib in liver cancer cells. Curr. Cancer Therapy Rev. 16 (4), 337–343.

Abdelhafez, O.H., Ali, T.F.S., Fahim, J.R., Desoukey, S.Y., Ahmed, S., Behery, F.A., et al., 2020. Anti-inflammatory potential of green synthesized silver nanoparticles of the soft coral Nephthea Sp. Supported by metabolomics analysis and docking studies. Int. J. Nanomedicine 15, 5345–5360.

Abdolmajid, E., Kharazi, H., Chalaki, M., Khojasteh, M., Haghighat, S., Attar, F., et al., 2019. Titanium oxide nanoparticles fabrication, hemoglobin interaction, white blood cells cytotoxicity, and antibacterial studies. J. Biomol. Struct. Dyn. 37 (11), 3007–3017.

Adaramola, B., Onigbinde, A., 2016. Effect of extraction solvent on the phenolic content, flavonoid content and antioxidant capacity of clove bud. IOSR J. Pharm. Biol. Sci. 11 (3), 33–38.

Afreen, A., Ahmed, R., Mehboob, S., Tariq, M., Alghamdi, H.A., Zahid, A.A., et al., 2020. Phytochemical-assisted biosynthesis of silver nanoparticles from *Ajuga bracteosa* for biomedical applications. Mater. Res. Express 7 (7), 075404.

Ahmad, T., Bustam, M.A., Irfan, M., Moniruzzaman, M., Asghar, H.M.A., Bhattacharjee, S., 2019. Mechanistic investigation of phytochemicals involved in green synthesis of gold nanoparticles using aqueous *Elaeis guineensis* leaves extract: role of phenolic compounds and flavonoids. Biotechnol. Appl. Biochem. 66 (4), 698–708.

Aisen, P., Wessling-Resnick, M., Leibold, E.A., 1999. Iron metabolism. Curr. Opin. Chem. Biol. 3 (2), 200–206.

Ajiboye, T.O., Mohammed, A.O., Bello, S.A., Yusuf, I.I., Ibitoye, O.B., Muritala, H.F., Onajobi, I.B., 2016. Antibacterial activity of *Syzygium aromaticum* seed: studies on oxidative stress biomarkers and membrane permeability. Microb. Pathog. 95, 208–215.

Ajitha, B., Reddy, Y.A.K., Lee, Y., Kim, M.J., Ahn, C.W., 2019. Biomimetic synthesis of silver nanoparticles using *Syzygium aromaticum* (clove) extract: catalytic and antimicrobial effects. Appl. Organomet. Chem. 33 (5), e4867.

Alexander, J.W., 2009. History of the medical use of silver. Surg. Infect. (Larchmt.) 10 (3), 289–292.

Anvarinezhad, M., Javadi, A., Jafarizadeh-Malmiri, H., 2020. Green approach in fabrication of photocatalytic, antimicrobial, and antioxidant zinc oxide nanoparticles – hydrothermal synthesis using clove hydroalcoholic extract and optimization of the process. Green Processes Synth. 9 (1), 375–385.

Aziz, W.J., Jassim, H.A., 2018. A new paradigm shift to prepare copper nanoparticles using biological synthesis and evaluation of antimicrobial activity. Plant Archives 18 (2), 2020–2024.

Azizi-Lalabadi, M., Ehsani, A., Divband, B., Alizadeh-Sani, M., 2019. Antimicrobial activity of titanium dioxide and zinc oxide nanoparticles supported in 4A zeolite and evaluation the morphological characteristic. Sci. Rep. 9 (1), 1–10.

Bagyalakshmi, B., Priyadarshini, S.L., Balamurugan, A., 2019. Anticancer activity of bee venom against lung cancer cell line (A549 cells) enhanced by iron oxide nanoparticles synthesized from *Syzygium aromaticum*. J. Drug Deliv. Ther. 9 (3-s), 248–254.

Bai, X., Wang, Y., Song, Z., Feng, Y., Chen, Y., Zhang, D., Feng, L., 2020. The basic properties of gold nanoparticles and their applications in tumor diagnosis and treatment. Int. J. Mol. Sci. 21 (7), 2480.

Banerjee, S., 2005. Clove (*Syzygium aromaticum* L.), a potential chemopreventive agent for lung cancer. Carcinogenesis 27 (8), 1645–1654.

Barua, S., Banerjee, P.P., Sadhu, A., Sengupta, A., Chatterjee, S., Sarkar, S., et al., 2017. Silver nanoparticles as antibacterial and anticancer materials against human breast, cervical and oral cancer cells. J. Nanosci. Nanotechnol. 17 (2), 968–976.

Bastos, C.A.P., Faria, N., Wills, J., Malmberg, P., Scheers, N., Rees, P., Powell, J.J., 2020. Copper nanoparticles have negligible direct antibacterial impact. NanoImpact 17, 100192.

Batiha, G.E.S., Beshbishy, A.M., Tayebwa, D.S., Shaheen, H.M., Yokoyama, N., Igarashi, I., 2019. Inhibitory effects of *Syzygium aromaticum* and *Camellia sinensis* methanolic extracts on the growth of Babesia and Theileria parasites. Ticks Tick-Borne Dis. 10 (5), 949–958.

Behnam, M.A., Emami, F., Sobhani, Z., Dehghanian, A.R., 2018. The application of titanium dioxide (TiO2) nanoparticles in the photo-thermal therapy of melanoma cancer model. Iran. J. Basic Med. Sci. 21 (11), 1133–1139.

Bogdanović, U., Lazić, V., Vodnik, V., Budimir, M., Marković, Z., Dimitrijević, S., 2014. Copper nanoparticles with high antimicrobial activity. Mater. Lett. 128, 75–78.

Chang, P.R., Yu, J., Ma, X., Anderson, D.P., 2011. Polysaccharides as stabilizers for the synthesis of magnetic nanoparticles. Carbohydr. Polym. 83 (2), 640–644.

Clement, J.L., Jarrett, P.S., 1994. Antibacterial silver. Metal-based drugs 1 (5–6), 467–482.

Cortés-Rojas, D.F., de Souza, C.R.F., Oliveira, W.P., 2014. Clove (*Syzygium aromaticum*): a precious spice. Asian Pac. J. Trop. Biomed. 4 (2), 90–96.

Dauthal, P., Mukhopadhyay, M., 2016. Noble metal nanoparticles: plant-mediated synthesis, mechanistic aspects of synthesis, and applications. Ind. Eng. Chem. Res. 55 (36), 9557–9577.

de Maciel, M.V.O.B., da Almeida, A.R., Machado, M.H., de Melo, A.P.Z., da Rosa, C.G., de Freitas, D.Z., et al., 2019. *Syzygium aromaticum* L. (clove) essential oil as a reducing agent for the green synthesis of silver nanoparticles. Open J. Appl. Sci. 09 (02), 45–54.

Deshpande, R., Bedre, M.D., Salimath, B., Sawle, B.D., Yalagatti, M.S., Abbaraju, V., 2012. Microwave-assisted rapid biosynthesis of stable silver nanoparticles using clove buds (*Syzygium aromaticum*) solution. Nano life 02 (02), 1250012.

Dube, P., Meyer, S., Madiehe, A., Meyer, M., 2020. Antibacterial activity of biogenic silver and gold nanoparticles synthesized from *Salvia africana-lutea* and *Sutherlandia frutescens*. Nanotechnology 31 (50), 505607.

El Ghallab, Y., Al Jahid, A., Jamal Eddine, J., Ait Haj Said, A., Zarayby, L., Derfoufi, S., 2019. *Syzygium aromaticum* L.: phytochemical investigation and comparison of the scavenging activity of essential oil, extracts and eugenol. Orient Pharm. Exp. Med. 20 (2), 153–158.

Etheridge, M.L., Campbell, S.A., Erdman, A.G., Haynes, C.L., Wolf, S.M., McCullough, J., 2013. The big picture on nanomedicine: the state of investigational and approved nanomedicine products. Nanomedicine 9 (1), 1–14.

Gao, H., Yang, H., Wang, C., 2017. Controllable preparation and mechanism of nano-silver mediated by the microemulsion system of the clove oil. Res. Phys. 7, 3130–3136.

Gawande, M.B., Goswami, A., Felpin, F.X., Asefa, T., Huang, X., Silva, R., et al., 2016. Cu and Cu-based nanoparticles: synthesis and applications in catalysis. Chem. Rev. 116 (6), 3722–3811.

Gour, A., Jain, N.K., 2019. Advances in green synthesis of nanoparticles. Artif. Cells, Nanomed. Biotechnol. 47 (1), 844–851.

Hameed, A., Khan, I., Azam, A., Naz, S.S., Khan, A., Saleem, M., Islam, N.U., 2014. Stability and enzyme inhibition activities of au nanoparticles using an aqueous extract of clove as a reducing and stabilizing agent. J. Chem. Soc. Pak. 36 (3), 542–547.

Hasanpoor, M., Fakhr Nabavi, H., Aliofkhazraei, M., 2017. Microwave-assisted synthesis of alumina nanoparticles using some plants extracts. J. Nanostruct. 7 (1), 40–46.

Hu, X., Zhang, Y., Ding, T., Liu, J., Zhao, H., 2020. Multifunctional gold nanoparticles: a novel nanomaterial for various medical applications and biological activities. Front. Bioeng. Biotechnol. 8, 990.

Iravani, S., Korbekandi, H., Mirmohammadi, S.V., Zolfaghari, B., 2014. Synthesis of silver nanoparticles: chemical, physical and biological methods. Res. Pharm. Sci. 9 (6), 385–406.

Jain, A., Wadhawan, S., Kumar, V., Mehta, S.K., 2018. Colorimetric sensing of Fe^{3+} ions in aqueous solution using magnesium oxide nanoparticles synthesized using green approach. Chem. Phys. Lett. 706, 53–61.

Jain, A., Wadhawan, S., Kumar, V., Mehta, S.K., 2019. pH-sensing strips based on biologically synthesized Ly-MgO nanoparticles. ACS Omega 4 (26), 21647–21657.

Jain, S., Jain, A., Kachhawah, P., Devra, V., 2015. Synthesis and size control of copper nanoparticles and their catalytic application. Trans. Nonferrous Met. Soc. Chin. 25 (12), 3995–4000.

Jirovetz, L., Buchbauer, G., Stoilova, I., Stoyanova, A., Krastanov, A., Schmidt, E., 2006. Chemical composition and antioxidant properties of clove leaf essential oil. J. Agric. Food Chem. 54 (17), 6303–6307.

Kanagesan, S., Hashim, M., Tamilselvan, S., Alitheen, N.B., Ismail, I., Hajalilou, A., Ahsanul, K., 2013. Synthesis, characterization, and cytotoxicity of iron oxide nanoparticles. Adv. Mater. Sci. Eng. 2013.

Kanneganti, A., Kumar, K.A., Vivek, V., 2014. Green synthesis of molybdenum trioxide nanoparticles using *Syzygium aromaticum* flower buds extract. Int. J. Sci. Res. Dev. 2 (4), 116–117.

Kargozar, S., Mozafari, M., 2018. Nanotechnology and nanomedicine: start small, think big. Mater. Today: Proc. 5 (7), 15492–15500.

Kaur, H., Kaur, S., Singh, M., 2013. Biosynthesis of silver nanoparticles by natural precursor from clove and their antimicrobial activity. Biologia (Poland) 68 (6), 1048–1053.

Kaur, K., Kaushal, S., Rani, R., 2019. Chemical composition, antioxidant and antifungal potential of clove (*Syzygium aromaticum*) essential oil, its major compound and its derivatives. J. Essent. Oil Bear. Plants 22 (5), 1195–1217.

Khalafi, T., Buazar, F., Ghanemi, K., 2019. Phycosynthesis and enhanced photocatalytic activity of zinc oxide nanoparticles toward organosulfur pollutants. Sci. Rep. 9 (1), 1–10.

Khan, I., Saeed, K., Khan, I., 2019. Nanoparticles: properties, applications and toxicities. Arab. J. Chem. 12 (7), 908–931.

Konop, M., Damps, T., Misicka, A., Rudnicka, L., 2016. Certain aspects of silver and silver nanoparticles in wound care: a mini review. J. Nanomater. 2016.

Kumar, V., Singh, K., Panwar, S., Mehta, S.K., 2017. Green synthesis of manganese oxide nanoparticles for the electrochemical sensing of p-nitrophenol. Int. Nano Lett. 7 (2), 123–131.

Lakhan, M.N., Chen, R., Shar, A.H., Chand, K., Shah, A.H., Ahmed, M., et al., 2020. Eco-friendly green synthesis of clove buds extract functionalized silver nanoparticles and evaluation of antibacterial and antidiatom activity. J. Microbiol. Methods 173, 105934.

Lakshmeesha, T.R., Kalagatur, N.K., Mudili, V., Mohan, C.D., Rangappa, S., Prasad, B.D., et al., 2019. Biofabrication of zinc oxide nanoparticles with *Syzygium aromaticum* flower buds extract and finding its novel application in controlling the growth and mycotoxins of *fusarium graminearum*. Front. Microbiol. 10 (JUN), 1244.

Le Ouay, B., Stellacci, F., 2015. Antibacterial activity of silver nanoparticles: a surface science insight. Nano Today 10 (3), 339–354.

Lee, W., Yeop, J., Heo, J., Yoon, Y.J., Park, S.Y., Jeong, J., et al., 2020. High colloidal stability ZnO nanoparticles independent on solvent polarity and their application in polymer solar cells. Sci. Rep. 10 (1), 18055.

Liu, J.Z., Yan, P.X., Yue, G.H., Kong, L.B., Zhuo, R.F., Qu, D.M., 2006. Synthesis of doped ZnS one-dimensional nanostructures via chemical vapor deposition. Mater. Lett. 60 (29–30), 3471–3476.

Maheswari, P., Ponnusamy, S., Harish, S., Muthamizhchelvan, C., Hayakawa, Y., 2020. Syntheses and characterization of *Syzygium aromaticum*, *Elettaria cardamomum* and *Cinnamomum verum* modified TiO2 and their biological applications. Mater. Sci. Semicond. Process. 105, 104724.

Majoumouo, M.S., Sibuyi, N.R.S., Tincho, M.B., Mbekou, M., Boyom, F.F., Meyer, M., 2019. Enhanced antibacterial activity of biogenic silver nanoparticles synthesized from *Terminalia mantaly* extracts. Int. J. Nanomedicine 14, 9031–9046.

Mallikarjuna, K., John Sushma, N., Narasimha, G., Manoj, L., Deva Prasad Raju, B., 2014. Phytochemical fabrication and characterization of silver nanoparticles by using pepper leaf broth. Arab. J. Chem. 7 (6), 1099–1103.

Martin, R., Buchwald, S.L., 2008. Palladium-catalyzed suzuki-miyaura cross-coupling reactions employing dialkylbiaryl phosphine ligands. Acc. Chem. Res. 41 (11), 1461–1473.

Meeker, H.G., Linke, H.A., 1988. The antibacterial action of eugenol, thyme oil, and related essential oils used in dentistry. Compendium (Newtown, Pa.) 9 (1).

Mishra, M.K., Narayan, A., Singh, R.K., Yadav, R.S., Nidhi, N.K., Pandey, A.C., 2013. Study of structural and optoelectronic properties of ZnO Codoped with Ca and Mg. Indian J. Mater. Sci. 2013, 1–6.

Mohindroo, J.J., Garg, U.K., Sharma, A.K., 2016. Optical properties of stabilized copper nanoparticles. AIP Conf. Proc. 1728 (1), 020534.

Mondal, P., Purkait, M.K., 2018. Green synthesized iron nanoparticles supported on pH responsive polymeric membrane for nitrobenzene reduction and fluoride rejection study: optimization approach. J. Clean. Prod. 170, 1111–1123.

Monguchi, Y., Ichikawa, T., Sajiki, H., 2017. Recent development of palladium-supported catalysts for chemoselective hydrogenation. Chem. Pharm. Bull. 65 (1), 2–9.

Nasrollahzadeh, M., Sajjadi, M., Iravani, S., Varma, R.S., 2020. Trimetallic nanoparticles: greener synthesis and their applications. Nanomaterials 10 (9), 1–27.

Nowak, K., Ogonowski, J., Jaworska, M., Grzesik, K., 2012. Clove oil—properties and applications science technique. Chem. Aust. 66 (2), 149–152.

Osaili, T.M., Albiss, B.A., Al-Nabulsi, A.A., Alromi, R.F., Olaimat, A., Al-Holy, M., et al., 2019. Effects of metal oxide nanoparticles with plant extract on viability of foodborne pathogens. J. Food Saf. 39 (5), e12681.

Ovais, M., Khalil, A.T., Islam, N.U., Ahmad, I., Ayaz, M., Saravanan, M., et al., 2018. Role of plant phytochemicals and microbial enzymes in biosynthesis of metallic nanoparticles. Appl. Microbiol. Biotechnol. 102 (16), 6799–6814.

Parlinska-Wojtan, M., Depciuch, J., Fryc, B., Kus-Liskiewicz, M., 2018. Green synthesis and antibacterial effects of aqueous colloidal solutions of silver nanoparticles using clove eugenol. Appl. Organomet. Chem. 32 (4), e4276.

Parveen, A., Kulkarni, N., Yalagatti, M., Abbaraju, V., Deshpande, R., 2018. In vivo efficacy of biocompatible silver nanoparticles cream for empirical wound healing. J. Tissue Viability 27 (4), 257–261.

Pattanayak, M., Mohapatra, D., Nayak, P.L., 2013. Green synthesis and characterization of zero valent iron nanoparticles from the leaf extract of *Syzygium aromaticum* (clove). Middle-East J. Sci. Res. 18 (5), 623–626.

Pavan, G., Nagarajugoud, M., Kalyani, S., Shireesha, P., Madhuri, M., Srikar, N., Abhilash, M., 2016. Novel approaches of condiment extract synthesis of nickel nanoparticles for antimicrobial activities: a green expertise. Int. J. Adv. Res. 4 (10), 842–846.

Pham, N.D., Duong, M.M., Le, M.V., Hoang, H.A., Pham, L.K.O., 2019. Preparation and characterization of antifungal colloidal copper nanoparticles and their antifungal activity against *Fusarium oxysporum* and *Phytophthora capsici*. C. R. Chim. 22 (11 – 12), 786–793.

Que, E.L., Domaille, D.W., Chang, C.J., 2008. Metals in neurobiology: probing their chemistry and biology with molecular imaging. Chem. Rev. 108 (5), 1517–1549.

Raghunandan, D., Bedre, M.D., Basavaraja, S., Sawle, B., Manjunath, S.Y., Venkataraman, A., 2010. Rapid biosynthesis of irregular shaped gold nanoparticles from macerated aqueous extracellular dried clove buds (*Syzygium aromaticum*) solution. Colloids Surf. B Biointerfaces 79 (1), 235–240.

Raghunandan, D., Ravishankar, B., Sharanbasava, G., Mahesh, D.B., Harsoor, V., Yalagatti, M.S., et al., 2011. Anti-cancer studies of noble metal nanoparticles synthesized using different plant extracts. Cancer Nanotechnol. 2 (1–6), 57–65.

Rahchamani, S.Z., Dizaji, H.R., Ehsani, M.H., 2015. Study of structural and optical properties of ZnS zigzag nanostructured thin films. Appl. Surf. Sci. 356, 1096–1104.

Rajesh, K.M., Ajitha, B., Reddy, Y.A.K., Suneetha, Y., Reddy, P.S., 2018. Assisted green synthesis of copper nanoparticles using *Syzygium aromaticum* bud extract: physical, optical and antimicrobial properties. Optik 154, 593–600.

Ramos, A.P., Cruz, M.A.E., Tovani, C.B., Ciancaglini, P., 2017. Biomedical applications of nanotechnology. Biophys. Rev. 9 (2), 79–89.

Rao, K.J., Paria, S., 2015. Mixed phytochemicals mediated synthesis of multifunctional Ag-Au-Pd nanoparticles for glucose oxidation and antimicrobial applications. ACS Appl. Mater. Interfaces 7 (25), 14018–14025.

Sameen, A., Fathima, S., Ramlal, S., Kumar, S., Khanum, F., 2014. Nanopackaging of silver using spice extract and their characterization. Sci. Technol. Arts Res. J. 3 (3), 52.

Santhoshkumar, T., Rahuman, A.A., Jayaseelan, C., Rajakumar, G., Marimuthu, S., Kirthi, A.V., et al., 2014. Green synthesis of titanium dioxide nanoparticles using P*sidium guajava* extract and its antibacterial and antioxidant properties. Asian Pac. J. Trop. Med. 7 (12), 968–976.

Sarkar, A., Maity, S., Chakraborty, P., Chakraborty, S.K., 2016. Synthesize of ZnO nano structure for toxic gas sensing application. Prog. Comput. Sci. 92, 199 206.

Sathishkumar, M., Rajamanickam, A.T., Saroja, M., 2018. Characterization, antimicrobial activity and photocatalytic degradation properties of pure and biosynthesized zinc sulfide nanoparticles using plant extracts. J. Mater. Sci. Mater. Electron. 29 (16), 14200–14209.

Sathishkumar, M., Saroja, M., Venkatachalam, M., 2017. Green synthesis, characterization and antimicrobial activity of ZnS using *Syzygium aromaticum* extracts. Int. J. ChemTech Res. 10 (9), 443–449.

Saxena, N., Praneeth, N.V.S., Rao, K.J., Paria, S., 2018. Organization of palladium nanoparticles into fractal patterns for highly enhanced catalytic activity and anode material for direct borohydride fuel cells applications. ACS Appl. Energy Mater. 1 (5), 2164–2175.

Sebastian, A., Nangia, A., Prasad, M.N.V., 2018. Green synthesis of iron nanoparticles from selected plant materials of Peninsular India. Proc. Natl. Acad. Sci., India, Sect. A 88 (2), 195–203.

Shafira, K.F., Azad, A.K., Labu, Z.K., Helal Uddin, A.B.M., 2020. Extraction and quantification of eugenol from clove buds using HPLC. Curr. Chromatogr. 7 (1), 17–23.

Shanthi, K., Sreevani, V., Vimala, K., Kannan, S., 2017. Cytotoxic effect of palladium nanoparticles synthesized from *Syzygium aromaticum* aqueous extracts and induction of apoptosis in cervical carcinoma. Proc. Natl. Acad. Sci. India Sect. B Biol. Sci. 87 (4), 1101–1112.

Sharma, C., Ansari, S., Ansari, M.S., Satsangee, S.P., Srivastava, M.M., 2020. Single-step green route synthesis of Au/Ag bimetallic nanoparticles using clove buds extract: enhancement in antioxidant bio-efficacy and catalytic activity. Mater. Sci. Eng. C 116, 111153.

Sharma, G., Kumar, A., Sharma, S., Naushad, M., Prakash Dwivedi, R., ALOthman, Z.A., Mola, G.T., 2019. Novel development of nanoparticles to bimetallic nanoparticles and their composites: a review. J. King Saud Univ. – Sci. 31 (2), 257–269.

Sharma, M., Pathak, M., Roy, B., 2017. Green synthesis of gold nanoparticles using *Cinnamomum verum*, *Syzygium aromaticum* and *Piper nigrum* extract. Asian J. Chem. 29 (8), 1693–1696.

Sil, S., Chaturvedi, D., Krishnappa, K.B., Kumar, S., Asthana, S.N., Umapathy, S., 2014. Density functional theoretical modeling, electrostatic surface potential and surface enhanced Raman spectroscopic studies on biosynthesized silver nanoparticles: observation of 400 pM sensitivity to explosives. J. Phys. Chem. A 118 (16), 2904–2914.

Singh, A.K., Talat, M., Singh, D.P., Srivastava, O.N., 2010. Biosynthesis of gold and silver nanoparticles by natural precursor clove and their functionalization with amine group. J. Nanopart. Res. 12 (5), 1667–1675.

Sujin Jeba Kumar, T., Akshay, V.R., Vasundhara, M., Arumugam, M., 2020. Biosynthesis of multiphase iron nanoparticles using *Syzygium aromaticum* and their magnetic properties. Colloids Surf. A Physicochem. Eng. Asp. 603, 125241.

Tekin, V., Kozgus Guldu, O., Dervis, E., Yurt Kilcar, A., Uygur, E., Biber Muftuler, F.Z., 2019. Green synthesis of silver nanoparticles by using eugenol and evaluation of antimicrobial potential. Appl. Organomet. Chem. 33 (7), e4969.

Thenmozhi, T., 2020. Functionalization of iron oxide nanoparticles with clove extract to induce apoptosis in MCF-7 breast cancer cells. 3 Biotech 10 (2), 82.

Thenmozhi, T., Rasajna, N., Rajesh, M., Sushma, N.J., 2019. Biosynthesis and characterization of iron oxide nanoparticles *via Syzygium aromaticum* extract and determination of its cytotoxicity against human breast cancer cell lines. J. Nanosci. Technol. 5 (1), 587–592.

Tsuji, J., 2004. Palladium Reagents and Catalysts: New Perspectives for the 21st Century, first ed. John Wiley & Sons, Ltd.

Ur Mashwani, Z.R., Khan, M.A., Khan, T., Nadhman, A., 2016. Applications of plant terpenoids in the synthesis of colloidal silver nanoparticles. Adv. Colloid Interface Sci. 234, 132–141.

Varalakshmi, V., Suganiya, S., Mala, R., 2015. Fabrication and characterization of hybrid sponge for healing of infectious burn wound. Recent Pat. Nanotechnol. 9 (3), 212–221.

Varghese, R.E., Ragavan, D., Sivaraj, S., Gayathri, D., Kannayiram, G., 2017. Anti-inflammatory activity of *Syzygium aromaticum* silver nanoparticles: in vitro and in silico study. Asian J. Pharm. Clin. Res. 10 (11), 370–373.

Venkatesan, P., Santhanalakshmi, J., 2010. Designed synthesis of Au/Ag/Pd trimetallic nanoparticle-based catalysts for sonogashira coupling reactions. Langmuir 26 (14), 12225–12229.

Venugopal, K., Rather, H.A., Rajagopal, K., Shanthi, M.P., Sheriff, K., Illiyas, M., Rather, R.A., Manikandan, E., Uvarajan, S., Bhaskar, M., Maaza, M., 2017. Synthesis of silver nanoparticles (Ag NPs) for anticancer activities (MCF 7 breast and A549 lung cell lines) of the crude extract of *Syzygium aromaticum*. J. Photochem. Photobiol. B 167, 282–289.

Vijayaraghavan, K., Nalini, S.P.K., Prakash, N.U., Madhankumar, D., 2012. Biomimetic synthesis of silver nanoparticles by aqueous extract of *Syzygium aromaticum*. Mater. Lett. 75, 33–35.

Wang, L., Xu, X., Yuan, X., 2010. Preparation and photoluminescent properties of doped nanoparticles of ZnS by solid-state reaction. JOL 130 (1), 137–140.

Wang, Z., Bai, X., 2020. One-pot synthesis of bio-supported Pd nanoparticles by using clove leaf and their catalytic performance for Suzuki coupling reaction. J. Mol. Struct. 1219, 128538.

Yin, I.X., Zhang, J., Zhao, I.S., Mei, M.L., Li, Q., Chu, C.H., 2020. The antibacterial mechanism of silver nanoparticles and its application in dentistry. Int. J. Nanomedicine 15, 2555–2562.

Yuwono, M., Siswandono, Hafid, A.F., Poernomo, A.T., Agil, M., Indrayanto, G., Ebel, S., 2002. Eugenol. In: Analytical Profiles of Drug Substances and Excipients. vol. 29. Elsevier Inc, pp. 149–177.

CHAPTER

Toxicity of clove (*Syzygium aromaticum*) extract

39

M. Joyce Nirmala*, P.J. Shiny*, Uppada Sumanth Raj, Narasa Saikrishna, and R. Nagarajan
Department of Chemical Engineering, Indian Institute of Technology Madras, Chennai, Tamil Nadu, India

1. Introduction

1.1 A brief description of clove

In recent years, medicinal plants and herbs have come into the limelight because of their natural origin and milder side effects compared to synthetic products. More than half of the world depends on traditional systems of medicine for primary health care. These natural medicines possess various pharmacological properties, such as bactericidal, virucidal, fungicidal, anti-inflammatory, antimicrobial, spasmolytic, sedative, analgesic, and local anesthetic activities.

Significant progress in the use of medicinal herbs has been achieved with the extraction of phytoconstituents such as glycosides, saponins, flavonoids, steroids, tannins, alkaloids, and terpenes (Batiha et al., 2020).

Dried, unopened, brown-colored flower buds of *Syzygium aromaticum* are commonly called cloves. They belong to the Myrtaceae family indigenous to the Maluku Islands of Indonesia. They are also farmed in many places around the world. Cloves are like tiny, round-headed nails, between 10 and 17.5 mm long and blackish-brown in color. Clove buds, leaves, and stems are used for different purposes in the pharmaceutical, food, flavoring, and fragrance industries. They have a wide range of applications in various fields. Clove is an ancient spice that has a history of usage in the field of Ayurveda. It is used in the perfume industry because of its aroma. Clove is considered one of the natural preservatives in many food industries to replace chemical substitutes, particularly in meat processing, because of its antioxidant and antimicrobial properties (Cortés-Rojas et al., 2014). Clove is known as the champion of antioxidants due to abundant flavonoids, hydroxy benzenes, phenylpropanoids, and other reductive substances.

Clove and its extracts have been used as medicine since ancient times. It is especially effective in the treatment of oral and dental problems. Clove essential oil is used in the treatment of burns and wounds. It is also used as a pain reliever for inflammatory disorders, tooth infections, and toothaches. It is a traditional warming and stimulating agent. Clove has been used for centuries to treat stomach disorders, vomiting, nausea, and liver and bowel disorders (Mbaveng and Kuete, 2017).

*Dr. M. Joyce Nirmala and Dr. P.J. Shiny share equal contribution.

Clove (*Syzygium aromaticum*). https://doi.org/10.1016/B978-0-323-85177-0.00007-0

1.2 Biological properties of the clove extract

Syzygium aromaticum extracts, formulated primarily using solvents, ethanol, methanol, and water, have been proved to have biological activities such as antibacterial, antioxidant, antifungal, antiviral, and anticarcinogenic. Some common applications and side effects caused by cloves are listed in Table 1.

1.2.1 Antioxidant activity

Compared to seeds, fruits, and vegetables, herbs and spices contain higher amounts of phenolics. Among the spices, higher amounts of phenolics and antioxidant compounds are found in cloves. Clove is a rich source of phenolic compounds such as flavonoids, hydroxybenzoic acids, and hydroxycinnamic acids. The main bioactive compound in clove is eugenol. Phenolic acids such as caffeic, ferulic, ellagic, and salicylic acids are found, while gallic acid is present in high concentrations. Flavonoids such as kaempferol, quercetin and its derivatives are present in lower concentrations (Cortés-Rojas et al., 2014). Studies suggest that the antioxidant activity of clove is comparable to BHT (butylated hydroxyanisole), a common synthetic food preservative. At concentrations of 20, 40, and 60 μg/mL, ethanol and aqueous clove extracts displayed inhibitions of up to 95% when tested as a metal chelant, superoxide radical DPPH scavenger. Because of their oxidative stability and shelf life, they have tremendous potential as food-grade preservatives and antioxidants. Controlled release of such antioxidants has been achieved by capsulation and encapsulation methods. Clove essential oil has antioxidant potential to reduce oxidative stress and revert short-term or long-term memory loss.

1.2.2 Antimicrobial activity

Several spice plants, such as garlic, mint, cinnamon, clove, etc., were tested for antimicrobial activity. Aqueous clove extract at 3% has shown complete bactericidal effect against common foodborne pathogens such as *Escherichia coli*, *Staphylococcus aureus*, and *Bacillus cereus*. Formulations containing eugenol and carvacrol have been effective in inhibiting the growth of microorganisms on the surface of the food. Eugenol is the main compound that is responsible for the antifungal activity of clove oil. Formulations containing eugenol have proven to be efficient anti-infective and antifungal medicines. These natural medicines' solubility and delivery are improved with the encapsulation process.

Table 1 Bioactive components present in clove and their benefits and side effects.

Components of clove	Benefits	Side effects
Eugenol	Protects the liver	Kidney and intestine problem
Quercetin	Treats oral diseases	Dilution of blood
Gallic acid	Fights lung cancer	Burning
Kaempferol	Improves digestion	Itching
Crategolic acid	Keeps diabetes in check	Allergies
Beta-caryophyllene	Enhances bone strength	Negative impact on sexual arousal
Oleanolic acid	Strengthens the immune system	
Stigmasterol	Strong antibacterial agent	
Eugenitin		
Bicornin		

Aqueous *Syzygium aromaticum* extracts, when combined with acyclovir, have shown antiviral efficacy against herpes simplex type 1 (HSV-1) virus and influenza A virus. Studies reveal that extracts of tannins, ellagic acid, gallic acid, flavonoids, and their isolated glycosides from alcoholic and aqueous clove buds have antithrombotic, antiprotozoal, anti-inflammatory, gastroprotective, and aphrodisiac properties. *Syzygium aromaticum* methanol extracts have shown antipiroplasmic efficacy against the multiplication of piroplasma parasites and antiplasmodial efficacy against chloroquine-resistant *Plasmodium falciparum* strain in vitro (Batiha et al., 2020; Cortés-Rojas et al., 2014).

These aqueous and alcoholic clove extracts are reported to have antipyretic, appetite-inducing, hypnotic, anxiolytic, antiemetic, analgesic, decongestant, antimicrobial, antiepileptic, myorelaxant, anti-inflammatory, antispasmodic, and acaricidal effects (Batiha et al., 2020).

2. Extraction types and active compounds identified

2.1 Aqueous extract of clove flower buds

Aqueous extraction is usually defined as the extraction of active components or materials from plants, buds, or leaves using distilled or deionized water as a solvent. Maceration, percolation, and pressurized liquid extraction are a few well-known examples of such types of extraction.

Eugenol is one of the main compounds extracted from cloves using aqueous extraction. In this extraction process, aliquots of clove buds are dried at 37°C for 24 h. They are blended, and the powder obtained is suspended in sterile saline solution. The powder is then centrifuged at 2000 rpm for about 3 min and the extract is collected. The yield of the extract obtained from 100 g of dried powder is approximately 40% (Chniguir et al., 2019).

Hydrodistillation is a commonly used aqueous method for the extraction of essential oils. In this method, dried clove buds are powdered and taken in a 500 mL volumetric flask and subjected to hydrodistillation for 4–6 h. The volatile distillate is collected and saturated with sodium chloride. To this distillate, petroleum ether or a suitable organic solvent is added. After some time, the hydro- and ether layers get separated. The distillate is then dehydrated by using anhydrous sodium sulfate. Finally, the sample is heated in a water bath at 60°C for recovering ether and the extract. The yield of extract obtained from 100 g of dried powder is approximately 11.5%. The concentration of eugenol obtained is 50.5%–53.5% (Wenqiang et al., 2007). Another study reported that the extraction yield depends on the particle size of the sample. By reducing the particle size of powdered cloves, extraction yield increases (Khajeh et al., 2004). In studies for analyzing the antimicrobial activity of aqueous clove extract against food-borne pathogen bacteria, the process of distillation with water is repeated thrice to procure a pure compound (Chaula et al., 2019; Puangpronpitag et al., 2009).

2.2 Solvent extraction

Solvent extraction is one of the most common and comprehensive methods used in the preparation of extracts. The solvent analysis is critical for determining the selectivity, solubility, and safety of solvents. Soluble residues that cause undesirable flavor changes in the food are a big hindrance to this method (Wenqiang et al., 2007). Organic solvents, primarily alcoholic, such as ethanol, methanol, phenol, diethyl ether, and *n*-hexane, are widely used to prepare *Syzygium aromaticum* extracts. The initial method of preparation is similar to the process of aqueous extraction. The selected buds of *Syzygium aromaticum* are washed, disinfected, and allowed to dry. Using a mechanical blender, the dried *Syzygium aromaticum* buds are then powdered into a homogeneous mixture. For a sufficient number of days, the desired amount of solvent and the powdered sample are left as a mixture to enable soaking. With the aid of a sterilized glass rod or magnetic stirrer, this mixture is frequently stirred. At the end of extraction, the mixture is filtered through filter paper, usually Whatman No. 1, to obtain a clear filtrate, and concentration methods are then employed. Soxhlet apparatus was used to prepare 50% ethanol extract for the determination of aphrodisiac activity in male mice.

This method has wide applications for the extraction of eugenol and essential oils from aromatic herbs. In this solvent extraction process, to obtain eugenol from the clove, the clove buds are ground and wrapped in filter paper. This filter paper is placed in an extraction thimble and inserted into a reflux flask in the Soxhlet apparatus containing a suitable organic solvent (Quan et al., 2004). The obtained extract is concentrated using a rotary vacuum evaporator at 50°C.

Several modifications have also been made to the conventional solvent extraction process and have led to higher efficiency than the traditional method. As an example, instead of Soxhlet extraction, batch extraction is used as an attractive alternative. This alternative method uses a reactor equipped with an agitator with four blades and a motor with 1200 rpm (Khalil et al., 2017).

Different methods of concentrating the extract have been employed depending on the application. For evaluating the antibacterial effect of clove extract against uropathogens, the ethanolic filtrate was concentrated using a water bath at 40°C, whereas in in vitro studies such as for anticandidal potency against vaginal candidiasis, a rotary evaporator is used for filtrate concentrations (Tajuddin et al., 2003; Faujdar et al., 2020).

2.3 Supercritical extraction

Supercritical fluid extraction (SFE) uses fluids with solubility similar to liquid and diffusivity similar to gas. They can dissolve a wide range of natural products. SFE enables plant material to be processed at low temperatures, limiting thermal degradation and avoiding toxic solvents. Due to its attractive merits, such as low critical temperature, colorless nature, inertness, nontoxic nature, nonflammability, and low polarity, supercritical carbon dioxide (S-CO_2) has been widely used in SFE and has proven to be ideal for the extraction of nonpolar natural products.

Eugenol, eugenyl acetate, and caryophyllene are the main compounds extracted via this process. SFE is conducted in a jacketed stainless steel vessel. First, the extractor is filled with finely sieved, weighed samples of clove and CO_2 of weight 10 times more than the sample weight. The vessel is sealed to ensure no leakage. It is subjected to high pressure and high temperature that are optimal. After a sufficient period of time, when equilibrium is obtained in the extractor, a sample is taken for analysis. This method yielded around 12%–13% of extracts to the dry weight of clove samples. About 86.7% and 13.3% of eugenol and eugenyl acetate are obtained, respectively, under optimum conditions (Yazdani et al., 2005).

In another method, the cleaned, dried *Syzygium aromaticum* buds are ground into fine particles with the help of an electric grinder or in a crushing chamber. In a standard SFE setup, 1 h before the extraction, the cooling and heating baths are switched on to ensure that the device reaches the desired temperature. The extractor is filled with the sample material, and the desired pressure is applied with the aid of a pump. The CO_2 flow rate is determined by a gas flow meter after the system is put under static pressure. The extract is collected in a collection vial after its passage through a micro-metering valve, followed by an evaluation of the yield. This procedure is followed twice to ensure that the results are definitive (Johner et al., 2018; Chatterjee and Bhattacharjee, 2012).

2.4 Identification of chemical compounds

Among spices and herbs, clove was a major source of phenolic molecules such as hydroxybenzoic acids, flavonoids, hydroxycinnamic acids, and eugenol. Among them, the major bioactive compound in clove is eugenol. The flavonoids in clove are kaempferol and quercetin. Phenolic acids present are ferulic, caffeic, ellagic, and salicylic acids (Cortés-Rojas et al., 2014). Another important compound in clove essential oil is α-humulin that is present in concentrations up to 2.1%. β-Pinene, limonene, farnesol, benzaldehyde, 2-heptanone, and ethyl hexanoate are other volatile compounds present in clove essential oil in lower concentrations (Mohammed et al., 2016). GC-MS analysis revealed the presence of 36 CEO components isolated by hydrodistillation, including major compounds such as eugenol and ethyl hexanoate, α-humulene calacorene, humulene, and calamen-one (Batiha et al., 2020).

Eugenol is a phenolic compound of the phenylpropanoid class and the main component of the clove. It makes up 45%–90% of the essential oil. It is used as a preservative in the food industry, primarily because of its antioxidant properties and as a flavoring agent for food and cosmetics. The findings of a case study revealed that the ingestion of 5–10 mL of clove oil by a 2-year-old boy resulted in pain and crying and that conditions worsened 3 h after ingestion. The acidosis was marked, and the patient was in a deep coma. During 8 h, he suffered from a generalized seizure, and within 24 h, liver function failure and also disseminated intravascular coagulopathy were also noted (Nejad et al., 2017). The concentration of eugenol in human dental pulp fibroblasts of primary teeth at which toxicity occurs has been investigated. The toxicity test findings showed toxic effects on dental pulp fibroblasts even at very low concentrations of eugenol (0.06 μM) (Escobar-García et al., 2016).

3. Toxicity studies

This section summarizes findings regarding the toxicity of clove extracts from toxicological studies conducted in recent years. Some examples of such studies on clove extracts, along with their mode of administration, as well as results, are listed in Table 2.

3.1 Toxic effect of clove extract at cellular level

Extracts of *Syzygium aromaticum* flower buds have demonstrated toxic behavior against different cell lines resulting in cytotoxicity. Ethanolic clove extract has been shown to improve the cytotoxic potency of gemcitabine in human cervical line cancer cells. Gemcitabine was noted for its extremely cytotoxic nature against cancer and normal cells, though comparatively less than *Syzygium aromaticum*. The effect of ethanol extracts was studied on HeLa, a human cervical carcinoma cell line, and lymphocytes in another research work. Cell viability testing was conducted to verify the optimum state and composition of *Syzygium aromaticum* extract and gemcitabine. The cells were subjected to different concentrations of gemcitabine (1–100 mM) or clove extract (0.5–5 mg/mL)/gemcitabine (5 mM) alone, clove extract (2, 3, and 4 mg/mL) alone, and their combinations for 24 h accompanied by MTT. Likewise, gemcitabine or clove extract (0.5–8 mg/mL) was tested on lymphocytes for 24 h to assess cell viability.

Table 2 Toxicological studies available on clove extracts.

Formulation	Concentration	Exposure	Test animal	Toxicity	References
Powder	3 g for 21 days	Oral	Rabbit	Renal damage, possible trigger for renal fibrosis	Nwaopara et al. (2008)
Aqueous extract	LD50 Intraperitoneal—263 mg/kg Oral routes—2500 mg/kg	Oral	Albino rats	Affects body organs, their enzymes as well as various functions	Agbaje et al. (2009)
Oil extract	360 mg/kg/day clove oil	Oral	Rats	Oxidative stress—decreased level of plasma GSH and low activity of glutathione peroxidase, liver damage	Adeyemi et al. (2018)
Extract	4 mg/kg for 7 days	Oral	Rat	No toxicity	Moghimian et al. (2017)
Methanol extract	600 mg/kg LD50—6803.3 mg/kg Safe dose: 1000 mg/kg 600 mg/kg	Oral	Wistar rat	Decrease in body weight Dose-dependent increase in serum creatinine, sodium, and potassium with increasing concentration	Usman et al. (2018)
Oil extract	LD50 18.2 and 21.7 mg/mL at 96 h	Direct exposure	*Danio rerio* and *Poecilia reticulata*	Sensitive at higher concentrations	Doleželová et al. (2011)
Ethanolic extract	4 mg, 20 mg, and 100 mg/kg BW for 56 days	Oral	Male golden hamsters	20 and 100 mg of the extract completely inhibited the testicular activity	Choi et al. (2014)
Ethanol and acetone extract	LC_{50}, acetone extract—0.6% Ethanol extract—0.36%	Direct exposure	On-field slug, *Limax flavus*	A rise in the mortality rate with an increase in the concentration, highly teratogenic leading to sterility	Mobarak et al. (2015)

The findings showed a dose-dependent decrease in cell viability when HeLa cells were treated with 50% ethanolic clove extract at a dose of 0.5–2 mg/mL for a period of 24 h. However, an improvement in cell viability for treatment with clove extract at higher doses (>2 mg/mL) was also observed. This may be due to the overexpression of Bcl-2 or Bcl-x, which could protect against the release of mitochondrial cytochrome *c*, caspase activation, and DNA fragmentation caused by chemotherapy. The percentage of cell death using a mixture of 5 mM gemcitabine and 2 mg/mL of *Syzygium aromaticum* extract was considerably higher (~65%) than cell death when each of the compounds was used alone. A combination of 3 mgmL aromatic extract and 5 mM gemcitabine has been observed to cause a similar effect that resulted in 77% cell death, which was more than the cell death of either of them alone (Aisha et al., 2011).

3.2 Toxic effect of clove extract on reproduction in mice

Syzygium aromaticum buds are widely known as a traditional medicine for treating male sexual disorders. A study was on the impact of *Syzygium aromaticum* on Parkes (P) strain mice. The flower buds were extracted in hexane as solvent (olive oil as a vehicle) and used to understand the effect on reproduction behavior in male mice. A total of 32 mice of equal weight were randomly distributed into four groups, with Group I as vehicle control, i.e., olive oil for 35 days. Groups II, III, and IV were orally administered with the lipid-soluble components of *Syzygium aromaticum* flower bud at doses of 15, 30, and 60 mg/kg body weight, respectively, for 35 days. Several male reproductive endpoints were statistically analyzed using biochemical estimates, sperm analyses, and fertility tests. The findings show that this treatment did not induce changes in the bodyweight of the treated animals, indicating that the treatment did not have a systemic toxic effect on the mice. A slight increase in sperm motility in mice treated with a lower dose of *Syzygium aromaticum* than in controls could be attributed to a change in the working environment of the epididymis as there has also been an increase in the concentration of epididymal sialic acid in the treated mice. On the other hand, higher doses of *Syzygium aromaticum* greatly lowered the motility and number of spermatozoa in the epididymis, as well as the amount of sialic acid in the duct; the adverse effect of higher doses of the clove extract in sperm morphology is also apparent in the substantial increase in morphologically irregular spermatozoa in the treated mice relative to controls. A marked drop in sperm count in cauda epididymis of mice treated with higher doses of *Syzygium aromaticum* is likely to be attributed to the suppressive effect of *Syzygium aromaticum* hexane extract on spermatogenesis. There was a small rise in the weight of the seminal vesicle and fructose concentration and the litter in females sired by males was administered with a lower dose of the hexane extract. However, at higher doses of the extract, these parameters were greatly decreased in treated mice than in controls. As a result, low fertility, as concluded in this research, at higher doses of *Syzygium aromaticum* extract, may be due to the adverse effects of the hexane extract on spermatogenesis, epididymis, and seminal vesicle. These results indicate that the male reproductive indices are biphasic in response to the *Syzygium aromaticum* hexane extract with beneficial effects in lower doses (15 mg) and adverse effects at higher concentrations (30, 60 mg) (Mishra and Singh, 2008).

In another study by Dehghani et al. (2012), 40 male mice were treated with 250, 500, and 1000 mg/kg/day of water/alcohol extract of *Syzygium aromaticum* for a period of 34 days. The number of sperms, their motility, and the level of sex hormones (testosterone) were found to be reduced in the treated animals. A significant increase in the concentration of estradiol concentration was observed compared to the control. However, spermatogenesis was significantly affected.

3.3 Sub-chronic assessment of Wistar rats

A recent study examined the safety of a standardized polyphenolic extract of *Syzygium aromaticum*, commonly known as Clovinol, as tested by oral acute and sub-chronic toxicity studies on adult Wistar rats. The test material, Clovinol, was prepared by grinding the pure dried clove buds into powder, which is then extracted with three to five times excess of a mixture consisting of ethanol and water (70:30 v/v) under ambient conditions, followed by purification and spray drying to a fine clovinal powder. Both male and female Wistar rats weighing around 170 g were used as test subjects for the toxicological studies. Acute and sub-chronic toxicological studies were performed, and clovinol in distilled water was orally administered. In the acute toxic study, 40 rats were classified into four classes with a similar weight of five animals (170 g) per sex. The controls in Group I were nontreated, and clovinol was administered in Groups II, III, and IV at 1.0, 2.0, and 5.0 g/kg body weight, respectively. Post 10, 30, 60, 120, 240, and 360 min doses and again, for the second time, twice daily for mortality and once daily for clinical indications over the 14-day study duration, both animals have been monitored for mortality, clinical and behavioral indication. The classification remains the same for sub-chronic assessment except for Group I, which was treated with 1 mL of water; Groups II, III, and IV were treated orally with Clovinol for 90 days at 0.25, 0.5, and 1.0 g/kg body weight, respectively. During a 14-day study period, no mortality or adverse effect was observed, and no clinical abnormal signs could be detected on oral administration of Clovinol at 1, 2, and 5 g/kg body weights. The weight and intake of food remained the same. Furthermore, the fact that all animals survived until a necropsy, and the lack of any major changes in various parameters, including changes in body weight, food and water intake, hematological parameters, biochemical parameters, organ toxicity, and organ tissue histopathology, demonstrated the safety of clovinol after repeated oral administration for 90 days in the sub-chronic assessment. This study established that the phenolic-rich ethanolic extract of clove was safe up to 1000 mg/kg body weight/day (NOAEL—no observed adverse effect level) (Vijayasteltar et al., 2016; Mohammed et al., 2016).

Another study by Issac et al. (2015) also reports that the acute and sub-chronic toxicity of alovinol does not exhibit any significant changes compared to the control Wistar rats. There were no behavioral/clinical changes, and no mortality was observed in the sub-acute toxicity studies at 0.5, 1.0, and 2.5 g/kg body weight. Acute toxicity studies reveal no mortality or any adverse reactions up to 5 g/kg body weight of rats.

3.4 Toxic effect of clove extract on aquatic species

Clove oil is used in hatcheries for immobilizing fish, artificial reproduction, surgery, and suppressing senses. It can be considered an effective natural and local anesthetic. Anesthesia causes loss of consciousness, mitigates pain, and gives a calming effect when used. Clove oil has been reported to be a safe anesthetic for fish irrespective of species, with a few exceptions. It is essential to administer the right concentration of clove oil for anesthesia. The recommended concentrations are dependent on the size of the fish, species of the fish, exposure time, and temperature (Roubach et al., 2005).

When low concentrations (approximately 48 mg/L) of clove oil have been used on fish for a long time, alterations in tissues such as "oedema" and "telangiectasis appearance" have been observed. However, some studies have proven that clove extract can be used as an alternative to clove oil. It is considered a safe anesthetic because it ensures increased shelf life. There are no specific laws and regulations on the use of anesthesia, which is a matter of concern.

A lower concentration of anesthesia helps fish maintain equilibrium, swimming activity, and breathing. While handling and transporting, light anesthesia helps to reduce stress (Da Costa Castro et al., 2008). When fish are not given anesthetic during transportation, it causes an increase in ammonia concentration through excretion. Long-term exposure to this will affect fish gills. In these cases, lower concentrations of clove extract have the potential to act as anesthetic agents. A 17.5 mg/L of clove extract is considered an optimum concentration to prevent damage to fish gills (Balamurugan et al., 2016).

3.5 Other toxic effects

A recent study investigated the impact of *Syzygium aromaticum* flower on testicular function in male golden hamsters, which have a photoperiod-inhibited reproductive behavior, such as winter weather. Ethanol extract of *Syzygium aromaticum* was used on adult male golden hamsters (*Mesocricetus auratus*). Untreated or control hamsters were divided into long photoperiod (LP) and short photoperiod (SP) cycles and were supplied drinking water during their study period. The experimental animals were kept in SP and administered orally with *Syzygium aromaticum* extract for 56 consecutive days. They were provided with 4, 20, or 100 mg of *Syzygium aromaticum* extract per kg of body weight/day, respectively, at levels termed SPL, SPM, and SPH (L—low, M—moderate, H—high). This study demonstrated total inhibition of testicular function in some male golden hamsters preserved in SP. SP normally inhibits the reproductive activity of the animals. Still, SP did not fully impair the sexual activity due to the lack of time during the present experiment's investigation process, resulting in partial regression. The administration of *Syzygium aromaticum* resulted in full involution in some animals treated with moderate and large doses of *Syzygium aromaticum* extract. This suggests that the animal was sensitive enough to respond to the *Syzygium aromaticum* extract administered dose. The golden hamsters living in LP have a continuous large mass of testicles, indicating their effect on functional activity. It is well known that the development of melatonin is related to the time cycle and is thought to be an obvious involuting component in the sexual activity of golden hamsters. Despite the imperfect influence of SP on the development of animals, the secretion of melatonin may be correlated with the complete regression of the testes caused by *Syzygium aromaticum* extract. The present investigation reflects the possible regressive role of *Syzygium aromaticum* extract in the sexual function of male golden hamsters. The finding is based on decreased testicular volume and decreased accessory sex organs such as the seminal vesicle and epididymis (Choi et al., 2014). Different forms of clove tested for their toxic effect on red ants are represented in Fig. 1.

The relative toxicity of *Syzygium aromaticum* extract on-field slug, *Limax flavus*, was tested under laboratory and field conditions with ethanol and acetone extracts. The plant extract was prepared by

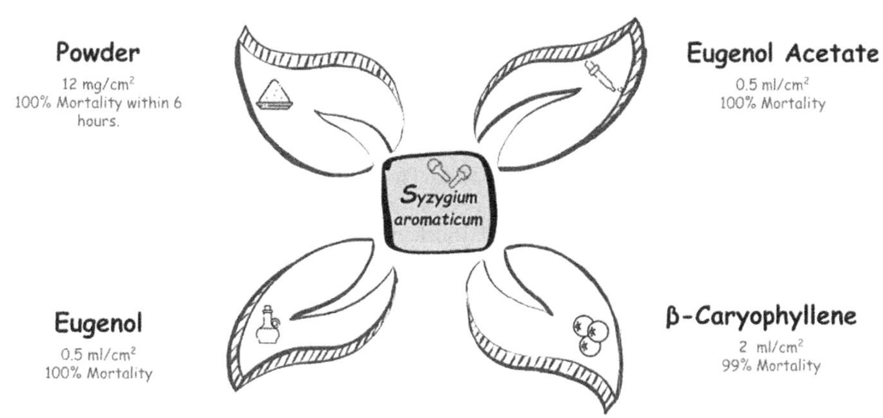

FIG. 1

Toxicity effect of different forms of *Syzygium aromaticum* on red ants.

grinding the dried buds of *Syzygium aromaticum*, followed by 95% solvent extraction using a homogenizer followed by drying. Results indicated a rise in the mortality rate with an increase in the concentration and complete mortality at the tested concentration of 2% for both extracts. The LC_{50} of acetone extract was found to be 0.6% and of the ethanol extract was 0.36%. The study indicated that the ethanol extract caused a significant reduction in population density and was highly teratogenic, leading to sterility (Mobarak et al., 2015).

4. Conclusion

This chapter discusses the toxicity of different types of clove extracts. Clove and its extracts are widely studied for their beneficial properties, while the data on its toxicity is limited. Clove is considered binate due to its vast benefits and properties when taken at the recommended dosage. Clove extracts are toxic when ingested at higher doses, mostly due to the active component, eugenol. Clove is commonly used as an aphrodisiac, and its capacity to stimulate sexual activity has been shown in Wistar rats and mice. Although a lower dose of hexane extract of clove increased testosterone production, higher doses were observed to cause a decrease, disrupting mice spermatogenesis. This toxic nature of clove extracts can be used for cell-targeted treatment and finds use to control pests such as red ants and soil slugs. While the study on the characteristics of cloves and their extracts is still underway, extreme caution should be exercised when used. The various properties and activities of clove extracts are still far from fully understood, and longer term human studies should be pursued in vitro and in vivo.

Acknowledgments

We gratefully acknowledge the support of Pratiksha Trust, Bangalore toward this research.

References

Adeyemi, J.A., Arowolo, O.K., Olawuyi, S.T., Alegbeleye, D., Ogunleye, A., Bamidele, O.S., Adedire, C.O., 2018. Effect of co-administration of Green tea (Camellia sinensis) on clove- (Syzygium aromaticum) induced hepatotoxicity and oxidative stress in Wistar rats. Indian J. Physiol. Pharmacol. 62 (2), 195–201. http://www.ijpp.com/IJPP%20archives/2018_62_2/195-201.pdf.

Agbaje, E.O., Adeneye, A.A., Daramola, A.O., 2009. Biochemical and toxicological studies of aqueous extract of Syzigium aromaticum (L.) Merr. & Perry (Myrtaceae) in rodents. Afr. J. Tradit. Complement. Altern. Med. 6 (3), 241–254. http://www.bioline.org.br/pdf?tc09033.

Aisha, A.F.A., Nassar, Z.D., Siddiqui, M.J., Abu-Salah, K.M., Alrokayan, S.A., Zhari, I., Shah, A.M.A.M., 2011. Evaluation of antiangiogenic, cytotoxic and antioxidant effects of Syzygium aromaticum L. extracts. Asian J. Biol. Sci., 282–290. https://doi.org/10.3923/ajbs.2011.282.290.

Balamurugan, J., Ajith Kumar, T.T., Prakash, S., Meenakumari, B., Balasundaram, C., Harikrishnan, R., 2016. Clove extract: a potential source for stress free transport of fish. Aquaculture 454, 171–175. https://doi.org/10.1016/j.aquaculture.2015.12.020.

Batiha, G.E.S., Alkazmi, L.M., Wasef, L.G., Beshbishy, A.M., Nadwa, E.H., Rashwan, E.K., 2020. Syzygium aromaticum l. (myrtaceae): traditional uses, bioactive chemical constituents, pharmacological and toxicological activities. Biomol. Ther. 10 (2). https://doi.org/10.3390/biom10020202.

Chatterjee, D., Bhattacharjee, P., 2012. Supercritical carbon dioxide extraction of eugenol from clove buds. Food Bioprocess Technol. 6 (10), 2587–2599.

Chaula, D., Laswai, H., Chove, B., Dalsgaard, A., Mdegela, R., Jacobsen, C., Hyldig, G., 2019. Effect of clove (Syzygium aromaticum) and seaweed (Kappaphycus alvarezii) water extracts pretreatment on lipid oxidation in sun-dried sardines (Rastrineobola argentea) from Lake Victoria, Tanzania. Food Sci. Nutr. 7 (4), 1406–1416. https://doi.org/10.1002/fsn3.975.

Chniguir, A., Zioud, F., Marzaioli, V., El-Benna, J., Bachoual, R., 2019. Syzygium aromaticum aqueous extract inhibits human neutrophils myeloperoxidase and protects mice from LPS-induced lung inflammation. Pharm. Biol. 57 (1), 56–64. https://doi.org/10.1080/13880209.2018.1557697.

Choi, D., Roh, H.S., Kang, D.W., Lee, J.S., 2014. The potential regressive role of *Syzygium aromaticum* on the reproduction of male golden hamsters. Dev. Reprod. 18, 57–64.

Cortés-Rojas, D.F., de Souza, C.R.F., Oliveira, W.P., 2014. Clove (Syzygium aromaticum): a precious spice. Asian Pac. J. Trop. Biomed. 4 (2), 90–96. https://doi.org/10.1016/S2221-1691(14)60215-X.

Da Costa Castro, A.L., De Farias Diniz, A., Martins, I.Z., Vendel, A.L., De Oliveira, T.P.R., De Lucena Rosa, I.M., 2008. Assessing diet composition of seahorses in the wild using a non destructive method: Hippocampus reidi (Teleostei: Syngnathidae) as a study-case. Neotrop. Ichthyol. 6 (4), 637–644. https://doi.org/10.1590/s1679-62252008000400012.

Dehghani, F., Heshmatpour, A., Panjehshahin, M.R., Talaei-Khozani, T., 2012. Toxic effects of water/alcoholic extract of Syzygium aromaticum on sperm quality, sex hormones and reproductive tissues in male mouse. IUFS J. Biol. 71 (2), 95–102.

Doleželová, P., Mácová, S., Plhalová, L., Pištěková, V., Svobodová, Z., 2011. The acute toxicity of clove oil to fish Danio rerio and Poecilia reticulata. Acta Vet. Brno 80 (3), 305–308. https://doi.org/10.2754/avb201180030305.

Escobar-García, M., Rodríguez-Contreras, K., Ruiz-Rodríguez, S., Pierdant-Pérez, M., Cerda-Cristerna, B., Pozos-Guillén, A., 2016. Eugenol toxicity in human dental pulp fibroblasts of primary teeth. J. Clin. Pediatr. Dent. 40 (4), 312–318. https://doi.org/10.17796/1053-4628-40.4.312.

Faujdar, S.S., Bisht, D., Sharma, A., 2020. Antibacterial activity of *Syzygium aromaticum* (clove) against uropathogens producing ESBL, MBL, and AmpC beta-lactamase: are we close to getting a new antibacterial agent? J. Family Med. Prim. Care 9 (1), 180–186.

Issac, A., Gopakumar, G., Kuttan, R., Maliakel, B., Krishnakumar, I.M., 2015. Safety and anti-ulcerogenic activity of a novel polyphenol-rich extract of clove buds (Syzygium aromaticum L). Food Funct. 6 (3), 842–852. https://doi.org/10.1039/c4fo00711e.

Johner, J.C.F., Hatami, T., Carvalho, P.I.N., Meireles, M.A.A., 2018. Impact of grinding procedure on the yield and quality of the extract from clove buds using supercritical fluid extraction. Open Food Sci. J., 1–7. https://doi.org/10.2174/1874256401810010001.

Khajeh, M., Yamini, Y., Sefidkon, F., Bahramifar, N., 2004. Comparison of essential oil composition of Carum copticum obtained by supercritical carbon dioxide extraction and hydrodistillation methods. Food Chem. 86 (4), 587–591. https://doi.org/10.1016/j.foodchem.2003.09.041.

Khalil, A.A., Rahman, U.U., Khan, M.R., Sahar, A., Mehmood, T., Khan, M., 2017. Essential oil eugenol: sources, extraction techniques and nutraceutical perspectives. RSC Adv. 7 (52), 32669–32681. https://doi.org/10.1039/c7ra04803c.

Mbaveng, A.T., Kuete, V., 2017. Syzygium aromaticum. In: Medicinal Spices and Vegetables From Africa: Therapeutic Potential Against Metabolic, Inflammatory, Infectious and Systemic Diseases. Elsevier Inc, pp. 611–625, https://doi.org/10.1016/B978-0-12-809286-6.00029-7.

Mishra, R.K., Singh, S.K., 2008. Safety assessment of Syzygium aromaticum flower bud (clove) extract with respect to testicular function in mice. Food Chem. Toxicol. 46 (10), 3333–3338. https://doi.org/10.1016/j.fct.2008.08.006.

Mobarak, S.A., Kandil, R.A., Kadel, S.M., 2015. Effect of clove plant extract on land slug and their reproductive system under laboratory and field conditions. J. Am. Sci. 11 (12), 1–6.

Moghimian, M., Abtahi-Evari, S.H., Shokoohi, M., Amiri, M., Soltani, M., 2017. Effect of Syzygium aromaticum (clove) extract on seminiferous tubules and oxidative stress after testicular torsion in adult rats. Physiol. Pharmacol. 21 (4), 343–350. http://phypha.ir/ppj/article-1-1257-en.pdf.

Mohammed, K.A.K., Abdulkadhim, H.M., Noori, S.I., 2016. Chemical composition and antibacterial effects of clove (*Syzygium aromaticum*) flowers. Int. J. Curr. Microbiol. Appl. Sci. 5, 483–489.

Nejad, M.S., Özgüneş, H., Başaran, N., 2017. Pharmacological and toxicological properties of eugenol. Turkish J. Pharm. Sci. 14 (2), 201–206.

Nwaopara, A.O., Odike, M.A.C., Inegbenebor, U., Nwaopara, S.O., Ewere, G.I., 2008. A comparative study on the effects of excessive consumption of ginger, clove, red pepper and black pepper on the histology of the kidney. Pak. J. Nutr. 7 (2), 287–291. https://doi.org/10.3923/pjn.2008.287.291.

Puangpronpitag, D., Niamsa, N., Sittiwet, C., 2009. Anti-microbial properties of clove (Eugenia caryophyllum Bullock and Harrison) aqueous extract against food-borne pathogen bacteria. Int. J. Pharmacol. 5 (4), 281–284. https://doi.org/10.3923/ijp.2009.281.284.

Quan, L., Li, S., Tian, S., Xu, H., Lin, A., Gu, L., 2004. Determination of organochlorine pesticides residue in ginseng root by orthogonal array design Soxhlet extraction and gas chromatography. Chromatographia 59 (1–2), 89–93.

Roubach, R., Gomes, L.C., Leão Fonseca, F.A., Val, A.L., 2005. Eugenol as an efficacious anaesthetic for tambaqui, Colossoma macropomum (Cuvier). Aquacult. Res. 36 (11), 1056–1061. https://doi.org/10.1111/j.1365-2109.2005.01319.x.

Tajuddin, Ahmad, S., Latif, A., Qasmi, I.A., 2003. Aphrodisiac activity of 50% ethanolic extracts of *Myristica fragrans* Houtt. (nutmeg) and *Syzgium aromaticum* (L) Merr and Perry (clove) in male mice: a comparative study. BMC Complement Altern. Med. 3 (6).

Usman, H.B., Ndatsu, Y., Abubakar, A., Abubakar, H., 2018. Kidney biomarkers as clue from toxicological studies of methanol extract of Syzygium aromaticum seed on Wister rats. Lapai J. Appl. Nat. Sci. 3 (1), 191–199.

Vijayasteltar, L., Nair, G.G., Maliakel, B., Kuttan, R., Krishnakumar, I.M., 2016. Safety assessment of a standardized polyphenolic extract of clove buds: subchronic toxicity and mutagenicity studies. Toxicol. Rep. 3, 439–449. https://doi.org/10.1016/j.toxrep.2016.04.001.

Wenqiang, G., Shufen, L., Ruixiang, Y., Shaokun, T., Can, Q., 2007. Comparison of essential oils of clove buds extracted with supercritical carbon dioxide and other three traditional extraction methods. Food Chem., 1558–1564. https://doi.org/10.1016/j.foodchem.2006.04.009.

Yazdani, F., Mafi, M., Farhadi, F., Tabar-Heidar, K., Aghapoor, K., Mohsenzadeh, F., Darabi, H.R., 2005. Supercritical CO_2 extraction of essential oil from clove bud: effect of operation conditions on the selective isolation of eugenol and eugenyl acetate. Z. Naturforsch. Sect. B J. Chem. Sci. 60 (11), 1197–1201. https://doi.org/10.1515/znb-2005-1113.

CHAPTER

Antiviral properties of clove (*Syzygium aromaticum*)

40

Raya Soltane[a,b], Mtat Dalila[c], Abdelwaheb Dhahri[d], Ahlam Alasiri[a], Rami Adel Pashameah[a], Nuha Almulla[a], Amani Chrouda[e], and Ahmed Mostafa[f]

Department of Basic Sciences, Adham University College, Umm Al-Qura University, Makkah, Saudi Arabia[a] Department of Biology, Faculty of Sciences, Tunis El Manar University, Tunis, Tunisia[b] Department of Chemistry, Preparatory Institute for Engineering Studies of Nabeul (IPEIN), University of Carthage, Tunis, Tunisia[c] Department of Polymer Materials Engineering, University of Lyon, UMR CNRS 5223, Lyon, Villeurbanne, France[d] Department of Chemistry, College of Science Al-Zulfi, Majmaah University, Al-Majmaah, Saudi Arabia[e] Center of Scientific Excellence for Influenza Viruses, Water Pollution Research Department, Environmental and Climate Change Research Institute, National Research Centre, Giza, Egypt[f]

1. Introduction

Viral diseases are of great concern to human health worldwide due to the quick emergence of drug-resistant viral strains. The development of effective treatments to combat human viral infections relies on identifying the new therapeutic agents against viral infections at low noncytotoxic concentrations in vitro and in vivo. Additionally, herbal therapeutic products have been documented as an essential source for discovering novel therapeutic molecules that have been used to inhibit and control viral infections (Astani et al., 2011).

For thousands of years, certain medical plants have been used in folk medicine; in particular, traditional medicines using natural products revealed to have antiviral compounds in vitro (Locher et al., 1996). For example, encephalomyocarditis virus and human immunodeficiency virus by *Opuntia streptacantha* extracts have been reported (Ahmad et al., 1996). At the same time, thyme and standardized *Echinacea purpurea* extract showed remarkable antiviral activity against viruses that cause respiratory infections (Pleschka et al., 2009; Lenz et al., 2018).

Syzygium aromaticum, known as clove, is a traditional spice belonging to the family of Myrtaceae that is original to the Maluku Islands in Indonesia, but has newly been cultivated in diverse places worldwide (Batiha et al., 2019). Numerous therapeutic usages of *S. aromaticum* have been documented. It has been used for food preservation and possesses many pharmacological activities. It is generally used in medicine as an antiseptic against infectious diseases due to the antimicrobial activities against oral microbes. Clove is moreover used in the food industry due to its antimicrobial activities for increasing shelf life.

In light of the documented broad antiviral activity of *S. aromaticum* extracts and its molecular constituents described in the scientific literature (Fig. 1), we will discuss the antiviral activity potency of *S. aromaticum* in the present chapter.

Clove (*Syzygium aromaticum*). https://doi.org/10.1016/B978-0-323-85177-0.00005-7

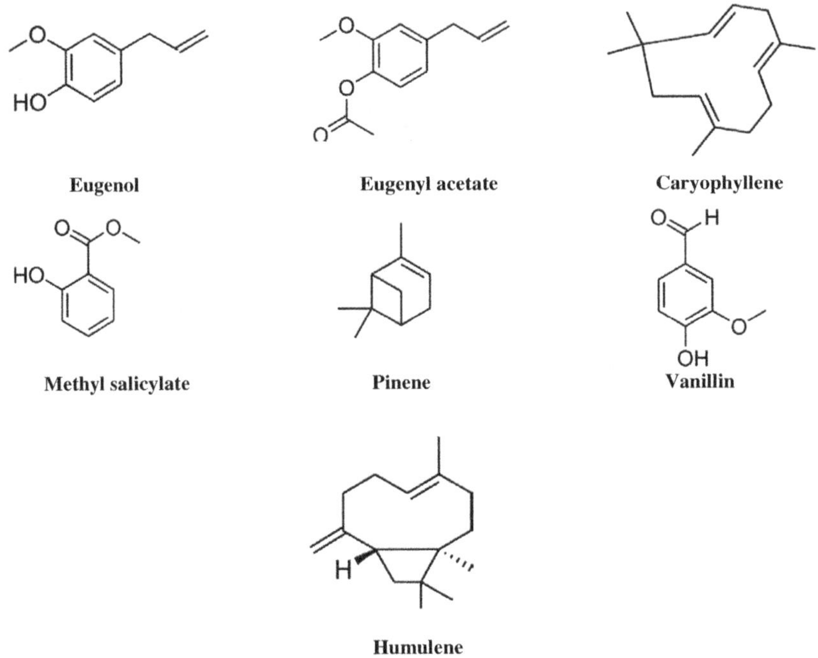

FIG. 1

Active ingredients in *S. aromaticum* extracts.

2. Antiherpes simplex virus (HSV) activity of *S. aromaticum*

S. aromaticum has been reported to have antiherpes simplex virus (HSV) activity combined with acyclovir (Kurokawa et al., 1995). Another research revealed that the antiviral efficacy of eugeniin, the compound isolated from the extracts of *S. aromaticum*, was identified as the anti-HSV compound. Five μg/mL concentration was active against HSV. The inhibitory action of eugeniin is on the viral DNA synthesis by acting as a selective inhibitor of the HSV-1 DNA polymerase (Kurokawa et al., 1998). Eugeniin also inhibited the development of acyclovir-phosphonoacetic acid-resistant HSV-1, thymidine kinase-deficient HSV-1, and wild HSV-2.

Eugenol proved to be active in an in vivo model of HSV-1 infection. Eugenol treatment considerably tardy the growth of herpetic keratitis in the cornea of HSV-1 infected mice, demonstrating that this composite can help treat ocular herpetic infections.

Eugenol may not be beneficial for internal uses as the antiviral effect is at least partly due to the virucidal effect, but taking into account the results of in vivo tests plus its stability at room temperature. It could be used for topical cures, principally in mixture with other drugs; then, its antiviral activity is slightly comparable to other drugs such as acyclovir.

3. Antiviral activity of green silver nanoparticles produced using *Syzygium aromaticum* aqueous buds extract

Nanobiotechnology can transform a broad group of applications like drug delivery (Elechiguerra et al., 2005). *S. aromaticum* extract was used as a reducing and stabilizing agent for the synthesis of silver nanoparticles. In current investigations, the antiviral activity of clove buds silver nanoparticles was inspected in vitro. The study evaluated the antiviral properties of clove buds extract modified silver nanoparticles against Newcastle disease virus (NDV) infection.

S. aromaticum modified AgNP provides the increased efficacy as an inhibitory agent against viral infection. The buds extract of extensively available therapeutic plant *S. aromaticum* could be used for the biosynthesis of silver nanoparticles with significant antiviral activity at higher concentrations throughout both in vitro and in vivo trials (Mehmood et al., 2020).

S. aromaticum produces stable silver nanoparticles, reducing aqueous silver ions shown by the dark brown solution after the $AgNO_3$ aqueous solution was added to the bud's extract. A study estimated the antiviral activity of *S. aromaticum* against NDV infection. The in vitro activity was measured by performing the spot assay and micro hemagglutination test. It was observed that stock solution of the clove extract silver nanoparticles (CESN) and 1:2 dilutions prevent the visible clusters/agglomerates of Chicken Red Blood Cells (Mehmood et al., 2020). Decrease in concentration directly relates to the decrease in antiviral activity and cytotoxicity CESN has revealed promising efficiency as an antiviral agent and can be used as an active antiviral alternative by minimalizing its cytotoxic effects.

4. Antiviral activity of eugenol

The essential bioactive agent in the clove is eugenol (4-allyl-1-hydroxy-2-methoxybenzene). It is the main constituent of the oil of cloves and is present in the essential oils of many plants, counting cinnamon, basil, and nutmeg. However, it also contains the essential vegetal sources of phenolic compounds, i.e., flavonoids, hydroxybenzoic acids, hydroxycinnamic acids, and hydroxyphenylpropens (Neveu et al., 2010). Numerous studies have reported eugenol counting therapeutic properties in dentistry (Woody and Davis, 1992), antimicrobial activities (Didry et al., 1994), inhibitory effects on lipid peroxidation (Nagababu and Laksmaiah, 1992), and peripheral vasodilator properties.

Eugenol has revealed encouraging activity against feline calicivirus (Aboubakr et al., 2016), tomato yellow leaf curl virus (Sun et al., 2016), influenza A virus (IAV) (Dai et al., 2013), Herpes Simplex virus type 1, Herpes Simplex virus 2, four airborne phages (Turgeon et al., 2016) as well as larvicidal activity against *Aedes aegypti* (Pandey et al., 2013). Dai et al. (2013) showed that eugenol inhibits autophagy and IAV duplication by interfering with ERK, p38MAPK, and IKK/NF-κB signaling pathways.

4.1 *In vitro* and *in vivo* antiviral activity of eugenol against human herpesvirus

Eugenol has been revealed to have antiviral activities against HSV-1 and HSV-2 (Benencia and Courrèges, 2000). It was repeatedly tested for antiviral activity against HSV-1 and HSV-2 viruses. In vitro, the replication of several enveloped DNA and RNA viruses, including HSV was studied (Reichling et al., 2009). On the other hand, various studies have revealed that eugenol induced glutathione *S*-transferase (GST) in rat liver in vivo (Palamara et al., 1995), leading to interference with the in vitro replication of HSV-1 due to the antiviral activity of the formed glutathione. Inhibitory concentration 50% values for the anti-HSV effects of eugenol were 25.6 and 16.2 mg/mL for HSV-1 and HSV-2 correspondingly, 250 mg/mL being the maximum quantity at which cytotoxicity was verified.

Eugenol was documented to have a virucidal effect against HSV and produced a complete reduction of HSV cuts after topically applied to the infected area by the virus (Tragoolpua and Jatisatienr, 2007). It was found to be safe at a wide range of tested concentrations. Advanced pharmacological studies showed that eugenol has powerful inhibitory properties against replicating either RNA or DNA viruses (Mancini et al., 2009).

4.2 Eugenol inhibitor of the Ebola virus *in vitro*

Since the Ebola virus (EBOV) epidemic in West Africa, there has been considerable effort to develop drugs to treat the EBOV. However, there is no FDA-approved treatment so far. Eugenol is much smaller and structurally unlike any compound that has been previously known as an inhibitor of EBOV, then it can offer novel mechanistic visions.

4.3 Inhibition of dengue virus protease by eugenol

Dengue virus (DENV) infection is a severe public health problem for several countries. There is still no specific antiviral treatment existing against these infections (Saleem et al., 2019). Novel use of clove as a larvicidal agent is an exciting approach to combat Dengue. Eugenol has been evaluated for its larvicidal activities against *Aedes aegypti* (Waliwitiya et al., 2009), with larvicidal activities as low as 44.5 ppm.

Eugeniin was also found to act as a potent inhibitor of the protease of the Dengue virus. The IC_{50} values of eugeniin against the proteases of DENV serotype-2 and serotype-3 were 94.7 nM and 7.5 µM, respectively. Therefore, considering the importance of DENV protease for the viral duplication cycle, eugeniin was suggested as a promising treatment in the context of anti-DENV therapeutics progress.

5. Antiviral activities of clove oil

Essential oils are well-identified inhibitors of microorganisms (Burt, 2004). Monoterpenes in clove essential oil such as eugenol, thymol, β-pinene, and limonene show the greatest important role as antiviral agents. They can directly inhibit the virus contagion mask viral structures, which are essential for adsorption or entry into the host cell (Schnitzler and Astani, 2014). It can dissolve or block the entry of the viral particle and denature its proteins or nucleic acids (DNA or RNA). In addition, it might display a role in denaturing elicitor proteins formed by these pathogens and ending the cell death program induction in the infected plant cell. The credible antimicrobial action for clove oil is related to eugenol, which contains around 68.6% of clove oil content (Iftikhar et al., 2013).

Clove essential oil nanoemulsion has revealed that good inhibitory potential against potato at 5% and 10% significantly reduced potato leafroll virus infection in potato plants, and also, the effect was found to be concentration-dependent. In all cases, clove oil nanoemulsion had a necessary inhibitory consequence on potato virus-Y (PVY) infection either in vitro or in vivo trials; it condensed the quantity of local lesions formed by PVY on Ch. quinoa and significantly increased the ratio of inhibition.

Hence, the presence of clove oil in the nano form helped increase the action of the inhibitory consequence where the use of nanoparticles offers a sustained and slow release of dynamic elements (Bilia et al., 2014). Incorporating eugenol oil nanoemulsion into culture intermediate was very effective and low cost to produce banana bunchy top virus-free banana plants and beneficial to reduce the frequency of fungal contamination (Mokbel et al., 2017).

6. Antiviral effects of *Syzygium aromaticum* against severe acute respiratory syndrome coronavirus 2 (SARS-CoV-2)

Since the epidemic of the coronavirus disease 2019 (COVID-19), scientists have been trying to study numerous active composites found in plants that have the potential to inhibit the proliferation of SARS-CoV-2, the cause of COVID-19. Treatment for the disease is mainly supportive. Scientists worldwide have been vigorously trying to develop efficient antiviral agents for the treatment of SARS-CoV-2.

The search for plant-based antivirals against the SARS-CoV-2 is promising, as numerous plants have been revealed to possess antiviral activities against beta coronaviruses (beta-CoVs). In China,

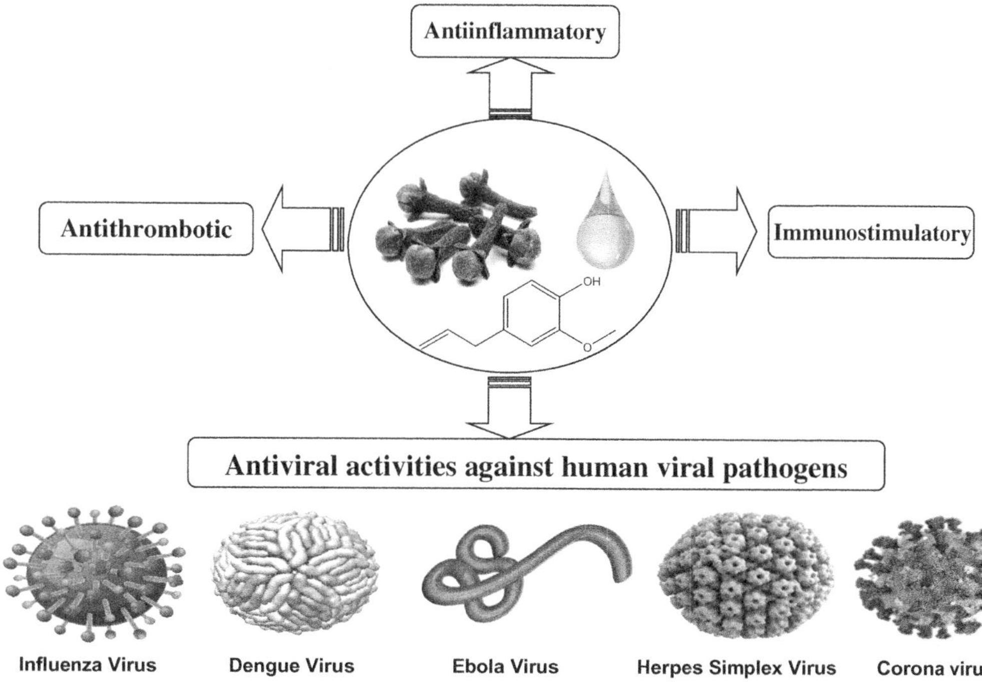

FIG. 2

Biological and antiviral effects of clove and its bioactive constituent.

traditional herbal medicine has been regularly used in combination with conventional medicine to treat SARS. There is evidence showing that herbal medicine is effective (Lin et al., 2003). It has been shown that natural plants contain antiviral activities to other coronaviruses (McCutcheon et al., 1995), and the mechanism of action of these herbal products is mainly through inhibition of viral duplication (Jassim and Naji, 2003).

Essential oils and their compound elements have been proposed for SARS-CoV-2 infections due to their lipophilic nature. These oils have the potential to enter the viral envelope leading to membrane disruption. Essential oils and their constituent phytochemicals disturb viral duplication and benefit the host respiratory system via mucus lysis and bronchodilation.

Silva et al. (2020) used molecular docking methods to screen the anti-SARS-CoV-2 efficacies of eugenol against the viral-specific proteins. Docking scores assumed that these composites have binding affinities toward SARS-CoV-2 spike protein, main protease (M^{pro}), RNA-dependent RNA polymerase, and human ACE-2 proteins (Silva et al., 2020).

7. Conclusion

In the present chapter, we brought together the antiviral effects of clove and their constituents (Fig. 2). Eugenol is the active element of the clove, and the FDA considered it a nontoxic constituent. Eugenol showed to be effective in an in vivo model of HSV-1 infection. The fast-developing fields of pharmacological and nanotechnology are bound to impact the designing of clove extract formulations. This will also negate the few cytotoxic properties of these compounds. The current chapter suggests that there will be merit in return to natural medicine to treat antiviral diseases. *S. aromaticum* can be considered a potential drug candidate; but new clinical and toxicological studies are necessary, as various toxicity issues in animals have been documented.

References

Aboubakr, H.A., Nauertz, A., Luong, N.T., Agrawal, S., El-Sohaimy, S.A., Youssef, M.M., et al., 2016. In vitro antiviral activity of clove and ginger aqueous extracts against feline Calicivirus, a surrogate for human norovirus. J. Food Prot. 79 (6), 1001–1012.

Ahmad, A., Davies, J., Randall, S., Skinner, G.R.B., 1996. Antiviral properties of extract of *Opuntia streptacantha*. Antivir. Res. 30, 75–80.

Astani, A., Reichling, J., Schnitzler, P., 2011. Screening for antiviral activities of isolated compounds from essential oils. Evid. Based Complement. Alternat. Med. 2011, 1–8.

Batiha, G.E.S., Beshbishy, A.A., Tayebwa, D.S., Shaheen, M.H., Yokoyama, N., Igarashi, I., 2019. Inhibitory effects of *Syzygium aromaticum* and *Camellia sinensis* methanolic extracts on the growth of *Babesia* and *Theileria* parasites. Ticks Tick. Borne Dis. 10, 949–958.

Benencia, F., Courrèges, M.C., 2000. In vitro and in vivo activity of eugenol on human herpesvirus. Phytother. Res. 14, 495–500.

Bilia, A.R., Guccione, C., Isacchi, B., Righeschi, C., Firenzuoli, F., Bergonzi, M.C., 2014. Essential oils loaded in nanosystems: a developing strategy for a successful therapeutic approach. In: Gori, L. (Ed.), Evidence-Based Complementary and Alternative Medicine, pp. 1–14.

Burt, S., 2004. Essential oils: their antibacterial properties and potential applications in food—a review. Int. J. Food Microbiol. 94, 223–253.

Dai, J.P., Zhao, X.F., Zeng, J., Wan, Q.Y., Yang, J.C., Li, W.Z., et al., 2013. Drug screening for autophagy inhibitors based on the dissociation of Beclin1-Bcl2 complex using BiFC technique and mechanism of eugenol on anti-influenza a virus activity. PLoS One 8 (4), e61026.

Didry, N., Dubreuil, L., Pinkas, M., 1994. Activity of thymol, carvacrol, cinnamaldehyde and eugenol on oral bacteria. Pharm. Acta Helv. 69, 25–28.

Elechiguerra, J.L., Burt, J.L., Morones, J.R., Camacho-Bragado, A., Gao, X., Lara, H.H., Yacaman, M.J., 2005. Interaction of silver nanoparticles with HIV-1. J. Nanobiotechnol. 3 (1), 6.

Iftikhar, S., Shahid, A.A., Javed, S., Nasir, I.A., Tabassum, B., Haider, M.S., 2013. Essential oils and latices as novel antiviral agent against *Potato leaf roll virus* and analysis of their phytochemical constituents responsible for antiviral activity. J. Agric. Sci. 5 (7), 167–188.

Jassim, S.A., Naji, M.A., 2003. Novel antiviral agents: a medicinal plant perspective. J. Appl. Microbiol. 95, 412–427.

Kurokawa, M., Nagasaka, K., Hirabayashi, T., Uyama, S., Sato, H., Kageyama, T., 1995. Efficacy of traditional herbal medicines in combination with acyclovir against herpes simplex virus type 1 infection in vitro and in vivo. Antivir. Res. 27 (1–2), 19–37.

Kurokawa, M., Hozumi, T., Basnet, P., Nakano, M., Kadota, S., Namba, T., 1998. Purification and characterization of eugeniin as an anti-herpesvirus compound from Geum japonicum and *Syzygium aromaticum*. J. Pharmacol. Exp. Ther. 284 (2), 728–735.

Lenz, E., Müller, C., Mostafa, A., Dzieciolowski, J., Kanrai, P., Dam, S., Cwientzek, U., Prenner, L.-N., Pleschka, S., 2018. Authorised medicinal product Aspecton® Oral Drops containing thyme extract KMTv24497 shows antiviral activity against viruses which cause respiratory infections. J. Herb. Med. 13, 26–33.

Lin, L., Han, Y., Yang, Z.M., 2003. Clinical observation on 103 patients of severe acute respiratory syndrome treated by integrative traditional Chinese and Western medicine. Zhongguo Zhong Xi Yi Jie He Za Zhi 23, 409–413.

Locher, C.P., Witvrow, M., De Bethune, M.P., et al., 1996. Antiviral activity of Hawaiian medicinal plants against human immunodeficiency virus type-1 (HIV-1). Phytomedicine 2, 259–264.

Mancini, D.A.P., Torres, R.P., Pinto, J.R., Mancini-Filho, J., 2009. Inhibition of DNA virus: herpes-1 (HSV-1) in cellular culture replication through an antioxidant treatment extracted from rosemary spice. Braz. J. Pharm. Sci. 45 (1), 127–133.

McCutcheon, A.R., Roberts, T.E., Gibbons, E., Ellis, S.M., Babiuk, L.A., Hancock, R.E., Towers, G.H., 1995. Antiviral screening of British Columbian medicinal plants. J. Ethnopharmacol. 49, 101–110.

Mehmood, Y., Farooq, Y., Yousaf, H., Riaz, H., Mahmood, K.R., Nawaz, A., Abid, Z., Gondal, M., Shamshad Malik, N., Barkat, K., Khalid, I., 2020. Antiviral activity of green silver nanoparticles produced using aqueous buds extract of *Syzygium aromaticum*. Pak. J. Pharm. Sci. 33, 839–845.

Mokbel, S.A., Khalil, A.A., El-Shazly, M.A., 2017. Efficiency of eugenol oil nanoemulsion against *Banana bunchy top virus* and contamination with fungi in plant tissue culture. Arab J. Biotech. 20 (1), 33–50.

Nagababu, E., Laksmaiah, N., 1992. Inhibitor effect of eugenol on non-enzymatic lipid peroxidation in rat liver mitochondria. Biochem. Pharmacol. 43, 2393–2400.

Neveu, V., Perez-Jiménez, J., Vos, F., Crespy, V., Du, C.L., Mennen, L., Knox, C., Eisner, R., Cruz, J., Wishart, D., Scalbert, A., 2010. Phenol-explorer: an online comprehensive database on polyphenol contents in foods. Database. (bap 024) 1–9.

Palamara, A.T., Perno, C.F., Ciriolo, M.R., et al., 1995. Evidence for antiviral activity of glutathione: in vitro inhibition of herpes simplex virus type 1 replication. Antivir. Res. 27, 237–253.

Pandey, S.K., Tandon, S., Ahmad, A., Singh, A.K., Tripathi, A.K., 2013. Structure-activity relationships of monoterpenes and acetyl derivatives against *Aedes aegypti* (Diptera: Culicidae) larvae. Pest Manag. Sci. 69 (11), 1235–1238.

Pleschka, S., Stein, M., Schoop, R., Hudson, J.B., 2009. Anti-viral properties and mode of action of standardized *Echinacea purpurea* extract against highly pathogenic avian influenza virus (H5N1, H7N7) and swine-origin H1N1 (S-OIV). Virol. J. 6, 197.

Reichling, J., Schnitzler, P., Suschke, U., Saller, R., 2009. Essential oils of aromatic plants with antibacterial, antifungal, antiviral, and cytotoxic properties—an overview. Forsch. Komplementmed. 16, 79–90.

Saleem, H.N., Batool, F., Mansoor, H.J., Shahzad-ul-Hussan, S., Saeed, M., 2019. Inhibition of dengue virus protease by eugeniin, isobiflorin, and biflorin isolated from the flower buds of *Syzygium aromaticum* (cloves). ACS Omega 4, 1525–1533.

Schnitzler, P., Astani, A., 2014. Antiviral activity of monoterpenes beta-pinene and limonene against herpes simplex virus *in vitro*. Iran. J. Microbiol. 6 (3), 149–155.

Silva, J., Figueiredo, P., Byler, K., Setzer, W., 2020. Essential oils as antiviral agents. Potential of essential oils to treat SARS-CoV-2 infection: an in-silico investigation. Int. J. Mol. Sci. 21, 3426.

Sun, W.J., Lv, W.J., Li, L.N., Yin, G., Hang, X., Xue, Y., et al., 2016. Eugenol confers resistance to tomato yellow leaf curl virus (TYLCV) by regulating the expression of SlPer1 in tomato plants. New Biotechnol. 33 (3), 345–354.

Tragoolpua, Y., Jatisatienr, A., 2007. Anti-herpes simplex virus activities of *Eugenia caryophyllus* (Spreng) bullock and S.G. Harrison and essential oil, eugenol. Phytother. Res. 21 (12), 1153–1158.

Turgeon, N., Michel, K., Ha, T.L., Robine, E., Moineau, S., Duchaine, C., 2016. Resistance of aerosolized bacterial viruses to four germicidal products. PLoS One 11 (12), 0168815.

Waliwitiya, R., Kennedy, C.J., Lowenberger, C.A., 2009. Larvicidal and oviposition-altering activity of monoterpenoids, trans-anethole and rosemary oil to the yellow fever mosquito *Aedes aegypti* (Diptera: Culicidae). Pest Manag. Sci. 65, 241–248.

Woody, T.L., Davis, R.D., 1992. The effect of eugenol-containing and eugenol-free temporary cements on microleakage in resin bonded restorations. Oper. Dent. 17, 175–180.

Index

Note: Page numbers followed by *f* indicate figures and *t* indicate tables.

A

Acaricidal activity, 477
Activated partial thromboplastin time (APTT) assay, 45
Active compounds
  aqueous extract of clove flower buds, 665
  identification of chemical compounds, 667
  solvent extraction, 665–666
  supercritical extraction, 666–667
Acute toxicity, of alovinol, 670
Adaptive neuro fuzzy inference system (ANFIS), 548–549, 552*t*
  vs. RSM and ANN, 552
α-Humulene, 501
Alchermes liqueur, clove in, 143*f*
Alzheimer's disease (AD), 627–628
Analysis of variance (ANOVA), 534
Anesthetic activity, 478
ANFIS. *See* Adaptive neuro fuzzy inference system (ANFIS)
ANN. *See* Artificial neural network (ANN)
ANOVA. *See* Analysis of variance (ANOVA)
Antibacterial activity, 474–475
  of CBEO, 577
  of CEO, 501–506, 502–506*t*
  of clove extracts, 621–622
  storage time on, 566–567
Anticancer actions of clove-AgNPs, 649
Anticancer effects, clove in vitro and in vivo
  aqueous fraction, 320–331
  liposoluble fraction of, 315–320
Anticholinesterase (ACh) agent, 627–628
Anticonvulsant activity, of clove extracts, 626
Antidiabetic activity, of clove extracts, 596–597, 625–626
Antifungal activity, 476
  of CEO, 507, 507–508*t*
  of clove extracts, 623–624
  of clove oil, 431
  of eugenol, 577
Antiherpes simplex virus (HSV) activity of clove, 676
Anti-inflammatory activity, 477
  of clove-AgNPs, 648
  of clove extracts, 596, 623
Antimicrobial activity, 640, 646–648, 653–654
  of CFOil, 423
  of clove extract, 664–665
  of ethanolic extract, 611–612
Antimicrobial-based edible coatings, 431
Antimicrobial essential oils, 518
Antimicrobial nanoemulsions, 541
Antimicrobial properties, of clove extracts, 608–611, 609–610*t*
Antimycobacterial activity, of clove extracts, 625
Antioxidant properties
  of CBEO, 576–577
  of clove extracts, 608–611, 609–610*t*
Antiplasmodial activity, of clove extracts, 625
Antiradical effect, of clove extracts, 594–595
Anti-stress effect, of clove extracts, 597
Antiviral activity
  of clove against SARS-CoV-2, 679–680
  of clove extracts, 622
  of clove oil, 678
  of eugenol, 677–678
    dengue virus protease by, 678
    Ebola virus in vitro, 678
    against human herpesvirus, 677–678
  of green silver nanoparticles, 677
Antiviral properties, of clove, 675
  antiherpes simplex virus activity of, 676
  clove oil, 678–679
  eugenol, 677–678
    dengue virus protease by, 678
    Ebola virus in vitro, 678
    in vitro and in vivo antiviral activity of, 677–678
  green silver nanoparticles, 677
  against SARS-CoV-2, 679–680
Anti-yeast study, 592–594
Ants, urban pests, 336, 339*t*
  mortality, of RIFA and TFA, 337–338
  repellency effect, 338
Applications, of clove, 427
  buds, clove
    air freshener, 31
    animal feeds, 31
    eugenol extraction, source for, 32
    food, 25–27
    kretek cigarettes, 30–31
    traditional medicine, 27–30
  of CFOil
    in edible oil blends, 424
    in meat products, 424
  of clove essential oil, 508, 509*t*
    in edible coating, 430–432
    in edible oil blends, 428

Applications, of clove *(Continued)*
in meat products, 428–430
clove volatile oil (CVO)
β-caryophyllene, 149
in biopesticides, 145–148
eugenol, 149
food preservative, 151–155
in food processing, 141–142
in medicines, 142–145
in perfumery, 145
pharmaceuticals, 155–158
safety, 150–151
tissue histology, clearing agents in, 148
of microencapsulated clove oil, 428
Aquatic species, clove extract on, 670
Aqueous extract, of clove flower buds, 665, 677
Aqueous extract of cloves (AEC), 320–329, 331
cytotoxic analysis, cancer cells, 323*t*
Aromatic cloves
period of flowering, 11
varieties of, 11
Artificial neural network (ANN), 490–491, 544–547, 547–548*f*
vs. RSM and ANFIS, 552
Atomizer, 519

B

BBD. *See* Box-Behnken design (BBD)
β-Caryophyllene, 40, 374–375, 501, 592
Bioactive compounds, in clove, 523, 525–527, 607, 664, 667
applications, of clove extracts, 413
chemical formula and names, 406–408*t*
extraction solvents, 409
microwave-assisted extraction, 412
phytochemical composition, 405
solvent extraction (maceration), 409–410
Soxhlet extraction, 410–411
supercritical fluid extraction, 412–413
ultrasound-assisted extraction, 411
Biological activities
and applications of CEO, 508, 509*t*
of clove oil, 577–578
of encapsulated CEO, 522–524
health-promoting activities of clove extracts, 620–628
antibacterial activity, 621–622
anticancer activity, 624–625
anticonvulsant activity, 626
antidiabetic activity, 625–626
antifungal activity, 623–624
anti-inflammatory activity, 623
antimycobacterial and antiplasmodial activities, 625
antioxidant activity, 621
antiviral activity, 622
hepatoprotective activity, 626–627
immunomodulation effects, 628
neuroprotection effect, 627–628
skin protective effect, 627
Biological properties, of clove extract, 664–665, 664*t*
antimicrobial activity, 664–665
antioxidant activity, 664
Biopesticides
fungicides, 146–147
insecticides, 147
insect repellants, 147–148
mechanisms, 148
Box-Behnken design (BBD), 542
Brake-specific fuel consumption (BSFC), 451
Brake thermal efficiency (BTE), 451
Brine shrimp lethality test (BSLT) assay, 186
Buds, clove
application
air freshener, 31
animal feeds, 31
eugenol extraction, source for, 32
food, 25–27
kretek cigarettes, 30–31
traditional medicine, 27–30
benzoic acid in, 23*f*
budding stage-1 of, 16*f*
budding stage-3 of, 16*f*
chemical compositions of, 21–23, 23–24*f*
chemical constituents, 11–13
cinnamic acid in, 23*f*
description of, 18–19
flavonoids in, 25*f*
full-budding stage of, 17*f*
full-flowering stage of, 17*f*
harvesting, at different stages, 11
in Indonesia, 13, 14*t*
phenylpropanoids in, 23*f*
production, in Pacific Ocean, 14–15
red color clove fruits, 18*f*
terpenoids and sterol in, 26–27*f*
various qualities, 11
young stage of, 16*f*

C

Candida albicans, 57
Canjica, Brazilian dessert sweet, 27, 29*f*
Cannabinoid receptor 2 (CB2), 374–375
Carboxymethyl cellulose-clove oil-polyvinyl alcohol films, 431
Catalytic activity, of PdNPs, 651
CBEO. *See* Clove bud essential oil (CBEO)
CCl_4-induced toxicity, in animals, 423–424

Cellular level, clove extract at, 667–668
Cell viability testing, 667–668
Central composite design (CCD), 493
CEO. *See* Clove essential oil (CEO)
CESN. *See* Clove extract silver nanoparticles (CESN)
CFOil. *See* Clove fixed oil (CFOil)
Chitosan (CS)-Myristic acid (MA) nanogel, 523
Chromobacterium violaceum, 153
Chuanxiong oil, 368
Clove *(Syzygium aromaticum)*, 586
  acetophenones in, 28*f*
  antiviral properties of, 675
    antiherpes simplex virus activity of, 676
    clove oil, 678–679
    eugenol, 677–678
    green silver nanoparticles, 677
    against SARS-CoV-2, 679–680
  Bandrek, Indonesia, 25, 28*f*
  biological and antiviral effects of, 679*f*
  buds (*see* Buds, clove)
  chemical constituents of
    clove bud essential oil, 462–463
    clove leaves essential oil, 463
    clove stem oil, 463
  chemistry, 2–6
  chromone glycosides in, 28*f*
  clove essential oil, functionality of, 461, 466–474*t*, 474*f*
    acaricidal activity, 477
    anesthetic activity, 478
    antibacterial activity, 474–475
    anticancer activity, 478
    antifungal activity, 476
    anti-inflammatory activity, 477
    antioxidant activity, 475–476
    herbicidal activity, 479
    larvicidal activity, 479
    leishmanicidal activity, 478
    nematicidal activity, 478–479
    structure, of different compounds in CEO, 464–465*t*
  composition of, 589–592, 590–591*t*
  distribution, 461
  extraction of, 586–588
  fiber, fragment of, 20*f*
  flowers, development stages of, 15–18
  functionality of, 2–6
    anticancer effect, 595
    antidiabetic effect, 596–597
    anti-inflammatory effect, 596
    antimicrobial effect, 592–594
    antioxidant and antiradical effect, 594–595
    anti-stress effect, 597
    applications, 599
    encapsulation applications of clove extract, 598–599
    food practices, 597–598
    testicular function effect, 597
  garden, in mountain of Java, 12*f*
  genetics, 20–21
  global export and import quantities, 4*f*
  ketone in, 28*f*
  market, 3–6
  medicinal properties, 461
  morphology, 462
  oil gland, 19*f*
  pollen grain, 21*f*
  spiral vessel, 20*f*
  techno-applications, 2–6
  trees, 12*f*
    with flowers, at full-budding stage, 13*f*
  unopened flower buds, 461
  UNSDG goals and applications, of medical plants, 1–6
Clove bud essential oil (CBEO)
  biological activities of, 577–578
  chemistry of, 575
  extraction of, 573–574
  multidisciplinary applications of, 578
  storage quality of, 575–577
  storage stability of, 575
Clove bud extract (CBE), 261–263
Clove bud powder (CBP), 198–199
Clove essential oil (CEO), 2, 455, 501
  advantages, pharmaceutical applications, 370*t*
  anticancer activity, 485–486
  antiviral activities of, 678
  applications, in food industries, 485–486
  aqueous pharmaceutical formulations, 363–364
  bioactive molecule, in pharmaceutical formulations
    anticancer and antiproliferative properties, 374–375
    antiinflammatory, antioxidant and antinociceptive properties, 369–372
    antimicrobial applications, 372–374
  as bio-additives, biodiesel-diesel fuel, 447
    antioxidant effect of, 447
    DIRJEN MIGAS specification, 451
    eugenol, 445
    extraction of, 446
    flavonoids, 445
    fuel additives, 448
    monomeric phenol compounds, 448
    phenolic acids, 445
    physical and chemical properties, 446
    plant parts, 445
  as bio-additives, biodiesel-diesel fuel high oxygen content, 445
  comparison of, 495–498, 495–497*t*

Clove essential oil (CEO) *(Continued)*
cost, 455
cyclodextrins, 389–390
electrospun nanofibers, 386–387
encapsulation, 517–518
biologic activities, 522–524
emulsification, 519–520
entrapment in liposomes, 521–522
in food matrices, 524–526
methods, 518–522
spray drying or lyophilization (freeze-drying), 518–519
encapsulation, nanostructured systems, 377–380
nanoemulsions (NEs), 378–380
engine performance
brake-specific fuel consumption (BSFC), 451
brake thermal efficiency (BTE), 451
combustion behavior of, 450
exhaust gas temperature (EGT), 452
ignition delay, 452
terpene compounds, 450
eugenol, 363–367
exhaust emissions
carbon monoxide (CO) and smoke, 453–454
nitrogen oxides (NOx), 453
particulate matters and unburnt hydrocarbon (HC), 454
smoke emissions, 452
extraction methods, 486–494, 487–490*t*
hydrodistillation, 490–491
microwave-assisted extraction (MAE), 494
ohmic heating assisted hydrodistillation (OAHD), 494
steam distillation, 486–490
supercritical fluid extraction (SFE), 491–492
ultrasound-assisted extraction, 492–494
health-promoting biological activities
antibacterial activities, 501–506, 502–506*t*
antifungal activities, 507, 507–508*t*
biological activities and applications, 508, 509*t*
high oxygen content, 450
hydrogel-based nanomaterials, 387–388
lipid-based nanoparticles
advantages, for pharmaceutical applications, 381*t*
differential scanning calorimetry (DSC), 382–383
liposomes, 380–381
medium-chain triglycerides (MCT), 382–383
nanostructured lipid carriers (NLC), 381–383
pore-forming toxins (PFTs), 381
solid lipid nanoparticles (SLN), 381–382
liposomes, 521–522
nanoemulsion, 541–542, 559–560, 561*t*, 678–679
emulsification process, 560–562
morphology, 562–563
optimal conditions of, 552–553
optimization techniques of, 542–549, 544*t*, 545–548*f*
rheological parameters, 567*t*
RSM *vs*. ANN *vs*. ANFIS, 552, 552*t*
stability studies of storage period, 563–567, 564*f*
techniques of, 549–551, 549–551*t*
permeation enhancer, for transdermal delivery, 368, 369*t*
pharmacological properties, 485–486
plant's secondary metabolism, 363
polymeric nanoparticles (PNPs), 383–386
biomedical activity, 385
bovine serum albumin conjugated to folic acid nanoparticles (BSA-CFNP), 386
encapsulation phenomena, 385
eugenol-loaded chitosan nanoparticles, 385
monodispersity, 385
nanocapsules, 383–385
nanospheres, 383–385
natural and synthetic polymers, 384*t*
in situ polymerization, 385–386
storage quality of, 573
biological activities, 577–578
clove bud essential oil (CBEO), 573–575
food products supplemented with CBEO, 575–577
multidisciplinary applications, 578
storage stability, 575
terpenes, 363–364
topical drug delivery, pharmaceutical formulations, 363–364
penetration enhancer, 366–369
poorly soluble drugs, 365–366
vector-borne diseases, 485–486
vesicle-derived carriers, 388–389
wound healing properties, 376–377
Clove extracts (CE), 223–224
food applications of, 607–608
antioxidant and antimicrobial properties of, 608–611, 609–610*t*
fish products, 612–613
meat and meat products, 611–612
proximate composition of, 608*t*
health-promoting activities of, 619–620
biological activities, 620–628
ethnobotanical knowledge and traditional uses, 620
health benefits, 628–629
toxicity of
biological properties, 664–665, 664*t*
description, 663
studies, 667–672, 668*t*, 671*f*
types and active compounds, 665–667
Clove extract silver nanoparticles (CESN), 677
Clove fixed oil (CFOil), 421–422, 427
composition and functionality of

antimicrobial activity of, 423
antiradical potential of, 423
in edible oil blends, 424
fatty acid profile of, 422, 422*t*
hepatoprotective effect of, 423–424
lipid classes and subclasses of, 422
in meat products, 424
tocol profile of, 423
food applications of
baked product and vegetable plant, 430
in edible coating, 430–432
in edible oil blends, 428
in meat products, 428–430
microencapsulated clove oil, 428
Clove flower buds, aqueous extract of, 665
Clove-mediated metallic nanoparticles, 640
AgNPs, 646–649, 649*f*
AuNPs, 649–650
clove phytochemicals in synthesis, 641–655
FeNPs, 650–651, 651*f*
mono-MNPs, 653–654
multi-MNPs, 654–655
PdNPs, 651–652, 652*f*
ZnSNPs/ZnONPs, 653
Clove oil-loaded nanoemulsion (CLN), 187
Clove oleoresins (CO)
antifungal activity, 60
application, 54–55
animal feeds, 60
biopesticides, 59
fish farming, 60
food preservatives, 56–57
pharmaceuticals, 57–59
safety, 55
biocide for biofilm removal, on outdoor stone monuments, 60
characteristics of, 50–51
compositions, 50–51
detection of compounds in, 53–54
from different plant varieties, 51–52
dyeing silk fabrics in, 60
extraction, 52–53
flavoring agent and adjuvant, in food products, 50
methanolic CO, 60
supercritical fluid extraction (SFE), using CO_2, 49–50
Clove phytochemicals, 641–655
AgNPs, 646–649, 649*f*
AuNPs, 649–650
FeNPs, 650–651, 651*f*
mono-MNPs, 653–654
multi-MNPs, 654–655
PdNPs, 651–652, 652*f*
ZnSNPs/ZnONPs, 653
Clove volatile oil (CVO)
aging process, of plants, 129–130
applications
β-caryophyllene, 149
in biopesticides, 145–148
eugenol, 149
food preservative, 151–155
in food processing, 141–142
in medicines, 142–145
in perfumery, 145
pharmaceuticals, 155–158
safety, 150–151
tissue histology, clearing agents in, 148
buds, 119*f*
characteristics and compositions, 118–127, 120–123*t*, 125–127*t*
composition detection, 140–141
different cultivation areas, 132
extraction and storage
alternative methods, 139–140
oil distillation, 135–137
oil storage, 140
pre- and post-distillation processes, 137–139
leaves, 124*f*
plant organs, different maturity of, 130–131
plant varieties, 128–129
post-harvesting processes, 133–134
pre- and post-distillation processes, 137–139
Clovinol, 669–670
Cockroaches, urban pests, 336, 339*t*
essential oils, 339
mortality
fumigant application, 341
surface contact application, 340–341
topical application, 340
repellency effect, 341, 342*t*
synthetic insecticides, 339
Code of Federal Regulations (CFR), 50
Compression ignition (CI) engines, 447
Confocal laser scanning microscopy (CLSM), 388–389
Couette emulsification process, 567
COVID-19 therapy, 1
Cyclodextrins (CDs), 389–390

D

Dengue virus protease, by eugenol, 678
Dental analgesic, clove oil, 144*f*
Deodorized aqueous extract of clove bud, 612
2,2-Diphenyl-l-1-picrylhydrazil (DPPH•), 370
1,1-Diphenyl-2-picrylhydrazyl (DPPH$^•$), 437–438

Dried clove buds, 15, 18, 19*f*, 209
Dripping method, 520
Droplet extrusion principle, 520
Droplet size, storage period on, 563–564, 564*f*
Dynamic light scattering (DLS), 91–92
Dynamic viscosity, 561, 563, 566

E

Ebola virus in vitro, eugenol inhibitor of, 678
Electron donor functional groups, 641
Emulsification process, 560–562
 encapsulation by, 519–520
 RSM of clove oil encapsulation by, 544*t*
Encapsulation
 applications, of clove extract, 598–599
 of clove essential oil, 517–518
  biologic activities of, 522–524
  in food matrices, 524–526
  methods, 518–522
 efficiency, 519
 by nanoemulsions, 560
Entrapment in liposomes, 521–522
Essential oils (EOs). *See* Clove essential oil (CEO)
Ethanol extract of clove (EEC), 315–316
Ethnobotanical knowledge of clove, 619–620
Ethosomes, 388
Eugenia caryophyllata bud extract (ECBE), 260–261
Eugeniin, 622, 678
Eugenol-based nutraceuticals
 nanoformulation-based nutraceuticals, 107–109
 pharmacology effects, 109–110, 110*t*
 phenolic compounds, 104
 source, extraction and chemistry of, 105–107
  hydrodistillation, 106
  microwave-assisted extraction (MWAE), 106
  solvent extraction, 105
  supercritical carbon dioxide (SC-CO_2) method, 106
  ultrasound-assisted extraction, 106–107
 structure of, 104*f*
 toxicological studies, 111, 111*t*
 traditional medicine systems, 103–104
Eugenol, clove and, 2, 575, 586, 588–589, 597–598, 628, 641–646, 665–667, 676
 acute toxicity, 290–293
 angiospermic plants, 68
 antibacterial activity of CEO and, 501–506, 502–506*t*
 antifungal activity of CEO and, 507, 507–508*t*
 antispasmodic activity, 69
 antiviral activity of, 677–678
  dengue virus protease by, 678
  Ebola virus in vitro, 678
  against human herpesvirus, 677–678
 bioactive compounds and toxic effects, 268–286
  β-caryophyllene, 270–281*t*
  caryophyllene oxide, 270–281*t*
  chavibetol, 270–281*t*
  fish, toxicity of clove against, 282–283
  α-humulene, 270–281*t*
  pests, toxicity of clove against, 283–286
 biological activities, 68
  and applications of CEO, 508, 509*t*
 chemical structure of, 607
 chloroform and basic aqueous solvents extraction, 68
 chronic toxicity of, 296–298
 clove oil, 68
 ecotoxicity, 299
 essential oils, 286
 genotoxicity, 298
 insect attractants, 68–69
 nonthermal techniques, 287
 phenylpropanes, 68
 phenylpropanoid, 286–287
 regulatory requirements, 299–300
 steam distillation, of cloves, 68
 subacute and sub-chronic toxicity of, 293–296, 294–295*t*
 therapeutic agent, in dentistry, 68
 tocicity and its effects, 288–290
 uses areas of, 287–288
Exhaust emissions
 carbon monoxide (CO) and smoke, 453–454
 nitrogen oxides (NOx), 453
 particulate matters and unburnt hydrocarbon (HC), 454
 smoke emissions, 452
Exhaust gas temperature (EGT), 452

F

FASN inhibitor-alcohol extract of clove (AEC), 626–627
Fatty acid profile of CFOil, 422, 422*t*
FCV. *See* Feline Calicivirus (FCV)
FDA. *See* Food and Drug Administration (FDA)
Federal Insecticide, Fungicide, and Rodenticide Act (FIFRA), 146
Feline Calicivirus (FCV), 622
Fish feeding, clove and its derivatives
 antibiotics, 258
 biological effects of, 259*f*
 chemical structure of, 260
 dietary supplements, 258
 health benefits and beneficial uses
  fish immunity, cloves effect on, 260–261
  fish performance, cloves effect on, 261–262
  semen quality, cloves effect on, 262–263
 pharmacological properties, 258–259
Flame ionization detector (FID), 141

Flavor and Extract Manufacturers Association (FEMA), 55, 150
Flos Caryophylli, 11
Food and Chemical Codex (FCC), 53
Food and Drug Administration (FDA), 2, 440, 508
Food applications
of CFoil
baked product and vegetable plant, 430
edible coating, 430–432
edible oil blends, 428
meat products, 428–430
microencapsulated clove oil, 428
of clove buds
animals, to treat tooth pain, 211
CEO, 211
culinary uses, 209
traditional medicinal uses, 209–211
of clove extracts, 607–608
antioxidant and antimicrobial properties of, 608–611, 609–610*t*
fish products, 612–613
meat and meat products, 611–612
proximate composition of, 608*t*
Food matrices, encapsulated CEO in, 524–526
Food polysaccharides, 43
Food preservatives
antibacterial activity against foodborne pathogens, 152–154
antifungal activity against foodborne pathogens, 154–155
antioxidant, 152
clove oleoresins (CO)
antimicrobial activity against foodborne pathogens, 56–57
antioxidant, 56
Food products, 517
supplemented with CBEO, 575–577
Frisian clove cheese, 27, 29*f*
Functionality, of clove
anticancer effect, 595
antidiabetic effect, 596–597
anti-inflammatory effect, 596
antimicrobial effect, 592–594
antioxidant and antiradical effect, 594–595
anti-stress effect, 597
applications, 599
of CFOil
antimicrobial activity, 423
antiradical potential, 423
edible oil blends, 424
fatty acid profile, 422, 422*t*
hepatoprotective effect, 423–424
lipid classes and subclasses, 422
meat products, 424
tocol profile, 423
encapsulation applications of clove extract, 598–599
food practices, 597–598
testicular function effect, 597

G

Galactomannan, 562
Gas chromatography-mass spectrometry (GCMS), 574
Gas chromatography triple quad time of flight (GC-Q-TOF), 217
Gelatin-chitosan-based edible films, 526
Gemcitabine, 667–668
Generally recognized as safe (GRAS), 50, 105, 151–152, 352, 421
Gray-based Taguchi method, 573–574
Green silver nanoparticles, antiviral activity of, 677
Green synthesis, of FeNPs, 650

H

Halloysite nanotubes (HNTs), 206
Health-promoting activities, of CEO, 619–620
antioxidant activity, 435–437
biological activities, 620–628
antibacterial activity, 501–506, 502–506*t*, 621–622
anticancer activity, 624–625
anticonvulsant activity, 626
antidiabetic activity, 625–626
antifungal activity, 507, 507–508*t*, 623–624
anti-inflammatory activity, 623
antimycobacterial and antiplasmodial activities, 625
antioxidant activity, 621
antiviral activity, 622
biological activities and applications, 508, 509*t*
hepatoprotective activity, 626–627
immunomodulation effects, 628
neuroprotection effect, 627–628
skin protective effect, 627
chemical composition of, 436–437
clove essential oil, 195–197
clove extracts, pharmacological activities
analgesic activities, 198
antidiabetic activities, 199
anti-inflammatory activities, 198
antimicrobial activities, 197
antioxidant activities, 197–198
hepatoprotective activities, 198
cold-pressed clove oil, 195–198
ethnobotanical knowledge and traditional uses, 620
extraction of, 437
health benefits, 628–629
microbicidal potential, 435–436
pharmacological activities of, 440
antibacterial activity, 438–439

Health-promoting activities, of CEO *(Continued)*
anticarcinogenic activity, 438
antidiabetic activities, 439–440
antiinflammatory activity, 439
antioxidant/free radical scavenging activity, 438
antithrombotic activity, 439
toxicity doses, 440
Heat reflux extraction, 574
Hepatoprotective activity
of clove extracts, 626–627
effect of CFOil, 423–424
Herbicidal activity, 479
Heterogeneous catalytic activity, 652
Hexane extracts, of clove buds, 597
High molecular weight polysaccharide (HMWPS), 44–45
High-pressure homogenization (HPH), 559–560
Human herpesvirus, in vitro and in vivo antiviral activity, 677–678
Hydrocarbon (HC) emission, 454
Hydrodistillation, 106, 490–491, 665
Hydrophilic-lipophilic balance (HLB), 379

I

Ibuprofen delivery, 368
Immunomodulation, 260
effects, of clove extracts, 628
Integrated pest management (IPM), 335
International Fragrance Association (IFRA), 295–296
In vitro antiviral activity, of eugenol, 677–678
In vivo antiviral activity, of eugenol, 677–678

J

Joint FAO/WHO Expert Committee on Food Additives (JECFA), 55, 151

K

Kai Palo, 25, 29*f*
Kan Plu toothpaste, clove oil, 143*f*

L

Larvicidal activity, 479
Leaves, clove
application
eugenol extraction, source of, 40
traditional medicine, 40
characteristics, 38
chemical composition of, 39–40
development stages, in Madagascar, 38, 38*f*
oil distillation, household industry, 37
oil glands, 38
varieties of, 38
volatiles and nonvolatiles chemical constituents, 37
Leishmanicidal activity, 478
Limax flavus, 671–672
Lipid classes, of CFOil, 422
Lipopolysaccharide (LPS) effects, 376
Liposome-encapsulated clove oil, 522
Liposomes, entrapment in, 521–522
Liquid chromatography-tandem mass spectrometry (LC-MS/MS), 436
Listeria monocytogenes, 424, 428–429, 475, 576
Low molecular weight polysaccharide (LMWPS), 44–45
Lyophilization (freeze-drying), 520
encapsulation using, 518–519

M

Maceration extraction technique, 586–587
Malondialdehyde (MDA), 356–357
Maslinic acid (MA), 320, 323*f*
Mass spectrometer (MS), 141
MATLAB software, 549
MCT. *See* Medium-chain triglycerides (MCT)
Meat and meat products
CFOil in, 424
of clove extracts, 611–614
clove oil in, 428–430
Meat industry, clove extracts applications in
antibacterial activity, of clove and clove EO, 353, 353*t*
antioxidant activity, 354–357, 355–356*t*
chemical composition and mechanisms of action
eugenol, 350–352
quality control, with CEO, 352
essential oils, 347
flower buds, 347–348
microbial contamination and oxidation, 347
oil production, 349–350
perspectives, 357–358
safety and quality of, 347
thermal and nonthermal technologies, 347
Medium-chain triglycerides (MCT), 563–564
Metal nanoparticles (MNPs), 641, 642–646*t*, 647*f*, 651
Methanolic oleoresins, 52
Methicillin-resistant Staphylococcus aureus (MRSA), 373
Methyleugenol, 371
Microencapsulated clove oil, 428
Microwave-assisted extraction (MWAE), 106, 494
bioactive compounds extraction, clove, 412
Minimum bactericidal concentration (MBC), 474–475
Minimum inhibitory concentration (MIC), 474–475
Molecular docking methods, 680
Mono-MNPs, 653–654
Monoterpenes, in clove essential oil, 678
MTT assay, 186–187

Multi-MNPs, 654–655
Multiphase clove-FeNPs, 651*f*

N

Nanobiotechnology, 677
Nanocomposites, clove chemical constituents
  antimicrobial packages, 204
  biodegradable materials, natural polymers, 203–204
  phytochemical compounds, 204
Nano-emulsification method, 520
Nanoemulsions (NEs), clove oil, 378–380, 520, 542, 559–560
  antimicrobial agent, in food, 96, 96*t*
  biofunctional properties, 82–83
  characterization of
    morphological characterization, 92
    particle charge, 92
    particle size, 91–92
  for delivery, clove oil, 83–84
  emulsifiers/surfactants, in emulsion system
    high molecular weight surfactants, 88–89
    low molecular weight surfactants, 88
  encapsulation of, 83–84
  and eugenol, 81–82
  in food matrix, 83
  functionalities, 93–95, 95*t*
  morphology, 562–563
  PDI, 565
  stability of
    flocculation and coalescence, 90
    gravitational separation, 89–90
    Ostwald ripening, 91
  techniques of EO, 549–551, 549–551*t*
Nanoformulation-based nutraceuticals, 107–109
Nanostructured lipid carriers (NLC), 381–383
Nanotechnology, 640
National Institute on Aging in the National Institutes of Health, 627–628
National Toxicology Program (NTP), 150
Natural antimicrobials, 541
Nematicidal activity, 478–479
Neurons, 544–546
Neuroprotection effect, of clove extracts, 627–628
Neutral lipids (NL), 422
Newcastle disease virus (NDV) infection, 677
Niosomes, 388
NL. *See* Neutral lipids (NL)
Nodes, 544–546
Nonsteroidal antiinflammatories drugs (NSAIDs), 371
Nonvolatile components, 436
No observed adverse effect level (NOAEL), 268
Nuclear Magnetic Resonance (NMR) spectroscopy, 50

O

Octopaminergic system, 343
Oedema tissue, 670
Ohmic heating assisted hydrodistillation (OAHD), 494
Oil-in-water (O/W) clove nanoemulsion, 541, 559–560
  clove oil nanoemulsions, 560, 561*t*
  emulsification process, 560–562
  morphology, 562–563
  rheological parameters, 567*t*
  stability studies of storage period, 563–567
    on antibacterial activity, 566–567
    on droplet size, 563–564, 564*f*
    on dynamic viscosity, 566
    on PDI, 565
    on zeta potential, 565–566
Oleanolic acid (OA), 320, 322*f*
Olmesartan, 379–380
Optimization techniques, of CEO, 542–549
  adaptive neuro fuzzy inference system, 548–549
  artificial neural network (ANN), 544–547, 547–548*f*
  response surface methodology (RSM), 543–544, 544*t*, 545–546*f*
ORAC. *See* Oxygen radical absorption capacity (ORAC)
Oral delivery of lipophilic drugs, 549–551
Organization for Economic Co-operation and Development (OECD), 150
Overall extraction curves (OEC), 492
Oxygen radical absorption capacity (ORAC), 257–258, 627–628

P

PdNPs, 651–652, 652*f*
Pediculocidal activity, 629
Peng-Robinson EoS, 492
Pharmaceuticals
  clove oleoresins (CO)
    antimalarial agents, 58
    antiviral agents, 59
    cancer drugs, 57–58
    dental formularies and related products, 57
    diabetes drugs, 58
    hangover supplements, 59
    hepatoprotective supplements, 59
    immunomodulators, 58
  clove volatile oil (CVO)
    animal feeds, 158–159
    antibiotics, for urinary tract infections, 157
    bacteria-resistant fabric, 158
    cancer drugs, 156–157
    cosmetics, 158
    dental formularies, 155–156
    diabetes drugs, 157
    ear preparation drugs, 156

Pharmaceuticals *(Continued)*
fish anesthesia, 159–160
mosquito control, 160
neuroprotective agents, 158
pediculicidal shampoo, 155
Phase-inversion composition, 549–551
Phenolics, clove
biological activities, 215–216, 223–226
clove buds, 215
extraction process, 216–217
structure and composition
phenolic acids and flavonoids, 217–220
tannins, 220–221
volatile phenolics, 221–222
varieties of, 215
Phosphatidylcholine, 521
Phospholipid liposome, 521*f*
Phospholipids, 521
PI3K/Akt/mTOR signaling pathway, 320
Plant phytochemicals, 640
Platelet-activating factor (PAF), 439
Polarized optical microscopy, 365
Polydispersity index (PDI), storage period on, 565
Polylactic acid (PLA), 205
Polymeric nanoparticles (PNPs)
clove essential oil (CEO), 383–386
biomedical activity, 385
bovine serum albumin conjugated to folic acid nanoparticles (BSA-CFNP), 386
encapsulation phenomena, 385
eugenol-loaded chitosan nanoparticles, 385
monodispersity, 385
nanocapsules, 383–385
nanospheres, 383–385
natural and synthetic polymers, 384*t*
in situ polymerization, 385–386
Polysaccharides, clove
biological activities, 45
flowering bud different stages, of clove, 43
organoleptic and therapeutic properties, 43
plant-based polysaccharides, 43
structure and composition, 44–45
textural properties, of food products, 43
Polyunsaturated fatty acids (PUFA), 259
Porphyromonas gingivalis, 57
Poultry feed, clove in
antibiotics, 235–236
dietary supplementation, 238–242
essential oil, 242–244
in diet of poultry, 244–247
fatty acids, 238*t*, 239*f*
food industry, clove extract in, 236
phyto-additive
as alternative of, 236–238
in drinking water, 247
phytochemicals, clove for intestinal health, 247–248
stress hormones synthesis, 235
Psychrophilic bacteria, 153

Q

Quadratic models, 543
Quantitative structure-activity relationship (QSAR), 144–145

R

Radical scavenging activity, 424
Red ants, 671–672, 671*f*
Respiratory burst activity (RBA), 260–261
Response surface methodology (RSM), 543–544, 543–544*t*, 546
vs. ANN and ANFIS, 552
Rhipicephalus microplus, 477
RIFA, invasive omnivorous pest, 336
Ripening inhibitor, 541
RMSE. *See* Root means square error values (RMSE)
RNA, 677–678
Root means square error values (RMSE), 552
RSM. *See* Response surface methodology (RSM)

S

Salmonella enteritidis, 424
Sardines *(Rastrineobola argentea)*, 613
SARS-CoV-2. *See* Severe acute respiratory syndrome coronavirus 2 (SARS-CoV-2)
SCG-nanoemulsion, 566–567
Secondary metabolites, of clove
bioactive compounds, chemical structures of, 178*f*
biological activities
anesthetic effect, 187
antibacterial activity, 184–185
anticancer activity, 186–187
antifungal activity, 185–186
anti-inflammatory activity, 186
antioxidant activity, 185
herbicidal activity, 188
insecticidal activity, 188
nematicidal activity, 187
economic value of, 176–177
geographical distribution, 176*f*
phenolic compounds, chemical structure of, 179*f*
phytochemicals, flower buds, 181–184*t*
phytochemistry, 177–183
Severe acute respiratory syndrome coronavirus 2 (SARS-CoV-2), 679–680
SFE. *See* Supercritical fluid extraction (SFE)

Shelf life, 524–527
Sikotok, aromatic clove, 11
Silence information regulator 1 (SIRT1) pathway, 627–628
Siputih, aromatic clove, 11
Skin protective effect, of clove extracts, 627
Solid lipid nanoparticles (SLN), 381–382
Soxhlet extraction, 52–53
  bioactive compounds extraction, clove, 410–411
  of clove, 666
  technique of, 586–587
Soy lecithin, 521
Spectrophotometry, 154
Spice oleoresins, 49
Spray drying or lyophilization (freeze-drying), encapsulation using, 518–519
Stability, 541–542, 549–553
  of O/W nanoemulsions, 552–553
  studies of clove nanoemulsion, 563–567
    on antibacterial activity, 566–567
    on droplet size, 563–564, 564*f*
    on dynamic viscosity, 566
    on PDI, 565
    on zeta potential, 565–566
Steam distillation methods, 486–490, 574
Storage quality, of clove oil, 573
  biological activities, 577–578
  clove bud essential oil (CBEO), 573–575
  food products supplemented with CBEO, 575–577
  multidisciplinary applications, 578
  stability, 575
Sub-chronic toxicity, of alovinol, 670
Supercritical carbon dioxide (SC-CO_2) method, 531–532
  apparatus of, 532–533, 533*f*
  chemical characteristics, 536–538, 538*t*
  eugenol-based nutraceuticals, 106
  optimization of, 533–536
    conditions of essential oil extraction, 536
    extraction pressure effect, 535
    extraction temperature effect, 535–536
Supercritical fluid extraction (SFE), 52–53, 491–492, 536–537, 574, 666–667
  bioactive compounds extraction, clove, 412–413
Supercritical fluid extraction assisted by cold pressing (SFEAP), 491–492
Syzygium aromaticum. *See* Clove

T

Tannins, in clove buds, 24*f*
TBARS. *See* Thiobarbituric acid reactive substances (TBARS)
Telangiectasis appearance, 670
Testicular function effect, 597
Thiobarbituric acid reactive substances (TBARS), 612
Threshold of toxicological concern (TTC), 150
Thrombin time (TT) assay, 45
TiO_2NPs, 654
Tocol profile, of CFOil, 423, 428
Tolak Angin, Indonesian herbal supplement, 30, 30*f*
Total flavonoid content (TFC), 437
Total phenolic compounds (TPC), 222
Toxicity, of clove extract
  biological properties of, 664–665, 664*t*
    antimicrobial activity, 664–665
    antioxidant activity, 664
  description of, 663
  studies of, 667–672, 668*t*
    aquatic species, 670
    cellular level, 667–668
    other toxic effects, 671–672, 671*f*
    reproduction in mice, 669
    sub-chronic assessment of Wistar rats, 669–670
  types and active compounds
    aqueous extract of clove flower buds, 665
    identification of chemical compounds, 667
    solvent extraction, 665–666
    supercritical extraction, 666–667
Traditional Chinese medicinal herb (TCM), 315
Transmission electron microscopy (TEM) analysis, 648
Tropical fire ant (TFA), 336
Tuni, aromatic clove, 11
Tunisian clove essential oil, 536–538, 538*t*, 567

U

Ultrasonic emulsification, 520
Ultrasound-assisted extraction (UAE), 52, 588
  bioactive compounds extraction, clove, 411
  eugenol-based nutraceuticals, 106–107
Ultrasound-assisted supercritical carbon dioxide (USC-CO_2) method, 411, 493–494, 574
United Nations Sustainable Development Goals (UNSDG), 1–6
United States Department of Agriculture (USDA), 627–628
Urban pests, clove
  ants, clove products against, 336, 339*t*
    mortality, of RIFA and TFA, 337–338
    repellency effect, 338
  bioactive compounds, mode of action of, 342–343
  cockroaches, clove products against, 336, 339*t*
    essential oils, 339
    mortality, 340–341
    repellency effect, 341
    synthetic insecticides, 339
  control methods, 335–336
  household insect pests, 335

Urban pests, clove *(Continued)*
  public health pests, 335
USDA. *See* United States Department of Agriculture (USDA)

V

van der Waals 1 (vdW1) mixing rule, 492
van der Waals 2 (vdW2) mixing rule, 492
Varroa destructor, 477
Viruses, 675, 677–678
Volatile organic compounds (VOCs), 222

W

WHO Expert Committee on Food Additives, 421
World Health Organization (WHO), 1, 440, 508, 625
Wound breaking strength (WBS), 376–377

Z

Zanzibar, aromatic clove, 11
Zeta potential, 561, 563, 565–567
  storage period on, 565–566
Zinc phthalocyanine (ZnPc), 365–366
ZnSNPs/ZnONPs, 653

Made in United States
Orlando, FL
04 March 2026

79035423R00400